FROM NEURON TO BRAIN

FROM NEURON TO BRAIN

FOURTH EDITION

John G. Nicholls
International School for Advanced Studies, Trieste, Italy

A. Robert Martin
Emeritus, University of Colorado School of Medicine

Bruce G. Wallace
University of Colorado School of Medicine

Paul A. Fuchs
The Johns Hopkins University School of Medicine

Sinauer Associates, Inc. • Publishers
Sunderland, Massachusetts • U.S.A.

ABOUT THE COVER

Functional MRI mapping of the human visual system. The map shows the activated (colored) areas bilaterally in the lateral geniculate nucleus and primary visual cortex during full-field visual stimulation. (Reproduced with permission from Wei Chen et al., 1999. *Proc. Natl. Acad. Sci. USA*, 96: 2430–2434.)

From Neuron to Brain, Fourth Edition

For information, address Sinauer Associates, 23 Plumtree Road, Sunderland, MA 01375 U.S.A.
Fax: 413-549-4300
Email: publish@sinauer.com; Internet: www.sinauer.com

Library of Congress Cataloging-in-Publication Data
From neuron to brain / John G. Nicholls ... [et al.].—4th ed.
p. cm.
Previous ed. Cataloged under: Nicholls, John G.
Includes bibliographical references and index.
ISBN 978-0-87893-439-3
1. Neurophysiology. 2. Brain. 3. Neurons. I. Nicholls, John G. From neuron to brain. II. Nicholls, John G.

QP355.2.K83 2000
573.8—dc21 00-036529

8 7 6 Printed in China

This book is dedicated to the memory of our friend and colleague, Steve Kuffler.

In a career that spanned 40 years, Stephen Kuffler made experiments on fundamental problems and laid paths for future research to follow. A feature of his work is the way in which the right problem was tackled using the right preparation. Examples are his studies on denervation, stretch receptors, efferent control, inhibition, GABA and peptides as transmitters, integration in the retina, glial cells, and the analysis of synaptic transmission. What gave papers by Stephen Kuffler a special quality were the clarity, the beautiful figures, and the underlying excitement. Moreover, he himself had done *every* experiment he described. Stephen Kuffler's work exemplified and introduced a multidisciplinary approach to the study of the nervous system. At Harvard he created the first department of neurobiology, in which he brought together people from different disciplines who developed new ways of thinking. Those who knew him remember a unique combination of tolerance, firmness, kindness, and good sense with enduring humor. He was the J. F. Enders University professor at Harvard, and was associated with the Marine Biological Laboratory at Woods Hole. Among his many honors was his election as a foreign member of the Royal Society.

PREFACE TO THE FOURTH EDITION

The aims of this new edition of *From Neuron to Brain* remain similar to those of the first edition, which was written 25 years ago. In the preface to that book the declared purpose was "...to describe how nerve cells go about their business of transmitting signals, how the signals are put together, and how out of this integration higher functions emerge. This book is directed to the reader who is curious about the workings of the nervous system but does not necessarily have a specialized background in biological sciences. We illustrate the main points by selected examples...".

Our style of writing has remained the same. In spite of the wealth of new facts, techniques, and concepts that have necessitated complete revision of the text and figures, our emphasis is on experiments, and on the way they are carried out. As before, references are given throughout the text to back up the statements that we make, and Appendices are provided at the end that we hope will help readers unfamiliar with the nervous system to deal with essential but somewhat dry facts and definitions.

A major task has been to keep the book up-to-date and manageable in size while retaining a narrative approach, in which we follow a line from the original inception of a new idea to an account of research being done today. As classical material becomes absorbed into common knowledge, its priority becomes lower, and inevitably we have had to change the emphasis in many chapters and delete descriptions of certain elegant but older experiments. Largely in response to comments from our readers, elements of the format and presentation have been changed for this edition. In particular, we have shortened the lengths of paragraphs, introduced more headings, and clarified the illustrations, partly by the introduction of full color and partly by a deliberate attempt to avoid overcrowding.

At the time that the first edition appeared, there were few other books on neurobiology. The picture is very different now with the availability of excellent, comprehensive texts, overviews, and monographs. We hope that this book will continue to present a readable and coherent account of how cellular and molecular approaches can provide insights into the workings of the brain. Rather than a description of all that is now known, we have tried to convey the excitement, beauty, and elegance of studies that use the whole range of techniques now at hand for progressing from gene expression and molecular events in neurons to the highest functions of the brain.

It is still appropriate to close by recalling the memory of Stephen Kuffler, who was an author of the first edition. The preface to the Second Edition of *From Neuron to Brain* ended thus: "The pleasure and satisfaction that we might hope to feel in re-creating a book that has seemed to fill a need has been diminished by the death of our friend and colleague, Steve Kuffler. We have tried to produce a book he would not have minded keeping his name on."

John G. Nicholls   Bruce G. Wallace

A. Robert Martin   Paul A. Fuchs

October, 2000

Acknowledgments

We are grateful to the numerous colleagues who have encouraged us and influenced our thinking. We particularly thank Drs. Ken Muller and Randall House who read all of the chapters. For reading individual chapters we thank Drs. W. Adams, H. Arechiga, J. Eugenin, J. Fernandez, F. Fernandez de Miguel, M. Lepre, J. Luque, A. Ribera, T. Shallice, and R. Wehner.

We wish to thank our colleagues who provided original illustrations from published and unpublished work that are acknowledged in the figure legends. We are particularly grateful for discussion with and for original plates provided by Drs. W. Almers, H. Arechiga, W. Betz, A. Brown, J. Black, R. Boch, F. Bonhoeffer, T. Bonhoeffer, A. M. Butt, W. Chen, M. Cohen, R. Costanzo, A. M. Craig, M. Crair, N. Davis, J. Driver, D. Furness, U. Garcia, A. W. Gehring, C. Gilbert, N. Hadjikhani, J. Hildebrand, V. Honrubia, J. Horton, M. Hubener, A. Kaneko, N. Kanwisher, S. Kierstead, J. Kinnamon, C.-P. Ko, S. R. Levinson, J. Lichtman, R. Llinás, U. J. McMahan, R. Michaels, K. Muller, E. Newman, M. M. Poo, M. Rijntjes, S. Roper, M. Ruegg, P. Sargent, E. Shooter, P. Shrager, P. Sterling, M. Stryker, M. Tessier-Lavigne, W. Thompson, R. Tootell, Y. Tsukamoto, K. Uĝurbil, D. Vaney, R. Vassar, S. Waxman, R. Wehner, A. Young, and H. Young.

We also thank the editors of the *Journal of Physiology*, the *Journal of Neurophysiology*, the *Journal of Neuroscience*, and *Neuron*, from which many of the illustrations were adapted.

JGN wishes to express his appreciation and thanks to Ms. I. Wittker for her secretarial help in the early phases of the book.

As in the previous editions, we feel extremely fortunate to have Andy Sinauer as our publisher, adviser, and friend; he saw the book through from its inception to the finish and was always available for dealing with any problems and for advice. His skill, unfailing taste, tact, and insight have made the collaboration a pleasure for us. We are also grateful to Kerry Falvey who, with great patience and courtesy, edited our text and weeded out inconsistencies. We also thank Christopher Small, Production Manager, and Jefferson Johnson and Joan Gemme, who paged the book. We owe a great debt to Imagineering Scientific and Technical Artworks, who translated our crude drawings and graphs into the elegant and consistently pleasing art that adds an essential dimension to the book.

From the Preface to the First Edition

Our aim is to describe how nerve cells go about their business of transmitting signals, how these signals are put together, and how out of this integration higher functions emerge. This book is directed to the reader who is curious about the workings of the nervous system but does not necessarily have a specialized background in biological sciences. We illustrate the main points by selected examples, preferably from work in which we have first-hand experience. This approach introduces an obvious personal bias and certain omissions.

We do not attempt a comprehensive treatment of the nervous system, complete with references and background material. Rather, we prefer a personal and therefore restricted point of view, presenting some of the advances of the past few decades by following the thread of development as it has unraveled in the hands of a relatively small number of workers. A survey of the table of contents reveals that many essential and fascinating fields have been left out: subjects like the cerebellum, the auditory system, eye movements, motor systems, and the corpus callosum, to name a few. Our only excuse is that it seems preferable to provide a coherent picture by selecting a few related topics to illustrate the usefulness of a cellular approach.

Throughout, we describe experiments on single cells or analyses of simple assemblies of neurons in a wide range of species. In several instances the analysis has now reached the molecular level, an advance that enables one to discuss some of the functional properties of nerve and muscle membranes in terms of specific molecules. Fortunately, in the brains of all animals that have been studied there is apparent a uniformity of principles for neurological signaling. Therefore, with luck, examples from a lobster or a leech will have relevance for our own nervous systems. As physiologists we must pursue that luck, because we are convinced that behind each problem that appears extraordinarily complex and insoluble there lies a simplifying principle that will lead to an unraveling of the events. For example, the human brain consists of over 10,000 million cells and many more connections that in their detail appear to defy comprehension. Such complexity is at times mistaken for randomness; yet this is not so, and we can show that the brain is constructed according to a highly ordered design, made up of relatively simple components. To perform all its functions it uses only a few signals and a stereotyped repeating pattern of activity. Therefore, a relatively small sampling of nerve cells can sometimes reveal much of the plan of the organization of connections, as in the visual system.

We also discuss "open-ended business," areas that are developing and whose direction is therefore uncertain. As one might expect, the topics cannot at present be fitted into a neat scheme. We hope, however, that they convey some of the flavor that makes research a series of adventures.

From Neuron to Brain expresses our approach as well as our aims. We work mostly on the machinery that enables neurons to function. Students who become interested in the nervous system almost always tell us that their curiosity stems from a desire to understand perception, consciousness, behavior, or other higher functions of the brain. Knowing of our preoccupation with the workings of isolated nerve cells or simple cell systems, they are frequently surprised that we ourselves started with similar motivations, and they are even more surprised that we have retained those interests. In fact, we believe we are working toward that goal (and in that respect probably do not differ from most of our colleagues and predecessors). Our book aims to substantiate this claim and, we hope, to show that we are pointed in the right direction.

Stephen W. Kuffler   John G. Nicholls
August, 1975

About the Authors

John G. Nicholls A. Robert Martin Bruce G. Wallace Paul A. Fuchs

John G. Nicholls

is Professor of Biophysics at the International School for Advanced Studies in Trieste. He was born in London in 1929 and received a medical degree from Charing Cross Hospital and a PhD in physiology from the Department of Biophysics at University College London, where he did research under the direction of Sir Bernard Katz. He has worked at University College London, at Oxford, Harvard, Yale and Stanford Universities and at the Biocenter in Basel. With Stephen Kuffler, he made experiments on neuroglial cells and wrote the first edition of this book. He is a Fellow of the Royal Society, a member of the Mexican Academy of Medicine, and the recipient of the Venezuelan Order of Andres Bello. He has given laboratory and lecture courses in neurobiology at Woods Hole and Cold Spring Harbor, and in universities in Argentina, Australia, Brazil, Chile, China, India, Israel, Malaysia, Mexico, Nigeria, the Philippines, Sri Lanka, Uruguay and Venezuela. His work concerns regeneration of the nervous system after injury, which he studied first in an invertebrate, the leech, and recently in immature mammalian spinal cord.

A. Robert Martin

is Professor Emeritus in the Department of Physiology at the University of Colorado School of Medicine. He was born in Saskatchewan in 1928 and majored in mathematics and physics at the University of Manitoba. He received a Ph.D. in Biophysics in 1955 from University College, London, where he worked on synaptic transmission in mammalian muscle under the direction of Sir Bernard Katz. From 1955 to 1957 he did postdoctoral research in the laboratory of Herbert Jasper at the Montreal Neurological Institute, studying the behavior of single cells in the motor cortex. He has taught at McGill University, the University of Utah, Yale University, and the University of Colorado Medical School, and has been a visiting professor at Monash University, Edinburgh University, and the Australian National University. His research has contributed to the understanding of synaptic transmission, including the mechanisms of transmitter release, electrical coupling at synapses, and properties of postsynaptic ion channels.

Bruce G. Wallace

is Professor of Physiology and Biophysics at the University of Colorado School of Medicine. He was born in Plainfield, New Jersey in 1947 and majored in biophysics at Amherst College. He received a Ph.D. in neurobiology in 1974 from Harvard University, where he worked with Edward Kravitz on transmitter biochemistry. From 1974 to 1977 he did postdoctoral research at Stanford University with John Nicholls, studying the function and regeneration of synapses in the leech nervous system. He has taught at Stanford and the University of Colorado Medical Schools. He is recognized for his research on the molecular mechanisms of synapse formation, including studies done in collaboration with U. J. McMahan that led to the identification of agrin and its role in regulating the differentiation of postsynaptic specializations.

Paul A. Fuchs

is Professor of Otolaryngology-Head and Neck Surgery, Professor of Biomedical Engineering, and Professor of Neuroscience in the Center for Hearing Sciences at The Johns Hopkins University School of Medicine. He was born in St. Louis, Missouri in 1951 and majored in biology at Reed College. He received a Ph.D. in Neuro- and Biobehavioral Sciences in 1979 from Stanford University where he investigated presynaptic inhibition at the crayfish neuromuscular junction under the direction of Donald Kennedy and Peter Getting. From 1979 to 1981 he did postdoctoral research with John Nicholls at Stanford University, examining synapse formation by leech neurons. From 1981 to 1983 he studied the efferent inhibition of cochlear hair cells with Robert Fettiplace and Andrew Crawford at Cambridge University. He has taught at the University of Colorado and The Johns Hopkins University Medical Schools. His research has focused on the role of voltage- and ligand-gated ion channels in cochlear hair cell function. This work has revealed details of functional differentiation, and identified a mechanism of cholinergic inhibition in hair cells.

Brief Table of Contents

Table of Contents

PART 1 Introduction 1

PART 2 Signaling in the Nervous System 23

PART 4 DEVELOPMENT OF THE NERVOUS SYSTEM 477

CHAPTER 25. CRITICAL PERIODS IN VISUAL AND AUDITORY SYSTEMS 549

PART 5 CONCLUSION 573

CHAPTER 26. OPEN QUESTIONS 575

APPENDIX A. CURRENT FLOW IN ELECTRICAL CIRCUITS A–1

APPENDIX B. METABOLIC PATHWAYS FOR THE SYNTHESIS AND INACTIVATION OF LOW-MOLECULAR-WEIGHT TRANSMITTERS B–1

APPENDIX C. STRUCTURES AND PATHWAYS OF THE BRAIN C–1

GLOSSARY G–1

BIBLIOGRAPHY BB–1

INDEX I–1

PART 1 INTRODUCTION

1 Principles of Signaling and Organization

THIS INTRODUCTION PROVIDES A FRAMEWORK for approaching chapters that deal with signaling, development, and functions of the nervous system in detail. Readers who are curious about the brain but unfamiliar with neurobiology often face certain difficulties in coming to grips with the subject. For example, the terminology of neurobiology, derived from anatomy, electricity, biochemistry, and molecular biology, is disconcerting. But because of the elaborate structure of the nervous system and the specialized features of neural signaling, it is unavoidable.

To describe signals in nerve cells (neurons) and to correlate such signals with our perception of the outside world, we have chosen the retina. The orderly structure of the retina, where the initial steps that lead to vision occur, makes it possible to follow signals literally from neuron to brain.

Information is transmitted in nerve cells by electrical signals. An essential task is to decode the content of the information they transmit. This decoding depends on where nerve fibers arise and where they go. Signals in the optic nerve carry visual information from the retina; similar impulses in a sensory nerve in the fingertip convey information about touch. Within the brain, individual nerve cells receive inputs from thousands of fibers. By integrating this information the cell creates a new message. This message can convey a complex meaning, such as the presence of a vertical bar of light in one's field of vision, or the movement of a tactile stimulus along one's finger.

One simplifying feature in the retina and throughout the nervous system is that nerve cells with similar properties are grouped together in layers or clusters. Another is that the brain uses stereotyped electrical signals to process information. The signals consist of changes in voltage produced by electrical currents flowing across cell membranes. Neurons use only two types of electrical signals: localized graded potentials, which spread over short distances, and action potentials, which are conducted rapidly over long distances.

Neurons transmit information to their targets by releasing chemical transmitter molecules at specialized junctions (known as synapses). The transmitter reacts with specific chemoreceptor molecules in the membrane of the target cell. This interaction gives rise to a localized graded potential that excites or inhibits the cell depending on the transmitter and the receptors involved. The efficacy of synaptic transmission is modified by use, hormones, and drugs.

During development, neurons depend on molecular signals derived from other cells. These signals determine the shape and position of the neuron, its survival, its transmitter, and the targets to which it connects. Once mature, most nerve cells cannot divide. Molecules in the environment of a neuron influence its capacity for repair after injury.

The central nervous system is an unresting assembly of cells that continually receives information, analyzes it, perceives it, and makes decisions. The brain can also take the initiative and produce coordinated, effective muscle contractions for walking, swallowing, or singing. To carry out the many aspects of behavior and to control directly or indirectly the whole of the body, the nervous system possesses an immense number of lines of communication provided by the nerve cells (**neurons**). Neurons are the fundamental units, or building blocks, of the brain. Our task is to learn the meaning behind their signaling.

One aim of this book is exemplified by its title: Throughout the following chapters we attempt to explain behavior and complex functions of the brain in terms of the underlying activity of nerve cells. A second aim is to understand the cellular and molecular mechanisms by which those signals arise. A third aim concerns the way in which structures and connections that subserve functions are established during development, become modified by experience, and may repair themselves after injury. In this chapter we summarize key concepts and essential background material.

Signaling in Simple Neuronal Circuits

Events that occur during the performance of simple reflexes can be followed and analyzed in detail. For example, when the tendon below the knee is tapped by a small hammer (the doctor does this to see whether you have syphilis and to play for time), thigh muscles are stretched and electrical impulses spread through sensory nerve fibers to the spinal cord, where motor cells are excited, producing impulses and contractions in muscles. The end result is that the leg extends and the knee is straightened. Such simplified circuits are essential for regulating muscular contractions that control movement of a limb. In simple reflex behavior like this, in which a stimulus leads to a defined output, the roles played by the signals in just two types of cells, sensory and motor, can be readily analyzed.

Complex Neuronal Circuitry in Relation to Higher Functions

Analyzing signaling in complex pathways that involve literally millions of neurons is far more difficult than analyzing simple reflexes. The transmission of information to the brain for perception of a sound, a touch, an odor, or a visual image requires the sequential activation of neuron after neuron, as does the execution of a simple voluntary movement. Serious problems in the analysis of signaling and circuitry arise from the dense packing of nerve cells, the intricacy of their connections, and the profusion of cell types. The brain is unlike an organ such as the liver, which consists of stereotyped populations of cells. If you have discovered how one region of the liver functions, you know a great deal about the liver as a whole. Knowledge of the cerebellum, however, gives you no idea about what the retina or any other part of the central nervous system does.

Despite the immense complexity of the nervous system, it is now possible to understand many aspects of the way in which neurons can act as the building blocks for perception. For example, by recording the activity of neurons in the pathway from the eye to the brain, we can follow signals first in cells that specifically respond to light and then onward, step by step through successive relays.

A remarkable feature of the way the visual system works is its ability to pick out contrasting patterns, colors, and movement over an immense range of light intensities. As you read this page, signaling within the eye itself ensures that the black letters stand out from the white page in either the dim light of a room or the brilliant sunshine on a beach (not perhaps the ideal place to concentrate on this book). Specific connections in the brain create a single image in the mind's eye, even though the two eyes are situated apart on the head and scan somewhat different regions of the outside world. Moreover, mechanisms exist to ensure that this image stays still (even though our eyes are continually making small movements) and to provide accurate information about the distance of the page so that you can turn it with your fingers.

How do the connections of nerve cells enable such phenomena to occur? Although we are still unable to provide complete explanations, much is now known about how these attributes of vision stem from well-defined neuronal circuits in the eye and in the initial relays within the brain. Numerous questions remain, however, about the relation between

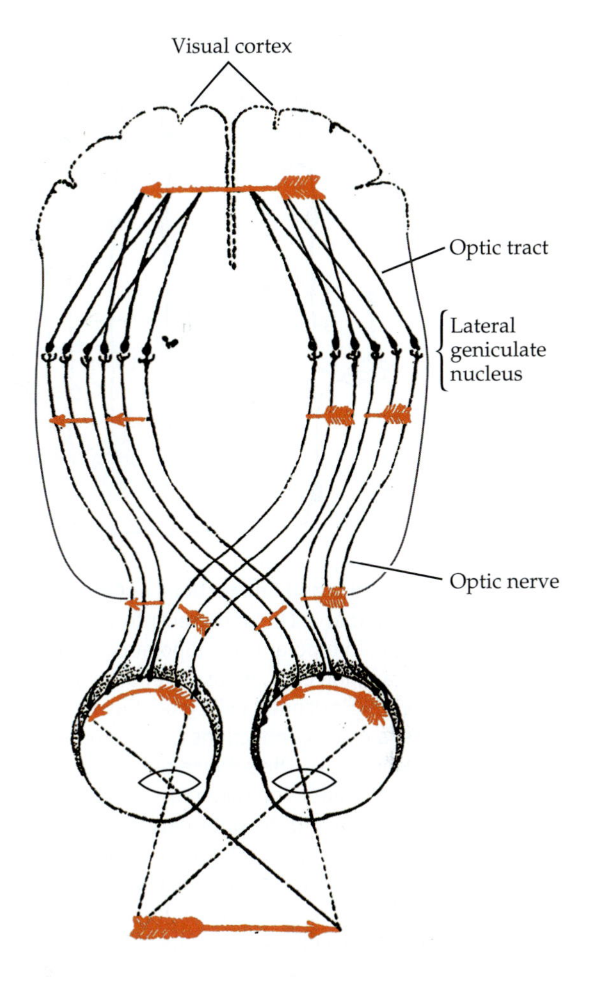

FIGURE 1.1 Pathways from the Eyes to the Brain through the optic nerve and the optic tract. The interposed relay is the lateral geniculate nucleus. Arrows indicate how images are reversed by the lens and how the specific crossing of axons causes the right visual field to be represented in the left brain, and vice versa. The figure has been modified from an original by Ramón y Cajal, which dates from 1892 ([1909–1911] Eng. trans. 1995).

neuronal properties and behavior. Thus, to read a page, you must maintain the posture of your body, head, and arms. The brain must further ensure that the eyeball remains moist through continuous secretion of tears, that breathing continues, and that untold other involuntary and unconscious bodily tasks are carried out. Such problems, which involve the seeing of a complete picture and the coordinated movements of the body as a whole, are still beyond the scope of this book.

The following discussion deals with questions of how neurons are organized and how electrical signals arise, propagate, and spread from nerve cell to nerve cell. The retina serves to illustrate general principles throughout the nervous system.

ORGANIZATION OF THE RETINA

Analysis of the visual world depends on the information coming from the retina, the initial stage of processing that sets the limits for our perception. Figure 1.1 shows pathways from the eye to higher centers in the brain. The image that falls on the retina is reversed but is otherwise a faithful representation of the external world. How can this picture be translated into our visual experience by way of electrical signals that are initiated in the retina and then travel through the optic nerves?

Shapes and Connections of Neurons

Figure 1.2 shows the various types of cells and their arrangement in the retina. Light entering the eye passes through the layers of transparent cells to reach the photoreceptors. The signals that leave the eye through the optic nerve fibers of ganglion cells provide the entire input for all of our vision.

Ramón y Cajal, about 1914

The drawing of Figure 1.2 was made by Santiago Ramón y Cajal[1] before the turn of the twentieth century. Ramón y Cajal was one of the greatest students of the nervous system, selecting examples from a wide range of animals with an almost unfailing instinct for the essential. An important generalization is that the shape and position of a neuron, as well as the origin and destination of its signals in the neural network, supply valuable clues to its function.

In Figure 1.2 it is apparent that the cells in the retina, like those elsewhere in the central nervous system (CNS), are densely packed. Early anatomists had to tease nervous tissue apart to see individual cells. Staining methods that impregnate all neurons are virtually useless for investigating cell shapes and connections because a structure like the retina appears as a dark blur of intertwined cells and processes. The electron micrograph of Figure 1.3 shows that the extracellular space surrounding neurons and their supporting cells is restricted to clefts only about 25 nanometers wide. Most of Ramón y Cajal's pictures were made with the Golgi staining method, which by an unknown mechanism stains just a few neurons at random out of the whole population, yet those few cells are stained in their entirety.

The schematic presentation in Figure 1.2 gives an idea of the orderly arrangement of the retinal neurons. It is possible to distinguish the photoreceptors, bipolar cells, and ganglion cells. The lines of transmission are from input to output, from photoreceptors through to ganglion cells. In addition, two other types of cells, the horizontal and amacrine cells, make connections linking the pathways.

[1]Ramón y Cajal, S. [1909–1911] 1995. *Histology of the Nervous System*, 2 vols. Translated by Neely Swanson and Larry Swanson. Oxford University Press, New York.

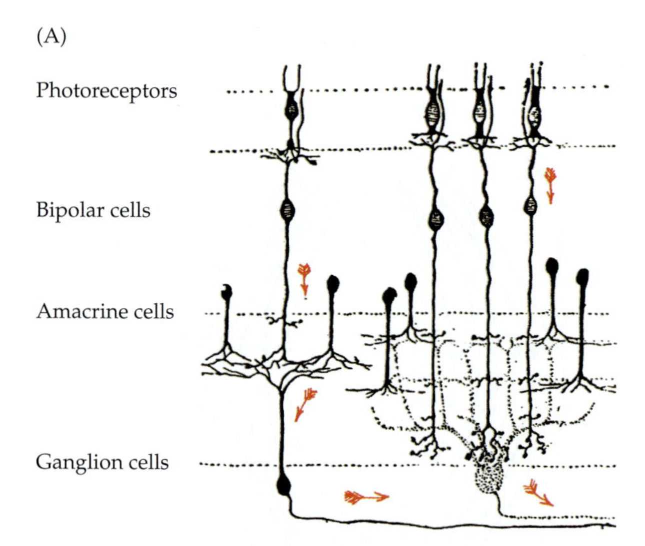

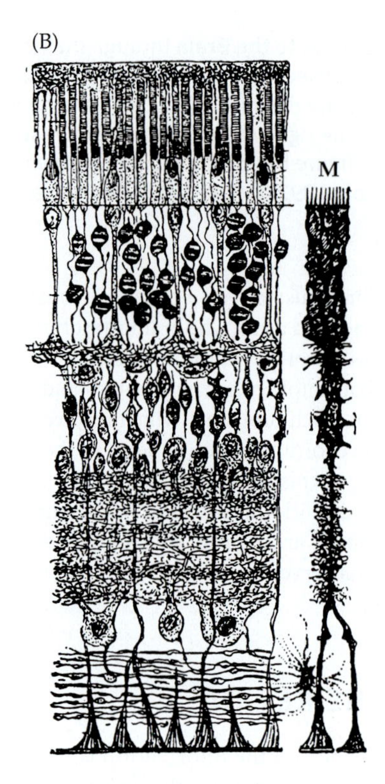

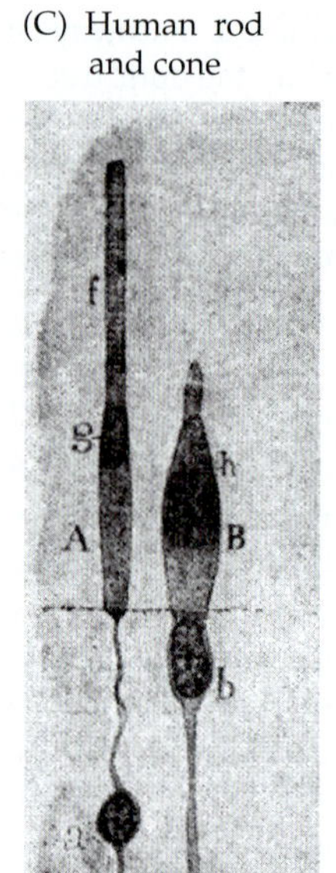

FIGURE 1.2 Structure and Connections of Cells in the Mammalian Retina. The photoreceptors (rods and cones) connect to bipolar cells. Bipolar cells in turn connect to ganglion cells, whose axons constitute the optic nerve. Horizontal cells (not shown) and amacrine cells make connections that are predominantly horizontal. (A) The scheme proposed by Ramón y Cajal for the direction taken by signals as they pass from receptors to the optic nerve fibers. This scheme still holds in general, but essential new pathways and feedback groups have been discovered since Ramón y Cajal's time. (B) Ramón y Cajal's depiction of the cellular elements of the retina and their orderly arrangement. The Müller cell (M) shown on the right is a satellite glial cell. (C) Drawings of a human rod (left) and cone (right) isolated from the retina. Light passes through the retina (in these drawings from bottom to top) to be absorbed by the outer segment (top) of the photoreceptor. There it produces a signal that spreads to the terminal to influence the next cell in line. By recording electrically from each cell in the retinal circuit, we can follow signals step by step and understand how the meaning of the signals changes. (After Ramón y Cajal, 1995.)

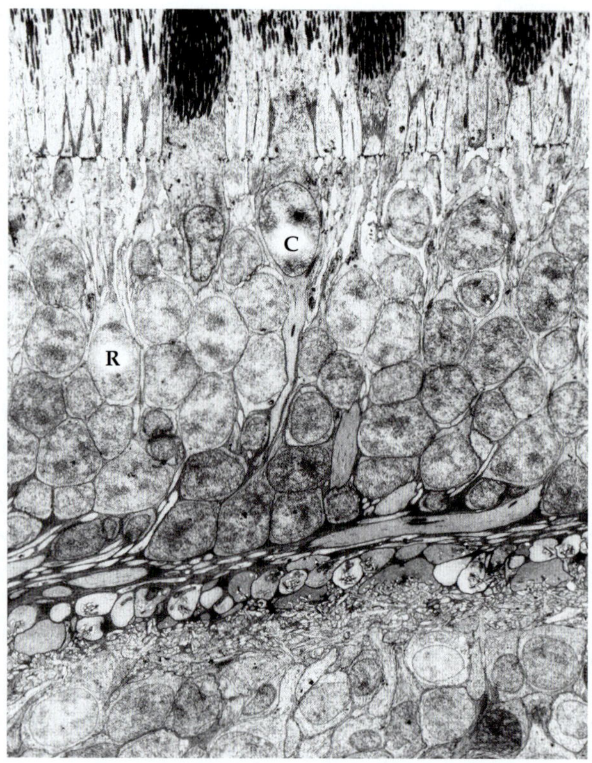

FIGURE 1.3 Dense Packing of Neurons in Monkey (Macaque) Retina. This electron micrograph shows a characteristic feature of the central nervous system: The cell membranes are separated by narrow, fluid-filled clefts. The photoreceptors and their processes can be followed to the outer plexiform layer where their terminals contact bipolar and horizontal cells. One cone (C) and one rod (R) are labeled. (Micrograph kindly provided by P. Sterling and Y. Tsukamoto.)

One idealized goal of neurobiology, implicit in Ramón y Cajal's drawings, is to understand how each cell contributes to the picture of the world we see.

Cell Body, Dendrites, and Axons

The ganglion cell shown in Figure 1.4 illustrates features shared by neurons throughout the central and peripheral nervous systems. The **cell body** contains the nucleus and other intracellular organelles common to nonneuronal as well as neuronal cells. The long process that leaves the cell body to form connections with target cells is known as the **axon**. The term **dendrite** applies to branches upon which incoming fibers make connections and that act as receiving stations for excitation or inhibition. In addition to the ganglion cell, Figure 1.4 shows other representative neurons. The terms for describing neuronal structures, particularly dendrites, are somewhat ambiguous, but they are still convenient and widely used.

Not all neurons conform to the simple plan of the cells shown in Figure 1.4. Certain nerve cells do not have axons; others have axons onto which incoming connections are made. Still others have dendrites that can conduct impulses and form connections with target cells. While ganglion cells conform to the caricature of a stereotyped neuron with dendrites, a cell body, and an axon, others in the retina do not. For example, photoreceptors (see Figure 1.2C) have no obvious dendrites. Activity in photoreceptors does not arise through input from another neuron but from an external stimulus, illumination. In the retina another exception to the usual picture is the absence of axons in photoreceptors.

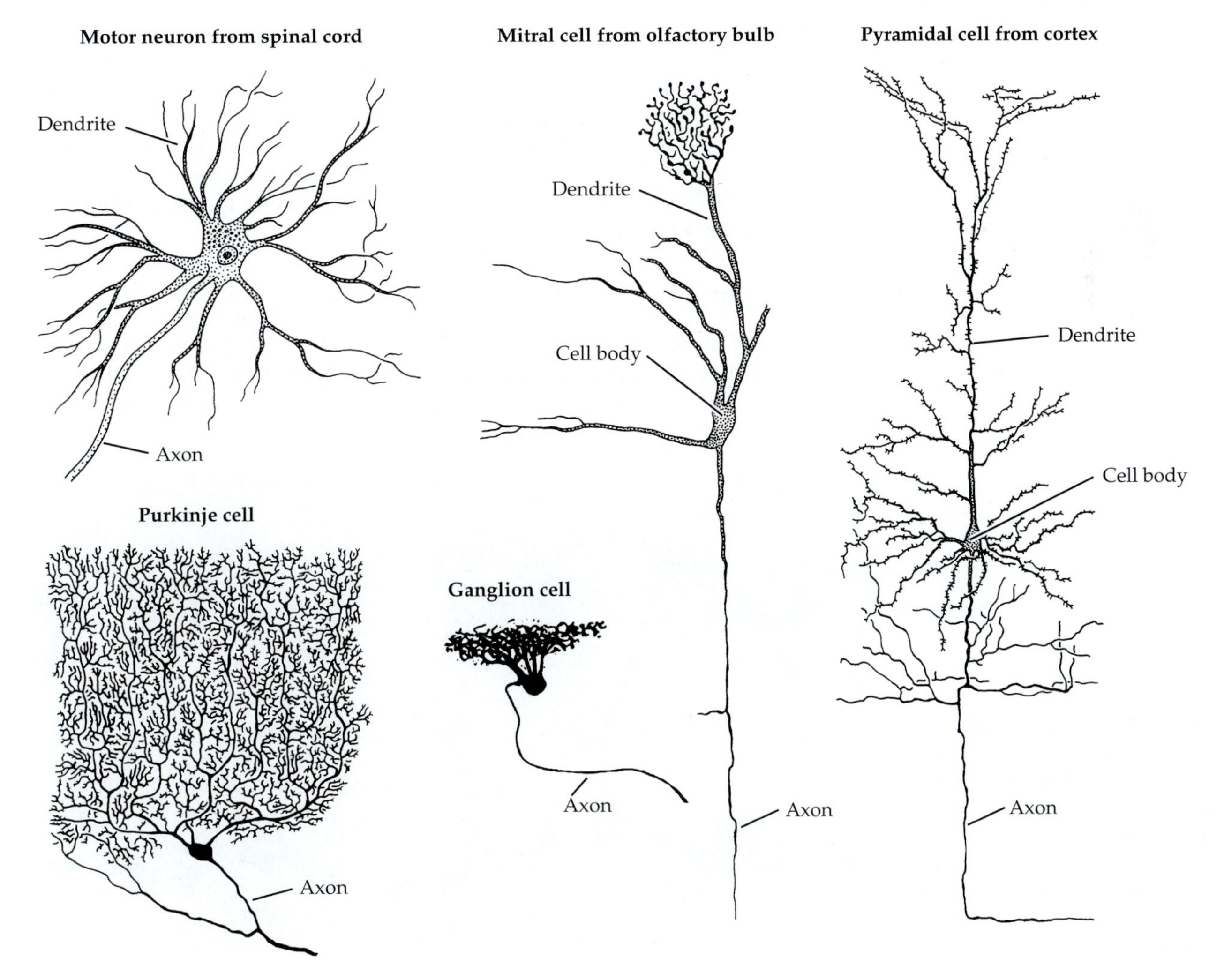

FIGURE 1.4 Shapes and Sizes of Neurons. Neurons have branches (the dendrites) on which other neurons form synapses, and axons that in turn make connections with other neurons. The motor neuron, drawn by Deiters in 1869, was dissected from a mammalian spinal cord. The other cells, stained by the Golgi method, were drawn by Ramón y Cajal. The pyramidal cell is from the cortex of a mouse, the mitral cell from the olfactory bulb (a relay station in the pathway concerned with smell) of a rat, the Purkinje cell from human cerebellum, and the ganglion cell from mammalian retina (animal not specified). (After Ramón y Cajal, 1995.)

Techniques for Identifying Neurons and Tracing Their Connections

Although the Golgi technique is still widely used in spite of its antiquity, many newer techniques have facilitated the functional identification of neurons and synaptic connections. Molecules that label a neuron in its entirety can be injected through a fine pipette that records electrical signals at the same time. Fluorescent markers such as Lucifer yellow are visible as they spread through the fine processes of a living cell. Alternatively, markers such as the enzyme horseradish peroxidase (HRP) or biocytin can be injected; after fixation they can be made to give rise to a dense product or a bright fluorescence. Neurons can also be stained by horseradish peroxidase that is applied extracellularly; the enzyme is taken up and transported to the cell body. Fluorescent carbocyanine dyes placed close to neurons dissolve in cell membranes and diffuse over the entire surface. These procedures are valuable for tracing the origins and destinations of axons from one region of the nervous system to another.

Antibodies have been made to characterize specific neurons, dendrites, axons, and synapses by selectively labeling intracellular or membrane components. Figure 1.5 shows a specific group of bipolar cells labeled by an antibody to the enzyme phosphokinase C. Antibody techniques also provide valuable tools for following the migration and differentiation of nerve cells during development. A complementary approach for characterizing neurons is by in situ hybridization: Specific tagged probes label the messenger RNA of a neuron that codes for a channel, receptor, enzyme, or structural element.

Nonneuronal Cells

The distinctive cell labeled M in Figure 1.2B represents of a class of nonneuronal cells present in the retina. Such cells are known as **glial cells**. Unlike neurons, they do not have axons or dendrites and are not directly connected to nerve cells. Glial cells are abundant throughout the nervous system. They play a number of roles in relation to neuronal signaling. For example, the axons of ganglion cells that run in the optic nerve conduct impulses rapidly because they are surrounded by an insulating lipid sheath called **myelin**. Myelin is formed by glial cells that wrap themselves around axons during development. Retinal glial cells are known as Müller cells.

Grouping of Cells According to Function

A remarkable feature of the retina is the distribution of cells according to function (see Figure 1.2). The photoreceptors and the horizontal, bipolar, amacrine, and ganglion cells all have their cell bodies and synapses situated in well-defined layers. Such layering is

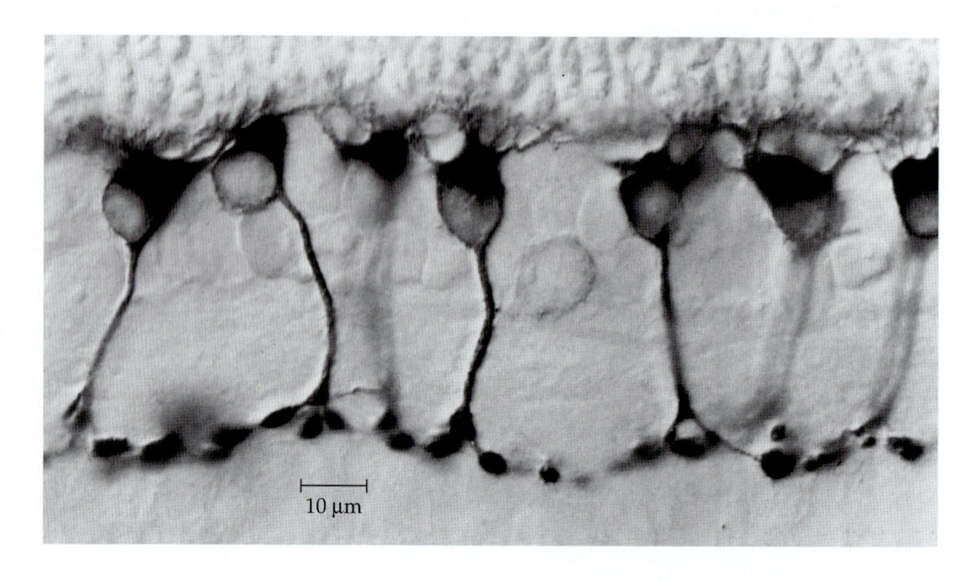

FIGURE 1.5 Population of Rod Bipolar Cells Stained by an Antibody against the enzyme phosphokinase C. Only bipolar cells that contain the enzyme are stained. Above are photoreceptors; below are ganglion cells. (Photograph kindly provided by H. M. Young and D. I. Vaney, University of Queensland.)

found throughout the brain. For example, the structure in which optic nerve fibers end (the lateral geniculate nucleus) consists of six layers of cells that can readily be distinguished by the naked eye. In many regions of the nervous system, cells with similar functions are clustered in obvious circumscribed groups known as **nuclei** (not to be confused with the cell nucleus) or **ganglia** (not to be confused with ganglion cells of the retina).

Subtypes of Cells in Relation to Function

The simplified description of Figure 1.2 omits certain features of the organization of retinal cells. There are many distinctive types of ganglion cells, horizontal cells, bipolar cells, and amacrine cells, each with characteristic morphology, transmitters, and physiological properties. For example, the photoreceptors fall into two easily recognizable classes—the rods and the cones—which perform different functions. The elongated rods are extremely sensitive to small changes in illumination. As you read this page, the ambient light is too bright for the rods, which function only in dim light after a prolonged period in darkness. The cones respond to visual stimuli in bright ambient light. Moreover, the cones are further subdivided into red-, green-, or blue-sensitive photoreceptors. The amacrine cells provide an extreme example of cell type diversity: More than 20 types can be recognized by structural and physiological criteria.

The retina thereby illustrates a profound problem of contemporary neurobiology. It is not known why so many types of amacrine cells are needed or what distinctive function each performs. It is a sobering reflection that throughout the central, peripheral, and enteric (gut) nervous systems, the function of the vast majority of nerve cells is not known. At the same time, this ignorance serves to point out that many basic concepts remain to be discovered, or even imagined.

Convergence and Divergence of Connections

The arrows in Figure 1.2A indicate a through line of transmission from receptors to ganglion cells. In reality, the picture is again far more complex. For example, there is a dramatic reduction in numbers from receptors to ganglion cells. Outputs from more than 100 million receptors converge to provide input to 1 million ganglion cells, the axons of which make up the optic nerve leading to higher centers. Many (but not all) ganglion cells therefore receive inputs from a large number of photoreceptors (**convergence**) by way of interposed cells. Similarly, the axon of a single ganglion cell branches extensively to supply many target cells (**divergence**) (see Figure 1.13).

In addition, contrary to the simple scheme of Figure 1.2A, arrows should also point sideways to indicate interactions among cells in the same layer (**lateral** connections) and even in the opposite direction—for example, backward from horizontal cells toward photoreceptors (**recurrent** connections). Such convergent, divergent, lateral, and recurrent connections are consistent features of pathways elsewhere in the nervous system. Thus, the simple step-by-step processing of signals is dramatically influenced by parallel and feedback interactions.

SIGNALING IN NERVE CELLS

To analyze events in the outside world or within our bodies, and to transmit information from cell to cell, nerve cells use electrical and chemical signals. The distances can be long, as from a toe to the spinal cord. Once again, the various signals are well represented in the retina. At the time that Ramón y Cajal drew the arrows in Figure 1.2A, almost no information was available about these signals, which makes his achievement all the more remarkable.

The progression can now be followed step by step as light falls on photoreceptors to generate electrical signals, which then influence bipolar cells. From bipolar cells, signals are conveyed to the ganglion cells, and from there onward to higher centers in the brain that give rise to our perception of the outside world. The following sections deal with the properties of these signals and the way in which they convey information.

Classes of Electrical Signals

The electrical signals generated by nerve cells fall into two main classes. First, **localized graded potentials** (see Figure 1.8) are generated by an extrinsic stimulus such as light falling on a photoreceptor in the eye, sound waves deforming a hair cell in the ear, or a touch stimulus pressing on a sensory nerve ending in the skin. Highly similar in their electrical characteristics, but with very different origins, are signals generated at synapses, the junctions between nerve cells and their targets (synapses will be discussed later in the chapter). As we shall see, all these signals are graded and localized to the site of origin and their spread depends on the passive properties of nerve cells.

Action potentials constitute the second major category (see Figure 1.9). Action potentials are initiated by localized graded potentials. Unlike local potentials, they propagate rapidly over long distances—for example, from the eye to higher centers along ganglion cell axons in the optic nerve, or from a motor cell in the spinal cord to a muscle in the leg. Again, unlike localized graded potentials, action potentials occurring in a neuron are fixed in amplitude and duration, like the dots in Morse code. It is important to realize that these action potentials traveling along the optic nerve fibers are not epiphenomena. They are the only signals that provide the brain with information about the visual world.

Signal transmission through the retina can be represented by the following simplified scheme:

Light → localized graded signal in photoreceptor → localized graded signal in bipolar cell → localized graded potential in ganglion cell → action potential in ganglion cell → conduction to higher centers

Universality of Electrical Signals

An important feature of electrical signals is that they are virtually identical in all nerve cells of the body, whether they carry commands for movement, transmit messages about colors, shapes, or painful stimuli, or interconnect various portions of the brain. A second important feature of signals is that they are so similar in different animals that even a sophisticated investigator (like one of the authors of this book) is unable to tell with certainty whether a photographic record of an action potential is derived from the nerve fiber of whale, mouse, monkey, worm, tarantula, or professor. In this sense action potentials can be considered to be stereotyped units. They are the universal coins for the exchange of information in all nervous systems that have been investigated. In the brain the great number of cells (probably 10^{10} to 10^{12} neurons) and the diversity of connections, rather than the types of signals, are what account for the complexity of the tasks that can be undertaken.

This idea was expressed in 1868 by the German physicist-biologist Hermann von Helmholtz. Starting from first principles, long before the facts as we know them were available, Helmholtz reasoned as follows[2]:

> The nerve fibers have often been compared with telegraphic wires traversing a country, and the comparison is well fitted to illustrate the striking and important peculiarity of their mode of action. In the network of telegraphs we find everywhere the same copper or iron wires carrying the same kind of movement, a stream of electricity, but producing the most different results in the various stations according to the auxiliary apparatus with which they are connected. At one station the effect is the ringing of a bell, at another a signal is moved, at a third a recording instrument is set to work.... In short, every one of the... different actions which electricity is capable of producing may be called forth by a telegraphic wire laid to whatever spot we please, and it is always the same process in the wire itself which leads to these diverse consequences.... All the difference which is seen in the excitation of different nerves depends only upon the difference of the organs to which the nerve is united and to which it transmits the state of excitation.

In fact, as will be shown in Chapter 6, minor differences in amplitude and time course are apparent in action potentials of different neurons. The statement that they are all the same is akin to saying that all oak trees are the same.

[2]Helmholtz, H. von. 1889. *Popular Scientific Lectures*. Longmans, London.

Techniques for Recording Signals from Neurons with Electrodes

For certain problems it is essential to record from a single neuron or even a single channel in a neuron; for others, an overview of activity in many neurons is required. The following description summarizes briefly key techniques for recording neuronal activity discussed throughout this book.

Recordings of action potentials were first made from peripheral nerves with **extracellular** electrodes. Currents flowing between a pair of silver wires stimulated nerve fibers while a second pair of wires at a distance recorded the action potentials. Within the central nervous system, recordings from a neuron or a group of neurons are made with an extracellular electrode, which consists of a single insulated wire or a glass capillary filled with salt solution (Figure 1.6A).

With an **intracellular** microelectrode we can directly measure the potential across the membrane (the potential difference between the inside and the outside of the cell), as well as excitation, inhibition, and impulse initiation. A glass microelectrode, which is filled with a salt solution and has a tip 0.1 μm in diameter or smaller, is inserted into the cell with the aid of a micromanipulator (Figure 1.6B). Microelectrodes are also used for passing electrical currents or for injecting molecules into the cytoplasm.

An alternative is to measure the membrane potential by a procedure known as **whole-cell patch recording** (Figure 1.6C). A larger pipette with a polished tip of approximately 1 μm is applied to the surface of the cell, where it fuses with the membrane to form a tight seal. After the membrane within the pipette tip has been ruptured, the fluid in the pipette makes contact with intracellular fluid.

Noninvasive Techniques for Recording Neuronal Activity

Using optical recording techniques, we can follow signaling in suitable preparations without using electrodes. Specially fabricated fluorescent dyes that bind to the cell membrane change their light absorbance during action potentials. Other noninvasive techniques, known as **positron emission tomography** (**PET**) and **magnetic resonance imaging** (**MRI**), allow us to determine which regions of the human, awake brain are active when stimuli are presented or movements are initiated. The MRI image in Figure 1.7 shows the location of activity moving from the eye to the cortex in response to a visual stimulus.

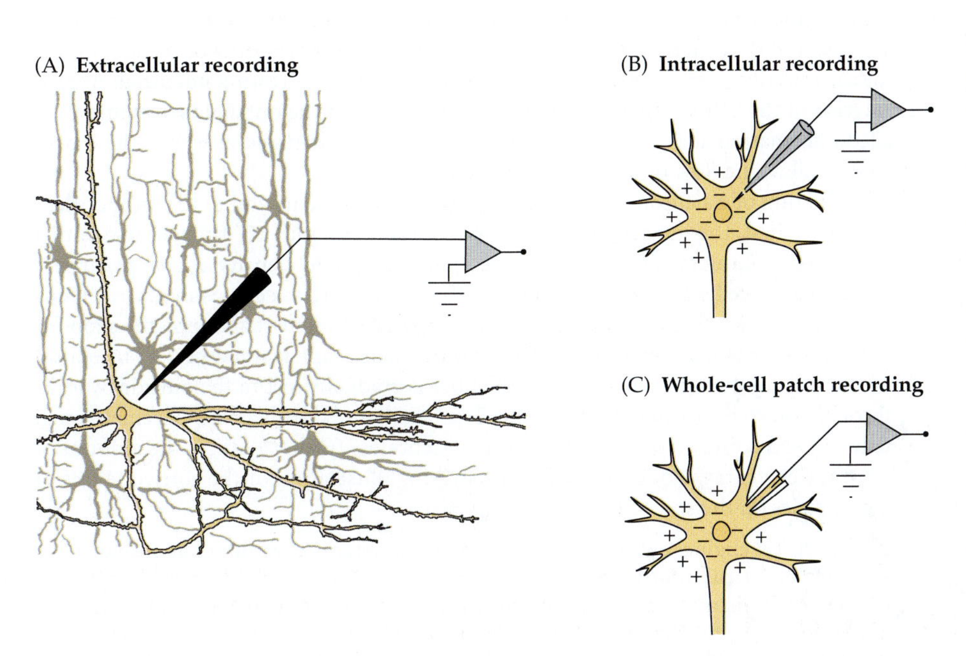

FIGURE 1.6 Electrical Recording Techniques. (A) The tip of a fine wire electrode is located close to a nerve cell in the cortex. (The wire above the tip is insulated.) Extracellular recording allows one to record from a single cell or from a group of cells. (B) Intracellular recordings are made with a fluid-filled glass capillary that has a tip of less than 0.1 μm in diameter, which is inserted into a neuron across the cell membrane. At rest, there is a potential difference of about 70 mV, the inside negative with respect to the outside. This difference is known as the resting potential. (C) Intracellular recordings are also made with patch electrodes. A patch electrode has a larger tip than that of an intracellular microelectrode; the tip makes an extremely tight seal with the cell membrane. If the membrane remains intact, the currents that flow as a single ion channel in the membrane opens or closes can be recorded. Alternatively, as shown here, the cell membrane can be ruptured to allow the diffusion of molecules between the pipette and the intracellular fluid of the cell (whole-cell patch clamp).

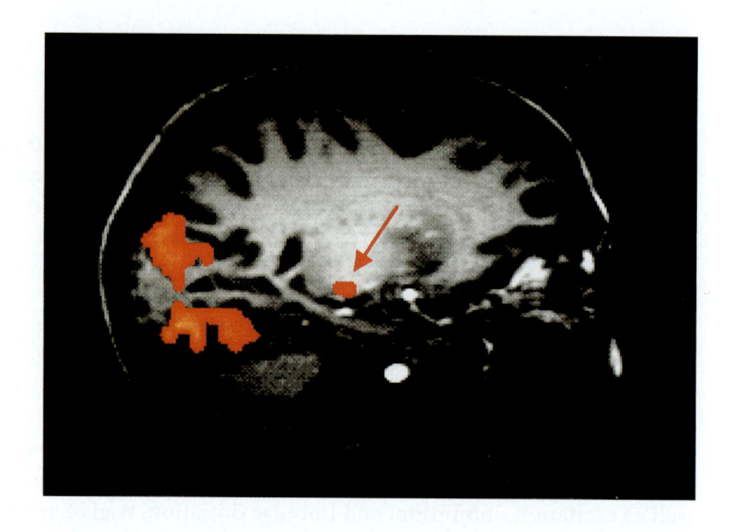

FIGURE 1.7 Magnetic Resonance Imaging of Living Brain. The subject was presented with visual stimuli that caused activity to be generated in the lateral geniculate nucleus (labeled with the red arrow) deep within the brain (see Figure 1.1 and Chapter 20) and the visual cortex. The color represents the level of activity, red corresponding to high and blue to low. (After Uğurbil et al., 1999; image kindly provided by K. Uğurbil.)

Overall measures of the averaged activity of the eye and brain are provided by the electroretinogram and electroencephalogram. They are mainly used to diagnose disorders of function.

Spread of Local Graded Potentials and Passive Electrical Properties of Neurons

Implicit in the wiring diagram of Ramón y Cajal (see Figure 1.2A) is the idea that changes in illumination of the retina influence the activity of the photoreceptors and eventually the fibers leaving the eye. For this influence to take effect, signals must spread not only from cell to cell, but along a cell, from one end to the other. How, for example, does the electrical signal generated at one end of the bipolar cell, in contact with the photoreceptor, spread along its length to reach the terminal that is close to the ganglion cell?

To answer this question, it is useful to consider the relevant structural components that carry the signals. A neuron such as a bipolar cell can be considered to be a long tube filled with a watery solution of salts (dissociated into positively and negatively charged ions) and proteins, separated from the extracellular solution by a membrane. The intracellular and extracellular solutions are of the same ionic strength but of different ionic composition. The cell membrane is relatively impermeable to ions on either side. Ions move through specific **ion channels** formed by proteins that span the membrane. Electrical and chemical stimuli cause the various channels for sodium, potassium, calcium, and chloride to open or close.

As a result of the differences in ionic concentrations on the two sides of the membrane and of the selective permeability of the channels, there exists a **resting potential:** At rest, the interior of the cell is negatively charged with respect to the outside (see Figure 1.6). Detailed information about the molecular structure of ion channels, as well as the way in which they allow the flow of ions, is presented in Chapter 2.

The structure of a neuron in the retina or elsewhere limits its ability to conduct electrical signals. First, the intracellular fluid, or axoplasm, is about 10^7 times worse than a metal wire as a conductor of electricity. One reason is that the density of charge carriers (ions) in the intracellular fluid is much lower than that of free electrons in a wire; in addition, the mobility of ions is lower. Second, the movement of currents along the axon for any great distance is hampered by the fact that the membrane is not a perfect insulator. Consequently, any current flowing along the fiber is gradually lost to the outside by leakage through ion channels in the membrane. The fact that nerve fibers are extremely small (usually not exceeding 20 μm in diameter in vertebrates) further reduces the amount of current they can carry. Hodgkin provided a striking illustration of the consequences of these properties on the spread of electrical signals[3]:

> If an electrical engineer were to look at the nervous system he would see at once that signaling electrical information along the nerve fibers is a formidable problem. In our nerves the diameter of the axis cylinder varies between about 0.1 μm and 20 μm. The inside of the fiber contains ions and is a reasonably good conductor of electricity. However, the fiber is so small that its longitudinal resistance is exceedingly high.

[3]Hodgkin, A. L. 1964. *The Conduction of the Nervous Impulse*. Liverpool University Press, Liverpool, England.

FIGURE 1.8 Localized Graded Potentials. Intracellular recordings are made from (A) a bipolar cell and (B) a ganglion cell with microelectrodes. (A) When light is absorbed by the photoreceptors, it gives rise to a signal that in turn produces a localized graded response in the bipolar cell. The resting potential across the membrane is reduced (the trace moves in an upward direction). This effect is known as a depolarization. The size of the signal in the bipolar cell depends on the intensity of illumination, hence the term "graded." The depolarization spreads to the far end of the bipolar cell passively. As it spreads, it becomes smaller in amplitude owing to the poor conducting properties of neurons. At the terminal of the bipolar cell, the depolarization causes the release of the chemical transmitter. (B) The transmitter produces a local graded potential in the ganglion cell. Because it is localized, the potential cannot spread for more than 1 mm (at most) along the axon. Whereas the bipolar cell is short enough for a local potential to spread to its endings, the ganglion cell has an axon several centimeters long. In these illustrations the local potentials were recorded from the cell bodies and were produced by transmitters acting on the dendrites. (A after Kaneko and Hashimoto, 1969; B after Baylor and Fettiplace, 1977.)

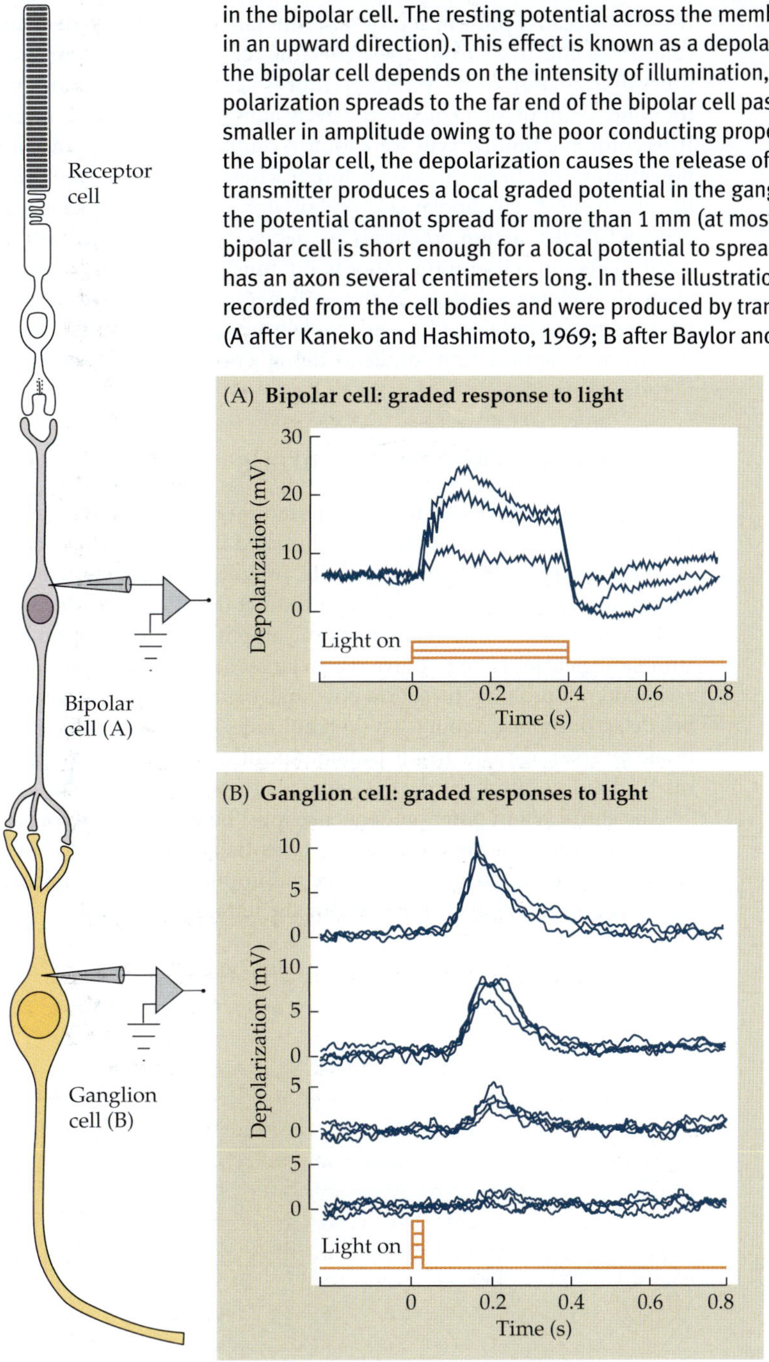

A simple calculation shows that in a 1 μm fiber containing axoplasm with a resistivity of 100 ohm-cm, the resistance per unit length is about 10^{10} ohm per cm. This means that the electrical resistance of a meter's length of small nerve fiber is about the same as that of 10^{10} miles of 22-gauge copper wire, the distance being roughly ten times that between the earth and the planet Saturn.

Passively conducted electrical signals, then, are severely attenuated, and limited to a short length of nerve fiber, 1 to 2 mm at most. In addition, when such a signal is brief, its time course may be severely distorted and its amplitude further attenuated by the electrical capacitance of the cell membrane. Nevertheless, localized potentials (Figure 1.8) are essential for the initiation and conduction of propagated signals.

Spread of Potential Changes in Bipolar Cells and Photoreceptors

It is only because photoreceptor and bipolar cells are so short that local graded signals can spread effectively from one end of the cell to the other. The electrical signal that signifies illumination of the photoreceptor is generated in the outer segment of the rod or the cone. From there it spreads passively along the cell to its terminal on the bipolar cell. If receptor and bipolar cells were longer, centimeters or even millimeters in length, local potentials would fizzle out long before reaching the terminal and would not be able to influence the next cell. Bipolar cells and photoreceptors thereby constitute exceptions to the general rule that action potentials are necessary to carry information along the length of a cell. Ganglion cells, on the other hand, must generate action potentials to send signals along their elongated axons in the optic nerve. The electrical recordings made from the neurons shown in Figure 1.8 were made from their cell bodies. The local potentials originate from synaptic actions on dendrites at a distance and spread passively to the recording site.

Properties of Action Potentials

One essential feature of the action potential is that it is a triggered, explosive, all-or-nothing event. An action potential is initiated in a ganglion cell by the signals impinging on it from bipolar and amacrine cells, provided their effects reach a critical level (the **threshold**). The action potential has a distinct threshold, and once initiated, its amplitude and duration are not determined by the amplitude and duration of the stimulus. Larger stimulating currents do not give rise to larger action potentials, and stimuli of longer duration do not prolong the action potential. Figure 1.9 shows that the action potential is a brief electrical pulse about 0.1 V in amplitude. At its peak, the potential across the membrane reverses sign (the inside becomes positive). The potential lasts for about 0.001 s (1 ms) and moves rapidly along the nerve fiber from one end to the other.

The entire action potential sequence must be completed before another action potential can be initiated. After each action potential, there is a period of enforced silence (the **refractory period**) during which a second impulse cannot be initiated. The frequency of repeated action potentials is limited by the refractory period.

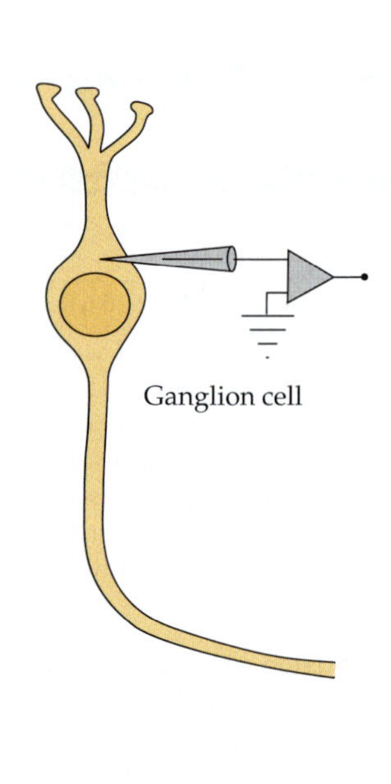

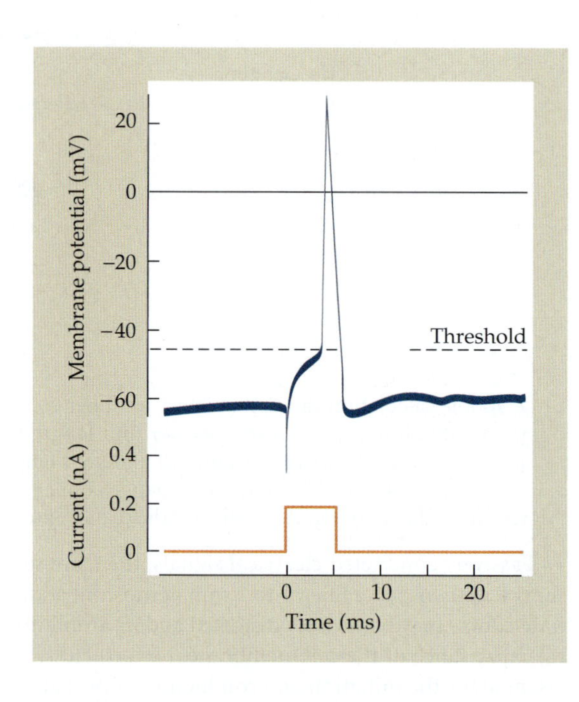

FIGURE 1.9 Action Potential recorded from a retinal ganglion cell with an intracellular microelectrode. When the stimulus, in this case current injected into the cell through the microelectrode, causes a depolarizing response that exceeds the threshold, the all-or-nothing action potential is initiated. During the action potential the inside of the neuron becomes positive. The action potential propagates along the axon of the ganglion cell to its terminal, where it causes transmitter to be released. (After Baylor and Fettiplace, 1977.)

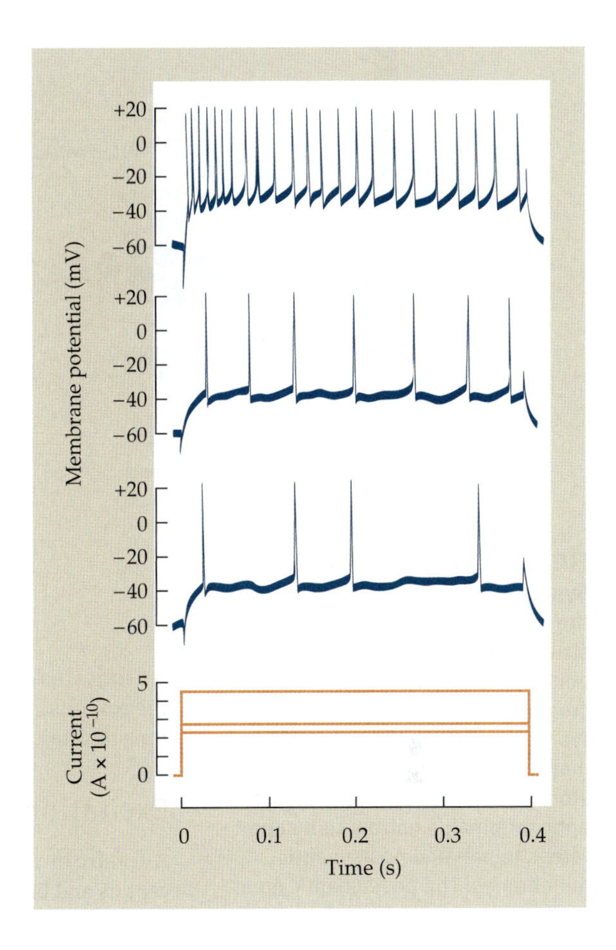

FIGURE 1.10 Frequency as a Signal of Intensity in a retinal ganglion cell. Depolarizing current passed through the microelectrode produces local potentials. Larger currents produce larger local potentials and higher frequencies of firing. (After Baylor and Fettiplace, 1979.)

Propagation of Action Potentials along Nerve Fibers

The impulse itself causes electrical currents to spread passively ahead of itself along the axon. Although the resulting local potential falls off steeply with distance, it nevertheless exceeds threshold. Hence, the action potential provides an electrical stimulus to the next region of the axon to be invaded. In this way, the impulse is reborn unchanged or regenerated as it propagates along the axon. The fastest action potentials travel in the largest fibers at a speed of about 120 m/s (430 km/h, or 270 miles/h) and are therefore capable of conveying information rapidly over a long distance.

Action Potentials as the Neural Code

Given that the action potential is fixed in amplitude, how is information about the intensity of a stimulus conveyed? Intensity is coded by frequency of firing. A more effective visual stimulus produces a greater depolarization and, as a consequence, higher frequencies of firing in ganglion cells (Figure 1.10). This generalization was first made by Adrian,[4] who showed that the frequency of action potential firing in a sensory nerve in the skin is a measure of the intensity of the stimulus. In addition, Adrian observed that stronger stimuli applied to the skin give rise to activity in a larger number of sensory fibers.

Synapses: The Sites for Cell-to-Cell Communication

Photoreceptors influence the bipolar cells, which influence the ganglion cells, and so on, through the progression that leads to visual perception. The structure at which one cell hands its information to the next is known as a **synapse**. The mechanisms of synaptic transmission constitute a major theme in modern neurobiology. Through synaptic interactions, a neuron such as the ganglion cell takes account of signals arising from many photoreceptors, thereby creating its own new message.

Chemically Mediated Synaptic Transmission

Figure 1.11A shows the highly organized structure at which a photoreceptor makes synaptic connections onto two bipolar cells. The **presynaptic** terminal of the photoreceptor is separated from the bipolar cells by a cleft that contains extracellular fluid. This space is too wide to be traversed by currents generated in the photoreceptor. Instead the photoreceptor terminal releases a **neurotransmitter** that is stored in presynaptic **vesicles**. The transmitter, the amino acid glutamate in this case, diffuses across the cleft to interact with specific protein molecules known as **receptors**, embedded in the membranes of the **postsynaptic** bipolar cells. (This terminology should not lead to confusion: The term "receptor" as used here means a chemoreceptor molecule and is not the same concept as that of a "sensory receptor" cell that responds to external stimuli—e.g., a photoreceptor.) The transmitters in a neuron and the receptors on its surface can be identified and visualized by a variety of techniques, including antibody labeling.

Activation of the receptor molecules in a bipolar cell by glutamate sets up graded local potentials that spread to its terminals. The more transmitter released, the higher the concentration in the cleft, the larger the number of activated receptors, and the larger the local potential. Such events occur rapidly, with a delay of about 1 ms. Essential features of

[4]Adrian, E. D. 1946. *The Physical Background of Perception*. Clarendon, Oxford, England.

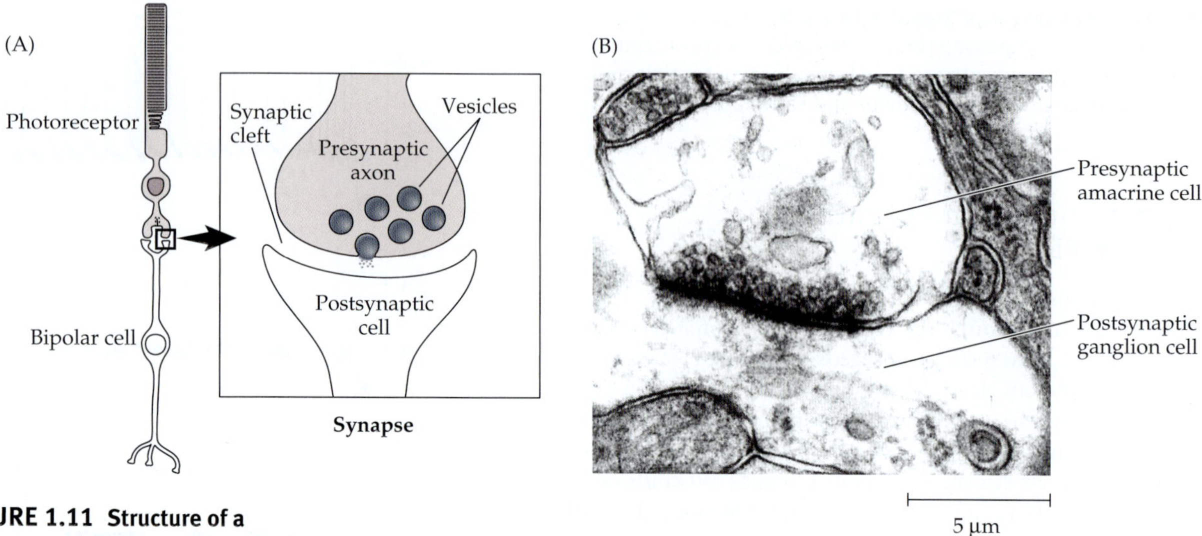

FIGURE 1.11 Structure of a Synapse. (A) These drawings show the principal features of synaptic structures made by a photoreceptor on a bipolar cell. (B) This electron micrograph shows the appearance of a typical synapse in the retina of a macaque monkey. The vesicles that store transmitter are in the presynaptic terminal; a narrow cleft separates the pre- and postsynaptic membranes. The postsynaptic membrane is densely stained. Transmitter released from the presynaptic amacrine cell diffuses across the cleft to interact with receptors on the postsynaptic ganglion cell. (Micrograph kindly provided by Y. Tsukamoto and P. Sterling.)

synaptic transmission were first revealed by Katz, Kuffler, and their colleagues,[5] who used the responses of receptors in muscle as a bioassay with extremely high sensitivity and time resolution for measuring transmitter release.

Excitation and Inhibition

A feature of synaptic transmission, as exemplified by the interactions between photoreceptors and bipolar cells in the retina, is that the transmitter released by a presynaptic terminal can excite or inhibit the next cell, depending on the receptors that cell possesses. For example, one class of glutamate receptor localized to certain bipolar cells reacts with glutamate to cause an excitatory signal. This signal spreads passively to the bipolar cell terminals at the other end of the cell, where it causes liberation of transmitter. Other classes of bipolar cells contain different glutamate receptors that produce a signal of the opposite sign when activated by glutamate. Again, the electrical signal spreads along the bipolar cell, but in this case it suppresses the release of transmitter. Excitatory and inhibitory synaptic potentials in ganglion cells are shown in Figure 1.12.

In neurons throughout the nervous system, combined excitatory and inhibitory inputs determine whether or not the threshold for action potential initiation will be reached. For example, a ganglion cell receives both excitatory and inhibitory inputs. If the threshold is crossed, a message is transmitted to higher centers; if not, no message is sent. In motor cells in the spinal cord, for example, excitatory and inhibitory influences from different fibers determine whether or not a finger will be bent. A motor cell of this sort receives 10,000 or more incoming fibers (Figure 1.13A). These release transmitters that drive the membrane potential toward or away from the threshold for impulse initiation. An individual cell in the cerebellum receives more than 100,000 inputs.

Electrical Transmission

Although the most prevalent form of synaptic transmission involves the release of transmitter molecules, certain cells in the retina, and in the rest of the nervous system, are linked by specialized junctions. At such synapses **electrical transmission** occurs. The pre- and postsynaptic membranes are closely apposed and linked by channels that connect the intracellular fluids of the two cells. This close connection allows local potentials and even action potentials to spread directly from cell to cell without a chemical transmitter. Metabolites and dyes can also spread from cell to cell. One important example in the retina is provided by the horizontal cells, which are electrically coupled in this way. By virtue of this property, graded potentials can spread from one horizontal cell to the next with marked effect on the processing of visual information in the retina. Electrical

[5]Katz, B. 1966. *Nerve, Muscle, and Synapse*. McGraw-Hill, New York.

FIGURE 1.12 Excitation and Inhibition. Intracellular recordings from a ganglion cell showing excitatory and inhibitory synaptic potentials. (A) A ganglion cell is depolarized by continuous release of excitatory transmitter during retinal illumination. If the depolarizing synaptic potential is large enough, threshold is crossed and action potentials are initiated in the ganglion cell. (B) Illumination of a different group of photoreceptors causes inhibition. Hyperpolarization of the membrane makes it more difficult to initiate an action potential. (After Baylor and Fettiplace, 1979.)

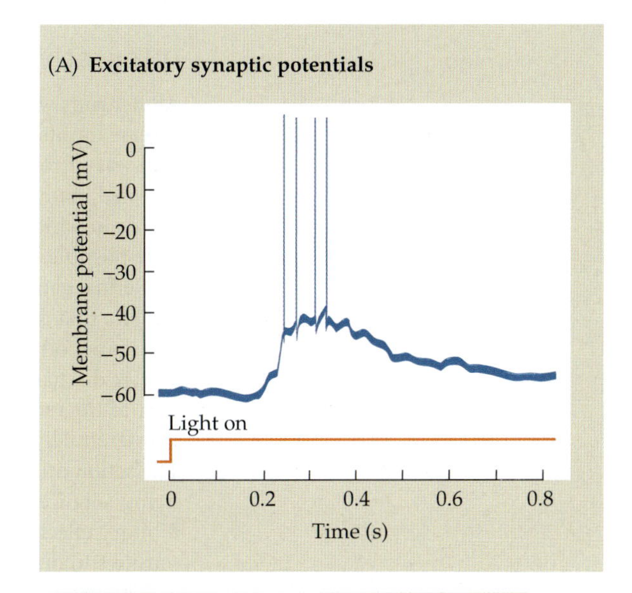

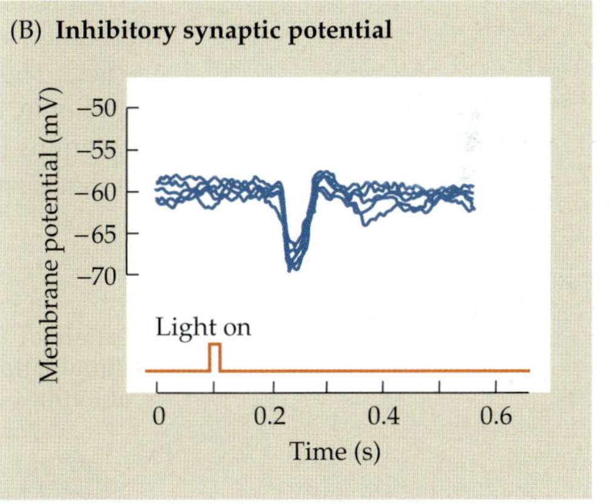

synapses are found elsewhere in the body—for example, between epithelial cells and the muscle fibers of gut and heart, and in the central nervous system, particularly in invertebrates.

Modulation of Synaptic Efficacy

Chemically mediated synaptic transmission shows great plasticity. Dramatic changes occur in the amount of transmitter that is released by a signal such as an action potential or a local potential that invades a presynaptic terminal. The photoreceptors in the retina provide an example: The amount of the transmitter glutamate that is released by a rod or a cone in response to a standard light stimulus can be increased or decreased by feedback to the terminal from horizontal cells. The horizontal cells themselves are influenced by other photoreceptors. This feedback loop plays a critical role in the way the eye adapts to different levels of illumination.

Other mechanisms that influence transmitter release depend on the history of impulse activity. During and after a train of impulses in a neuron, the amount of transmitter it releases can increase or decrease dramatically, depending on the frequency and duration of the activity. Modulation of efficacy can also be postsynaptic in origin. Long- and short-term plasticity are the focus of intense contemporary research.

FIGURE 1.13 Multiple Connections of Individual Neurons. (A) Approximately 10,000 presynaptic axons converge to form endings that are distributed over the surface of a motor neuron in the spinal cord. This drawing is based on a reconstruction made from electron micrographs. (B) This drawing shows the divergence of the axon of a single horizontal cell that branches extensively to supply many postsynaptic target cells. (A from Poritsky, 1969; B after Fisher and Boycott, 1974.)

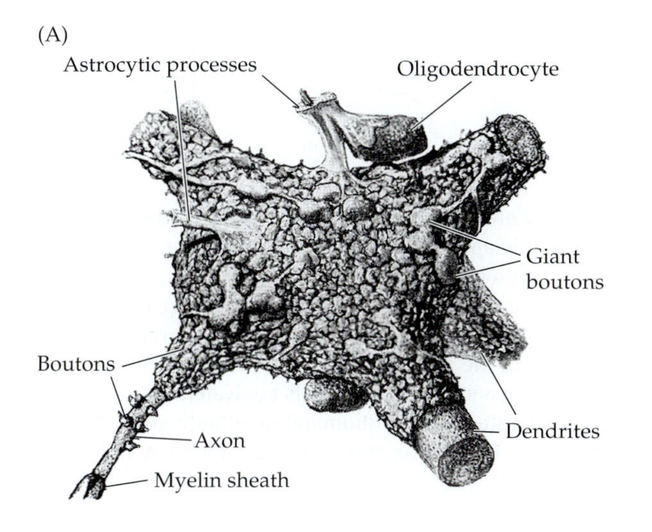

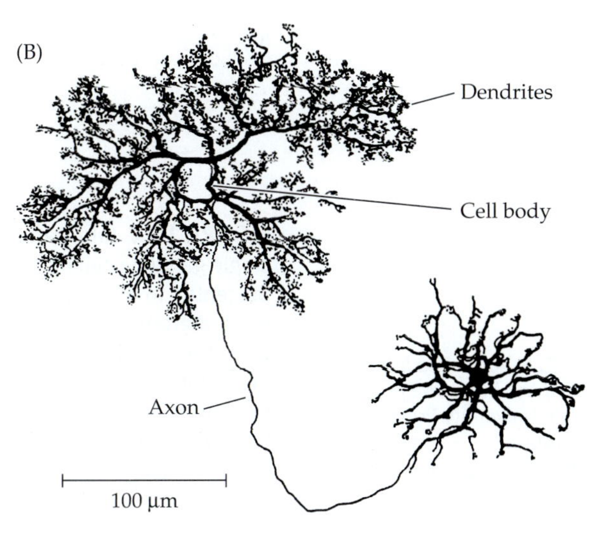

Integrative Mechanisms

All neurons within the central nervous system take account of influences arriving from diverse inputs to create their own new messages with new meanings. The term **integration** was introduced by Sherrington[6] (who also coined the word "synapse"). Sherrington revealed many of the essential concepts that permeate modern neurobiology by experiments in which he measured the contractions of muscles, before electrical recordings were possible.

Retinal ganglion cells once again provide an excellent example of integration. Kuffler[7] was the first to show that ganglion cells respond best to a small light spot or dark spot that shines on a few receptors in a particular region of the retina. Such a spot gives rise to a brisk discharge of action potentials (Figure 1.14A). A larger spot shone over the same part of the retina is far less effective: This is because a different group of receptors circumferentially arranged around the others is influenced by the change in illumination. The action of these photoreceptors on bipolar cells gives rise to inhibition of ganglion cell firing (Figure 1.14B). Summation of the excitatory effect of a small spot and the inhibitory effect from the surrounding region means that the ganglion cell is relatively insensitive to diffuse light (Figure 1.14C).

The meaning of the signal in a ganglion cell has thereby become more complex than information about "light" or "dark" and refers to a contrasting pattern of light in the visual field. This complex signal arises because each ganglion cell is influenced, albeit indirectly, not by one photoreceptor but by many. For a particular ganglion cell, the specific connections through bipolar, horizontal, and amacrine cells determine the pattern of light that is required for it to discharge action potentials.

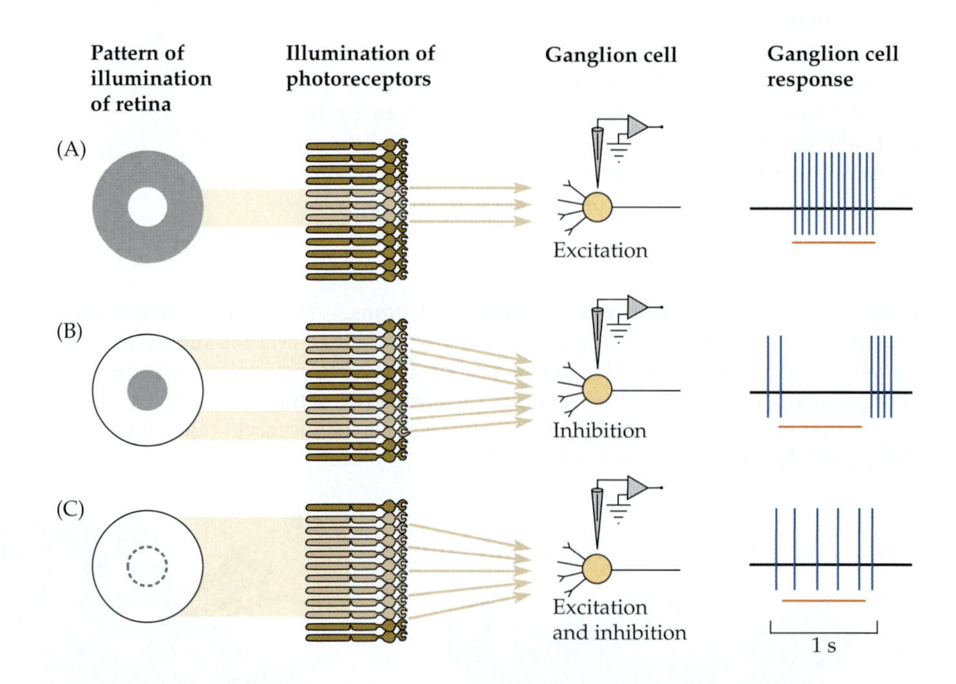

FIGURE 1.14 Integration by Ganglion Cells. Extracellular recordings made from a single ganglion cell in the retina of a lightly anesthetized cat while patterns of light were presented to the eye (see Figure 1.15). (A) A small spot of light presented to a centrally located group of photoreceptors gives rise to excitation and a brisk discharge of action potentials. (B) Light presented as a ring, or annulus, to illuminate a circumferential group of photoreceptors gives rise to inhibition of the ganglion cell, which prevents the cell from firing. Removal of the inhibition at the end of illumination is equivalent to excitation, which gives rise to a burst of action potentials. (C) Illumination of both groups of receptors causes integration of excitation and inhibition and a weak discharge of action potentials. (After Kuffler, 1953.)

[6] Sherrington, C. S. 1906. *The Integrative Action of the Nervous System.* Reprint, Yale University Press, New Haven, CT, 1966.

[7] Kuffler, S. W. 1953. *J. Neurophysiol.* 16: 37–68.

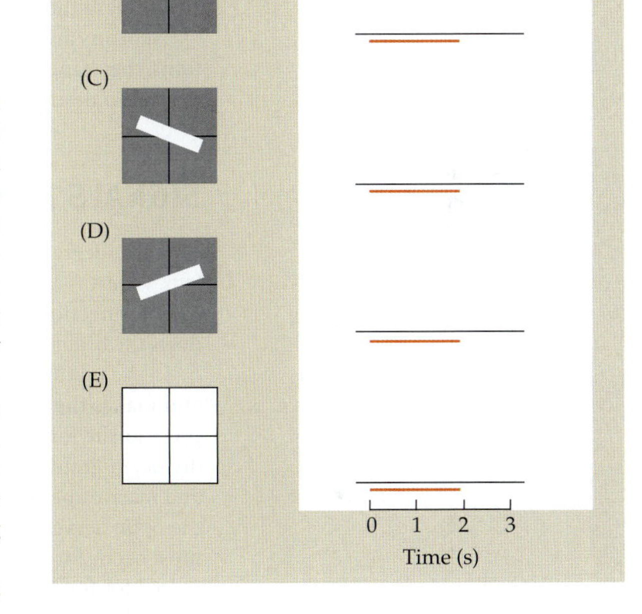

FIGURE 1.15 Information Conveyed by Action Potentials. Extracellular recordings from a neuron in the cerebral cortex of a lightly anesthetized cat. Action potentials in this cell indicate that a bar of light that is almost vertical shines on one particular part of the visual field. The small drawings to the left of the graphs show how visual stimuli such as bars or edges with different orientations and positions are presented to the eye. (A) The cortical cell fires a burst of action potentials when the light stimulus consists of a vertical bar of light in one particular part of the visual field. (B–E) Bars with different orientations or diffuse light fail to evoke action potentials. The cortical cell integrates information arriving by way of relays from a large number of photoreceptors, some of which (corresponding to those illuminated by the vertical bar) give rise to excitation on the cortical cell, the others giving rise to inhibition. (After Hubel and Wiesel, 1959.)

Complexity of the Information Conveyed by Action Potentials

Three relays beyond the retina, even more sophisticated information about the visual world is given by the action potentials in cortical nerve cells.[8] Cortical neurons do not respond simply to light or dark on the retina. Instead, whether or not action potentials occur depends on the pattern of retinal illumination, which is specific for each type of cortical cell. For example, one type of cortical cell responds selectively to a bar of light with a specific orientation (vertical, oblique, or horizontal), moving in a particular direction in a particular part of the visual field (Figure 1.15). The firing of such a cell is not influenced by diffuse light or by a bar of an inappropriate orientation or one moving in the wrong direction. Hence, its action potentials provide precise information about the visual stimulus to higher centers in the brain. This increase in the meaning attributed to a stereotyped action potential is explained by the precise connections of lower-order cells to the cortical cell and the way in which the cortical cell integrates incoming signals by summation of localized graded potentials.

The transformation of information can be simply summarized as follows:

> A signal in a photoreceptor indicates a change in light intensity in that area of the field of vision.
> A signal in a ganglion cell indicates the presence of contrast.
> A signal in a cortical neuron indicates the presence of an oriented bar of light.

Complex integration occurs in other sensory systems. Thus, the position and direction of a mechanical stimulus moving along a fingertip act as selective stimuli for particular cells in the region of the cerebral cortex that are concerned with tactile stimuli moving in one direction.

Two important conclusions about signaling in the nervous system are (1) that nerve cells act as the building blocks for perception and (2) that the abstract significance of the message can be extremely complex, depending on the inputs a neuron receives.

CELLULAR AND MOLECULAR BIOLOGY OF NEURONS

Like other types of cells, neurons possess the cellular machinery for metabolic activity and for synthesizing membrane proteins (such as ion channels and receptors). Moreover, ion channels and receptors are distributed to precise locations in the cell membrane. Channels specific for sodium or for potassium are present in membranes of ganglion cell

[8] Hubel, D. H., and Wiesel, T. N. 1977. *Proc. R. Soc. Lond. B* 198: 1–59.

axons at discrete, well-defined sites. These channels enable the action potential to be initiated and conducted.

The presynaptic terminals of optic nerve fibers (like those of photoreceptors and bipolar cells, and indeed like presynaptic nerve terminals in general) contain in their membranes specific channels through which calcium ions can flow. Calcium entry is what triggers the release of transmitters. Each type of neuron synthesizes, stores, and releases its characteristic transmitter(s). As for other membrane proteins, the receptors for specific transmitters are located at well-defined postsynaptic sites under the presynaptic terminals. In addition, separate membrane proteins, known as pumps and transporters, maintain the constancy of the internal milieu of the cell.

A major specialization in the cell biology of neurons, compared to other types of cells, arises from the presence of the axon. Since axons do not have the machinery for protein synthesis, essential molecules must be carried to the nerve terminals by a process known as **axonal transport**, often over long distances. All the molecules required for maintenance of structure and function, as well as the appropriate membrane channels, must travel from the cell body in this way; similarly, molecules taken up at the ending are carried back to the cell body by the same mechanism.

Neurons are different from most other cells in a second way: With very few exceptions, they cannot divide after differentiation. As a result, in an adult animal, neurons that have been destroyed cannot usually be replaced.

SIGNALS FOR DEVELOPMENT OF THE NERVOUS SYSTEM

The high degree of organization in a structure such as the retina poses a fascinating problem. Whereas a computer requires a brain to wire it, the brain must establish and tune its own connections. What seems so puzzling is how the proper assembly of the parts endows the brain with its extraordinary properties.

In the mature retina, each cell type is situated in the correct layer—or even sublayer—and makes the correct connections with the appropriate targets. This arrangement is a prerequisite for function. For ganglion cells to develop, for example, precursor cells must divide, migrate, differentiate into the appropriate shapes with the appropriate properties, and receive specific synapses. The axons must find their way over long distances through the optic nerve to end in the appropriate layer of the next relay station. Similar processes must occur for the various divisions of the nervous system so that complex structures required for function are formed.

Study of the mechanisms by which highly complex structures such as the retina are formed is a key problem in modern neurobiology. An understanding of how intricate wiring diagrams are established in development may provide clues about the functional properties and about the genesis of functional disorders. (If you know how an electrical circuit has been wired, you may be able to understand what the components are doing, and consequently you may be able to repair it.) Key molecules play a crucial role in differentiation, outgrowth, pathfinding, synapse formation, and survival of neurons. Such molecules are now being identified at an ever increasing rate, and their mechanisms of action are being studied. Interestingly, molecular signals that give rise to the outgrowth of axons and formation of connections can be regulated by electrical signals. Activity plays a role in determining the pattern of connections.

Genetic approaches have made it possible to identify genes that control the differentiation of entire organs, such as the eye as a whole. Gehring[9] and his colleagues have studied the expression of a gene in the fruit fly (*Drosophila*), known as *eyeless*, that controls the development of the eyes. After deletion of this gene, eyes fail to develop. Homologous genes in mice and humans (known as *small eye* and *aniridia*, respectively) share extensive sequence identity. If the fly *eyeless* gene or the mammalian homologue of the gene is introduced and overexpressed, the fly develops multiple ectopic eyes over its antennae, wings, and legs (Figure 1.16). The gene can therefore orchestrate the formation of an en-

[9]Halder, G., Callaerts, P., and Gehring, W. J. 1995. *Science* 267: 1788–1792.

(A)

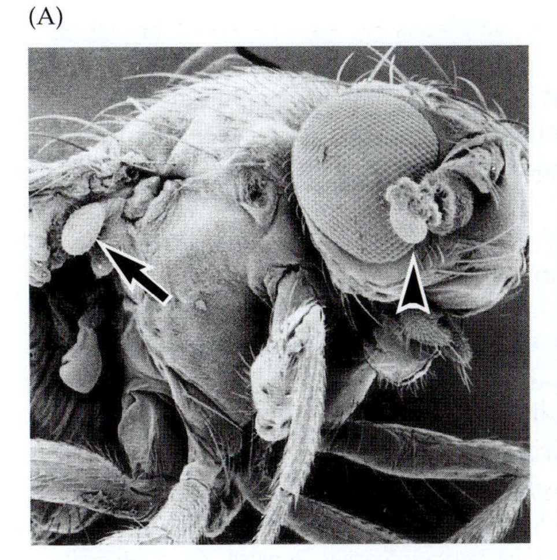

(B)

FIGURE 1.16 Genetic Influences on Development of the Eye in the fruit fly, *Drosophila*. A gene known as *eyeless* controls development of the eye in the fruit fly. After deletion of this gene, eyes fail to appear. Overexpression leads to the development of ectopic eyes that are morphologically normal. (A) This scanning electron micrograph shows such ectopic eyes on the antenna (arrowhead) and on the wing (arrow). (B) Here the wing eye is shown at higher magnification. A gene with strikingly similar sequence homology in the mouse also leads to the formation of ectopic eyes in the fly if it is overexpressed. (After Halder, Callaerts, and Gehring, 1995; micrographs kindly provided by W. Gehring.)

tire eye, in a mouse or a fly, even though the eyes themselves have completely different structures and properties.

REGENERATION OF THE NERVOUS SYSTEM AFTER INJURY

Not only does the nervous system wire itself when it is developing, but it can also restore certain connections after injury (something your computer cannot do!). For example, axons in an arm can grow back after the nerve has been injured so that function can be restored; the hand can once again be moved, and sensation returns. Similarly, in a frog or a fish or an invertebrate, lesions in the central nervous system are followed by axon regeneration and functional recovery. After the optic nerve of a frog or a fish has been cut, fibers grow back to the brain and the animal can see again. However, this ability is not the case in the adult mammalian central nervous system: Regeneration does not occur. The molecular signals that cause this failure and its biological significance for the functioning of the nervous system are not known.

SUMMARY

- Neurons are connected to each other in a highly specific manner.
- At synapses, information is transmitted from cell to cell.
- In relatively simple circuits, such as the retina, it is possible to trace connections and understand the meaning of signals.
- Neurons within the eye and the brain act as building blocks for perception.
- Signals in neurons are highly stereotyped and similar in all animals.
- Action potentials conduct unfailingly over long distances.
- Local graded potentials depend on passive electrical properties of nerve cells and spread only over short distances.
- Owing to the peculiar structure of neurons, specialized cellular mechanisms are required for axonal transport of proteins and organelles to and from the cell body.
- During development, neurons migrate to their final destinations and become connected to their targets.
- Molecular cues provide guidance for growing axons.

Suggested Reading

All the experiments and concepts described in this introductory chapter are treated in more detail and fully referenced in later chapters. The following sources represent key reviews for understanding how neurobiology has developed over the years.

Adrian, E. D. 1946. *The Physical Background of Perception.* Clarendon, Oxford, England.

Helmholtz, H. 1962/1927. *Helmholtz's Treatise on Physiological Optics.* J. P. C. Southhall (ed.). Dover, New York.

Hodgkin, A. L. 1964. *The Conduction of the Nervous Impulse.* Liverpool University Press, Liverpool, England.

Hubel, D. H. 1988. *Eye, Brain and Vision.* Scientific American Library, New York.

Katz, B. 1966. *Nerve, Muscle, and Synapse.* McGraw-Hill, New York.

Ramón y Cajal, S. [1909–1911] 1995. *Histology of the Nervous System*, 2 vols. Translated by Neely Swanson and Larry Swanson. Oxford University Press, New York.

Sherrington, C. S. 1906. *The Integrative Action of the Nervous System.* Reprint, Yale University Press, New Haven, CT, 1961.

PART 2 SIGNALING IN THE NERVOUS SYSTEM

Membrane Proteins and Their Function

Properties of Neurons and Glia

Communication between Excitable Cells

2 Ion Channels and Signaling

ELECTRICAL SIGNALS THAT ARE NECESSARY FOR NEURAL FUNCTION are mediated by the flow of ions through aqueous pores in the nerve cell membrane. These pores are formed by transmembrane proteins called ion channels. It is possible to record and measure ion currents through single channels. In this chapter we discuss the functional properties of ion channels, such as their specificity for one ion species or another, and how they are regulated to produce the required types of signaling. Individual channels are selective for either cations or anions, and some cationic channels can be highly selective for a single ionic species, such as sodium. Channels fluctuate between open and closed states, often with a characteristic open time. Their contribution to the flow of ionic current across the cell membrane is determined by the relative amount of time they spend in the open state. Channel opening is regulated by a variety of mechanisms. Some of these mechanisms are physical, such as changes in membrane tension or membrane potential; others are chemical, involving the binding of activating molecules (ligands) to sites on the extracellular mouth of the channel, or the binding of particular ions or molecules to the inner mouth.

An important property of channels, in addition to the kinetics of opening and closing, is the relative ability of an open channel to pass ionic current. One way in which ions can pass through an open channel is by simple diffusion; another is by interacting with internal binding sites, hopping from one to the next through the pore. In either case, movement through the channel is passive, driven by concentration gradients and by the electrical potential gradient across the membrane.

The amount of current flowing through an open channel down an electrochemical gradient depends on the permeability of the channel to the ion species involved. The magnitude of the current also depends on the concentration of the ions at either mouth of the channel. These two factors, permeability and concentration, determine the channel conductance.

In Chapter 1 we discussed how the transfer of information in the nervous system is mediated by two types of electrical signals in nerve cells: graded potentials that are localized to specific regions of the nerve cell membrane, and action potentials that are propagated along the entire length of a neuronal process. These signals are superimposed on a steady electrical potential across the cell membrane called the resting membrane potential. Depending on cell type, nerve cells at rest have steady membrane potentials ranging from about –30 mV to almost –100 mV, the negative sign meaning that the inside of the membrane is negative with respect to the outside.

Signaling in the nervous system is mediated by changes in the membrane potential: In sensory receptors, an appropriate stimulus, such as touch, sound, or light, causes local **depolarization** (making the membrane potential less negative) or **hyperpolarization** (membrane potential more negative). Similarly, neurotransmitters at synapses act by depolarizing or hyperpolarizing the postsynaptic cell. Action potentials, which are large, brief pulses of depolarization, propagate along axons to carry information from one place to the next in the nervous system.

All such changes in membrane potential are produced by the movement of ions across the nerve cell membrane. For example, inward movement of positively charged sodium ions reduces the net negative charge on the inner surface of the membrane or, in other words, causes depolarization. Conversely, outward movement of positively charged potassium ions results in an increase in net negative charge, causing hyperpolarization, as does inward movement of negatively charged chloride ions.

How do ions move across the cell membrane, and how is their movement regulated? The major pathway for rapid movement of ions into and out of the cell is through **ion channels**, which are protein molecules that span the membrane and form pores through which ions can pass. Ion currents are regulated by controlling the opening and closing of such channels. Knowledge of the functional behavior of ion channels has provided an essential advance in our understanding of how electrical signals are generated.

PROPERTIES OF ION CHANNELS

The Nerve Cell Membrane

Cell membranes consist of a fluid mosaic of lipid and protein molecules. As shown in Figure 2.1A, the lipid molecules are arranged in a bilayer about 6 nm thick, with their polar, hydrophilic heads facing outward and their hydrophobic tails extending to the middle of the layer. The lipid is sparingly permeable to water, and virtually impermeable to ions. Embedded in the lipid bilayer are protein molecules—some on the extracellular

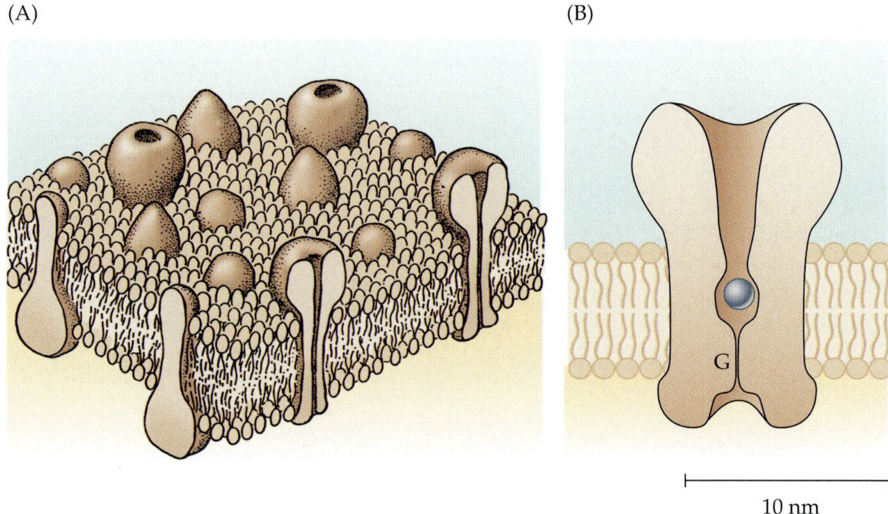

FIGURE 2.1 Cell Membrane and Ion Channel. (A) The cell membrane is composed of a lipid bilayer embedded with proteins. Some of the proteins traverse the lipid layer, and some of these membrane-spanning proteins form membrane channels. (B) This schematic representation shows a membrane channel in cross section, with a central water-filled pore and channel "gate" (G). The gate opens and closes irregularly; the probability of opening may be regulated by the membrane potential, by the binding of a ligand to the channel, or by other biophysical or biochemical conditions. A sodium ion, surrounded by a single shell of water molecules, is shown to scale in the pore for size comparison.

side, some facing the cytoplasm, and some spanning the membrane. Many of the membrane-spanning proteins form ion channels. Ions such as potassium, sodium, calcium, or chloride move through such channels passively, driven by concentration gradients and by the electrical potential across the membrane.

Other membrane-spanning proteins function as transport molecules (**pumps** and **transporters**) that move substances across the membrane against their electrochemical gradients. Transport molecules maintain the ionic composition of the cytoplasm by pumping back across the cell membrane ion species that have moved down their electrochemical gradients into or out of the cell. They also perform the important function of carrying metabolic substances such as glucose and amino acids across cell membranes. Properties of transport molecules are discussed in Chapter 4.

What Does an Ion Channel Look Like?

The molecular composition of ion channels and their configuration in the cell membrane are discussed in detail in Chapter 3, but it is useful at this point to have some idea of the general physical features of an ion channel protein. These are illustrated by Figure 2.1B. The protein spans the membrane and has a central water-filled pore open to both the intracellular and the extracellular spaces. On each side the pore widens to form a vestibule (or mouth), and the restricted region within the plane of the membrane contains a **gate** that can open and close to regulate the passage of ions.

The size of the protein varies considerably from one channel type to the next, and some have additional structural features. Figure 2.1B represents a channel of medium dimensions. A sodium ion, hydrated by a single shell of water molecules, is shown in the pore for size comparison.

Channel Selectivity

Membrane channels vary considerably in their selectivity: Some are permeable to cations, some to anions. Some cation channels are selective for a single ion species. For example, some allow permeation of sodium almost exclusively, others of potassium, still others of calcium. Others are relatively nonspecific, allowing the passage of even small organic cations. Anion channels involved in signaling tend to have low specificity, but they are referred to as "chloride channels" because chloride is the major permeant anion in biological solutions. In addition, some channels (called connexons) connect adjacent cells and allow the passage of most inorganic ions and many small organic molecules. Connexons are discussed in Chapter 7.

Open and Closed States

Although for simplicity we must represent protein molecules as static structures, they are never still. Because of their thermal energy, all large molecules are inherently dynamic. At room temperature, chemical bonds stretch and relax, and twist and wave around their equilibrium positions. Although individual movements are of the order of only 10^{-12} m in magnitude, with frequencies approaching 10^{13} Hz, such atomic trembling can underlie much larger and slower changes in conformation of the molecule. The reason is that numerous rapid motions of the atoms occasionally allow groups to slide by one another in spite of mutual repulsive interactions that would otherwise keep them in place. Such a transition, once achieved, can last for many milliseconds, or even seconds. Hemoglobin provides an example: The binding sites for oxygen to the heme groups are buried inside the molecule, and not immediately accessible. Oxygen binding, and its subsequent escape, can be accomplished only by the dynamic formation of transient access pathways to the heme pocket; thus, the molecule "breathes" in order to perform its function.[1]

In ion channel proteins, molecular transitions occur between open and closed states, with transitions between states being virtually instantaneous. If we examine the behavior of any given channel, we find that open times vary randomly: Sometimes the channel opens for only a millisecond or less, sometimes for very much longer (see Figure 2.4).

[1]Karplus, M., and Petsko, G. A. 1990. *Nature* 347: 631–639.

However, each channel has its own characteristic **mean open time** (τ) around which these durations of opening fluctuate.

Some channels in the resting cell membrane open frequently; the probability of finding such channels in the open state is relatively high. Most of these are potassium and chloride channels associated with the resting membrane potential. The remainder are predominantly in the closed state, and the probability that an individual channel will open is low. When such channels are **activated** by an appropriate stimulus, the probability of openings increases sharply. On the other hand, channels that open frequently at rest may be **deactivated** by a stimulus; that is, their frequency of opening is decreased. An important point to remember is that activation or deactivation of a channel means an increase or decrease in the *probability of channel opening*, not an increase or decrease in mean open time of the channel.

In addition to activation and deactivation, two other factors regulate current flow through ion channels. One is that certain channels can enter a conformational state in which activation no longer occurs, even though the activating stimulus is still present. In channels that respond to depolarization of the cell membrane, this condition is called **inactivation**; in channels that respond to chemical stimuli, the condition is known as **desensitization.** The second mechanism is **open channel block.** This occurs when, for example, a large molecule (such as a toxin) binds to a channel and physically occludes the pore. Another example is the block of some cation channels by magnesium ions, which do not themselves permeate the channel, but bind in its inner mouth and prevent the permeation of other cations.

Modes of Activation

Figure 2.2 summarizes the modes of channel activation. Some channels respond specifically to physical changes in the nerve cell membrane. Prominent in this group are the **voltage-activated** channels. An example is the voltage-sensitive sodium channel, which is responsible for the regenerative depolarization that underlies the rising phase of the action potential (Chapter 6). Also in this group are **stretch-activated** channels, which respond to mechanical distortion of the cell membrane. Stretch receptors are found, for example, in mechanoreceptors in the skin (Chapter 17).

(A) **Channels activated by physical changes in the cell membrane**

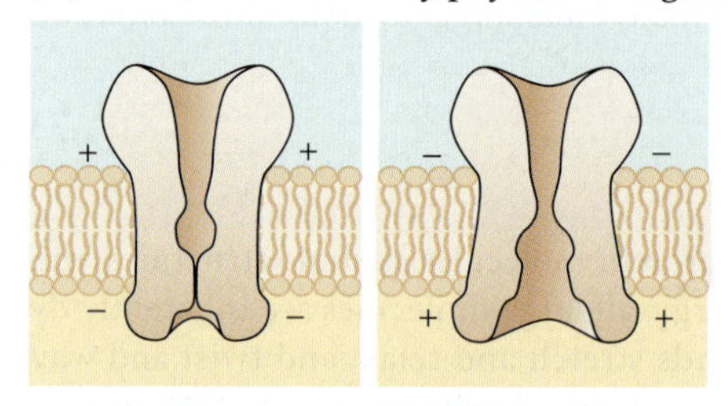

Voltage-activated

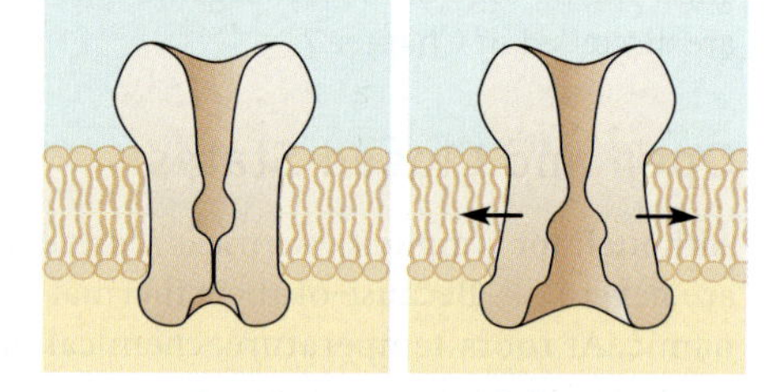

Stretch-activated

(B) **Channels activated by ligands**

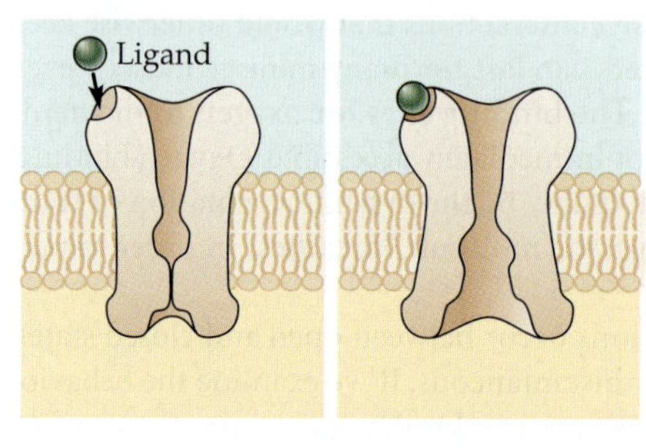

Extracellular activation

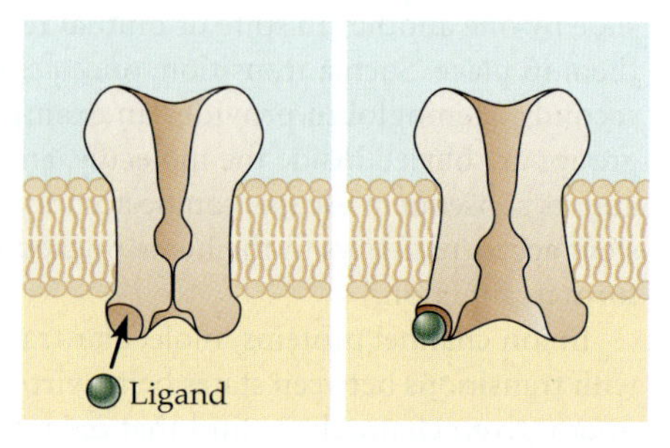

Intracellular activation

FIGURE 2.2 Modes of Channel Activation. The probability of channel opening is influenced by a variety of stimuli. (A) Some channels respond to changes in the physical state of the membrane, specifically changes in membrane potential (voltage-activated) and mechanical distortion (stretch-activated). (B) Ligand-activated channels respond to chemical agonists, which attach to binding sites on the channel protein. Neurotransmitters, such as glycine and acetylcholine, act on extracellular binding sites. Included among a wide variety of intracellular ligands are calcium ions, subunits of G proteins, and cyclic nucleotides.

Other channels are activated when chemical agonists attach to binding sites on the channel protein. These **ligand-activated** channels are further divided into two subgroups, depending on whether the binding sites are extracellular or intracellular. Channels that respond to extracellular activation include, for example, cation channels in the postsynaptic membranes of skeletal muscle that are activated by the neurotransmitter acetylcholine, released from presynaptic nerve terminals (Chapter 9). This activation allows sodium to enter the cell, causing muscle depolarization.

Ligand-activated channels responding to intracellular stimuli include channels that are sensitive to local changes in the concentration of specific ions. For example, calcium-activated potassium channels are activated by local increases in intracellular calcium and, in many cells, play a role in repolarizing the membrane during termination of the action potential. Other intracellular ligands include the cyclic nucleotides: Cyclic GMP, for example, is responsible for the activation of sodium channels in retinal rods, thereby playing an important role in visual transduction (Chapter 19).

These classifications are not rigid: For example, calcium-activated potassium channels are also voltage-sensitive, and some voltage-activated channels are sensitive to intracellular ligands.

MEASUREMENT OF SINGLE-CHANNEL CURRENTS

Patch Clamp Recording

Experimental tools have been devised to measure ion fluxes through individual channels, first indirectly by analysis of membrane noise,[2,3] and later by direct observation, using patch clamp recording methods.[4,5] These methods provide direct answers to questions of obvious physiological interest about channels; for example, how much current does a single channel carry? how long does a channel stay open? how do its open and closed times depend on voltage? on the activating molecule?

The development of the patch clamp by Erwin Neher, Bert Sakmann, and their colleagues has contributed enormously to our knowledge of the functional behavior of membrane channels. In patch clamp recording, the tip of a small glass pipette (with an internal diameter of about 1 μm) is sealed to the membrane of a cell. Under ideal conditions, with slight suction on the pipette, a seal resistance of greater than 10^9 ohms (hence the term "gigaohm seal") forms around the rim of the pipette tip between the cell membrane and the glass (Figure 2.3A–B). When the pipette is connected to an appropriate amplifier, small currents across the patch of membrane inside the pipette tip can be recorded (Figure 2.3F). The high-resistance seal ensures that such currents flow through the amplifier rather than escaping through the rim of the patch. The recorded events consist of rectangular pulses of current, reflecting the opening and closing of single channels. In other words, we can observe in real time the activity of single protein molecules in the membrane.

Erwin Neher (left) and Bert Sakmann (right)

In their simplest form the current pulses appear irregularly, with nearly fixed amplitudes and variable durations (Figure 2.4A). In some cases, however, records of current are more complex: For example, channels may exhibit open states with more than one current level, as in Figure 2.4B, where the open channels often close to smaller "substate" levels. In addition, channels may display complicated kinetics. For example, channel openings may occur in bursts (Figure 2.4C).

In summary, patch clamp techniques offer two advantages for studying the behavior of channels: First, the isolation of a small patch of membrane allows us to observe the activity of only a few channels, rather than the thousands that may be active in an intact cell. Second, the very high resistance of the seal enables us to observe extremely small currents. As a result, we are able to obtain accurate measures of the amplitudes of single-channel currents and to analyze the kinetic behavior of the channels.

Recording Configurations with Patch Electrodes

Patch clamp methods permit other recording configurations. Having made a seal to form a cell-attached patch, we can then pull the patch from the cell to form an inside-out patch

[2]Katz, B., and Miledi, R. 1972. *J. Physiol.* 224: 665–699.

[3]Anderson, C. R., and Stevens, C. F. 1973. *J. Physiol.* 235: 665–691.

[4]Neher, E., Sakmann, B., and Steinbach, J. H. 1978. *Pflügers Arch.* 375: 219–228.

[5]Hamill, O. P., et al. 1981. *Pflügers Arch.* 391: 85–100.

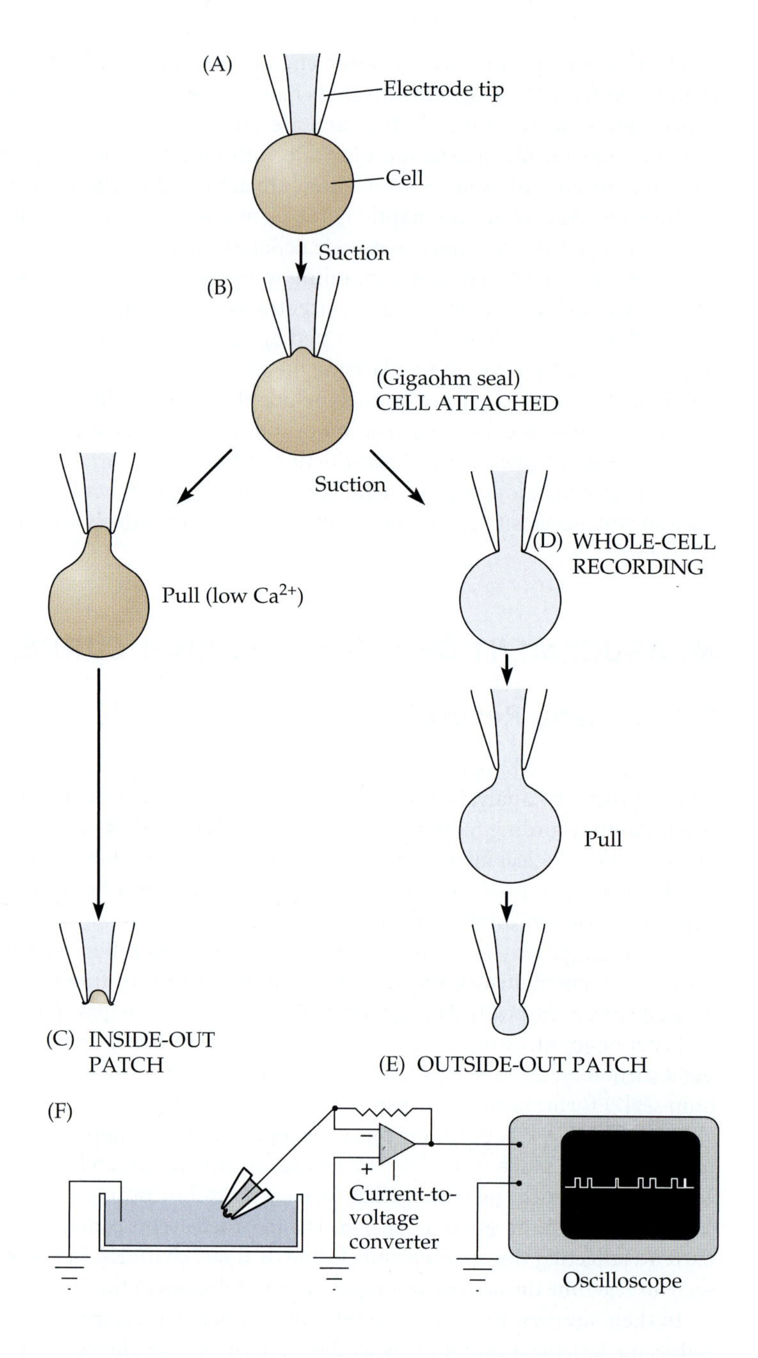

FIGURE 2.3 Patch Clamp Recording. (A–E) Patch configurations, represented schematically. The electrode forms a seal on contact with the cell membrane (A), which is converted to a gigaohm seal by gentle suction (B). Records may then be made from the patch of membrane within the electrode tip (cell-attached patch). Pulling away from the cell results in the formation of a cell-free vesicle, whose outer membrane can then be ruptured to form an inside-out patch (C). Alternatively, the membrane within the electrode tip may be ruptured by further suction to obtain a whole-cell recording (D) or, by pulling, to obtain an outside-out patch (E). (F) Recording arrangement. The patch electrode is connected to an amplifier that converts channel currents to voltage signals. The signals are then displayed on an oscilloscope trace or computer screen so that amplitudes and durations of single-channel currents can be measured. (A–E after Hamill et al., 1981.)

(see Figure 2.3C), with the cytoplasmic face of the patch membrane facing the bathing solution. Alternatively, with slight additional suction we can rupture the membrane inside the patch to provide access to the cell cytoplasm (Figure 2.3D). In this condition currents are recorded from the entire cell (whole-cell recording). Finally, we may first obtain a whole-cell recording and then pull the electrode away from the cell to form a thin neck of membrane that separates and seals to form an outside-out patch (see Figure 2.3E). Each of these configurations has an advantage, depending on the type of channel we are studying and the kind of information we wish to obtain. For example, if we wish to apply a variety of chemical ligands to the outer membrane of the patch, then an outside-out patch is most convenient.

One feature of whole-cell recording with a patch pipette is that substances can exchange between the cell cytoplasm and the pipette. This exchange (sometimes referred to as dialysis) can be useful in providing a method for changing intracellular ion concentrations to those predetermined when the pipette is filled. On the other hand, particularly if

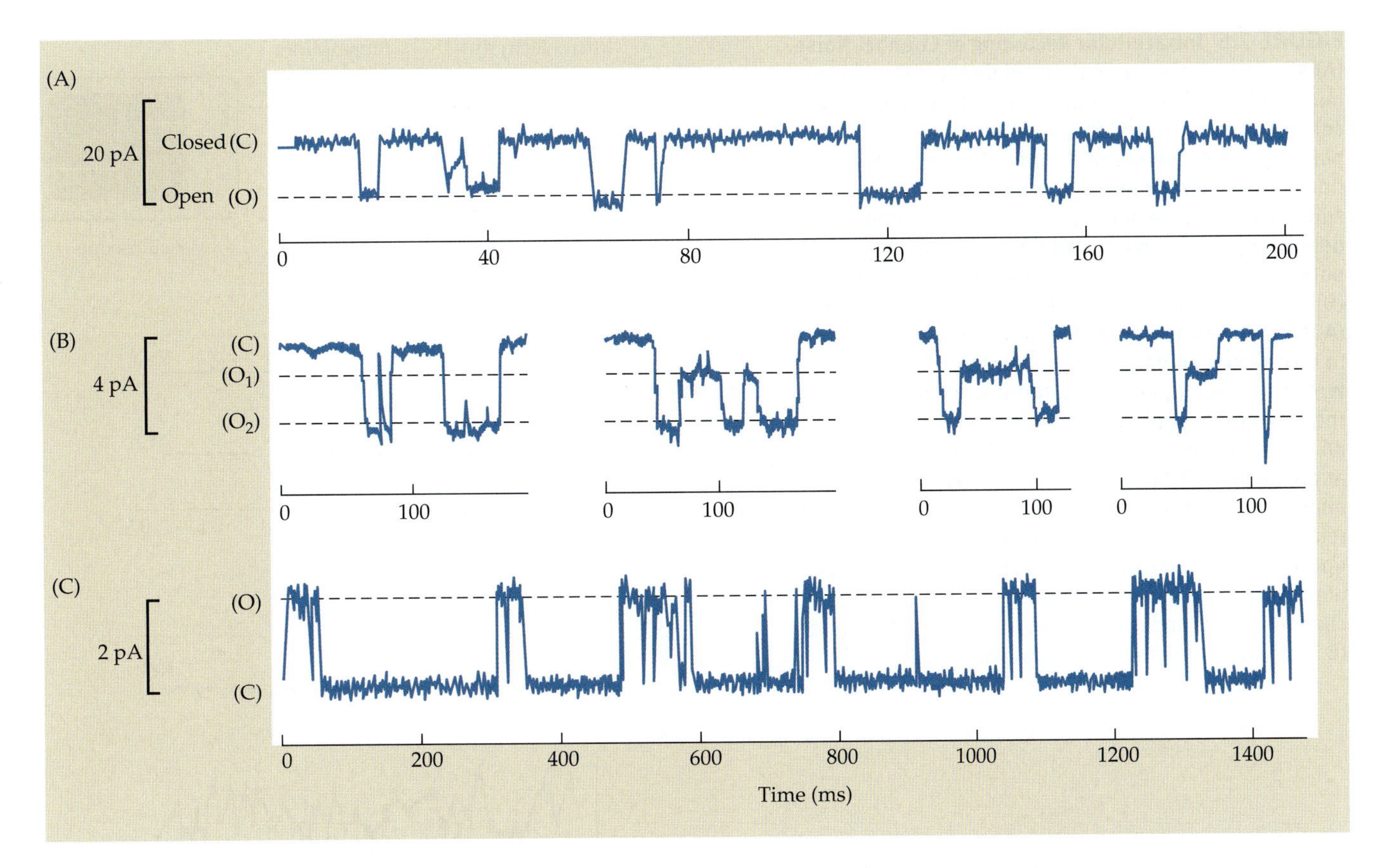

FIGURE 2.4 Examples of Patch Clamp Recordings. (A) Glutamate-activated channel currents recorded in a cell-attached patch from locust muscle occur irregularly, with a single amplitude and varied open times. Downward deflections indicate current flowing into the cell. (B) Acetylcholine-activated currents from single channels in an outside-out patch from cultured embryonic rat muscle reach a maximum amplitude of about 3 pA and relax to a substate current of about 1.5 pA. Downward deflections indicate inward current. (C) Pulses of outward current through glycine-activated chloride channels in an outside-out patch from cultured chick spinal cord cells are interrupted by fast closing and reopening transitions to produce bursts. (A after Cull-Candy, Miledi, and Parker, 1980; B after Hamill and Sakmann, 1981; C from A. I. McNiven and A. R. Martin, unpublished.)

the cell is small, important cytoplasmic constituents can be lost rapidly into the pipette solution. Such loss can be avoided by using a perforated patch.[6] The patch pipette is loaded with a pore-forming substance, such as the antibiotic nystatin, and a seal to the cell is formed. After a delay, pores form in the patch, allowing whole-cell currents to be recorded.

Intracellular Recording with Microelectrodes

Before patch clamp techniques were developed, properties of membrane channels were deduced from experiments in which glass microelectrodes were used to record membrane potential or membrane current from whole cells. The introduction of the glass microelectrode for intracellular recording from single cells by Ling and Gerard[7] in 1949 was at least as important as the introduction of patch clamp recording three decades later. The technique provided a method for accurate measurements of resting membrane potentials, action potentials, and responses to synaptic activation in muscle fibers and neurons (Chapter 1).

The intracellular recording technique is illustrated in Figure 2.5A. A sharp glass micropipette, with a tip diameter of less than 0.5 μm and filled with a concentrated salt solution (e.g., 3 *M* KCl), serves as an electrode and is connected to an amplifier to record the potential at its tip. When the pipette is pushed against the cell membrane, penetration into the cytoplasm is signaled by the sudden appearance of the resting potential. If the penetration is successful, the membrane seals around the outer surface of the pipette, so that the cytoplasm remains isolated from the extracellular fluid.

Intracellular Recording of Channel Noise

In the early 1970s, Katz and Miledi[2] did pioneering experiments on frog muscle fibers, in which they used intracellular recording techniques to examine the characteristics of the "noise" produced by acetylcholine (ACh) at the neuromuscular junction. At this

[6]Horn, R., and Marty, A. 1988. *J. Gen. Physiol.* 92: 145–149.

[7]Ling, G., and Gerard, R. W. 1949. *J. Cell Comp. Physiol.* 34: 383–396.

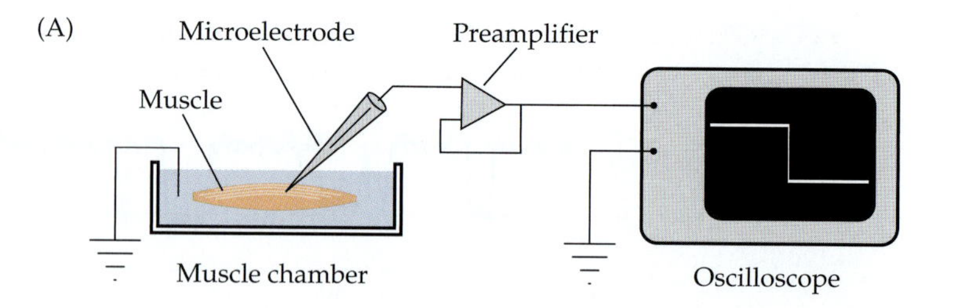

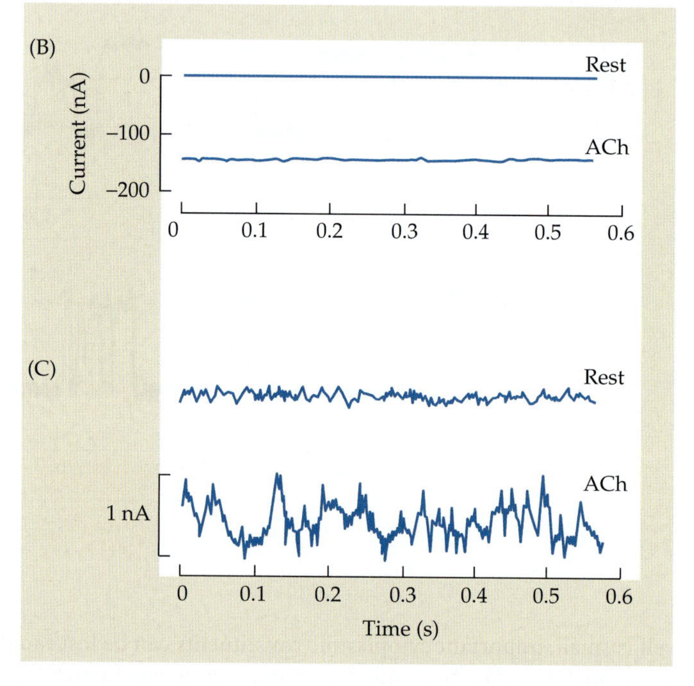

FIGURE 2.5 Intracellular Recording of Channel Noise. (A) Arrangement for recording membrane potentials of muscle fibers with a microelectrode. The electrode is connected to a preamplifier, and the signals are displayed on an oscilloscope or computer screen. Penetration of the electrode into a fiber is marked by the sudden appearance of the resting potential (downward deflection on the screen). After penetration, changes in potential due to channel activation can be measured. (B) Intracellular records of the effect of acetylcholine (ACh). In this experiment, additional circuitry was used to record membrane current (rather than membrane potential). At rest (upper trace), there is no current across the membrane; application of ACh produces about 130 nA of inward current (lower trace). (C) Traces in B shown at greater amplification. The baseline shows little fluctuation at rest; the inward current produced by ACh shows relatively large fluctuations ("noise"), due to random opening and closing of ACh-activated channels. Analysis of the increased noise yields values for the single-channel current and the mean open time of the channels. (B and C after Anderson and Stevens, 1973.)

synapse, ACh released from the presynaptic nerve terminal opens ligand-gated channels in the postsynaptic membrane that allow cations to enter, thus depolarizing the muscle (Chapter 9). Katz and Miledi applied ACh directly to the synaptic region and observed that the resulting depolarization was "noisy"; that is, during the depolarization, fluctuations in the electrical recording were larger than the normal baseline fluctuations at rest. This increase in noise was due to the random opening and closing of the ACh-activated channels. In other words, application of ACh resulted in the opening of a large number of channels, and this number fluctuated in a random way as ACh molecules bombarded the membrane.

By applying noise analysis techniques, Katz and Miledi were able to obtain information about the behavior of the individual channels activated by ACh. Subsequently, similar experiments were done on the same preparation by Anderson and Stevens[3] in which they used measurements of membrane current to deduce the size and duration of ionic currents through single channels (Figure 2.5B).

Although noise analysis involves moderately complex algebra, the underlying principles are straightforward: First, if the single-channel currents are relatively large, the noise will be large as well. Second, channels that open for a relatively long time will produce only low frequency noise; channels that open only briefly will produce higher frequency noise. Examination of the amplitude and the frequency composition of the noise produced by ACh-activated channels at the neuromuscular junction showed that about 10 million ions per second flowed through an open channel, and that the mean open time (τ) of the channel was 1 to 2 ms.

Although largely supplanted by patch clamp methods, noise analysis is still useful for studying membrane channels in cells that are not amenable to patch clamping—for example, in some parts of the intact central nervous system.[8] In addition, noise analysis is a relatively quick method of obtaining information about the properties of a population of channels, and is useful for relating whole-cell currents to identified channel types. It can-

[8]Gold, M. R., and Martin, A. R. 1983. *J. Physiol.* 342: 99–117.

not, however, provide detailed information about single-channel behavior, such as the existence of complicated kinetics or multiple conductance states.

Channel Conductance

The kinetic behavior of a channel—that is, the durations of its closed and open states—can provide information about the steps involved in channel opening and closing, and the rate constants associated with these steps. The channel current, on the other hand, is a direct measure of how rapidly ions move through a channel. The current depends not only on channel properties but also on the transmembrane potential. Consider, for example, the outside-out membrane patch shown in Figure 2.6, which contains a single spontaneously active channel that is permeable to potassium. The solutions in both the patch pipette and the bath contain 150 m*M* potassium. Potassium ions move in both directions through the open channel, but because the concentrations are equal, there is no net movement in either direction. Thus, no current is seen (Figure 2.6B).

Fortunately, the patch clamp recording system has an important feature that has not yet been mentioned: We can apply voltage to the pipette solution and thus produce a voltage difference across the membrane patch. When a potential of +20 mV is applied

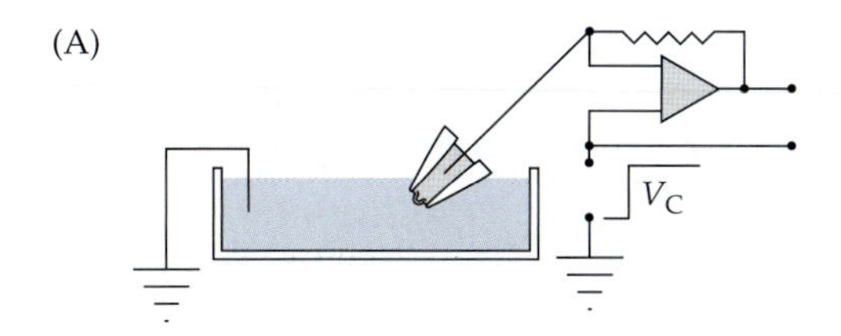

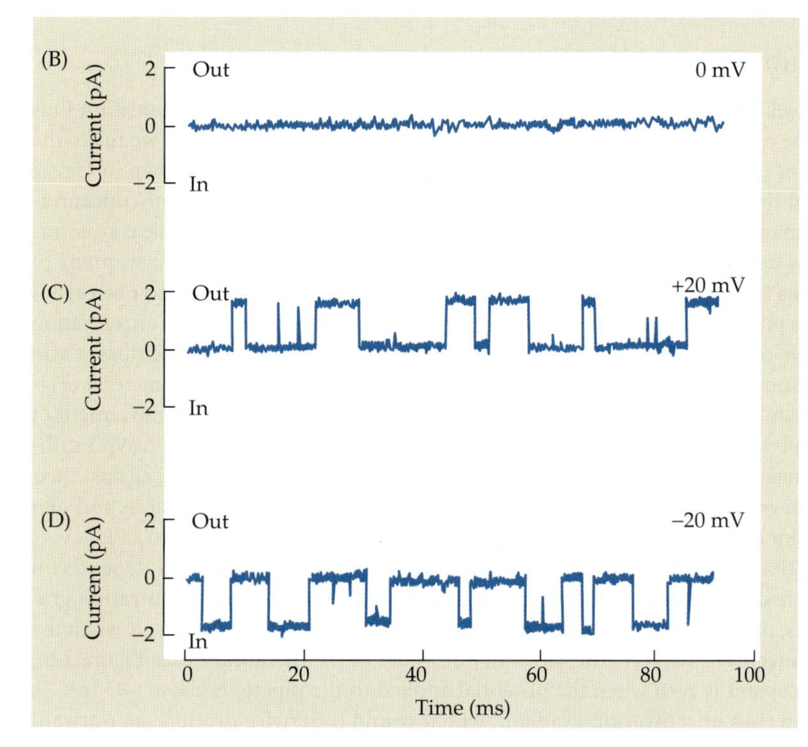

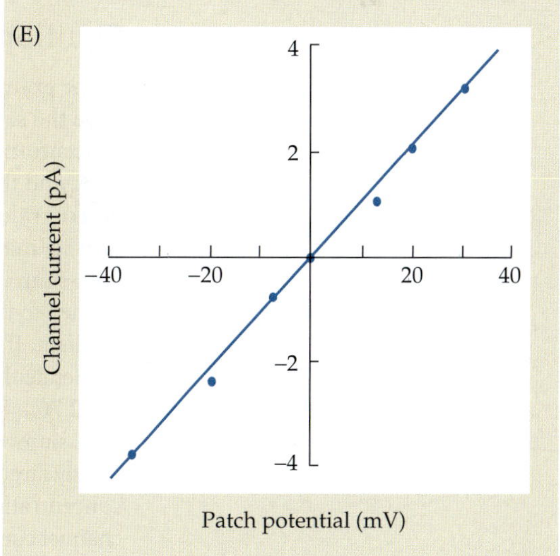

FIGURE 2.6 Effect of Potential on Currents through a single, spontaneously active potassium channel in an outside-out patch, with 150 m*M* potassium in both the electrode and the bathing solution. (A) The recording system. The output from the patch clamp amplifier is proportional to the current across the patch. The potential across the patch is equal to the potential (V_C) applied to the input of the amplifier as shown. Positive charge flowing out of the electrode is defined as positive current. (B) When no potential is applied to the patch, no channel currents are seen because there is no net flux of potassium through the channels. (C) Application of +20 mV to the electrode results in an outward current (upward deflections) of about 2 pA through the channels. (D) A –20 mV potential results in inward channel currents (downward deflections) of the same amplitude as in C. (E) Channel currents as a function of applied voltage. The slope of the line is the channel conductance (γ). In this case, γ = 110 pS (picosiemens).

(Figure 2.6C), each channel opening results in a pulse of outward current because positively charged potassium ions are driven outward through the channel by the electrical gradient between the pipette solution and the bath. On the other hand, when the inside is made negative by 20 mV (Figure 2.6D), current flows the other way, through the open channel into the pipette.

The effect of voltage on the size of the current is plotted in Figure 2.6E. The relationship is linear: The current (I) through the channel is proportional to the voltage (V) applied to it,

$$I = \gamma V$$

Many readers will recognize this as Ohm's law (Appendix A); those who do not should not worry. The constant of proportionality, γ, is called the **channel conductance**. For a particular applied voltage, a high-conductance channel will carry a lot of current, a low-conductance channel only a little.

Conductance has the units of siemens (S). In nerve cells, the potential across membrane channels is usually expressed in units of millivolts ($1 \text{ mV} = 10^{-3} \text{ V}$), currents through single channels in picoamperes ($1 \text{ pA} = 10^{-12} \text{ A}$), and conductances of single channels in picosiemens ($1 \text{ pS} = 10^{-12} \text{ S}$). In Figure 2.6E, a potential of +20 mV produced a current of about 2.2 pA, so the channel conductance ($\gamma = I/V$) was 2.2 pA/20 mV = 110 pS.

Conductance and Permeability

The conductance of a channel depends on two factors: The first is the ease with which ions can pass through the open channel; this is an intrinsic property of the channel known as the **channel permeability**. The second is the concentration of the ions in the region of the channel. Clearly, if there are no potassium ions in the inside or outside solution, there can be no current flow through an open potassium channel, no matter how large its permeability or how great a potential is applied. If only a few potassium ions are present, then for a given permeability and a given potential the channel current will be smaller than when potassium ions are present in abundance. One way to think of these relations is as follows:

Open channel → permeability

Permeability + ions → conductance

Equilibrium Potential

In the examples of channel current just considered, the concentrations of potassium ions were the same on both sides of the membrane patch. What happens when we make the concentrations different? Imagine that we make an outside-out patch as shown in Figure 2.7A, and that the potassium concentration in the bath is 3 m*M* (similar to the concentration outside many cells), and in the electrode 90 m*M* (similar to the cytoplasmic concentration in many cells). Now when the channel opens, there will be a net movement of potassium ions through the channel from the pipette to the bath, even when no potential is applied to the pipette (Figure 2.7B); potassium ions simply move down their concentration gradient. If we make the pipette positive, the potential gradient across the membrane will accelerate the outward potassium ion movement, and the channel current will increase (Figure 2.7C). If, on the other hand, we make the pipette negative, the outward movement of potassium will be retarded, and the channel current will decrease (Figure 2.7D). With sufficiently large negativity, potassium ions will flow *inward* across the membrane, against their concentration gradient (Figure 2.7E). If we make a number of such observations and plot channel current against applied voltage, we get a result like that shown in Figure 2.7F.

Figure 2.7F illustrates that the potassium current through the channel depends on both the electrical potential across the membrane and the potassium concentration gradient—that is, on the **electrochemical gradient** for potassium. Unlike the result when the potassium concentrations were the same on both sides of the membrane (see Figure 2.6), the channel current is zero when the potential applied to the pipette is about –85 mV. In this condition the concentration gradient, which would otherwise produce an outward

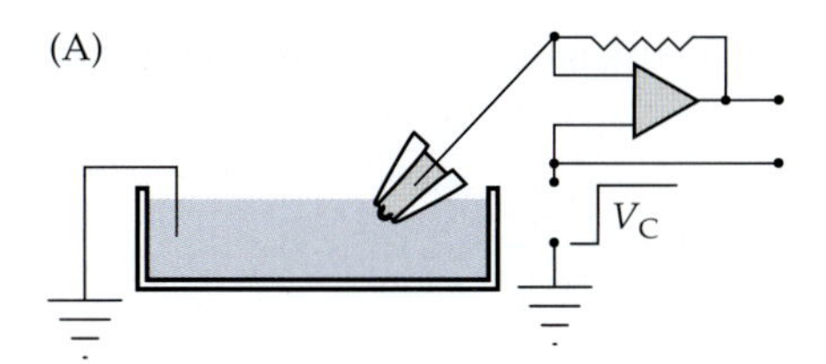

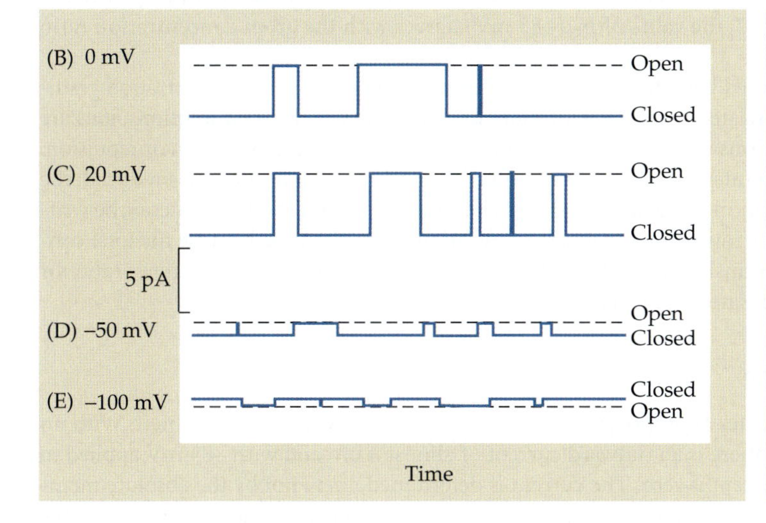

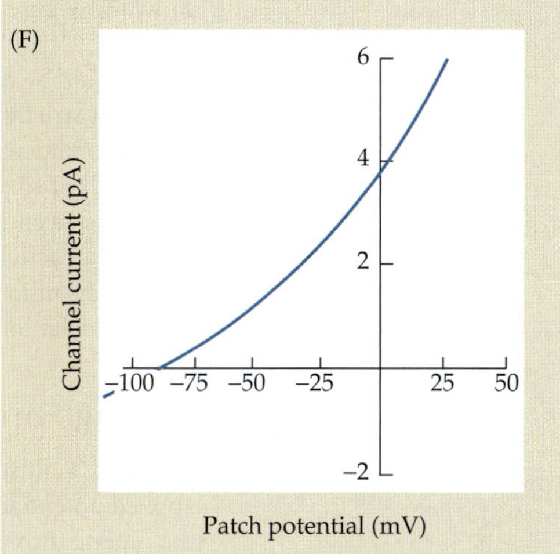

FIGURE 2.7 Reversal Potential for Potassium Currents in a hypothetical experiment using an outside-out patch with the concentration of potassium in the recording pipette ("intracellular" concentration) 90 m*M* and in the bathing solution ("extracellular" concentration) 3 m*M*. (A) Recording arrangement as in Figure 2.6. (B) With no potential applied to the pipette, the flux of potassium from the electrode to the bath along its concentration gradient produces outward channel currents. (C) When a potential of +20 mV is applied to the pipette, outward currents increase in amplitude. (D) Application of –50 mV to the pipette reduces outward currents. (E) At –100 mV, currents are reversed. (F) The current–voltage relation indicates zero current at –85 mV, which is the potassium equilibrium potential (E_K).

flux of potassium through the channel, is balanced exactly by the electrical potential gradient tending to move ions through the channel in the opposite direction. The difference in potential that just balances the potassium concentration difference is called the **potassium equilibrium potential** (E_K). The equilibrium potential depends only on the ion concentrations on either side of the membrane—not on the properties of the channel or on the mechanism of ion permeation through the channel.

The Nernst Equation

Exactly how large a potential is required to balance a given potassium concentration difference across the membrane? One guess might be that E_K would be proportional simply to the difference between the inside concentration, $[K]_i$, and the outside concentration, $[K]_o$; this is not quite right. Rather the required potential depends on the difference between the *logarithms* of the concentrations:

$$E_K = k\left(\ln[K]_o - \ln[K]_i\right)$$

The constant k is given by RT/zF, where R is the thermodynamic gas constant, T the absolute temperature, z the valence of the ion (in this case +1), and F the Faraday (the number of coulombs of charge in 1 mole of a monovalent ion). The answer, then, is

$$E_K = \frac{RT}{zF}\left(\ln[K]_o - \ln[K]_i\right)$$

which is the same as

$$E_K = \frac{RT}{zF}\ln\frac{[K]_o}{[K]_i}$$

This is the Nernst equation for potassium. RT/zF has the dimensions of volts and is equal to about 25 mV at room temperature (20°C). It is sometimes more convenient to use the logarithm to the base 10 (log) of the concentration ratio, rather than the natural logarithm (ln). Then RT/zF must be multiplied by ln(10), or 2.31, which gives a value of 58 mV. In summary,

$$E_K = 25 \ln \frac{[K]_o}{[K]_i} = 58 \log \frac{[K]_o}{[K]_i}$$

At mammalian body temperature (37°C), 58 mV increases to about 61 mV. For the cell shown in Figure 2.7, the value of E_K (–85 mV) agrees with the given concentration ratio (1:30).

It should be noted here that the rate of diffusion of an ion down a concentration gradient is not strictly related to its concentration. In all but the very weakest solutions, ions are subject to interactions with one another—for example, electrostatic attraction or repulsion. The result of such interactions is that the effective concentration of the ion is reduced; the effective concentration is called the **activity**. The Nernst equation should, in theory, be written with an activity ratio rather than a concentration ratio. However, because the total concentrations of ions inside and outside the cell are similar (Chapter 5), the activity ratio for any particular ion is not significantly different from its concentration ratio.

Driving Force

Figure 2.7F illustrates an important point about current flow through channels: With no applied potential there is an outward current of almost 4 pA, and with –85 mV applied to the pipette the current is zero. The current is determined, then, not by the absolute membrane potential (V_m), but rather by the difference between the membrane potential and the equilibrium potential for the ion in question, in this case potassium (E_K). This difference, $V_m - E_K$, is the **driving force** for the movement of ions through the channel. Referring again to Figure 2.7F, when the membrane potential is zero the driving force is +85 mV.

Nonlinear Current–Voltage Relations

A second feature of the current–voltage relation in Figure 2.7F is that, unlike the one in Figure 2.6E, it is not linear. As we move away from the equilibrium potential in the depolarizing direction (i.e., toward zero), the current changes more rapidly than if we move in the hyperpolarizing direction. This is because of the dependence of conductance on concentration. The concentration of potassium inside the pipette is much higher than in the external solution. As a result more ions are available to carry outward current than to carry inward current. The farther we move away from the equilibrium potential, the more prominent the effect becomes, so the current–voltage relation has a marked upward curvature, even though, in this example, the channel permeability is quite independent of voltage.

Nonlinear current–voltage relations also occur in some channels because the channels themselves rectify; that is, their permeability is voltage-dependent, so they allow ions to move through their pores in one direction much more readily than in the other. One example is a voltage-sensitive potassium channel, called the inward rectifier channel, that allows potassium ions to move into the cell when the membrane potential is negative with respect to the potassium equilibrium potential, but little or no outward movement when the difference in potential is reversed. The current–voltage relation of this channel is similar to that shown in Box 2.1.

Ion Permeation through Channels

How do ions actually pass through channels? One way could be by diffusion through a water-filled pore. Diffusion formed the basis of early ideas about ion permeation, but for most channels it does not provide an adequate description of the permeation process. This is because the channels themselves interact with the ions. For example, because they are charged, ions in solution are always accompanied by closely apposed water molecules (see Figure 2.1). In the case of cations, the water molecules are oriented so that their oxygens, which carry net negative charges, lie closest to the ions. If the pore is relatively nar-

Box 2.1 Measuring Channel Conductance

Investigators often state that a particular channel has a particular conductance, say 100 pS. Because conductance depends on concentration, such a statement tells us little unless we know the ionic conditions under which the measurement was made. For example, if we consider a potassium channel with a potassium concentration of 150 m*M* on either side of the membrane, then in a more physiological environment, with only 5 m*M* potassium on the extracellular side, the channel conductance can be expected to be five to ten times smaller.

A second problem arises when the current–voltage relation is not linear, as in Figure 2.7, when the ion concentrations on either side of the channel are not symmetrical, or as in the illustration shown here, when the channel is heavily rectifying. We have defined channel conductance, γ, as the slope of the current–voltage relation, $\gamma = I/V$, but the definition becomes ambiguous in the circumstances described here because the slope is not constant.

Experimenters have chosen two ways of specifying conductance in such cases. One way is to hold the voltage across the membrane patch at a particular value and measure the resulting current through the channel. In the illustration, the equilibrium potential for the current is –75 mV; when the patch is held at –25 mV, the channel current is 0.6 pA. Thus, a driving force of 50 mV produces a current of 0.6 pA, and the channel conductance is 0.6 pA/50 mV = 12 pS. This is the slope of the brown line, which is known as **chord conductance.** Note that it is necessary to say at what membrane potential the measurement was made; the chord conductance at –25 mV is not the same as the chord conductance at 0 mV.

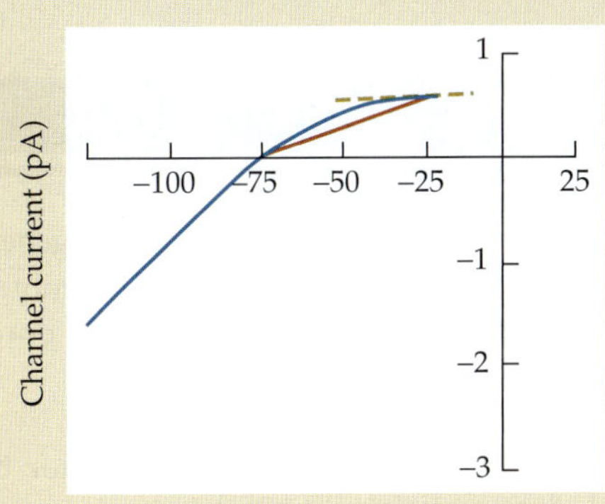

The second way of specifying conductance is to measure the slope of the current–voltage relation at the point of interest. This is the slope of the dashed line and is called the **slope conductance,** which in this example is about 3 pS at –25 mV. The measurement tells us that although the channel is passing a substantial current at –25 mV, it will not pass proportionately more if the driving force is increased.

In summary, the conductance characteristics of a channel can be specified completely only by showing a complete current–voltage relation and specifying the ionic conditions under which it was obtained. Otherwise, single numbers given in the literature for channel conductance usually imply chord conductance. Perhaps the most useful convention for comparing one type of channel to another would be to state the slope conductance at the equilibrium potential, with specified ion concentrations.

row, an ion must acquire a certain amount of energy to escape from its associated waters of hydration and squeeze through the neck of the channel (Chapter 3). Once in the channel, the ion may be attracted to or repelled by electrostatic charges lining the channel wall, or it may be bound to sites from which it must escape to continue its journey. Such interactions affect both ion selectivity and the rate of ion flux through the channel. Channel models that deal with ion permeation in this way are called Eyring rate theory models.[9] In general, such models are more successful than simple diffusion models in describing channel selectivity and conductance.

Significance of Ion Channels

The array of channel types described in this chapter enables neurons to receive signals from the external world or from other neurons, to carry signals over long distances, to modulate the activity of other neurons and effector organs, and to alter their own signaling properties in response to internal metabolic changes. All the complexities of perception and analysis of signals depend ultimately on ion channel activity, as does the generation of complex motor outputs from the nervous system.

An important point to remember is that all the ion fluxes that underlie signaling are due to ions moving passively through open channels along concentration and potential gradients. In other words, the neuron makes use of standing electrochemical gradients to

[9]Johnson, F. H., Eyring, H., and Polissar, M. J. 1954. *The Kinetic Basis of Molecular Biology*. Wiley, New York.

generate ion movements, and hence to generate electrical signals. Such fluxes eventually would dissipate the gradients, but this does not happen because cells use metabolic energy to maintain the ionic composition of the cytoplasm. Specialized mechanisms underlying the active transport of ions are discussed in Chapter 4.

Summary

- Electrical signals in the nervous system are generated by the movement of ions across the nerve cell membrane. These ionic currents flow through the aqueous pores of membrane proteins known as ion channels.

- Channels vary in their selectivity: Some cation channels allow only sodium, potassium, or calcium to pass; others are less selective. Anion channels are relatively nonselective for smaller anions, but they pass mainly chloride ions because of the relative abundance of chloride in the extracellular and intracellular fluids.

- Channels fluctuate between open and closed states. Each channel has a characteristic mean open time. When channels are activated their probability of opening increases. Deactivation reduces opening frequency. Channels may also be inactivated or blocked.

- Channels can be classified by their mode of activation: stretch-activated, voltage-activated, or ligand-activated.

- Ions move through channels passively in response to concentration and electrical gradients across the membrane.

- The net flux of ions through a channel due to a concentration gradient can be reduced by an opposing electrical gradient. The electrical potential that reduces the net flux to exactly zero is called the equilibrium potential for that ionic species. The relation between equilibrium potential and the concentration gradient is given by the Nernst equation.

- The driving force for the movement of an ion across the membrane is the difference between its equilibrium potential and the membrane potential. Ionic current flow through a single channel depends on the driving force for the ion in question and on the conductance of the channel for that ion. The conductance depends, in turn, on the intrinsic ionic permeability of the channel and on the inside and outside ion concentrations.

SUGGESTED READING

Hille, B. 1992. *Ion Channels in Excitable Membranes*, 2nd Ed. Sinauer, Sunderland, MA, pp. 291–314.

Pun, R. Y. K., and Lecar, H. 1998. Patch clamp techniques and analysis. In N. Sperelakis (ed.), *Cell Physiology Source Book*, 2nd Ed. Academic Press, San Diego, CA, pp. 391–405.

3 STRUCTURE OF ION CHANNELS

THE MOLECULAR STRUCTURE OF ION CHANNELS can be resolved and related to their function by a variety of experimental methods. These include biochemical isolation of channel proteins, molecular cloning to determine amino acid sequences of the proteins, site-directed mutagenesis to alter the sequences in selected locations, and expression of channel proteins in host cells, such as *Xenopus* oocytes, to examine channel function. In addition, high-resolution electron microscopy, and electron and X-ray diffraction, make it possible to determine the physical conformation of channels.

These combined experimental approaches have been applied most extensively to a ligand-activated channel, the nicotinic acetylcholine receptor. This receptor is composed of five separate subunits arranged around a central core. Two of these—the α subunits—contain receptors for the ligand. Each subunit contains four membrane-spanning regions (designated M1, M2, M3, and M4), connected by intracellular and extracellular loops. The five M2 regions have been identified as lining the pore and forming the gating structure of the channel. The acetylcholine receptor is representative of a genetic superfamily of ligand receptors that includes receptors for glycine, γ-aminobutyric acid, and serotonin.

Voltage-activated channels form another superfamily. Experimental evidence indicates that the voltage-activated sodium channel is a single large molecule with four repeating domains arranged around a central core, each with six transmembrane segments (designated S1–S6). In each domain the loop of amino acids between S5 and S6 dips into the center of the structure to contribute to the pore lining. Voltage-activated calcium channels have a similar structure. Voltage-activated potassium channels are similar in molecular configuration, but with one important difference: Instead of being single molecules, they are assembled from four separate subunits.

Amino acid sequences and partial structural information have been obtained for a number of other families of receptors and channels. With few exceptions, all are made up of an array of subunits, each subunit containing at least two transmembrane regions, and each with a potential pore-forming loop.

For the nervous system to function properly, neurons must perform a widely varied repertoire of electrical behavior. Thus, an impulse generated by one neuron may suppress the electrical activity of dozens of its neighbors, travel long distances to produce excitation of other neuronal groups, and influence the responsiveness of still other target neurons in a variety of more subtle ways. All of this behavior depends on the activation or deactivation of ion channels, thereby regulating the flow of ion currents across the nerve cell membranes.

An essential task is to understand how ion channels perform their function. This understanding depends on knowledge of the molecular structure of the channel proteins. Amino acid sequences have been determined for a large number of ion channels, and their configurations in the plasma membrane have been postulated. In this chapter we

Box 3.1 Cloning Receptors and Channels

The application of molecular genetic techniques to study of the nervous system has triggered extraordinarily rapid progress in the identification and characterization of proteins in neurons and their synaptic targets. What makes this approach so powerful is that the techniques are relatively straightforward, widely applicable, and (because of the intrinsic amplification of the procedures) exquisitely sensitive.

As the figure shows, the first step in isolating a cDNA clone for a receptor or channel is to obtain mRNA from a tissue in which the protein is synthesized. The chance of isolating the clone of interest depends on the relative abundance of the corresponding mRNA in the starting material. Thus, appropriate sources of mRNA for cloning acetylcholine receptor subunits would be electric organs of electric fish and eels or embryonic skeletal muscle. Reverse transcriptase enzymes are used to make cDNA copies of total or size-selected poly(A)+ mRNA. These cDNA copies are incorporated into vectors (specially constructed bits of DNA resembling viruses) that can grow and multiply in host bacteria. The collection of viral or plasmid vectors is called a library.

To screen such a cDNA library, host bacteria are exposed to the library under conditions such that no bacterium is "infected" with more than one vector. Colonies of bacteria carrying the cDNA of interest are identified by oligonucleotide probes that recognize the amplified cDNA or by antibody probes that recognize the protein synthesized from the cDNA. Bacteria harboring a vector bearing the appropriate cDNA insert can then be grown in large numbers and the cDNA insert isolated and sequenced. The sequence of nucleotides is translated into the corresponding sequence of amino acids, and the deduced amino acid sequence can be examined for features such as hydrophobic transmembrane domains, α-helical regions, sites for posttranslational modifications, or stretches of amino acids that are similar to those in other proteins.

A variety of enzymes in addition to reverse transcriptase are crucial for these and other manipulations: (1) restriction

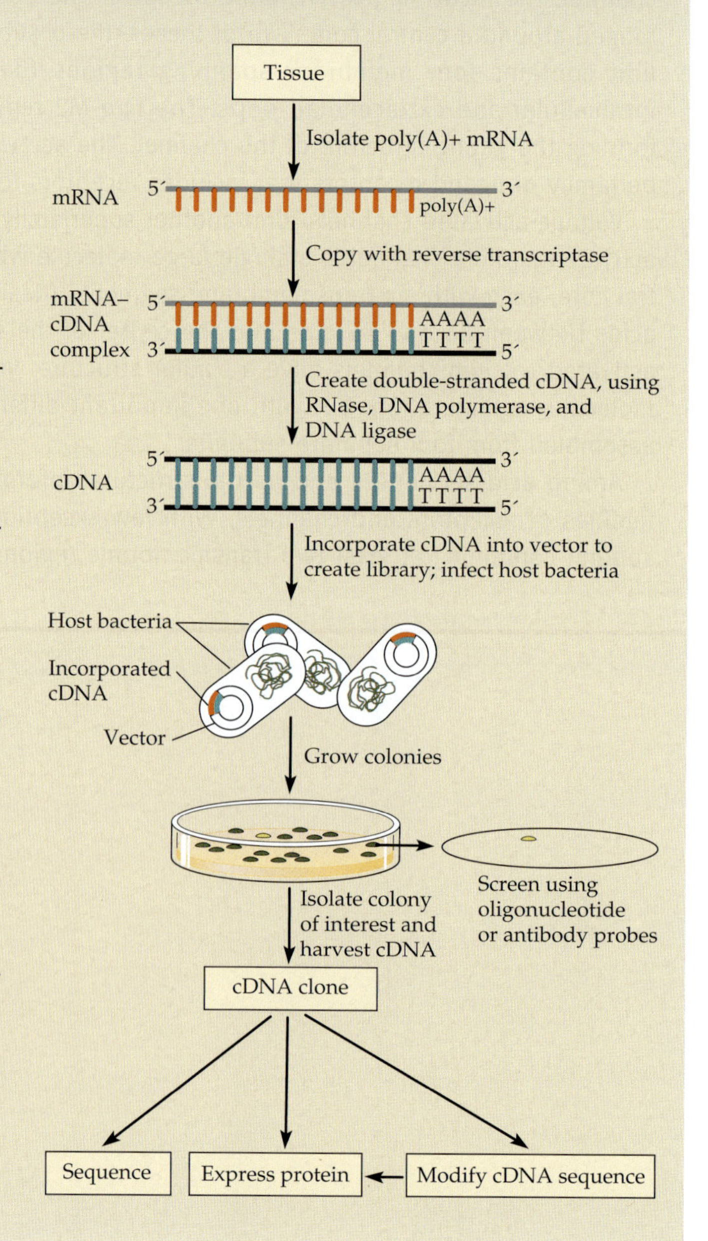

examine two types of channels in detail: The first is a group of ligand-activated channels represented by the nicotinic acetylcholine receptor. The second is a group of voltage-activated cation channels that allow the movement of sodium, potassium, and calcium across the cell membrane. These two channel types are the models on which other postulates of channel structure are based.

What does a channel molecule look like, and how does it work? In recent years remarkable progress has been made in deducing the principles of channel structure and function. Two experimental advances have been instrumental in obtaining this information. The first was the ability to isolate and sequence cDNA (complementary DNA) clones of channel proteins (Box 3.1) and thus to obtain the corresponding amino acid sequences. The techniques used in this process also provide the opportunity to alter bases in the DNA (**site-directed mutagenesis**) and thereby substitute one amino acid for another at selected locations in the protein.

The second advance was the development of techniques whereby cDNA clones can be used to express channel proteins in host cells such as *Xenopus* oocytes (see Box 3.3). In this way the functional properties of cloned channels can be measured. The combination of techniques makes it possible to tinker with specific regions of a channel molecule and then determine how such tinkering affects its function. For example, changing even a single amino acid can have a marked effect on the ion selectivity of a channel.

THE NICOTINIC ACETYLCHOLINE RECEPTOR

The first channel to be studied in detail was the nicotinic ACh receptor ("nAChR," or simply "AChR"). Note that for ligand-activated channels, the word "receptor" rather than "channel" is used routinely; this is because characterization of such molecules has relied primarily on the binding of activating molecules (agonists), antagonists, toxins, and antibodies, rather than on the specific channel properties of the protein. Nicotinic ACh receptors are expressed in postsynaptic membranes of vertebrate skeletal muscle fibers, in

endonucleases that recognize specific nucleotide sequences and cut double-stranded DNA exactly at those points; (2) exo- and endonucleases that selectively digest single- or double-stranded RNA or DNA; (3) ligase enzymes that couple DNA molecules together; and (4) polymerase enzymes that replicate them.

Once the cDNA clone for a protein has been isolated, similar recombinant DNA techniques can be used to test the function of its various domains. Parts of the cDNA sequence can be deleted, then the remaining sequence inserted back into an appropriate vector and a truncated version of the protein produced. Stretches of cDNA from different proteins can be spliced together, and the resulting hybrid protein expressed and tested for activity. Using specific oligonucleotide primers to initiate DNA synthesis, individual bases in the cDNA can be changed by site-directed mutagenesis to produce proteins in which a single amino acid has been altered.

Several techniques are used to express such modified transcripts. Bacterial hosts can often be used to produce large amounts of protein; however, many eukaryotic proteins are not properly processed or glycosylated in bacteria. This drawback can be overcome by injecting RNA synthesized from the cDNA into oocytes (see Box 3.3) or by transfecting cDNA vectors into eukaryotic cell lines. Then the proteins are synthesized by the host cells and can be incorporated into their membranes. Transfected cells can be fused with other cells, such as muscle cells, that are more convenient for analysis of the properties of the encoded protein.

Although the results of such experiments must be interpreted with care, the ability to make specific, discrete alterations in protein structure is a powerful tool for understanding how proteins carry out their functions. In addition, the screening of libraries with probes derived from cDNAs encoding proteins of known function allows one to search for and isolate clones for related proteins. Small differences in homologous proteins can be identified and characterized by using the polymerase chain reaction (PCR) technique to amplify selected regions of mRNAs. The sensitivity and specificity of PCR are so great that mRNA from a single neuron can be analyzed. Such experiments have led to the characterization of superfamilies of receptor and channel proteins and to the identification of a surprising number of isoforms of their subunits.

neurons throughout the nervous systems of invertebrates and vertebrates, and in the neuroeffector junctions of electric organs of a number of electric fish.

The receptors are activated by ACh released from presynaptic nerve terminals, and on activation they open to form channels through which cations can enter or leave the postsynaptic cell. They are designated "nicotinic" to reflect the fact that the actions of ACh are mimicked by nicotine, and to distinguish them from the very different ACh receptors (mAChRs) that can be activated by muscarine. Muscarinic ACh receptors are not ion channels; instead their activation sets in motion intracellular messenger systems that, in turn, affect ion channel activity.

Biochemical isolation and characterization of the nicotinic AChR were facilitated by the availability of a concentrated source of receptors: the remarkably dense assembly of synapses on electrocyte membranes in the electric organ of the ray *Torpedo*. The electrocytes of this and other strongly electric fish are modified muscle cells. Large numbers of them are arrayed anatomically such that when depolarized simultaneously they can produce, as a group, voltages approaching 100 V, sufficient to stun nearby prey in the surrounding water. After being extracted from the electrocyte membranes, AChR molecules were separated from other membrane proteins by means of their high affinity for α-bungarotoxin, a neurotoxin known to bind to channels with high specificity in intact electrocytes and in vertebrate muscle. The *Torpedo* AChR was found to consist of four glycoprotein subunits (α, β, γ, and δ), of about 40, 50, 60, and 65 kDa (kilodaltons), respectively. The isolated AChR was shown to retain the major functional properties of the native ionic channel when reincorporated into lipid vesicles.[1]

Physical Properties of the ACh Receptor

The size and orientation of the intact channel with respect to the lipid membrane has been determined by high-resolution electron microscope imaging and other physical techniques.[2–4] Its physical structure is shown in Figure 3.1. Five subunits—two α, one β, one γ, and one δ—form a circular array around a central pore. The molecule is about 8.5

[1]Tank, D. W., et al. 1983. *Proc. Natl. Acad. Sci. USA* 80: 5129–5133.

[2]Wise, D. S., Schoenborn, B. P., and Karlin, A. 1981. *J. Biol. Chem.* 256: 4124–4126.

[3]Unwin, N., Toyoshima, C., and Kubalek, E. 1988. *J. Cell Biol.* 107: 1123–1138.

[4]Toyoshima, C., and Unwin, N. 1988. *Nature* 336: 247–250.

FIGURE 3.1 The ACh Receptor. (A) The complete AChR consists of five subunits—two α, one β, one γ, and one δ—spaced radially in increments of about 72° around a central core. The α subunits contain receptor sites for acetylcholine. The position of the β and γ subunits may be reversed. (B) Longitudinal and transverse electron microscope images of cylindrical vesicles from postsynaptic membranes of *Torpedo*, showing closely packed ACh receptors. (C) Transverse section of the tube at higher magnification. (D) Further enlarged image of a single ACh receptor, showing its position and size relative to the membrane bilayer. Dense blob under the receptor is intracellular receptor–associated protein. (A based on Stroud and Finer-Moore, 1985, and Toyoshima and Unwin, 1988; B, C, and D kindly provided by N. Unwin.)

(A)

Extracellular

δ α γ β α

Intracellular

5 nm

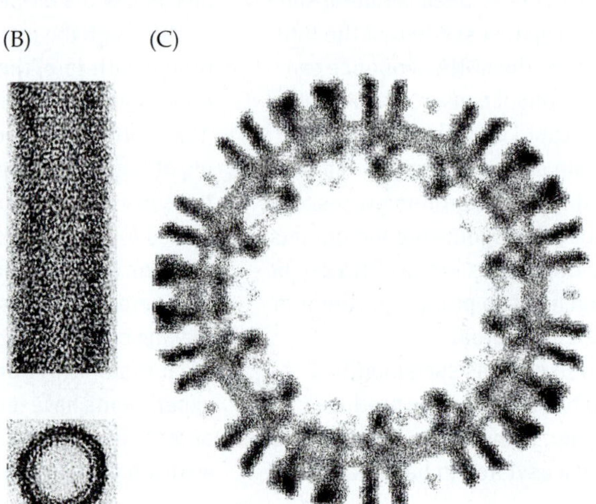

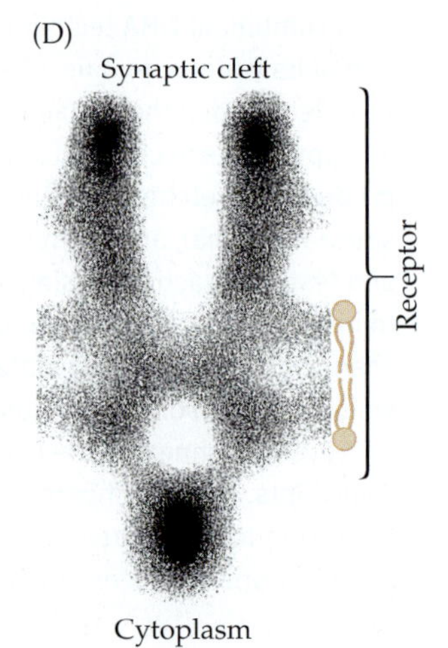

nm across at its widest point, which is in the extracellular domain, and about 11 nm long. The extracellular portion extends about 5 nm above the surface of the membrane. The central pore is about 0.7 nm in diameter, as predicted from earlier measurements of its selectivity to large cations.[5,6]

Amino Acid Sequence of AChR Subunits

After preliminary biochemical studies,[7] the cDNA for each subunit was cloned and sequenced.[8–10] Figure 3.2 shows the corresponding amino acid sequence for the α subunit from *Torpedo* (human and bovine subunits are slightly different). The sequences of the other three subunits are highly similar (homologous), with various amino acid insertions and deletions, and discussion of the structural configuration of any one is generally applicable to the others.

[5]Maeno, T., Edwards, C., and Anraku, M. 1977. *J. Neurobiol.* 8: 173–184.

[6]Dwyer, T. M., Adams, D. J., and Hille, B. 1980. *J. Gen. Physiol.* 75: 469–492.

[7]Raftery, M. A., et al. 1980. *Science* 208: 1454–1457.

[8]Noda, M., et al. 1982. *Nature* 299: 793–797.

[9]Noda, M., et al. 1983. *Nature* 301: 251–255.

[10]Noda, M., et al. 1983. *Nature* 302: 528–532.

Met Ile Leu Cys

–20 ... –10

Ser Tyr Trp His Val Gly Leu Val Leu Leu Leu Phe Ser Cys Cys Gly Leu Val Leu Gly

10 ... 20

Ser Glu His Glu Thr Arg Leu Val Ala Asn Leu Leu Glu Asn Tyr Asn Lys Val Ile Arg

30 ... 40

Pro Val Glu His His Thr His Phe Val Asp Ile Thr Val Gly Leu Gln Leu Ile Gln Leu

50 ... 60

Ile Ser Val Asp Glu Val Asn Gln Ile Val Glu Thr Asn Val Arg Leu Arg Gln Gln Trp

70 ... 80

Ile Asp Val Arg Leu Arg Trp Asn Pro Ala Asp Tyr Gly Gly Ile Lys Lys Ile Arg Leu

90 ... 100

Pro Ser Asp Asp Val Trp Leu Pro Asp Leu Val Leu Tyr Asn Asn Ala Asp Gly Asp Phe

110 ... 120

Ala Ile Val His Met Thr Lys Leu Leu Leu Asp Tyr Thr Gly Lys Ile Met Trp Thr Pro

130 ... 140

Pro Ala Ile Phe Lys Ser Tyr Cys Glu Ile Ile Val Thr His Phe Pro Phe Asp Gln Gln

150 ... 160

Asn Cys Thr Met Lys Leu Gly Ile Trp Thr Tyr Asp Gly Thr Lys Val Ser Ile Ser Pro

170 ... 180

Glu Ser Asp Arg Pro Asp Leu Ser Thr Phe Met Leu Ser Gly Glu Trp Val Met Lys Asp

190 ... 200

Tyr Arg Gly Trp Lys His Trp Val Tyr Tyr Thr Cys Cys Pro Asp Thr Pro Tyr Leu Asp

210 ... 220

Ile Thr Tyr His Phe Ile Met Gln Arg Ile Pro Leu Tyr Phe Val Val Asn Val Ile Ile

230 ... 240

Pro Cys Leu Leu Phe Ser Phe Leu Thr Gly Leu Val Phe Tyr Leu Pro Thr Asp Ser Gly

M1

250 ... 260

Glu Lys Met Thr Leu Ser Ile Ser Val Leu Leu Ser Leu Thr Val Phe Leu Leu Val Ile

M2

270 ... 280

Val Glu Leu Ile Pro Ser Thr Ser Ser Ala Val Pro Leu Ile Gly Lys Tyr Met Leu Phe

290 ... 300

Thr Met Ile Phe Val Ile Ser Ser Ile Ile Ile Thr Val Val Val Ile Asn Thr His His

M3

310 ... 320

Arg Ser Pro Ser Thr His Thr Met Pro Gln Trp Val Arg Lys Ile Phe Ile Asp Thr Ile

330 ... 340

Pro Asn Val Met Phe Phe Ser Thr Met Lys Arg Ala Ser Lys Glu Lys Gln Glu Asn Lys

350 ... 360

Ile Phe Ala Asp Asp Ile Asp Ile Ser Asp Ile Ser Gly Lys Gln Val Thr Gly Glu Val

370 ... 380

Ile Phe Gln Thr Pro Leu Ile Lys Asn Pro Asp Val Lys Ser Ala Ile Glu Gly Val Lys

390 ... 400

Tyr Ile Ala Glu His Met Lys Ser Asp Glu Glu Ser Ser Asn Ala Ala Glu Glu Trp Lys

410 ... 420

Tyr Val Ala Met Val Ile Asp His Ile Leu Leu Cys Val Phe Met Leu Ile Cys Ile Ile

M4

430

Gly Thr Val Ser Val Phe Ala Gly Arg Leu Ile Glu Leu Ser Gln Glu Gly

FIGURE 3.2 Amino Acid Sequence of the AChR Subunit. Sequences in color indicate hydrophobic regions (M1, M2, M3, and M4) that are capable of forming membrane-spanning α helices. In the M1 region, 16 of the 22 amino acids are hydrophobic (see Box 3.2). The other membrane-spanning regions are of similar composition. The initial underscored segment is the signal sequence. (After Numa et al. 1983.)

Higher-Order Chemical Structure

Although the primary structure of the subunits does not provide unique information about how the protein is arranged in the membrane, various models can be made, based on the characteristics of the amino acids in the sequence. As with any very large protein, segments of the molecule can be expected to fold into ordered secondary structures such as α helices or β sheets. These secondary structures themselves fold to produce a tertiary structure in each subunit. Finally five subunits (two α, one β, one γ, and one δ) join together to form the final quaternary structure—that is, the complete channel.

Models of the secondary and tertiary structures depend on a number of considerations—for example, identifying in the primary sequence extended runs of nonpolar (and hence hydrophobic) amino acid residues capable of forming α helices or other structures of sufficient length to span the membrane. In the original model proposed by Numa and his colleagues for the subunit structures,[9] four such regions were identified (M1–M4; see Figure 3.2), and the model shown in Figure 3.3 was postulated. You may find it useful to use the hydropathy indices (Box 3.2) of the amino acid residues in these regions to verify the validity of these conclusions. Recent evidence suggests that all the transmembrane regions except M2 may be β sheets rather than α helices (see Figure 3.4A).

How do we know which parts of the molecule are extracellular and which intracellular? To begin with, the NH_2 terminus is preceded by a relatively hydrophobic region of 24 amino acids (see Figures 3.2 and 3.5) that is taken to be the signal sequence necessary for insertion of the protein into the membrane. Consequently, the NH_2 terminus is assumed to be extracellular. Consistent with this arrangement is the fact that the two adjacent cysteine residues at positions 192 and 193 in the subunit are associated with the extracellular binding site for ACh. In addition, the initial extracellular segment constitutes about half of the entire molecule, which is in agreement with the mass distribution of the intact receptor (see Figure 3.1). Given an even number of membrane crossings, the COOH terminus is also extracellular. The general features of the model are in agreement with other observations. For example, when closely packed, ACh receptors aggregate in pairs (dimers) formed by disulfide cross-links between cysteine residues near the COOH termini of the δ subunits. These cross-links have been shown to be extracellular.[11,12]

[11]McCrea, P. D., Popot, J-L., and Engleman, D. M. 1987. *EMBO J.* 6: 3619–3626.

[12]DiPaola, M., Czajkowski, C., and Karlin, A. 1989. *J. Biol.Chem.* 264: 15457–15463.

[13]Miledi, R., Parker, I., and Sumikawa, K. 1983. *Proc. R. Soc. Lond. B* 218: 481–484.

Channel Structure and Function

An essential technique for relating channel function to structure is the expression of channels in *Xenopus* oocytes (Box 3.3) or other host cells by injection of the appropriate mRNA, or transfection with cDNA vectors.[13] In such experiments, electrical recording techniques

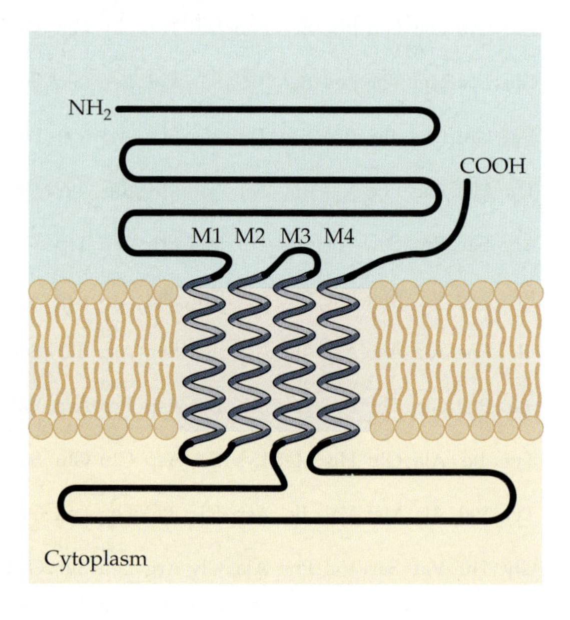

FIGURE 3.3 Model of the Tertiary Structure of an AChR Subunit, as proposed originally from amino acid sequence analysis. Regions M1 through M4 each form membrane-spanning helices, and both the carboxy (COOH) terminus and the amino (NH_2) terminus of the peptide lie in the extracellular space.

are used to measure the characteristics of single-channel currents or whole-cell currents (representing the behavior of the entire population of inserted channels). Oocytes normally do not express nicotinic ACh receptors or voltage-activated sodium channels in their membranes. Yet after the appropriate message has been injected, they not only express the protein subunits, but also miraculously assemble them to form functionally active channels.

Channel structure can be altered by site-directed mutagenesis. In this technique, mutant cDNAs are constructed, with mutations directed at a particular site in the channel protein,

Box 3.2 Classification of Amino Acids

Channel subunits, like all other peptides, are composed of amino acids, and the amino acid side chains are what determine many of the local chemical and physical properties of the subunits. The 20 amino acids fall into three groups: basic, acidic, and neutral, as shown below. The three-letter and one-letter abbreviations for each amino acid are included in parentheses. Acidic and basic amino acids are hydrophilic. The neutral amino acids are arranged according to hydropathy indices (numbers below each amino acid) proposed by Kyte and Doolittle,[14] starting with the most hydrophobic (positive numbers) and progressing to the most hydrophilic (negative numbers). Sections of a peptide are candidates for membrane-spanning regions if they contain sequences of hydrophobic amino acids that are capable of forming an α helix long enough to traverse the lipid bilayer (about 20 amino acid residues; see Figure 3.2) or a membrane-spanning β sheet (see Figure 3.4A).

[14]Kyte, J., and Doolittle, R. F. 1982. *J. Mol. Biol.* 157: 105–132.

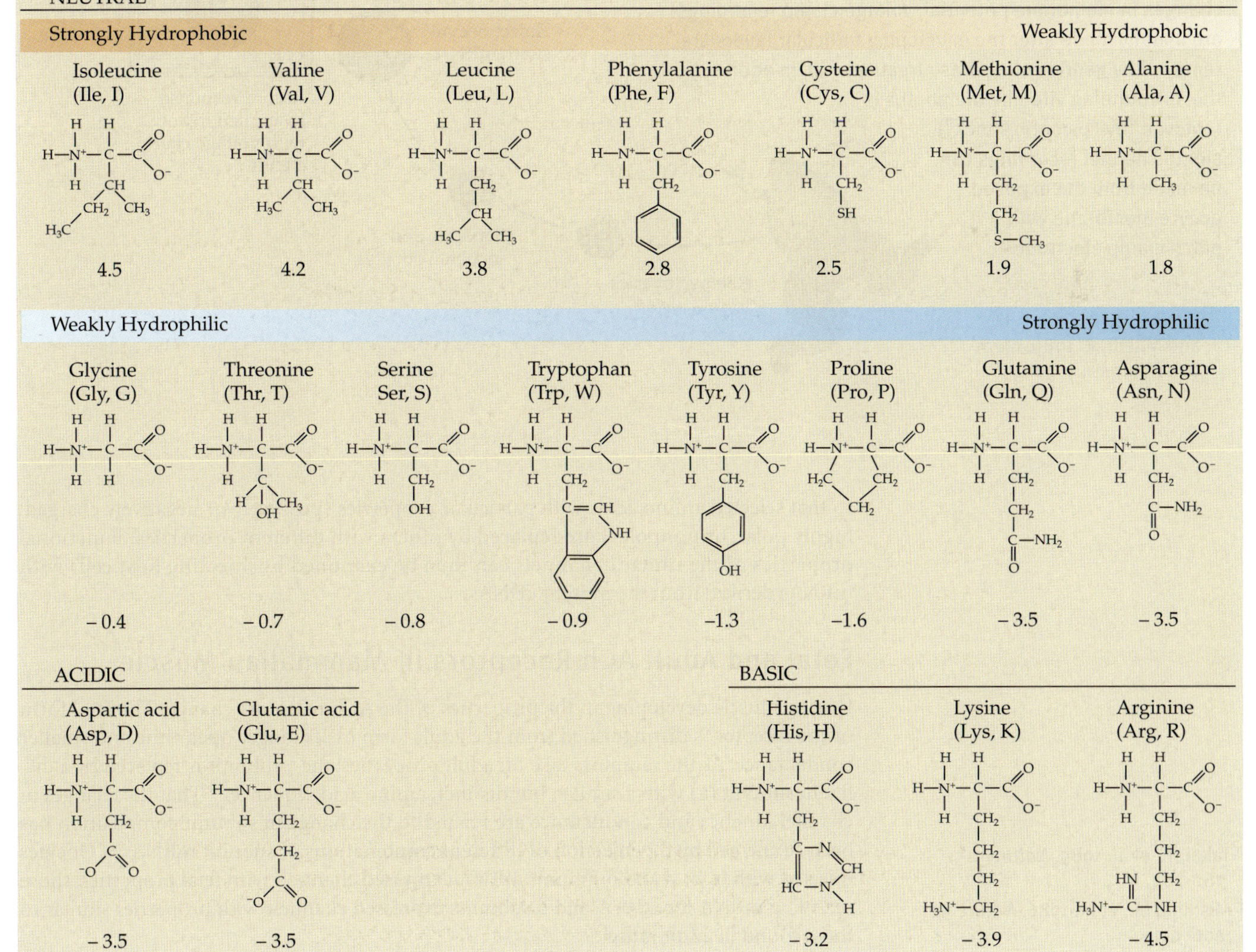

Box 3.3 Expression of Receptors and Channels in *Xenopus* Oocytes

The expression of proteins by messenger RNA in oocytes, as well as in other host cells, has been an indispensable tool for examining the properties of receptors and channels. The methods of oocyte preparation and mRNA isolation were first described in detail by Miledi and his colleagues.[15] The steps are shown here schematically.

As a first step in preparing mRNA for injection, an homogenate of brain (or other tissue of interest) is made in a solution that contains a protein denaturant to inactivate RNAse. After being injected, total brain poly(A)+ mRNA expresses a large variety of proteins. Partial mRNA purification—for example, by size separation—reduces this number; in addition, the remaining mRNAs, being more concentrated, express more of the desired protein. Using RNA derived from cDNA clones, it is possible to express only the desired receptor or channel. The separated oocytes are incubated overnight in a saline solution before being injected with the RNA, and after injection an additional 2 to 7 days are allowed for expression of the message.

To record responses of the oocytes to ligands, or to changes in membrane potential, voltage clamp recordings are usually made after the enveloping follicular layers are removed by gentle collagenase treatment. If, in addition, the surrounding vitelline membrane is removed (by "osmotic shock"), single-channel recordings can be made from the exposed oocyte membrane with patch clamp electrodes.

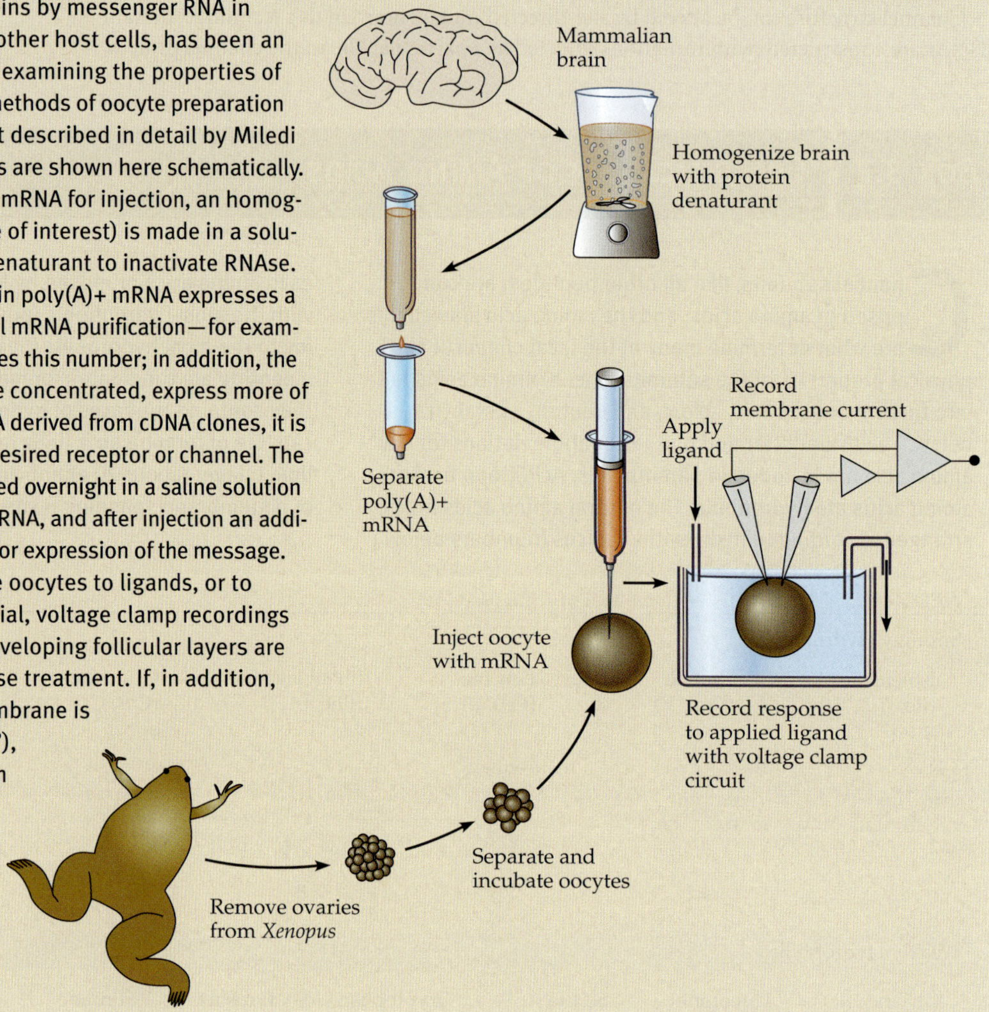

[15]Sumikawa, K., Parker, I., and Miledi, R. 1989. *Methods Neurosci.* 1: 30–45.

so that selected amino acids with particular properties (positively or negatively charged, highly polar, or nonpolar) are replaced by others with different properties. Functional properties of the mutant channels can then be examined by injecting host cells with mRNAs derived from the mutant cDNAs.

Fetal and Adult ACh Receptors in Mammalian Muscle

During muscle development, the properties of the ACh receptors change. The fetal form of the receptor is distinguished from the adult form by its longer open time and smaller conductance. As the receptors take on adult properties, the γ subunit is replaced by a different subunit (ε) with a similar, but distinct, amino acid sequence.[16] That the changes in channel kinetics and conductance are related to the change in subunit composition has been confirmed by the injection of different combinations of subunit mRNAs.[17] Oocytes injected with α, β, γ, and δ subunit mRNA expressed channels with fetal properties; those receiving mRNA for α, β, ε, and δ subunits expressed channels with properties similar to those found in adult muscle.

[16]Takai, T., et al. 1985. *Nature* 315: 761–764.

[17]Mishina, M., et al. 1986. *Nature* 321: 406–411.

Which AChR Subunits Line the Pore?

Mutagenesis and host cell expression have been used to obtain structural information about the pore lining. Once the tertiary structure of the subunits has been deduced we are still left in doubt about the precise topological arrangement of the transmembrane segments with respect to one another, and which of these contributes to formation of the pore. These characteristics are important because the pore structure can be expected to be the major determinant of channel ion selectivity and conductance. One approach to the problem is to make one or more mutations in the suspected pore-lining region and see whether channel selectivity or conductance is altered. Another property of the pore is its ability to bind certain molecules. For example, the molecule known as QX222 has been shown to bind to a site in the open AChR channel, thereby blocking ion current through the channel. Mutations of AChR subunits indicated that the M2 helix forms part of the wall of the open channel,[18] as indicated in Figure 3.4A. Specifically, M2 regions of mouse AChR α and δ subunits contain the following amino acid sequences, going from cytoplasm to extracellular fluid, with the numbers referring to the α subunit (see Figure 3.2):

243 250 255 261

α: **M T L S I S V L L S L T V F L L V I V**

δ: **T S V A I S V L L A Q S V L L L I S**

We might expect the relatively hydrophilic amino acids (see Box 3.2), such as the serines (S) and threonines (T), to be exposed to the aqueous pore, and the more hydrophobic isoleucines (I), for example, to be nestled against the membrane lipid or other parts of the protein. When the serines in the underscored positions were replaced by alanines (weakly hydrophobic), the conductance of the mutated channels was reduced by almost half. In addition, the binding affinity for QX222 was greatly reduced. These effects are consistent with the idea that the serine residues lie in the wall of the aqueous channel and contribute directly to its functional properties.

The residues highlighted in color in the α subunit sequence above are believed to contribute to the lining of the pore. These residues were identified in experiments by Karlin and his colleagues[19] in which each residue was mutated, one at a time, to cysteine. The mutant subunit was then expressed in oocytes together with wild-type β, γ, and δ subunits. Membrane currents produced by ACh were measured before and after exposure of the oocyte to the hydrophilic reagent MTSEA (methanethiosulfonate ethylammonium). The reagent reacts selectively with the cysteine sulfhydryl group (see Box 3.2), but will do so only if the substituted cysteine is at a water-accessible position on the subunit—that is, exposed to the aqueous pore. The reagent attenuated the responses to ACh only in channels with mutations of the residues highlighted in color. The pattern of exposure is consistent with a helix, interrupted in the middle by an extended structure containing residues 250, 251, and 252.

High-Resolution Imaging of the ACh Receptor

A powerful approach to determining channel topology is the use of high-resolution imaging, as shown in Figure 3.1D. Unwin[20,21] has extended the resolution of imaging of ACh receptors in tubular vesicles to better than 0.9 nm, thereby revealing many details of the internal structure of the protein. The five subunits are similar in physical appearance, except that each α subunit has a prominent pocket on its synaptic surface, presumably representing a binding site for ACh. The principal features of the subunit structure are summarized in Figure 3.4A. Three rods of high density, probably representing α helices, are found in each subunit around the extracellular vestibule, extending outward from, and nearly normal to, the plane of the membrane. Their physical positions are roughly as shown, but their sequential positions in the chain of amino acid residues are not known. Images from within the plane of the membrane indicate that each subunit contains only one helical structure (instead of the expected four), located in the pore region and possessing a marked kink in the middle.

[18]Leonard, R. J., et al. 1988. *Science* 242: 1578–1581.

[19]Akabas, M. H., et al. 1994. *Neuron* 13: 919–927.

[20]Unwin, N. 1993. *J. Mol. Biol.* 229: 1101–1124.

[21]Unwin, N. 1995. *Nature* 373: 37–43.

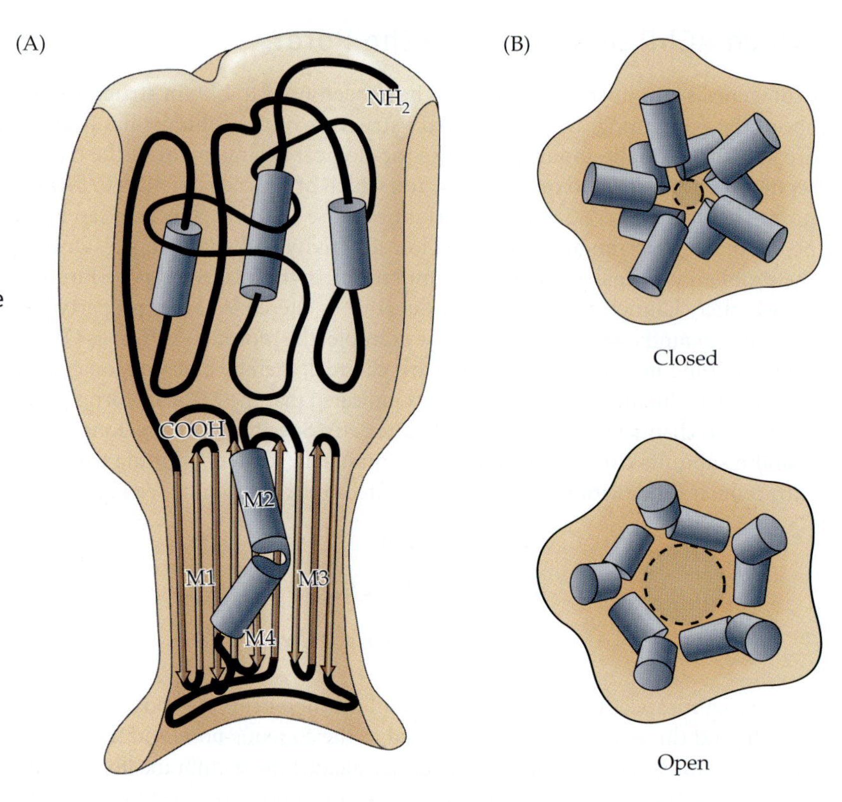

FIGURE 3.4 Proposed Model of the AChR Subunit Structure, based on electron microscope imaging. (A) Three helical elements lie within the extracellular region around the synaptic vestibule. Within the plane of the membrane, regions M1, M3, and M4 form a β sheet. M2 is a split α helix extending into the pore region. (B) Cross sections of the complete receptor, consisting of five subunits, in the plane of the membrane. In the closed receptor, the M2 helices extend into the core of the structure. An open pore is created by the rotation of each helix toward the channel wall. The model is not in agreement with biochemical experiments that suggest that the gate is closer to the cytoplasmic end of the pore.

In the complete receptor, the assembly of five central helices is encased by a star-shaped ring of dense material (Figure 3.4B). The interpretation of these observations is that the centrally located helices are the pore-forming M2 regions and the surrounding ring is a continuous antiparallel β sheet, formed by the remaining 15 transmembrane segments (three in each subunit). The M1, M3, and M4 segments are located arbitrarily in Figure 3.4A because their locations relative to each other are not known. The selection of three β strands per segment is also arbitrary, but is based on three considerations: (1) An odd number is needed if the carboxy terminus of each subunit is to remain on the extracellular side of the membrane; (2) three strands traversing the membrane thickness in each segment requires approximately 25 residues, which is about equal to the lengths assigned to the hydrophobic segments (see Figures 3.2 and 3.5); and (3) if the β strand spacing is about 0.5 nm, then the circumference of the star-shaped ring is about 20 nm, close to that seen in the electron diffraction images.

Open and Closed States of the ACh Receptor

Figure 3.4A represents one of the α subunits in the closed receptor, in which the kinked M2 region projects into the pore. To obtain images of open receptors, Unwin applied ACh within 5 ms before rapid freezing, thus trapping the receptors in the open state.[21] This was accomplished by spraying an ACh solution onto grids holding the tubular vesicles just before plunging the grids into ethane cooled by liquid nitrogen.

Comparison of images made with and without ACh exposure shows that ACh binding causes rotational displacement of the helical rods in the extracellular loops, and suggests that this motion is transmitted, in turn, to the M2 helices. Whatever the mechanism, the M2 helices themselves appear to rotate out of the pore region to lie tangentially along the surrounding β sheet (see Figure 3.4B), thereby forming an open channel. The precise location of the gate is uncertain. The structural evidence suggests that the point of closure is near the kink in the M2 helix, possibly at the leucine residue in position 251 (L251).[21] However, experiments by Karlin and his colleagues on mutant channels expressed in oocytes suggest that the gate is at the cytoplasmic end of the helix.[22] They found that water-soluble reagents penetrated the closed pore as far as threonine T244 when added to the outside solution, but barely entered at all when added to the cytoplasm.

[22]Wilson, G. G., and Karlin, A. 1998. *Neuron* 20: 1269–1281.

Diversity of Neuronal AChR Subunits

After the nicotinic AChR was sequenced, similar isolation and sequencing was carried out for subunits of nicotinic ACh receptors from autonomic ganglia and from vertebrate brain (neuronal nicotinic ACh receptors). Subunits analogous to the muscle receptor α subunit are identified by the presence of adjacent cysteine residues near the proximal end of the amino terminus (positions 192 and 193 in Figure 3.2). The remainder are designated as "β." Together with the α, β, γ, δ, and ε subunits from electric organs and muscle, these subunits form a **family** of common genetic origin.

To date, 11 subunits have been isolated from chicken and rat brain:[23,24] α2, 3, 4, 5, 6, 7, 8, and 9, and β2, 3, and 4. Injection into oocytes of mRNA for α2, α3, or α4 with β2 or β4 results in the formation of channels.[25] Channels formed from two or more types of subunits are said to be **heteromultimeric**. Expression of α7, 8, or 9 alone is sufficient for the formation of **homomultimeric** channels.[24] The existence of a large family of subunits available for channel formation suggests that they may be able to combine selectively to form a diverse variety of channel isotypes with different functional properties such as ion selectivity, conductance, and kinetics.

Subunit Composition of Neuronal ACh Receptors

How many subunits assemble to form a neuronal ACh receptor? This was determined by making mutations in chicken α and β subunits that altered channel conductance.[26] Coinjection into oocyte nuclei of normal and mutant α cDNA, together with native β cDNA, resulted in three different channel conductances. One of these was the same as the conductance of normal channels, another the same as that of channels obtained with only mutant cDNA. The third conductance presumably represented channels with one native and one mutant α subunit. Because there was one (and only one) such intermediate type, it was concluded that the channels contained two α subunits. When native and mutant β cDNAs were injected, the resulting channels showed four different conductances, indicating the presence of three non-α subunits. The deduced stoichiometry of the channel, then, was $(\alpha)_2(\beta)_3$.

A RECEPTOR SUPERFAMILY

GABA, Glycine, and 5-HT Receptors

Three types of anion (chloride) channels mediate synaptic inhibition in the nervous system. All three are activated by amino acids: the type A and type C receptors for γ-aminobutyric acid ($GABA_A$[27,28] and $GABA_C$[29] receptors) and the glycine receptor (glyR).[30] Each receptor type has a number of subunit isoforms, all closely homologous to the AChR subunits. For example, six α subunits, three β, three γ, one δ, and one ε have been identified for $GABA_A$ receptors. One of several receptors for serotonin, the $5\text{-}HT_3$ receptor, forms a cation channel with functional properties similar to those of the nicotinic AChR.[31] A $5\text{-}HT_3$ receptor subunit, also known as SERα1, has been cloned.[32] Like the GABA and glycine receptor subunits, SERα1 closely resembles the nicotinic AChR subunits in amino acid composition.

Although the ACh and $5\text{-}HT_3$ receptors are cation-selective, while the GABA and glycine receptors are anion-selective, the close similarity of the amino acid sequences of their subunits suggests that all four receptor families have a common genetic origin. Together, they are said to form a genetic **superfamily**.[33] Their relatively high degree of sequence homology suggests that their higher-order structures are likely to be similar as well. One indication of this structural similarity is shown in Figure 3.5, in which a hydropathy index for the amino acids in the sequence (see Box 3.2) is plotted against residue number for two $GABA_A$ receptor subunits from mammalian brain and for the *Torpedo* AChRα subunit. The three peptides have similar profiles, with four hydrophobic sequences between residues 220 and 500 indicating possible membrane-spanning regions. Similar hydropathy profiles are seen for glycine and $5\text{-}HT_3$ receptor subunits as well. Thus, the proposed tertiary structures for glycine, GABA, and $5\text{-}HT_3$ subunits are analogous to that of the AChR subunits (see Figure 3.3).

[23]Ortells, M. O., and Lunt, G. G. 1995. *Trends Neurosci.* 18: 121–127.

[24]Colquhoun, L. M., and Patrick, J. W. 1997. *Adv. Pharmacol.* 39: 191–220.

[25]McGehee, D. S., and Role, L. W. 1995. *Annu. Rev. Physiol.* 57: 521–546.

[26]Cooper, E., Couturier, S., and Ballivet, M. 1991. *Nature* 350: 235–238.

[27]Sieghart, W. 1995. *Pharmacol. Rev.* 47: 181–234.

[28]McKernan, R. M., and Whiting, P. J. 1996. *Trends Neurosci.* 19: 139–143.

[29]Bormann, J., and Feigenspan, A. 1995. *Trends Neurosci.* 18: 515–519.

[30]Kuhse, J., Betz, H., and Kirsch, J. 1995. *Curr. Opin. Neurobiol.* 5: 318–323.

[31]Yakel, J. L., and Jackson, M. B. 1988. *Neuron* 1: 615–621.

[32]Maricq, A. V., et al. 1991. *Science* 254: 432–437.

[33]Ortells, M. O., and Lunt, G. G. 1995. *Trends Neurosci.* 18: 121–127.

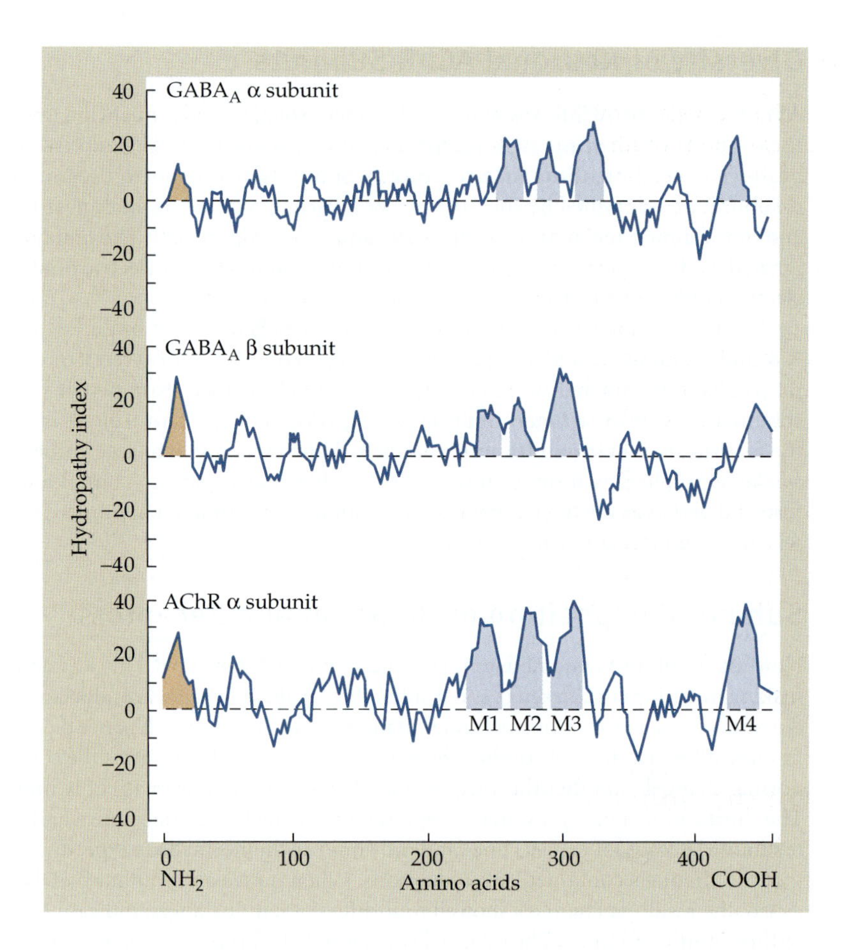

FIGURE 3.5 Hydropathy Indices for the amino acid sequences of the GABA$_A$ α and β subunits and the AChR α subunit. Indices are calculated by taking a moving sum of the individual indices of adjacent amino acids in the sequence (see Box 3.2). Regions in the positive range are hydrophobic, those in the negative range hydrophilic. Blue indicates principal hydrophobic regions; the four areas in similar positions toward the carboxy terminus correspond to the M1 through M4 regions of the ACh receptor. Brown regions are signal sequences. (Amino acid numbering includes the signal sequence on the amino terminus.) (After Schofield et al., 1987.)

Ion Selectivity of Ligand-Gated Channels

It might seem surprising that the same receptor superfamily should include both cation- and anion-selective channels. Given their apparent structural similarity, how is it that their selectivities are different? One feature of channels in general is that their extramembranous connecting loops form structures that extend considerable distances from the surfaces of the lipid bilayer (e.g., see Figure 3.1). Unwin[34] has pointed out that in the cation-selective channels, the loops lining the walls of the projecting channel vestibules have a net excess of negative charges; in anion-selective channels the excess is positive. Since the channel openings are about 2 nm in diameter and the effective radius for electrostatic interaction in physiological solutions is about 1 nm, excess charges in the vestibules can contribute considerably to the accumulation of counterions—for example, of chloride in the mouth of the glycine receptor. Such accumulation might contribute to the preference of a channel for anions over cations (or the reverse) and, in addition, enhance its conductance (remember that conductance depends on concentration).

VOLTAGE-ACTIVATED CHANNELS

Channels activated specifically by depolarization of the cell membrane include voltage-activated sodium channels responsible for the depolarizing phase of the nerve action potential, and voltage-activated potassium channels associated with membrane repolarization. Also included in this group are voltage-activated calcium channels, which in some tissues are responsible for action potential generation or prolongation, and which subserve many other functions, such as muscle contraction and release of neurotransmitters. Each of these three families of channels has a number of isotypes found in different species and in different parts of the nervous system, and like the nicotinic AChR and its homologues, they constitute a superfamily of common genetic origin.

[34]Unwin, N. 1989. *Neuron* 3: 665–676.

The Voltage-Activated Sodium Channel

The methods that were successful in characterizing the molecular structure of the AChR were applied with equal success to the voltage-activated sodium channel. The essential steps were biochemical extraction and isolation of the protein,[35–37] followed by isolation of cDNA clones and deduction of the amino acid sequence.[38] As with the AChR, an electric fish—this time the eel *Electrophorus electricus*—provided a rich source of material, and high-affinity toxins were available to facilitate isolation of the protein, principally tetrodotoxin (TTX) and saxitoxin (STX); both of these molecules block ion conduction in the native channels by occluding the pore of the open channel. Subsequently, sodium channels were isolated from brain and skeletal muscle. The sodium channel purified from eel consists of a single large (260 kDa) protein and is representative of a diverse family of structurally similar proteins.

In mammalian brain, the primary 260 kDa (α) subunit is accompanied by two additional subunits: β_1 (36 kDa) and β_2 (33 kDa). The β_1 subunit has been shown to increase the rate of inactivation of the channel.[39] Several different mRNAs have been identified for the α subunits of brain, producing various channel subtypes. At least two additional isotypes are expressed in mammalian skeletal muscle—one from adult muscle (RSkM1),[40] another that is characteristic of developing fetal muscle, or denervated muscle (RSkM2).[41] A third isotype is found in mammalian heart.[42] After translation, the channel protein is heavily glycosylated; about 30% of the mass of the mature eel channel consists of carbohydrate chains containing large amounts of sialic acid.

Amino Acid Sequence and Tertiary Structure of the Sodium Channel

The eel channel is composed of a sequence of 1832 amino acids, containing four successive domains (I–IV) of 300 to 400 residues each, with about 50% sequence homology from one to the next. Each domain is architecturally equivalent to one subunit of the AChR family of channel proteins. Unlike AChR subunits, however, the sodium channel domains are expressed together as a single protein. Within each domain are multiple hydrophobic or mixed hydrophobic and hydrophilic (amphipathic) sequences capable of forming transmembrane helices.

In the most commonly accepted model of the channel topology (Figure 3.6A), each domain has six such membrane-spanning segments (S1–S6). As with the AChR subunits, the domains are arranged radially around the pore of the channel (Figure 3.6D). Of particular interest is the S4 region, which is highly conserved in all four domains and has a positively charged arginine or lysine residue at every third position on the transmembrane helix. This feature occurs in all voltage-sensitive channels and is believed to be involved in voltage-sensitive activation (Chapter 6).

Voltage-Activated Calcium Channels

The family of voltage-activated calcium channels contains several subtypes that have been classified by their functional properties, such as sensitivity to membrane depolarization and persistence of activation (Table 3.1).[43,44] Channel isotypes have been cloned from skeletal, cardiac, and smooth muscle, and from brain. The amino acid sequence of the primary channel-forming subunit (α_1) is similar to that of the voltage-activated sodium channel.[45] In particular, the putative transmembrane regions, S1 to S6, are highly homologous to those of the sodium channel. Consequently, an analogous tertiary structure has been postulated (see Figure 3.6B).

Although the expression of α_1 subunits alone is sufficient for the formation of functional calcium channels in host cells, three accessory subunits are coexpressed in native cell membranes: $\alpha_2\delta$, a dimer with the extracellular α_2 portion linked to the membrane-spanning δ portion by a disulfide bridge; β, a membrane protein located intracellularly; and γ, an integral membrane protein with four putative transmembrane segments. Coexpression of various subunit combinations suggests that the $\alpha_2\delta$ and β subunits influence both channel conductance and kinetics, and that the γ subunit plays a role in the voltage sensitivity of the channel.[46]

[35]Miller, J., Agnew, W. S., and Levinson, S. R. 1983. *Biochemistry* 22: 462–470.

[36]Hartshorn, W. A., and Catterall, W. A. 1984. *J. Biol. Chem.* 259: 1667–1675.

[37]Barchi, R. L. 1983. *J. Neurochem.* 40: 1377–1385.

[38]Noda, M., et al. 1984. *Nature* 312: 121–127.

[39]Cormick, K. A., et al. 1998. *J. Biol. Chem.* 273: 3954–3962.

[40]Trimmer J. S., et al. 1989. *Neuron* 3: 33–49.

[41]Kallen, R. G., et al. 1990. *Neuron* 4: 233–342.

[42]Rogart, R. B., et al. 1989. *Proc. Natl. Acad. Sci. USA* 86: 8170–8174.

[43]Hofmann, F., Biel, M., and Flockerzi, V. 1994. *Annu. Rev. Neurosci.* 17: 399–418.

[44]Randall, A., and Tsien, R. W. 1995. *J. Neurosci.* 15: 2995–3012.

[45]Tanabe, T., et al. 1987. *Nature* 328: 313–318.

[46]Walker, D., and De Waard, M. 1998. *Trends Neurosci.* 21: 148–154.

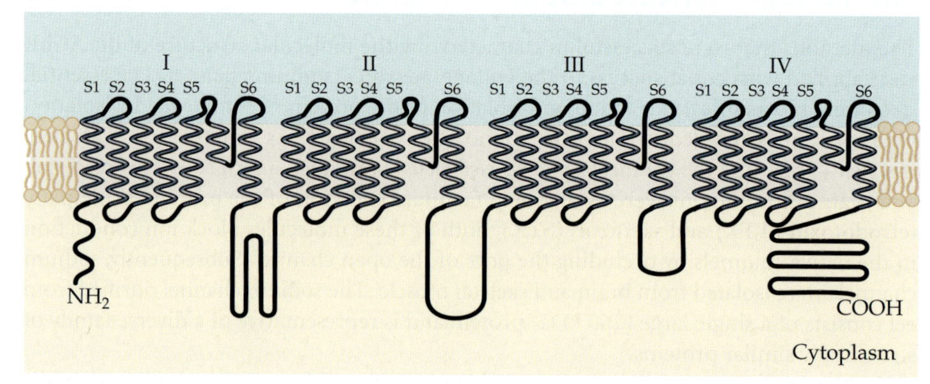

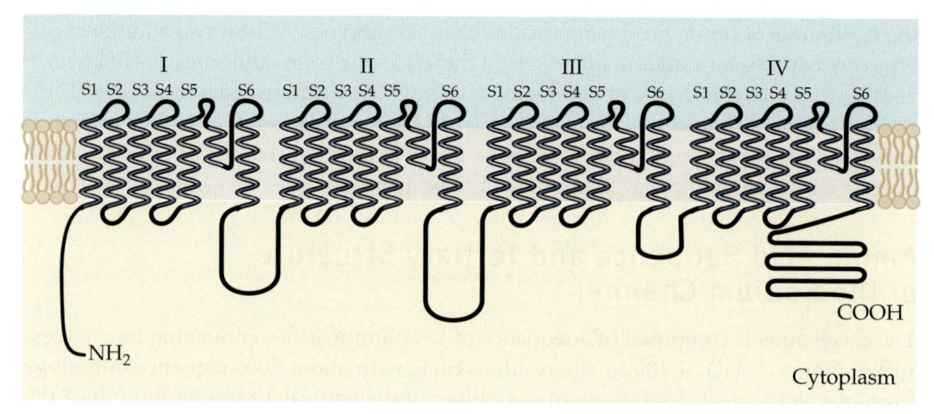

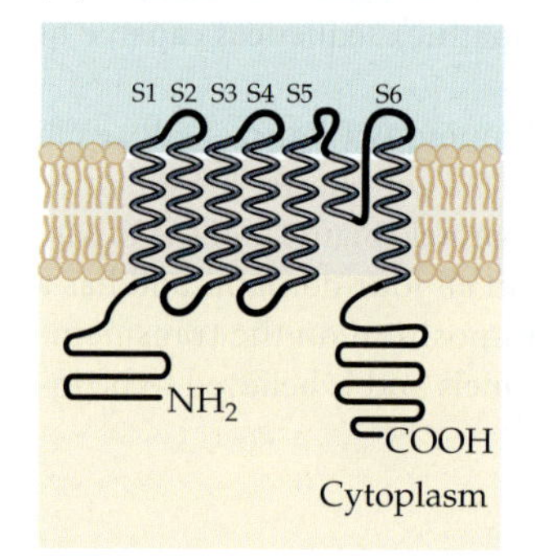

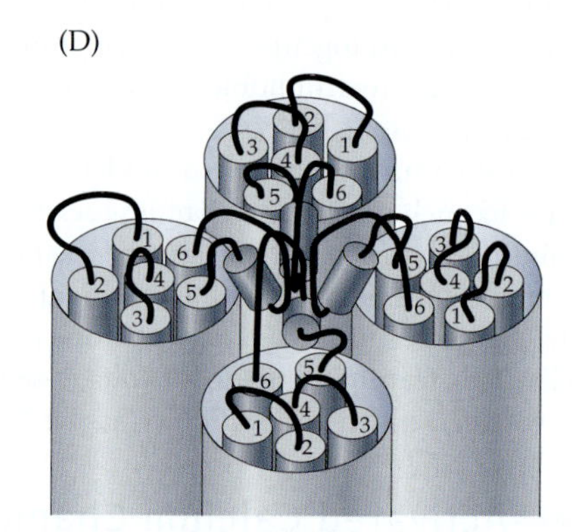

FIGURE 3.6 Voltage-Activated Channel Structure. (A) Voltage-activated sodium channel, represented as a single protein with four domains (I–IV) connected by intracellular loops. Each domain has six transmembrane segments (S1–S6), with a pore-forming structure between the fifth and sixth. (B) The structure of voltage-activated calcium channels is similar to that of the sodium channel. (C) The potassium channel subunit resembles a single domain of the sodium channel. (D) The proposed three-dimensional channel structure, with the domains in a circular array and the S5–S6 connectors dipping into the pore. The relative positions of S1 through S6 in each domain are not known and thus are shown arbitrarily.

Voltage-Activated Potassium Channels

Voltage-sensitive potassium channels play an important role in nerve excitation and conduction. A number of distinct genetic messages give rise to a diverse family of these proteins. The first potassium channel to be sequenced was from *Drosophila* and is called Shaker after a genetic mutant in which expression of the channel is defective.[47] When the mutant flies are anesthetized with ether (e.g., for counting), they go through a period of trembling, or "shaking," before becoming immobile. The mutation itself provided a different approach to cloning the channel, one that did not rely on prior identification of the protein. Genetic analysis indicated the approximate location of the *shaker* gene on the *Drosophila* genome. Overlapping genomic clones from normal and mutant flies were isolated from that region, and comparison of normal and mutant sequences led to identification of the *shaker* gene.

An unexpected finding was that the amino acid sequence of the protein is much shorter than that of voltage-sensitive sodium or calcium channels. It contained a single domain (see

[47]Papazian, D. M., et al. 1987. *Science* 237: 749–753.

Table 3.1
Functional classification of voltage-activated calcium channels and corresponding clones of α subunits

Type[a]	Threshold[b]	Inactivation	Clone	Alternative designation	Tissue source
T	LV	Yes	—	—	—
L	HV	No	S	CaCh1	Skeletal muscle
—	—	—	Ca, Cb[c]	CaCh2a, CaCh2b	Brain and heart
—	—	—	D	CaCh3	Brain
P	HV	Slow	A	CaCh4	Brain
Q	HV	Slow	—	—	—
N	HV	Yes	B	CaCh5	Brain
R	HV	No	E	CaCh6	Brain

[a] Designations T, L, and N originally meant **T**ransient, **L**ong-lasting, and **N**either T nor L; P refers to **P**urkinje cells.
[b] HV and LV indicate high-voltage and low-voltage thresholds for opening.
[c] Lowercase designations a and b indicate splice variants.

Figure 3.6C), similar to a segment of domain IV of the eel sodium channel. Experimental evidence indicates that the single protein subunits assemble to form multimeric ion channels in the membrane.[48] Four distinct subfamilies of potassium channel proteins have been cloned from *Drosophila*—called Shaker, Shab, Shaw, and Shal—all with mammalian counterparts (Table 3.2). Each subfamily is made up of various isoforms (e.g., Shaker1, Shaker2, and so on). Isoforms from the same subfamily, when expressed in host cells, combine to form heteromultimeric channels, but those from different subfamilies do not.[49]

Like sodium and calcium channels, voltage-activated potassium channels are expressed with accessory (β) subunits.[50] Two subfamilies have been identified: Kvβ 1.1 through Kvβ 1.3, and Kvβ 2.1. When expressed with the primary subunits, the β subunits affect the voltage sensitivity and inactivation properties of the channels.

How Many Subunits Make a Potassium Channel?

A comparison of the potassium channel structure to that of the closely related sodium and calcium channels suggests that the complete channel is formed by an assembly of four subunits. The potassium channel blocker charybdotoxin (CTX), in combination with mutants resistant to the toxin, was used to examine this question.[51] Wild-type and mutant channel subunits from *Drosophila* were expressed in various relative amounts in oocytes. When only wild-type mRNA was used, forming homomultimeric channels, potassium currents in the oocyte membrane were completely blocked by high concentrations of CTX. Mutant channels, however, were unaffected. Currents in oocytes injected with both mutant and wild-type mRNA were only partially blocked.

It is known that binding to a single subunit is sufficient to produce the block. Therefore, in oocytes injected with mixtures of wild-type and mutant mRNA, only homomultimeric channels that form solely from the mutant subunits should remain unaffected by

[48]Timpe, L. C., et al. 1988. *Nature* 331: 143–145.
[49]Salkoff, L., et al. 1992. *Trends Neurosci.* 15: 161–166.
[50]Jan, L. Y., and Jan, Y. N. 1997. *Annu. Rev. Neurosci.* 20: 91–123.
[51]MacKinnon, R. 1991. *Nature* 350: 232–238.

Table 3.2
Voltage-activated potassium channel subunit genes in *Drosophila*, and corresponding mammalian genes

Drosophila gene	Mammalian genes
Shaker 1 to 12	Kv 1.1 to 1.12
Shab 1 and 2	Kv 2.1 and 2.2
Shaw 1 to 4	Kv 3.1 to 3.4
Shal 1 and 2	Kv 4.1 and 4.2

the toxin. The fraction of such channels that would form from random association of wild-type and mutant subunits can be calculated from the ratio of the injected mRNAs (mutant/wild type), assuming equal potency of expression, and will vary depending on the number of subunits in the channel. For example, if there are four subunits and if 90% of the RNA is mutant, then the probability that the channels will be composed entirely of mutant subunits is ($[0.9]^4$), or 66%. The remainder (34%) will have at least one wild-type subunit and be subject to block by the toxin. Block of the current by CTX would be 27% if the channels were trimers, and 41% for pentamers. Block of about 34% was observed experimentally, providing direct evidence for a tetrameric structure.

Pore Formation in Voltage-Activated Channels

A consistent feature of all of the voltage-activated channel sequences is a moderately hydrophobic region in the extracellular S5–S6 loop. As in the experiments described previously for the M2 region of the AChR, mutations in this region of the Shaker potassium channel reduced the binding affinity for the blocking molecule TEA (tetraethylammonium) and altered the conductance properties of the channel.[52,53] It was concluded that this segment dips into the channel mouth to form the pore region.[54] This conclusion is supported by the X-ray diffraction studies that are discussed later in this chapter (see Figure 3.7). The S5–S6 loops form short helices that dip into the center of the channel; residues ascending from the lower ends of helices then form the upper pore lining (see Figure 3.6D).

Mutations in the pore region also affect ion selectivity. For example, in rat brain sodium channels, substitution of a positively charged glutamate residue for a negatively charged lysine in the pore region of the third domain causes the channel to take on many of the properties of a calcium channel.[55] Rather than being permeable only to sodium, the mutated channel has poor selectivity for monovalent cations, and at physiological ion concentrations most of the channel current is carried by calcium.

High-Resolution Imaging of a Potassium Channel

The structure of potassium channels from *Streptomyces lividans* ($K_{CS}A$ channels) has been examined with X-ray crystallography at a resolution of 3.2 Å.[56] The bacterial channels belong to a class of potassium channels whose subunits have only two transmembrane segments rather than six. Another example is the inward-rectifying potassium channel, which will be discussed later (see Figure 3.8). The two segments are structurally equivalent to S5 and S6 in the voltage-activated channels, and the amino acid sequence of the pore region in the S5–S6 loop is highly conserved across all potassium channel subunits, regardless of the number of membrane crossings.[55,57] The advantage of studying the bacterial potassium channel is that it can be produced in quantities sufficiently large for crystallization, thus enabling analysis by X-ray diffraction.

The $K_{CS}A$ channel is a tetramer. Figure 3.7 is a sectional view of the channel, showing its major structural features. Near the amino terminus of each subunit is an outer helix that spans the membrane from the cytoplasmic side of the membrane to the outer surface. The outer helix is followed by a short helix that points into the pore, and then an inner helix that returns to the cytoplasmic side. The links between the outer and short helices form four turrets that surround the external opening of the pore and contain the binding sites for TEA and other channel-blocking toxins.[55] In each subunit, the link between the central end of the short helix and the inner helix contributes to the pore structure. The four links combine to form a restricted passage responsible for the ion selectivity of the channel—the selectivity filter. A relatively large central cavity and a lower internal pore connect the selectivity filter with the cytoplasm.

Selectivity for potassium is achieved by both the size and the molecular composition of the selectivity filter. Its diameter is about 0.3 nm, and amino acids in its wall are oriented so that successive rings of four carbonyl oxygens, one from each subunit, are exposed to the pore. The pore diameter is adequate to accommodate a dehydrated potassium ion (about 0.27 nm in diameter), but stripping waters of hydration from the penetrating ion requires considerable energy (Chapter 2). This requirement is minimized by the exposed

[52]Yool, A. J., and Schwarz, T. L. 1991. *Nature* 349: 700–704.

[53]Yellen, G., et al. 1991. *Science* 251: 939–942.

[54]Miller, C. 1992. *Curr. Biol.* 2: 573–575.

[55]Heinemann, S. H., et al. 1992. *Nature* 356: 441–443.

[56]Doyle, D. A., et al. 1998. *Science* 280: 69–77.

[57]MacKinnon, R., et al. 1998. *Science* 280: 106–109.

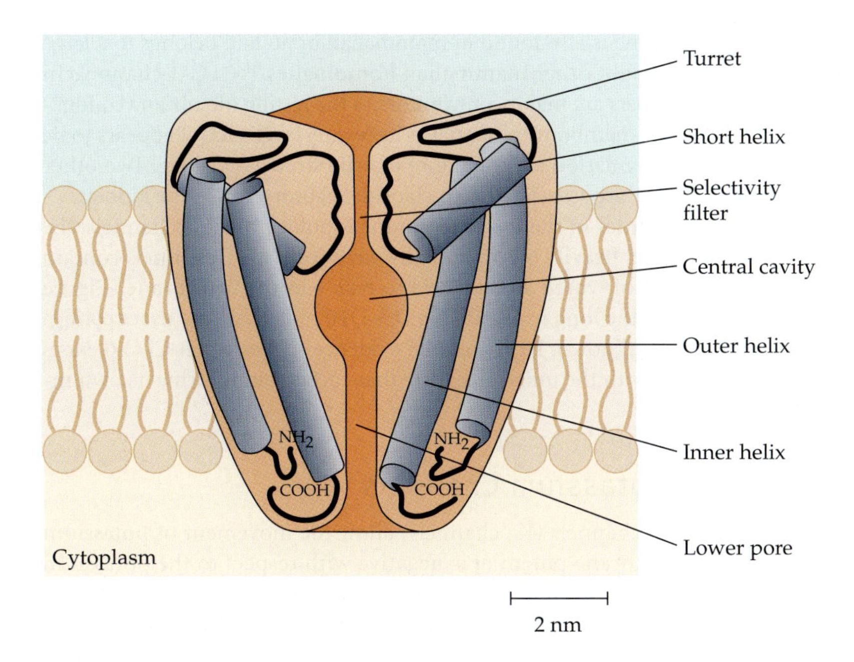

FIGURE 3.7 Potassium Channel Structure. Sectional view of a $K_{CS}A$ potassium channel, showing two of four subunits, one on either side of the central pore. Each subunit has two membrane-spanning helices and a short helix pointing into the pore. The connections between the outer helices and the short helices of the four subunits form four turrets that surround the pore entrance and contain binding sites for blocking molecules. The four connections between the short helices and the inner helices combine to form the selectivity filter, which allows the permeation of potassium, cesium, and rubidium but excludes smaller cations such as sodium and lithium. (After Doyle et al., 1998.)

oxygens, which provide effective substitutes for the oxygen atoms that normally surround the ion in the hydrated molecule. Smaller ions, such as sodium (diameter 0.19 nm) or lithium (0.12 nm), are excluded because they cannot make intimate contact with all four oxygens simultaneously, so they remain hydrated. Ions larger than cesium (0.33 nm in diameter) cannot penetrate the pore because of their size. This structural basis for ion selectivity is in accordance with traditional ideas about ion permeation through channels.[58]

The X-ray diffraction studies have yet to provide information about structural changes that accompany potassium channel gating. One clue to the location of the gate is that, in Shaker potassium channels, substances that have access to the inner pore from the cytoplasmic side of open channels can penetrate only a short distance when the channels are closed. Thus, the gate must close near the cytoplasmic end of the pore.[59]

OTHER CHANNELS

A number of other channels are important for neuronal function. The molecular structures of some of these have been defined. Among these is a family of subunits that form glutamate-activated channels, and another family that form channels activated by ATP. Representatives of a large family of voltage-activated chloride channels have been cloned, and the amino acid sequences of subunits that form voltage-sensitive inward-rectifying potassium channels have been determined. In addition, the structural arrangement of various channels activated by intracellular ligands is known.

A general theme is that an assembly of four or more subunit proteins (or, in the case of voltage-activated sodium and calcium channels, four domains of a single protein) is required to form an ion channel. The actual number of subunits is often not known directly, but it is usually assumed to be four for voltage-sensitive channels and for cation channels activated by intracellular ligands. Channels activated by extracellular ligands are generally assumed to contain five subunits. The following examples provide a flavor of the kinds of structural variations seen in such proteins, as well as their similarities to and differences from the ligand-activated and voltage-activated superfamilies already discussed.

Voltage-Activated Chloride Channels

Voltage-activated chloride channels were first cloned from the electric organ of *Torpedo*.[60] Known as CLC-0 channels, they are expressed in high density on the noninnervated face of cells and provide a low-resistance pathway for currents generated by electrical activity of the

[58]Mullins, L. J. 1975. *Biophys. J.* 15: 921–931.

[59]Liu, Y., et al. 1997. *Neuron* 19: 175–184.

[60]Jentsch, T. J., Steinmeyer, K., and Schwarz, G. 1990. *Nature* 348: 510–514.

innervated face. The CLC-0 gene is also found in mammalian brain and belongs to a large family that includes at least eight other mammalian homologues.[61] CLC-1 channels in mammalian skeletal muscle fibers are major contributors to the resting membrane conductance and serve to stabilize the membrane potential at its resting level. CLC-2 appears to be associated with cell volume regulation and therefore may be stretch-sensitive. Two other isotypes, CLC-K1 and CLC-K2, are associated with chloride reabsorption in the kidney.

There are several uncertainties in the transmembrane topology of CLC channels, the elucidation of which has relied heavily on hydropathy analyses.[62] CLC channels contain 13 hydrophobic domains, 11 of which are believed to reside in the membrane (Figure 3.8A). One unusual feature is the long hydrophobic D9–D10 region, whose exact configuration in the membrane is unknown. Experimental evidence suggests that CLC-0 exists in the membrane as a dimer, with the unusual feature that each subunit forms its own independent channel.[63]

Inward-Rectifying Potassium Channels

Inward-rectifying potassium channels (K_{ir} channels) allow the movement of potassium ions into cells when the membrane potential is negative with respect to the potassium equilibrium potential, but little outward potassium movement. There are at least five subfamilies of the channel ($K_{ir}1$–$K_{ir}5$) found in brain, heart, and kidney.[64,65] One family, $K_{ir}3$, forms channels that are activated by intracellular G proteins (Chapter 10). The block of outward potassium flux appears to be due to blockade of the channels by intracellular magnesium and/or by intracellular polyamines. The proposed subunit structure (Figure 3.8B) is similar to that of the $K_{CS}A$ channel, with only two membrane crossings. The magnesium-blocking site appears to be in M2, near the cytoplasmic end of the channel.

ATP-Activated Channels

Adenosine-5′-triphosphate (ATP) acts as a neurotransmitter in smooth muscle cells, in autonomic ganglion cells, and in neurons of the central nervous system.[66,67] Because ATP is a purine, its receptor molecules are known as purinergic receptors. One is the P2X receptor, a ligand-gated cation channel. Seven P2X subunits ($P2X_1$–$P2X_7$) have been cloned.[68] Their proposed tertiary structure (Figure 3.8C), with two membrane-spanning segments, is similar to that of the $K_{CS}A$ channel subunits.

Glutamate Receptors

Glutamate is the most important and prevalent excitatory neurotransmitter in the central nervous system, activating at least three cation channel types. The three types have distinct functional properties and are distinguished experimentally by their different sensitivities to glutamate analogues.[69] One, the NMDA receptor, responds selectively to *N*-methyl-D-aspartate. The other two are activated selectively by AMPA (α-amino-3-hydroxy-5-methyl-4-isoxazole propionic acid) and kainate, respectively. Because of this selectivity, the three chemical analogues are important experimental tools. Remember, however, that the native neurotransmitter for all three receptor types is glutamate, not one of the analogues.

So far, 16 cDNAs for glutamate receptor subunits have been identified by molecular cloning. Five of these—named NR1 and NR2A through NR2D—are involved in the formation of NMDA receptors. AMPA receptors are formed from another set, designated as GluRA through GluRD (or GluR1–GluR4), and kainate receptors are assembled from subunits KA1 and/or KA2, combined with GluR5, 6, or 7. Two homologous subunits, δ_1 and δ_2, remain unassigned to any particular receptor. The channels are believed to be pentamers, by analogy to the nicotinic ACh receptor, but there is no further similarity. The amino acid sequences of the subunits[70–72] are nearly twice as long as those of the superfamily containing the ACh, 5-HT_3, GABA, and glycine receptors, and there is little or no homology between the two groups.

The subunit proteins have four putative transmembrane segments, which were first thought to be arranged in a manner analogous to those of the AChR (see Figure 3.3). However, there is increasing evidence in favor of the arrangement shown in Figure

[61]Jentsch, T. J., et al. 1995. *J. Physiol.* 482P: 19S–25S.

[62]Schmidt-Rose, Y., and Jentsch, T. J. 1997. *Proc. Natl. Acad. Sci. USA* 94: 7633–7638.

[63]Middleton, R. E., Pheasant, D. J., and Miller, C. 1996. *Nature* 383: 337–340.

[64]Doupnik, C. A., Davidson, N., and Lester, H. A. 1995. *Curr. Opin. Neurobiol.* 5: 268–277.

[65]Nicholls, C. G., and Lopatin, A. N. 1997. *Annu. Rev. Physiol.* 59: 171–191.

[66]Burnstock, G. 1996. *Drug Dev. Res.* 39: 204–242.

[67]Burnstock, G. 1996. *P2-Purinoceptors: Localization, Function and Transduction Mechanisms*. Wiley, London.

[68]Soto, F., Garcie-Guzman, M., and Stühmer, W. 1997. *J. Membr. Biol.* 160: 91–100.

[69]Seeburg, P. H. 1993. *Trends Neurosci.* 16: 359–365.

[70]Moriyoshi, K., et al. 1991. *Nature* 354: 31–37.

[71]Keinän, K., et al. 1990. *Science* 249: 556–560.

[72]Egebjerg, J., et al. 1991. *Nature* 351: 745–748.

(A) **Voltage-activated chloride channel**

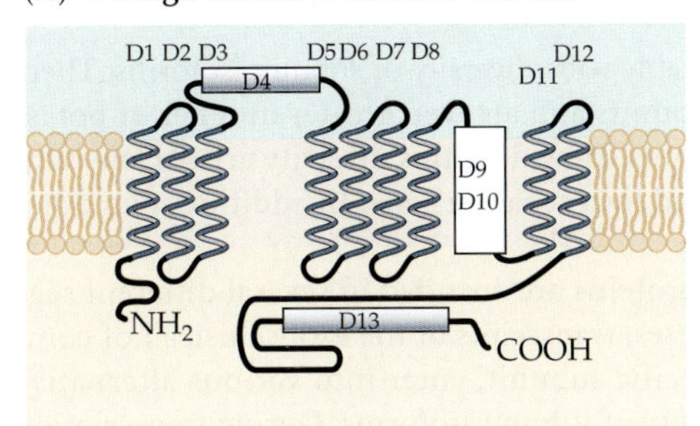

(B) **Inward rectifier subunit**

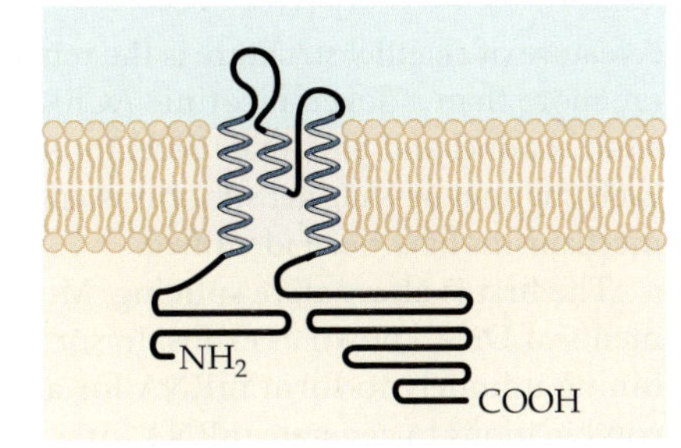

(C) **Purinergic receptor subunit**

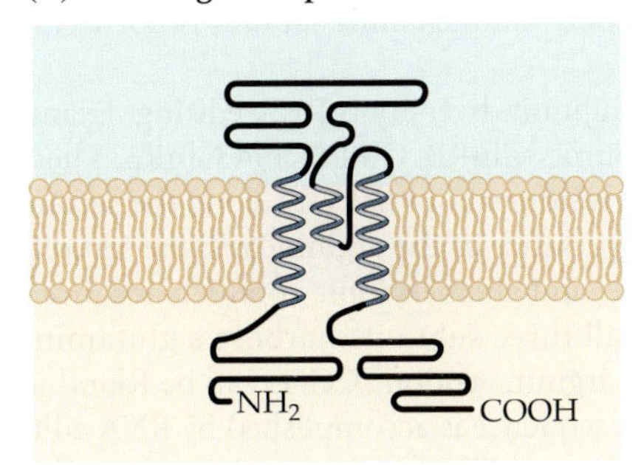

(D) **Glutamate receptor subunit**

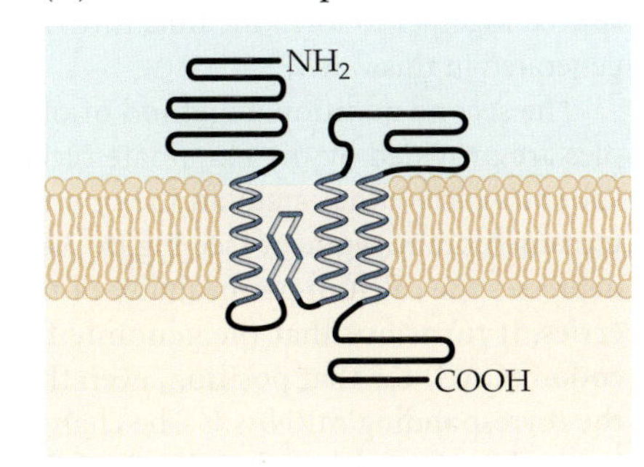

(E) **Cyclic nucleotide–gated channel subunit**

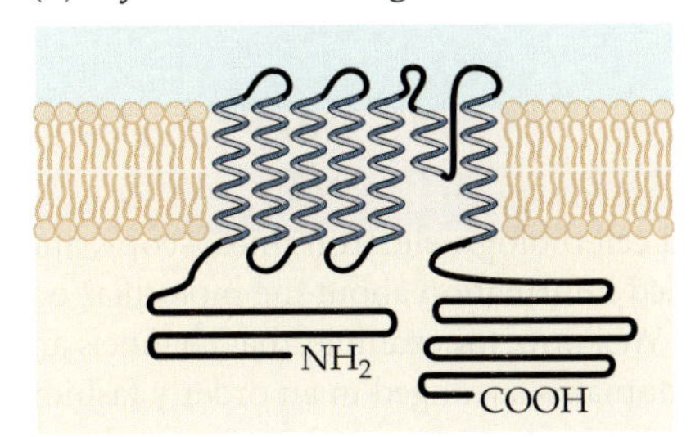

FIGURE 3.8 Examples of Distinctive Types of Channels. (A) The voltage-activated chloride channel protein is thought to have 11 membrane-spanning regions; the arrangement of segment D9–D10 is not known. The protein is believed to assemble in the membrane as a dimer to form a "double-barreled" channel structure. (B) Proposed secondary structure of a subunit of the inward-rectifying potassium channel, with two membrane-spanning regions and a pore-lining loop. The complete channel is assumed to be a tetramer. (C) The subunit structure of receptors for ATP appears similar to that of the inward rectifier. (D) Glutamate receptor subunits are believed to have three transmembrane helices and a pore-forming loop entering the channel from the cytoplasmic side. The complete receptor is thought to be a pentamer. (E) The subunit structure of channels activated by cyclic AMP or cyclic GMP is similar to that of voltage-activated potassium channels (see Figure 3.6C). Complete channels are assumed to be tetramers.

3.8D,[73] in which the second segment enters the membrane from the cytoplasmic face to form a hairpin loop contributing to the pore lining. Although unique to glutamate receptor subunits, this configuration is supported by several lines of evidence, including mutation experiments in which glycosylation sites were introduced into the native protein.[74] Because glycosylating enzymes occur only in the lumen of the endoplasmic reticulum, glycosylation of the modified protein can be expected to occur only at extracellular sites. As expected, glycosylation occurred on the amino terminus, but otherwise only on the loop between the third and fourth segments, consistent with the model.

Channels Activated by Cyclic Nucleotides

Channels in retinal and olfactory epithelial cells are influenced by the intracellular concentration of cyclic AMP or cyclic GMP.[75,76] These are cation channels, with varying selectivities for potassium, sodium, or calcium. The amino acid sequence of the channel subunits has some homology with the subunits of the voltage-sensitive channels, and their proposed structure (Figure 3.8E) includes six membrane-spanning regions inferred from hydropathy indices. The S4 region has a sequence of regularly spaced charged residues, but these are fewer in number than found in the voltage-sensitive family (usually four rather than six or seven). Consistent with activation by intracellular ligands, most of the mass of the channel is on the cytoplasmic side of the membrane. The subunits are believed to form ion channels by assembling in the membrane as tetramers.

[73] Dani, J. A., and Mayer, M. I. 1995. *Curr. Opin. Neurobiol.* 5: 310–317.

[74] Hollmann, M., Maron, C., and Heinemann, S. 1994. *Neuron* 13: 1331–1343.

[75] Finn, J. T., Grunwald, M. E., and Yau, K-W. 1996. *Annu. Rev. Physiol.* 58: 395–426.

[76] Zagotta, W. N., and Siegelbaum, S. A. 1996. *Annu. Rev. Neurosci.* 19: 235–263.

DIVERSITY OF SUBUNITS

A feature of channel structure is the remarkably wide diversity of subunit isoforms. There are more than a dozen nicotinic AChR subunits and an even greater number of potassium channel and glutamate receptor subunits. How does this diversity arise? Generally, each channel or channel subunit is encoded by a separate gene. In addition, two other mechanisms have been identified.

The first is **alternative splicing**. Most proteins are encoded in several different segments of DNA known as exons. In some cases, transcripts of the exons, instead of combining uniquely to form mRNA for a specific subunit, enter into various alternative combinations to generate mRNA for a variety of subunit isoforms. During transcription an unknown regulatory mechanism determines which of the alternative RNAs are to be used. The remainder are excised from the transcript and the desired RNA segments spliced together to form the final mRNA. *Shaker* potassium channel subunit isoforms are generated in this way.

The second additional method of obtaining subunit diversity is **RNA editing**. Examples are provided by the glutamate receptor subunits GluRB, GluR5, and GluR6. These subunits carry either a glutamine or an arginine residue about halfway along the second transmembrane segment (see Figure 3.8D). The presence of the arginine residue severely depresses the calcium permeability of the channel and alters its ionic conductance properties. It turns out that the genomic DNA for all three subunits harbors a glutamine codon (CAG) for that position, even though an arginine codon (CGG) can be found in the corresponding mRNAs.[77] The change in base sequence is accomplished by RNA editing in the cell nucleus. Virtually all of the message for GluRB is edited in this way, as well as some of the message for GluR5 and GluR6. In GluR6, additional editing from A to G is found in the first transmembrane segment.[78]

CONCLUSION

Modern methods of biochemistry, molecular and cell biology, electron microscopy, and electron and X-ray diffraction have provided detailed information about the molecular organization and structure of channels and receptors. We know, for example, that channels are formed by four or more polypeptide subunits, or domains, arranged in an orderly fashion around a central core. Each subunit or domain contains, in turn, two to six transmembrane regions joined by extra- and intracellular connecting loops. The combined assembly constitutes the essential pore-forming structure. Channels that are relatively selective, such as the voltage-dependent channels, are usually (perhaps always) tetramers; larger, less selective, ligand-activated channels are pentamers. As an extension of this principle, the largest channels—gap junctions—have a hexameric structure (Chapter 7).

The functional roles of the individual transmembrane regions of the subunit proteins are still relatively obscure, but some details are known. For example, there is strong evidence that the M2 region in the AChR superfamily of subunits forms the lining of the pore, and probably the gating structure as well. X-ray diffraction has revealed the structural basis of ion selectivity in potassium channels and, by implication, in channels having similar primary sequences in their pore regions, including the voltage-dependent channels, inward-rectifying channels, cyclic nucleotide–gated channels, and channels activated by ATP. One persistent feature of channel modeling is the assumption that the transmembrane segments are α helices. Structural studies support this idea for the $K_{CS}A$ channel but not for the nicotinic AChR: All but the M2 segment appear to be made up of β sheets.

The extramembranous loops subserve a number of functions, a principal one being to provide binding sites for extracellular or intracellular molecules that regulate channel functions. In addition, charge concentration in the extramembranous vestibules may help determine ion selectivity and increase channel conductance. Many details remain to be worked out, but with the techniques now at hand we can expect to see continued rapid advancement in our knowledge of the molecular basis of nervous system function.

[77]Köhler, M., et al. 1993. *Neuron* 10: 491–500.

[78]Kyte, J., and Doolittle, R. F. 1982. *J. Mol. Biol.* 157: 105–132.

Summary

■ Nicotinic acetylcholine receptors from the electric organ of *Torpedo* consist of five subunits (two α, and three others, designated β, γ, and δ) arranged around a central pore. The α subunits contain the binding sites for ACh.

■ In each subunit of the AChR the string of amino acids folds to form four membrane-spanning segments (M1–M4) joined by intracellular and extracellular loops. The M2 region has been shown to be associated with the pore lining of the channel.

■ Eleven distinct nicotinic AChR subunit isoforms have been isolated from nervous tissues, eight having ACh-binding sites (and therefore designated as α_2–α_9), and three others (β_2–β_4). Most α–β combinations can form channels in host cells.

■ Subunits of receptors for γ-aminobutyric acid (GABA), glycine, and serotonin (5-HT) are analogous in structure to the family of ACh receptor subunits. Together these four receptor families form a superfamily of common genetic origin.

■ The voltage-activated sodium channel from eel electric organs is a single molecule of about 1800 amino acids, within which are four repeating domains (I–IV). The domains are architecturally equivalent to the subunits of other channels; within each there appear to be six membrane-spanning regions (S1–S6) connected by intracellular and extracellular loops. The eel channel is representative of a diverse family of channel isotypes present in muscle and brain.

■ The family of voltage-activated calcium channel proteins is analogous in structure to the voltage-activated sodium channel. Voltage-activated potassium channels are structurally similar, but with an important genetic difference: The four repeating units are expressed as individual subunits, not as repeating domains of a single molecule. There are at least 20 distinct potassium channel subunit genes, grouped into four subfamilies. Together, the voltage-activated sodium, calcium, and potassium channels constitute a genetic superfamily.

■ Subunits of other ligand-activated and voltage-activated channel types vary considerably in size and amino acid composition. Some resemble voltage-activated channel subunits, but most differ markedly from members of both the voltage-activated and the AChR superfamilies. Some have only two or three membrane-spanning regions, others as many as 10.

SUGGESTED READING

Akabas, M. H., Kauffman, C., Archdeacon, P., and Karlin, A. 1994. Identification of acetylcholine receptor channel-lining residues in the entire M2 segment of the α subunit. *Neuron* 13: 919–927.

Doyle, D. A., Cabral, J. M., Pfeutzner, A. K., Gulbis, J. M., Cohen, S. L., Chait, B. T., and McKinnon, R. 1998. The structure of the potassium channel: Molecular basis of K^+ conduction and selectivity. *Science* 280: 69–77.

Hille, B. 1992. *Ionic Channels in Excitable Membranes*, 2nd Ed. Sinauer Associates, Sunderland, MA, pp. 236–258, 423–444.

Levinson, S. R. 1998. Structure and mechanism of voltage-gated ion channels. In N. Sperelakis (ed.), *Cell Physiology Source Book*, 2nd Ed. Academic Press, San Diego, CA, pp. 406–428.

McGehee, D. S., and Role, L. W. 1995. Physiological diversity of nicotinic acetylcholine receptors expressed by vertebrate neurons. *Annu. Rev. Physiol.* 57: 521–546.

Miller, C. 1992. Hunting for the pore of voltage-gated channels. *Curr. Biol.* 2: 573–575.

Unwin, N. 1993. Nicotinic acetylcholine receptor at 9 Å resolution. *J. Mol. Biol.* 229: 1101–1124.

Unwin, N. 1995. Acetylcholine receptor imaged in the open state. *Nature* 373: 37–43.

4 Transport across Cell Membranes

IONS THAT ENTER OR LEAVE A NEURON THROUGH CHANNELS do so passively, driven by electrical and chemical concentration gradients. In the face of such movements, concentrations in the cytoplasm are kept constant by active transport mechanisms that use energy to move ions back across the membrane against their electrochemical gradients. Constant intracellular ion concentrations are necessary to maintain both the resting membrane potential and the ability to generate electrical signals.

Primary active transport uses energy provided by the hydrolysis of ATP. The most prevalent example of this kind of transport is sodium–potassium exchange. The molecule that is responsible for transport is sodium–potassium ATPase, which carries three sodium ions out of the cell and two potassium ions in for every molecule of ATP hydrolyzed. Each cycle of the pump results in the net movement of one ionic charge across the membrane, so the pump is said to be electrogenic. Additional examples are ATPases responsible for calcium transport out of the cytoplasm: Plasma membrane calcium ATPase pumps calcium out of the cell, and endoplasmic and sarcoplasmic reticulum ATPases pump calcium from the cytoplasm into intracellular compartments.

Secondary active transport uses the movement of sodium down its electrochemical gradient to transport other ions across the membrane, either in the same direction (cotransport) or in the opposite direction (ion exchange). An example is sodium–calcium exchange, in which the influx of three sodium ions is used to transport a single calcium ion out of the cell. Like all other transport systems, this exchange system is reversible: Depending on the chemical and electrical gradients for the two ions, it can be made to run in a "forward" or "backward" direction. A second sodium–calcium exchange system, found in retinal cells, transports a single calcium ion, plus a potassium ion, out of the cell in exchange for four sodium ions. Sodium influx is also used to transport chloride and bicarbonate across the cell membrane. All such mechanisms provide pathways for sodium to enter the cell down its electrochemical gradient, and therefore depend on sodium–potassium ATPase to maintain that gradient.

Transport of neurotransmitters is essential for neuronal function. Ion-coupled transport is responsible for concentrating transmitters in synaptic vesicles within the cytoplasm of presynaptic terminals, as well as for recovering transmitter molecules after their release into the synaptic cleft.

A number of transport ATPases and ion exchangers have been isolated and cloned, and proposals have been made about their configurations in the membrane. All appear to have 10 to 12 transmembrane segments and are assumed to form channel-like structures through which substances are moved by alternate exposure of binding sites to the extracellular and intracellular spaces.

In Chapter 2 we discussed how electrical signals in nerve cells are generated by the movement of ions through channels in the plasma membrane. For example, inward movement of positively charged sodium ions reduces the net negative charge on the inner surface of the membrane or, in other words, causes membrane depolarization. Inward movement of negatively charged chloride ions, on the other hand, causes hyperpolarization. All ion movements through channels are passive, driven by the concentration and electrical gradients across the cell membrane. Clearly, if such ion movements continued without compensation, the system eventually would run down and both the concentration gradients and the cell membrane potential would disappear.

Ions that leak into or out of the cell at rest or during electrical activity are recovered by a variety of active transport mechanisms that move ions back across the membrane against their electrochemical gradients. **Primary active transport** is driven directly by metabolic energy, specifically by the hydrolysis of ATP. **Secondary active transport** uses energy provided by the flux of an ion (usually sodium) down its established electrochemical gradient to transport other ions across the cell membrane, either in the same direction (**cotransport**), or in the opposite direction (**ion exchange**).

THE SODIUM–POTASSIUM EXCHANGE PUMP

Most excitable cells have resting membrane potentials in the range of –60 to –90 mV, while the equilibrium potential for sodium (E_{Na}), is usually of the order of +50 mV. Thus, there is a large electrochemical gradient tending to drive sodium into the cell, and sodium does, in fact, enter through various pathways in the cell membrane. In addition, the equilibrium potential for potassium (E_K) is more negative than the resting potential, so potassium ions continually move out of the cell. To maintain the viability of the cell, sodium must be transported back out and potassium back in, against their electrochemical gradients. This perpetual task is carried out by the **sodium–potassium exchange pump**, which transports three sodium ions out across the cell membrane for every two potassium ions carried inward.

Early studies of sodium–potassium exchange, done by Hodgkin and Keynes and their colleagues on the giant axon of the squid,[1,2] showed clearly that the source of energy for the transport process is hydrolysis of ATP. During the same period Skou[3] demonstrated that an ATPase isolated from crab nerve had many of the biochemical properties that would be expected of a sodium–potassium exchange pump. Specifically, just as both sodium and potassium are required for the pump to work, the enzyme was stimulated by the simultaneous presence of sodium and potassium. Moreover, both sodium–potassium exchange and activity of the enzyme were inhibited by ouabain.

These observations led to the conclusion that the enzyme **sodium–potassium ATPase** is itself the transport molecule. It is one of a family of "P-type" ATPases (so called because they form a phosphorylated intermediate) that derive energy for ion translocation from hydrolysis of ATP. Other members of the same family include calcium ATPases, which transport calcium out of the cytoplasm of cells, and hydrogen–potassium ATPase, whose best-known function is the secretion of vast quantities of acid into the lumen of the stomach.

Biochemical Properties of Sodium–Potassium ATPase

The biochemical properties of sodium–potassium ATPase have been known for a number of years.[4] The stoichiometry of cation binding to the enzyme is as expected from the transport characteristics: An average of three sodium and two potassium ions are bound for each molecule of ATP hydrolyzed. The requirement for sodium is remarkably specific. It is the only substrate accepted for net outward transport; conversely, it is the only monovalent cation *not* accepted for inward transport. Thus, lithium, ammonium, rubidium, cesium, and thallium are all able to substitute for potassium in the external solution but not for sodium in the internal solution. The requirement for external potassium is not absolute. In its absence the pump will extrude sodium at about 10% of capacity in an "uncoupled" mode.

[1]Hodgkin, A. L., and Keynes, R. D. 1955. *J. Physiol.* 128: 28–60.

[2]Caldwell, P. C., et al. 1960. *J. Physiol.* 152: 561–590.

[3]Skou, J. C. 1957. *Biochim. Biophys. Acta* 23: 394–401.

[4]Skou, J. C. 1988. *Methods Enzymol.* 156: 1–25.

The transport system is blocked specifically by digitalis glycosides (drugs used for treating congestive heart failure), particularly ouabain and strophanthidin. Although they block sodium and potassium transport by ATPase, these drugs have no effect on the passive movements of sodium and potassium through membrane channels.

Experimental Evidence that the Pump Is Electrogenic

Because sodium–potassium ATPase transports unequal numbers of sodium and potassium ions, each cycle of the pump results in the outward movement of one positive charge across the membrane. For this reason the pump is said to be **electrogenic**. The electrogenic nature of the pump was tested experimentally in squid axons,[1,2,5] and in snail neurons by Thomas.[6,7] Snail neurons are sufficiently large to permit the insertion of several micropipettes through the cell membrane into the cytoplasm without damaging the cell. To examine how internal sodium concentration, pump current, and membrane potential are related, Thomas used two intracellular pipettes to deposit ions in the cell—one filled with sodium acetate, the other with lithium acetate (Figure 4.1A). A third intracellular pipette was used as an electrode to record membrane potential. A fourth pipette was used as a current electrode for voltage clamp experiments (Chapter 6), and yet a fifth, made of sodium-sensitive glass, to monitor the intracellular sodium concentration. To inject sodium, the sodium-filled pipette was made positive with respect to the lithium pipette. Thus, current flow in the injection system was between the two pipettes, with none of the injected current flowing across the cell membrane.

The result of sodium injection is shown in Figure 4.1B. After a brief injection, the cell became hyperpolarized by about 20 mV, because of increased pump activity. The potential recovered gradually over several minutes, as the excess sodium was extruded. Injection of lithium (by making the lithium pipette positive) produced no hyperpolarization.

Several lines of evidence showed that the potential change after sodium injection was due to the action of a sodium pump. For example, the hyperpolarization could be greatly reduced or abolished by addition of the transport inhibitor ouabain to the bathing solution (Figure 4.1C). Similarly, sodium injection had little effect on the potential when potassium was absent from the external solution; reintroduction of potassium after sodium injection, however, resulted in immediate hyperpolarization (Figure 4.1D).

Quantitative estimates of the pump rate and the exchange ratio were obtained by voltage clamp experiments. This technique (Chapter 6) provided a means of measuring ionic current across the membrane, while the membrane potential was held constant ("clamped"). At the same time, intracellular sodium concentration was monitored. Sodium injection produced a transient rise in intracellular sodium, accompanied by an outward surge of current whose amplitude and duration followed the sodium concentration change (Figure 4.1E). The net charge carried out of the cell, calculated by measuring the area under the membrane current, amounted to only about one-third of the charge injected in the form of sodium ions. This result was consistent with the idea that for every three sodium ions pumped out of the cell, two potassium ions are carried inward.

Mechanism of Ion Translocation

The general sequence of events believed to underlie translocation of sodium and potassium by sodium–potassium ATPase is illustrated in Figure 4.2. Sodium- and potassium-binding sites, located within a channel-like structure, are exposed alternately to the intracellular and extracellular solutions. The cyclic conformational changes are driven by phosphorylation and dephosphorylation of the protein and are accompanied by changes in binding affinity for the two ions. Inward-facing sites have a low affinity for potassium and a high affinity for sodium (Figure 4.2A). Binding of three sodium ions causes a conformational change that leads, in turn, to ATP binding, followed by phosphorylation of the enzyme (Figure 4.2B). Phosphorylation then produces a further conformational change that exposes the ion-binding sites to the extracellular solution (Figure 4.2C). The outward-facing sites have low sodium and high potassium affinities, so that the sodium ions are replaced by two potassium ions (Figure 4.2D). Potassium binding leads to dephosphorylation of the enzyme (Figure 4.2E) and a return to the starting conformation (Figure 4.2F). The potassium ions are then released into the cytoplasm.

[5]Baker, P. F., et al. 1969. *J. Physiol.* 200: 459–496.

[6]Thomas, R. C. 1969. *J. Physiol.* 201: 495–514.

[7]Thomas, R. C. 1972. *J. Physiol.* 220: 55–71.

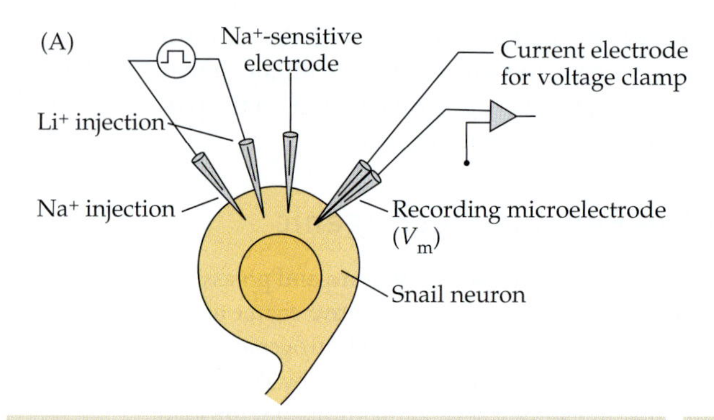

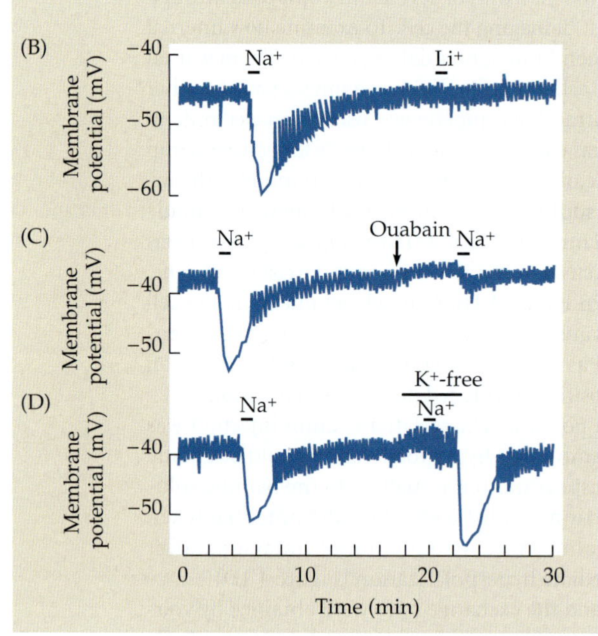

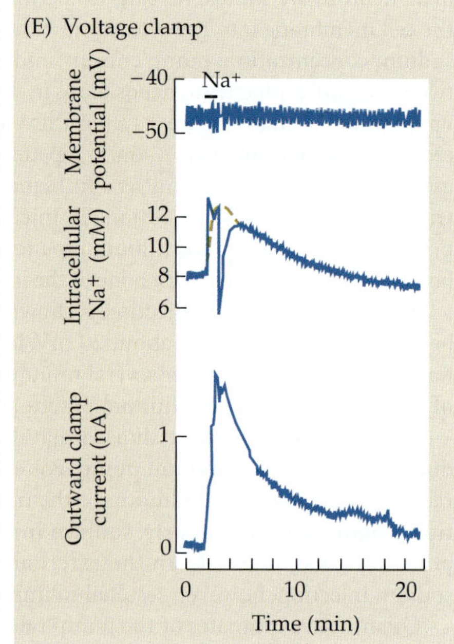

FIGURE 4.1 Effects of Sodium Injection into a snail neuron on intracellular sodium concentration, membrane potential, and membrane current. (A) Two micropipettes are used to inject either sodium or lithium. A sodium-sensitive electrode measures the intracellular sodium concentration. Another electrode measures membrane potential or, alternatively, is used in combination with a current-passing electrode to hold the membrane potential steady while measuring membrane current (shown in E). (B) Hyperpolarization of the membrane following intracellular injection of sodium. (The small rapid deflections are spontaneously occurring action potentials, reduced in size because of the poor frequency response of the pen recorder.) Injection of lithium does not produce hyperpolarization. (C) After application of ouabain (20 μg/ml), which blocks the sodium pump, hyperpolarization by sodium injection is greatly reduced. (D) Removal of potassium from the extracellular solution blocks the pump, so sodium injection produces no hyperpolarization until potassium is restored. (E) Measurements of membrane current, with the membrane potential clamped at −47 mV. Sodium injection results in an increase in intracellular sodium concentration, and an outward current across the cell membrane. Sharp deflections on the sodium concentration record are artifacts from the injection system; the time course of the concentration change is indicated by the dashed line. (After Thomas, 1969.)

CALCIUM PUMPS

Changes in intracellular calcium play an important role in many neuronal functions, such as release of neurotransmitters at synapses, activation of ion channels in the cell membrane, and regulation of a number of cytoplasmic enzymes. In muscle, calcium plays a primary role in the initiation of contraction. All these functional roles are associated

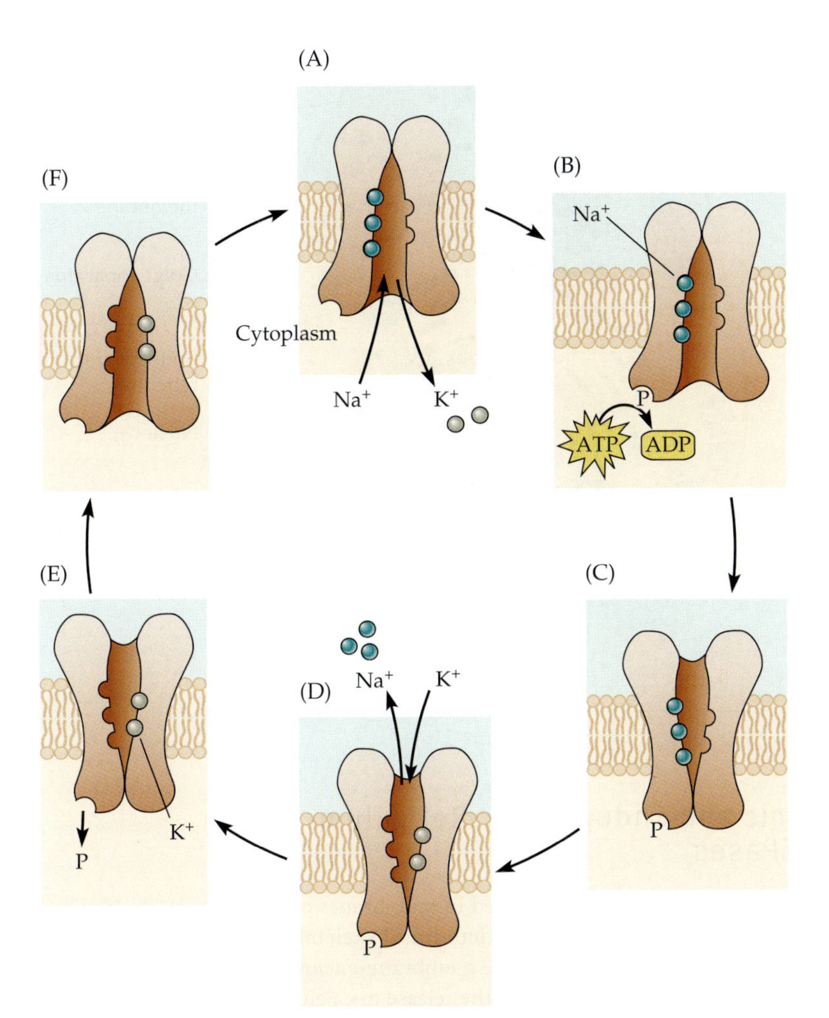

FIGURE 4.2 Ion Translocation by Na–K ATPase. (A) Inward-facing binding sites have a high affinity for sodium and a low affinity for potassium. Previously bound potassium ions are released and three sodium ions are bound. (B) Sodium binding is followed by ATP binding and phosphorylation of the enzyme. (C) The phosphorylated enzyme undergoes a conformational change such that its binding sites face the extracellular solution. (D) Outward-facing sites have a low affinity for sodium and high affinity for potassium, and they bind two potassium ions. (E) Potassium binding leads to dephosphorylation. (F) Dephosphorylation is followed by a return to the starting conformation.

with transient increases in cytoplasmic calcium concentration, so it is important that the resting concentration be kept low. Otherwise the various mechanisms would be activated continuously rather than in response to specific stimuli.

Cytoplasmic calcium concentration is altered in two ways: by calcium entry or exit through the plasma membrane and by release from or sequestration into intracellular compartments (Figure 4.3),[8] principally the endoplasmic reticulum (in muscle, the sarcoplasmic reticulum) and mitochondria.[9] Cytoplasmic concentrations of free calcium have been measured by injecting molecules such as aequorin[10] or FURA2[11] that absorb or emit light in the presence of ionized calcium, or by transfecting cells with genetically engineered calcium-sensitive fluorescent protein complexes.[12,13] The absorption or fluorescence, which is dependent on calcium concentration, is then monitored with highly sensitive optical techniques. In most neurons the resting concentration of free calcium ranges from 10 to 100 n*M*. The extracellular concentration of free calcium in vertebrate interstitial fluid is in the range of 2 to 5 m*M*.

Maintenance of the low intracellular concentration requires that calcium be transported outward across the plasma membrane against a very substantial electrochemical gradient. In addition, transport systems within the cell maintain high concentrations of calcium within intracellular compartments. Calcium concentrations in the endoplasmic reticulum can reach 400 μ*M*,[12,14] and in the sarcoplasmic reticulum of muscle they may go as high as 10 m*M*.[15] Transport of calcium out of the cytoplasm across the plasma membrane, as well as into intracellular compartments, is accomplished by **calcium ATPase.** An additional mechanism for the transport of calcium across the plasma membrane is discussed later in this chapter.

[8]Simpson, P. B., Challiss, R. A. J., and Nahorski, S. R. 1995. *Trends Neurosci.* 18: 299–306.

[9]Babcock, D. F., et al. 1997. *J. Cell Biol.* 136: 833–844.

[10]Baker, P. F., Hodgkin, A. L., and Ridgeway, E. B. 1971. *J. Physiol.* 218: 709–755.

[11]Tsien, R. Y. 1988. *Trends Neurosci.* 11: 419–424.

[12]Miyawaki, A., et al. 1997. *Nature* 388: 882–887.

[13]Persechini, A., Lynch, J. A., and Romoser, V. A. 1997. *Cell Calcium* 22: 209–216.

[14]Golovina, V. A., and Blaustein, M. P. 1997. *Science* 275: 1643–1648.

[15]Edes, I., and Kranias, E. G. 1995. In *Cell Physiology Source Book*. Academic Press, New York, pp. 156–165.

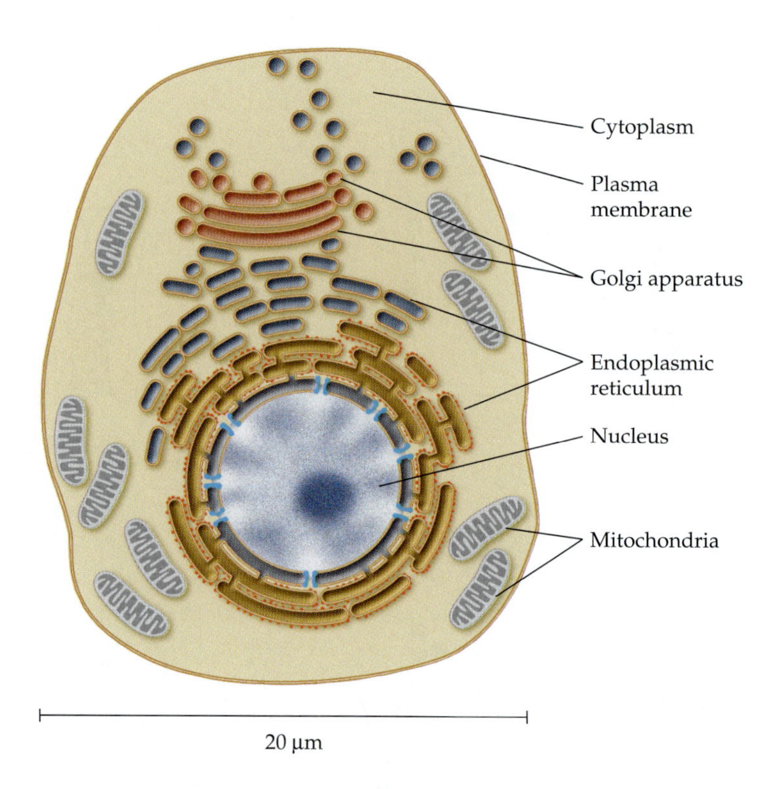

FIGURE 4.3 Intracellular Compartments. Within the cytoplasm of animal cells are numerous intracellular compartments. Some of these compartments—principally the endoplasmic reticulum and mitochondria—sequester calcium.

Sarcoplasmic and Endoplasmic Reticulum Calcium ATPases

One family of calcium ATPases is concentrated in membranes of the endoplasmic reticulum of neurons and in the sarcoplasmic reticulum of skeletal muscle. These enzymes transport calcium from the cytoplasm into the membrane-enclosed intracellular compartments. In muscle, contraction is triggered by the release of calcium from the sarcoplasmic reticulum into the myoplasm. The high density of calcium ATPase in the membranes of the sarcoplasmic reticulum enables rapid recovery of free calcium from the cytoplasm, thereby promoting muscle relaxation.

The transport cycle is analogous to that described in Figure 4.2 for sodium–potassium ATPase: It begins with the attachment of two calcium ions to high-affinity sites ($K_{m(Ca)}$ ≈100 n*M*)[16] facing the cytoplasm. The enzyme is then phosphorylated and undergoes a conformational change resulting in the release of the calcium ions to the reticular compartment. Release of the bound calcium is followed by dephosphorylation and return to the starting molecular configuration.

Plasma Membrane Calcium ATPase

Calcium ATPase is also found in the plasma membranes of all cells. The plasma membrane enzyme is similar in structure and function to its endoplasmic and sarcoplasmic reticular counterparts, but it differs in some details.[17] The intracellular binding site has a high affinity for calcium,[18] but only one calcium ion is bound during the transport cycle. In nerve and muscle the enzyme is only sparsely distributed in the plasma membrane, so its transport capacity is relatively low. Nonetheless, it manages to compensate for calcium influx into resting cells.

[16]Schatzmann, H. J. 1989. *Annu. Rev. Physiol.* 51: 473–485.

[17]Carafoli, E. 1994. *FASEB J.* 8: 993–1002.

[18]DiPolo, R., and Beaugé, L. 1979. *Nature* 278: 271–273.

SODIUM–CALCIUM EXCHANGE

Many ion transport mechanisms make use of an entirely different principle for the uphill transfer of ions across the cell membrane: Instead of relying on hydrolysis of ATP, ion movement is coupled to the inward flux of sodium down its electrochemical gradient. Sodium, entering the cell downhill through transport molecules, provides the energy required to carry other ions uphill against their electrochemical gradients. A simple example is the 1:1 sodium–hydrogen exchange mechanism that contributes to the maintenance of intracellular pH. Protons are carried out of the cell against their electrochemical gradients in exchange for inward movement of sodium. Hydrogen, calcium, potassium, chloride, and bicarbonate are all transported in this way. These secondary active transport systems account for a measurable fraction of the sodium entry into the resting cell. Ultimately, of course, the mechanisms depend on sodium–potassium ATPase to maintain the sodium gradient by pumping sodium ions back out of the cell. In some instances, potassium ions, moving out of the cell down their electrochemical gradient, also contribute energy to secondary active transport processes.

The NCX Transport System

At least two sodium–calcium exchange systems are found in plasma membranes. The most widely distributed of these, now known as the NCX (Na–Ca exchange) transport system, was first studied in cardiac muscle,[19] crab nerve,[20] and squid axon.[21] The transport molecule carries one calcium ion outward for each group of three sodium ions entering the cell.[22] The NCX exchanger has a lower affinity for calcium ($K_{1/2(Ca)} \approx 1.0\ \mu M$)[23,24] than does calcium ATPase, but because the exchange molecules occur at a much higher density, their transport capacity is about 50 times greater. The exchange system is called into play in excitable cells when calcium influx due to electrical activity overwhelms the transport ability of the ATPase.

The experiment shown in Figure 4.4 illustrates the operation of the sodium–calcium exchange mechanism in a squid axon. The intracellular concentration of ionized calcium was measured by aequorin luminescence. In the steady state, the influx of calcium along its electrochemical gradient is balanced by outward transport through the ion exchanger (Figure 4.4A). At the beginning of the experiment (Figure 4.4B), the intracellular calcium concentration is relatively high, as the axon is bathed in high (112 m*M*) calcium. When the extracellular calcium concentration is reduced, the passive influx decreases. As a result the intracellular calcium concentration falls, increasing the driving force for calcium entry until the passive influx again equals the rate of extrusion. Reducing extracellular sodium concentration, on the other hand, increases intracellular calcium concentration because the reduced driving force for sodium entry reduces the rate at which the exchanger extrudes calcium. Intracellular calcium concentration then rises until calcium influx is reduced by the same amount. Replacement of sodium by lithium (which does not enter the exchanger) results in a further increase of intracellular calcium.

Reversal of Na–Ca Exchange

Ion exchange mechanisms can be made to run backward by altering one or more of the ionic gradients involved in the exchange. An interesting feature of the NCX family of exchangers is that such reversal can occur readily under physiological conditions, in which case calcium *enters* through the system and sodium is extruded. The direction of transport is determined simply by whether the energy provided by the entry of three sodium ions is greater than or less than the energy required to extrude one calcium ion. One factor determining this energy balance is the membrane potential of the cell. Such dependence on potential arises because the exchange is not electrically neutral: Each forward cycle of the transport molecule results in a net transfer of one positive charge inward across the membrane. As a result, forward transport is facilitated by membrane hyperpolarization and impeded, or even reversed, by depolarization. It is useful to note that nonneutral ion exchangers, although voltage-sensitive, are not electrogenic; unlike pumps, they make use of standing electrochemical gradients rather than contributing to them.

The energy dissipated by sodium entry (or required for sodium extrusion) is simply the product of the charge moved across the membrane and the driving force for such movement or, in other words, the charge multiplied by the difference between the sodium

[19]Reuter, H., and Seitz, N. 1968. *J. Physiol.* 195: 451–470.

[20]Baker, P. F., and Blaustein, M. P. 1968. *Biochim. Biophys. Acta* 150: 167–179.

[21]Baker, P. F., et al. 1969. *J. Physiol.* 200: 431–458.

[22]Caputo, C., Bezanilla, F., and DiPolo, R. 1989. *Biochim. Biophys. Acta* 986: 250–256.

[23]Rasgado-Flores, H., Santiago, E. M., and Blaustein, M. P. 1989. *J. Gen. Physiol.* 93: 1219–1241.

[24]Philipson, K. D., et al. 1996. *Ann. N.Y. Acad. Sci.* 779: 20–28.

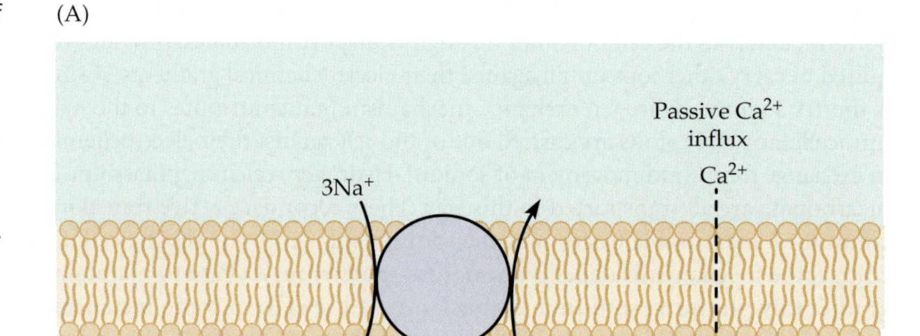

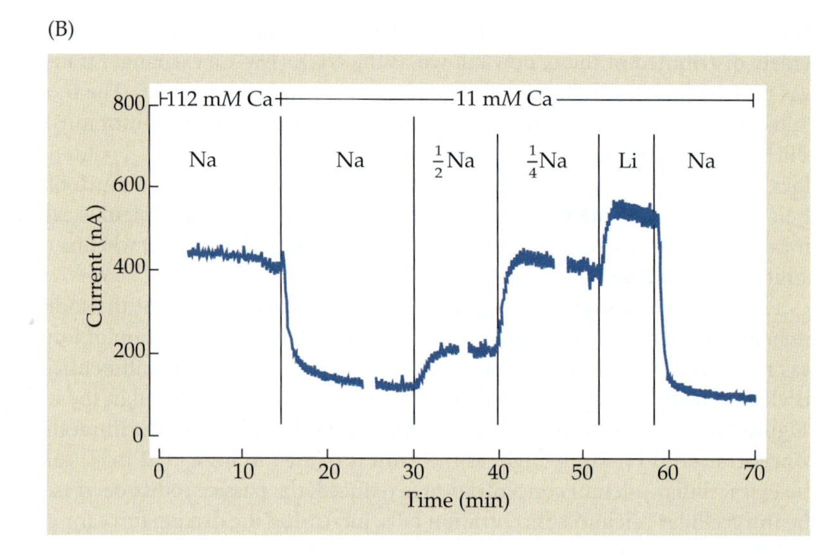

FIGURE 4.4 Transport of Calcium Ions out of a squid axon. (A) Scheme for sodium–calcium exchange. The influx of three sodium ions down their electrochemical gradient is coupled to the extrusion of one calcium ion. The calcium concentration reaches a steady state when outward transport through the exchanger is equal to the inward Ca^{2+} leak. (B) Effect of changes in extracellular calcium and sodium on intracellular calcium concentration. Changes in intracellular calcium concentration are measured by changes in the luminescence of injected aequorin, indicated by nanoamperes of current from the photodetector. Increased readings mean increased concentrations of intracellular free calcium. Reducing the extracellular calcium concentration from 112 m*M* to 11 m*M* reduces the intracellular concentration. Reducing extracellular sodium reduces outward calcium transport, and hence increases intracellular concentration. Lithium ions do not substitute for sodium in the transport system. (From Baker, Hodgkin, and Ridgeway, 1971.)

equilibrium potential (E_{Na}) and the membrane potential (V_m). For three sodium ions, this is $3(E_{Na} - V_m)$. Similarly, for a single (divalent) calcium ion the energy is $2(E_{Ca} - V_m)$. At some value of membrane potential the energies will be exactly equal and no exchange will occur. If we call this the reversal potential, V_r, then

$$3(E_{Na} - V_r) = 2(E_{Ca} - V_r)$$

and (by rearrangement)

$$V_r = 3E_{Na} - 2E_{Ca}$$

At more negative potentials, sodium moves into the cell and calcium is transported out. At more positive potentials, the sodium and calcium fluxes are reversed.

Now suppose a nerve cell has internal sodium and calcium concentrations of 15 m*M* and 100 n*M*, respectively, and is bathed in a solution containing 150 m*M* sodium and 2 m*M* Ca. These are reasonable physiological values for mammalian cells. Using the Nernst equation (Chapter 2), the equilibrium potential for sodium is +58 mV and for calcium is +124 mV. Ion movement through the exchanger is zero at a membrane potential (V_r) of –74 mV. This value is in the range of resting membrane potentials for many cells, so in any given cell, ion movements through the exchanger may be in one direction or the other, depending on membrane potential, and on whether or not there has been previous sodium or calcium accumulation. In heart muscle cells, entry of calcium through the sodium–calcium exchange system during the action potential can contribute to activation of the contractile process,[25,26] and, upon repolarization, calcium extrusion through the same system is important for relaxation.[27]

[25]LeBlanc, N., and Hume, J. R. 1990. *Science* 248: 372–376.

[26]Lipp, P., and Niggli, E. 1994. *J. Physiol.* 474: 439–446.

[27]Bridge, J. H. B., Smolley, J. R., and Spitzer, K. W. 1990. *Science* 248: 376–378.

Sodium–Calcium Exchange in Retinal Rods

It is evident that the NCX type of exchanger would be a poor system for extruding calcium from cells with small resting potentials; rather than being extruded, calcium would be accumulated until a relatively high steady-state concentration was reached in the cytoplasm. An example of such a cell is the mammalian retinal rod, which has a resting membrane potential of the order of –40 mV (Chapter 19). These cells contain a second kind of sodium–calcium exchange molecule, known as RetX, in their plasma membranes.[28–30] The RetX exchanger is given an additional "boost" over the NCX exchanger by two differences in stoichiometry: (1) There are four (rather than three) sodium ions entering for each calcium ion extruded, and (2) additional energy is available from the extrusion of one potassium ion, which, like sodium, moves down its electrochemical gradient during the exchange. The reversal potential for sodium–potassium–calcium exchange is as follows:

$$V_r = 4E_{Na} - E_K - 2E_{Ca}$$

Using the previous assumptions about E_{Na} (58 mV) and E_{Ca} (124 mV), and assuming that $E_K = -90$ mV, then $V_r = +74$ mV. Clearly, transport through the RetX exchanger is unlikely to reverse.

Another way of viewing the exchange system is to ask what value of E_{Ca} would be needed for V_r to reach –40 mV, the resting potential of the cell. Using the same equation, we find that the answer is 181 mV, which with 2 m*M* extracellular calcium is equivalent to an intracellular concentration of 1 n*M*. In other words, at a membrane potential of –40 mV it is energetically possible for the exchanger to reduce the cytoplasmic calcium concentration to 1 n*M*. It is interesting to note that under the same circumstances the NCX system could only reduce the intracellular calcium concentration to 383 n*M*.

[28]Schnetkamp, P. P., Basu, D. K., and Szerencsei, R. T. 1989. *Am. J. Physiol.* 257: C153–C157.

[29]Cervetto, L., et al. 1989. *Nature* 337: 740–743.

[30]Schnetkamp, P. P. 1995. *Cell Calcium* 18: 322–330.

[31]Inagaki, C., Hara, M., and Zeng, H. T. 1996. *J. Exp. Zool.* 275: 262–268.

CHLORIDE TRANSPORT

Intracellular chloride concentration is closely regulated in all cells. Such regulation is particularly important in neurons because direct synaptic inhibition depends on the maintenance of low intracellular chloride (Chapter 9). Although a chloride-sensitive ATPase has been reported in cultured neurons from rat brain,[31] indicating that there might be a primary active transport mechanism for chloride, the bulk of chloride transport across the plasma membrane of nerve cells is by three secondary transport mechanisms: chloride–bicarbonate exchange, which carries chloride outward across the plasma membrane; outward potassium–chloride cotransport; and inward sodium–potassium–chloride cotransport. These mechanisms are summarized in Figure 4.5.

Chloride–Bicarbonate Exchange

Chloride–bicarbonate exchange (see Figure 4.5A) is found in most cell types. Because bicarbonate is an important buffer for hydrogen ions in the cell cytoplasm, this ex-

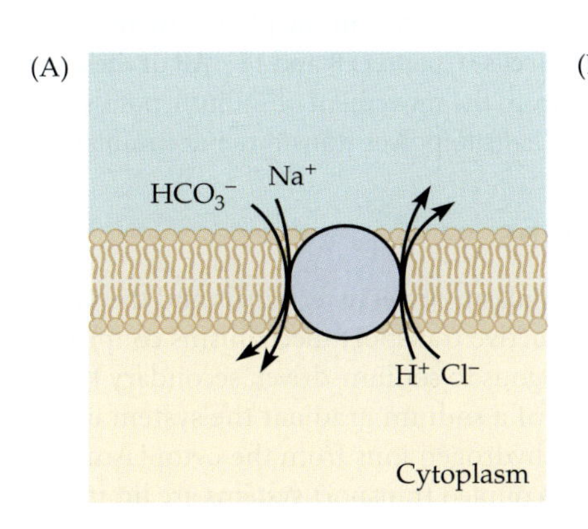

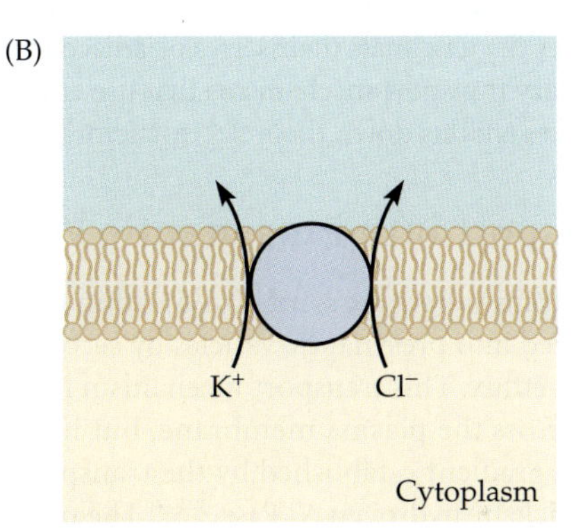

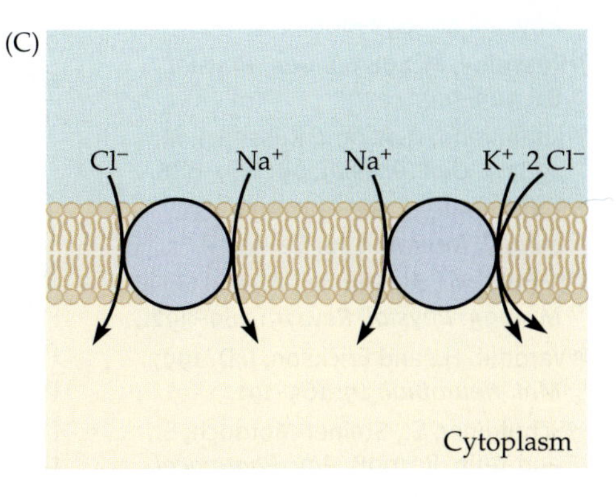

FIGURE 4.5 Mechanisms of Chloride Transport. (A) Chloride–bicarbonate exchange operates in parallel with sodium–hydrogen exchange to regulate intracellular pH. (B) Potassium–chloride cotransport uses the outward electrochemical gradient for potassium to transport chloride out of the cell. (C) In some neurons, inward chloride transport is mediated by two independent mechanisms, both using the electrochemical gradient for sodium. One is sodium–chloride cotransport, the other sodium–potassium–chloride cotransport, with a stoichiometry of 1:1:2. Note that all the chloride transport systems are electrically neutral.

change system, operating in parallel with sodium–hydrogen exchange, contributes to the regulation of intracellular pH. The exchange mechanism has been studied by Thomas,[32] who used pH-sensitive microelectrodes to measure intracellular hydrogen ion concentrations in snail neurons. He acidified the cytoplasm by injecting it with HCl or exposing it to CO_2 and measured the rate of recovery to normal. Recovery was prolonged when extracellular bicarbonate concentration was reduced or when intracellular chloride was depleted. Furthermore, recovery was virtually abolished when sodium was removed from the extracellular bathing solution. Thus, recovery involved inward movement of sodium and bicarbonate in exchange for chloride. The exchange mechanism is inhibited by 4-aceto-4′-isothiocyanostilbene-2,2′-disulfonic acid (SITS) and a related compound called DIDS. Both chloride and bicarbonate are carried against their electrochemical gradients; the energy required for their transport is obtained from the passive inward movement of sodium down its electrochemical gradient.

Potassium–Chloride Cotransport

Another transport mechanism for chloride is the cotransport of chloride and potassium outward across the cell membrane (see Figure 4.5B).[33] The system is insensitive to SITS and DIDS but is blocked by furosemide and bumetanide, substances known to block chloride transport in some tissues, such as renal tubular cells. Since ion transport by the system is insensitive to extracellular sodium concentration, it appears likely that energy required for outward chloride transport is supplied solely by the outward movement of potassium down its electrochemical gradient.

Inward Chloride Transport

In many cells, such as skeletal muscle fibers, tubular cells of the kidney, and squid axons, chloride is actively accumulated. Inward chloride transport is dependent on both sodium and potassium concentrations in the extracellular bathing medium, is insensitive to DIDS, and is blocked by furosemide and bumetanide. Two such transport systems have been shown to coexist in kidney cells, one with the Na:K:Cl stoichiometry of 1:1:2 and the second with a Na:Cl stoichiometry of 1:1 (Figure 4.5C).[34] In sqid axon, Russell[35,36] has demonstrated sodium- and potassium-dependent inward chloride transport with a Na:K:Cl stoichiometry of 1:2:3. Whether this indicates the existence of yet a third inward transport system, or whether it represents the two systems seen in kidney cells operating in parallel, is not clear. Inward chloride transport coupled to inward sodium and potassium movements has also been reported in spinal cord cells of tadpoles.[37]

TRANSPORT OF NEUROTRANSMITTERS

In addition to transporting inorganic ions, nerve cells have mechanisms for accumulating substances involved in synaptic transmission. Neurotransmitters are transported into organelles within the cytoplasm of presynaptic nerve terminals, where they are stored ready for release. After release, transmitters such as norepinephrine, serotonin, GABA, glycine, and glutamate are recovered from the synaptic cleft by transporters in the plasma membranes of either the terminals themselves or adjacent glial cells (Chapters 8 and 13). All of these are secondary transport mechanisms that use energy from the movement of sodium, potassium, or hydrogen ions down their electrochemical gradients to power transmitter accumulation.

Transport into Presynaptic Vesicles

Neurotransmitters are synthesized within the cytoplasm of nerve terminals and then concentrated into presynaptic vesicles by secondary active transport mechanisms coupled to proton efflux. This transport mechanism is analogous to sodium-driven secondary transport across the plasma membrane, but instead of a sodium gradient the system uses a proton gradient established by the transport of hydrogen ions from the cytoplasm into the vesicle by hydrogen ATPase.[38–40] The proton-coupled transport systems are limited in number: A vesicular monoamine transporter is responsible for the uptake of norepineph-

[32]Thomas, R. C. 1977. *J. Physiol.* 273: 317–338.

[33]Payne, J. A. 1997. *Am J. Physiol.* 273: C1516–C1525.

[34]Kaplan, M. R., et al. 1996. *Annu. Rev. Physiol.* 58: 649–668.

[35]Russell, J. M. 1983. *J. Gen. Physiol.* 81: 909–925.

[36]Altamirano, A. A., and Russell, J. M. 1987. *J. Gen. Physiol.* 89: 669–686.

[37]Rohrbough, J., and Spitzer, N. C. 1996. *J. Neurosci.* 16: 82–91.

[38]Schuldiner, S., Shirvan, A., and Linial, M. 1995. *Physiol. Rev.* 75: 369–392.

[39]Varoqui, H., and Erickson, J. D. 1997. *Mol. Neurobiol.* 15: 165–191.

[40]Schuldiner, S., Steiner-Mordoch, S., and Yelin, R. 1998. *Adv. Pharmacol.* 42: 223–227.

rine, dopamine, histamine, and serotonin (5-HT). Likewise, GABA and glycine are concentrated by a common transport system. ACh has its own transport system, as does glutamate. The proposed stoichiometry for the uptake systems is shown in Figure 4.6. Because the exchanges are not electrically neutral, all are dependent on the potential across the vesicle membrane as well as on the pH gradient. Glutamate transport is unusual in that it includes cotransport of chloride into the vesicle.[41]

Transmitter Uptake

Neurotransmitters are also transported into neurons and adjacent glial cells (Chapters 8 and 13). In general, such recovery serves two purposes: (1) The transmitter is removed from the extracellular space in the region of the synapse. This removal helps terminate its synaptic action and prevents its diffusion to other synaptic regions.[42,43] (2) Transmitter molecules recovered by the nerve terminal can be packaged again for re-release. All uptake mechanisms use the electrochemical gradient for sodium to carry transmitter substances across the plasma membrane into the cytoplasm.

There are two major categories of transmitter uptake systems (Figure 4.7).[44] In one, sodium influx and potassium efflux are coupled to the uptake process. This system is used for the uptake of neutral and acidic amino acids, including glutamate.[45] Inward transport of one glutamate ion is coupled to the influx of two or three sodium ions and the efflux of one potassium ion. In addition, there is either outward transport of one hydroxyl ion or inward transport of a proton, leading to acidification of the cytoplasm.[46,47]

The second uptake system couples sodium influx to transmitter uptake and influx of chloride. This system transports GABA, glycine, norepinephrine, dopamine, and serotonin into the cell. In each cycle one or two sodium ions enter the cell, accompanied by one transmitter molecule and a single chloride ion.[48]

One interesting aspect of the indirectly coupled transport mechanisms for transmitter uptake is that they are not, as a rule, electrically neutral. Thus, the direction of transport can be reversed by membrane depolarization, possibly within the physiological range of membrane potentials.[49] Indeed, outward transport of GABA by this mechanism has been demonstrated in catfish retinal cells.[50] The transport system, then, not only mediates uptake of transmitter, but might also function as a mechanism for transmitter release. Reversal of transmitter uptake can also have deleterious effects. After brain damage by stroke or trauma, outward transport of glutamate from depolarized glial cells can lead to the accumulation of cytotoxic amounts of glutamate around neurons in the damaged area, and thereby to further cell death.[51]

One neurotransmitter, ACh, is recycled in a different way. ACh is synthesized from acetyl coenzyme A and choline (Chapter 13). After ACh is released from the presynaptic terminal, its postsynaptic action is terminated by an enzyme (acetylcholinesterase; Chapter 9) that hydrolyzes it to acetate and choline. Approximately half of the choline is recov-

[41]Hartinger, J., and Jahn, R. 1993. *J. Biol. Chem.* 268: 23122–23127.

[42]Barbour, B., et al. 1994. *Neuron* 12: 1331–1343.

[43]Tong, G., and Jahr, C. E. 1994. *Neuron* 13: 1195–1203.

[44]Kanner, B. I. 1994. *J. Exp. Biol.* 196: 237–249.

[45]Kanai, Y. 1997. *Curr. Opin. Cell Biol.* 4: 565–572.

[46]Bouvier, M., et al. 1992. *Nature* 360: 471–474.

[47]Hediger, M. A., et al. 1995. *J. Physiol.* 482P: 7S–17S.

[48]Mager, S., et al. 1996. *J. Neurosci.* 16: 5405–5414.

[49]Attwell, D., Barbour, B., and Szatkowski, M. 1993. *Neuron* 11: 401–407.

[50]Cammack, J. N., and Schwartz, E. A. 1993. *J. Physiol.* 472: 81–102.

[51]Takahashi, M., et al. 1997. *J. Exp. Biol.* 200: 401–409.

(A) **Monoamines and acetylcholine**

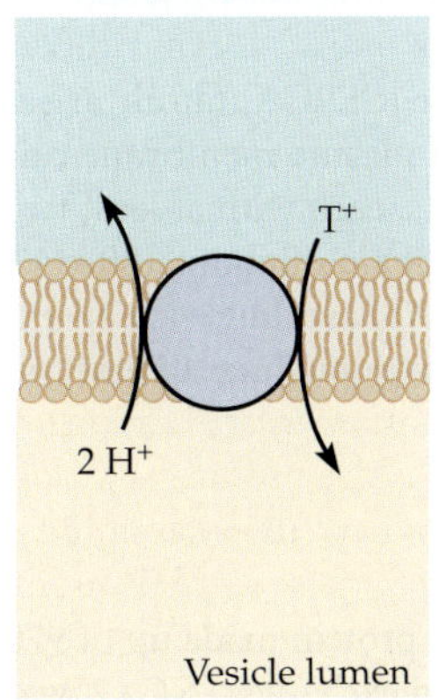

(B) **GABA and glycine**

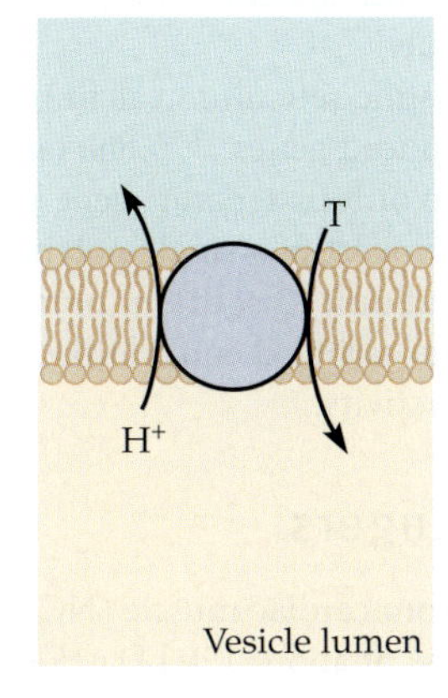

(C) **Glutamate**

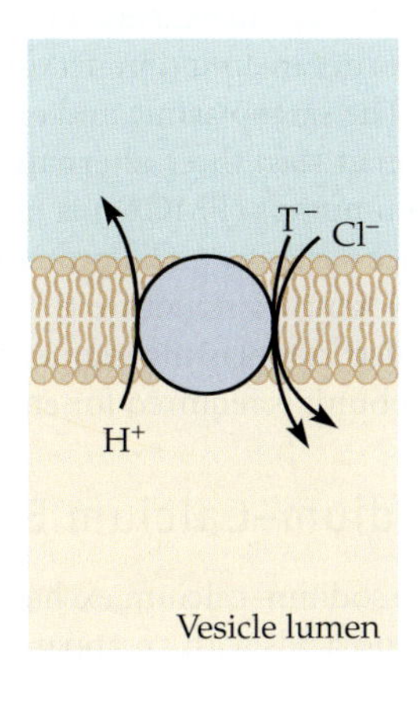

FIGURE 4.6 Transport of Neurotransmitters into Synaptic Vesicles. Hydrogen ATPase transports protons into the vesicle from the cytoplasm, creating an electrochemical gradient for proton efflux. Proton efflux through the secondary transport molecule provides the energy for accumulation of the neurotransmitter (T) in the vesicle. The proposed stoichiometry is 2:1 for monoamine and ACh transport (A), 1:1 for glycine, GABA, and glutamate transport (B and C). Glutamate transport is accompanied by chloride entry into the vesicle.

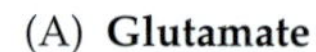

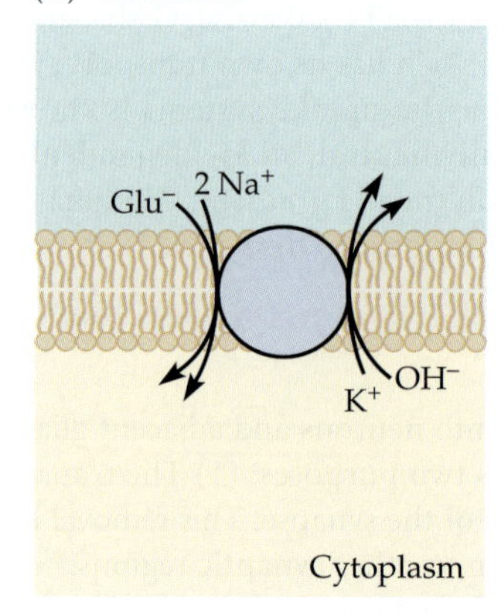

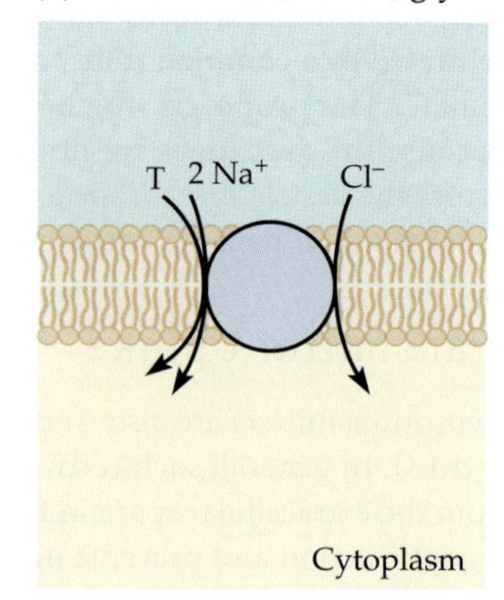

FIGURE 4.7 Uptake of Neurotransmitters. (A) Glutamate uptake is coupled to the influx of two sodium ions and efflux of one potassium ion, and it is accompanied by the extrusion of one OH^- (or one HCO_3^-) ion. (B) In the GABA, glycine, and monoamine (norepinephrine, dopamine, and serotonin) uptake systems, recovery of the transmitter (T) is coupled to the influx of two sodium ions and accompanied by the uptake of a single chloride ion. Choline from the hydrolysis of acetylcholine is recovered into the nerve terminal with the same stoichiometry as the GABA, glycine, and monoamine systems.

ered into the nerve terminal by a high-affinity ($K_m = 2\ \mu M$) uptake mechanism and reused for ACh synthesis. Like the monoamine, GABA, and glycine transport systems, choline uptake is dependent on extracellular sodium and chloride.[52]

MOLECULAR STRUCTURE OF TRANSPORTERS

So far we have dealt with functional properties of transporters with no reference to their molecular structure. As with membrane channels (Chapter 2), each functional group is represented by a specific transport protein or, more commonly, a family of proteins. Many of these have been isolated and cloned, and deductions have been made about their configurations in the membrane. Their structures are summarized in the discussion that follows and in Figure 4.8. Most appear to have 10 to 12 transmembrane segments, and they are assumed to form channel-like structures through which substances are moved by alternate exposure of binding sites to the extracellular and intracellular spaces.

ATPases

The molecular structure of sodium–potassium ATPase is known in some detail.[53,54] It is assembled from two subunits, α and β. The α subunit, with an apparent molecular mass of about 100 kDa, is responsible for the enzymatic activity of the pump and contains all the substrate-binding sites. The smaller (35 kDa) β subunit has a number of extracellular glycosylation sites and is necessary for pump function, but its precise role is not clear. Both the α and β subunits have been sequenced.[55,56] Their proposed structure is shown in Figure 4.8A. The nucleotide-binding and phosphorylation sites have been localized to the large cytoplasmic region of the α subunit between the fourth and fifth membrane-spanning segments. The β subunit contains only one putative membrane-spanning region; the bulk of the peptide is in the extracellular space. There are three α subunit isoforms (α_1, α_2, and α_3), all expressed in the nervous system. Two (β_1 and β_2) of three known β subunit isoforms are found in nervous tissue.

The sarcoplasmic and endoplasmic reticulum calcium ATPase (SERCA) family arises from at least three alternatively spliced genes.[15,57] The family of plasma membrane calcium pumps (PMCAs) is made up of four separate gene products, each with at least two splice variants.[58] The proteins all consist of a single polypeptide chain of about 100 kDa, analogous in structure to the α subunit of sodium–potassium ATPase, but with an extended cytoplasmic segment at the carboxyl end. Unlike sodium–potassium ATPase, no β subunit is required for enzyme activity.

Sodium–Calcium Exchangers

The sodium–calcium exchanger from cardiac muscle (NCX1) is a protein made up of 970 amino acids with an apparent mass of about 120 kDa.[59] A second exchanger, NCX2, was

[52]Cooper, J. R., Bloom, F. E., and Roth, R. H. (eds.). 1996. *The Biochemical Basis of Neuropharmacology*. Oxford University Press, New York, pp. 194–225.

[53]Pressley, T. A. 1996. *Miner. Electrolyte Metab.* 22: 264–271.

[54]Chow, D. C., and Forte, J. G. 1995. *J. Exp. Biol.* 198: 1–17.

[55]Kawakami, K., et al. 1985. *Nature* 316: 733–736.

[56]Noguchi, S., et al. 1986. *FEBS Lett.* 196: 315–320.

[57]Anderson, J. P., and Vilsen, B. 1995. *FEBS Lett.* 359: 101–106.

[58]Filoteo, A. G., et al. 1997. *J. Biol. Chem.* 272: 23741–23747.

[59]Nicoll, D. A., Longoni, S., and Philipson, K. D. 1990. *Science* 250: 562–565.

(A) **Sodium–potassium ATPase**

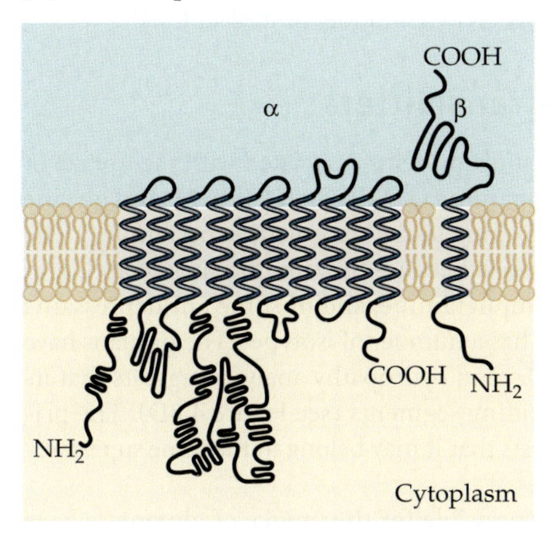

(B) **Sodium–calcium exchanger**

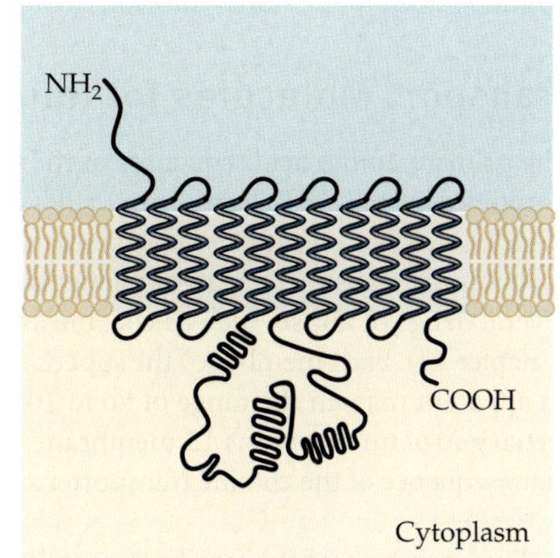

(C) **Potassium–chloride cotransporter**

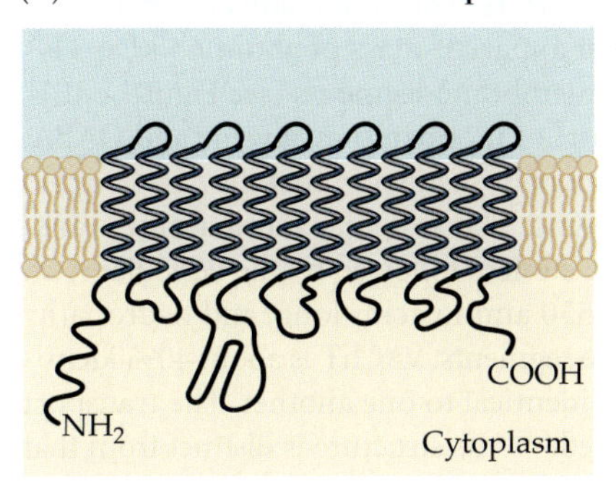

(D) **Monoamine transporter**

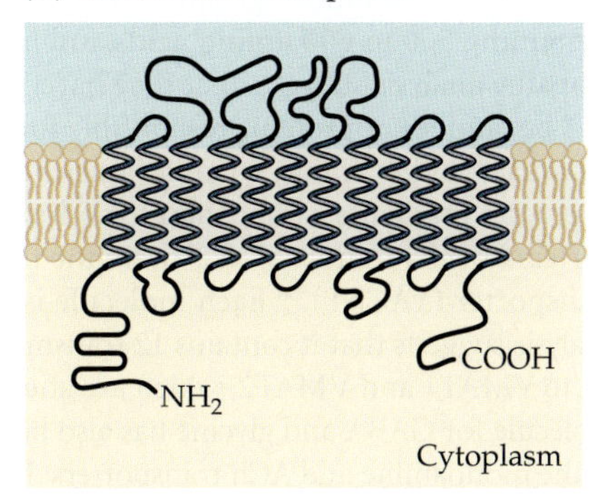

(E) Glutamate transporter

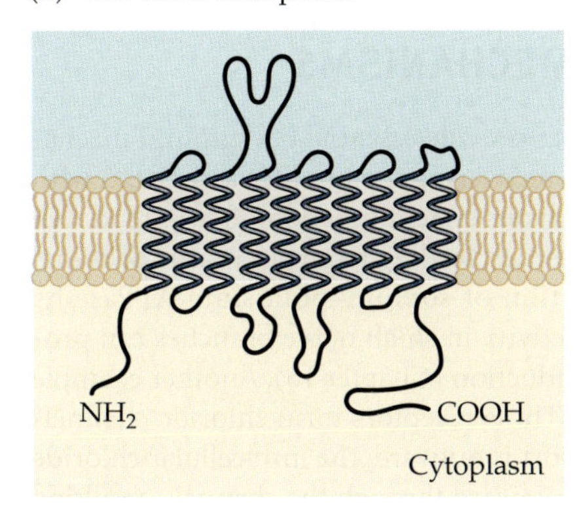

FIGURE 4.8 Proposed Molecular Configurations of Transport Molecules. Various transport molecules have been cloned and their structure in the membrane deduced from hydropathy analyses. (A) Sodium–potassium ATPase consists of an α subunit, with 8 to 10 transmembrane segments, and a smaller β subunit that spans the membrane only once. Calcium ATPases have a similar structure. (B) Sodium–calcium exchangers have 11 transmembrane segments. (C) Potassium–chloride cotransporters, anion exchange molecules, and sodium–potassium–chloride transporters all share the same membrane configuration, characterized by 12 transmembrane segments. (D) The superfamily of transporters for the uptake of monoamines, GABA, and glycine has a motif of 12 transmembrane segments. (E) Glutamate transport molecules are smaller, with 10 transmembrane segments.

originally cloned from rat brain, and both clones have a number of splicing isoforms.[60] Hydropathy analysis and studies with antibodies suggest that each of these exchangers contains 11 transmembrane segments, with a large intracellular loop between segments 5 and 6 (see Figure 4.8B).[61] RetX is somewhat larger, with 1199 amino acids and an apparent mass of about 130 kDa.[62] It has very little sequence homology to the NCX molecules, but a similar hydropathy profile suggests that it has the same membrane topology.

Other Ion Transporters

Two potassium–chloride cotransporters (KCC1 and KCC2) have been identified in rat brain.[63,64] Each of these proteins has about 1100 amino acids and an apparent molecular mass of about 120 kDa. Their primary structures suggest that they contain 12 transmembrane segments, bounded by amino- and carboxy-terminal regions in the cytoplasm (see

[60] Philipson, K. D., et al. 1996. *Ann. N.Y. Acad. Sci.* 779: 20–28.

[61] Cook, O., Low, W., and Rahamimoff, H. 1998. *Biochim. Biophys. Acta* 1371: 40–52.

[62] Reiländer, H., et al. 1992. *EMBO J.* 11: 1689–1695.

[63] Payne, J. A., Stevenson, T. J., and Donaldson, L. F. 1996. *J. Biol. Chem.* 271: 16245–16252.

[64] Gillen, C., et al. 1996. *J. Biol. Chem.* 271: 16237–16244.

Figure 4.8C). Molecules of the anion exchange (AE) family have similar structures,[65] as do the sodium–potassium–chloride cotransport (NKCC) family of molecules.[34]

Transport Molecules for Neurotransmitters

The primary amino acid sequences of the transport proteins associated with the uptake of norepinephrine (NET), serotonin (SERT), dopamine (DAT), GABA (GAT), and glycine (GLYT) into axon terminals suggest that they belong to the same gene superfamily.[66] The monoamine transporters are of particular interest because they are the sites of action of several drugs of abuse, such as cocaine and amphetamine, and of some antidepressants (Chapter 13). Each member of the superfamily has a number of isotypes. The proteins have an apparent mass in the range of 80 to 100 kDa, and hydropathy analysis suggests that its tertiary structure contains 12 membrane-spanning segments (see Figure 4.8D). The primary sequence of the choline transporter suggests that it may belong to the same superfamily.[67]

Five members of the family of proteins responsible for the uptake of glutamate have been isolated.[68] These transport proteins are structurally distinct from the superfamily containing monoamine, glycine, and GABA transporters. They are relatively small, each containing 500 to 600 amino acids and having an apparent mass of about 65 kDa. Hydropathy analysis suggests that they have 10 transmembrane segments (see Figure 4.8E).

The families of proteins responsible for transport of monoamines, glycine and GABA, glutamate, and ACh into synaptic vesicles are functionally distinct in that they are coupled to proton gradients rather than to sodium gradients. The vesicular monoamine transporters (VMAT1 and VMAT2) were the first to be cloned, followed by the vesicular ACh transporter (VAChT).[38] Each molecule is 520 to 530 amino acids long, and hydropathy analysis suggests that it contains 12 transmembrane segments. VAChT is about 40% identical to VMAT1 and VMAT2, which are about 65% identical to one another. The transport molecule for GABA and glycine has also been cloned.[69,70] Its structure is distinct from that of the monoamine and ACh transporters: It has only 10 putative transmembrane segments.

SIGNIFICANCE OF TRANSPORT MECHANISMS

Primary and secondary active transport molecules provide essential background mechanisms for maintaining cell homeostasis and for recovering metabolites associated with synaptic transmission. However, their roles in nervous system function often extend well beyond such relatively mundane housekeeping duties, leading to an active role of transport in cell signaling. For example, the activation of sodium–potassium ATPase by sodium accumulation during action potential activity in small nerve branches can produce a transient hyperpolarization and block conduction (Chapter 15). Another example is related to the activation of $GABA_A$ receptors. These receptors form chloride channels (Chapter 3). In cells in which net chloride transport is outward, the intracellular chloride concentration is relatively low and chloride flows inward through the channels, resulting in hyperpolarization. In cells in which inward chloride transport is dominant, however, intracellular chloride concentration is relatively high and chloride flows out of the cell, producing depolarization.[37] Thus, the action of a neurotransmitter depends directly on the nature of chloride transport in the cell.

Perhaps of more general significance are the neurotransmitter uptake systems. Prompt removal of transmitters from synaptic clefts can prevent prolonged activation or inactivation of postsynaptic receptors, and recycling maintains the supply of transmitters in presynaptic terminals. Thus, the transport mechanisms play an important role in regulating the dynamic behavior of nervous system synapses.

In summary, it is useful to think of channels as mediating electrical signaling, and transport molecules as maintaining the background conditions under which such signaling occurs. We should remember, however, that the two types of molecules often interact in more complicated ways to regulate nervous system function.

[65]Alper, S. L. 1991. *Annu. Rev. Physiol.* 53: 549–564.

[66]Nelson, N., and Lill, H. 1994. *J. Exp. Biol.* 196: 213–228.

[67]Mayser, W., Schloss, P., and Betz, H. 1992. *FEBS Lett.* 305: 31–36.

[68]Palacin, M., Estévez, R., Bertran, J., and Zorzano, A. 1998. *Physiol Rev.* 78: 969–1054.

[69]McIntire, S. L., et al. 1997. *Nature* 389: 870–876.

[70]Sagne, C., et al. 1997. *FEBS Lett.* 417: 177–183.

SUMMARY

■ A number of membrane proteins are involved in transporting substances into and out of cells. One example is sodium–potassium ATPase, which transports three sodium ions outward across the cell membrane and two potassium ions inward for each molecule of ATP hydrolyzed. The transport system maintains intracellular sodium and potassium concentrations at constant levels even though the ions leak into and out of the cells.

■ Calcium concentrations in the cell cytoplasm are kept low by two classes of calcium ATPases. One, plasma membrane calcium ATPase, transports calcium out of the cell. The other, endoplasmic and sarcoplasmic reticulum ATPases, concentrate calcium into intracellular compartments.

■ Another mechanism for calcium transport is sodium–calcium exchange: Sodium, entering the cell down its electrochemical gradient, provides energy for outward transport of calcium. This is an example of secondary transport, which relies on maintenance of the sodium gradient by sodium–potassium ATPase. In most neurons the transport molecule exchanges three sodium ions for one calcium ion. Under some physiological conditions, the exchanger can run in reverse. In retinal rods the transport molecule carries one calcium ion and one potassium ion out of the cell in exchange for the entry of four sodium ions.

■ There are two major mechanisms for extrusion of chloride from cells: One is chloride–bicarbonate exchange, which is important for intracellular pH regulation and depends on the sodium electrochemical gradient for its operation. The second is potassium–chloride cotransport, which relies on the electrochemical gradient for outward potassium movement across the cell membrane. In some cells chloride is accumulated rather than extruded. Chloride accumulation relies on the sodium electrochemical gradient and is accompanied by inward movement of potassium.

■ Within presynaptic nerve terminals, neurotransmitters are concentrated into synaptic vesicles in exchange for protons. The proton gradient across the vesicle membrane is maintained by hydrogen ATPase.

■ Neurotransmitters released from nerve terminals are recovered by secondary transport mechanisms that rely on the electrochemical gradient for sodium. There are two main types of transmitter uptake systems: The system that transports glutamate couples uptake with sodium influx and potassium efflux, and the system that transports other transmitters couples uptake with sodium influx and chloride influx.

■ The amino acid sequences of most transport molecules are known, and hydropathy analysis has been used to predict their conformation in the membrane. Most have 10 to 12 transmembrane segments and are assumed to form porelike structures through which ions move by alternate exposure of binding sites to the extracellular and intracellular spaces.

SUGGESTED READING

Altamirano, A. A., and Russell, J. M. 1987. Coupled Na/K/Cl efflux. "Reverse" unidirectional fluxes in squid giant axons. *J. Gen. Physiol.* 89: 669–686.

Baker, P. F., Hodgkin, A. L., and Ridgeway, E. B. 1971. Depolarization and calcium entry in squid giant axons. *J. Physiol.* 218: 709–755.

Kanner, B. I. 1994. Sodium-coupled neurotransmitter transport: Structure, function and regulation. *J. Exp. Biol.* 196: 237–249.

Schatzmann, H. J. 1989. The calcium pump of the surface membrane of the sarcoplasmic reticulum. *Annu. Rev. Physiol.* 51: 473–485.

Schnetkamp, P. P. 1995. Calcium homeostasis in vertebrate retinal rod outer segments. *Cell Calcium* 18: 322–330.

Schuldiner, S., Shirvan, A., and Linial, M. 1995. Vesicular neurotransmitter transporters: From bacteria to humans. *Physiol. Rev.* 75: 369–392.

Simpson, P. B., Challiss, R. A. J., and Nahorski, S. R. 1995. Neuronal Ca^{2+} stores: Activation and function. *Trends Neurosci.* 18: 299–306.

Skou, J. C. 1988. Overview: The Na,K pump. *Methods Enzymol.* 156: 1–25.

Thomas, R. C. 1969. Membrane currents and intracellular sodium changes in a snail neurone during extrusion of injected sodium. *J. Physiol.* 201: 495–514.

5 IONIC BASIS OF THE RESTING POTENTIAL

AT REST, A NEURON HAS A STEADY ELECTRICAL POTENTIAL across its plasma membrane, the inside being negative with respect to the outside. In relation to the extracellular fluid, the neuron has a high intracellular potassium concentration and low intracellular concentrations of sodium and chloride, so that potassium tends to diffuse out of the cell and sodium and chloride tend to diffuse in. The tendency for potassium and chloride to diffuse down their concentration gradients is opposed by the membrane potential. In a model cell permeable only to potassium and chloride, the concentration gradients and the membrane potential can be balanced exactly so that there is no net flux of either ion across the membrane. The membrane potential is then equal to the equilibrium potential for both potassium and chloride.

In the model cell, changing the extracellular potassium concentration changes the potassium equilibrium potential, and hence the membrane potential. In contrast, changing the extracellular chloride concentration eventually leads to a change in intracellular chloride. As a result the chloride equilibrium potential and the membrane potential are unchanged.

Real cells are also permeable to sodium. At rest, sodium ions constantly move into the cell, reducing the internal negativity of the membrane. As a result, potassium, being no longer in equilibrium, leaks out. If there were no compensation, these fluxes would lead to changes in the internal concentrations of sodium and potassium. However, the concentrations are maintained by the sodium–potassium exchange pump, which transports sodium out and potassium in across the cell membrane in a ratio of 3 sodium to 2 potassium. The resting membrane potential depends on the potassium equilibrium potential, the sodium equilibrium potential, the relative permeabilities of the cell membrane to the two ions, and the pump ratio. At the resting potential, the passive fluxes of sodium and potassium are exactly matched by the rates at which they are transported in the opposite direction. Because the sodium–potassium exchange pump transports more positive ions outward than inward across the membrane, it makes a direct contribution of several millivolts to the membrane potential.

The chloride equilibrium potential may be positive or negative with respect to the resting membrane potential, depending on chloride transport processes. Although the chloride distribution plays little role in determining the resting membrane potential, a substantial chloride permeability is important in some cells for electrical stability.

Electrical signals are generated in nerve cells and muscle fibers primarily by changes in permeability of the cell membrane to ions such as sodium and potassium. Increases in permeability allow ions to move inward or outward across the cell membrane down their electrochemical gradients. As we discussed in Chapter 2, permeability increases are due to activation of ion channels. Ions moving through the open channels change the charge on the cell membrane, and hence change the membrane potential. In order to understand how signals are generated, it is necessary to understand the nature of the standing ionic gradients across the cell membrane, and how these influence the resting membrane potential.

A MODEL CELL

It is useful to begin with the model cell shown in Figure 5.1. This cell contains potassium, sodium, chloride, and a large anion species, and it is bathed in a solution of sodium and potassium chloride. Other ions present in real cells, such as calcium or magnesium, are ignored for the moment, as their direct contributions to the resting membrane potential are negligible. The extracellular and intracellular ion concentrations in the model cell are similar to those found in frogs. In birds and mammals, ion concentrations are somewhat higher; in marine invertebrates such as the squid, very much higher (see Table 5.1). The model cell membrane is permeable to potassium and chloride, but not to sodium or to the internal anion. There are three major requirements for such a cell to remain in a stable condition:

1. The intracellular and extracellular solutions must each be electrically neutral. For example, a solution of chloride ions alone cannot exist; their charges must be balanced by an equal number of positive charges on cations such as sodium or potassium (otherwise electrical repulsion would literally blow the solution apart).
2. The cell must be in osmotic balance. If not, water will enter or leave the cell, causing it to swell or shrink, until osmotic balance is achieved. Osmotic balance is achieved when the total concentration of solute particles inside the cell is equal to that on the outside.
3. There must be no net movement of any particular ion into or out of the cell.

Ionic Equilibrium

How are the concentrations of the permeant ions maintained in the model cell, and what electrical potential is developed across the cell membrane? Figure 5.1 shows that the two ions are distributed in reverse ratio: Potassium is more concentrated on the inside of the cell, chloride on the outside. Imagine first that the membrane is permeable only to potassium; the question that arises immediately is why potassium ions do not diffuse out of the cell until the concentrations on either side of the cell membrane are equal. The answer is that they cannot because as they diffuse outward, positive charges accumulate on

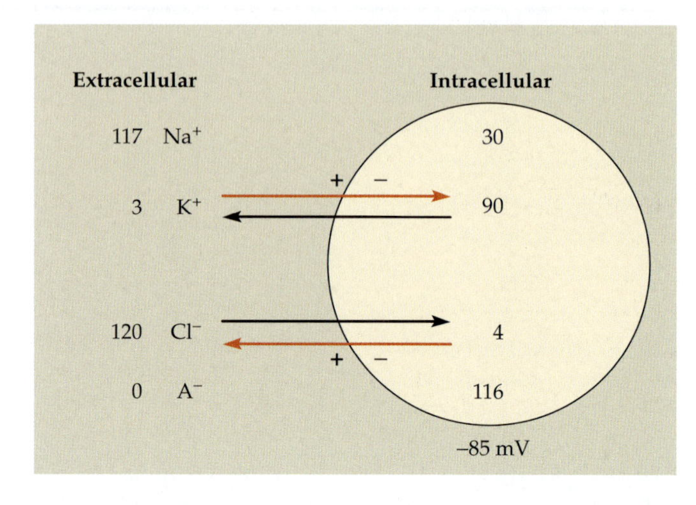

FIGURE 5.1 Ion Distributions in a Model Cell. The cell membrane is impermeable to Na^+ and to the internal anion (A^-), and permeable to K^+ and Cl^-. The concentration gradient for K^+ tends to drive it out of the cell (black arrow); the potential gradient tends to attract K^+ into the cell (orange arrow). In a cell at rest the two forces are exactly in balance. Concentration and electrical gradients for Cl^- are in the reverse directions. Ion concentrations are expressed in millimolar (m*M*).

the outer surface of the membrane and an excess of negative charges is left on the inner surface. As a result, a difference in potential develops across the membrane, the inside being negative with respect to the outside. The electrical gradient slows the efflux of positively charged potassium ions, and when the potential becomes sufficiently large, further net efflux of potassium is stopped. This is the potassium equilibrium potential (E_K). At E_K the effects of the concentration gradient and the potential gradient on ion flux through the membrane balance one another exactly. Individual potassium ions still enter and leave the cell, but no *net* movement occurs. The potassium ion is in **equilibrium**.

The conditions for potassium to be in equilibrium across the cell membrane are the same as those described in Chapter 2 for maintaining zero net flux through an individual channel in a membrane patch. There, a concentration gradient was balanced by a potential applied to the patch pipette. The important difference here is that the ion flux itself produces the required transmembrane potential. In other words, equilibrium in the model cell is automatic and inevitable. Recall from Chapter 2 that the potassium equilibrium potential is given by the Nernst equation:

$$E_K = \frac{RT}{zF} \ln \frac{[K]_o}{[K]_i} = 58 \log \frac{[K]_o}{[K]_i}$$

where $[K]_o$ and $[K]_i$ are the external and internal potassium ion concentrations. For the cell shown in Figure 5.1, E_K is 58 log (1/30) = –85 mV.

Suppose now that, in addition to potassium channels, the membrane has chloride channels. Because for an anion $z = -1$, the equilibrium potential for chloride is:

$$E_{Cl} = -58 \log \frac{[Cl]_o}{[Cl]_i}$$

or (from the properties of logarithmic ratios):

$$E_{Cl} = 58 \log \frac{[Cl]_i}{[Cl]_o}$$

In our model cell, the chloride concentration ratio is again 1:30, and E_{Cl} is also –85 mV. As with potassium, the membrane potential of –85 mV balances exactly the tendency for chloride to move down its concentration gradient, in this case *into* the cell.

In summary, the tendency for potassium ions to leave the cell and for chloride ions to diffuse inward are both opposed by the membrane potential. Because the concentration ratios for the two ions are of exactly the same magnitude (1:30), their equilibrium potentials are exactly the same. Since potassium and chloride are the only two ions that can move across the membrane and both are in equilibrium at –85 mV, the model cell can exist indefinitely with no net gain or loss of ions.

Electrical Neutrality

The charge separation across the membrane of our model cell, produced by outward movement of potassium and inward movement of chloride, means that there is an excess of anions inside the cell and of cations outside. This appears to violate the principle of electrical neutrality that we started with, but in fact does not. Potassium ions diffusing outward collect as excess cations against the outer membrane surface, leaving excess anions closely attracted to the inner surface. Both the potassium ions and the counterions they leave behind are, in effect, removed from the intracellular bulk solution, leaving it neutral. Similarly, chloride ions diffusing inward add to the collection of excess anions on the inner surface and leave counterions in the outer charged layer, so that the extracellular solution remains neutral as well. The outer layer of cations and inner layer of anions, of equal and opposite charge, are not in free solution, but are held to the membrane surface by mutual attraction. Thus, the membrane acts as a capacitor, separating and storing charge.

This does not mean that any given anion or cation is locked in position against the membrane. Ions in the charged layer interchange freely with those in the bulk solution.

The point is that although the identities of the ions in the layer are constantly changing, their total number remains constant, and the bulk solution remains neutral.

Another question we might ask about charge separation is whether the number of ions accumulated in the charged layer represents a significant fraction of the total number of ions in the cell. The answer is that it does not. If we consider our cell to have a radius of 25 μm, then at a concentration of 120 m*M* there are roughly 4×10^{12} cations and an equal number of anions in the cytoplasm. At a membrane potential of –85 mV, the amount of charge separated by the membrane is about 5×10^{11} univalent ions/cm^2 (Chapter 7). Our cell has a surface area of about 8×10^{-5} cm^2, so there are approximately 4×10^7 negative ions collected at the inner surface of the membrane, or 1/100,000 the number in free solution. Thus, the movements of potassium and chloride ions required to establish the membrane potential have no significant effect on intracellular ion concentrations.

The Effect of Extracellular Potassium and Chloride on Membrane Potential

In neurons, and in many other cells, the resting membrane potential is sensitive to changes in extracellular potassium concentration but is relatively unaffected by changes in extracellular chloride. To understand how this comes about it is useful to consider the consequences of such changes in the model cell. We will assume throughout this discussion that the volume of the extracellular fluid is infinitely large. Thus, movements of ions and water into or out of the cell have no significant effect on extracellular concentrations. Figure 5.2A shows the changes in intracellular composition and membrane potential that result from increasing extracellular potassium from 3 to 6 m*M*. This is done by replacing 3 m*M* NaCl with 3 m*M* KCl, thereby keeping the osmolarity unchanged, with a total

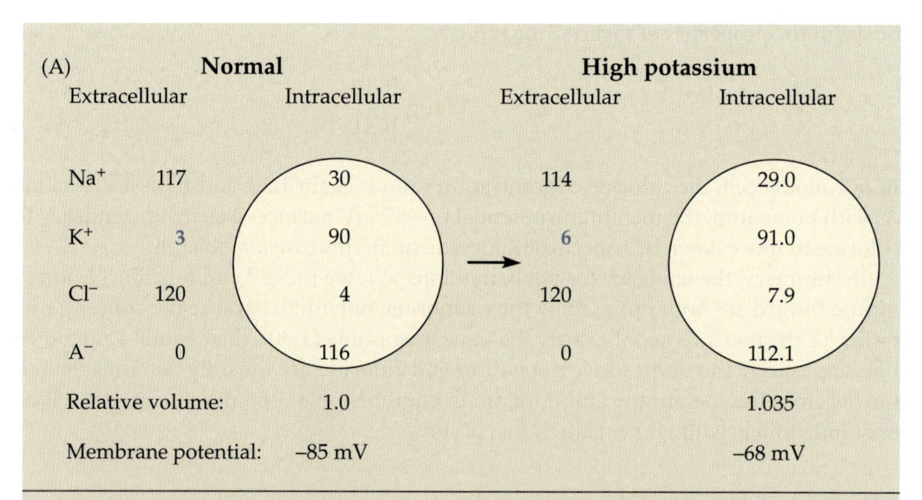

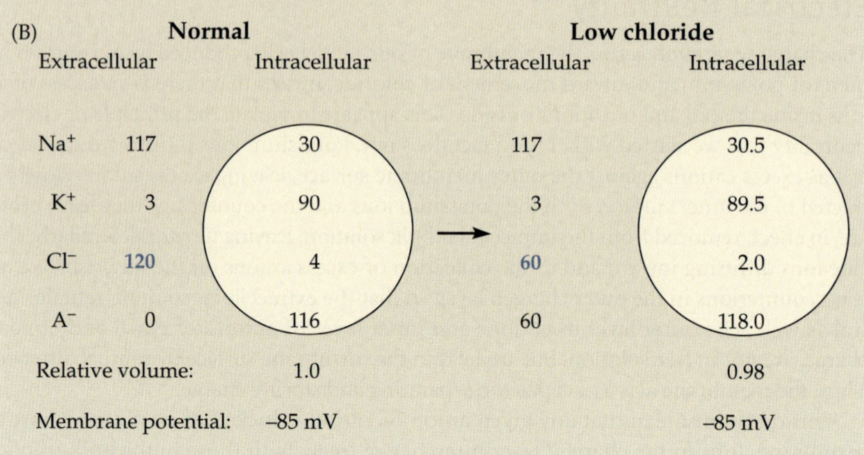

FIGURE 5.2 Effects of Changing Extracellular Ion Composition on intracellular ion concentrations and on membrane potential. (A) Extracellular K^+ concentration is doubled and, to keep osmolarity constant, Na^+ concentration is reduced. (B) Half the extracellular Cl^- is replaced by an impermeant anion, A^-. Ion concentrations are in millimolar (m*M*), and extracellular volumes are assumed to be very large with respect to cell volumes, so fluxes into and out of the cell do not change extracellular concentrations.

solute concentration of 240 m*M*. The increase in extracellular potassium reduces the concentration gradient for outward potassium movement, while initially leaving the electrical gradient unchanged. As a result there will be a net inward movement of potassium ions. As positive charges accumulate on its inner surface, the membrane is depolarized. This, in turn, means that chloride ions are no longer in equilibrium, and they move into the cell as well. Potassium and chloride entry continues until a new equilibrium is established, with both ions at a new concentration ratio consistent with the new membrane potential, in this example –68 mV.

Potassium and chloride entry is accompanied by the entry of water to maintain osmotic balance, resulting in a slight increase in cell volume. When the new equilibrium is reached, intracellular potassium has increased in concentration from 90 to 91 m*M*, intracellular chloride has increased in concentration from 4 to 7.9 m*M*, and the cell volume has increased by 3.5%.

At first glance it seems that more chloride than potassium has entered the cell, but think what the concentrations would be if the cell did *not* increase in volume: The concentrations of both ions would be greater than the indicated values by 3.5%. Thus, the intracellular chloride concentration would be about 8.2 m*M* (instead of 7.9 m*M*), and intracellular potassium would be about 94.2 m*M*, both 4.2 m*M* higher than in the original solution. In other words, we can think first of potassium and chloride entering in equal quantities (except for the trivial difference required to change the charge on the membrane), and then of water following to achieve the final concentrations shown in the figure.

Similar considerations apply to changes in extracellular chloride concentration, but with a marked difference: When the new steady state is finally reached, the membrane potential is essentially unchanged. The consequences of a 50% reduction in extracellular chloride concentration are shown in Figure 5.2B, in which 60 m*M* of chloride in the solution bathing the cell is replaced by an impermeant anion. Chloride leaves the cell, depolarizing the membrane toward the new chloride equilibrium potential (–68 mV). Potassium, being no longer in equilibrium, leaves as well. As in the previous example, potassium and chloride leave the cell in equal quantities (accompanied by water). Because the intracellular concentration of potassium is high, the fractional change in concentration produced by the efflux is relatively small. However, the efflux of chloride causes a sizable fractional change in the intracellular chloride concentration, and hence in the chloride equilibrium potential. As chloride continues to leave the cell, the equilibrium potential returns toward its original value. The process continues until the chloride and potassium equilibrium potentials are again equal and the membrane potential is restored.

MEMBRANE POTENTIALS IN SQUID AXONS

The idea that the resting membrane potential is the result of an unequal distribution of potassium ions between the extracellular and intracellular fluids was first proposed by Bernstein[1] in 1902. He could not test this hypothesis directly, however, because there was no satisfactory way of measuring membrane potential. It is now possible to measure membrane potential accurately, and to see whether changes in external and internal potassium concentrations produce the potential changes predicted by the Nernst relation.

The first such experiments were done on giant axons that innervate the mantle of the squid. The axons are up to 1 mm in diameter,[2] and their large size permits the insertion of recording electrodes into their cytoplasm to measure transmembrane potential directly (Figure 5.3A). Further, they are remarkably resilient and continue to function even when their axoplasm has been squeezed out with a rubber roller and replaced with an internal perfusate (Figure 5.3B and C). Thus, their internal, as well as external, ion composition can be controlled. A. L. Hodgkin, who together with A. F. Huxley initiated many experiments on squid axon (for which they later received the Nobel prize), has said,[3]

> It is arguable that the introduction of the squid giant nerve fiber by J. Z. Young in 1936 did more for axonology than any other single advance during the last forty years. Indeed a distinguished neurophysiologist remarked recently at a congress dinner (not, I thought, with the utmost tact), "It's the squid that really ought to be given the Nobel Prize."

[1]Bernstein, J. 1902. *Pflügers Arch.* 92: 521–562.

[2]Young, J. Z. 1936. *Q. J. Microsc. Sci.* 78: 367–386.

[3]Hodgkin, A. L. 1973. *Proc. R. Soc. Lond. B* 183: 1–19.

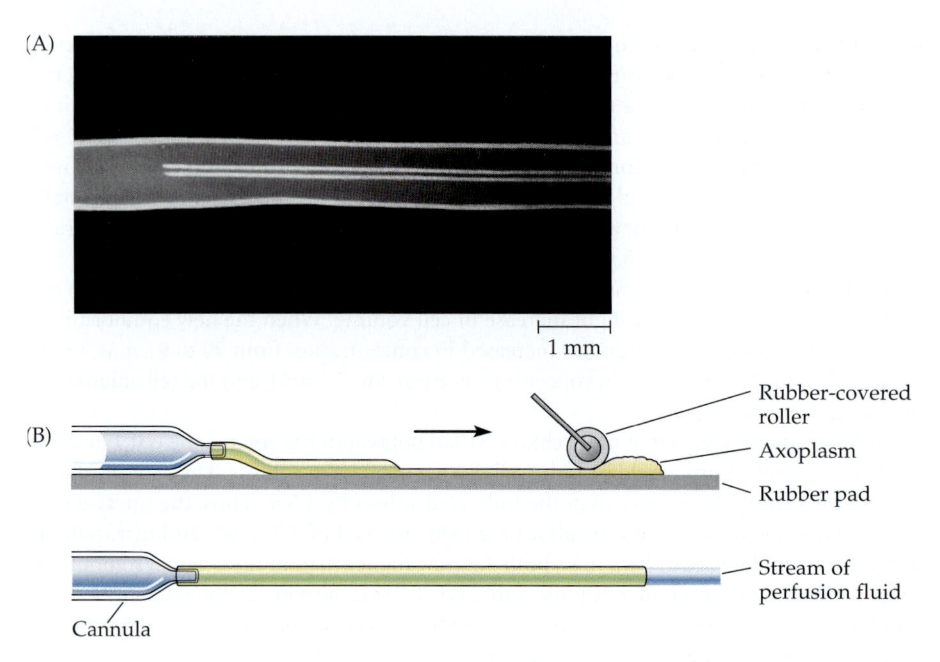

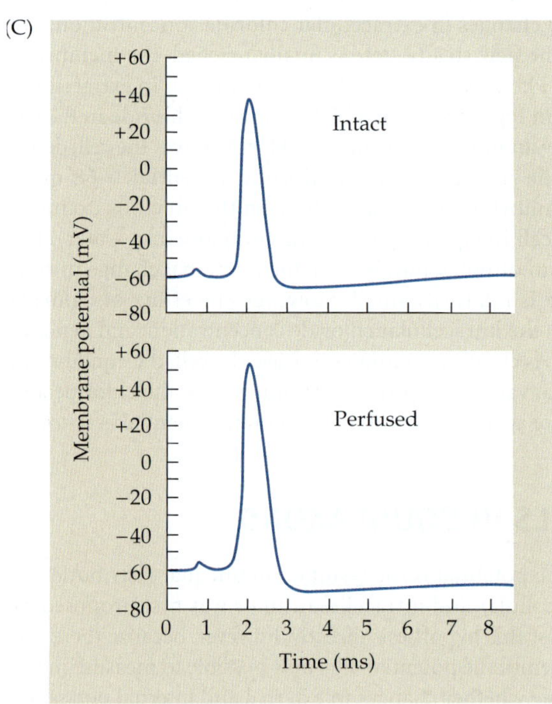

FIGURE 5.3 Recording from a Squid Axon. (A) Isolated giant axon of the squid, with axial recording electrode inside. (B) Extrusion of axoplasm from the axon, which is then cannulated and perfused internally. (C) Comparison of records before (intact) and after perfusion shows that the resting and action potentials are unaffected by removal of the axoplasm. (A from Hodgkin and Keynes, 1956; B and C after Baker, Hodgkin, and Shaw, 1962.)

The concentrations of some of the major ions in squid blood and in the axoplasm of the squid nerves are given in Table 5.1 (several ions, such as magnesium and internal anions, are omitted). Experiments on isolated axons are usually done in seawater, with the ratio of intracellular to extracellular potassium concentrations 40:1. If the membrane potential (V_m) were equal to the potassium equilibrium potential, it would be −93 mV. In fact, the measured membrane potential is considerably less negative (about −65 to −70 mV). On the other hand, the membrane potential is more negative than the chloride equilibrium potential, which is about −55 mV.

Bernstein's hypothesis was tested by measuring the resting membrane potential and comparing it with the potassium equilibrium potential at various extracellular potassium concentrations. (As with our model cell, such changes would be expected to produce no significant change in internal potassium concentration.) From the Nernst equation

Table 5.1
Concentrations of ions inside and outside freshly isolated axons of squid

	Concentration (m*M*)		
Ion	Axoplasm	Blood	Seawater
Potassium	400	20	10
Sodium	50	440	460
Chloride	60	560	540
Calcium	0.1 μ*M*[a]	10	10

Source: After Hodgkin, 1964.
[a] Ionized intracellular calcium from Baker, Hodgkin, and Ridgeway, 1971.

(Chapter 2), changing the concentration ratio by a factor of 10 should change the membrane potential by 58 mV at room temperature. The results of such an experiment on squid axon, in which the external potassium concentration was changed, are shown in Figure 5.4. The external concentration is plotted on a logarithmic scale on the abscissa and the membrane potential on the ordinate. The expected slope of 58 mV per 10-fold change in extracellular potassium concentration is realized only at relatively high concentrations (straight line), with the slope becoming less and less as external potassium is reduced. This result indicates that the potassium ion distribution is not the only factor contributing to the membrane potential.

The Effect of Sodium Permeability

From the experiments on squid axon we can conclude that the hypothesis made by Bernstein in 1902 is almost correct: The membrane potential is strongly but not exclusively dependent on the potassium concentration ratio. How do we account for the deviation from the Nernst relation shown in Figure 5.4? Simply by abandoning the notion that the membrane is impermeable to sodium. A real cell membrane has, in fact, a permeability to sodium that ranges between 1 and 10% of its permeability to potassium.

To consider the effect of sodium permeability, we begin with our model cell and, for the moment, ignore any movement of chloride ions. The membrane potential is at the potassium equilibrium potential, so there is no net movement of potassium across the membrane. If we now make the cell permeable to sodium, both the concentration gradi-

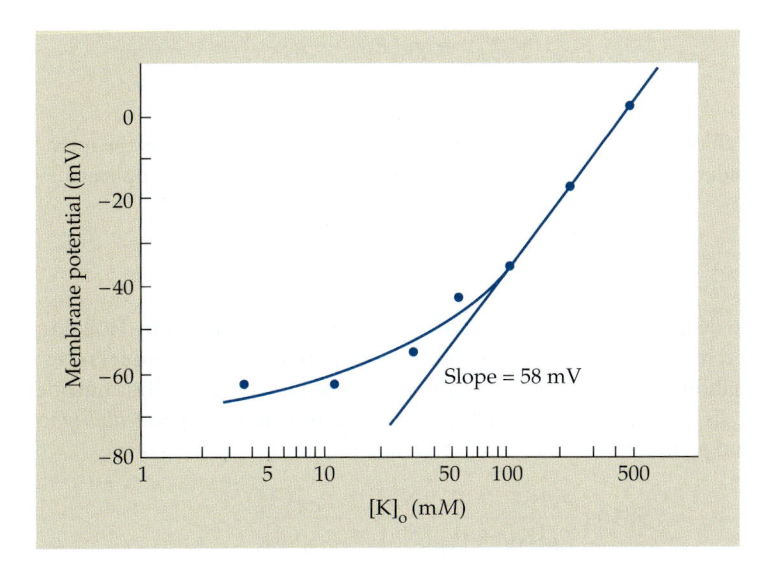

FIGURE 5.4 Membrane Potential versus External Potassium Concentration in squid axon, plotted on a semilogarithmic scale. The straight line is drawn with a slope of 58 mV per 10-fold change in extracellular potassium concentration, according to the Nernst equation. Because the membrane is also permeable to sodium, the points deviate from the straight line, especially at low potassium concentrations. (After Hodgkin and Keynes, 1955.)

ent and the membrane potential tend to drive sodium into the cell. As sodium ions enter, the accumulation of positive charge depolarizes the membrane. As a result, potassium is no longer in equilibrium and potassium ions leave the cell. As the depolarization progresses, the driving force for sodium influx decreases and that for potassium efflux increases. The process continues until the influx of sodium is exactly balanced by the efflux of potassium. At that point there is no further charge accumulation, and the membrane potential remains constant. In summary, the membrane potential lies between the potassium and sodium equilibrium potentials, and is the potential at which the sodium and potassium currents are exactly equal and opposite.

Chloride ions participate in the process as well, but as we have already seen, there is ultimately an adjustment in intracellular chloride concentration in the model cell so that the chloride equilibrium potential matches the new membrane potential. As the cation fluxes gradually reach a balance, the intracellular chloride concentration increases until there is no net chloride flux across the membrane.

The Constant Field Equation

To determine the exact membrane potential in our model cell we have to consider the individual ion currents across the membrane. The inward sodium current (i_{Na}) depends on the driving force for sodium ($V_m - E_{Na}$) (Chapter 2), and on the membrane conductance for sodium (g_{Na}). The conductance depends on the average number of sodium channels that are open in the resting membrane: The more open channels, the greater the conductance. So the sodium current is:

$$i_{Na} = g_{Na}(V_m - E_{Na})$$

This is also true for potassium and chloride:

$$i_K = g_K(V_m - E_K)$$

$$i_{Cl} = g_{Cl}(V_m - E_{Cl})$$

If we assume that chloride is in equilibrium, so that $i_{Cl} = 0$, then for the membrane potential to remain constant, the potassium and sodium currents must be equal and opposite:

$$g_K(V_m - E_K) = -g_{Na}(V_m - E_{Na})$$

It is useful to examine this relation in more detail. Suppose g_K is much larger than g_{Na}. Then, if the currents are to be equal, the driving force for potassium efflux must be much smaller than that for sodium entry. In other words, the membrane potential must be much closer to E_K than to E_{Na}. Conversely, if g_{Na} is relatively large, the membrane potential will be closer to E_{Na}.

By rearranging the equation we arrive at an expression for the membrane potential:

$$V_m = \frac{g_K E_K + g_{Na} E_{Na}}{g_K + g_{Na}}$$

If, for some reason, chloride is not at equilibrium, then chloride currents across the membrane must be considered as well, and the equation becomes slightly more complicated:

$$V_m = \frac{g_K E_K + g_{Na} E_{Na} + g_{Cl} E_{Cl}}{g_K + g_{Na} + g_{Cl}}$$

These ideas were developed originally by Goldman,[4] and independently by Hodgkin and Katz.[5] However, instead of considering equilibrium potentials and conductances, they derived an equation for membrane potential in terms of ion concentrations outside the cell ($[K]_o$, $[Na]_o$, $[Cl]_o$) and inside the cell ($[K]_i$, etc.), and membrane *permeability* to each ion (p_K, p_{Na}, and p_{Cl}):

$$V_m = 58 \log \frac{p_K[K]_o + p_{Na}[Na]_o + p_{Cl}[Cl]_i}{p_K[K]_i + p_{Na}[Na]_i + p_{Cl}[Cl]_o}$$

[4]Goldman, D. E. 1943. *J. Gen. Physiol.* 27: 37–60.

[5]Hodgkin, A. L., and Katz, B. 1949. *J. Physiol.* 108: 37–77.

As before, if chloride is in equilibrium, the chloride terms disappear. This equation is sometimes called the GHK equation after its originators and is known also as the **constant field equation** because one of the assumptions made in arriving at the expression was that the voltage gradient (or "field") across the membrane is uniform. It is entirely analogous to the conductance equation and makes the same predictions: When the permeability to potassium is very high relative to the sodium and chloride permeabilities, the sodium and chloride terms become negligible and the membrane potential approaches the equilibrium potential for potassium: $V_m = 58 \log ([K]_o/[K]_i)$. Increasing sodium permeability causes the membrane potential to move toward the sodium equilibrium potential.

The constant field equation provides us with a useful general principle to remember. The membrane potential depends on the relative conductances (or permeabilities) of the membrane to the major ions, and on the equilibrium potentials for those ions. In real cells the resting permeabilities to potassium and chloride are relatively high, so the resting membrane potential is close to the potassium and chloride equilibrium potentials. When sodium permeability is increased, as during an action potential (Chapter 6) or an excitatory postsynaptic potential (Chapter 9), the membrane potential moves toward the sodium equilibrium potential.

The Resting Membrane Potential

As useful as the constant field equation is, it does not provide us with an accurate description of the resting membrane potential. This is because the requirement for zero net current across the membrane is not, in itself, adequate: Our third requirement for the cell to remain in a stable condition—namely, that *each* individual ionic current must be zero—is not satisfied. As a result, the cell will gradually fill up with sodium and chloride and lose potassium. In real cells, intracellular sodium and potassium concentrations are kept constant by sodium–potassium ATPase (Chapter 4). To counteract the constant influx of sodium and the efflux of potassium, the pump transports a matching amount of each ion in the opposite direction (Figure 5.5). Thus, metabolic energy is used to maintain the cell in a **steady state**.

In order to have a more complete and accurate description of the resting membrane potential, we must consider both the passive ion fluxes and the activity of the pump. Again, we first consider the currents carried by passive fluxes of sodium and potassium across the membrane:

$$i_{Na} = g_{Na}(V_m - E_{Na})$$

$$i_K = g_K(V_m - E_K)$$

We no longer assume that the sodium and potassium currents are equal and opposite, but if we know how they are related we can, as before, obtain an equation for the membrane potential in terms of the sodium and potassium equilibrium potentials and their relative conductances. This is where the pump comes in. Because it keeps intracellular sodium and potassium concentrations constant by transporting the ions in the ratio of 3 Na to 2K (Chapter 4), it follows that the passive ion fluxes must be in the same ratio: $i_{Na}/i_K =$

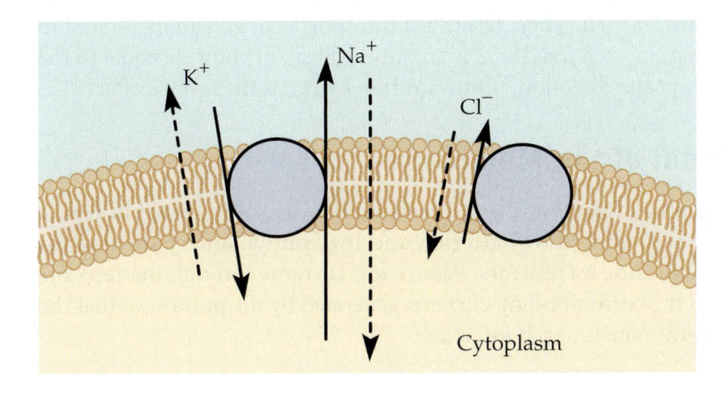

FIGURE 5.5 Passive Ion Fluxes and Pumps in a Steady State. Net passive ion movements across the membrane are indicated by dashed arrows, transport systems by solid arrows and circles. Lengths of arrows indicate the relative magnitudes of net ion movements. Total flux is zero for each ion. For example, the net inward leak of Na^+ is equal to the rate of outward transport. Na:K transport is coupled with a ratio of 3:2. In any particular cell, Cl^- transport may be outward (as shown) or inward.

3/2. So we can write

$$\frac{i_{Na}}{i_K} = \frac{g_{Na}(V_m - E_{Na})}{g_K(V_m - E_K)} = -1.5$$

The ratio is negative because the sodium and potassium currents are flowing in opposite directions. By rearranging, we get

$$V_m = \frac{1.5g_K E_K + g_{Na}E_{Na}}{1.5g_K + g_{Na}}$$

This equation is similar to the expression derived previously for the model cell, and it makes the same kinds of predictions: The membrane potential depends on the relative magnitudes of g_K and g_{Na}. The difference is that the potassium term is multiplied by a factor of 1.5. Because of this factor, the membrane potential is closer to E_K than would otherwise be the case. Thus, the driving force for sodium entry is increased, and that for potassium efflux is reduced. As a result the Na/K passive fluxes are in a ratio of 3:2 rather than 1:1.

In summary, the real cell differs from the model cell in that the resting membrane potential is the potential at which the passive influx of sodium is 1.5 times the passive efflux of potassium, rather than the potential at which the two fluxes are equal and opposite. The passive inward and outward currents are determined by the equilibrium potentials and conductances for the two ions; the required ratio of 3:2 is determined by the transport characteristics of the pump.

The problem of finding an expression for the resting membrane potential of real cells, taking into account the transport activity, was first considered by Mullins and Noda,[6] who used intracellular microelectrodes to study the effects of ionic changes on membrane potential in muscle. Like Goldman and Hodgkin and Katz, they derived an expression for membrane potential in terms of permeabilities and concentrations. The result is equivalent to the equation we have just derived using conductances and equilibrium potentials:

$$V_m = 58 \log \frac{rp_K[K]_o + p_{Na}[Na]_o}{rp_K[K]_i + p_{Na}[Na]_i}$$

where r is the absolute value of the transport ratio (3:2). The equation provides an accurate description of the resting membrane potential, provided all the other permeant ions (e.g., chloride) are in a steady state.

Chloride Distribution

How do these considerations apply to chloride? As for all other ions, there must be no net chloride current across the resting membrane. As already discussed (see Figure 5.2B), chloride is able to reach equilibrium simply by an appropriate adjustment in internal concentration, without affecting the steady-state membrane potential. In many cells, however, there are transport systems for chloride as well (Chapter 4). In squid axon and in muscle, chloride is transported actively into the cells; in many nerve cells active transport is outward (see Figure 5.5). The effect of inward transport is to add an increment to the equilibrium concentration such that there is an outward leak of chloride equal to the rate of transport in the opposite direction.[7] Outward transport has the reverse effect.

An Electrical Model of the Membrane

For those attuned to electrical diagrams, these considerations are summarized in Figure 5.6. E_{Na}, E_K, and E_{Cl} are represented by batteries, and the conductance pathways for sodium, potassium, and chloride by resistors. Passive ion currents through the resistors are equal and opposite to the corresponding currents generated by the pumps, so that the net current across the membrane for each ion is zero.

[6]Mullins, L. J., and Noda, K. 1963. *J. Gen. Physiol.* 47: 117–132.

[7]Matthews, G., and Wickelgren, W. O. 1979. *J. Physiol.* 293: 393–414, Appendix (by A. R. Martin).

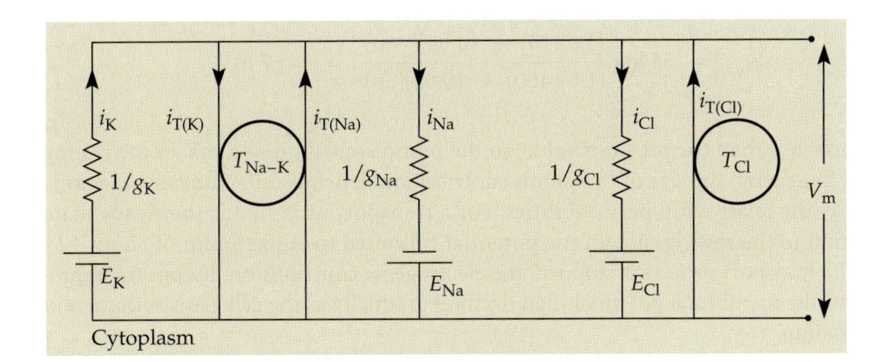

FIGURE 5.6 Electrical Model of the Steady-State Cell Membrane shown in Figure 5.5. E_K, E_{Na}, and E_{Cl} are the Nernst potentials for the individual ions. The individual ion conductances are represented by resistors (having a resistance of $1/g$ for each ion). The individual ion currents (i_K, i_{Na}, and i_{Cl}) are equal and opposite to the currents ($i_{T(K)}$, $i_{T(Na)}$, and $i_{T(Cl)}$) supplied by the sodium–potassium exchange pump (T_{Na-K}), and the chloride pump (T_{Cl}), so the net flux of each ion across the membrane is zero. The resulting membrane potential is V_m.

Predicted Values of Membrane Potential

How do these considerations explain the relation between potassium concentration and membrane potential shown in Figure 5.4? The answer becomes evident if we use real numbers in the equations. In squid axon, the permeability constants for sodium and potassium are roughly in the ratio 0.04:1.0.[5] We can use these relative values, together with the ion concentrations given in Table 5.1, to calculate the resting membrane potential in seawater:

$$V_m = 58 \log \frac{(1.5)(10) + (0.04)(460)}{(1.5)(400) + (0.04)(50)} = -73 \text{ mV}$$

Now we can see quantitatively why, when extracellular potassium is altered, the membrane potential fails to follow the Nernst potential for potassium, as shown in Figure 5.4. If, in the numerator of the equation, we look at the magnitude of the term involving extracellular potassium concentration ($1.5 \times 10 = 15$) and the term that involves extracellular sodium concentration ($0.04 \times 460 = 18.4$), we see that potassium contributes only about 45% to the total. Because of this, doubling the external potassium concentration does not double the numerator (as would happen in the Nernst equation), and as a consequence, the effect on the membrane potential of changing the extracellular potassium concentration is less than would be expected if potassium were the only permeant ion. When the external potassium concentration is raised to a high enough level (100 m*M* in Figure 5.4), the potassium term becomes sufficiently dominant for the relation to approach the theoretical limit of 58 mV per 10-fold change in concentration. This effect is enhanced by a factor discussed in Chapter 3: Many potassium channels are voltage-activated and open when the membrane is depolarized by increasing extracellular potassium. Because of the increased permeability to potassium, the relative contribution of sodium to the membrane potential is further reduced.

In general, nerve cells have resting potentials on the order of –70 mV. In some cells, such as vertebrate skeletal muscle,[8] the resting potential can be –90 mV or larger, reflecting a low ratio of sodium permeability to potassium permeability. Glial cells in particular have a very low permeability to sodium, so that their resting potentials are nearly identical to the potassium equilibrium potential (Chapter 8). Other cells, such as leech ganglion cells[9] and receptors in the retina,[10] have relatively high membrane permeabilities to sodium and resting membrane potentials as small as –40 mV.

Contribution of the Sodium–Potassium Pump to the Membrane Potential

The sodium–potassium transport system is **electrogenic** because each cycle of the pump results in the net outward transfer of one positive ion, thereby contributing to the excess negative charge on the inner face of the membrane. How large is this contribution? An easy way to find out is to calculate what the membrane potential would be if the pump were *not* electrogenic, or, in other words, if $r = 1$. Repeating the previous calculation with this condition yields the following:

[8]Fatt, P., and Katz, B. 1951. *J. Physiol.* 115: 320–370.

[9]Nicholls, J. G., and Baylor, D. A. 1968. *J. Neurophysiol.* 31: 740–756.

[10]Baylor, D. A., and Fuortes, M. G. F. 1970. *J. Physiol.* 207: 77–92.

$$V_m = 58 \log \frac{(1.0)(10) + (0.04)(460)}{(1.0)(400) + (0.04)(50)} = -67 \text{ mV}$$

This is 6 mV less than the previous value, so the pump contributes –6 mV to the resting potential. In general, the size of the pump contribution depends on a number of factors, particularly the relative ion permeabilities. For a transport ratio of 3:2, the steady-state contribution to the resting membrane potential is limited to a maximum of about –11 mV.[11] If the transport process is stopped, the electrogenic contribution disappears immediately, and the membrane potential then declines gradually as the cell gains sodium and loses potassium.

Ion Channels Associated with the Resting Potential

The resting permeabilities of membranes to sodium, potassium, and chloride have been determined in many nerve cells. It is a curious fact, however, that none of the channels underlying these resting permeabilities have been precisely identified in any specific cell. Candidates for potassium channels active in the resting membrane vary from one cell to the next. Among these are channels activated by intracellular cations: sodium-activated and calcium-activated potassium channels. In addition, many nerve cells have "M" potassium channels that are open at rest and closed by intracellular messengers (Chapter 16). Although it is unlikely that a large fraction of voltage-activated potassium channels ("delayed rectifier" and "A" channels) are open at rest, only 0.1 to 1% of the total number would be required to account for a substantial fraction of the resting conductance.[12]

The specific sources of the resting sodium permeability of nerve cells are also uncertain. A part can be attributed to the movement of sodium through potassium channels, most of which have a sodium/potassium permeability ratio that ranges between 1% and 3%.[13] In addition, both inward sodium and outward potassium fluxes may occur through cation channels that show little selectivity for potassium over sodium.[14,15] An additional sodium influx is through sodium-dependent secondary active transport systems (Chapter 4). Finally, tetrodotoxin has been shown to block a small fraction of the resting sodium conductance, indicating a contribution by voltage-activated sodium channels.[9]

Chloride channels of the CLC family (Chapter 3) are widely distributed in nerve and muscle. The presence of chloride channels is important in that they serve to stabilize the membrane potential (see the next section). The channels also interact with chloride transport systems to determine intracellular chloride concentrations.[16,17] When CLC channel expression is low—for example, in embryonic neurons in the hippocampus—E_{Cl} is positive with respect to the resting membrane potential because of inward transport and accumulation of chloride in the cytoplasm. In adult neurons, the expression of CLC channels increases the chloride conductance of the membrane so that excess accumulation cannot occur, and E_{Cl} becomes equal to the membrane potential. In central nervous system neurons, chloride channels can account for as much as 10% of the resting membrane conductance.[18]

[11]Martin, A. R., and Levinson, S. R. 1985. *Muscle Nerve* 8: 354–362.

[12]Edwards, C. 1982. *Neuroscience* 7: 1335–1366.

[13]Hille, B. 1992. *Ionic Channels of Excitable Membranes*, 2nd Ed. Sinauer Associates, Sunderland, MA, p. 352.

[14]Yellen, G. 1982. *Nature* 296: 357–359.

[15]Chua, M., and Betz, W. J. 1991. *Biophys. J.* 59: 1251–1260.

[16]Staley, K., et al. 1996. *Neuron* 17: 543–551.

[17]Mladinić, M., et al. 1999. *Proc. R. Soc. Lond. B* 266: 1207–1213.

[18]Gold, M. R., and Martin, A. R. 1983. *J. Physiol.* 342: 99–117.

CHANGES IN MEMBRANE POTENTIAL

It is important to keep in mind that the discussion of resting membrane potential is always in reference to *steady-state* conditions. For example, we have said that changing the extracellular chloride concentration has little effect on membrane potential because the intracellular chloride concentration accommodates to the change. This is true in the long run, but the intracellular adjustment takes time, and while it is occurring there will indeed be a transient effect.

The steady-state potential is the baseline upon which all changes in membrane potential are superimposed. How are such changes in potential produced? In general, transient changes, such as those that mediate signaling between cells in the nervous system, are the result of transient changes in membrane permeability. As we already know from the constant field equation, an increase in sodium permeability (or a decrease in potassium permeability) will move the membrane potential toward the sodium equilibrium potential, producing depolarization. Conversely, an increase in potassium permeability will produce hyperpolarization. Another ion of importance in signaling is calcium. Intracellular calcium concentration is very low (Chapter 4), and in most cells E_{Ca} is greater than +150 mV. Thus, an increase in calcium permeability results in calcium influx and depolarization.

The role of chloride permeability in the control of membrane potential is of particular interest. As we have noted, chloride makes little contribution to the resting membrane potential. Instead, intracellular chloride concentration adjusts to the potential and is modified by whatever chloride transport mechanisms are operating in the cell membrane. The effect of a transient increase in chloride permeability can be either hyperpolarizing or depolarizing, depending on whether the chloride equilibrium potential is negative or positive with respect to the resting potential. This, in turn, depends on whether intracellular chloride is depleted or concentrated by the transport system. In either case, the change in potential is usually relatively small. Even so, the increased chloride permeability can be important for the regulation of signaling because it tends to hold the membrane potential near the chloride equilibrium potential and thus attenuates changes in potential that are produced by other influences.

This stabilization of the membrane potential is important for controlling the excitability of many cells, such as skeletal muscle fibers, that have a relatively high chloride permeability at rest. In such cells a transient influx of positive ions causes less depolarization than would otherwise be the case because it is countered by an influx of chloride through already open channels. Chloride channel mutations that reduce chloride conductance are responsible for several muscle diseases. The diseased muscles are hyperexcitable (myotonic) because of a loss of the normal stabilizing influence of a high chloride conductance.[19,20]

[19]Barchi, R. L. 1997. *Neurobiol. Dis.* 4: 254–264.

[20]Cannon, S. C. 1996. *Trends Neurosci.* 19: 3–10.

SUMMARY

■ Nerve cells have a high intracellular concentration of potassium and low intracellular concentrations of sodium and chloride, so that potassium tends to diffuse out of the cell, and sodium and chloride tend to diffuse in. The tendency for potassium and chloride to diffuse down their concentration gradients is opposed by the electrical potential across the cell membrane.

■ In a model cell permeable only to potassium and chloride, the concentration gradients can be balanced exactly by the membrane potential, so that there is no net flux of either ion across the membrane. The membrane potential is then equal to the equilibrium potential for both potassium and chloride.

■ Changing the extracellular potassium concentration changes the potassium equilibrium potential, and hence the membrane potential. Changing the extracellular chloride concentration, on the other hand, leads ultimately to a change in intracellular chloride, so that the chloride equilibrium potential and the membrane potential differ from their original values only transiently.

■ The plasma membranes of real cells are permeable to sodium, as well as to potassium and chloride. As a result, there is a constant passive influx of sodium into the cell, and an efflux of potassium. These fluxes are balanced exactly by active transport of the ions in the opposite directions, in the ratio of 3 sodium to 2 potassium. Under these circumstances, the membrane potential depends on the sodium equilibrium potential, the potassium equilibrium potential, the relative conductance of the membrane to the two ions, and the pump ratio.

■ Because the sodium–potassium exchange pump transports more positive ions outward than inward across the membrane, it makes a direct contribution of several millivolts to the membrane potential.

■ The chloride equilibrium potential may be positive or negative with respect to the resting membrane potential, depending on chloride transport processes. Although the chloride distribution plays little role in determining the resting membrane potential, a high chloride permeability is important for electrical stability.

SUGGESTED READING

Hodgkin, A. L., and Katz, B. 1949. The effect of sodium ions on the electrical activity of the giant axon of the squid. *J. Physiol.* 108: 37–77. (The constant field equation is derived in Appendix A of this paper.)

Junge, D. 1992. *Nerve and Muscle Excitation*, 3rd Ed. Sinauer Associates, Sunderland, MA, Chapters 1–3.

Mullins, L. J., and Noda, K. 1963. The influence of sodium-free solutions on membrane potential of frog muscle fibers. *J. Gen. Physiol.* 47: 117–132.

6 Ionic Basis of the Action Potential

THE IONIC MECHANISMS RESPONSIBLE FOR GENERATING action potentials have been described quantitatively by using the voltage clamp method to measure membrane currents. From such measurements it is possible to determine which components of the currents are carried by different ion species, and to deduce the magnitude and time course of the underlying changes in ionic conductances. Such experiments have shown that depolarization increases sodium conductance and, more slowly, potassium conductance. The activation of sodium conductance is transient, being followed by inactivation. The increase in potassium conductance persists for as long as the depolarizing pulse is maintained. The dependence of sodium and potassium conductances on membrane potential and their sequential timing account quantitatively for the amplitude and time course of the action potential, as well as for other phenomena, such as threshold and refractory period.

Patch clamp experiments have been used to examine the behavior of individual sodium and potassium channels associated with the action potential. The behavior of the channels is consistent with previous voltage clamp experiments on whole cells: Depolarization increases the probability that sodium and potassium channels will open. For both ion channels the increase in this probability follows the same time course as that of the corresponding voltage clamp currents. For example, sodium channels open most frequently near the beginning of a depolarizing pulse and openings then become less frequent as inactivation develops.

Other cation channels can be involved in action potential generation. In some cells, voltage-activated calcium channels are responsible for the rising phase of the action potential, and repolarization can involve activation of a variety of potassium channel types.

A. L. Hodgkin, 1949

SODIUM AND POTASSIUM CURRENTS

In Chapter 5 we showed that the resting potential is determined mainly by the potassium concentration ratio (as postulated by Bernstein in 1902), but it is influenced as well by the concentration ratio of sodium and, to a lesser extent, by that of chloride. At the same time that Bernstein proposed his hypothesis about the nature of the resting potential, Overton[1] made the important discovery that sodium ions are necessary for nerve and muscle cells to produce action potentials, and he suggested (somewhat hesitantly) that the action potential might come about by sodium entering the cell. Further clarification of this idea came with experiments on the squid axon.

In 1939, Hodgkin and Huxley[2] showed that at the peak of the action potential there was an **overshoot** during which the membrane potential became transiently positive on the inside. This suggested that sodium was indeed involved, because sodium entry across the membrane would continue beyond zero membrane potential until the sodium equilibrium potential (E_{Na}) was reached. Ten years later, Hodgkin and Katz[3] showed that reducing external sodium concentration, and hence E_{Na}, produced corresponding reductions in the overshoot (Figure 6.1). They concluded that the action potential was the result of a large, transient increase in the sodium permeability of the membrane. We now know that this permeability increase is due to the opening of a large number of voltage-activated sodium channels.

What about the falling phase of the action potential? One might expect that the membrane potential would return to the resting level if the sodium channels simply closed. Indeed, this is one factor involved. If nothing else occurred, however, the return in most cells would be much slower than that observed experimentally. This is because the overall resting permeability of the membrane is relatively small, and consequently the loss of the accumulated positive charge through resting potassium and chloride channels would take several, or even tens of, milliseconds. The return to normal is very rapid because of a second large increase in membrane permeability—this time due to the opening of voltage-activated potassium channels. The membrane potential, having raced toward E_{Na}, now returns with almost equal rapidity toward E_K. The increase in potassium permeability can last for several milliseconds, so that in many cells the membrane is actually *hyperpolarized* beyond its normal resting potential for a time (see Figure 6.1).

To summarize, the action potential is the result of a sudden, large increase in sodium permeability of the membrane. The resulting inrush of sodium and accumulation of positive charge on the inner surface of the membrane drives the potential toward E_{Na}. Repolarization is accomplished by a subsequent large increase in potassium permeability, and loss of the accumulated positive charge, carried now by the efflux of potassium ions as the membrane returns toward E_K. Explanation of the mechanisms underlying generation of the action potential leads directly to understanding impulse propagation, discussed in Chapter 7.

A. F. Huxley, 1974

[1]Overton, E. 1902. *Pflügers Arch.* 92: 346–386.

[2]Hodgkin, A. L., and Huxley, A. F. 1939. *Nature* 144: 710–711.

[3]Hodgkin, A. L., and Katz, B. 1949. *J. Physiol.* 108: 37–77.

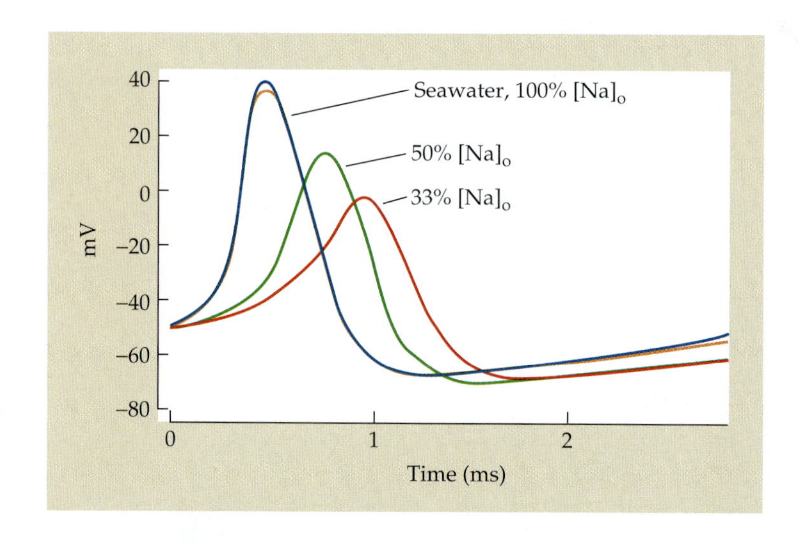

FIGURE 6.1 Role of Sodium in Action Potential Generation. Action potentials recorded from a squid axon bathed in seawater (blue), in solutions containing 50% (green) and 33% normal sodium (red), and then returned to seawater (orange). (After Hodgkin and Katz, 1949.)

How Many Ions Enter and Leave during an Action Potential?

If the interior of the nerve gains sodium during the rising phase of the action potential and loses potassium during the falling phase, then it follows that the sodium and potassium concentrations in the cytoplasm must change. The magnitude of the concentration change can be determined in two ways: by calculation and by direct measurement.

Calculation of the relation between membrane potential and charge separation by the membrane is discussed in detail in Chapter 7. At a membrane potential of −67 mV, about 4×10^{11} negative charges/cm^2 are collected on the inner surface of the membrane. At the peak of the action potential (+40 mV) these negative charges are replaced by 2.4×10^{11} positive charges, requiring an influx of 6.4×10^{11} sodium ions/cm^2. This is equivalent to about 10^{-12} mol/cm^2. Experimental measurements of radioactive sodium entering and radioactive potassium leaving the fiber during action potential activity[4] gave values in the range of 3×10^{-12} to 4×10^{-12} mol/cm^2.The values are higher than those calculated theoretically, largely because the calculation takes no account of the fact that the sodium and potassium fluxes overlap in time. Thus, the actual amount of sodium influx is greater than that required to charge the membrane to the peak of the action potential because potassium efflux (carrying charge in the opposite direction) begins before the peak is reached.

How does the sodium influx affect concentration? A 1 cm segment of squid axon, 1 mm in diameter, has a surface area of 0.31 cm^2, so that an influx of 3.5×10^{-12} mol/cm^2 into the segment amounts to about 10^{-12} mol of sodium. The same length of axon has a volume of 7.85×10^{-12} l, and contains (at 50 mmol/l) 4×10^{-7} mol of sodium, so that the influx changes the sodium concentration by only about 2.5 parts in a million. The same potassium efflux represents only about 3 parts in 10 million of the potassium content of the fiber.

Action potential activity in very small nerve processes can produce changes in intracellular ion concentrations that are more significant than those seen in large squid axons. For example, a nerve terminal 1 μm in diameter and 100 μm long has a surface area of 3×10^{-6} cm^2 and a volume of 8×10^{-14} l. During an action potential, an influx of 3.5×10^{-12} mol/cm^2, like that seen in squid axon, would result in about 10^{-17} mol of sodium accumulating in the terminal. At an intracellular concentration of 20 m*M*, the terminal contains about 1.5×10^{-15} mol, so a single impulse increases the intracellular sodium concentration by 0.7%. A rapid burst of 50 impulses would, in theory, increase the intracellular sodium concentration by 35%, with a corresponding reduction in intracellular potassium. Sodium influx accelerates the activity of the sodium–potassium exchange pump (Chapter 4) so that the concentrations are restored rapidly to their resting values.

Positive and Negative Feedback during Conductance Changes

The main feature underlying the ion currents associated with the action potential is that both the sodium and potassium conductances are voltage-dependent: The probability that the channels will open increases with depolarization. Depolarization increases the membrane conductance to sodium and, with a delay, to potassium as well. The effect on sodium conductance is **regenerative**: A small depolarization increases the number of open sodium channels; the resulting entry of sodium down its electrochemical gradient produces still more depolarization, opening more sodium channels, leading to still more rapid sodium entry, and so on (Figure 6.2A). This process of cumulative self-enhancement is known as **positive feedback.** In contrast, the voltage dependence of potassium conductance is self-limiting and involves **negative feedback** (Figure 6.2B). Depolarization increases the number of open potassium channels, resulting in the efflux of potassium down its electrochemical gradient. Rather than reinforcing the depolarization, the efflux leads to repolarization and return of the potassium conductance to its resting level.

Measuring Conductance

The ideas we have discussed so far were proposed by Hodgkin, Huxley, and Katz[5] and developed in detail by Hodgkin and Huxley,[6–9] who carried out and analyzed elegant electrophysiological experiments on the giant axon of the squid. They showed experimentally

[4]Keynes, R. D., and Lewis, P. R. 1951. *J. Physiol.* 114: 151–182.

[5]Hodgkin, A. L., Huxley, A. F., and Katz, B. 1952. *J. Physiol.* 116: 424–448.

[6]Hodgkin, A. L., and Huxley, A. F. 1952. *J. Physiol.* 116: 449–472.

[7]Hodgkin, A. L., and Huxley, A. F. 1952. *J. Physiol.* 116: 473–496.

[8]Hodgkin, A. L., and Huxley, A. F. 1952. *J. Physiol.* 116: 497–506.

[9]Hodgkin, A. L., and Huxley, A. F. 1952. *J. Physiol.* 117: 500–544.

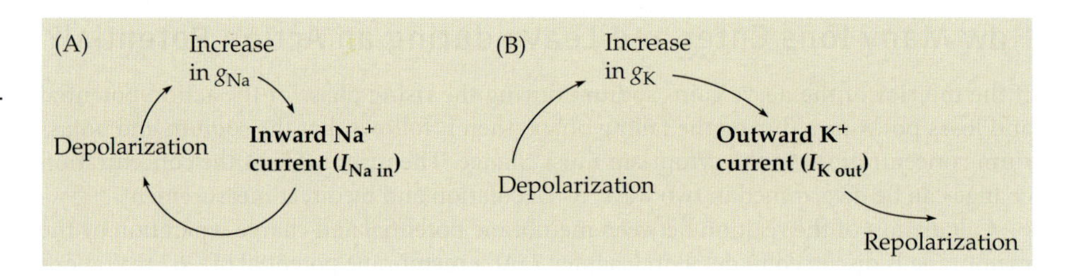

FIGURE 6.2 Effects of Increasing Conductances (g_{Na} and g_K) on membrane potential. (A) Sodium entry reinforces depolarization. (B) Potassium efflux leads to repolarization.

that changes in sodium and potassium conductances occurred, and that the changes were timed appropriately and were of the correct magnitude to account exactly for the magnitude and time course of the action potential.

What kinds of experiments were done to arrive at these conclusions? At first thought, it appears simple to obtain the appropriate measurements of conductance of the membrane to sodium (g_{Na}) or potassium (g_K). All that is needed is to measure the amount of current (I) flowing inward or outward across the membrane at various levels of potential (V_m), since

$$g_{Na} = \frac{I_{Na}}{(V_m - E_{Na})}$$

$$g_K = \frac{I_K}{(V_m - E_K)}$$

However, there are two problems to be solved before this approach becomes practical. The first is that current flowing across the membrane will change the membrane potential; this, in turn, will alter the membrane conductances. The solution was to devise a method for rapidly setting the membrane potential to any desired level and then holding it at that level while measuring the magnitude and time course of the membrane current. Because the voltage is fixed for the period of observation, the observed current will represent accurately the underlying changes in membrane conductance. The second problem is to separate the ionic components of the current so that their individual characteristics can be assessed. This has been accomplished in a number of ways, including the replacement of sodium with impermeant ions and, later, the use of selective toxins and poisons.

VOLTAGE CLAMP EXPERIMENTS

The **voltage clamp** was devised by Cole and his colleagues[10,11] and developed further by Hodgkin, Huxley, and Katz.[6] The experimental arrangement is described in Box 6.1. All we need to know to understand the experiments themselves is that the method permits us to set the membrane potential of the cell almost instantaneously at any level and hold it there ("clamp" it), while at the same time recording the current flowing across the membrane. Figure 6.3A shows an example of the currents that occur when the membrane potential is stepped suddenly from its resting value (in this example –65 mV) to a depolarized level (–9 mV). The current produced by the voltage step consists of three phases: (1) a brief outward surge lasting only a few microseconds, (2) an early inward current, and (3) a late outward current.

Capacitative and Leak Currents

The initial brief surge of current is the capacitative current, which occurs because the step from one potential to another alters the charge on the membrane capacitance. If the clamp amplifier is capable of delivering a large amount of current, then the membrane can be charged rapidly and this current will last only a very short time. Once the new potential is reached, there is no more capacitative current. In practice the surge of capacitative current lasts only about 20 μs and is followed by a small steady outward current.

The steady outward current is through the resting membrane conductances and is known as leak current. Leak current is carried largely by potassium and chloride ions,

Box 6.1 The Voltage Clamp

The figure illustrates an experimental arrangement for voltage clamp experiments on squid axons. The axon is bathed in seawater, and into one end two fine silver wires are inserted longitudinally. One of the wires provides a measure of the potential inside the fiber with respect to that of the seawater (which is grounded) or, in other words, a measure of the membrane potential (V_m). It is also connected to one input of the voltage clamp amplifier. The other input is connected to a variable voltage source, which can be set by the person doing the experiment; the value to which it is set is thus known as the **command potential**. The voltage clamp amplifier delivers current from its output whenever there is a voltage difference between the inputs. The output current flows across the cell membrane between the second fine silver wire and the seawater (arrows); it is measured by the voltage drop across a small series resistor.

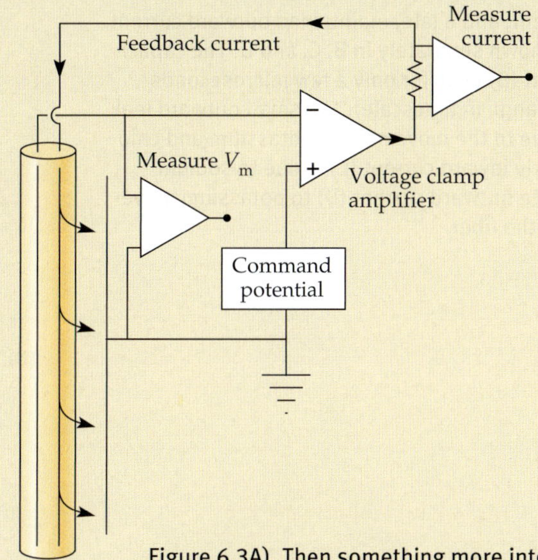

The circuit is arranged so that the output current tends to cancel any voltage difference between the two inputs, and it works as follows: Suppose that the resting potential of the fiber is –70 mV and the command potential is set to –70 mV as well. Because the voltages at the two inputs of the amplifier are equal, there will be no output current. If the command potential is stepped to, say, –65 mV, then because of the 5 mV difference between the inputs, the amplifier delivers positive current into the axon and across the cell membrane. The current produces a voltage drop across the membrane, driving V_m to –65 mV and removing the voltage difference between the two inputs. In this way the membrane potential is kept equal to the command potential. If the circuitry is properly designed, the change in V_m is achieved within a few microseconds.

Now suppose that the command potential is stepped from –70 to –15 mV. We would expect that the amplifier would deliver positive current to the axon to drive V_m to –15 mV. This is indeed what happens, but only transiently (see Figure 6.3A). Then something more interesting occurs. The depolarization to –15 mV produces an increase in sodium conductance, and there is a consequent flow of sodium ions *inward* across the membrane. In the absence of the clamp, this would tend to depolarize the membrane still further (toward the sodium equilibrium potential); with the clamp in place, however, the amplifier provides just the correct amount of negative current to hold the membrane potential constant. In other words, the current provided by the amplifier is exactly equal to the current flowing across the membrane. Here, then, is the great power of the voltage clamp: In addition to holding the membrane potential constant, it provides an exact measure of the membrane current required to do so. Voltage clamp measurements can now be made in small nerve cells by using the whole-cell method of patch clamp recording (Chapter 2).

varies linearly with voltage displacement from rest in either direction, and lasts throughout the duration of the voltage step. However, during most of the response it is obscured by much larger ion currents.

Currents Carried by Sodium and Potassium

Hodgkin and Huxley showed that the second and third phases of the current were due first to the entry of sodium and then to the exit of potassium across the cell membrane. In addition, they were able to deduce the relative size and time course of the separate currents. One convenient way was to abolish the sodium current by replacing most of the extracellular sodium by choline (an impermeant cation). With an appropriate reduction in extracellular sodium concentration, the sodium equilibrium potential could be made equal to the potential during the depolarizing step (–9 mV in Figure 6.3A). Consequently there was no net current through the activated sodium channels. This left only the potassium current, shown

[10]Marmont, G. 1940. *J. Cell. Comp. Physiol.* 34: 351–382.

[11]Cole, K. S. 1968. *Membranes, Ions and Impulses*. University of California Press, Berkeley.

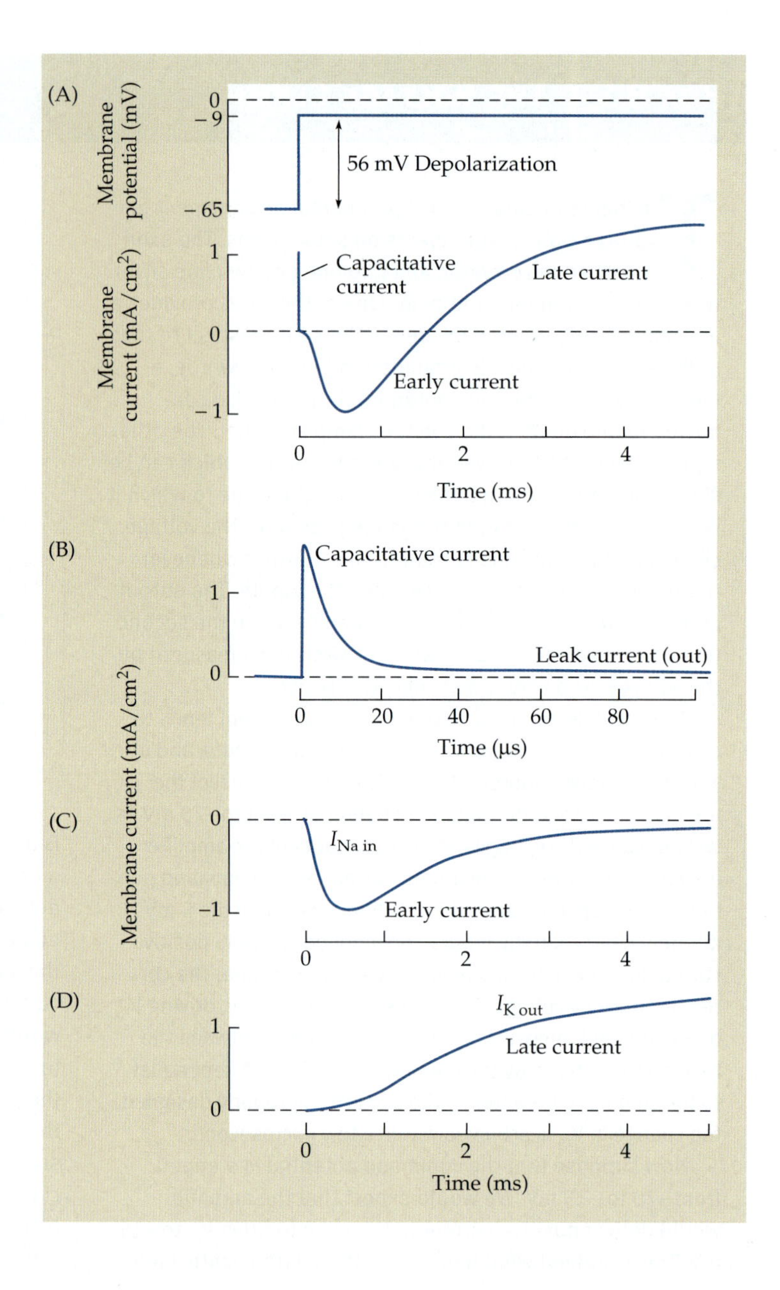

FIGURE 6.3 Membrane Currents produced by depolarization. (A) Currents measured by a voltage clamp during a 56 mV depolarization of a squid axon membrane. The currents (lower trace) consist of a brief positive capacitative current, an early transient phase of inward current, and a late, maintained outward current. These are shown separately in B, C, and D. The capacitative current (B) lasts for only a few microseconds (note the change in timescale). The small outward leak current is due to the movement of potassium and chloride. The early inward current (C) is due to sodium entry, the late outward current (D) to potassium movement out of the fiber.

in Figure 6.3D. Subtraction of the potassium current from the total ion current (Figure 6.3A) then revealed the magnitude and time course of the sodium current (Figure 6.3C).

Selective Poisons for Sodium and Potassium Channels

Since the original experiments of Hodgkin and Huxley, convenient pharmacological methods have been found for blocking sodium and potassium currents selectively. Tetrodotoxin (TTX) and its pharmacological companion saxitoxin (STX) have been particularly useful for blocking sodium channels (Chapter 3). TTX is a virulent poison, concentrated in the ovaries and other organs of puffer fish, whose potent effects have given rise to the Chinese proverb "To throw away life eat blowfish." Kao[12] has reviewed the fascinating history of TTX, beginning with the discovery of its effects by the Chinese emperor Shun Nung (2838–2698 B.C.), who personally tasted 365 drugs while compiling a pharmacopoeia and lived (for an amazingly long time) to tell the tale. STX is synthesized by marine dinoflagellates and concentrated by filter-feeding shellfish, such as the Alaskan butter clam *Saxidomus*. Its virulence competes with that of TTX: Ingestion of a single clam (cooked or not) can be fatal.

[12]Kao, C. T. 1966. *Pharmacol. Rev.* 18: 977–1049.

The great advantage of TTX for neurophysiological studies is that its action is highly specific. Working with squid axons, Moore, Narahashi, and colleagues showed that it blocks the voltage-activated sodium conductance selectively at concentrations of less than 1 μ*M*.[13] When a TTX-poisoned axon is subjected to a depolarizing voltage step, no inward sodium current is seen, but only the outward potassium current (Figure 6.4A and B). The potassium current is unchanged in amplitude and time course by the poison. Application of TTX to the inside of the membrane by adding it to an internal perfusing solution has no effect. The actions of STX are indistinguishable from those of TTX. Both toxins appear to bind to the same site in the outer mouth of the channel through which sodium ions move, thereby physically blocking ion current through the channel.[14]

Just as TTX and STX block sodium channels selectively, a number of substances have been found that have similar effects on the voltage-activated potassium channels associated with the action potential. For example, in squid axons and in frog myelinated axons Armstrong, Hille, and others have shown that voltage-activated potassium currents are blocked by tetraethylammonium (TEA, in concentrations greater than 10 m*M*) (see Figure 6.4C).[15] In squid axon, TEA must be added to the internal solution, and exerts its action at the inner mouth of the potassium channel; in other preparations, such as the frog node of Ranvier, TEA is effective at an external site as well. Other compounds, such as 4-aminopyridine (4-AP) and 3,4-diaminopyridine (DAP), block potassium currents when applied in millimolar concentrations to either the inside or the outside of the membrane.

Dependence of Ion Currents on Membrane Potential

Having established that the early and late currents are due to sodium influx, followed by potassium efflux, Hodgkin and Huxley then determined how the magnitude and time course of the currents depend on membrane potential. Currents produced by various levels of depolarization from a holding potential of –65 mV are shown in Figure 6.5A. First, a step hyperpolarization to –85 mV (the bottom trace in Figure 6.5A) produces only a small inward current, as would be expected from the resting properties of the membrane. As already shown in Figure 6.3, moderate depolarizing steps each produce an early inward

[13]Narahashi, T., Moore, J. W., and Scott, W. R. 1964. *J. Gen. Physiol.* 47: 965–974.

[14]Hille, B. 1970. *Prog. Biophys. Mol. Biol.* 21: 1–32.

[15]Armstrong, C. M., and Hille, B. 1972. *J. Gen. Physiol.* 59: 388–400.

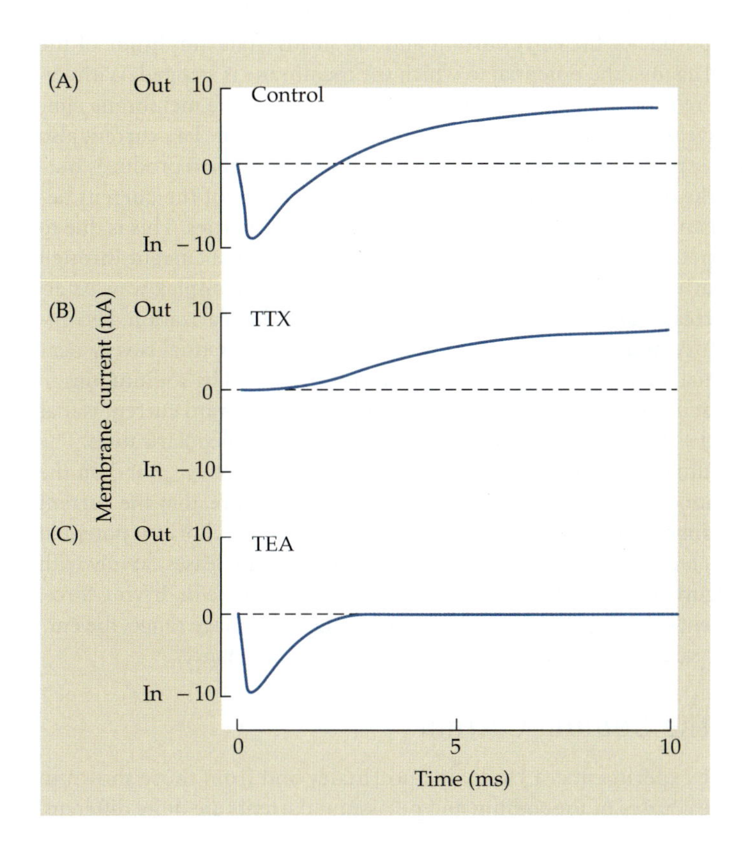

FIGURE 6.4 Pharmacological Separation of Membrane Currents into sodium and potassium components. Membrane currents were produced by clamping of the membrane potential to 0 mV in a frog myelinated nerve. (A) Control record in normal bathing solution. (B) The addition of 300 n*M* tetrodotoxin (TTX) causes the sodium current to disappear while the potassium current remains. (C) The addition of tetraethylammonium (TEA) blocks the potassium current, leaving the sodium current intact. (After Hille, 1970.)

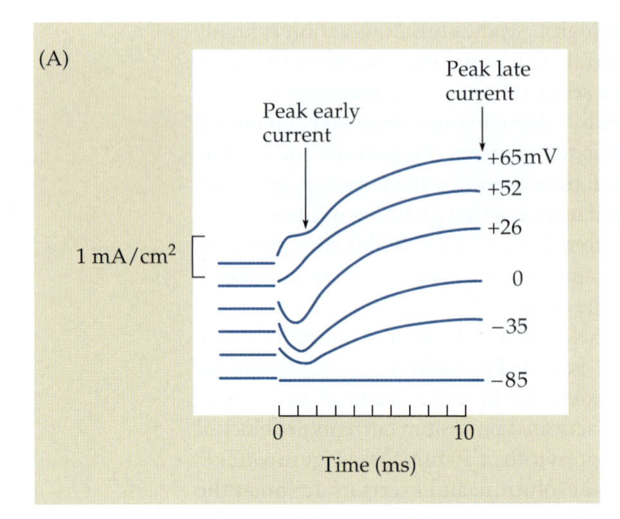

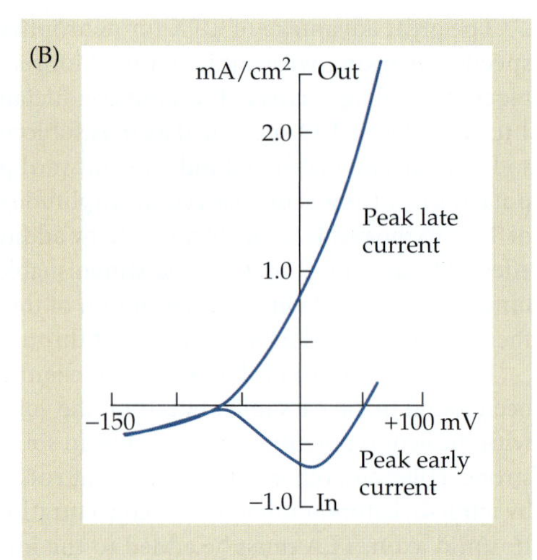

FIGURE 6.5 Dependence of Early and Late Currents on Potential. (A) Currents produced by voltage steps from a holding potential of –65 mV to a hyperpolarized level (–85 mV) and to successively increasing depolarized levels as indicated. The late potassium current increases as the depolarizing steps increase. The early sodium current first increases, then decreases with increasing depolarization; it is absent at +52 mV and reversed in sign at +65 mV. (B) Peak currents plotted against the potential to which the membrane is stepped. Late outward current increases rapidly with depolarization. Early inward current first increases in magnitude, then decreases, reversing to outward current at about +55 mV (the sodium equilibrium potential). (After Hodgkin, Huxley, and Katz, 1952.)

current followed by a sustained outward current. With greater depolarizations the early current becomes smaller, at about +52 mV it is absent, and as the depolarizing step is increased still further it reverses, becoming outward.

The current–voltage relations for the early and late currents are shown in Figure 6.5B, in which the peak amplitude of the early current and the steady-state amplitude of the late current are plotted against the potential to which the membrane is stepped. With hyperpolarizing steps there is no separation of early and late currents; the membrane simply responds as a passive resistor, with the expected inward current. The late current also behaves as one would expect of a resistor in the sense that depolarization produces outward current, but as the depolarization is increased the magnitude of the current becomes much greater than expected from the resting membrane properties. This is due to the voltage-activated potassium conductance, which allows additional current through the membrane. The early inward current behaves in a much more complex way. As already noted, it first increases and then decreases with increasing depolarization, becoming zero at about +52 mV and then reversing in sign. The reversal potential is very near the equilibrium potential for sodium, as expected for a current carried by sodium ions.

One point of interest in the current–voltage relation for the early inward current is that between about –50 and +10 mV the current increases with increasing depolarization. The magnitude of the sodium current depends on the sodium conductance (g_{Na}) and on the driving force for sodium entry ($V_m - E_{Na}$). One might expect, therefore, that the current will *decrease* as the membrane potential moves toward the sodium equilibrium potential and the driving force is reduced. However, the sodium conductance increases rapidly with depolarization (see Figure 6.7), and this increase outweighs the decrease in driving force. Thus, the sodium current—$i_{Na} = g_{Na}(V_m - E_{Na})$—increases. In this voltage range, the current–voltage relation is said to have a region of "negative slope conductance."

Inactivation of the Sodium Current

It is apparent from the experiments of Hodgkin and Huxley and from those shown in Figure 6.4 that the time courses of the sodium and potassium currents are quite different.

The potassium current is much delayed compared to the onset of the sodium current, but once developed it remains high throughout the duration of the step. The sodium current, on the other hand, rises much more rapidly but then decreases to zero, even though the membrane is still depolarized. This decline of the sodium current is called **inactivation.**

Hodgkin and Huxley studied the nature of the inactivation process in detail. In particular, they investigated the effect of hyperpolarizing and depolarizing prepulses on the peak amplitude of the sodium current produced by a subsequent depolarizing step. Records from such an experiment are shown in Figure 6.6. In Figure 6.6A the membrane is stepped from a holding potential of –65 to –21 mV, producing a peak sodium current of about 1 mA/cm^2. When the step is preceded by a hyperpolarizing prepulse of –13 mV, the peak sodium current is increased (Figure 6.6B). Depolarizing prepulses, on the other hand, cause a decrease in the sodium current (Figure 6.6C and D). The effects of hyperpolarizing and depolarizing prepulses are time-dependent; brief pulses of only a few milliseconds duration have little effect. In the experiment shown here, the prepulses are of sufficient duration (30 ms) for the effects to reach their maximum.

The results are shown quantitatively in Figure 6.6E, in which the peak sodium current is plotted against the potential during the prepulse. The peak current after a prepulse is expressed as a fraction of the control current. With a depolarizing prepulse to about –30

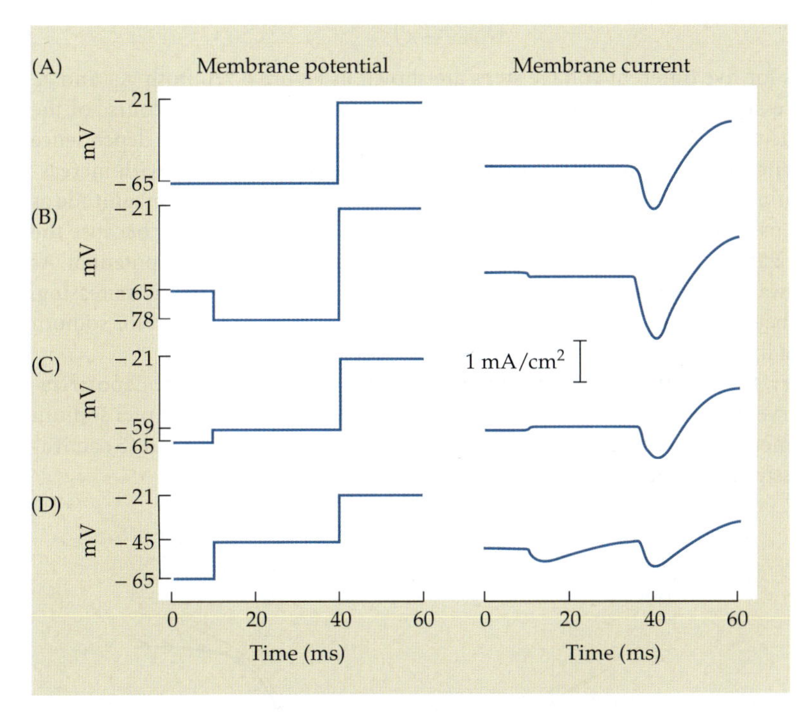

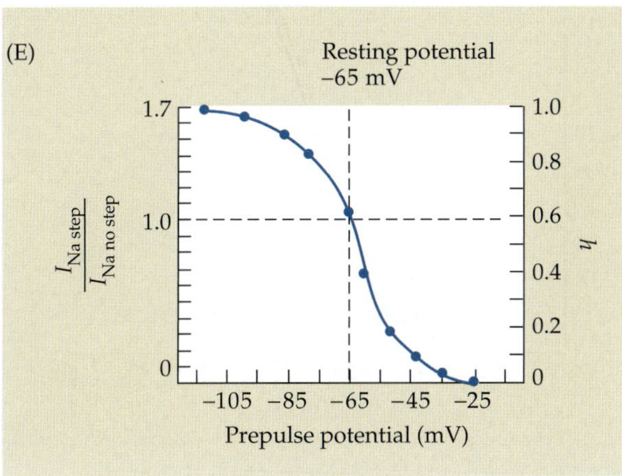

FIGURE 6.6 Effect of Membrane Potential on Sodium Currents. (A) A depolarizing step from –65 to –21 mV produces inward sodium current, followed by outward potassium current. (B) When the depolarizing step is preceded by a 30 ms hyperpolarizing step, the sodium current is increased. Prior depolarizing steps (C and D) reduce the size of the inward current. (E) The fractional increase or reduction of the sodium current as a function of membrane potential during the preceding conditioning step. The maximum current with a hyperpolarizing step to –105 mV is about 1.7 times larger than the control value. A depolarizing step to –25 mV reduces the subsequent response to zero. Full range of the sodium current is scaled from zero to unity by the *h* ordinate.

mV, the subsequent sodium current was reduced to zero; that is, inactivation was complete. Hyperpolarizing prepulses to –95 mV or beyond increased the sodium current by a maximum of about 70%. Hodgkin and Huxley represented this range of sodium currents from zero to their maximum value with a single parameter (h), varying between zero (complete inactivation) to 1 (no inactivation), as indicated on the right-hand ordinate of Figure 6.6E. In these experiments there was about 40% inactivation at the resting potential. Subsequent experiments have shown that all neurons show some degree of sodium channel inactivation at rest.

Sodium and Potassium Conductances as Functions of Potential

Having measured the magnitude and time course of sodium and potassium currents as a function of the membrane potential, V_m, and having determined the equilibrium potentials E_{Na} and E_K, Hodgkin and Huxley were then able to deduce the magnitude and time courses of the sodium and potassium conductance changes, using the relations noted earlier:

$$g_{Na} = \frac{I_{Na}}{(V_m - E_{Na})}$$

$$g_K = \frac{I_K}{(V_m - E_K)}$$

The results for five different voltage steps are shown in Figure 6.7A. Both g_{Na} and g_K increase progressively with increasing membrane depolarization. The time course of the sodium conductance is similar to that of the sodium current, but its voltage dependence is quite different (see Figure 6.5). The conductance increases progressively with increasing depolarization, whereas the current first increases and then decreases in magnitude as the voltage steps increase in amplitude. The current decreases progressively because the larger depolarizations come progressively closer to the sodium equilibrium potential. As a result, the inward current decreases, even though the sodium conductance is increasing. The relations between peak conductance and membrane potential are shown for sodium and potassium in Figure 6.7B. The curves are remarkably similar.

In summary, the results obtained by Hodgkin and Huxley indicated that depolarization of the nerve membrane leads to three distinct processes: (1) activation of a sodium conductance mechanism, (2) subsequent inactivation of the mechanism, and (3) activation of a potassium conductance mechanism.

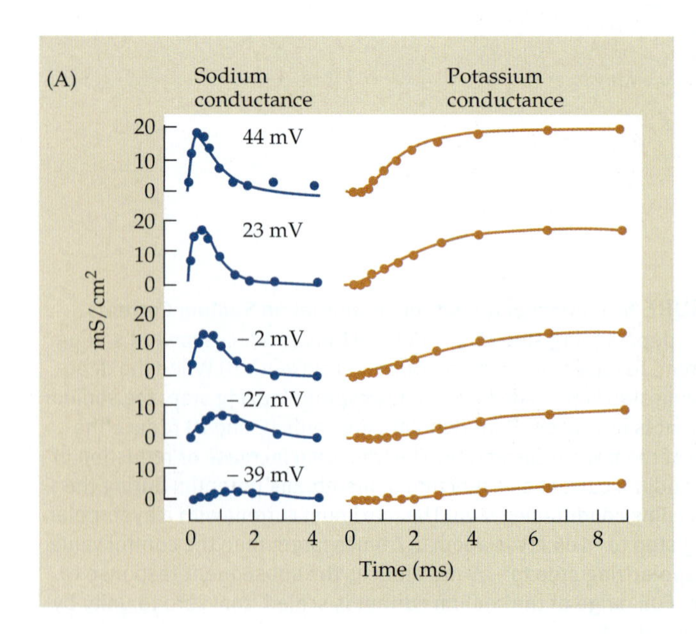

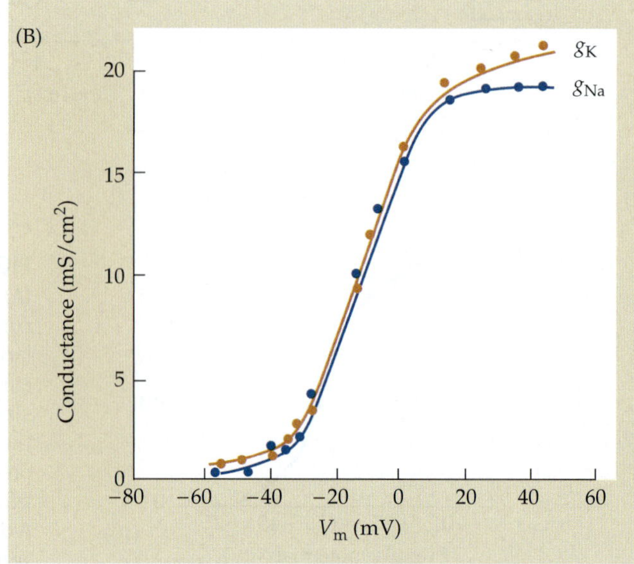

FIGURE 6.7 Sodium and Potassium Conductances. (A) Conductance changes produced by voltage steps from –65 mV to the indicated potentials. Peak sodium conductance and steady-state potassium conductance both increase with increasing depolarization. (B) Peak sodium conductance and steady-state potassium conductance plotted against the potential to which the membrane is stepped. Both increase steeply with depolarization between –20 and +10 mV. (After Hodgkin and Huxley, 1952b.)

Quantitative Description of Potassium and Sodium Conductances

After obtaining the experimental results, Hodgkin and Huxley proceeded to develop a mathematical description of the precise time courses of the sodium and potassium conductance changes produced by the depolarizing voltage steps. To deal first with the potassium conductance, one might imagine that the effect of a sudden change in membrane potential would be to provide a driving force for the movement of one or more charges in the voltage-activated potassium channel that would then lead to channel opening. If a single process were involved, the change in the overall potassium conductance might be expected to be governed by ordinary first-order kinetics; that is, its rise after the onset of the voltage step would be exponential.

Instead the onset of the potassium conductance change was found to be S-shaped, with a marked delay (see Figure 6.7A). Because of this delay, and because the potassium conductance increase occurred during depolarization but not hyperpolarization (i.e., it rectified), it was called the **delayed rectifier.** Hodgkin and Huxley were able to account for the S-shaped onset of the conductance by assuming that the opening of each potassium channel required the activation of four first-order processes—for example, the movement of four charged particles in the membrane. In other words, the S-shaped time course of activation could be fitted by the product of four exponential functions. The increase in potassium conductance for a given voltage step, then, was described by the relation

$$g_K = g_{K(max)} n^4$$

where $g_{K(max)}$ is the maximum conductance reached for the particular voltage step and n is a rising exponential function varying between zero and unity, given by $n = 1 - e^{-t/\tau_n}$.

The dependence of $g_{K(max)}$ on voltage is shown in Figure 6.7. The exponential time constant, τ_n, is also voltage-dependent: The increase in conductance becomes more rapid with larger depolarizing steps. At 10°C τ_n ranges from about 4 ms for small depolarizations to 1 ms for depolarization to zero.

The time course of the rise in sodium conductance, also S-shaped, was fitted by an exponential raised to the third power. In contrast, the fall in sodium conductance due to inactivation was consistent with a simple exponential decay process. For a given voltage step the overall time course of the sodium conductance change was the product of the activation and inactivation processes:

$$g_{Na} = g_{Na(max)} m^3 h$$

where $g_{Na(max)}$ is the maximum level to which g_{Na} would rise if there were no inactivation, and $m = 1 - e^{-t/\tau_m}$. The inactivation process is a falling, rather than a rising, exponential and is given by $h = e^{-t/\tau_h}$. As with the potassium conductance, $g_{Na(max)}$ is voltage-dependent, as are the activation and inactivation time constants. The activation time constant, τ_m, is much shorter than that for potassium, having a value at 10°C on the order of 0.6 ms near the resting potential, decreasing to about 0.2 ms at zero potential. The inactivation time constant, τ_h, on the other hand, is similar in magnitude to τ_n.

Reconstruction of the Action Potential

Once the empirical expressions were obtained for sodium and potassium conductances as a function of voltage and time, Hodgkin and Huxley were able to predict the entire time course of the action potential, and of the underlying conductance changes. Starting with a depolarizing step to just above threshold, they calculated what the subsequent potential changes would be at successive intervals of 0.01 ms. Thus, during the first 0.01 ms after the membrane had been depolarized to, say, –45 mV, they calculated how g_{Na} and g_K would change, what increments of I_{Na} and I_K would result, and then the effect of the net current on V_m. Knowing the change in V_m at the end of the first 0.01 ms, they then repeated the calculations for the next time increment, and so on all through the rising and falling phases of the action potential (a laborious exercise to undertake in the days before computers, or even electronic calculators, were available).

The calculations duplicated with remarkable accuracy the naturally occurring action potential in the squid axon. Calculated and observed action potentials produced by brief depolarizing pulses at three different stimulus strengths are compared in Figure 6.8A. To appreciate fully the magnitude of this accomplishment, it is necessary to keep in mind that the calculations used to duplicate the action potential were based on measurements of current that were made under completely artificial conditions with the membrane potential clamped first at one value, then at another.

The mechanisms of action potential generation are summarized in Figure 6.8B, which shows the calculated magnitude and time course of a propagated action potential in a squid axon, together with the calculated changes in sodium and potassium conductance.

Threshold and Refractory Period

In addition to describing the action potential, Hodgkin and Huxley were able to explain in terms of ionic conductance many other properties of excitable axons, such as **thresh-**

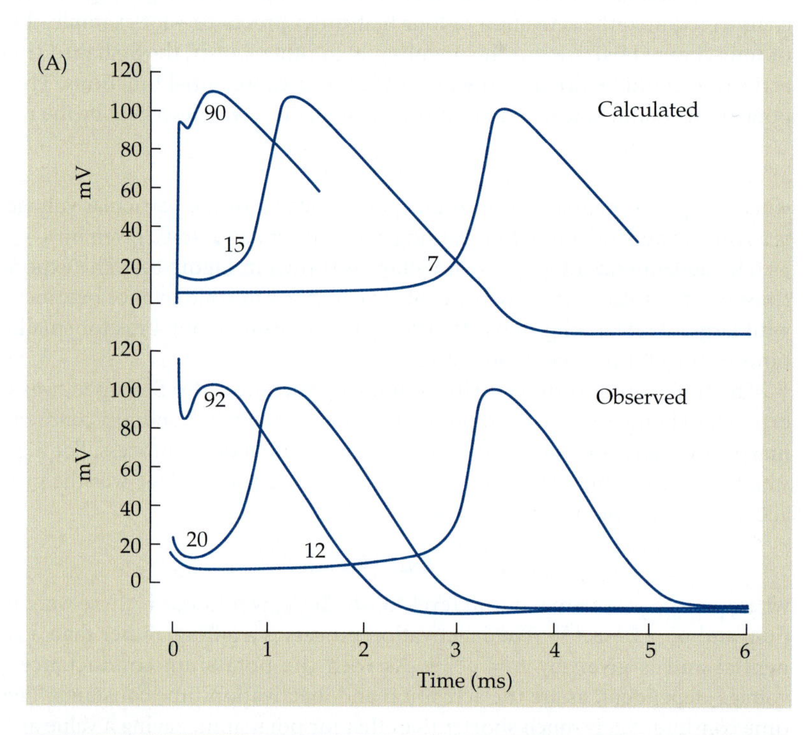

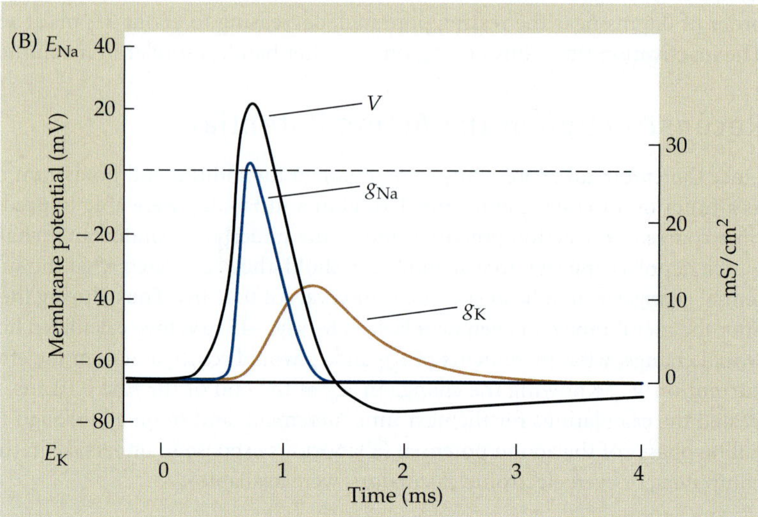

FIGURE 6.8 Reconstruction of the Action Potential. (A) Calculated action potentials produced by brief depolarizations of three different amplitudes (upper panel) are compared with those recorded under the same conditions (lower panel). (B) Relation between conductance changes (g_{Na} and g_K) and the action potential (V), calculated for a propagated action potential in a squid axon. (After Hodgkin and Huxley, 1952d.)

old and **refractory period**. Further, their findings have been found to be applicable to a wide variety of other excitable tissues.

How do the findings predict the threshold membrane potential at which the impulse takes off, especially when it might seem that a discontinuity like threshold would require a discontinuity in g_{Na} or g_K? The phenomenon can be understood if we imagine passing current through the membrane to depolarize it just to threshold, and then turning the current off. Because the membrane is depolarized, there will be an increase in outward current over that at rest (through potassium and leak channels). We will also have activated some sodium channels, increasing inward sodium current. At threshold the inward and outward currents are exactly equal and opposite, just as they are at rest. However, there is an important difference: The balance of currents is now unstable. If an extra sodium ion enters the cell, the depolarization is increased, g_{Na} increases, and more sodium enters. The outward current can no longer keep up with the sodium influx, and the regenerative process explodes. If, on the other hand, an extra potassium ion leaves the cell, the depolarization is decreased, g_{Na} decreases, sodium current decreases, and the excess outward current causes repolarization. As the membrane potential approaches its resting level, the potassium current decreases until it again equals the resting inward sodium current. Depolarization above threshold results in an increase in g_{Na} sufficient for inward sodium movement to swamp outward potassium movement immediately. Subthreshold depolarization fails to increase g_{Na} sufficiently to override the resting potassium conductance.

And how is the refractory period explained? Two changes develop during an action potential that make it impossible for the nerve fiber to produce a second action potential immediately: (1) Inactivation of sodium channels is maximal during the falling phase of the action potential and requires several more milliseconds to be removed. During this time few if any channels are available to contribute to an increase in g_{Na}. (2) Because of activation of potassium channels, g_K is very large during the falling phase of the action potential and decreases slowly back to its resting level. During this time a very large increase in g_{Na} is required to initiate any regenerative depolarization. These two factors result in an **absolute refractory period** lasting throughout the falling phase of the action potential during which no amount of externally applied depolarization can initiate a second regenerative response. This is followed by a **relative refractory period**, during which the threshold gradually returns to normal as sodium channels recover from inactivation and potassium channels close.

It was an extraordinary achievement for Hodgkin and Huxley to have provided rigorous quantitative explanations of such complex biophysical properties of membranes. Although subsequent observations on single channels have provided a new depth to our understanding of the underlying molecular mechanisms, by no stretch of the imagination would single-channel studies on their own, without the previous voltage clamp experiments and insights, have been able to account for how a nerve cell generates and conducts impulses. The older work has become enriched, rather than supplanted, by the new.

Gating Currents

Hodgkin and Huxley suggested that sodium channel activation was associated with the translocation of charged structures, or particles, within the membrane. Such charge movements would be expected to contribute to the capacitative current produced by a depolarizing voltage step. After a number of technical difficulties were resolved, such gating currents were finally seen.[16,17]

How is the gating current separated from the usual capacitative current expected with a step change in membrane potential (e.g., see Figure 6.3)? Briefly, currents associated merely with charging and discharging the membrane capacitance should be symmetrical. That is, they should be of the same magnitude for depolarizing steps as for hyperpolarizing steps. On the other hand, currents associated with sodium channel activation should appear upon depolarization of, say, 50 mV from a holding potential of –70 mV, but not upon hyperpolarization. In other words, if the channels are already closed, there should be no gating current upon further hyperpolarization. Similarly, gating currents associated with channel closing might be expected at the termination of a brief depolarizing pulse

[16]Armstrong, C. M., and Bezanilla, F. 1974. *J. Gen. Physiol.* 63: 533–552.

[17]Keynes, R. D., and Rojas, E. 1974. *J. Physiol.* 239: 393–434.

but not after a hyperpolarizing pulse. One experimental way of recording gating currents, then, is to sum the currents produced by two identical voltage steps of opposite polarity. The asymmetry due to gating currents is shown in parts a and b of Figure 6.9A. The current at the beginning of the depolarizing pulse is larger than that produced by the hyperpolarizing pulse because of the additional charge movement associated with gating of the sodium channel. When the two currents are added (part c of Figure 6.9A), the net result is the gating current (or "asymmetry current") alone.

An example of gating current in a squid axon, obtained by cancellation of the capacitative current, is shown in Figure 6.9B. Voltage-sensitive potassium currents were blocked with TEA. A step depolarization of perfused squid axon produced an outward gating current, followed by an inward sodium current. The sodium current was much smaller than usual because extracellular sodium concentration was reduced to 20% of normal. The gating current is shown alone in Figure 6.9C, after tetrodotoxin was added to the solution to eliminate the sodium current entirely (note the change in scale). The evidence that asymmetry currents observed in this way are, in fact, associated with sodium channel activation has been summarized by Armstrong.[18]

Activation and Inactivation of Single Channels

Patch clamp techniques have now provided detailed information about the way in which single sodium channels respond to depolarization. One experiment using this technique is illustrated in Figure 6.10. The records are from a cell-attached patch on a cultured rat muscle fiber.[19] To remove inactivation of sodium channels, a steady command potential was applied to the electrode, hyperpolarizing the patch membrane to −100 mV or so. On successive trials, a 40 mV depolarizing pulse was applied to the electrode for about 20 ms (part a of Figure 6.10B). In about one-third of the trials no sodium channels were activated. In the remainder, one or more single-channel currents appeared during the pulse, occurring most frequently near the onset of depolarization (part b of Figure 6.10B). The mean channel current was 1.6 pA. Assuming the sodium equilibrium potential to be +30 mV, the driving potential for sodium entry was about −90 mV; thus, the single-channel conductance was about 18 pS. This is comparable to sodium channel conductances mea-

[18]Armstrong, C. M. 1981. *Physiol. Rev.* 61: 644–683.

[19]Sigworth, F. J., and Neher, E. 1980. *Nature* 287: 447–449.

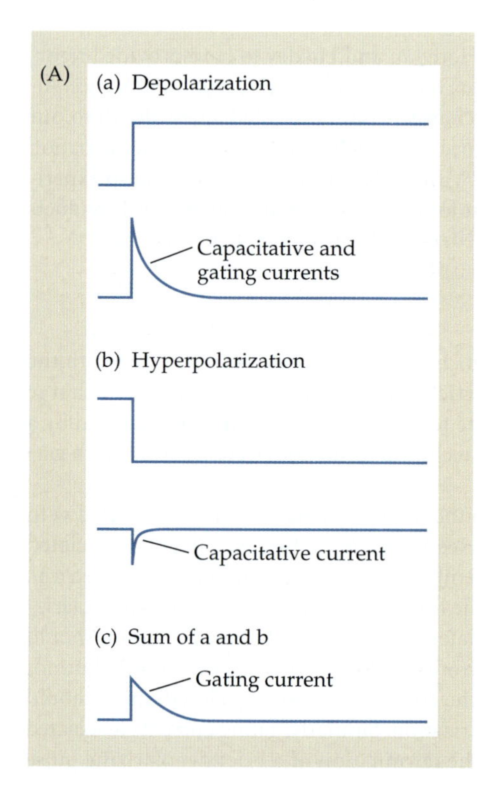

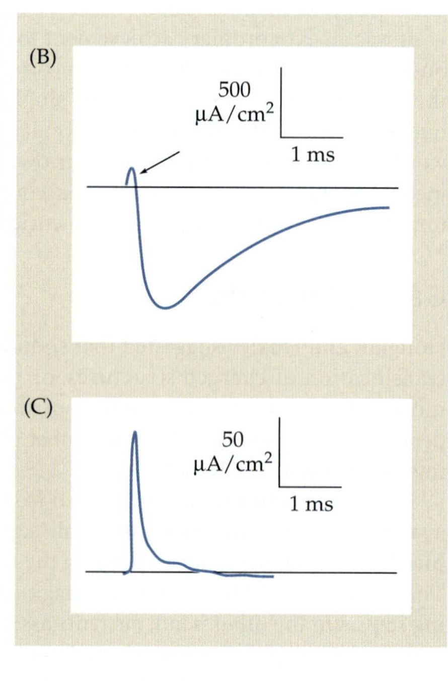

FIGURE 6.9 Sodium Channel Gating Current. (A) Method of separating gating current from capacitative current. A depolarizing pulse (a) produces capacitative current in the membrane, plus gating current. A hyperpolarizing pulse of the same amplitude (b) produces capacitative current only. When the responses to a hyperpolarizing and a depolarizing pulse are summed (c), capacitative currents cancel out and only gating current remains. (B) Record of current from a squid axon in response to a depolarizing pulse, after cancellation of capacitative current. Inward sodium current was reduced by lowering extracellular sodium to 20% of normal. The small outward current (arrow) preceding the inward current is the sodium channel gating current. (C) Response to depolarization from the same preparation after adding TTX to the bathing solution, recorded at higher amplification. Only the gating current remains. (B and C after Armstrong and Bezanilla, 1977.)

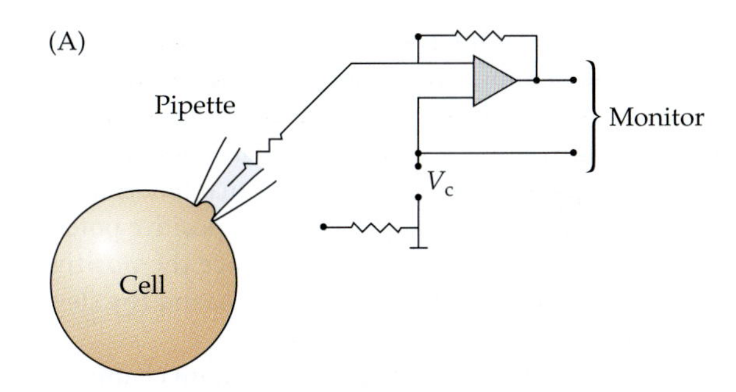

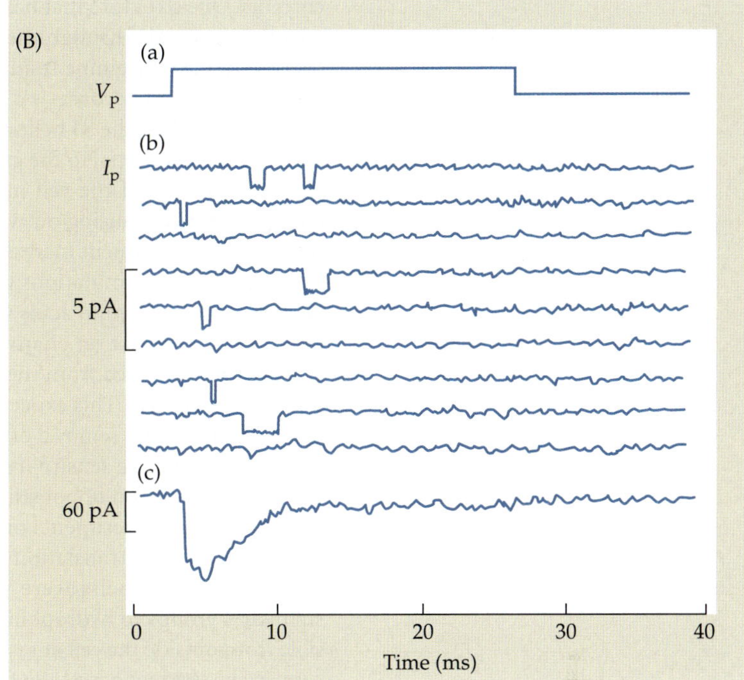

FIGURE 6.10 Sodium Channel Currents recorded from cell-attached patch on a cultured rat muscle cell. (A) Recording arrangement. V_c = the command potential applied to the membrane patch. (B) Repeated depolarizing voltage pulses applied to the patch, with the waveform shown in (a), produce single-channel currents (downward deflections) in the nine successive records shown in (b). The sum of 300 such records (c) shows that channels open most often in the initial 1 to 2 ms after the onset of the pulse, after which the probability of channel opening declines with the time constant of inactivation. (After Sigworth and Neher, 1980.)

sured in a variety of other cells. When 300 of the individual traces were summed (part c of Figure 6.10B), the result was an inward current that followed the same time course as that expected from the whole-cell sodium current.

A major point of interest in Figure 6.10 is that the mean channel open time (0.7 ms) is short relative to the overall time course of the summed current. Specifically, the time constant of decay of the summed current (about 4 ms) does not reflect the length of time that individual channels remain open. Instead, it indicates a slow decay in the probability of channel opening. The processes of activation (m^3) and inactivation (h) represent first an increase and then a decrease in the *probability* that a channel will open for a brief period. Their product (m^3h) describes the time course of the overall probability change. The probability increases rapidly near the beginning of the pulse, reaches a peak, and then decreases with time. In any given trial an individual channel may open immediately after the onset of the pulse, at any subsequent time during the pulse, or not at all.

A second point of interest is illustrated by referring again to Figure 6.9: The movement of charge is virtually complete in 0.5 ms, before sodium current reaches its peak—that is, before most of the channels have opened. This means that the conformational change in the channel protein associated with the charge movement is distinct from the conformational change associated with channel opening itself. Thus, it is not accurate to think of the charges as being connected with some kind of "handle" that opens a gate in the channel directly. Instead the charge movement simply increases the probability that the channel structure will enter the open (or the inactivated) state.

MOLECULAR MECHANISMS OF ACTIVATION AND INACTIVATION

Gating of Voltage-Activated Channels

Structural studies of voltage-activated channels suggest that channel gating occurs near the cytoplasmic end of the pore (Chapter 2). One question not yet answered is how the gate itself is coupled to changes in membrane potential. For voltage-activated gating to occur, there must be charged elements within the channel protein that are displaced by membrane depolarization. It is this charge displacement that is responsible for the gating currents. One structure that has attracted particular interest in this regard is the S4 helix, which, as it winds through the plane of the membrane, contains a string of positively charged lysine or arginine residues located at every third position (Figure 6.11). This feature, which is highly conserved within the superfamily of voltage-activated channels, has led to the idea that the S4 helices comprise the voltage-sensing elements that link changes in membrane potential to the gating mechanism.[20] Thus, application of a positive potential to the inside of the cell membrane (depolarization) would displace the positive charges outward, causing outward movement of the helix (Figure 6.11B), and (by steps unknown) a consequent increase in the probability of channel opening.

To test this idea, mutations were directed at S4 regions of rat brain sodium channel.[21] Neutral or acidic amino acids were substituted for one or more of the basic residues to determine the effect on channel activation. It would be expected that when positive charges were removed from the helix a greater voltage change would then be required to produce a response. This expectation was realized for substitutions near the cytoplasmic end of the helix, but removal of charge from the extracellular half had the reverse effect—an increased voltage sensitivity. Similar equivocal results have been obtained with mutants in the S4 region of potassium A-channels.[22]

Biochemical experiments on mutated channels also support the idea that activation is accompanied by translation of the S4 segment.[23,24] In these experiments residues at either end of the helix were replaced by cysteine. The accessibility of the cysteine sulfhydryl groups to hydrophilic reagents was then tested (Chapter 3). Residues inaccessible from outside the cell at rest became accessible when the membrane was depolarized; conversely, residues accessible from the inside at rest became inaccessible upon depolarization, suggesting outward movement of the helix.

In summary, although the experimental evidence is incomplete, it seems safe to assume that outward movement of the S4 helix is the first step in the gating process. At rest, the internal negativity holds the helix toward the cytoplasmic end of the channel protein. When depolarization occurs, the reduction in internal negativity allows the helix to move outward. The movement of S4 then sets in motion additional conformational changes that ultimately allow the gate to open.

Sodium Channel Inactivation

Hodgkin and Huxley's experiments with prepulses suggested that inactivation was a distinct phenomenon, separable from the activation process. A subsequent experimental observation supporting this idea was that pronase, a mixture of proteolytic enzymes, when perfused through the inside of a squid axon, led to a delay in onset of inactivation and, eventually, to its abolition.[25] The enzyme was ineffective when applied in the same concentration to the outer surface. It appeared, then, that pronase had degraded a portion of the cytoplasmic end of the sodium channel associated specifically with the inactivation

[20]Sigworth, F. J. 1994. *Q. Rev. Biophys.* 27: 1–40.

[21]Stühmer, W., et al. 1989. *Nature* 239: 597–603.

[22]Papazian, D. M., et al. 1991. *Nature* 349: 305–349.

[23]Yang, N., George, A. L., and Horn, R. 1996. *Neuron* 16: 113–122.

[24]Larsson, H. P., et al. 1996. *Neuron* 16: 387–397.

[25]Armstrong, C. M., Bezanilla, F., and Rojas, E. 1973. *J. Gen. Physiol.* 62: 375–391.

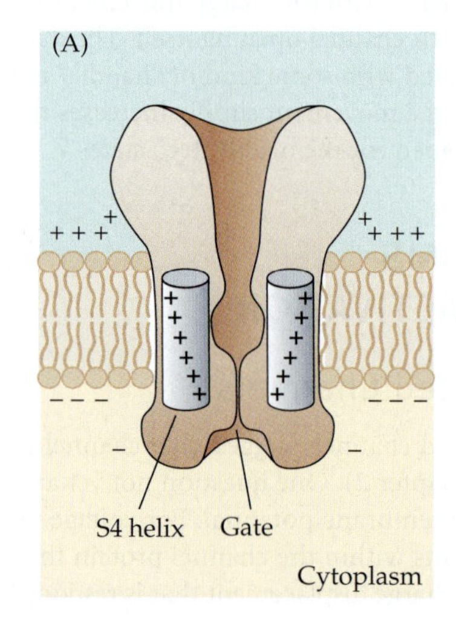

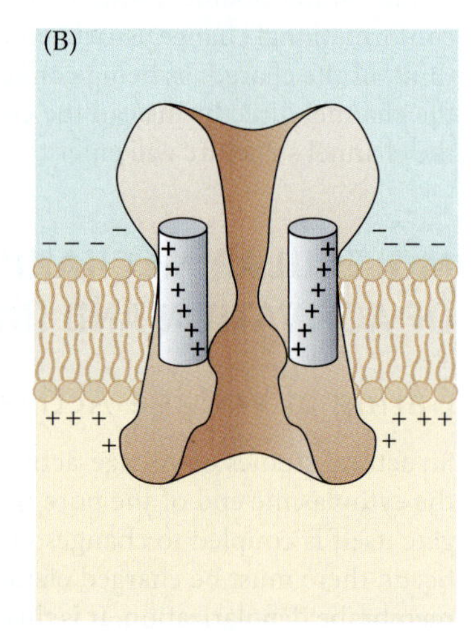

FIGURE 6.11 Proposed Shift of S4 Helices by membrane depolarization. Charged S4 helices are represented in two of the four domains of a voltage-activated sodium channel. (A) At the resting potential the helices are held against the inner end of the channel and the channel-gating elements are closed. (B) Depolarization causes outward movement of the positively charged helices, allowing the gate to open.

process. This led Armstrong and Bezanilla to propose a "ball and chain" model whereby an intracellular blocking particle (the ball), tethered to the cytoplasmic end of the channel by a flexible link (the chain), swings in to block the channel during inactivation.[26]

Experiments with sodium channels have identified the intracellular loop of amino acids between domains III and IV as being closely involved in the inactivation process. The loop is about 45 residues in length, and it is envisioned as a hairpin that swings into the inner vestibule of the channel to block the pore. In experiments with rat brain channels expressed in oocytes, three adjacent amino acid residues in the middle of the loop have been identified as essential for inactivation to occur.[27,28] When they were removed or replaced by site-directed mutagenesis, inactivation was severely attenuated or abolished. Similar experiments also identified groups of glycine and proline residues at either end of the loop involved in inactivation. These are assumed to be "hinge" regions that allow the hairpin to flip into the vestibule.[29]

Sodium channel activation and inactivation are also affected by a group of lipid-soluble toxins, including veratridine, an alkaloid from plants of the lily family, and batrachotoxin from the skin of South American frogs. They virtually eliminate inactivation so that the channels remain open indefinitely.[30] In addition, the voltage dependence of activation is shifted so that the channels are open at the normal resting potential.

Inactivation of Potassium A-Channels

The identification of an intracellular structure associated with inactivation was made first on potassium A-channels from *Drosophila* (Chapter 3) that, unlike delayed rectifier channels in squid axons, inactivate during maintained depolarization. Experiments on this channel provided evidence that a particular intracellular string of amino acids is associated with inactivation, and revived the "ball-and-chain" model proposed earlier for sodium channel inactivation. The model is illustrated in Figure 6.12. The ball is a clump of amino acids and the chain a string of residues tethering it to the main channel structure. Upon depolarization, the ball binds to a site in the inner vestibule of the channel, thereby blocking the pore.

This model of inactivation was tested in potassium A-channels by examining the behavior of channels formed in oocytes from mutant subunits (recall that the A-channel is a tetramer rather than a single polypeptide). Mutations and deletions were made in the 80 or so amino acids between the amino terminus and the first (S1) membrane helix.[31] Channels formed by mutant subunits with deletion of residues 6 through 46 showed virtually no inactivation, suggesting that some or all of these residues were involved in the normal inactivation process. When a synthetic peptide matching the first 20 amino acids in the N-terminal chain was simply added to the solution bathing the cytoplasmic face of the membrane, inactivation was restored with a linear dose dependence over the concentration range of 0 to 100 μM.[32] This amazing observation provides unusually strong support for the idea that in potassium A-channel subunits, the first 20 or so amino acid residues constitute a blocking particle responsible for inactivation of the channel. Because

[26]Armstrong, C. M., and Bezanilla, F. 1977. *J. Gen. Physiol.* 70: 567–590.

[27]West, J. W., et al. 1992. *Proc. Natl. Acad. Sci. USA* 89: 10910–10914.

[28]Kallenberger, S., et al. 1997. *J. Gen. Physiol.* 109: 589–605.

[29]Kallenberger, S., et al. 1997. *J. Gen. Physiol.* 109: 607–617.

[30]Catterall, W. A. 1980. *Annu. Rev. Pharmacol. Toxicol.* 20: 15–43.

[31]Hoshi, T., Zagotta, W. N., and Aldrich, R. W. 1990. *Science* 250: 533–550.

[32]Zagotta, W. N., Hoshi, T., and Aldrich, R. W. 1990. *Science* 250: 568–571.

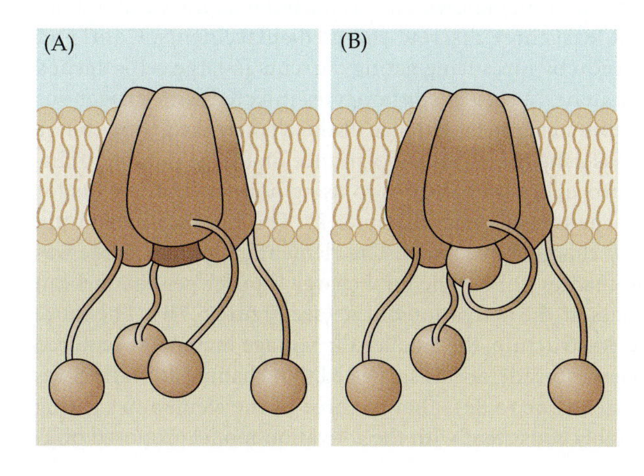

FIGURE 6.12 Ball-and-Chain Model of Inactivation of a voltage-activated potassium channel. The figure shows the complete channel, with one ball-and-chain element tethered to each of the four channel subunits. (A) Gating elements at the inner (cytoplasmic) end of the channel are open. (B) One of the four inactivation balls enters the inner vestibule to block the open channel.

it involves the N-terminal structure, this type of inactivation in potassium channels is often referred to as N-type inactivation. Some potassium channels also display a slower C-type inactivation, originally suspected to involve the carboxy terminus but later found to be related to structures near the outer mouth of the pore.[33,34]

Kinetic Models of Channel Activation and Inactivation

From their observations that the time courses of activation of the sodium and potassium currents were best fitted by exponential functions raised to the third and fourth powers (m^3 and n^4), Hodgkin and Huxley suggested that activation could be explained by the independent displacement of three or four charged particles in the membrane. For example, we might imagine that a voltage step has to produce displacements of the S4 helices in all four domains of the potassium channel before the channel can open. Similarly, we might suppose that at least three such displacements are necessary for sodium channel activation. Further, we might suppose that in the sodium channel one or more of the displacements also leads ultimately to inactivation. A parallel model of this kind has been proposed by Keynes.[35,36]

The idea that four separate events (such as S4 helix displacements) are necessary for channel opening gives rise to the possibility of 16 different channel states: no displacement (one state), one displacement in any one of four domains (four possible states), two displacements in any two domains (six possible states), three displacements in any three (four possible states), and displacements in all four (one state). If the steps are independent and kinetically identical, then this reduces to five states: no displacement, displacement in any one domain, in any two, in any three, and in all four. On this basis, the transition from closed to open can be represented as follows:

$$C_4 \leftrightarrow C_3 \leftrightarrow C_2 \leftrightarrow C_1 \leftrightarrow O$$

where C_4 represents the state of the channel at rest, C_3 and so on represent a series of closed states into which the channel can be driven by depolarization, and O is the open state. For sodium channels we must add the inactivation process. Measurements of both macroscopic and single-channel currents suggest that the sodium channel can be inactivated whether or not it has opened previously.[37,38] Thus, inactivation (I) can occur both from the open state and from one or more of the closed states:

$$\begin{array}{ccccccccc} C_4 & \leftrightarrow & C_3 & \leftrightarrow & C_2 & \leftrightarrow & C_1 & \leftrightarrow & 0 \\ & & & & & & & \searrow\!\!\nwarrow \;\; \swarrow\!\!\nearrow & \\ & & & & & & & I & \end{array}$$

Many variations of this kind of model have been proposed with more or fewer steps and with more than one inactivated state.[39,40] They differ from the original model of Hodgkin and Huxley in the sense that activation and inactivation are envisioned as sharing a number of sequential events, rather than proceeding in parallel as independent processes. In addition, although progression through any number of the steps may depend on membrane potential, the final steps leading to activation and inactivation need not themselves be voltage-dependent.[41,42]

How many states really exist? This is not known for certain, but it appears that sodium channel activation involves at least three discrete charge displacements. Conti and Stühmer[43] reached this conclusion by measuring gating currents in large cell-attached membrane patches ("macro patches") on *Xenopus* oocytes injected with exogenous mRNA coding for rat brain sodium channels. The size of the elementary gating charge movement was deduced by measuring the mean and variance of a large number of individual gating currents. The procedure is analogous to using noise measurements for determining the size of single-channel currents (Chapter 2): The variance/mean ratio gives the size of the elementary charge movements. The elementary gating charge was calculated to be 2.3 electronic charges (2.3e). The total charge transfer per channel can be estimated from the steepness of the activation curve (see Figure 6.7B): The more charges there are on the affected structure, the smaller the voltage increment required to change its conformation. These considerations suggested that channel activation was associated with a charge transfer of 6e to 8e—that is, three of the elementary gating charges. This finding is remarkably consistent with the activation model proposed orig-

[33]Hoshi, T. W., Zagotta, N., and Aldrich, R. W. 1991. *Neuron* 7: 547–556.

[34]Choi, K. L., Aldrich, R. W., and Yellen, G. 1991. *Proc. Natl. Acad. Sci. USA* 88: 5092–5095.

[35]Keynes, R. D. 1990. *Proc. R. Soc. Lond. B* 240: 425–432.

[36]Keynes, R. D., and Elinder, F. 1998. *Proc. R. Soc. Lond. B* 265: 263–270.

[37]Bean, B. P. 1981. *Biophys. J.* 35: 595–614.

[38]Aldrich, R. W., and Stevens, C. F. 1983. *Cold Spring Harb. Symp. Quant. Biol.* 48: 147–153.

[39]Hille, B. 1992. *Ionic Channels of Excitable Membranes*, 2nd Ed. Sinauer Associates, Sunderland, MA, Chapter 18.

[40]Patlach, J. 1991. *Physiol. Rev.* 71: 1047–1080.

[41]Aldrich, R. W., and Stevens, C. F. 1987. *J. Neurosci.* 7: 418–431.

[42]Cota, G., and Armstrong, C. M. 1989. *J. Gen. Physiol.* 94: 213–232.

[43]Conti, F., and Stühmer, W. 1989. *Eur. Biophys. J.* 17: 53–59.

inally by Hodgkin and Huxley. Conti and Stühmer note that it is attractive to identify the charge transfers with structural transitions in three of the four sodium channel domains, and to ascribe inactivation to interaction of these three with the fourth. Such findings, of course, take no account of additional state transitions that are electrically silent.

In summary, the kinetic models suggest that depolarization initiates a series of stepwise conformational changes that lead eventually to channel opening, with one or more alternate steps leading to inactivation. Although we can imagine in a very general way how such structural changes might occur in the protein, it is difficult to specify them precisely (Chapter 3). An initial step has been made in relating inactivation to particular groups of amino acids in the sodium channel and in potassium channel subunits. Further correlations will no doubt be forthcoming when the molecular anatomy is understood in greater detail.

Properties of Channels Associated with the Action Potential

The conductance of the voltage-activated sodium channel has been determined directly by patch clamp measurements to be about 20 pS in cultured rat muscle fibers (see Figure 6.10) and 14 pS in rat spinal motoneurons.[44] The density of sodium channels has been determined in a number of tissues by measuring the density of TTX-binding sites. Using tritiated tetrodotoxin, Levinson and Meves[45] estimated that in squid axon an average of 553 molecules were bound to each square micrometer of membrane. Values in other tissues have been found to range from a low of 2 molecules/μm^2 in neonatal rat optic nerve[46] to 2000 molecules/μm^2 at the node of Ranvier in rabbit sciatic nerve.[47] Sodium channel densities in skeletal muscle have been measured by depolarizing small areas of the membrane with a focal extracellular pipette and measuring inward sodium currents through the underlying membrane. The currents varied from one patch to the next, indicating variations in channel density. Sodium channels were found to be most concentrated near the end-plate region, and to decrease in density with distance from the end plates, reaching a low of about 10% of their maximum density near the tendons.[48] In addition, sodium channels in the muscle fiber membrane were found to be distributed in clusters rather than uniformly.[49]

The conductance of the potassium channels underlying the late current has been measured in cut-open squid axon.[50] Delayed rectifier channels were found to have conductances of 10, 20, and 40 pS, with the 20 pS channel predominating. Potassium channels in frog muscle, like sodium channels, are distributed in clusters.[36] However, the sodium and potassium channel clusters are not co-localized. Delayed rectifier channels appear to be totally absent at nodes of Ranvier in rabbit myelinated nerve, since depolarization produces no late outward current.[51] During the action potential, repolarization is achieved by a large leak current after rapid inactivation of the sodium channels.

Other Potassium Channels Contributing to Repolarization

In addition to potassium channels associated with the delayed rectifier current, neurons have a number of voltage-activated potassium channel types,[52] some of which can contribute to action potential repolarization. One voltage-activated channel is the potassium **A-channel**, which is activated rapidly by depolarization. The contribution of A-channels to action potential repolarization is minimal for two reasons: They inactivate rapidly, and in most cells activation occurs only after a preceding hyperpolarization; that is, they are usually inactivated at rest. Two other voltage-sensitive potassium channels, **M-channels** (Chapter 16) and **S-channels**, are similar to delayed rectifier channels in that they open in response to depolarization. M-channels have the additional feature of being inactivated by acetylcholine through *muscarinic* ACh receptors (hence their name). S-channels are open at the resting membrane potential and inactivated by *serotonin*.

Calcium-activated potassium channels can also contribute to action potential repolarization.[53] During the action potential, calcium ions enter the cell through voltage-activated calcium channels (see the next section). In many cells this inward calcium cur-

[44]Safronov, V. B., and Vogel, W. 1995. *J. Physiol.* 487: 91–106.

[45]Levinson, S. R., and Meves, H. 1975. *Phil. Trans. R. Soc. Lond. B* 270: 349–352.

[46]Waxman, S. G., et al. 1989. *Proc. Natl. Acad. Sci. USA* 86: 1406–1410.

[47]Ritchie, J. M. 1986. *Ann. N.Y. Acad. Sci.* 479: 385–401.

[48]Beam, K. G., Caldwell, J. H., and Campbell, D. T. 1985. *Nature* 313: 588–590.

[49]Almers, W., Stanfield, P., and Stühmer, W. 1983. *J. Physiol.* 336: 261–284.

[50]Lanno, I., Webb, C. K., and Bezanilla, F. 1988. *J. Gen. Physiol.* 92: 179–196.

[51]Chiu, S. Y., et al. 1979. *J. Physiol.* 292: 149–166.

[52]Mathie, A., Wooltorton, J. R., and Watkins, C. S. 1998. *Gen. Pharmacol.* 30: 13–44.

[53]Vergara, C., et al. 1998. *Curr. Opin. Neurobiol.* 8: 321–329.

rent stimulates an increase in potassium conductance that contributes to repolarization and produces a subsequent hyperpolarization. The calcium-activated potassium channels have at least three subtypes with very large (200 pS), intermediate (30 pS), and small (10 pS) conductances. Their presence can be demonstrated experimentally by raising intracellular calcium—for example, by injection from an intracellular micropipette.[54] Following such an injection the membrane conductance of the cell increases rapidly and the resting membrane potential approaches the equilibrium potential for potassium. The resistance and potential then return to their control levels as the excess calcium is removed from the cytoplasm by internal buffering mechanisms and outward transport. Still other potassium channels are activated by intracellular sodium.[55,56] Their activation by sodium influx during the action potential may contribute to repolarization in some cells.[57]

THE ROLE OF CALCIUM IN EXCITATION

Calcium Action Potentials

The membranes of nerve and muscle fibers contain a variety of voltage-activated calcium channels (see Chapter 3 for calcium channel classifications and properties). Calcium ions enter the cell through such channels during the action potential, and this entry plays a key role in a variety of processes (Chapters 9 through 12). For example, a transient increase in intracellular calcium during the action potential is responsible for secretion of chemical transmitters by neurons and for contraction of muscle fibers.

In some muscle fibers and some neurons, calcium currents become sufficiently large to contribute significantly to, or even be solely responsible for, the rising phase of the action potential. Because g_{Ca} increases with depolarization, the process is a regenerative one, entirely analogous to that discussed for sodium. The participation of calcium in the action potential process was first studied in invertebrate muscle fibers by Fatt and Ginsborg[58] and subsequently by Hagiwara.[59] Calcium action potentials occur in cardiac muscle, in a wide variety of invertebrate neurons, and in neurons in the vertebrate autonomic and central nervous systems.[60] Such action potentials occur in nonneural cells as well, including a number of endocrine cells and some invertebrate egg cells. The voltage-dependent calcium currents can be blocked by adding millimolar concentrations of cobalt, manganese, or cadmium to the extracellular bathing solution. Barium can substitute for calcium as the permeant ion; magnesium, on the other hand, cannot. A particularly striking example of the coexistence of sodium and calcium action potentials in the same cell is found in the mammalian cerebellar Purkinje cell, which generates sodium action potentials in its cell body and calcium action potentials in the branches of its dendritic tree.[61,62]

Calcium Ions and Excitability

Calcium ions also affect excitation: A reduction in extracellular calcium increases the excitability of nerve and muscle cells; conversely, increasing extracellular calcium decreases excitability. Frankenhaeuser and Hodgkin[63] used voltage clamp experiments to examine these effects in the squid axon and found that when extracellular calcium was reduced, the voltage dependence of sodium channel activation was shifted so that smaller depolarizing pulses were required to reach threshold and to produce sodium currents equivalent to those in normal solution. The reduction in depolarizing pulse amplitudes was constant throughout the range of excitation and depended on calcium concentration. A fivefold reduction in extracellular calcium resulted in a 10 to 15 mV reduction in the required depolarization.

Frankenhaeuser and Hodgkin suggested that the effect might be associated with screening by calcium ions of negative charges fixed to the outer surface of the membrane. Such charges at the membrane surface can arise, for example, by glycosylation of membrane proteins with carbohydrate chains that include negatively charged sialic acid. The eel sodium channel itself has over 100 sialic acid residues.[64] As long as the charges were screened, the potential gradient across the membrane would be the same as the measured

[54]Meech, R. W. 1974. *J. Physiol.* 237: 259–277.

[55]Partridge, L. D., and Thomas, R. C. 1976. *J. Physiol.* 254: 551–563.

[56]Martin, A. R., and Dryer, S. E. 1989. *Q. J. Exp. Physiol.* 74: 1033–1041.

[57]Koh, D-S., Jonas, P., and Vogel, W. 1994. *J. Physiol.* 479(Pt.2): 183–197.

[58]Fatt, P., and Ginsborg, B. L. 1958. *J. Physiol.* 142: 516–543.

[59]Hagiwara, S., and Byerly, L. 1981. *Annu. Rev. Neurosci.* 4: 69–125.

[60]Hagiwara, S. 1983. *Membrane Potential-Dependent Ion Channels in Cell Membrane. Phylogenetic and Developmental Approaches.* Raven, New York.

[61]Llinás, R., and Sugimori, M. 1980. *J. Physiol.* 305: 197–213.

[62]Ross, W. N., Lasser-Ross, N., and Werman, R. 1990. *Proc. R. Soc. Lond. B* 240: 173–185.

[63]Frankenhaeuser, B., and Hodgkin, A. L. 1957. *J. Physiol.* 137: 218–244.

[64]Miller, J. A., Agnew, W. S., and Levinson, S. R. 1983. *Biochemistry* 22: 462–470.

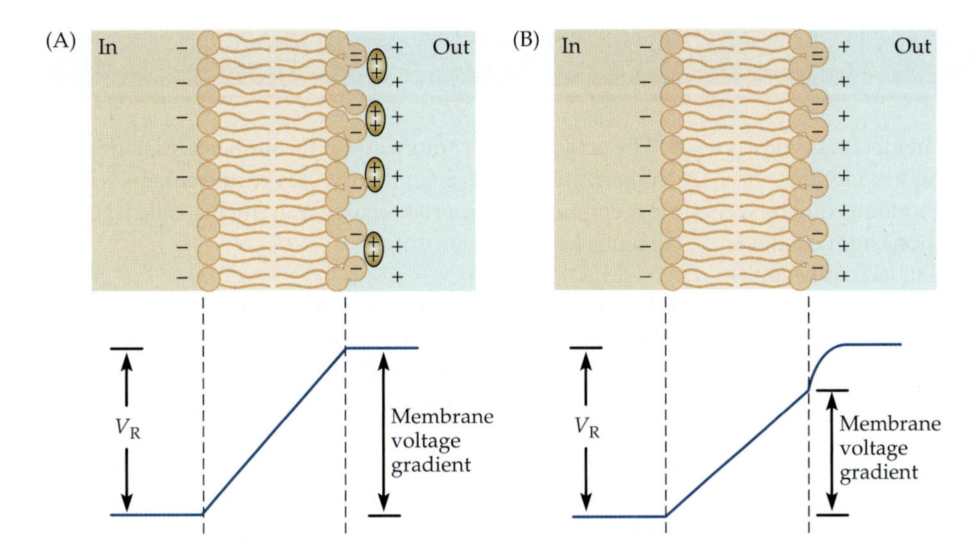

FIGURE 6.13 Effect of Surface Charge on Membrane Potential, proposed to explain the effects of calcium on action potential threshold. (A) The membrane structure includes negatively charged elements on the outer surface whose charge is neutralized by divalent cations. The resting membrane potential, V_R, produced by ionic charge separation, is determined by the composition of the intracellular and extracellular fluids. (B) When the fixed negative surface charges are unscreened (e.g., by removing calcium from the extracellular solution) the resting potential is unchanged, but the shape of the potential profile is altered by the surface negativity, reducing the potential gradient across the membrane.

resting potential (Figure 6.13A). Upon removal of calcium the unscreened charges would add an increment of negative potential to the outer surface, thereby reducing the potential gradient across the membrane (Figure 6.13B). This idea introduces a new concept with regard to membrane potential: The potential between the intracellular and extracellular solutions is determined by extracellular and intracellular ion concentrations and ion permeabilities, as discussed in Chapter 5. However, the shape of the potential gradient may depend on charged molecules fixed to the membrane surface. This can have a pronounced effect on voltage-sensitive components in the membrane, since they sense only the voltage gradient in their immediate vicinity.

One problem with the surface charge idea is that reducing the potential gradient across the membrane by removing extracellular calcium should affect not only activation but also inactivation of sodium channels, as well as activation of potassium channels. A reduction in the depolarization required for sodium channel inactivation and potassium channel activation should lead to a *decrease* in excitability. In fact, for reasons that are not clear, calcium removal has a much smaller effect on these parameters than on sodium activation.[64,65] It is perhaps not surprising that local changes in the potential gradient across the membrane should have varying effects on different channel molecules, or even on different regions of the same molecule, depending on the location of the voltage-sensitive element relative to the surface charges. Whatever the reason for these disparities, the net effect of calcium is to stabilize the membrane, maintaining a margin of safety between the resting membrane potential and the threshold for action potential initiation.

[65]Hille, B. 1968. *J. Gen. Physiol.* 51: 221–236.

SUMMARY

- The action potential in most nerve cell membranes is produced by a transient increase in sodium conductance that drives the membrane potential toward the sodium equilibrium potential, followed by an increase in potassium conductance that returns the membrane potential to its resting level.

- The increases in conductance occur because sodium and potassium channels in the membrane are voltage-dependent: Their opening probability increases with depolarization.

- Voltage clamp experiments on squid axons have provided detailed information about the voltage dependence and time course of the conductance changes. When the cell membrane is depolarized, the sodium conductance is activated rapidly, then inactivated. Potassium conductance is activated with a delay and remains high as long as the depolarization is maintained.

- The time course and voltage dependence of the sodium and potassium conductance changes account precisely for the amplitude and time course of the action potential, as well as other phenomena, such as activation threshold and refractory period.

- The activation of sodium and potassium conductances by depolarization requires, in theory, charge movements within the membrane. Appropriate charge movements, called gating currents, have been measured.

- Patch clamp experiments on voltage-activated sodium and potassium channels are consistent with voltage clamp experiments and reveal new details about the process of excitation. For example, sodium channels open for a relatively short time, and the probability that they open during a depolarizing step first increases and then decreases, corresponding to activation and inactivation of sodium conductance in the whole cell. Various kinetic models have been proposed for channel activation and inactivation.

- Calcium plays an important role in excitation. In some cells calcium influx, rather than sodium influx, is responsible for the rising phase of the action potential. In addition, membrane excitability is controlled by extracellular calcium concentration. As extracellular calcium decreases, excitability increases.

SUGGESTED READING

Armstrong, C. M., and Hille, B. 1998. Voltage-gated ion channels and electrical excitability. *Neuron* 20: 371–380.

Frankenhaeuser, B., and Hodgkin, A. L. 1957. The action of calcium on the electrical properties of squid axons. *J. Physiol.* 137: 218–244.

Hille, B. 1992. *Ionic Channels of Excitable Membranes*, 2nd Ed. Sinauer Associates, Sunderland, MA, Chapters 2–5.

Hodgkin, A. L., Huxley, A. F., and Katz, B. 1952. Measurement of current-voltage relations in the membrane of the giant axon of *Loligo*. *J. Physiol.* 116: 424–448.

Hodgkin, A. L., and Huxley, A. F. 1952. Currents carried by sodium and potassium ion through the membrane of the giant axon of *Loligo*. *J. Physiol.* 116: 449–472.

Hodgkin, A. L., and Huxley, A. F. 1952. The components of the membrane conductance in the giant axon of *Loligo*. *J. Physiol.* 116: 473–496.

Hodgkin, A. L., and Huxley, A. F. 1952. The dual effect of membrane potential on sodium conductance in the giant axon of *Loligo*. *J. Physiol.* 116: 497–506.

Hodgkin, A. L., and Huxley, A. F. 1952. A quantitative description of membrane current and its application to conduction and excitation in nerve. *J. Physiol.* 117: 500–544.

7 NEURONS AS CONDUCTORS OF ELECTRICITY

ACTION POTENTIALS PROPAGATE ALONG AXONS by the longitudinal spread of current. As each region of the membrane generates an all-or-nothing impulse, it depolarizes and excites the adjacent, not yet active region. This depolarization gives rise to a new regenerative impulse. To understand impulse propagation, as well as synaptic transmission and integration, one has to know how electrical currents spread passively along a nerve.

As current spreads along a nerve axon or dendrite, it becomes attenuated with distance. This attenuation depends on a number of factors, principally the diameter and membrane properties of the fiber. Longitudinal current spreads farther along a fiber with large diameter and high membrane resistance. The electrical capacitance of the membrane influences the time course of the electrical signals and usually their spatial spread as well. To estimate how far a subthreshold potential change will spread, one needs to know the geometry and membrane characteristics of the neuron, and, in addition, the time course of the potential change.

The axons of many vertebrate nerve cells are covered by a high-resistance, low-capacitance myelin sheath. This sheath acts as an effective insulator and forces currents associated with the nerve impulse to flow through the membrane at intervals where the sheath is interrupted (nodes of Ranvier). The impulse jumps from one such node to the next, and thereby its conduction velocity is increased. Myelinated nerves occur in pathways in the nervous system where speed of conduction is important.

Electrical activity can also pass between neurons through specialized regions of close membrane apposition called gap junctions. Pathways for current flow in such regions are provided by intercellular channels called connexons.

PASSIVE ELECTRICAL PROPERTIES OF NERVE AND MUSCLE MEMBRANES

The permeability properties of nerve cell membranes and the way in which these properties produce regenerative electrical responses have been discussed in the preceding chapters. In this chapter we describe in more detail how currents spread along nerve fibers to produce local graded potentials.

The **passive electrical properties** of neurons, specifically the resistance and capacitance of the nerve cell membrane, and the resistance of the cytoplasm, play a major role in signaling. At sensory end organs they are the link between the stimulus and the production of impulses; along axons they allow the impulse to spread and propagate; at synapses they enable the postsynaptic neuron to add and subtract synaptic potentials that arise from numerous converging inputs—some close to the cell body, others on remote dendritic sites. To understand phenomena such as impulse initiation and propagation and interactions between synaptic inputs, it is necessary to know how electrical signals spread along nerve processes. The discussion that follows deals primarily with the spread of current along nerve fibers of uniform diameter—that is, along cylindrical conductors. Further, throughout the discussion we will assume that in the absence of regenerative action potentials the nerve fiber membranes are indeed passive—that is, that changes in potential below action potential threshold do not activate any voltage-sensitive channels that would change the membrane resistance. The concepts we develop are also applicable in principle to more complex structures, such as arborizations of axon terminals or branched dendritic trees with nonuniform electrical properties.[1,2] Such complex structures play an important role in nervous system function, but quantitative treatments of their electrical properties require more complicated analyses.

Nerve and Muscle Fibers as Cables

A cylindrical nerve fiber has the same formal components as an undersea cable—namely, a central or core conductor and an insulating sheath surrounded by a conducting medium. However, the two systems are quantitatively very dissimilar. In a cable, the core conductor is usually copper, which has a very high conductance, and the surrounding insulating sheath is neoprene, plastic, or some other material of very high resistance. In addition, the sheath is usually relatively thick, so it has a very low capacitance (Appendix A). Voltage applied to one end of such a cable will spread an immense distance because the resistance to longitudinal current flow along the copper conductor is relatively low and virtually no current is lost through the insulating sheath. In a nerve fiber, on the other hand, the core conductor is a salt solution similar in concentration to that bathing the nerve and (compared with copper) is a poor conductor. Furthermore, the plasma membrane of the fiber is a relatively poor insulator and, being thin, has a relatively high capacitance. A voltage signal applied to one end of a nerve fiber, then, will fail to spread very far for two reasons: (1) The core material has a low conductance; that is, the resistance to current flow down the fiber is high. (2) Current that starts off flowing down the axoplasm is lost progressively along the fiber by outward leakage through the poorly insulating plasma membrane.

The analysis of current flow in cables was developed by Lord Kelvin for application to transatlantic telephone transmission and refined by Oliver Heaviside in the late nineteenth century. Heaviside was the first to consider the effect of resistive leak through the insulation, equivalent to the membrane resistance in nerve, and made many other contributions to cable theory, including the concept of what he called impedance. Cable theory was first applied coherently to nerve fibers by Hodgkin and Rushton,[3] who used extracellular electrodes to measure the spread of applied current along lobster axons. Later, intracellular electrodes were used in a number of nerve and muscle fibers for similar studies.

Here we consider how the spread of current along a cylindrical axon is affected by the resistive properties of the membrane and of the axoplasm. The main requirement is that we keep in mind Ohm's law: A given amount of current, i, passed through a resistor, r, produces a voltage $v = ir$ (Appendix A). Later in this chapter we consider the additional effects of membrane capacitance on the magnitude and time course of the current spread.

[1]Rall, W. 1967. *J. Neurophysiol.* 30: 1138–1168.

[2]Lev-Tov, A., et al. 1983. *J. Neurophysiol.* 50: 399–412.

[3]Hodgkin, A. L., and Rushton, W. A. H. 1946. *Proc. R. Soc. Lond. B* 133: 444–479.

Flow of Current in a Cable

One way to gain an intuitive feeling about how current spreads in a cable is to think of the spread of heat along a metal rod surrounded by insulation and immersed in a conducting material (such as water). If one end of the rod is heated continuously, heat spreads along the rod and, as it spreads, is lost to the surrounding medium. At progressively greater distances from the heated end, the temperature becomes progressively lower; as the temperature decreases with distance, the rate at which heat is lost decreases as well. Assuming that the surrounding medium is a good heat conductor, the distance over which the heat spreads depends primarily on (1) the conductivity of the rod and (2) the effectiveness of the insulation in preventing heat loss.

Flow of electrical current in a cable can be described in similar terms. A voltage applied to one end of the cable causes current to flow along the core, some of which is lost to the surrounding medium through the insulation. At progressively greater distances from the end, less and less current remains. The distance along which the current spreads depends on the conductivity of the core and the effectiveness of the insulation in preventing current loss. Low-resistance insulation allows all the current to leak out before it can spread very far. A larger-resistance insulation allows the current to flow a greater distance along the core.

In an axon, current is carried by the flow of ions: If we inject current from a microelectrode into a nerve fiber (e.g., a lobster axon), as shown in Figure 7.1A, positive charges flowing into the axoplasm from the tip of the microelectrode repel other cations and attract anions. By far the most abundant small ion in the axoplasm is potassium, which therefore carries most of the current. The current flows longitudinally along the axon,

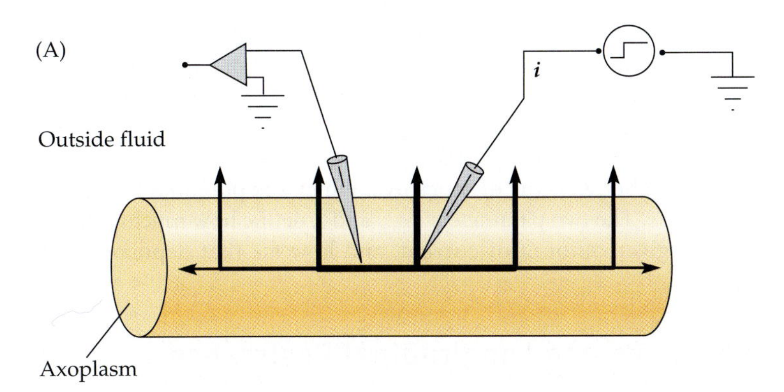

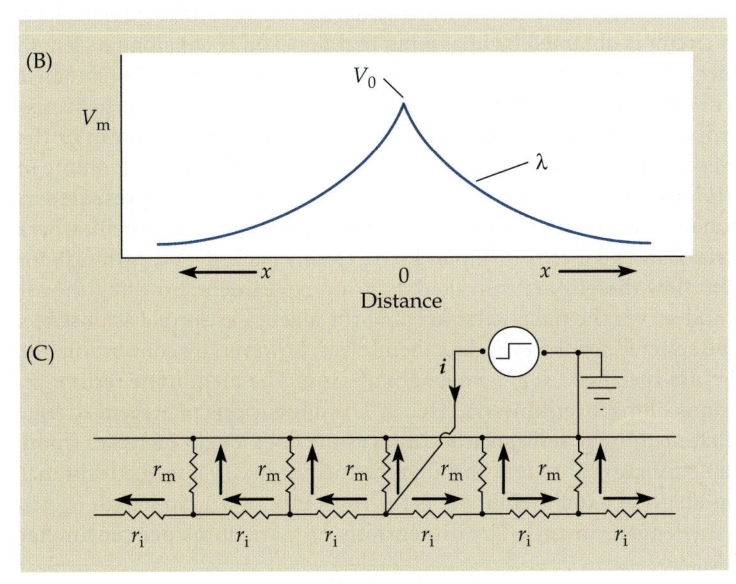

FIGURE 7.1 Pathways for Current Flow in an axon. (A) Current flow (i) across the membrane produced by a steady injection of current from a microelectrode. Thickness of the arrows indicates current density at various distances from the point of injection. The second electrode (left) records membrane potential at various distances from the current electrode. (B) Potential (V_m) measured along the axon as a function of distance (x) from the point of current injection. Decay of voltage is exponential, with a length constant λ. (C) Equivalent electrical circuit, assuming zero resistance in the external fluid and ignoring membrane capacitance. r_i = longitudinal resistance of axoplasm per unit length; r_m = membrane resistance of a unit length.

and as it spreads away from the electrode some is lost by ion movements through the membrane. The distance the potential spreads along the axon depends on the resistance of the cell membrane relative to that of the axoplasm. A low-resistance membrane with high ionic conductances allows current to leak out before it can spread very far; a larger-resistance membrane allows a greater portion of the current to spread laterally before escaping to the external solution.

Input Resistance and Length Constant

In Figure 7.1A the relative amounts of current flowing across the membrane at various distances from the current electrode are indicated roughly by the thickness of the arrows. The potential change produced across the membrane at any given distance is proportional to the current flow across the membrane at that point, in accordance with Ohm's law. Two questions arise from an experiment of this kind: (1) For a given amount of current injected into the pipette, how much voltage change will be produced at the electrode? (2) How far will this voltage change spread along the fiber? These questions can be answered by measuring the potential change with a second micropipette, which we can insert at various positions along the fiber, as indicated in Figure 7.1A. The results of such measurements are shown in Figure 7.1B. The current produces a change in potential that is greatest at the point of injection and falls off with distance on either side.

The decrease in potential with distance from the current electrode is exponential, so the potential (V_x) at any distance x on either side is given by

$$V_x = V_0 e^{-x/\lambda}$$

The peak potential change, V_0, is proportional to the size of the injected current. The constant of proportionality is known as the **input resistance** of the fiber, r_{input}. It is the average resistance presented by the fiber to the flow of current through the axoplasm and surface membrane to the extracellular solution. Thus, if the amount of current injected is i, then

$$V_0 = ir_{input}$$

The **length constant** of the fiber, λ, is the distance over which the potential falls to $1/e$ (37%) of its maximum value. The two parameters r_{input} and λ define how much depolarization is produced by a given amount of current, and how far that depolarization spreads along the fiber.

Membrane Resistance and Longitudinal Resistance

We can think of a cable as a series of resistive elements, r_m and r_i, connected in a chain (Figure 7.1C). The circuit is obtained by imagining that the axon is cut along its length into a series of short cylinders. The membrane resistance, r_m, represents the resistance across the cylinder wall; the longitudinal resistance, r_i, represents the internal resistance along the axoplasm from the midpoint of one cylinder to the midpoint of the next. Because nerves in a recording chamber are normally bathed in a large volume of fluid, the *extracellular* longitudinal resistance along the cylinders is represented as zero. This approximation is not always adequate in the central nervous system, where nerve axons, dendrites, and glial cells (Chapter 8) are closely packed, and pathways for extracellular current flow thereby are restricted. For our experiment, however, the assumption is valid and serves the purpose of keeping the algebra as simple as possible. Any length could be selected for the cylinders themselves; however, by convention, the resistances r_m and r_i are specified for a 1 cm length of axon. The membrane resistance, r_m, has the dimensions ohms × centimeters (Ωcm). The dimensions may seem strange until one realizes that membrane resistance *decreases* as the fiber length increases (more channels are available for current to leak through the membrane). Thus, the resistance in ohms of a given length of axon membrane is the resistance of a 1 cm length (r_m, in Ωcm) divided by the length (in cm). The dimensions of r_i are ohms per centimeter (Ω/cm), as expected.

Calculating Membrane Resistance and Internal Resistance

The length constant of the fiber depends on both r_m and r_i:

$$\lambda = \left(\frac{r_m}{r_i}\right)^{1/2}$$

This expression has the dimensions of centimeters as required, and it fulfills the intuitive expectation that the distance over which the potential change spreads should increase with increasing membrane resistance (which prevents loss of current across the membrane) and decrease with increasing internal resistance (which resists current flow along the core of the fiber).

Similarly, the input resistance depends on both parameters:

$$r_{input} = 0.5(r_m r_i)^{1/2}$$

Again the expression has the required dimensions (ohms), and tells us that the input resistance increases with both membrane resistance and internal resistance. The factor 0.5 arises because the axon extends in both directions from the point of current injection; each half has an input resistance equal to $(r_m r_i)^{1/2}$.

Knowing these relations, we can determine the resistive properties of the membrane and axoplasm from experiments such as that shown in Figure 7.1. After we have measured r_{input} and λ experimentally, it is a simple matter to rearrange the equations to calculate r_m and r_i:

$$r_m = 2r_{input}\lambda$$

$$r_i = \frac{2r_{input}}{\lambda}$$

Specific Resistance

The calculated values of r_m and r_i specify the resistive characteristics of a cylindrical segment of the axon 1 cm in length. They do not, however, provide precise information about the resistive properties of the membrane itself, or of the axoplasmic material, because these depend on the size of the fiber. All else being equal, a 1 cm length of small fiber should have a higher membrane resistance than the same length of larger fiber, simply because the smaller fiber has less membrane surface. On the other hand, the smaller fiber might contain a much higher density of ion channels and thereby have about the same resistance per unit length.

To compare one *membrane* to another we need to know the specific resistance (R_m) of each, which is the resistance of 1 cm^2 of membrane and has the units Ωcm^2. A 1 cm length of axon of radius a has an area of $2\pi a$ cm^2. Its membrane resistance (r_m) is obtained by dividing R_m by the membrane area: $r_m = R_m/2\pi a$. Turning the equation around, we get

$$R_m = 2\pi a r_m$$

R_m is important because it is independent of geometry and therefore enables us to compare the membrane of one cell with that of another of quite different size or shape.

In most neurons, R_m is determined primarily by the resting permeabilities to potassium and chloride (Chapter 5); these vary considerably from one cell to the next. The average value for R_m reported by Hodgkin and Rushton for lobster axons was about 2000 Ωcm^2; in other preparations, measurements range from less than 1000 Ωcm^2 for membranes with a large number of channels through which ions can leak to more than 50,000 Ωcm^2 for membranes with relatively few such channels.

The specific resistance (R_i) of the axoplasm is the internal longitudinal resistance of a 1 cm length of axon 1 cm^2 in cross-sectional area. It is also independent of geometry and is a measure of how freely ions migrate through the intracellular space. To calculate R_i from r_i for a cylindrical axon, recall that the resistance along the core of a cylinder decreases as the cross-sectional area increases. Therefore, the resistance of a 1 cm length of axon (r_i) is obtained by dividing R_i by the cross-sectional area of the axon: $r_i = R_i/\pi a^2$. Again we can turn the equation around, to get

$$R_i = r_i \pi a^2$$

R_i has the dimensions Ωcm. In squid nerve, it has a value of about 30 Ωcm at 20°C, or about 10^7 times that of copper. This value is expected from the ionic composition of squid axoplasm.[3] In mammals, where the ion concentration in the cytoplasm is lower, the specific resistance is higher, about 125 Ωcm at 37°C; in frogs, with still lower ion concentration, the specific resistance is about 250 Ωcm at 20°C.

The Effect of Diameter on Cable Characteristics

Given a specific resistance, R_i, for the axoplasm and a specific membrane resistance, R_m, how are the cable parameters r_{input} and λ influenced by fiber diameter? The answer can be obtained from the relations presented in the preceding paragraphs. Beginning with input resistance, we know that $r_{input} = 0.5(r_m r_i)^{1/2}$ and that $r_m = R_m/2\pi a$ and $r_i = R_i/\pi a^2$. Putting these relations together, we get

$$r_{input} = 0.5\left(\frac{R_m R_i}{2\pi^2 a^3}\right)^{1/2}$$

Thus, as the fiber radius (a) increases, the input resistance decreases, varying inversely with the $3/2$ power of the radius.

Using the same approach, we find the length constant:

$$\lambda = \left(\frac{aR_m}{2R_i}\right)^{1/2}$$

Other properties being equal, λ increases with the square root of the fiber radius. We can use this relation to compare various fibers, assuming in each case that the specific membrane resistance is 2000 Ωcm^2: A squid axon of 1 mm diameter with a specific internal resistance of 30 Ωcm would have a length constant of almost 13 mm. Because of its smaller diameter and larger specific internal resistance, a 50 μm diameter frog muscle fiber would have a length constant of only 1.4 mm, and a 1 μm diameter mammalian nerve fiber a length constant of 0.3 mm.

In summary, the cable parameters r_{input} and λ determine the size of a signal generated in a nerve process, and how far the signal will spread. For example, other properties being equal, an excitatory synaptic potential (Chapter 9) will be larger in a small dendritic process (larger r_{input}) than in a large one. On the other hand, in the larger dendrite the potential will spread farther toward the cell body (larger λ). However, r_{input} and λ depend not only on fiber size but also on the resistive properties of the cytoplasm and plasma membrane. It is reasonably safe to assume that the specific resistance of the cytoplasm is the same in all cells in any animal class. Guesses about specific membrane resistance are much less certain; values may vary between one cell and the next by a factor of 50 or more.

Membrane Capacitance

In addition to allowing the passage of ionic currents, the cell membrane accumulates ionic charges on its inner and outer surfaces (Chapter 5). Electrically, the charge separation means that the membrane has the properties of a capacitor. In general, a capacitor consists of two conducting sheets, called plates, separated by a layer of insulating material; in manufactured capacitors the conductors are usually metallic foil, and the insulator is mica or a plastic such as Mylar. In the case of a nerve cell the conductors are the two layers of fluid that lie against either side of the membrane, and the insulating material is the membrane itself. When a capacitor is charged by connecting a battery to the two plates, it accumulates an excess of positive charges on one plate, leaving an equal excess of negative charges on the other. Its capacitance (C) is defined by how much charge (Q) it will accumulate for each volt of potential (V) applied to it; specifically, $C = Q/V$. C has the units coulombs per volt, or farads (F). The closer together the plates are, the greater their ability to separate and store charge. Because the cell membrane is only about 5 nm thick, it is capable of storing a relatively large amount of charge. Typically nerve cell membranes have a capacitance on the order of 1 μF/cm² (1 μF = 10^{-6} F). Turning the equation around, the charge stored in a capacitor is given by $Q = CV$. Thus, if a cell has a resting potential of –80 mV, the amount of excess negative charge at the inner surface of the membrane will be $(1 \times 10^{-6}) \times (80 \times 10^{-3}) = 8 \times 10^{-8}$ coulombs/cm², which is equivalent to 5×10^{11} univalent ions (0.8 pmol) for each square centimeter of membrane.

The current flowing into or out of a capacitor can be deduced from the relation between charge and voltage, and remembering that current (i, in amperes) is the rate of change of charge with time, that is: amperes = coulombs/s. Thus, since $Q = CV$, we can write

$$i = \frac{dQ}{dt} = C\frac{dV}{dt}$$

The rate of change of the voltage on the capacitor is directly proportional to the current flow. If the current i is constant, then the voltage will change at a constant rate $dV/dt = i/C$.

The relations between current and voltage in circuits with resistors and capacitors in parallel are illustrated in Figure 7.2. When a rectangular current pulse of amplitude i is applied to a resistor (R), this produces a voltage pulse across the resistor of amplitude $V = iR$ (Figure 7.2A). If the same pulse is applied to a capacitor (C), the voltage on the capacitor builds up at a rate $dV/dt = i/C$ (Figure 7.2B). When the two elements R and C are combined in parallel (Figure 7.2C), all of the initial current goes to charge the capacitor at a rate of i/C; however, as soon as voltage starts to develop across the capacitor, current is driven through the resistor as well. As the voltage increases further, more current is diverted

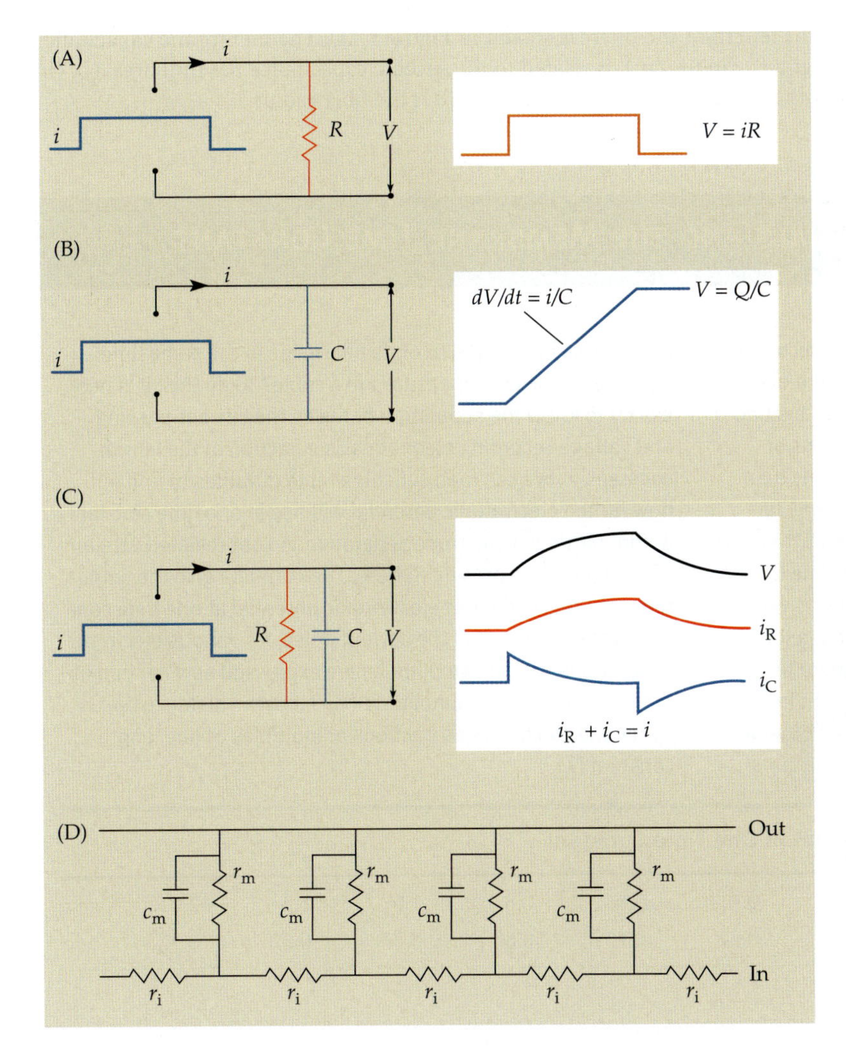

FIGURE 7.2 Effect of Capacitance on Time Course of potentials. (A) Potential (V) produced by a rectangular pulse of current (i) in a purely resistive network. Voltage is proportional to, and has the same time course as, the applied current. (B) In a purely capacitative network the *rate of change* of voltage is proportional to the applied current. (C) In a combined RC network the initial surge of current is into the capacitor (i_C); by the end of the pulse all of the current flows through the resistor (i_R). Voltage rises to the final value iR exponentially with time constant $\tau = RC$. After termination of the current pulse, the capacitance discharges through the resistance with the same time constant and i_C and i_R are equal and opposite. (D) Electrical model of a cable as in Figure 7.1C, but with membrane capacitance per unit length (c_m) added.

through the resistor, and the rate at which the capacitor charges is decreased. Eventually all of the applied current flows through the resistor, producing a potential $V = iR$, and the capacitor is fully charged to the same potential. When the pulse is terminated, the charge stored in the capacitor leaks away through the resistor and the voltage returns to zero.

Time Constant

The rise and fall of the potential in Figure 7.2C are described by exponential functions. The rising phase during the pulse is described by

$$V = iR(1 - e^{-t/\tau})$$

where t is the time from the beginning of the pulse. It turns out that the exponential time constant, τ, is given by the product of the resistance and capacitance in the circuit: $\tau = RC$. It is the time for the potential to rise to a fraction ($1 - 1/e$, or 63%) of its final value. The fall in voltage is again exponential, with the same time constant. Just as the voltage rises and falls exponentially, so must the resistive current, i_R. On the rising phase, then, the resistive current starts at zero and rises exponentially toward its final value i. Conversely, the capacitative current, i_C, starts at i and falls with the same time constant. After termination of the pulse, because external current is no longer being applied, the only current flowing across the resistor is that flowing out of the capacitor. Consequently, the resistive and capacitative currents must be equal and opposite as shown.

The circuit just described, with a resistor and capacitor in parallel, can be used to represent a spherical nerve cell with an axon and dendrites so small that they make only negligible contributions to the electrical properties of the cell. In the equivalent circuit for an axon or muscle fiber, however, the membrane capacitance, like the resistance, is distributed along the length of the fiber, as shown in Figure 7.2D. The membrane capacitance per unit length, c_m (in µF/cm), is related to the specific capacitance per unit area, C_m (in µF/cm^2), by the expression $c_m = 2\pi a C_m$, where a is the fiber radius.

Box 7.1 Electrotonic Potentials and the Membrane Time Constant

The electrotonic potentials shown in Figure 7.3, recorded at various distances along an axon from the point of current injection, do not rise and fall exponentially. Instead their waveforms are described by complicated functions of both time and distance. The potentials rise more rapidly than exponentials near the current electrode and more slowly farther away. Consequently, the membrane time constant, $\tau = R_m C_m$, cannot be obtained simply by measuring the time for an electrotonic potential to rise to 63% of its final value, as with an exponential voltage change. At the point of current injection, the potential rises to 84% of its maximum amplitude in one time constant. At a distance of one length constant from the point of current injection the potential does reach 63% of its final value in one time constant, but two length constants away it reaches only 37% of its final value in the same time. How, then, does one measure τ in a cable? To do this, it is necessary to know the separation between the current-passing and voltage-recording electrodes as a fraction of the length constant, λ. We can then consult a table of values to find out how far the electrotonic potential will rise in one time constant at that particular electrode separation. An abbreviated version of such a table is presented here. The numbers give the amplitude (V_τ) reached by an electrotonic potential at one time constant after the onset of the current pulse, for various electrode separations (d). The amplitudes are expressed as a fraction of the final steady-state amplitude (V_∞) and the distance between the electrodes as fractions or multiples of one length constant (λ).

Amplitude Reached by an Electrotonic Potential in One Time Constant

Separation (d/λ)	0	0.2	0.4	0.6	0.8	1.0	1.5	2.0
Amplitude (V_τ/V_∞)	0.84	0.81	0.77	0.73	0.68	0.63	0.50	0.37

Source: Calculated from Hodgkin and Rushton, 1946.

The membrane time constant of a spherical cell or of a fiber ($\tau_m = R_m C_m$) is independent of cell or fiber size. This is because an increase in radius (and hence in membrane surface area) causes not only an increase in capacitance, but also a corresponding decrease in resistance, so the product remains constant. Because C_m has been found to be approximately the same (1 $\mu F/cm^2$) for all nerve and muscle membranes, τ provides a convenient measure for the specific membrane resistance of a cell. The time constant is the third parameter specifying the behavior of an axon, the other two being the input resistance and the length constant. Time constants in nerve and muscle cells typically range from 1 to 20 ms.

Capacitance in a Cable

How does the time constant affect current flow in a cable? As with the simple *RC* circuit (see Figure 7.2C), the rise and fall of the potential change produced by a rectangular current pulse are slowed by the presence of the capacitance. The effects are more complicated, however, because current no longer flows into a single capacitor; instead each segment of the circuit with its capacitative and resistive elements interacts with the others. Because of this interaction, the rising and falling phases of the potential changes are not exponential, and the growth and decline of the potentials become increasingly prolonged as records are made farther and farther from the point of current injection (Figure 7.3). Because the rate of rise depends on the distance between the recording electrode and the point of current injection, the membrane time constant cannot be obtained by measuring the time to rise to 63% of its final value, except in the unique case when the separation is one length constant (Box 7.1).

Again, the effect of membrane capacitance can be explained in terms of ion movements. When positive current is injected into the axon, intracellular ions (mostly potassium) on the inside spread longitudinally along the fiber. Some ions accumulate to change the charge on the membrane capacitance, and some flow out through the membrane resistance. (Negatively charged chloride ions are, at the same time, moving in the opposite direction.) Eventually the membrane potential reaches a new steady state with the distributed capacitances fully charged to their new potential and a steady ionic current through the membrane. The time required to reach the steady state is determined by the membrane time constant.

A further consequence of the membrane capacitance is that brief signals do not spread as far as signals of long duration. For sufficiently long pulses, when the potential can reach a steady state, the capacitance is fully charged and the spatial distribution of the potential is determined by the resistances of the membrane and cytoplasm; that is, $V_x = V_0 e^{-x/\lambda}$, as described earlier. However, for brief events, such as synaptic potentials, the current flow giving rise to the signal may end before the membrane capacitances become fully charged. This has the effect of reducing the spread of the potential along the fiber. In other words, for brief signals the effective length constant is less than for those of longer duration. In addition, such signals are distorted as they spread along the fiber, the peaks becoming more rounded and occurring progressively later with increasing distance.

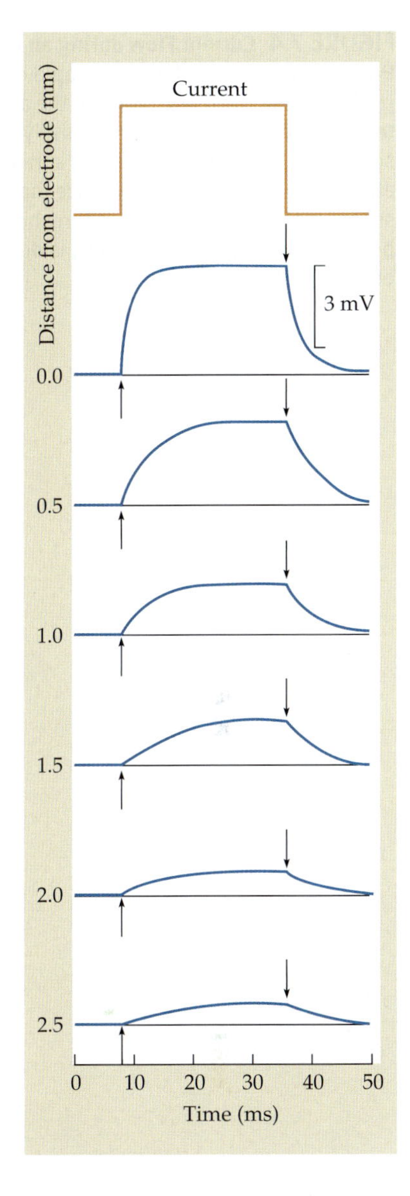

FIGURE 7.3 Spread of Potential along a lobster axon, recorded with a surface electrode. A rectangular current pulse is applied at 0 mm, producing an electrotonic potential. With increasing distance from the site of current injection, the rise time of the potential change is slowed and the height of the plateau attenuated. (After Hodgkin and Rushton, 1946.)

PROPAGATION OF ACTION POTENTIALS

Action potential propagation along a nerve fiber depends on the passive spread of current ahead of the active region to depolarize the next segment of membrane to threshold. To illustrate the nature of the current flow involved in impulse generation and propagation, we can imagine the action potential frozen at an instant in time and plot its spatial distribution along the axon as shown in Figure 7.4. The distance occupied depends on its duration and conduction velocity. For example, if the action potential duration is 2 ms and it is conducted at 10 m/s (10 mm/ms), then the potential will be spread over a 20 mm length of axon (almost an inch). Near the leading edge of the action potential, where the membrane potential has reached threshold, there is a rapid influx of sodium ions along their electrochemical gradient, depolarizing the cell membrane. Just as when current is injected through a microelectrode, the inward current spreads longitudinally in the fiber away from the active region. Ahead of the active region this current depolarizes a new seg-

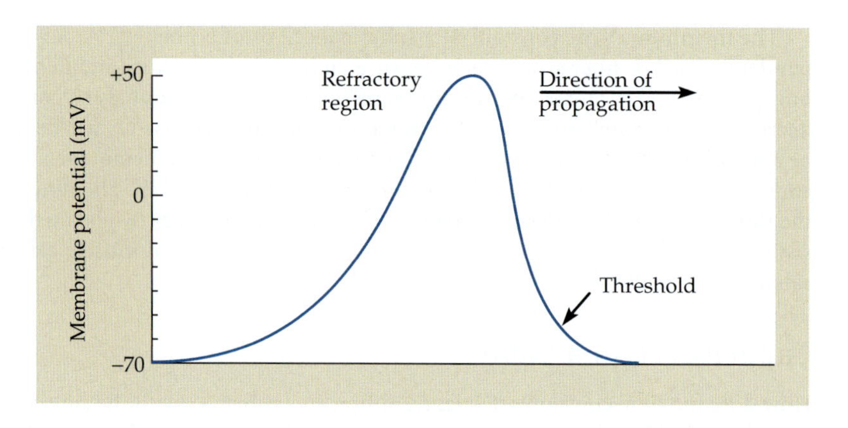

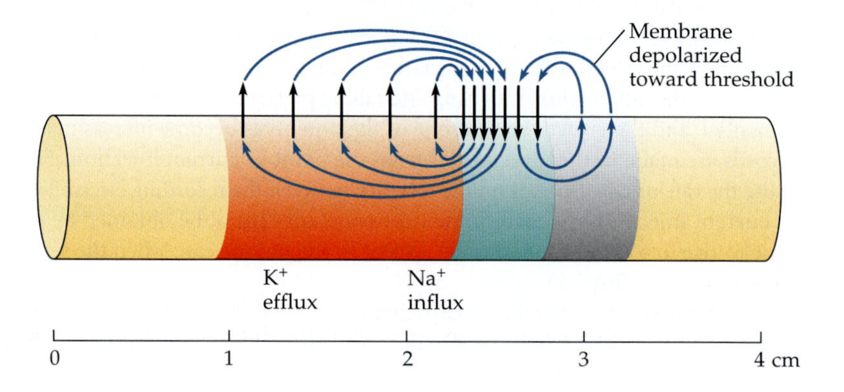

FIGURE 7.4 Current Flow during an Action Potential at an instant in time. Rapid depolarization during the rising phase of the action potential is due to the influx of positively charged sodium ions. The positive current spreads ahead of the impulse to depolarize the adjacent segment of membrane toward threshold. Repolarization on the falling phase is due to the efflux of potassium ions.

ment of membrane toward threshold. Behind the peak of the action potential the potassium conductance is high and current flows out through potassium channels, restoring the membrane potential toward its resting level.

Normally, impulses arise at one end of an axon and travel to the other. However, there is no inherent directionality to propagation. Impulses produced at a neuromuscular junction in the middle of a muscle fiber travel away from the junction in both directions toward the tendons. Except in unusual circumstances, however, an action potential, once initiated, cannot double back on itself, reversing its direction of propagation. This is because of the refractory period. In the refractory region, indicated in Figure 7.4, the sodium conductance is largely inactivated and the potassium conductance is high, so a backward-conducting regenerative response cannot occur. As the action potential leaves the region, the membrane potential returns to its resting value, sodium channel inactivation is removed, potassium conductance returns to normal, and excitability recovers.

Conduction Velocity

The conduction velocity of the action potential depends on how quickly and how far ahead of the active region the membrane capacitance is discharged to threshold by the spread of positive charge. This, in turn, depends on the amount of current generated in the active region and on the cable properties of the fiber. If the membrane time constant, $\tau_m = R_m C_m$, is small, the membrane will depolarize to threshold quickly, speeding conduction. If the length constant, $\lambda = (r_m/r_i)^{1/2}$, is large, the depolarizing current will spread a correspondingly large distance ahead of the active region, and the conduction velocity will be high.

How do these factors relate to fiber size? As already noted, the time constant is independent of size. The length constant, on the other hand, increases with the square root of the fiber diameter. Large fibers, then, conduct more rapidly than small fibers. A more detailed theoretical approach indicates that in unmyelinated fibers, such as squid axons, the velocity of propagation should vary directly with the square root of the fiber diameter.[4]

[4]Hodgkin, A. L. 1954. *J. Physiol.* 125: 221–224.

Myelinated Nerves and Saltatory Conduction

In the vertebrate nervous system the larger nerve fibers are myelinated. Myelin is formed in the periphery by Schwann cells and in the CNS by oligodendrocytes (Chapter 8). The cells wrap themselves tightly around axons, and with each wrap the cytoplasm between the membrane pair is squeezed out so that the result is a spiral of tightly packed membranes (Chapter 8). The number of wrappings (lamellae) ranges from a low of between 10 and 20 to a maximum of about 160.[5] A wrapping of 160 lamellae means that there are 320 membranes in series between the plasma membrane of the axon and the extracellular fluid. Thus, the effective membrane resistance is increased by a factor of 320, and the membrane capacitance is reduced to the same extent. In terms of dimensions, myelin usually occupies 20 to 40% of the overall diameter of the fiber. The myelin sheath is interrupted periodically by nodes of Ranvier, exposing patches of axonal membrane. The internodal distance is usually about 100 times the external diameter of the fiber, and it ranges from 200 μm to 2 mm.

The effect of the myelin sheath is to restrict membrane current flow largely to the node because ions cannot flow easily into or out of the high-resistance internodal region and the internodal capacitative currents are very small as well. As a result, excitation jumps from node to node, thereby greatly increasing the conduction velocity. Such impulse propagation is called **saltatory conduction** (from the Latin *saltare*, "to jump, leap, or dance"). Saltatory conduction does not mean that the action potential occurs in only one node at time. While excitation is jumping from one node to the next on the leading edge of the action potential, many nodes behind are still active. Myelinated axons not only conduct more rapidly than unmyelinated ones but also are capable of firing at higher frequencies for more prolonged periods of time. These capabilities may be related to an additional consequence of myelination—namely, that during impulse propagation fewer sodium and potassium ions enter and leave the axon because regenerative activity is restricted to the nodes. Consequently, less metabolic energy is required by the nerve cell to maintain the appropriate intracellular ion concentrations.

Experiments that demonstrated saltatory conduction were first made in 1941 by Tasaki[6] and later by Huxley and Stampfli,[7] who recorded current flow at nodes and internodes. Such an experiment on a single myelinated axon is illustrated in Figure 7.5. The nerve is placed in three saline pools, the central pool being narrow and separated from the others by air gaps of very high resistance. Electrically, the pools are connected by the external recording circuitry as shown, so that during impulse propagation currents that would otherwise be interrupted by the air gap flow instead into or out of the central pool through the resistor (R). The voltage drop across the resistor provides a measure of the magnitude and direction of the currents.

In the first experiment (Figure 7.5A) the central pool contains a node of Ranvier. Upon stimulation of the nerve, current first flows outward through the node and back toward the region of oncoming excitation (upward deflection) as the node is depolarized to threshold. This is followed by inward current at the node (downward deflection) when threshold is reached and an action potential is generated. When the central pool contains an internode (Figure 7.5B), there is no inward current, but only two small peaks of capacitative and resistive current flowing from the central pool toward the regions of excitation as the impulse first approaches in compartment 1 and then travels onward in compartment 3. Experiments such as this confirmed that there is no inward current, and hence no regenerative activity, in the internodal region.

Sophisticated techniques for recording saltatory conduction in undissected mammalian axons in situ have been developed by Bostock and Sears.[8] With such techniques it is possible to measure both inward currents at the nodes and longitudinal internodal currents and thereby to estimate accurately the positions of the nodes and distances between them.

Conduction Velocity in Myelinated Fibers

Conduction velocities of myelinated fibers vary from a few meters per second to more than 100 m/s. The world speed record is held by myelinated axons of the shrimp, which conduct at speeds in excess of 200 m/s (447 miles/h).[9] In the vertebrate nervous system, peripheral nerves have been classified into groups according to conduction velocity[10] and function

[5]Arbuthnott, E. R., Boyd, I. A., and Kalu, K. U. 1980. *J. Physiol.* 308: 125–157.

[6]Tasaki, I. 1959. In *Handbook of Physiology*, Section 1, Vol. 1, Chapter 3. American Physiological Society, Bethesda, MD, pp. 75–121.

[7]Huxley, A. F., and Stampfli, R. 1949. *J. Physiol.* 108: 315–339.

[8]Bostock, H., and Sears, T. A. 1978. *J. Physiol.* 280: 273–301.

[9]Xu, K., and Terakawa, S. 1999. *J. Exp. Biol.* 202: 1979–1989.

[10]Gasser, H. S., and Erlanger, J. 1927. *Am. J. Physiol.* 80: 522–547.

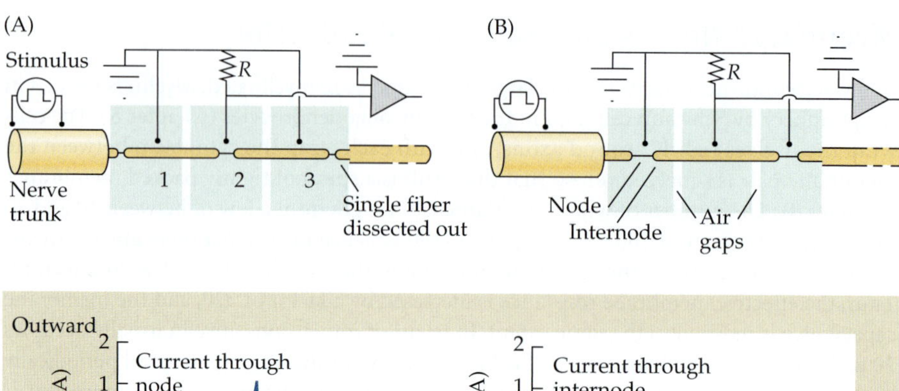

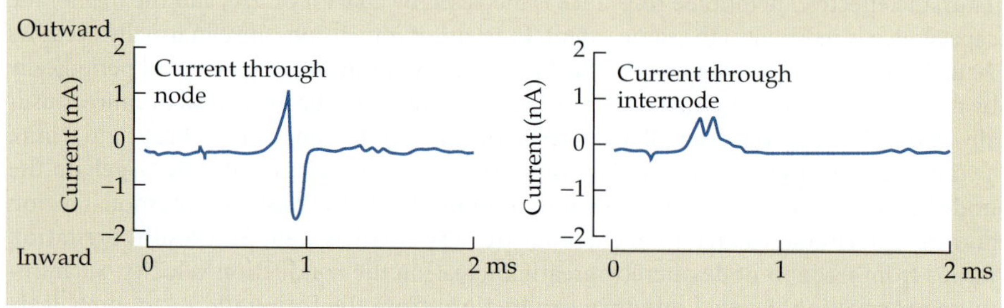

FIGURE 7.5 Current Flow through a Myelinated Axon. A single myelinated axon passes through two air gaps that create three compartments not linked by extracellular fluid. During the propagated action potential, currents into and out of the center compartment (2) flow through the resistor (*R*); the voltage drop across the resistor provides a measure of the current. (A) A node of Ranvier is in compartment 2. Initially, as the action potential approaches and the node is being depolarized, current flows through the resistor from compartment 2 to compartment 1 (upward deflection); when threshold is reached at the node, a large inward flux follows and the current is reversed. (B) An internode is in the center compartment and there is only outward current flow from the compartment, with no inward current, as the action potential first approaches and then leaves the internodal segment. (After Tasaki, 1959.)

(Box 7.2). Theoretical calculations suggest that in myelinated fibers, conduction velocity should be proportional to the diameter of the fiber.[11] Boyd and his colleagues have shown that in mammals, large myelinated fibers (greater than 11 μm in diameter) have a conduction velocity in meters per second that is equal to approximately six times their outside diameter in micrometers; for smaller fibers the constant of proportionality is about 4.5.[12]

One theoretical point of interest is the best thickness for the myelin sheath for optimal conduction velocity, given a particular outer diameter. Obviously the increase in membrane resistance in the myelinated region will be greater with a thick sheath than with a thin one. On the other hand, as the myelin thickness increases, the cross-sectional area of the axoplasm must decrease, thereby increasing internal longitudinal resistance. The first effect would be expected to increase conduction velocity, the second to decrease it. It turns out that the optimal compromise between these opposing effects is achieved when the axon diameter is about 0.7 times the overall fiber diameter. As already noted, the observed ratio in mammalian peripheral nerve ranges between 0.6 and 0.8.

The calculated optimum internodal length for conduction is also approximately that found in reality—namely, about 100 times the external fiber diameter. Greater internodal distances would allow the excitation to spread farther, tending to increase conduction velocity. On the other hand, as the internodal distance increases, current flow from one node to the next decreases because of the greater longitudinal resistance. As a result, the nodal depolarization produced by activity in a preceding node would be smaller and rise more slowly, tending to decrease conduction velocity. Because of these opposing factors, modest variations in internodal length have little effect on conduction velocity. With a very large internodal distance, of course, depolarization from activity in a preceding node would no longer reach threshold, and conduction would be blocked.

[11] Rushton, W. A. H. 1951. *J. Physiol.* 115: 101–122.

[12] Arbuthnott, E. R., Boyd, I. A., and Kalu, K. U. 1980. *J. Physiol.* 308: 125–157.

Box 7.2 Classification of Nerve Fibers in Vertebrates

If we stimulate a peripheral nerve electrically at one end and record from it some distance away, the resulting record has a series of peaks. The peaks occur because of the dispersion of nerve impulses that travel at different velocities and therefore arrive at the recording electrode at different times after the stimulus. For example, a record taken from a rat sciatic nerve with 50 mm between the stimulating and recording electrodes might look like the figure shown here (the rapid deflection at the beginning is an artifact due to current spread from the stimulating electrode).

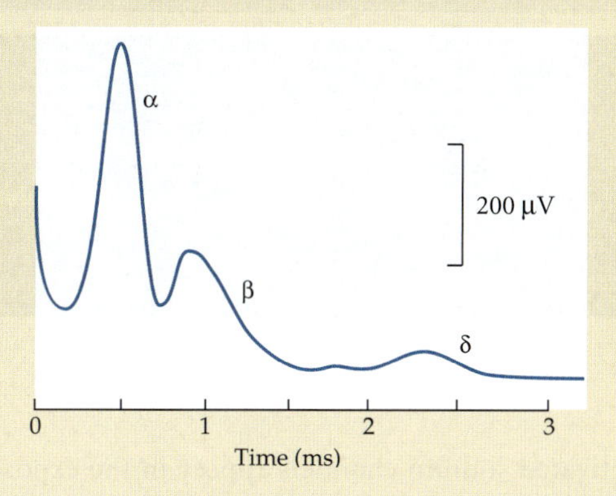

Vertebrate nerve fibers were classified into groups on the basis of differences in conduction velocity, combined with differences in function. Unfortunately, two such classifications developed simultaneously. In the first system, group A refers to myelinated fibers in peripheral nerve; in mammals, these conduct at velocities ranging from 5 to 120 m/s. Group A fibers were further subdivided according to conduction velocity into α (80–120 m/s), β (30–80 m/s), and δ (5–30 m/s). These conduction velocity peaks are indicated in the record shown here. Group B consists of myelinated fibers in the autonomic nervous system, which have conduction velocities in the lower part of the A-fiber range. Group C contains unmyelinated fibers, which conduct very slowly (less than 2 m/s). The term "γ fibers" is reserved for motor nerves supplying muscle spindles (Chapter 17), which have conduction velocities that span the β and lower part of the α range.

The second nomenclature applies to sensory fibers arising in muscle: Group I, corresponding to Aα; Group II (Aβ); Group III (Aδ). Group I afferent fibers were further classified into two separate groups depending on whether they conveyed information from muscle spindles (Ia) or from sensory receptors in tendons (Ib).

Distribution of Channels in Myelinated Fibers

In myelinated fibers, voltage-sensitive sodium channels are highly concentrated in the nodes of Ranvier, with potassium channels more concentrated under the paranodal sheath.[13] The properties of the axon membrane in the paranodal region normally covered by myelin were first examined by Ritchie and his colleagues.[14] To do this, the myelin was loosened by enzyme treatment or osmotic shock. Voltage clamp studies were then made of currents in the region of the node and compared with those obtained before the treatment. Such experiments showed that in rabbit nerve, nodes of Ranvier normally display only inward sodium current upon excitation. Repolarization occurs not through an increase in potassium conductance (as in other cells considered so far) but instead by rapid sodium inactivation and current flow through a relatively large resting conductance. When the axon membrane adjacent to the nodes (the paranodal region) was exposed, excitation then produced a delayed outward potassium current, with no increase in inward current, indicating that the newly exposed membrane contained delayed rectifier channels but not sodium channels. Later immunocytochemical studies confirmed that in rat myelinated nerve, voltage-sensitive potassium channels were confined to the paranodal region (Figure 7.6A).[15,16] It is interesting that the nodal regions of *Xenopus* axons contain *sodium-activated* potassium channels.[17] Such channels might be activated by sodium influx during the rising phase of the action potential and thereby speed repolarization.

Channels in Demyelinated Axons

Mammalian axons that have been demyelinated chronically by exposure to diphtheria toxin can develop continuous conduction through a demyelinated region,[7] implying that

[13]Vabnick, I., and Shrager, P. 1998. *J. Neurobiol.* 37: 80–96.

[14]Chiu, S. Y., and Ritchie, J. M. 1981. *J. Physiol.* 313: 415–437.

[15]Wang, H., et al. 1993. *Nature* 365: 75–79.

[16]Rasband, M. N., et al. 1998. *J. Neurosci.* 18: 36–47.

[17]Koh, D. S., Jonas, P., and Vogel, W. 1994. *J. Physiol.* 479(Pt.2): 183–197.

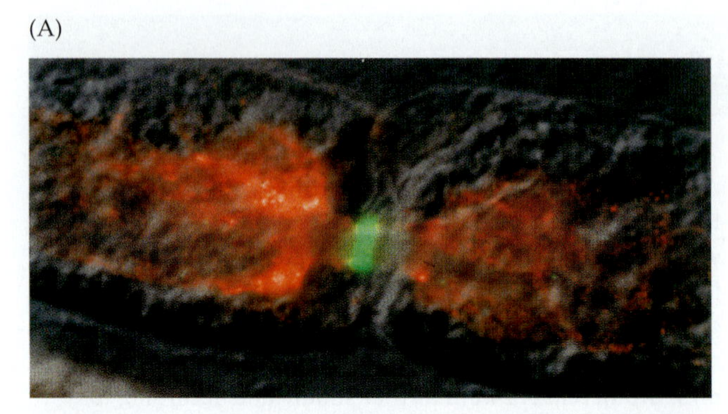

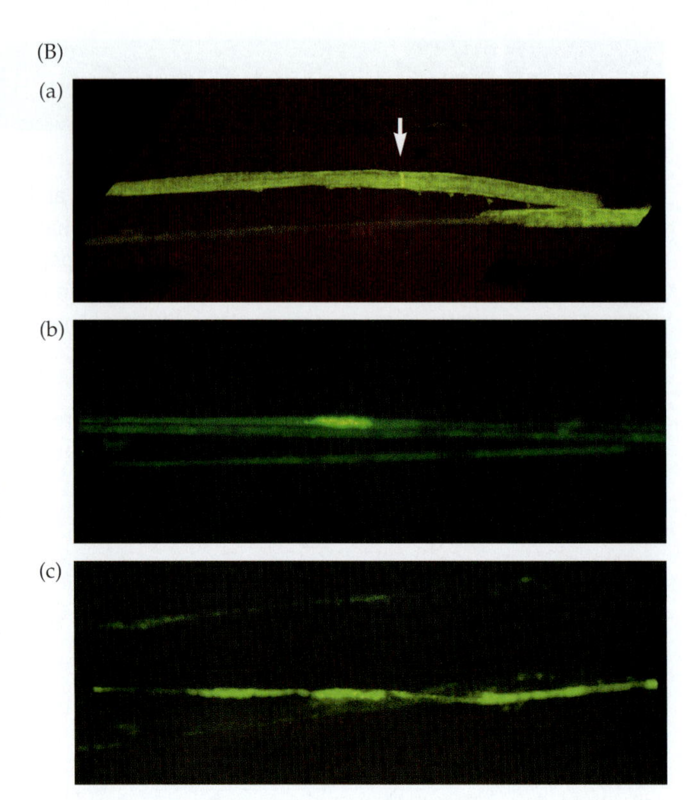

FIGURE 7.6 Distribution of Sodium and Potassium Channels in myelinated axons. (A) In rat sciatic nerve, sodium channels (green) are tightly clustered in the node of Ranvier, and potassium channels (red) are sequestered in the paranodal region. Note the sharp decrease in axon diameter within the node. (B) Disruption of sodium channel distribution after demyelination of goldfish lateral-line nerve. In the myelinated axon (a), sodium channels are concentrated at the node (arrow). Fourteen days after the beginning of demyelination (b), sodium channels appear in irregular patches. At 21 days (c), more patches have appeared, distributed along the length of the nerve. (A after Rasband et al. 1998, kindly provided by P. Shrager; B after England et al. 1996, kindly provided by S. R. Levinson.)

after demyelination, voltage-activated sodium channels appear in the exposed axon membrane. In other experiments, labeling of demyelinated nerves with antibodies to sodium channels shows that channels disappear from clusters in the former nodal regions and that new channels are distributed along previously myelinated regions (Figure 7.6B).[18] Voltage-activated potassium channels are redistributed as well.[14,19] Upon remyelination the normal clustering of sodium and potassium channels in newly formed nodes and paranodal regions is restored.

Geometry and Conduction Block

The simple uniform cable is an idealized structure resembling an unmyelinated axon, but not a neuron in its entirety, with its cell body, elaborate dendritic arborization, and numerous axonal branches. The complex geometry of the neuron provides many possibilities for block of action potential propagation. Specifically, propagation may fail wherever there is an abrupt expansion of membrane area. In such a situation the active membrane may not be able to provide enough current to depolarize the larger membrane area to threshold. For example, where an axon divides into two branches, the active segment of the single axon must contribute sufficient current to activate both branches. Under normal circumstances an impulse will usually be unimpeded, but after repeated firing, block may occur at the branch point. Other factors contribute to such block: In leech sensory cells, block can occur because of persistent hyperpolarization induced by increased electrogenic activity of the sodium pump (Chapter 15), and because of long-lasting increases in potassium permeability, both of which increase the amount of current required for depolarization to threshold.[20–22]

In myelinated peripheral nerve the safety factor for conduction is about 5; that is, the depolarization produced at a node by excitation of a preceding node is approximately five times larger than necessary to reach threshold. Again, this safety factor is reduced where branching occurs. Similarly, when the myelin sheath terminates—for example, near the end of a motor nerve—the current from the last node is then distributed over a large area

[18]England, J. D., Levinson, S. R., and Shrager, P. 1996. *Microsc. Res. Tech.* 34: 445–451.

[19]Bostock, H., Sears, T. A., and Sherratt, R. M. 1981. *J. Physiol.* 313: 301–315.

[20]Yau, K. W. 1976. *J. Physiol.* 263: 513–538.

[21]Gu, X. N., Macagno, E. R., and Muller, K. J. 1989. *J. Neurobiol.* 20: 422–434.

[22]Baccus, S. A. 1998. *Proc. Natl. Acad. Sci. USA* 95: 8345–8350.

Box 7.3 Stimulating and Recording with External Electrodes

For much physiological work in the central and peripheral nervous systems, extracellular electrodes are used to stimulate or to record from axons of various diameters. When two extracellular electrodes are used to stimulate a nerve trunk (A in the figure), much of the current is diverted through the extracellular fluid; the remainder enters individual axons under the positive electrode, flows along the axoplasm, and exits under the negative electrode. Corresponding electrical circuits are shown in B and C. Current enters and leaves the axon through the input resistances, and between the regions of entry and exit it flows along an internal longitudinal resistance. Under the positive electrode there is a voltage drop across the membrane, producing a local hyperpolarization (the outside is made more positive). Current flow toward the negative electrode produces an additional voltage gradient along the core of the axon. Current leaving the axon under the negative electrode causes a local depolarization. These three voltage gradients must sum to equal the total voltage drop in the extracellular solution between the electrodes.

The voltage that must be applied to bring the membrane under the negative electrode to threshold depends on fiber diameter: Large fibers require less stimulating voltage than small fibers, because of geometrical factors. As fiber size increases, the internal longitudinal resistance (which varies inversely with the square of the diameter) decreases more rapidly than the input resistance (which varies inversely with the $3/2$ power of the diameter). As a result, less of the applied voltage is dissipated along the core of the fiber, and a greater voltage drop is produced across the membrane. B and C illustrate this principle numerically. The diagrams represent the pathways for current flow through two unmyelinated fibers, between a pair of stimulating electrodes 2 cm apart. Plausible values for input resistances and longitudinal resistance for a 40 μm fiber are given in part B. A potential of 500 mV between the electrodes produces a depolarization of 23 mV. For a fiber with half the diameter (20 μm) (part C) the input resistances are increased by a factor of 2.8, the longitudinal resistance is increased by a factor of 4, and the depolarization is reduced to 16 mV.

Extracellular electrodes can also be used to record action potential activity in nerve trunks. This is possible because during action potential propagation, longitudinal currents in the surrounding extracellular fluid create potential gradients along the nerve fiber. Because of their larger membrane area and lower core resistance, larger fibers generate more current, and hence larger extracellular gradients, than smaller fibers. As a result, they produce larger signals at the recording electrodes.

The relation between size and threshold for stimulation by external electrodes is fortunate for physiological and clinical purposes. For example, threshold and conduction velocity can be tested in motor nerves, which are relatively large, without exciting pain fibers, which are very much smaller. Just as the largest fibers are the easiest to stimulate, they are often the most difficult to block—for example, by cooling or by local anesthetics. Again, this means that pain fibers can be blocked with anesthetic without interfering with conduction in larger sensory and motor fibers. But this relation between size and block of conduction does not always hold: Because of additional geometrical effects, block by localized pressure affects large axons first, then smaller axons as the pressure is increased.

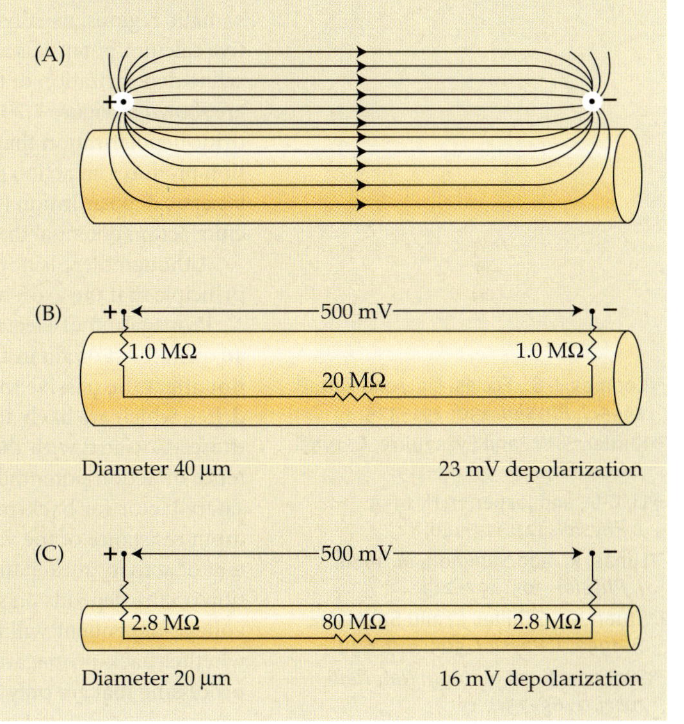

of unmyelinated nerve terminal membrane and, as a consequence, provides less overall depolarization than would occur at a node. It is perhaps for this reason that the last few internodes before an unmyelinated terminal are shorter than normal: so that more nodes can contribute to depolarization of the terminal.[23]

[23]Quick, D. C., Kennedy, W. R., and Donaldson, L. 1979. *Neuroscience* 4: 1089–1096.

CONDUCTION IN DENDRITES

Apart from considerations of geometry, some regions of the neuron have a lower threshold for action potential initiation than others. This was first observed in spinal motoneurons by J. C. Eccles and his colleagues.[24] They found that upon depolarization action potentials were initiated first in the initial segment of the axon (between the cell body and the first internode) and then propagated both outward along the axon and back into the soma (cell body) and dendrites of the cell. At about the same time, Kuffler and Eyzaguirre found that depolarization of the dendrites in the crayfish stretch receptor initiated action potentials in or near the cell body, rather than in the dendrites themselves.[25] Observations of this kind led to the idea that dendrites were generally unexcitable and served only to transmit signals passively from dendritic synapses to the initial segment of the axon. This idea arose in spite of numerous observations to the contrary. For example, earlier extracellular recordings of electrical activity within the mammalian motor cortex by Li and Jasper gave clear indication of action potentials traveling upward along pyramidal cell dendrites from their cell bodies to the cortical surface, with a conduction velocity of about 3 m/s.[26]

Dendritic action potentials are now known to occur in a variety of neurons, mediated by regenerative sodium and calcium currents. Cerebellar Purkinje cells, in addition to producing sodium action potentials in their somatic regions, generate calcium action potentials in their dendrites.[27] As shown in Figure 7.7A, calcium action potentials generated in a dendrite spread effectively into the soma. Somatic action potentials, on the other hand, are not propagated into the dendrites, but spread passively a short distance into the dendritic tree.

Like Purkinje cells, cortical pyramidal cells exhibit sodium action potentials in their somatic regions, usually arising in the initial segment of the axon. In addition, regenerative calcium potentials are observed in the distal dendrite.[28,29] Responses of a pyramidal cell to depolarization of the distal dendrite by activation of excitatory synapses (Chapter 9) are shown in Figure 7.7B. Modest synaptic activation (part a of Figure 7.7B) produces dendritic depolarization that is attenuated as it spreads passively to the soma. The depolarization produces an action potential in the soma that spreads back into the dendrite. Stronger synaptic depolarization (part b of Figure 7.7B) results in direct activation of a dendritic calcium action potential that precedes the action potential generated in the soma.

Although there is now ample evidence of regenerative activity in dendrites, the general principle that the axon hillock is usually the most excitable region of the cell still holds.[30]

Propagation of electrical signals in a dendrite is clearly much more complex than in an axon. First of all, in the axon the assumption that subthreshold potential changes do not affect the passive membrane properties is more or less reasonable. Not so in dendrites, which are likely to contain a variety of voltage-dependent channels in addition to those associated with the action potential. Still more complexity is added by the coexistence of action potentials and synaptic potentials in the dendritic tree. For example, the safety factor for back-propagation of action potentials from the soma depends on the input resistance of the various branches; the input resistances, in turn, depend on the extent of activity at excitatory and inhibitory synapses. Thus, whether or not back-propagation occurs depends on synaptic activity.[31–33] At the same time, synaptic channels that are voltage-dependent will behave differently from one moment to the next, depending on whether back-propagation has occurred.[34] These factors add new dimensions to signal processing that are only beginning be understood.

[24]Coombs, J. S., Eccles, J. C., and Fatt, P. 1955. *J. Physiol.* 130: 291–325.

[25]Kuffler, S. W., and Eyzaguirre, C. 1955. *J. Gen. Physiol.* 39: 87–119.

[26]Li, C-L., and Jasper, H. H. 1953. *J. Physiol.* 121: 117–140.

[27]Llinás, R., and Sugimori, M. 1980. *J. Physiol.* 305: 197–213.

[28]Stuart, G., Schiller, J., and Sakmann, B. 1997. *J. Physiol.* 505: 617–632.

[29]Svoboda, K., et al. 1999. *Nat. Neurosci.* 2: 65–73.

[30]Stuart, G., et al. 1997. *Trends Neurosci.* 20: 125–131.

[31]Tsubokawa, H., and Ross, W. N. 1996. *J. Neurophysiol.* 76: 2896–2906.

[32]Sandler, V. M., and Ross, W. N. 1999. *J. Neurophysiol.* 81: 216–224.

[33]Larkum, M. E., Zhu, J. J., and Sakmann, B. 1999. *Nature* 398: 338–341

[34]Markram, H., et al. 1997. *Science* 275: 213–215.

PATHWAYS FOR CURRENT FLOW BETWEEN CELLS

In most circumstances electrical signals cannot pass directly from one cell to the next. Certain cells, however, are **electrically coupled**. The properties and functional role of electrical synapses are discussed in Chapter 9. Here we describe special intercellular structures that are required for electrical continuity between cells. The necessity for such specialized structures to mediate the spread of current from one cell to the next is shown in Box 7.4.

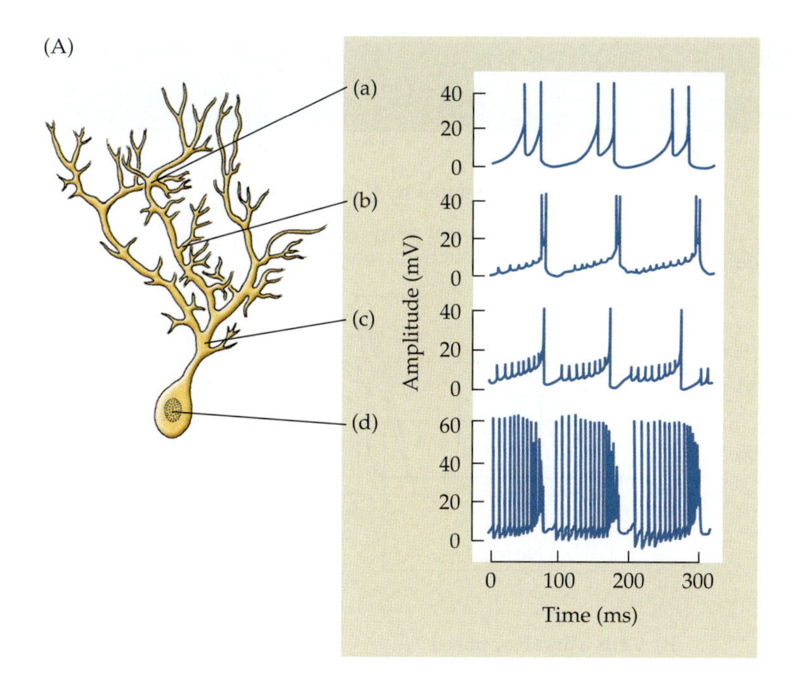

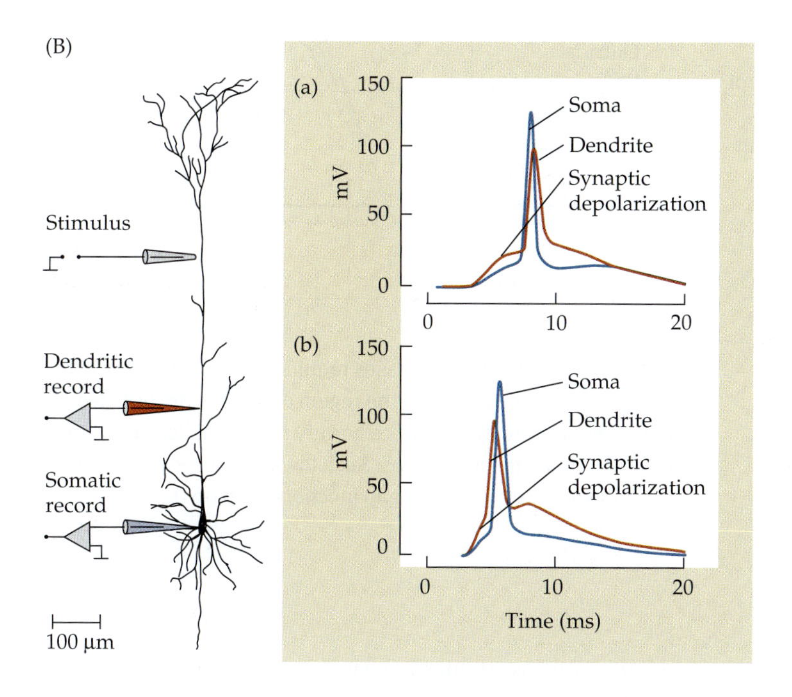

FIGURE 7.7 Spread of Action Potentials in Dendrites. (A) Records from a cerebellar Purkinje cell, obtained by impaling the cell at the indicated locations and passing a depolarizing current through the electrode. Near the end of the dendritic tree (a), depolarization produces long-duration calcium action potentials. In the cell soma (d), a steady depolarizing current produces high-frequency sodium action potentials, interrupted periodically by calcium action potentials. At intermediate locations (b and c), depolarization produces calcium action potentials in the dendrite. Accompanying sodium action potentials, generated in the soma, spread passively into the dendritic tree and die out after a short distance. (B) Conduction in a cortical pyramidal cell. The cortical cell dendrite is depolarized by activating distal excitatory synapses. (a) Moderate depolarization of the dendrite is attenuated as it spreads to the soma, where it initiates an action potential. The action potential then spreads back into the dendrite. (b) Larger depolarization produces a calcium action potential in the dendrite that precedes action potential initiation in the soma. (A from Llinás and Sugimori, 1980; B after Stuart, Schiller, and Sakmann, 1997.)

Structural Basis for Electrical Coupling: The Gap Junction

At sites of electrical coupling the intercellular current flows through **gap junctions**.[35] The gap junction is a region of close apposition of two cells characterized by aggregates of particles distributed in hexagonal arrays in each of the adjoining membranes (Figure 7.8). Each particle, called a **connexon**, is made up of six protein subunits arranged in a circle about 10 nm in diameter, with a 2 nm diameter central core.[36,37] Identical particles in the apposing cells are exactly paired to span the 2 or 3 nm gap in the region of contact. The core provides the pathway for the flow of small ions and molecules between cells. The conductance of an individual channel connecting adjacent cells (i.e., two connexons in series) is on the order of 100 pS.[38]

[35]Loewenstein, W. 1981. *Physiol. Rev.* 61: 829–913.

[36]Goodenough, D. A., Goliger, J. A., and Paul, D. 1996. *Annu. Rev. Biochem.* 65: 475–502.

[37]Perkins, G., Goodenough, D., and Sosinsky, G. 1997. *Biophys. J.* 72: 533–544.

[38]Neyton, J., and Trautmann, A. 1985. *Nature* 317: 331–335.

Box 7.4 Current Flow between Cells

The figure shown here indicates pathways for current flow between the ends of two cylindrical neuronal processes, labeled "a" and "b" in the diagrams. In the first example (A), the ends are separated by a gap (as occurs, for example, at chemical synapses). Current passed into one of the processes (a) exits along the length of the process and out the end. How is the current flow distributed quantitatively? This can be determined by considering the resistances involved. The cylindrical conductor is a cable extending in one direction only. Its input resistance is:

$$r_{input} = \left(\frac{R_m R_i}{2\pi^2 a^3}\right)^{1/2}$$

The resistance of the circular end is

$$r_e = \frac{R_m}{\pi a^2}$$

Suppose that for both cells $R_m = 2000\ \Omega cm^2$, $R_i = 100\ \Omega cm$, and $a = 10\ \mu m$ (all plausible values). Then r_{input} will be 3.2 MΩ, and r_e 637 MΩ. Because the end resistance is 200 times larger than the input resistance, only 0.5% of the current injected into the cylinder will flow outward through the end. Further, all of the current leaving the end will flow laterally out of the cleft, which has a low resistance, rather than entering the end of the second process (b).

What if we butt the two ends of the processes tightly together, as shown in B? The resistance to outward current flow along process a will be unchanged. The resistance to current flow into process b will be 1274 MΩ (r_c, the resistance of the two end membranes in series) plus 3.2 MΩ (the input resistance of process b). As a result, current flowing through the two ends and into process b is 0.025% of the total. Accordingly, an applied current of 10 nA would depolarize cell a by 31.7 mV, cell b by only about 79 µV. Clearly, significant coupling between the two processes requires not only that current be prevented from escaping the region of apposition, but also that the intercellular resistance be very much smaller than that normally found in membranes. Such low-resistance intercellular pathways are created by gap junctions (see text).

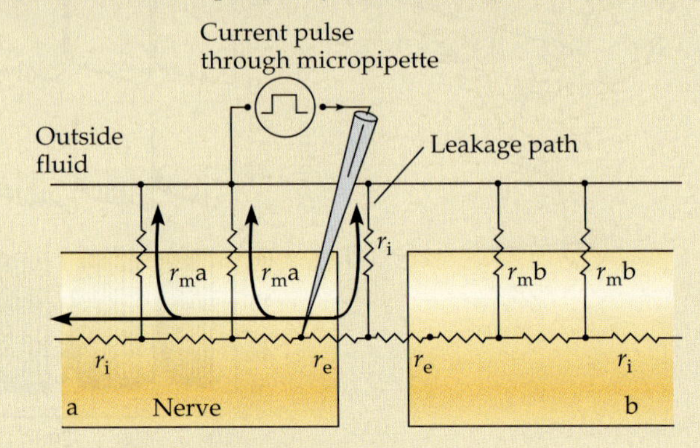

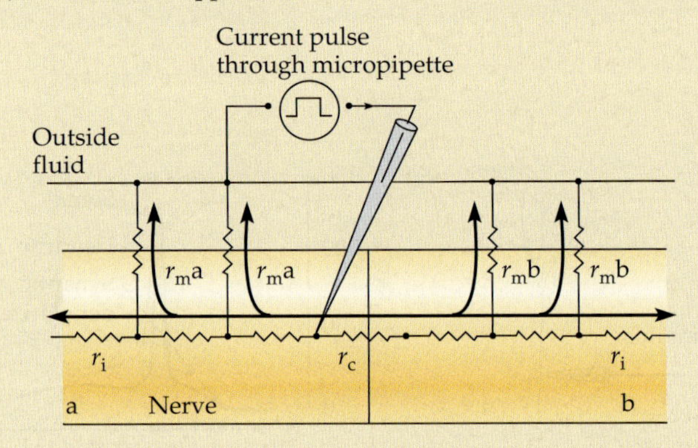

A variety of connexon subunit proteins (**connexins**) have been isolated and sequenced, ranging in mass from 26 to 56 kDa.[36,39] Each is named according to its deduced mass—for example, connexin32 (32 kDa) is found in rat liver, connexin43 (43 kDa) in heart muscle, and so on. Hydropathy plots (Chapter 3) suggest that the connexins contain four membrane-spanning helices. Antibody-binding studies are consistent with this model and indicate that the amino terminus (and hence the carboxy terminus as well) resides in the cytoplasm. Only one, or perhaps a few, specific connexins form gap junctions in any given tissue, but functional intercellular coupling can occur between connexins of different types—for example, when message for connexin32 is injected into one cell and for connexin43 in the other.[40] Gap junctions can be formed artificially by the injection of mRNA coding for connexin into pairs of apposed *Xenopus* oocytes.[41] A major unsolved problem is how connexons in the juxtaposed cells align to make a channel without, at the same time, forming pores between the cytoplasm and the extracellular solution.

[39] Larsen, W. J., and Veenstra, R. D. 1998. In *Cell Physiology Source Book*, 2nd Ed. Academic Press, New York, pp. 467–480.

[40] Swensen, K. I., et al. 1989. *Cell* 57: 145–155.

[41] Werner, R., et al. 1985. *J. Membr. Biol.* 87: 253–268.

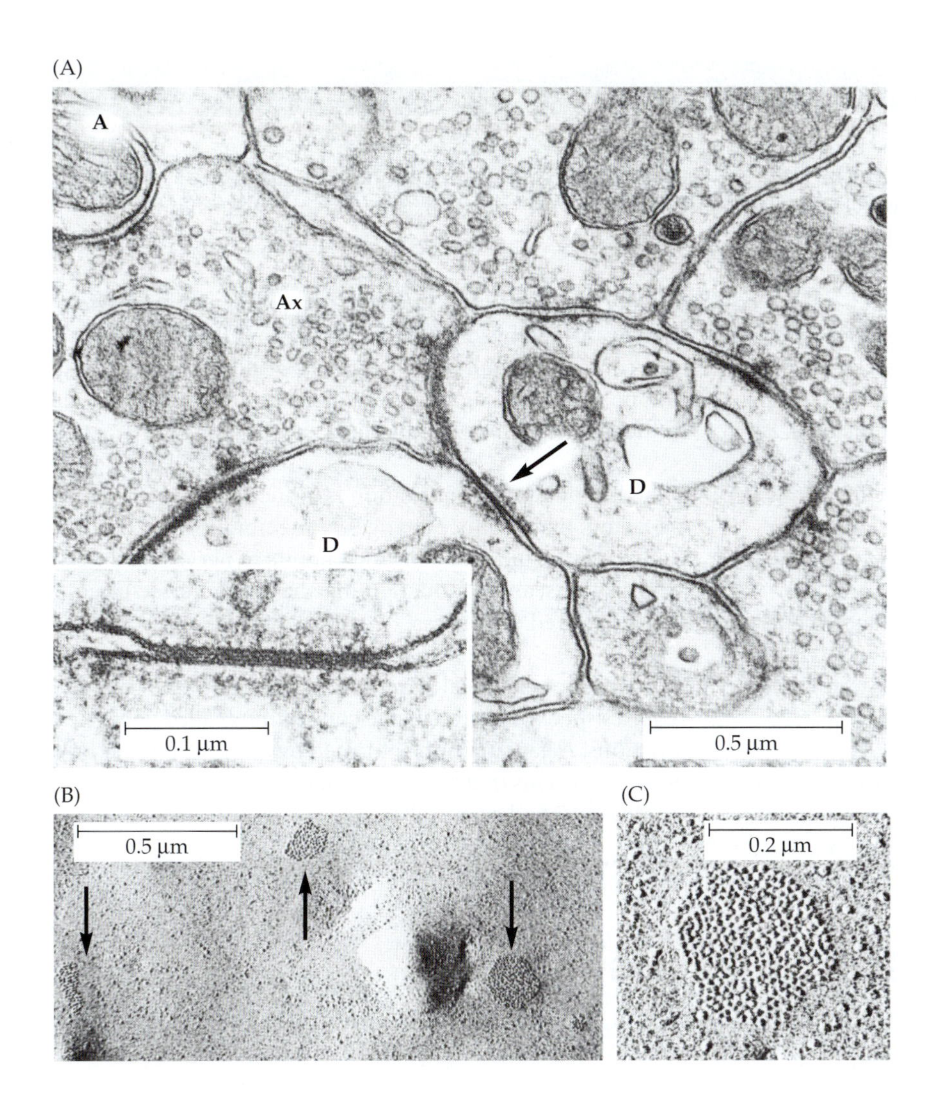

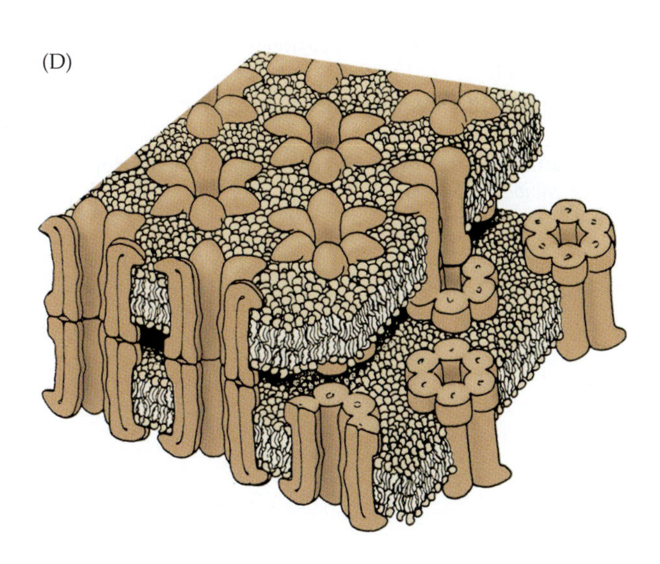

FIGURE 7.8 Gap Junctions between neurons. (A) Two dendrites (labeled D) in the inferior olivary nucleus of the cat are joined by a gap junction (arrow), shown at higher magnification in the inset. The usual space between the cells is almost obliterated in the contact area, which is traversed by cross-bridges. (B) Freeze-fracture through the presynaptic membrane of a nerve terminal that forms gap junctions with a neuron in the ciliary ganglion of a chicken. A broad area of the cytoplasmic fracture face is exposed, showing clusters of gap junction particles (arrows). (C) Higher magnification of one such cluster. Each particle in the cluster represents a single connexon. (D) Gap junction region, showing individual connexons bridging the gap between the lipid membranes of two apposed cells. (A from Sotelo, Llinás, and Baker, 1974; B and C from Cantino and Mugnani, 1975; D after Makowski et al., 1977.)

Summary

■ The spread of local graded potentials in a neuron, and the propagation of action potentials along a nerve fiber, depend on the electrical properties of the cytoplasm and the cell membrane.

■ When a steady current is injected into a cylindrical fiber, the size of a local graded potential is determined by the input resistance of the fiber, r_{input}, and the distance over which it spreads is determined by the length constant of the fiber, λ.

■ The input resistance and length constant are determined, in turn, by the specific resistances of the cell membrane (R_m) and axoplasm (R_i), and by the fiber diameter.

■ Cell membranes, in addition to having a resistance, have a capacitance. The effect of the membrane capacitance, C_m, is to slow the rise and decay of signals. The magnitude of this effect is determined by the membrane time constant, $\tau = R_m C_m$.

■ Propagation of an action potential along a fiber depends on the passive spread of current from the active region into the next segment of membrane. The conduction velocity depends on the time constant and length constant of the membrane.

■ Large nerve fibers in vertebrates are wrapped in myelin sheaths formed by glial cells or Schwann cells. The myelin is interrupted by regularly spaced gaps, or nodes. During action potential propagation, excitation jumps from one node to the next (saltatory conduction).

■ Action potential propagation is influenced by geometrical factors that produce changes in membrane area: Propagation may be blocked at branch points in nerve terminal arborizations, and conduction may have a preferred direction in tapered dendrites.

■ Transfer of electrical signals from one cell to the next requires special low-resistance connections called gap junctions. A gap junction is formed by a collection of connexons, proteins that form aqueous channels between the cytoplasms of adjacent cells.

SUGGESTED READING

Arbuthnott, E. R., Boyd, I. A., and Kalu, K. U. 1980. Ultrastructural dimensions of myelinated peripheral nerve fibres in the cat and their relation to conduction velocity. *J. Physiol.* 308: 125–157.

Goodenough, D. A., Goliger, J. A., and Paul, D. 1996. Connexins, connexons, and intercellular communication. *Annu. Rev. Biochem.* 65: 475–502.

Perkins, G., Goodenough, D., and Sosinsky, G. 1997. Three-dimensional structure of the gap junction connexon. *Biophys. J.* 72: 533–544.

Rushton, W. A. H. 1951. A theory of the effects of fibre size in medullated nerve. *J. Physiol.* 115: 101–122.

Vabnick, I., and Shrager, P. 1998. Ion channel redistribution and function during development of the myelinated axon. *J. Neurobiol.* 37: 80–96.

8 Properties and Functions of Neuroglial Cells

NERVE CELLS IN THE CENTRAL AND THE PERIPHERAL NERVOUS SYSTEM are surrounded by satellite cells. These consist of Schwann cells in the periphery and neuroglial cells in the CNS. In this chapter we discuss the structure and properties of the satellite cells, their interactions with neurons, and open questions regarding their functions.

Neuroglial cells make up about one-half of the volume of the brain and greatly outnumber neurons. The main classes of neuroglial cells are oligodendrocytes, astrocytes, and radial glial cells. Microglial cells constitute a separate population of wandering phagocytotic cells in the nervous system. Neurons and glial cells are densely packed. Their membranes are separated from each other by narrow fluid-filled extracellular spaces that are about 20 nm wide. Glial cell membranes, like those of neurons, contain channels for ions, receptors for transmitters, ion transport pumps, and amino acid transporters. In addition, glial cells are linked to each other by low-resistance gap junctions that permit direct passage of ions and small molecules. Glial cells, which have more negative resting potentials than neurons, do not generate action potentials.

An essential function of oligodendrocytes and Schwann cells is to form myelin around axons and speed up conduction of the nerve impulse. Glial cells and Schwann cells also guide growing axons to their targets. Microglial cells invade regions of damage or inflammation and phagocytose debris.

By virtue of the close apposition of glial and neuronal membranes, dynamic interactions occur between the two types of cells. Thus, neurons release K^+ into narrow extracellular spaces during the conduction of impulses, thereby raising its concentration and depolarizing the glial membrane. Glial cells influence the composition of fluid surrounding neurons by taking up K^+, as well as transmitters that accumulate after neuronal activity. Glial cells secrete transmitters, nutrients, and trophic molecules into extracellular space. It is hard to estimate quantitatively the part played by these mechanisms in the normal functioning of neurons.

Nerve cells in the brain are intimately surrounded by satellite cells called **neuroglial cells, glial cells,** or **glia.** It has been estimated that they outnumber neurons by at least 10 to 1 and make up about one-half of the bulk of the nervous system. From the time of their discovery, the problem of what the glial cells do has posed a challenging question for neurobiologists. In spite of the prevalence of glial cells, the physiological activities of the nervous system are often discussed in terms of neurons only, as if glia did not exist. This chapter emphasizes experiments dealing with the physiological properties of glial cells and their functional interactions with neurons.

Historical Perspective

Glial cells were first described in 1846 by Rudolf Virchow, who thought of them as "nerve glue" and gave them their name. Excerpts from a paper by Virchow give the flavor of his thinking:[1]

> Hitherto, considering the nervous system, I have only spoken of the really nervous part of it. But. . . it is important to have a knowledge of that substance. . . which lies between the proper nervous parts, holds them together and gives the whole its form. . . [this] has induced me to give it a new name, that of neuroglia. . . . Experience shows us that this very interstitial tissue of the brain and spinal marrow is one of the most frequent sites of morbid change. . . . Within the neuroglia run the vessels, which are therefore nearly everywhere separated from the nervous substance by a slender intervening layer, and are not in immediate contact with it.

In subsequent years, neuroglial cells were studied intensively by neuroanatomists and pathologists, who knew them to be the most common source of tumors in the brain. This is not so surprising, because certain glial cells—unlike the vast majority of neurons—can still divide in the mature animal. Among early speculations about glial cell function in relation to neurons were structural support, secretion of trophic molecules, and the prevention of "cross talk" by current spread during conduction of nerve impulses.[2] A nutritive role of neuroglia was proposed by Golgi in about 1883.[3] He wrote:

> Neuroglia. . . serves to distribute nutrient substances. . . (and). . . is different from ordinary connective tissue by virtue of its morphological and chemical characteristics and its different embryological origin.

Golgi's ideas seemed so reasonable and had such force that they were not decisively tested for many years.

Appearance and Classification of Glial Cells

A distinct structural feature of neuroglial cells compared with neurons is the absence of axons. A representative picture of mammalian neuroglial cells is shown in Figure 8.1. In the vertebrate CNS glial cells are usually subdivided into several distinctive classes.[4,5]

Astrocytes make contacts with capillaries and neurons. There are two principal subgroups: (1) fibrous astrocytes, which contain filaments and are prevalent among bundles of myelinated nerve fibers in the white matter of the brain; (2) protoplasmic astrocytes, which contain less fibrous material and are abundant in the gray matter around nerve cell bodies, dendrites, and synapses.

Oligodendrocytes are predominant in the white matter, where they form myelin around larger axons (Chapter 7).

Radial glial cells play an essential role in the developing mammalian central nervous system. They stretch through the thickness of the spinal cord, retina, cerebellum, or cerebral cortex to the surface, forming elongated filaments along which developing neurons migrate to their final destinations. Cells in adult CNS resembling radial glia are the Bergmann cells in the cerebellum and Müller cells in the retina.

Ependymal cells that line the inner surfaces of the brain, in the ventricles, are usually classified as glial cells.

Microglial cells are distinct from neuroglial cells in structure, properties, and lineage.[6,7] They resemble macrophages in the blood and probably arise from them.

[1]Virchow, R. 1959. *Cellularpathologie.* Hirschwald, Berlin. (Excerpts are from pp. 310, 315, and 317.)

[2]Ramón y Cajal, S. [1909–1911] 1995. *Histology of the Nervous System,* Vol. 1. Oxford University Press, New York.

[3]Golgi, C. 1903. *Opera Omnia,* Vols. 1 and 2. U. Hoepli, Milan, Italy.

[4]Kettenmann, H., and Ransom, B. R. (eds.). 1995. *Neuroglia.* Oxford University Press, New York.

[5]Ransom, B. R. 1991. *Ann. N.Y. Acad. Sci.* 633: 19–26.

[6]Kreutzberg, G. W. 1996. *Trends Neurosci.* 19: 312–318.

[7]Brown, H. C., and Perry, V. H. 1998. *Glia* 23: 361–373.

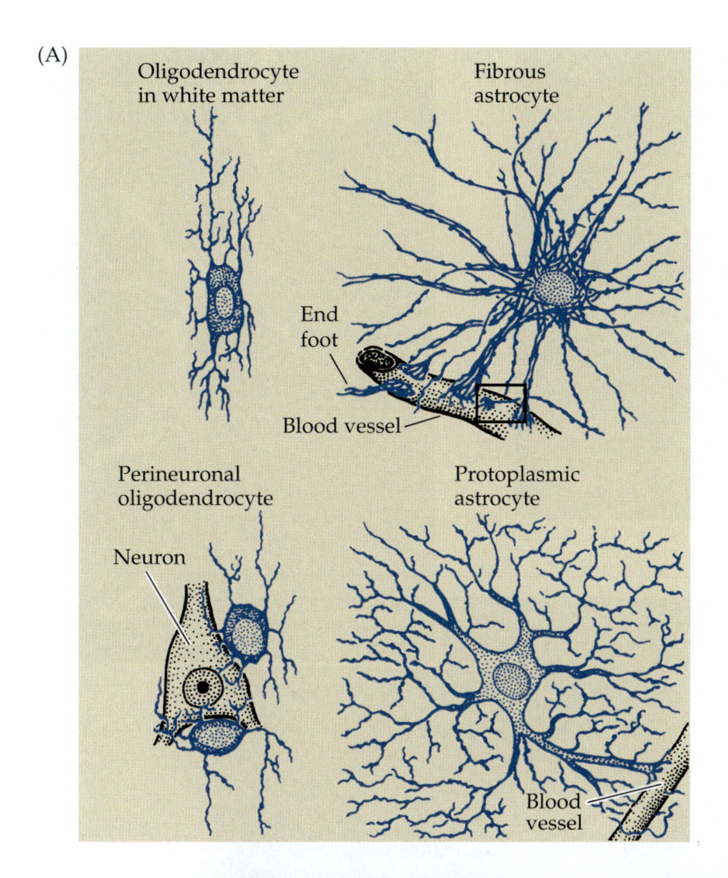

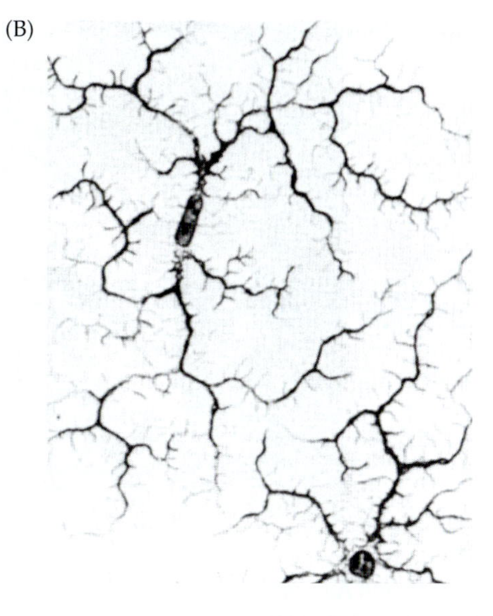

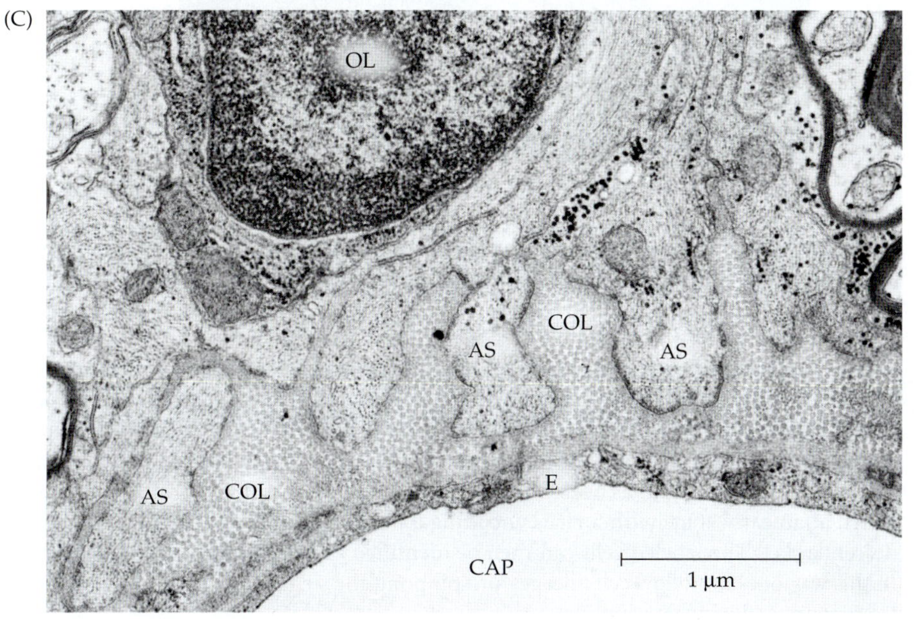

FIGURE 8.1 Neuroglial Cells in Mammalian Brain. (A) Oligodendrocytes and astrocytes, stained with silver impregnation, represent the principal types of neuroglial cells in vertebrate brain. They are closely associated with neurons. (B) Microglial cells are small, wandering, macrophage-like cells. (C) Electron micrograph of glial cells in rat optic nerve. In the lower portion is the lumen of a capillary (CAP) lined with endothelial cells (E). The capillary is surrounded by end feet formed by processes of fibrous astrocytes (AS). Between the end feet and the endothelial cells is a space filled with collagen fibers (COL). In the upper portion is part of a nucleus of an oligodendrocyte (OL), and to the right are axons surrounded by myelin wrapping. (A and B after Penfield, 1932, and Del Rio-Hortega, 1920; C from Peters, Palay, and Webster, 1991).

In vertebrate peripheral nerves and ganglia, **Schwann cells** are analogous to glial cells. They form myelin around the fast-conducting axons. Schwann cells also enclose smaller axons (less than 1 μm in diameter) without forming a myelin sheath.

The term **satellite cell** is used in this chapter to refer collectively to nonneuronal cells in general; it includes glial cells in the CNS and Schwann cells in the periphery.

The various types of satellite cells can be distinguished by injecting them with labels, such as dyes in living preparations, or by immunological techniques (Figure 8.2). Antibodies have been prepared that bind specifically to astrocytes, to oligodendrocytes, to microglia,

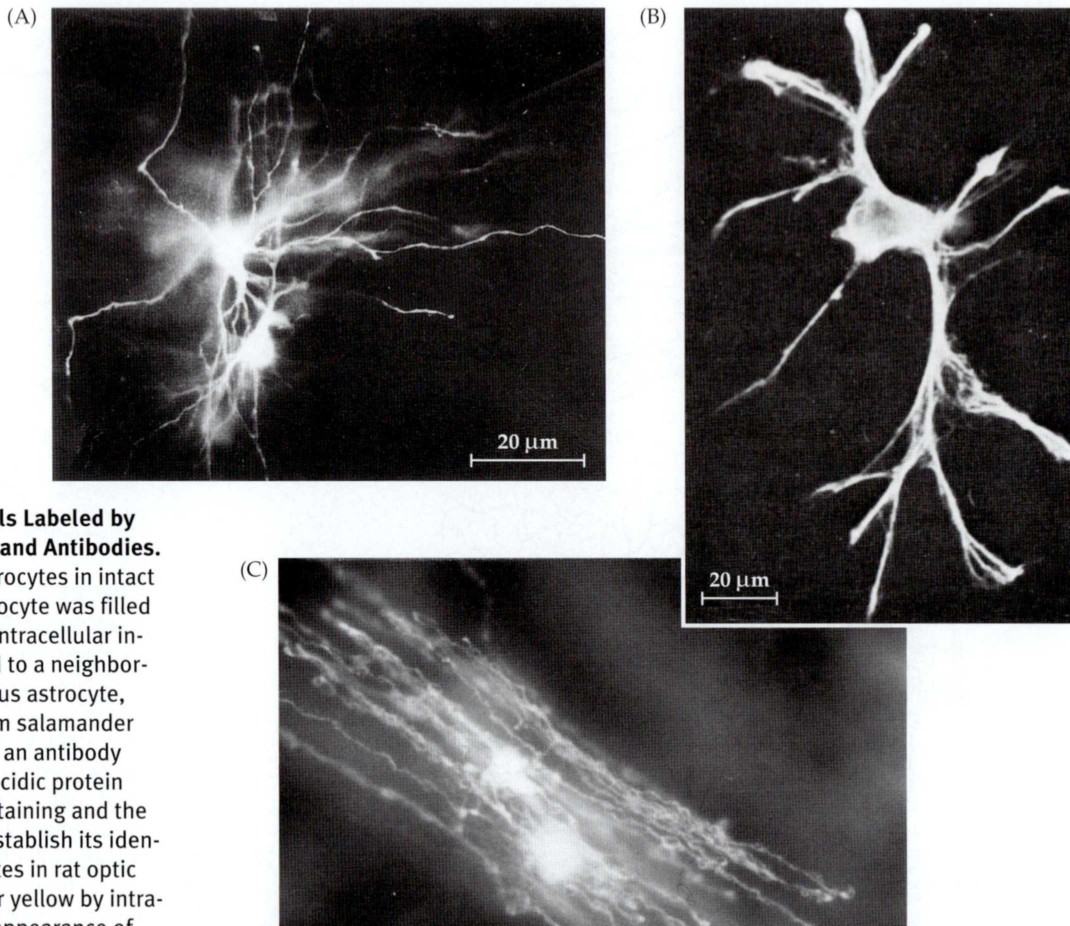

FIGURE 8.2 Glial Cells Labeled by Intracellular Injection and Antibodies. (A) Dye coupling of astrocytes in intact rat optic nerve. An astrocyte was filled with Lucifer yellow by intracellular injection. The dye spread to a neighboring astrocyte. (B) Fibrous astrocyte, freshly dissociated from salamander optic nerve, labeled by an antibody against glial fibrillary acidic protein (GFAP). The antibody staining and the shape unequivocally establish its identity. (C) Oligodendrocytes in rat optic nerve filled with Lucifer yellow by intracellular injection. The appearance of the longitudinal processes that run symmetrically in parallel is characteristic for oligodendrocytes. (A and C after Butt and Ransom, 1993, photos kindly provided by A. M. Butt; B from Newman, 1986.)

or to Schwann cells.[8] For example, fibrous astrocytes can be stained with an antibody against a protein known as GFAP[9] (glial fibrillary acidic protein) (see Figure 8.2B).

Like the neurons in the peripheral and central nervous systems, glial cells and Schwann cells have different embryological origins: Glial cells in the central nervous system are derived from precursor cells that line the neural tube constituting the inner surface of the brain; Schwann cells arise from the neural crest (Chapter 23). In animals such as the leech, glial cell development can be observed directly.[10] Precursor cells differentiate into glial cells that surround neuronal cell bodies, axons, and synapses. In vertebrate embryos, glial development has been observed by labeling precursor cells. A few cells are infected at an early stage with a virus encoding a marker gene that is handed on to their descendants.[11] The labeled cells can then be identified as astrocytes or oligodendrocytes. In this way one can follow cell lineages and pinpoint the stages at which glial cells diverge from neurons during development.

[8]Martini, R., and Schachner, M. 1997. *Glia* 19: 298–310.

[9]Bignami, A., and Dahl, D. 1974. *J. Comp. Neurol.* 153: 27–38.[10]Stent, G. S., et al. 1992. *Int. Rev. Neurobiol.* 33: 109–193.

[10]Stent G. S., et al. 1992. *Int. Rev. Neurobiol.* 33: 109-193.et al., 1992

[11]Luskin, M. B. 1998. *J. Neurobiol.* 36: 221–233.

Structural Relations between Neurons and Glia

A glance at electron micrographs of brain tissue brings home the close packing of neurons and glia. Figure 8.3 shows an example from the cerebellum of a rat. The section is filled with neurons and glial cells, which can be distinguished by a number of criteria. Glial processes tend to be thin, at times less than 1 μm thick. Only around the glial nuclei are there larger volumes of glial cytoplasm. The extracellular space is restricted to narrow clefts, about 20 nm wide, that separate all cell boundaries. No special connections, such as gap junctions, are seen between neurons and glia in adult CNS, and physiological tests

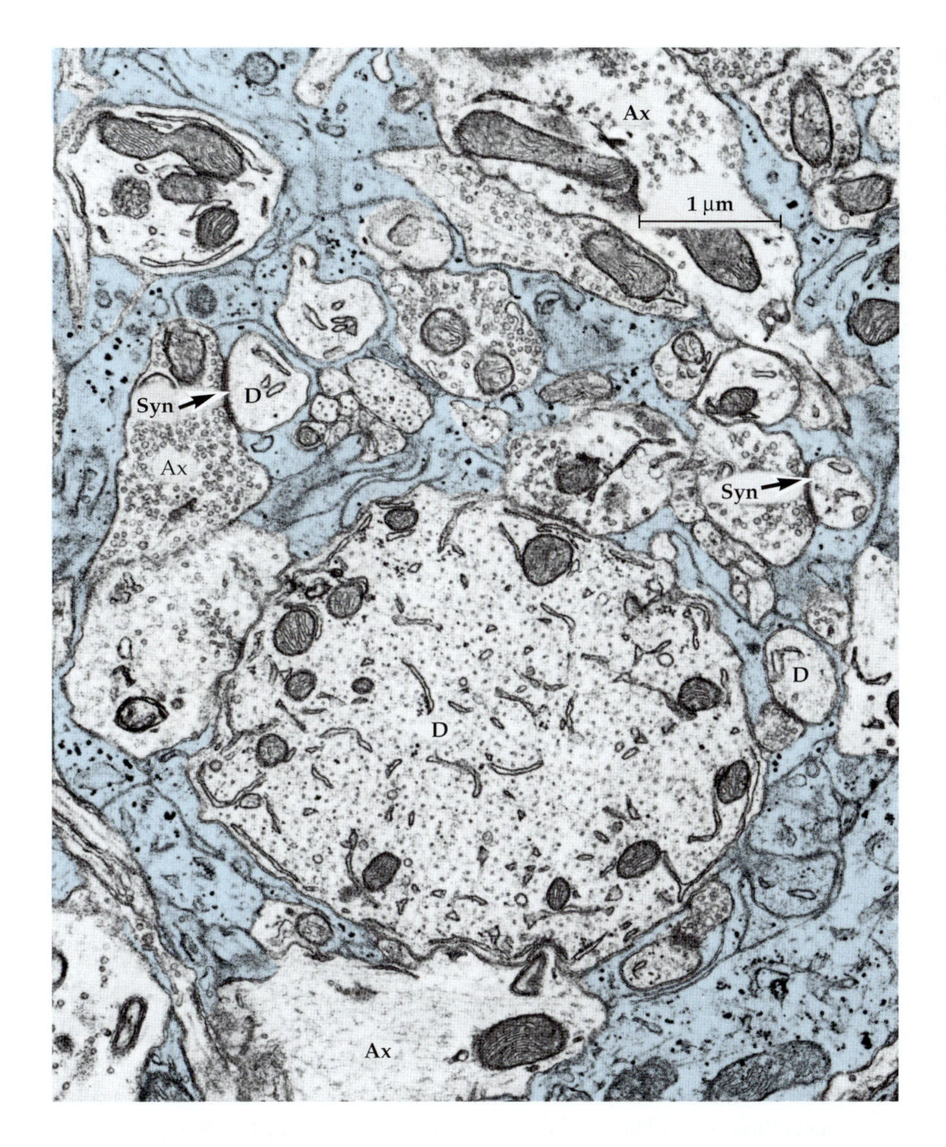

FIGURE 8.3 Neurons and Glial Processes in Rat Cerebellum. The glial contribution is lightly colored. The neurons and glial cells are always separated by clefts about 20 nm wide. The neural elements are dendrites (D) and axons (Ax). Two synapses (Syn) are marked by arrows. (After Peters, Palay, and Webster, 1991.)

fail to reveal direct low-resistance pathways between them. Such pathways do, however, link glial cells to one another through gap junctions (Chapter 7).[12,13] The relation between glial cells, neurons, capillaries, and extracellular space is diagrammed in Figure 8.4.

PHYSIOLOGICAL PROPERTIES OF NEUROGLIAL CELL MEMBRANES

For technical reasons, the membrane properties of glial cells were first analyzed in the leech CNS. The glial cells in a leech ganglion are large and transparent (see Figure 15.2). They appear under the dissecting microscope as spaces between nerve cells, and they can be recorded from by sharp microelectrodes or by whole-cell patch electrodes.[12,14,15] After one has established its physiological properties, the glial cell can be injected with a fluorescent marker, such as Lucifer yellow, and its form observed in the living preparation. Once leech glial cells had been described, it became practicable to record from and label amphibian and mammalian glial cells, which were found to share many key properties with them. Far from being a roundabout approach, this turned out to be a shortcut to the study of glia in the vertebrate CNS.[16]

Glial cells have resting potentials greater (more negative inside) than those of neurons. The largest membrane potentials recorded from neurons are about –75 mV, whereas the

[12]Kuffler, S. W., and Potter, D. D. 1964. *J. Neurophysiol.* 27: 290–320.

[13]Kuffler, S. W. 1967. *Proc. R. Soc. Lond. B* 168: 1–21.

[14]Butt, A. M., and Ransom, B. R. 1993. *J. Comp. Neurol.* 338: 141–158.

[15]Ransom, B. R., and Sontheimer, H. 1992. *J. Clin. Neurophysiol.* 9: 224–251.

[16]Kuffler, S. W., and Nicholls, J. G. 1966. *Ergeb. Physiol.* 57: 1–90.

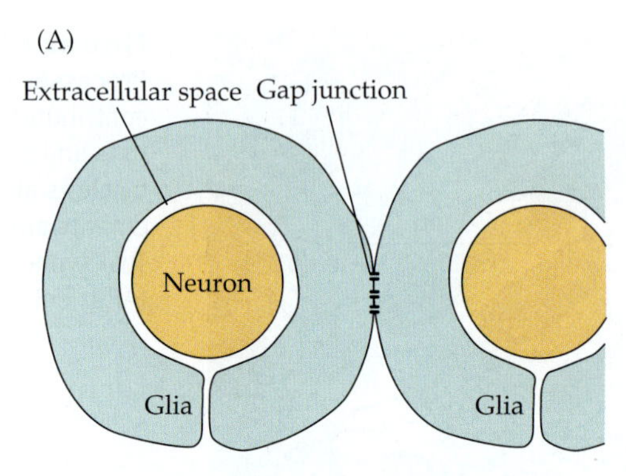

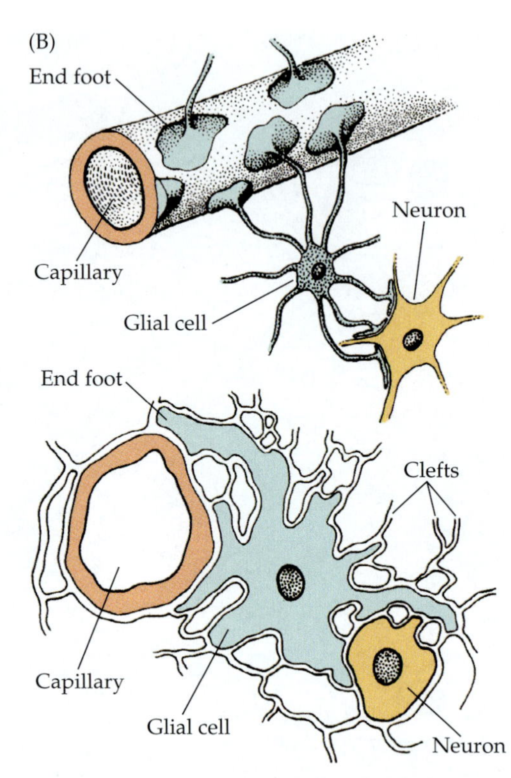

FIGURE 8.4 Neurons, Glia, Extracellular Space, and Blood. (A) Neuronal–glial and glial–glial relationships. While neurons are always separated from glia by continuous clefts, the interiors of glial cells are linked by gap junctions. (B) Relations of capillaries, glia, and neurons as seen by light and electron microscopy. The pathway for diffusion from the capillary to the neuron is through the aqueous intercellular clefts. Cell dimensions are not in proportion. (After Kuffler and Nicholls, 1966.)

values for glial cells approach –90 mV. Another distinguishing feature of glial cells is the absence of conducted action potentials. In only a few examples of cultured glial cells have regenerative responses been reported.

The glial membrane behaves like a potassium electrode; that is, its behavior follows the Nernst equation in solutions containing different potassium concentrations (Chapter 5). Ions other than potassium make only a small contribution to the resting membrane potential.[17] Figure 8.5 shows a series of membrane potential measurements plotted against $[K]_o$ on a logarithmic scale. The solid line is the theoretical slope of 59 mV per tenfold change in concentration predicted by the Nernst equation (at 24°C). This behavior differs significantly from that of most neurons, whose responses to potassium do not follow the Nernst equation in the physiological range of 2 to 4 m*M* $[K]_o$ (Chapter 5).

The distribution of potassium channels has been explored over the surface of Müller glial cells and astrocytes isolated from the retina and optic nerves of many species, such as frogs and tadpoles,[18] salamanders,[19] and rabbits.[20] The potassium sensitivity is distributed in a characteristic pattern, being highest over the end feet and lower over the soma of the Müller cell. Figure 8.6 shows an isolated salamander Müller cell and its responses to a high concentration of potassium applied to different regions of the surface from a pipette. At early stages of development, potassium channels are distributed more uniformly over Müller cells. For example, the high concentration at the end feet occurs as tadpoles become transformed into adult frogs.[18] The possible significance is discussed later in this chapter.

[17]Kuffler, S. W., Nicholls, J. G., and Orkand, R. K. 1966. *J. Neurophysiol.* 29: 768–787.

[18]Rojas, L., and Orkand, R. K. 1999. *Glia* 25: 199–203.

[19]Newman, E., and Reichenbach, A. 1996. *Trends Neurosci.* 19: 307–312.

[20]Brew, H., et al. 1986. *Nature* 324: 466–468.

[21]Rose, C. R., Ransom, B. R., and Waxman, S. G. 1997. *J. Neurophysiol.* 78: 3249–3258.

Ion Channels, Pumps, and Receptors in Glial Cell Membranes

Glial cells and Schwann cells in culture display a variety of ion channels and pumps in their membranes:

1. As already shown, potassium conductances predominate.[16]
2. Voltage-activated sodium and calcium channels are present in the membranes of Schwann cells and astrocytes.[21] The overall ratio of potassium permeability to sodium permeability of Müller cells has been estimated to be approximately 100:1. As men-

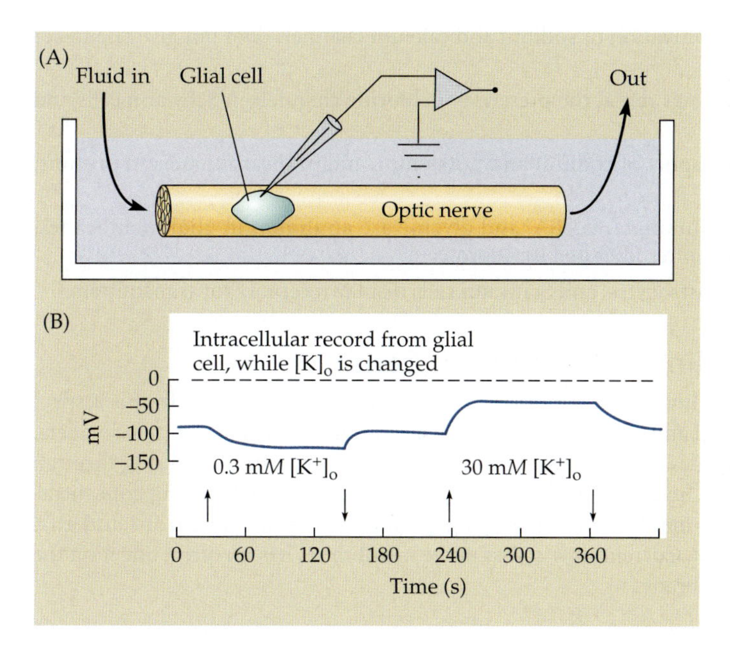

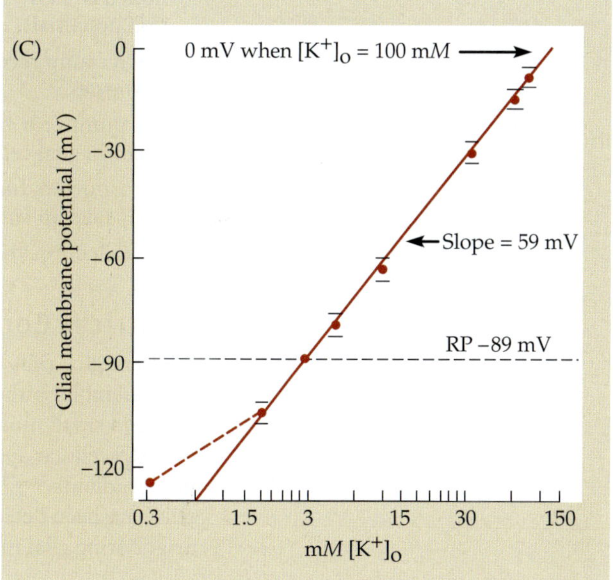

FIGURE 8.5 Dependence of Glial Membrane Potential on potassium concentration. (A) System for recording from a glial cell in mud puppy optic nerve while changing external potassium concentration. (B) Reducing the potassium concentration from the normal 3.0 m*M* to 0.3 m*M* hyperpolarizes the membrane; increasing the potassium concentration to 30 m*M* depolarizes by 59 mV. (C) The relation between potassium concentration and membrane potential predicted by the Nernst equation (solid line) accurately fits the experimental results, except at very low extracellular potassium concentrations. Neurons are less sensitive than glia to small changes in potassium concentration in the physiological range. RP = resting potential. (After Kuffler, Nicholls, and Orkand, 1966.)

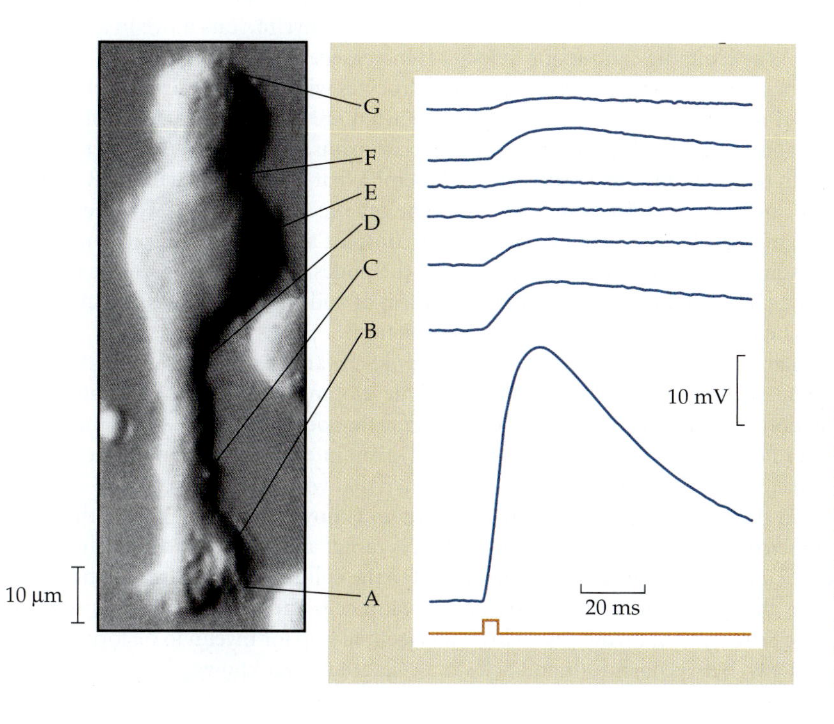

FIGURE 8.6 Responses to Potassium of a Müller Glial Cell isolated from salamander retina. Recordings were made with an intracellular electrode while potassium was applied to different sites. A is the end foot, and G is the distal part of the cell. The sensitivity to potassium is much greater at the end foot, suggesting a higher concentration of potassium channels in that region. (After Newman, 1987; micrograph kindly provided by E. Newman.)

tioned already, the activation of sodium and calcium channels does not give rise to action potentials.

3. Patch clamp recordings reveal the presence of chloride channels in Schwann cells and astrocytes.[22]
4. Ion pumps for transport of sodium and potassium, and of bicarbonate and protons, occur in glial cells.[23]
5. Transporters for glutamate, GABA, and glycine are abundant in glial membranes: They take up transmitter liberated by neurons.[24]
6. Oligodendrocytes, astrocytes, and Schwann cells display receptors for transmitters.[25]

Electrical Coupling between Glial Cells

Adjacent glial cells are linked to each other by gap junctions (Chapter 7). In this respect they resemble epithelial and gland cells, as well as smooth muscle and heart muscle fibers. Ions and small molecules are exchanged directly between cells without passing through the extracellular space, and such interconnections may be used for reducing concentration gradients.[26–28] As mentioned earlier, no gap junctions between neurons and glial cells have been detected. Current flow across nerve membranes has no direct effect on the neighboring glial membrane.

FUNCTIONS OF NEUROGLIAL CELLS

Over the years, almost every nervous system task for which no other obvious explanation has been found has been attributed to glial cells. In the following sections we discuss first the functions of glial cells that have been well established, and then questions about their functional role that require further elucidation.

Myelin and the Role of Neuroglial Cells in Axonal Conduction

One important function of oligodendrocytes and Schwann cells is to produce the myelin sheath around axons—a high-resistance covering akin to insulating material around wires (Chapter 7). The myelin is interrupted at the nodes of Ranvier (Figure 8.7), which occur at regular intervals.[29] At nodes within the CNS, a characteristic feature is the presence of astrocytic processes that contact the axon.[30] Since the ionic current associated with the conducted nerve impulse cannot flow across the myelin, ions move in and out at the nodes. As a result, the conduction velocity is increased.

The association of Schwann cells or oligodendrocytes with axons to form myelin raises a number of interesting problems. For example, what are the genetic or environmental factors that enable glial cells to select the appropriate axons, to surround them at the right time, and to maintain myelin sheaths around them? What are the characteristics of neurological disorders of myelin caused by disease or genetic abnormalities? For the formation of the myelin sheath during development, complex and precise interactions occur between axons and satellite cells. The spacing of the nodes, the seals between the two cell types at the paranodal areas, and the distribution of sodium and potassium channels must be matched and positioned for rapid conduction.

The dynamic interactions between neurons and Schwann cells have been analyzed by experiments made in tissue culture. Schwann cell development, as well as myelination and remyelination of axons, in culture parallel the events that occur in vivo.[31] At the molecular level, key proteins that play a part in Schwann cell–axon interactions have been identified.[8] For example, Shooter and his colleagues have shown that when Schwann cells are cultured in a dish on their own, a peripheral myelin protein (known as PMP22) is synthesized. Under these conditions, the turnover of PMP22 is rapid, and it is degraded in the endoplasmic reticulum. When neurons are introduced to the culture, as in Figure 8.8, the fate of the protein changes. After contact between Schwann cells and axons, PMP22 is translocated to the Schwann cell membrane. This is an essential step for myelin to be formed. The signals that pass between neurons and Schwann cells are not yet known.[32,33]

[22]Ritchie, J. M. 1987. *J. Physiol.* (Paris) 82: 248–257.

[23]Astion, M. L., Chavatal, A., and Orkand, R. K. 1991. *Glia* 4: 461–468.

[24]Szatkowski, M., Barbour, B., and Attwell, D. 1990. *Nature* 348: 443–446.

[25]Porter, J. T., and Mccarthy, K. D. 1998. *Prog. Neurobiol.* 51: 439–455.

[26]Loewenstein, W. R. 1999. *The Touchstone of Life.* Oxford University Press, Oxford, England.

[27]Rose, C. R., and Ransom, B. R. 1996. *J. Physiol.* 491: 291–305.

[28]Zahs, K. R., and Newman, E. A. 1997. *Glia* 20: 10–22.

[29]Bunge, R. P. 1968. *Physiol. Rev.* 48: 197–251.

[30]Black, J. A., and Waxman, S. G. 1988. *Glia* 1: 169–183.

[31]Waxman, S.G. 1997. *Curr. Biol.* 7: 406–410.

[32]Notterpek, L., Shooter, E. M., and Snipes, G. J. 1997. *J. Neurosci.* 17: 4190–4200.

[33]Tobler, A. R., et al. 1999. *J. Neurosci.* 19: 2027–2036.

(A) **Schematic diagram of arrangement of myelin**

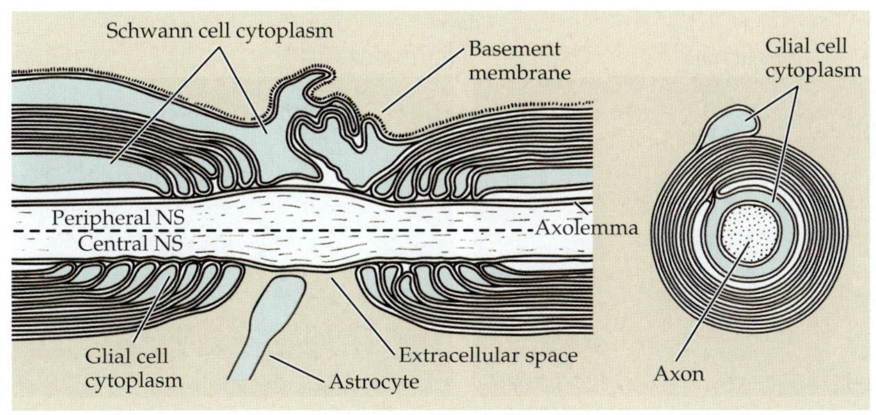

(B)

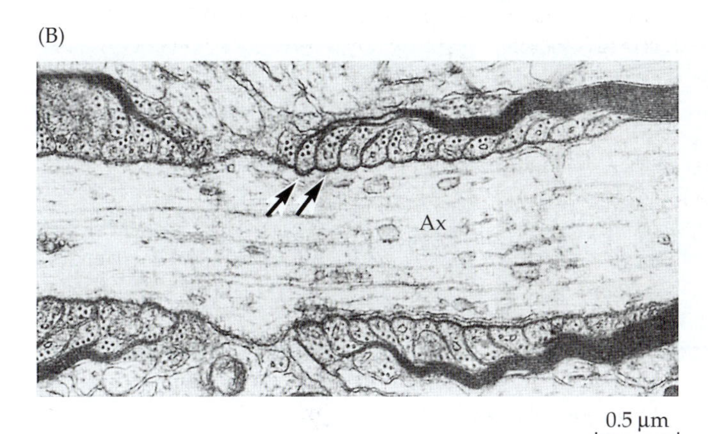

(C)

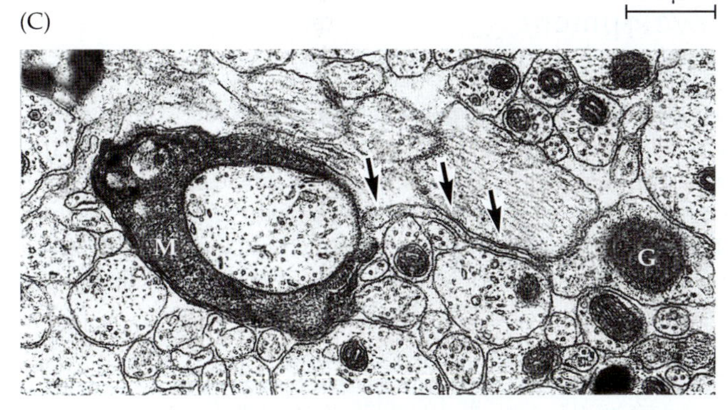

FIGURE 8.7 Myelin and Nodes of Ranvier. Oligodendrocytes and Schwann cells form the wrapping of myelin around axons. (A) At the nodes of Ranvier, like the one shown here on the left, the myelin is interrupted and the axon is exposed. The upper half of the nodal region, with a loose covering of processes, is typical of the arrangement in peripheral nerves. The lower part is representative of a node within the central nervous system. Here an astrocytic finger comes into close apposition with the nodal membrane. To the right is a transverse section through a myelin-covered axon. (B) Electron micrograph of a nodal region in a myelinated fiber in rat CNS. At the edge of the node is a specialized close contact area between the membrane of the axon (Ax) and the membrane of the myelin wrapping (arrows). (C) Cross section of a myelinated axon at a node that is contacted by a process (marked with arrows) from a perinodal astrocytic glial cell (G). Myelin (M) is absent at the site of contact between the astrocyte and the node. (A after Bunge, 1968; B from Peters, Palay, and Webster, 1991; C from Sims et al., 1985; micrograph kindly provided by J. Black and S. Waxman.)

The exact amount of PMP22 that is produced is critical for proper myelination; with over- or underexpression of PMP22, disorders occur. Figure 8.9 shows that changes in a single amino acid of PMP22 (e.g., from leucine to proline) result in "trembler" mice, which exhibit deficient myelination and serious neurological defects. Hereditary human neuropathies arise in families with the same mutation.

A number of experiments have provided evidence that glial cells can influence the clustering of sodium channels in myelinated nerve fibers. Changes occur in the distribution of ion channels in nodes, paranodal areas, and internodes as axons become myelinated, demyelinated, and remyelinated.[34,35] The process resembles the clustering of transmitter receptors at postsynaptic sites brought about by innervation. Astrocytic fingers in the nodal region themselves show intense labeling with saxitoxin (a toxin that binds to sodium channels) (Chapter 6), indicating a high density of sodium channels in the glial membrane.[36] It has been proposed that transfer of sodium channels might occur from astrocytes to nodes of Ranvier,[37] but there is no direct evidence for this interesting speculation.

[34]Kaplan, M. R., et al. 1997. *Nature* 386: 724–728.

[35]Rasband, M. N., et al. 1999. *J. Neurosci.* 19: 7516–7528.

[36]Ritchie, J. M., et al. 1990. *Proc. Natl. Acad. Sci. USA* 87: 8290–9294.

[37]Shrager, P., Chiu, S. Y., and Ritchie, J. M. 1985. *Proc. Natl. Acad. Sci. USA* 82: 948–952.

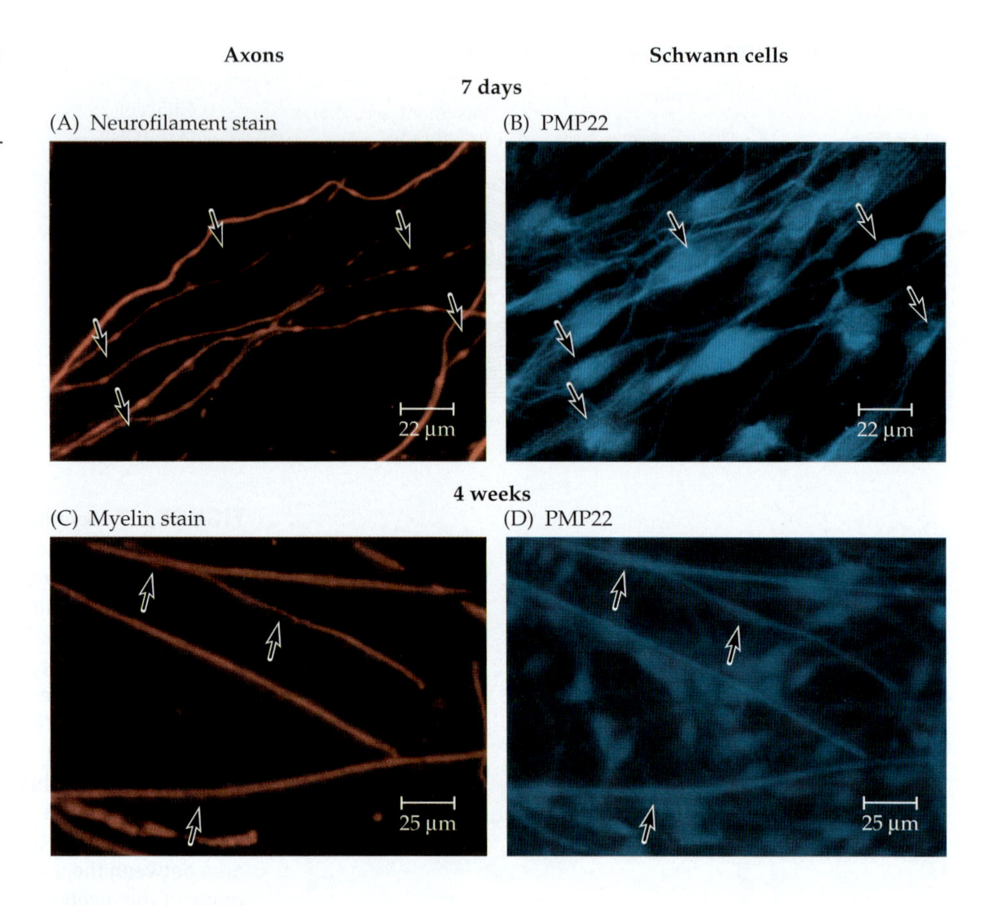

FIGURE 8.8 Localization of Myelin Protein (PMP22) in short- and long-term myelinating cultures of axons and Schwann cells. Changes in distribution of myelin protein PMP22 induced by coculture of axons with Schwann cells. (A, B) 1-week-old coculture of neurons (A) and Schwann cells (B) doubly stained with monoclonal antineurofilament and polyclonal PMP22 antiserum. Arrows point to Schwann cells that are in contact with neuronal processes. At this early stage the glial and neuronal proteins have different distributions, with PMP22 mainly in Schwann cell bodies. (C, D) After 4 weeks in medium that promotes myelination, PMP22 becomes colocalized with myelin segments (stained by antibody Po). Arrows point to axons (C) and to the cell bodies of elongated Schwann cells (D), with uniform PMP22 staining over the cell membrane. (After Pareek et al., 1997; photos kindly provided by E. Shooter.)

Glial Cells, CNS Development, and Secretion of Growth Factors

Essential aspects of development that involve satellite cells are described in Chapter 23. Here we emphasize certain key features of the functional role played by glia, Schwann cells, and microglia. For example, glial and Schwann cells secrete molecules such as nerve growth factor and laminin; these molecules promote the outgrowth of neurites, both in culture and in the animal.[38,39] In addition, a neurite-promoting protein of glia known as GDN (glial-derived nexin) has been identified. Monard and his colleagues[40] have sug-

[38]Heuman, R. 1987. *J. Exp. Biol.* 132: 133–150.

[39]van der Laan, L. J., et al. 1997. *J. Neurosci. Res.* 50: 539–548.

[40]Bleuel, A., et al. 1995. *J. Neurosci.* 15: 759–761.

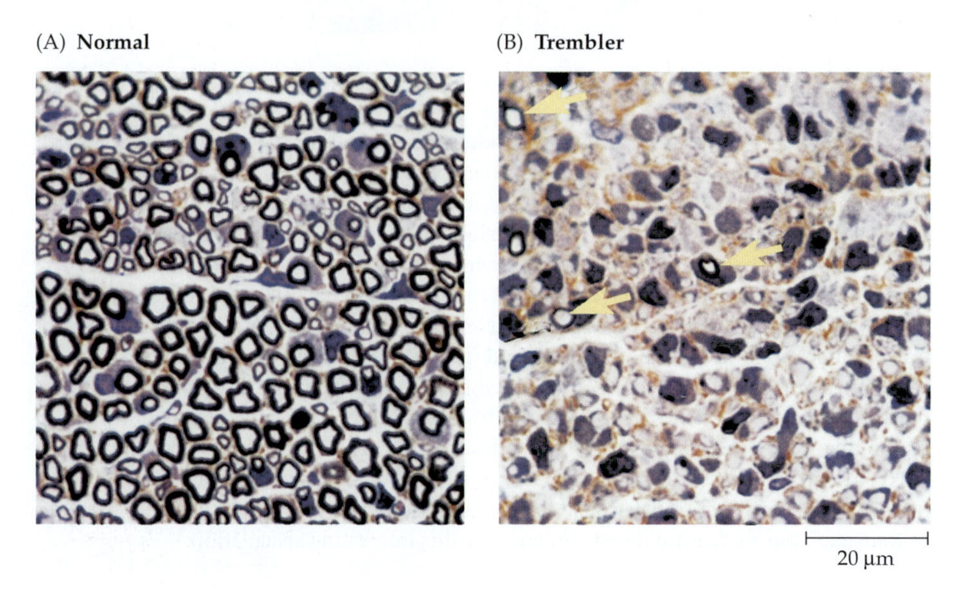

FIGURE 8.9 Deficient Myelination in "Trembler" Mutant Mice with a genetic defect in a myelin protein, PMP22. Morphological appearance of sciatic nerves in normal (A) and mutant trembler (B) mice, aged 10 days. Note the marked differences in axon caliber and myelin thickness between normal and trembler mice (indicated by arrows in B) in microscopic sections at equivalent magnifications. Also note the severity of dysmyelination. A single amino acid mutation from leucine to proline produces trembler neuropathy in mice and in humans. (After Notterpek, Shooter, and Snipes, 1997.)

gested that GDN, which is a potent protease inhibitor, could prevent degradation of extracellular matrix molecules that are essential for growth. Glial cells can also act as repellents that inhibit neurite outgrowth.[41] For example, a protein known as NI-35/250 is present in myelin and oligodendrocytes. This molecule stops neuronal growth cones from advancing and causes their collapse. Growth-inhibiting proteins could also help to delineate tracts and boundaries in the spinal cord: Growing fibers would be stopped from entering or wandering out of specified pathways inappropriately. The possible effects of these proteins on regeneration after injury are discussed in Chapter 24.

During CNS development, glial cells have been shown to play a role in the aggregation of neurons into well-defined nuclei. Prospective nuclei and structures that develop in situ and in culture are first outlined by glial cells.[42,43] Thus, before the neurons themselves have arrived in the somatosensory cortex of the developing mouse, the well-defined "barrels" (Chapter 18) have already been outlined by glia that take up their places at early stages.

A mechanism by which radial glial cells guide neuronal migration during development has been demonstrated in experiments by Rakic,[44] Hatten,[45,46] and their colleagues (see also Chapter 23). During development of the cerebral cortex, hippocampus, and cerebellum in monkeys and humans, nerve cells migrate to their destinations along radial glial processes. A time-lapse sequence of the movement of a hippocampal neuron along a radial glial cell is shown in Figure 8.10. Migrating neurons recognize surface molecules on the glial processes specific for their neuronal type; for example, hippocampal radial glial cells provide guidance cues for hippocampal but not cortical neurons. The mature neurons send out axons and form connections with their targets.

[41]Caroni, P., and Schwab, M. E. 1988. *J. Cell Biol.* 106: 1281–1288.

[42]Willbold, E., et al. 1995. *Eur. J. Neurosci.* 7: 2277–2284.

[43]Cooper, N. G. F., and Steindler, D. A. 1986. *Brain Res.* 380: 341–348.

[44]Rakic, P. 1981. *Trends Neurosci.* 4: 184–187.

[45]Hatten, M. E. 1999. *Annu. Rev. Neurosci.* 22: 511–539.

[46]Zheng, C., Heintz, N., and Hatten, M. E. 1996. *Science* 272: 417–419.

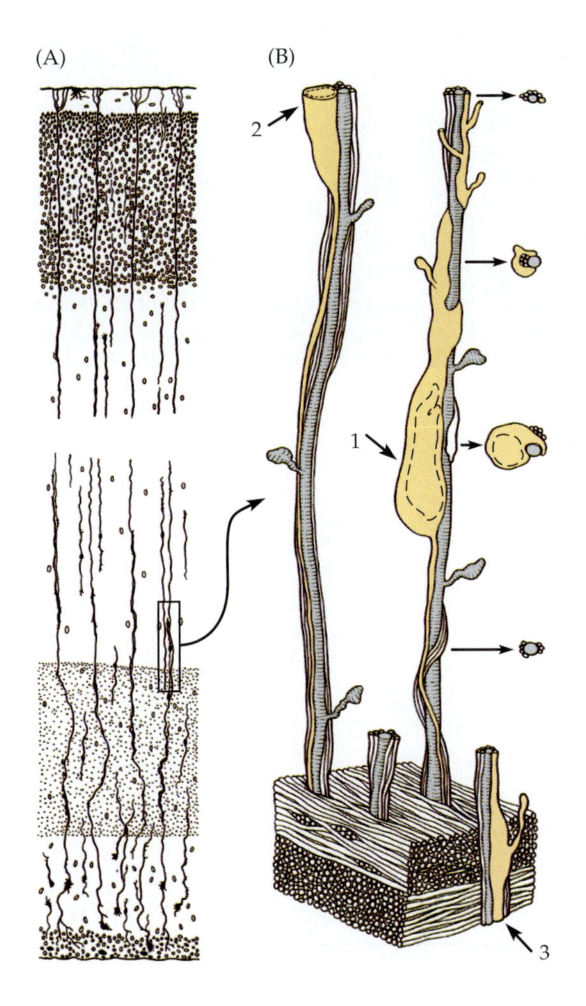

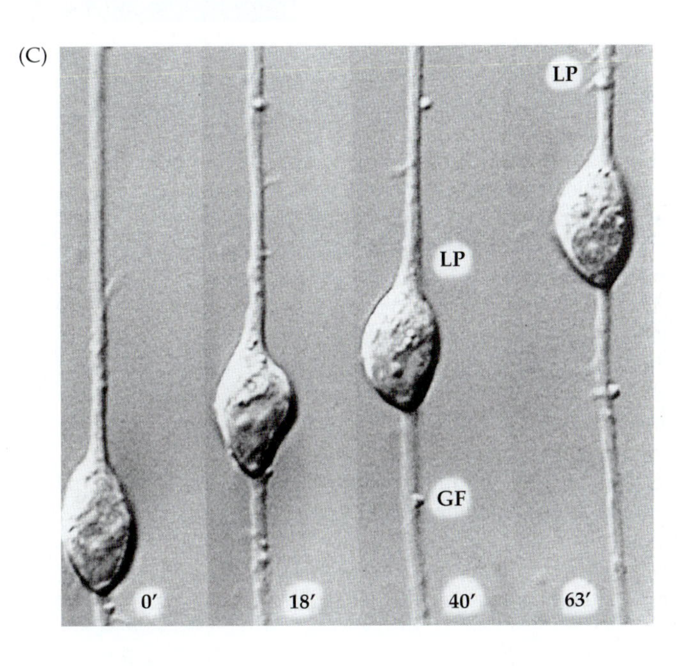

FIGURE 8.10 Neurons Migrating along Radial Glia during development. (A) Camera lucida drawing of the occipital lobe of developing cortex of a monkey fetus at mid-gestation. Radial glial fibers run from the ventricular zone below to the surface of the developing cortex above. (B) Three-dimensional reconstruction of migrating neurons. The migrating cell (1) has a voluminous leading process that follows the radial glia, using it as guideline. Cell 2, which has migrated farther, retains a process still connected to the radial glia. Cell 3 is beginning to send a process along the radial glia before migrating. (C) Migration of a hippocampal neuron along a radial glial fiber (GF) in vitro. As time progresses, the leading process (LP) moves farther up, with the neuronal cell body following. Times indicated at the bottom represent real time, in minutes, taken from video photography. (A and B after Rakic, 1988; C from Hatten, 1990.)

Role of Microglial Cells in CNS Repair and Regeneration

Astrocytes, microglia, and Schwann cells react to neuronal injury by replication.[6,7] They participate in the removal of debris and in scar formation. As a first step, resident microglial cells and macrophages that invade damaged CNS from blood at the site of an injury divide and scavenge debris from dying cells. In the cockroach, during repair, CNS macrophage-like cells give rise to the new glial cells that ensheathe neurons and regulate the fluid environment.[47]

In the leech, the role of microglial cells in regeneration has been studied by Muller and his colleagues.[48] (As an aside, it may be mentioned that it was in the CNS of the leech that such wandering cells first were given the name "microglia," by del Rio-Hortega.[49]) Normally microglial cells are evenly distributed in leech ganglia and in the bundles of axons that link them (Figure 8.11). Immediately after damage to the CNS, microglial cells migrate to the site of the lesion, at a rate of about 300 µm/h. There they accumulate and phagocytose damaged tissue. By in situ hybridization and by antibody staining, microglial cells have been shown to produce laminin at the lesion site.[50] Laminin (Chapter 24) is an extracellular matrix molecule that promotes neurite outgrowth in culture and in vivo, in the leech and in vertebrates.

[47]Smith, P. J., Howes, E. A., and Treherne, J. E. 1987. *J. Exp. Biol.* 132: 59–78.

[48]McGlade-McCulloh, E., et al. 1989. *Proc. Natl. Acad. Sci. USA* 86: 1093–1097.

[49]Del Rio-Hortega, P. 1920. *Trab. Lab. Invest. Biol. Madrid* 18: 37–82.

[50]Luebke, A. E., Dickerson, E. M., and Muller, K. J. 1993. *Soc. Neurosci. Abstr.* 19: 1084.

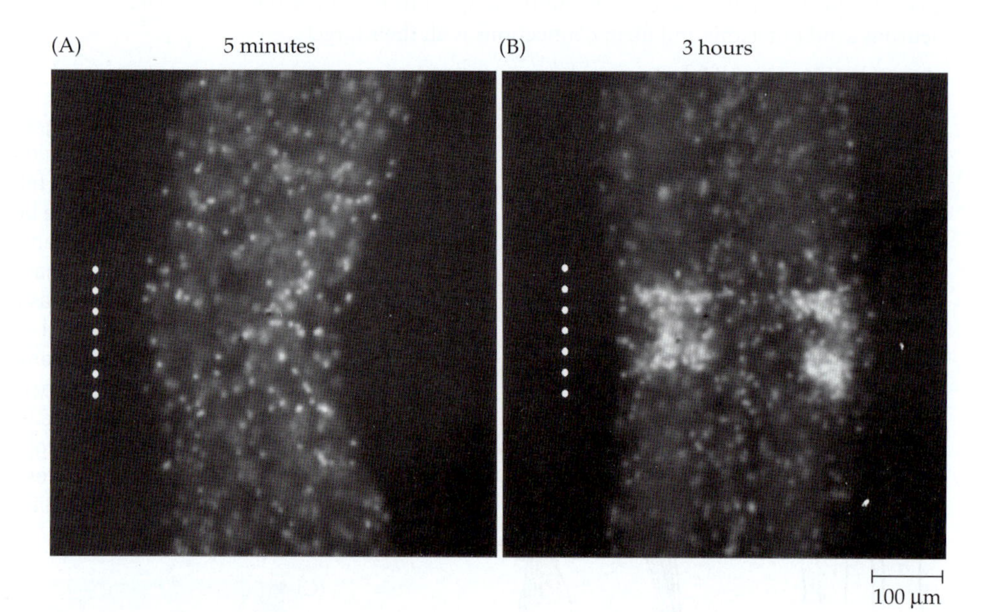

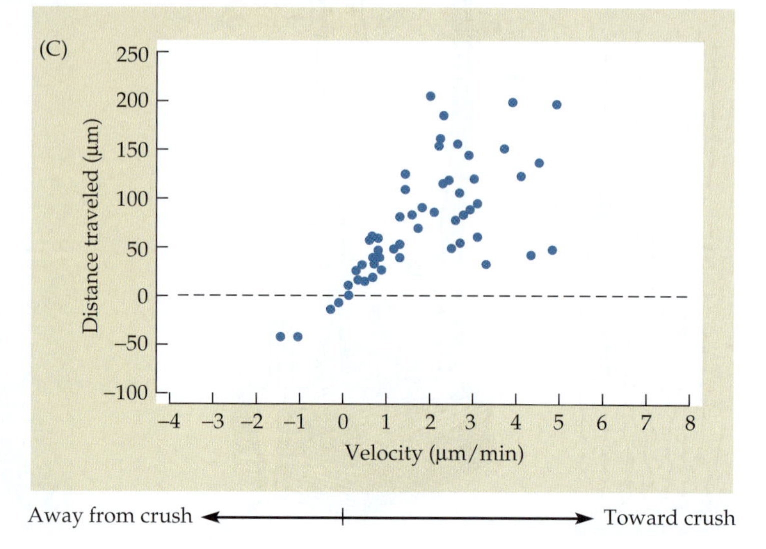

FIGURE 8.11 Migration of Microglial Cells in Injured CNS. (A) Microglia in the leech CNS were stained with a fluorescent nuclear dye (Hoechst 33342). The bundle of axons linking ganglia had been crushed 5 minutes earlier. The extent of the crush is indicated by the dotted line. The nuclei of microglial cells were still evenly distributed at this time. (B)Three hours after the injury, microglial cells had accumulated at the crush site. There they produced the growth-promoting molecule laminin. (C) Velocities and distances traveled by microglial cells as they moved toward a lesion in leech CNS. Microglial cells were tracked by video-microscopy at 10-min intervals in injured leech preparations. In uninjured preparations, microglial cells make only short random movements. (A and B after Chen et al., 2000, micrographs kindly provided by K. J. Muller; C after McGlade-McCulloh et al., 1989.)

Schwann Cells as Pathways for Outgrowth in Peripheral Nerves

The way in which Schwann cells can guide axons and promote outgrowth has been demonstrated in a series of experiments by Thompson and his colleagues.[51–53] They took advantage of the visibility and accessibility of motor nerve terminals on skeletal muscle end plates (Chapter 24). Figure 8.12A provides a diagrammatic summary of these experiments. In adult rats the soleus muscle in the leg was partially denervated. In confirmation of earlier studies it was shown that uninjured motor axons sprouted to occupy denervated end plates. Under these circumstances an axon can sprout to innervate up to five times the usual number of muscle fibers. Axon growth was visualized by an antibody against neurofilaments (red in Figure 8.12). Schwann cells were stained with a different specific antibody (blue in Fugure 8.12). Direct observation showed that the first outgrowth consisted of Schwann cells that extended from denervated muscle fibers to reach the intact axons (part c

[51]Son, Y. J., and Thompson, W. J. 1995. *Neuron* 14: 125–132.

[52]Son, Y. J., and Thompson, W. J. 1995. *Neuron* 14: 133–141.

[53]Son, Y. J., Trachtenberg, J. T., and Thompson, W. J. 1996. *Trends Neurosci.* 19: 280–285.

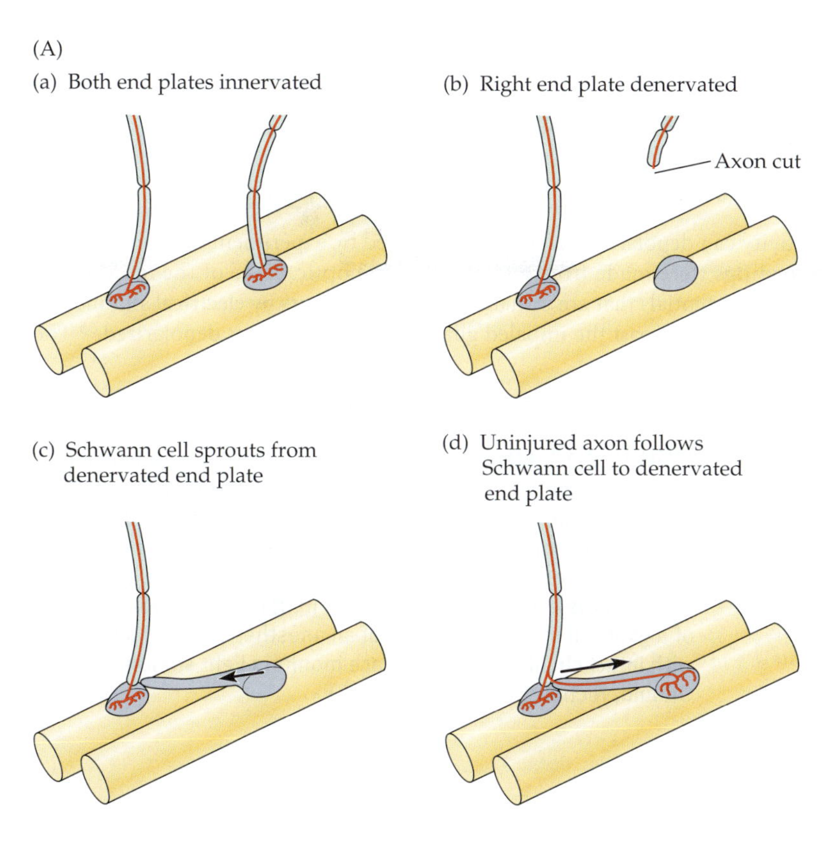

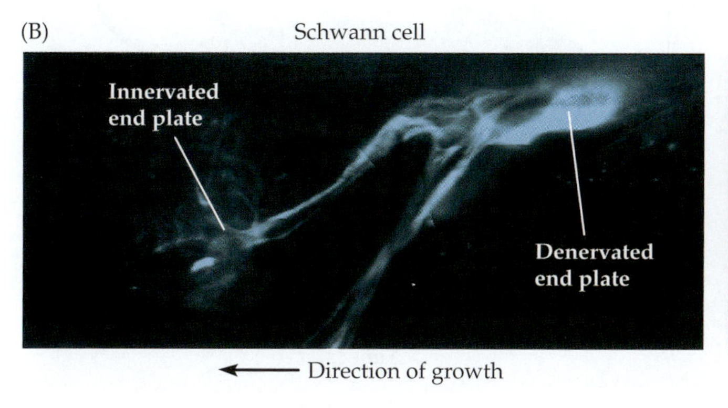

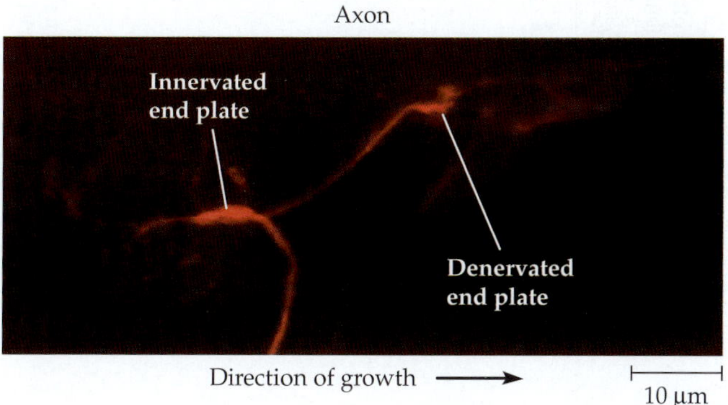

FIGURE 8.12 Role of Schwann Cells in Guiding Axons to denervated motor end plates of skeletal muscle fibers. (A) Schematic representation of the effects of partial denervation of a rat muscle. (a) At normal nerve–muscle synapses, an axon and its Schwann cell (shown in blue) are closely apposed. (b) The axon innervating the right-hand muscle fiber is cut, leading to degeneration of the nerve terminal. (c) In response to this denervation, the terminal Schwann cells on the denervated muscle fiber (blue) grow processes, one of which reaches the nerve terminal at the adjoining muscle fiber. (d) An axonal sprout is induced from the uninjured nerve terminal. It grows along the Schwann cell process to the denervated end plate, which it reinnervates. (B) Growth of axonal sprouts (labeled with antineurofilament antibody) to denervated synapses along processes extended by Schwann cells (labeled with a monoclonal antibody, 4E2, which is specific for Schwann cell bodies and processes). Three days after partial denervation, a neurofilament-labeled nerve sprout has grown from the innervated junction (red) to a denervated junction (blue) by following the Schwann cell process that had grown earlier. The innervated and denervated motor end plates were identified by the patterns of staining of axons and Schwann cells. (After Son and Thompson, 1995; micrographs kindly provided by W. Thompson.)

of Figure 8.12A). Only later did those axons sprout and follow the track laid down by Schwann cells. Implantation of Schwann cells next to an uninjured axon also promoted sprouting without a denervated target.

In other experiments a peripheral nerve was cut; as expected, after a short delay axons grew out from the proximal stump. Once again, however, the first step was the outgrowth of Schwann cell processes that provided the substrate on which the axons could subsequently grow toward their targets.

A Cautionary Note

Although experiments made at neuromuscular synapses clearly implicate Schwann cells in directing neurite outgrowth during regeneration, it seems prudent to limit the scope of such generalizations. For example, in the CNS of the leech, lesioned axons grow back to re-form their original connections after all the large glial cells surrounding them have been killed. Abundant connections and synapses are formed in embryonic CNS in which glial cells are sparsely distributed. Moreover, in the complete absence of neuroglial cells, synapses with an array of normal properties form rapidly in culture. It therefore seems unlikely that "synapses might neither form nor function if there were no glia."[54]

EFFECTS OF NEURONAL ACTIVITY ON GLIAL CELLS

Potassium Accumulation in Extracellular Space

That nerve activity can depolarize glial cells is illustrated by experiments shown in Figure 8.13. The recordings were made from a glial cell in the optic nerve of the mud puppy (*Necturus*). Action potentials that are initiated in the nerve fibers by electrical stimulation or by flashes of light travel past the impaled glial cell, which becomes depolarized.[55] The

[54]Pfrieger, F. W., and Barres, B. A. 1996. *Curr. Opin. Neurobiol.* 6: 615–621, p. 619.

[55]Orkand, R. K., Nicholls, J. G., and Kuffler, S. W. 1966. *J. Neurophysiol.* 29: 788–806.

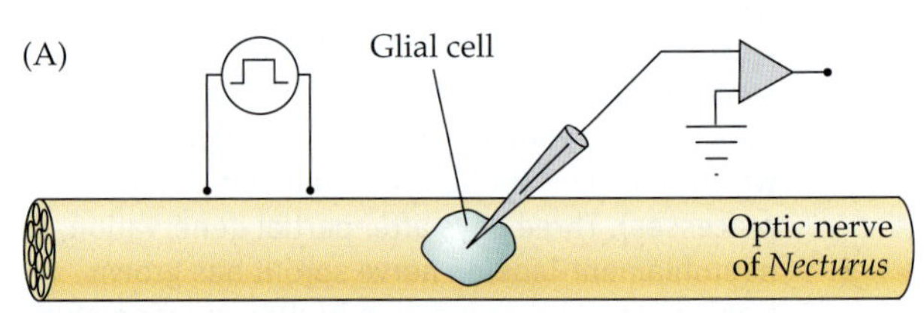

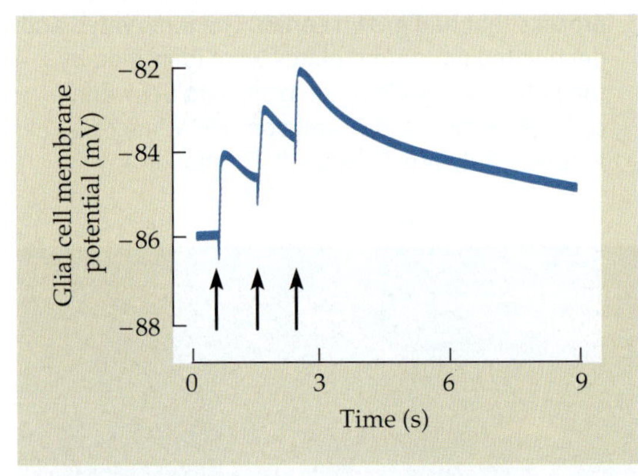

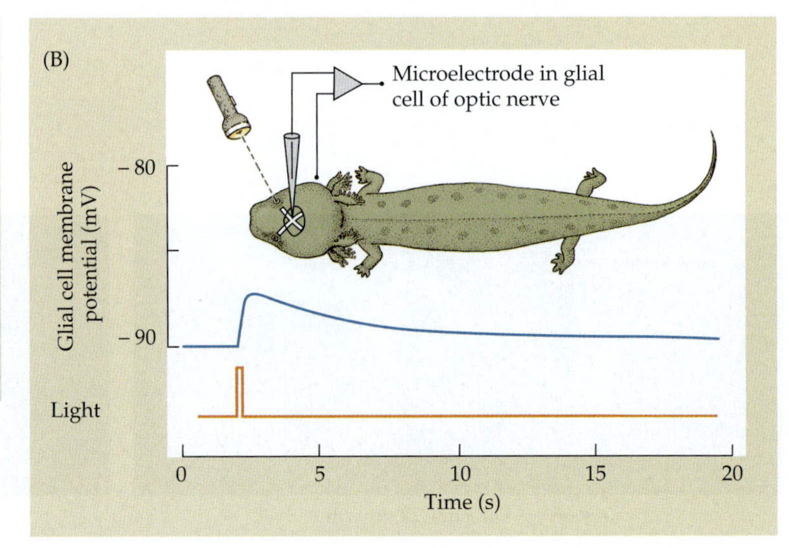

FIGURE 8.13 Effect of Action Potentials on Glial Cells in mud puppy optic nerve. (A) Synchronous impulses evoked by electrical stimulation of nerve fibers cause glial cells to become depolarized. The amplitude of the potentials depends on the number of axons activated and on the frequency of stimulation. (B) Illumination of the eye with a 0.1 s flash of light causes depolarization of a glial cell in the optic nerve of an anesthetized mud puppy, with intact circulation. Lower trace monitors light stimulus. (After Orkand, Nicholls, and Kuffler, 1966.)

depolarization is graded. Similarly, in the mammalian cortex, glial cells become depolarized depending on the number of nerve fibers activated and on the frequency when neurons in their vicinity are activated by stimulation of neural tracts, peripheral nerves, the surface of the cortex, or sensory input.[56]

The cause of glial depolarization is potassium efflux from axons. When potassium accumulates in the intercellular clefts, it changes the $[K]_o/[K]_i$ ratio and alters the membrane potential of glial cells. Potassium-sensitive glass electrodes have been used to measure accumulation of potassium in the extracellular spaces of the CNS during neuronal activity.[57]

Changes of membrane potential in glial cells indicate the level of impulse traffic in their environment. Potassium signaling between neurons and glia is different from specific synaptic activity. Synaptic actions are confined to specialized regions on neuronal cell bodies and dendrites and may be excitatory or inhibitory. In contrast, signaling by potassium is not confined to structures containing receptors but occurs anywhere the glial cell is exposed to potassium. Neurons exposed to increased external K concentrations become less depolarized than glia because the neuronal response deviates from the Nernst equation in the physiological range (Chapter 5).

Current Flow and Potassium Movement through Glial Cells

Currents flow between regions of a cell that are at different potentials. Nerve cells, of course, use this as the mechanism for conduction: Current flows between inactive regions of an axon and the part that is occupied by a nerve impulse. Since glial cells are linked to each other by low-resistance connections, their conducting properties are similar to those of a single elongated cell. Consequently, if several glial cells become depolarized by increased potassium concentrations in their environment, they draw current from the unaffected cells. Similarly, an elongated Müller cell that extends through the thickness of the retina generates current when the potassium concentration increases locally over part of its surface (see Figures 8.6 and 8.14). Inward current in the region of raised $[K]_o$, carried by potassium ions, spreads to other regions of the glial cell, and through gap junctions to other glial cells. The currents generated by glial cells contribute to recordings made from the eye or the skull with extracellular electrodes, known as the electroretinogram (ERG) and the electroencephalogram (EEG). Such measures of overall activity are valuable for the clinical diagnosis of pathological conditions.

Spatial Buffering of Extracellular Potassium Concentration by Glia

One obvious property of glial cells is to separate and group neuronal processes. As a result, the potassium concentration increases around some neurons while others in a separate compartment are protected. An attractive concept is that glial cells regulate the potassium concentration in intercellular clefts, a process known as "spatial buffering."[16,58] According to this hypothesis, glial cells act as conduits for uptake of potassium from the clefts to preserve the constancy of the environment. Since glial cells are coupled to each other, potassium enters in one region and leaves at another, as already described (Figure 8.14). That

[56]Ransom, B. R., and Goldring, S. 1973. *J. Neurophysiol.* 36: 869–878.

[57]Dietzel, I., Heinemann, U., and Lux, H. D. 1989. *Glia* 2: 25–44.

[58]Karowski, C. J., Lu, H-K., and Newman, E. A. 1989. *Science* 224: 579–580.

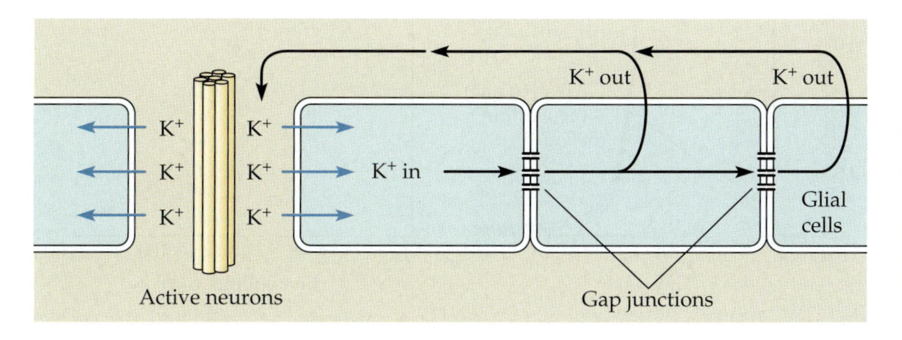

FIGURE 8.14 Potassium Currents in Glial Cells. The glial cells in the diagram are linked by gap junctions. Potassium released by active axons in one region depolarizes the glial cell and enters, causing current flow and outward movement of potassium elsewhere in the glial tissue. The concept of spatial buffering of potassium has been postulated as a mechanism for influencing neuronal function by glial cells.

potassium will move through glial cells as a consequence of potassium buildup is inevitable. It is, however, not simple to estimate quantitatively how much potassium actually moves or how much these movements reduce the extracellular potassium concentration. For such calculations, numerous assumptions about geometry, conductance, diffusion, and active transport of potassium into neurons and glial cells must be made.[59]

Effects of Transmitters on Glial Cells

Transmitters such as GABA, glutamate, glycine, and acetylcholine act on glial membranes to produce depolarizing or hyperpolarizing responses.[25,60–62] Figure 8.15 shows activation of $GABA_A$ receptors by GABA in retinal Müller cells.[63] These receptors are similar to those of neurons in many but not all respects. The physiological functions of the neurotransmitter receptors in glia have not been defined.

Conversely, glial cells play a role in transmitter uptake in the CNS, under normal and pathological conditions. The extracellular concentration of a transmitter such as glutamate, norepinephrine, or glycine that has been released at synapses is reduced in part by diffusion, but mainly by uptake into neurons and into glial cells.[64–66] As in neurons, glutamate transport in glial cells is coupled to inward movement of sodium along its electrochemical gradient (Chapter 4). In the absence of a removal mechanism, excessively high levels of external glutamate can activate NMDA receptors in neurons, and this in turn can lead to calcium entry and cell death. Quantitative estimates indicate that glial cell transport plays a key role in preventing such excessive rises in extracellular glutamate concentration. It has been shown that transgenic mice lacking the astrocytic glutamate transporter GLT-1 develop epilepsy and increased susceptibility to convulsants.[25]

[59]Odette, L. L., and Newman, E. A. 1988. *Glia* 1: 198–210.

[60]Conti, F., et al. 1997. *Mol. Neurobiol.* 14: 1–18.

[61]Reichelt, W., et al. 1997. *Neuroreport* 8: 541–544.

[62]Lohr, C., and Deitmar, J. W. 1997. *J. Exp. Biol.* 200: 2565–2573.

[63]Qian, H., et al. 1996. *Proc. R. Soc. Lond. B* 263: 791–796.

[64]Tanaka, K., et al. 1997. *Science* 276: 1699–1702.

[65]Takahashi, M., et al. 1997. *J. Exp. Biol.* 200: 401–409.

[66]Matsui, K., Hosoi, N., and Tachibana, M. 1999. *J. Neurosci.* 19: 6755–6766.

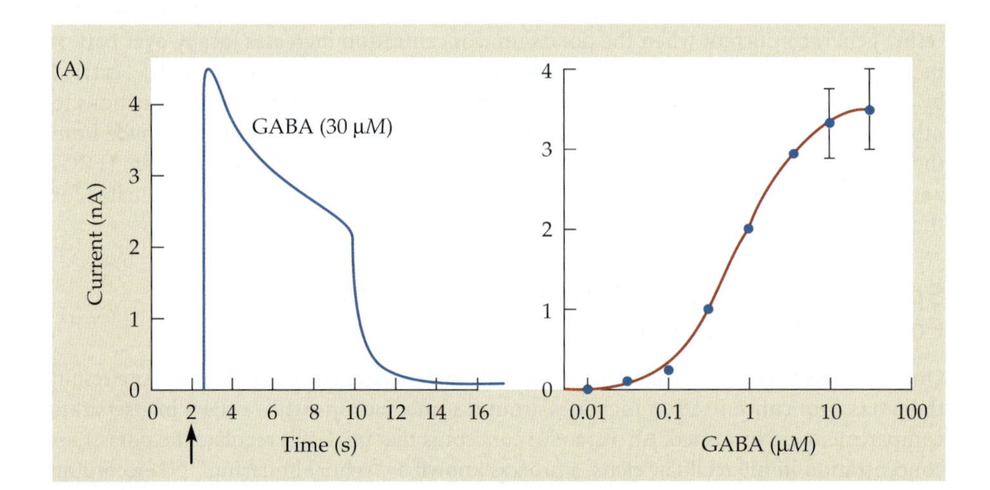

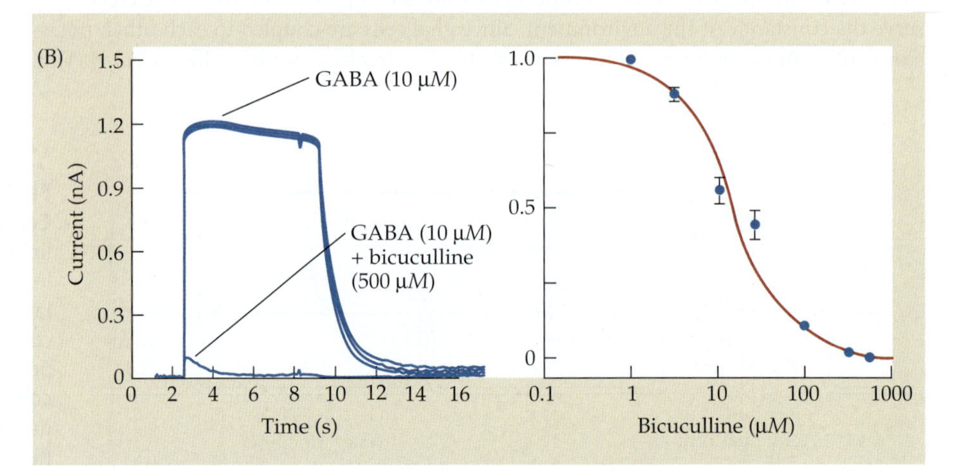

FIGURE 8.15 Responses of Glial Cells to GABA. Responses of Müller glial cells in skate retina to GABA. (A) Current induced by GABA (30 μ*M*) in a glial cell voltage-clamped at 0 mV. The dose–response relation for the peak of the GABA current is shown on the right (error bars indicate standard error of the mean). (B) The effect of GABA was blocked by bicuculline, a $GABA_A$ antagonist. (After Qian et al., 1996.)

Release of Transmitters by Glial Cells

If glial cells themselves become depolarized by raised extracellular potassium or by glutamate, or if intracellular sodium concentration is increased, their membranes transport glutamate out of the glial cell into the extracellular space.[67] This mechanism is similar to that for reversed transport described in Chapter 4. Figure 8.16 illustrates an experiment showing currents associated with glutamate release by glial cells. Such reverse transport can exacerbate the deleterious effects of brain injury. Injured and dying nerve cells release glutamate and K^+ and depolarize glial cells, which in turn release more glutamate.

One of the best-established demonstrations of transmitter release by a satellite cell occurs during regeneration in the peripheral nervous system. At denervated motor end plates, Schwann cells come to occupy the sites vacated by motor nerve terminals. There they release ACh, giving rise to miniature potentials in muscle.[68]

Calcium Waves in Glial Cells

In networks of glial cells in culture or in situ, transient increases in intracellular calcium concentration arise by release from intracellular stores (Figure 8.17). Using fluorescent indicators one can observe such oscillatory waves of increased concentration (or "calcium spikes") as they propagate from glial cell to glial cell through gap junctions.[69,70] The calcium waves can be triggered by depolarization, by transmitters, or by mechanical stimulation. They resemble the calcium waves seen in neuronal networks[71] (Chapter 23). Newman and his colleagues have shown that propagating intracellular calcium waves can trigger glutamate release by glial cells in the retina; this in turn can influence neuronal firing patterns.[70] The physiological significance of these waves is not known.

[67]Billups, B., and Attwell, D. 1996. *Nature* 379: 171–174.

[68]Dennis, M., and Miledi, R. 1974. *J. Physiol.* 237: 431–452.

[69]Newman, E. A., and Zahs, K. R. 1997. *Science* 275: 844–847.

[70]Newman, E. A., and Zahs, K. R. 1998. *J. Neurosci.* 18: 4022–4028.

[71]Verkhratsky, A., Orkand, R. K., and Kettenmann, H. 1998. *Physiol. Rev.* 78: 99–139.

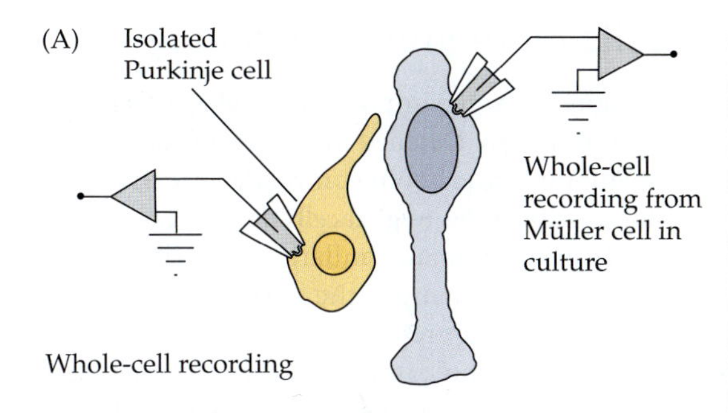

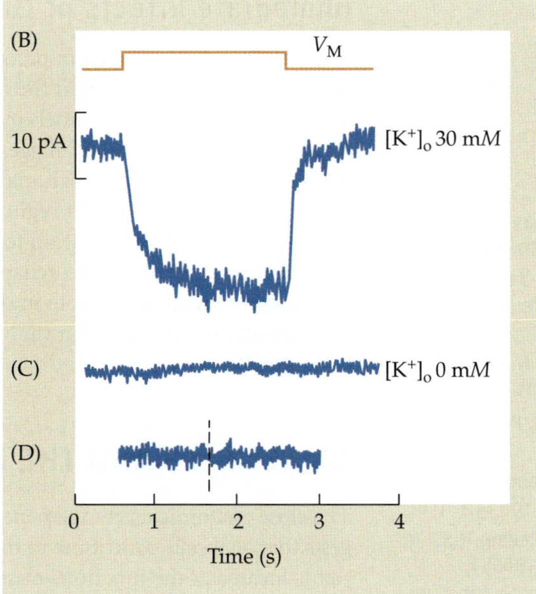

FIGURE 8.16 Release of Glutamate by Glial Cells. Release of glutamate generated by reversal of the glutamate uptake carrier in a Müller cell. (A) Depolarization-induced release of glutamate from a Müller cell (right) is monitored by recording glutamate-elicited currents from an adjacent Purkinje cell (left). The Purkinje cell acts as a detector with high sensitivity and time resolution. (B) Depolarization of the Müller cell from –60 to +20 mV (top trace) elicits an inward current in the nearby Purkinje cell. The Purkinje cell current is generated by activation of its glutamate receptors. The response to glutamate disappears when the Purkinje cell is moved away from the Müller cell (C) or when extracellular K^+ is omitted (D). In fluid containing 0 mM K^+, reverse glutamate transport by the Müller cell is blocked. (After Billups and Attwell, 1996.)

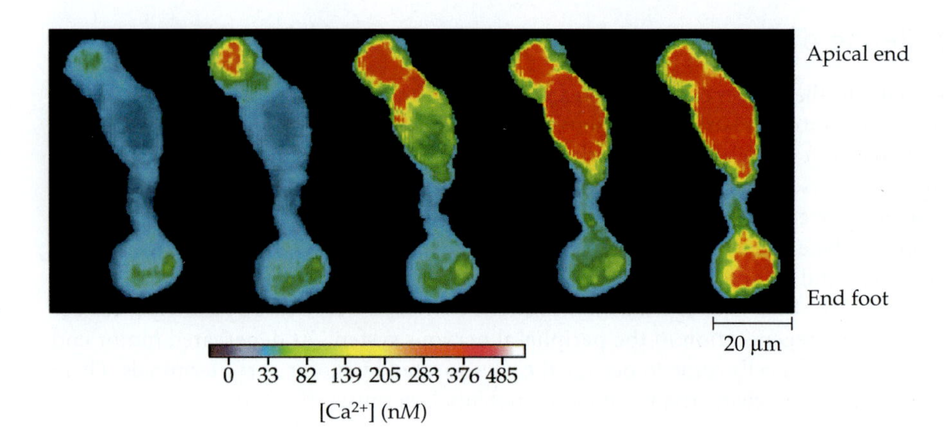

FIGURE 8.17 Calcium Wave in a Dissociated Salamander Müller Cell. The Ca^{2+} wave, elicited by addition of 100 *nM* ryanodine in the absence of extracellular Ca^{2+}, begins at the apical end of the cell and travels toward the cell end foot. The intracellular Ca^{2+} concentration was imaged using the Ca^{2+}-indicator dye fura-2. Images were obtained at 7 s intervals. With glial cells that are coupled by gap junctions, the calcium waves can spread from cell to cell in a continuous progression. Similar waves are initiated by physiological stimuli. They resemble those seen in neurons during development. (After Keirstead and Miller, 1997; image kindly provided by S. Keirstead.)

Transfer of Metabolites from Glial Cells to Neurons

From studies made in honeybees and vertebrates, evidence has accumulated that glial cells can supply nutrients to neurons.[72–75] In aggregates of photoreceptors and Müller cells isolated from guinea pig retina it has been shown that lactate, derived from glycogen in glial cells, accumulates in extracellular fluid. Under conditions of increased neuronal activity, measurements of intracellular and extracellular concentrations of lactate indicate that it is transferred from glia to photoreceptor neurons. The demonstration of such transfer in culture suggests a similar role for glial cells within intact central nervous systems under conditions of high neuronal activity or anoxia.

Immediate Effects of Glial Cells on Neuronal Signaling

The experiments described in preceding sections show the importance of glial cells in development, regeneration, and the formation of myelin. Detailed information is now also available about the ion channels in their membranes and about control of the extracellular fluid composition. Many suggestions for possible roles of glial cells depend on experiments made in culture, which may not apply in the animal. Similarly, glial cells vary in their properties in different regions of the CNS. For example, the optic nerve, while it presents a favorable preparation for studying glia,[76] is not representative of CNS tissue in general, since it contains no synapses. As yet, there is no compelling evidence that glial cells play a crucial, direct role in dynamic activities such as the stretch reflex, or the responses of a complex cell in the visual cortex under normal physiological conditions (Chapter 20).

GLIAL CELLS AND THE BLOOD–BRAIN BARRIER

The close anatomical arrangement of glial cells, capillaries, and neurons in the brain suggests that glial cells contribute to the blood–brain barrier (Box 8.1). The blood–brain barrier is located at the junctions of specialized endothelial cells that line brain capillaries.[77] The role of glial cells has been shown by growing endothelial cells and astrocytes in culture.[78,79] Grown on their own, the endothelial cells are occasionally attached to each other. The presence of astrocytes, however, triggers the formation of complete bands of tight junctions resembling those seen in vivo. These junctions, which completely occlude the intercellular spaces between endothelial cells, account for the impermeability of brain capillaries. Molecules must go through rather than between endothelial cells. Conversely, the presence of endothelial cells from brain capillaries in culture causes distinctive assemblies of membrane particles to appear in astrocytes. These interactions are specific for astrocytes and brain-capillary endothelial cells. No comparable results occur with fibro-

[72]Tsacopoulos, M., and Magistretti, P. J. 1996. *J. Neurosci.* 16: 877–885.
[73]Tsacopoulos, M., Poitry-Yamate, C. L., and Poitry, S. 1997. *J. Neurosci.* 17: 2383–2390.
[74]Coles, J. A., and Abbott, N. J. 1996. *Trends Neurosci.* 19: 358–362.
[75]Ransom, B. R., and Fern, R. 1997. *Glia* 21: 134–141.
[76]Ransom, B. R., and Orkand, R. K. 1996. *Trends Neurosci.* 19: 352–358.
[77]Brightman, M. W., and Reese, T. S. 1969. *J. Cell Biol.* 40: 668–677.
[78]Tao-Cheng, J. H., Nagy, Z., and Brightman, M. W. 1987. *J. Neurosci.* 7: 3293–3299.
[79]Tao-Cheng, J. H., Nagy, Z., and Brightman, M. W. 1990. *J. Neurocytol.* 19: 143–153.

blasts or endothelial cells from peripheral vessels. Methods for promoting the uncoupling of endothelial cells to produce an increase in capillary permeability are now being investigated. They could provide a method for circumventing the blood–brain barrier (see the next section) and allowing therapeutically useful substances to enter the brain that would otherwise not be able to do so.[80]

[80]Rubin, L. L., et al. 1992. *Ann. N.Y. Acad. Sci.* 633: 420–425.

Box 8.1 The Blood–Brain Barrier

A homeostatic system controls the fluid environment in the brain and prevents fluctuations in its composition. This constancy seems particularly important in a system where the activity of so many cells is integrated and where small variations may upset the balance of delicately poised excitatory and inhibitory influences. Within the brain there are three fluid compartments: (1) the blood supplied to the brain through a dense network of capillaries, (2) the cerebrospinal fluid (CSF) that surrounds the bulk of the nervous system and is contained in the internal cavities (ventricles), and (3) the fluid in the intercellular clefts (Figure I).[81]

The **blood–brain barrier** depends on specialized properties of the endothelial cells of capillaries in the brain, which are far less permeable than those supplying organs in the periphery. Proteins, ions, and hydrophilic molecules cannot pass through; lipophilic molecules (such as alcohol) and gases can. A second essential component is the choroid plexus: Specialized epithelial cells surround the choroid plexus capillaries and secrete CSF (Figure IIA). CSF itself is almost devoid of protein, containing only about 1/200 of the amount present in blood plasma. Proteins, electrolytes, transmitters, and a variety of drugs, including penicillin, injected directly into the bloodstream act rapidly on peripheral tissues such as muscle, heart, or glands—but they have little or no effect on the CNS. When administered by way of the CSF, however, the same substances exert a prompt and strong action.[77,82,83]

Within the brain, ions and small particles reach neurons by passing through the narrow 20 nm intercellular clefts

[81]Saunders, N. R., Habgood, M. D., and Dziegielewska, K. M. 1999. *Clin. Exp. Pharmacol. Physiol.* 26: 11–19.

[82]Reese, T. S., and Karnovsky, M. J. 1967. *J. Cell Biol.* 34: 207–217.

[83]Saunders, N., and Dziegielewska, K. M. 1997. *News Physiol. Sci.* 12: 21–31.

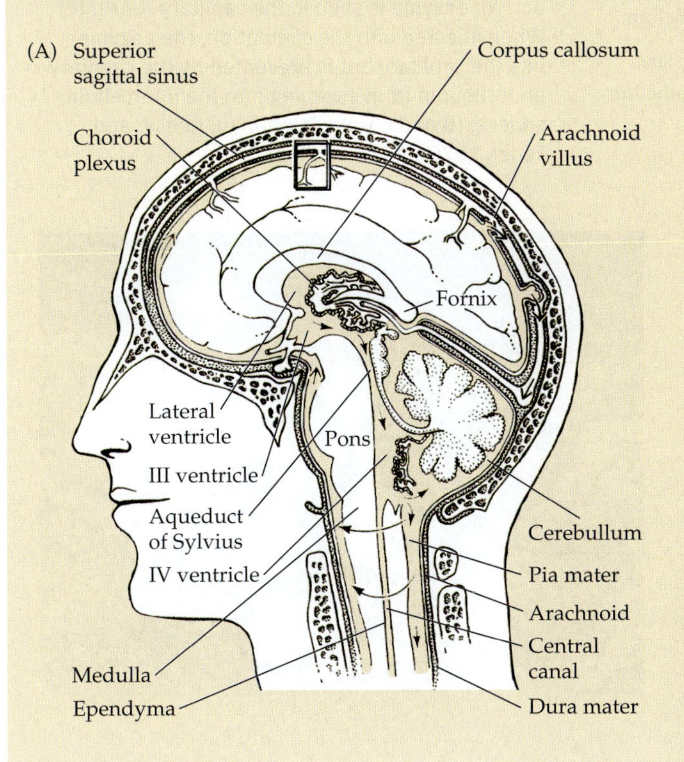

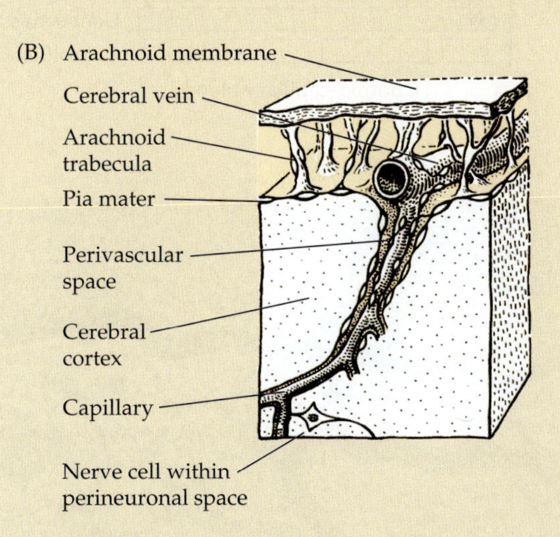

FIGURE I Distribution of Cerebrospinal Fluid and its relation to larger blood vessels and to structures surrounding the brain. (A) All spaces containing CSF communicate with each other. (B) CSF is drained into the venous system through the arachnoid villi.

Box 8.1 *(continued)*

and not through glia. In Figure IIB, after injection of microperoxidase into CSF, electron-dense molecules deposited by peroxidase reaction are lined up in clefts and fill extracellular spaces. This result shows that large molecules can pass between the ependymal cells that line the ventricles and through intercellular clefts. In contrast, the junctions between endothelial cells lining the blood capillaries in the brain provide a barrier. Tracers injected into CSF do not enter the capillaries. Figure IIC shows the opposite result: When enzyme was injected into the circulation, brain capillaries filled with enzyme but none entered the intercellular spaces. During development the barrier is already present, but the range of substances that can enter the CSF is qualitatively and quantitatively different. For example, the CSF has a high protein concentration during development (up to 10%) and contains molecules not detected in adult CSF.[84,85] Knowledge of blood–brain barrier properties is important for understanding pharmacological actions of drugs and their effects on the body.

[84] Balslev, Y., Saunders, N. R., and Mollgard, K. 1997. *J. Neurocytol.* 26: 133–148.

[85] Saunders, N. R., Habgood, M. D., and Dziegielewska, K. M. 1999. *Clin. Exp. Pharmacol. Physiol.* 26: 85–91.

(A)
Brain-capillary blood
Basement membrane
Endothelium
Pericapillary glial cells
Neurons
Gap junction
Subependymal glial cells
Ependyma
Circumferential junction
CSF
Choroid epithelium
Choroid plexus
Connective tissue
Choroid endothelium
Choroid-capillary blood

(B)

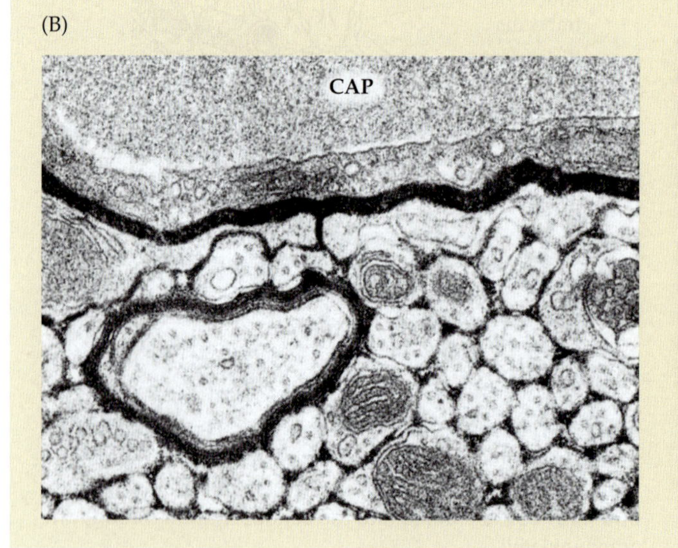

(C)

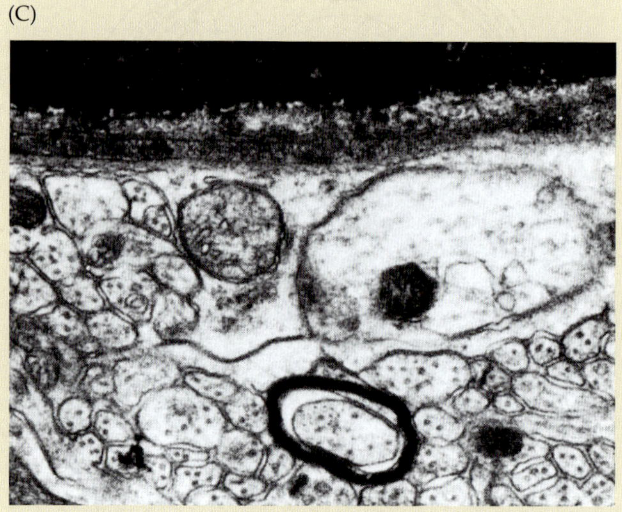

FIGURE II Pathways for Diffusion in the Brain. (A) Schematic presentation of cells involved in the exchange of materials between blood, CSF, and intercellular spaces. Molecules are free to diffuse through the endothelial cell layer that lines capillaries in the choroid plexus, which are not tightly linked. They are, however, restrained by circumferential tight junctions between the choroid epithelial cells, which secrete CSF. There are no barriers between the bulk fluid of CSF and the various cell layers, such as ependyma, glia, and neurons. The endothelial cells lining brain capillaries are joined by circumferential junctions. This prevents free diffusion of molecules out of the blood into the brain tissue or from the brain into the blood. (B) Demonstration in the mouse that the enzyme microperoxidase diffuses freely from cerebrospinal fluid into the intercellular spaces of the brain, which are filled with the dark reaction product. No enzyme is seen in the capillary (CAP). (C) When injected into the circulation, the enzyme fills the capillary but is prevented by the capillary endothelium from escaping into the intercellular spaces. (B and C from Brightman, Reese, and Feder, 1970.)

Astrocytes and Blood Flow through the Brain: A Speculation

We can now take account of three facts to suggest a possible role for astrocytes in the mammalian brain. First, they envelop the brain capillaries with their end feet. (Indeed, it was this feature that led Golgi and so many others to suggest that they provide materials to the neurons.) Second, as we show in Chapters 21 and 22, activity localized to a particular region of the brain causes a dramatic increase in the blood flow through that region, as measured by positron emission tomography (PET), magnetic resonance imaging (MRI), or optical recording. Third, glial cells register the overall level of activity of the neurons in their vicinity. Paulson and Newman[86] have put forward the attractive idea that end feet of depolarized astrocytes might feed back onto capillaries and cause localized vasodilatation. Through glial signaling, active neurons would be supplied with extra oxygen and glucose. As for mechanisms, we can speculate that raised concentrations of potassium, hydrogen ions, or nitric oxide (NO) liberated from glial end feet could influence capillary endothelial cells. Paulson and Newman's proposal is reminiscent of Golgi's original idea, which it reproduces but with signals flowing in the opposite direction. Instead of glial cells carrying nutrients through their cytoplasm from blood to neurons, neuronal activity leads to highly localized vasodilatation and increased blood flow just where it is needed.

GLIAL CELLS AND IMMUNE RESPONSES OF THE CNS

Until recently it was generally accepted that the tissues of the central nervous system were not patrolled by the surveillance mechanisms of the immune system. The blood–brain barrier, the absence of a lymphatic system, and the comparative ease with which grafts can be accepted all suggest the absence of immune responses to foreign antigens. Thus, CNS functions are not disrupted by the massive allergic reactions to a bee sting or poison ivy. Astrocytes in culture and in situ, however, have been shown to react with T lymphocytes, whose activity they can either stimulate or suppress. Evidence has now accumulated to show that microglia and activated T lymphocytes do enter the brain and can mediate acute inflammation of brain tissue.[87,88] The role played by glial cells in interactions between the immune system and the nervous system represents an interesting and challenging problem.

[86]Paulson, O. B., and Newman, E. A. 1987. *Science* 237: 896–898.

[87]Neumann, H., et al. 1996. *Eur. J. Neurosci.* 8: 2582–2590.

[88]Neumann, H., et al. H. 1998. *Proc. Natl. Acad. Sci. USA* 95: 5779–5784.

SUMMARY

- Glial cells in the brain and Schwann cells in the periphery envelop neurons.
- Oligodendrocytes have short processes and myelinate larger axons.
- Astrocytes surround brain capillaries.
- Schwann cells myelinate peripheral axons and produce trophic molecules.
- Microglial cells remove debris after damage and are involved in inflammatory responses of the nervous system.
- Glial cells have more negative resting potentials than neurons and do not produce action potentials.
- Glial cells are electrically coupled to each other but not to neurons.
- Glial cell membranes contain ion channels for sodium, potassium, and calcium, as well as receptors, pumps, and transporters.
- Glial cells play roles in development, in regeneration, and in homeostatic control of the fluid environment of neurons.

SUGGESTED READING

General Reviews

Kettenmann, H., and Ransom, B. R. (eds.). 1995. *Neuroglia*. Oxford University Press, New York.

Kuffler, S. W., and Nicholls, J. G. 1966. The physiology of neuroglial cells. *Ergeb. Physiol.* 57: 1–90.

Newman, E., and Reichenbach, A. 1996. The Müller cell: A functional element of the retina. *Trends Neurosci.* 19: 307–312.

Paulson, O. B., and Newman, E. A. 1987. Does the release of potassium from astrocyte endfeet regulate cerebral blood flow? *Science* 237: 896–898.

Porter, J. T., and Mccarthy, K. D. 1998. Astrocytic neurotransmitter receptors in-situ and in-vivo. *Prog. Neurobiol.* 51: 439–455.

Ransom, B. R., and Orkand, R. K. 1996. Glial-neuronal interaction in non-synaptic areas of the brain: Studies in the optic nerve. *Trends Neurosci.* 19: 352–358.

Ransom, B. R., and Sontheimer, H. 1992. The neurophysiology of glial cells. *J. Clin. Neurophysiol.* 9: 224–251.

Saunders, N. R., Habgood, M. D., and Dziegielewska, K. M. 1999. Barrier mechanisms in the brain. I. Adult brain. *Clin. Exp. Pharmacol. Physiol.* 26: 11–19.

Original Papers

Araque, A., Li, N., Doyle, R. T., and Haydon, P. G. 2000. SNARE protein-dependent glutmate release from astrocytes. *J. Neurosci.* 20: 666–673.

Kuffler, S. W., and Potter, D. D. 1964. Glia in the leech central nervous system: Physiological properties and neuron-glia relationship. *J. Neurophysiol.* 27: 290–320.

Newman, E. A., and Zahs, K. R. 1998. Modulation of neuronal activity by glial cells in the retina. *J. Neurosci.* 18: 4022–4028.

Pareek, S., Notterpek, L., Snipes, G. J., Naef, R., Sossin, W., Laliberte, J., Iacampo, S., Suter, U., Shooter, E. M., and Murphy, R. A. 1997. Neurons promote the translocation of peripheral myelin protein 22 into myelin. *J. Neurosci.* 17: 7754–7762.

Paulson, O. B., and Newman, E. A. 1987. Does the release of potassium from astrocyte endfeet regulate cerebral blood flow? *Science* 237: 896–898.

Rasband, M. N., Peles, E., Trimmer, J. S, Levinson, S. R., Lux, S. E., and Shrager, P. 1999. Dependence of nodal sodium channel clustering on paranodal axoglial contact in the developing CNS. *J. Neurosci.* 19: 7516–7528.

Son, Y. J., and Thompson, W. J. 1995. Nerve sprouting in muscle is induced and guided by processes extended by Schwann cells. *Neuron* 14: 133–141.

Takahashi, M., Billups, B., Rossi, D., Sarantis, M., Hamann, M., and Attwell, D. 1997. The role of glutamate transporters in glutamate homeostasis in the brain. *J. Exp. Biol.* 200: 401–409.

Tsacopoulos, M., Poitry-Yamate, C. L., and Poitry, S. 1997. Ammonium and glutamate released by neurons are signals regulating the nutritive function of a glial-cell. *J. Neurosci.* 17: 2383–2390.

Zheng, C., Heintz, N., and Hatten, M. E. 1996. CNS gene encoding astrotactin, which supports neuronal migration along glial fibers. *Science* 272: 417–419.

9 PRINCIPLES OF DIRECT SYNAPTIC TRANSMISSION

SYNAPSES ARE POINTS OF CONTACT between nerve cells and their targets where signals are handed on from one cell to the next. At electrical synapses, current flows from the presynaptic nerve terminal into the postsynaptic cell to alter its membrane potential. Electrical transmission is prevalent in the nervous systems of invertebrates and also occurs at synapses in the mammalian CNS. At chemical synapses, the arrival of the action potential in the presynaptic nerve terminal causes neurotransmitter molecules to be released. At direct chemical synapses, the transmitter binds to ionotropic receptors in the membrane of the postsynaptic cell that are themselves ion channels. As a result, the conformation of the receptor changes, the channel opens, ions flow, and the membrane potential changes. At indirect chemical synapses, metabotropic receptors and intracellular second messengers are involved (Chapter 10).

The channels opened at excitatory synapses allow cations to enter, driving the membrane potential toward threshold. At inhibitory synapses, transmitters open channels that are permeable to anions, tending to keep the membrane potential negative to threshold. At both excitatory and inhibitory synapses, the direction of current flow is determined by the balance of concentration and electrical gradients acting on the permeant ions.

Synapses between motor nerves and muscle fibers provided important preparations for understanding the mechanisms of direct chemical synaptic transmission. In the mammalian central nervous system, directly mediated excitatory and inhibitory transmission occurs at synapses where acetylcholine, glutamate, γ-aminobutyric acid, serotonin, and purines are released to activate ionotropic receptors.

More than one type of transmitter may be released at a single chemical synapse, and many transmitters act both rapidly, by binding to and opening ion channels directly, and more slowly through indirect mechanisms.

Information is transmitted from one neuron to another, or from a neuron to an effector cell such as a muscle fiber, at a specialized point of contact, the **synapse**. In this chapter we consider the principles of **direct synaptic transmission**. Direct, or "fast," synapses include electrical synapses, at which transmission is mediated by the flow of current from the pre- to the postsynaptic cell. More common are direct chemical synapses, at which a neurotransmitter is released from the axon terminal and binds to receptors on the target cell that are ion channels.

Subsequent chapters describe how chemical neurotransmitters influence target cells indirectly by binding to receptors that trigger intracellular signaling cascades (Chapter 10), how neurotransmitters are released (Chapter 11), how neurotransmitters are synthesized and stored within axon terminals (Chapter 13), and how the efficacy of transmission can be dramatically altered by repetitive activity (Chapter 12). Because of the variety and complexity of synaptic interactions, it is useful to begin by reviewing the history of some of the fundamental ideas.

NERVE CELLS AND SYNAPTIC CONNECTIONS

It was not always clear that the two main components of the synapse—presynaptic terminal and postsynaptic cell—are morphologically distinct. In the second half of the nineteenth century there was vigorous debate between proponents of the **cell theory**, who considered that neurons were independent units, and those who thought that nerve cells were a **syncytium**, interconnected by protoplasmic bridges. Not until the late nineteenth century did it become generally accepted that nerve cells are independent units. It remained for electron microscopy to obtain definitive evidence that each neuron is surrounded completely by its own plasma membrane. Even so, electron microscopy and other modern techniques eventually revealed that some neurons are, in fact, connected by channels, called connexons, that permit direct intercellular flow of ions and other small molecules (Chapter 7).

The disagreement about synaptic structure was accompanied by a parallel disagreement about function. In 1843, Du Bois-Reymond showed that flow of electrical current was involved in both muscle contraction and nerve conduction, and it required only a small extension of this idea to conclude that transmission of excitation from nerve to muscle was also due to current flow (Figure 9.1A).[1] Du Bois-Reymond himself favored an alternative explanation: the secretion by the nerve terminal of an excitatory substance that then caused muscle contraction (Figure 9.1B). However, the idea of animal electricity had such a potent hold on people's thinking that it was more than 100 years before contrary evidence finally overcame the assumption of electrical transmission between nerve and muscle and, by extension, between nerve cells in general.

[1]Du Bois-Reymond, E. 1848. *Untersuchungen über thierische Electricität*. Reimer, Berlin.

FIGURE 9.1 Electrical and Chemical Synaptic Transmission. (A) At electrical synapses, current flows directly from one cell to another through connexons, intercellular channels that cluster to form gap junctions. (B) At a chemical synapse, depolarization of the presynaptic nerve terminal triggers the release of neurotransmitter molecules, which interact with receptors on the postsynaptic neuron, causing excitation or inhibition.

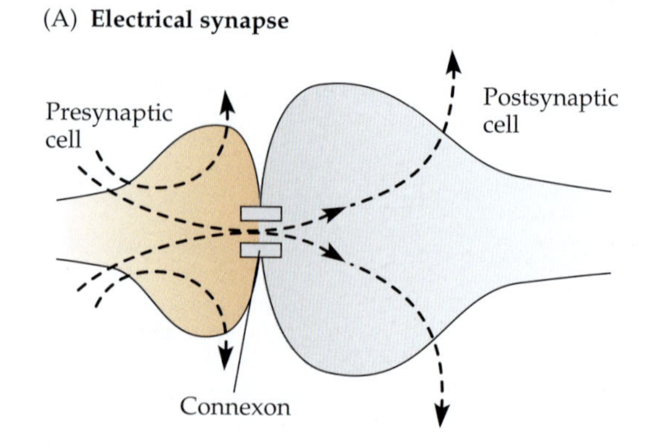

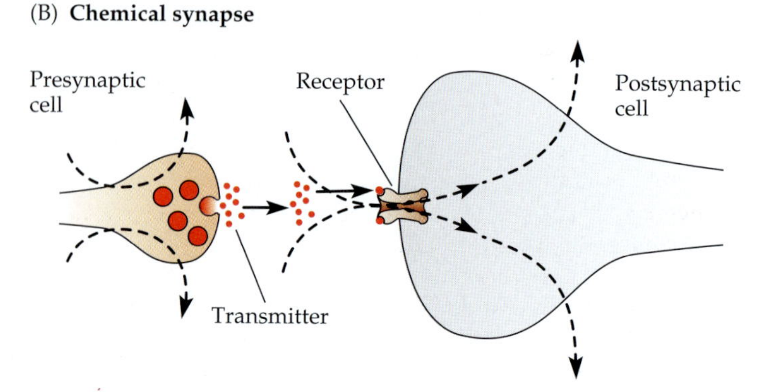

Chemical Synaptic Transmission in the Autonomic Nervous System

One reason that the idea of chemical synaptic transmission seemed unattractive is the speed of signaling between nerve cells or between nerve and muscle. The fraction of a second that intervenes between stimulation of a motor axon and contraction of the corresponding muscle did not appear to provide sufficient time for a chemical neurotransmitter to be released from the nerve terminal and interact with receptors on the postsynaptic target to cause excitation. This difficulty did not exist in the autonomic nervous system that controls glands and blood vessels. There the effects of nerve stimulation are slow and prolonged, waxing and waning over the course of many seconds (Chapter 16).

In 1892, Langley proposed that synaptic transmission in mammalian autonomic ganglia was chemical, rather than electrical, based on the observation that transmission through the mammalian ciliary ganglion was blocked selectively by nicotine.[2] About a decade later, Elliot pointed out that an extract from the adrenal gland, adrenaline (epinephrine), mimicked the action of sympathetic nerve stimulation when applied directly to the target tissues and suggested that it might be secreted by nerve terminals as a transmitter.[3] It was not until 1921, however, that Otto Loewi did a direct and simple experiment that established the chemical nature of transmission at autonomic synapses between the vagus nerve and the heart.[4] He perfused the heart of a frog and stimulated the vagus nerve, thereby slowing the heartbeat. When the fluid from the inhibited heart was transferred to a second unstimulated heart, it too began to beat more slowly. Apparently stimulation of the vagus nerve had caused an inhibitory substance to be released into the perfusate. Loewi and his colleagues demonstrated in subsequent experiments that the substance was mimicked in every way by acetylcholine (ACh).

It is an amusing sidelight that Loewi had the idea for his experiment in a dream, wrote it down in the middle of the night, but could not decipher his writing the next morning. Fortunately the dream returned, and this time Loewi took no chances; he rushed to the laboratory and performed the experiment. Later he reflected:

> On mature consideration, in the cold light of morning, I would not have done it. After all, it was an unlikely enough assumption that the vagus should secrete an inhibitory substance; it was still more unlikely that a chemical substance that was supposed to be effective at very close range between nerve terminal and muscle be secreted in such large amounts that it would spill over and, after being diluted by the perfusion fluid, still be able to inhibit another heart.[4]

Subsequently, in the early 1930s, the role of ACh in synaptic transmission in ganglia in the autonomic nervous system was firmly established by Feldberg and his colleagues.[5] Highlights of such experiments and ideas from the beginning of the twentieth century are contained in the writings of Dale, for several decades one of the leading figures in British physiology and pharmacology.[6] Among his many contributions are the clarification of the action of acetylcholine at synapses in autonomic ganglia and the establishment of its role in neuromuscular transmission.

Henry Dale (left) and Otto Loewi, mid-1930s. (Kindly provided by Lady Todd and W. Feldberg.)

Chemical Synaptic Transmission at the Vertebrate Skeletal Neuromuscular Junction

In 1936, Dale and his colleagues demonstrated that stimulation of motor nerves to vertebrate skeletal muscle caused the release of ACh.[7] In addition, when ACh was injected into arteries supplying the muscle, it caused a large synchronous contraction of the muscle fibers. Later, electrical recording techniques were used to characterize the change in muscle fiber membrane potential evoked by stimulation of the motor axon and demonstrate that it could be mimicked by application of ACh. It was also shown that the responses to both nerve stimulation and direct application of ACh were antagonized by curare, a South American Indian arrow poison that blocks the ACh receptor, and potentiated by eserine, a drug that prevents the hydrolysis of ACh by the enzyme acetylcholinesterase. Such experiments served to establish firmly the idea of chemical synaptic transmission at nerve–muscle synapses. As we shall see, elaborate pre- and postsynaptic specializations allow chemical synaptic transmission to take place on a millisecond timescale. Thus, the

[2]Langley, J. N., and Anderson, H. K. 1892. *J. Physiol.* 13: 460–468.

[3]Elliot, T. R. 1904. *J. Physiol.* 31: (Proc.) xx–xxi.

[4]Loewi, O. 1921. *Pflügers Arch.* 189: 239–242.

[5]Feldberg, W. 1945. *Physiol. Rev.* 25: 596–642.

[6]Dale, H. H. 1953. *Adventures in Physiology*. Pergamon, London.

[7]Dale, H. H., Feldberg, W., and Vogt, M. 1936. *J. Physiol.* 86: 353–380.

long-championed hypothesis of electrical transmission, which had been accepted almost universally for about 100 years on inadequate evidence, was finally forsaken for good experimental reasons, only to be shown valid after all at a different kind of synapse.

ELECTRICAL SYNAPTIC TRANSMISSION

Identification and Characterization of Electrical Synapses

In 1959, Furshpan and Potter, using intracellular microelectrodes to record from nerve fibers in the abdominal nerve cord of the crayfish, discovered an **electrical synapse** between neurons that mediated the animal's escape reflex (Figure 9.2A).[8] They demonstrated that an action potential in a lateral giant fiber led, by direct intercellular current flow, to depolarization of a giant motor fiber leaving the cord (Figure 9.2B). The depolarization was sufficient to initiate an action potential in the postsynaptic fiber. The electrical coupling was in one direction only; depolarization of the postsynaptic fiber did not lead to presynaptic depolarization (Figure 9.2C). In other words, the synapse **rectified.**

Unlike the crayfish giant synapse, most electrical synapses do not exhibit rectification, but conduct equally well in both directions. We now know that the morphological spe-

[8]Furshpan, E. J., and Potter, D. D. 1959. *J. Physiol.* 145: 289–325.

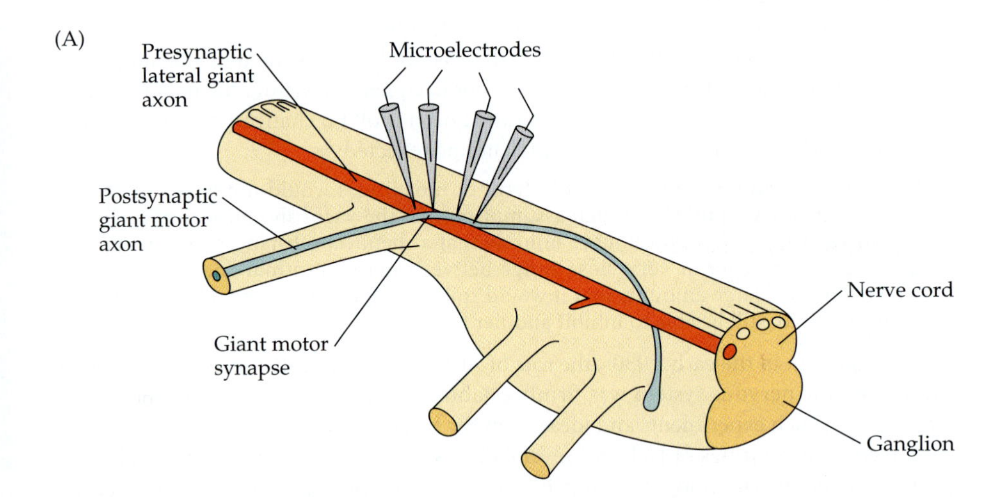

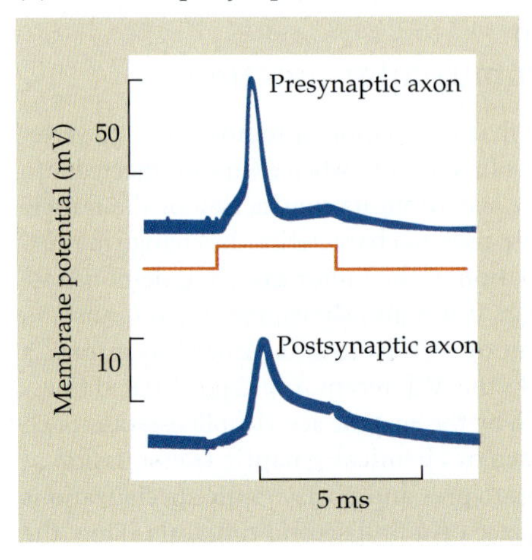

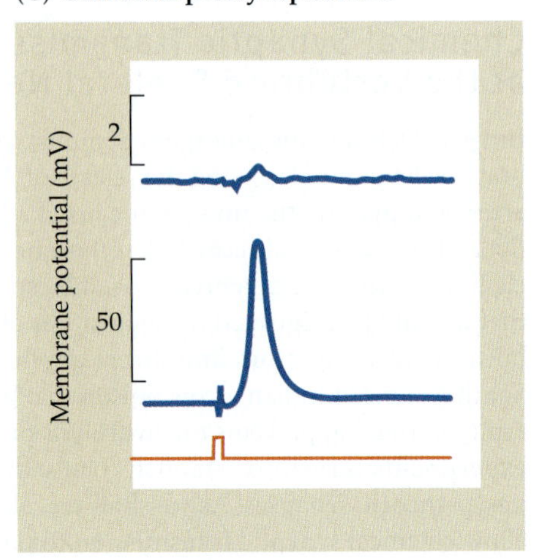

FIGURE 9.2 Electrical Synaptic Transmission at a Giant Synapse in the crayfish central nervous system. (A) The experimental preparation. The presynaptic lateral giant axon makes an electrical synapse with the postsynaptic giant motor axon in the abdominal nerve cord. (B) Depolarization of the presynaptic axon spreads immediately to the postsynaptic fiber. In this case each cell reaches threshold and fires an action potential. (C) When the postsynaptic axon is stimulated directly to give an action potential, depolarization spreads poorly from the postsynaptic to the presynaptic axon. The synapse is said to rectify. (After Furshpan and Potter, 1959.)

cialization for electrical coupling at the crayfish giant synapse and other electrical synapses is the **gap junction**.[9] Gap junctions are formed by an assembly of connexons, which allows current to pass from one cell to the next (Chapter 7).

Electrical transmission has been demonstrated at a wide variety of synapses,[10,11] such as those between motoneurons in the spinal cord of the frog,[12] sensory neurons in the mesencephalic nucleus of the rat,[13] pyramidal cells in the rat hippocampus,[14] and horizontal cells in the zebra fish retina.[15] In the leech, electrical coupling between pairs of sensory neurons has the remarkable property that depolarization spreads readily from either cell to the other, but hyperpolarization spreads poorly; that is, the electrical connections are doubly rectifying.[16,17]

The degree of electrical coupling between cells is usually expressed as a **coupling ratio**; a ratio of 1:4 means that one-fourth of the presynaptic voltage change appears in the postsynaptic cell. For cells to be strongly electrically coupled, the resistance of the junction between the cells must be very low and there must be a reasonable match between the sizes of the presynaptic and postsynaptic elements (Chapter 7).

Electrical and chemical transmission often coexist at a single synapse. Such combined electrical and chemical synapses were first found between cells of the avian ciliary ganglion, where a chemical synaptic potential (produced by ACh) is preceded by an electrical coupling potential (see Figure 9.3).[18] Similar synapses occur widely in vertebrates—for example, onto spinal interneurons of the lamprey[19] and spinal motoneurons of the frog.[20] Postsynaptic cells also may receive separate chemical and electrical synaptic inputs from different sources. For example, in leech ganglia (Chapter 15) motor neurons receive three distinct types of synaptic input from sensory neurons signaling three different modalities; one input is chemical, one electrical, and one combined electrical and chemical.[21]

Synaptic Delay at Chemical and Electrical Synapses

One characteristic of electrically mediated synaptic transmission is the absence of a **synaptic delay**. At chemical synapses there is a pause of approximately 1 ms between the arrival of an impulse in the presynaptic terminal and the appearance of an electrical potential in the postsynaptic cell. The delay is due to the time taken for the terminal to release transmitter (Chapter 11). At electrical synapses there is no such delay; current spreads instantaneously from one cell to the next.

The presence of both electrical and chemical transmission at the same synapse provides a convenient means of comparing the two modes of transmission. This is illustrated in Figure 9.3, which shows intracellular records from a cell in the ciliary ganglion of the chick. Stimulation of the preganglionic nerve leads to an action potential in the postsynaptic cell, with very short latency (Figure 9.3A). When the cell is hyperpolarized slightly (Figure 9.3B), the action potential arises at a later time, revealing an early, brief depolarization that, because the cell has been hyperpolarized, is now subthreshold. This depolarization is an electrical coupling potential, produced by current flow from the presynaptic nerve terminal into the cell. Further hyperpolarization (Figure 9.3C) blocks the initiation of the action potential altogether, revealing the underlying chemical synaptic potential. The cells, then, have the property that, under normal conditions, initiation of a postsynaptic action potential by chemical transmission is preempted by electrical coupling. In this example the coupling potential precedes the chemical synaptic potential by about 2 ms, giving us a direct measure of the synaptic delay. Additional experiments on these cells have shown that the electrical coupling is bidirectional; that is, the synapses do not rectify.

There are several advantages to electrical transmission. One is that electrical synapses are more reliable than chemical synapses; transmission is less likely to fail because of synaptic depression or to be blocked by neurotoxins.[22,23] A second advantage is the greater speed of electrical transmission. Speed is important in rapid reflexes involving escape reactions, in which the saving of a millisecond may be crucial for surviving attack by a predator. Other functions include the synchronization of electrical activity of groups of cells[24,25] and intercellular transfer of key molecules such as calcium, ATP, and cAMP.[26] Dopamine has been shown to modulate the activity of gap junctions between cells in the retina.[27,28] Thus, gap junctions do not function merely as passive connections, but can be dynamic components of neuronal circuits.

[9]Bennett, M. V. 1997. *J. Neurocytol.* 26: 349–366.
[10]Loewenstein, W. 1981. *Physiol. Rev.* 61: 829–913.
[11]Llinás, R. 1985. In *Gap Junctions*. Cold Spring Harbor Laboratory, Cold Spring Harbor, NY, pp. 337–353.
[12]Grinnell, A. D. 1970. *J. Physiol.* 210: 17–43.
[13]Baker, R., and Llinás, R. 1971. *J. Physiol.* 212: 45–63.
[14]MacVicar, B. A., and Dudek, F. E. 1981. *Science* 213: 782–785.
[15]McMahon, D. G. 1994. *J. Neurosci.* 14: 1722–1734.
[16]Baylor, D. A., and Nicholls, J. G. 1969. *J. Physiol.* 203: 591–609.
[17]Acklin, S. E. 1988. *J. Exp. Biol.* 137: 1–11.
[18]Martin, A. R., and Pilar, G. 1963. *J. Physiol.* 168: 443–463.
[19]Rovainen, C. M. 1967. *J. Neurophysiol.* 30: 1024–1042.
[20]Shapovalov, A. I., and Shiriaev, B. I. 1980. *J. Physiol.* 306: 1–15.
[21]Nicholls, J. G., and Purves, D. 1972. *J. Physiol.* 225: 637–656.
[22]Brodin, L., et al. 1994. *J. Neurophysiol.* 72: 592–604.
[23]Dityatev, A. E., and Clamann, H. P. 1996. *J. Neurophysiol.* 76: 3451–3459.
[24]Draguhn, A., et al. 1998. *Nature* 394: 189–192.
[25]Brivanlou, I. H., Warland, D. K., and Meister, M. 1998. *Neuron* 20: 527–539.
[26]Loewenstein, W. R. 1999. *The Touchstone of Life*. Oxford University Press, New York.
[27]DeVries, S. H., and Schwartz, E. A. 1989. *J. Physiol.* 414: 351–375.
[28]Hampson, E. C. G. M., Vaney, D. I., and Weiler, R. 1992. *J. Neurosci.* 12: 4911–4922.

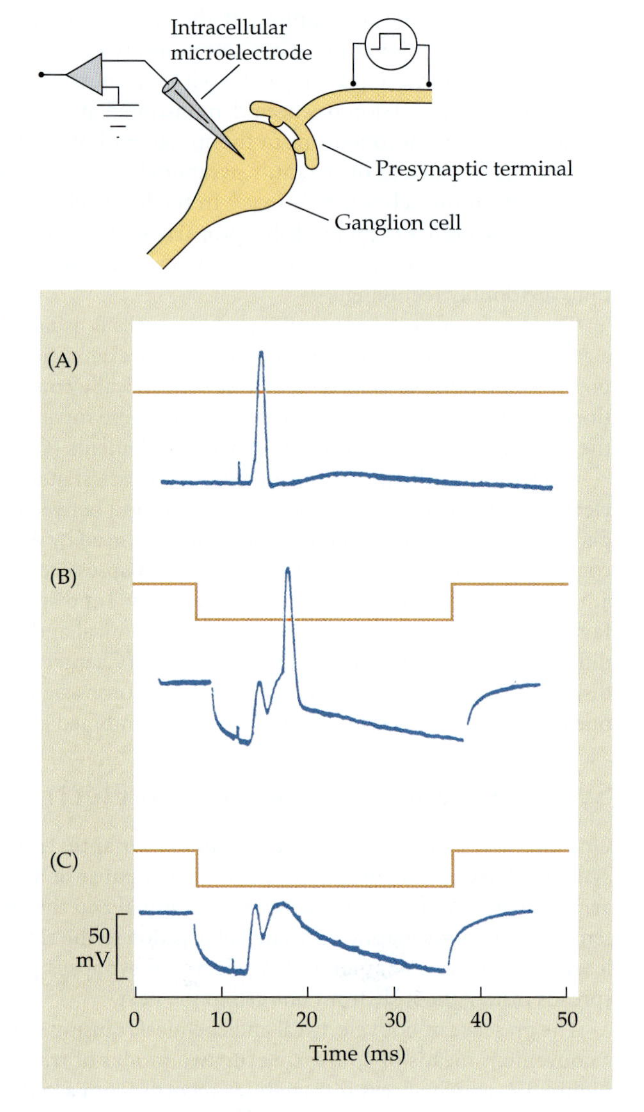

FIGURE 9.3 Electrical and Chemical Synaptic Transmission in a chick ciliary ganglion cell, recorded with an intracellular microelectrode. (A) Stimulation of the preganglionic nerve produces an action potential in the ganglion cell (lower trace). (B) When the ganglion cell is hyperpolarized by passing current through the recording electrode (upper trace), the cell reaches threshold later, revealing an earlier, transient depolarization. This depolarization is an electrical synaptic potential (coupling potential), caused by current flow into the ganglion cell from the presynaptic terminal. In A, the electrical synaptic potential depolarized the ganglion cell to threshold, initiating an action potential. (C) Slightly greater hyperpolarization prevents the ganglion cell from reaching threshold, exposing a slower chemical synaptic potential. The chemical synaptic potential follows the coupling potential with a synaptic delay of about 2 ms at room temperature. (After Martin and Pilar, 1963.)

CHEMICAL SYNAPTIC TRANSMISSION

Certain obvious questions arise when one considers the elaborate scheme for **chemical synaptic transmission**, which entails the secretion of a specific chemical by a nerve terminal and its interaction with postsynaptic receptors (see Figure 9.1B). How does the terminal liberate the chemical? Is there a special feature of the action potential mechanism that causes secretion? How is the interaction of a transmitter with its postsynaptic receptor rapidly converted into excitation or inhibition? The release process will be considered in detail in Chapter 11; the present discussion is concerned with the question of how transmitters act on the postsynaptic cell at direct chemical synapses.

Many of the pioneering studies of chemical synaptic transmission were done on relatively simple preparations, such as the skeletal neuromuscular junction of the frog. At the time, this particular preparation had the advantage that the neurotransmitter (ACh) had been definitively identified. (Many years later was it shown that ATP is co-released with acetylcholine by motor nerve endings to act as a second transmitter.[29])

Synaptic Structure

Chemical synapses are complex in structure. Figure 9.4 illustrates the morphological features of the neuromuscular junction of the frog. Individual axons branch from the incom-

[29]Silinsky, E. M., and Redman, R. S. 1996. *J. Physiol.* 492: 815–822.

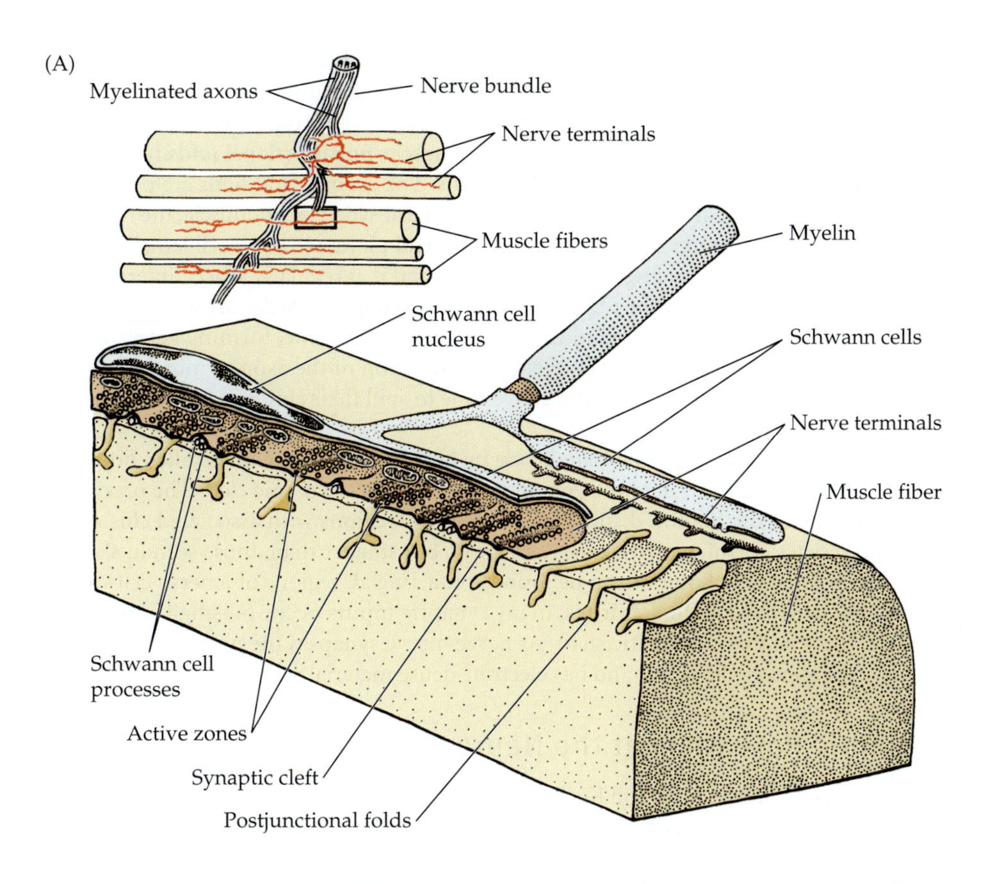

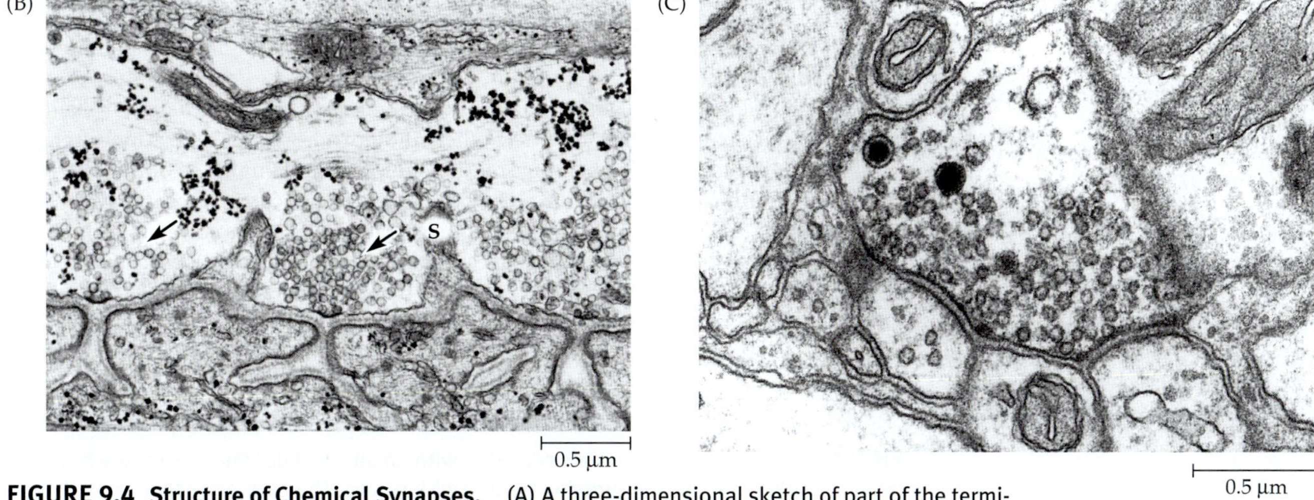

FIGURE 9.4 Structure of Chemical Synapses. (A) A three-dimensional sketch of part of the terminal arbor of a motor axon at the frog skeletal neuromuscular junction. The low-power view (inset) shows several skeletal muscle fibers and their innervation (the region depicted in more detail is indicated by the box). Synaptic vesicles are clustered in the nerve terminal in special regions opposite the openings of the postjunctional folds. These regions, called active zones, are the sites of transmitter release into the synaptic cleft. Fingerlike processes of Schwann cells extend between the terminal and the postsynaptic membrane, separating active zones. (B) Electron micrograph of a longitudinal section through a portion of the neuromuscular junction. In the nerve terminal, clusters of vesicles lie over thickenings in the presynaptic membrane—the active zones (arrows). Schwann cell processes (S) separate the clusters. In the muscle, postjunctional folds open into the synaptic cleft directly under the active zone. The band of fuzzy material in the cleft, which follows the contours of the postjunctional folds, is the synaptic basal lamina. (C) Electron micrograph of synapses in the central nervous system of the leech. As at the frog neuromuscular junction, clusters of synaptic vesicles are focused on dense regions of the presynaptic membrane, forming active zones, and are juxtaposed to postsynaptic densities. (B kindly provided by U. J. McMahan; C kindly provided by K. J. Muller.)

ing motor nerve, lose their myelin sheath, and give off terminal branches that run in shallow grooves on the surface of the muscle. The **synaptic cleft** between the terminal and the muscle membrane is about 30 nm wide. Within the cleft is the **basal lamina**, which follows the contours of the muscle fiber surface. On the muscle, **postjunctional folds** radiate into the muscle fiber from the cleft at regular intervals. The grooves and folds are peculiar to skeletal muscle and are not a general feature of chemical synapses. In muscle, the region of postsynaptic specialization is known as the **motor end plate**. Schwann cell lamellae cover the nerve terminal, sending fingerlike processes around it at regularly spaced intervals.

Stephen Kuffler, John Eccles, and Bernard Katz (left to right) in Australia, about 1941.

Within the cytoplasm of the terminal, clusters of **synaptic vesicles** are associated with electron-dense material attached to the presynaptic membrane, forming **active zones**. Synaptic vesicles are sites of ACh storage; upon excitation of the axon terminal, they fuse with the presynaptic membrane at the active zone to spill their contents into the synaptic cleft by **exocytosis** (Chapter 11).[30]

Synapses on nerve cells are usually made by nerve terminal swellings called **boutons**, which are separated from the postsynaptic membrane by a synaptic cleft. The presynaptic membrane of the bouton displays electron-dense regions with associated clusters of synaptic vesicles, forming active zones similar to, but smaller than, those seen in skeletal muscle (Figure 9.4C). Boutons can be found contacting all regions of a nerve cell—dendrites, cell body, axon. On dendrites, many synaptic inputs occur on small **spines** that project from the main dendritic shaft. At nerve–nerve synapses the postsynaptic membrane often appears thickened and has electron-dense material associated with it.

Synaptic Potentials at the Neuromuscular Junction

Early studies by Eccles, Katz, and Kuffler used extracellular recording techniques to study the **end plate potential** (EPP) in muscle.[31,32] The EPP is the depolarization of the end plate region of the muscle fiber following motor nerve excitation, produced by ACh released from the presynaptic nerve terminals. Synaptic potentials similar to the EPP are seen in nerve cells. A synaptic potential that excites a postsynaptic cell is usually referred to as an **excitatory postsynaptic potential** (**EPSP**), and one that inhibits as an **inhibitory postsynaptic potential** (IPSP).

Normally the amplitude of the EPP in a skeletal muscle fiber is much greater than that needed to initiate an action potential. When an appropriate concentration of curare (about 1 μM) is added to the bathing solution, however, the amplitude of the EPP is reduced to below threshold, so that it no longer evokes (and becomes obscured by) the action potential (Figure 9.5). The effect of curare is graded; if the concentration is increased far enough, the end plate potential disappears entirely as all the receptors become blocked.

The intracellular microelectrode[33] was used by Fatt and Katz[34] to study in detail the time course and spatial distribution of the EPP in muscle fibers treated with curare. They stimulated the motor nerve and recorded the EPP intracellularly at various distances from

[30]Heuser, J. E. 1989. *Q. J. Exp. Physiol.* 74: 1051–1069.

[31]Eccles, J. C., and O'Connor, W. J. 1939. *J. Physiol.* 97: 44–102.

[32]Eccles, J. C., Katz, B., and Kuffler, S. W. 1942. *J. Neurophysiol.* 5: 211–230.

[33]Ling, G., and Gerard, R. W. 1949. *J. Cell. Comp. Physiol.* 34: 383–396.

[34]Fatt, P., and Katz, B. 1951. *J. Physiol.* 115: 320–370.

FIGURE 9.5 Synaptic Potentials recorded with an intracellular microelectrode from a mammalian neuromuscular junction treated with curare. The curare concentration in the bathing solution was adjusted so that the amplitude of the synaptic potential was near threshold and so on occasion evoked an action potential in the muscle fiber. (From Boyd and Martin, 1956.)

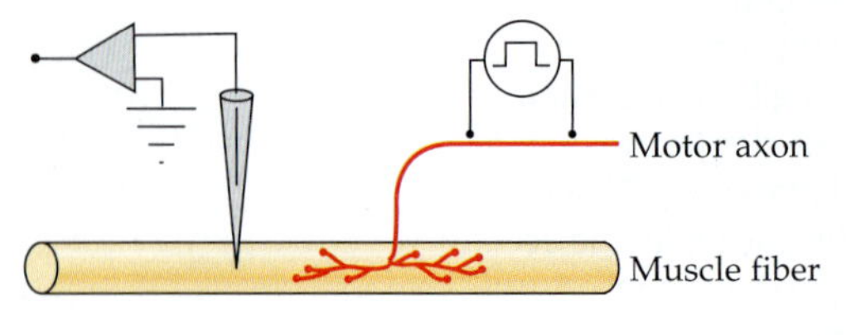

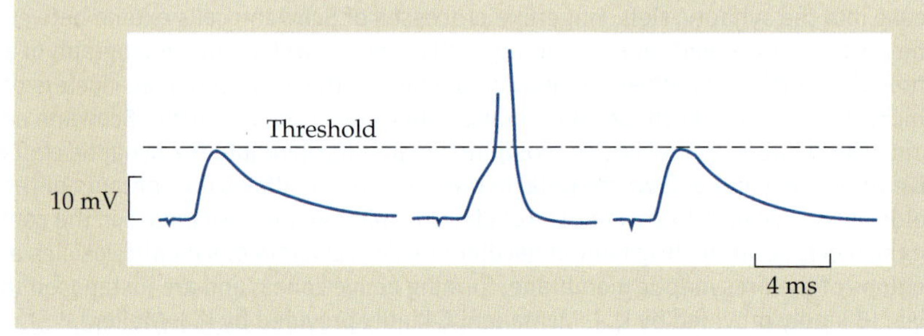

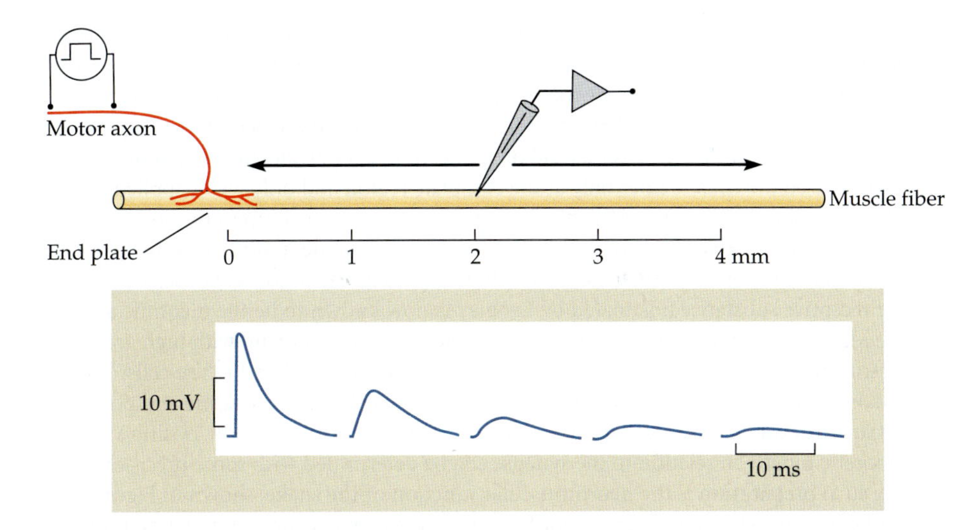

FIGURE 9.6 Decay of Synaptic Potentials with Distance from the end plate region of a muscle fiber. As the distance from the end plate increases, synaptic potentials recorded by an intracellular electrode decrease in size and rise more slowly. (After Fatt and Katz, 1951.)

the end plate (Figure 9.6). At the end plate the depolarization rose rapidly to a peak and then declined slowly over the next 10 to 20 ms. As they moved the recording microelectrode farther and farther away from the end plate, the EPP amplitude became progressively smaller and its time to peak progressively longer. Fatt and Katz showed that after reaching its peak, the EPP decayed at a rate that was consistent with the time constant of the muscle fiber membrane, and that the decrement in EPP peak amplitude with distance from the end plate was predicted by the muscle fiber cable properties. Accordingly, they concluded that the end plate potential is generated by a brief surge of current that flows into the muscle fiber locally at the end plate and causes a rapid depolarization. The potential then decays passively, spreading beyond the end plate in both directions as it dies away.

Mapping the Region of the Muscle Fiber Receptive to ACh

The existence of special properties of skeletal muscle fibers in the region of innervation has been known since the beginning of the twentieth century. For example, Langley[35] assumed the presence of a "receptive substance" around motor nerve terminals, based on the finding that this region of the muscle fiber was particularly sensitive to various chemical agents, such as nicotine. Shortly after the introduction of the glass microelectrode for intracellular recording, microelectrodes were also used for discrete application of ACh (and later other drugs as well) to the end plate region of muscle.[36] The technique is illustrated in Figure 9.7A.

[35]Langley, J. N. 1907. *J. Physiol.* 36: 347–384.

[36]Nastuk, W. L. 1953. *Fed. Proc.* 12: 102.

FIGURE 9.7 Mapping the Distribution of ACh Sensitivity by Ionophoresis at the frog neuromuscular junction. (A) An ACh-filled pipette is placed close to the neuromuscular junction, and ACh is ejected from the tip by a brief, positive, voltage pulse (ionophoresis). An intracellular microelectrode is used to record the response from the muscle fiber. (B) Responses to small ionophoretic pulses of ACh applied at different distances from the axon terminal (indicated by the blue dots in [A]). The amplitude and rate of rise of the response decrease rapidly as ACh is applied farther from the terminal. (After Peper and McMahan, 1972.)

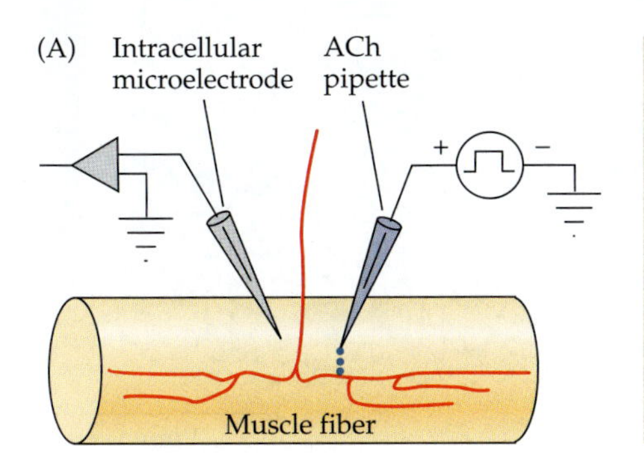

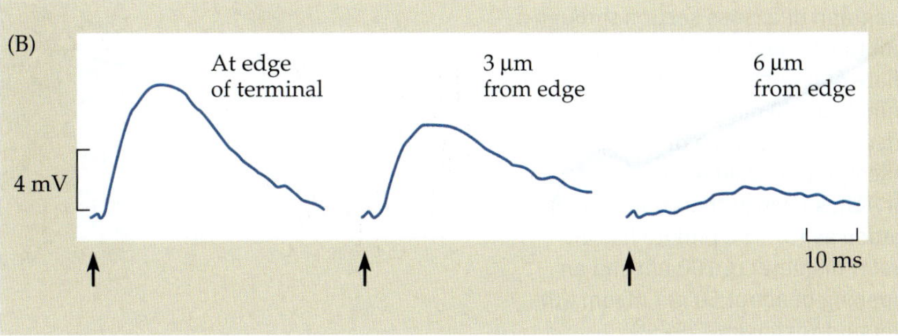

A microelectrode is inserted into the end plate of a muscle fiber for recording membrane potential while an ACh-filled micropipette is held just outside the fiber. To apply ACh, a brief positive voltage pulse is applied to the top of the pipette, causing a spurt of positively charged ACh ions to leave the pipette tip. This method of ejecting charged molecules from pipettes is known as ionophoresis. Using this method of application, del Castillo and Katz showed that ACh depolarized the muscle fiber only at the end plate region and only when applied to the outside of the fiber.[37] When the ACh-filled pipette is in close apposition to the end plate region, the response to ionophoresis is rapid (Figure 9.7B). Movement of the pipette by only a few micrometers results in a reduction in amplitude and slowing of the response.

The receptive substance postulated by Langley is now known to be the nicotinic acetylcholine receptor. The technique of ionophoresis made it possible to map with high accuracy the distribution of postsynaptic ACh receptors in muscle fibers[38] and nerve cells.[39] This method is particularly useful with thin preparations in which the presynaptic and postsynaptic structures can be resolved with interference contrast optics[40] and the position of the ionophoretic pipette in relation to the synapse can be determined with some precision.

One such preparation is the neuromuscular junction of the snake, shown in Figure 9.8. The end plates in snake muscle are about 50 μm in diameter, resembling in their compactness those seen in mammals. Each axon terminal consists of 50 to 70 terminal swellings, analogous to synaptic boutons, from which transmitter is released. The swellings rest in craters sunk into the surface of the muscle fiber. An electron micrograph of such a synapse is shown in Figure 9.8B, again illustrating the characteristic features observed at all chemical synapses. Figure 9.8C shows an electron micrograph of a typical ionophoretic micropipette. The opening is about 50 nm, similar in size to a synaptic vesicle.

The sharp delineation of sensitivity to ACh at the snake neuromuscular junction can be demonstrated dramatically in muscle fibers in which the motor nerve terminal has been removed by bathing the muscle in a solution of the enzyme collagenase, which frees the terminal without damaging the muscle fiber.[41,42] The process of lifting off the terminals is shown in Figure 9.9A. Each of the boutons leaves behind a circumscribed crater lined with the exposed postsynaptic membrane. This is shown in more detail in Figure 9.9B, in which an ACh-filled micropipette points at an empty crater. If the tip of the pipette is placed on the postsynaptic membrane, 1 pC (picocoulomb) of charge passed through the pipette releases enough ACh to cause, on the average, a 5 mV depolarization. The sensitivity of the membrane is then said to be 5000 mV/nC (Figure 9.9C). In contrast, at a distance of about 2 μm, just outside the crater, the same amount of ACh applied to the extrasynaptic membrane produces a response that is 50 to 100 times smaller. Along the rims of the craters the sensitivity fluctuates over a wide range.

[37]del Castillo, J., and Katz, B. 1955. *J. Physiol.* 128: 157–181.

[38]Miledi, R. 1960. *J. Physiol.* 151: 24–30.

[39]Dennis, M. J., Harris, A. J. and Kuffler, S. W. 1971. *Proc. R. Soc. Lond. B* 177: 509–539.

[40]McMahan, U. J., Spitzer, N. C., and Peper, K. 1972. *Proc. R. Soc. Lond. B* 181: 421–430.

[41]Betz, W. J., and Sakmann, B. 1973. *J. Physiol.* 230: 673–688.

[42]Kuffler, S. W., and Yoshikami, D. 1975. *J. Physiol.* 244: 703–730.

Other Techniques for Determining the Distribution of ACh Receptors

A second way to determine the distribution of ACh receptors is to use α-bungarotoxin, a snake toxin that binds selectively and irreversibly to nicotinic ACh receptors. The distri-

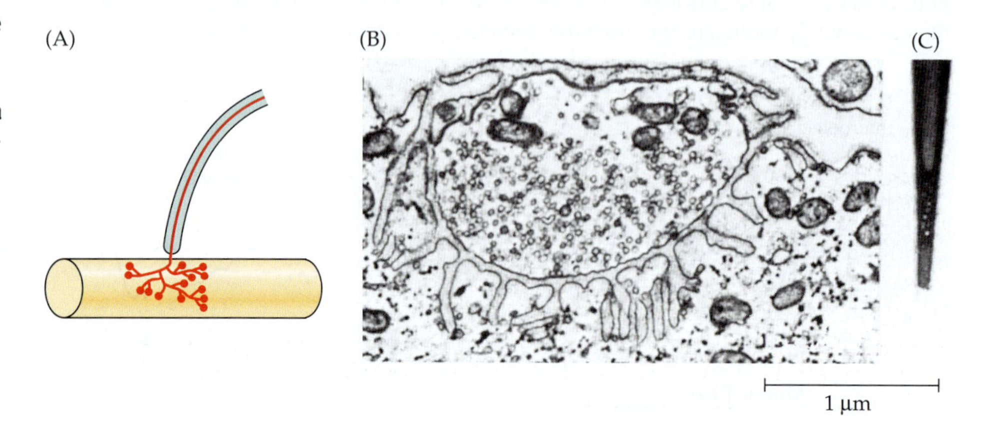

FIGURE 9.8 Skeletal Neuromuscular Junction of the Snake. (A) An end plate on a skeletal muscle of a snake. The axon terminates in a cluster of boutons. (B) Electron micrograph of a cross section through a bouton. Synaptic vesicles, which mediate ACh release from the nerve terminal, are 50 nm in diameter. (C) Electron micrograph of the tip of a micropipette used for ionophoresis of ACh, shown at the same magnification as (B). The pipette has an outer diameter of 100 nm and an opening of about 50 nm. (From Kuffler and Yoshikami, 1975.)

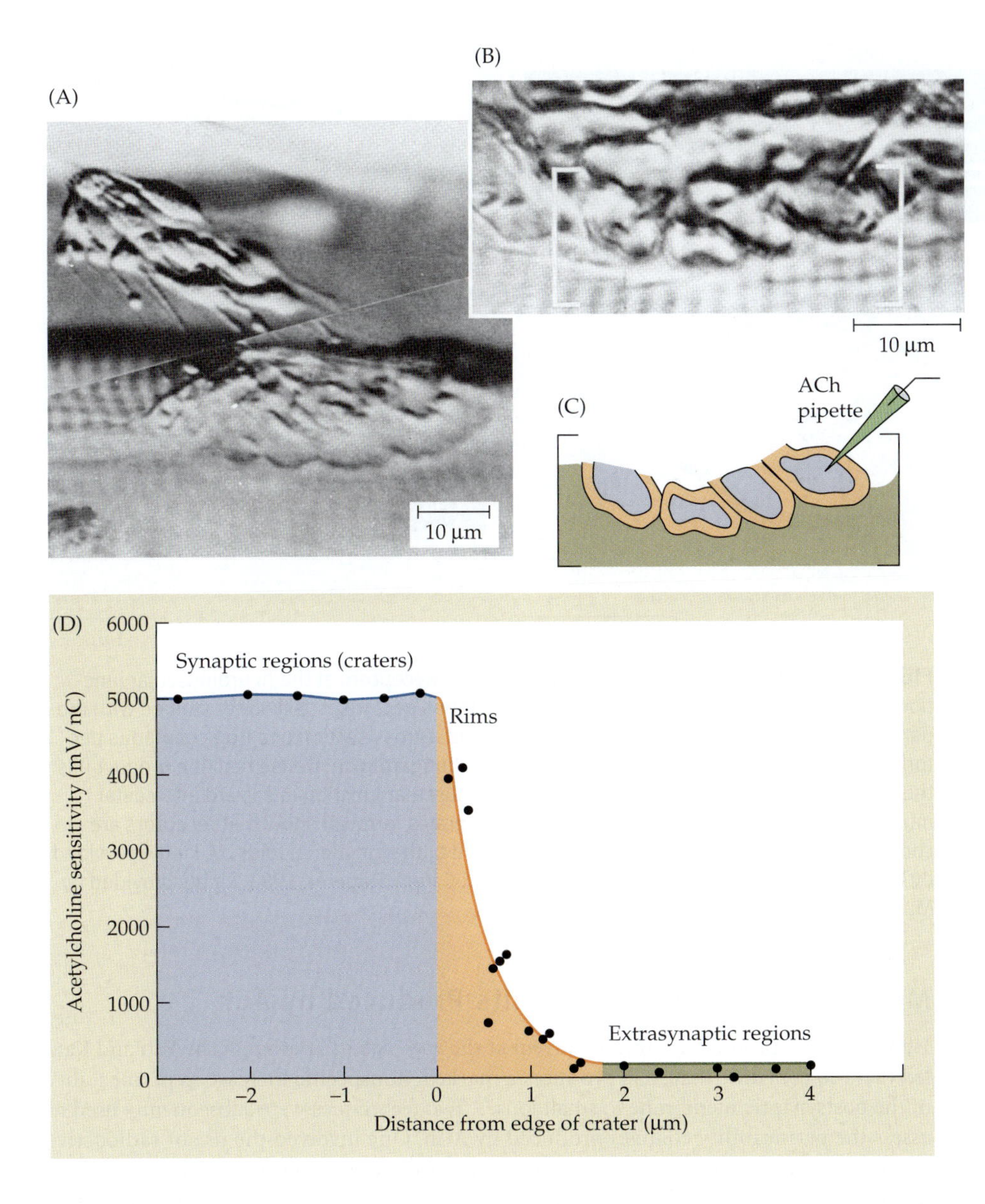

FIGURE 9.9 Acetylcholine Receptor Distribution at the skeletal neuromuscular junction of the snake. (A) Light micrograph showing the removal of the nerve terminal from the neuromuscular junction in a muscle treated with collagenase. (B) Light micrograph of the postsynaptic surface of the muscle cell exposed by removal of the nerve terminal. An ACh-filled pipette (entering from upper right; compare with part C) points to a crater that had been occupied by a terminal bouton. (C) Drawing of the area bracketed in B, showing the position of four craters formerly occupied by terminal boutons. Blue areas represent postsynaptic membrane within the craters, orange areas are the crater rims, and green areas are extrasynaptic regions. (D) Distribution of ACh sensitivity. The craters have a uniformly high sensitivity to ACh (5000 mV/nC); the sensitivity declines steeply at the rims of the craters; extrasynaptic regions have a uniformly low ACh sensitivity (100 mV/nC). (After Kuffler and Yoshikami, 1975.)

bution of bound toxin can be visualized using histochemical techniques. For example, fluorescent markers can be attached to α-bungarotoxin and the distribution of receptors visualized by fluorescence microscopy[43] (Figure 9.10A); or the enzyme horseradish peroxidase (HRP) can be linked to α-bungarotoxin and its dense reaction product visualized in the electron microscope[44] (Figure 9.10B). Such techniques confirm that receptors are highly restricted to the membrane immediately beneath the axon terminal. Even more precise, quantitative estimates of the concentration of ACh receptors can be obtained using radioactive α-bungarotoxin and autoradiography (Figure 9.10C).[45] By counting the number of silver grains exposed in the emulsion, the density of receptors can be determined. In muscle the density is highest along the crests and upper third of the junctional folds (about $10^4/\mu m^2$); the density in extrasynaptic regions is much lower (about $5/\mu m^2$).[46] Transmitter receptors are highly concentrated in the postsynaptic membrane at synapses throughout the central and peripheral nervous systems.

Another way to apply neurotransmitters and other substances to nerve and muscle fiber membranes is by **pressure ejection**. Brief pulses of pressure are applied to the top of a pipette to drive solution from the tip. The method has an advantage over ionophoresis in that the substance to be applied need not carry a net charge. Substances can also be applied to cells or membrane patches using fast flow techniques that allow extremely rapid changes between solutions of known composition.[47]

[43]Ravdin, P., and Axelrod, D. 1977. *Anal. Biochem.* 80: 585–592.

[44]Burden, S. J., Sargent, P. B., and McMahan, U. J. 1979. *J. Cell Biol.* 82: 412–425.

[45]Fertuck, H. C., and Salpeter, M. M. 1974. *Proc. Natl. Acad. Sci. USA* 71: 1376–1378.

[46]Salpeter, M. M. 1987. In *The Vertebrate Neuromuscular Junction*. Alan R. Liss, New York, pp. 1–54.

[47]Heckmann, M., and Dudel, J. 1997. *Biophys. J.* 72: 2160–2169.

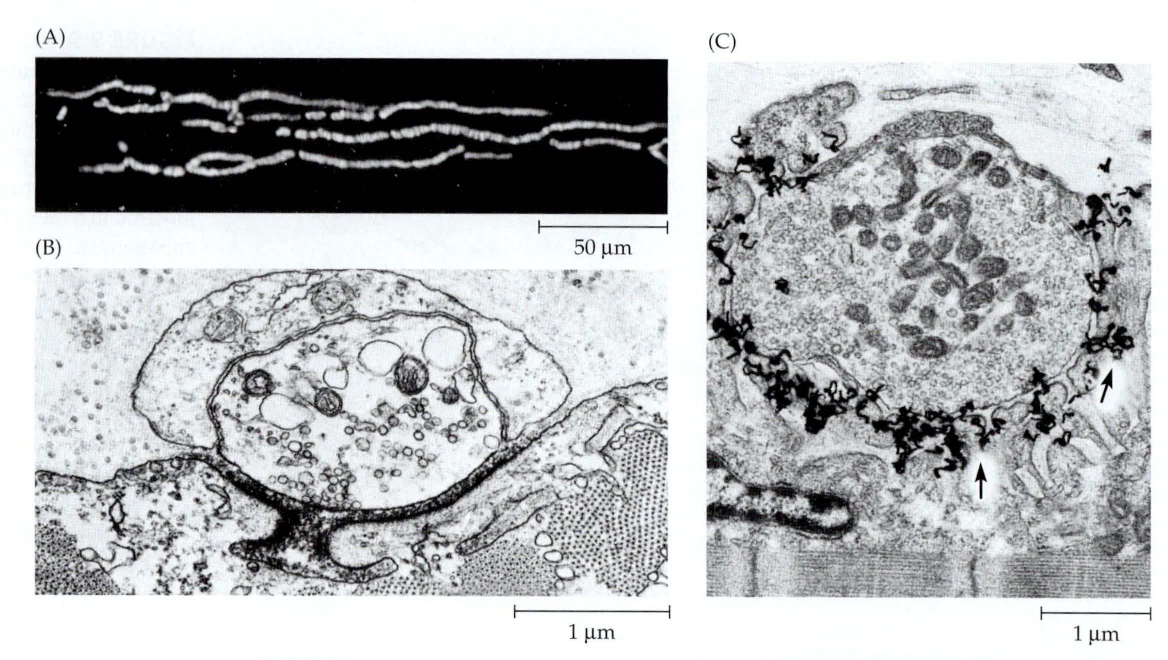

FIGURE 9.10 Visualizing the Distribution of ACh Receptors at the neuromuscular junction. (A) Fluorescence micrograph of a frog cutaneous pectoris muscle fiber stained with rhodamine α-bungarotoxin. (B) Electron micrograph of a cross section of a frog cutaneous pectoris neuromuscular junction labeled with HRP-α-bungarotoxin. Dense reaction product fills the synaptic cleft. (C) Autoradiograph of a neuromuscular junction in a lizard intercostal muscle labeled with [^{125}I]-α-bungarotoxin. Silver grains (arrows) show that receptors are concentrated at the tops and along the upper third of the junctional folds. (A kindly provided by W. J. Betz; B kindly provided by U. J. McMahan; C from Salpeter, 1987, kindly provided by M. M. Salpeter.)

Measurement of Ionic Currents Produced by ACh

How does ACh produce an inward current at the end plate? Experiments by Fatt and Katz led them to conclude that ACh produces a marked, nonspecific increase in permeability of the postsynaptic membrane to small ions.[33] Two techniques were subsequently used to assess the permeability changes produced by ACh. One involved the use of radioactive isotopes, which showed that the permeability of the postsynaptic membrane was increased to sodium, potassium, and calcium, but not to chloride.[48] This experiment provided convincing evidence concerning the ion species involved but did not reveal the details of the conductance changes, their timing, or their voltage dependence. This information was provided by voltage clamp experiments, first performed by A. and N. Takeuchi, who used two microelectrodes to voltage-clamp the end plate region of muscle fibers.[49] The experimental arrangement is shown in Figure 9.11A. Two microelectrodes were inserted into the end plate region of a frog muscle fiber—one for recording membrane potential (V_m), the other for injecting current to clamp the membrane potential at the desired level. The nerve was then stimulated to release ACh, or, in later experiments, ACh was applied directly by ionophoresis. Subsequently, similar experiments were carried out by Magleby and Stevens[50] in muscle fibers treated with hypertonic glycerol, which has the advantage of preventing muscle fibers from contracting when depolarized, but the disadvantage of leaving the muscle fibers in an artificially depolarized state.

Figure 9.11B illustrates results from such a glycerol-treated muscle fiber. With the muscle membrane potential clamped at –40 mV, nerve stimulation produced an inward current, which would have caused a depolarization if the fiber had not been voltage-clamped. At more negative holding potentials, the end plate current increased in amplitude. When the membrane was depolarized, the end plate current decreased in amplitude. With further depolarization the current reversed direction and was outward.

[48]Jenkinson, D. H., and Nicholls, J. G. 1961. *J. Physiol.* 159: 111–127.
[49]Takeuchi, A., and Takeuchi, N. 1959. *J. Neurophysiol.* 22: 395–411.
[50]Magleby, K. L., and Stevens, C. F. 1972. *J. Physiol.* 223: 151–171.

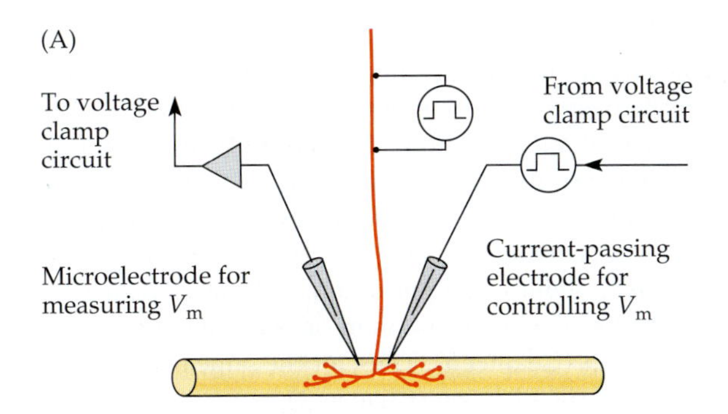

FIGURE 9.11 Reversal Potential for Synaptic Currents measured by voltage clamp recording. (A) Scheme for voltage clamp recording at the motor end plate. (B) Synaptic currents recorded at membrane potentials between –120 and +38 mV. When the muscle membrane potential is clamped below 0 mV, synaptic current flows into the muscle. Such inward current would depolarize the muscle if it were not voltage-clamped. When the end plate potential is clamped above 0 mV, synaptic current flows out of the cell. (C) Plot of peak end plate current as a function of membrane potential. The relation is nearly linear, with the reversal potential close to 0 mV. (After Magleby and Stevens, 1972.)

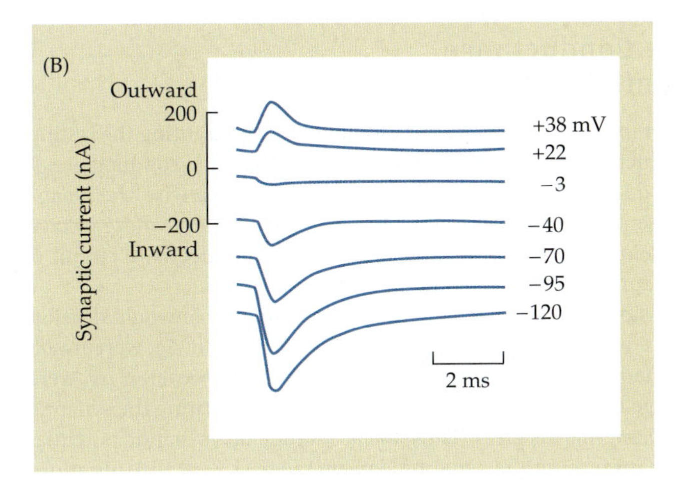

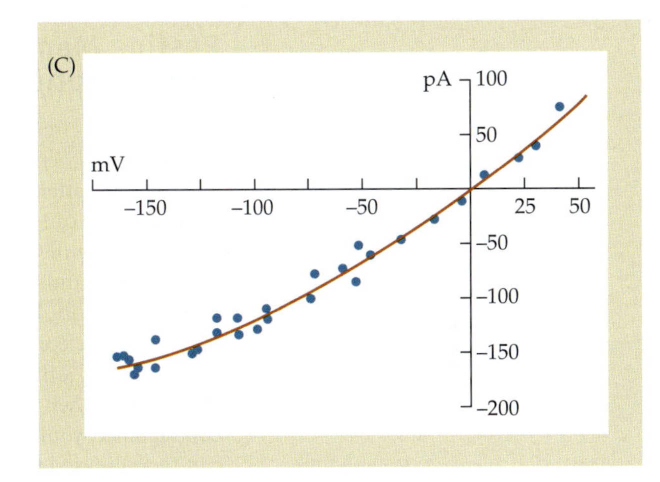

Figure 9.11C shows a plot of the peak amplitude of the end plate current as a function of holding potential. The current changed from inward to outward near zero membrane potential. Accordingly, zero is called the **reversal potential**, V_r. In earlier experiments on intact muscle fibers, A. and N. Takeuchi estimated the reversal potential to be about –15 mV.[51]

Significance of the Reversal Potential

The reversal potential for the end plate current gives us information about the ionic currents flowing through the channels activated by ACh in the postsynaptic membrane. For example, if the channels were permeable exclusively to sodium, then current through the channels would be zero at the sodium equilibrium potential (about +50 mV). The other major ions, potassium and chloride, have equilibrium potentials near –90 mV, the normal resting membrane potential (Chapter 5); the calcium equilibrium potential is approximately +120 mV. None of the ions has an equilibrium potential in the range of 0 to –15 mV. What ions, then, are involved in the response? Consistent with the results of radioactive tracer experiments, A. and N. Takeuchi showed that changing the concentrations of sodium, potassium, or calcium in the bathing solution resulted in changes in the reversal potential, but changes in extracellular chloride did not.[33,52] They concluded that the effect of ACh was to produce a general increase in *cation* permeability.

Relative Contributions of Sodium, Potassium, and Calcium to the End Plate Potential

ACh opens channels in the end plate membrane that, at the normal resting potential, allow sodium and calcium ions to leak inward and potassium ions outward along their electrochemical gradients. Because the calcium conductance of the channels is small, the contribution of calcium to the overall synaptic current can be ignored, as can that of other

[51] Takeuchi, A., and Takeuchi, N. 1960. *J. Physiol.* 154: 52–67.
[52] Takeuchi, N. 1963. *J. Physiol.* 167: 128–140.

cations, such as magnesium. (It should be noted that the low calcium *conductance* is due to its low extracellular and intracellular concentrations; calcium *permeability* is about 20% of the sodium permeability.) The equivalent electrical circuit is shown in Figure 9.12A. The resting membrane consists of the usual sodium, potassium, and chloride channels. It is in parallel with ACh-activated channels for sodium and potassium, Δg_{Na} and Δg_K. The Takeuchis calculated that for a reversal potential $V_r = -15$ mV, the ratio of the sodium to potassium conductance changes, $\Delta g_{Na}/\Delta g_K$, is about 1.3 (Box 9.1). The channel opened by ACh is, in fact, nearly equally permeable to sodium and potassium.[53] However, taking the extracellular and intracellular solutions together, there are more sodium than potassium ions available to move through the channels (Chapter 5). Thus, for the same permeability change, the sodium conductance change is slightly larger (Chapter 2).

Resting Membrane Conductance and Synaptic Potential Amplitude

The electrical circuit shown in Figure 9.12A can be simplified by representing the resting membrane as a single conductance, g_{rest} (equal to the sum of all the ionic conductances), and a single battery, V_{rest} (equal to the resting membrane potential). Likewise, the synaptic membrane can be represented by a single conductance Δg_s and a battery whose voltage is equal to the reversal potential V_r (Figure 9.12B). A feature of this electrical circuit is that the amplitude of a synaptic potential depends on both Δg_s and g_{rest}.

For simplicity, let us consider the steady-state membrane potential that would develop if the synaptic conductance were activated for a long period of time. If Δg_s were much larger than g_{rest}, then the membrane potential would approach V_r. However, if Δg_s were equal to g_{rest}, then the change in membrane potential produced by activating the synaptic conductance would be only one-half as great. Thus, the amplitude of a synaptic potential can be increased by either increasing the synaptic conductance (i.e., activating more synaptic channels) or decreasing the resting conductance. Indeed, a reduction in mem-

[53]Adams, D. J., Dwyer, T. M., and Hille, B. 1980. *J. Gen. Physiol.* 75: 493–510.

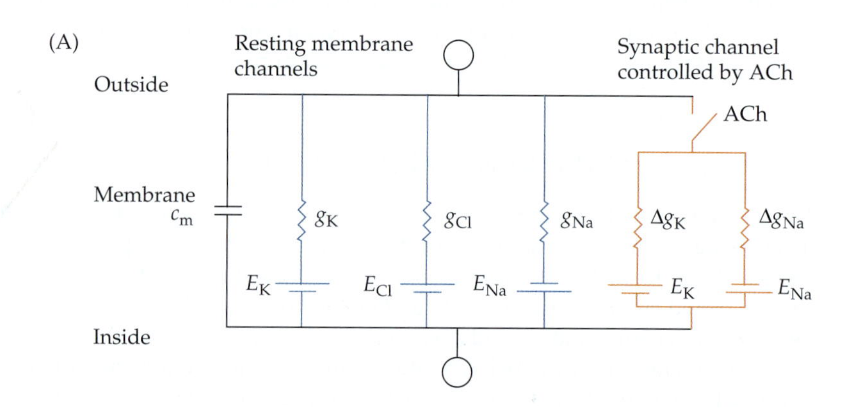

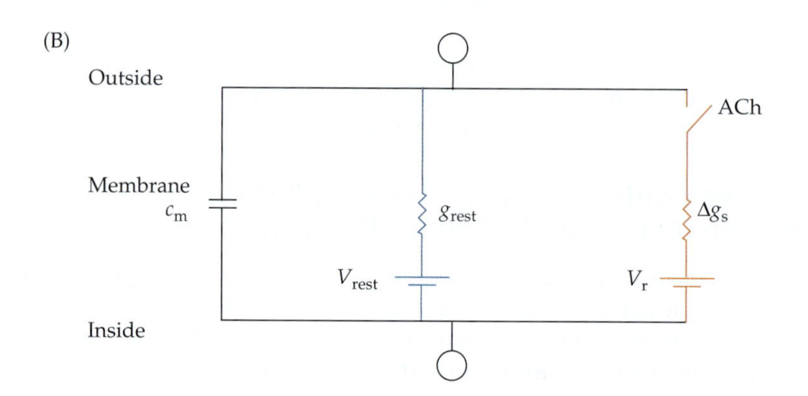

FIGURE 9.12 Electrical Model of the Postsynaptic Membrane with channels activated by ACh in parallel with the resting membrane channels and with the membrane capacitance, c_m. (A) The synaptic channel opened by ACh is electrically equivalent to two independent pathways for sodium and potassium. The resting membrane has channels for potassium, chloride, and sodium. (B) The synaptic channel can be represented as a single pathway with conductance Δg_s and a battery equal to the reversal potential V_r. The resting membrane can be represented as a single pathway with conductance g_{rest} and a battery equal to V_{rest}.

Box 9.1 Electrical Model of the Motor End Plate

How did A. and N. Takeuchi calculate the ratio of sodium to potassium conductance for the channels opened by ACh? They proposed an electrical model of the muscle cell membrane similar to that shown in Figure 9.12A. Although ACh receptors do not form separate pathways for sodium and potassium, the two ions move through the channel independently. Therefore, the synaptic conductance and reversal potential can be represented by separate conductances (Δg_{Na} and Δg_K) and driving potentials (E_{Na} and E_K) for sodium and potassium. Accordingly, separate expressions can be written for the sodium and potassium currents (ΔI_{Na} and ΔI_K):

$$\Delta I_{Na} = \Delta g_{Na}(V_m - E_{Na})$$

$$\Delta I_K = \Delta g_K(V_m - E_K)$$

These equations provide a means of determining the relative conductance changes to sodium and potassium produced by ACh, once the reversal potential (V_r) is determined. Since the Takeuchis considered only *changes* in current resulting from the action of ACh, they could ignore the resting membrane channels. The net synaptic current is zero at the reversal potential; therefore, at this potential the inward sodium current is exactly equal and opposite to the outward potassium current. So when $V_m = V_r$,

$$\Delta g_{Na}(V_r - E_{Na}) = -\Delta g_K(V_r - E_K)$$

It follows that

$$\frac{\Delta g_{Na}}{\Delta g_K} = \frac{-(V_r - E_K)}{(V_r - E_{Na})}$$

We can rearrange the equations regarding synaptic sodium and potassium currents to predict the reversal potential when the relative conductances are known:

$$V_r = \frac{\Delta g_{Na}E_{Na} + \Delta g_K E_K}{(\Delta g_{Na} + \Delta g_K)}$$

Thus, the reversal potential is simply the average of the individual equilibrium potentials, weighted by the relative conductance changes. This relationship can be extended to include any number or variety of ions, so it is applicable at any synapse where transmitters produce a change in conductance of the postsynaptic membrane to one or more ions. This relationship was found to predict how changes in E_{Na} and E_K, produced by changes in extracellular concentrations of sodium and potassium, affected the reversal potential at the neuromuscular junction.[51]

Such predictions were accurate only for small changes in extracellular sodium and potassium, however, because channel conductance is determined in part by ion concentration (Chapters 2 and 5). Therefore, the effect of a large change in sodium, potassium, or calcium concentration on reversal potential is predicted accurately only if the resulting change in conductance is taken into account. Alternatively, the analysis can be made in terms of permeabilities, using the constant field equation developed by Goldman, Hodgkin, and Katz (Chapter 5).[54,55]

[54]Ritchie, A., and Fambrough, D. M. 1975. *J. Gen. Physiol.* 65: 751–767.

[55]Lassignal, N. L., and Martin, A. R. 1977. *J. Gen. Physiol.* 70: 23–36.

brane conductance is an important mechanism for modulating synaptic strength. For example, certain inputs to autonomic ganglion cells in the bullfrog close potassium channels, thereby increasing the amplitude of excitatory synaptic potentials produced by other inputs to the cell (Chapter 16).

Kinetics of Currents through Single ACh Receptor Channels

To what extent does the time course of the end plate current reflect the behavior of individual ACh channels? For example, do individual channels open and close repetitively during the end plate current, with the probability of channel opening declining with time? Or do individual channels open only once, so that the time course of the current is determined by how long channels remain open?

Definitive answers to such questions came only with the advent of patch clamp techniques, by which the behavior of individual channels could be observed directly (Chapter 2).[56] When ACh was applied continuously, ACh channels were shown to open instantaneously in an all-or-nothing fashion, and then close at a rate that matched exactly the rate of decay of the end plate current.[57] This idea can be summarized by the following

[56]Neher, E., and Sakmann, B. 1976. *Nature* 260: 799–802.

[57]Dionne, V. E., and Leibowitz, M. D. 1982. *Biophys. J.* 39: 253–261.

scheme, indicating the interaction between the transmitter molecule A (for "agonist") and the postsynaptic receptor molecule R:

$$2A + R \rightleftharpoons A_2R \underset{\alpha}{\overset{\beta}{\rightleftharpoons}} A_2R^*$$

Two ACh molecules combine with the channel (one on each α subunit; Chapter 3), which then undergoes a change in conformation from the closed (A_2R) to the open (A_2R^*) state. The transitions between the open and closed states are characterized by the rate constants α and β, as indicated. Now consider the time course of the end plate current, as illustrated in Figure 9.13. ACh arriving at the postsynaptic membrane opens a large number of channels almost simultaneously. Because ACh is lost rapidly from the synaptic cleft (because of hydrolysis by cholinesterase and diffusion), each channel opens only once. As the channels close, the synaptic current declines. Thus, the time course of decay of the end plate current reflects the rate at which individual ACh channels close. Channels close at the rate $\alpha \times [A_2R^*]$; that is, many channels close very quickly, and fewer and fewer channels close at longer and longer times (see Figure 9.13). As with all independent or random events, the open times are distributed exponentially, and the **mean open time** (τ) is equal to the time constant of the decay of the end plate current, $1/\alpha$.

Patch clamp experiments have revealed a number of details of channel activation that were previously undetectable. For example, local anesthetics such as procaine prolong the falling phase of the end plate current (in addition to blocking sodium channels and therefore action potentials). Patch clamp experiments showed that this prolongation occurs because procaine causes the current through open ACh channels to "flicker" on and off, apparently as a consequence of the anesthetic molecule itself moving rapidly into and out of the pore of the open channel and thereby intermittently obstructing current flow.[58–60] Subsequent experiments revealed that procaine also inhibits channel opening by binding to sites on the receptor *outside* the pore.[61] Indeed, channel properties can be modified by a wide variety of drugs, including cocaine,[62] barbiturates,[63] steroids,[64] and general anesthetics,[65] which bind to sites both within and outside the pore.

The properties of acetylcholine receptors change during development. There is a fetal form of the acetylcholine receptor, which has a low conductance and a long and variable open time, and an adult form, which has a higher conductance and shorter open time.[66,67] The switch from embryonic to adult receptors is caused by a change in subunit composition (Chapters 3 and 23), and splice variants of one of the embryonic subunits may account for variation in mean channel open time early in development.[68]

[58]Neher, E., and Steinbach, J. H. 1978. *J. Physiol.* 277: 153–176.

[59]Galzi, J-L., et al. 1991. *Annu. Rev. Pharmacol.* 31: 37–72.

[60]Lester, H. A. 1992. *Annu. Rev. Biophys. Biomol. Struct.* 21: 267–292.

[61]Niu, L., and Hess, G. P. 1993. *Biochemistry* 32: 3831–3835.

[62]Niu, L., Abood, L. G., and Hess, G. P. 1995. *Proc. Natl. Acad. Sci. USA* 92: 12008–12012.

[63]Dilger, J. P., et al. 1997. *J. Gen. Physiol.* 109: 401–414.

[64]Bouzat, C., and Barrantes, F. J. 1996. *J. Biol. Chem.* 271: 25835–25841.

[65]Dilger, J. P., Liu, Y., and Vidal, A. M. 1995. *Eur. J. Anaesthesiol.* 12: 31–39.

[66]Mishina, M., et al. 1986. *Nature* 321: 406–411.

[67]Grassi, F., et al. 1998. *J. Physiol.* 508: 393–400.

[68]Herlitze, S., et al. 1996. *J. Physiol.* 492: 775–787.

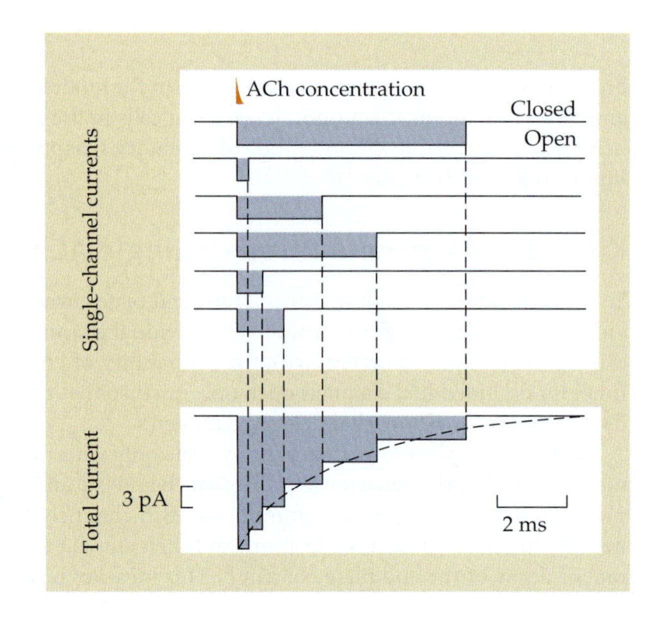

FIGURE 9.13 Total End Plate Current Is the Sum of Individual Channel Currents. Current flow through six individual channels is depicted in the top panel. Channels open instantaneously in response to ACh. ACh is rapidly hydrolyzed, preventing any further channel openings. Channel open times are distributed exponentially. The individual channel currents sum to give the total end plate current (lower panel). The time constant of the decay of the total current is equal to the mean open time of the individual channels.

DIRECT SYNAPTIC INHIBITION

The principles that underlie direct chemical synaptic excitation at the neuromuscular junction also apply to **direct chemical inhibitory synapses.** Whereas excitation occurs by opening channels in the postsynaptic membrane whose reversal potential is *positive* to threshold, direct chemical synaptic inhibition is achieved by opening channels whose reversal potential is *negative* to threshold. Direct chemical synaptic inhibition occurs by activating channels permeable to chloride, an anion that typically has an equilibrium potential at or near the resting potential. Pioneering studies of direct chemical synaptic inhibition were made on the crustacean neuromuscular junction,[69,70] the crayfish stretch receptor,[71] and spinal motoneurons of the cat.[72]

Reversal of Inhibitory Potentials

Spinal motoneurons are inhibited by sensory inputs from antagonistic muscles, by way of inhibitory interneurons in the spinal cord. The effect of activation of inhibitory inputs can be studied by an experiment similar to that illustrated in Figure 9.14A. The motoneuron is impaled with two micropipettes—one to record potential changes, the other to pass current through the cell membrane. At the normal resting potential (about –75 mV), stimulation of the inhibitory inputs causes a slight hyperpolarization of the cell—the **inhibitory postsynaptic potential (IPSP)** (Figure 9.14B). When the membrane is depolarized by passing positive current into the cell, the amplitude of the IPSP is increased. When the cell is hyperpolarized to –82 mV, the inhibitory potential is very small and reversed in sign, and at –100 mV the reversed inhibitory potential is increased in amplitude. The reversal potential in this experiment is thus about –80 mV.

[69]Dudel, J., and Kuffler, S. W. 1961. *J. Physiol.* 155: 543–562.

[70]Takeuchi, A., and Takeuchi, N. 1967. *J. Physiol.* 191: 575–590.

[71]Kuffler, S. W., and Eyzaguirre, C. 1955. *J. Gen. Physiol.* 39: 155–184.

[72]Coombs, J. S., Eccles, J. C. and Fatt, P. 1955. *J. Physiol.* 130: 326–373.

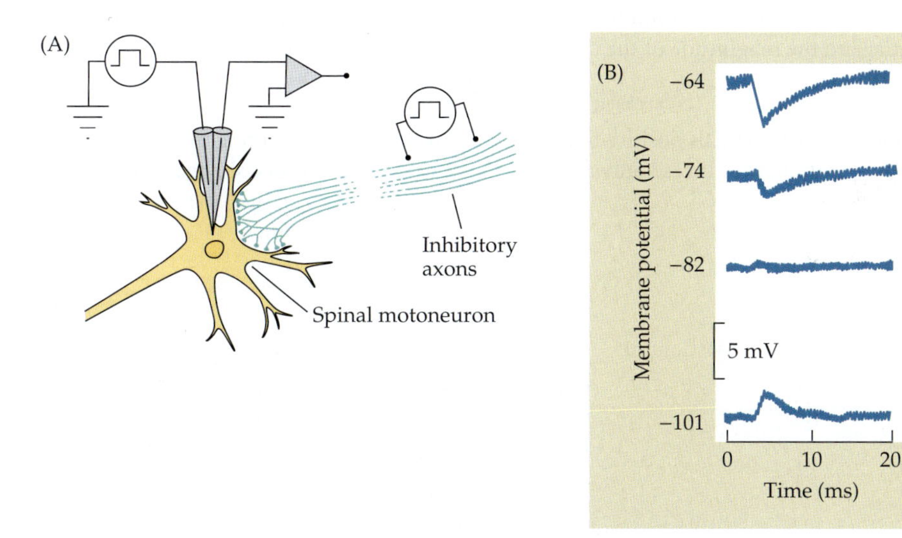

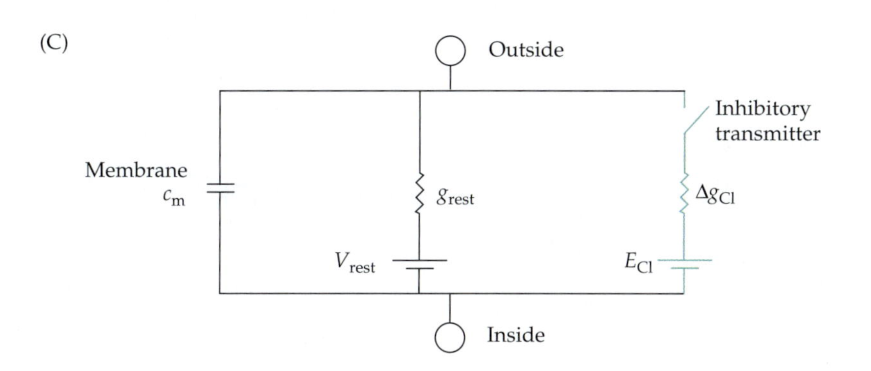

FIGURE 9.14 Direct Inhibitory Chemical Synaptic Transmission. (A) Scheme for intracellular recording from a cat spinal motoneuron and stimulation of inhibitory synaptic inputs. The membrane potential of the motoneuron is set to different levels by passing current through a second intracellular microelectrode. (B) Intracellular records of synaptic potentials evoked at membrane potentials between –64 and –101 mV. The reversal potential is between –74 and –82 mV. (C) Electrical model of the motoneuron membrane with chloride channels activated by the inhibitory transmitter, Δg_{Cl}, in parallel with the resting membrane channels, g_{rest}, and the membrane capacitance, c_m. (A and B after Coombs, Eccles, and Fatt, 1955.)

Inhibitory channels are permeable to anions, with permeabilities roughly correlated with the hydrated radius of the penetrating ion.[72,73] In physiological circumstances, the only small anion present in any quantity is chloride. Thus, in spinal motoneurons injection of chloride into the cell from a micropipette shifts the chloride equilibrium potential, and hence the reversal potential for the IPSP, in the positive direction. In other preparations, changes in extracellular chloride have been shown to produce corresponding changes in the chloride equilibrium potential and the IPSP reversal potential, but such experiments often give ambiguous results. This is because changes in extracellular chloride concentration lead eventually to proportionate changes in intracellular concentration as well (Chapter 5), so any change in chloride equilibrium potential is only transient.

One way around this difficulty is to remove chloride entirely, as shown in Figure 9.15. The records are from a reticulospinal cell in the brainstem of the lamprey, in which inhibitory synaptic transmission is mediated by glycine.[74] Membrane potential was recorded with an intracellular microelectrode. A second electrode was used to pass brief hyperpolarizing current pulses into the cell; the resulting changes in potential provided a measure of the cell's input resistance. Finally, a third micropipette was used to apply glycine to the cell close to an inhibitory synapse, using brief pressure pulses. Glycine application resulted in a slight hyperpolarization, with a marked reduction in input resistance (Figure 9.15A), as would be expected if glycine activated a large number of chloride channels. To test this idea, chloride was removed from the bathing solution and replaced by the impermeant ion isethionate. As a result, intracellular chloride was also removed, by efflux through chloride channels open at rest. After 20 min, glycine application produced no detectable change in membrane potential or input resistance (Figure 9.15B), indicating that no ions other than chloride pass through the inhibitory channels. The restoration of normal extracellular chloride concentration (Figure 9.15C) resulted in restoration of the response.

Given that the inhibitory response involves an increase in chloride permeability, the reversal potential for the inhibitory current will be equal to the chloride equilibrium potential and the magnitude of the current will be given by

$$\Delta i_{\text{inhibitory}} = \Delta i_{\text{Cl}} = \Delta g_{\text{Cl}}(V_{\text{m}} - E_{\text{Cl}})$$

At membrane potentials positive to E_{Cl} the current is outward, resulting in membrane hyperpolarization. In this case outward current is carried by an influx of negatively charged

[73]Hille, B. 1992. *Ionic Channels of Excitable Membranes*, 2nd Ed. Sinauer, Sunderland, MA.

[74]Gold, M. R., and Martin, A. R. 1983. *J. Physiol.* 342: 99–117.

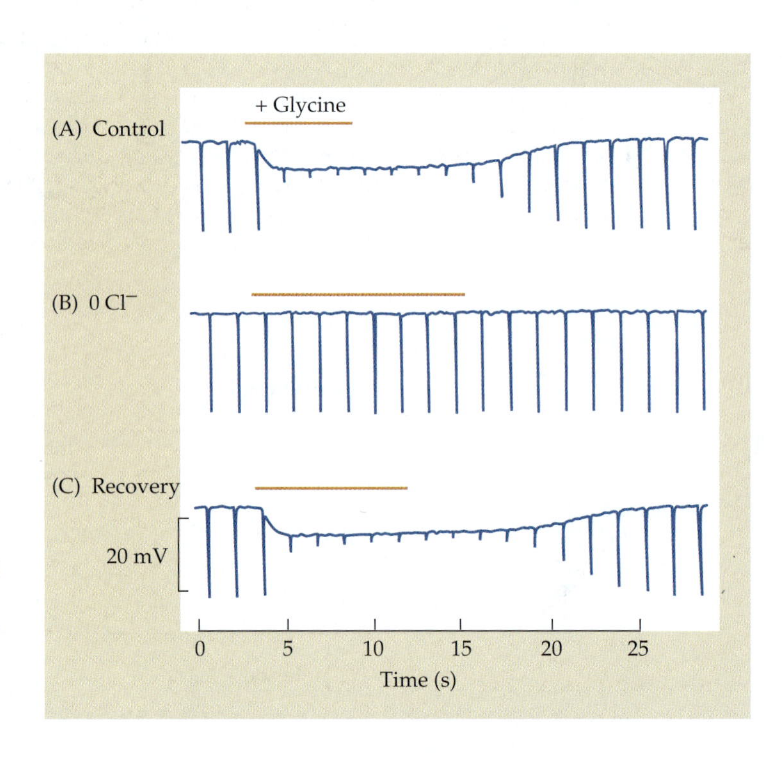

FIGURE 9.15 Inhibitory Response to Glycine Depends on Chloride. Intracellular microelectrode recordings from a neuron in the brainstem of the lamprey. (A) Resting membrane potential is –63 mV. Brief downward voltage deflections are produced by 10 nA current pulses from a second intracellular microelectrode; their amplitude indicates membrane resistance. On application of glycine (bar), the cell is hyperpolarized by about 7 mV and membrane resistance is reduced drastically. (B) After 20 min in chloride-free bathing solution, the response to glycine is abolished. (C) Five minutes after return to normal chloride solution the response has recovered. (From Gold and Martin, 1983.)

chloride ions. At membrane potentials negative to E_{Cl}, inhibition causes an efflux of chloride ions, resulting in depolarization. The equivalent circuit is shown in Figure 9.14C.

Early in postnatal development of the mammalian central nervous system, GABA and glycine paradoxically depolarize and thereby excite neurons in the hippocampus.[75] This effect is due not to differences in the properties of the channels opened by GABA and glycine, but to a difference in the regulation of intracellular chloride that results in a change in the chloride equilibrium potential.

Stephen W. Kuffler in 1975

Presynaptic Inhibition

So far we have defined excitatory and inhibitory synapses on the basis of the effect of the transmitter on the postsynaptic membrane—that is, on whether the postsynaptic permeability change is to cations or to anions. However, a number of early experiments indicated that in some instances it was difficult to account for inhibition in terms of postsynaptic permeability changes alone.[76,77] The paradox was resolved by the discovery of an additional inhibitory mechanism, **presynaptic inhibition**, described in the mammalian spinal cord by Eccles and his colleagues[78] and at the crustacean neuromuscular junction by Dudel and Kuffler.[69] Presynaptic inhibition results in a reduction in the amount of transmitter released from excitatory nerve terminals.[69,79,80]

As shown in Figure 9.16, the action of the inhibitory nerve at the crustacean neuromuscular junction is exerted not only on the muscle fibers, but also on the excitatory terminals. The presynaptic effect is brief, reaching a peak in a few milliseconds and declining to zero after a total of 6 to 7 ms. For the maximum inhibitory effect to occur, the impulse must arrive in the inhibitory presynaptic terminal several milliseconds before the action potential arrives in the excitatory terminal. The importance of accurate timing is shown in Figure 9.16, where parts A and B show the excitatory and inhibitory potentials following separate stimulation of the corresponding nerves. In Figure 9.16C both nerves are stimulated, but the action potential in the inhibitory nerve follows that in the excitatory nerve by 1.5 ms, arriving too late to exert any effect. In Figure 9.16D, on the other hand, the action potential in the inhibitory nerve precedes that in the excitatory nerve and causes a marked reduction in the size of the excitatory postsynaptic potential.

The presynaptic effect, like that on the postsynaptic membrane, is mediated by γ-aminobutyric acid (GABA), and is associated with a marked increase in chloride per-

[75]Mladinic, M., et al. 1999. *Proc. R. Soc. Lond. B* 266: 1207–1213.

[76]Fatt, P., and Katz, B. 1953. *J. Physiol.* 121: 374–389.

[77]Frank, K., and Fuortes, M. G. F. 1957. *Fed. Proc.* 16: 39–40.

[78]Eccles, J. C., Eccles, R. M., and Magni, F. 1961. *J. Physiol.* 159: 147–166.

[79]Kuno, M. 1964. *J. Physiol.* 175: 100–112.

[80]Rudomin, P. and Schmidt, R. F. 1999. *Exp. Brain Res.* 129: 1–37.

FIGURE 9.16 Presynaptic Inhibition in a crustacean muscle fiber innervated by one excitatory and one inhibitory axon. (A) Stimulation of the excitatory axon (E) produces a 2 mV EPSP. (B) Stimulation of the inhibitory axon (I) produces a depolarizing IPSP of about 0.2 mV. (C) If the inhibitory stimulus follows the excitatory one by a short interval, there is no effect on the EPSP. (D) If the inhibitory stimulus precedes the excitatory one by a few milliseconds, the EPSP is almost abolished. The importance of precise timing indicates that the inhibitory nerve is having a presynaptic effect, reducing the amount of excitatory neurotransmitter that is released. (After Dudel and Kuffler, 1961.)

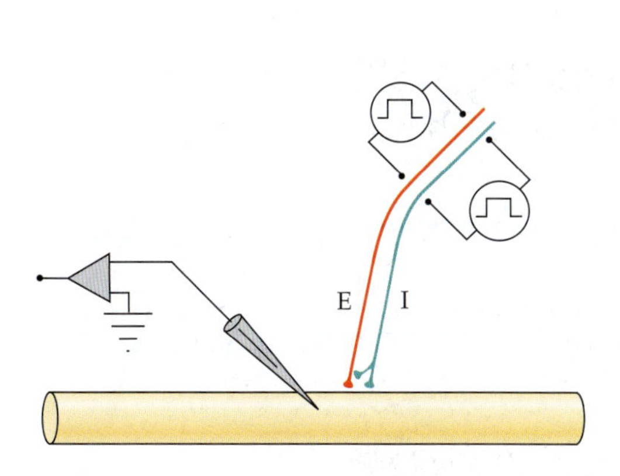

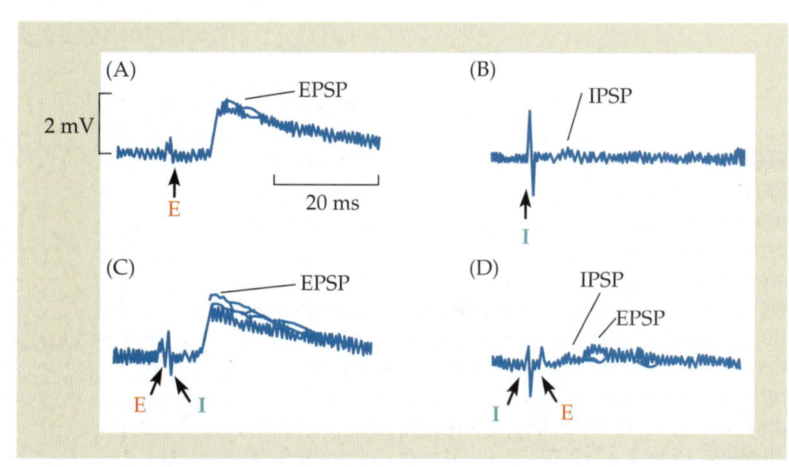

meability in the presynaptic terminals.[81,82] An explanation that has been suggested is that when chloride permeability is high, the depolarizing effect of sodium influx during the rising phase of the action potential is canceled in part by an accompanying influx of chloride. As a result, the presynaptic action potential is smaller in amplitude, and its effectiveness in releasing transmitter is reduced. At many mammalian synapses, presynaptic inhibition has been shown to be due to inhibition of voltage-dependent calcium channels in the axon terminal.[83]

In the nervous system in general, presynaptic and postsynaptic inhibition serve quite different functions. Postsynaptic inhibition reduces the excitability of the cell itself, rendering it relatively less responsive to all excitatory inputs. Presynaptic inhibition is much more specific, aimed at a particular input and leaving the postsynaptic cell free to go about its business of integrating information from other sources.[84] Presynaptic inhibition implies that inhibitory axons make synaptic contact with axon terminals. Such axo-axonic synapses have been demonstrated directly by electron microscopy at the crustacean neuromuscular junction[85] and at numerous locations in the mammalian central nervous system.[86] Moreover, inhibitory nerve terminals themselves can be influenced presynaptically;[87] the requisite ultrastructural arrangement has been reported at inhibitory synapses on crayfish stretch receptors.[88] There is also evidence for **presynaptic excitation**, synaptic inputs that enhance the release of transmitter from presynaptic nerve terminals.[89]

Desensitization

The response to a neurotransmitter often decreases during repeated or prolonged application, a phenomenon termed **desensitization**. It was described in detail at the neuromuscular junction by Katz and Thesleff, who showed that with prolonged application of ACh the depolarizing response of the muscle fiber steadily declines (Figure 9.17).[90] Desensitization is an intrinsic molecular property of the ACh receptor;[91] however, the rate of receptor desensitization and recovery is modulated by phosphorylation.[92–94] Under normal physiological conditions desensitization does not play a significant role in the response of skeletal muscle to ACh released from axon terminals.[95] However, in muscles that have been poisoned with inhibitors of cholinesterase, such as the organophosphorus compounds used as insecticides and nerve gases, the persistence of ACh in the synaptic cleft is sufficient to cause desensitization and block synaptic transmission.[96]

Receptors for glutamate and GABA also desensitize.[97] At synapses in the central nervous system where glutamate and GABA are released as direct chemical transmitters, desensitization of the postsynaptic receptors occurs even under normal physiological conditions and appears to play an important role in determining the amplitude and time course of postsynaptic potentials.[95,98]

[81]Takeuchi, A., and Takeuchi, N. 1966. *J. Physiol.* 183: 433–449.

[82]Fuchs, P. A., and Getting, P. A. 1980. *J. Neurophysiol.* 43: 1547–1557.

[83]Wu, L.-G. and Saggau, P. 1997. *Trends Neurosci.* 20: 204–212.

[84]Lomeli, J., et al. 1998. *Nature* 395: 600–604.

[85]Atwood, H. L., and Morin, W. A. 1970. *J. Ultrastruct. Res.* 32: 351–369.

[86]Schmidt, R. F. 1971. *Ergeb. Physiol.* 63: 20–101.

[87]Nicholls, J. G., and Wallace, B. G. 1978. *J. Physiol.* 281: 157–170.

[88]Nakajima, Y., Tisdale, A. D., and Henkart, M. P. 1973. *Proc. Natl. Acad. Sci. USA* 70: 2462–2466.

[89]Radcliffe, K. A., et al. 1999. *Ann. N.Y. Acad. Sci.* 868: 591–610.

[90]Katz, B., and Thesleff, S. 1957. *J. Physiol.* 138: 63–80.

[91]Auerbach, A., and Akk, G. 1998. *J. Gen. Physiol.* 112: 181–197.

[92]Huganir, R. L., and Greengard, P. 1990. *Neuron* 5: 555–567.

[93]Hardwick, J. C., and Parsons, R. L. 1996. *J. Neurophysiol.* 76: 3609–3616.

[94]Paradiso, K., and Brehm, P. 1998. *J. Neurosci.* 18: 9227–9237.

[95]Jones, M. V., and Westbrook, G. L. 1996. *Trends Neurosci.* 19: 96–101.

[96]Magleby, K. L., and Pallotta, B. S. 1981. *J. Physiol.* 316: 225–250.

[97]Dudel, J., Adelsberger, H., and Heckmann, M. 1997. *Invertebr. Neurosci.* 3: 89–92.

[98]Trussell, L. O. 1999. *Annu. Rev. Physiol.* 61: 477–496.

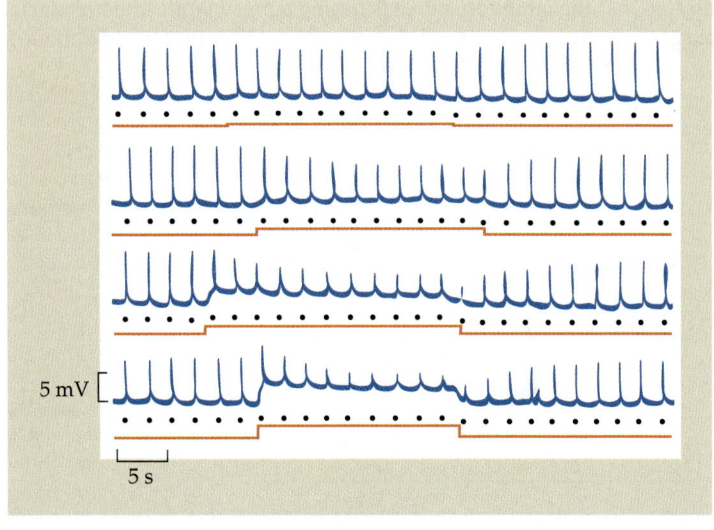

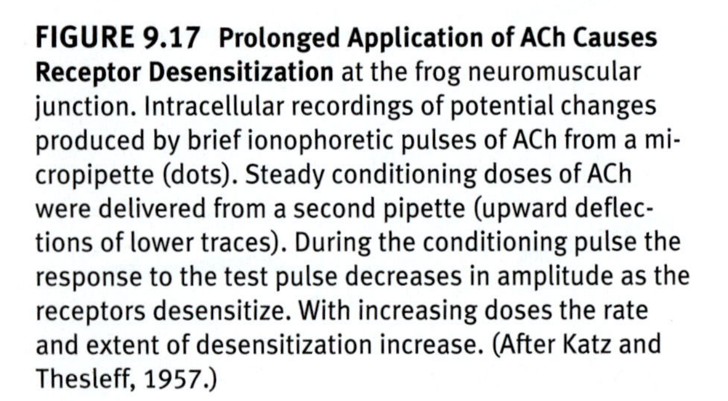

FIGURE 9.17 Prolonged Application of ACh Causes Receptor Desensitization at the frog neuromuscular junction. Intracellular recordings of potential changes produced by brief ionophoretic pulses of ACh from a micropipette (dots). Steady conditioning doses of ACh were delivered from a second pipette (upward deflections of lower traces). During the conditioning pulse the response to the test pulse decreases in amplitude as the receptors desensitize. With increasing doses the rate and extent of desensitization increase. (After Katz and Thesleff, 1957.)

Receptors Mediating Direct and Indirect Chemical Transmission

Direct chemical synaptic transmission is mediated by ion channels in the postsynaptic membrane that are activated by binding the neurotransmitter released by the presynaptic cell. Such ligand-activated ion channels are also referred to as **ionotropic neurotransmitter receptors.** In the mammalian central nervous system three major transmitters act at direct chemical synapses. Two of these, GABA and glycine, activate receptors that are anion channels and so are inhibitory, at least in the adult (Chapter 13). Glutamate, the most prevalent excitatory transmitter in the mammalian central nervous system, acts on several different types of cation-selective ionotropic receptors. Other important transmitters in the mammalian CNS that activate cation-selective ionotropic receptors include acetylcholine, serotonin, and purines.

All the directly acting neurotransmitters mentioned here, as well as transmitters such as dopamine, histamine, norepinephrine, and neuropeptides, also influence postsynaptic cells by a quite different mechanism, **indirect chemical transmission**. Typically, indirect chemical synaptic transmission is mediated by **metabotropic receptors**, postsynaptic receptors that produce an intracellular second messenger. The second messenger, in turn, influences the activity of ion channels, causing excitation or inhibition, and often affects other intracellular targets as well. Indirect chemical synaptic transmission is discussed in detail in Chapter 10.

Summary

- Signaling between nerve cells and their targets can occur by chemical or electrical synaptic transmission.

- Electrical synaptic transmission is mediated by the direct flow of current from cell to cell.

- At chemical synapses a neurotransmitter released from the presynaptic terminal activates receptors in the postsynaptic membrane; the time required for transmitter release imposes a minimum synaptic delay of approximately 1 ms.

- Direct chemical synaptic transmission occurs when the postsynaptic receptor activated by a neurotransmitter is itself an ion channel. Such ligand-activated ion channels are called ionotropic transmitter receptors.

- At direct excitatory synapses, such as the vertebrate skeletal neuromuscular junction, the neurotransmitter (in this case ACh) opens cation-selective channels, allowing sodium, potassium, and calcium ions to flow down their electrochemical gradients.

- The relative permeability of a channel for various ions determines the reversal potential; at excitatory synapses the reversal potential is more depolarized than the threshold for action potential initiation.

- Direct chemical synaptic inhibition occurs when a neurotransmitter opens anion-selective channels, which allow chloride ions to flow down their electrochemical gradient. The reversal potential for such currents is the chloride equilibrium potential (E_{Cl}); inhibition occurs if E_{Cl} is hyperpolarized to threshold.

- Many transmitter receptors desensitize; that is, their response decreases during repeated or prolonged transmitter application.

SUGGESTED READING

General Reviews

Galzi, J-L., Revah, F., Bessis, A., and Changeux, J-P. 1991. Functional architecture of the nicotinic acetylcholine receptor: From electric organ to brain. *Annu. Rev. Pharmacol.* 31: 37–72.

Hille, B. 1992. *Ionic Channels of Excitable Membranes*, 2nd Ed. Sinauer, Sunderland, MA.

Jones, M. V., and Westbrook, G. L. 1996. The impact of receptor desensitization on fast synaptic transmission. *Trends Neurosci.* 19: 96–101.

Lester, H. A. 1992. The permeation pathway of neurotransmitter-gated ion channels. *Annu. Rev. Biophys. Biomol. Struct.* 21: 267–292.

Llinás, R. 1985. Electrotonic transmission in the mammalian central nervous system. In M. V. L. Bennett and D. C. Spray (eds.), *Gap Junctions.* Cold Spring Harbor Laboratory, Cold Spring Harbor, NY, pp. 337–353.

Salpeter, M. M. (ed.). 1987. *The Vertebrate Neuromuscular Junction.* Alan R. Liss, New York.

Original Papers

Coombs, J. S., Eccles, J. C., and Fatt, P. 1955. The specific ionic conductances and the ionic movements across the motoneuronal membrane that produce the inhibitory post-synaptic potential. *J. Physiol.* 130: 326–373.

del Castillo, J., and Katz, B. 1955. On the localization of end-plate receptors. *J. Physiol.* 128: 157–181.

Dudel, J., and Kuffler, S. W. 1961. Presynaptic inhibition at the crayfish neuromuscular junction. *J. Physiol.* 155: 543–562.

Fatt, P., and Katz, B. 1951. An analysis of the end-plate potential recorded with an intra-cellular electrode. *J. Physiol.* 115: 320–370.

Furshpan, E. J., and Potter, D. D. 1959. Transmission at the giant motor synapses of the crayfish. *J. Physiol.* 145: 289–325.

Kuffler, S. W., and Yoshikami, D. 1975. The distribution of acetylcholine sensitivity at the post-synaptic membrane of vertebrate skeletal twitch muscles: Iontophoretic mapping in the micron range. *J. Physiol.* 244: 703–730.

Magleby, K. L., and Stevens, C. F. 1972. The effect of voltage on the time course of end-plate currents. *J. Physiol.* 223: 151–171.

Martin, A. R., and Pilar, G. 1963. Dual mode of synaptic transmission in the avian ciliary ganglion. *J. Physiol.* 168: 443–463.

Neher, E., and Sakmann, B. 1976. Single channel currents recorded from membrane of denervated frog muscle fibres. *Nature* 260: 799–801.

Paradiso, K., and Brehm, P. 1998. Long-term desensitization of nicotinic acetylcholine receptors is regulated via protein kinase A-mediated phosphorylation. *J. Neurosci.* 18: 9227–9237.

Takeuchi, A., and Takeuchi, N. 1960. On the permeability of the end-plate membrane during the action of transmitter. *J. Physiol.* 154: 52–67.

Takeuchi, A., and Takeuchi, N. 1966. On the permeability of the presynaptic terminal of the crayfish neuromuscular junction during synaptic inhibition and the action of γ-aminobutyric acid. *J. Physiol.* 183: 433–449.

Takeuchi, A., and Takeuchi, N. 1967. Anion permeability of the inhibitory post-synaptic membrane of the crayfish neuromuscular junction. *J. Physiol.* 191: 575–590.

10 Indirect Mechanisms of Synaptic Transmission

NEUROTRANSMITTERS BIND TO METABOTROPIC RECEPTORS that influence ion channels and pumps indirectly through membrane-associated or cytoplasmic second messengers. At many synapses in the central and autonomic nervous systems, transmission occurs solely by such indirect mechanisms. At other synapses indirect mechanisms modulate direct transmission.

Indirect synaptic transmission is often mediated by G protein–coupled receptors. G proteins, so called because they bind guanine nucleotides, are trimers of three subunits: α, β, and γ. When a G protein is activated by its receptor, the α and $\beta\gamma$ subunits dissociate. The free subunits bind to and modulate the activity of intracellular targets. Some G protein subunits bind to ion channels, producing relatively brief effects. For example, when ACh binds to its muscarinic receptor on cells in the atrium of the heart, a G protein is activated. The freed $\beta\gamma$ subunit binds to and opens a potassium channel, thereby slowing the heart.

A second mechanism of G protein action is through activation of enzymes that produce intracellular second messengers. An example is the activation of a G protein in cardiac muscle cells by binding of norepinephrine to β-adrenergic receptors. The activated α and $\beta\gamma$ subunits stimulate the enzyme adenylyl cyclase. The resulting increase in cyclic AMP activates another enzyme, cAMP-dependent protein kinase, which modifies the activity of channels and enzymes through phosphorylation. Such responses may last for seconds, minutes, or hours, often persisting long after the transmitter interaction with the receptors has stopped. These mechanisms provide an enormous signal amplification.

Potassium and calcium channels are prime targets for such indirect transmitter action. Indirect action can cause channels to be opened, closed, or changed in their voltage sensitivity. Thus, indirectly acting transmitters open potassium channels in heart atrial cells, inhibit N-type calcium channels in sympathetic neurons, and increase the probability that calcium channels will open in response to depolarization in cardiac muscle cells. Changes in channel activation in axon terminals modify transmitter release; in postsynaptic cells such changes alter spontaneous activity and the responses to synaptic inputs.

In addition to direct mechanisms of synaptic transmission, discussed in Chapter 9, transmitters act indirectly by binding to receptors that are not themselves ion channels. Such receptors, in turn, modify the activity of ion channels, ion pumps, or receptor proteins (Figure 10.1). Indirect mechanisms are also referred to as "slow" or "second messenger linked." Many indirectly acting receptors produce their effects through interaction with GTP-binding proteins, or G proteins,[1] and are referred to as **metabotropic receptors.**

At some synapses, such as those between sympathetic neurons and muscle fibers in the heart[2] or between photoreceptors and certain bipolar cells in the retina,[3] transmission may occur solely through indirectly coupled receptors. Alternatively, indirectly acting transmitters may influence the efficacy of direct synaptic transmission (Chapter 16), a process referred to as **neuromodulation.** In this chapter we first describe the structure and function of metabotropic receptors and G proteins. Then we consider examples of the synaptic interactions they mediate. We will see that metabotropic receptors, acting through G proteins, influence a remarkable number of intracellular signaling mechanisms that can profoundly affect neuronal activity for seconds, minutes, hours, or even longer.

METABOTROPIC RECEPTORS AND G PROTEINS

Structure of Metabotropic Receptors

G protein–coupled metabotropic receptors comprise a superfamily of membrane proteins characterized by seven transmembrane domains, with an extracellular amino terminus and an intracellular carboxy terminus (Figure 10.2).[4] More than a thousand different metabotropic receptors have been identified. Some are activated by acetylcholine (muscarinic cholinergic receptors); others bind norepinephrine (α- and β-adrenergic receptors); still others respond to GABA, 5-HT, dopamine, glutamate, purines, or neuropeptides; and some are activated by light (rhodopsin) (Chapter 19), by odorants (Chapter 17), or by proteases.

[1]Dunlap, K., Holz, G. G., and Rane, S. G. 1987. *Trends Neurosci.* 10: 244–247.

[2]McDonald, T. F., et al. 1994. *Physiol. Rev.* 74: 365–507.

[3]de la Villa, P., Kurahashi, T., and Kaneko, A. 1995. *J. Neurosci.* 15: 3571–3582.

[4]Ji, T. H., Grossmann, M., and Ji, I. 1998. *J. Biol. Chem.* 273: 17299–17302.

(A) **Direct transmitter action**

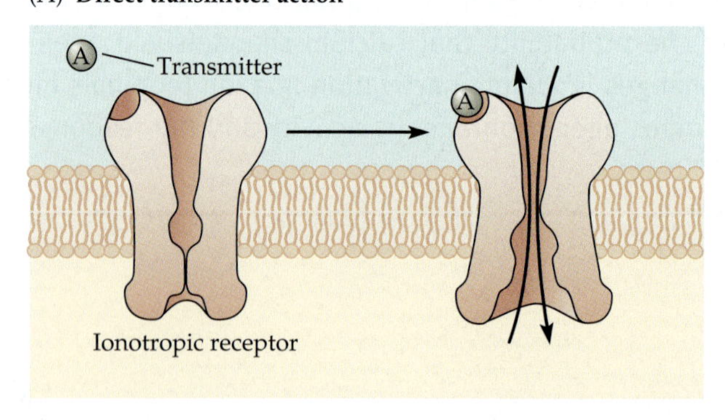

(B) **Indirect transmitter action**

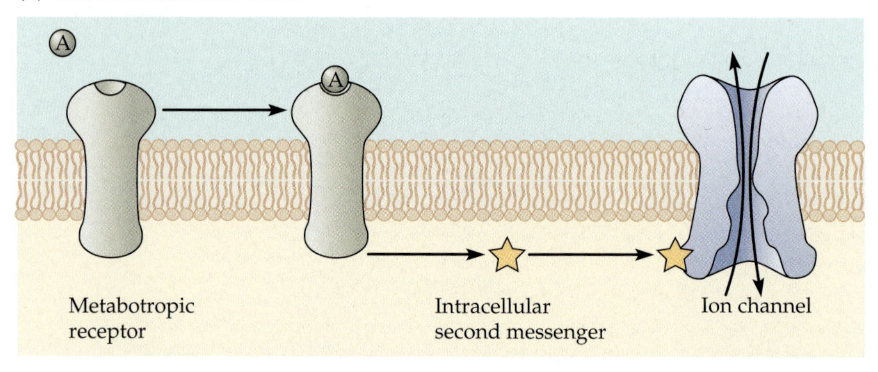

FIGURE 10.1 Direct and Indirect Transmitter Action. (A) At direct chemical synapses the transmitter binds to an ionotropic receptor. Ionotropic receptors are ligand-activated ion channels. (B) Indirectly acting transmitters bind to metabotropic receptors. Metabotropic receptors are not themselves ion channels, but rather activate intracellular second messenger signaling pathways that influence the opening and closing of ion channels.

(A)

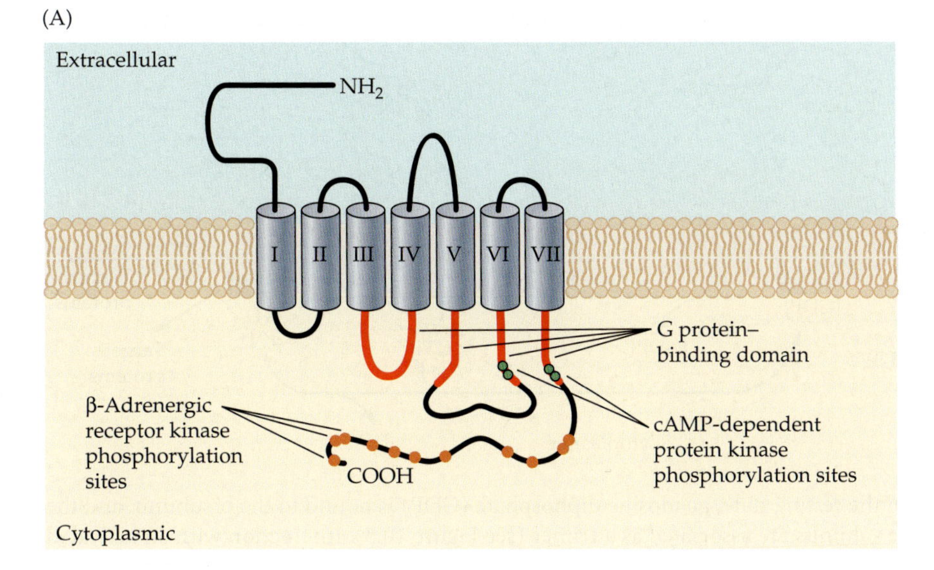

(B) **Receptor for amines, nucleotides, eicosanoids**

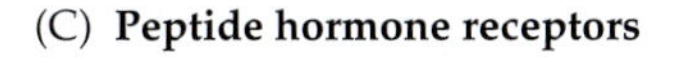

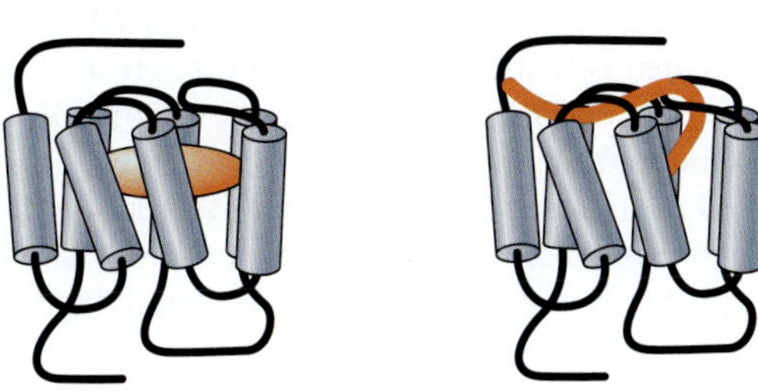

(D) **Neurotransmitter (glutamate, GABA) receptor**

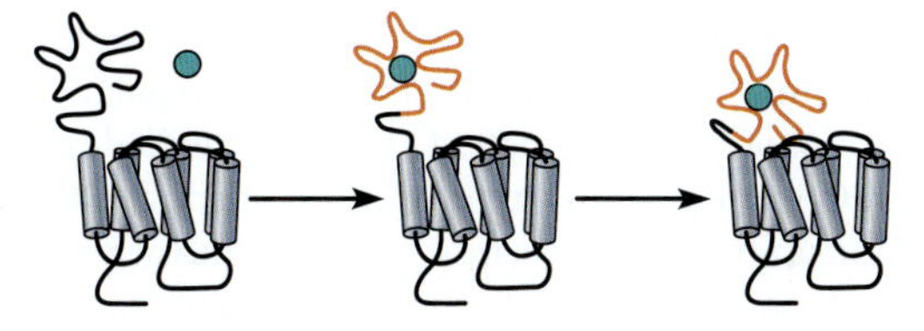

FIGURE 10.2 Metabotropic or G Protein–Coupled Transmitter Receptors. (A) Indirectly coupled transmitter receptors have seven transmembrane domains, an extracellular amino terminus, and an intracellular carboxy terminus. The second and third cytoplasmic loops, together with the amino-terminal region of the intracellular tail, mediate binding to the appropriate G protein. Phosphorylation of sites on the third cytoplasm loop and the carboxy terminus by second messenger–related kinases, such as cAMP-dependent protein kinase, causes receptor desensitization. Phosphorylation of sites on the carboxy terminus by G protein coupled receptor kinases (GRKs), such as β-adrenergic receptor kinase (βARK), causes receptor desensitization, binding of the protein arrestin, and termination of the response. (B) Portions of the transmembrane domains form the ligand-binding sites of metabotropic receptors that bind amines, nucleotides, and eicosanoids. (C) Ligands bind to the outer portions of the transmembrane domains of peptide hormone receptors. (D) The amino-terminal tail forms the ligand-binding domain of metabotropic receptors for glutamate and GABA. (After Ji, Grossmann, and Ji, 1998.)

Biochemical, structural, and molecular genetic experiments have indicated several distinct modes of ligand binding, each of which ultimately produces a similar rearrangement of the α-helical regions that form the transmembrane core of the receptor (see Figure 10.2). Portions of the second and third cytoplasmic loops, together with the membrane-proximal region of the carboxy tail, mediate binding to and activation of the appropriate G protein.[5,6]

G Protein Structure and Function

G proteins, so named because they bind guanine nucleotides, are trimers made up of three subunits: α, β, and γ (Figure 10.3).[7] Multiple forms are known for each G protein subunit (20 for α, 6 for β, 12 for γ), providing a bewildering array of possible permutations. G proteins are grouped into four main classes according to the structure and targets of their α subunits: G_s stimulates adenylyl cyclase, G_i inhibits adenylyl cyclase, G_q couples to phospholipase C, and G_{12} has unknown targets. The G_i family includes G_t (transducin), which activates cGMP phosphodiesterase (Chapter 19), and two G_o isoforms, which bind to ion channels. However, a particular G protein can couple to more than one effector, and different G proteins can modulate the activity of the same ion channel.[8]

[5]Wess, J. 1997. *FASEB J.* 11: 346–354.

[6]Wess, J. 1998. *Pharmacol. Ther.* 80: 231–264.

[7]Hamm, H. E. 1998. *J. Biol. Chem.* 273: 669–672.

[8]Ross, E. M. 1989. *Neuron* 3: 141–152.

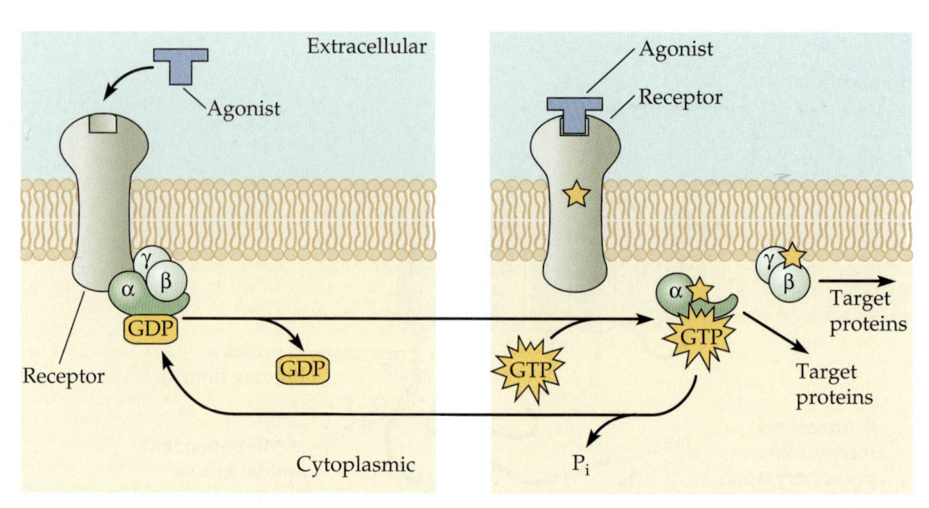

FIGURE 10.3 Indirectly Coupled Transmitter Receptors Act through G Proteins. G proteins are trimers of α, β, and γ subunits. Activation of a metabotropic receptor by agonist binding (indicated by a star) promotes the exchange of GTP for GDP on the α subunit of the G protein. This activates the α subunit and the βγ complex, causing them to dissociate from the receptor and from one another. The free activated α–GTP subunit and βγ complex each interact with target proteins. Hydrolysis of GTP to GDP and inorganic phosphate (P_i) by the endogenous GTPase activity of the α subunit leads to association of the αβγ complex, terminating the response.

In the resting state, guanosine diphosphate (GDP) is bound to the α subunit, and the three subunits are associated as a trimer (see Figure 10.3). Interaction with an activated receptor allows guanosine triphosphate (GTP) to replace GDP on the α subunit, resulting in dissociation of the α and βγ subunits. (The β and γ subunits remain together under physiological conditions.) Free α and βγ subunits bind to and modulate the activity of target proteins.[7,9,10] The free α subunit has intrinsic GTPase activity that results in the hydrolysis of the bound GTP to GDP. This causes the reassociation of the α and βγ subunits into the G protein complex and terminates their activity. The lifetime of the activated G protein subunits is modulated by proteins called GTPase-activating proteins, or GAPs, which influence the rate at which GTP bound to the α subunit is hydrolyzed.[11]

The details of the interactions of the G protein subunits with each other and with their targets are beginning to be understood using molecular biological[5] and X-ray crystallographic techniques.[12–14] In addition, probes have been developed for identifying responses mediated by G proteins (Box 10.1).

Desensitization

As is the case for many ionotropic receptors (Chapter 9), the responsiveness of G protein–coupled receptors may desensitize—that is, wane in the face of persistent stimulation. Two mechanisms mediate desensitization.[15] In one, a second messenger–regulated kinase, such as cAMP-dependent protein kinase, phosphorylates G protein–coupled receptors, blocking their interaction with G proteins. This kind of mechanism can be relatively nonselective: More than one transmitter may activate cAMP-dependent protein kinase, cAMP-dependent protein kinase may phosphorylate more than one type of G protein–coupled receptor, and both active and inactive receptors may be phosphorylated. A second cellular mechanism mediating rapid, agonist-specific desensitization involves two other protein families. The first step is selective phosphorylation of *activated* receptors by a specific G protein–coupled receptor kinase, or GRK, such as β-adrenergic receptor kinase (βARK). Then, a protein of the arrestin family binds to the phosphorylated receptor, blocking its ability to activate G proteins and thereby terminating the response.

DIRECT MODULATION OF CHANNEL FUNCTION BY G PROTEINS

G proteins have been implicated in the regulation of at least a dozen different potassium, sodium, or calcium channels by nearly 70 different receptors.[16] G proteins couple receptor activation to modulation of channel activity either directly, by binding to the channel itself, or indirectly, by regulating a second messenger pathway. The multiplicity of signaling cascades makes it possible for a single transmitter first to excite and then to inhibit a postsynaptic cell, or to produce other patterns of responses in various types of neurons.

[9]Clapham, D. E., and Neer, E. J. 1997. *Annu. Rev. Pharmacol. Toxicol.* 37: 167–203.

[10]Gautam, N., et al. 1998. *Cell. Signalling* 10: 447–455.

[11]Berman, D. M., and Gilman, A. G. 1998. *J. Biol. Chem.* 273: 1269–1272.

[12]Lambright, D. G., et al. 1996. *Nature* 379: 311–319.

[13]Tesmer, J. J. G., et al. 1997. *Science* 278: 1907–1916.

[14]Sunahara, R. K., et al. 1997. *Science* 278: 1943–1947.

[15]Lefkowitz, R. J. 1998. *J. Biol. Chem.* 273: 18677–18680.

[16]Birnbaumer, L., Abramowitz, J., and Brown, A. M. 1990. *Biochim. Biophys. Acta* 1031: 163–224.

Box 10.1 Identifying Responses Mediated by G Proteins

Several tests can be used to identify responses mediated by G proteins. For example, activation of the α subunit requires that bound GDP be replaced by GTP. Accordingly, G protein–mediated events have an absolute requirement for cytoplasmic GTP and will be blocked by intracellular perfusion with solutions lacking GTP. Two analogues of GTP—GTPγS and Gpp(NH)p—are useful because they cannot be hydrolyzed by the endogenous GTPase activity of the α subunit. Like GTP, they can replace GDP on the α subunit and activate it. However, because they cannot be hydrolyzed, these analogues activate the α subunit permanently. Thus, intracellular perfusion with either of these analogues enhances and greatly prolongs agonist-induced activation of G protein–mediated responses, and may even initiate responses in the absence of agonist. Likewise, AlF_4^- persistently activates G proteins. On the other hand, GDPβS, an analogue of GDP, binds strongly to the GDP site on the α subunit and resists replacement by GTP. Thus, GDPβS inhibits G protein–mediated responses by maintaining the αβγ complex in the inactive state.

Two bacterial toxins are useful for characterizing G protein–mediated processes. Each is an enzyme that catalyzes the covalent attachment of ADP-ribose to an arginine residue on the α subunit. Cholera toxin acts on α_s, irreversibly activating it; pertussis toxin acts on members of the α_i family, irreversibly blocking their activation and so inhibiting responses mediated by the corresponding G proteins.

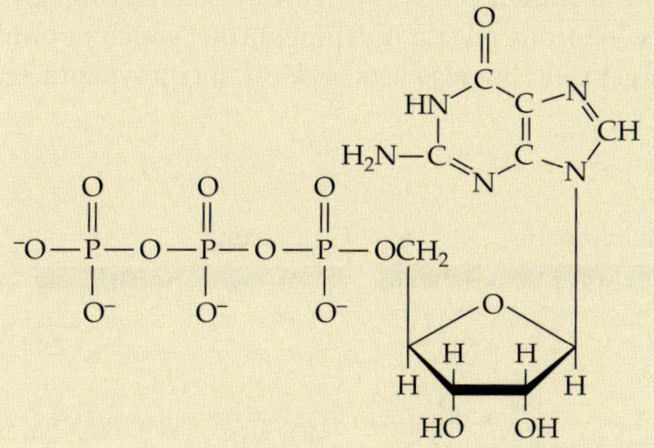

GTP
Guanosine 5′-triphosphate

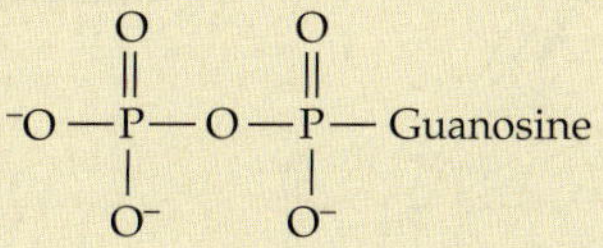

GDP
Guanosine 5′- diphosphate

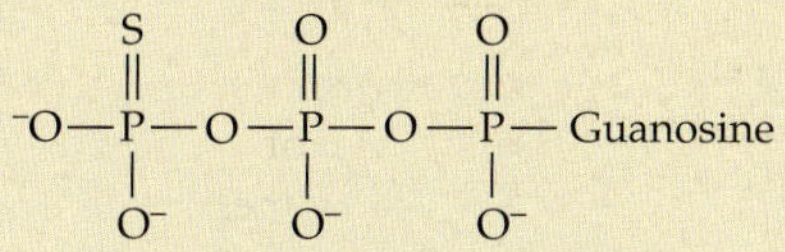

GTPγS
Guanosine 5′-*O*-[γ-thio] triphosphate

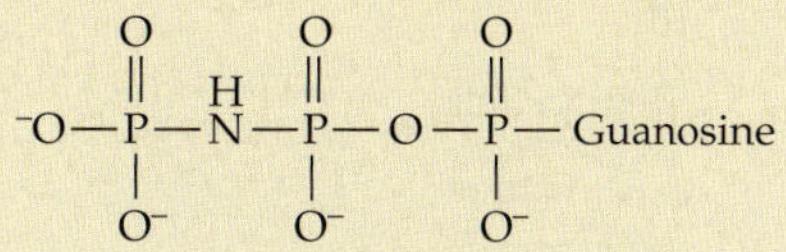

Gpp(NH)p
Guanosine 5′- [β,γ-imido] triphosphate

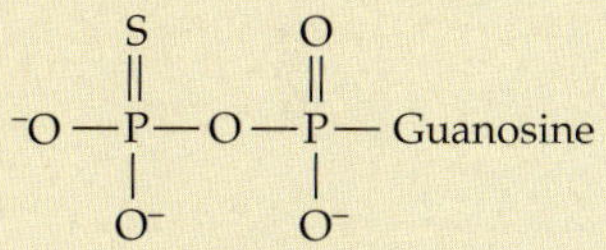

GDPβS
Guanosine 5′-*O*-[β -thio] diphosphate

G Protein Activation of Potassium Channels

The effect of acetylcholine on slowing the heartbeat, first studied by Loewi some 70 years ago (Chapter 9), is now known to be mediated by a metabotropic receptor, the **muscarinic acetylcholine receptor.** (Muscarinic receptors are so named because of their selective activation by the drug muscarine; ionotropic ACh receptors are activated selectively by nicotine and so are called nicotinic [Chapters 3 and 9].) Activation of muscarinic receptors in the heart opens potassium channels,[17] causing hyperpolarization.

Results of experiments by Breitwieser, Szabo, Pfaffinger, Trautwein, Hille, and their colleagues indicated that activation of muscarinic receptors is coupled to opening of potassium channels by G proteins.[18,19] For example, they found that intracellular GTP is required,[20] that activation of potassium channels by muscarinic agonists is greatly prolonged by intracellular application of a nonhydrolyzable analogue of GTP known as Gpp(NH)p,[21] and that muscarinic activation of potassium channels is blocked by pertussis toxin,[20] which inactivates G_i proteins (see Box 10.1). An important clue to the mechanism of potassium channel activation in cardiac muscle cells came from experiments by

[17] Sakmann, B., Noma, A., and Trautwein, W. 1983. *Nature* 303: 250–253.

[18] Brown, A. M., and Birnbaumer, L. 1990. *Annu. Rev. Physiol.* 52: 197–213.

[19] Szabo, G., and Otero, A. S. 1990. *Annu. Rev. Physiol.* 52: 293–305.

[20] Pfaffinger, P. J., et al. 1985. *Nature* 317: 536–538.

[21] Breitwieser, G. E., and Szabo, G. 1985. *Nature* 317: 538–540.

Clapham and his colleagues using inside-out excised membrane patches. They showed that when pure recombinant βγ subunit was applied to the intracellular side of the patch, potassium channels opened (Figure 10.4).[22] This finding demonstrated that the βγ subunit, rather than the α subunit, is responsible for potassium channel activation. Subsequent experiments with a cloned muscarinic potassium channel (GIRK1) indicated that βγ subunits interact directly with potassium channels (see Figure 10.4B).[23,24]

Using muscle cells dissociated from the atrium of the heart, Soejima and Noma found that potassium channel activity in cell-attached patches was increased when muscarinic agonists were added to the patch pipette solution, but not when agonists were added to the bath (Figure 10.5).[25] Thus, activated βγ subunits appear unable to traverse the region of the pipette–membrane seal to influence channels on the other side. This so-called membrane-delimited, or direct, mechanism of G protein action reflects the limited range over which βγ subunits can act.

[22]Wickman, K. D., et al. 1994. *Nature* 368: 255–257.

[23]Reuveny, E., et al. 1994. *Nature* 370: 143–146.

[24]Huang, C-L., et al. 1995. *Neuron* 15: 1133–1143.

[25]Soejima, M., and Noma, A. 1984. *Pflügers Arch.* 400: 424–431.

[26]Lipscombe, D., Kongsamut, S., and Tsien, R. W. 1989. *Nature* 340: 639–642.

G Protein Inhibition of Calcium Channels

A second example of direct interaction between G proteins and ion channels comes from studies by Tsien and his colleagues on transmitter release by neurons isolated from sympathetic ganglia of adult frogs.[26] These neurons release norepinephrine, which not only activates receptors on postsynaptic target cells, but also acts back on the presynaptic ter-

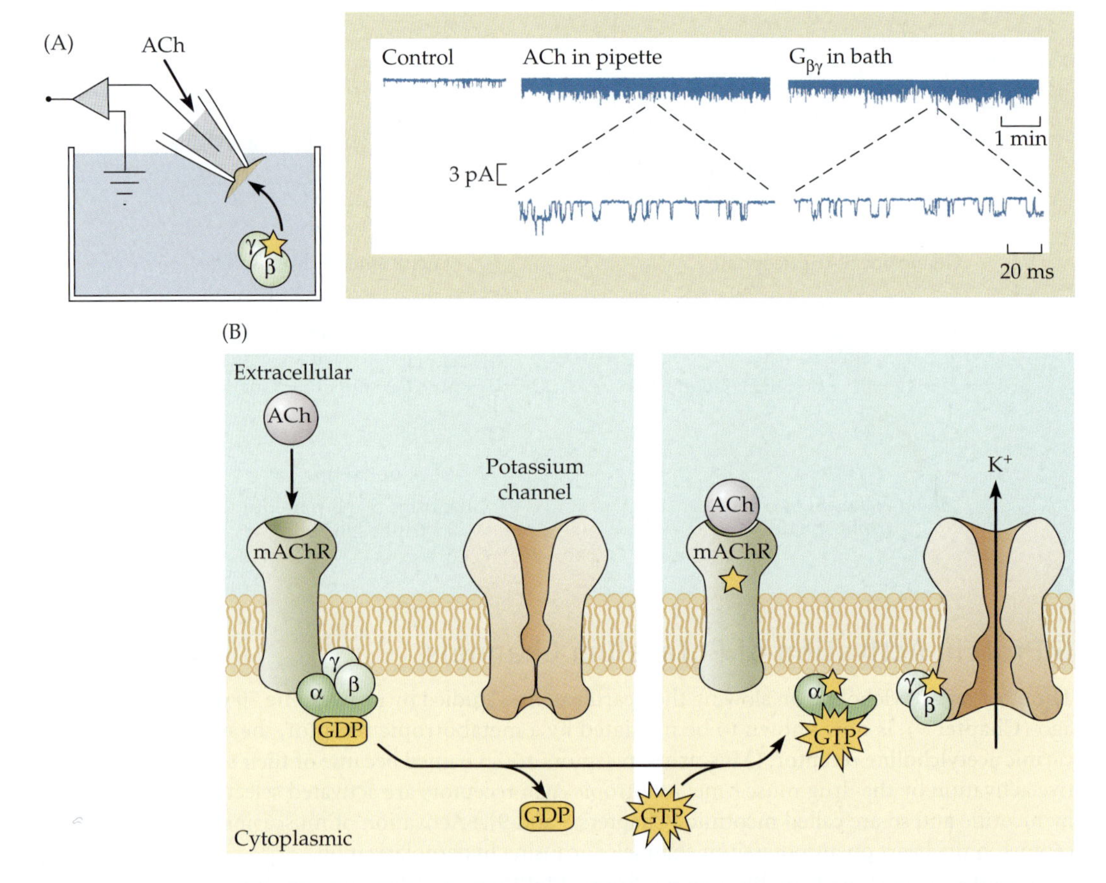

FIGURE 10.4 Direct Modulation of Channel Function by G Proteins. (A) Application of the $G_{\beta\gamma}$ complex to the intracellular surface of an isolated patch of membrane from a rat atrial muscle cell ($G_{\beta\gamma}$ in bath) results in an increase in potassium channel current similar to that seen when acetylcholine is added to the extracellular side of the patch (ACh in pipette). (B) Schematic representation of events in an intact cell. Binding of ACh to muscarinic receptors (mAChR) activates a G protein (indicated by a star); activated βγ complex binds directly to and opens a potassium channel. (A after Wickman et al., 1994.)

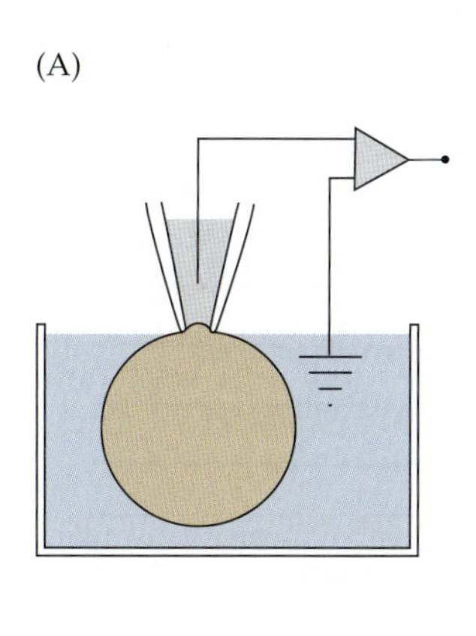

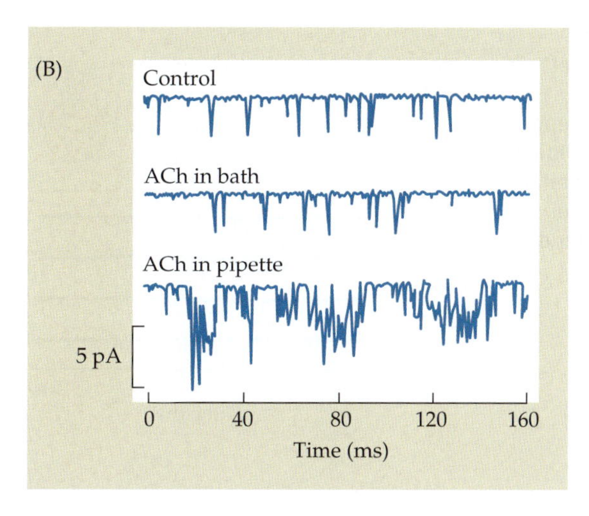

FIGURE 10.5 Direct, or Membrane-Delimited, Effects of G Proteins operate over short distances. (A) Effects of ACh were assayed by cell-attached patch clamp recording. Acetylcholine could be perfused into either the patch pipette or the bath. (B) Recordings of single-channel currents before and during addition of ACh. Compared with the control, channel activity increased only when ACh was added to the patch pipette. (After Soejima and Noma, 1984.)

minals themselves, via α_2-adrenergic receptors (Figure 10.6A). Activation of such **autoreceptors** decreases transmitter release (Figure 10.6B) by reducing the probability of opening of N-type calcium channels (Figure 10.7). (Calcium channels have been grouped into classes [L, T, N, P, Q, and R] that differ in their kinetic and pharmacological properties [Chapter 3]. Release of transmitter from sympathetic neurons is controlled by calcium entering through N-type channels.[27]) Experiments with nonhydrolyzable derivatives of GTP indicate that the response to norepinephrine is mediated by G proteins. Inhibition

[27]Hirning, L. D., et al. 1988. *Science* 239: 57–61.

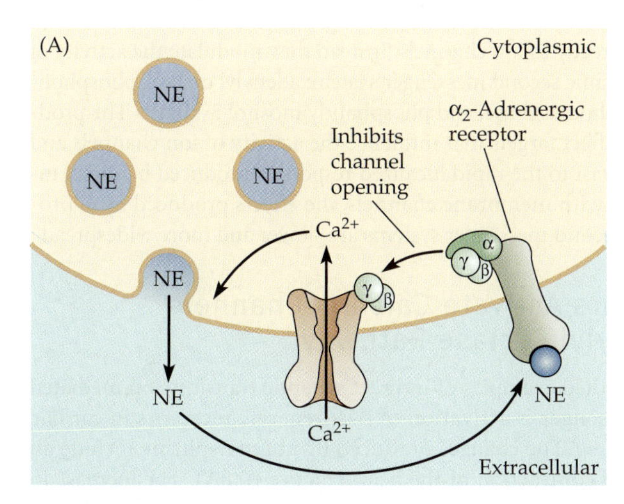

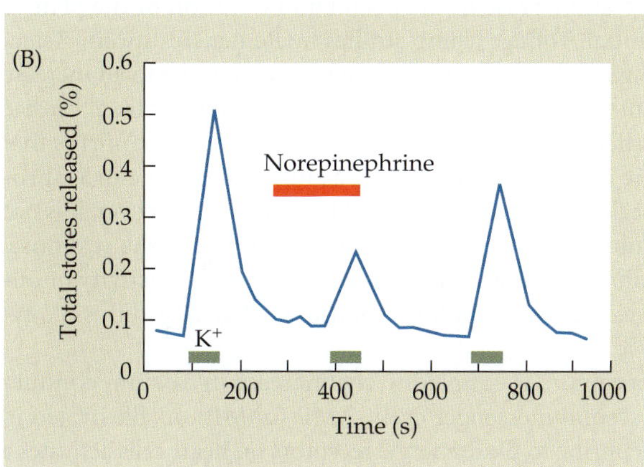

FIGURE 10.6 Presynaptic Autoreceptors Reduce Transmitter Release. (A) Norepinephrine (NE) released from sympathetic neurons combines with α_2-adrenergic receptors in the terminal membrane (called autoreceptors), activating a G protein. The activated $\beta\gamma$ complex binds to calcium channels, decreasing calcium influx and so limiting further transmitter release. (B) Norepinephrine reduces the release of transmitter from sympathetic ganglia. Ganglia were loaded with radioactive norepinephrine and then enclosed in a perfusion chamber. Transmitter release was evoked by depolarization with a solution containing 50 m*M* potassium (green bars). Addition of 30 μ*M* unlabeled norepinephrine to the perfusion solution (red bar) reduced the amount of radiolabeled transmitter released in response to potassium-induced depolarization. (B after Lipscombe, Kongsamut, and Tsien, 1989.)

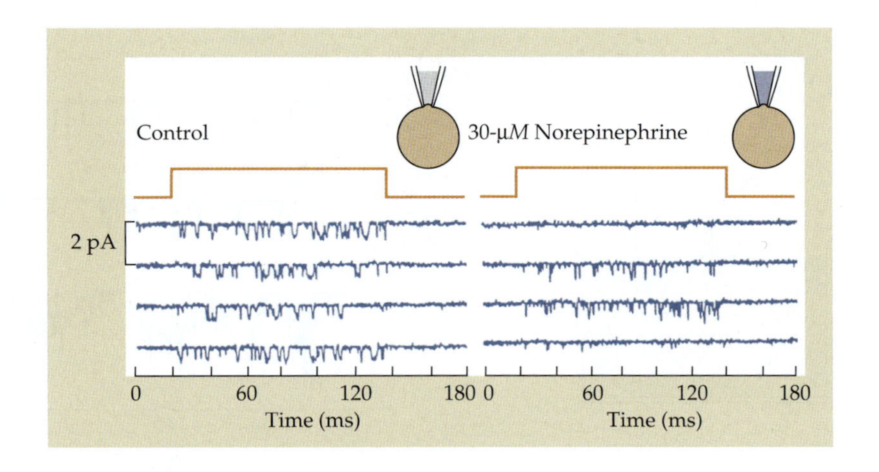

FIGURE 10.7 Norepinephrine Inhibits Calcium Channel Activity. Single-channel currents were recorded in cell-attached patches; channels were activated with a depolarizing pulse (top trace). When norepinephrine was included in the patch electrode, the unitary currents did not change in size, but channel openings were less frequent and of shorter duration. (After Lipscombe, Kongsamut, and Tsien, 1989.)

of calcium channel activity in cell-attached patches is observed when norepinephrine is present in the recording pipette, but not when added to the extracellular solution, suggesting a direct interaction between G proteins and calcium channels. Results of experiments in which α or βγ subunits were overexpressed or injected into cells indicated that βγ subunits inhibit N-type calcium channels.[28,29] Such autoinhibitory effects provide a rapid and localized mechanism for regulation of transmitter release.[27]

G PROTEIN ACTIVATION OF CYTOPLASMIC SECOND MESSENGER SYSTEMS

Many G proteins do not bind directly to ion channels. Instead they modulate the activity of an enzyme involved in a cytoplasmic second messenger system: adenylyl cyclase, phospholipase C, phospholipase A_2, phosphodiesterase, and phosphatidylinositol 3-kinase. The products of these enzymes, in turn, affect targets that influence the activity of ion channels and other cellular processes. In contrast to the rapid localized responses produced by direct interaction of G protein subunits with membrane channels, the effects produced by G proteins that activate cytoplasmic second messenger systems are slower and more widespread.

β-Adrenergic Receptors Activate Calcium Channels via a G Protein–Adenylyl Cyclase Pathway

One of the most thoroughly studied examples of indirect synaptic transmission mediated by an intracellular second messenger is activation of β-adrenergic receptors in cardiac muscle cells by norepinephrine.[2,30] The changes produced by norepinephrine include an increase in the rate and force of contraction of the heart (Figure 10.8A). The increase in contractile force is due in part to an increase in the height and duration of the plateau phase of the cardiac action potential. Voltage clamp studies by Reuter, Trautwein, Tsien, and others indicate that the change in the size of the action potential is due to an increase in the underlying calcium current (Figure 10.8B). Single-channel recording from cardiac muscle cells, using the cell-attached mode of the patch clamp technique, confirms that stimulation with a β-adrenergic receptor agonist, such as norepinephrine or isoproterenol, produces an increase in calcium channel activity (Figure 10.9). Moreover, it is not necessary that the agonist be added to the pipette solution to observe the response. Adding isoproterenol to the medium bathing the cell causes an increase in activity of calcium channels *within* the patch, a diagnostic test for responses mediated by diffusible cytoplasmic second messengers.[31,32]

Activation of β-adrenergic receptors is coupled to the increase in calcium conductance through the intracellular second messenger cyclic AMP (cAMP). As illustrated in Figure 10.10, binding of norepinephrine to β-adrenergic receptors on heart cells activates a G protein, G_s, releasing its α and βγ subunits. In this case, both the α and the βγ subunits

[28]Ikeda, S. R. 1996. *Nature* 380: 255–258.

[29]Herlitze, S., et al. 1996. *Nature* 380: 258–262.

[30]Tsien, R. W. 1987. In *Neuromodulation: The Biochemical Control of Neuronal Excitability*. Oxford University Press, New York, pp. 206–242.

[31]Reuter, H., et al. 1983. *Cold Spring Harb. Symp. Quant. Biol.* 48: 193–200.

[32]Tsien, R. W., et al. 1983. *Cold Spring Harb. Symp. Quant. Biol.* 48: 201–212.

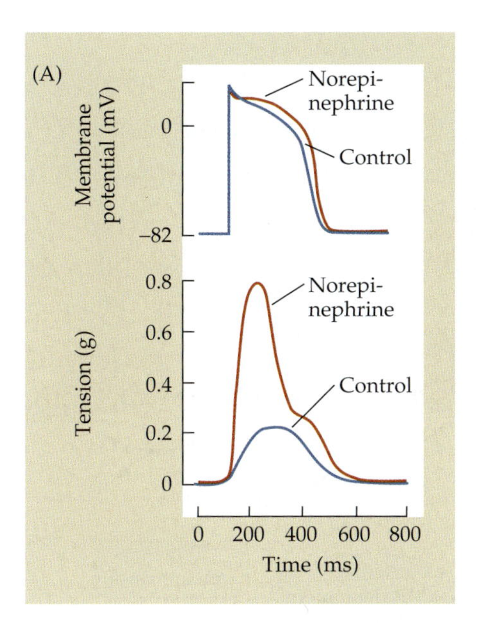

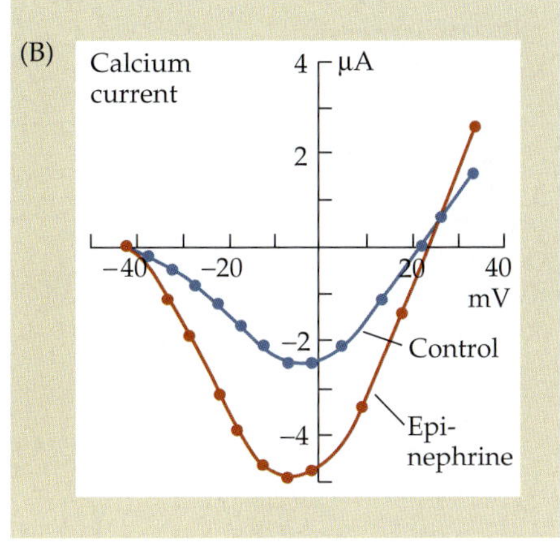

FIGURE 10.8 Activation of β-Adrenergic Receptors in Cardiac Muscle Increases Calcium Current. (A) The increase in calcium current produced by activation of β-adrenergic receptors, in this case by addition of 10^{-6} *M* norepinephrine, increases action potential amplitude and duration and the tension produced by cardiac muscle cells. (B) Current–voltage relationship of calcium current in a myocardial cell measured under voltage clamp conditions in the absence and presence of 0.5 μ*M* epinephrine, a β-adrenergic receptor agonist. (A after Reuter et al., 1983; B after Reuter, 1974.)

bind to and activate the enzyme adenylyl cyclase; the stimulatory effect of the α subunit is much greater than that of the βγ subunit. (There is considerable diversity in the effects of different G_{α} and $G_{\beta\gamma}$ subunits on various isoforms of adenylyl cyclase.[33,34]) Adenylyl cyclase converts ATP to cyclic AMP, a readily diffusible intracellular second messenger, which activates another enzyme, cAMP–dependent protein kinase. The catalytic subunits of this protein kinase mediate the transfer of phosphate from ATP to the hydroxyl groups of serine and threonine residues in a variety of enzymes and channels, thereby modifying their activity.

Several lines of evidence are consistent with this scheme for β-adrenergic stimulation of calcium conductance in heart muscle.[2] The experimental approach takes advantage of well-established features of cyclic AMP acting as an intracellular second messenger, as outlined in Box 10.2. For example, calcium channel activity is increased by forskolin, by membrane-permeable derivatives of cyclic AMP, by inhibitors of phosphodiesterase, and by direct intracellular injection of cyclic AMP itself. Similarly, intracellular injection of the catalytic subunit of cyclic AMP–dependent protein kinase leads to an increase in calcium current, while injection of excess regulatory subunit or inhibitors of protein kinase block adrenergic stimulation of calcium currents. ATPγS, an analogue of ATP, augments adrenergic activation of calcium channels by forming stably phosphorylated proteins, while intracellular injection of protein phosphatases prevents or reverses adrenergic stimulation of calcium currents by rapidly removing protein phosphate residues.

[33] Taussig, R., and Gilman, A. G. 1995. *J. Biol. Chem.* 270: 1–4.

[34] Tang, W-J., and Gilman, A. G. 1991. *Science* 254: 1500–1503.

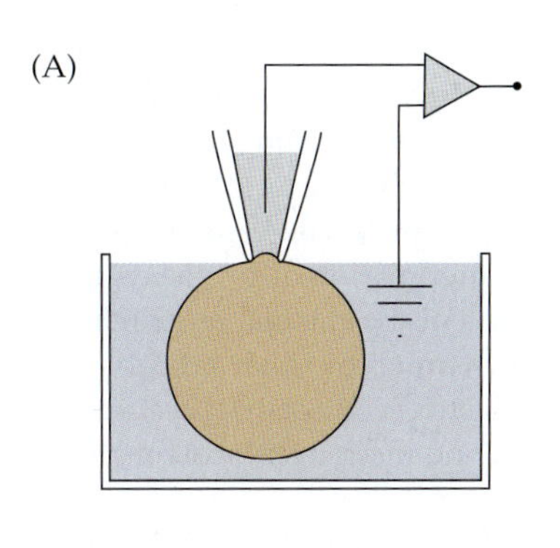

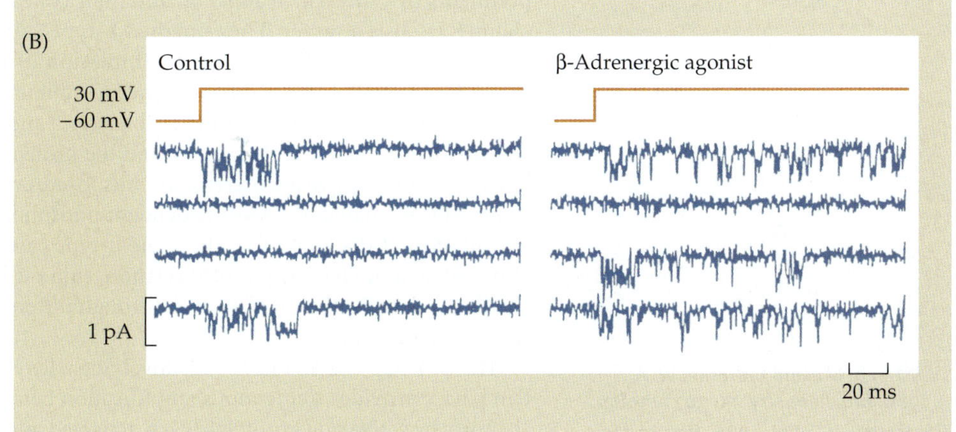

FIGURE 10.9 β-Adrenergic Agonists Cause an Increase in Calcium Channel Activity during a depolarizing pulse. (A) Recordings are from a voltage-clamped cell-attached patch. (B) Consecutive records of the activity of a patch containing two calcium channels. Addition of 14 μ*M* isoproterenol, a β-adrenergic agonist, to the bath causes an increase in the probability that the calcium channel will open when the cell is depolarized. (After Tsien, 1987.)

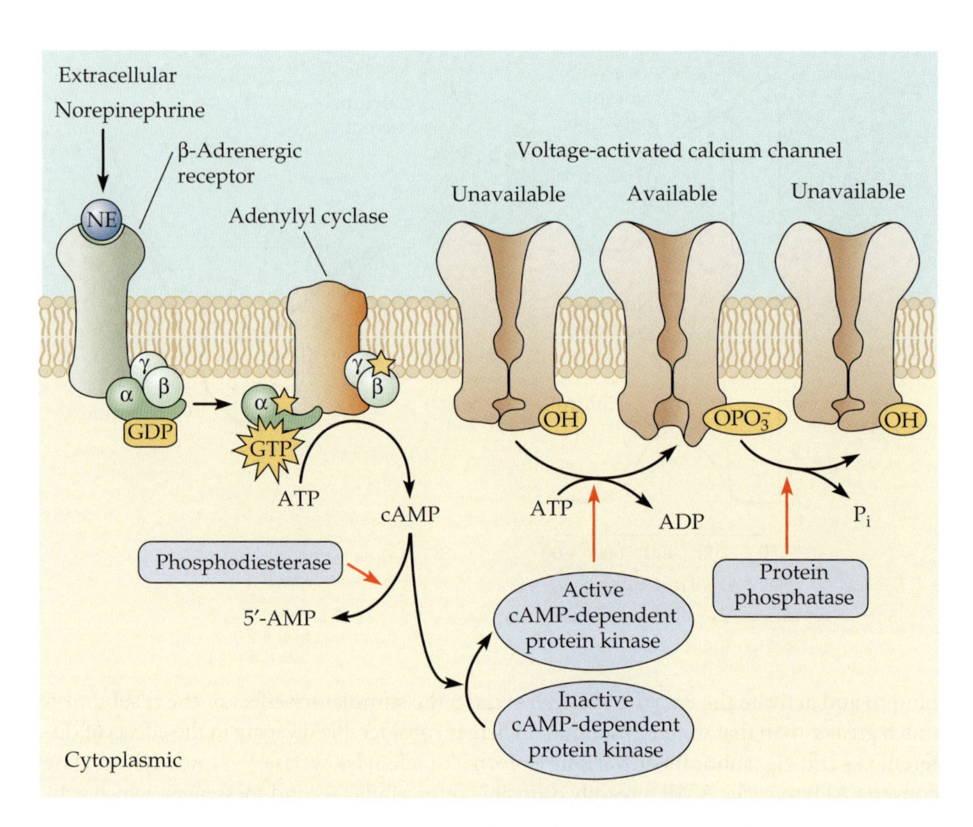

FIGURE 10.10 β-Adrenergic Receptors Act through the Intracellular Second Messenger Cyclic AMP to increase calcium channel activity. Binding of norepinephrine to β-adrenergic receptors activates, through a G protein, the enzyme adenylyl cyclase. Adenylyl cyclase catalyzes the conversion of ATP to cyclic AMP. As the concentration of cyclic AMP increases, it activates cAMP-dependent protein kinase, an enzyme that phosphorylates proteins on serine and threonine residues (—OH). The response to norepinephrine is terminated by the hydrolysis of cyclic AMP to 5′-AMP and the removal of protein phosphate residues by protein phosphatases. In cardiac muscle cells, norepinephrine causes phosphorylation of voltage-activated calcium channels, converting them to a form that can be opened by depolarization (available).

Regulation of Calcium Channel Activity by Other Signaling Pathways

In addition to norepinephrine, other neurotransmitters and hormones modulate the activity of calcium channels in cardiac muscle.[2] Many of these agents act through metabotropic receptors and G proteins to stimulate or inhibit adenylyl cyclase; others affect different intracellular second messenger systems. An example is the effect of acetylcholine on muscarinic receptors, which activates a G_i protein that *inhibits* the activity of adenylyl cyclase. This reduces the concentration of cyclic AMP and so decreases calcium channel activity.

Modulation of Calcium Channel Activity by Phosphorylation

The experiments described here clearly demonstrate that the effects of β-adrenergic stimulation on calcium currents are mediated by an increase in cyclic AMP and activation of protein kinase, but they do not identify the protein or proteins being phosphorylated. Experiments by Catterall, Trautwein, and their colleagues demonstrated that the calcium channel itself is a target. They purified L-type calcium channels (the type of calcium channels affected by β-adrenergic stimulation of heart muscle) from skeletal muscle using specific, high-affinity dihydropyridine inhibitors. The purified channels were incorporated into lipid vesicles[35] or planar bilayers[36] and exposed to active cAMP-dependent protein kinase (Figure 10.11). As a result the protein was phosphorylated and the probability of channel opening increased. Thus, β-adrenergic modulation of L-type calcium channels is mediated by cAMP-dependent phosphorylation of the channel protein itself. Modulation of the activity of the same L-type calcium channels by other hormones is also mediated by channel phosphorylation, either through effects on adenylyl cyclase and cAMP-dependent protein kinase, or through different second messenger–protein kinase signaling pathways.[2]

The two-step enzymatic cascade involving adenylyl cyclase and cAMP-dependent protein kinase provides tremendous amplification compared to direct opening or closing of channels by activated G proteins. Each activated adenylyl cyclase can catalyze the synthesis of many molecules of cyclic AMP and thereby activate many protein kinase molecules,

[35]Curtis, B. M., and Catterall, W. A. 1986. *Biochemistry* 25: 3077–3083.

[36]Flockerzi, V., et al. 1986. *Nature* 323: 66–68.

Box 10.2 Cyclic AMP as a Second Messenger

Experiments by Sutherland, Krebs, Walsh, Rodbell, Gilman, and their colleagues, initially aimed at understanding how the hormones epinephrine and glucagon elicit breakdown of glycogen in the liver, led to the discovery of cyclic AMP and the concept of intracellular second messengers.[37–39] They showed that binding of the hormone to its receptor activates a G protein, which, in turn, stimulates the enzyme adenylyl cyclase. Adenylyl cyclase catalyzes the synthesis of cyclic AMP from ATP. The increase in cyclic AMP concentration activates cAMP-dependent protein kinase, an enzyme that phosphorylates its target proteins on serine and threonine residues. Cyclic AMP is subsequently degraded by phosphodiesterase to AMP, and the phosphate residues on the target proteins are removed by protein phosphatases (see Figure 10.10).

Some tests used to determine if the response to a transmitter or hormone is mediated by cyclic AMP depend on activating adenylyl cyclase or elevating cyclic AMP directly. For example, intracellular injection of cyclic AMP and addition of membrane-permeable derivatives of cyclic AMP, such as 8-bromo-cyclic AMP or dibutyryl-cyclic AMP, mimic cAMP-mediated responses. Similarly, direct activation of adenylyl cyclase by forskolin mimics the response. Inhibitors of phosphodiesterase, such as the methylxanthines theophylline and caffeine, either mimic or enhance the response, depending on the endogenous level of cyclase activity. Other procedures test the involvement of cAMP-dependent protein kinase (also known as protein kinase A). This enzyme is composed of two regulatory and two catalytic subunits. In the absence of cyclic AMP the four subunits exist as a complex, the regulatory subunits blocking the activity of the catalytic subunits. When cyclic AMP binds to the regulatory subunits, the complex dissociates, freeing active catalytic subunits. Thus, intracellular injection of purified catalytic subunit will mimic responses mediated by increased cyclic AMP, while injection of excess regulatory subunits will be inhibitory. Additional inhibitors of this enzyme have been developed, including H-8 (which also inhibits several other protein serine kinases), specific peptide inhibitors, and derivatives of ATP that cannot be used by the kinase as a source for phosphate residues. These inhibitors block responses mediated by cyclic AMP.

On the other hand, treatments that inhibit protein phosphatases augment and prolong responses mediated by cyclic AMP. These include injection of specific phosphatase inhibitors and of ATPγS, an analogue of ATP that can be used as a cosubstrate by cAMP-dependent protein kinase, forming phosphoproteins with thiophosphate linkages, which are resistant to hydrolysis by protein phosphatases.

[37]Sutherland, E. W. 1972. *Science* 177: 401–408.
[38]Schramm, M., and Selinger, Z. 1984. *Science* 225: 1350–1356.
[39]Gilman, A. G. 1987. *Ann. Rev. Biochem.* 56: 615–649.

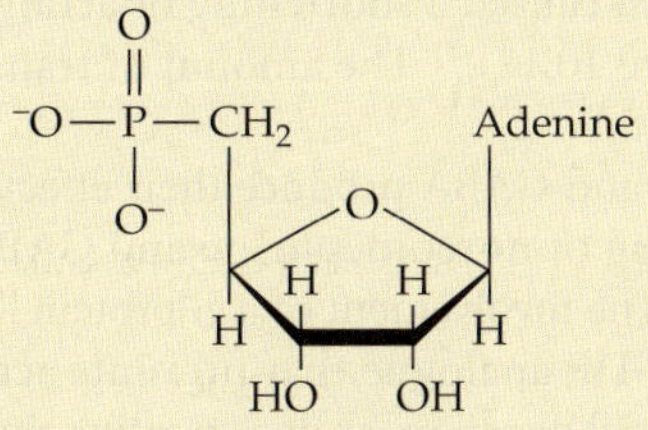

AMP
Adenosine 5′-monophosphate

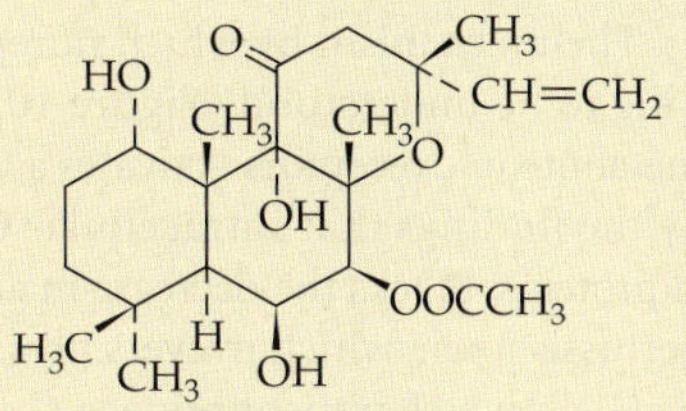

Forskolin

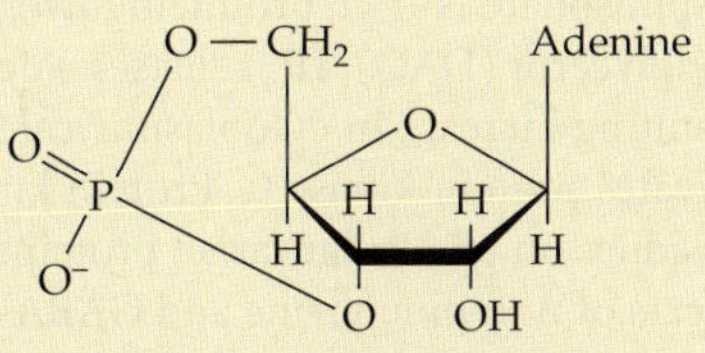

Cyclic AMP (cAMP)
Adenosine 3′, 5′-monophosphate

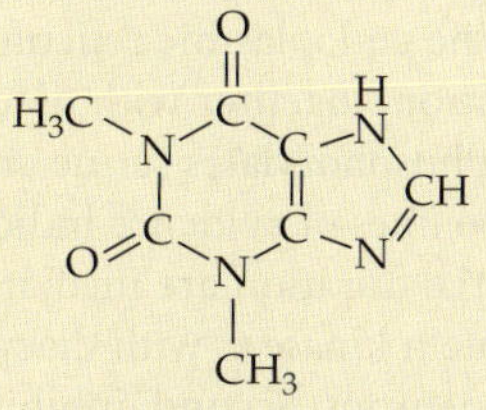

Theophylline
1, 3-Dimethylxanthine

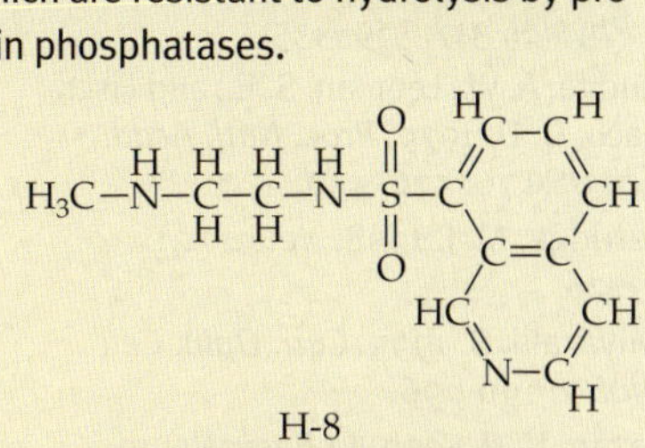

H-8
N-2-(methylamino)ethyl]-5-isoquinolinesulfonamide

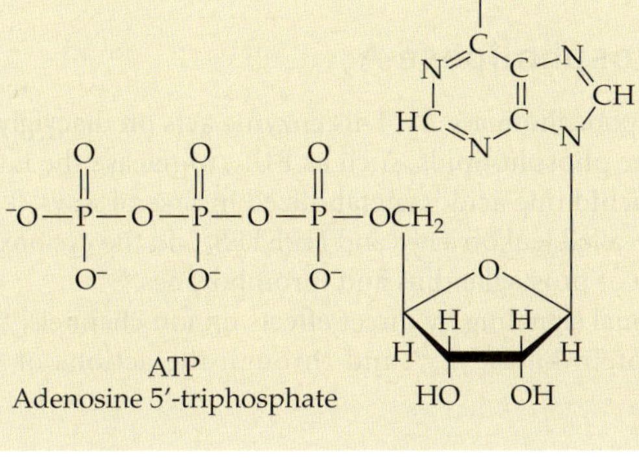

ATP
Adenosine 5′-triphosphate

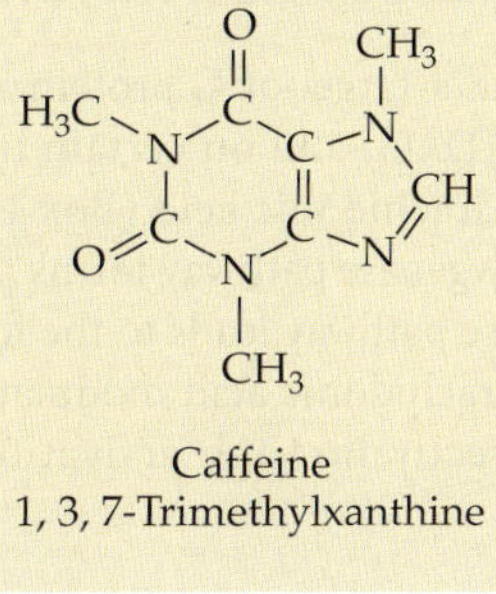

Caffeine
1, 3, 7-Trimethylxanthine

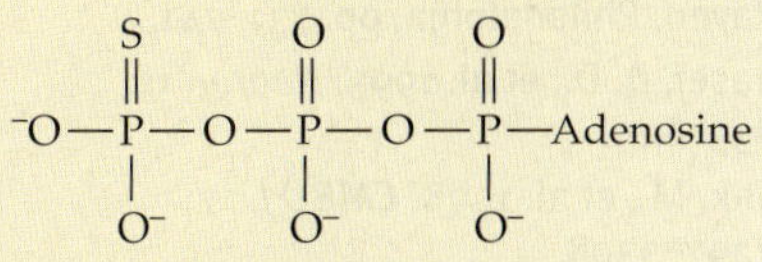

ATPγS
Adenosine 5′-*O*-[γ-thio] triphosphate

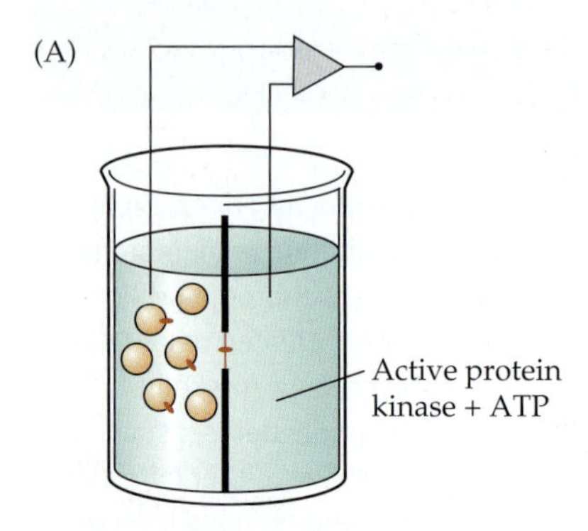

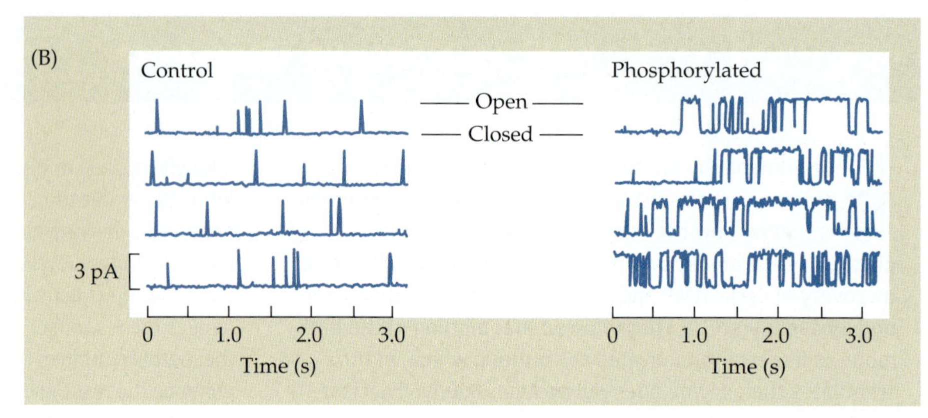

FIGURE 10.11 Phosphorylation of Calcium Channels Increases Their Probability of Opening. (A) Purified calcium channels, incorporated into small lipid vesicles (liposomes), are added to the solution on one side of a phospholipid bilayer. When liposomes fuse into the bilayer membrane, single-channel events can be recorded. (B) Phosphorylation increases the probability of channel opening. Single-channel records before and after adding ATP and the catalytic subunit of cAMP–dependent protein kinase to the solution bathing one side of the bilayer. Biochemical studies demonstrate that the calcium channel itself becomes phosphorylated. (After Flockerzi et al., 1986.)

and each activated kinase can phosphorylate many proteins. Thus, the activity of many molecules of a target protein at widespread sites may be modulated by the occupation of a few receptors. Moreover, cAMP-dependent protein kinase can phosphorylate a variety of proteins and so modulate a broad spectrum of cellular processes.

G Protein Activation of Phospholipase C

Modulation of the calcium current flowing during a presynaptic action potential is one mechanism by which synaptic efficacy is altered.[40] For example, when Fischbach and his colleagues made intracellular recordings from neurons dissociated from chick dorsal root ganglia and grown in cell culture, they found that GABA, norepinephrine, serotonin, and the peptides enkephalin and somatostatin each caused a shortening of action potential duration by decreasing calcium current (Figure 10.12).[41] The amount of transmitter released was reduced by a corresponding amount.[42]

The mechanism by which norepinephrine and GABA produce their effects is beginning to be understood (Figure 10.13). Binding of norepinephrine and GABA to their metabotropic receptors activates a G_i protein. The involvement of a G protein is indicated by the findings that intracellular GDPβS, a GDP analogue that prevents activation of G proteins, blocks the decrease in action potential duration, as does treating the cells with pertussis toxin, which irreversibly inactivates α subunits of the G_i family.

G_i activates phospholipase C (see Figure 10.13). Phospholipase C hydrolyzes the membrane lipid phosphatidylinositol 4,5-bisphosphate (PIP_2), producing two products: inositol 1,4,5-trisphosphate (IP_3) and diacylglycerol (DAG). IP_3 releases calcium ions from the endoplasmic reticulum.[43,44] The resulting increase in cytoplasmic calcium concentration, together with diacylglycerol, activates protein kinase C. Protein kinase C, in turn, phosphorylates serine and threonine residues on particular target proteins.

Two lines of evidence indicate that the effects of norepinephrine and GABA on action potential duration are mediated by protein kinase C[40] (Box 10.3). First, direct activation of protein kinase C with diacylglycerol analogues causes a decrease in calcium current in these neurons. Second, inhibitors of protein kinase C block the decrease in calcium current produced both by diacylglycerol analogues and by norepinephrine and GABA.

G Protein Activation of Phospholipase A_2

Another target of G protein action is phospholipase A_2. This enzyme acts on diacylglycerol (DAG) and on certain membrane phospholipids, such as PIP_2, to release the fatty acid arachidonic acid (Box 10.4). Arachidonic acid is metabolized in several ways: The lipoxygenase pathway forms products called leukotrienes and HPETES, and the cyclooxygenase pathway leads to the formation of prostaglandins and thromboxanes.[45]

Arachidonic acid modulates neuronal signaling by direct effects on ion channels,[46,47] indirectly through activation of protein kinase C,[48] and through the actions of its

40Dolphin, A. C. 1990. *Annu. Rev. Physiol.* 52: 243–255.

41Dunlap, K., and Fischbach, G. D. 1981. *J. Physiol.* 317: 519–535.

42Mudge, A. W., Leeman, S. E., and Fischbach, G. D. 1979. *Proc. Natl. Acad. Sci. USA* 76: 526–530.

43Berridge, M. J. 1998. *Neuron* 21: 13–26.

441Miyazaki, S. 1995. *Curr. Opin. Cell Biol.* 7: 190–196.

45Bazan, N. G. 1999. In *Basic Neurochemistry: Molecular, Cellular and Medical Aspects*, 6th Ed. Lippincott-Raven, Philadelphia, pp. 731–741.

46Fraser, D. D., et al. 1993. *Neuron* 11: 633–644.

47Fink, M., et al. 1998. *EMBO J.* 17: 3297–3308.

48Majewski, H., and Iannazzo, L. 1998. *Prog. Neurobiol.* 55: 463–475.

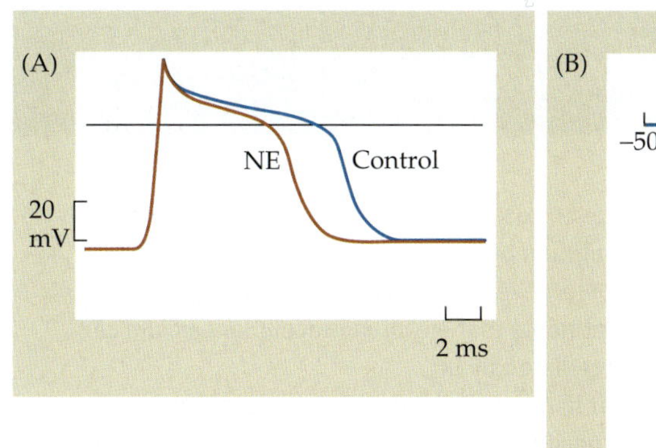

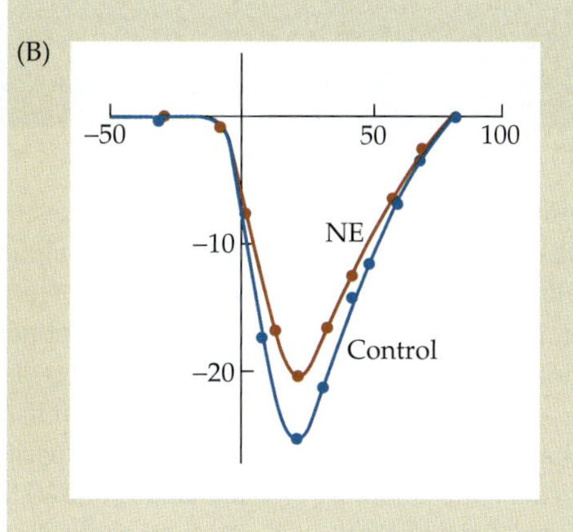

FIGURE 10.12 Norepinephrine Decreases the Duration of Action Potentials in chick dorsal root ganglion cells in culture. (A) Norepinephrine (NE; 10^{-5} *M*) causes a decrease in the duration of the action potential recorded with an intracellular microelectrode. (B) Current–voltage relationship for calcium current recorded from cells bathed in solutions containing tetrodotoxin and tetraethylammonium, and 10 m*M* calcium, with or without 10^{-4} *M* norepinephrine. Norepinephrine reduces the voltage-activated calcium current, which accounts for the decreased duration of the action potential. (After Dunlap and Fischbach, 1981.)

metabolites.[49] In *Aplysia* sensory neurons, for example, arachidonic acid is produced in response to the peptide FMRFamide and is metabolized along the lipoxygenase pathway to 12-HPETE. 12-HPETE binds to particular S-current potassium channels and increases their probability of opening (Figure 10.14).[50,51] This effect, together with other changes that hyperpolarize the postsynaptic motor cell, inhibits sensorimotor neuron transmission.[52,53]

Signaling via Nitric Oxide and Carbon Monoxide

Nitric oxide (NO), a water- and lipid-soluble gas produced from arginine by nitric oxide synthase, acts as a transmitter by diffusing from the cytoplasm of one cell into neighboring cells and activating guanylyl cyclase.[54,55] NO was first characterized as an important regulator of blood pressure, mediating the vasodilatation caused by acetylcholine.[56] The interaction of ACh with muscarinic receptors on vascular endothelial cells leads to activation of phospholipase C, formation of IP_3, and release of calcium from intracellular stores (Figure 10.15).

[49]Piomelli, D. 1994. *Crit. Rev. Neurobiol.* 8: 65–83.

[50]Piomelli, D., et al. 1987. *Nature* 328: 38–43.

[51]Buttner, N., Siegelbaum, S. A., and Volterra, A. 1989. *Nature* 342: 553–555.

[52]Pieroni, J. P., and Byrne, J. H. 1992. *J. Neurosci.* 12: 2633–2647.

[53]Belkin, K. J., and Abrams, T. W. 1993. *J. Neurosci.* 13: 5139–5152.

[54]Ignarro, L. J. 1990. *Pharmacol. Toxicol.* 67: 1–7.

[55]Schmidt, H. H., and Walter, U. 1994. *Cell* 78: 919–925.

[56]Furchgott, R. F., and Zawadzki, J. V. 1980. *Nature* 288: 373–376.

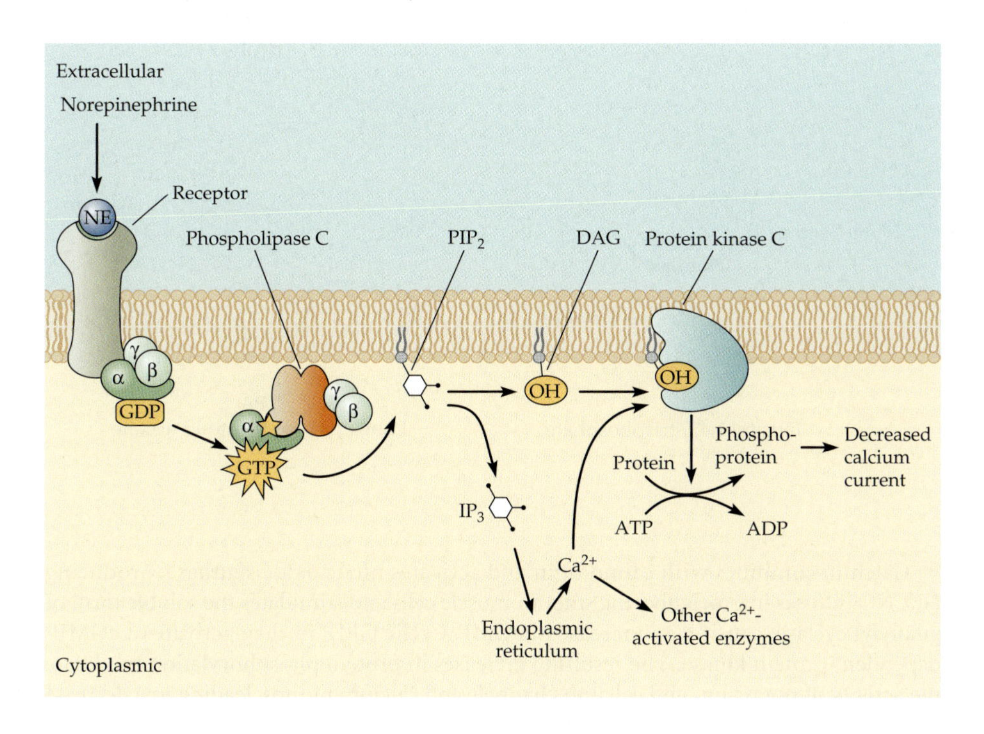

FIGURE 10.13 Norepinephrine Acts through the Intracellular Second Messengers Diacylglycerol and IP_3 to reduce calcium current in chick dorsal root ganglion cells. The binding of norepinephrine to its receptor activates, through a G protein, the enzyme phospholipase C. This enzyme hydrolyzes the phospholipid phosphatidylinositol 4,5-bisphosphate (PIP_2), releasing two intracellular second messengers: diacylglycerol (DAG) and inositol 1,4,5-trisphosphate (IP_3). IP_3 releases calcium from the endoplasmic reticulum into the cytoplasm. DAG and calcium together activate protein kinase C. Protein kinase C catalyzes increased protein phosphorylation. In chick neurons, this causes a decrease in calcium current.

Box 10.3 Diacylglycerol and IP_3 as Second Messengers

Activation of receptors coupled through G proteins to the enzyme phospholipase C leads to hydrolysis of the membrane lipid phosphatidylinositol 4,5-bisphosphate (PIP_2), producing two intracellular second messengers: diacylglycerol (DAG) and inositol 1,4,5-trisphosphate (IP_3) (see Figure 10.13).

IP_3 is water soluble, diffuses through the cytoplasm, and binds to receptors on the endoplasmic reticulum that allow calcium ions, sequestered therein, to be released. Calcium then acts as a "third" messenger to regulate the activity of calcium-dependent proteins in the cell.

DAG is hydrophobic and remains associated with the membrane, where it activates protein kinase C, a protein serine kinase. In unstimulated cells, protein kinase C is found in the cytoplasm in an inactive form; in the presence of calcium and DAG, it becomes associated with the membrane and is activated. The active kinase is capable of phosphorylating a variety of proteins on serine and threonine residues.

No simple tests exist to reliably identify responses mediated by IP_3. Inhibition by agents that interfere with phosphatidylinositol turnover, such as lithium, can suggest a role for IP_3, and treatments that elevate intracellular calcium can mimic effects mediated by IP_3.

On the other hand, the role of protein kinase C can be assessed using specific activators and inhibitors of this enzyme. Analogues of DAG that directly stimulate protein kinase C include synthetic diacylglycerols such as OAG and tumor-promoting phorbol esters such as TPA. Phospholipid analogues such as sphingosine, and protein serine kinase inhibitors such as H-7, inhibit protein kinase C and thus block responses mediated by this enzyme.

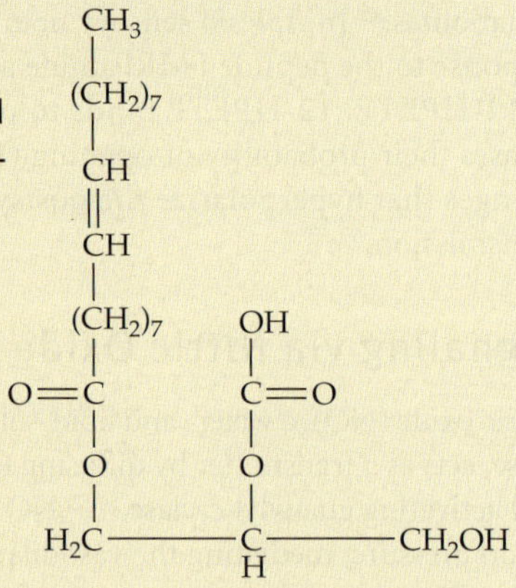

R Usually stearate ($CH_3(CH_2)_{16}COO^-$)

R′ Usually arachidonate ($CH_3(CH_2)_4(CH{=}CHCH_2)_4(CH_2)_2COO^-$)

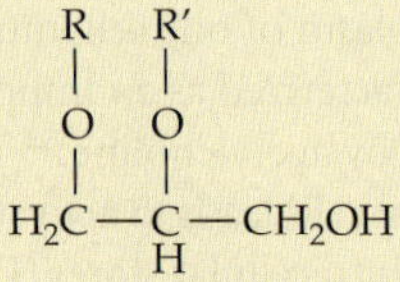

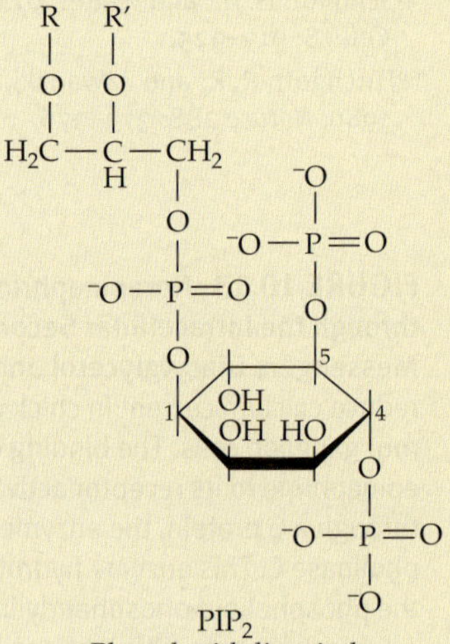

PIP_2
Phosphatidylinositol 4,5-bisphosphate

DAG
Diacylglycerol

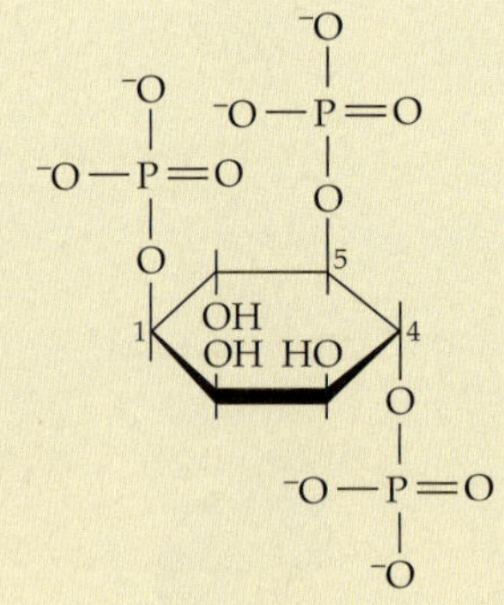

IP_3
Inositol 1,4,5-trisphosphate

OAG
1-Oleoyl-2-acetylglycerol

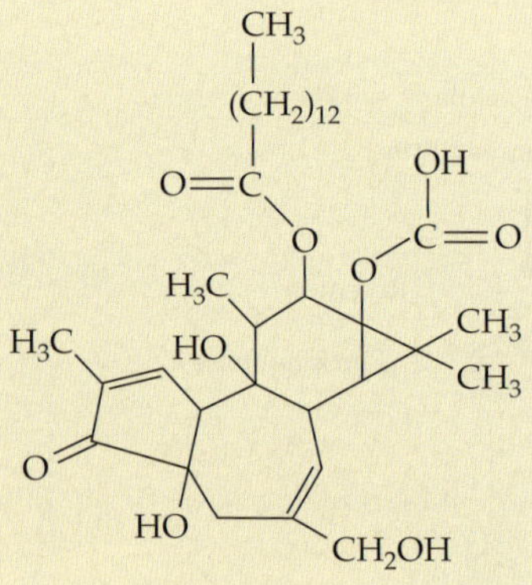

TPA
Phorbol 12-myristate 13-acetate

Calcium combines with calmodulin and activates nitric oxide synthase, producing NO. NO diffuses into neighboring smooth muscle cells and stimulates the soluble form of guanylyl cyclase, causing an increase in cGMP. Cyclic GMP in turn activates a cGMP-dependent protein kinase. The resulting increases in protein phosphorylation modulate the activity of potassium and calcium channels and calcium pumps, leading to a decrease in intracellular calcium concentration, which causes relaxation. NO is inactivated within

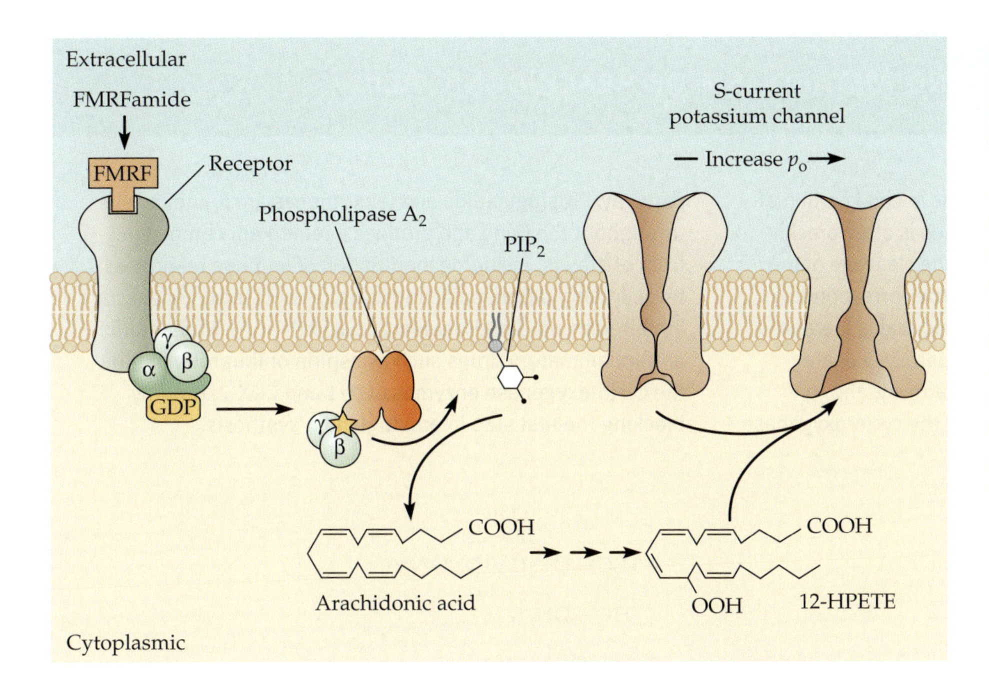

FIGURE 10.14 The Neuropeptide FMRFamide Acts through the Intracellular Second Messenger Arachidonic Acid to increase potassium channel activity in sensory neurons in the CNS of *Aplysia*. Binding of FMRFamide to its receptor activates, through a G protein, the enzyme phospholipase A_2. This enzyme degrades membrane phospholipids, such as PIP_2, forming arachidonic acid. Arachidonic acid is metabolized along the lipoxygenase pathway to form 12-HPETE, which binds to S-current potassium channels and increases their probability of opening (p_o).

seconds by reaction with superoxides and by formation of complexes with heme-containing proteins such as hemoglobin.

Elucidation of the NO signaling pathway led to an understanding of the action of nitroglycerin, long used to treat the chest pains associated with insufficient flow of blood to the heart (angina pectoris). Nitroglycerin acts as an NO donor and is particularly effective in relaxing coronary arteries. This increases the flow of blood to the heart and relieves the pain.

In brain, nitric oxide synthase is found associated with an ionotropic glutamate receptor, the NMDA receptor (Chapter 3).[57,58] Calcium ions entering through activated NMDA receptors bind calmodulin, and the calcium–calmodulin complex activates nitric oxide synthase.[59] The major effect of NO in brain, as in smooth muscle, is to stimulate

[57]Bredt, D. S., and Snyder, S. H. 1989. *Proc. Natl. Acad. Sci. USA* 86: 9030–9033.

[58]Garthwaite, J., et al. 1989. *Eur. J. Pharmacol.* 172: 413–416.

[59]Snyder, S. H., Jaffrey, S. R., and Zakhary, R. 1998. *Brain Res. Brain Res. Rev.* 26: 167–175.

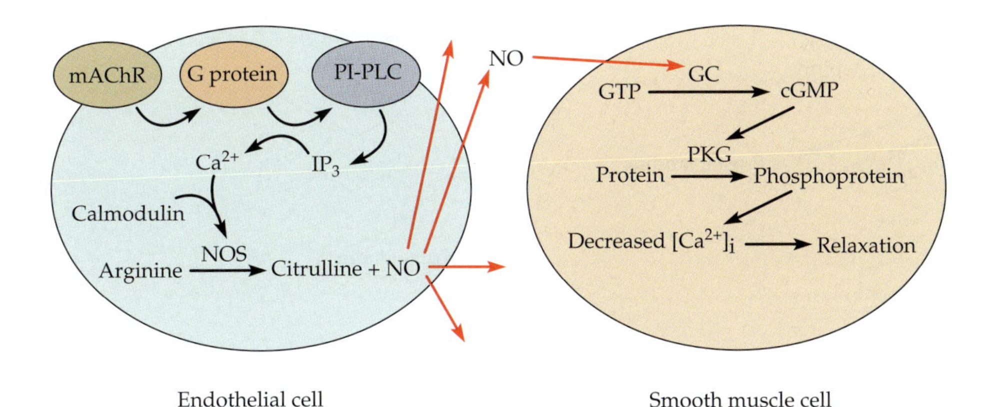

FIGURE 10.15 Paracrine Signaling by Release of Nitric Oxide. ACh binds to muscarinic receptors (mAChR) on vascular endothelial cells, activating phosphotidylinositide-specific phospholipase C (PI-PLC). PI-PLC forms inositol trisphosphate (IP_3), which releases calcium from intracellular stores. Calcium, together with calmodulin, activates nitric oxide synthase (NOS), producing nitric oxide (NO). NO diffuses into neighboring smooth muscle cells and stimulates guanylyl cyclase (GC), increasing cGMP. Cyclic GMP activates cGMP-dependent protein kinase (PKG). The resulting increases in protein phosphorylation lead to a decrease in intracellular calcium concentration, causing relaxation. NO is rapidly degraded, so that it affects only nearby cells—hence the term "paracrine."

Box 10.4 Formation and Metabolism of Arachidonic Acid

In mammals, the fatty acyl chain on the second carbon of the glycerol backbone of phospholipids is often arachidonate. Phospholipase A_2 catalyzes the cleavage of the fatty acyl chain at this position, yielding the corresponding lysolipid and arachidonic acid. Arachidonic acid is metabolized to leukotrienes and HPETEs (hydroperoxyeicosatetraenoic acids) along the lipoxygenase pathway, and to prostaglandins and thromboxanes along the cyclo-oxygenase pathway. Prostaglandins and leukotrienes are found in cells throughout the body and produce a remarkably broad spectrum of effects, including modulation of immune responses, fever, inflammation, pain, and apoptosis.[42] Lipoxygenases are inhibited by nordihydroguaiaretic acid (NDGA). Nonsteroidal anti-inflammatory drugs such as aspirin or ibuprofen inhibit the cyclo-oxygenase enzymes COX-1 and COX-2, thereby blocking the first step in prostaglandin synthesis.

Phosphatidylinositol 4,5-bisphosphate —Phospholipase A_2→ Lysophosphatidylinositol 4,5-bisphosphate + Arachidonic acid

Arachidonic acid → Lipoxygenase pathway (inhibited by NDGA) → Leukotrienes (Leukotriene B_4)

Arachidonic acid → Cyclooxygenase pathway (inhibited by Aspirin) → Prostaglandins (Prostaglandin A_2)

[60]Daniel, H., Levenes, C., and Crepel, F. 1998. *Trends Neurosci.* 21: 401–407.

[61]Savchenko, A., Barnes, S., and Kramer, R. H. 1997. *Nature* 390: 694–698.

[62]Hawkins, R. D., Zhuo, M., and Arancio, O. 1994. *J. Neurobiol.* 25: 652–665.

[63]Park, J. H., Straub, V. A., and O'Shea, M. 1998. *J. Neurosci.* 18: 5463–5476.

[64]Wang, R. 1998. *Can. J. Physiol. Pharmacol.* 76: 1–15.

guanylyl cyclase.[55] NO has been implicated in a variety of synaptic interactions in both invertebrates and vertebrates.[60–63]

Carbon monoxide (CO) has properties similar to those of NO.[59,64] CO is produced by the enzyme heme oxygenase, which is found in vascular endothelial cells and can be activated by protein kinase C–mediated phosphorylation. CO, a lipid- and water-soluble gas like NO, diffuses into neighboring smooth muscle cells and stimulates cGMP production, causing vasodilatation. In addition, both nitric oxide synthase and heme oxygenase are found in neurons in the walls of the gut, and both NO and CO have been shown to contribute to neurally induced relaxation of intestinal smooth muscle. A specific form of heme oxygenase, HO2, is concentrated in brain and has a distribution similar to that of soluble guanylyl cyclase, suggesting that CO may also have a role in signaling in the CNS.

A striking feature of signaling by NO and CO is that these compounds cannot be stored in synaptic vesicles for release at specific sites in axon terminals, juxtaposed to specialized postsynaptic membranes bearing appropriate receptors. Rather, they diffuse indiscriminately from the site at which they are produced into neighboring cells, their spread being limited only by their short lifetimes. This type of signaling, intermediate between neurotransmission and the release of hormones into the bloodstream by endocrine organs, has been termed **paracrine.** Clearly, specificity in the effects of such paracrine signals depends on the distribution and properties of enzymes activated or inhibited by NO and CO.

Modulation of Potassium and Calcium Channels by Indirectly Coupled Receptors

An important generalization is that potassium and calcium channels are prime targets for modulation by transmitters acting through indirectly coupled receptors. Changes in channel activation can influence the resting potential, spontaneous activity, the response to other excitatory or inhibitory inputs, and the amount of calcium that enters during an action potential. Such effects play a major role in signaling in the nervous system. It is impossible to predict what effects indirect mechanisms will have on channel activity in a particular cell. For example, norepinephrine and GABA decrease calcium current in chick sensory cells through activation of protein kinase C,[40] while GABA, but not norepinephrine, acts through the cyclic AMP pathway to lengthen the duration of calcium action potentials in lamprey sensory neurons by inhibiting calcium-activated potassium channels.[65] The complexity extends to the level of single cells: The activity of rat superior cervical ganglion neurons is modulated by at least nine transmitters acting through five G protein–coupled pathways that influence two calcium channels and a potassium channel.[66]

CALCIUM AS AN INTRACELLULAR SECOND MESSENGER

The concentration of calcium within cells can be raised by influx through ligand- or voltage-activated channels (Chapter 3) and through the activity of calcium pumps and exchangers (Chapter 4). In addition, calcium can be released from intracellular storage sites by calcium influx, sodium influx, and IP_3.[67,68] Depletion of intracellular calcium stores, in turn, opens another class of calcium channels in the surface membrane, called i_{crac} channels for the calcium release–activated calcium current they carry.[69,70]

With the development of optical methods for the measurement of calcium concentration within cells came two remarkable findings. One was the extent to which increases in calcium concentration were often confined to particular regions of the cell, creating calcium microdomains.[71–74] Thus, the effects of changes in intracellular calcium depend on the subcellular distribution of calcium, of the calcium-activated proteins, and of their target enzymes and ion channels. A second finding was that physiological signals often give rise to dynamic changes in calcium concentration, such as oscillations, spikes, sparks, and waves, rather than a sustained increase.[44,68,75,76] Different spatiotemporal calcium profiles, in turn, activate specific intracellular biochemical pathways.[77]

Calcium-Mediated Rapid Synaptic Inhibition

Synaptic inhibition between efferent auditory nerve fibers and cochlear hair cells in the chick provides an example in which calcium, entering the hair cell through an ionotropic receptor, acts as an intracellular second messenger to activate another ion channel.[78] At this synapse acetylcholine, released from the efferent nerve terminals onto the postsynaptic membrane of the hair cells, activates ionotropic neuronal acetylcholine receptors that allow entry of calcium and other cations. The effect would ordinarily be excitatory, as at the neuromuscular junction. In hair cells, however, the incoming calcium opens calcium-activated potassium channels, thereby producing inhibition (Figure 10.16A).

[65]Leonard, J. P., and Wickelgren, W. O. 1986. *J. Physiol.* 375: 481–497.

[66]Hille, B. 1994. *Trends Neurosci.* 17: 531–536.

[67]Lipscombe, D., et al. 1988. *Neuron* 1: 355–365.

[68]Tsien, R. W., and Tsien, R. Y. 1990. *Annu. Rev. Cell Biol.* 6: 715–760.

[69]Lewis, R. S. 1999. *Adv. Second Messenger Phosphoprotein Res.* 33: 279–307.

[70]Putney, J. W., Jr., and McKay, R. R. 1999. *BioEssays* 21: 38–46.

[71]Ross, W. N., Arechiga, H., and Nicholls, J. G. 1988. *Proc. Natl. Acad. Sci. USA* 85: 4075–4078.

[72]Bacskai, B. J., et al. 1995. *Neuron* 14: 19–28.

[73]Tucker, T., and Fettiplace, R. 1995. *Neuron* 15: 1323–1335.

[74]Llinás, R., Sugimori, M., and Silver, R. B. 1995. *J. Physiol. (Paris)* 89: 77–81.

[75]Meyer, T., and Stryer, L. 1991. *Annu. Rev. Biophys. Biophys. Chem.* 20: 153–174.

[76]Cheng, H., et al. 1996. *Am. J. Physiol.* 270: C148–159.

[77]DeKoninck, P., and Schulman, H. 1998. *Science* 279: 227–230.

[78]Fuchs, P. A., and Murrow, B. W. 1992. *J. Neurosci.* 12: 800–809.

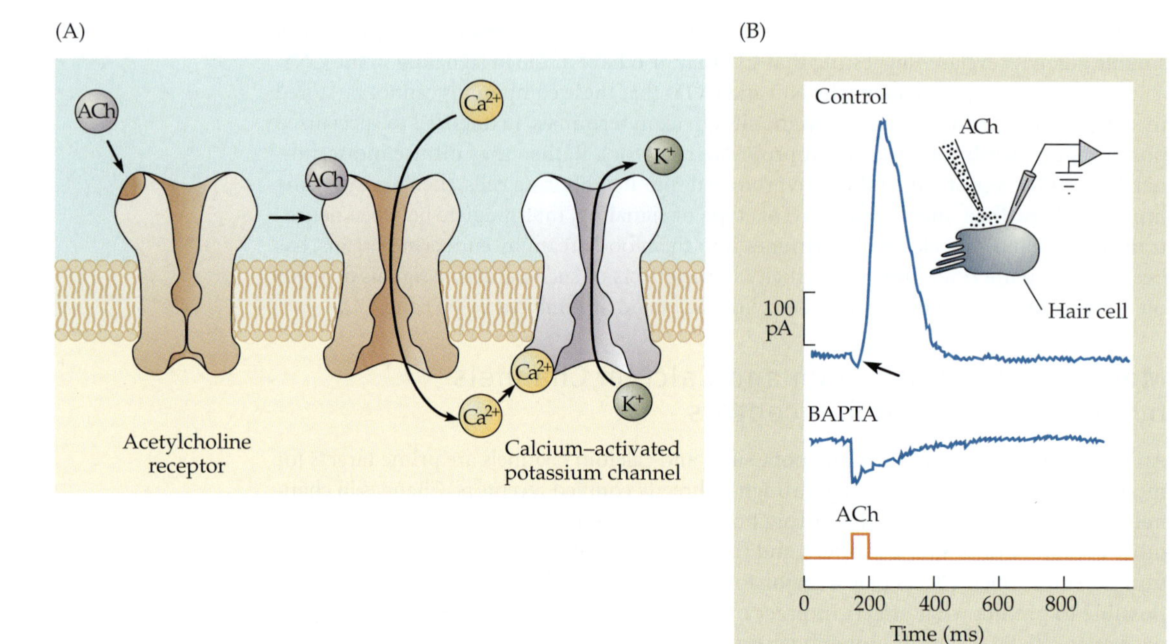

FIGURE 10.16 Inhibition by ACh-Activated Cation Channels in hair cells from the chick cochlea. (A) In chick hair cells ACh binds to ionotropic receptors that allow cations, including calcium, to flow into the cell. Intracellular calcium causes calcium-activated potassium channels to open, leading to outward potassium current and hyperpolarization. (B) In a whole-cell recording (inset), application of ACh near the base of a hair cell produces a small, transient inward current (arrow) followed by a large outward current. In the intact cell, the outward current would be inhibitory. If the calcium chelator BAPTA is added to the recording electrode, and hence to the cell cytoplasm, ACh application produces only inward current. No outward current is seen because incoming calcium ions are chelated and so prevented from activating potassium channels. (Records kindly provided by P. A. Fuchs)

As illustrated in Figure 10.16B, patch clamp electrodes were used to obtain records of whole-cell currents from the hair cells. Application of a brief pulse of ACh solution from a micropipette produced a large outward (inhibitory) potassium current, preceded by a small, brief inward current. When the same experiment was repeated with the calcium chelator BAPTA in the recording pipette (and hence in the cytoplasm of the cell), the outward current was blocked, revealing a substantial inward cation current. The incoming calcium, being chelated, could no longer act as an intracellular second messenger to open calcium-activated potassium channels. A similar mechanism of inhibition may occur in neurons in the brain.[79]

Complexity of Calcium Signaling Pathways

Intracellular calcium regulates potassium, cation-selective, and chloride channels in the plasma membrane, as well as the membrane-bound phospholipases C and A_2 (Figure 10.17).[80] These enzymes, which are involved in the production of the intracellular second messengers IP_3, diacylglycerol, and arachidonic acid, are also regulated by G proteins, as described earlier. Within the cytoplasm, calcium activates three major targets: protein kinase C, calmodulin, and a calcium-dependent protease called calpain.

Calpains are a group of proteases that regulate the cytoskeleton, as well as a number of membrane proteins.[81,82] Calmodulin is a ubiquitous protein having four calcium-binding sites.[83,84] When these sites are occupied, calmodulin activates calcium/calmodulin-dependent protein kinases, adenylyl cyclase, cyclic nucleotide phosphodiesterase, a protein phosphatase called calcineurin, and nitric oxide synthase.

[79]Wong, L. A., and Gallagher, J. P. 1991. *J. Physiol.* 436: 325–346.

[80]Marty, A. 1989. *Trends Neurosci.* 12: 420–424.

[81]Sorimachi, H., Ishiura, S., and Suzuki, K. 1997. *Biochem. J.* 328: 721–732.

[82]Johnson, G. V., and Guttmann, R. P. 1997. *BioEssays* 19: 1011–1018.

[83]Vogel, H. J. 1994. *Biochem. Cell Biol.* 72: 357–376.

[84]Ghosh, A., and Greenberg, M. E. 1995. *Science* 268: 239–247.

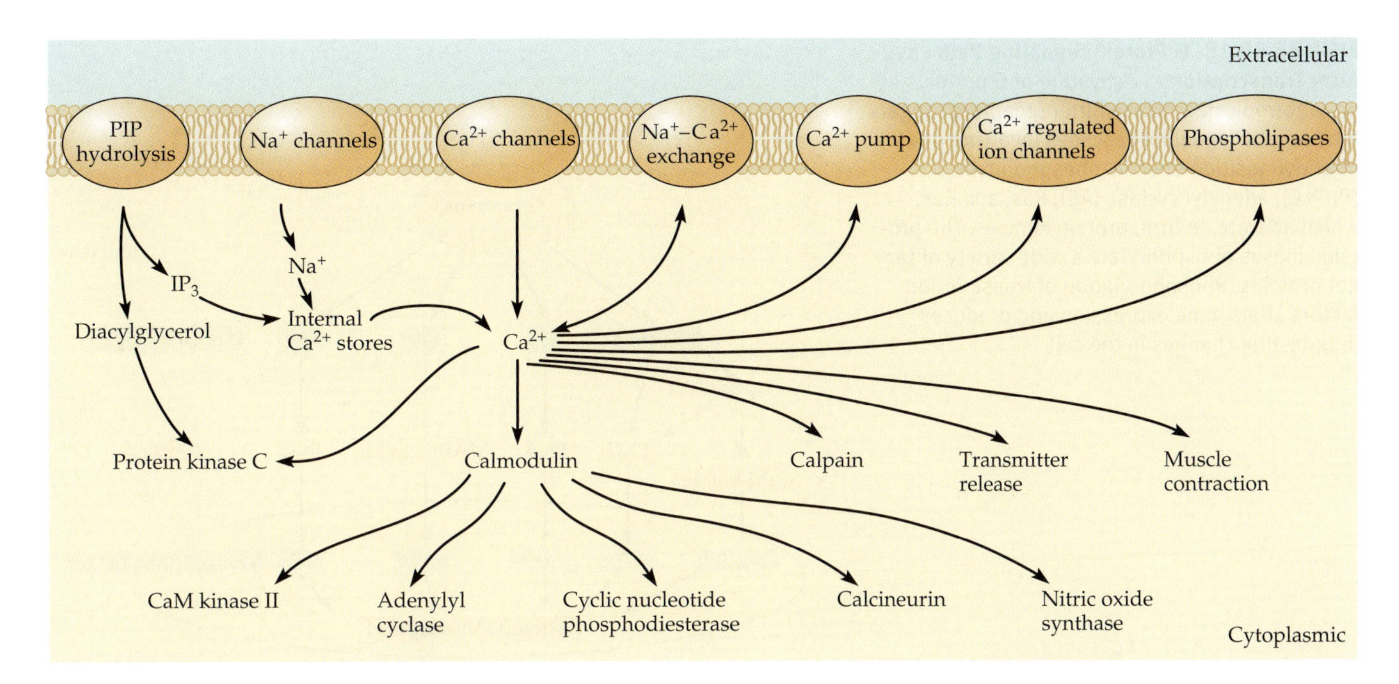

FIGURE 10.17 Calcium as an Intracellular Second Messenger. The concentration of calcium in the cytoplasm is regulated by influx through membrane channels, by the activity of calcium pumps and exchangers in the plasma membrane, by sequestration in internal stores such as the endoplasmic reticulum, and by release from internal storage sites by sodium influx, calcium influx, and IP_3. Calcium, in turn, regulates membrane and cytosolic proteins, including ion channels, exchangers, pumps, phospholipases, protein kinase C, calmodulin, and calpain. (After Kennedy, 1989.)

PROLONGED TIME COURSE OF INDIRECT TRANSMITTER ACTION

Synaptic interactions mediated by indirect mechanisms typically develop more slowly and last much longer than those mediated by direct mechanisms. At the skeletal neuromuscular junction, only one or two milliseconds are required for acetylcholine to be released, diffuse across the synaptic cleft, and bind to and open ionotropic acetylcholine receptors. These events are much too fast to be mediated by enzymes such as adenylyl cyclase or phospholipase C, which take many milliseconds to catalyze the synthesis of a single molecule of cyclic AMP or the hydrolysis of a membrane lipid. Even activation of a membrane channel by binding of a G protein subunit to the channel itself tends to have a time course of seconds, reflecting the lifetime of the activated α subunit.[85] Responses mediated by enzymatic production of diffusible cytoplasmic second messengers such as cyclic AMP or IP_3 are slower still, lasting seconds to minutes, reflecting the slow time course of changes in second messenger concentration.

Yet experience tells us that changes in signaling in the nervous system can last a lifetime. How can such long-lasting changes in synaptic efficacy be produced? One answer comes from the properties of several of the protein kinases discussed in this chapter. These enzymes are themselves targets for phosphorylation. For example, when activated by calcium, CaM kinase II phosphorylates itself.[86] If several of the subunits are phosphorylated, the properties of the enzyme change: It becomes constitutively active and no longer requires the presence of the calcium–calmodulin complex for activity. This provides one mechanism by which a transient increase in calcium concentration can be translated into long-lasting activation of the kinase, which in turn can cause sustained changes in the activity of its other target proteins.

For changes to persist for days or longer, protein synthesis is usually required. Many of the second messenger systems described in this chapter have been shown to produce changes in protein synthesis (Figure 10.18).[87–89] Such changes typically occur as a result of activation of one or more protein phosphorylation signaling cascades, which lead to phosphorylation of transcription factors and, consequently, altered gene expression. The most rapid effects that have been measured occur in expression of immediate early genes, such as c-*fos*, c-*jun*, and *zif*/268I, which encode inducible transcription factors of the Fos, Jun, and Krox families.[90] Upon translation, these proteins enter the nucleus, where they regulate further gene expression, ultimately producing metabolic or structural changes that permanently alter the response of the cell.

[85]Casey, P. J., et al. 1988. *Cold Spring Harb. Symp. Quant. Biol.* 53: 203–208.

[86]Miller, S. G., and Kennedy, M. B. 1986. *Cell* 44: 861–870.

[87]Gutkind, J. S. 1998. *J. Biol. Chem.* 273: 1839–1842.

[88]Schulman, H., and Hyman, S. E. 1999. In *Fundamental Neuroscience.* Academic Press, New York, pp. 269–316.

[89]Heist, E. K., and Schulman, H. 1998. *Cell Calcium* 23: 103–114.

[90]Tischmeyer, W., and Grimm, R. 1999. *Cell. Mol. Life Sci.* 55: 564–574.

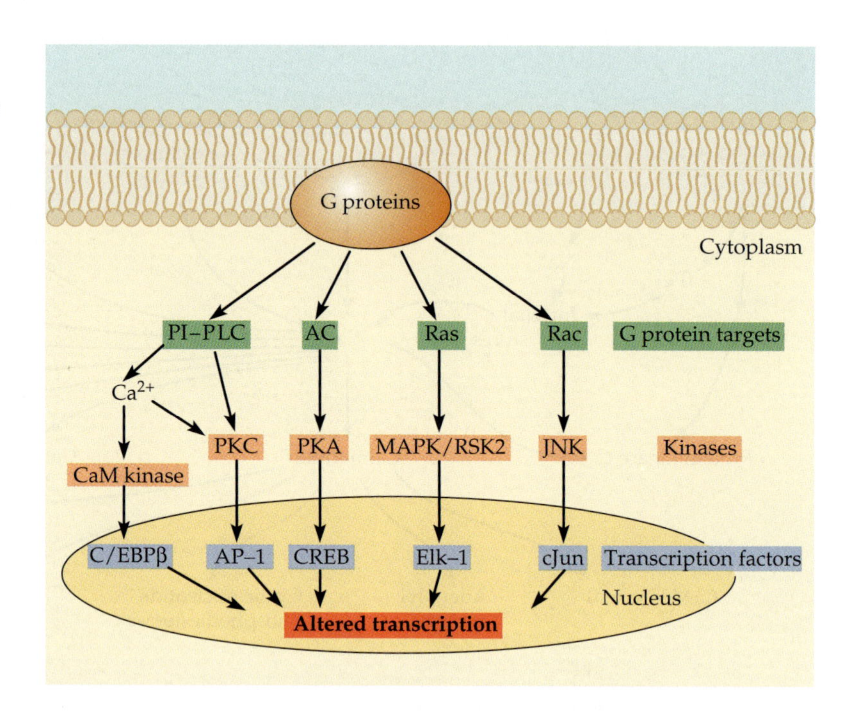

FIGURE 10.18 G Protein Signaling Pathways Alter Transcription. Activation of G proteins by metabotropic neurotransmitter receptors triggers intracellular signaling cascades involving phosphatidylinositide-specific phospholipase C (PI-PLC), adenylyl cyclase (AC), Ras, and Rac, which activate, in turn, protein kinases. The protein kinases phosphorylate a wide variety of target proteins. Phosphorylation of transcription factors alters gene expression and produces long-lasting changes in the cell.

SUMMARY

- Neurotransmitters activate metabotropic receptors in target cells. Metabotropic receptors are not themselves ion channels, rather they modify the activity of ion channels, ion pumps, or other receptor proteins by indirect mechanisms.

- Metabotropic receptors produce their effects through G proteins. Examples of metabotropic receptors include muscarinic acetylcholine receptors; α- and β-adrenergic receptors; some of the receptors for GABA, 5-HT, dopamine, and glutamate; and receptors for neuropeptides, for light, and for odorants.

- G proteins are αβγ heterotrimers. In the resting state, GDP is bound to the α subunit, and the three subunits are associated as a trimer. When activated by a metabotropic receptor, GDP is replaced with GTP, the α and βγ subunits dissociate, and the free subunits activate one or more intracellular targets. The activity of G protein subunits is terminated by hydrolysis of GTP to GDP by the endogenous GTPase activity of the α subunit, and the recombination of α and βγ subunits into a trimer.

- Some G protein βγ subunits bind directly to ion channels and stimulate or inhibit their activity. Other G protein α or βγ subunits activate adenylyl cyclase, phospholipase C, or phospholipase A_2, generating intracellular second messengers that can have widespread effects.

- Indirectly acting transmitters influence the activity of potassium and calcium channels. Changes in channel activity in turn influence the resting potential, spontaneous activity, the response to other inputs, or the amount of calcium entering during an action potential, and thereby the amount of transmitter release.

- Changes in intracellular calcium or calcium–calmodulin concentration regulate ion channels, phospholipases C and A_2, protein kinase C, calpain, adenylyl cyclase, cyclic nucleotide phosphodiesterase, and nitric oxide synthase. Both the distribution of changes in intracellular calcium, which can be highly localized, and their dynamics (calcium waves and oscillations) are important determinants of calcium action.

- Transmitter actions mediated by indirect mechanisms have time courses that vary from milliseconds to years. Rapid effects are produced by direct changes in ion channel activity, effects of intermediate duration by activation and phosphorylation of enzymes and other proteins, and very long-lasting effects by regulation of protein synthesis.

SUGGESTED READING

General Reviews

Berridge, M. J. 1998. Neuronal calcium signaling. *Neuron* 21: 13–26.

Clapham, D. E., and Neer, E. J. 1997. G protein βγ subunits. *Annu. Rev. Pharmacol. Toxicol.* 37: 167–203.

Hamm, H. E. 1998. The many faces of G protein signaling. *J. Biol. Chem.* 273: 669–672.

Hille, B. 1994. Modulation of ion-channel function by G-protein-coupled receptors. *Trends Neurosci.* 17: 531–535.

Kaczmarek, L. K., and Levitan, I. B. (eds.). 1987. *Neuromodulation: The Biochemical Control of Neuronal Excitability*. Oxford University Press, New York.

Levitan, I. B., and Kaczmarek, L. K. 1997. *The Neuron: Cell and Molecular Biology*, 2nd Ed. Oxford University Press, New York.

Piomelli, D. 1994. Eicosanoids in synaptic transmission. *Crit. Rev. Neurobiol.* 8: 65–83.

Schmidt, H. H., and Walter, U. 1994. NO at work. *Cell* 78: 919–925.

Schulman, H., and Hyman, S. E. 1999. Intracellular signaling. In M. J. Zigmond, F. E. Bloom, S. C. Landis, J. L. Roberts, and L. R. Squire (eds.), *Fundamental Neuroscience*. Academic Press, New York, pp. 269–316.

Wess, J. 1998. Molecular basis of receptor/G-protein-coupling selectivity. *Pharmacol. Ther.* 80: 231–264.

Original Papers

Dunlap, K., and Fischbach, G. D. 1981. Neurotransmitters decrease the calcium conductance activated by depolarization of embryonic chick sensory neurones. *J. Physiol.* 317: 519–535.

Flockerzi, V., Oeken, H-J., Hofmann, F., Pelzer, D., Cavalié, A. and Trautwein, W. 1986. Purified dihydropyridine-binding site from skeletal muscle t-tubules is a functional calcium channel. *Nature* 323: 66–68.

Lambright, D. G., Sondek, J., Bohm, A., Skiba, N. P., Hamm, H. E., and Sigler, P. B. 1996. The 2.0Å crystal structure of a heterotrimeric G protein. *Nature* 379: 311–319.

Sakmann, B., Noma, A., and Trautwein, W. 1983. Acetylcholine activation of single muscarinic K^+ channels in isolated pacemaker cells of the mammalian heart. *Nature* 303: 250–253.

Sunahara, R. K., Tesmer, J. J. G., Gilman, A. G., and Sprang, S. R. 1996. Crystal structure of the adenylyl cyclase activator $G_{s\alpha}$. *Science* 278: 1943–1947.

Tesmer, J. J. G., Sunahara, R. K., Gilman, A. G., and Sprang, S. R. 1997. Crystal structure of the catalytic domains of adenylyl cyclase in a complex with $G_{s\alpha}$·GTPγS. *Science* 278: 1907–1916.

Wickman, K. D., Iñiguez-Lluhi, J. A., Davenport, P. A., Taussig, R., Krapivinsky, G. B., Linder, M. E., Gilman, A. G., and Clapham, D. E. 1994. Recombinant G-protein βγ-subunits activate the muscarinic-gated atrial potassium channel. *Nature* 368: 255–257.

11 Transmitter Release

THE STIMULUS FOR NEUROTRANSMITTER RELEASE is depolarization of the nerve terminal. Release occurs as a result of calcium entry into the terminal through voltage-activated calcium channels. Invariably a delay of about 0.5 ms intervenes between presynaptic depolarization and transmitter release. Part of the delay is due to the time taken for calcium channels to open; the remainder is due to the time required for calcium to cause transmitter release.

Transmitter is secreted in multimolecular packets (quanta), each containing several thousand transmitter molecules. In response to an action potential, anywhere from 1 to as many as 300 quanta are released almost synchronously from the nerve terminal, depending on the type of synapse. At rest, nerve terminals release quanta spontaneously at a slow rate, giving rise to spontaneous miniature synaptic potentials. There is also at rest a continuous, nonquantal leak of transmitter from nerve terminals.

One quantum of transmitter corresponds to the contents of one synaptic vesicle and comprises several thousand molecules of a low-molecular-weight transmitter. Release occurs by the process of exocytosis, during which the synaptic vesicle membrane fuses with the presynaptic membrane and the contents of the vesicle are released into the synaptic cleft. The components of the vesicle membrane are then retrieved by endocytosis, sorted in endosomes, and recycled into new synaptic vesicles.

A number of questions arise concerning the way in which presynaptic neurons release transmitter. Experimental answers to such questions require a highly sensitive, quantitative, and reliable measure of the amount of transmitter released, with a time resolution in the millisecond range. In many of the experiments described in this chapter, this measure is provided by the membrane potential of the postsynaptic cell. Once again, the vertebrate neuromuscular junction, where ACh is known to be the transmitter, offers many advantages. However, to obtain more complete information about the release process, it is useful to be able to record from the presynaptic endings as well; for example, such recordings are needed to establish how calcium and membrane potential affect transmitter release. The presynaptic terminals at vertebrate skeletal neuromuscular junctions are typically too small for electrophysiological recording (but see Morita and Barrett, 1990[1]); however, this can be done at a number of synapses, such as the giant fiber synapse in the stellate ganglion of the squid,[2] giant terminals of goldfish retinal bipolar cells,[3] and calyciform synapses in the avian ciliary ganglion[4] and the rodent brainstem.[5] Moreover, new techniques allow transmitter release to be monitored by means that do not require recording from the postsynaptic cell. In this chapter we discuss electrophysiological and morphological experiments that characterize the release process. The proteins that mediate transmitter release are described in Chapter 13.

CHARACTERISTICS OF TRANSMITTER RELEASE

Axon Terminal Depolarization and Release

The stellate ganglion of the squid was used by Katz and Miledi to determine the precise relation between the membrane potential of the presynaptic terminal and the amount of transmitter release.[6] The preparation and the arrangement for recording from the presynaptic terminal and the postsynaptic fiber simultaneously are shown in Figure 11.1A. When tetrodotoxin (TTX) was applied to the preparation, the presynaptic action potential gradually decreased in amplitude over the next 15 min (Figure 11.1B). The postsynaptic action potential also decreased in amplitude, but then abruptly disappeared because the excitatory postsynaptic potential (EPSP) failed to reach threshold. From this point on the size of the synaptic potential could be used as a measure of the amount of transmitter released.

When the amplitude of the excitatory postsynaptic potential is plotted against the amplitude of the failing presynaptic impulse, as in Figure 11.1C, the synaptic potential is seen to decrease rapidly as the presynaptic action potential amplitude falls below about 75 mV, and at amplitudes less than about 45 mV there are no postsynaptic responses. Tetrodotoxin has no effect on the sensitivity of the postsynaptic membrane to transmitter, so the fall in synaptic potential amplitude indicates a reduction in the amount of transmitter released from the presynaptic terminal. Thus, there is a threshold for transmitter release at about 45 mV depolarization, after which the amount released, and hence the EPSP amplitude, increases rapidly with presynaptic action potential amplitude.

Katz and Miledi used an additional procedure to explore the relation further: They placed a second electrode in the presynaptic terminal, through which they applied brief (1–2 ms) depolarizing current pulses, thereby mimicking a presynaptic action potential. The relationship between the amplitude of the artificial action potential and that of the synaptic potential was the same as the relation obtained with the failing action potential during TTX poisoning (see Figure 11.1C). This result indicates that the normal fluxes of sodium and potassium ions responsible for the action potential are not necessary for transmitter release; depolarization is the trigger.

Synaptic Delay

One characteristic of the transmitter release process evident in Figure 11.1B is the **synaptic delay**, the time between the onset of the presynaptic action potential and the beginning of the synaptic potential (Chapter 9). In these experiments on the squid giant synapse, which were done at about 10°C, the delay was 3 to 4 ms. Detailed measurements at the frog neuromuscular junction show a synaptic delay of 0.5 ms at room temperature

[1]Morita, K., and Barrett, E. F. 1990. *J. Neurosci.* 10: 2614–2625.

[2]Bullock, T. H., and Hagiwara, S. 1957. *J. Gen. Physiol.* 40: 565–577.

[3]Heidelberger, R., and Matthews, G. 1992. *J. Physiol.* 447: 235–256.

[4]Martin, A. R., and Pilar, G. 1963. *J. Physiol.* 168: 443–463.

[5]Borst, J. G. G., and Sakmann, B. 1996. *Nature* 383: 431–434.

[6]Katz, B., and Miledi, R. 1967. *J. Physiol.* 192: 407–436.

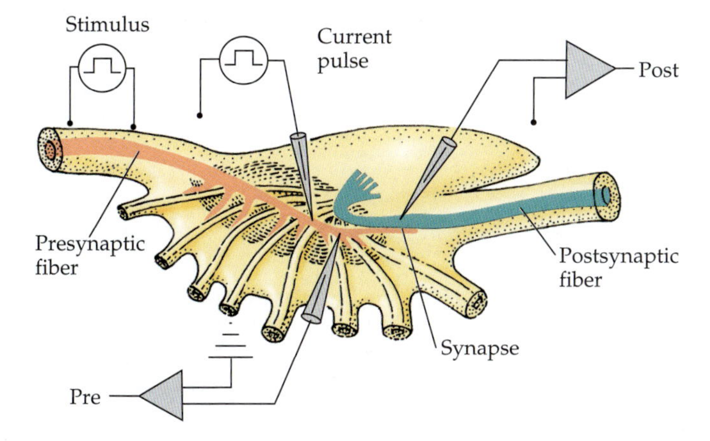

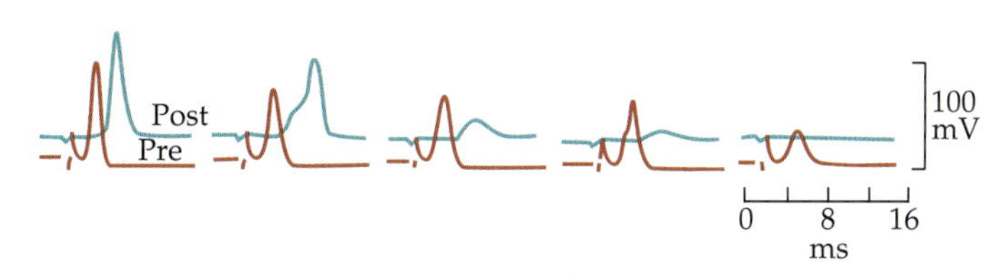

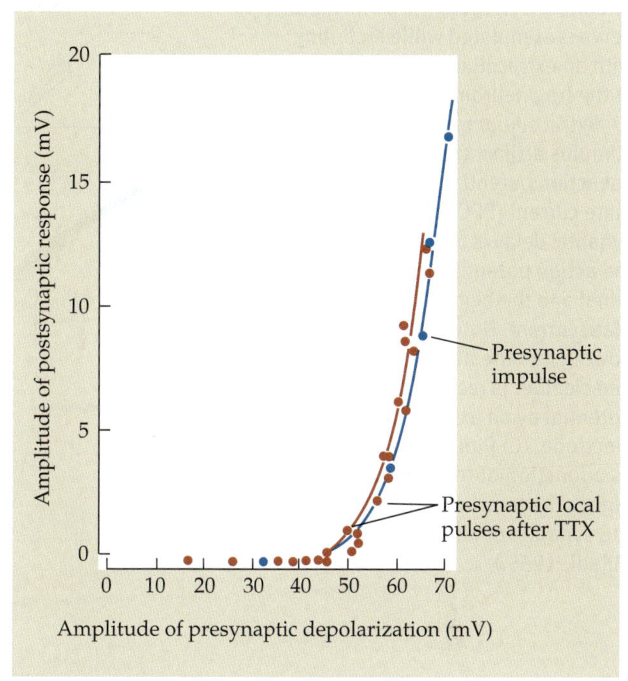

FIGURE 11.1 Presynaptic Impulse and Postsynaptic Response at a squid giant synapse. (A) Sketch of the stellate ganglion of the squid, illustrating the two large axons that form a chemical synapse. Both axons can be impaled with microelectrodes as shown. (B) Simultaneous recordings from the presynaptic axons (brown records) and postsynaptic axons (blue records) during the development of conduction block by TTX. As the amplitude of the presynaptic action potential decreases, so does the size of the postsynaptic potential. (Note that the two largest presynaptic action potentials evoke postsynaptic action potentials.) (C) The relation between the amplitude of the presynaptic action potential and the postsynaptic potential. Blue circles represent results in B; brown circles represent results obtained by applying depolarizing current pulses to the presynaptic terminals after complete TTX block. (A after Bullock and Hagiwara, 1957; B and C after Katz and Miledi, 1977c.)

Ricardo Miledi

(Figure 11.2).[7] The time is too long to be accounted for by diffusion of ACh across the synaptic cleft (a distance of 50 nm), which should take no longer than about 50 μs. When ACh is applied to the junction ionophoretically from a micropipette, delays of as little as 150 μs can be achieved. Furthermore, synaptic delay is much more sensitive to temperature than would be expected if it were due to diffusion. Cooling the frog nerve–muscle preparation to 2°C increases the delay to as long as 7 ms (Figure 11.2B), whereas the delay in the response to ionophoretically applied ACh is not perceptibly altered. Thus, the delay is largely in the transmitter release mechanism.[8]

Evidence That Calcium Is Required for Release

Calcium has long been known as an essential link in the process of synaptic transmission. When its concentration in the extracellular fluid is decreased, release of ACh at the neuromuscular junction is reduced and eventually abolished.[9,10] The importance of calcium for release has been established at synapses, irrespective of the nature of the transmitter. (One exception is the release of GABA from horizontal cells in the fish retina; see Chapter 4.[11]) The role of calcium has been generalized further to other secretory processes, such as liberation of hormones by cells of the pituitary gland, release of epinephrine from

[7] Katz, B., and Miledi, R. 1965. *J. Physiol.* 181: 656–670.

[8] Parnas, H., Segel, L., Dudel, J., and Parnas, I. 2000. *Trends Neurosci.* 23: 60–68.

[9] del Castillo, J., and Stark, L. 1952. *J. Physiol.* 116: 507–515.

[10] Dodge, F. A., Jr., and Rahamimoff, R. 1967. *J. Physiol.* 193: 419–432.

[11] Schwartz, E. A. 1987. *Science* 238: 350–355.

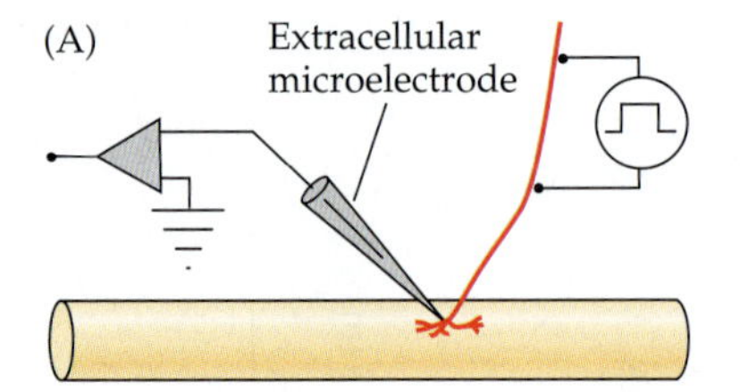

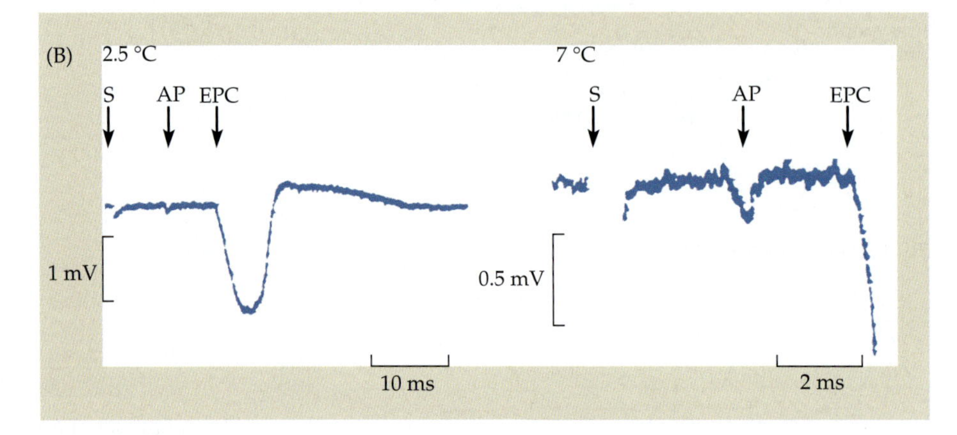

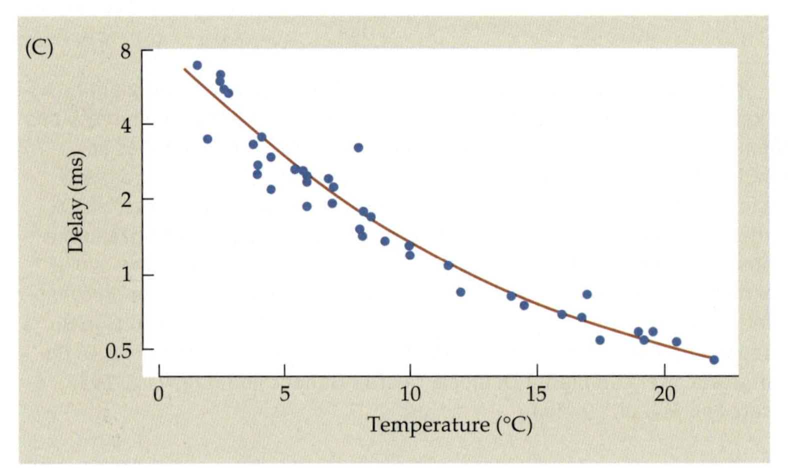

FIGURE 11.2 Synaptic Delay at a Chemical Synapse. (A) The motor nerve is stimulated while recording with an extracellular microelectrode at the frog neuromuscular junction. (B) Extracellular recordings of the stimulus artifact (S), the axon terminal action potential (AP), and the endplate current (EPC) at 2.5 and 7°C. The synaptic delay is the time between the action potential in the nerve terminal and the beginning of the endplate current. Note that the current flowing into the nerve terminal or the muscle fiber is recorded as a negative potential by an extracellular microelectrode. (C) Plot of synaptic delay as a function of temperature; the higher the temperature, the briefer the synaptic delay. (After Katz and Miledi, 1965.)

the adrenal medulla, and secretion by salivary glands.[12,13] As discussed in the next section, evoked transmitter release is preceded by calcium entry into the terminal and is antagonized by ions that block calcium entry, such as magnesium, cadmium, nickel, manganese, and cobalt. Transmitter release can be reduced, then, either by removing calcium from the bathing solution or by adding a blocking ion. For transmitter release to occur, calcium must be present in the bathing solution at the time of depolarization of the presynaptic terminal.[14]

Measurement of Calcium Entry into Presynaptic Nerve Terminals

Subsequent experiments have indicated that the calcium conductance of the membrane is increased by depolarization and that calcium enters with each action potential. Using voltage clamp techniques, Llinás and his colleagues measured the magnitude and time course of the calcium current produced by presynaptic depolarization at the squid giant synapse. An example is shown in Figure 11.3A. The sodium and potassium conductances associated with the action potential were blocked by TTX and TEA (tetraethylammonium) so that only the voltage-activated calcium channels remained.

[12]Penner, R., and Neher, E. 1988. *J. Exp. Biol.* 139: 329–345.

[13]Kasai, H. 1999. *Trends Neurosci.* 22: 88–93.

[14]Katz, B., and Miledi, R. 1967. *J. Physiol.* 189: 535–544.

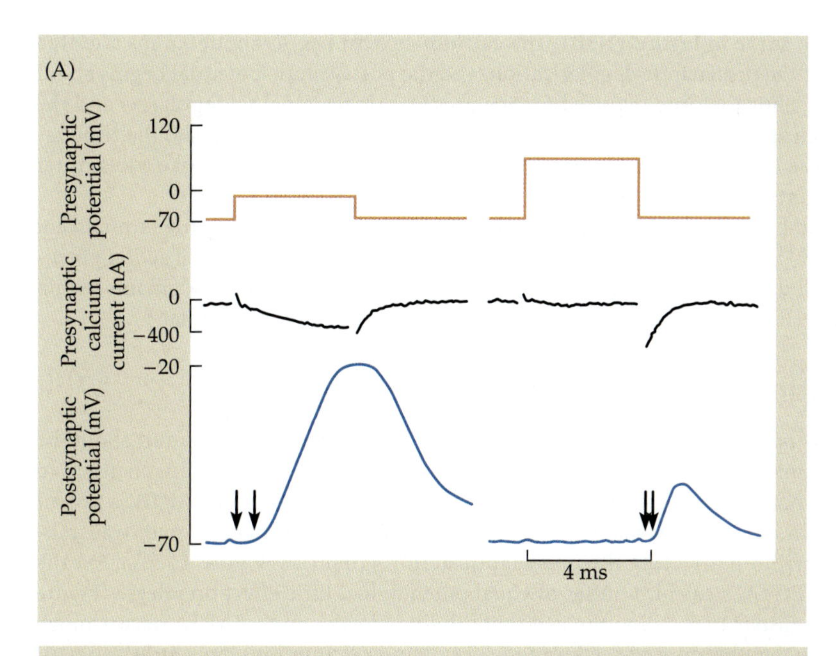

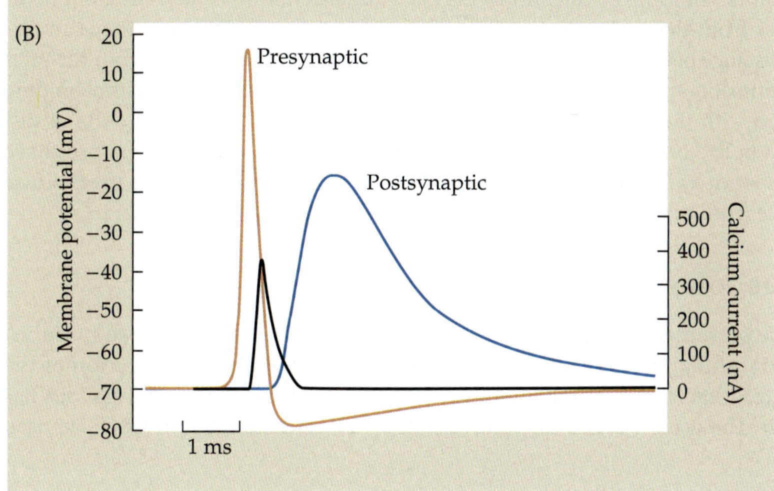

FIGURE 11.3 Presynaptic Calcium and Transmitter Release at the squid giant synapse. (A,B) The presynaptic terminal is voltage-clamped and treated with TTX and TEA to abolish voltage-activated sodium and potassium currents. (A) Records show potentials applied to the presynaptic fiber (upper trace), presynaptic calcium current (middle trace), and EPSP in the postsynaptic fiber (lower trace). A voltage pulse from –70 to –18 mV (left panel) results in a slow inward calcium current and, after a delay of about 1 ms (arrows), an EPSP. A larger depolarization, to +60 mV (right panel), suppresses calcium entry. At the end of the pulse, a surge of calcium current is followed within about 0.2 ms (arrows) by an EPSP. (B) If a voltage change identical in shape to a normal action potential is produced by the voltage clamp (labeled Presynaptic), then the EPSP is indistinguishable from that seen normally (labeled Postsynaptic). The black curve gives the magnitude and time course of the calcium current. The synaptic delay between the beginning of presynaptic depolarization and the beginning of postsynaptic response is due in part to the time required to open calcium channels and in part to the time for calcium entry to trigger transmitter release. (After Llinás, 1982.)

Depolarizing the presynaptic terminal to –18 mV (upper record in left panel) produced an inward calcium current in the terminal that increased slowly in magnitude to about 400 nA (middle record), and a large synaptic potential in the postsynaptic cell (lower record). When the terminal was depolarized to +60 mV, approximating the calcium equilibrium potential, the calcium current was suppressed during the pulse (see right panel) and no synaptic potential was seen. This demonstrates that depolarization of the terminal is not sufficient on its own to trigger release; calcium entry must also occur. On repolarization, there was a brief calcium current, as calcium flowed in through channels opened during the depolarization, accompanied by a small postsynaptic potential.

The effect of an artificial action potential is shown in Figure 11.3B. A presynaptic action potential, recorded before addition of TTX and TEA to the preparation, was “played back” through the voltage clamp circuit to produce exactly the same voltage change in the terminal. The postsynaptic potential is indistinguishable from that produced by a normal presynaptic action potential, confirming that the sodium and potassium currents that normally accompany the action potential are not necessary for transmitter release.

The voltage clamp technique also enabled Llinás and his colleagues to measure the magnitude and time course of the calcium current produced by the artificial action potential (black curve in Figure 11.3B). The calcium current begins about 0.5 ms after the beginning of the presynaptic depolarization, and the postsynaptic potential begins about 0.5 ms later. Thus the time required for the presynaptic terminal to depolarize and the calcium channels to open accounts for the first half of the synaptic delay; the time required for the calcium concentration to rise within the terminal and evoke transmitter release accounts for the remainder.

Llinás and his colleagues also visualized calcium entry directly, using the luminescent dye aequorin.[15,16] They showed that as a result of a brief train of presynaptic action potentials, intracellular calcium concentration reached 100 to 200 μ*M* in microdomains within the terminal (Figure 11.4), considered to correspond to active zones (Chapter 9).

Localization of Calcium Entry Sites

Experiments on the squid giant synapse have provided additional information about the role of calcium in release and, in particular, about the proximity of calcium channels to the sites of transmitter secretion.[17] In these experiments, injection of BAPTA, a potent calcium buffer, into the presynaptic terminal resulted in a severe attenuation of transmitter release, without affecting the presynaptic action potential (Figure 11.5A). On the other hand, EGTA, a calcium buffer of equal potency, had little effect on release (Figure 11.5C). This disparity is due to the fact that binding of calcium to BAPTA occurs much more rapidly than binding of calcium to EGTA. Thus, calcium ions have little opportunity to diffuse from their site of entry before being bound by BAPTA, but they can traverse some distance before being captured by EGTA (Figure 11.5B and D). From the rates of calcium diffusion and binding to EGTA, it can be calculated that the calcium-binding site associated with the release process must lie within 100 nm or less of the site of calcium entry. On the other hand, similar experiments at some neuronal synapses have shown an effect of EGTA on release, suggesting that in these cells calcium may diffuse some distance from calcium channels to sites that trigger or modulate release.[5]

Role of Depolarization in Release

The evidence presented so far indicates that transmitter release is triggered by an increase in intracellular calcium concentration brought about by depolarization of the presynaptic terminal and opening of voltage-activated calcium channels. This idea has been tested by the use of "caged calcium," a buffer that releases calcium when exposed

[15]Llinás, R. 1982. *Sci. Am.* 247(4): 56–65.

[16]Llinás, R., Sugimori, M., and Silver, R. B. 1992. *Science* 256: 677–679.

[17]Adler, E. M., et al. 1991. *J. Neurosci.* 11: 1496–1507.

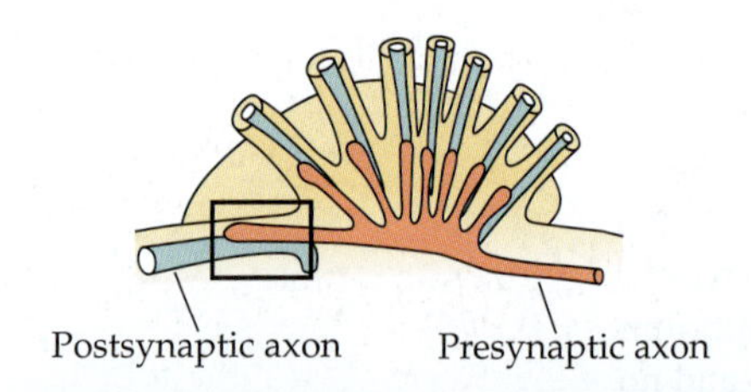

(A)

(B)

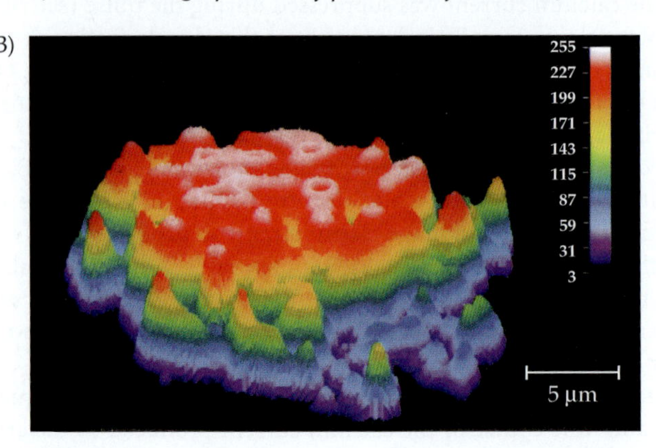

FIGURE 11.4 Microdomains of Calcium within the Presynaptic Terminal at the Squid Giant Synapse. (A) Distribution of calcium within the presynaptic axon terminal at rest, determined by intracellular injection of a calcium-sensitive dye (box in illustration at left shows the region imaged). (B) A brief train of presynaptic action potentials results in the appearance of microdomains of high calcium concentration within the axon terminal. (After Llinás, Sugimori, and Silver, 1992; micrographs kindly provided by R. Llinás.)

to intense ultraviolet light.[18–20] When caged calcium was injected into the presynaptic terminal of the squid giant synapse or the crayfish neuromuscular junction, illumination of the terminal evoked the release of transmitter (Figure 11.6). Transmitter release resembling that seen after a presynaptic action potential occurred under conditions in which the intracellular calcium concentration was increased to approximately 100 μM. Similar experiments on terminals of bipolar cells from the goldfish retina gave comparable results.[21]

At the same time, certain properties of the release process cannot be accounted for solely on the basis of calcium flux into the nerve terminal. These have been investigated at neuromuscular synapses of crayfish and frogs by Parnas and his colleagues, who have shown that when intracellular calcium is held at a constant, elevated level and further calcium entry is blocked, depolarization causes release.[8] The site at which depolarization acts is not yet known; one suggestion is that depolarization influences voltage-sensitive autoreceptors in the presynaptic nerve terminal membrane.[22]

[18]Hochner, B., Parnas, H., and Parnas, I. 1989. *Nature* 342: 433–435.

[19]Delany, K. R., and Zucker, R. S. 1990. *J. Physiol.* 426: 473–498.

[20]Zucker, R. S. 1993. *J. Physiol. (Paris)* 87: 25–36.

[21]Heidelberger, R., et al. 1994. *Nature* 371: 513–515.

[22]Linial, M., Ilouz, N., and Parnas, H. 1997. *J. Physiol.* 504: 251–258.

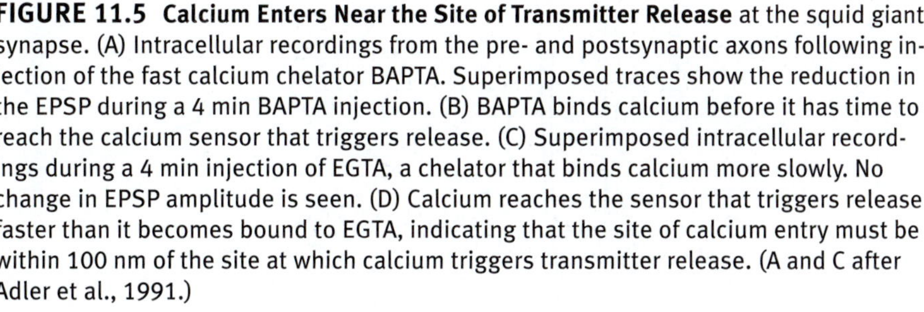

FIGURE 11.5 Calcium Enters Near the Site of Transmitter Release at the squid giant synapse. (A) Intracellular recordings from the pre- and postsynaptic axons following injection of the fast calcium chelator BAPTA. Superimposed traces show the reduction in the EPSP during a 4 min BAPTA injection. (B) BAPTA binds calcium before it has time to reach the calcium sensor that triggers release. (C) Superimposed intracellular recordings during a 4 min injection of EGTA, a chelator that binds calcium more slowly. No change in EPSP amplitude is seen. (D) Calcium reaches the sensor that triggers release faster than it becomes bound to EGTA, indicating that the site of calcium entry must be within 100 nm of the site at which calcium triggers transmitter release. (A and C after Adler et al., 1991.)

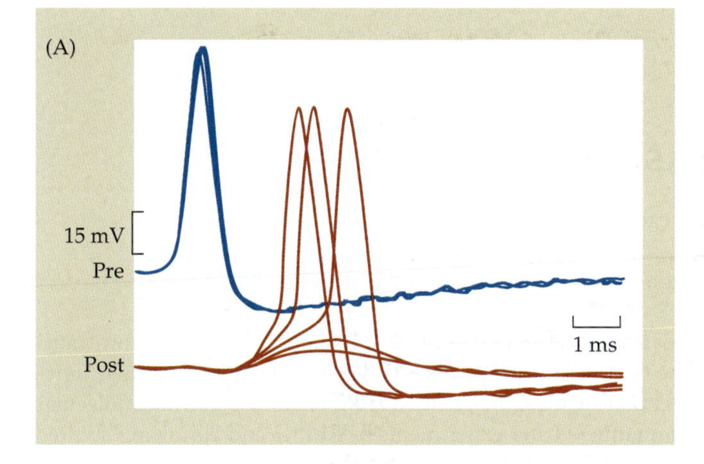

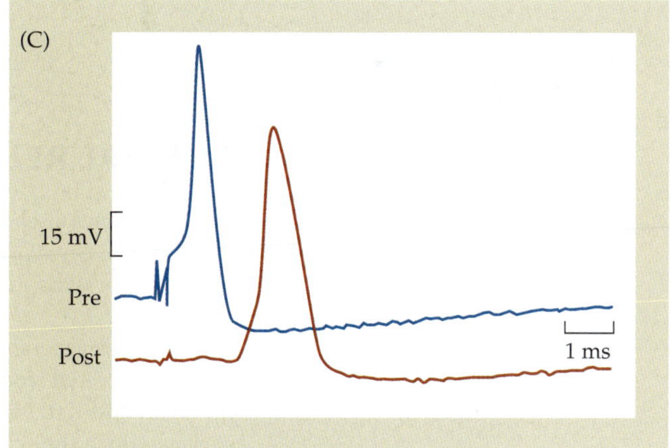

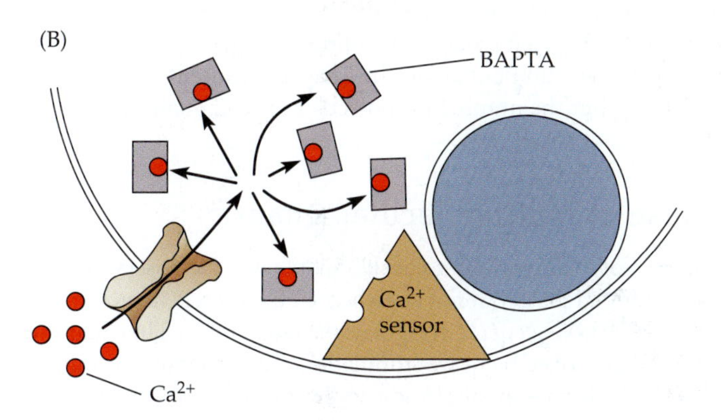

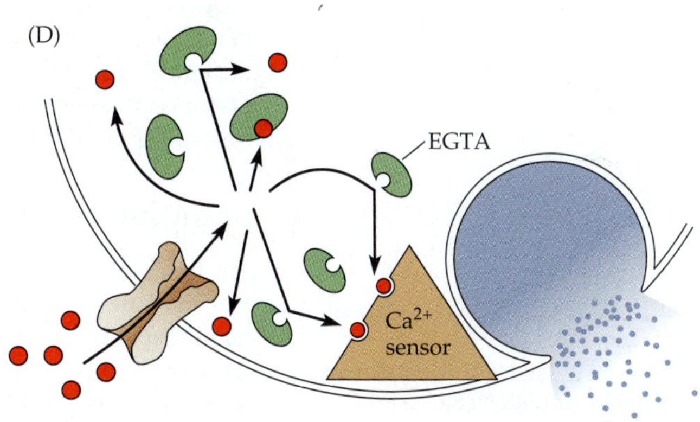

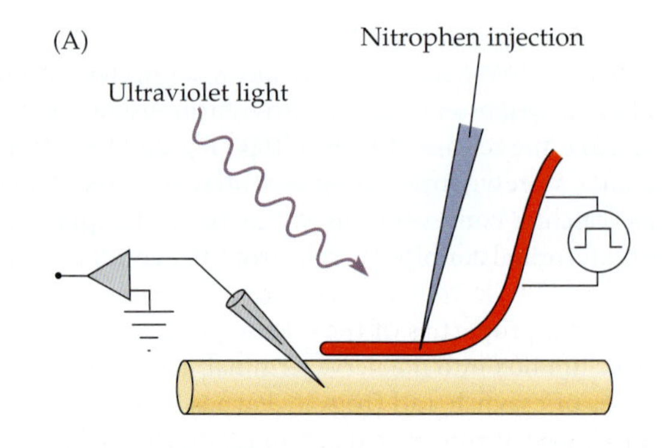

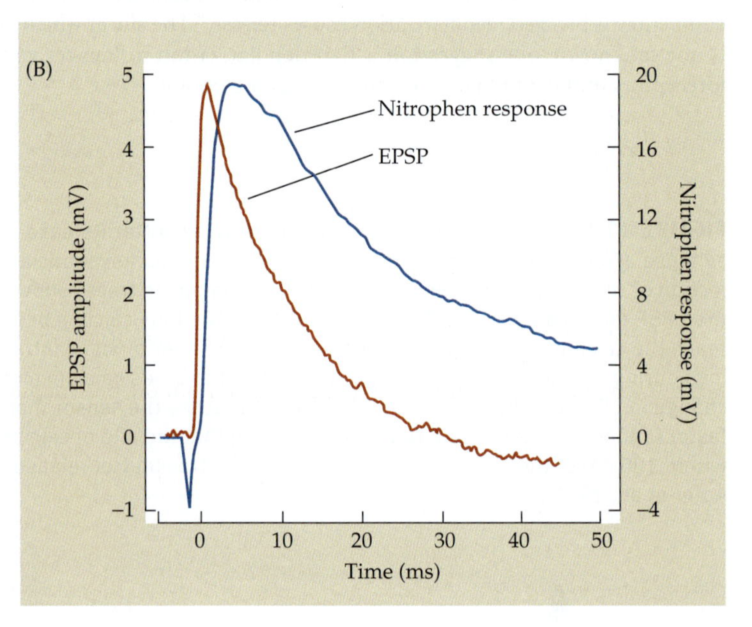

FIGURE 11.6 An Increase in Intracellular Calcium Is Sufficient to Trigger Rapid Transmitter Release at the squid giant synapse. (A) Nitrophen, a form of "caged calcium," is injected into the presynaptic terminal. Transmitter release is monitored by recording intracellularly from the postsynaptic axon. (B) Intracellular records show the postsynaptic response to nerve stimulation (EPSP) and to release of calcium from nitrophen by a flash of ultraviolet light (nitrophen response). An abrupt increase in intracellular calcium causes an increase in transmitter release that is nearly as rapid as that produced by a presynaptic action potential. The decay of the nitrophen response is slower and incomplete because the photolyzed nitrophen buffers calcium to a concentration higher than the normal level at rest. (B after Zucker, 1993.)

QUANTAL RELEASE

So far the general scheme can be summarized as follows:

presynaptic depolarization → calcium entry → transmitter release

Now that this general framework has been established, it remains to be shown how transmitter is secreted from the terminals. In experiments on the frog neuromuscular junction, Fatt and Katz showed that ACh can be released from terminals in multimolecular packets, which they called **quanta**.[23] Later experiments by Kuffler and Yoshikami showed that each quantum corresponds to approximately 7000 molecules of ACh.[24] Quantal release then means that only 0, 7000, 14,000, or so on molecules will be released at a time, not 4250 or 10,776. In general, at any given synapse the number of quanta released from the nerve terminal in response to an action potential (the **quantum content** of a synaptic response) may vary considerably, but the number of molecules in each quantum (**quantum size**) is fixed (with a variance of about 10%).

Spontaneous Release of Multimolecular Quanta

The first evidence for packaging of ACh in multimolecular quanta was the observation by Fatt and Katz[23] that at the motor end plate, but not elsewhere in the muscle fiber, spontaneous depolarizations of about 1 mV occurred irregularly (Figure 11.7). They had the same time course as the potentials evoked by nerve stimulation. The spontaneous miniature potentials (MEPPs) were decreased in amplitude and eventually abolished by in-

[23]Fatt, P., and Katz, B. 1952. *J. Physiol.* 117: 109–128.

[24]Kuffler, S. W., and Yoshikami, D. 1975. *J. Physiol.* 251: 465–482.

creasing concentrations of the ACh receptor antagonist curare, and they were increased in amplitude and time course by acetylcholinesterase inhibitors such as prostigmine (Figure 11.7C). These two pharmacological tests indicated that the potentials were produced by the spontaneous release of discrete amounts of ACh from the nerve terminal and ruled out the possibility that they might be due to single ACh molecules. Subsequently, patch electrode recordings demonstrated directly that the amount of current that flows through an individual ACh receptor will produce a potential change in the muscle fiber of approximately 1 μV (Chapter 2). Thus, a spontaneous miniature potential is produced by the opening of more than a thousand ACh receptors.

Additional evidence confirmed in a variety of different ways that the spontaneous miniature potentials are indeed due to multimolecular packets of ACh liberated by the nerve terminal. For example, depolarization of the nerve terminal by passing a steady current through it causes an increase in frequency of the spontaneous activity, whereas muscle depolarization has no effect on frequency.[25] Botulinum toxin, which blocks release of ACh in response to nerve stimuli, also abolishes the spontaneous activity.[26] Shortly after denervation of a muscle, as the motor nerve terminal degenerates, the miniature potentials disappear.[27] Surprisingly, after an interim period, spontaneous potentials reappear in denervated frog muscle; these arise because of ACh released from Schwann cells that have engulfed segments of the degenerating nerve terminals by phagocytosis.[28]

Nonquantal Release

In addition to being released by the motor nerve terminal in the form of individual quanta, ACh leaks continuously from the cytoplasm into the extracellular fluid. In other words, there is a steady nonquantal "ooze" of ACh from the presynaptic terminal.[29,30] Indeed, the amount of ACh that leaks from the nerve terminal in this way is about 100 times greater than that released in the form of spontaneous quanta. This can be determined by comparing the total amount of ACh released from a muscle, measured biochemically, to the amount released as quanta, calculated from MEPP frequency and the total number of end plates in the muscle.

[25]del Castillo, J., and Katz, B. 1954. *J. Physiol.* 124: 586–604.

[26]Brooks, V. B. 1956. *J. Physiol.* 134: 264–277.

[27]Birks, R., Katz, B., and Miledi, R. 1960. *J. Physiol.* 150: 145–168.

[28]Reiser, G., and Miledi, R. 1989. *Brain Res.* 479: 83–97.

[29]Katz, B., and Miledi, R. 1977. *Proc. R. Soc. Lond. B* 196: 59–72.

[30]Vyskocil, F., Nikolsky, E., and Edwards, C. 1983. *Neuroscience* 9: 429–435.

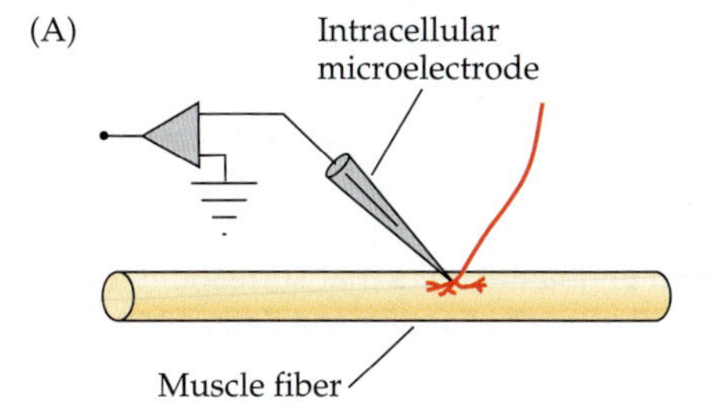

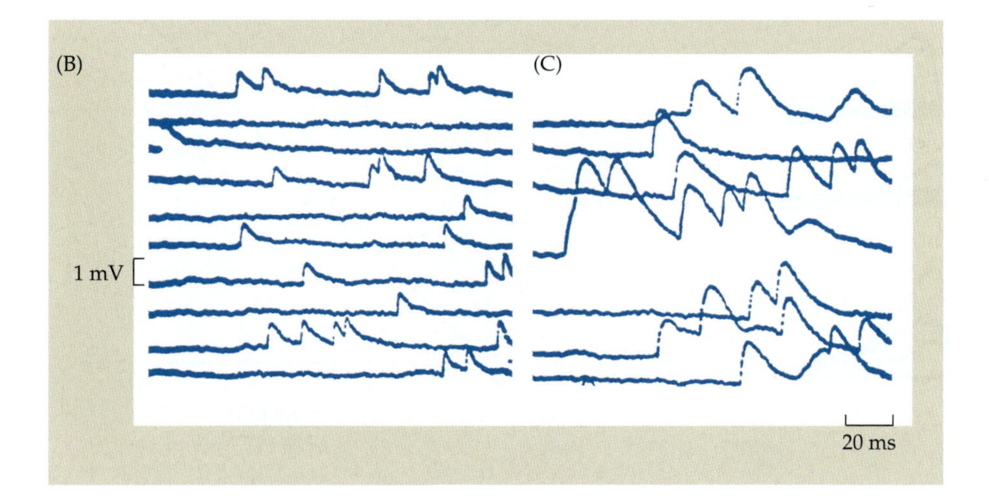

FIGURE 11.7 Miniature Synaptic Potentials occur spontaneously at the frog neuromuscular junction. (A) Intracellular recording from a muscle fiber in the region of the motor end plate. (B) Spontaneous miniature synaptic potentials are about 1 mV in amplitude and are confined to the end-plate region of the muscle fiber. (C) After addition of prostigmine, which prevents acetylcholinesterase from hydrolyzing ACh, miniature synaptic potentials are increased in amplitude and duration, but the frequency at which they occur is unchanged. This indicates that each miniature is due to a quantal packet of ACh, rather than to a single ACh molecule. (After Fatt and Katz, 1952.)

Under normal circumstances, the leak of ACh from the presynaptic terminal does not produce a postsynaptic response; the amount of cholinesterase in the synaptic cleft is sufficient to hydrolyze such a dribble of ACh before any receptors in the postsynaptic membrane are activated. Its postsynaptic effect can be detected only when cholinesterase is inhibited. In contrast, the simultaneous release of 7000 molecules of ACh in a quantum locally overwhelms the enzyme, allowing ACh to reach its postsynaptic receptors and cause a MEPP (Chapter 13).

Fluctuations in the End-Plate Potential

A typical synaptic potential at the skeletal neuromuscular junction depolarizes the postsynaptic membrane by 50 to 70 mV. Which mechanism of ACh release—ooze or quanta—underlies such a synaptic potential? Fatt and Katz observed that when synaptic transmission had been reduced by lowering the extracellular calcium concentration and adding extracellular magnesium, the responses to stimulation fluctuated in a stepwise manner, as shown in Figure 11.8A.[23] Some stimuli produced no response at all, a failure of transmission. Some stimuli produced a response of about 1 mV in amplitude, similar in size and shape to a spontaneous miniature potential; others evoked responses that appeared to be two, three, or four times the size of the spontaneous miniature potential.

This remarkable observation led Fatt and Katz to propose the **quantum hypothesis**: that the single quantal events observed to occur spontaneously also represented the building blocks for the synaptic potentials evoked by stimulation. Normally the end-plate potential is made up of about 200 quantal units, and variations in its size are not obvious. In low calcium concentrations, the quantal *size* remains the same, but the quantum *content* is small—perhaps 1, 2, or 3 quanta—and fluctuates randomly from trial to trial, resulting in stepwise fluctuations in the amplitude of the end-plate potential.

FIGURE 11.8 The End-Plate Potential Is Composed of Quantal Units That Correspond to Spontaneous Miniature Potentials. Presynaptic release of ACh at a frog neuromuscular junction was reduced by lowering the calcium concentration in the bathing solution. (A) Sets of intracellular records, each showing two to four superimposed responses to nerve stimulation. The amplitude of the end-plate potential (EPP) varies in a stepwise fashion; the smallest response corresponds in amplitude to a spontaneous miniature potential (MEPP). (B) Comparison of the mean quantal content (m) of the EPP determined in two ways: by applying the Poisson distribution, $m = \ln(N/n_0)$, and by dividing the mean EPP amplitude by the mean MEPP amplitude. Agreement of the two estimates supports the hypothesis that the EPP is composed of quantal units that correspond to spontaneous MEPPs. (A after Fatt and Katz, 1952; B after del Castillo and Katz, 1954.)

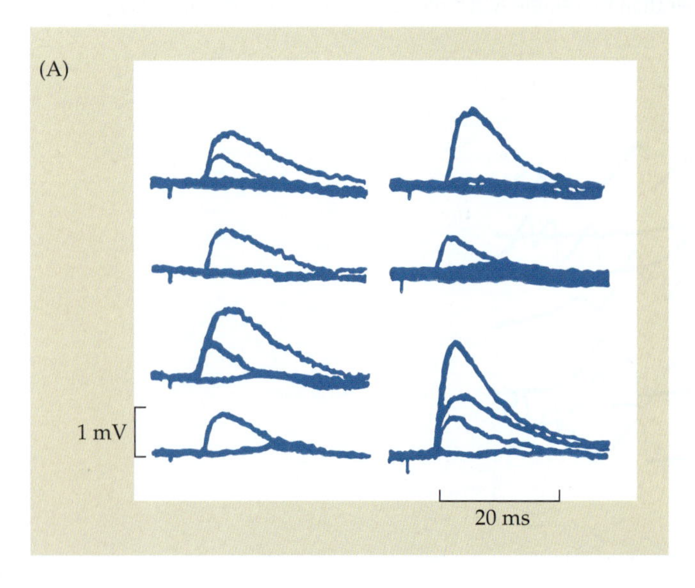

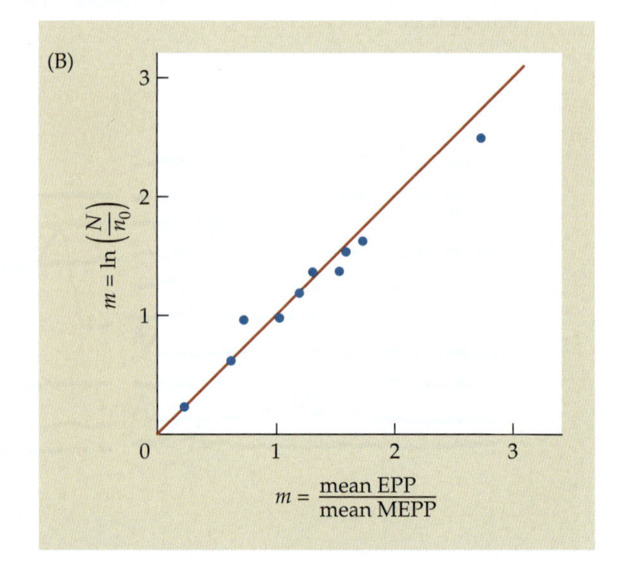

Statistical Analysis of the End-Plate Potential

Bernard Katz, 1950

Del Castillo and Katz realized that to test the quantum hypothesis adequately required a statistical analysis.[31] Accordingly, they proposed that the motor nerve terminal contains thousands of quantal packets of acetylcholine (n), each of which has a probability (p) of being released in response to a nerve impulse, and that quanta are released independently, that is, the release of one has no influence on the probability of release of the next. Then, in a large number of trials the mean number of quanta released per trial, m, would be given by np, and the number of times the response consisted of 0, 1, 2, 3, 4, . . . , or x quanta would be given by the **binomial distribution**. However, del Castillo and Katz had no way of measuring n or p experimentally, so they could not use the binomial distribution to test the hypothesis that the end-plate potential is made up of units of the same size as the spontaneous miniature potentials. In order to deal with this difficulty, they reasoned as follows:

> Under normal conditions, p may be assumed to be relatively large, that is a fairly large part of the synaptic population responds to an impulse. However, as we reduce the Ca and increase the Mg concentration, the chances of responding are diminished and we observe mostly complete failures with an occasional response of one or two units. Under these conditions, when p is very small, the number of units x which make up the e.p.p. in a large series of observations should be distributed in the characteristic manner described by Poisson's law.

The **Poisson distribution** is an approximation to the binomial distribution when p is very small. The crucial difference is that to predict a Poisson distribution it is not necessary to know either n or p. The experimenter needs to measure only their product, m, the mean number of quanta released per trial. Thus, for a Poisson distribution the expected number of responses containing x quanta is given by

$$n_x = N\left(\frac{m^x}{x!}\right)e^{-m}$$

One of the best-known applications of the Poisson distribution was an analysis of the number of Prussian cavalry officers killed each year by a horse kick. There were a large number of such officers (n), and the probability of any one of them being killed (p) was very low. In some years there were "failures"—no one was killed; in other years, one or perhaps two were killed. Over a long period, the number of years in which zero, one, two, or three officers were killed was described closely by the Poisson equation, using only the mean number of "successful" kicks per year (m) to determine the theoretically expected distribution.

Another convenient example for which the Poisson equation would predict the distribution of events is a nickel slot machine used for gambling. The unit size is fixed at 5 cents, the machine contains a large number of nickels, and the probability of any one nickel being released is very low and independent of other nickels. If the mean number of nickels paid out per play is known, then during a long period of play the Poisson equation will accurately predict the number of times there is no payoff, the number of times the player receives one nickel, or two, and so on. Again, the important feature of the Poisson equation is that the characteristics of the distribution depend only on m.

To test if the fluctuations in the end-plate potential at low calcium concentration are adequately described by the Poisson distribution, then, it is necessary only to have a measure of m, the mean number of units released per trial. This is obtained by dividing the mean amplitude of the evoked potentials by the unit size, the mean amplitude of the spontaneous miniature potential:

$$m = \frac{\text{mean amplitude of evoked potentials}}{\text{mean amplitude of miniature potential}}$$

For the slot machine the corresponding calculation for determining m would be the average amount of money paid out per trial (likely not to be very much, say 1.5 cents/trial) divided by the size of the unit (5 cents/unit), giving $m = 0.3$ unit/trial. If the end-plate potential amplitudes are distributed according to the Poisson equation, then m can also be determined from the number of failures, n_0. When $x = 0$ in the Poisson equation, $n_0 = Ne^{-m}$ (since both m^0 and $0! = 1$). Rearranging this result gives

[31] del Castillo, J., and Katz, B. 1954. *J. Physiol.* 124: 560–573.

$$m = \ln\left(\frac{N}{n_0}\right)$$

Del Castillo and Katz bathed a neuromuscular junction in a solution containing low calcium and high magnesium concentrations and recorded a large number of end-plate potentials evoked by nerve stimulation, as well as a large number of spontaneous miniature potentials. When they calculated m in these two entirely different ways, they found the estimates in excellent agreement, providing strong support for the quantum hypothesis (Figure 11.8B).

A more stringent test of the quantum hypothesis is to predict the entire distribution of response amplitudes, using only m and the mean amplitude of the unit potential (Figure 11.9). To do this, m is calculated from the ratio of the mean evoked potential to the mean spontaneous miniature potential, as before. Then the number of expected responses containing 0, 1, 2, 3, . . . units is calculated. To account for the slight variation in size of the unit, the expected number of responses containing one unit is distributed about the mean unit size with the same variance as the spontaneous events (Figure 11.9, inset). Similarly, the predicted number of responses containing 2, 3, or more units are distributed about their means with proportionately increasing variances. The individual distributions are then summed to give the theoretical distribution shown by the continuous curve. The agreement with the experimentally observed distribution (bars) provides additional support for the quantum hypothesis.

At many synapses the probability of release is sufficiently high that the Poisson distribution is not applicable. Under such conditions the binomial distribution itself must be used. For the binomial distribution the number of units capable of responding can be high or low, as can the probability of release. All that is required is that quanta be released independently. As before, if we take the mean quantal release (m) to be the product of the number of units capable of responding (n) and the average probability of release (p), then the relative occurrence of multiple events predicted by the binomial distribution is

$$n_x = N[n!/(n-x)!x!]\,p^x q^{n-x}$$

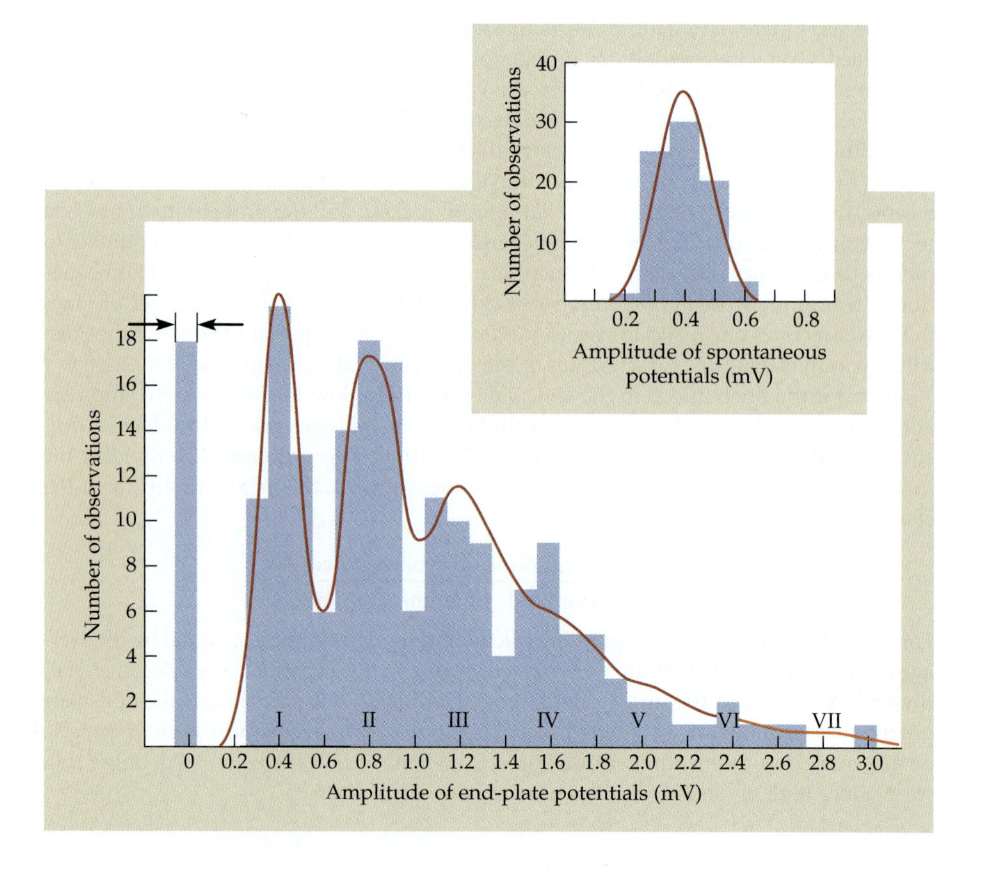

FIGURE 11.9 Amplitude Distribution of end-plate potentials at a mammalian neuromuscular junction in high (12.5 m*M*) magnesium solution. The histogram shows the number of end-plate potentials observed at each amplitude. The peaks of the histogram occur at 0 mV (failures) and at one, two, three, and four times the mean amplitude of the spontaneous miniature end-plate potentials (inset), indicating responses comprising 1, 2, 3, and 4 quanta. The solid line represents the theoretical distribution of end-plate potential amplitudes calculated according to the Poisson equation and allowing for the spread in amplitude of the quantal size. The arrows indicate the predicted number of failures.(From Boyd and Martin, 1956.)

where n_x is the number of responses containing x quanta, N is the number of trials, and $q = 1 - p$. Adherence of the release process to binomial statistics was first demonstrated at the crayfish neuromuscular junction.[32]

In summary, there is now ample evidence that transmitter is released in packets, or quanta.[33,34] Accurate determination of quantum size and quantum content are important in establishing the site of action of treatments that modulate synaptic transmission (Chapters 10, 12, and 16). In general, presynaptic modulatory effects change the amount of transmitter released by changing quantum content, not quantum size. On the other hand postsynaptic modulatory influences change the sensitivity of the postsynaptic cell to transmitter and alter quantum size, not the number of quanta released. When the release probability (p) is very low, as in a low-calcium medium, the Poisson distribution provides a good means of analyzing fluctuations. When the probability of release is high, binomial statistics are required to describe the distribution of responses. In addition, binomial statistics can provide information as to whether changes in the amount of transmitter released arise from changes in the number of available quanta or in the probability of their release.

Quantum Content at Neuronal Synapses

One striking feature of the vertebrate nervous system is the reduction in mean quantum content as one moves from the neuromuscular junction, where there is little integration (m = 200–300), to autonomic ganglia (m = 2–20),[35,36] to synapses in the central nervous system (at which m can be as low as 1),[37,38] where postsynaptic cells are concerned with integrating a myriad of incoming signals. At the synapse between a primary afferent fiber from a muscle spindle and a spinal motoneuron, for example, the mean quantum content is about 1.[39] This does not mean, however, that transmission fails most of the time, as would be expected for a Poisson distribution. Rather, release conforms to binomial statistics, with a high probability ($p \sim 0.9$) and low number of available quanta ($n \sim 1$). The evidence at most CNS synapses favors quantal release as the mechanism of synaptic transmission. However, it is often difficult to apply a simple Poisson or binomial statistical analysis. The complexity and diversity of synaptic connections in the CNS has led to the introduction of more sophisticated statistical treatments.[38,40,41]

Number of Molecules in a Quantum

Although it was clear from the experiments of Katz, Fatt, and del Castillo that at the neuromuscular junction one quantum contained more than one acetylcholine molecule, the question of how many molecules were in a quantum remained. The first accurate determination was made by Kuffler and Yoshikami, who used very fine pipettes for ionophoresis of ACh onto the postsynaptic membrane of snake muscle.[24] By careful placement of the pipette, they were able to produce a response to a brief pulse of ACh that mimicked almost exactly the spontaneous miniature potential (Figure 11.10). To measure the number of molecules released by the pipette, ACh was released by repetitive pulses into a small (about 0.5 μl) droplet of saline under oil (Figure 11.11). The droplet was then applied to the end plate of a snake muscle fiber and the resulting depolarization measured. The response was compared with responses to droplets of exactly the same size containing known concentrations of ACh. In this way the concentration of ACh in the test droplet was determined and the number of ACh molecules released per pulse was calculated. The pulse of ACh required to mimic a spontaneous miniature potential contained approximately 7000 molecules.

Number of Channels Activated by a Quantum

Given that a quantum of ACh consists of about 7000 molecules, one might expect that only a few thousand of these would actually combine with postsynaptic receptors at the neuromuscular junction, the remainder being lost to diffusion out of the cleft or hydrolysis by cholinesterases. This expectation is correct. The number of receptors activated by a quantum can be determined by comparing the conductance change that occurs during a miniature potential with that produced by a single ACh-activated channel.[42] Measurements of miniature end-plate currents by voltage clamp in frog muscle indicate a peak

[32]Johnson, E. W., and Wernig, A. 1971. *J. Physiol.* 218: 757–767.

[33]Rahamimoff, R., and Fernandez, J. M. 1997. Pre- and postfusion regulation of transmitter release. *Neuron* 18: 17–27.

[34]Stjärne, L., et al. 1994. *Molecular and Cellular Mechanisms of Neurotransmitter Release*. Raven, New York.

[35]Blackman, J. G., and Purves, R. D. 1969. *J. Physiol.* 203: 173–198.

[36]Martin, A. R., and Pilar, G. 1964. *J. Physiol.* 175: 1–16.

[37]Redman, S. 1990. *Physiol. Rev.* 70: 165–198.

[38]Edwards, F. A., Konnerth, A., and Sakmann, B. 1990. *J. Physiol.* 430: 213–249.

[39]Kuno, M. 1964. *J. Physiol.* 175: 81–99.

[40]Walmsley, B., Alvarez, F. J., and Fyffe, R. E. W. 1998. *Trends Neurosci.* 21: 81–88.

[41]Bekkers, J. M. 1994. *Curr. Opin. Neurobiol.* 4: 360–365.

[42]Magleby, K. L., and Weinstock, M. M. 1980. *J. Physiol.* 299: 203–218.

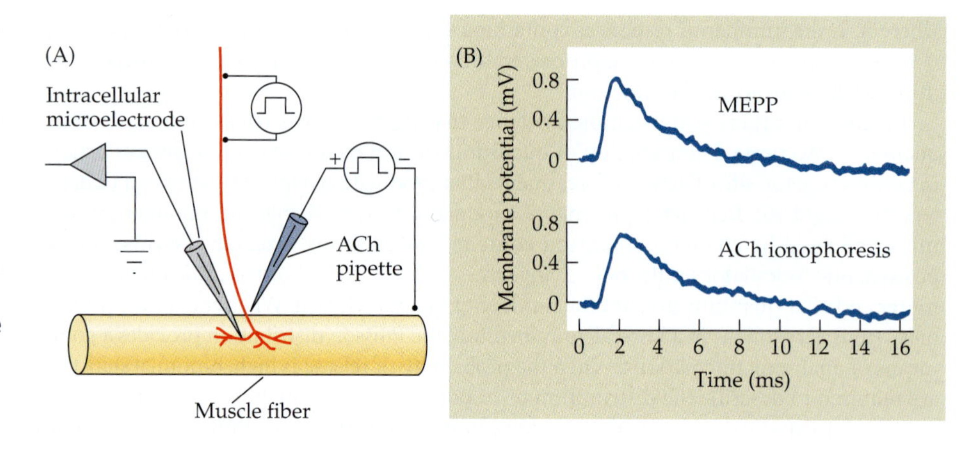

FIGURE 11.10 The Number of ACh Molecules in a Quantum is determined by mimicking a spontaneous miniature end-plate potential with an ionophoretic pulse of ACh. (A) An intracellular microelectrode records spontaneous miniature end-plate potentials (MEPPs) and the response to ionophoretic application of ACh. (B) A MEPP is mimicked almost exactly by an ionophoretic pulse of ACh. The rate of rise of the ionophoretic ACh pulse is slightly slower because the ACh pipette is further from the postsynaptic membrane than is the nerve terminal. (B after Kuffler and Yoshikami, 1975a.)

conductance change on the order of 40 nS. A single frog ACh receptor has a conductance of about 30 pS. Thus, a miniature end-plate potential is produced by about 1300 open channels. (This corresponds to 2600 molecules of ACh, since it takes 2 molecules of ACh to open a channel; see Chapters 9 and 13.) This is similar to the number calculated by Katz and Miledi, who estimated the contribution of a single channel to the end-plate potential from noise measurements.[43] A similar value for the number of channels opened by a quantum of transmitter was obtained at glycine-mediated inhibitory synapses in lamprey brainstem cells.[44] Lower values are observed at other synapses. For example, at synapses on hippocampal cells a quantal response corresponds to activation of 15 to 65 channels.[38,45]

Why are there such differences among synapses? A little thought leads to the conclusion that the number of postsynaptic receptors activated by a quantum of transmitter released from a single presynaptic bouton must be tailored to the size of the cell. In large cells with low input resistances, such as skeletal muscle fibers or lamprey Müller cells, a large number of receptors must be activated for the effect of a quantum to be significant. Activation of the same number of receptors on a very small cell, on the other hand, would overwhelm all other conductances, depolarizing the cell to a potential near zero if the synapse were excitatory, or locking its membrane potential firmly at the chloride equilibrium potential if the effect were inhibitory.

How is the match between cell size and the number of receptors activated by a quantum achieved? Is the number of molecules in a quantum reduced, or is the number of available postsynaptic receptors lower? Precise values for the number of molecules of

[43]Katz, B., and Miledi, R. 1972. *J. Physiol.* 244: 665–699.

[44]Gold, M. R., and Martin, A. R. 1983. *J. Physiol.* 342: 85–98.

[45]Jonas, P., Major, G., and Sakmann, B. 1993. *J. Physiol.* 472: 615–663.

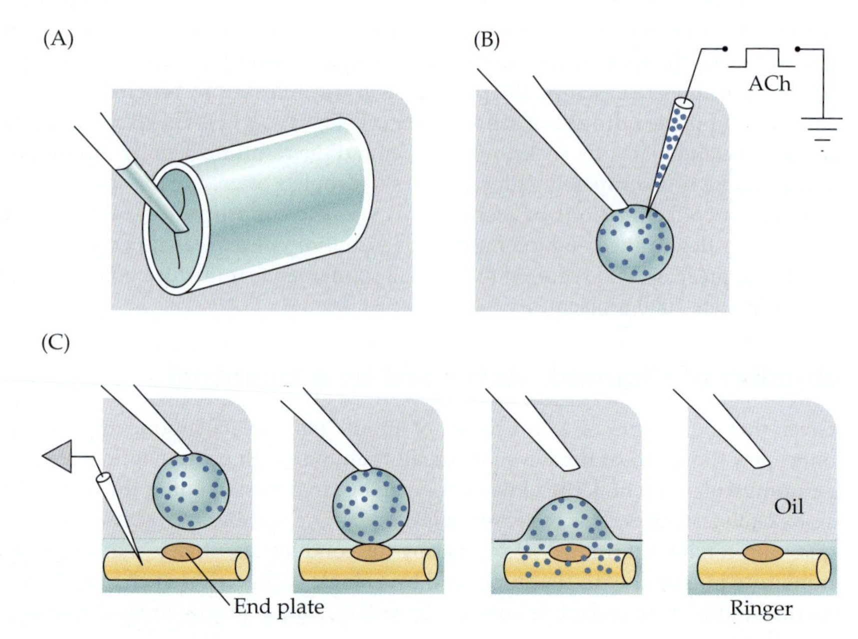

FIGURE 11.11 Assay of ACh Ejected from a Micropipette by ionophoresis. (A) A droplet of fluid is removed from the dispensing capillary under oil. (B) ACh is injected into the droplet by a series of ionophoretic pulses, each identical to that used to mimic a spontaneous miniature potential (see Figure 11.10B). (C) After its volume is measured, the ACh-loaded droplet is touched against the oil–Ringer interface at the end plate of a snake muscle, discharging its contents into the aqueous phase. The depolarization of the end plate is measured (not shown) and compared with that produced by droplets with known ACh concentration. Once the concentration in the test droplet is determined, the amount of ACh released per pulse from the electrode can be calculated. (After Kuffler and Yoshikami, 1975b.)

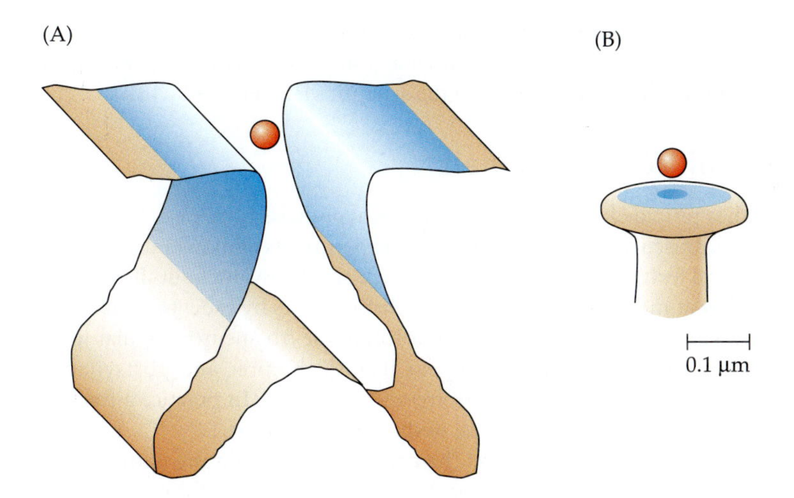

FIGURE 11.12 The Area of Postsynaptic Membrane Relative to the Size of a Synaptic Vesicle. (A) At the frog neuromuscular junction ACh receptors are packed at high density (~10,000/μm²) over a large postsynaptic area (shaded blue). Accordingly, receptors outnumber ACh molecules, and the size of the quantal event varies with the variation in the number of molecules per quantum. (B) At a typical hippocampal synapse, postsynaptic receptors are packed less densely (~2800/μm²) over a very small area (0.04 μm²). As a result, the number of transmitter molecules in a quantum is sufficient to saturate the available receptors, and quantal events show very little fluctuation in amplitude.

transmitter in a CNS synaptic vesicle are not available. However, the reported estimate for glutamate-containing vesicles is 4000,[46] the same order of magnitude as the number of ACh molecules in vesicles at the neuromuscular junction. On the other hand, analysis of quantal fluctuations at excitatory and inhibitory synapses on hippocampal cells suggests that the number of available postsynaptic receptors is much lower than at the neuromuscular junction.[38,45] Quantal events at hippocampal synapses activate only about 15 to 65 channels. The amplitude of these quantal events shows remarkably little variance, suggesting that the number of molecules released in a single quantum is always more than sufficient to activate all the available receptors. Conversely, at the neuromuscular junction an increase in the number of transmitter molecules in a quantum will result in a larger quantal event.[47,48] The difference in available postsynaptic receptors deduced from quantal fluctuations is consistent with the difference in synaptic morphology (Figure 11.12): At the neuromuscular junction, receptors are packed at high density (~10,000/μm²) throughout a large expanse of postsynaptic membrane, providing an essentially limitless sea of receptors for each quantum of transmitter (Chapter 13). At a typical hippocampal synapse the estimated postsynaptic receptor density is lower (~2800/μm²),[49] and the area occupied by postsynaptic membrane is very small (0.04 μm²);[50] thus, fewer than 100 postsynaptic receptors may be present.

Changes in Mean Quantal Size at the Neuromuscular Junction

Although the size of spontaneous miniature potentials at any particular synapse tends to remain constant, exceptions occur under certain circumstances. For example, during development and regeneration of motor nerve terminals, the amplitudes of spontaneous miniature potentials, rather than being distributed normally, are skewed into the baseline noise; that is, there are large numbers of very small spontaneous potentials.[51,52] Conversely, spontaneous synaptic potentials larger than the usual miniature potentials are seen occasionally.[53] In some instances these appear to be due to the spontaneous release of two or more quanta simultaneously; in others, their size shows no clear relation to normal quantal amplitude. Finally, in some myoneural diseases that afflict humans, such as myasthenia gravis, spontaneous miniature and evoked synaptic potentials are reduced in amplitude owing to a reduction in the number of receptors in the postsynaptic membrane.[54]

VESICLE HYPOTHESIS OF TRANSMITTER RELEASE

Shortly after del Castillo, Fatt, and Katz demonstrated by electrophysiological methods that transmitter release was quantal, the first electron micrographs of the neuromuscular junction revealed that axon terminals contain many small membrane-bound synaptic vesicles (Figure 11.13A).[55,56] Thus, the **vesicle hypothesis** of transmitter release was sug-

[46]Villanueva, S., Fiedler, J., and Orrego, F. 1990. *Neuroscience* 37: 23–30.

[47]Hartzell, H. C., Kuffler, S. W., and Yoshikami, D. 1975. *J. Physiol.* 251: 427–463.

[48]Salpeter, M. M. 1987. In *The Vertebrate Neuromuscular Junction*. Alan R. Liss, New York, pp. 1–54.

[49]Harris, K. M., and Landis, D. M. M. 1986. *Neuroscience* 19: 857–872.

[50]Schikorski, T., and Stevens, C. F. 1997. *J. Neurosci.* 17: 5858–5867.

[51]Denis, M. J., and Miledi, R. 1974. *J. Physiol.* 239: 571–594.

[52]Erxleben, C. and Kriebel, M. E. 1988. *J. Physiol.* 400: 659–676.

[53]Vautrin, J., and Kriebel, M. E. 1991. *Neuroscience* 41: 71–88.

[54]Drachman, D. B. 1994. *New England J. Med.* 330: 1797–1810.

gested: that a quantum of transmitter corresponds to the contents of one vesicle and that release occurs by a process of **exocytosis**, in which a vesicle fuses with the presynaptic plasma membrane and releases its contents into the synaptic cleft (Figure 11.13B).

Ultrastructure of Nerve Terminals

Ultrastructural studies provided support for the vesicle hypothesis of release. Many were first made at the neuromuscular junction. Subsequent experiments demonstrated that the principal morphological features of chemical synapses are similar throughout the nervous system, suggesting that at most chemical synapses, release occurs by exocytosis of transmitter-containing vesicles. A schematic view of a portion of the frog neuromuscular junction is shown in Figure 11.14, as it might appear if both the pre- and postsynaptic membranes were split open by the technique of freeze-fracturing. (In practice a fracture would occur in one membrane or the other, not both at the same time.)

The upper portion of Figure 11.14 shows the presynaptic membrane with vesicles lined up on the cytoplasmic side. Some are represented in the process of exocytosis. The exposed surface of the cytoplasmic half of the presynaptic membrane shows intramembranous particles protruding from the fracture face; matching pits are seen on the fracture face of the outer portion of the presynaptic membrane. Similar particles and pits occur on the fracture faces of the postsynaptic membrane. In addition, in the presynaptic membrane, sites of exocytosis are visualized as large indentations in the cytoplasmic portion of the membrane and fractured vesicle "stalks" in the outer portion.

A conventional transmission electron micrograph of a horizontal section through an active zone is shown in Figure 11.15A. It is equivalent to looking down onto an active zone from the nerve terminal cytoplasm. An orderly row of synaptic vesicles is lined up along either side of the band of dense material. Figure 11.15B shows a corresponding image of the fracture face of the cytoplasmic portion of the presynaptic membrane. This is equivalent to viewing the same region from the synaptic cleft. Rows of particles, each about 10 nm in diameter, flank the active zone on each side. Severed stalks, believed to indicate exocytic openings, appear more laterally. As described earlier, electrophysiological experiments in which calcium buffers were injected into presynaptic terminals indicated a close association be-

[55]Reger, J. F. 1958. *Anat. Rec.* 130: 7–23.
[56]Birks, R., Huxley, H. E., and Katz, B. 1960. *J. Physiol.* 150: 134–144.

FIGURE 11.13 Release of Neurotransmitter by Synaptic Vesicle Exocytosis. High-power electron micrographs of frog neuromuscular junctions. (A) A cluster of synaptic vesicles within the presynaptic terminal contacts an electron-dense region of the presynaptic membrane, forming an active zone. (B) A single stimulus was applied to the motor nerve in the presence of 4-aminopyridine, a drug that greatly increases transmitter release by prolonging the action potential, and the tissue was frozen within milliseconds. Vesicles docked at the active zone have fused with the presynaptic membrane and released their contents into the synaptic cleft by exocytosis. (A, micrograph kindly provided by U. J. McMahan; B from Heuser, 1977).

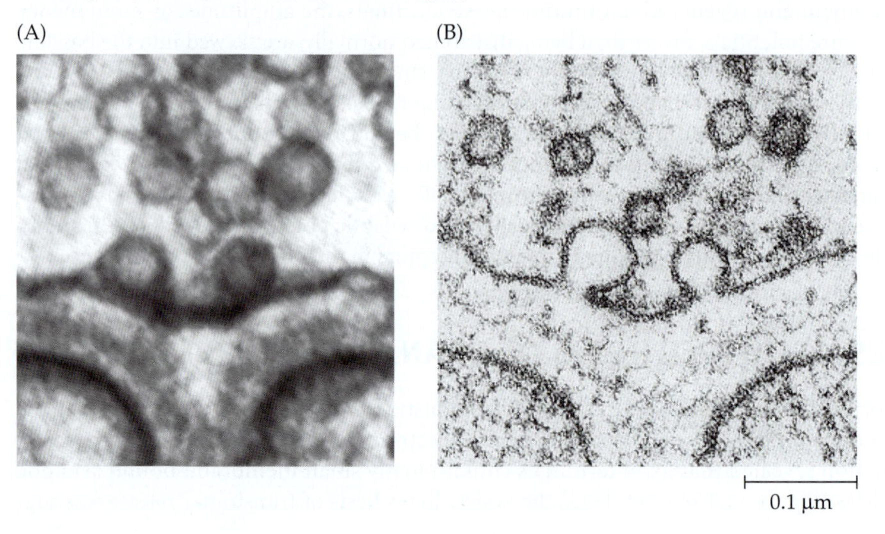

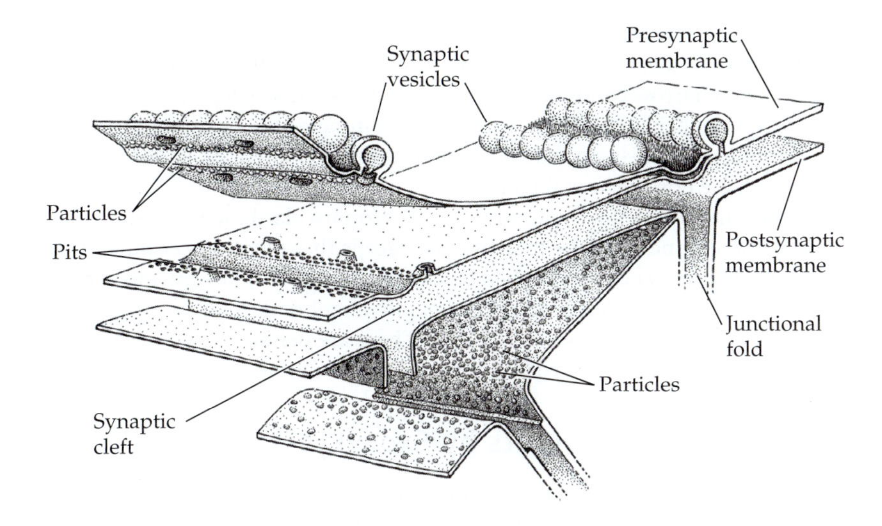

FIGURE 11.14 Synaptic Membrane Structure at the frog neuromuscular junction. Three-dimensional view of presynaptic and postsynaptic membranes, with each membrane split along its intramembranous plane as might occur in freeze-fracture. The cytoplasmic half of the presynaptic membrane at the active zone shows on its fracture face protruding particles whose counterparts are seen as pits on the fracture face of the outer membrane leaflet. Vesicles fusing with the presynaptic membrane give rise to pores and protrusions on the two fracture faces. The fractured postsynaptic membrane in the region of the folds shows a high concentration of particles on the fracture face of the cytoplasmic leaflet; these are ACh receptors. (Kindly provided by U. J. McMahan.)

tween calcium channels and release sites. Thus, at least some of the pits seen in Figure 11.15B might correspond to the voltage-activated calcium channels that trigger exocytosis.

Results obtained with toxin-binding studies at the neuromuscular junction of the frog and the mouse are consistent with this idea.[57–59] ω-Conotoxin, which blocks neuromus-

[57]Robitaille, R., Adler, E. M., and Charlton, M. P. 1990. *Neuron* 5: 773–779.

[58]Cohen, M. W., Jones, O. T., and Angelides, K. J. 1991. *J. Neurosci.* 11: 1032–1039.

[59]Sugiura, Y., et al. 1995. *J. Neurocytol.* 24: 15–27.

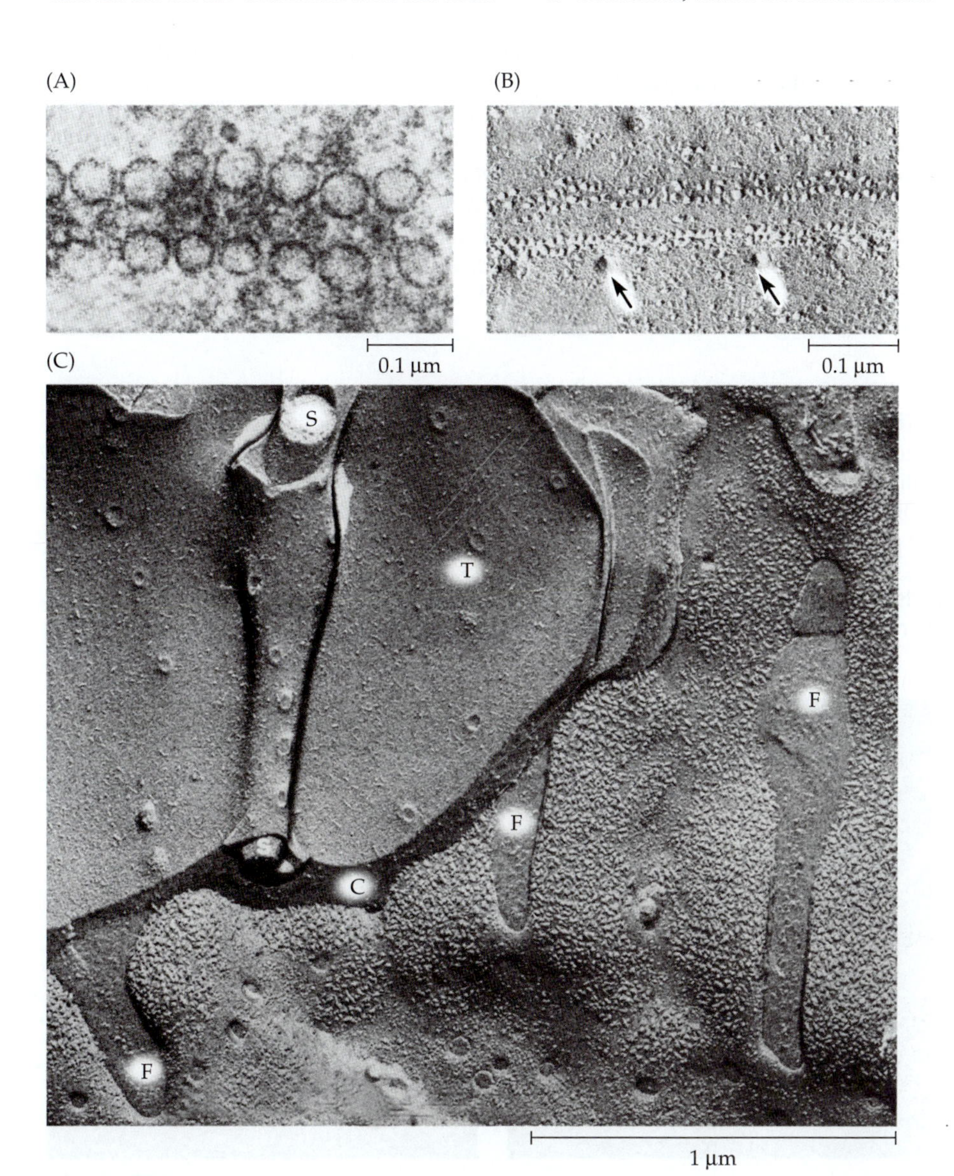

FIGURE 11.15 Structure of the Frog Neuromuscular Junction. (A) Transmission electron micrograph of a section through the nerve terminal parallel to an active zone, showing two lines of vesicles. (B) Fracture face of the cytoplasmic half of the presynaptic membrane in an active zone. The active zone region is delineated by particles about 10 nm in diameter and flanked by pores (arrows) caused by fusion of synaptic vesicles with the membrane. (C) Low-power view of fractured synaptic region. The fracture passes first through the presynaptic terminal membrane (T), showing the fracture face of the outer leaflet, then crosses the synaptic cleft (C) to enter the postsynaptic membrane. On the fracture face between the folds (F) one sees on the cytoplasmic leaflet aggregates of ACh receptors. A Schwann cell process (S) passes between the nerve terminal and the muscle. (A from Couteaux and Pecot-Déchavassine, 1970; B and C from Heuser, Reese, and Landis, 1974.)

cular transmission irreversibly by binding to presynaptic calcium channels,[60] was coupled to a fluorescent molecule. Upon microscopic examination, the fluorescence was found to be concentrated in narrow bands at 1 μm intervals, the same spacing as that of the active zones in the terminal (Figure 11.16). Combined staining of the postjunctional ACh receptors with fluorescent α-bungarotoxin showed that the presynaptic bands were in spatial register with the postjunctional folds.

A low-power freeze-fracture image of a frog neuromuscular junction is shown in Figure 11.15C. At the upper left, the first fracture face is that of the outer portion of the presynaptic membrane. The fracture then breaks across the synaptic cleft and exposes the face of the cytoplasmic portion of the postsynaptic membrane. Clusters of particles are seen along the sides of the postsynaptic folds. These are believed to correspond to the ACh receptors that are concentrated in this region of the end plate (Chapter 9).[61–63]

Release of Vesicle Contents by Exocytosis

A prediction of the hypothesis that neurotransmitter release occurs by vesicle exocytosis is that stimulation will release the total soluble contents of synaptic vesicles. This prediction was first tested not in neurons but for adrenal medullary cells, from which chromaffin granules could be purified and their contents analyzed.[64] Chromaffin granules are organelles analogous to but much larger than synaptic vesicles; they contain epinephrine, norepinephrine, ATP, the synthetic enzyme dopamine β-hydroxylase, and proteins called chromogranins. All of these components are released in response to stimulation of the adrenal medulla, and they appear in the perfusate in exactly the same proportions as are found within the purified granules.

A good correspondence exists between vesicle contents and release from neurons as well, although it is difficult to isolate pure populations of synaptic vesicles from nerve terminals in order to determine their contents. For example, small synaptic vesicles in sympathetic neurons contain norepinephrine and ATP; the larger dense-core vesicles contain, in addition, dopamine β-hydroxylase and chromogranin A. Stimulation of sympathetic axons results in the release of all of these vesicle constituents.[65] Similarly, vesicles isolated from cholinergic neurons contain ATP as well as ACh, and both are released by stimulation of cholinergic nerves.[66]

[60]Olivera, B. M., et al. 1994. *Annu. Rev. Biochem.* 63: 823–867.

[61]Heuser, J. E., Reese, T. S., and Landis, D. M. D. 1974. *J. Neurocytol.* 3: 109–131.

[62]Peper, K., et al. 1974. *Cell Tissue Res.* 149: 437–455.

[63]Porter, C. W., and Barnard, E. A. 1975. *J. Membr. Biol.* 20: 31–49.

[64]Kirshner, N. 1969. *Adv. Biochem. Psychopharmacol.* 1: 71–89.

[65]Smith, A. D., et al. 1970. *Tissue Cell* 2: 547–568.

[66]Silinsky, E. M., and Redman, R. S. 1996. *J. Physiol.* 492: 815–822.

FIGURE 11.16 Distribution of Calcium Channels at the neuromuscular junction. Mouse neuromuscular junctions double-labeled with α-bungarotoxin (red) and with antibodies to calcium channels (green) and observed by confocal laser microscopy. Pseudocolor images of calcium channels (A), ACh receptors (B), and superimposed images (C). The position of the calcium channels matches that of the active zones in the nerve terminal, concentrated in narrow bands that are in register with the openings of the junctional folds, as marked by ACh receptors (arrowheads). (After Sugiura et al. 1995; micrographs courtesy of C.-P. Ko.)

(A)

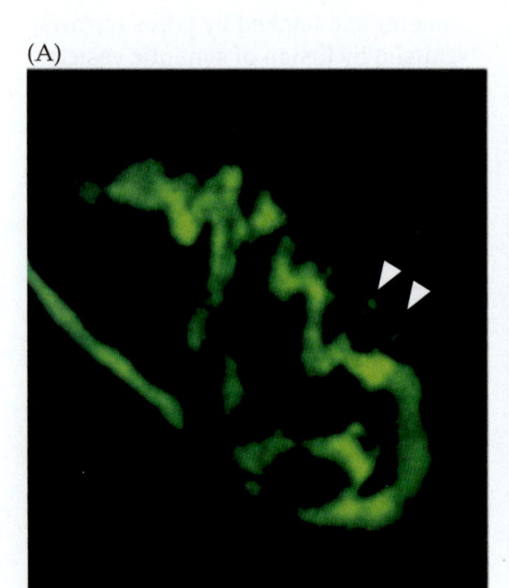

(B)

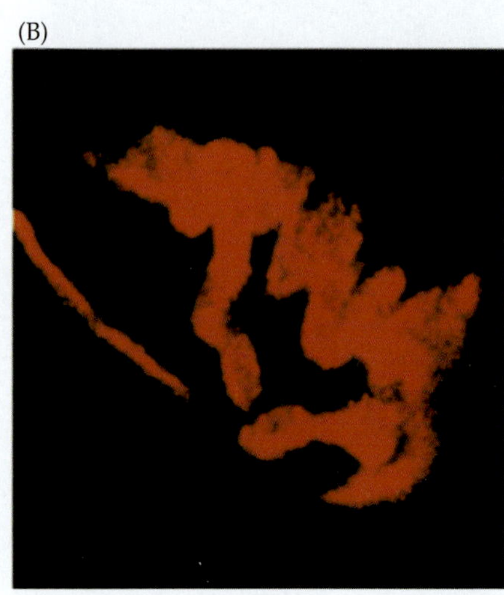

(C)

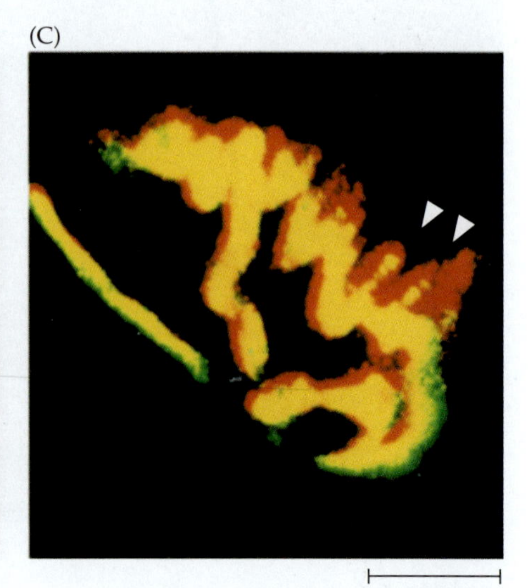

The idea that one quantum of transmitter corresponds to the contents of one synaptic vesicle has been examined quantitatively for cholinergic neurons. Vesicles purified from the terminals of the cholinergic electromotor neurons in the electric organ of the marine ray *Narcine brasiliensis* (a relative of *Torpedo californica*) were found to contain about 47,000 molecules of ACh.[67] If synaptic vesicles at the frog neuromuscular junction had the same intravesicular ACh concentration, then, making allowance for their smaller size, they would contain 7000 molecules of ACh. This is in excellent agreement with electrophysiological estimates of the number of ACh molecules in a quantum.[24]

[67]Wagner, J. A., Carlson, S. S., and Kelly, R. B. 1978. *Biochemistry* 17: 1199–1206.

[68]Heuser, J. E., et al. 1979. *J. Cell Biol.* 81: 275–300.

Morphological Evidence for Exocytosis

An important experimental innovation developed by Heuser and Reese and their colleagues[68] enabled frog muscle to be quick-frozen within milliseconds after a single shock to the motor nerve and then to be prepared for freeze-fracture. With such an experiment it was possible to obtain scanning electron micrographs of vesicles caught in the act of fusing with the presynaptic membrane and to determine with some accuracy the time course of such fusion. To do this, the muscle is mounted on the undersurface of a falling plunger, with the motor nerve attached to stimulating electrodes. As the plunger falls, a stimulator is triggered, shocking the nerve at a selected interval before the muscle smashes into a copper block cooled to 4 K with liquid helium. An essential part of the experiment is that the duration of the presynaptic action potential is increased by addition of 4-aminopyridine (4-AP) to the bathing solution. This treatment greatly increases the magnitude and duration of quantal release evoked by a single shock and hence the number of vesicle openings seen in the electron micrographs (Figure 11.17A and B).

Two important observations were made: First, the maximum number of vesicle openings occurred when stimulation preceded freezing by 3 to 5 ms. This corresponded to the peak of the postsynaptic current recorded from curarized, 4-AP–treated muscles in separate experiments. In other words, the maximum number of vesicle openings coincided in time with the peak postsynaptic conductance change determined physiologically. Second, the number of vesicle openings increased with 4-AP concentration, and the increase was related linearly to the estimated increase in quantum content of the end-plate potentials by 4-AP, again obtained from separate physiological experiments

(A)

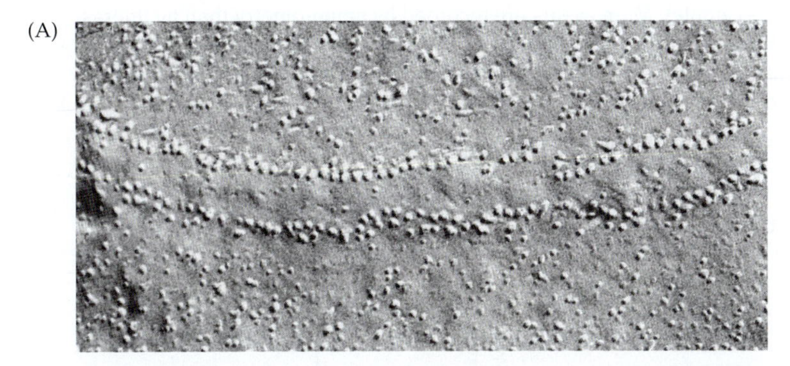

(B)

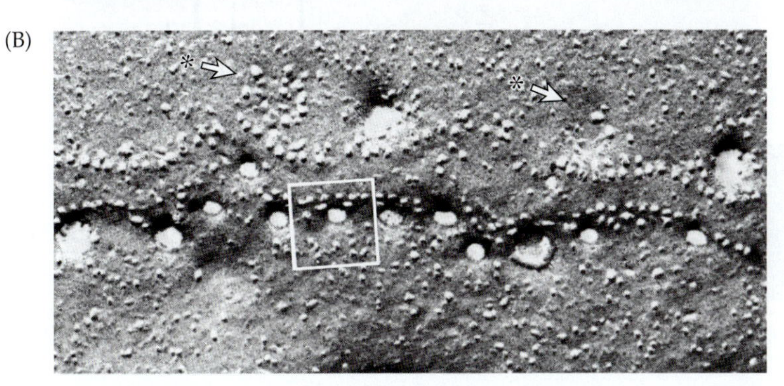

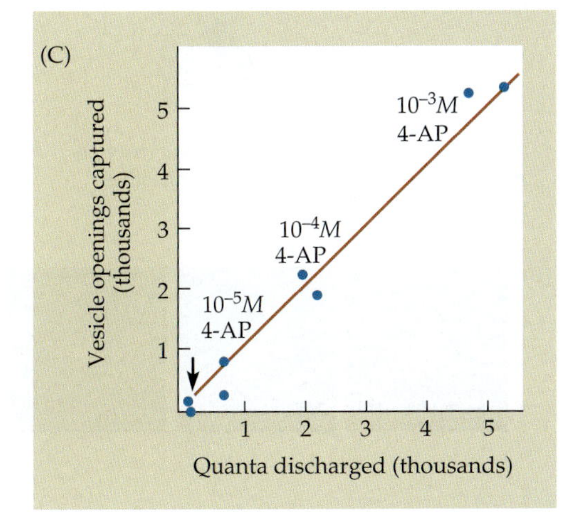

FIGURE 11.17 Vesicle Exocytosis Corresponds to Quantal Release. (A) Freeze-fracture electron micrograph of the cytoplasmic half of the presynaptic membrane in a frog nerve terminal (as if observed from the synaptic cleft). The region of the active zone appears as a slight ridge delineated by membrane particles (about 10 nm in diameter). (B) Similar view of a terminal that was frozen just at the time the nerve began to discharge large numbers of quanta (5 ms after stimulation). "Holes" (box) are sites of vesicle fusion; shallow depressions (indicated by asterisks and arrows) mark where vesicles have collapsed flat after opening. (C) Comparison of the number of vesicle openings (counted in freeze-fracture images) and the number of quanta released (determined from electrophysiological recordings). The diagonal line is the 1:1 relationship expected if each vesicle that opened released 1 quantum of transmitter. Transmitter release was varied by adding different concentrations of 4-AP (arrow indicates control, without 4-AP). (From Heuser et al., 1979; micrographs kindly provided by J. E. Heuser.)

(Figure 11.17C). Thus, vesicle openings were correlated both in number and in time course with quantal release. In later experiments, Heuser and Reese[69] characterized the time course of vesicle openings in greater detail, showing that openings first increase during a 3 to 6 ms period after stimulation and then decrease over the next 40 ms.

Exocytosis has also been observed in living cells by evanescent-wave microscopy, a fluorescence microscopy technique that greatly reduces background fluorescence by exciting only a 300-nm-thick layer of cytosol. For these experiments, peptide-containing vesicles in neuronlike PC12 cells and chromaffin granules in adrenomedullary cells were labeled with a fluorescent dye. The release of catecholamines was measured by amperometry, a very sensitive method in which a carbon fiber microelectrode is used to detect transmitters by the current they produce when they are oxidized. Using evanescent-wave microscopy, individual fluorescent vesicles could be seen to dock at the plasma membrane and then disappear as they released their fluorescent contents by exocytosis (Figure 11.18).[70,71] Each time a fluorescent vesicle disappeared, the release of a quantum of transmitter was detected by amperometry. These optical techniques are also being used to track the movements of vesicles within cells, as they approach the plasma membrane and dock prior to fusion.[72,73]

In summary, there is now much evidence that synaptic vesicles are the morphological correlate of the quantum of transmitter, each vesicle containing a few thousand transmitter molecules. Vesicles can release their contents by exocytosis both spontaneously at a low rate (producing miniature synaptic potentials) and in response to presynaptic depolarization. This view is not universally held,[74] but other mechanisms proposed for quantal release, such as calcium-activated quantal gates in the presynaptic membrane, have

[69]Heuser, J. E., and Reese, T. S. 1981. *J. Cell Biol.* 88: 564–580.

[70]Lang, T., et al. 1997. *Neuron* 18: 857–863.

[71]Steyer, J. A., Horstmann, H., and Almers, W. 1997. *Nature* 388: 474–478.

[72]Steyer, J. A., and Almers, W. 1999. *Biophys. J.* 76: 2262–2271.

[73]Oheim, M., et al. 1999. *Philos. Trans. R. Soc. Lond. B* 354: 307–318.

[74]Dunant, Y., and Israel, M. 1998. *Neurochem. Res.* 23: 709–718.

FIGURE 11.18 Exocytosis Observed in Living Cells. (A) Chromaffin cells growing on a glass coverslip in cell culture were labeled with a fluorescent dye, which becomes concentrated in chromaffin vesicles. Individual vesicles docked at the plasma membrane were visualized by evanescent-wave microscopy. At the same time, release of catecholamines was detected by amperometry. (B) High-power images of a single chromaffin vesicle at 2 s intervals after the cell was stimulated with high potassium. The spot disappears abruptly and permanently as the vesicle undergoes exocytosis and releases its fluorescent contents. (C) The time course of exocytosis in response to an increase in extracellular potassium concentration, as recorded by amperometric detection of catecholamine release and the disappearance of fluorescent spots. Note the coincidence of release and spot disappearance (the arrows mark one example). More events are recorded by amperometry than by fluorescence because the amperometric electrode detects exocytosis over a large part of the cell, while only a small portion of the cell surface is imaged by evanescent-wave microscopy. (After Steyer, Horstmann, and Almers, 1997; micrographs kindly provided by W. Almers.)

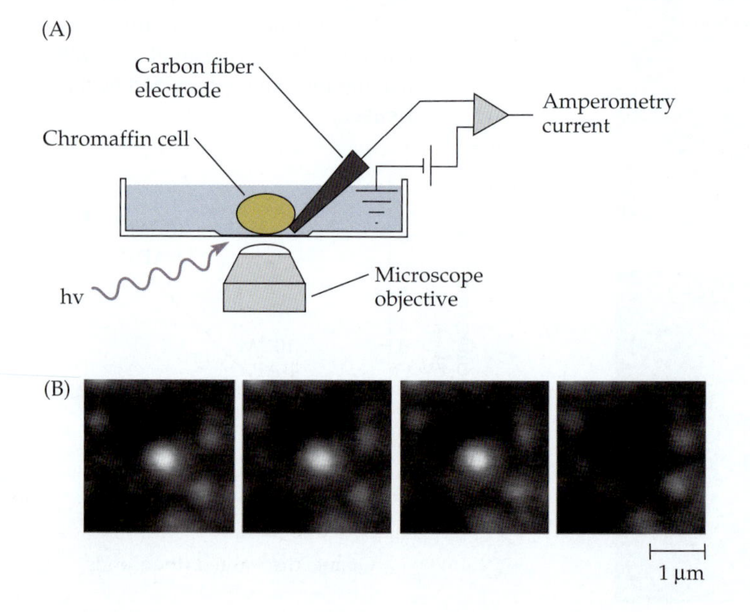

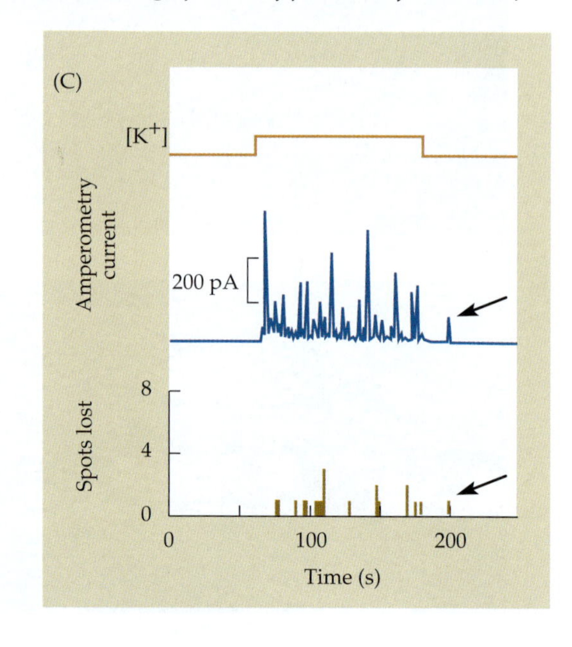

less extensive experimental support. As mentioned earlier, there is evidence that at some specialized synapses in the retina, depolarization can release transmitter through transport proteins in the presynaptic membrane, a mechanism that is nonquantal, not mediated by vesicle exocytosis, and not dependent on calcium influx.[75,76]

Recycling of Vesicle Components

What happens to a depleted synaptic vesicle after it has released its transmitter store? Does it meld with and become incorporated into the presynaptic membrane, or does it remain distinct and pinch back off into the cytoplasm as soon as its contents have been discharged? At neuromuscular, ganglionic, and CNS synapses, periods of intense stimulation have been shown to deplete synaptic vesicles and increase the surface area of the axon terminal, indicating that after releasing their contents, empty vesicles flatten out and become part of the terminal membrane (Figure 11.19).[77–79]

Heuser and Reese[80] found that components of the vesicle membrane are retrieved and recycled into new synaptic vesicles. They studied recycling of vesicles in frog motor nerve terminals by stimulating nerve–muscle preparations in the presence of horseradish peroxidase (HRP), an enzyme that catalyzes the formation of an electron-dense reaction product. When electron micrographs of terminals fixed after short periods of electrical stimulation were examined, HRP was found primarily in coated vesicles around the outer margins of the synaptic region, suggesting that these vesicles had been formed from the terminal membrane by endocytosis and, in the process, had captured HRP from the extracellular space (Figure 11.20). HRP also appeared, after a delay, in synaptic vesicles. Synaptic vesicles loaded in this way with HRP could then be depleted of the enzyme by stimulation in HRP-free medium, an experimental result supporting the idea that the recaptured membrane and enclosed HRP were recycled into the vesicle population from which release occurs.

The composition of synaptic vesicle membrane differs from that of the terminal plasma membrane; nevertheless, the appropriate vesicle membrane components are re-

[75]Cammack, J. N., and Schwartz, E. A. 1993. *J. Physiol.* 472: 81–102.

[76]Cammack, J. N., Rakhilin, S. V., and Schwartz, E. A. 1994. *Neuron* 13: 949–960.

[77]Ceccarelli, B., and Hurlbut, W. P. 1980. *Physiol. Rev.* 60: 396–441.

[78]Dickinson-Nelson, A., and Reese, T. S. 1983. *J. Neurosci.* 3: 42–52.

[79]Wickelgren, W. O., et al. 1985. *J. Neurosci.* 5: 1188–1201.

[80]Heuser, J. E., and Reese, T. S. 1973. *J. Cell Biol.* 57: 315–344.

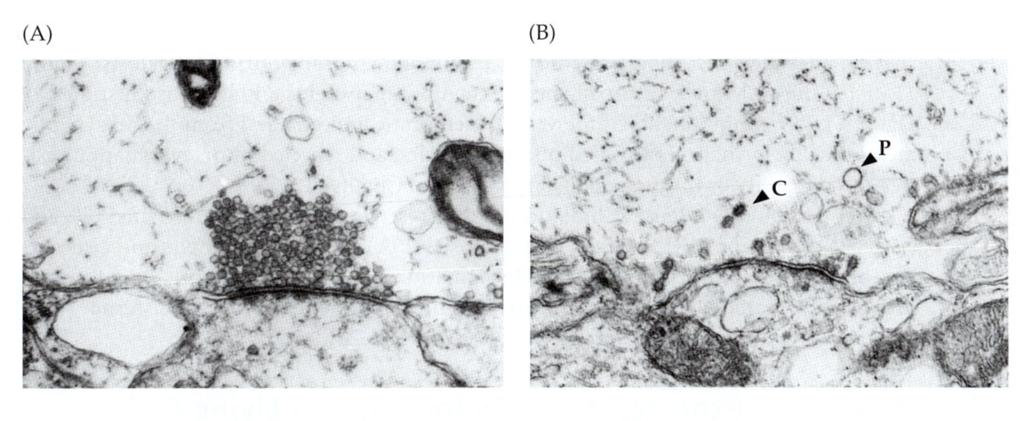

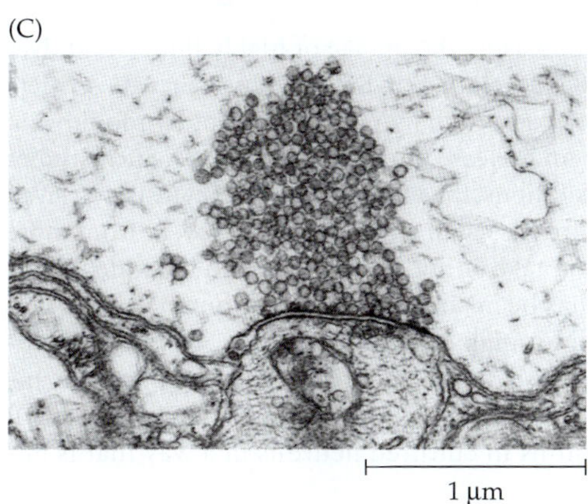

FIGURE 11.19 Stimulation Causes a Reversible Depletion of Synaptic Vesicles in lamprey giant axons. (A) Control synapse fixed after 15 min in saline. Synaptic vesicles are clustered at the presynaptic membrane. (B) Synapse fixed after stimulation of the spinal cord for 15 min at 20 Hz. Note the depletion of synaptic vesicles, the presence of coated vesicles (C) and pleomorphic vesicles (P), and the expanded presynaptic membrane. (C) Synapse fixed 60 min after cessation of stimulation. Note similarities to the control synapse in A. (Micrographs kindly provided by W. O. Wickelgren.)

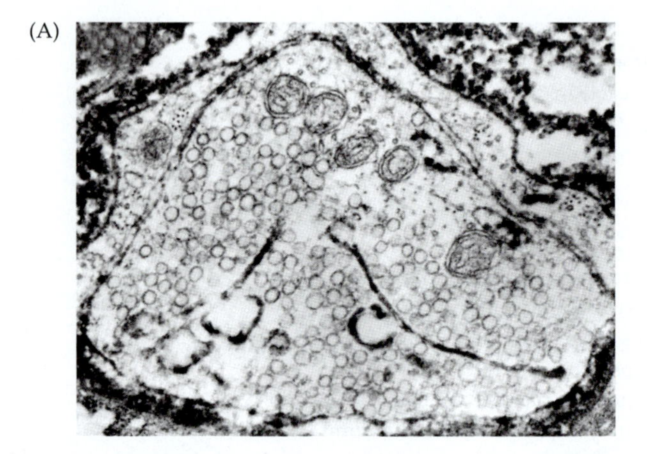

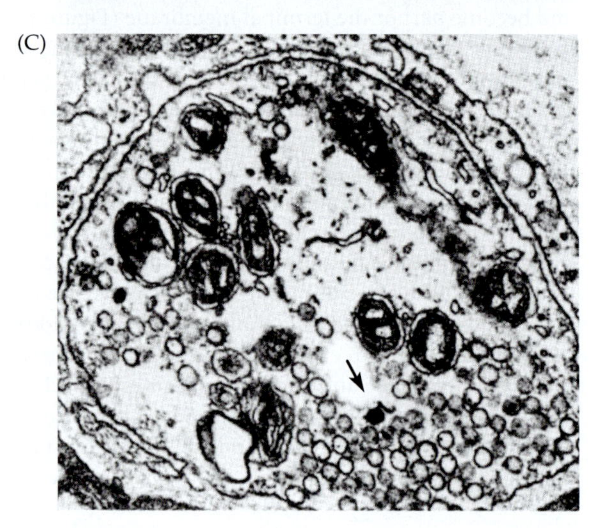

FIGURE 11.20 Recycling of Synaptic Vesicle Membrane. Electron micrographs of cross sections of frog neuromuscular junctions stained with horseradish peroxidase (HRP). (A) The nerve was stimulated for 1 min in saline containing HRP; electron-dense reaction product can be seen in the extracellular space and in cisternae and coated vesicles. (B) The nerve was stimulated for 15 min in HRP, then allowed to recover for 1 h while the HRP was washed out of the muscle. Many synaptic vesicles contain HRP reaction product, indicating that they have been formed from membrane retrieved by endocytosis. (C) The axon terminal was loaded with HRP and rested as in B, then stimulated a second time and allowed to recover an additional hour. Few vesicles are labeled (arrow), indicating that the recaptured membrane and enclosed HRP were recycled into the vesicle population from which release occurs. (From Heuser and Reese, 1973; micrographs kindly provided by J. E. Heuser.)

covered after their incorporation into the terminal membrane during exocytosis (Figure 11.21).[81] The cycle for recovery of specific membrane proteins and lipids is illustrated in Figure 11.22.[82] Components of the vesicle membrane are retrieved from the presynaptic membrane by formation of coated pits and vesicles, and recycled into new synaptic vesicles either directly or through endosomes (see also Chapter 13).

After particularly intense stimulation in the presence of HRP, large, uncoated pits and cisternae containing HRP are seen (see Figure 11.20A). Such uncoated pits and cisternae appear to represent bulk, nonselective invaginations of excess presynaptic membrane.[81] Presumably, coated vesicles then remove specific components from such cisternae to be recycled, directly or through endosomes (see Figure 11.22).

Monitoring Exocytosis and Endocytosis in Living Cells

Analogous experiments have now been done using the uptake of highly fluorescent dyes to mark recycled vesicles.[83] This technique, developed by Betz and his colleagues, offers the advantage that vesicle recycling can be observed in living preparations by monitoring the stimulation-dependent accumulation and release of dye (Figure 11.23). Such studies reveal that under normal physiological conditions, the entire cycle of exocytosis, retrieval, and re-formation of synaptic vesicles requires less than a minute;[84–86] recovery from more intense stimulation is slower.[79] Moreover, uptake of fluorescent dyes into individual boutons of cultured hippocampal neurons has been shown to occur in quantal steps that have the magnitude expected for uptake into single synaptic vesicles.[87] Such quantal uptake occurred within seconds of evoked release, suggesting that unitary exocytic and endocytic events are closely coupled. This technique makes it possible to analyze quantal release from individual presynaptic boutons in cultured neurons in a way that is not

[81]Valtorta, F., et al. 1988. *J. Cell Biol.* 107: 2717–2727.

[82]Miller, T. M., and Heuser, J. E. 1984. *J. Cell Biol.* 98: 685–698.

[83]Cochilla, A. J., Angleson, J. K., and Betz, W. J. 1999. *Annu. Rev. Neurosci.* 22: 1–10.

[84]Betz, W. J., and Bewick, G. S. 1993. *J. Physiol.* 460: 287–309.

[85]Ryan, T. A., et al. 1993. *Neuron* 11: 713–724.

[86]Teng, H., et al. 1999. *J. Neurosci.* 19: 4855–4866.

[87]Ryan, T. A., Reuter, H., and Smith, S. J. 1997. *Nature* 388: 478–482.

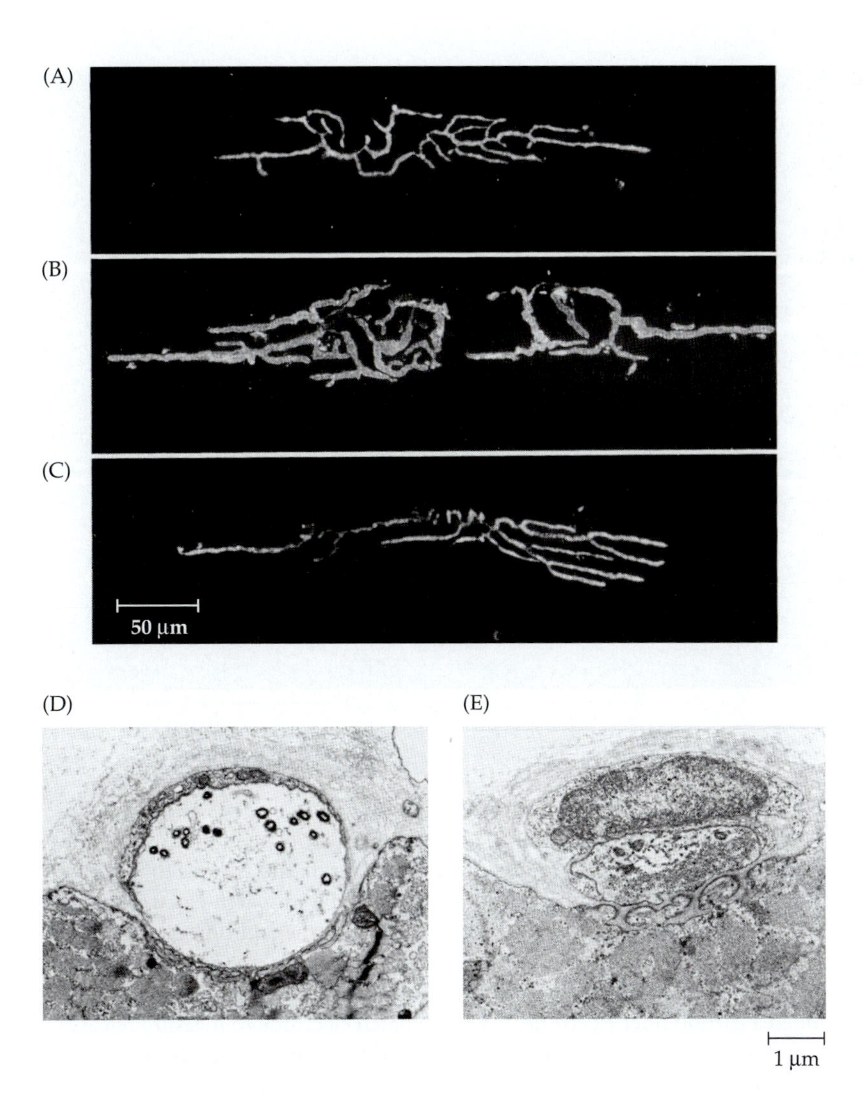

FIGURE 11.21 Recycling of Specific Synaptic Vesicle Membrane Proteins. (A–C) Fluorescence micrographs of frog neuromuscular junctions labeled with antibodies to synaptophysin (a vesicle membrane protein) and fluorescein-conjugated second antibodies. (D,E) Electron micrographs of cross sections of neuromuscular junctions. (A) Normal junction. The axon terminal membrane must be made permeable with detergent in order for antibodies to reach synaptophysin. (B,D) Muscle was treated with α-latrotoxin, which causes quantal transmitter release in the absence of calcium. This was done in calcium-free medium, which blocks endocytosis. Under these conditions, axon terminals are depleted of synaptic vesicles, appear distended, and stain without being made permeable. This indicates that synaptic vesicles have fused with the terminal membrane during exocytosis while retrieval of vesicle membrane was blocked, leaving synaptophysin exposed on the surface. (C,E) Muscle was treated with α-latrotoxin in normal saline. Terminals have a normal appearance and can be stained only after being made permeable. Under these conditions the vesicle population is maintained by active recycling while more than two times the initial store of quanta is released. Thus, despite the active turnover of synaptic vesicles, no detectable synaptophysin remains on the terminal surface, demonstrating the specificity and efficiency of synaptic vesicle membrane retrieval. (From Valtorta et al., 1988; micrographs kindly provided by F. Valtorta.)

complicated by changes in the postsynaptic response. Because recording electrodes are not necessary, such optical methods have proved useful in studies of presynaptic function and long-term plasticity.

A promising approach to the study of vesicle fusion is the use of dissociated nonneuronal secretory cells, such as mast cells and chromaffin cells, in which exocytosis of large,

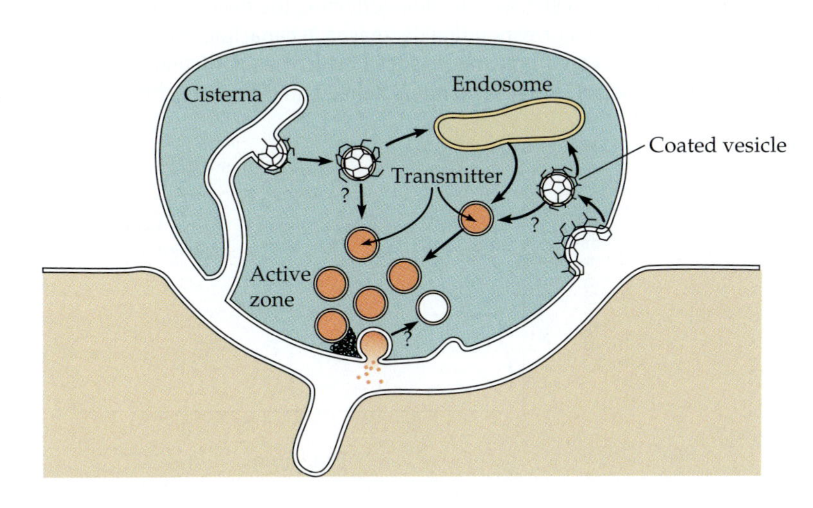

FIGURE 11.22 Proposed Pathways for Membrane Retrieval during vesicle recycling. After exocytosis, clathrin-coated vesicles selectively recapture synaptic vesicle membrane components. New synaptic vesicles are formed from coated vesicles, either directly or through endosomes. After intense stimulation, retrieval occurs from uncoated pits and cisternae. The new synaptic vesicles formed from recycled membrane are filled with transmitter and can be released by stimulation.

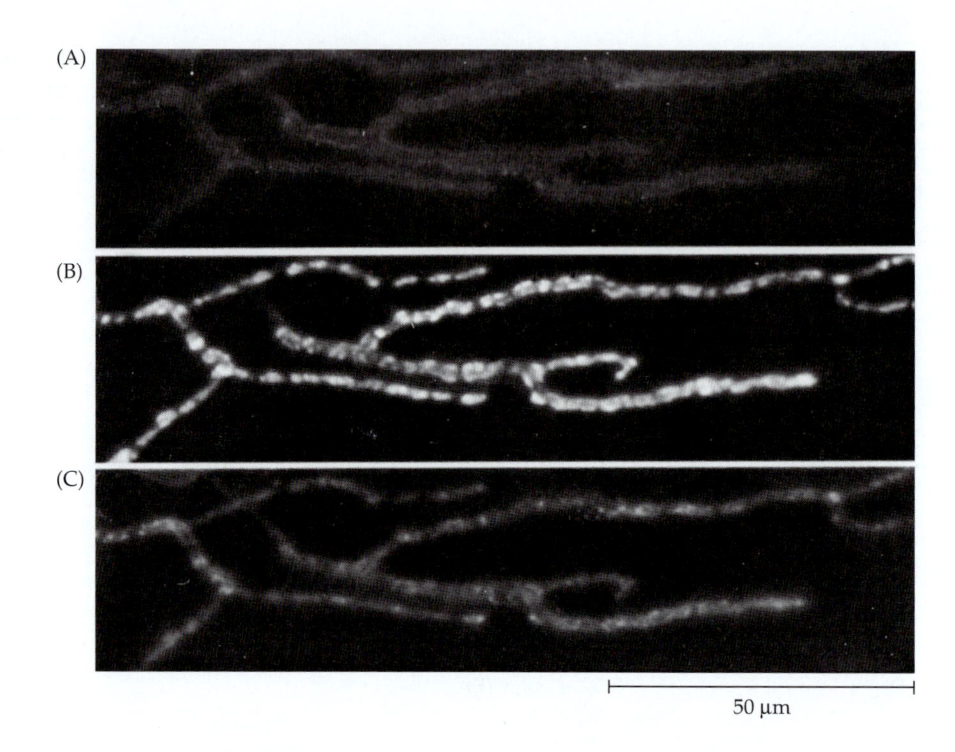

FIGURE 11.23 Activity-Dependent Uptake and Release of fluorescent dye by axon terminals at the frog neuromuscular junction. Fluorescence micrographs of axon terminals in a cutaneous pectoris muscle. (A) Muscle was bathed for 5 min in fluorescent dye (2 μ*M* FM1-43) and washed for 30 min. Only small amounts of dye remain associated with the terminal membrane. (B) The same muscle was then bathed in dye for 5 min while the nerve was stimulated (10 Hz) and washed for 30 min. The fluorescent patches are clusters of synaptic vesicles that were filled with dye during recycling. (C) The same muscle was then stimulated at 10 Hz for 5 min and washed for 30 min. Stimulation released most of the dye. (Micrographs kindly provided by W. J. Betz.)

dense-cored secretory granules can be followed simultaneously by light microscopy, electrophysiological recording, and amperometry, which detects the amines that they release.[88,89] The fusion of single granules can been recorded with patch electrodes as an increase in the electrical capacitance of the cell, arising from the addition of the granule membrane to the cell surface; membrane retrieval results in a capacitance decrease (Figure 11.24).[90] In experiments on chromaffin cells, Almers, Alvarez de Toledo, and their colleagues added a carbon fiber electrode inside the patch electrode to measure release of the catecholamines contained within the granules.[91] Release was usually detected coincidentally with an increase in capacitance, as expected for exocytosis and incorporation of the granule membrane into the plasma membrane (Figure 11.25). However, approximately 15% of release events were associated with transient, incomplete increases in capacitance. In such cases, exocytosis apparently occurred through a small temporary

[88]Penner, R., and Neher, E. 1989. *Trends Neurosci.* 12: 159–163.

[89]Angleson, J. K., and Betz, W. J. 1997. *Trends Neurosci.* 20: 281–287.

[90]Fernandez, J. M., Neher, E., and Gomperts, B. D. 1984. *Nature* 312: 453–455.

[91]Albillos, A., et al. 1997. *Nature* 389: 509–512.

[92]Lim, N. F., Nowycky, M. C., and Bookman, R. J. 1990. *Nature* 344: 449–451.

[93]von Gersdorff, H., and Matthews, G. 1997. *J. Neurosci.* 17: 1919–1927.

FIGURE 11.24 Release and Retrieval of Vesicle Membrane monitored by changes in membrane capacitance. Increases in cell capacitance measured with the whole-cell patch pipette recording technique occur in a stepwise fashion reflecting the fusion of individual vesicles with the plasma membrane. Corresponding decreases in capacitance are seen during vesicle retrieval. The recordings are from a rat mast cell, which has particularly large secretory vesicles (800 nm in diameter). (After Fernandez, Neher, and Gomperts, 1984.)

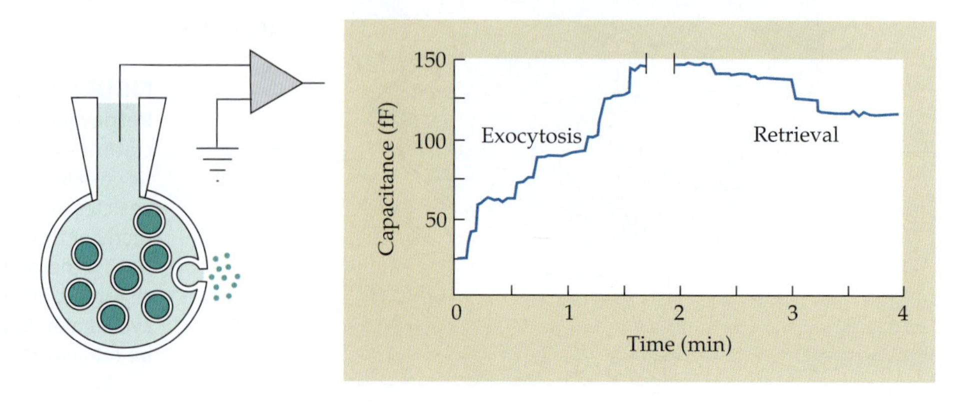

"fusion pore" that then closed, allowing the granule to pinch back off without ever becoming incorporated into the surface membrane (Figure 11.25D). Such "kiss and run" events may allow release of small molecules, such as catecholamines, which can be rapidly replenished, while retaining large proteins, which if lost would have to be replaced by synthesis of entirely new granules at the Golgi apparatus.

Changes in capacitance associated with the release of multiple quanta have been measured from individual nerve terminals isolated from the vertebrate CNS.[92,93] It is not yet possible to record changes in capacitance associated with the fusion of single synaptic vesicles. Although complete fusion of synaptic vesicles into the presynaptic membrane clearly occurs during periods of intense stimulation, it remains to be determined whether or not under more normal physiological conditions "kiss and run" vesicle recycling might occur in nerve terminals.[94,95]

[94]Fesce, R., et al. 1994. *Trends Cell Biol.* 4: 1–4.

[95]Klingauf, J., Kavalali, E. T., and Tsien, R. W. 1998. *Nature* 394: 581–585.

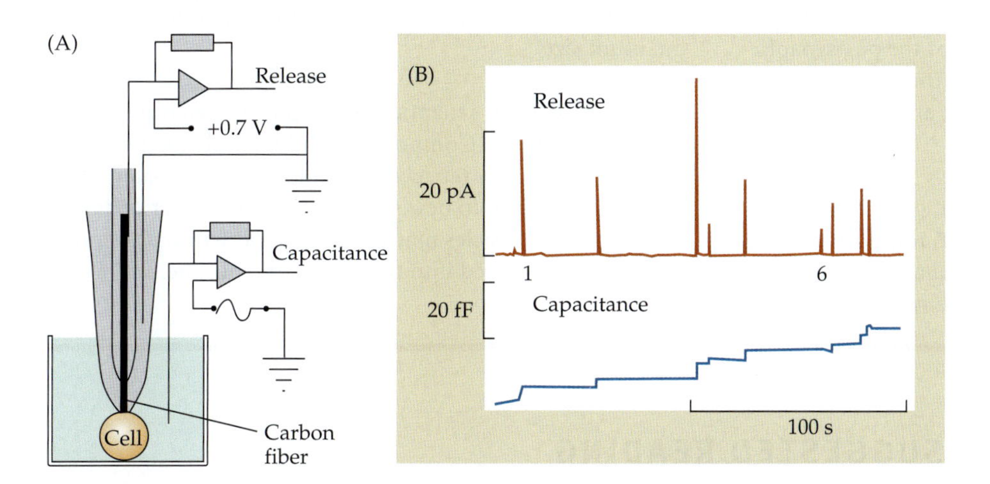

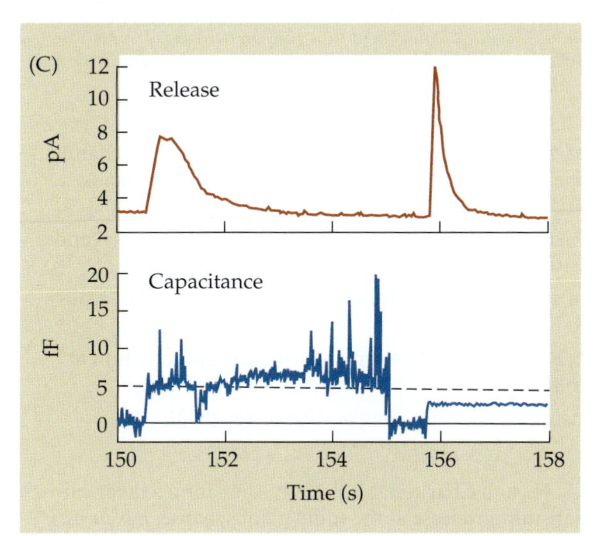

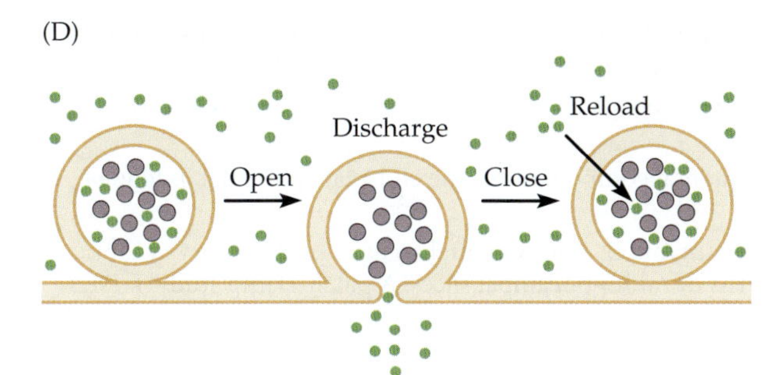

FIGURE 11.25 Coincident Increases in Membrane Capacitance and Release of Catecholamines from chromaffin cells. (A) A carbon fiber electrode inside the patch pipette measures catecholamine release by amperometry, while at the same time the electrode is used to measure capacitance within the patch. (B) Simultaneous recording of catecholamine release (top trace) and capacitance (bottom trace). All exocytic events detected by catecholamine release coincide with increases in capacitance. (C) The sixth and seventh exocytic events in part B, displayed on an expanded scale. The sixth exocytic event coincides with a transient, flickering increase in capacitance that lasts about 5 s. The seventh exocytic event coincides with an abrupt and long-lasting increase in capacitance. (D) Transient increases in capacitance may correspond to exocytosis through a temporary fusion pore that rapidly closes, allowing the vesicle to pinch back off into the cytoplasm without ever becoming incorporated into the plasma membrane. Under such circumstances small molecules may be released while larger proteins are retained in the vesicle. (After Albillos et al., 1997.)

SUMMARY

- When an axon terminal is depolarized, voltage-activated calcium channels open, increase the intracellular calcium concentration, and cause transmitter release.

- Transmitter is released in multimolecular packets, or quanta, that arise when transmitter-containing synaptic vesicles fuse with the plasma membrane and release their contents by exocytosis. There is also a continuous, non-quantal leak of transmitter from axon terminals at rest.

- The synaptic delay between the beginning of the presynaptic depolarization and the beginning of the postsynaptic potential is due to the time required for the nerve terminal to depolarize, calcium channels to open, and increased intracellular calcium to cause exocytosis.

- Exocytosis occurs at a slow rate at rest, producing spontaneous miniature synaptic potentials. In response to an action potential, anywhere from 1 to 300 quanta are released nearly simultaneously, depending on the synapse.

- Synaptic vesicles contain several thousand molecules of transmitter. The number of postsynaptic receptors activated by a quantum of transmitter varies considerably from about 15 to 1500, depending on the synapse.

- The distribution of amplitudes of spontaneous miniature and evoked postsynaptic potentials can be analyzed by statistical methods to determine the quantum size and quantum content of the response. Neuromodulatory influences that act presynaptically tend to influence quantum content; those that act postsynaptically tend to influence quantum size.

- After exocytosis, synaptic vesicles may flatten out into the plasma membrane. Components of the vesicle membrane are then specifically retrieved by endocytosis of coated vesicles and recycled into new synaptic vesicles. Under certain circumstances, vesicles may pinch back off without ever becoming incorporated into the surface membrane.

SUGGESTED READING

General Reviews

Cochilla, A. J., Angleson, J. K., and Betz, W. J. 1999. Monitoring secretory membrane with FM1-43 fluorescence. *Annu. Rev. Neurosci.* 22: 1–10.

Llinás, R. 1982. Calcium in synaptic transmission. *Sci. Am.* 247(4): 56–65.

Olivera, B. M., Miljanich, G. P., Ramachandran, J., and Adams, M. E. 1994. Calcium channel diversity and neurotransmitter release: The omega-conotoxins and omega-agatoxins. *Annu. Rev. Biochem.* 63: 823–867.

Parnas, H., Segel, L., Dudel, J., and Parnas, I. 2000. Autoreceptors, membrane potential and the regulation of transmitter release. *Trends Neurosci.* 23: 60–68.

Redman, S. 1990. Quantal analysis of synaptic potentials in neurons of the central nervous system. *Physiol. Rev.* 70: 165–198.

Walmsley, B., Alvarez, F. J., and Fyffe, R. E. W. 1998. Diversity of structure and function at mammalian central synapses. *Trends Neurosci.* 21: 81–88.

Zucker, R. S. 1993. Calcium and transmitter release. *J. Physiol. (Paris)* 87: 25–36.

Original Papers

Adler, E. M., Augustine, G. J., Duffy, S. N., and Charlton, M. P. 1991. Alien intracellular calcium chelators attenuate neurotransmitter release at the squid giant synapse. *J. Neurosci.* 11: 1496–1507.

Betz, W. J., and Bewick, G. S. 1993. Optical monitoring of transmitter release and synaptic vesicle recycling at the frog neuromuscular junction. *J. Physiol.* 460: 287–309.

Boyd, I. A., and Martin, A. R. 1956. The end-plate potential in mammalian muscle. *J. Physiol.* 132: 74–91.

del Castillo, J., and Katz, B. 1954. Quantal components of the end-plate potential. *J. Physiol.* 124: 560–573.

Edwards, F. A., Konnerth, A., and Sakmann, B. 1990. Quantal analysis of inhibitory synaptic transmission in the dentate gyrus of rat hippocampal slices: A patch-clamp study. *J. Physiol.* 430: 213–249.

Fatt, P., and Katz, B. 1952. Spontaneous subthreshold potentials at motor nerve endings. *J. Physiol.* 117: 109–128.

Fernandez, J. M., Neher, E., and Gomperts, B. D. 1984. Capacitance measurements reveal stepwise fusion events in degranulating mast cells. *Nature* 312: 453–455.

Heuser, J. E., and Reese, T. S. 1973. Evidence for recycling of synaptic vesicle membrane during transmitter release at the frog neuromuscular junction. *J. Cell Biol.* 57: 315–344.

Heuser, J. E., Reese, T. S., Dennis, M. J., Jan, Y., Jan, L., and Evans, L. 1979. Synaptic vesicle exocytosis captured by quick freezing and correlated with quantal transmitter release. *J. Cell Biol.* 81: 275–300.

Katz, B., and Miledi, R. 1967. A study of synaptic transmission in the absence of nerve impulses. *J. Physiol.* 192: 407–436.

Katz, B., and Miledi, R. 1967. The timing of calcium action during neuromuscular transmission. *J. Physiol.* 189: 535–544.

Kuffler, S. W., and Yoshikami, D. 1975. The number of transmitter molecules in a quantum: An estimate from iontophoretic application of acetylcholine at the neuromuscular synapse. *J. Physiol.* 251: 465–482.

Llinás, R., Sugimori, M., and Silver, R. B. 1992. Microdomains of high calcium concentration in a presynaptic terminal. *Science* 256: 677–679.

Miller, T. M., and Heuser, J. E. 1984. Endocytosis of synaptic vesicle membrane at the frog neuromuscular junction. *J. Cell Biol.* 98: 685–698.

Ryan, T. A., Reuter, H., and Smith, S. J. 1997. Optical detection of a quantal presynaptic membrane turnover. *Nature* 388: 478–482.

Steyer, J. A., Horstmann, H., and Almers, W. 1997. Transport, docking and exocytosis of single secretory granules in live chromaffin cells. *Nature* 388: 474–478.

12 SYNAPTIC PLASTICITY

THE EFFICACY OF TRANSMISSION AT A SYNAPSE is not fixed, but can vary as a consequence of patterns of ongoing activity. Short trains of presynaptic action potentials can produce either facilitation of transmitter release from the presynaptic terminal that persists for several hundred milliseconds, or depression of release lasting for seconds, or a combination of both. A second phase of facilitation, called augmentation, can also last for seconds. Longer-lasting trains of presynaptic action potentials produce post-tetanic potentiation (PTP), an increase in transmitter release that lasts for tens of minutes. A persistent increase in calcium concentration in the presynaptic terminal underlies these changes in release.

At many synapses repetitive activity can produce not only short-term changes, but also alterations in synaptic efficacy that last for hours, or even days. The two phenomena of this type are known as long-term potentiation (LTP) and long-term depression (LTD). LTP is mediated by an increase in calcium concentration in the postsynaptic cell that sets in motion a series of second messenger systems that recruit additional receptors into the postsynaptic membrane and also increase receptor sensitivity. LTD appears to occur in response to smaller increases in postsynaptic calcium concentration and is accompanied by a reduction in the number and sensitivity of postsynaptic receptors.

Other forms of LTP and LTD appear to involve presynaptic mechanisms. Both LTP and LTD have been postulated to be substrates for various forms of learning and memory formation, but current evidence for this idea is inconclusive.

So far we have discussed the transmission of excitatory and inhibitory signals at directly activated synapses in terms of a single action potential arriving in the presynaptic nerve terminal, causing depolarization of the terminal, calcium entry, and transmitter release, followed by postsynaptic depolarization or hyperpolarization, depending on the properties of the postsynaptic receptors. Under such circumstances, except for statistical variations, the responses at any given synapse are relatively regular and stereotyped. During everyday activity, however, synapses in the nervous system are not usually activated by the arrival of an occasional presynaptic action potential. Instead, constant streams of action potentials invade the terminals, sometimes regularly at clocklike intervals, sometimes in bursts of varying frequency and duration.

Such ongoing activity can have marked effects on the efficacy of synaptic transmission. For example, when a brief train of stimuli is applied to a presynaptic nerve, during the train the amplitude of the resulting postsynaptic potentials may either increase (synaptic **facilitation**) or decrease (synaptic **depression**). Such changes continue after the activity itself has ended, and they have been classified according to the duration over which they persist (Figure 12.1). Facilitation appears instantly and continues throughout the stimulus train. After the stimulation has ceased, test pulses applied to the nerve show that the increase in synaptic efficacy persists for several hundred milliseconds. A slower phase of facilitation, lasting for several seconds, is called **augmentation**.

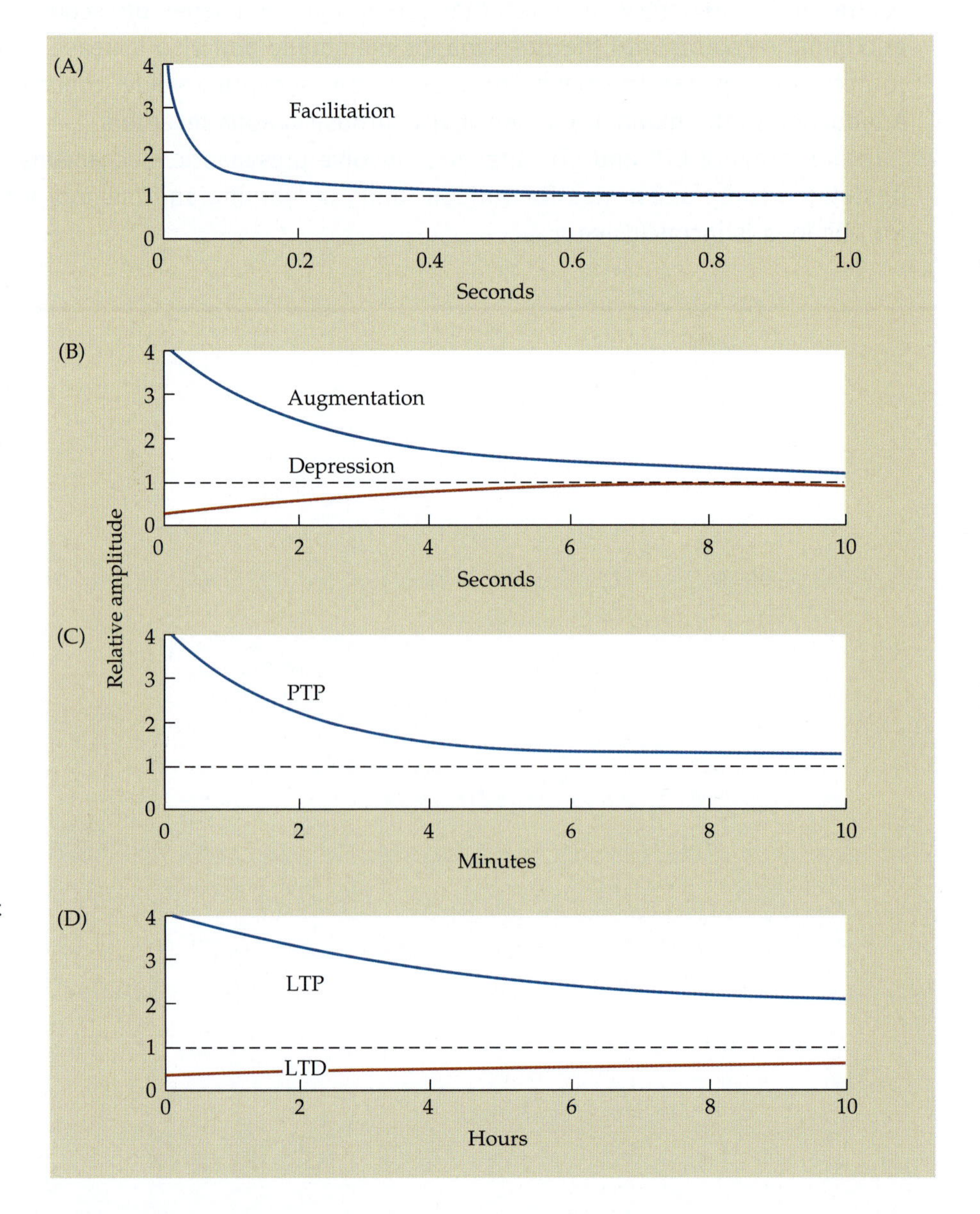

FIGURE 12.1 Time Courses of Activity-Induced Changes in synaptic transmission. (A) The main component of facilitation decays over a period of about 100 ms, with a smaller phase persisting for several hundred ms. (B) Recovery from depression is complete, and augmentation is largely dissipated, by 10 s. (C) PTP lasts for more than 10 min, and (D) LTP and LTD last well beyond 10 h.

When repetitive activity results in synaptic depression, recovery also occurs over a period of several seconds. A relatively long, high-frequency train of stimuli (commonly called a tetanus because such a train of stimuli applied to a muscle, or to its motor nerve, produces a tetanic muscle contraction) usually results in synaptic depression, but it is followed a few seconds later by an increase in synaptic potential amplitude that can last for tens of minutes. This is called **posttetanic potentiation (PTP)**. Repetitive activation of synapses in the central nervous system can produce still more persistent changes in synaptic transmission called **long-term potentiation (LTP)** and **long-term depression (LTD)**. These can last for many hours, or even days.

As we examine the mechanisms underlying these phenomena, a common thread emerges: Much of the regulation of synaptic transmission is mediated by changes in intracellular calcium concentration, either in the presynaptic nerve terminal or the postsynaptic cell, or in both. Thus, synaptic regulation depends ultimately on mechanisms that control the uptake and extrusion of calcium across the plasma membrane and its mobilization from and sequestration into subcellular stores. These mechanisms, which range from direct activation of calcium channels (Chapter 9) to complex second messenger systems (Chapter 10), provide means of regulating synaptic efficacy over time spans ranging from milliseconds to hours.

SHORT-TERM CHANGES IN SIGNALING

Short-term changes in the amplitude of synaptic potentials have been studied most extensively at synapses in the peripheral nervous system, such as those in vertebrate skeletal muscle or in autonomic ganglia. Nonetheless, such changes have been demonstrated throughout the central nervous system as well. Here we will discuss facilitation, augmentation, depression, and PTP at the frog neuromuscular junction and at synapses in the ciliary ganglion of the chicken. They are "short term" in that they normally last for periods ranging from a few tens of milliseconds (facilitation) to tens of minutes (PTP). In special circumstances, however, PTP in such preparations can last for hours.

Facilitation and Depression of Transmitter Release

At many synapses the most immediate effect of repetitive stimulation is synaptic facilitation. This is illustrated in Figure 12.2A, which shows end-plate potentials recorded from a frog neuromuscular junction, produced by a short train of impulses to the motor nerve. The amplitudes of the potentials (measured from the starting point of each rising phase) increase progressively during the train. Furthermore, the effect outlasts the stimulus train, so that the response to a test stimulus 230 ms later is still larger than the first response in the sequence. Facilitation can be separated into two components: The largest one decays with a time constant of about 50 ms; a smaller component has a time constant of decay of about 250 ms.[1] Facilitation has been shown to be due to an increase in the mean number of quanta of transmitter (m; see Chapter 11) released by the presynaptic terminal.[2–4] Further statistical analyses at the crayfish neuromuscular junction have led to the suggestion that this increase in m is due, in turn, to an increase in p, the probability of transmitter release, rather than in n, the number of available quanta.[5] Facilitation can more than double the amount of transmitter released from the nerve terminal.

Transmitter release can also be subject to synaptic depression if the number of quanta released by a train of stimuli is large. In the experiment shown in Figure 12.2A, the end-plate potentials were reduced in amplitude by lowering calcium in the bathing solution; that is, the initial quantum content of the potentials was low (10 or less). A similar experiment on a muscle in high calcium concentration is shown in Figure 12.2B. In this experiment the quantal release is very much greater, but the responses have been reduced in amplitude by blocking the postjunctional ACh receptors with curare. During repetitive stimulation the responses become progressively smaller in amplitude. As with facilitation, depression outlasts the stimulus train (not shown).

Depression of the end-plate potential, like facilitation, is presynaptic in origin. Its mechanism is not completely clear, but the fact that a large quantal release of transmitter is needed to produce depression suggests that one factor is depletion of vesicles from the

[1]Mallart, A., and Martin, A. R. 1967. *J. Physiol.* 193: 679–697.
[2]del Castillo, J., and Katz, B. 1954. *J. Physiol.* 124: 574–585.
[3]Dudel, J., and Kuffler, S. W. 1961. *J. Physiol.* 155: 543–562.
[4]Kuno, M. 1964. *J. Physiol.* 175: 100–112.
[5]Wernig, A. 1972. *J. Physiol.* 226: 751–759.

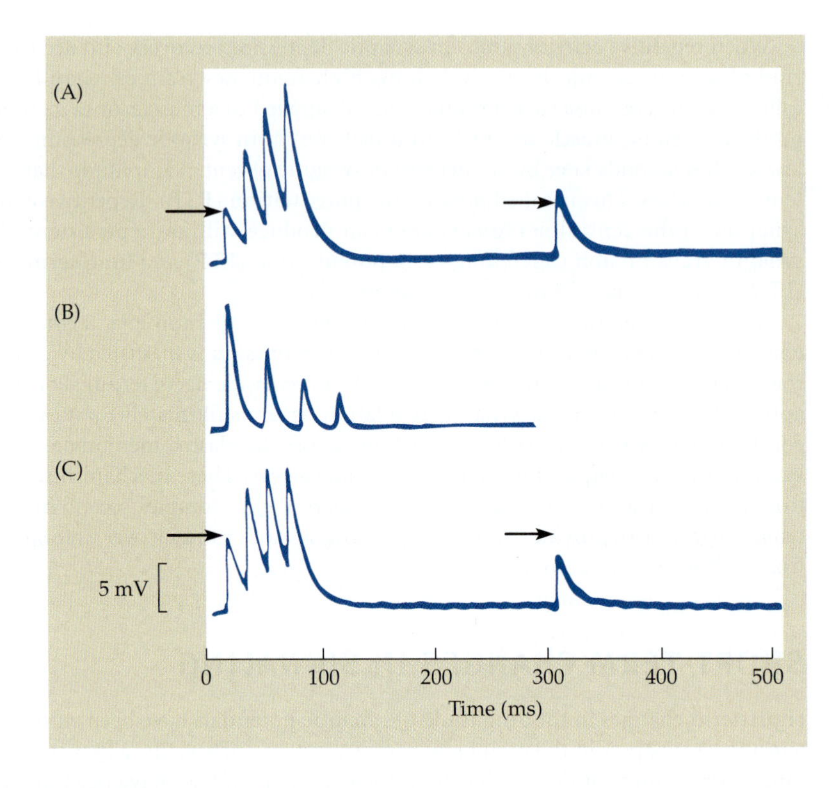

FIGURE 12.2 Facilitation and Depression at the vertebrate neuromuscular junction. (A) Muscle bathed in low calcium solution to reduce the quantum content of the response. Amplitudes of the end-plate potentials increase progressively during a train of four impulses. The response to a test pulse 230 ms later is still facilitated (arrows indicate initial amplitude). (B) Similar experiment with a curarized preparation in high calcium solution. The response amplitudes decrease progressively during the train. (C) Interaction between facilitation and depression in normal calcium concentration. There is no further facilitation after the second response because of the onset of depression, and the test response at 230 ms is depressed. (A and C after Mallart and Martin, 1968; B after Lundberg and Quilisch, 1953.)

nerve terminal during the conditioning train.[6] A second factor involves modulation of ACh release from the presynaptic terminal by co-release of ATP (Chapter 13). ATP is hydrolyzed to adenosine, which then acts on the presynaptic terminal to reduce the quantal content of the response.[7] After a long period of repetitive stimulation, depression can be severe, reducing the amplitude of the synaptic potential to less than 20% of its initial value.

In summary, transmitter release from presynaptic terminals is subject to two relatively short-term modifications. The first, facilitation, appears to be due to an increase in efficacy of release of quanta from the presynaptic terminal. The second, depression, may be associated with both depletion of quanta and reduced release efficacy. Figure 12.2C illustrates how these two effects can interact. During the train of impulses the initial effect of facilitation outweighs that of depression, and the responses increase in amplitude. Later, when the test pulse is given, facilitation has partially worn off and is overridden by depression, which has a more prolonged time course.

Role of Calcium in Facilitation

Experimental evidence obtained by Katz and Miledi[8] suggested that facilitation of transmitter release by the second of two action potentials arriving in the nerve terminal was related to a residue of calcium left over from the first. Several other possible mechanisms have been ruled out. For example, it has been shown at synapses in the chick ciliary ganglion that facilitation is not due to an increase in amplitude or duration of the second action potential in the presynaptic terminal.[9] Similarly, at synapses between cultured leech neurons, facilitation of the second of two responses occurs with no increase in presynaptic calcium *entry*.[10] Thus, the idea that facilitation is related to residual calcium seems the most viable. With repetitive stimulation, continuing calcium accumulation would lead to a progressive increase in transmitter release, as in Figure 12.2A. Various theoretical approaches used to examine the relation between intracellular calcium kinetics and the time course of facilitation have been reviewed by Parnas, Parnas, and Segel.[11]

Augmentation of Synaptic Transmission

After the time courses of facilitation and depression had been characterized, Magleby and Zengel discovered an additional effect of repetitive stimulation that they called augmen-

[6]Mallart, A., and Martin, A. R. 1968. *J. Physiol.* 196: 593–604.

[7]Redman, R. S., and Silinsky, E. M. 1994. *J. Physiol.* 477: 117–127.

[8]Katz, B., and Miledi, R. 1968. *J. Physiol.* 195: 481–492.

[9]Martin, A. R., and Pilar, G. 1964. *J. Physiol.* 175: 16–30.

[10]Stewart, R. R., Adams, W. B., and Nicholls, J. G. 1989. *J. Exp. Biol.* 144: 1–12.

[11]Parnas, H., Parnas, I., and Segel, L. A. 1990. *Int. Rev. Neurobiol.* 32: 1–50.

tation.[12,13] Augmentation is an increase in synaptic potential amplitude produced by repetitive stimulation that comes on more slowly than facilitation and decays over a much longer time period, having a time constant of decay of 5 to 10 s (see Figure 12.1B). Like facilitation, augmentation is due to an increase in transmitter release from the presynaptic nerve terminals. At the frog neuromuscular junction, augmentation and facilitation together can increase synaptic potential amplitude by a factor of more than five.

Posttetanic Potentiation

Posttetanic potentiation (PTP) is similar to facilitation and augmentation in that it refers to an increase in transmitter release from the presynaptic nerve terminal due to prior stimulation. It is different in that its onset is considerably delayed, so that it reaches its maximum several seconds after stimulation has ceased, and it last for tens of minutes. An example is shown in Figure 12.3, this time from an experiment on a cell in a ciliary ganglion from a chicken, treated with curare to reduce the amplitude of the excitatory postsynaptic potential (EPSP). In addition, in order to prevent the synaptic potential from triggering an action potential, the cell was hyperpolarized (long downward deflection in Figure 12.3A) before stimulating the presynaptic nerve. The first upward deflection in each trace is an electrical coupling potential; the second, slower, depolarization is the EPSP, produced by release of ACh from the presynaptic terminal (Chapter 9). It is the EPSP that is

[12]Magleby, K. L., and Zengel, J. E. 1976. *J. Physiol.* 257: 449–470.

[13]Zengel, J. E., and Magleby, K. L. 1982. *J. Gen. Physiol.* 80: 582–611.

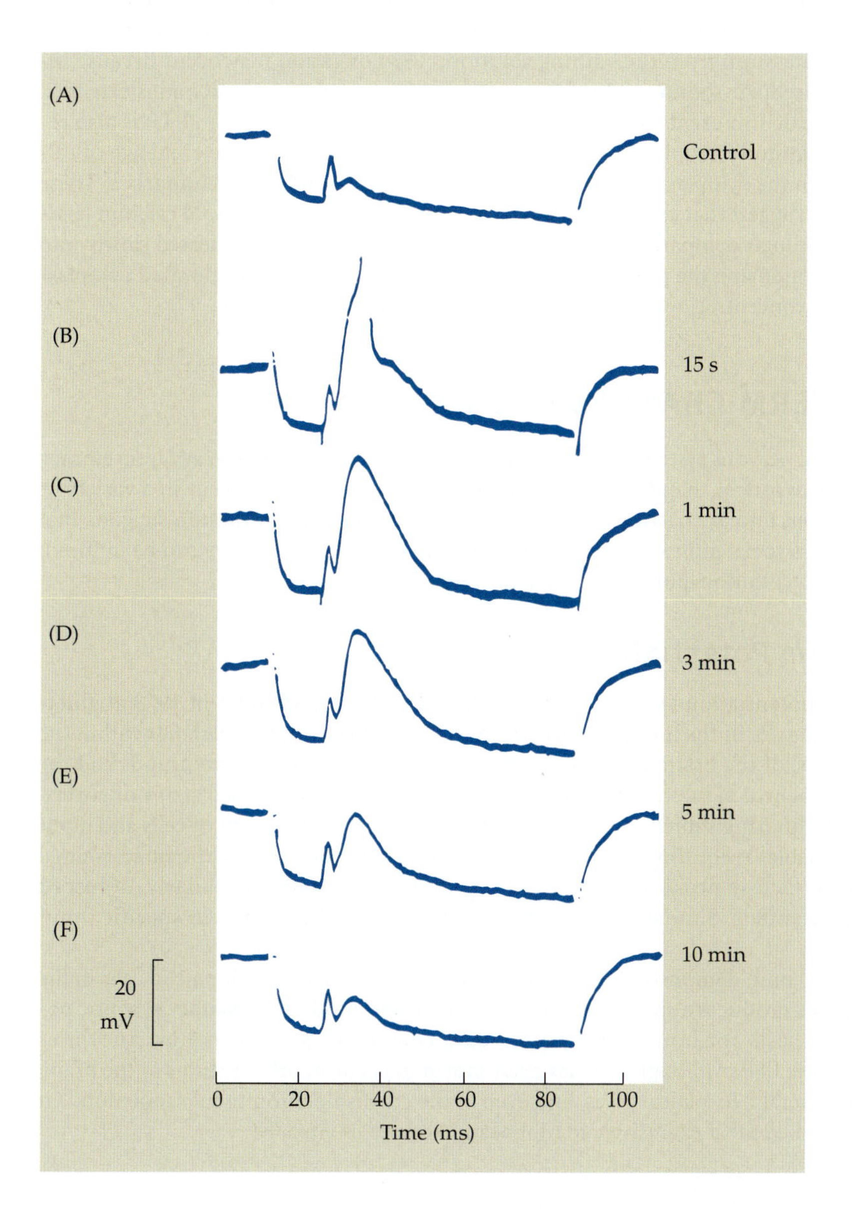

FIGURE 12.3 Posttetanic Potentiation of the excitatory postsynaptic potential (EPSP) in a chick ciliary ganglion cell, produced by preganglionic nerve stimulation. Potentials were recorded with an intracellular microelectrode. To prevent action potential initiation, the EPSP amplitude was reduced with curare, and a hyperpolarizing pulse was applied through the recording electrode before each stimulus. (A) The control record shows electrical coupling potential (brief depolarization) followed by a small EPSP. (B) The response recorded 15 s after the end of a train of 1500 stimuli applied to the preganglionic nerve. The EPSP amplitude is more than six times greater than the control, giving rise to an action potential. The amplitude of the coupling potential is unchanged. (C–F) Test stimuli at 1, 3, 5, and 10 min after the tetanus show the slow decline of potentiation, with the EPSP in the last record still more than twice the control value. (From Martin and Pilar, 1964.)

of interest. Initially the EPSP was only about 4 mV in amplitude (because of curarization). The presynaptic nerve was then stimulated at 100 pulses/s for 15 s (1500 stimuli), which caused a transient depression of the EPSP (not shown). Fifteen seconds later (Figure 12.3B), however, a single test stimulus produced an EPSP well over 20 mV in amplitude—so large, in fact, that it exceeded threshold and produced an action potential! The EPSP amplitudes produced by subsequent test shocks then declined (Figure 12.3C–F), but the response was still twice the pretetanic amplitude 10 min after the end of the tetanic stimulation.

Like facilitation, PTP is presynaptic in origin; that is, it is due to increased release of quanta from the presynaptic terminal and is accounted for by an increase in intracellular calcium. The exact mechanism underlying the phenomenon is obscure, but experiments at the neuromuscular junction of the frog have shown that it depends on calcium entry into the nerve terminal during the conditioning train of stimuli. For example, if calcium is removed from the bathing solution while the stimulation is applied, then no potentiation occurs.[14] On the other hand, sodium entry is not necessary, since PTP can be produced by trains of artificial depolarizing pulses applied to the nerve terminal in the presence of tetrodotoxin (TTX).[15] In that circumstance, the magnitude of the potentiation is increased with increasing extracellular calcium concentration, and in very high calcium (83 m*M*), potentiation after 500 shocks lasts for more than 2 hours.

Although sodium is not necessary for potentiation, sodium entry nevertheless contributes to its duration. At the rat neuromuscular junction, potentiation is prolonged by treatments that block extrusion of sodium by Na–K ATPase, such as adding ouabain or removing potassium from the bathing solution.[16] Prolongation may occur because increased intracellular sodium reduces the rate at which accumulated calcium is extruded by sodium–calcium exchange. At the crayfish neuromuscular junction, PTP is also reduced in magnitude and duration by treatments that interfere with the exchange of calcium between the cytoplasm and intracellular stores, for example, mitochondria.[17] These experiments suggest that calcium influx during the tetanus results in rapid calcium loading of intracellular compartments. The accumulated calcium is then released slowly into the cytoplasm during the posttetanic period, thereby maintaining an elevated cytoplasmic calcium concentration.

LONG-TERM CHANGES IN SIGNALING

In the central nervous system, repetitive activity can produce changes in synaptic efficacy that last much longer than those seen at peripheral synapses. They occur in a variety of brain locations and are particularly intriguing because their long duration suggests that they may be associated in some way with memory. Two types of change can be induced: long-term potentiation and long-term depression.

Long-Term Potentiation

Long-term potentiation (LTP) was first described by Bliss and Lømo in 1973 at glutamatergic synapses in the hippocampal formation.[18] This structure, which lies within the temporal lobe of the brain, consists of two regions, known as the hippocampus and the dentate gyrus, which in cross section appear as interlocking C-shaped strips of cortex, plus the neighboring subiculum (Figure 12.4). Its orderly arrangement of cells and input pathways enables recording electrodes to be inserted into the brain of the intact animal and placed in close proximity to known cell types, or even intracellularly, to record synaptic potentials. Similarly, stimulating electrodes can be located in specific input pathways.

Bliss and Lømo demonstrated that high-frequency stimulation of inputs to cells in the dentate gyrus produces a subsequent increase in the amplitude of excitatory synaptic potentials that lasts for hours, or even for days (Figure 12.5). This is now known as **homosynaptic LTP**. Although LTP has been shown to occur in other regions of the brain, including several neocortical areas, and even at the crayfish neuromuscular junction,[19] it has been studied most extensively in hippocampal slices in vitro.[20,21]

[14] Rosenthal, J. L. 1969. *J. Physiol.* 203: 121–133.
[15] Weinreich, D. 1970. *J. Physiol.* 212: 431–446.
[16] Nussinovitch, I., and Rahamimoff, R. 1988. *J. Physiol.* 396: 435–455.
[17] Tang, Y-G., and Zucker, R. S. 1997. *Neuron* 18: 483–491.
[18] Bliss, T. V. P., and Lømo, T. 1973. *J. Physiol.* 232: 331–356.
[19] Baxter, D. A., Bittner, G. D., and Brown, T. H. 1985. *Proc. Natl. Acad. Sci. USA* 82: 5978–5982.
[20] Andersen, P., et al. 1977. *Nature* 266: 736–737.
[21] Malenka, R. C., and Nicoll, R. A. 1999. *Science* 285: 1870–1874.

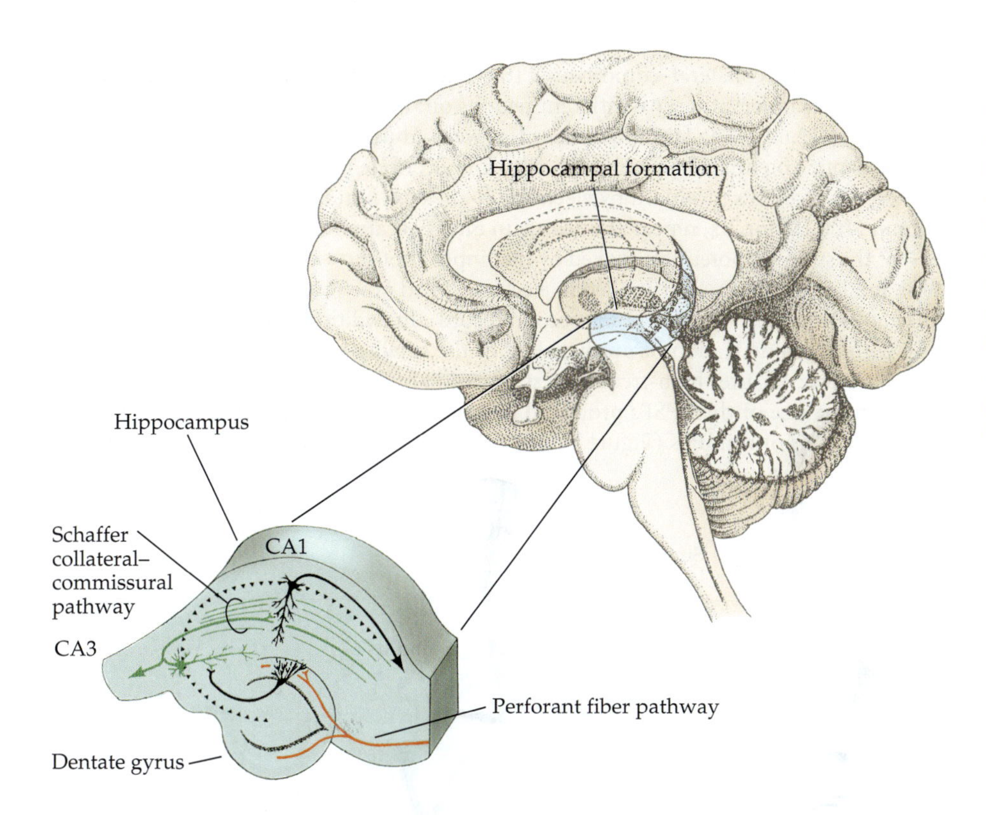

FIGURE 12.4 The Hippocampal Formation lies buried in the temporal lobe and consists of two interlocking C-shaped strips of cortex, the dentate gyrus and the hippocampus, together with the neighboring subiculum. Pyramidal cells in area CA1 are innervated by the Schaffer collateral–commissural pathway. Granule cells in the dentate gyrus are innervated by the perforant fiber pathway from the subiculum and, in turn, innervate CA3 cells.

Associative LTP in Hippocampal Pyramidal Cells

Experiments by T. H. Brown and his colleagues revealed that repetitive activity at one synaptic input to a cell could influence whether or not another input to the same cell was potentiated by repetitive activity. [22] This is called **associative LTP**. An example is shown in Figure 12.6. Intracellular recordings were made from a pyramidal cell in area CA1 of the hippocampus, and two extracellular stimulating electrodes were located in the input path-

[22]Barrionuevo, G., and Brown, T. H. 1983. *Proc. Natl. Acad. Sci. USA* 80: 7347–7351.

FIGURE 12.5 Long-Term Potentiation (LTP) in the hippocampus of an anesthetized rabbit. (A) Synaptic responses to perforant fiber pathway stimulation were recorded from granule cells in the dentate gyrus. (B) Brief tetanic stimuli (15/s for 10 s) were given at times marked by the arrows. Each tetanus caused an increase in the amplitude of the synaptic response (red circles), eventually lasting for hours. In the same cell, responses to stimulation of a control pathway not receiving tetanic stimulation (blue circles) were unchanged. (After Bliss and Lømo, 1973.)

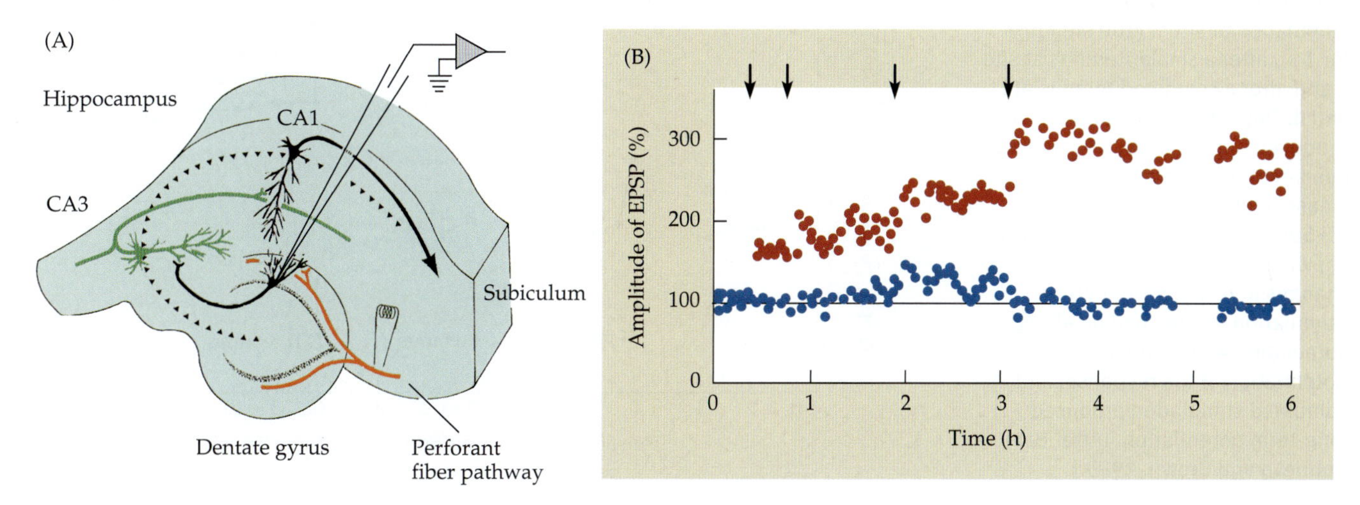

way (the Schaffer collateral–commissural tract) in such a way as to stimulate subpopulations of axons innervating two different regions on the dendritic arbor of the pyramidal cell (Figure 12.6A). The stimulus intensities were adjusted so that electrode I evoked a large synaptic potential in a pyramidal cell, while electrode II evoked a much smaller one.

Records of EPSPs produced by electrode II are shown in Figure 12.6B. Brief trains of stimuli (100 Hz for 1 s, repeated once again after 5 s) applied to electrode I resulted in LTP of the synaptic potentials evoked by that input (not shown), as in the experiments of Bliss and Lømo. The stimuli applied to electrode I had no effect on the smaller EPSP produced by electrode II (Figure 12.6B, second record). In addition, high-frequency stimulation with electrode II did not produce LTP of the smaller response (Figure 12.6B, third record). However, after high-frequency stimulation of I and II together, there was an increase in the size of the EPSPs produced by electrode II (Figure 12.6B, fourth record), lasting for tens of minutes (Figure 12.6C). This was called associative LTP because the

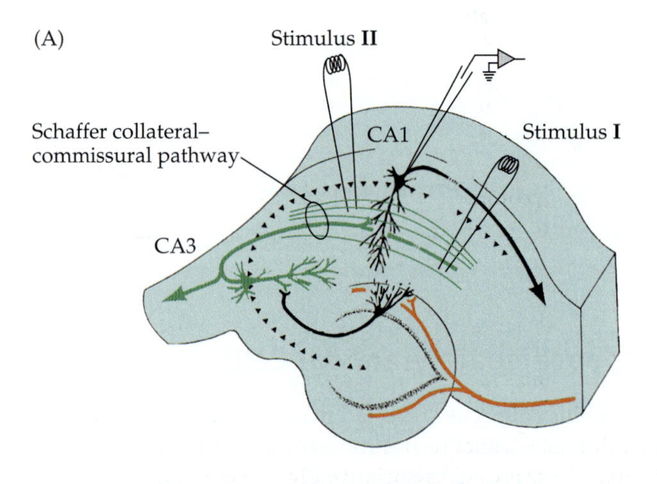

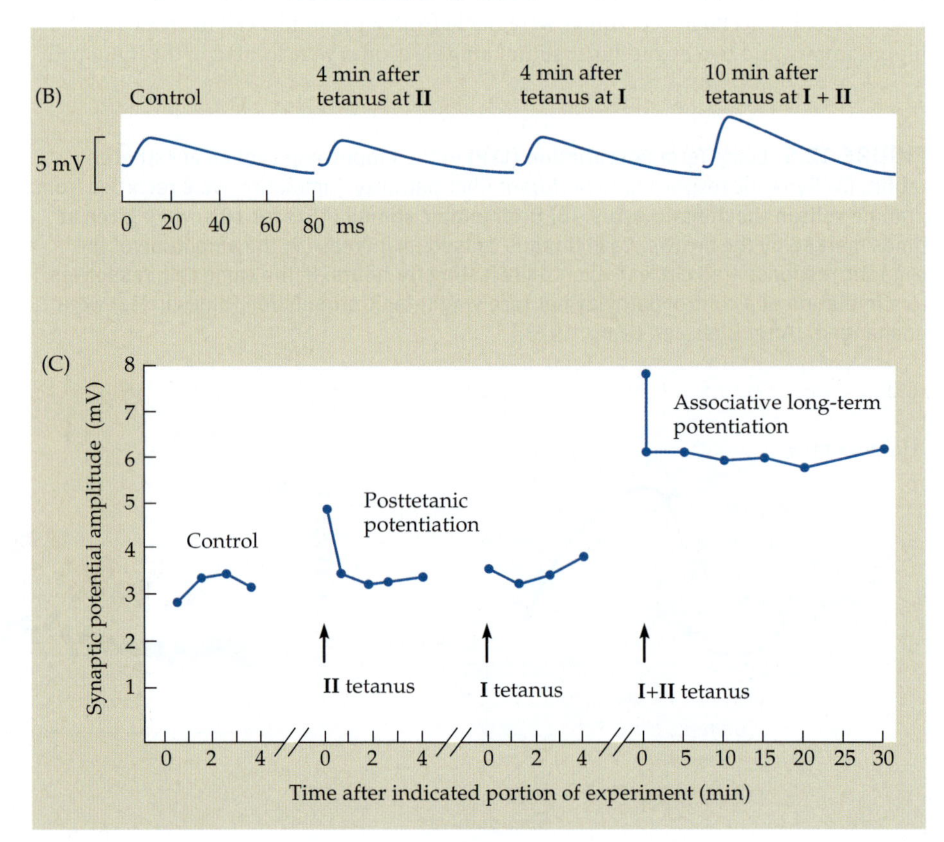

FIGURE 12.6 Associative LTP in a rat hippocampal slice. (A) Intracellular records were made from a CA1 pyramidal cell while stimulating two distinct groups of presynaptic fibers in the Schaffer collateral–commissural pathway (stimulus I and stimulus II). The stimuli were adjusted so that responses to stimulation at site I were five times greater than those to stimulation at site II. (B) Averaged responses to stimulation at site II in the control condition, after tetanic stimulation at site II (100 stimuli/s for 1 s), after a similar tetanus at site I, and after a combined tetanus at I and II. Only the combined tetanus produced potentiation; test shock 10 min later indicated a twofold increase in response amplitude. (C) Summary of the results in part B, showing the time course of the changes in response amplitude. Stimulation at site I had no effect, stimulation at II produced a brief potentiation of the response, and combined stimulation produced long-term potentiation. (After Barrionuevo and Brown, 1983.)

prolonged increased response to input II was produced only when repetitive stimulation of that input was associated with simultaneous stimulation of input I.

Mechanisms Underlying the Induction of LTP

A complete picture of the mechanisms underlying LTP induction has yet to emerge. There is general agreement, however, that one important factor is an increase in calcium concentration in the postsynaptic cell. In the CA1 pyramidal cell this is accomplished by entry of calcium through NMDA-type glutamate receptors (Chapter 3). NMDA receptors form cation channels with the unusual characteristic that they are blocked at normal resting potentials. Block is due to occupation of the channels by magnesium ions from the extracellular solution, which are removed when the receptors are depolarized.[23,24] Most glutamate-sensitive cells express both NMDA and non-NMDA (AMPA) receptors in their postsynaptic membrane,[25] so that both are activated by glutamate released from excitatory presynaptic terminals.

NMDA receptors have a relatively high calcium conductance, but calcium entry is dependent on membrane depolarization sufficient to remove magnesium block of the channels. Thus, in the experiment demonstrating associative LTP, repetitive activation of input II alone could not produce LTP because the EPSPs produced by cation influx through AMPA receptors were not sufficiently large to unblock NMDA receptors and permit calcium entry. However, when stimulation was accompanied by larger depolarizations due to activation of input I, the NMDA receptors were unblocked, permitting calcium entry and causing LTP. The involvement of NMDA receptors is indicated by the fact that NMDA antagonists block the induction of LTP but do not prevent LTP if applied after it has been induced.[26,27]

Two lines of evidence support the idea that LTP is induced by a postsynaptic increase in calcium concentration: First, it has been demonstrated that intracellular calcium concentration does indeed increase during stimulation, and when such an increase is prevented by prior injection of calcium buffer into the cell, LTP is prevented.[28,29] Second, elevation of postsynaptic calcium concentration by other means leads to a prolonged increase in EPSP amplitude.[30] Thus, the source of calcium appears to be unimportant, and at some synapses LTP can be mediated by calcium entry though voltage-activated calcium channels, at others by calcium release from intracellular stores.[29]

Increased intracellular calcium can activate a large number of intracellular biochemical pathways. Two pathways important for the induction of LTP involve calcium/calmodulin-dependent protein kinase II (CaMKII) and cAMP-dependent protein kinase.[21,31] CaMKII is found in high concentrations in the postsynaptic densities of dendritic spines, and intracellular injection of inhibitors of CaMKII prevents the induction of LTP.[32,33] Induction of LTP is also deficient in mice after genetic deletion of a critical CaMKII subunit.[34] Similarly, inhibition of cAMP-dependent protein kinase reduces LTP.[29,32,33]

LTP Expression

After the discovery of LTP, the question arose whether the increased amplitude of the potentiated synaptic potentials was due to an increase in transmitter release from presynaptic nerve terminals, or to an increase in the transmitter sensitivity of the postsynaptic membrane. It was attractive to think that LTP, like facilitation, augmentation, and PTP, reflected an increase in the quantum content of the synaptic response, and early experiments showed that was the case. An example is shown in Figure 12.7, where the amplitude distribution of the synaptic potentials is plotted before and after potentiation. In this and other experiments, statistical analysis of the amplitude distributions indicated that the mean quantum content of the potentiated synaptic potential was increased.[21,35,36] However, in other cases the same type of analysis gave the reverse result: The increase appeared to be in the size of the individual quanta, not in the number of quanta in the response.[37]

The finding of an increase in quantum content of the potentiated response on the one hand and an increase in quantal size on the other was taken as indicating two different mechanisms of expression of LTP: presynaptic (more quanta released) and postsynaptic

[23]Nowak, L., et al. 1984. *Nature* 307: 462–465.

[24]Mayer, M. L., Westbrook, G. L., and Guthrie, P. B. 1984. *Nature* 309: 261–263.

[25]Takumi, Y., et al. 1999. *Ann. N.Y. Acad. Sci.* 868: 474–481.

[26]Collingridge, G. L., Kehl, S. J., and McClennan, H. 1983. *J. Physiol.* 334: 33–46.

[27]Muller, D., Joly, M., and Lynch, G. 1988. *Science* 242: 1694–1697.

[28]Lynch, G., et al. 1983. *Nature* 304: 719–721.

[29]Yeckel, M. F., Kapur, A., and Johnston, D. 1999. *Nature Neurosci.* 2: 525–633.

[30]Malenka, R. C., et al. 1988. *Science* 242: 81–84.

[31]Schulman, H. 1995. *Curr. Opin. Neurobiol.* 5: 375–381.

[32]Malenka, R. C., et al. 1989. *Nature* 340: 554–557.

[33]Malinow, R., Schulman, H., and Tsien, R. W. 1989. *Science* 245: 862–866.

[34]Silva, A. J., et al. 1992. *Science* 257: 501–506.

[35]Malinow, R., and Tsien, R. W. 1990. *Nature* 346: 177–180.

[36]Bekkers, J. M., and Stevens, C. F. 1990. *Nature* 346: 724–729.

[37]Reid, C. A., and Clements, J. D. 1999. *J. Physiol.* 518(pt. 1): 121–130.

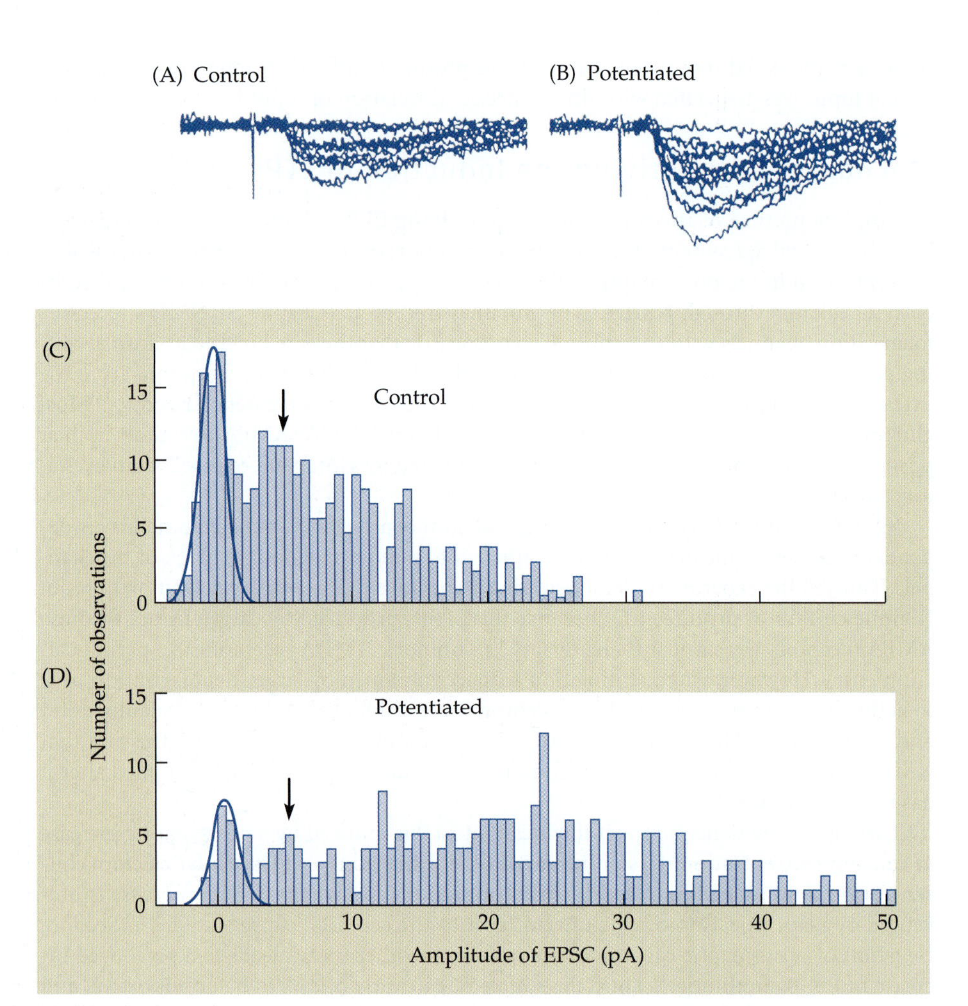

FIGURE 12.7 Change in Quantum Content of synaptic responses during LTP. Records from a rat hippocampal slice. (A,B) Sixteen superimposed whole-cell records of excitatory synaptic currents (EPSCs) in a CA1 pyramidal cell (A) before and (B) after a conditioning train of stimuli. Note the quantal steps in the current amplitudes. After conditioning, the fraction of failures is decreased and there are many more multiquantal responses. (C,D) Distribution of current amplitudes (C) before and after (D) conditioning. The normal curve is fitted to baseline noise (failures); arrows indicate mean current produced by a single quantum. After potentiation, the number of failures is reduced and the mean current is increased in amplitude by a factor of almost three, while the single-quantum current is unchanged. (After Malinow and Tsien, 1990.)

(larger response to each quantum). Current evidence favors the idea that the observed increases in quantal size and in quantum content are both mediated by changes in the postsynaptic membrane.

Silent Synapses

How can a change in the postsynaptic cell appear as an increase in the number of quanta in the synaptic response? If one thinks of the postsynaptic membrane as a static structure, then the only way that the quantum content of a synaptic potential could increase would be by an increase in the number of quanta released from the presynaptic terminal. Suppose, however, that a number of presynaptic excitatory boutons overlie postsynaptic regions on dendritic spines that contain only a few functional AMPA receptors, or perhaps none at all. Under resting conditions, release of a quantum of glutamate from such a bouton would produce little or no response; such a synapse would be "silent."[21] Now suppose that after the induction of LTP, glutamate receptors of the AMPA type were inserted into the postsynaptic membranes of the silent synapses. These would now respond to quanta released from the presynaptic terminal, and the quantum content of the response would increase (Figure 12.8).

Up-Regulation of Receptors

There is now substantial evidence for the idea that AMPA receptors are up-regulated during the expression of LTP, the most direct being the demonstration that AMPA receptor subunits are delivered to dendritic spines after repetitive stimulation accompanied by NMDA receptor activation.[38] The AMPA receptor subunit GluR1 (Chapter 3) was tagged with green fluorescent protein (GFP) and expressed transiently in hippocampal CA1 pyramidal cells. When cell dendrites were examined with laser scanning microscopy and

[38]Shi, S. H., et al. 1999. *Science* 284: 1811–1816.

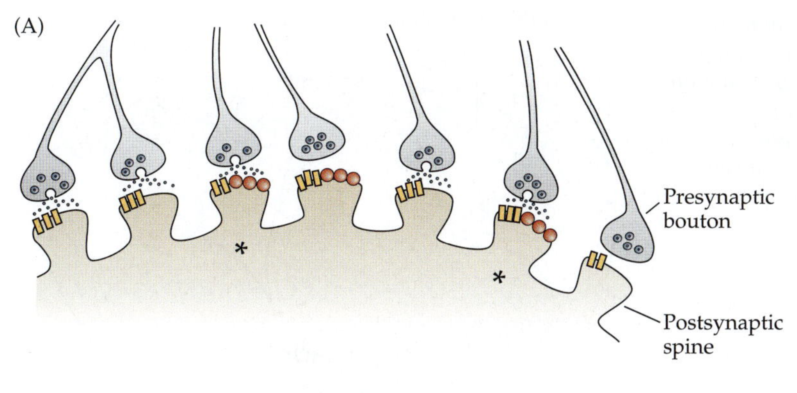

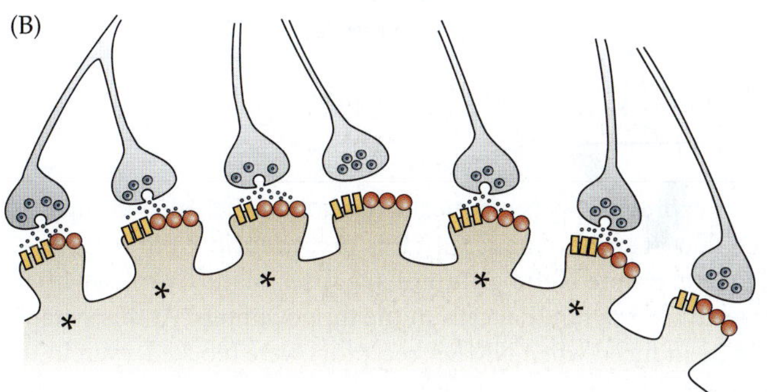

FIGURE 12.8 Proposed Mechanism for Increased Quantum Content during LTP. (A) Release of five quanta of glutamate from presynaptic boutons (indicated by omega figures) activates only two postsynaptic spines (asterisks) because many spines contain no AMPA receptors (red circles) and are "silent." Thus, the quantum content of the response is two, even though five quanta are released. NMDA receptors (yellow rectangles) are activated only if depolarization is sufficient to remove magnesium block. (B) During LTP, AMPA receptors are inserted into the postsynaptic membranes of the spines and the quantum content of the response is increased to five, with no change in the number of quanta released.

with electron microscopy, most of the protein (GluR1-GFP) in the dendrites was found in intracellular compartments, and only about half of the dendritic spines showed fluorescence. After stimulation, tagged receptors were delivered rapidly to dendritic spines and, in addition, to clusters on the dendrites. Almost all of the spines were fluorescent, even those nearly devoid of label before stimulation. The results suggest that many of the excitatory dendritic spines are silent and receive a full complement of AMPA receptors after repetitive stimulation.

Additional evidence for up-regulation of AMPA receptors has been summarized by Malenka and Nicoll.[21] For example, it has been shown by immunohistochemistry that all Schaffer collateral–commissural synapses contain NMDA receptors, but that only a fraction of these contain co-localized AMPA receptors.[39] Correspondingly, electrophysiological experiments reveal many synapses in CA1 pyramidal cells that are activated only by NMDA; these acquire AMPA responses during LTP.[40,41] The idea that such acquisition is due to the insertion of new receptors is supported by the observation that LTP is reduced by postsynaptic injection of compounds that interfere with membrane fusion.[42]

Still other experiments have demonstrated that after induction of LTP, new spines appear on the CA1 pyramidal cell dendrites.[43] Potentiation of the synaptic response by about 80% was accompanied by roughly a 13% increase in measured spine density.

The current view of factors underlying the induction and expression of LTP is summarized in Figure 12.9. Many dendritic spines (perhaps about half) contain primarily NMDA receptors that are unresponsive to glutamate at the resting potential. When the dendrite is depolarized sufficiently, repetitive synaptic activation of the receptors results in calcium entry. The incoming calcium binds to calmodulin. Calcium–calmodulin (CaM) activates CaMKII, which then autophosphorylates, converting it to a form that remains active after the calcium concentration has returned to basal levels. CaMKII has two effects on synaptic transmission: (1) It can phosphorylate AMPA receptors in the membrane, increasing their channel conductance and thereby increasing quantal size. (2) It can facilitate mobilization of reserve AMPA receptors from the cytoplasm into the plasma membrane, so that more postjunctional sites are available for activation by quantal packages of glutamate released from the nerve terminal.

Although the events described here summarize the basic mechanism currently proposed for LTP, the scheme is complicated by a variety of proposed modulating influences, involving more than 100 other molecules and receptors. Those are discussed elsewhere.[21,44]

[39]Takumi, Y., et al. 1999. *Nature Neurosci.* 2: 618–624.

[40]Liao, D., Hessler, N. A., and Malinow, R. 1995. *Nature* 375: 400–404.

[41]Isaac, J. T. R., Nicoll, R. A., and Malenka, R. C. 1995. *Neuron* 15: 427–434.

[42]Lledo, P. M., et al. 1998. *Science* 279: 339–403.

[43]Engert, F., and Bonhoeffer, T. 1999. *Nature* 399: 66–70.

[44]Sanes, J. R., and Lichtman, J. W. 1999. *Nature Neurosci.* 2: 597–604.

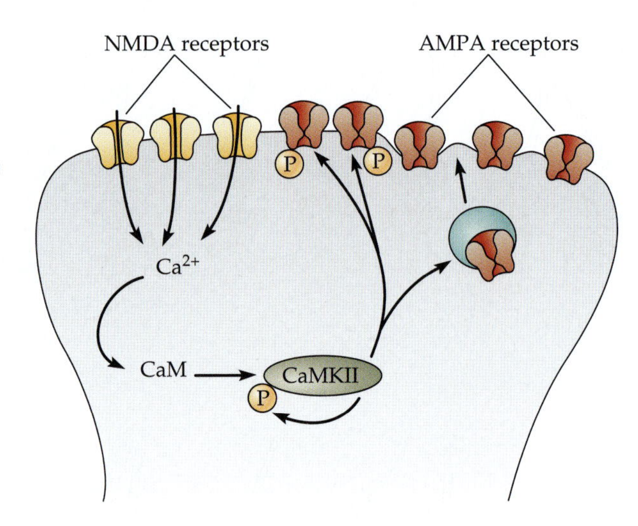

FIGURE 12.9 Proposed Mechanism for LTP. Activation of NMDA receptors allows calcium entry into the spine, activating CaMKII, which undergoes autophosphorylation, thereby maintaining its own activity after calcium concentration has returned to normal. CaMKII phosphorylates AMPA receptors already present in the postsynaptic membrane and/or promotes the insertion of new receptors from a reserve pool. (After Malenka and Nicoll, 1999.)

Presynaptic LTP

Although the idea that LTP is expressed postsynaptically has good experimental support, there is lingering evidence that at some synapses LTP is presynaptic, or at least has a presynaptic component. One example is the group of synapses made by mossy fibers from dentate granule cells onto CA3 pyramidal cells in the hippocampus. At this synapse it was found that LTP could be induced when NMDA receptors were blocked, even in the absence of any postsynaptic response at all,[45] suggesting that the phenomenon was entirely presynaptic. However, subsequent experiments indicated that the LTP was nonetheless accompanied by an increase in postsynaptic calcium concentration.[29] Both the increase in calcium concentration and the LTP depended on the activation of metabotropic glutamate receptors (mGluR) (Chapter 10) and were blocked by inhibitors of cAMP-dependent protein kinase, suggesting that the increased calcium concentration was due to release from intracellular stores.

One might conclude, then, that LTP at the mossy fiber–CA3 pyramidal cell synapse is postsynaptic after all, mediated through metabotropic receptors. However, one observation appears inconsistent with such a conclusion: LTP at this synapse is accompanied by a decrease in facilitation. At most central synapses, when two stimuli are applied presynaptically—say, 50 ms apart—the postsynaptic response to the second shock is facilitated. The mechanism underlying this paired-pulse facilitation is the same as at the neuromuscular junction—namely, an increase in the number of quanta of transmitter released from the presynaptic terminal. Because facilitation decreases when the mean quantum content increases (see Figure 12.2), the decrease in facilitation that accompanies LTP would suggest an increase in quantal release from the terminal. In any case, the fact that facilitation is reduced suggests that a presynaptic change has occurred. This criterion has been used as indication of presynaptic LTP at a number of other synapses.[46–49]

The expression of LTP presynaptically in response to the activation of postsynaptic metabotropic receptors implies the existence of a retrograde message across the synapse. One candidate for the messenger is nitric oxide (NO), which can be synthesized in the postsynaptic cell and diffuses rapidly into the surrounding tissue (Chapter 10).

Long-Term Depression

Long-term depression (LTD) of synaptic transmission, the inverse of LTP, was first reported to occur at the Schaffer collateral input to CA1 pyramidal cells[50] and has since been studied in a number of other areas, including hippocampal area CA3, the dentate gyrus, various cortical regions, and the cerebellum.[51] **Homosynaptic LTD** is a prolonged depression of synaptic transmission produced by previous repetitive activity in the same pathway (Figure 12.10A). It can be induced by a variety of stimulus protocols—for example, prolonged low-frequency stimulation (1–5 stimuli/s for 5–15 min), low-frequency stimulation with paired pulses, or brief high-frequency stimulation (50–100 stimuli/s for 1–5 s). In Schaffer collaterals, homosynaptic LTD is blocked by NMDA receptor antagonists,[52,53] as well as by pyramidal cell hyperpolarization and postsynaptic injection of cal-

[45]Zalutsky, R. A., and Nicoll, R. A. 1990. *Science* 248: 1619–1624.

[46]Maren, S., and Fanselow, M. S. 1995. *J. Neurosci.* 15: 7548–7564.

[47]Rogan, M. T., and LeDoux, J. E. 1995. *Neuron* 15: 127–136.

[48]Castro-Alamancos, M. A., and Calcagnotto, M. E. 1999. *J. Neurosci.* 19: 9090–9097.

[49]Voronin, L. L., Rossokhin, A. V., and Sokolov, M. V. 1999. *Neurosci. Behav. Physiol.* 29: 347–354.

[50]Lynch, G. S., Dunwiddie, T., and Gribkoff, V. 1977. *Nature* 266: 737–739.

[51]Linden, D. J., and Conner, J. A. 1995. *Annu. Rev. Neurosci.* 18: 319–357.

[52]Dudek, S. M., and Bear, M. F. 1992. *Proc. Natl. Acad. Sci. USA* 89: 4363–4367.

[53]Mulkey, R. M., and Malenka, R. C. 1992. *Neuron* 9: 967–975.

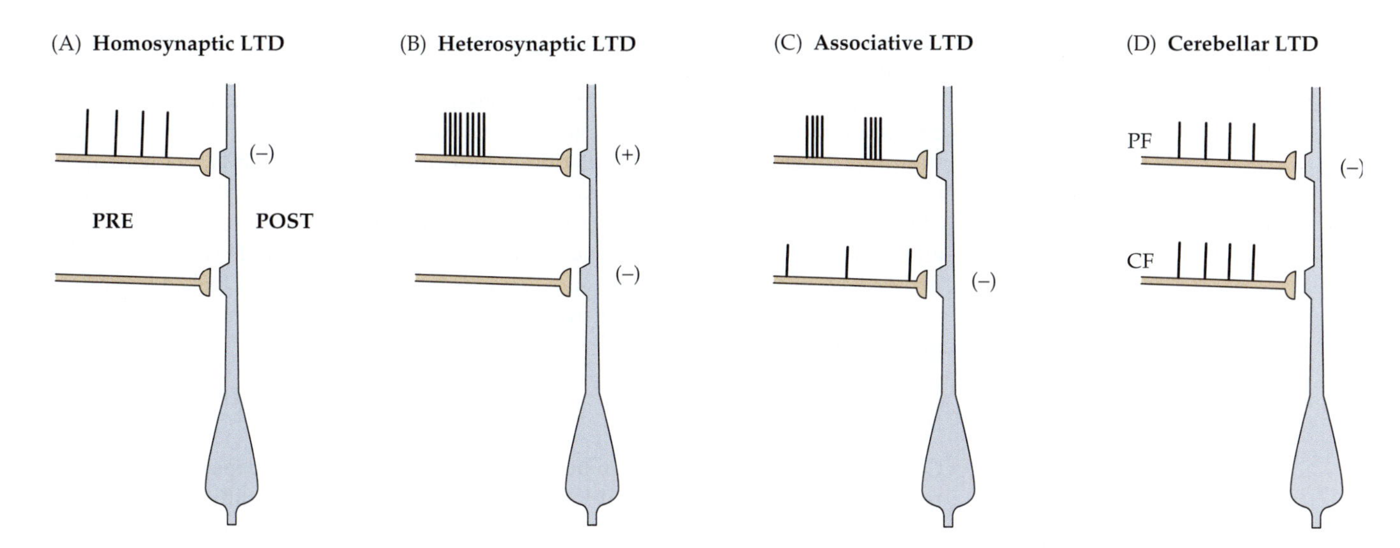

FIGURE 12.10 Types of Long-Term Depression, classified according to stimulus conditions. Symbols indicate potentiation (+) or depression (–) of the synaptic response after the conditioning stimuli. (A) Homosynaptic LTD is produced by prolonged low-frequency stimulation of the same afferent pathway. (B) Heterosynaptic LTD is produced by tetanic stimulation of a neighboring pathway, which may itself be potentiated after the stimulus train. (C) Associative LTD is produced by low-frequency stimulation of the test pathway, together with brief out-of-phase tetani applied to the conditioning pathway. (D) LTD in the cerebellum is produced by coordinate low-frequency stimulation of the climbing-fiber (CF) and parallel-fiber (PF) inputs to Purkinje cells. (After Linden and Connor, 1995.)

cium chelators.[48] However, in other brain locations NMDA antagonists have no effect; instead, induction of LTD appears to involve metabotropic glutamate receptors.[46]

Heterosynaptic LTD is a prolonged depression of synaptic transmission produced by previous activity in a different afferent pathway to the same cell (Figure 12.10B). This form of LTD was first reported as a correlate of homosynaptic LTP induced in Schaffer collateral input to CA1 pyramidal cells;[45] that is, induction of LTP in one pathway resulted in depression of transmission at nearby synapses. Later experiments, using chronic recordings from perforant path–dentate gyrus synapses, showed that the depression could persist for days.[54] In acute preparations, heterosynaptic LTD lasts for hours.[55] The phenomenon requires external calcium[56] and is accompanied by a rise in postsynaptic calcium concentration.[57] In the hippocampus it involves NMDA receptor activation,[46] but it can also be induced by postsynaptic depolarization without activation of NMDA receptors.[52] In the dentate gyrus it is blocked by L-type calcium channel blockers.[58]

Associative LTD has been reported to occur with stimulus protocols similar to those used for associative LTP. Combined weak and strong stimulation of two inputs results in depression of the weakly stimulated input (Figure 12.10C). One difference between the protocols for associative LTP and associative LTD is that during induction of LTD the two stimuli are delivered out of phase. As in associative LTP, postsynaptic depolarizing pulses can substitute for the strong synaptic stimulation.[59] Overall, however, experiments made to demonstrate associative LTD in the hippocampus have yielded inconsistent results, and in some cases have required special stimulating protocols—for example, prior "priming" by so-called theta (5 stimuli/s) stimulation.[43]

LTD in the Cerebellum

One important site of LTD is the cerebellar cortex. There Purkinje cells receive excitatory input from two sources (Chapter 22): Parallel fibers, arising from granule cells, diverge widely to form a large number of synapses on secondary and tertiary dendrites; climbing fibers, from the inferior olivary nucleus, make strong synaptic connections on the soma and proximal dendritic tree. Parallel fibers use glutamate as a transmitter, and their synapses contain both mGluR and AMPA receptors. The transmitter used by the climbing fibers has not been definitely identified. No NMDA receptors are found in the adult cerebellum.[46,60,61]

Cerebellar LTD was first studied by Ito and his colleagues.[62] They applied a series of low-frequency (1–4/s) paired stimuli to parallel-fiber and climbing-fiber pathways (Figure 12.10D) for about 5 min. Subsequent responses to parallel-fiber stimulation were depressed for several hours. In addition, when application of glutamate to the dendritic field was paired with climbing-fiber stimulation, subsequent responses to glutamate were depressed, suggesting that the phenomenon was mediated by postsynaptic changes. Reliable demonstration of cerebellar LTD proved somewhat difficult in vivo, but it was

[54]Krug, M., et al. 1985. *Brain Res.* 360: 264–272.

[55]Colbert, C. M., Burger, B. S., and Levy, W. B. 1992. *Brain Res.* 571: 159–161.

[56]Christofi, G., et al. 1993. *J. Neurophysiol.* 69: 219–229.

[57]Barry, M. F., et al. 1996. *Hippocampus* 6: 3–8.

[58]Christie, B. R., and Abraham, W. C. 1994. *Neurosci. Lett.* 167: 41–45.

[59]Debanne, D., and Thompson, S. M. 1996. *Hippocampus* 6: 9–16.

[60]Levenes, C., Daniel, H., and Crepel, F. 1998. *Prog. Neurobiol.* 55: 79–91.

[61]Daniel, H., Levenes, C., and Krepel, F. 1998. *Trends Neurosci.* 21: 401–407.

[62]Ito, M., Sakurai, M., and Tongroach, P. 1982. *J. Physiol.* 324: 113–134.

established as an unambiguous phenomenon in cerebellar slice preparations[63] and in culture.[64]

It was also shown in both slice preparations and cerebellar cultures that Purkinje cell depolarization, producing calcium action potentials in the dendrites (Chapter 7), could substitute for climbing-fiber stimulation in inducing LTD.[65,66] However, neither climbing-fiber stimulation nor depolarization alone is effective in inducing LTD; coactivation of glutamate receptors, either by parallel-fiber stimulation or by direct application of glutamate, is required. In the case of parallel-fiber stimulation, LTD is input specific: That is, only the stimulated inputs are depressed. LTD induction is blocked by postsynaptic calcium chelators,[67] and there is a large calcium accumulation following climbing-fiber stimulation.[68]

Induction of LTD

Conditions under which LTD can be produced vary considerably, depending on which type is being studied and in what location. As a consequence, it is difficult to come up with a consistent view of the factors underlying its induction. Various schemes have been summarized in a number of review articles.[51,60,61] One consistent feature is that LTD, like LTP, depends on postsynaptic accumulation of calcium. In the hippocampus, calcium entry appears to be primarily through NMDA receptors, although heterosynaptic LTD can be induced by depolarization alone without receptor activation and is attenuated by L-type calcium channel blockers. This suggests that whereas local depolarization and calcium accumulation through NMDA receptors produces LTP at the activated input, spread of depolarization to adjacent synaptic regions produces LTD by calcium entry through voltage-gated calcium channels. In other brain regions, free calcium concentration may be elevated by release from intracellular stores by IP_3 (inositol 1,4,5-trisphosphate) after activation of metabotropic glutamate receptors (Chapter 10). In cerebellar Purkinje cells, where NMDA receptors are absent, calcium entry is through voltage-sensitive calcium channels that generate dendritic action potentials.

Why does calcium accumulation produce LTP in some circumstances and LTD in others? At the moment there is no clear answer to that question, except that the difference appears to be one of concentration: A relatively large increase results in LTP, a smaller increase in LTD. In accordance with this idea is the observation that a stimulus barely able to induce homosynaptic LTP in CA1 cells with normal extracellular calcium concentration produces LTD when extracellular calcium is reduced.[53]

Second Messenger Systems Mediating LTD

The second messenger processes that couple moderate increases in intracellular calcium to LTD are not at all clear. As with LTP, calmodulin inhibitors have been shown to block homosynaptic LTD,[69] and LTD has been reported to be absent in CaMKII mutant mice. In addition, homosynaptic LTD is blocked by postsynaptic injection of protein phosphatase inhibitors.

In the cerebellum, induction of LTD occurs only if both mGluR and AMPA receptors are activated. Since climbing-fiber stimulation can result in massive calcium entry through voltage-activated channels, it is unlikely that the essential role of mGluR activation in Purkinje cells is the release of calcium from intracellular stores by IP_3. Instead it appears to involve the activation of protein kinase C (PKC), as indicated by the observation that cerebellar LTD is blocked by PKC inhibitors. Sodium influx through AMPA receptors seems to play a role in the induction of LTD, since substitution of lithium or cesium for extracellular sodium has been reported to block LTD.[70]

Several studies in cerebellar slice preparations have implicated nitric oxide (NO) production in the induction and maintenance of LTD; however, other studies have failed to substantiate any role for NO. A major difficulty in including NO in the second messenger pathway is that NO synthase, necessary for NO production, is notably absent in Purkinje cells.

[63] Sakurai, M. 1987. *J. Physiol.* 394: 463–480.

[64] Hirano, T. 1990. *Neurosci. Lett.* 119: 141–144.

[65] Crepel, F., and Jaillard, D. 1991. *J. Physiol.* 432: 123–141.

[66] Hirano, T. 1990. *Neurosci. Lett.* 119: 145–147.

[67] Sakurai M. 1990. *Proc. Natl. Acad. Sci. USA* 87: 3383–3385.

[68] Ross, W. N., and Werman, R. 1987. *J. Physiol.* 389: 319–336.

[69] Mulkey, R. M., Herron, C. E., and Malenka, R. C. 1993. *Science* 261: 1051–1055.

[70] Linden, D. J., Smeyne, M., and Connor, J. A. 1993. *Neuron* 10: 1093–1100.

LTD Expression

The expression of LTD involves a reduction in postsynaptic sensitivity. In addition to the reduced sensitivity to applied glutamate already noted, during LTD the amplitudes of miniature excitatory synaptic potentials are often decreased.[71–73] These changes are accompanied by dephosphorylation of the GluR1 subunit of the AMPA receptors,[74] implying a reduction in single-channel conductance. Furthermore, during LTD in cultured hippocampal cells there is a decrease in the number of AMPA receptors clustered in the postsynaptic membrane.[67] Thus, the mechanisms underlying LTD expression appear to be the exact reverse of those responsible for LTP. As with LTP, basic features discussed here are subject to a wide variety of modulating influences.

In addition, as with LTP, there is indication of presynaptic involvement: In many cases the decrease in synaptic potential amplitude is accompanied by an increase in paired-pulse facilitation.[75,76]

Significance of Changes in Synaptic Efficacy

The phenomena discussed here provide the nervous system with built-in mechanisms to regulate synaptic efficacy over a timescale ranging from tens of milliseconds to several days. How these mechanisms are utilized in nervous system function, and their relative significance, are to a large extent unknown. Of interest are the two classes of potentiation. PTP, which is presynaptic in origin, can last for tens of minutes in the ciliary ganglion, and for hours at the vertebrate neuromuscular junction when there is high calcium accumulation in the presynaptic terminals. Such high calcium concentrations could easily be achieved in small presynaptic boutons with a large surface-to-volume ratio. Perhaps part of the confusion in attempting to classify LTP as presynaptic or postsynaptic arises from the fact that at some synapses there may be considerable temporal overlap of PTP and LTP, making them difficult to separate, at least in experiments lasting only a few hours.

It is a matter of basic belief among those who study nervous system function that learning and memory involve long-term changes in synaptic efficacy, and for this reason the mechanisms underlying LTP and LTD are of particular interest. This interest is strengthened further because both phenomena exhibit a characteristic postulated by Donald Hebb to be required for associative learning[77]—namely, that increases in synaptic strength should occur when the presynaptic and postsynaptic elements are coactive. In modern jargon, synapses that exhibit such a property are known as Hebbian, and it is sometimes assumed that if the requirement has been satisfied, then learning has occurred.

A number of correlations have been shown between spatial learning in intact animals and LTP in hippocampal slices.[78–80] For example, both can be blocked by antagonists of NMDA receptors or metabotropic glutamate receptors and by inhibitors of calcium/calmodulin-dependent protein kinase. However, the nature of the behavioral deficit associated with the block is not always clear. For example, rats under NMDA antagonists have general sensorimotor disturbances that interfere with negotiating a water maze (suggesting that learning ability has been compromised), but they can learn it readily if they are first made familiar with the general requirements of the task.[81] Thus, NMDA receptor–mediated LTP does not seem to be an essential requirement. Similar ambiguities have been encountered with gene deletions: Some that eliminate LTP produce a deficit in the spatial learning ability; others do not.[82]

There is growing evidence for the idea that LTP in the amygdala might be a substrate for aversive ("fear") conditioning. Rats trained to associate foot shock with an auditory tone exhibit an exaggerated auditory startle reflex, and cells in the amygdala show an LTP-like increase in their synaptic response to electrical stimulation of the auditory pathway from the medial geniculate nucleus.[83,84] Conversely, induction of LTP at the same synapses by electrical stimulation results in an increase in the response to auditory stimuli.[68] Both effects are blocked by NMDA receptor antagonists.[85,86] In conclusion, while an unequivocal relation between LTP and spatial learning tasks has not been established, LTP may play a role in more discrete learning tasks, such as classical conditioning.

[71]Oliet, S., Malenka, R. C., and Nicoll, R. A. 1996. *Science* 271: 1294–1297.

[72]Carrol, R. C., et al. 1999. *Nature Neurosci.* 2: 454–460.

[73]Murashima, M., and Hirano, T. 1999. *J. Neurosci.* 19: 7326–7333.

[74]Lee, H-K., et al. 1998. *Neuron* 21: 1151–1162.

[75]Berretta, N., and Cherubini, E. 1998. *Eur. J. Neurosci.* 10: 2957–2963.

[76]Domenici, M. R., Berretta, N., and Cherubini, E. 1998. *Proc. Natl. Acad. Sci. USA* 95: 8310–8315.

[77]Hebb, D. O. 1949. *The Organization of Behavior.* Wiley, New York.

[78]Izquierdo, I., and Medina, J. H. 1995. *Neurobiol. Learn. Mem.* 63: 19–32.

[79]Martinez, J. L., Jr., and Derrick, B. E. 1996. *Annu. Rev. Psychol.* 47: 173–203.

[80]Elgersma, Y., and Silva, A. J. 1999. *Curr. Opin. Neurobiol.* 9: 209–213.

[81]Cain, D. P. 1998. *Neurosci. Biobehav. Rev.* 22: 181–193.

[82]Holscher, C. 1999. *J. Neurosci. Res.* 58: 62–75.

[83]Rogan, M. T., Staubil, U. V., and LeDoux, J. E. 1997. *Nature* 390: 604–607.

[84]McKernan, M. G., and Shinnick-Gallagher, P. 1997. *Nature* 390: 607–611.

[85]Miserendino, M. J., et al. 1990. *Nature* 345: 716–718.

[86]Fanselow, M. S., and Kim, J. J. 1994. *Behav. Neurosci.* 108: 210–212.

Summary

- Short periods of synaptic activation can result in facilitation, depression, or augmentation of transmitter release, or a combination of these effects.
- Facilitation decays gradually over a few hundred milliseconds; synaptic depression and augmentation persist for several seconds.
- Facilitation is related to a persistent increase in cytoplasmic calcium concentration in the presynaptic terminal.
- Longer periods of repetitive stimulation result in posttetanic potentiation (PTP) of transmitter release, which can last for tens of minutes and, like facilitation, is mediated by an increase in presynaptic terminal calcium concentration.
- In various parts of the central nervous system, repetitive stimulation can result in long-term potentiation (LTP) or long-term depression (LTD) of synaptic strength.
- The change in synaptic efficacy during LTP or LTD may be homosynaptic, involving only the stimulated input, or heterosynaptic, affecting adjacent synapses on the same dendrite; in addition, heterosynaptic effects may be associative, requiring the coordinate activation of both synapses.
- LTP is produced by an increase in calcium concentration in the postsynaptic cell, and appears to involve both the insertion of new receptors into the postsynaptic membrane and an increase in receptor sensitivity.
- LTD also requires an increase in postsynaptic calcium concentration and appears to be mediated by a decrease in receptor number and sensitivity.
- Both LTP and LTD can also involve changes in transmitter release from the presynaptic terminal.
- Although there are some correlations between LTP and LTD and behavioral tasks involving learning, no unequivocal relation between these long-term synaptic changes and memory formation have been established.

SUGGESTED READING

General Reviews

Daniel, H., Levenes, C., and Krepel, F. 1998. Cellular mechanisms of cerebellar LTD. *Trends Neurosci.* 21: 401–407.

Linden, D. J., and Conner, J. A. 1995. Long-term synaptic depression. *Annu. Rev. Neurosci.* 18: 319–357.

Malenka, R. C., and Nicoll, R. A. 1999. Long-term potentiation—A decade of progress? *Science* 285: 1870–1874.

Original Papers

Barrionuevo, G., and Brown, T. H. 1983. Associative long-term potentiation in hippocampal slices. *Proc. Natl. Acad. Sci. USA* 80: 7347–7351.

Bliss, T. V. P., and Lømo, T. 1973. Long-lasting potentiation of synaptic transmission in the dentate of the anesthetized rabbit following stimulation of the perforant path. *J. Physiol.* 232: 331–356.

delCastillo, J., and Katz, B. 1954. Statistical factors involved in neuromuscular facilitation and depression. *J. Physiol.* 124: 574–585.

Ito, M., Sakurai, M., and Tongroach, P. 1982. Climbing fibre induced depression of both mossy fibre responsiveness and glutamate sensitivity of cerebellar Purkinje cells. *J. Physiol.* 324: 113–134.

Malinow, R., and Tsien, R. W. 1990. Presynaptic enhancement shown by whole-cell recordings of long-term potentiation in hippocampal slices. *Nature* 346: 177–180. [12]

Mallart, A., and Martin, A. R. 1967. Analysis of facilitation of transmitter release at the neuromuscular junction of the frog. *J. Physiol.* 193: 679–697.

Shi, S. H., Hayashi, Y., Petralia, R. S., Zaman, S. H., Wenthold, R. J., Svoboda, K., and Malinow, R. 1999. Rapid spine delivery and redistribution of AMPA receptors after synaptic NMDA receptor activation. *Science* 284: 1811–1816.

Weinrich, D. 1970. Ionic mechanisms of post-tetanic potentiation at the neuromuscular junction of the frog. *J. Physiol.* 212: 431–446.

Zengel, J. E., and Magleby, K. L. 1982. Augmentation and facilitation of transmitter release. A quantitative description at the frog neuromuscular junction. *J. Gen. Physiol.* 80: 582–611.

13 Cellular and Molecular Biochemistry of Synaptic Transmission

AT CHEMICAL SYNAPSES, NEURONS RELEASE NEUROPEPTIDES and low-molecular-weight transmitters, such as acetylcholine. Low-molecular-weight transmitters are synthesized in the axon terminal. A number of mechanisms ensure that their supply is adequate to meet the demands of release; these include storage of transmitters in synaptic vesicles, rapid changes in the activity of enzymes mediating transmitter synthesis, and long-term changes in the number of enzyme molecules in the terminal. Neuropeptides are synthesized and incorporated into vesicles in the cell body, then shipped down the axon for storage and release.

Synaptic vesicles and other organelles move by fast axonal transport toward the terminal (anterograde transport) and back to the cell body (retrograde transport). Slow axonal transport moves cytoplasmic proteins and components of the axonal cytoskeleton from the cell body toward the terminal.

Synaptic vesicle exocytosis occurs by a mechanism for vesicle trafficking and fusion that is highly conserved from yeast to flies to humans, described by the SNARE hypothesis. In nerve terminals, proteins on the synaptic vesicle bind to proteins on the presynaptic membrane to form a complex that docks the vesicle, ready for release. When calcium enters the terminal, it binds to a protein in this complex, thereby triggering membrane fusion and exocytosis.

Neurotransmitter receptors are concentrated in the membrane of the postsynaptic cell immediately beneath the nerve terminal. Cytoskeletal and membrane-associated proteins immobilize transmitter receptors and create a subsynaptic scaffold to which intracellular signaling proteins are recruited.

The final step in chemical synaptic transmission is removal of the transmitter from the synaptic cleft. Low-molecular-weight transmitters are either degraded after release or taken up into glial cells or axon terminals where they are repackaged into vesicles and released again. Neuropeptides are removed by diffusion. Drugs that interfere with transmitter degradation or uptake can have profound effects on signaling, indicating that such removal processes play an important role in synaptic function.

Progress in elucidating the biochemical basis of synaptic transmission was initially slow. Sixty years after the idea of chemical synaptic transmission was first proposed by Elliot, in 1904,[1] only three compounds—acetylcholine, norepinephrine, and γ-aminobutyric acid—had been unequivocally identified as neurotransmitters, and their receptors were simply classified on the basis of pharmacology: nicotinic or muscarinic for acetylcholine, α- or β-adrenergic for norepinephrine. The last few decades have witnessed an explosion in biochemical research on the nervous system, fueled by rapid conceptual and technological advances in biochemistry and genetics. As a result, over 50 compounds have been shown to be neurotransmitters. The protein structure and molecular mechanisms of action of neurotransmitter receptors have been established (Chapters 2 and 3), and the ways in which transmitters mediate or modulate synaptic transmission have been characterized (Chapters 9 and 10).

In this chapter we focus on the proteins and organelles that mediate transmitter synthesis, storage, release, and removal (Figure 13.1). Although many were identified in neural tissues, some were first characterized in very different contexts. Remarkable examples are the proteins mediating transmitter release, which were identified on the basis of their role in transport in the Golgi apparatus and secretion from yeast cells.

NEUROTRANSMITTERS

The Identification of Transmitters

Pioneering experiments by Langley, Loewi, Feldberg, and Dale identified acetylcholine as the transmitter at the vertebrate skeletal neuromuscular junction and in autonomic ganglia (Chapter 9). Definitive experiments by Cannon, von Euler, and Peart identified norepinephrine as the transmitter released by sympathetic neurons,[2] and those of Florey, Kuffler, Kravitz, and their colleagues established γ-aminobutyric acid as an inhibitory transmitter at the crustacean neuromuscular junction.[3] Such studies involved readily accessible peripheral synapses where it is possible to stimulate selectively the presynaptic axon, to record intracellularly from the postsynaptic cell, to apply chemicals to the postsynaptic membrane through a micropipette, and to collect transmitter from release sites. Thus it could be shown that a molecule suspected to be the transmitter was synthesized by the presynaptic cell, stored in the axon terminal, released by nerve stimulation, and mimicked the effects of the endogenous transmitter by physiological and pharmacological tests.

In contrast to such peripheral structures, regions of the central nervous system contain different types of neurons intermingled with each other. This greatly complicates transmitter identification. Fortunately, cells that use the same transmitter are sometimes grouped together, with axons traveling in bundles to end in well-defined regions. Hence it becomes possible to stimulate identified populations of axons and to collect the transmitter they release. Such experiments can be made in the intact CNS using, for example, microdialysis[4,5] or a push–pull cannula[6,7] to perfuse synaptic sites. Alternatively, slices of CNS tissue, 100 μm or so in thickness, can be cut in order to stimulate particular fiber tracts while recording from postsynaptic cells. Such slices can be maintained in vitro while transmitters are applied through a micropipette and collected from the perfusate[8] or detected by extraordinarily sensitive amperometric techniques.[9]

[1]Elliot, T. R. 1904. *J. Physiol.* 31: (Proc.) xx–xxi.

[2]von Euler, U. S. 1956. *Noradrenaline.* Charles Thomas, Springfield, IL.

[3]Hall, Z. W., Hildebrand, J. G. and Kravitz, E. A. 1974. *Chemistry of Synaptic Transmission.* Chiron Press, Newton, MA.

[4]Ding, R., Asada, H., and Obata, K. 1998. *Brain Res.* 800: 105–113.

[5]Miele, M., Boutelle, M. G., and Fillenz, M. 1996. *J. Physiol.* 497: 745–751.

[6]Perschak, H. and Cuenod, M. 1990. *Neuroscience* 35: 283–287.

[7]Myers, R. D., Adell, A., and Lankford, M. F. 1998. *Neurosci. Biobehav. Rev.* 22: 371–387.

[8]Vollenweider, F. X., Cuenod, M. and Do, K. Q. 1990. *J. Neurochem.* 54: 1533–1540.

[9]Jaffe, et al. 1998. *J. Neurosci.* 18: 3548–3553.

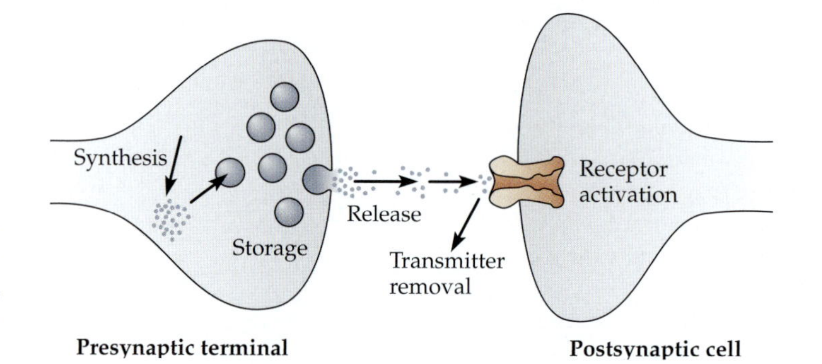

FIGURE 13.1 Chemical Synaptic Transmission. At chemical synapses, neurotransmitters are synthesized, stored in synaptic vesicles, and released by exocytosis. Transmitters diffuse across the synaptic cleft, activate receptors on the postsynaptic cell, and then are removed by diffusion, uptake, or degradation.

Additional techniques for identifying transmitters at CNS synapses and visualizing the distribution of cells that release them have been developed and are described in Chapter 14. Many of these rely on detecting the expression of proteins that mediate different aspects of chemical synaptic transmission.

Neurotransmitters as Messengers

The same neurotransmitter may be released at many synapses in different regions of the nervous system; the message conveyed by its release is determined by the precise way in which neurons are interconnected. For example, the release of acetylcholine from motor nerve terminals in a gastrocnemius muscle causes extension of the foot, whereas acetylcholine released at certain synapses in the CNS may enable you to remember a scene from your childhood. On the other hand, some neuropeptides are found in only a very small number of cells, which subserve a specific function. In such cases, diffuse application of the neuropeptide may elicit a characteristic behavior, determined by the distribution and properties of its receptors on central neurons. For example, infusing neuropeptide Y into the cerebrospinal fluid causes overeating and obesity.[10]

Transmitter Molecules

Transmitters can be divided into three groups. One group contains the "classic" neurotransmitters, low-molecular-weight substances that are stored in vesicles in nerve terminals. This group includes acetylcholine, norepinephrine, epinephrine, dopamine, 5-hydroxytryptamine (5-HT, or serotonin), histamine, adenosine triphosphate (ATP), and the amino acids γ-aminobutyric acid (GABA), glutamate, and glycine (Figure 13.2). A second group of transmitters stored in vesicles consists of the neuropeptides (Table 13.1), more than 40 of which have been identified in the mammalian CNS. Nitric oxide (NO) and carbon monoxide (CO) constitute a third group of transmitters that are highly lipid- and water-soluble gases. Accordingly, they cannot be stored in cells. It seems highly likely that more transmitter substances remain to be discovered.

Many substances that were first characterized as neurotransmitters in invertebrates or at peripheral synapses in vertebrate nervous systems have been shown to act in the mammalian CNS. It is remarkable that the same transmitters are

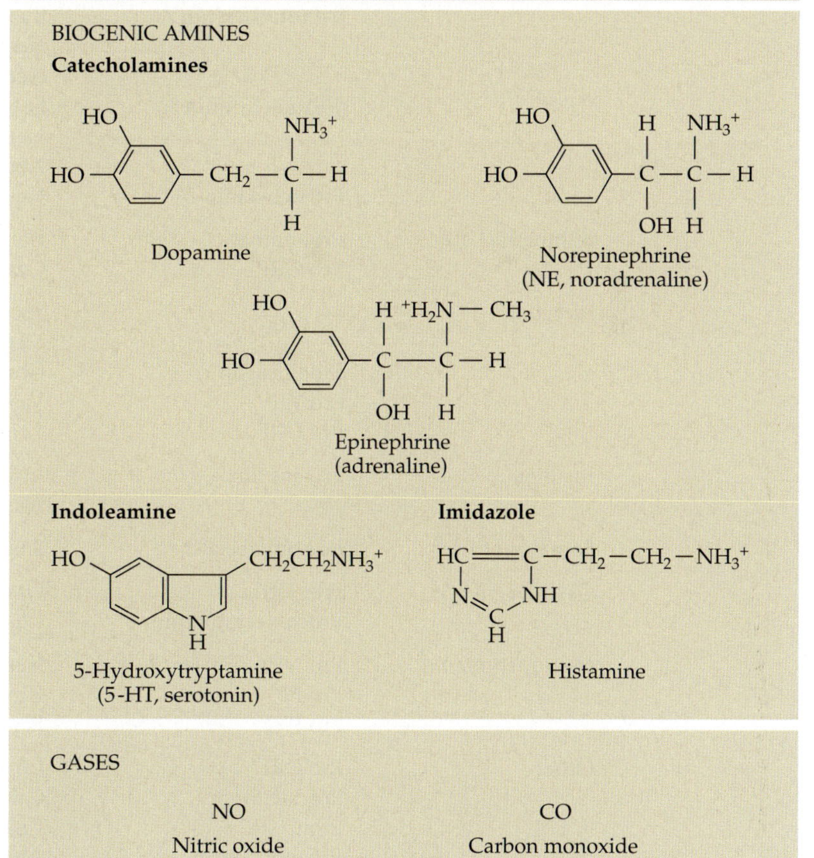

FIGURE 13.2 Low-Molecular-Weight Neurotransmitters.

[10]Zakrzewska, K. E., et al. 1999. *Endocrinology* 140: 3183–3187.

Table 13.1
Neuropeptides in mammalian nervous systems

Family	Precursor	Neuropeptide
Opioid	Pro-opiomelanocortin (POMC)	Corticotropin (ACTH)
		β-Lipotropin
		α-MSH
		α-Endorphin
		β-Endorphin
		γ-Endorphin
	Pro-enkephalin	Met-enkephalin
		Leu-enkephalin
	Prodynorphin	α-Neoendorphin
		β-Neoendorphin
		Dynorphin A
		Dynorphin B (rimorphin)
		Leumorphin
Neurohypophyseal	Provasopressin	Vasopressin
		Neurophysin II
	Pro-oxytocin	Oxytocin
		Neurophysin I
Tachykinins	α-Protachykinin A	Substance P
	β-Protachykinin A	Substance P
		Neurokinin A
		Neuropeptide K
	γ-Protachykinin A	Substance P
		Neurokinin A
		Neuropeptide γ
	Protachykinin B	Neurokinin B
Bombesin/GRP	Probombesin	Bombesin
	Pro GRP	Gastrin releasing peptide (GRP)
Secretins	—	Secretin
	—	Motilin
	Proglucagon	Glucagon
	Pro VIP	Vasoactive intestinal peptide (VIP)
	Pro GRF	Growth hormone–releasing factor (GRF)
Insulins	Pro-insulin	Insulin
	—	Insulin-like growth factors
Somatostatins	Prosomatostatin	Somatostatin
Gastrins	Progastrin	Gastrin
	Procholecystokinin	Cholecystokinin (CCK)
Neuropeptide Y	Pro NPY	Neuropeptide Y (NPY)
	Pro PP	Pancreatic polypeptide (PP)
	Pro PYY	Peptide YY (PYY)
Other	Pro CRF	Corticotropin-releasing factor (CRF)
	Procalcitonin	Calcitonin
	Pro CGRP	Calcitonin gene-related peptide (CGRP)
	Proangiotensin	Angiotensin
	Probradykinin	Bradykinin
	Pro TRH	Thyrotropin-releasing hormone (TRH)
	—	Neurotensin
	—	Galanin
		Luteinizing hormone–releasing hormone (LHRH)

found so widely distributed in the animal kingdom, from leeches and insects to lampreys and mammals.

NEUROTRANSMITTER SYNTHESIS

Where are transmitter molecules synthesized, and how are transmitter stores maintained and replenished? Are transmitters shipped to the nerve terminal ready-made, or are they assembled there from precursors provided by the cell body? The answers to such questions are different for different transmitters. Classic low-molecular-weight transmitters are produced within the axon terminal from common cellular metabolites and are incorporated into small synaptic vesicles (50 nm in diameter) for storage and release. NO and CO are synthesized within terminals as well. However, since they cannot be packaged in vesicles, NO and CO immediately diffuse out of nerve terminals to act on their targets (Chapter 10). Neuropeptide transmitters, on the other hand, are synthesized in the cell body, packaged in large dense-core vesicles (100–200 nm in diameter), and shipped down the axon.

An added complication is that many, perhaps all, neurons release more than one transmitter, typically a low-molecular-weight transmitter and one or more neuropeptides.[11,12] Figure 13.3 shows the localization of 5-HT and the neuropeptides FLRFamide and allatostatin within individual neurons in the CNS of the moth.[13] At many synapses such cotransmitters have been shown to act synergistically (Chapter 16).[11,14] Low-molecular-weight transmitters are released by single impulses, whereas trains of impulses are often required to release neuropeptide cotransmitters.

There is also considerable evidence that many neurons release more than one low-molecular-weight transmitter, an idea that would have been greeted with ridicule only a few years ago. For example, some spinal interneurons release both glycine and GABA; each inhibits the postsynaptic target through distinct receptors.[15] Other spinal neurons

[11]Lundberg, J. M. 1996. *Pharmacol. Rev.* 48: 113–178.

[12]Bondy, C. A., et al. 1989. *Cell. Mol. Neurobiol.* 9: 427–446.

[13]Homberg, U. and Hildebrand, J. G. 1989. *J. Comp. Neurol.* 288: 243–253.

[14]Whim, M. D., Church, P. J., and Lloyd, P. E. 1993. *Mol. Neurobiol.* 7: 335–347.

[15]Jonas, P., Bischofberger, J., and Sandkuhler, J. 1998. *Science* 281: 419–424.

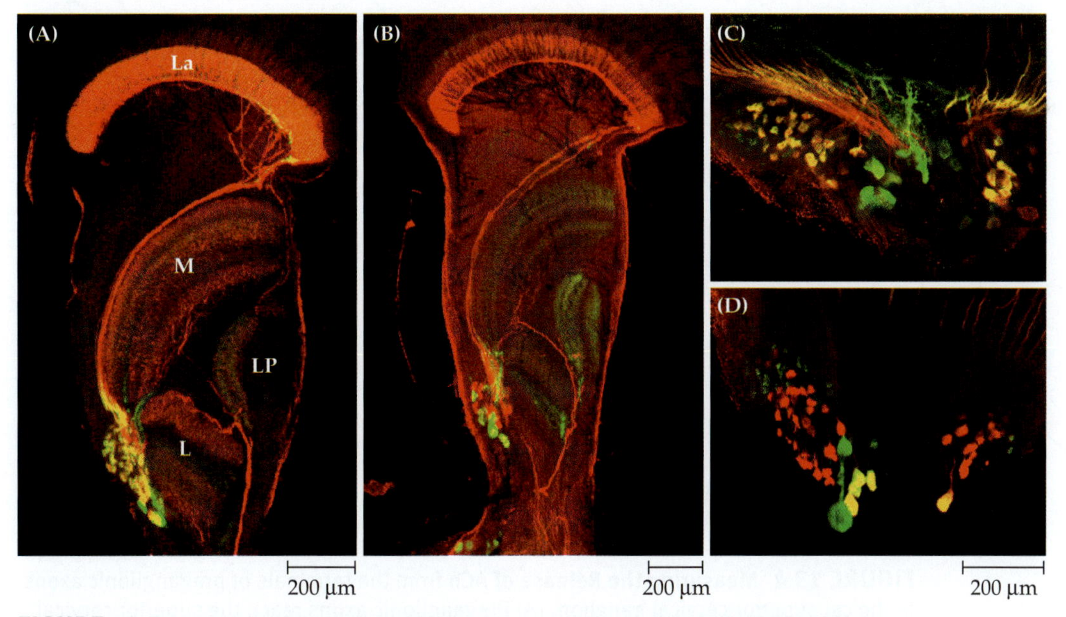

FIGURE 13.3 5-Hydroxytryptamine, FLRFamide, and Allatostatin in Individual Nerve Cells of the moth, *Manduca*. (A, C) Fluorescence confocal micrographs of a section of the moth optic lobe labeled with antisera to 5-HT (red) and to the neuropeptide FLRFamide (green). (A) Longitudinal section. Cells in the OL2 cluster (lower left) project mostly to the lamina (La). (C) Frontal view of cell cluster. Most cells contain both 5-HT and FLRFamide, and so appear yellow. (B, D) Micrographs of a section labeled with antisera to 5-HT (red) and to the neuropeptide allatostatin (green). (B) Longitudinal section. Cells containing 5-HT and/or allatostatin project to the lamina, the medulla (M), the lobula (L), and the lobula plate (LP). (D) Frontal view of cell cluster. Colocalization of 5-HT and allatostatin occurs in a small subpopulation of cells. (Micrographs kindly provided by N. T. Davis and J. G. Hildebrand.)

co-release ATP as an excitatory transmitter and GABA as an inhibitory transmitter.[16] A remarkable finding is that many so-called monoaminergic neurons in the vertebrate central nervous system (those containing dopamine, serotonin, or norepinephrine) also contain glutaminase, the enzyme mediating glutamate synthesis in axon terminals that release glutamate (discussed later in this section), suggesting that they release both transmitters.[17,18] For example, neurons in the raphe nuclei release both serotonin and glutamate from a uniform set of synapses, which contain two different vesicle types. Ventral midbrain dopamine-containing neurons make two different types of terminals. Many contain both glutamate, which produces in their targets fast excitatory postsynaptic potentials, and dopamine, which diffuses over a wider area to evoke slower modulatory actions. Others contain only glutamate.

Synthesis of ACh

One of the first thorough investigations of how transmitters are accumulated in nerve terminals and how transmitter stores are maintained during periods of activity was made by Birks and MacIntosh in their studies of acetylcholine in the terminals of preganglionic axons in the superior cervical ganglion of the cat (Figure 13.4A and B; see also Chapter 16).[19] They cannulated the carotid artery and the jugular vein, perfused the ganglion with solutions containing anticholinesterase, and analyzed the perfusate for acetylcholine. A small amount of acetylcholine was continually released from the ganglion at rest, amounting to 0.1% of the total contents each minute (Figure 13.4C). The fact that the level of acetylcholine in the ganglion remained constant meant that acetylcholine was synthesized continually at rest. (Subsequently it was shown that the ongoing rate of synthesis of acetylcholine at rest, determined by monitoring the incorporation of radioactively

[16]Jo, Y. H., and Schlichter, R. 1999. *Nature Neurosci.* 2: 241–245.

[17]Kaneko, T., et al. 1990. *Brain Res.* 507: 151–154.

[18]Sulzer, D., et al. 1998. *J. Neurosci.* 18: 4588–4602.

[19]Birks, R. I., and MacIntosh, F. C. 1961. *J. Biochem. Physiol.* 39: 787–827.

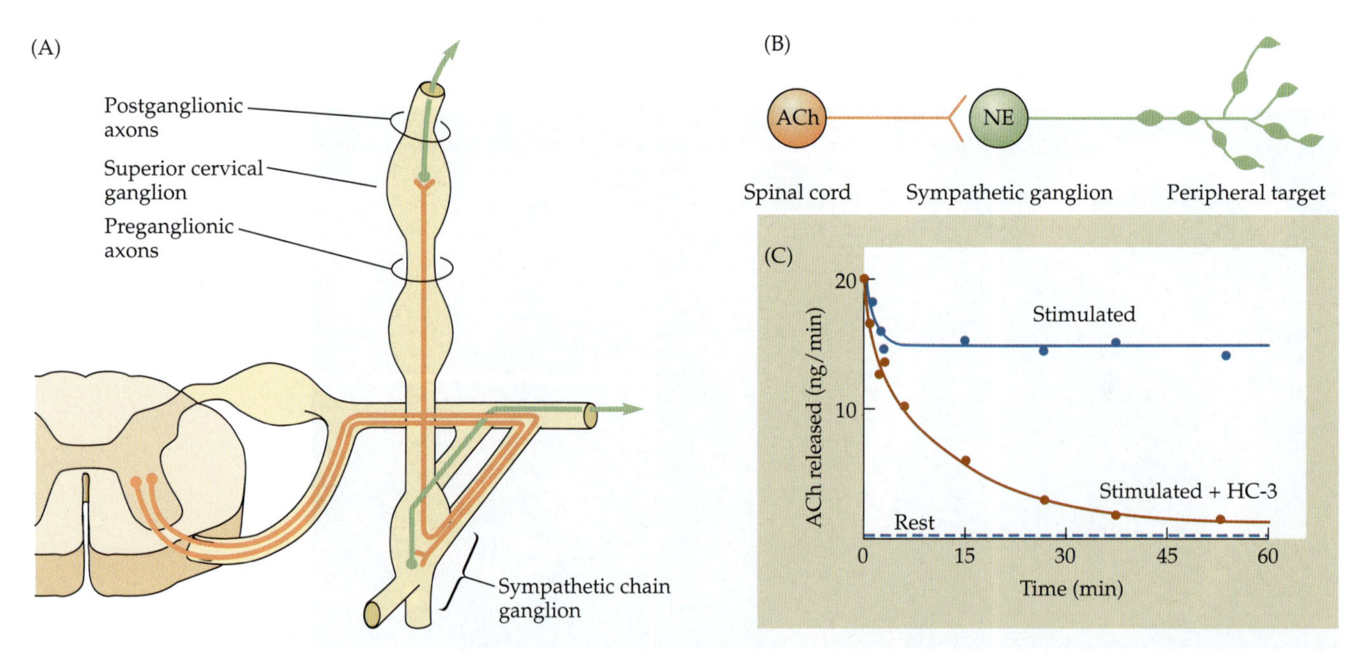

FIGURE 13.4 Measuring the Release of ACh from the terminals of preganglionic axons in the cat superior cervical ganglion. (A) Preganglionic axons reach the superior cervical ganglion from more posterior ganglia in the sympathetic chain. (B) Preganglionic neurons, whose cell bodies lie in the spinal cord, release acetylcholine as a transmitter at synapses in sympathetic ganglia. Ganglion cells release norepinephrine (NE) from varicosities along their processes in the periphery. (C) Release of ACh from a cat sympathetic ganglion perfused with oxygenated plasma containing 3×10^{-5} *M* eserine to inhibit acetylcholinesterase. In control medium, preganglionic stimulation at 20/s causes a sustained, 100-fold increase in the rate of ACh release, compared to release at rest. Release decreases rapidly during stimulation in the presence of 2×10^{-5} M hemicholinium (HC-3), which blocks choline uptake. (After Birks and MacIntosh, 1961.)

labeled choline into acetylcholine, is so high that an amount equal to the entire store of acetylcholine is degraded and resynthesized within the axon terminals every 20 min.[20])

Birks and MacIntosh then stimulated the preganglionic nerve with long trains of impulses and found that the quantity of acetylcholine released from the ganglion increased 100-fold, so that an amount corresponding to 10% of the original content was released each minute (see Figure 13.4C). Remarkably, this rate of release was maintained for over an hour with no change in the level of acetylcholine in the ganglion. Thus, during an hour of stimulation an axon terminal can release an amount of acetylcholine equal to many times its original content without having its stores depleted.

The only exogenous ingredient the nerve terminals need to maintain their stores of acetylcholine under such conditions is choline, which is taken up from the surrounding fluid via an active transport process (Figure 13.5). The requirement for extracellular choline was demonstrated both by perfusing the preparation with solutions lacking choline and by blocking choline uptake into the axon terminals with hemicholinium (HC-3). In both cases the level of acetylcholine in the ganglion and the amount released by stimulation fell rapidly (see Figure 13.4C).

How is acetylcholine synthesis controlled to meet the demands of release? Our understanding of the mechanisms regulating acetylcholine synthesis and storage in cholinergic nerve terminals is surprisingly limited. The enzymatic reactions are summarized in Figure 13.5 and shown in detail in Appendix B. Acetylcholine is synthesized from choline and acetyl CoA (acetyl coenzyme A) by the enzyme choline acetyltransferase and is hydrolyzed to choline and acetate by acetylcholinesterase. Both enzymes are found in the cytosol. Because the reaction catalyzed by choline acetyltransferase is reversible, one factor controlling the level of acetylcholine is the **law of mass action**. For example, a fall in acetylcholine concentration caused by release would favor net synthesis until equilibrium was reestablished. However, the regulatory mechanisms at work within cholinergic axon terminals are more complex than this. For example, under resting conditions the accumulation of acetylcholine is limited by ongoing hydrolysis by intracellular acetylcholinesterase; inhibition of acetylcholinesterase within nerve terminals causes their content of acetylcholine to increase.[19,20] Thus, the level to which acetylcholine accumulates represents a steady state between ongoing synthesis and degradation. This is a common feature

[20]Potter, L. T. 1970. *J. Physiol.* 206: 145–166.

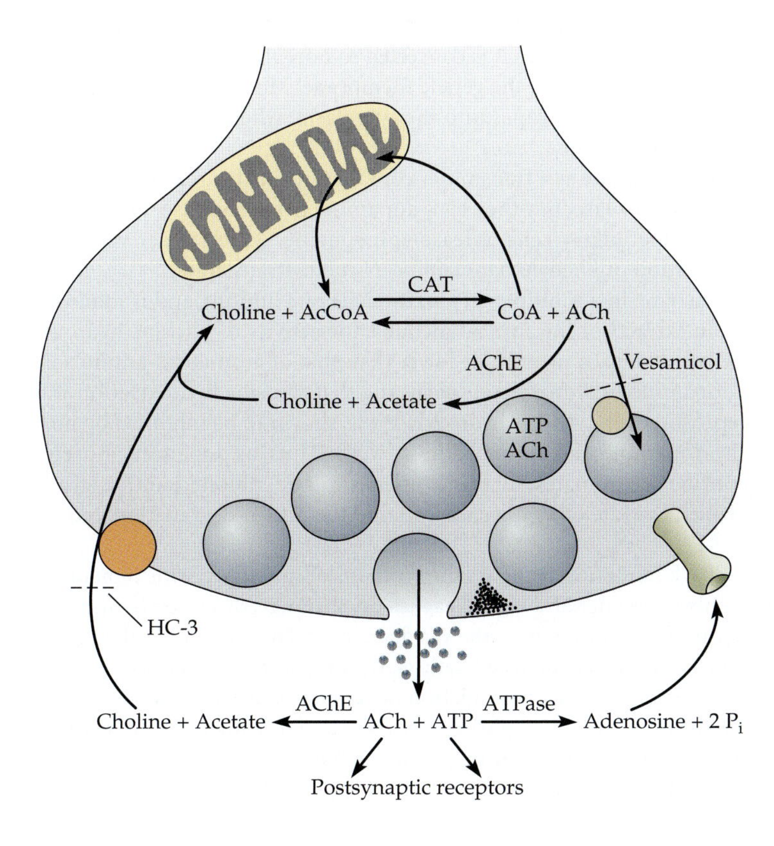

FIGURE 13.5 Pathways of Acetylcholine Synthesis, Storage, Release, and Degradation. Acetylcholine is synthesized from choline and acetyl coenzyme A (AcCoA) by choline acetyltransferase (CAT) and is degraded by acetylcholinesterase (AChE). AcCoA is synthesized primarily in mitochondria; choline is supplied by a high-affinity active transport system that can be inhibited by hemicholinium (HC-3). ACh is packaged into vesicles together with ATP for release by exocytosis. Transport of ACh into vesicles is blocked by vesamicol. Vesicular ACh is protected from degradation. After release, ACh is degraded by extracellular AChE to choline and acetate. About half of the choline transported into cholinergic axon terminals comes from the hydrolysis of ACh that has been released. At some synapses, ATP combines with postsynaptic receptors. ATP is hydrolyzed by extracellular ATPases to adenosine and phosphate (P_i); adenosine can combine with presynaptic receptors to modulate release.

of the metabolism of low-molecular-weight transmitters. Although it seems wasteful, such constant turnover may be an unavoidable consequence of the mechanisms that ensure an adequate supply of transmitter is always available.

Much of the ACh in nerve terminals is sequestered in synaptic vesicles, whereas ACh synthesis and degradation occur in the cytosol. Thus, to have an effect on the rate of synthesis, the release of acetylcholine must reduce the cytoplasmic concentration of ACh. This presumably occurs by the movement of cytoplasmic ACh into newly formed vesicles. Similar interplay between cytoplasmic synthesis and vesicular storage and release is another common feature of the metabolism of low-molecular-weight transmitters.

In cholinergic nerve terminals in the central nervous system the supply of choline, the supply of the cosubstrate acetyl CoA (made in mitochondria), and the activity of choline acetyltransferase have also been shown to regulate the rate of acetylcholine synthesis.[21,22]

Synthesis of Dopamine and Norepinephrine

Another mechanism by which the rate of synthesis of substances in cells is controlled is **feedback inhibition**, in which the rate-limiting step in a biosynthetic pathway is inhibited by the final product. A good example comes from studies by von Euler, Axelrod, Udenfriend, and their colleagues on the synthesis, storage, and release of norepinephrine (NE) in sympathetic neurons and in secretory cells of the adrenal medulla.[23] Adrenal medullary cells resemble sympathetic neurons in many ways: They share the same embryonic origin, they are innervated by cholinergic axons that originate in the central nervous system, and they release a catecholamine in response to stimulation. (The term **catecholamine** is used to designate collectively the substances DOPA, dopamine, norepinephrine, and epinephrine, all of which contain a catechol nucleus—a benzene ring with two adjacent hydroxyl groups—and an amino group; Appendix B.) Mammalian sympathetic neurons release norepinephrine (those in frog release epinephrine); adrenal medullary cells release epinephrine as well as norepinephrine.

Norepinephrine is synthesized from the common cellular metabolite tyrosine in a series of three steps: Tyrosine is converted to DOPA by the enzyme tyrosine hydroxylase, DOPA to dopamine by aromatic L-amino acid decarboxylase, and dopamine to norepinephrine by dopamine β-hydroxylase (Figure 13.6; see also Appendix B). The conversion of tyrosine to DOPA and DOPA to dopamine occurs in the cytoplasm. Dopamine is then transported into synaptic vesicles, where it is converted to norepinephrine by dopamine β-hydroxylase, which is associated with the vesicle membrane. Much of the norepinephrine is stored within vesicles; some escapes into the cytoplasm, where it is susceptible to degradation by monoamine oxidase.

Neurons that release dopamine as a transmitter contain tyrosine hydroxylase and aromatic L-amino acid decarboxylase, but they lack dopamine β-hydroxylase. Other neurons, as well as adrenal medullary cells, release epinephrine, which is derived from norepinephrine by the action of phenylethanolamine-*N*-methyltransferase.

Typically the first enzyme in a multiple-step pathway is rate limiting and is inhibited by the final product. In extracts of the adrenal medulla, the activity of tyrosine hydroxylase was found to be two orders of magnitude lower than that of aromatic L-amino acid decarboxylase and dopamine β-hydroxylase, suggesting that tyrosine hydroxylation was the rate-limiting step. Moreover, tyrosine hydroxylase was shown to be inhibited by norepinephrine (and by dopamine and epinephrine as well). Thus, as dopamine, norepinephrine, or epinephrine accumulates, further synthesis will be inhibited until a steady state is reached, at which the rate of synthesis is equal to the rate of degradation and release (Figure 13.6).

Evidence that feedback inhibition regulates the synthesis of norepinephrine in neurons came from experiments by Weiner and his colleagues on terminals of sympathetic axons innervating the smooth muscles of a duct, the vas deferens.[24] They measured the rate of norepinephrine synthesis in the terminals by bathing the preparation in radioactively labeled precursors and monitoring the accumulation of radioactively labeled norepinephrine. They found that the rate of norepinephrine synthesis was more than threefold greater if the first enzymatic step was bypassed by providing DOPA rather than tyrosine as the precursor. This confirmed that the conversion of tyrosine to DOPA was rate limiting.

[21]Jope, R. 1979. *Brain Res. Rev.* 1: 313–344.

[22]Parsons, S. M., et al. 1987. *Ann. N.Y. Acad. Sci.* 493: 220–233.

[23]Axelrod, J. 1971. *Science* 173: 598–606.

[24]Weiner, N., and Rabadjija, M. 1968. *J. Pharmacol. Exp. Ther.* 160: 61–71.

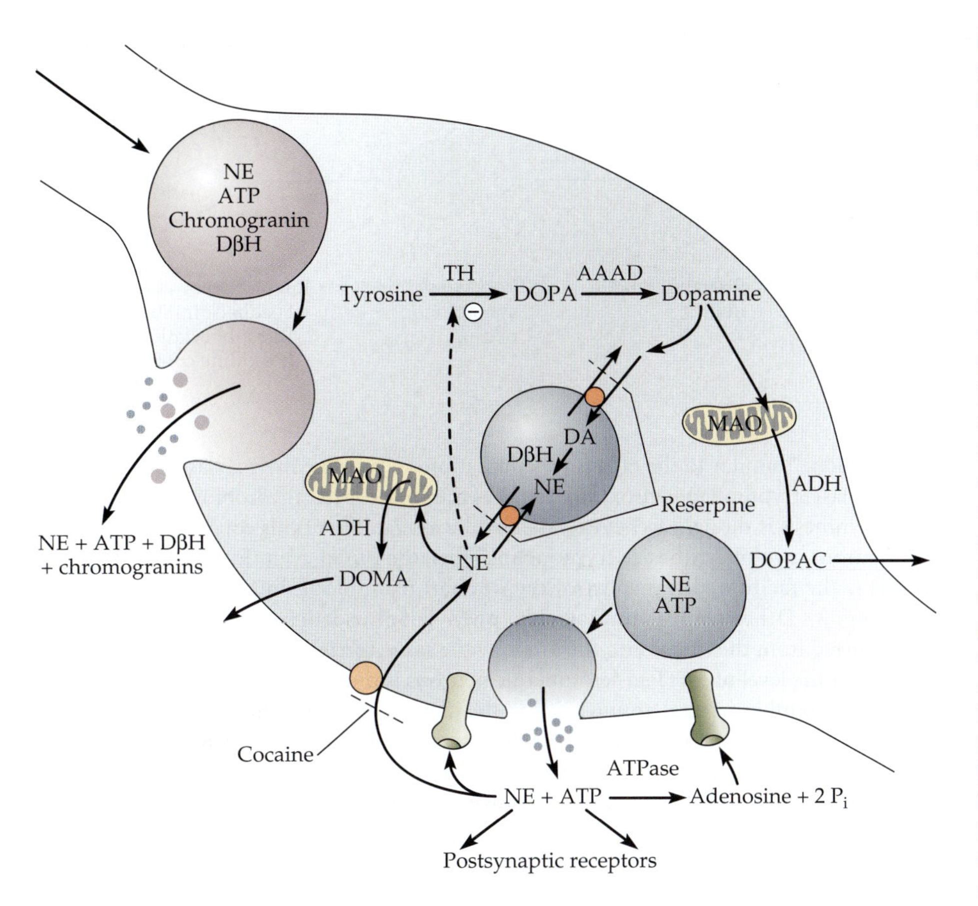

FIGURE 13.6 Pathways of Norepinephrine Synthesis, Storage, Release, and Uptake. Tyrosine is converted to DOPA by tyrosine hydroxylase (TH). DOPA is converted to dopamine (DA) by aromatic L-amino acid decarboxylase (AAAD). Dopamine is transported into vesicles, where it is converted to norepinephrine (NE) by dopamine β-hydroxylase (DβH). Norepinephrine inhibits TH, thus regulating synthesis by feedback inhibition. Transport of dopamine and norepinephrine into vesicles is blocked by reserpine. Vesicles also contain ATP (large dense-core vesicles contain soluble DβH and chromogranins as well). All soluble components of vesicles are released together. NE, ATP, adenosine, and peptides derived from chromogranins can bind to pre- or postsynaptic receptors. After release, norepinephrine is transported back into the varicosity by an uptake mechanism that is blocked by cocaine. Norepinephrine in the cytoplasm can be repackaged into vesicles for release. Within the varicosity, monoamine oxidase (MAO) and aldehyde dehydrogenase (ADH) degrade norepinephrine to 3,4-dihydroxymandelic acid (DOMA) and dopamine to 3,4-dihydroxyphenylacetic acid (DOPAC).

To test the idea that the rate-limiting step was controlled by feedback inhibition, they varied the concentration of norepinephrine in the cytoplasm in two ways. First, taking advantage of the fact that sympathetic axon terminals have a specific transport mechanism for NE, they added it to the bathing fluid, which caused an increase in NE concentration in the terminals. This decreased the rate at which norepinephrine was synthesized from tyrosine. Conversely, nerve stimulation, which lowers the concentration of NE in the cytoplasm, increased the rate of conversion of tyrosine to NE almost twofold. No such increase was seen, however, if NE was added to the bath during nerve stimulation. Apparently uptake from the medium was sufficient to maintain the level of norepinephrine in the axon terminals and so limit its biosynthesis.

Additional factors affect catecholamine synthesis (Figure 13.7). When axon terminals are stimulated to release norepinephrine, tyrosine hydroxylase acquires a higher affinity for its cofactor tetrahydrobiopterin (Appendix B) and becomes less sensitive to inhibition by NE.[25] These changes are associated with a reversible phosphorylation of the enzyme by kinases activated by the influx of calcium ions.[26,27] An additional factor that regulates tyrosine hydroxylase activity is the concentration of tetrahydrobiopterin, which is synthesized from guanosine triphosphate.[28] Thus, a variety of mechanisms act to ensure that the rate of synthesis of norepinephrine meets the demands of release.

Synthesis of 5-HT

Serotonin is synthesized from tryptophan. The first step, conversion of tryptophan to 5-hydroxytryptophan (5-HTP) by the enzyme tryptophan hydroxylase, is rate limiting (Figure 13.8; see also Appendix B).[29] 5-HTP is decarboxylated to serotonin (5-HT) by aromatic L-amino acid decarboxylase, the same enzyme that converts DOPA to dopamine. Stimulation of neurons releasing 5-HT causes an increase in the rate of conversion of tryptophan to 5-hydroxytryptophan. It has been suggested that this is due to changes in the properties of tryptophan hydroxylase caused by calcium-dependent phos-

[25]Joh, T. H., Park, D. H., and Reis, D. J. 1978. *Proc. Natl. Acad. Sci. USA* 75: 4744–4748.

[26]Zigmond, R. E., Schwarzchild, M. A., and Rittenhouse, A. R. 1989. *Annu. Rev. Neurosci.* 12: 415–461.

[27]Nagatsu, T. 1995. *Essays Biochem.* 30: 15–35.

[28]Nagatsu, T., and Ichinose, H. 1999. *Mol. Neurobiol.* 19: 79–96.

[29]Boadle-Biber, M. C. 1993. *Prog. Biophys. Mol. Biol.* 60: 1–15.

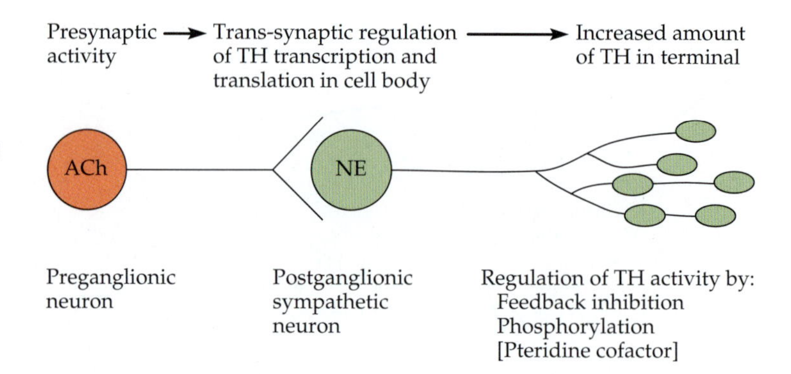

FIGURE 13.7 Regulation of Tyrosine Hydroxylase in sympathetic neurons. The expression of tyrosine hydroxylase (TH) is influenced by the activity of the presynaptic neuron, a process referred to as trans-synaptic regulation. This determines the amount of TH present in the cell and nerve terminal. Within nerve terminals, there is local control of tyrosine hydroxylase activity.

phorylation,[30] similar to the effects of stimulation on tyrosine hydroxylase. Like tyrosine hydroxylase, tryptophan hydroxylase requires the cofactor tetrahydrobiopterin, and serotonin synthesis is thought to be regulated by the availability of this cofactor.

Neurons cannot synthesize tryptophan. Thus, the initial event leading to 5-HT synthesis is the facilitated transport of tryptophan from blood into cerebrospinal fluid (Chapter 8). Other neutral amino acids (phenylalanine, leucine, and methionine) are transported from the blood into the brain by the same carrier. Thus, an important determinant of the level of 5-HT in serotonergic neurons is the relative amount of tryptophan compared to other neutral amino acids in the diet. As a result, behaviors associated with 5-HT function (Chapter 14) are particularly sensitive to dietary influences.[31] For example, volunteers fed a low-protein diet for a day and then given a tryptophan-free amino acid mixture showed an increase in aggressive behavior[32] and changes in sleep cycle.[33]

Synthesis of GABA

GABA is synthesized from glutamate by the enzyme glutamic acid decarboxylase (GAD). This reaction was first characterized as part of the so-called GABA shunt, a series of reactions by which α-ketoglutarate could be converted to succinate (Figure 13.9). The GABA shunt was originally considered to be a brain-specific pathway for glucose metabolism that bypassed part of the Krebs cycle (hence the term "shunt"). The discovery that GABA is the major inhibitory transmitter in the brain,[34] together with the finding that glutamic acid decarboxylase is found only in neurons releasing GABA, suggests that the GABA shunt is not of general importance in glucose metabolism.

Kravitz and his colleagues showed that in crustacean inhibitory neurons, physiological levels of GABA inhibit glutamic acid decarboxylase, indicating that feedback inhibition regulates the accumulation of GABA.[35] Several additional regulators of GABA synthesis have been identified in the mammalian brain, including ATP, inorganic phosphate, and the cofactor pyridoxal phosphate.[36] Two forms of glutamic acid decarboxylase are present in brain:[37] GAD_{67} has a high affinity for pyridoxal phosphate and so may be constitutively active; GAD_{65} has a lower affinity, and its activity may be rapidly regulated by cofactor availability. Mutant mice lacking GAD_{65} have normal behavior and levels of GABA but are slightly more susceptible to seizures. GAD_{67} knockouts show a substantial reduction in brain GABA and die shortly after birth of severe cleft palate.[38]

[30]Hamon, M., et al. 1981. *J. Physiol. (Paris)* 77: 269–279.

[31]Sandyk, R. 1992. *Int. J. Neurosci.* 67: 127–144.

[32]Moeller, F. G., et al. 1996. *Psychopharmacology* 126: 96–103.

[33]Voderholzer, U., et al. 1998. *Neuropsychopharmacology* 18: 112–124.

[34]Roberts, E. 1986. In *Benzodiazepine/GABA Receptors and Chloride Channels: Structural and Functional Properties*. Alan R. Liss, New York, pp. 1–39.

[35]Hall, Z. W., Bownds, M. D., and Kravitz, E. A. 1970. *J. Cell Biol.* 46: 290–299.

[36]Martin, D. L. 1987. *Cell. Mol. Neurobiol.* 7: 237–253.

[37]Erlander, M. G., et al. 1991. *Neuron* 7: 91–100.

[38]Asada, H., et al. 1997. *Proc. Natl. Acad. Sci. USA* 94: 6496–6499.

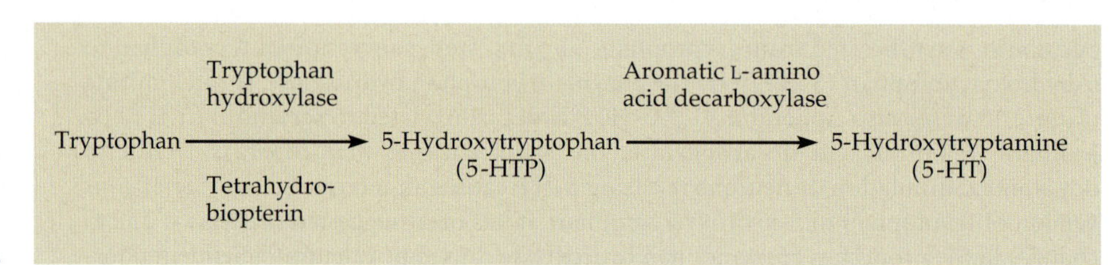

FIGURE 13.8 Synthesis of 5-HT.

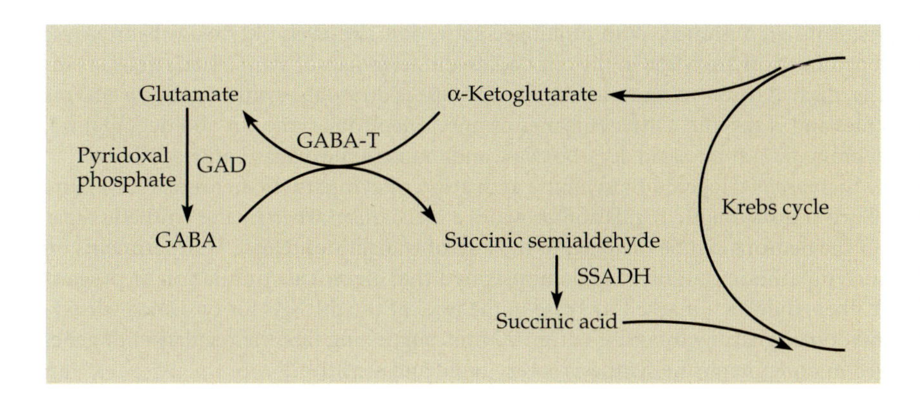

FIGURE 13.9 GABA Synthesis and Metabolism. GABA is synthesized from glutamate by the enzyme glutamic acid decarboxylase (GAD), which requires pyridoxal phosphate as a cofactor. Glutamate is synthesized from α-ketoglutarate by the enzyme GABA α-oxoglutarate transaminase (GABA-T) or from glutamine (see Figure 13.10). GABA is metabolized to succinic acid by GABA-T and succinic semialdehyde dehydrogenase (SSADH).

Synthesis of Glutamate

Glutamate is the major excitatory transmitter in the brain. There is more than one pathway for glutamate synthesis in cells. In neurons, glutamate destined for release as a transmitter is derived primarily from glutamine by a phosphate-activated form of the enzyme glutaminase (Figure 13.10).[39] A brain-specific inorganic phosphate transporter is localized selectively to terminals of glutamatergic neurons, and so is in a position to regulate glutamate synthesis.[40] Much of the glutamate released by neurons is taken up by glial cells and converted to glutamine. Glutamine, in turn, is released from the glial cells, taken up by neurons, and converted back to glutamate.

Short- and Long-Term Regulation of Transmitter Synthesis

The regulatory mechanisms described so far operate rapidly to change the rate of synthesis within nerve terminals. In addition to such short-term effects, there are long-term regulatory mechanisms. A good example comes from the response of the sympathetic nervous system to prolonged exposure of an animal to stress. When the body is stressed, sympa-

[39] Laake, J. H., et al. 1999. *Neuroscience* 88: 1137–1151.

[40] Bellocchio, E. E., et al. 1998. *J. Neurosci.* 18: 8648–8659.

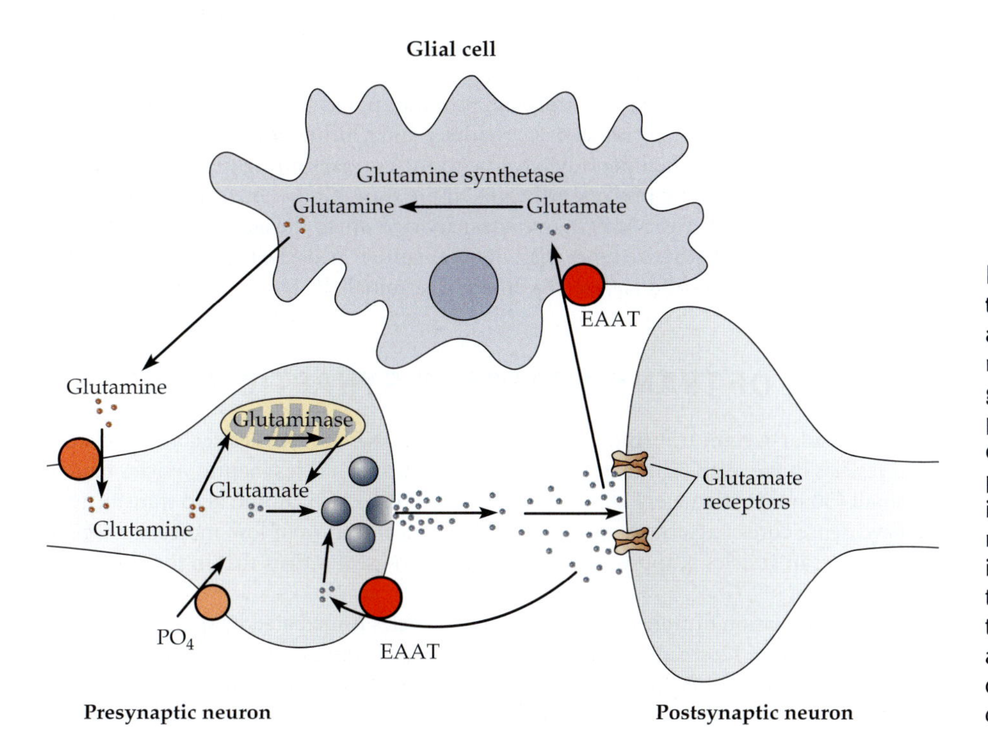

FIGURE 13.10 Pathways for Glutamate Synthesis, Storage, Release, and Uptake in glutamatergic neurons. Glutamate is synthesized from glutamine within mitochondria by a phosphate-dependent form of the enzyme glutaminase. An inorganic phosphate (PO_4) transporter is localized to glutamatergic terminals. After release, some glutamate is taken up into presynaptic terminals; most is taken up by glial cells and converted to glutamine, which is then released and taken up into nerve terminals for conversion to glutamate. EAAT = excitatory amino acid transporter.

thetic neurons are activated. With prolonged activation, the levels of tyrosine hydroxylase and dopamine β-hydroxylase in the cell bodies and terminals of sympathetic neurons increase as much as three- to fourfold.[41,42] The increase is due to the synthesis of new enzyme molecules and is specific. Other enzymes of norepinephrine synthesis and degradation, such as aromatic L-amino acid decarboxylase and monoamine oxidase, are not affected.

The increase is triggered by synaptic activation of sympathetic neurons (see Figure 13.7). Such **trans-synaptic regulation** provides a mechanism whereby the synthetic capability of the neurons can be matched to the rate of transmitter release.[43] Experiments on human sympathetic ganglia have demonstrated that electrical stimulation of preganglionic fibers induces a marked increase in the level of the mRNAs for tyrosine hydroxylase and dopamine β-hydroxylase within 20 min, suggesting that the regulation of genes involved in norepinephrine synthesis is very rapid and sensitive.[44]

Synthesis of Neuropeptides

Regulation of the stores of peptide transmitters is complicated by the separation between the sites of synthesis and release. Peptides are synthesized on ribosomes, which are found in neuronal cell bodies but not in axons or nerve terminals. This arrangement has two consequences: First, the rate of synthesis of peptides is regulated in the cell body, and the peptides must then be moved to the terminal by axonal transport (which will be discussed shortly). This is a slow process compared to the rapid local control of the synthesis and storage of low-molecular-weight transmitters within the axon terminal. Second, the amount of a peptide available for release is limited to the amount on hand in the terminal. However, the binding of peptides to their receptors occurs at a much lower concentration (in the range of 10^{-10} to 10^{-8} *M*) than the binding of low-molecular-weight transmitters, such as acetylcholine, to their receptors (10^{-7} to 10^{-4} *M*), and the mechanisms by which they are removed from the synaptic cleft are generally slower. Moreover, neuropeptide receptors, like other metabotropic receptors, act indirectly through intracellular pathways that can provide tremendous amplification (Chapter 10). As a consequence, only a few molecules of a peptide are needed to influence a postsynaptic target, so the demands of release can be met by the supply of molecules transported from the cell body.

Peptides are synthesized as part of larger precursor proteins, which often contain the sequences for more than one biologically active peptide[45,46] (Table 13.1 and Figure 13.11). The initial steps in neuropeptide precursor synthesis are those typical of secreted proteins: synthesis in the endoplasmic reticulum, signal peptide cleavage, processing in the Golgi apparatus, and incorporation into large (100–200 nm) dense-core vesicles. Later steps are unique to neurons and endocrine cells. They are catalyzed by specific endoproteases that cleave the precursor protein into the appropriate peptide molecules, exopeptidases that remove C-terminal basic residues, and a bifunctional amidating enzyme that converts the C-terminal peptidyl glycine to the corresponding peptide amide (see Figure 13.11B). Proteolytic processing begins in the *trans*-Golgi network and continues within large dense-core vesicles as they are transported down the axon and stored in the terminal. Some cells synthesize more than one transmitter peptide; these can be differentially sorted into vesicles and targeted to different terminals.[47]

STORAGE OF TRANSMITTERS IN SYNAPTIC VESICLES

Low-molecular-weight transmitters such as ACh and NE are synthesized and packaged into vesicles in the axon terminal. In electron micrographs, such synaptic vesicles tend to be small (50 nm in diameter) and can appear clear (e.g., ACh, amino acid transmitters) or have dense cores (e.g., biogenic amines). The concentration of low-molecular-weight transmitters in vesicles is approximately 0.5 *M*, much greater than that in the surrounding cytoplasm.

The accumulation of transmitters in synaptic vesicles is mediated by specific transport proteins (Figure 13.12; see also Chapter 4). Four vesicular transporters have been identified: a monoamine transporter for all biogenic amines (called VMAT), a transporter for GABA and glycine, a transporter for acetylcholine (called VAChT), and one for glutamate.[48–50]

[41]Thoenen, H., Mueller, R. A., and Axelrod, J. 1969. *Nature* 221: 1264.

[42]Thoenen, H., Otten, U., and Schwab, M. 1979. In *The Neurosciences: Fourth Study Program*. MIT Press, Cambridge, MA, pp. 911–928.

[43]Comb, M., Hyman, S. E., and Goodman, H. M. 1987. *Trends Neurosci.* 10: 473–478.

[44]Schalling, M., et al. 1989. *Proc. Natl. Acad. Sci. USA* 86: 4302–4305.

[45]Mains, R. E., and Eipper, B. A. 1999. In *Basic Neurochemistry: Molecular, Cellular, and Medical Aspects*, 6th Ed. Lippincott-Raven, Philadelphia, pp. 363–382.

[46]Sossin, W. S., Fisher, J. M., and Scheller, R. H. 1989. *Neuron* 2: 1407–1417.

[47]Paganetti, P., and Scheller, R. H. 1994. *Brain Res.* 633: 53–62.

[48]Schuldiner, S., Shirvan, A., and Linial, M. 1995. *Physiol. Rev.* 75: 369–392.

[49]Varoqui, H., and Erickson, J. D. 1997. *Mol. Neurobiol.* 15: 165–191.

[50]Reimer, R. J., Fon, E. A., and Edwards, R. H. 1998. *Curr. Opin. Neurobiol.* 8: 405–412.

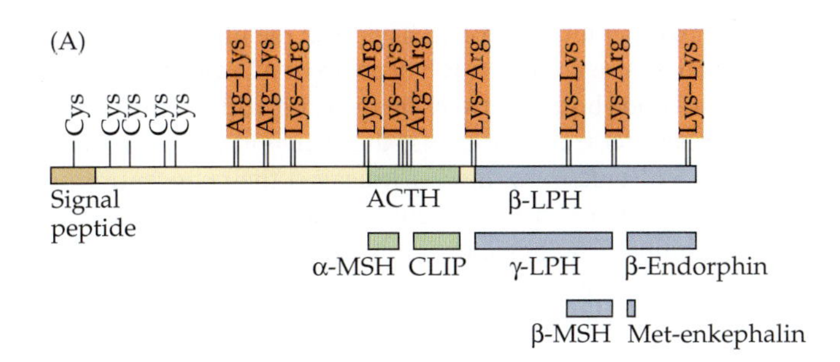

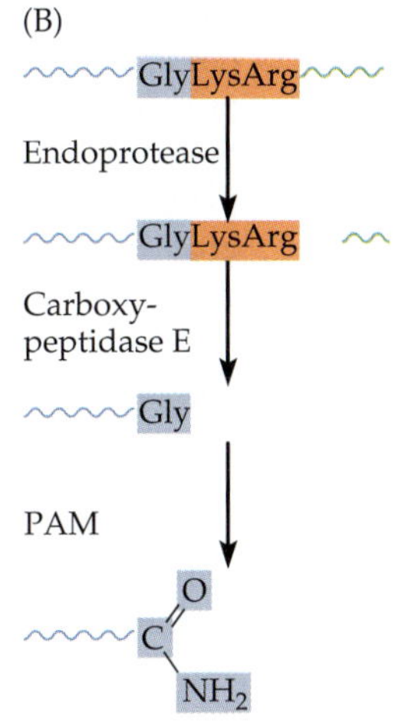

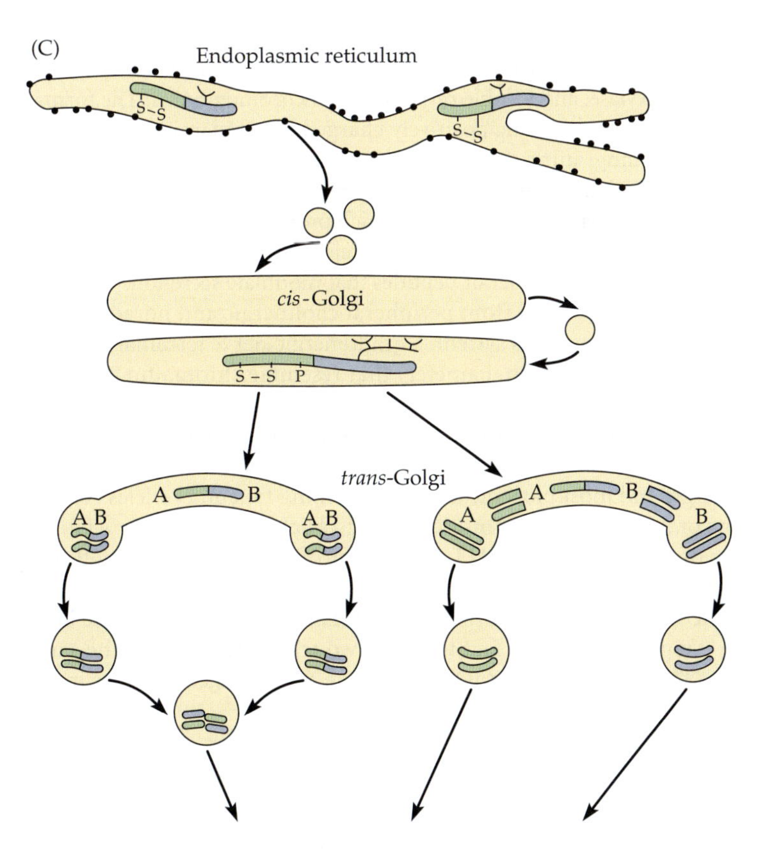

FIGURE 13.11 Synthesis of Neuropeptides. (A) Structure of bovine pro-opiomelanocortin. The locations of known peptide components are shown by colored boxes. Paired basic amino acid residues—common targets for processing enzymes—are indicated. (B) Processing of neuropeptide precursors usually begins with cleavage on the carboxy-terminal side of the recognition site by an endoprotease. The basic residues are trimmed by carboxypeptidase E. If the peptide ends in glycine, the enzyme peptidyl glycine α-amidating monooxygenase (PAM) converts the carboxy terminus to an amide. (C) Neuropeptide precursors are directed into the lumen of the endoplasmic reticulum by a signal sequence. In the endoplasmic reticulum, disulfide bonds are formed and N-linked glycosylation occurs. The propeptide is then transported through the Golgi apparatus, where further modifications, such as sulfation and phosphorylation, take place. Two packaging schemes are illustrated. On the left, a propeptide is packaged into vesicles budding from the Golgi; as the vesicle matures, the propeptide is cleaved, resulting in two peptides (A and B) packaged in the same vesicle. On the right, a propeptide is cleaved within the Golgi, followed by sorting of peptides into separate vesicles. (After Sossin, Fisher, and Scheller, 1989.)

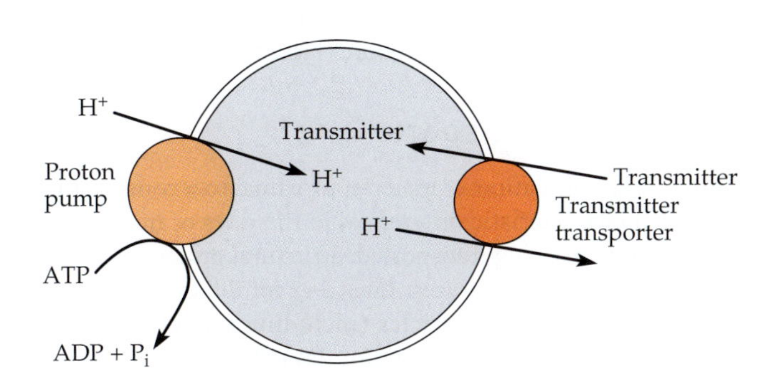

FIGURE 13.12 Transport of Transmitters into Synaptic Vesicles Is Driven by a Proton Electrochemical Gradient. An ATP-powered pump transports protons into synaptic vesicles, making the vesicle interior acidic and positive relative to the cytoplasm. Neurotransmitters are carried into the vesicles by specific transporters, energetically coupled to the proton electrochemical gradient.

Each of the transporters is an integral membrane protein having 12 membrane-spanning domains. Synaptic vesicle membranes also contain a H^+-dependent ATPase that pumps protons into the vesicle, making the inside both positively charged and acidic compared to the cytoplasm.[51] The vesicular transmitter transporters use the energy in the voltage and pH gradients to move transmitter molecules into the vesicle against their concentration gradient. Because of their small size, synaptic vesicles probably contain only a single copy of the proton pump, which is a large complex composed of at least 12 subunits.

Vesicular transmitter transporters do not have the same specificity as postsynaptic receptors. Consequently, molecules that do not activate postsynaptic receptors sometimes accumulate in synaptic vesicles and are released from axon terminals as false transmitters.[52,53]

The first vesicles to be purified and analyzed biochemically were from the adrenal medulla. These large vesicles (200–400 nm in diameter) are called chromaffin granules because of their tendency to stain with chromium salts. In addition to catecholamines, chromaffin granules contain high concentrations of ATP, a soluble form of the synthetic enzyme dopamine β-hydroxylase, and soluble proteins called chromogranins. The formation of multimolecular complexes among positively charged catecholamines, negatively charged ATP, and the chromogranins appears to aid in packing and storing the catecholamines at concentrations that might otherwise be hyperosmotic.[54] In addition, release of catecholamines is accompanied by release of ATP, which produces effects of its own and so acts as a cotransmitter.[55,56] Likewise, one of the proteins, chromogranin A, has been shown to serve as a precursor for a number of peptides that modulate secretion.[56]

Synaptic vesicles have been purified from peripheral cholinergic and noradrenergic terminals, and from the central nervous system. Noradrenergic nerve terminals contain large dense-core vesicles (70–200 nm in diameter) that resemble chromaffin granules, containing chromogranins, the soluble form of dopamine β-hydroxylase, and neuropeptide transmitters. The more numerous, small synaptic vesicles in catecholamine-containing nerve terminals, as well as those in cholinergic nerve terminals, contain little soluble protein. Both cholinergic and most biogenic amine-containing synaptic vesicles contain high concentrations of ATP.[57,58] Serotonin-containing vesicles contain little ATP but have large amounts of a high-affinity 5-HT–binding protein.[59] Uptake of ATP into amine- and ACh-containing vesicles has been characterized,[54,60] but the ATP transporter has not been identified. The observation that some spinal cord neurons release both GABA and glycine,[15] together with the fact that the two transmitters share the same vesicular transporter,[61,62] suggests that GABA and glycine are packaged together in the same vesicle.

AXONAL TRANSPORT

Proteins found in axon terminals must have been shipped there from the cell body, where they are synthesized. The first evidence for movement of material along axons came from experiments by Weiss and his colleagues, who ligated peripheral nerves and described the ballooning out of axons just proximal to the site of constriction and the subsequent movement of the accumulated material along the axons after the constriction was removed.[63] These effects suggested that normally there is a continuous bulk movement of axoplasm along the axon at the rate of 1–2 mm/day, which was given the name "axoplasmic flow." This idea was buttressed by later experiments using radioactively labeled amino acids to follow the movement of proteins from neuronal cell bodies along peripheral and central axons.[64] Such movement has even been observed in single axons in cell culture (Figure 13.13).[65]

Rate and Direction of Axonal Transport

Measurement of the time course of accumulation of material proximal to a constriction or in axon terminals demonstrated characteristic differences in the rates of movement within the broad spectrum of components being transported. Structural proteins, such as tubulin and neurofilament proteins, move at the slowest rates, 1–2 mm/day; membrane-enclosed organelles, such as mitochondria and vesicles (including synaptic vesicles packed with transmitter), move much faster (up to 400 mm/day).[66] Such rapid move-

[51]Stevens, T. H., and Forgac, M. 1997. *Annu. Rev. Cell Dev. Biol.* 13: 779–808.

[52]Kopin, I. J. 1968. *Annu. Rev. Pharmacol.* 8: 377–394.

[53]Luqmani, Y. A., Sudlow, G., and Whittaker, V. P. 1980. *Neuroscience* 5: 153–160.

[54]Johnson, R. G., Jr. 1988. *Physiol. Rev.* 68: 232–307.

[55]Burnstock, G. 1995. *J. Physiol. Pharmacol.* 46: 365–384.

[56]Kupfermann, I. 1991. *Physiol. Rev.* 71: 683–732.

[57]Dowdall, M. J., Boyne, A. F., and Whittaker, V. P. 1974. *Biochem. J.* 140: 1–12.

[58]De Potter, W. P., Smith, A. D., and De Schaepdryver, A. F. 1970. *Tissue Cell* 2: 529–546.

[59]Tamir, H., et al. 1994. *J. Neurochem.* 63: 97–107.

[60]Stadler, H., and Kiene, M-L. 1987. *EMBO J.* 6: 2217–2221.

[61]Dumoulin, A., et al. 1999. *J. Cell Sci.* 112: 811–823.

[62]Sagne, C., et al. 1997. *FEBS Lett.* 417: 177–183.

[63]Weiss, P., and Hiscoe, H. B. 1948. *J. Exp. Zool.* 107: 315–395.

[64]Droz, B., and Leblond, C. P. 1963. *J. Comp. Neurol.* 121: 325–346.

[65]Koehnle, T. J., and Brown, A. 1999. *J. Cell Biol.* 144: 447–458.

[66]Grafstein, B., and Forman, D. S. 1980. *Physiol. Rev.* 60: 1167–1283.

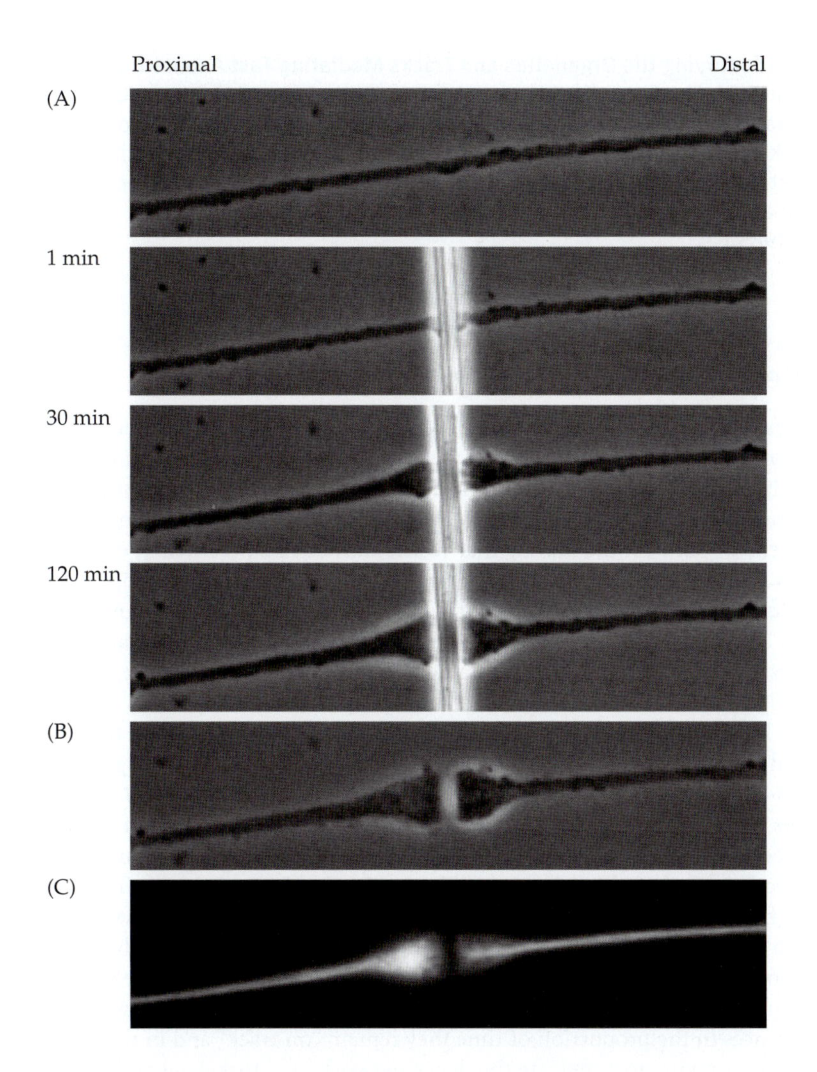

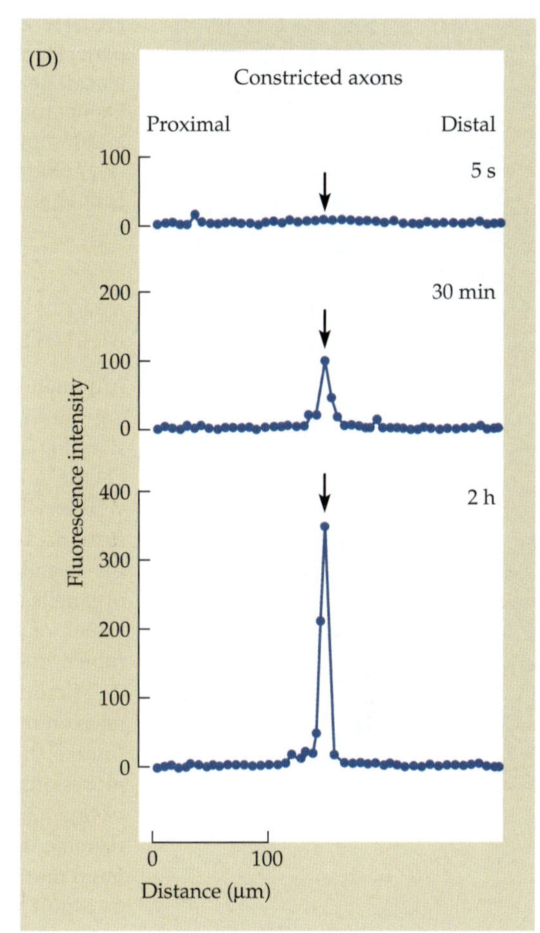

FIGURE 13.13 Slow Axonal Transport is demonstrated by the accumulation of cytoskeletal components at the site of axonal constriction. A single axon from a cultured rat dorsal root ganglion neuron was constricted by pressure from a glass fiber. (A) Phase-contrast images show the axon immediately before constriction and after 1, 30, and 120 min. (B) Two h after constriction the cell was fixed, and the glass fiber removed. (C) Fluorescence micrograph of the axon labeled with antineurofilament protein antibodies. (D) Graphs of fluorescence intensity as a function of distance along axons constricted for 5 s, 30 min, or 2 h. The time course of accumulation of neurofilament protein at the site of the constriction (arrow) indicated that the average transport rate was approximately 3 mm/day. (After Koehnle and Brown, 1999; micrographs kindly provided by A. Brown.)

ment could not be accounted for by bulk flow of cytoplasm, and the general term **axonal transport** was adopted.

Some proteins and organelles move toward the axon terminal (**anterograde transport**), and others from the terminal to the cell body (**retrograde transport**).[66,67] Retrograde transport of membrane-enclosed organelles returns material to the cell body for recycling or degradation, and has been shown to be crucial for the movement of trophic molecules such as nerve growth factor from axon terminals back to cell bodies (Chapter 23).[68]

Neuroanatomists have developed tracers, such as horseradish peroxidase, fluorescently labeled beads, and even viruses, that are carried in anterograde and retrograde directions by axonal transport. Using such tracers, it is possible to map synaptic connections, even over long distances, by visualizing individual axons, their terminal arborizations, and their cell bodies.[69,70]

[67]Vallee, R. B., and Bloom, G. S. 1991. *Annu. Rev. Neurosci.* 14: 59–92.

[68]Bartlett, S. E., Reynolds, A. J., and Hendry, I. A. 1998. *Immunol. Cell Biol.* 76: 419–423.

[69]Kuypers, H. G. J. M., and Ugolini, G. 1990. *Trends Neurosci.* 13: 71–75.

[70]Teune, T. M., et al. 1998. *J. Comp. Neurol.* 392: 164–178.

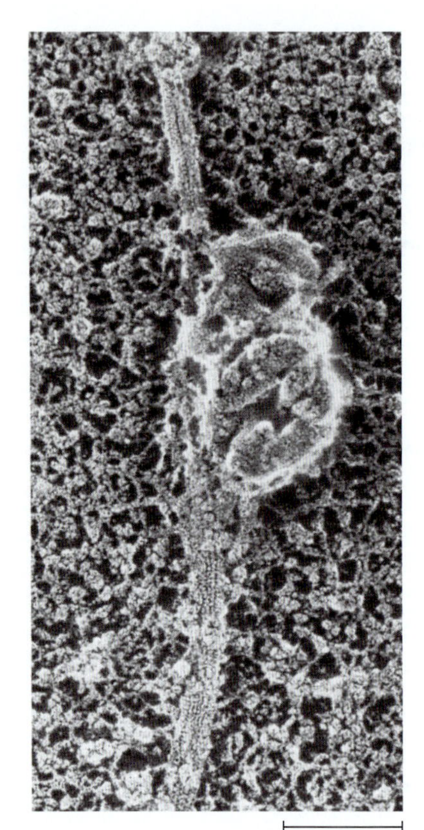

FIGURE 13.14 Identifying the Organelles and Tracks Mediating Fast Axonal Transport. Electron micrograph of a vesicle attached to a microtubule in extruded squid axoplasm. Before fixation this organelle was observed by light microscopy moving along a filamentous track at a rate corresponding to fast axonal transport. The electron micrograph shows that the organelle is a synaptic vesicle, and the track is a microtubule. A layer of granular and finely filamentous material coats the glass substrate. (From Schnapp et al., 1985.)

Microtubules and Fast Transport

Although early experiments demonstrated that axonal transport required metabolic energy and relied on intact microtubules, for 30 years little progress was made in understanding its mechanism. Then two technological advances triggered very rapid progress: (1) the development of microscopic techniques that allowed direct visualization of single vesicles within cells[71,72] and (2) the finding that vesicle movements persisted in cell-free systems, such as extruded squid axoplasm.[73] Studies by Reese, Sheetz, Schnapp, Vale, Block, and their colleagues have demonstrated that transport occurs by the attachment of organelles, such as mitochondria and vesicles, to microtubules. Mechanochemical enzymes, or motors, hydrolyze ATP and use the energy to carry organelles along the microtubule "track" (Figure 13.14).[74,75]

Microtubules have an inherent polarity; in axons the "plus" end points toward the distal axon terminal. Anterograde transport is powered by kinesin, which moves organelles toward the plus end; retrograde transport is powered by cytoplasmic dynein, which moves organelles toward the minus end (Figure 13.15).[76] Specific receptors on the surface of organelles mediate the attachment of either kinesin or cytoplasmic dynein and thus regulate the direction of organelle movement (Figure 13.16).[77] Remarkably, a single kinesin motor has been shown to pull an organelle along at speeds equivalent to fast axonal transport;[78] each molecule of ATP hydrolyzed produces a "step" of approximately 8 nm, corresponding to the distance from one αβ tubulin dimer to the next along the microtubule protofilament.[79,80] Differences in the rate of transport of different components arise from differences in the proportion of time they remain "on track" and in the resistance they encounter trying to penetrate the dense network of cytoskeletal and cross-bridging elements within the axon.

Mechanism of Slow Axonal Transport

Soluble proteins of intermediary metabolism and cytoskeletal elements, such as microtubules and neurofilaments, move from the cell body toward the axon terminal by slow transport.[66] There is considerable debate about whether microtubules and neurofilaments move as intact polymers[81] or if polymerized filaments are stationary and tubulin and neurofilament monomers or oligomers are transported.[82] What is clear is that diffusion cannot account for the axonal transport of cytoskeletal proteins and that an active process is involved.

TRANSMITTER RELEASE AND VESICLE RECYCLING

The quantal release of transmitter by vesicle exocytosis (Chapter 11) raises a number of fundamental questions: How is the distribution of synaptic vesicles within nerve terminals organized? What mechanisms recruit vesicles to the active zone? How does calcium influx trigger exocytosis?

Sorting of Vesicles within the Nerve Terminal

It appears that not all vesicles are created equal. Studies using radioactive precursors to follow transmitter synthesis, incorporation into vesicles, and release suggest that there

[71]Inoue, S. 1981. *J. Cell Biol.* 89: 346–356.

[72]Allen, R. D., Allen, N. S., and Travis, J. L. 1981. *Cell Motil.* 1: 291–302.

[73]Brady, S. T., Lasek, R. J., and Allen, R. D. 1982. *Science* 218: 1129–1131.

[74]Vale, R. D., and Fletterick, R. J. 1997. *Annu. Rev. Cell Dev. Biol.* 13: 745–777.

[75]Vallee, R. B., and Gee, M. A. 1998. *Trends Cell Biol.* 8: 490–494.

[76]Hirokawa, N. 1998. *Science* 279: 519–552.

[77]Sheetz, M. P. 1999. *Eur. J. Biochem.* 262: 19–25.

[78]Howard, J., Hudspeth, A. J., and Vale, R. D. 1989. *Nature* 342: 154–158.

[79]Svoboda, K., et al. 1993. *Nature* 365: 721–727.

[80]Mandelkow, E., and Hoenger, A. 1999. *Curr. Opin. Cell Biol.* 11: 34–44.

[81]Baas, P. W., and Brown, A. 1997. *Trends Cell Biol.* 7: 380–384.

[82]Hirokawa, N., et al. 1997. *Trends Cell Biol.* 7: 382–388.

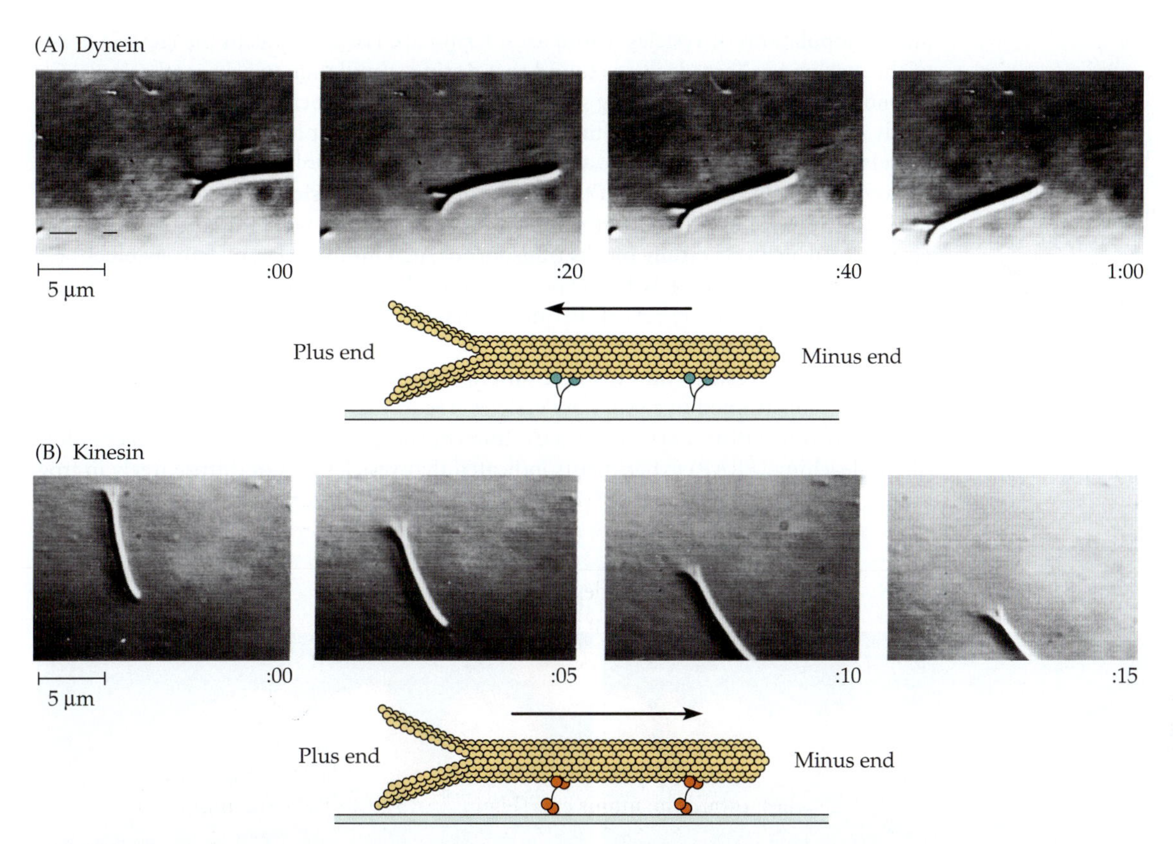

FIGURE 13.15 The Molecular Motors Dynein and Kinesin Propel Microtubules in Opposite Directions. (A,B) Sequential images of the movement of microtubule fragments on purified fast-transport "motors." Time is indicated in minutes. Purified cytoplasmic dynein (A) or kinesin (B) was adsorbed to a coverslip, and fragments of microtubules were added. When the fragments contacted the surface, they were propelled toward their frayed (distal or +) end on dynein and toward their compact (proximal or –) end on kinesin, as illustrated. (After Paschal and Vallee, 1987; micrographs kindly provided by R. Vallee.)

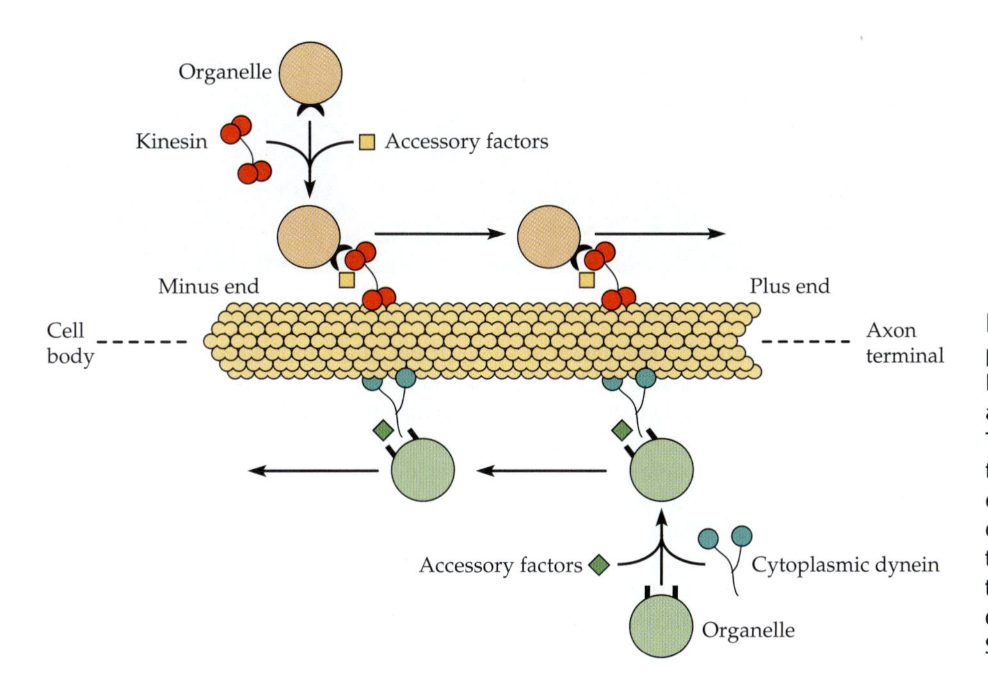

FIGURE 13.16 Fast Axonal Transport Powered by Kinesin and Dynein. In the axon, microtubules are stationary and have a polarity: The plus end points toward the axon terminal, the minus end toward the cell body. Kinesin and cytoplasmic dynein, together with accessory factors, attach to organelles and propel them toward the axon terminal and cell body, respectively. (After Vallee, Shpetner, and Paschal, 1989.)

are subpopulations of vesicles within axon terminals. For example, in the electric organ of the marine ray *Torpedo* it was found that during stimulation, newly synthesized ACh was not spread uniformly among synaptic vesicles, but was localized to those vesicles recently formed by recycling.[83,84] Studies on mammalian motor and sympathetic axon terminals have shown that newly synthesized transmitter molecules are preferentially released.[20,85] Such results suggest that a subpopulation of vesicles recycles rapidly, while most of the vesicles are held in reserve. This scheme fits with the observation that vesamicol, which specifically blocks transport of ACh into vesicles, selectively blocks release of newly synthesized ACh.[22] Experiments in which vesicles are labeled with fluorescent dyes as they recycle reveal distinct pools of vesicles in axon terminals in some preparations,[86] but not others.[87]

A division of vesicles into distinct subpopulations requires that the movements of vesicles be constrained in some way; vesicles free to move by Brownian motion would mix uniformly within nerve terminals. Indeed, so-called fluorescence recovery after photobleaching (FRAP) experiments indicated that vesicles do not diffuse freely in frog motor nerve terminals (Figure 13.17).[88] In these experiments, vesicles were labeled with

[83]Zimmermann, H., and Denston, C. R. 1977. *Neuroscience* 2: 695–714.

[84]Zimmermann, H., and Denston, C. R. 1977. *Neuroscience* 2: 715–730.

[85]Kopin, I. J., et al. 1968. *J. Pharmacol. Exp. Ther.* 161: 271–278.

[86]Kuromi, H., and Kidokoro, Y. 1998. *Neuron* 20: 917–925.

[87]Henkel, A. W., Lübke, J. and Betz, W. J. 1996. *Proc. Natl. Acad. Sci. USA* 93: 1918–1923.

[88]Henkel, A. W., et al. 1996. *J. Neurosci.* 16: 3960–3967.

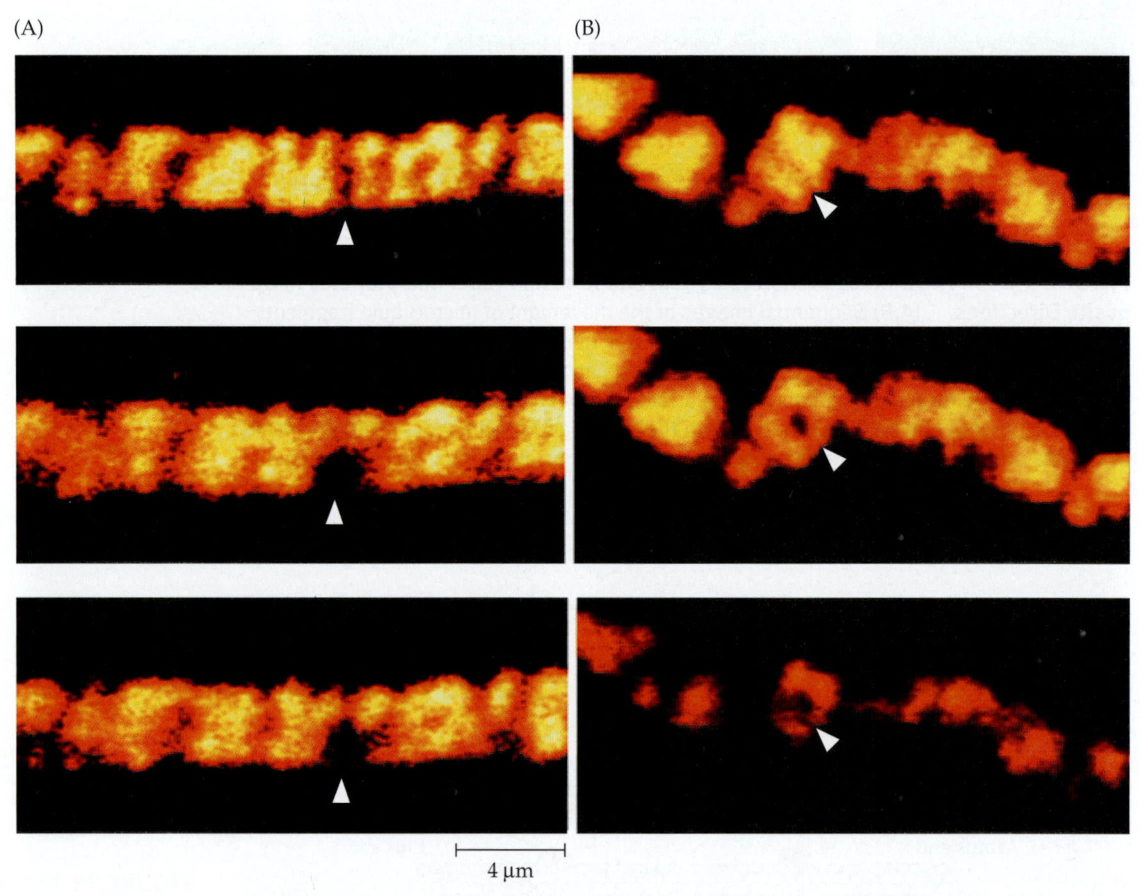

FIGURE 13.17 Restricted Movement of Synaptic Vesicles within Nerve Terminals. Fluorescence micrographs of frog nerve terminals in which synaptic vesicles have been labeled with a fluorescent dye. (A) Resting nerve terminal before (top), immediately after (middle), and 28 min after (bottom) bleaching the dye in a small spot among the vesicle clusters with intense illumination. The bleached region (arrowhead) is nearly unchanged after 28 min, indicating little spontaneous movement of dye-containing vesicles. (B) As in part A, except after bleaching the terminal was stimulated at 10 Hz for 5 min. Although all the vesicle clusters dim as vesicles undergo exocytosis and release their dye, no increase in fluorescence occurs in the bleached region, indicating no net movement of dye-containing vesicles into this area. (After Henkel et al. 1996; micrographs kindly provided by W. J. Betz.)

a fluorescent dye, and a small region in a cluster of labeled vesicles was bleached. The bleached spot remained, even in the face of ongoing release, indicating that the vesicles were not free to move laterally.

Greengard and his colleagues have identified a family of proteins in axons and axon terminals, called synapsins, that cross-link vesicles to the cytoskeleton, preventing their movement.[89,90] Interestingly, the binding of synapsins to vesicles is regulated by phosphorylation; addition of phosphate groups to synapsins dissociates them from vesicles. One role that restricted vesicle movement might play in regulating transmitter release is illustrated by experiments by Llinás and his colleagues[91] in which dephosphorylated synapsin I, injected directly into a presynaptic axon terminal at the squid giant synapse, caused a decrease in the amount of transmitter released by nerve stimulation. When this was followed by injection of a specific protein kinase that is able to phosphorylate synapsin I, the amount of transmitter released by stimulation increased. Thus, under normal conditions, the activity of endogenous kinases and phosphatases might modulate the amount of transmitter released from axon terminals by regulating phosphorylation of synapsins.

Conserved Mechanisms for Vesicle Trafficking

How are vesicles brought to the plasma membrane and induced by calcium entry to liberate their cargo into the synaptic cleft? Experiments aimed at characterizing the components of nerve terminals[92] led to the discovery that several proteins localized to synapses were homologous to proteins required for secretion in yeast[93] and for vesicle-mediated transport within the Golgi complex.[94] The realization that the same basic machinery underlies all membrane trafficking, from yeast to mammals, laid the foundation for rapid progress. Genetic, physiological, and biochemical approaches are beginning to reveal the complex sequence of steps by which transmitter exocytosis occurs.[95,96] The first step is the formation of a complex between proteins on the synaptic vesicle membrane and proteins on the presynaptic membrane at the active zone. This complex holds the vesicle in position (the vesicle is said to be "docked") and mediates membrane fusion in response to calcium influx. The proteins mediating docking and fusion are called SNARES, short for SNAP receptors, because they were first identified as receptors for another protein required for secretion in yeast called soluble NSF attachment protein, or SNAP. Accordingly, the scheme for vesicle binding and fusion is referred to as the **SNARE hypothesis** (Box 13.1).

Synaptotagmin and the Calcium Dependence of Neurotransmitter Release

A protein known as synaptotagmin appears to be the calcium sensor for release.[96] Synaptotagmin binds calcium in the appropriate concentration range and displays calcium-dependent changes in its interactions with vesicle and plasma membrane proteins. In synaptotagmin knockouts, there is little or no evoked release, although spontaneous release is increased. Thus, synaptotagmin is thought to act as a brake to prevent fusion from occurring in the absence of calcium, then to accelerate fusion and exocytosis when calcium influx occurs.

Calcium channels and the plasma membrane SNARE proteins syntaxin and SNAP-25 bind to one another, which may play a role in the co-localization of calcium channels and sites of transmitter release.[97] Calcium channels and synaptotagmin also bind to one another, which could bring synaptotagmin into position to sense the microdomain of increased calcium concentration near the open channel.[98]

Bacterial Neurotoxins Target the SNARE Complex

Identification of the components of the SNARE complex led to an understanding of the mechanism of action of the very potent neurotoxins produced by *Clostridium* bacteria: tetanus toxin and the botulinum toxins.[99] A single toxin molecule is sufficient to prevent transmitter release from an entire nerve terminal. Each toxin is composed of disulfide-

[89]De Camilli, P., and Greengard, P. 1986. *Biochem. Pharmacol.* 35: 4349–4357.

[90]Hilfiker, S., et al. 1999. *Philos. Trans. R. Soc. Lond. B* 354: 269–279.

[91]Llinás, R., et al. 1985. *Proc. Natl. Acad. Sci. USA* 82: 3035–3039.

[92]Bennett, M. K., and Scheller, R. H. 1994. *Curr. Opin. Neurobiol.* 4: 324–329.

[93]Ferro-Novick, S., and Jahn, R. 1994. *Nature* 370: 191–193.

[94]Rothman, J. E. 1994. *Nature* 372: 55–63.

[95]Gerst, J. E. 1999. *Cell. Mol. Life Sci.* 55: 707–734.

[96]Bajjalieh, S. M. 1999. *Curr. Opin. Neurobiol.* 9: 321–328.

[97]Rettig, J., et al. 1996. *Proc. Natl. Acad. Sci. USA* 93: 7363–7368.

[98]Seagar, M., et al. 1999. *Philos. Trans. R. Soc. Lond. B* 354: 289–297.

[99]Pellizzari, R., et al. 1999. *Philos. Trans. R. Soc. Lond. B* 354: 259–268.

Box 13.1 The SNARE Hypothesis

A fundamental concept of how vesicle targeting and fusion occurs is the SNARE hypothesis.[95,96] The idea is that the selective docking of a transport vesicle with the appropriate target membrane occurs through the formation of a complex between a *vesicle* membrane protein (**v-SNARE**) and the corresponding *target* membrane protein (**t-SNARE**). (The term "SNARE" is short for SNAP receptor, because the SNARE proteins were first identified as receptors for another protein involved in exocytosis, called α-SNAP.) Once formed, the SNARE complex leads, either directly or indirectly, to membrane fusion. Three families of SNARE proteins have been identified: the VAMP/synaptobrevin family of v-SNAREs, and two families of t-SNAREs, the syntaxin family and the SNAP-25 family. Proteins of the synaptotagmin family are considered to be either a second family of v-SNAREs or important regulators of SNARE function.

Outlined here and in the figure below is a scheme for the recruitment of synaptic vesicles to the presynaptic membrane, docking at active zones, and calcium-evoked exocytosis.[92,93] Many proteins have been omitted for the sake of clarity, and changes will be required as more information becomes available.

During **preparation**, the initial step leading to exocytosis, the proteins *N*-ethylmaleimide-sensitive factor (NSF) and soluble NSF attachment protein (α-SNAP) act on synaptobrevin, syntaxin, and SNAP-25. (Note that α-SNAP and SNAP-25 are different, unrelated proteins.) Preparation causes preexisting SNARE complexes to dissociate, activates the SNARE proteins (indicated by a star in the figure), and removes negative regulators of exocytosis, such as n-sec1. Preparation requires ATP hydrolysis by NSF and readies vesicles and target membranes for subsequent steps.

Next, the synaptic vesicle protein rab3, a member of the family of low-molecular-weight GTPases, promotes reversible vesicle attachment to the presynaptic membrane (**tethering**). Tethering allows formation of the SNARE complex consisting of synaptobrevin, syntaxin, and SNAP-25. This brings the vesicle into a docked position in the active zone, immediately adjacent to the plasma membrane and calcium channels. **Docking** is an irreversible step that involves some degree of membrane fusion.

A further **priming** reaction is required before a docked vesicle can be released. At some point during docking or priming, synaptotagmin is recruited to the SNARE complex. Calcium influx then triggers the completion of **fusion** of the vesicle and presynaptic membranes, a step that requires GTP. A fusion pore is created, which rapidly widens to allow transmitter to escape into the synaptic cleft.

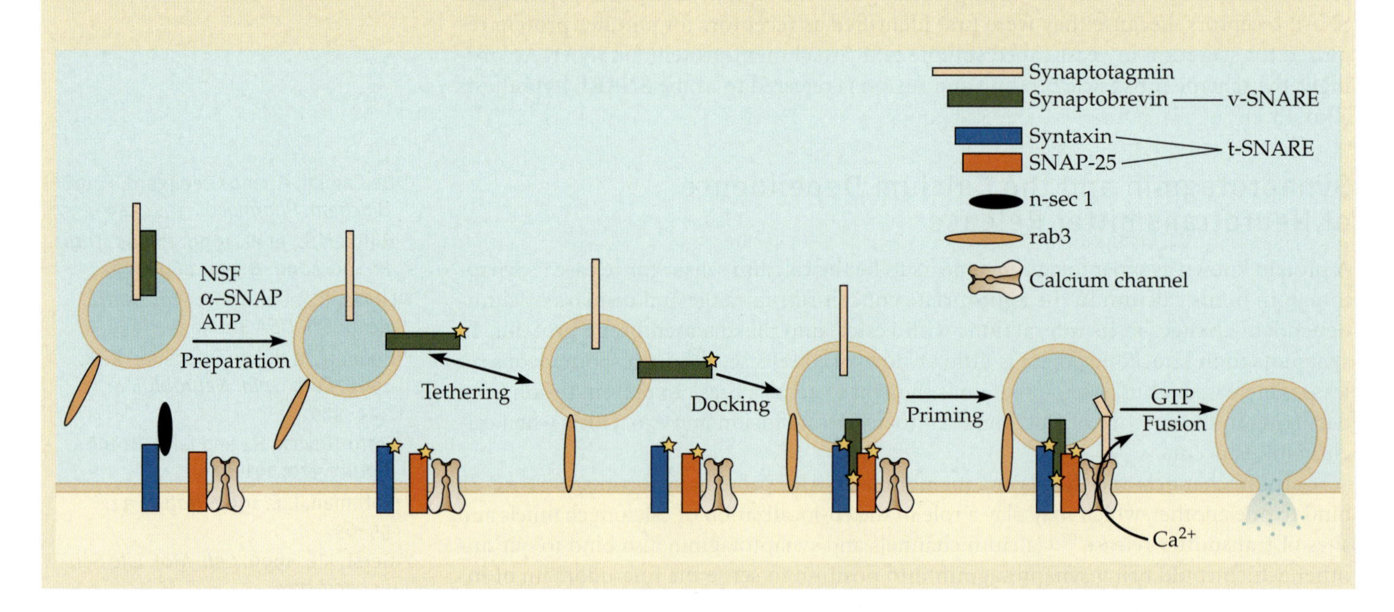

linked heavy and light chains. The heavy chain binds to the plasma membrane and translocates the toxin to the inside of the nerve terminal. The disulfide bond is then cleaved, and the free light chain acts as an endoprotease that specifically degrades a component of the SNARE complex (Figure 13.18).

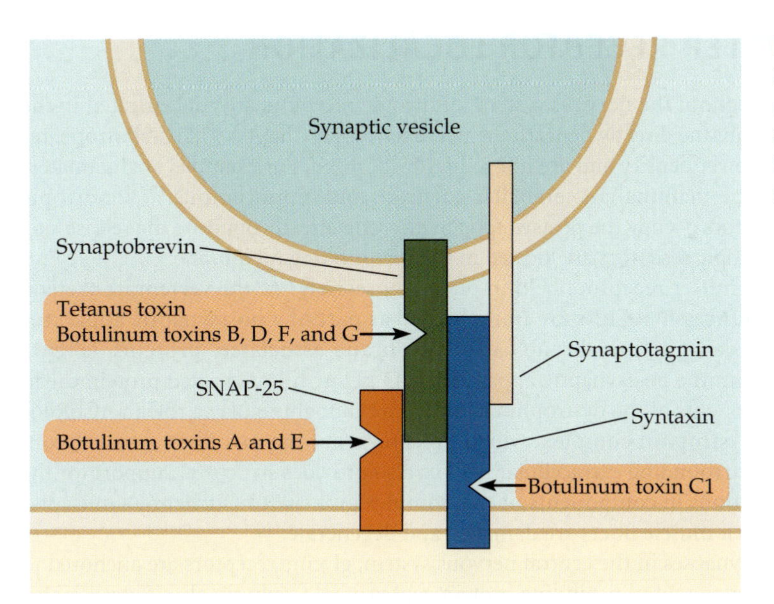

FIGURE 13.18 Site of Action of Tetanus and Botulinum Toxins. These toxins are proteases that degrade components of the SNARE complex. Botulinum toxin C1 acts on syntaxin, botulinum toxins A and E degrade SNAP-25, and botulinum toxins B, D, F, and G and tetanus toxin target synaptobrevin.

Recovery of Synaptic Vesicle Membrane Components by Endocytosis

As described in Chapter 11, after exocytosis the components of the synaptic vesicle membrane are retrieved and recycled. The best-characterized pathway for recycling is the classic endocytic pathway, in which vesicles flatten completely into the presynaptic membrane following exocytosis and components are retrieved by endocytosis via clathrin-coated pits (Figure 13.19).[100] In addition to clathrin, this pathway requires adapter proteins to select the appropriate constituents for recycling, and dynamin, a calcium-sensitive GTPase, to complete the pinching off of coated vesicles from the plasma membrane.[101] Two pathways have been suggested for the recycling of components retrieved by endocytosis. Endocytic vesicles may either lose their coat to re-form synaptic vesicles directly, or pass through an endosomal compartment from which synaptic vesicles are formed. Recycling via endocytosis is generally thought to take 30 s to 1 min;[102,103] however, experiments at snake neuromuscular junctions suggest that a much faster time course is possible.[104]

An alternative proposal is that vesicles pinch off immediately after releasing their contents, without flattening into the plasma membrane (see Figure 13.19).[105] Such "kiss and run" exocytosis was proposed to account for the very rapid recycling that occurs at some synapses—for example, at central synapses where small presynaptic boutons containing relatively few vesicles release quanta of transmitter at high frequency.

[100]Miller, T. M., and Heuser, J. E. 1984. *J. Cell Biol.* 98: 685–698.

[101]Robinson, M. S. 1994. *Curr. Opin. Cell Biol.* 6: 538–544.

[102]Betz, W. J., and Bewick, G. S. 1993. *J. Physiol.* 460: 287–309.

[103]Ryan, T. A., et al. 1993. *Neuron* 11: 713–724.

[104]Teng, H., et al. 1999. *J. Neurosci.* 19: 4855–4866.

[105]Kavalali, E., Klingauf, J., and Tsien, R. W. 1999. *Philos. Trans. R. Soc. Lond. B* 354: 337–346.

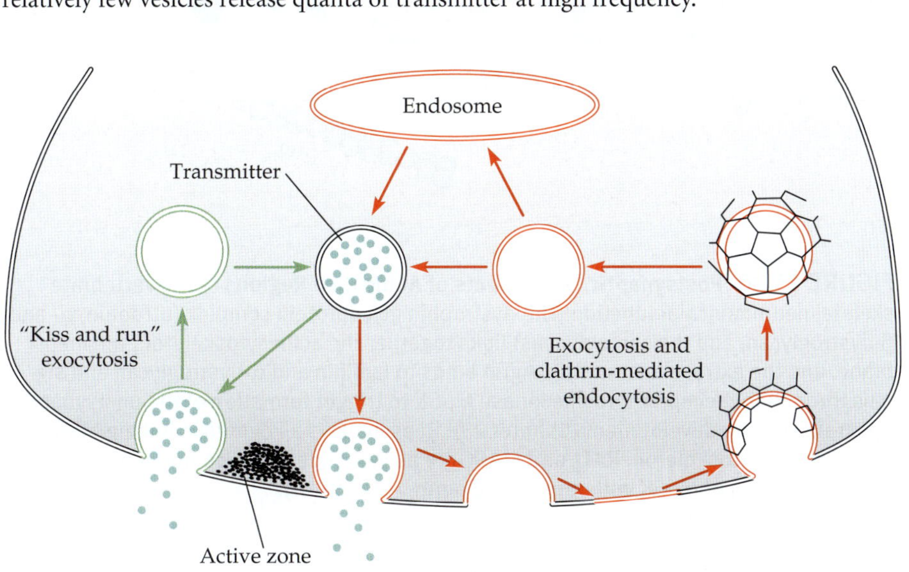

FIGURE 13.19 Two Pathways for Vesicle Membrane Recycling after exocytosis. In the best-characterized pathway, vesicles flatten into the plasma membrane after exocytosis, and components are recycled via clathrin-mediated endocytosis and formation of new vesicles. At some synapses there is evidence for "kiss and run" exocytosis, in which vesicles rapidly pinch off after exocytosis without merging with the plasma membrane.

TRANSMITTER RECEPTOR LOCALIZATION

At synapses throughout the nervous system ionotropic receptors are concentrated in the postsynaptic membrane directly beneath the nerve terminal (Chapter 9); metabotropic receptors occur at lower density and are not so highly localized. For example, at glutamatergic synapses in the mammalian cerebellar cortex[106] and hippocampus,[107] ionotropic AMPA-type receptors occupy the postsynaptic membrane directly opposite the release site, whereas metabotropic receptors are located in the surrounding membrane.

How are transmitter receptors held in place so precisely? At the vertebrate skeletal neuromuscular junction, AChRs are immobilized as part of a postsynaptic apparatus formed by cytoskeletal, membrane, and membrane-associated proteins (Figure 13.20).[108,109] Within the postsynaptic apparatus, a 43 kD AChR-associated protein called rapsyn and components of the dystrophin complex are thought to play a role in AChR localization. The dystrophin complex, which links together the myofiber cytoskeleton, membrane, and surrounding extracellular matrix, also provides structural support for the muscle cell. Mutations in components of this complex give rise to Duchenne's muscular dystrophy, in which muscle fibers are damaged and degenerate.[110]

At inhibitory synapses in the central nervous system, glycine receptors are anchored to the cytoskeleton by the tubulin-binding protein gephyrin.[111] Gephyrin also is essential for localization of $GABA_A$ receptors in postsynaptic membranes, although direct interactions between gephyrin and $GABA_A$ receptor subunits have not been demonstrated. Gephyrin does interact with several intracellular components that mediate responses to activity and trophic factors. Such interactions are thought to play a central role in the assembly and stabilization of postsynaptic specializations at inhibitory synapses.

At excitatory synapses in the central nervous system, three families of proteins that interact with glutamate receptors have been identified (Figure 13.21).[112,113] Proteins in each of the families have one or more PDZ domains, which are conserved regions that mediate protein–protein interactions. NMDA-type glutamate receptors bind to proteins of the PSD-95 family, which are major components of the postsynaptic density. AMPA-type

[106]Nusser, Z., et al. 1994. *Neuroscience* 61: 421–427.

[107]Luján, R., et al. 1996. *Eur. J. Neurosci.* 8: 1488–1500.

[108]Apel, E. D., and Merlie, J. P. 1995. *Curr. Opin. Neurobiol.* 5: 62–67.

[109]Sanes, J. R., and Lichtman, J. W. 1999. *Annu. Rev. Neurosci.* 22: 389–442.

[110]Petrof, B. J. 1998. *Mol. Cell. Biochem.* 179: 111–123.

[111]Kneussel, M., and Betz, H. 2000. *J. Physiol.* 525: 1–9.

[112]O'Brien, R. J., Lau, L. F., and Huganir, R. L. 1998. *Curr. Opin. Neurobiol.* 8: 364–369.

[113]Sheng, M. and Lee, S. H. 2000. *Nature Neurosci.* 3: 633–635.

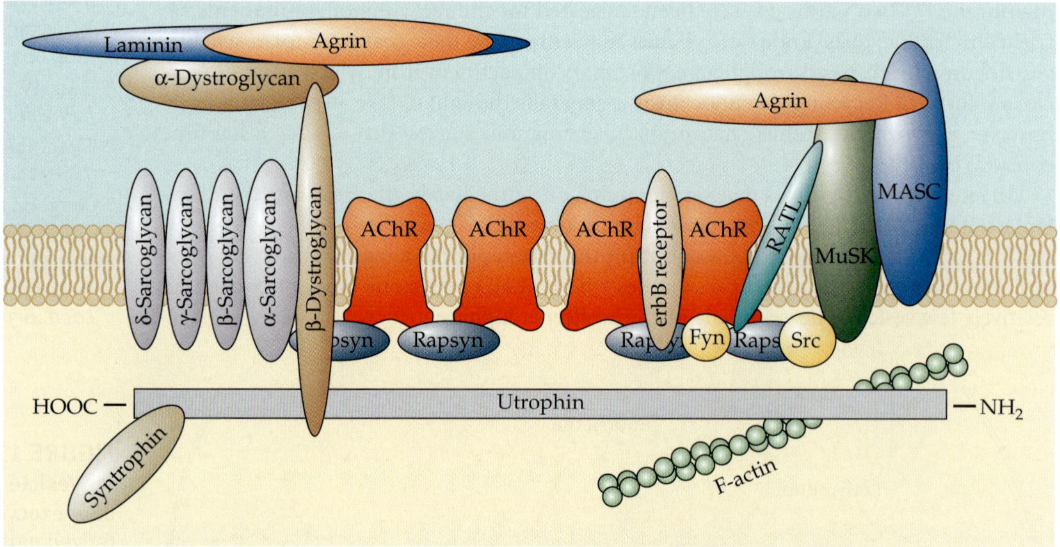

FIGURE 13.20 Postsynaptic Components of AChR-Rich Regions at the vertebrate skeletal neuromuscular junction. The dystrophin glycoprotein complex (utrophin, α- and β-dystroglycan, and the sarcoglycans) links together the actin cytoskeleton, the membrane, and the extracellular matrix. Agrin binds to laminin and α-dystroglycan and signals through the receptor tyrosine kinase MuSK to trigger formation of the postsynaptic apparatus during development (Chapter 23). Rapsyn plays a key role in linking MuSK and AChRs to the cytoskeleton. RATL and MASC are as yet unidentified components that mediate interaction of MuSK with rapsyn and agrin, respectively.

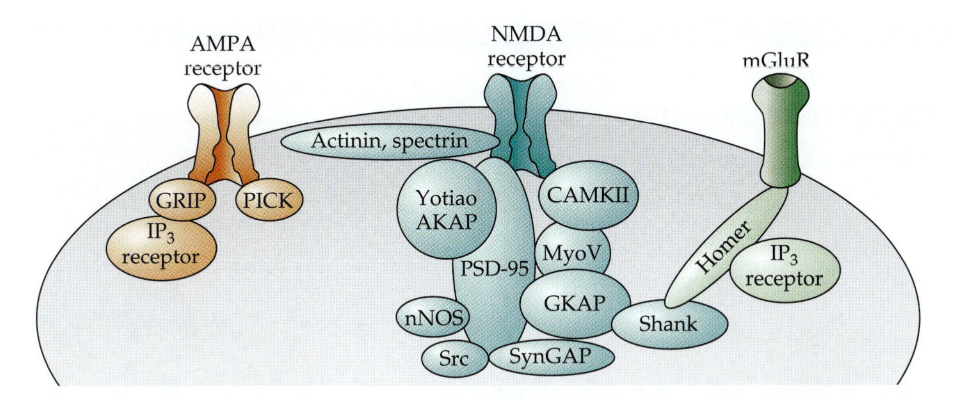

FIGURE 13.21 Glutamate Receptors Are Linked to a Postsynaptic Scaffold that includes proteins involved in intracellular signaling cascades. Members of the GRIP protein family link AMPA receptors to the IP_3 receptor. PSD-95 and its homologues connect NMDA receptors to Yotiao, nNOS, Src, SynGAP, and GKAP. CaMKII binds to NMDA receptors and to MyoV. The Homer protein family links metabotropic glutamate receptors to Shank and thereby to the NMDA receptor complex. (After Sheng and Lee, 2000.)

glutamate receptors bind to proteins of the GRIP family, and metabotropic glutamate receptors bind to members of the Homer protein family. Attention focused originally on the role such proteins play in localizing receptors to synaptic sites. It is becoming apparent, however, that these proteins also play an important role in creating an intracellular scaffold to which particular signaling proteins are recruited, including nitric oxide synthase, receptor tyrosine kinases Raf, MAP, and Rsk kinases, IP_3 receptors, and Ras-like small GTPases.[113] Thus, these proteins may determine not only receptor location, but also the consequences of receptor activation.

Presynaptic Receptors

Neurotransmitter receptors are also found on presynaptic nerve terminals. Such presynaptic receptors regulate transmitter release (Chapter 10) and may play a role in the release mechanism itself (Chapter 11).

REMOVAL OF TRANSMITTERS FROM THE SYNAPTIC CLEFT

The final step in chemical synaptic transmission is the removal of transmitter from the synaptic cleft. The mechanisms for transmitter removal include diffusion, degradation, and uptake into glial cells or nerve terminals.

Removal of ACh by Acetylcholinesterase

As described in Chapter 9, the action of ACh is terminated by the enzyme acetylcholinesterase (AChE), which hydrolyzes ACh to choline and acetate. Much of the choline is transported back into the nerve terminal and reused for ACh synthesis. At the vertebrate skeletal neuromuscular junction, acetylcholinesterase is bound to the synaptic basal lamina, that portion of the muscle fiber's sheath of extracellular matrix material that occupies the synaptic cleft and junctional folds (Figure 13.22).[114] There are 2600 catalytic subunits of AChE per square micrometer of synaptic basal lamina[115] (compared to the 10^4 AChRs per square micrometer in the postsynaptic membrane).

It might seem inefficient to have acetylcholinesterase situated between the axon terminal and the postsynaptic membrane, forcing molecules of ACh to traverse a minefield of degradative enzymes before having an opportunity to interact with their postsynaptic re-

[114]McMahan, U. J., Sanes, J. R., and Marshall, L. M. 1978. *Nature* 271: 172–174.

[115]Salpeter, M. M. 1987. In *The Vertebrate Neuromuscular Junction*. Alan R. Liss, New York, pp. 1–54.

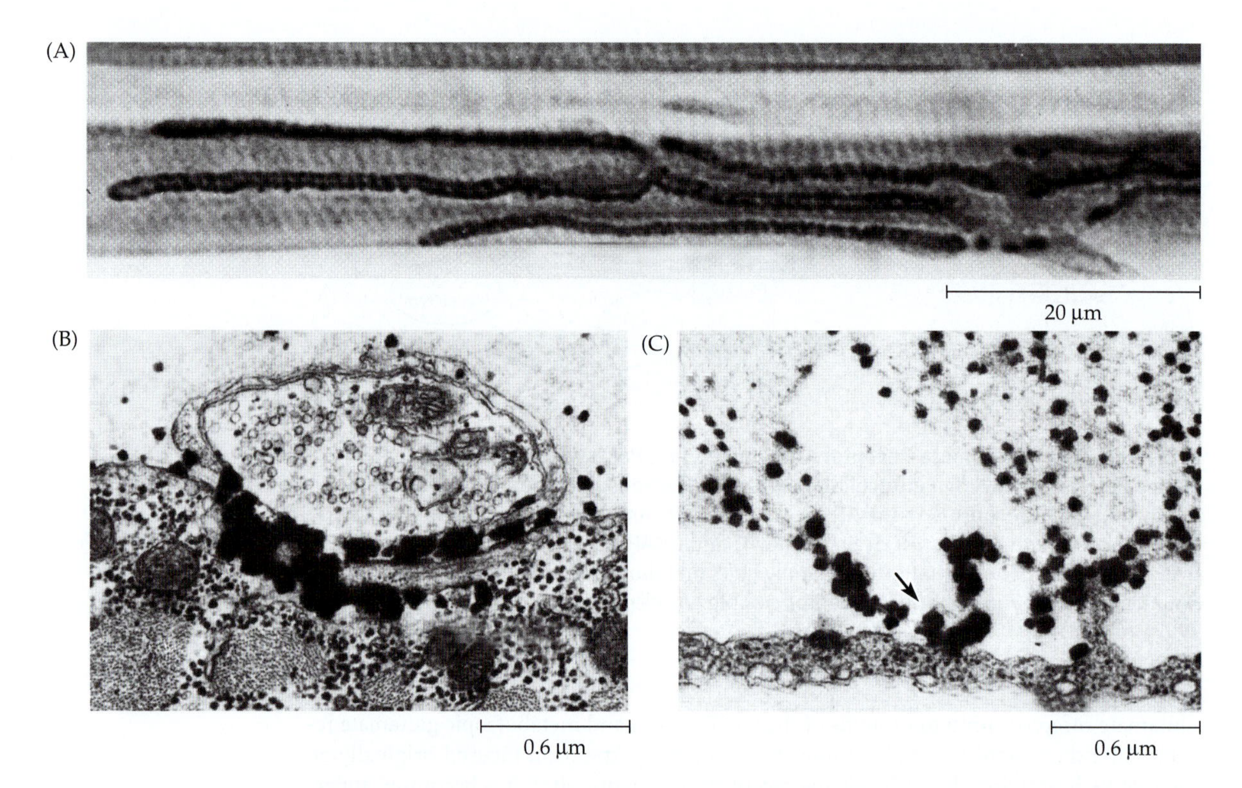

FIGURE 13.22 Acetylcholinesterase Is Concentrated in the synaptic basal lamina at the skeletal neuromuscular junction. (A) Light micrograph of a neuromuscular junction in a frog cutaneous pectoris muscle stained by a histochemical procedure for acetylcholinesterase. The dark reaction product lines the synaptic gutters and junctional folds. (B) Electron micrograph of a cross section of an axon terminal from a muscle stained for acetylcholinesterase as in part A. The electron-dense reaction product fills the synaptic cleft and the junctional folds. (C) Electron micrograph of a damaged muscle in which the nerve terminal, Schwann cell, and muscle fiber have degenerated and been phagocytized, leaving only empty basal lamina sheaths. The damaged muscle was stained for acetylcholinesterase; reaction product is associated with the synaptic basal lamina (arrow). (Micrographs kindly provided by U. J. McMahan.)

ceptors. However, if the dimensions of the cleft and the rates of ACh diffusion, binding, and hydrolysis are taken into consideration, a simple scheme emerges, called the **saturated disk**.[116] Following the release of one quantum, the concentration of ACh increases almost instantaneously (within microseconds) across the width of the cleft to a level high enough (0.5 m*M*) to saturate both ACh receptors and esterase within a disk approximately 0.5 μm in diameter centered on the release site. Binding of ACh to its receptors and to acetylcholinesterase is rapid compared to the rate at which acetylcholinesterase can hydrolyze ACh (it takes AChE 0.1 ms to hydrolyze one molecule of ACh). Therefore, the fraction of ACh molecules released that bind initially to postsynaptic receptors is determined by the ratio of receptors to esterase; this means that approximately 20% of the ACh molecules bind to AChE, and 80% to ACh receptors.

Binding causes a precipitous fall in ACh concentration. The concentration then remains low because AChE can hydrolyze ACh molecules much faster (10/ms) than they are released from receptors as the channels close ($\tau = 1$ ms; Chapter 9). Thus, by 0.1 ms or so after release, the concentration of ACh in the cleft has fallen to levels that make the probability of two ACh molecules being available to bind and open another receptor negligible.

Such an analysis predicts that inhibition of acetylcholinesterase should have a more pronounced effect on the duration of the synaptic potential than on its amplitude, which is the case; amplitude is increased 1.5- to 2-fold, and the duration is increased 3- to 5-fold.[117,118] Thus, the organization of the neuromuscular junction and the density and ki-

[116]Bartol, T. M., et al. 1991. *Biophys. J.* 59: 1290–1307.
[117]Fatt, P., and Katz, B. 1951. *J. Physiol.* 115: 320–370.
[118]Katz, B., and Miledi, R. 1973. *J. Physiol.* 231: 549–574.

netic properties of ACh receptors and AChE combine to produce a synapse that is capable of very rapid responses and efficient use of ACh.

Removal of ATP by Hydrolysis

Like ACh, the action of ATP is rapidly terminated by hydrolysis.[119] EctoATP diphosphohydrolase (ADPase or apyrase) hydrolyzes ATP to ADP, and ADP to AMP. AMP is converted to adenosine by ecto-5′-nucleotidase. Both enzymes are found on glial cells, as well as at synaptic sites on neurons.[120,121] At many synapses, adenosine modulates synaptic transmission by combining with receptors on pre- and postsynaptic cells.[122] Its action is terminated by uptake and by adenosine deaminase, which degrades adenosine to inosine.

Removal of Transmitters by Uptake

The actions of dopamine, norepinephrine, glutamate, 5-HT, glycine, and GABA are terminated by uptake of the transmitter into presynaptic nerve terminals, postsynaptic cells, or glial cells by specific transport proteins.[123,124] Transmitters transported into nerve terminals can be repackaged and released again. There are two major protein families of transmitter transporters; both use the energy provided by the movement of sodium down its electrochemical gradient to drive transmitter uptake (Chapter 4). The sodium- and chloride-dependent transporters for norepinephrine (NET), GABA (GAT), glycine (GLYT), dopamine (DAT), and 5-HT (SERT) each have 12 putative transmembrane domains, intracellular amino and carboxy termini, and several extracellular glycosylation sites. Extracellular sodium and chloride ions are required and are cotransported with the transmitter.

Sodium-dependent transporters for anionic and zwitterionic amino acids make up a second major neurotransmitter transporter family.[124] They are proteins with 10 transmembrane domains. Members of this family (called GLAST, GLT-1, EAAC1 in rats and EAAT1-5 in humans) terminate the action of glutamate and aspartate. The stoichiometry of transport varies among family members: Two to three sodium ions are transported together with each glutamate into the cell, one potassium ion is transported out, and either a proton is transported in or a hydroxyl (or bicarbonate) ion is transported out. In rats, the GLAST and GLT-1 glutamate transporters are found predominantly in glial cells; EAAC1 is localized to neurons.[125] Knockout experiments reveal that the majority of glutamate uptake is into glial cells, but that all three transporters must be present for normal function.[126]

The transporters that carry neurotransmitters into nerve terminals differ from those that carry them into synaptic vesicles. For example, the uptake of norepinephrine into the terminal is blocked by cocaine, but its uptake into vesicles is blocked by reserpine. An interesting exception is the drug fenfluramine, which inhibits transport of 5-HT by both the vesicular transporter and the plasma membrane transporter.[127] As a result, 5-HT leaks out of synaptic vesicles and accumulates in the cytoplasm, which results in release of 5-HT from the terminal by running the plasma membrane transporter in reverse.

The importance of such uptake mechanisms is revealed by the consequences of drugs that inhibit neurotransmitter transporters.[123] Tricyclic antidepressants such as desipramine are potent inhibitors of norepinephrine uptake, and some of the newer antidepressants—such as fluoxetine (Prozac), sertraline (Zoloft), and paroxetine (Paxil)—are particularly potent inhibitors of 5-HT uptake.[128] Failure to remove glutamate from the extracellular space in the CNS can result in excessive activation of glutamate receptors, leading to epilepsy, excitotoxic neurodegeneration, and eventually paralysis.[126]

The action of peptide transmitters is terminated by receptor desensitization and removal of the peptide from the extracellular fluid by diffusion and degradation.[129] There is no evidence for uptake of peptides into presynaptic terminals. Enzymes that hydrolyze specific peptides are beginning to be characterized.[130]

[119]Zimmermann, H., and Braun, N. 1996. *J. Auton. Pharmacol.* 16: 397–400.

[120]Wang, T. F., and Guidotti, G. 1998. *Brain Res.* 790: 318–322.

[121]Maienschein, V., and Zimmermann, H. 1996. *Neuroscience* 70: 429–438.

[122]Brundege, J. M., and Dunwiddie, T. V. 1997. *Adv. Pharmacol.* 39: 353–391.

[123]Amara, S. G., and Kuhar, M. J. 1993. *Annu. Rev. Neurosci.* 16: 73–93.

[124]Palacín, M., et al. 1998. *Physiol. Rev.* 78: 969–1054.

[125]Rothstein, J. D., et al. 1994. *Neuron* 13: 713–725.

[126]Rothstein, J. D., et al. 1996. *Neuron* 16: 675–686.

[127]Frazer, A., and Hensler, J. G. 1999. In *Basic Neurochemistry: Molecular, Cellular, and Medical Aspects*, 6th Ed. Lippincott-Raven, Philadelphia, pp. 263–292.

[128]Fuller, R. W. 1995. *Prog. Drug Res.* 45: 167–204.

[129]Grady, E., et al. 1997. *J. Invest. Dermatol. Symp. Proc.* 2: 69–75.

[130]Turner, A. J., and Tanzawa, K. 1997. *FASEB J.* 11: 355–364.

Summary

- Acetylcholine, norepinephrine, epinephrine, dopamine, 5-hydroxytryptamine, histamine, adenosine triphosphate, γ-aminobutyric acid, glycine, and glutamate are classic low-molecular-weight transmitters.

- Nitric oxide and carbon monoxide are low-molecular-weight transmitters that are lipid and water soluble and are not stored in cells.

- Neuropeptides form a third group of transmitters.

- Many neurons release more than one transmitter, typically a low-molecular-weight transmitter and one or more neuropeptides.

- Classic low-molecular-weight transmitters are synthesized in axon terminals, packaged in small synaptic vesicles, and stored for release. Feedback mechanisms control the number and activity of enzymes that catalyze transmitter synthesis so as to maintain an adequate supply of transmitter.

- Neuropeptides are synthesized in the cell body, processed and packaged in large dense-core vesicles in the Golgi apparatus, and transported to the axon terminal.

- Slow axonal transport moves soluble proteins and components of the cytoskeleton from the cell body to the axon terminal at a rate of 1 to 2 mm/day.

- Fast axonal transport moves vesicles and other organelles at speeds of up to 400 mm/day either toward the terminal (anterograde transport) or toward the cell soma (retrograde transport). Fast transport is mediated by molecular motors that move organelles along microtubules.

- A highly conserved mechanism for vesicle trafficking and membrane fusion, described by the SNARE hypothesis, mediates synaptic vesicle exocytosis. Synaptotagmin appears to be the calcium sensor for release.

- After transmitter release, components of the synaptic vesicle membrane are retrieved by endocytosis and recycled.

- A complex postsynaptic apparatus immobilizes transmitter receptors by linking them to the cytoskeleton, and it also provides an intracellular scaffold for signaling proteins that determine some of the consequences of receptor activation.

- The final step in chemical synaptic transmission is the removal of transmitter from the synaptic cleft by diffusion, degradation, or uptake. Prompt transmitter removal is important for normal synaptic function.

SUGGESTED READING

General Reviews

Bajjalieh, S. M. 1999. Synaptic vesicle docking and fusion. *Curr. Opin. Neurobiol.* 9: 321–328.

Cooper, J. R., Bloom, F. E., and Roth, R. H. 1996. *The Biochemical Basis of Pharmacology*, 7th Ed. Oxford University Press, New York.

Hilfiker, S., Pieribone, V. A., Czernik, A. J., Kao, H-T., Augustine, G. J., and Greengard, P. 1999. Synapsins as regulators of neurotransmitter release. *Philos. Trans. R. Soc. Lond. B* 354: 269–279.

Hirokawa, N. 1998. Kinesin and dynein superfamily proteins and the mechanism of organelle transport. *Science* 279: 519–526.

Kneussel, M., and Betz, H. 2000. Receptors, gephyrin and gephyrin-associated proteins: Novel insights into the assembly of inhibitory postsynaptic membrane specializations. *J. Physiol.* 525: 1–9.

Lundberg, J. M. 1996. Pharmacology of cotransmission in the autonomic nervous system: Integrative aspects on amines, neuropeptides, adenosine triphosphate, amino acids and nitric oxide. *Pharmacol. Rev.* 48: 113–178.

O'Brien, R. J., Lau, L. F., and Huganir, R. L. 1998. Molecular mechanisms of glutamate receptor clustering at excitatory synapses. *Curr. Opin. Neurobiol.* 8: 364–369.

Palacín, M., Estévez, R., Bertran, J., and Zorzano, A. 1998. Molecular biology of mammalian plasma membrane amino acid transporters. *Physiol. Rev.* 78: 969–1054.

Pellizzari, R., Rossetto, O., Schiavo, G., and Montecucco, C. 1999. Tetanus and botulinum neurotoxins: Mechanism of action and therapeutic uses. *Philos. Trans. R. Soc. Lond. B* 354: 259–268.

Robinson, M. S. 1994. The role of clathrin, adaptors, and dynamin in endocytosis. *Curr. Opin. Cell Biol.* 6: 538–544.

Sanes, J. R., and Lichtman, J. W. 1999. Development of the vertebrate neuromuscular junction. *Annu. Rev. Neurosci.* 22: 389–442.

Schuldiner, S., Shirvan, A., and Linial, M. 1995. Vesicular neurotransmitter transporters: From bacteria to humans. *Physiol. Rev.* 75: 369–392.

Sheng, M. and Lee, S. H. 2000. Growth of the NMDA receptor industrial complex. *Nature Neurosci.* 3: 633–635.

Siegel, G. J., Agranoff, B. W., Albers, R. W., Fisher, S. K., and Uhler, M. D. (eds.). 1999. *Basic Neurochemistry: Molecular, Cellular, and Medical Aspects*, 6th Ed. Lippincott-Raven, Philadelphia.

Vallee, R. B., and Bloom, G. S. 1991. Mechanisms of fast and slow axonal transport. *Annu. Rev. Neurosci.* 14: 59–92.

Original Papers

Birks, R. I., and MacIntosh, F. C. 1961. Acetylcholine metabolism of a sympathetic ganglion. *J. Biochem. Physiol.* 39: 787–827.

Brady, S. T., Lasek, R. J., and Allen, R. D. 1982. Fast axonal transport in extruded axoplasm from squid giant axon. *Science* 218: 1129–1131.

Howard, J., Hudspeth, A. J., and Vale, R. D. 1989. Movement of microtubules by single kinesin molecules. *Nature* 342: 154–158.

Jonas, P., Bischofberger, J., and Sandkuhler, J. 1998. Corelease of two fast neurotransmitters at a central synapse. *Science* 281: 419–424.

Kuromi, H., and Kidokoro, Y. 1998. Two distinct pools of synaptic vesicles in single presynaptic boutons in a temperature-sensitive *Drosophila* mutant, *shibire*. *Neuron* 20: 917–925.

McMahan, U. J., Sanes, J. R., and Marshall, L. M. 1978. Cholinesterase is associated with the basal lamina at the neuromuscular junction. *Nature* 271: 172–174.

Nusser, Z., Mulvihill, E., Streit, P., and Somogyi, P. 1994. Subsynaptic segregation of metabotropic and ionotropic glutamate receptors as revealed by immunogold localization. *Neuroscience* 61: 421–427.

Schnapp, B. J., Vale, R. D., Sheetz, M. P., and Reese, T. S. 1985. Single microtubules from squid axoplasm support bi-directional movement of organelles. *Cell* 40: 455–462.

Sulzer, D., Joyce, M. P., Lin, L., Geldwert, D., Haber, S. N., Hattori, T., and Rayport, S. 1998. Dopamine neurons make glutamatergic synapses in vitro. *J. Neurosci.* 18: 4588–4602.

14 Neurotransmitters in the Central Nervous System

THIS CHAPTER IS CONCERNED WITH THE FUNCTIONAL ROLE of particular transmitters within the central nervous system. Important insights into function come from studies of transmitter distribution, the effects of drugs that interfere with synthesis, storage, release, or action, and from mutations or knockout experiments targeting proteins that mediate these processes. Such studies also provide clues to biochemical mechanisms underlying nervous system dysfunction and suggest possible new therapies.

γ-Aminobutyric acid (GABA) mediates inhibitory interactions throughout the CNS by way of three classes of receptors. $GABA_A$ receptors are the most common. They respond to GABA with an increase in chloride conductance, and their activity is modulated by anticonvulsant drugs such as barbiturates and benzodiazepines. Glutamate is the major excitatory transmitter in the CNS. It interacts with two classes of ionotropic receptors that are distinguished by the types of agonists that activate them and by their relative permeability to calcium. GABA and glutamate also act on metabotropic receptors.

Acetylcholine acts as a transmitter in many brain regions through metabotropic muscarinic receptors. In addition, presynaptically located nicotinic receptors modulate CNS activity. Nuclei in the basal forebrain provide extensive and diffuse cholinergic innervation to the cortex and hippocampus. The basal forebrain cholinergic system is a conspicuous locus of degenerative changes in Alzheimer's disease, although neurons releasing other transmitters also are affected.

ATP acts on ionotropic or metabotropic receptors to modulate, or mediate directly, synaptic transmission in the CNS. Adenosine modulates CNS transmission by way of metabotropic receptors.

Considerable interest in substance P and the opioid peptides stems from the finding that they are involved in pain sensation. Substance P is released by primary afferent fibers that respond to noxious stimuli. Enkephalin, an endogenous opioid released by interneurons in the spinal cord, suppresses pain sensation by blocking the release of substance P from primary afferent terminals. Other opioid peptides act at synapses within the brain to alter our perception of pain. Both substance P and opioid peptides are involved in other neural functions in addition to pain sensation.

A remarkable feature of the distribution of norepinephrine, dopamine, epinephrine, serotonin, and histamine in the CNS is that relatively few neurons release these amines as transmitters. The arborizations of those cells are, however, extensive and broadly distributed, so that each neuron sends literally thousands of branches throughout the CNS. This morphology correlates with the physiological roles of amine-containing neurons in modulating synaptic activity in widespread areas of the CNS, so as to regulate global functions such as attention, arousal, the sleep–wake cycle, mood, and affect.

The complexity and plasticity of synaptic connections within the central nervous system provide the physical substrate for behavior. Thus, knowledge of the transmitters involved in synaptic function and their mechanism of action is central to our understanding of the brain. In addition, the synthetic and degradative pathways for each neurotransmitter offer potential targets for pharmacological intervention to treat the imbalances arising in disease states. Still greater promise for basic understanding and clinical intervention is offered by the rich variety of membrane receptors that have been identified for each transmitter type, using molecular cloning techniques. However, the new knowledge arising from molecular genetic studies would not be of use without also establishing the identity, mechanisms of action, and location of neurotransmitters in the central nervous system.

Material in this chapter recapitulates some of the information presented in Chapters 3 and 13. Here, however, we emphasize the types of neurons that contain different transmitters and their distribution in the central nervous system. Thus, we begin with a description of techniques for transmitter identification that have been useful for mapping the distribution of neurons releasing a particular transmitter, summarized in Figure 14.1. These include visualization of the transmitter itself; labeling one of the proteins mediating its action, synthesis, or degradation; and detecting mRNA transcripts for such proteins. Identification of candidate genes allows one to produce transgenic animals or to use antisense oligonucleotides to block specific mRNAs (Box 14.1). Ultimately, such informa-

Box 14.1 Molecular Methods and CNS Transmitters

Techniques of molecular biology can be used to modify synaptic function in vitro and, more remarkably, in living animals. Proteins mediating transmitter storage and release, pre- and postsynaptic transmitter receptors, and transporters that remove transmitter from the synaptic cleft all can be targeted by these techniques. **Antisense oligonucleotides** can be introduced into cells to combine with a specific messenger RNA (mRNA) and prevent its translation into protein, or make it subject to enzymatic degradation.[1] The chief challenges to this technique are to determine a sequence that blocks synthesis effectively and further, to establish that such effects are specific to that sequence. Scrambled, or "mis-sense," oligonucleotides are used as negative controls. In this way it is possible to assess the contribution of identified gene products to synaptic function in cells in vitro or in brain slice preparations.[2,3]

The function of identified gene products in the intact brain can be studied in **transgenic animals** using the technique of **homologous recombination**. Here an altered and usually nonfunctional variant of a target gene is synthesized and transfected into pluripotent embryonic stem cells. Identical sequences shared by the native and foreign genes allow recombination to occur, with the chance for integration of the nonfunctional transgene into some fraction of the embryonic cells. The recombinant stem cells are injected into a blastocyst that is implanted into a host mother. Transgenic offspring can be identified by a visible marker gene within the transgene (such as one coding for a difference in coat color), and confirmed by analysis of the DNA. The heterozygous transgenics are then bred to produce homozygous mice in which both copies of the target gene have been disrupted—a **knockout** mouse.

Perhaps it is no surprise to note that transgenic knockout mice do not always give decisive answers. In some cases the knockout gene proves to have been necessary for development or survival, so homozygous animals die. In other cases the knockout appears to have no discernible effect—possibly because of redundancy, with other gene products able to substitute. Nonetheless, a number of informative knockout mice have been made that have revealed the function of peptide transmitters or transmitter receptors in the CNS.[4] More consistent and useful results may become possible as cell type–specific promoters are incorporated into the transgene. These regulatory elements of DNA can direct the synthesis of the transgenic protein so that only particular cell types are affected. Another promising variant of this technique is to "knock in" a gene that has been modified to have a novel pharmacology, and which can be activated specifically.

[1]Yu, C., et al. 1993. *Dev. Genet.* 14: 296–304.
[2]Sumikawa, K., and Miledi, R. 1988. *Proc. Natl. Acad. Sci. USA* 85: 1302–1306.
[3]Ramirez-Latorre, J., et al. 1996. *Nature* 1996. 380: 347–351.
[4]Davies, R. W., Gallagher, E. J., and Savioz, A. 1994. *Prog. Neurobiol.* 42: 319–331.

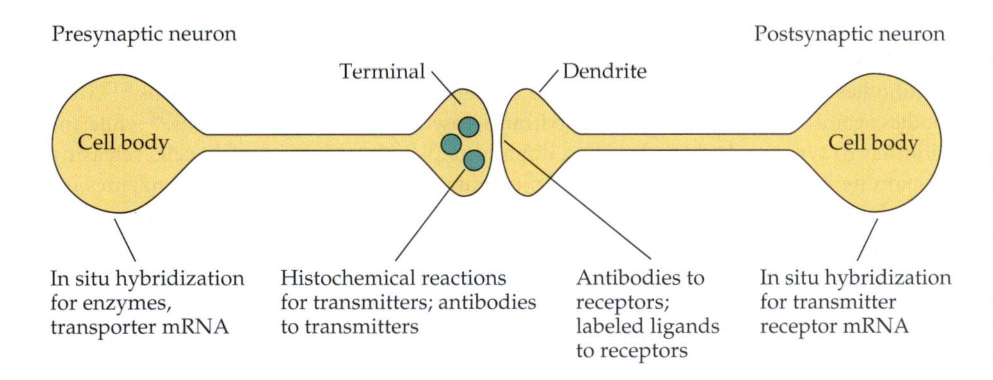

FIGURE 14.1 Methods for Identifying Neurotransmitters in the CNS. Labeled antibodies or nucleotide probes can be used to detect the expression of enzymes involved in synthetic and degradative pathways in the presynaptic neuron. The neurotransmitter itself can be detected by chemical reaction or by antibodies to a conjugated form of the transmitter. Specific uptake of radiolabeled transmitter can identify some neurons. Ligands or antibodies to postsynaptic receptors, or nucleotide probes for receptor mRNA, provide means to identify cells sensitive to a particular transmitter.

tion may suggest therapeutic interventions for treating central nervous system defects using genetic engineering and gene transfer techniques.

MAPPING TRANSMITTER DISTRIBUTION

One method for visualizing neurons that release different transmitters in the CNS is to map their distribution by histological techniques or by immunohistochemistry. The Falck and Hillarp[5] fluorescence method labels neurons containing biogenic amines such as dopamine, norepinephrine, and 5-hydroxytryptamine (5-HT, or serotonin). When condensed with formaldehyde, each of these substances emits light of a characteristic wavelength under ultraviolet illumination[6] (Figure 14.2). Specific antibodies have been made against small neurotransmitters, such as GABA (γ-aminobutyric acid), 5-HT, dopamine,[7] and neuropeptides. The antibodies are coupled, either directly or indirectly, to markers that can be detected by light, fluorescence, or electron microscopy. Cloning of neuropeptide cDNAs and in situ hybridization techniques have provided additional tools for identifying neurons that synthesize neuropeptides, and for studying the regulation of peptide expression.[8]

An alternative to labeling the transmitter itself is to make specific probes for enzymes mediating transmitter synthesis and degradation. Neurons contain high levels of

[5]Falck, B., et al. 1962. *J. Histochem. Cytochem.* 10: 348–354.

[6]de la Torre, J. C. 1980. *J. Neurosci. Methods* 3: 1–5.

[7]Wässle, H., and Chun, M. H. 1988. *J. Neurosci.* 8: 3383–3394.

[8]Mengod, G., Charli, J. L., and Palacios, J. M. 1990. *Cell Mol. Neurobiol.* 10: 113–126.

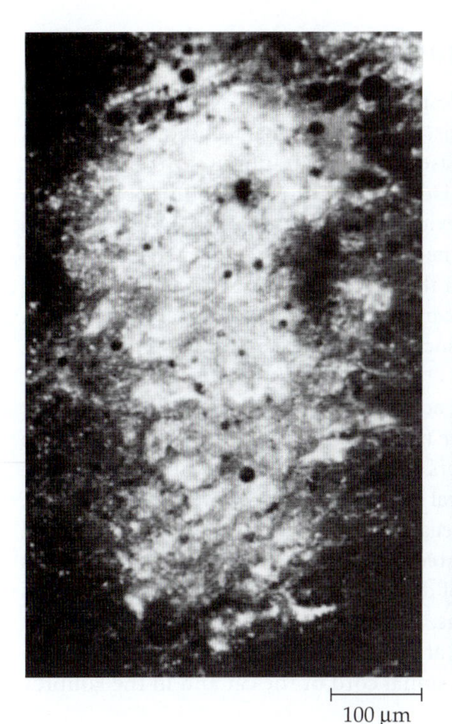

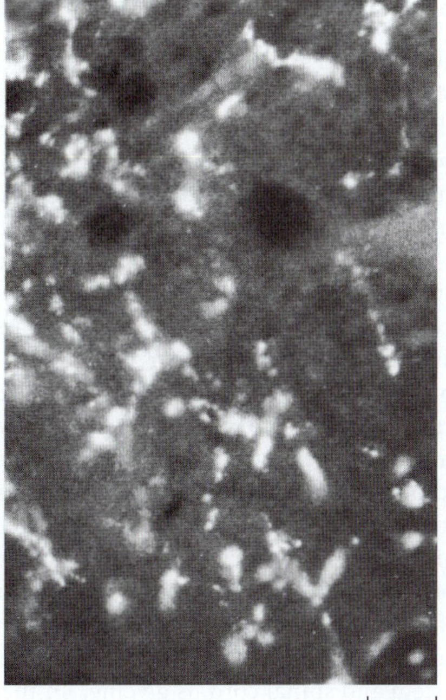

FIGURE 14.2 Visualization of Biogenic Amine-Containing Cells and their terminal arborizations by formaldehyde-induced fluorescence. (A) Norepinephrine-containing cells in the locus coeruleus. (B) The terminal arborizations of locus coeruleus cells in the hippocampus. (From Harik, 1984.)

enzymes catalyzing the synthesis of the transmitter they release. For example, cells having GABA as a transmitter possess high levels of the enzyme glutamic acid decarboxylase. Antibodies to the enzyme have been used to identify GABA-containing neurons.[9,10] Antibodies against the enzyme choline acetyltransferase label cholinergic neurons,[11] while antibodies to tyrosine hydroxylase and dopamine β-hydroxylase identify cells releasing dopamine and norepinephrine.[12] Complementary DNAs for many of the enzymes involved in transmitter synthesis have been cloned. The localization and regulation of expression of mRNA for these enzymes has been studied using in situ hybridization techniques.[13] Enzymes involved in transmitter degradation are less reliable indicators of transmitter identity, although for acetylcholine and GABA the distribution of their degradative enzymes has proven a useful marker under appropriate conditions.[14–16]

The chemical nature of synapses also can be identified using probes that recognize the receptors on postsynaptic cells. For instance, monoclonal antibodies against GABA receptors have been used to localize synapses where GABA is likely to be the transmitter.[17] Alternatively, the transmitter itself, or a specific agonist or antagonist, can be radioactively labeled and used to localize receptors.[18–20] Other techniques exploit mechanisms of transmitter inactivation. For example, cells that release biogenic amines or GABA have specific uptake systems for recapturing these transmitters. Thus, bathing tissue in radioactive norepinephrine, dopamine, 5-HT, or GABA can selectively label nerve terminals that release these transmitters.

The rapid growth of knowledge about the genes that encode neurotransmitter receptors has enabled a host of experimental strategies employing molecular genetic techniques. In situ hybridization locates the mRNA. Reverse-transcriptase polymerase chain reaction (RT-PCR) amplification of identified gene products can be performed on microdissected neural tissue, and even on single isolated cells.[21] Transgenic animals can be bred in which a candidate receptor gene is "knocked out" or otherwise altered (see Box 14.1). Behavioral and physiological studies can then assess the contribution of the transgene product to synaptic function. A similar kind of experiment can be performed on cells in vitro into which an antisense oligonucleotide is introduced.[22] Finally, naturally occurring mutations in neurotransmitter-related genes in humans and other species lead to behavioral changes that speak to the function of the encoded gene product. In the case of a neurotransmitter receptor, there may be anatomical or functional changes in a region of the brain, or even in particular neurons that were suspected to be sensitive to that neurotransmitter.

GABA and Glycine: Inhibitory Transmitters in the CNS

GABA is released at inhibitory synapses in widespread areas of the CNS. In an elegant series of studies at the level of individual neurons, Otsuka, Ito, Obata, and their colleagues established that cerebellar Purkinje cells release GABA as an inhibitory transmitter at synapses on cells in the brainstem.[23–25] The visual system provides another example in which both physiological and morphological evidence indicates that GABA is released as a transmitter (Chapters 19 and 20). Thus, in the retina, GABA is associated with certain types of horizontal and amacrine cells.[26–29] In the lateral geniculate nucleus, some cells contain the synthetic enzyme glutamic acid decarboxylase, and in the visual cortex GABA uptake and glutamic acid decarboxylase are associated with several distinct types of local-circuit inhibitory neurons.[30]

Local application of agents that block the action of GABA has effects on signaling: Bicuculline and picrotoxin change the receptive field organization of ganglion cells in the retina and complex cells in the cortex (Chapters 19 and 20).[31,32] A striking finding is the high proportion of GABA neurons in the central nervous system. In several cortical areas, for example, one out of every five neurons releases GABA as a transmitter.[33] The overall importance of inhibitory interactions mediated by GABA for signaling in the CNS is demonstrated by the finding that application of drugs that block GABA receptors, such as picrotoxin and penicillin, produces convulsions.[34]

A second transmitter mediating inhibition at CNS synapses, particularly those in the brainstem and spinal cord, is glycine. In the spinal cord of the cat and in the lamprey

[9]Matsuda, T., Wu, J-Y., and Roberts, E. 1973. *J. Neurochem.* 21: 159–166, 167–172.

[10]Hendrickson, A. E., et al. 1983. *J. Neurosci.* 3: 1245–1262.

[11]Wainer, B. H., et al. 1984. *Neurochem. Int.* 6: 163–182.

[12]Miachon, S., et al. 1984. *Brain Res.* 305: 369–374.

[13]Kawata, M., Yuri, K., and Sano, Y. 1991. *Comp. Biochem. Physiol. C* 98: 41–50.

[14]Shute, C. C. D., and Lewis, P. R. 1963. *Nature* 199: 1160–1164.

[15]Wallace, B. G., and Gillon, J. W. 1982. *J. Neurosci.* 2: 1108–1118.

[16]Nagai, T., et al. 1984. In *Classical Transmitters and Transmitter Receptors in the CNS.* Vol. 3 of *Handbook of Chemical Neuroanatomy.* Elsevier, Amsterdam, pp. 247–272.

[17]Hendry, S. H. C., et al. 1990. *J. Neurosci.* 10: 2438–2450.

[18]Kuhar, M. J., De Souza, E. B., and Unnerstall, J. R. 1986. *Annu. Rev. Neurosci.* 9: 27–59.

[19]Palacios, J. M., et al. 1990. *Ann. N.Y. Acad. Sci.* 600: 36–52.

[20]Palacios, J. M., et al. 1990. *Prog. Brain Res.* 84: 243–253.

[21]Kacharmina, J. E., Crino, P. B., and Eberwine, J. 1999. *Methods Enzymol.* 303: 3–18.

[22]Brussaard, A. B. 1997. *J. Neurosci. Methods* 71: 55–64.

[23]Obata, K. 1969. *Experientia* 25: 1283.

[24]Otsuka, M., et al. 1971. *J. Neurochem.* 18: 287–295.

[25]Obata, K., Takeda, K., and Shinozaki, H. 1970. *Exp. Brain Res.* 11: 327–342.

[26]Sterling, P. 1983. *Annu. Rev. Neurosci.* 6: 149–185.

[27]Lam, D. M. 1997. *Invest. Ophthalmol. Vis. Sci.* 38: 553–556.

[28]Wassle, H., et al. 1998. *Vision Res.* 38: 1411–1430.

[29]Fletcher, E. L., and Wassle, H. 1999. *J. Comp. Neurol.* 413: 155–167.

[30]Gilbert, C. D. 1983. *Annu. Rev. Neurosci.* 6: 217–247.

[31]Caldwell, J. H., and Daw, N. W. 1978. *J. Physiol.* 276: 299–310.

[32]Sillito, A. M. 1979. *J. Physiol.* 289: 33–53.

[33]Naegele, J. R., and Barnstable, C. J. 1989. *Trends Neurosci.* 12: 28–34.

[34]Olsen, R. W., and Leeb-Lundberg, F. 1981. In *GABA and Benzodiazepine Receptors.* Raven, New York, pp. 93–102.

CNS, glycine plays a key role as an inhibitory transmitter.[35,36] The increase in chloride conductance in target neurons that is produced by the stimulation of inhibitory pathways is accurately mimicked by application of glycine from a pipette. The effects of two different transmitters, such as GABA and glycine, both of which open chloride channels, can be distinguished with single-channel recording or noise analysis to determine the conductance and kinetics of the opened channels. In this way one can demonstrate that the characteristics of the channel opened by the natural transmitter during synaptic activity resemble those activated by exogeneously applied GABA or glycine. A simple and useful discrimination between the two transmitters can also be made by the use of agents that block inhibition, such as strychnine for glycine and picrotoxin or bicuculline for GABA.[37]

The ionotropic glycine receptor (glyR) is a member of the ligand-gated ion channel superfamily that includes the nicotinic ACh, GABA, and 5-HT receptors[38] (Chapter 3). The importance of glycinergic transmission is emphasized by the existence of mutations of the α subunit of the glycine receptor in mice and humans that result in motor and behavioral defects.[39,40]

GABA Receptors

Three classes of receptors for GABA have been identified in the CNS. Two of these, $GABA_A$ and $GABA_C$, are ionotropic receptors; the $GABA_B$ receptor is metabotropic. Four classes of $GABA_A$ receptor subunits have been identified (Chapter 3): α, β, γ, and δ.[41,42] Experimentally, channels can be assembled from α and β subunits, or from either alone. Multiple variants exist within each class of subunit, and different combinations of subunits produce receptors with different properties. Studies of the distribution of mRNAs encoding variant subunits demonstrate differences among brain regions, suggesting regional specificity in receptor subtype.[33] As seen in Figure 14.3, the α_1 and α_6, but not α_3, subtypes are highly expressed in the cerebellum. Some combinations of subunits (e.g., α_1, β_2, γ_2) appear to be distributed more widely than others.[43] It is postulated that differences in subunit composition contribute to the heterogeneous pharmacology displayed by native GABA receptors.[44]

[35] Martin, A. R. 1990. In *Glycine Neurotransmission.* Wiley, New York, pp. 171–191.

[36] Luddens, H., and Wisden, W. 1991. *Trends Pharmacol. Sci.* 12: 49–51.

[37] Matthews, G., and Wickelgren, W. O. 1979. *J. Physiol.* 293: 393–415.

[38] Zafra, F., Aragon, C., and Gimenez, C. 1997. *Mol. Neurobiol.* 14: 117–142.

[39] Shiang, R., et al. 1993. *Nature Genet.* 5: 351–358.

[40] Ryan, S. G., et al. 1994. *Nature Genet.* 7: 131–135.

[41] Johnston, G. A. 1996. *Pharmacol. Ther.* 69: 173–198.

[42] Vafa, B., and Schofield, P. R. 1998. *Int. Rev. Neurobiol.* 42: 285–332.

[43] Rabow, L. E., Russek, S. J., and Farb, D. H. 1995. *Synapse* 21: 189–274.

[44] Weiner, J. L., Gu, C., and Dunwiddie, T. V. 1997. *J. Neurophysiol.* 77: 1306–1312.

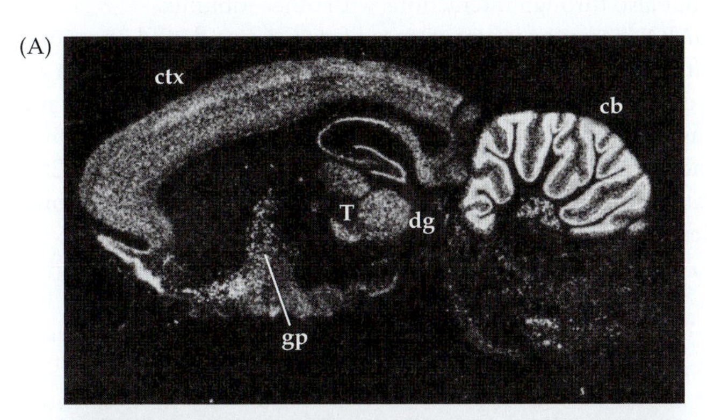

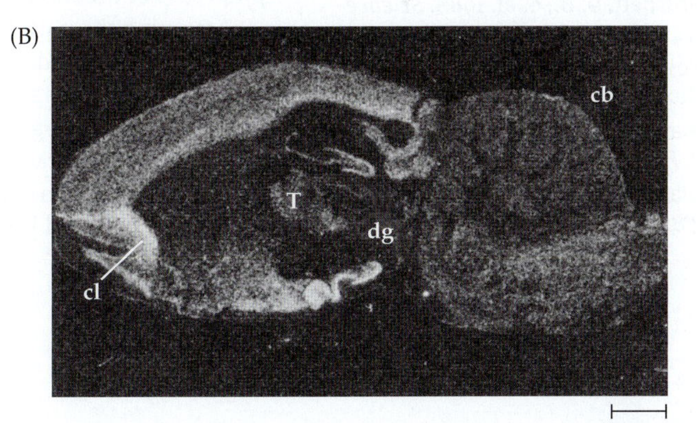

FIGURE 14.3 The Distribution of $GABA_A$ Receptor Subunit mRNAS in the rat brain. Light-microscopic autoradiographs of serial sections of rat brain following in situ hybridization with ^{35}S-labeled antisense oligonucleotide probes for the α_1 (A), α_3 (B), and α_6 (C) $GABA_A$ receptor subunits. cb = cerebellum, cl = claustrum, ctx = cortex, dg = dentate gyrus, gp = globus pallidus, gr = cerebellar granule cells, T = thalamus. (After Luddens and Wisden, 1991; autoradiographs kindly provided by W. Wisden.)

In the early 1980s, responses to GABA with a slow time course and insensitivity to bicuculline were observed on neurons in the CNS.[45] These slow responses were found to be activated by baclofen, which has no effect on $GABA_A$ receptors. The receptors mediating these effects are metabotropic $GABA_B$ receptors. A $GABA_B$ receptor gene has been cloned.[46,47] It codes for a seven-transmembrane G protein–coupled receptor related to metabotropic glutamate receptors (Chapter 10). Postsynaptic $GABA_B$ receptors activate an inward rectifier potassium channel (GIRK or Kir3.0).[48–50] Presynaptic $GABA_B$ receptors suppress the activity of voltage-gated calcium channels.[51]

Retinal bipolar cells have rapid GABA-gated chloride currents like those due to $GABA_A$ activation but that are insensitive to bicuculline (or baclofen). These have been termed $GABA_C$ receptors.[52] They are pharmacologically, genetically, and functionally distinct from $GABA_A$ receptors.[53] $GABA_C$ receptors are especially sensitive to the analogue of GABA called CACA (*cis*-4-aminocrotonic acid). Two related genes (*rho1* and *rho2*) code for a $GABA_C$-like receptor when expressed in *Xenopus* oocytes,[54] and the rho2 subunit is found throughout the brain. $GABA_A$ and $GABA_C$ subunits may coassemble to generate still another class of functionally distinct receptors.[55]

Modulation of $GABA_A$ Receptor Function by Benzodiazepines and Barbiturates

A distinctive feature of $GABA_A$ receptors is their regulation by allosteric modulation.[56] $GABA_A$ receptors exhibit binding sites for two classes of modulators: benzodiazepines and barbiturates (Figure 14.4). Benzodiazepines, which include diazepam (Valium) and chlordiazepoxide (Librium), are antianxiety drugs and muscle relaxants; barbiturates, such as phenobarbital and secobarbital, are anticonvulsants. Both increase GABA-induced chloride current—benzodiazepines by increasing the frequency of channel opening, and barbiturates by prolonging channel burst duration.[48] The α and β subunits each have binding sites for both GABA and barbiturates (see Figure 14.4); indeed, expression of either subunit alone can result in formation of a functional receptor.[57] Benzodiazepines bind to the γ_2 subunit, and the γ subunit must be coexpressed with α and β subunits to form a $GABA_A$ receptor that can be modulated by benzodiazepines.[58] The affinity and specificity of each binding site for its ligand is determined not only by the intrinsic properties of the subunit, but also through interactions with other subunits.

Assuming the native $GABA_A$ receptor is a pentamer of two α, two β, and one γ or δ subunit, the known diversity in subunits is sufficient to produce hundreds of different receptors. Determining the distribution and properties of various receptor subtypes and identifying the endogenous ligands, if any exist, for the barbiturate and benzodiazepine binding sites are areas of active research.[59] GABA (and glutamate) receptors in the CNS and retina also are modulated by zinc.[60,61] Zinc is present at high concentrations in some synaptic vesicles and is released during neuronal activity.

[45]Hill, D. R., and Bowery, N. G. 1981. *Nature* 290: 149–152.
[46]Kaupmann, K., et al. 1997. *Nature* 386: 239–246.
[47]Bettler, B., Kaupmann, K., and Bowery, N. 1998. *Curr. Opin. Neurobiol.* 8: 345–350.
[48]Sodickson, D. L., and Bean, B. P. 1996. *J. Neurosci.* 16: 6374–6385.
[49]Luscher, C., et al. 1997. *Neuron* 19: 687–695.
[50]Kaupmann, K., et al. 1998. *Proc. Natl. Acad. Sci. USA* 95: 14991–14996.
[51]Ikeda, S. R. 1996. *Nature* 380: 255–258.
[52]Drew, C. A., et al. 1984. *Neurosci. Lett.* 52: 317–321.
[53]Lukasiewicz, P. D. 1996. *Mol. Neurobiol.* 12: 181–194.
[54]Wang, T. L., Guggino, W. B., and Cutting, G. R. 1994. *J. Neurosci.* 14: 6524–6531.
[55]Enz, R., and Cutting, G. R. 1998. *Vision Res.* 38: 1431–1441.
[56]Costa, E. 1991. *Neuropsychopharmacology* 4: 225–235.
[57]Blair, L. A. C., et al. 1988. *Science* 242: 577–579.
[58]Pritchett, D. B., et al. 1988. *Science* 242: 577–579.
[59]Costa, E. 1998. *Annu. Rev. Pharmacol. Toxicol.* 38: 321–350.
[60]Mayer, M. L., and Vyklicky, L., Jr. 1989. *J. Physiol.* 415: 351–365.
[61]Qian, H., et al. 1997. *J. Neurophysiol.* 78: 2402–2412.

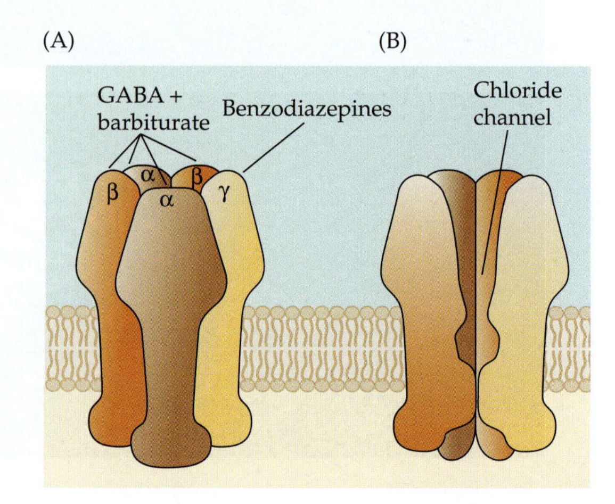

FIGURE 14.4 Hypothetical Model for the $GABA_A$ Receptor. (A) The receptor is drawn as an $\alpha_2\beta_2\gamma$ complex, by analogy with the nicotinic acetylcholine receptor; the actual number and arrangement of subunits in the native receptor is not known. GABA and barbiturates bind to the α and β subunits, each of which can form a functional homomultimeric receptor; benzodiazepines bind to the γ subunit. (B) Cross-sectional view of the receptor.

Glutamate Receptors in the CNS

A variety of physiological tests have supported the idea that glutamate, first established as an excitatory transmitter at the locust neuromuscular junction,[62] and in the squid giant synapse,[63] is released as the transmitter at excitatory synapses in the central nervous sytem.[64] As described in Chapter 3, ionotropic glutamate receptors can be grouped into two broad classes: NMDA (based on binding to *N*-methyl-D-aspartate) and non-NMDA, including the AMPA (binding α-amino-3-hydroxy-5-methyl-4-isoxazole propionic acid), or kainate, receptors.[65] NMDA receptors form calcium-permeable channels whose gating is subject to a number of important modulating factors. Magnesium ions block current flow through the NMDA receptor. This magnesium block can be relieved by strong depolarization (Chapter 12).[66–68] The activity of NMDA receptors also depends critically on the presence of extracellular glycine, which acts allosterically to enable channel opening by glutamate.[69]

The role of calcium influx through NMDA receptors in long-term changes in signaling at central synapses was described in Chapter 12. Excessive influx of calcium ions through NMDA receptors also has been implicated in the neurotoxicity of a variety of insults to the nervous system, including anoxia, hypoglycemia, and seizure. Under such conditions glutamate remains elevated for prolonged periods, persistently activating NMDA receptors and allowing intracellular calcium to reach cytotoxic levels. NMDA receptor antagonists can prevent such neuronal cell death.[70]

The ubiquitous importance of glutamate as a CNS neurotransmitter is revealed by autoradiographs of NMDA and AMPA receptor distribution (Figure 14.5). Both receptor types are widely expressed throughout the cerebral cortex and in many subcortical regions. At the same time, marked differences in relative levels are also found, such as in the cerebellum, where Purkinje cells express higher numbers of AMPA receptors than NMDA receptors.[71] Still more restricted patterns are found when the distribution of individual subunits is examined using in situ hybridization. Indeed, selective expression of AMPA receptor subtypes by neurons in nuclei of the auditory brainstem results in channels whose rapid gating is consistent with the role of these cells in precise temporal resolution for sound localization (Chapter 18).[72]

Nitric Oxide as a Transmitter in the CNS

The biological activity of nitric oxide (NO) was first recognized by its ability to cause relaxation of endothelial smooth muscle[73] (Chapter 10). A neuronal form of the NO synthesizing enzyme, NOS, is expressed in the CNS.[74–76] Excessive production of NO is neurotoxic and may contribute to the pathogenesis of neurological disorders.[77] Trans-

[62]Usherwood, P. N., and Machili, P. 1969. *Nature* 210: 634–636.

[63]Kawai, N., et al. 1983. *Brain Res.* 278: 346–349.

[64]Fonnum, F. 1984. *J. Neurochem.* 42: 1–11.

[65]Hollmann, M., and Heinemann, S. 1994. *Annu. Rev. Neurosci.* 17: 31–108.

[66]Nowak, L., et al. 1984. *Nature* 307: 463–465.

[67]Mayer, M. L., and Westbrook, G. L. 1987. *Prog. Neurobiol.* 28: 197–276.

[68]Johnson, J. W., and Ascher P. 1990. *Biophys. J.* 57: 1085–90.

[69]Johnson, J. W., and Ascher P. 1987. *Nature* 325: 529–531.

[70]Choi, D. W., Koh, J-Y., and Peters, S. 1988. *J. Neurosci.* 8: 185–196.

[71]Young, A. B., et al. 1995. In *Excitatory Amino Acids and Synaptic Transmission.* Academic Press, Harcourt, Brace and Co., New York, pp. 29–40.

[72]Trussell, L. 1997. *Curr. Opin. Neurobiol.* 7: 487–492.

[73]Palmer, R. M., Ferrige, A. G., and Moncada, S. 1987. *Nature* 327: 524–526.

[74]Bredt, D. S., et al. 1991. *Neuron* 7: 615–624.

[75]Hope, B. T., et al. 1991. *Proc. Natl. Acad. Sci. USA* 88: 2811–2814.

[76]Egberongbe, Y. I., et al. 1994. *Neuroscience* 59: 561–578.

[77]Dawson, V. L., and Dawson, T. M. 1998. *Prog. Brain Res.* 118: 215–229.

FIGURE 14.5 The Distribution of Glutamate Receptors in the CNS. (A) NMDA receptors in a parasagittal autoradiograph of rat brain labeled by tritiated glutamate (conditions chosen so that AMPA receptors were not labeled). (B) Tritiated AMPA binding sites. AMPA and NMDA receptors are widespread and generally overlap, as in cerebral cortex. However, differences also can be seen: for example, there are relatively few NMDA receptors in the molecular layer of cerebellum. C = cerebral cortex; G = granule cell layer of cerebellum; M = molecular layer of cerebellum. (After Young et al., 1995.)

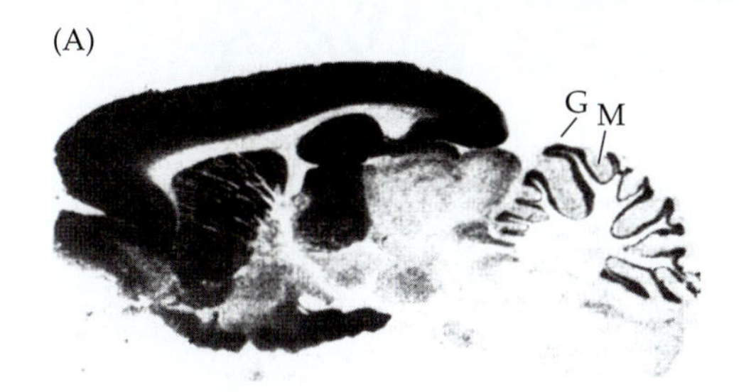

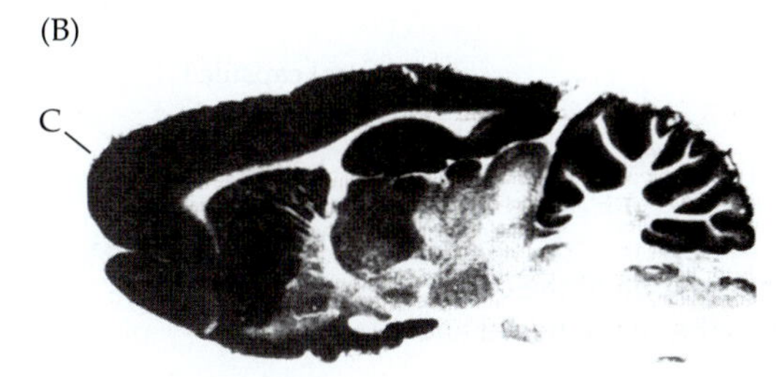

genic mice lacking neuronal NOS are resistant to cerebral ischemia,[78] and chemical lesion of dopaminergic neurons is prevented by inhibition of NOS.[79] NOS activity may be particularly relevant to cholinergic networks in the CNS. NOS-containing neurons in the cortex are innervated specifically by cholinergic basal forebrain neurons,[80] which are themselves rich in NOS.

Acetylcholine: Basal Forebrain Nuclei

The first compound to be identified as a transmitter in the CNS was acetylcholine, which is released at synapses made by collaterals of spinal motoneurons onto Renshaw cells.[81] This fast excitatory nicotinic synapse is, however, atypical. Subsequent studies of the effects of ACh in many brain regions have demonstrated a variety of responses that are mediated instead by indirectly coupled muscarinic receptors. These include increases in cation conductance, increases or decreases in various potassium conductances, and decreases in calcium conductance. Although nicotinic cholinergic receptors are expressed in the CNS (one compelling indicator of their existence is addiction to cigarette smoking), their synaptic functions have been more difficult to discern.

Cholinergic neurons are found in nuclei scattered throughout the brain, and cholinergic axons innervate most regions of the CNS. Prominent sources of cholinergic inputs to the cortex and the hippocampus are nuclei in the basal forebrain, especially the septal nuclei and the nucleus basalis (Figure 14.6). Cholinergic neurons in these nuclei have widespread and diffuse projections, innervating the cortex, hippocampus, amygdala, thalamus, and brainstem. Lesions of the nucleus basalis reduce the levels of choline acetyltransferase in the cortex by more than 50%.[82]

Cholinergic Neurons, Cognition, and Alzheimer's Disease

Animal and human studies suggest that cholinergic systems are important for learning, memory, and cognition.[83] Compounds that block muscarinic receptors, such as atropine and scopolamine, disrupt the acquisition and performance of learned behaviors, as can lesions of the nucleus basalis. Agents that inhibit acetylcholinesterase, such as physostigmine, can enhance performance in learning and memory tasks and can reverse some of the effects of basal forebrain lesions. However, such lesions inevitably involve cells releasing transmitters other than acetylcholine, and treatments aimed at enhancing performance of cholinergic neurons only partially reverse their effects. Thus, it seems certain that memory is not mediated exclusively by cholinergic neurons, but rather that they,

[78]Huang, Z., et al. 1994. *Science* 265: 1883–1885.

[79]Hantraye, P., et al. 1996. *Nature Med.* 2: 1017–1021.

[80]Vaucher, E., Linville, D., and Hamel, E. 1997. *Neuroscience* 79: 827–836.

[81]Eccles, J. C., Eccles, R. M., and Fatt, P. 1956. *J. Physiol.* 131: 154–169.

[82]Muir, J. L. 1997. *Pharmacol. Biochem. Behav.* 56: 687–696.

[83]Dunnett, S. B., and Fibiger, H. C. 1993. *Prog. Brain Res.* 98: 413–420.

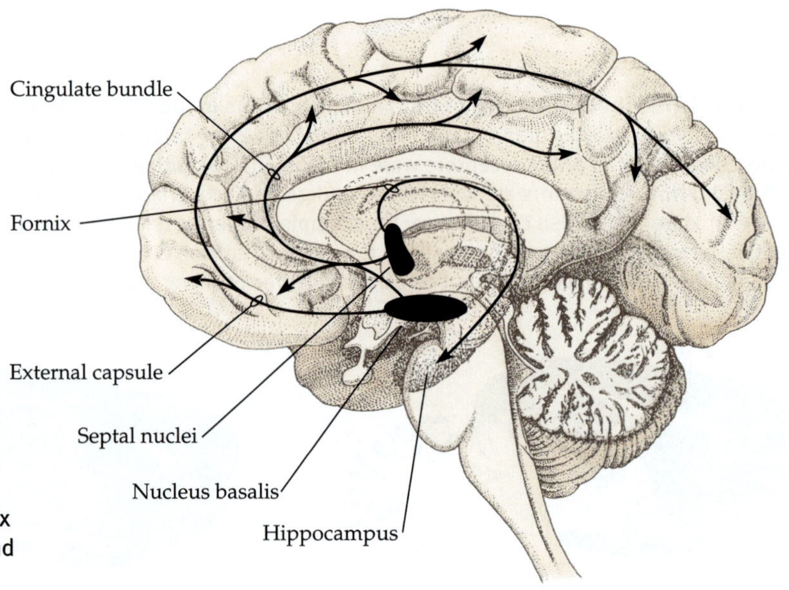

FIGURE 14.6 Cholinergic Innervation of the cortex and hippocampus by neurons in the septal nuclei and nucleus basalis.

along with neurons releasing other transmitters, provide modulatory input to cortical and hippocampal neurons.[84]

Interest in the role of basal forebrain cholinergic neurons in learning and memory was increased by the findings that declines in cognitive capacity with aging are paralleled by decreases in the levels of the synthetic enzyme choline acetyltransferase in the cortex and hippocampus and by the loss of cholinergic neurons in the basal forebrain.[85] Particularly pronounced are changes in patients with Alzheimer's disease, a progressive neurodegeneration causing a loss of memory and cognitive function. Alzheimer's disease is characterized by the accumulation in neurons throughout the CNS of insoluble aggregates (neurofibrillary tangles) of modified versions of proteins normally associated with the cytoskeleton. There is also a formation of extracellular senile plaques, composed of a core region filled with insoluble fibrils of improperly degraded amyloid proteins surrounded by degenerating neurites (Figure 14.7).[86,87] Such lesions occur in cholinergic neurons in the basal forebrain and in their axon terminals in the hippocampus. In addition, on autopsy, substantially reduced numbers of cholinergic neurons are found in the nucleus basalis. These changes were not reflected in the levels of muscarinic cholinergic receptors, but instead could be correlated with a loss of nicotine-binding sites.[88]

Among the 10 neuronal nicotinic receptor subunits that have been found in mammals, the most prevalent receptor combination in the CNS is thought to be a pentamer composed of two α_4 and three β_2 subunits. The role of these specific gene products in Alzheimer's disease is unknown, but it has been shown that β_2 knockout mice perform poorly in associative memory tasks and do not display the usual improvement upon nicotine administration.[89] Nicotinic receptors may act as presynaptic modulators of transmitter release in the CNS.[90] For example, low doses of nicotine enhance synaptic transmission in the hippocampus.[91] Finally, certain inherited neural disorders can be linked with mutations of nicotinic receptor subunits. Autosomal-dominant familial nocturnal frontal lobe epilepsy has been linked to two different mutations in the α_4 gene of Australian[92] and Norwegian families.[93]

The presence of neurofibrillary tangles in Alzheimer's disease is not limited to cholinergic neurons. Projection and local-circuit neurons releasing many other transmitters are involved, including neurons releasing norepinephrine, dopamine, serotonin, glutamate, GABA, somatostatin, neuropeptide Y, and substance P. Thus, there is no direct evidence that damage to cholinergic neurons in the basal forebrain is solely responsible for the cognitive decline in Alzheimer's disease, and attempts to alleviate such cognitive deficits with drugs aimed at enhancing performance of cholinergic neurons have achieved only limited effects.[94]

[84]Aigner, T. G. 1995. *Curr. Opin. Neurobiol.* 5: 155–160.

[85]Kasa, P., Rakonczay, Z., and Gulya, K. 1997. *Prog. Neurobiol.* 52: 511–535.

[86]Selkoe, D. J. 1991. *Neuron* 6: 487–498.

[87]Murrell, J., et al. 1991. *Science* 254: 97–99.

[88]Gotti, C., Fornasari, D., and Clementi, F. 1997. *Prog. Neurobiol.* 53: 199–237.

[89]Picciotto, M. R., et al. 1995. *Nature* 374: 65–67.

[90]McGehee, D. S., et al. 1995. *Science* 269: 1692–1696.

[91]Gray, R., et al. 1996. *Nature* 383: 713–716.

[92]Steinlein, O. K., et al. 1995. *Nature Genet.* 11: 201–203.

[93]Steinlein, O. K., et al. 1997. *Hum. Mol. Genet.* 6: 943–947.

[94]Lena, C., and Changeux, J. P. 1997. *Curr. Opin. Neurobiol.* 7: 674–682.

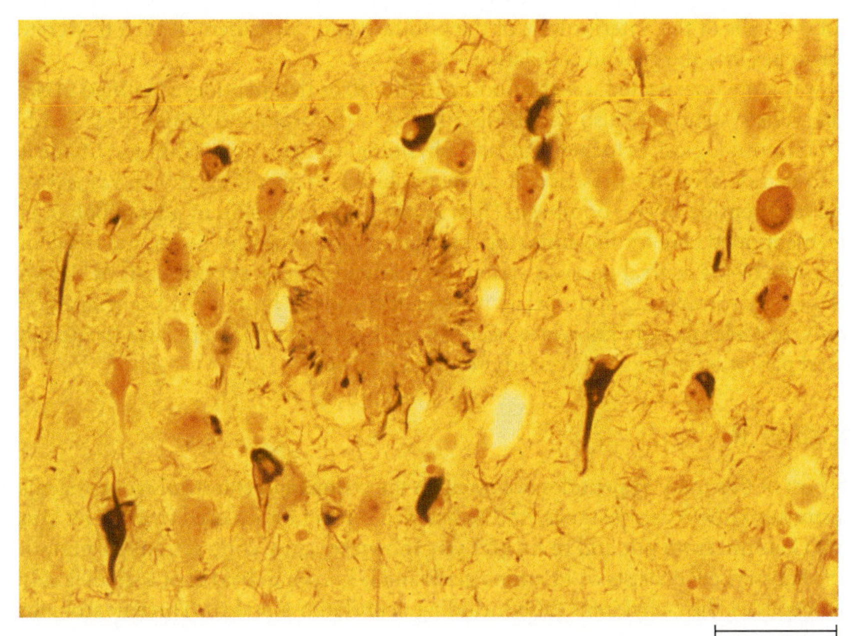

FIGURE 14.7 Neurofibrillary Tangles and Senile Plaques characteristic of Alzheimer's disease. Scattered among cytologically normal neurons in this section of the amygdala from an Alzheimer's patient are abnormal pyramidal cells filled with darkly staining neurofibrillary tangles, formed by the accumulation of bundles of paired helical filaments. In the center, a senile plaque consists of a large, compacted deposit of extracellular amyloid surrounded by a halo of dilated, structurally abnormal neurites. Modified Bielschowsky silver stain. (Micrograph kindly provided by D. J. Selkoe.)

ATP and Adenosine as CNS Transmitters

Adenosine triphosphate (ATP) is found in high concentration in synaptic vesicles and acts as a neurotransmitter through binding to one of a large (and growing) number of ionotropic or metabotropic **purinergic receptors**.[95] A family of seven genes encode ionotropic purinergic receptors, referred to as P2X receptors that form ligand-gated ion channels[96] (Chapter 3). Metabotropic P2Y receptors are seven-transmembrane, G protein–coupled receptors encoded by eight genes identified to date.[97] The purinergic receptors also include those that bind adenosine, a metabolite of ATP that acts as a modulator of transmission at many synapses.[98,99] Four genes are known to encode adenosine (P1) receptors, and these are all metabotropic, G protein–coupled receptors.[100]

ATP and adenosine originally were established as neurotransmitters acting on smooth muscle, and they are known to act throughout the peripheral nervous system, including effects on nociceptive afferents[101] (Chapter 17) and on cells of the inner ear.[102–104] Purinergic receptors also are found in the brain, where they provide a variety of modulatory actions that can influence behavior. For example, transgenic mice lacking one type of adenosine receptor normally expressed in basal ganglia showed reduced levels of exploratory activity; the male knockout mice showed higher levels of aggression toward intruders.[105] The first metabotropic ATP (P2Y) receptor to be cloned was isolated from the chick brain.[106] Several types of ionotropic (P2X) ATP receptors are distributed widely in the brain,[107,108] including some that are found on catecholaminergic neurons,[109] where they could mediate a presynaptic effect for ATP released from catecholamine-containing synaptic vesicles. In addition to serving as a synaptic modulator in the CNS, ATP is a direct synaptic transmitter in peripheral ganglia. Even in the CNS, neurons of the medial habenula have spontaneous and evoked synaptic currents that are blocked by suramin, the specific P2X receptor antagonist.[110,111]

PEPTIDE TRANSMITTERS IN THE CNS

It was in the gut that the first hormone—secretin—was discovered by Bayliss and Starling in 1902.[112] Since then, numerous additional intestinal hormones have been isolated and characterized. Many of the intestinal hormones, including secretin, gastrin, bradykinin, somatostatin, and cholecystokinin (CCK), were later shown to be peptides. These peptides are found in the terminals of autonomic axons that innervate the gut and in neurons of the enteric nervous system (Chapter 16). It had also been known since the 1950s that certain neurons within the brain could secrete peptide hormones into the local circulation. For example, nerve cells in the hypothalamus secrete releasing factors that reach the endocrine cells of the anterior lobe of the pituitary and cause them in turn to secrete other hormones into the general circulation.[113]

What was quite unexpected was the finding in the 1970s that peptides identified in the enteric nervous system were also widely distributed in the brain and spinal cord.[114] Advances in immunological, cytochemical, and physiological detection techniques made it possible to demonstrate the presence of cholecystokinin, bradykinin, gastrin, vasoactive intestinal polypeptide (VIP), bombesin (first isolated from the skin of a frog, *Bombina bombina*, mentioned here for its onomatopoetic name), and other gut hormones in widespread regions of the central nervous system. In many instances, peptides can be shown to be released by stimulating appropriate regions of the intact brain or in brain slices.[115] Conversely, peptides already known to occur in the hypothalamus were later located in the gut and pancreas, where they exerted profound effects.

Substance P

An early hint of this unity of peptides occurring in the central and enteric nervous systems is a transmitter known as substance P.[116] Substance P was first isolated in 1931 by Von Euler and Gaddum from gut and brain and was shown to cause contractions of smooth muscle.[117] Substance P consists of 11 amino acids and is known to be one of a small family of related peptides, the tachykinins (also including neurokinin A and B). Three neurokinin

[95] Burnstock, G. 1997. *Neuropharmacology* 36: 1127–1139.

[96] MacKenzie, A. B., Surprenant, A., and North, R. A. 1999. *Ann. N.Y. Acad. Sci.* 868: 716–729.

[97] Barnard, E. A., et al. 1996. *Ciba Found. Symp.* 198: 166–180.

[98] Ribeiro, J. A., et al. 1996. *Prog. Brain Res.* 109: 231–241.

[99] Brundege, J. M., and Dunwiddie, T. V. 1997. *Adv. Pharmacol.* 39: 353–391.

[100] Palmer, T. M., and Stiles, G. L. 1995. *Neuropharmacology* 34: 683–694.

[101] Burnstock, G., and Wood, J. N. 1996. *Curr. Opin. Neurobiol.* 6: 526–532.

[102] Glowatzki, E., et al. 1997. *Neuropharmacology* 36: 1269–1275.

[103] Heilbronn, E., Jarlebark, L., and Lawoko, G. 1995. *Neurochem. Int.* 27: 301–311.

[104] Housley, G. D. 1998. *Mol. Neurobiol.* 16: 21–48.

[105] Ledent, C., et al. 1997. *Nature* 388: 674–678.

[106] Webb, T. E., et al. 1993. *FEBS Lett.* 324: 219–225.

[107] Kanjhan, R., et al. 1999. *J. Comp. Neurol.* 407: 11–32.

[108] Shibuya, I., et al. 1999. *J. Physiol.* 514: 351–367.

[109] Vulchanova, L., et al. 1996. *Proc. Natl. Acad. Sci. USA* 93: 8063–8067.

[110] Edwards, F. A., Gibb, A. J., and Colquhoun, D. 1992. *Nature* 359: 144–147.

[111] Edwards, F. A., Robertson, S. J., and Gibb, A. J. 1997. *Neuropharmacology* 36: 1253–1268.

[112] Bayliss, W. M., and Starling, E. H. 1902. *J. Physiol.* 28: 325–353.

[113] Harris, G. W., Reed, M., and Fawcett, C. P. 1966. *Br. Med. Bull.* 22: 266–272.

[114] Krieger, D. T. 1983. *Science* 222: 975–985.

[115] Iversen, L. L., et al. 1980. *Proc. R. Soc. Lond. B* 210: 91–111.

[116] Maggio, J. E. 1988. *Annu. Rev. Neurosci.* 11: 13–28.

[117] Von Euler, U. S., and Gaddum, J. H. 1931. *J. Physiol.* 72: 74–87.

receptors have been identified (NK1, NK2, and NK3) that belong to the superfamily of G protein–coupled receptors.[118] Substance P is present in endings of small-diameter sensory axons concerned with nociception in the dorsal layers of the spinal cord, where it acts as a transmitter (Figure 14.8) (see also Chapter 18). When substance P, or its receptor (NK1), was eliminated in transgenic knockout mice, sensitivity to intense pain was reduced.[119,120] The NK1 knockout mice also showed reduced territorial aggression, which may be related to the normal expression of substance P and neurokinin receptors in nerve fibers and secretory cells of the hypothalamic–pituitary–adrenal axis.[121]

Opioid Peptides

Interest in brain peptides increased further in the mid-1970s as a result of two sets of experiments made by Kosterlitz, Hughes, Goldstein, Snyder, and their colleagues.[122–124] First, they found in the brain and in the gut receptors to which morphine and other derivatives of opium (opiates) bind with high specificity. Second, they identified peptides within the brain that have actions similar to those of opiates. First to be characterized were the enkephalins, which are pentapeptides; one enkephalin is known as met-enkephalin and the other as leu-enkephalin, depending on whether the carboxy-terminal amino acid is methionine or leucine. Other key findings were that opioid peptides (peptides with opiate activity) and their receptors were concentrated in regions of the brain known to be involved in the perception of pain, that stimulation of these regions of the brain could produce analgesia,[125] and that this analgesia was reversed by naloxone, a drug that blocks opiate receptors. Interest was spurred still further by the discovery of opioid neurons in the spinal cord whose axons ended on the substance P–containing terminals supposed to mediate pain sensation, and by the finding that opiates block the release of substance P from the sensory terminals[126] (see Figure 14.8).

A clue as to how enkephalin blocks release of substance P comes from studies of dorsal root ganglion neurons in culture.[127] When stimulated, these isolated neurons release substance P. Release is blocked by enkephalin, which causes a decrease in the duration of the action potential by binding to opiate receptors of the μ subtype and activating a calcium-dependent potassium channel.[128] Other opioid peptides bind to a second subtype of opiate receptors known as κ receptors. They reduce transmitter release by inhibiting

[118]Patacchini, R., and Maggi, C. A. 1995. *Arch. Int. Pharmacodyn. Ther.* 329: 161–184.

[119]Cao, Y. Q., et al. 1998. *Nature* 392: 390–394.

[120]De Felipe, C., et al. 1998. *Nature* 392: 394–397.

[121]Culman, J., and Unger, T. 1995. *Can. J. Physiol. Pharmacol.* 73: 885–891.

[122]Hughes, J., et al. 1975. *Nature* 258: 577–579.

[123]Teschemacher, H., et al. 1975. *Life Sci.* 16: 1771–1776.

[124]Pert, C. B., and Snyder, S. H. 1973. *Science* 179: 1011–1014.

[125]Fields, H. L., and Basbaum, A. I. 1978. *Annu. Rev. Physiol.* 40: 217–248.

[126]Jessel, T. M., and Iversen, L. L. 1977. *Nature* 268: 549–551.

[127]Mudge, A., Leeman, S., and Fischbach, G. 1979. *Proc. Natl. Acad. Sci. USA* 76: 526–530.

[128]Werz, M. A., and Macdonald, R. L. 1983. *Neurosci. Lett.* 42: 173–178.

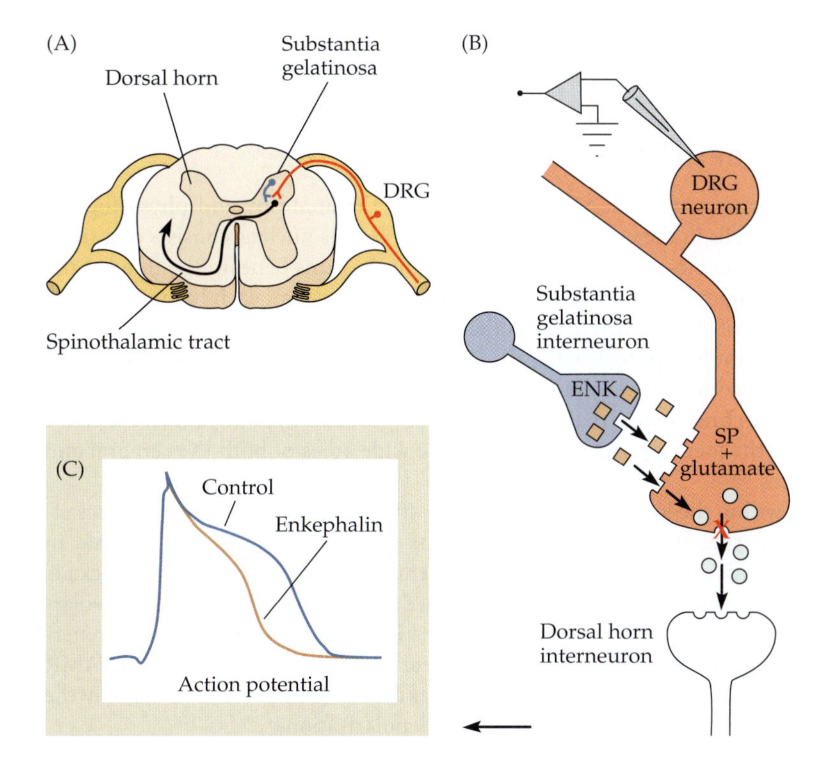

FIGURE 14.8 Pathway for Transmission of Pain Sensation in the Spinal Cord. (A, B) Dorsal root ganglion (DRG) cells responding to noxious stimuli release substance P (SP) and glutamate at their synapses with interneurons in the dorsal horn of the spinal cord. Interneurons containing enkephalin (ENK) in the substantia gelatinosa of the dorsal horn block transmission by inhibiting transmitter release from terminals of DRG cells. (C) Intracellular recordings from the dorsal root ganglion cell demonstrate that enkephalin acts by causing a decrease in the duration of the action potential. (C after Mudge, Leeman, and Fischbach, 1979.)

voltage-dependent calcium channels.[129] Both these and a third receptor type, the δ opioid receptor, form a small family of G protein–coupled opioid receptors. They act to decrease the activity of adenylyl cyclase and so lower the concentration of cAMP.[130]

Although only three opioid receptors are known thus far, numerous endogenous ligands have been identified. The amino-terminal sequences of all opioid peptides contain the so-called opioid motif, Tyr-Gly-Gly-Phe-[Met/Leu], followed by carboxy-terminal extensions resulting in peptides from 5 to 31 amino acids in length. β-Endorphin, for example, is found in the pituitary gland, the brain, the pancreas, and the placenta. This 31-residue peptide is generated from a large molecule that also acts as a precursor for other hormones, such as corticotropin (ACTH).[131] Dynorphin A is a 17–amino acid peptide whose pharmacological profile and anatomical distribution differ markedly from those of β-endorphin.[132]

Injection of enkephalins into the brain, either intraventricularly, or into the immediately surrounding nuclei (the "central gray"), not only mimics the analgesic and euphoric effects of opiates, but also produces other profound changes of behavior, such as muscular rigidity, suggesting that they act in regions of the central nervous system in addition to those subserving pain sensation.

REGULATION OF CENTRAL NERVOUS SYSTEM FUNCTION BY BIOGENIC AMINES

Within the mammalian central nervous system there is evidence that norepinephrine, dopamine, 5-hydroxytryptamine (5-HT, or serotonin), and histamine act as transmitters. These biogenic amines (Chapter 13) are found in pathways essential for sensory and motor performance, as well as for higher functions. However, out of the billions of neurons in the human brain, very few contain biogenic amines—such cells number only in the thousands. What is more, many of the cells containing these transmitters are clustered together in discrete regions of the brainstem. Neurons in these clusters or nuclei (which are shown schematically in Figures 14.9–14.12) extend axons to supply virtually all areas of the brain. In some cases these cells form synapses in which the presynaptic terminals are closely apposed to their postsynaptic targets; in other locations no obvious postsynaptic targets are seen. These anatomical characteristics suggest that an important function of amine-containing neurons is to modulate synaptic activity simultaneously in widespread areas of the central nervous system. Consistent with such a neuromodulatory role is the finding that biogenic amines act through indirectly coupled receptors (with one exception in the case of one type of serotonin receptor).[133–135]

Norepinephrine: The Locus Coeruleus

The **locus coeruleus** provides a good illustration of the anatomy and physiology of amine-containing cells in the central nervous system.[136] This nucleus consists of a small cluster of norepinephrine-containing cells that lies in the pons beneath the floor of the fourth ventricle (Figure 14.9). In the rat central nervous system, each locus coeruleus (one on either side of the brainstem) contains approximately 1500 cells; together these 3000 neurons account for approximately one-half of all norepinephrine-containing cells in the brain. These comparatively few cells have extensive projections, sending axons to the cerebellum, cerebral cortex, thalamus, hippocampus, and hypothalamus. Indeed, a single neuron in the locus coeruleus can innervate wide areas of both the cerebral and cerebellar cortices.[137]

Stimulation within the locus coeruleus or application of norepinephrine produces effects in central neurons that depend on the type of receptor activated. For example, in hippocampal pyramidal cells the most prominent effect of norepinephrine is a blockade of the slow calcium-activated potassium conductance that underlies the after-hyperpolarization following a train of action potentials.[138] The response is mediated by β-adrenergic receptors, which activate adenylyl cyclase, increasing the level of intracellular cAMP. The effect of blocking the slow after-hyperpolarization is to increase dramatically the number of action potentials elicited by prolonged depolarizations.

[129]Macdonald, R. L., and Werz, M. A. 1986. *J. Physiol.* 377: 237–249.

[130]Akil, H., et al. 1998. *Drug Alcohol Depend.* 51: 127–140.

[131]Mains, R. E., Eipper, B. A., and Ling, N. 1977. *Proc. Natl. Acad. Sci. USA* 74: 3014–3018.

[132]Watson, S. J., et al. 1978. *Nature* 275: 226–228.

[133]Cooper, J. R., Bloom, F. E., and Roth, R. H. 1996. *The Biochemical Basis of Pharmacology*, 7th Ed. Oxford University Press, New York.

[134]Fillenz, M. 1990. *Noradrenergic Neurons.* Cambridge University Press, Cambridge.

[135]Waterhouse, B. D., et al. 1991. *Prog. Brain Res.* 88: 351–362.

[136]Foote, S. L., Bloom, F. E., and Aston-Jones, G. 1983. *Physiol. Rev.* 63: 844–914.

[137]Swanson, L. W. 1976. *Brain Res.* 110: 39–56.

[138]Madison, D. V., and Nicoll, R. A. 1986. *J. Physiol.* 372: 245–259.

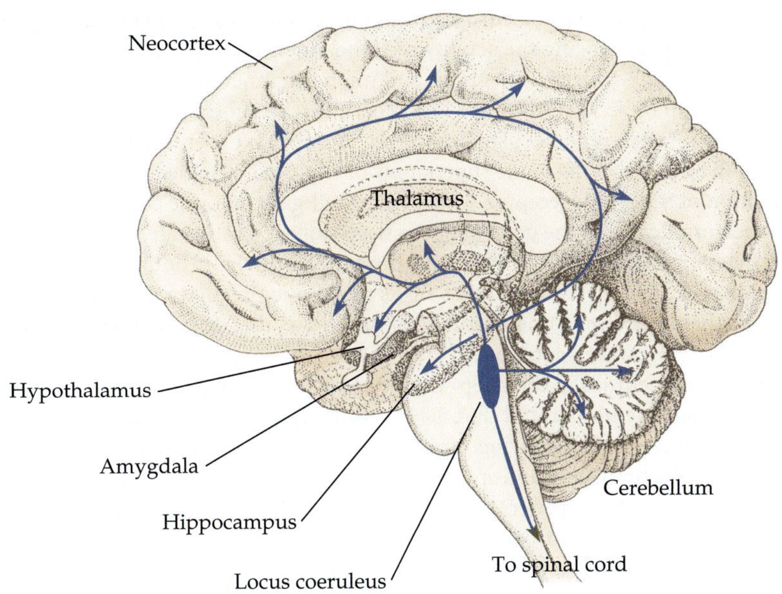

FIGURE 14.9 Projections of Norepinephrine-Containing Neurons in the Locus Coeruleus. The locus coeruleus lies in the pons just beneath the floor of the fourth ventricle. Neurons projecting from it innervate widespread regions of the brain and spinal cord.

The projections of the locus coeruleus form part of the **ascending reticular activating system**, a functionally defined projection from the brainstem reticular formation to higher brain centers. This pathway regulates attention, arousal, and circadian rhythms. For example, norepinephrine regulates circadian patterns of melatonin synthesis in the pineal gland in mammals.[139] Widespread projections from a few neurons, such as is the case for cells in the locus coeruleus, seem particularly well suited to carry out such a global function.

5-HT: The Raphe Nuclei

Like norepinephrine, 5-hydroxytryptamine (5-HT, also known as serotonin) is localized to a few nuclei in the brainstem.[140] These are the raphe nuclei, which lie directly along the midline of the brainstem from the midbrain to the medulla (Figure 14.10). (The term "raphe" comes from the French for "seam.") Nuclei in the medulla project to the spinal cord

[139]Stehle, J. H., et al. 1993. *Nature* 365: 314–320.

[140]Steinbusch, H. W. M. 1984. In *Classical Transmitters and Transmitter Receptors in the CNS*. Vol. 3 of *Handbook of Chemical Neuroanatomy*. Elsevier, New York, pp. 68–125.

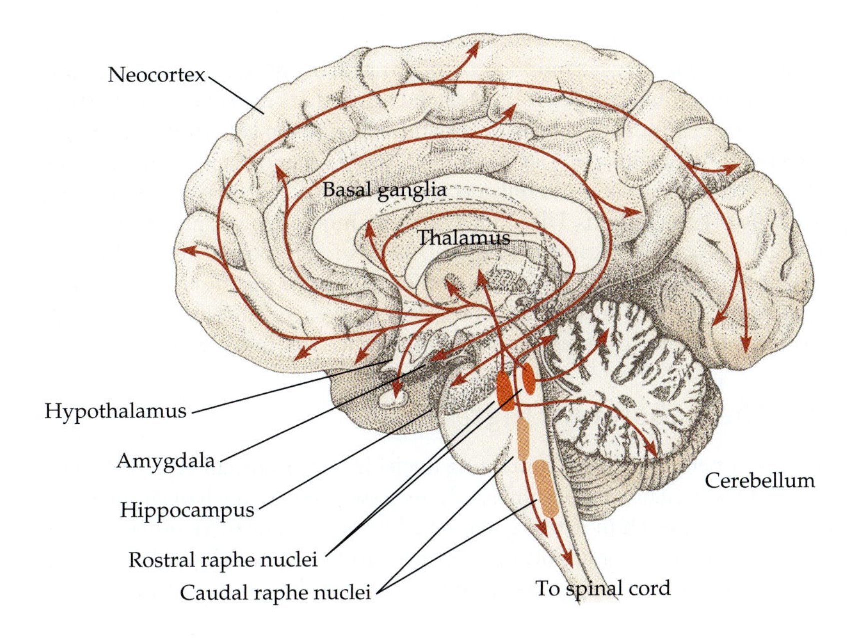

FIGURE 14.10 Neurons Containing 5-HT Form a Chain of Raphe Nuclei lying along the midline of the brainstem. More caudal nuclei innervate the spinal cord, more rostral nuclei innervate nearly all regions of the brain.

and modulate transmission in spinal cord pathways involved in the perception of pain (Chapter 18), as well as influencing the activity of spinal interneurons and motoneurons (see next paragraph). Raphe nuclei in the pons and midbrain innervate almost the entire brain, and together with projections from the locus coeruleus, they form part of the ascending reticular activating system. Molecular cloning techniques have identified 15 different 5-HT receptors. All of these belong to the superfamily of G protein–coupled receptors, with the single exception of the 5-HT_3 receptor, which is a ligand-gated choride channel.[141]

5-HT has been implicated in the control of the sleep–wake cycle. In cats, the depletion of 5-HT pharmacologically or by destruction of the raphe nuclei produces insomnia, which is reversed by administration of 5-HT or its metabolic precursor. This led to the "monoaminergic theory of sleep," one postulate of which was that sleep was triggered by the release of 5-HT.[142] However, subsequent studies have shown that this oversimplifies the role of serotonergic neurons in the sleep–wake cycle. For example, single-unit recordings in the raphe nuclei demonstrated that 5-HT–containing cells increase their firing rate during arousal, but fall silent during rapid eye movement (REM) sleep.[143] Indeed, serotonergic input to the brainstem and spinal cord increases the activity of motoneurons and elements of central pattern generators involved in locomotion and respiration.[144] 5-HT enhances the excitability of brainstem (trigeminal) motoneurons by reducing potassium conductances, as well as by increasing sodium conductance and a nonselective cation conductance.[145] In addition to general arousal, serotonergic neurons play a role in other complex behaviors, including aggression and the establishment of social dominance.[146] 5-HT also is implicated in cognitive performance through its modulation of cholinergic neurons.[147] LSD (lysergic acid diethylamide), which so dramatically modifies human perception, acts primarily on serotonergic transmission.[148]

Histamine: The Tuberomammillary Nucleus

Histamine was first identified in the 1920s as a natural constituent of the liver and lungs and in tissues throughout the body.[149] In the periphery, the predominant site of histamine storage and release is the mast cell. Histamine affects a variety of peripheral tissues and is involved in diverse physiological processes, including allergic reactions, response to injury, and regulation of gastric secretions. Histamine also acts as a neurotransmitter in the brain.[150,151] Three pharmacologically distinct receptor types bind histamine, two of which (H1 and H2) have been cloned.[152,153] These are G protein–coupled receptors, similar to those mediating the effects of the other biogenic amines. Histamine acts through the H1 receptor to depolarize cholinergic neurons of the nucleus basalis by reducing potassium conductance and activating a tetrodotoxin-insensitive sodium conductance.[154] In nodose ganglion cells (vagal afferents), H1 activation blocks a resting potassium conductance and the potassium channels that generate a slow, postspike after-hyperpolarization,[155] both effects leading to increased excitability.

The cell bodies of histaminergic neurons are concentrated in a small region of the hypothalamus, the tuberomammillary nucleus, and extend axons that reach almost all parts of the CNS[156] (Figure 14.11). Individual histamine neurons have axon collaterals that innervate several different brain regions.[157] Like neurons releasing other biogenic amines, histaminergic neurons arborize diffusely and only occasionally form classic synapses with clear pre- and postsynaptic specializations. Histaminergic axons innervate not only neurons, but glial cells, small blood vessels, and capillaries as well. Based on this morphology and the effects of drugs that influence histaminergic transmission, it appears that histamine neurons regulate general brain activities, such as the arousal state and energy metabolism, through indirect mechanisms mediated by receptors on neurons, astrocytes, and blood vessels.

Dopamine: The Substantia Nigra

There are four prominent dopamine-containing nuclei in the brainstem (Figure 14.12). One of these lies in the arcuate nucleus and sends processes into the median eminence of the hypothalamus, an area rich in peptide-releasing hormones. The three other clusters of dopamine cells lie in the midbrain; they project primarily to the basal ganglia, a group of nuclei in the center of the brain that are important for the control of movement

[141]Gerhardt, C. C., and van Heerikhuizen, H. 1997. *Eur. J. Pharmacol.* 334: 1–23.

[142]Jouvet, M. 1972. *Ergeb. Physiol.* 64: 166–307.

[143]Jacobs, B. L., and Fornal, C. A. 1991. *Pharmacol. Rev.* 43: 563–578.

[144]White, S. R., et al. 1996. *Prog. Brain Res.* 107: 183–199.

[145]Hsiao, C. F., et al. 1997. *J. Neurophysiol.* 77: 2910–2924.

[146]Raleigh, M. J., et al. 1991. *Brain Res.* 559: 181–190.

[147]Cassel, J-C., and Jeitsch, H. 1995. *Neuroscience* 69: 1–41.

[148]Aghajanian, G. K. and Marek, G. J. 1999. *Neuropsychopharmacology* 21 (2 suppl.): 16s–23s.

[149]Douglas, W. W. 1980. In *Goodman and Gilman's The Pharmacological Basis of Therapeutics*. Macmillan, New York, pp. 608–618.

[150]Prell, G. D., and Green, J. P. 1986. *Annu. Rev. Neurosci.* 9: 209–254.

[151]Wada, H., et al. 1991. *Trends Neurosci.* 14: 415–418.

[152]Gantz, I., et al. 1991. *Proc. Natl. Acad. Sci. USA* 88: 429–433.

[153]Yamashita, M., et al. 1991. *Proc. Natl. Acad. Sci. USA* 88: 11515–11519.

[154]Gorelova, N., and Reiner, P. B. 1996. *J. Neurophysiol.* 75: 707–714.

[155]Jafri, M. S., et al. 1997. *J. Physiol.* 503: 533–546

[156]Panula, P., et al. 1990. *Neuroscience* 34: 127–132.

[157]Kohler, C., et al. 1985. *Neuroscience* 16: 85–110.

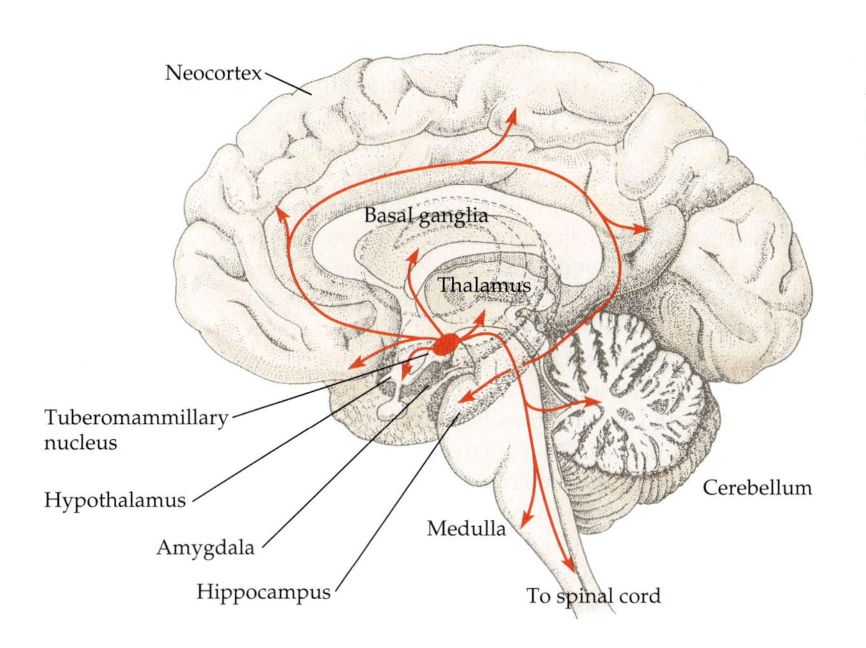

FIGURE 14.11 Histamine-Containing Neurons Are Localized to the Tuberomammillary Nucleus in the hypothalamus. These neurons have diffuse projections throughout the brain and spinal cord.

(Chapter 22). As is the case for other neurons releasing biogenic amines, a few cells with widely ramifying projections are involved. In the rat, for example, there are approximately 7000 dopamine cells in one of these midbrain clusters, the substantia nigra; each of these neurons, however, gives rise to an estimated 250,000 varicosities spread throughout its targets in the basal ganglia.[158]

Five G protein–coupled dopamine receptors have been cloned, divided into two functional groups by their effects on adenlyl cyclase: D1 and D5 increase enzyme activity; D2, D3, and D4 decrease enzyme activity.[159] D1 receptors are encoded by a gene with a single exon,[160,161] while alternative splicing of the D2 gene produces several additional subtypes.[162]

It is the progressive degeneration of one discrete group of dopaminergic neurons that is the most prominent feature of Parkinson's disease. In patients with degeneration of neurons in the pars compacta of the substantia nigra, nerve cells in the basal ganglia become deprived of their presynaptic dopaminergic inputs. As axon terminals disappear, the levels of dopamine in the basal ganglia fall and characteristic motor disorders appear: difficulty in initiating movement, muscular rigidity, and a tremor at rest.

[158]Yurek, D. M., and Sladek, J. R., Jr. 1990. *Annu. Rev. Neurosci.* 13: 415–440.

[159]Missale, C., et al. 1998. *Physiol. Rev.* 78: 189–225.

[160]Dearry, A., et al. 1990. *Nature* 347: 72–76.

[161]Sunahara, R. K., et al. 1990. *Nature* 347: 80–83.

[162]Monsma, F. J., et al. 1989. *Nature* 342: 926–929.

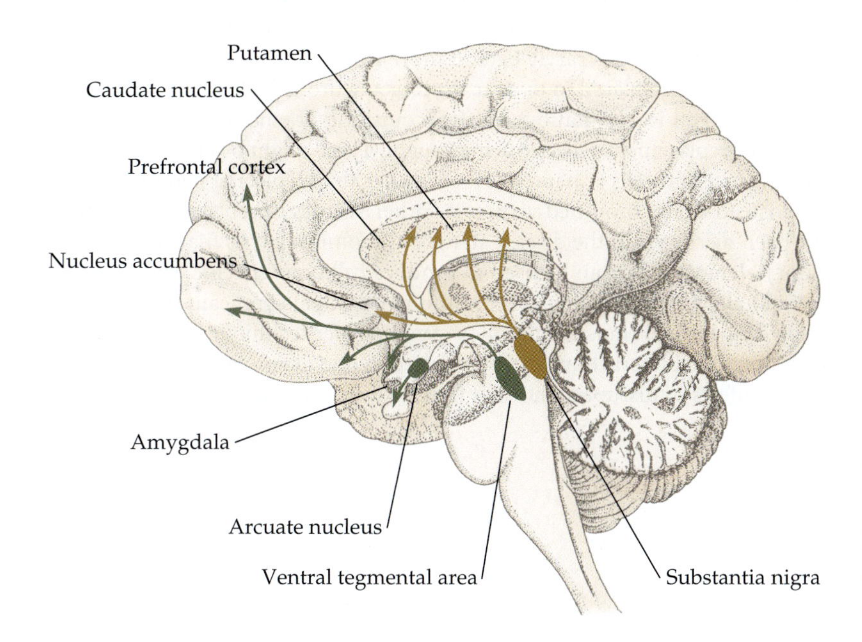

FIGURE 14.12 Neurons Containing Dopamine are found in nuclei in the hypothalamus and midbrain. Those in the arcuate nucleus project to the median eminence of the hypothalamus, forming the tuberoinfundibular system. Dopamine neurons in the substantia nigra project to the caudate nucleus and putamen (collectively called the striatum) of the basal ganglia, forming the nigrostriatal pathway. Dopamine neurons in the ventral tegmental area project to the nucleus accumbens, amygdala, and prefrontal cortex, forming the mesolimbic and mesocortical systems.

A triumph of neuropharmacology was the technique of replacement therapy for the treatment of Parkinson's disease. The idea was to attempt to alleviate the symptoms of the disease by restoring the level of dopamine in the basal ganglia. It was known that dopamine itself would not cross the blood–brain barrier, so the precursor for dopamine, L-DOPA, was given. Remarkably, patients receiving oral doses of L-DOPA usually showed improvement, and autopsies of patients who died from other causes while undergoing L-DOPA therapy had near normal concentrations of dopamine in their basal ganglia. Presumably the dopamine neurons that do remain in Parkinson's patients are able to synthesize and release sufficient transmitter if provided with extra precursor.

If one thinks in terms of neuromuscular transmission, where the precise timing and location of release is critically important, it is difficult to imagine how excess transmitter derived from a distant surviving terminal could substitute for the absence of transmitter release from a terminal lost to degeneration. It would be like trying to restore appropriate contractions of skeletal muscles in coordinated movements by diffuse application of acetylcholine. However, if one thinks in terms of *neuromodulation* mediated by indirectly coupled receptors, where effects have an inherently slow time course and specificity is determined by the nature and distribution of the receptors, then one can begin to appreciate how a generalized elevation in dopamine concentration could restore normal activity.

The axonal projection from the substantia nigra to the basal ganglia is not the only dopaminergic pathway in the central nervous system, however. Accordingly, as one might expect, a complication of giving L-DOPA to alleviate the symptoms of Parkinson's disease is that the balance of inputs to other areas is disrupted as well. Dopamine is released by neurons in areas of the brain controlling mood and affect; accordingly, patients undergoing L-DOPA therapy can experience psychiatric disturbances. Conversely, many of the drugs given to psychiatric patients, which are thought to act by blocking the interaction of dopamine with its receptor, produce Parkinson's-like motor side effects.[163]

Targeting Specific Synapses

The challenge for development of new drugs for systemic treatment of CNS disorders is to identify for a specific synapse not only the neurotransmitter released, but also those features that set that synapse apart from others using the same transmitter. The next step is to design a specific pharmacological agent to exploit that difference. The rich variety of postsynaptic transmitter receptors offers one avenue for finding drugs with such high selectivity. Transmitter transporters, presynaptic receptors, and proteins mediating transmitter synthesis, storage, and release offer additional potential targets.

A still-experimental alternative to systemic administration of drugs is to replace degenerating or defective cells by transplantation of appropriate embryonic neurons.[164] Neurons transplanted into the adult CNS have been shown to survive and make synaptic connections; however, their axons rarely extend more than a few millimeters (Chapter 24). Therefore, grafts must be placed in or near the target neurons. Other drawbacks arise from both practical and ethical limitations on the acquisition of embryonic tissue. Thus, alternative sources of replacement cells are being sought. These might be engineered from pluripotent embryonic stem cells (Chapter 23). Or, rather than replacing neurons, it may prove possible instead to use modified viruses or other vectors to replace genes whose functions have been disrupted in the disease state.[165] Genomic analysis may identify upstream regulatory elements that can be used to produce cell type–specific expression of proteins that could alleviate a pathological condition. Although significant problems remain to be resolved, neuronal transplantation or gene transfer techniques may provide a means by which functional interactions can be restored in one region of the CNS without the unwanted side effects of disrupting normal transmission elsewhere.

[163]Baldessarini, R. J., and Tarsy, D. 1980. *Annu. Rev. Neurosci.* 3: 23–41.

[164]Dunnett, S. B., et al. 1997. *Br. Med. Bull.* 53: 757–776.

[165]Horellou, P., and Mallet, J. 1997. *Mol. Neurobiol.* 15: 241–256.

SUMMARY

■ Neurons employing a particular transmitter in the CNS can usually be identified by the presence of the transmitter itself or enzymes involved in its synthesis. The expression of proteins involved in transmitter storage, degradation, and uptake are useful, although less reliable indicators.

■ GABA and glycine are the principal inhibitory neurotransmitters in the brain and spinal cord.

■ GABA acts on three classes of receptors: ionotropic $GABA_A$ and $GABA_C$ receptors and metabotropic $GABA_B$ receptors.

■ Glutamate is the principal excitatory neurotransmitter of the CNS.

■ There are three classes of glutamate receptors: NMDA and AMPA ionotropic receptors, and metabotropic receptors.

■ Calcium signals initiated by NMDA receptors can stimulate the production of nitric oxide (NO). NO acts as a diffusible neuromodulator.

■ Cholinergic neurons in the nucleus basalis and septal nuclei arborize extensively throughout the cortex. Neurodegenerative diseases that are characterized by loss of memory and cognitive function show losses of cholinergic neurons.

■ Neuropeptides are released as neurotransmitters in the CNS. Particular peptides are used in different areas of the nervous system and in circuits mediating diverse physiological functions.

■ Relatively few cells in the CNS release biogenic amines (norepinephrine, 5-HT, dopamine, and histamine) as neurotransmitters; the cell bodies of those that do are clustered in nuclei in the brainstem. Cells that release norepinephrine, 5-HT, and histamine have extensive arborizations.

■ Cells that release dopamine have more circumscribed projections. Those in the arcuate nucleus project to the median eminence of the hypothalamus to modulate release of hypothalamic peptides. Dopamine-containing cells in the substantia nigra project to the basal ganglia and influence motor activity; those in the ventral tegmental area project to the nucleus accumbens, amygdala, and prefrontal cortex, where they influence mood and affect.

SUGGESTED READING

General Reviews

Betz, H. 1990. Ligand-gated ion channels in the brain: The amino acid receptor superfamily. *Neuron* 5: 383–392.

Cooper, J. R., Bloom, F. E., and Roth, R. H. 1996. *The Biochemical Basis of Pharmacology*, 7th Ed. Oxford University Press, New York.

Costa, E. 1998. From $GABA_A$ receptor diversity emerges a unified vision of GABAergic inhibition. *Annu. Rev. Pharmacol. Toxicol.* 38: 321–350.

Davies, R. W., Gallagher, E. J., and Savioz, A. 1994. Reverse genetics of the mouse central nervous system: Targeted genetic analysis of neuropeptide function and reverse genetic screens for genes involved in human neurodegenerative disease. *Prog. Neurobiol.* 42: 319–331.

Dunnett, S. B., Kendall, A. L., Watts, C., and Torres, E. M. 1997. Neuronal cell transplantation for Parkinson's and Huntington's diseases. *Br. Med. Bull.* 53: 757–776.

Gotti, C., Fornasari, D., and Clementi, F. 1997. Human neuronal nicotinic receptors. *Prog. Neurobiol.* 53: 199–237.

Hollman, M., and Heinemann, S. 1994. Cloned glutamate receptors. *Annu. Rev. Neurosci.* 17: 31–108.

Kasa, P., Rakonczay, Z., and Gulya, K. 1997. The cholinergic system in Alzheimer's disease. *Prog. Neurobiol.* 52: 511–535.

Lena, C., and Changeux, J. P. 1997. Pathological mutations of nicotinic receptors and nicotine-based therapies for brain disorders. *Curr. Opin. Neurobiol.* 7: 674–682.

Maggio, J. E. 1988. Tachykinins. *Annu. Rev. Neurosci.* 11: 13–28.

Martin, A. R. 1990. Glycine- and GABA-activated chloride conductances in lamprey neurons. In O. P. Ottersen and J. Storm-Mathisen (eds.), *Glycine Neurotransmission*. Wiley, New York, pp. 171–191.

Nicoll, R. A., Malenka, R. C., and Kauer J. A. 1990. Functional comparison of neurotransmitter receptor subtypes in mammalian central nervous system. *Physiol. Rev.* 70: 513–565.

Sargent, P. B. 1993. The diversity of neuronal nicotinic acetylcholine receptors. *Annu. Rev. Neurosci.* 16: 403–443.

Schuman, E. V., and Madison, D. V. 1995. Nitric oxide and synaptic function. *Annu. Rev. Neurosci.* 17: 153–183.

Original Papers

Jonas, P., Bischofberger, J., and Sandkuhler, J. 1998. Corelease of two fast neurotransmitters at a central synapse. *Science* 281: 419–424.

Johnson, J. W. and Ascher, P. 1987. Glycine potentiates the NMDA response in cultured mouse brain neurons. *Nature* 325: 529–531.

Kaupmann, K., Huggel, K., Heid, J., Flor, P. J., Bischoff, S., Mickel, S. J., McMaster, G., Angst, C., Bittiger, H., Froestl, W., and Bettler, B. 1997. Expression cloning of $GABA_B$ receptors uncovers similarity to metabotropic glutamate receptors. *Nature* 386: 239–246.

Luscher, C., Jan, L. Y., Stoffel, M., Malenka, R. C., and Nicoll, R. A. 1997. G protein-coupled inwardly-rectifying K channels (GIRKs) mediate postsynaptic but not presynaptic transmitter actions in hippocampal neurons. *Neuron* 19: 687–695.

Mains, R. E., Eipper, B. A., and Ling, N. 1977. Common precursor to corticotropins and endorphins. *Proc. Natl. Acad. Sci. USA* 74: 3014–3018.

McGehee, D. S., Heath, M. J., Gelber, S., Devay, P., and Role, L. W. 1995. Nicotine enhancement of fast excitatory synaptic transmission in CNS by presynaptic receptors. *Science* 269: 1692–1696.

Nowak, L., et al. 1984. Magnesium gates glutamate-activated channels in mouse central neurones. *Nature* 307: 463–465.

Sommer, B., Keinanen, K., Verdoorn, T. A., Wisden, W., Burnashev, N., Herb, A., Kohler, M., Takagi, T., Sakmann, B., and Seeburg, P. H. 1990. Flip and flop: A cell-specific functional switch in glutamate-operated channels of the CNS. *Science* 249: 1580–1585.

Von Euler, U. S., and Gaddum, J. H. 1931. Unidentified depressor substance in certain tissue extracts. *J. Physiol.* 72: 74–87.

Yu, C., Brussaard, A. B., Yang, X., Listerud, M., and Role, L. W. 1993. Uptake of antisense oligonucleotides and functional block of acetylcholine receptor subunit gene expression in primary embryonic neurons. *Dev. Genet.* 14: 296–304.

PART 3 INTEGRATIVE MECHANISMS

Circuits Mediating Stereotyped Behavior

Acquisition and Analysis of Sensory Information

The Visual System

Initiation and Control of Movement

15 Cellular Mechanisms of Integration and Behavior in Leeches, Ants, and Bees

EXPERIMENTS MADE IN INVERTEBRATES have provided crucial insights into cellular and molecular mechanisms of signaling. Because of their simplified nervous systems and wide diversity, animals such as flies, bees, ants, worms, snails, lobsters, and crayfish offer advantages for studying how nerve cells integrate information to produce coordinated behavior.

One convenient example is the CNS of the leech. In this animal it is possible to identify a single neuron, measure its biophysical properties, trace its connections, and define its role in integrative actions such as reflexes, swimming, and avoidance behavior. The leech CNS consists of 21 stereotyped ganglia, each of which contains only about 400 neurons. Individual sensory cells, motoneurons, and interneurons can be recognized by visual inspection and by recording electrically. Each mechanosensory cell in the ganglion responds selectively to touch, pressure, or noxious stimulation and innervates a well-defined area of skin. Sensory cells transmit information to interneurons and motoneurons by electrical and chemical synapses. After repetitive stimulation at natural frequencies, transmission in sensory neurons becomes blocked at branch points where small axons feed into larger processes. This disconnects sensory neurons from some but not all of their targets temporarily. With circuits consisting of identified neurons, it is also possible to study how individual neurons form connections during development and during regeneration after injury.

Ants and bees exhibit complex behaviors that exemplify an important principle: Measurements of behavior provide insights into integrative mechanisms. Hence, one can work downward from behavior to brain to neuron. From analysis of the path taken by an ant or a bee, one can progress to the properties of the photoreceptors and the wiring that made it possible. Desert ants migrate over long distances while foraging for food. As an ant searches, it moves in a zigzag meandering path. Once the ant has found its food source, however, it turns toward its nest and marches directly there in a straight line. The CNS keeps track of turns to the left and right that the ant made along its initial trajectory, calculates the position from which it started, and directs it straight for home. Such integration of information depends on coordinates provided by polarized light from the sun. The desert ant's eyes contain specific groups of photoreceptors for polarized light. These photoreceptors supply information to the CNS, which calculates and keeps track of movements in space to create a new vector. Ants and bees exemplify the ability of invertebrate nervous systems containing relatively few neurons to make complex computations. Thanks to their diverse forms, invertebrates constitute appealing preparations for studying cellular mechanisms that underlie behavior.

Throughout this book, invertebrate neurons are used to illustrate fundamental mechanisms that are significant for understanding signaling. From the squid axon and giant synapse, for example, were derived principles of nearly universal validity for conduction of action potentials and for synaptic transmission. Cells of other invertebrates were used to illustrate passive electrical properties and the mechanisms responsible for synaptic inhibition.

The reasons for choosing invertebrate preparations are often technical. Certain problems can be solved more easily in invertebrate nervous systems. First, if the individual nerve cells are large and readily accessible, they can be recognized and studied by electrophysiological and molecular techniques. Second, although the behavior of invertebrates can be elaborate, it is highly stereotyped, reproducible, and readily analyzed. Third, because of the opportunities afforded by genetic manipulations, the fruit fly (*Drosophila melanogaster*) and the nematode worm *Caenorhabditis elegans*[1] have been essential for studies on the development of the nervous system. The mechanisms are so highly conserved throughout evolution that genes for receptors and genes for channels are strikingly similar in their sequences in flies, worms, and higher vertebrates (Chapters 13 and 23). In human beings, as in *Drosophila* and leeches, homeobox genes are responsible for the development of body segments, major divisions of the brain, and specialized organs (such as eyes).[2,3]

At the same time it is worth emphasizing that curiosity about the nervous systems of invertebrates goes beyond studying mechanisms that might apply to higher animals. The way in which a leech swims, an ant navigates, a bee dances, a cricket sings, or a fly flies are all problems of first-order interest.

FROM NEURONS TO BEHAVIOR AND VICE VERSA

Invertebrates provide the opportunity for following the thread from the birth of a single identified cell through the elaboration of its branching patterns and the formation of its connections. In a suitable animal with large neurons, one can analyze the biophysical properties of a single cell and observe how they are related to the coordinated behavior of the whole animal. Similarly, one can follow molecular events occurring in that cell as the animal modifies its behavior in response to outside influences and internal programs. Such analyses are possible because stereotyped behavioral responses in invertebrates are performed by relatively few neurons, whereas analogous responses in mammals require many thousands of neurons.

By contrast, in ants and bees individual neurons are smaller and harder to study physiologically. Nevertheless, by acute and quantitative observation of behavior one can infer and then define underlying cellular and integrative mechanisms. Thus, it has been possible to analyze quantitatively the sensory mechanisms that are used in highly complex tasks by an ant or a bee as it locates a distant target and then finds the shortest pathway back to the nest. A complementary approach to that of cellular neurobiology is provided by this type of analysis: from behavior, to brain, to neuron. In subsequent chapters that deal with sensory mechanisms, perception, and motor coordination in the mammalian CNS, we shall see again that it is behavior (e.g., the detection of where a sound has arisen in space) that provides the starting point for revealing cellular mechanisms.

Each type of system has its own set of advantages for approaching specific problems. The neural circuits for coordinated elementary units of behavior, such as postural reflexes, feeding, circadian rhythms, escape reactions, and swimming, have been traced in crayfish and snails.[4,5] Central nervous systems of insects have been used to study a variety of problems, including development and regeneration, flight, walking, navigation, and communication by sound.[6–9] The animals themselves and the scope of the problems are so varied that a comprehensive review is impossible. Indeed, monographs abound on *Aplysia*, leech, the marine snail *Hermissenda*, crayfish, and isolated invertebrate neurons in culture.[10–14]

We have singled out for fuller discussion the nervous systems of just three invertebrates. The leech, with its limited behavior, stereotyped CNS, and small numbers of neurons, provides preparations in which one can study the properties, connections, and functions of in-

[1]Hodgkin, J. 1999. *Cell* 98: 277–280.

[2]Boncinelli, E., Mallamaci, A., and Broccoli, V. 1998. *Adv. Genet.* 38: 1–29.

[3]Halder, G., Callaerts, P., and Gehring, W. J. 1995. *Science* 267: 1788–1792.

[4]Staras, K., Kemenes, G., and Benjamin, P. R. 1999. *J. Neurophysiol.* 81: 1261–1273.

[5]Garcia, U., et al. 1994. *Cell. Mol. Neurobiol.* 14: 71–88.

[6]Reichert, H., and Boyan, G. 1997. *Trends Neurosci.* 20: 258–264.

[7]Edwards, J. S. 1997. *Brain Behav. Evol.* 50(1): 8–12.

[8]Pearson, K. G. 1995. *Curr. Opin. Neurobiol.* 5: 786–791.

[9]Engel, J. E., and Hoy, R. R. 1999. *J. Exp. Biol.* 202: 2797–2806.

[10]Kandel, E. R. 1979. *Behavioral Biology of* Aplysia. W. H. Freeman, San Francisco.

[11]Muller, K. J., Nicholls, J. G., and Stent, G. S. (eds.). 1981. *Neurobiology of the Leech.* Cold Spring Harbor Laboratory, Cold Spring Harbor, NY.

[12]Alkon, D. 1987. *Memory Traces in the Brain.* Cambridge University Press, Cambridge.

[13]Atwood, H. L. (ed.). 1982. *Biology of Crustacea*, Vol. 3. Academic Press, New York.

[14]Beadle, D. J., Lees, G., and Kater, S. B. 1988. *Cell Culture Approaches to Invertebrate Neuroscience.* Academic Press, London.

dividual identified nerve cells in great detail. In ants and bees, analysis of the behavior itself is the starting point. This has revealed neuronal mechanisms that provide the animals with information about polarized light and magnetic fields in the outside world.

INTEGRATION BY INDIVIDUAL NEURONS IN THE CNS OF THE LEECH

Since the days of ancient Greece and Rome, leeches have been applied by physicians to patients suffering from diseases such as epilepsy, angina, tuberculosis, meningitis, and hemorrhoids—an unpleasant treatment that almost certainly did more harm than good to unfortunate victims. By the nineteenth century, use of the medicinal leech was so prevalent that the animal became almost extinct in western Europe, forcing Napoleon to import about 6 million leeches from Hungary in one year to treat his soldiers. This mania for leeching had one benefit for contemporary biology: The medicinal application of leeches stimulated basic research on their reproduction, development, and anatomy. In the late nineteenth century, founders of experimental embryology, such as Whitman, chose the leech to follow the fates of early embryonic cells. Similarly, its nervous system was extensively studied by a roster of distinguished anatomists, including Ramón y Cajal, Sanchez, Gaskell, Del Rio Hortega, Odurih, and Retzius.[15] Interest in the leech thereafter declined, to be rekindled in 1960 when Stephen Kuffler and David Potter[16] first applied modern neurophysiological techniques to its nervous system. This set the stage for extensive studies of its development,[17,18] biophysics and molecular biology,[19–22] circuitry, and behavior.[23–25]

Leech Ganglia: Semiautonomous Units

The body and the nervous system of the leech are rigorously segmented and consist of a number of repeating units (segments) that are similar throughout the length of the animal. Each segment is innervated by a stereotyped ganglion that is much like all the others. Even the specialized head and tail "brains" shown in Figure 15.1 consist of fused ganglia, in which many characteristic features of segmental ganglia are still recognizable.[26]

Each ganglion contains only about 400 nerve cells, which have distinctive shapes, sizes, positions, and branching patterns. A ganglion innervates a well-defined territory of the body by way of paired axon bundles (**roots**), and it communicates with neighboring and distant parts of the nervous system through another set of bundles (**connectives**). Integration thus occurs in a succession of clear-cut steps:

1. Each segmental ganglion receives information from a circumscribed body segment, the performance of which it regulates.
2. Neighboring ganglia influence each other by direct interconnections.
3. The coordinated operation of the whole nerve cord and the animal is governed by the "brains" at each end of the leech.

Perhaps the main appeal of the leech is the beauty of the ganglion as it appears under the microscope, with its neurons so recognizable and so familiar from segment to segment, from specimen to specimen, from species to species (Figure 15.2). As one looks at these limited aggregates of cells laid out in an orderly pattern, one cannot but marvel at how they, on their own, being the brain of the creature, are responsible for all its movements, hesitations, avoidance, mating, feeding, and sensations. In addition to the aesthetic pleasure provided by the preparation, there is the intellectual excitement of trying to solve the circuitry and logic of a finite, well-organized nervous system, one cell at a time.

Before one can work out how the animal performs its movements, it is necessary to know about the individual cells: their properties, connections, and functions.

Sensory Cells in Leech Ganglia

When one strokes, presses, or pinches the skin of a leech, a sequence of movements follows. One segment or more shortens abruptly, and the skin becomes raised into a series

[15]Payton, W. B. 1981. In *Neurobiology of the Leech*. Cold Spring Harbor Laboratory, Cold Spring Harbor, NY, pp. 27–34.

[16]Kuffler, S. W., and Potter, D. D. 1964. *J. Neurophysiol.* 27: 290–320.

[17]Stent, G. S., and Weisblat, D. 1982. *Sci. Am.* 246(1): 136–146.

[18]Weisblat, D. A., et al. 1999. *Curr. Top. Dev. Biol.* 46: 105–132.

[19]Calabrese, R. L. 1998. *Curr. Opin. Neurobiol.* 8(6): 710–717.

[20]Szczupak, L., et al. 1998. *J. Exp. Biol.* 201: 1895–1906.

[21]Catarsi, S., and Drapeau, P. 1996. *Cell. Mol. Neurobiol.* 16: 699–713.

[22]Kleinhaus, A. L., and Angstadt, J. D. 1995. *J. Neurobiol.* 27: 419–433.

[23]Yu, X., Nguyen, B., and Friesen, W. O. 1999. *J. Neurosci.* 19: 4634–4643.

[24]Thompson, W. J., and Stent, G. S. 1976. *J. Comp. Physiol.* 111: 309–333.

[25]Kristan, W. B. 1983. *Trends Neurosci.* 6: 84–88.

[26]Coggeshall, R. E., and Fawcett, D. W. 1964. *J. Neurophysiol.* 27: 229–289.

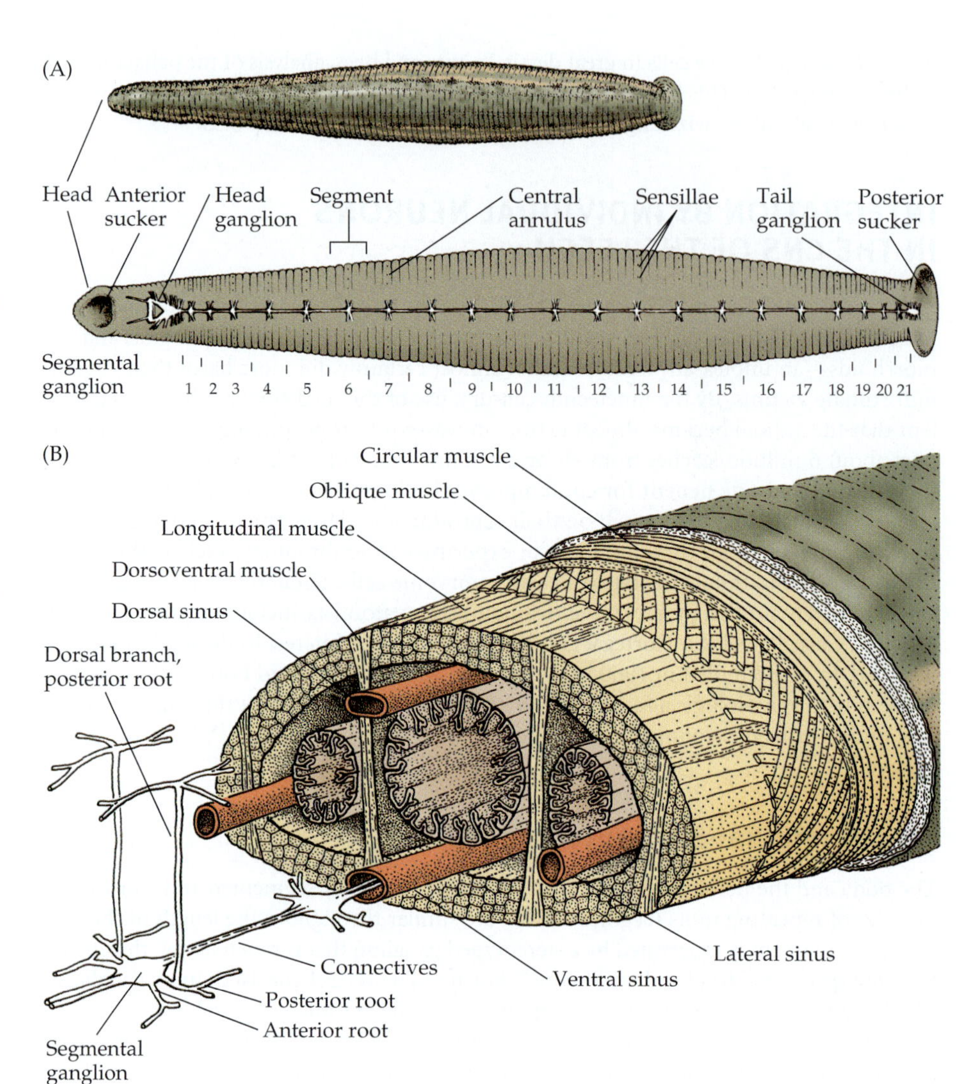

FIGURE 15.1 Central Nervous System of the Leech. (A) The CNS of the leech consists of a chain of 21 segmental ganglia, a head ganglion, and a tail ganglion. Over most of the body, five circumferential annuli make up each segment; the central annulus is marked by sensory end organs responding to light and touch (the sensillae). (B) The nerve cord lies in the ventral part of the body within a blood sinus. Ganglia, which are linked to each other by bundles of axons (the connectives), innervate the body wall by paired roots. The muscles are arranged in three principal layers: circular, oblique, and longitudinal. In addition, there are dorsoventral muscles that flatten the animal, and fibers immediately under the skin that raise it into ridges.

of distinct ridges. Subsequently the animal bends, writhes, or swims away. One can reliably identify the individual sensory and motor cells that mediate these reflexes according to their shapes, sizes, positions, and electrical characteristics.[27–29] Figure 15.2 shows the distribution of identified sensory cells, motoneurons, and interneurons in a leech ganglion. The 14 neurons labeled T, P, and N in Figure 15.2 are sensory and represent three sensory modalities. Each cell responds selectively to touch (T), pressure (P), or noxious (N) mechanical stimulation of the skin. Figure 15.3 illustrates the responses of sensory cells to various forms of cutaneous stimuli. T cells give transient responses to light touch of the skin surface (Figure 15.3A). Their sensory endings consist of small dilatations situated between epithelial cells on the surface of the skin. T cells adapt rapidly to a maintained step indentation and cease firing within a fraction of a second. The P cells respond only to a marked pressure or deformation of the skin and show a slowly adapting discharge (Figure 15.3B). The N cells require still stronger mechanical stimuli, such as radical deformation produced by pinching the skin with blunt forceps (Figure 15.3D). The modalities and responses of these neurons in leech resemble those of mechanoreceptors in the human skin (Chapter 18), which distinguish among touch, pressure, and noxious or painful stimuli. In the invertebrate, however, a single nerve cell does the job of many neurons supplying our own skin in a densely innervated region such as the fingertip.

Figure 15.4 shows that each sensory cell innervates a defined territory. The area is mapped by recording from a cell while applying mechanical stimuli to the skin or by la-

[27]Nicholls, J. G., and Baylor, D. A. 1968. *J. Neurophysiol.* 31: 740–756.

[28]Stuart, A. E. 1970. *J. Physiol.* 209: 627–646.

[29]Lockery, S. R., and Kristan, W. B., Jr. 1990. *J. Neurosci.* 10: 1816–1829.

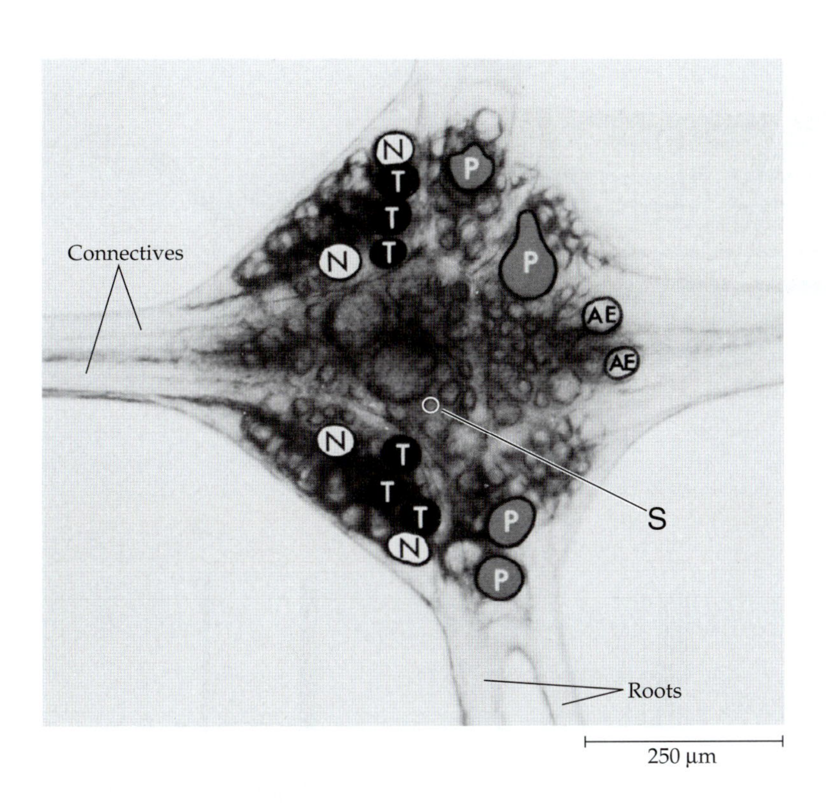

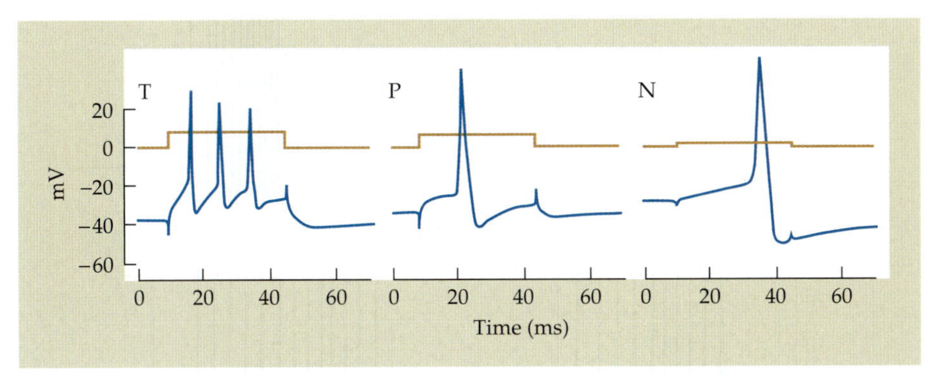

FIGURE 15.2 Ventral View of Leech Segmental Ganglion. Individual cells are clearly recognized. The three sensory cells responding to touch (T) and the pairs of cell types responding to pressure (P) or noxious (N) mechanical stimulation of the skin are labeled. Each type of cell gives distinctive action potentials, as shown by the traces below. Impulses in T cells are briefer and smaller than those in P or N cells. Current injected into cells through the microelectrode is monitored on the upper traces. The cells outlined in the posterior part of the ganglion are the annulus erector (AE) motoneurons. The S cell is a small unpaired neuron that is connected to its homologues in neighboring ganglia and plays a part in escape reflexes of the animal. (After Nicholls and Baylor, 1968; Stuart, 1970.)

beling the cell and its axons with a marker such as horseradish peroxidase.[30] The boundaries can be conveniently identified by landmarks, such as segmentation or the coloring of skin, so that one can predict reliably which cell will fire when a particular area is touched, pressed, or pinched. Thus, one of the touch-sensitive T cells innervates dorsal skin, another ventral skin, and a third laterally situated skin. Similarly, the two P sensory cells divide the skin into roughly equal dorsal and ventral areas. The elaborate and stereotyped branching pattern of a P sensory cell injected with horseradish peroxidase is shown in Figure 15.4. The T and N cells also send axons along the connectives to neighboring ganglia, and then out to minor receptive fields on either side of the major region of innervation.[31] The terminal branches of an individual cell each supply a circumscribed area of skin with no overlap between them, even though endings from other sensory neurons of the same modality may encroach on the same territory. It will be shown later that conduction is blocked in these fine branches. In a system that has such clear-cut boundaries in the periphery, it has also been possible to determine how the receptive fields become established during development and regeneration.[32,33]

Additional sensory cells that respond specifically to light, to chemical stimuli, to vibration, and to stretch of the body wall have been found in the head and in the periphery of the leech.[34,35]

[30]Blackshaw, S. E. 1981. *J. Physiol.* 320: 219–228.

[31]Yau, K. W. 1976. *J. Physiol.* 263: 513–538.

[32]Wang, H., and Macagno, E. R. 1997. *J. Neurosci.* 17: 2408–2419.

[33]Huang, Y., et al. 1998. *J. Comp. Neurol.* 397: 394–402.

[34]Blackshaw, S. E., and Nicholls, J. G. 1995. *J. Neurobiol.* 27: 267–276.

[35]Blackshaw, S. E., and Thompson, S. W. 1988. *J. Physiol.* 396: 121–137.

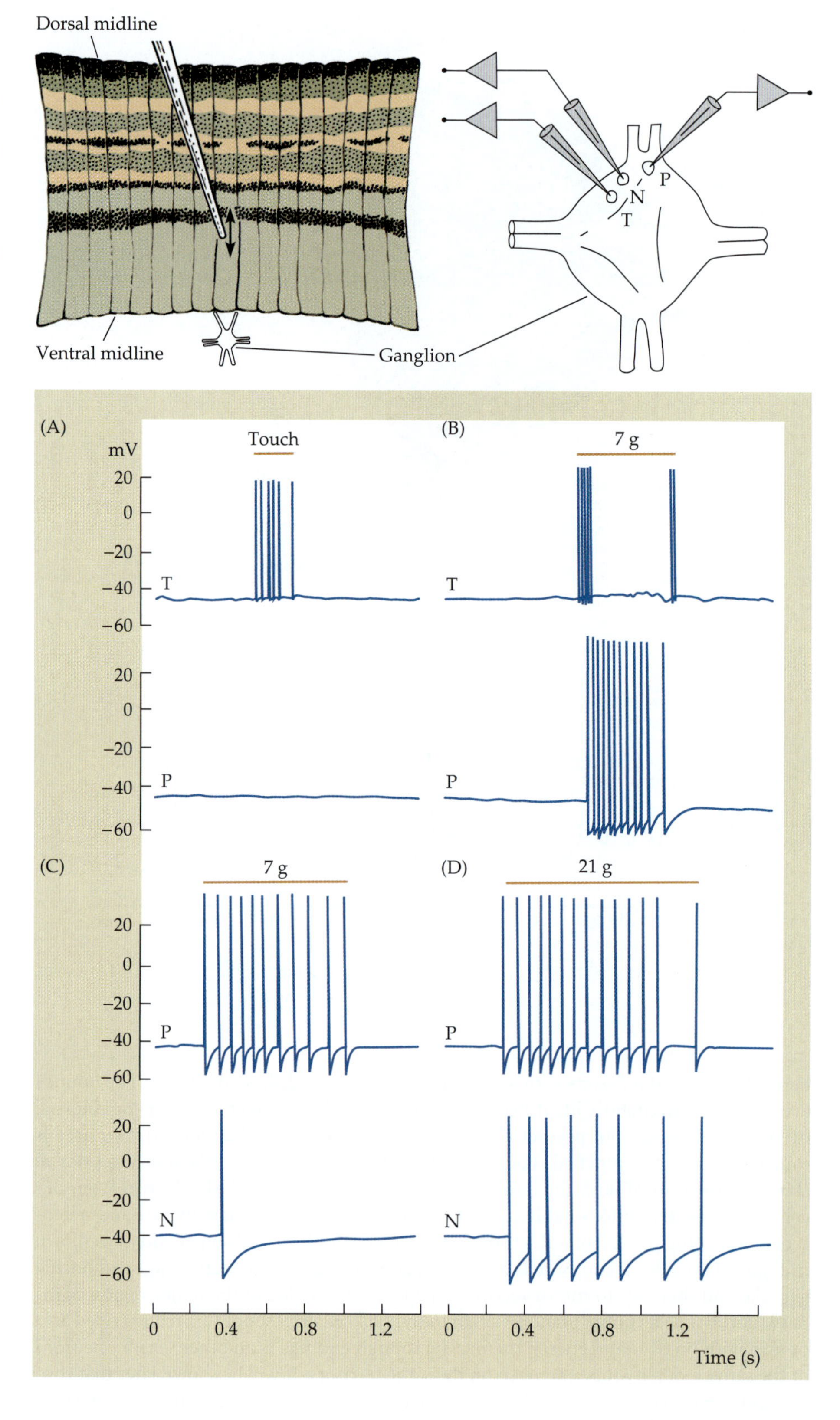

FIGURE 15.3 Responses to Skin Stimulation by leech sensory neurons. Intracellular records of T, P, and N sensory cells (see Figure 15.2). The preparation consists of a piece of skin and the ganglion that innervates it. Cells are activated by touching or pressing their receptive fields in the skin. (A) A T cell responds to light touching that is not strong enough to stimulate the P cell. (B) Stronger, maintained pressure evokes a prolonged discharge from the P cell and a rapidly adapting "on" and "off" response from the T cell. (C and D) Still stronger pressure is needed to activate the N cell. (After Nicholls and Baylor, 1968.)

Motor Cells

Individual motor cells in leech ganglia are shown in Figure 15.2. The criterion for showing that a cell is indeed motor is that each impulse in the cell gives rise to a conducted action potential in its axon leading to the muscle, and then to a synaptic potential in the

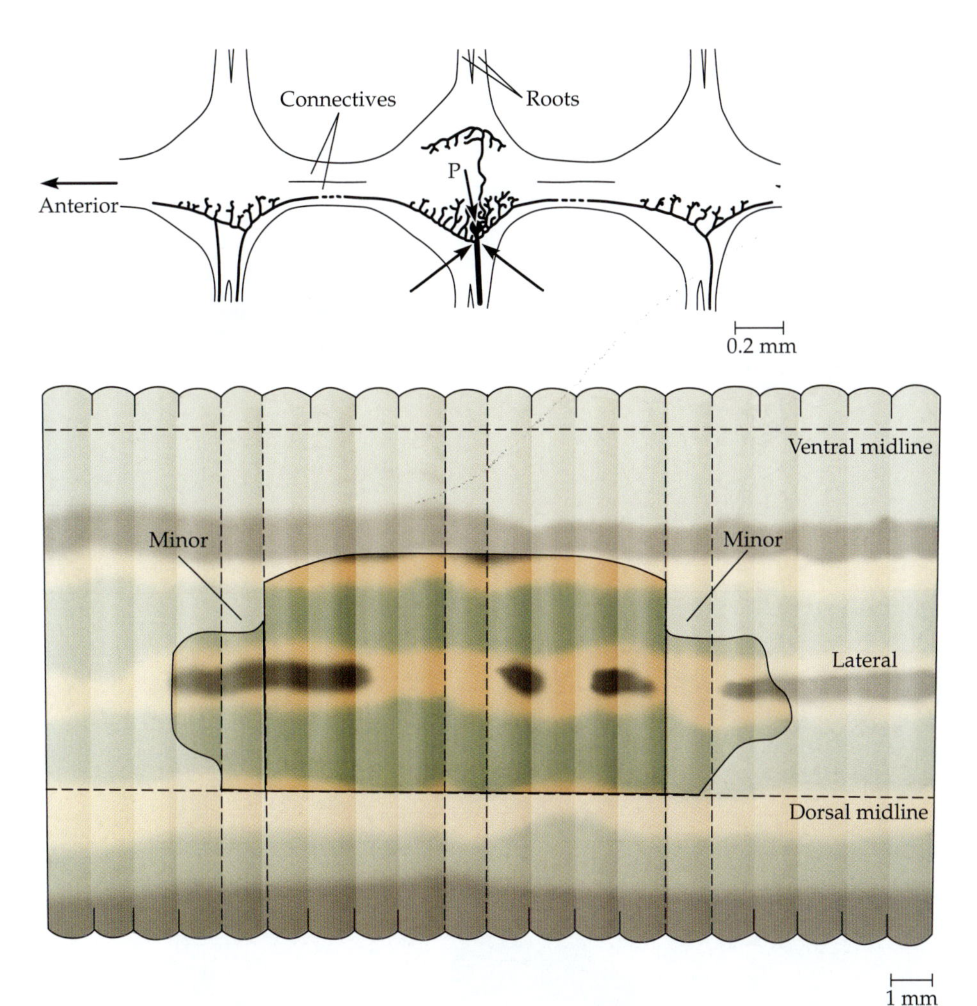

FIGURE 15.4 Receptive Field of a Pressure Sensory Cell. This P cell has an axon that runs out through the root of its own ganglion to supply the skin of the segment in which it is situated. Other axons of smaller diameter pass along the connectives to neighboring ganglia. These axons pass through the appropriate roots to innervate additional (minor) territories in adjacent segments. The second P cell (not shown) innervates similar territory situated more medially (i.e., closer to the ventral midline) but with similar longitudinal extent. There is considerable overlap between the receptive fields of the two cells. Hence, pressure applied to dorsal skin will activate the P cell shown in this figure. Pressure applied to ventral skin will activate the other P cell. Pressure applied on lateral skin will activate both P cells. The fact that axons with a small diameter supply the minor fields by the connectives and roots has implications for conduction. Although pressure applied anywhere within the receptive field will activate action potentials, conduction can become blocked where an axon of small diameter feeds into a larger-diameter axon, as at the points marked by arrows. (After Gu, 1991.)

muscle fiber. More than 20 pairs of motor cells supplying the various muscles that flatten, lengthen, shorten, and bend the body, as well as the motor cells that control the heart, have been identified in the segmental ganglia.[28] Muscles receive inhibitory and modulatory peptidergic inputs, as well as excitatory inputs that are mediated by acetylcholine. Deletion of a single cell can give rise to an obvious deficit in behavior—like a genetic "knockout" experiment for a single adult neuron.[36] For example, in each ganglion there is only one annulus erector motor cell on each side (labeled AE in Figure 15.2). Impulses in this cell cause the skin of the leech to be raised into ridges like a concertina (Figure 15.5). One AE motor cell can be killed by injecting it with a mixture of proteolytic enzymes (pronase) in an otherwise intact animal. When tests are made after the animal has been allowed to recover, the region of skin formerly innervated exclusively by the killed cell fails to become erect in response to appropriate sensory stimuli. The deficit is not permanent, however. Eventually branches of other intact AE cells come to supply the denervated territory.[37]

Connections of Sensory and Motor Cells

In the nervous systems of invertebrates, synapses between neurons are usually situated not on the cell bodies but on fine processes within a central region of the ganglion (the neuropil).[26,38–40] Synaptic potentials originating in the neuropil spread into the cell body, where they are recorded as excitatory and inhibitory potentials. Currents injected into the cell body can influence synaptic potentials and the release of transmitter. Despite its complexity, the neuropil is organized in an orderly manner. This fact was revealed by the experiments of Muller and McMahan, who first devised the technique of intracellular injection of horseradish peroxidase, using identified nerve cells in the leech.

The branching patterns of sensory and motor cells within the neuropil are characteristic, each cell displaying its own configuration. Examples of the typical ramifications of

[36]Bowling, D., Nicholls, J. G., and Parnas, I. 1978. *J. Physiol.* 282: 169–180.
[37]Blackshaw, S. E., Nicholls, J. G., and Parnas, I. 1982. *J. Physiol.* 326: 261–268.
[38]Muller, K. J., and McMahan, U. J. 1976. *Proc. R. Soc. Lond. B* 194: 481–499.
[39]French, K. A., and Muller, K. J. 1986. *J. Neurosci.* 6: 318–324.
[40]Macagno, E. R., Muller, K. J., and Pitman, R. M. 1987. *J. Physiol.* 387: 649–664.

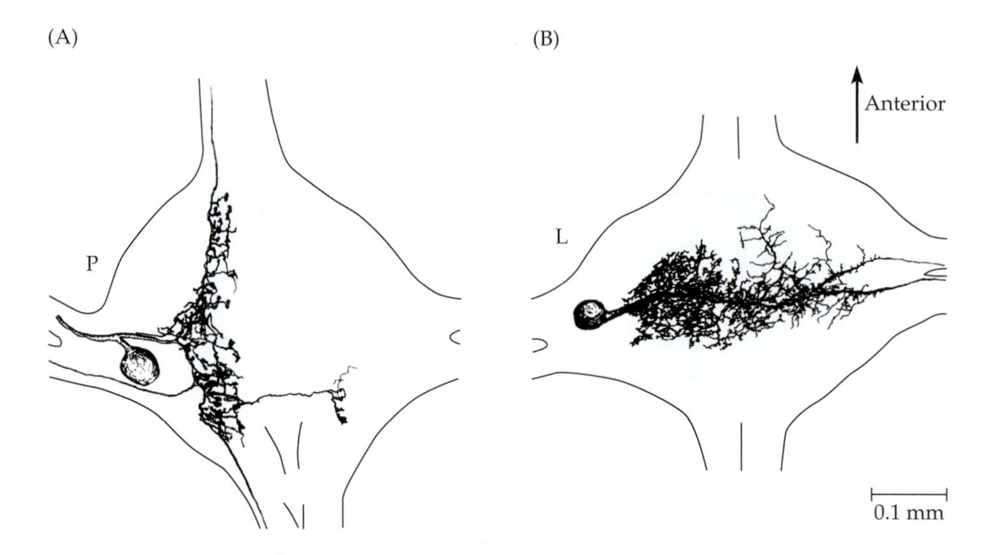

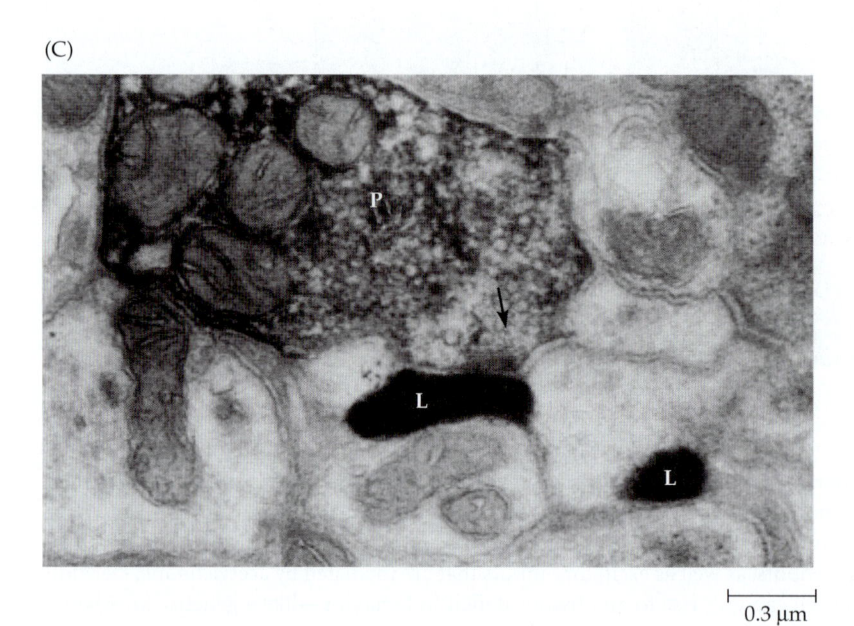

FIGURE 15.5 Structures of Pre- and Postsynaptic Cells labeled by intracellular injection of horseradish peroxidase. (A) The arborization of a P cell (shown in Figure 15.4) is profuse with numerous small varicosities. They represent sites of presynaptic endings that release transmitter. (B) The L motoneuron sends its axons out through contralateral roots. Its processes within the ganglion are smooth and represent postsynaptic sites upon which synapses are made. (C) A synapse (arrow) made by a P cell on an L motoneuron in the neuropil. Both cells were injected with horseradish peroxidase. (After Muller, 1981.)

identified neurons are shown in Figures 15.4 and 15.5. A single sensory cell supplies many postsynaptic targets, and its presynaptic endings are themselves supplied by numerous terminals arising from other neurons that modulate its release of transmitters. Electrical synapses appear as gap junctions between neurons with a narrowed space, 4 to 6 nm, separating the two membranes. Fluorescent dyes such as Lucifer yellow injected into one cell will usually, but not always, cross such electrical junctions and spread into other coupled cells.

The T, P, and N sensory cells responding to mechanical stimuli make excitatory connections on the L motoneurons (called "L" because they innervate longitudinal muscles) used for shortening the leech. Several lines of evidence, including electron microscopy, have shown that the connections are direct (i.e., that there are no known intermediary cells).[41] This fact is important because only if each constituent of a circuit and its properties are known can one pinpoint the sites at which any interesting modifications in signaling take place.

The mechanism of transmission onto the L motor cell is characteristically different for each of the sensory cells. The N cells act through chemical synapses (with only a hint of electrical coupling), the T cells through rectifying electrical synapses, and the P cells by combination of both mechanisms.[41] The same P and N cells also make direct chemical

[41]Nicholls, J. G., and Purves, D. 1972. *J. Physiol.* 225: 637–656.

synapses on another cell, the AE motoneuron. Transmitters used by leech neurons are acetylcholine, GABA, glutamate, dopamine, serotonin, and peptides.[42–44]

Short-Term Changes in Synaptic Efficacy

While an animal is swimming or moving on a surface, mechanical stimulation of the skin causes trains of impulses. Under these conditions of repetitive activity, transmission at chemical synapses shows considerable changes in strength (Chapter 12). Variable effectiveness of different chemical synapses permits sequential activation or inactivation of two different postsynaptic cells by a single presynaptic sensory neuron.[45] It can activate first one neuron and then the other. This type of differential effect adequately explains how pressing the skin of a leech leads first to shortening of the animal and then to erection of its annuli.

The shortening of the body wall occurs abruptly and is poorly maintained, whereas the annuli become erect more slowly and stay erect longer. This behavior correlates well with the synaptic potentials shown in Figure 15.6 that facilitate at different rates and trigger the two motor responses in sequence. When the N cell fires repetitively after natural stimulation, chemically mediated synaptic potentials are recorded in both the L and the AE motoneurons. These synaptic potentials undergo phases of facilitation and depression. The facilitation, however, is characteristically greater and longer lasting at synapses on the AE motoneuron. A number of indirect experiments suggest that the facilitation and depression are due to changes in the amount of transmitter released at presynaptic N cell terminals. Similar effects have been found in the target cells supplied by cortical presynaptic neurons.[46]

In the leech and in *Aplysia*,[47,48] there are delayed lines, through interneurons, that parallel the direct route. These serve to coordinate more complex directional movements evoked by mechanical stimuli. For example, Kristan and his colleagues have shown that pressure applied to one side of the leech causes the animal to bend. The direction of the bend depends on the position of the mechanical stimulus on the body wall.[49,50] The reflex causes the bend to be directed away from the stimulus. The relative frequency of firing of the four P neurons, which indicate the stimulus position, is analyzed by a network of 25 to 30 interneurons, so as to produce a bending in the appropriate direction (Figure 15.7). This is achieved by contraction of certain muscles and relaxation of others. The manner in which the specific motoneurons are activated depends on precise computational analysis by the network. Interestingly, as in the somatogastric ganglion of the lobster,[51–53] any specific interneuron may be involved in the calculation of a number of different movements.

Membrane Potential, Presynaptic Inhibition, and Transmitter Release

In the leech, the pumping action of the blood vessels that constitute its heart is controlled by motoneurons in the CNS. Blood circulation depends on peristaltic waves moving along the length of the animal and coordinated pumping mechanisms on the two sides of the animal. In the well-defined circuit of neurons controlling the heartbeat of the leech, the rhythmicity is established by cyclical modulation of the membrane potential of presynaptic terminals of interneurons.[24] This in turn modulates the efficacy with which they release transmitter. Presynaptic depolarization and hyperpolarization by only a few millivolts have significant effects.[54] Specifically, maintained depolarization of the presynaptic terminals at the synapse between the interneuron and a heart motoneuron enables an action potential to release more transmitter. Conversely, if the presynaptic terminal is hyperpolarized, the impulse releases less transmitter.

A shift in resting potential of as little as 5 millivolts (from –40 to –35 mV) can cause a threefold increase in the amount of transmitter released by an impulse. (The amplitude and the duration of the presynaptic action potential are not obviously altered by such small changes in resting potential.) The nerve terminal is periodically hyperpolarized as a result of inhibitory inputs from other identified interneurons. This markedly reduces the numbers of quanta of transmitter liberated by each impulse that invades. Here, then, is

[42]Thorogood, M. S., Almeida, V. W., and Brodfuehrer, P. D. 1999. *J. Comp. Neurol.* 405: 334–344.

[43]Bruns, D., and Jahn, R. 1995. *Nature* 377: 62–65.

[44]Nadim, F., and Calabrese, R. L. 1997. *J. Neurosci.* 17: 4461–4472.

[45]Muller, K. J., and Nicholls, J. G. 1974. *J. Physiol.* 238: 357–369.

[46]Reyes, A., et al. 1998. *Nature Neurosci.* 1(4): 279–285.

[47]Hickie, C., Cohen, L. B., and Balaban, P. M. 1997. *Eur. J. Neurosci.* 9: 627–636.

[48]Walters, E. T., and Cohen, L. B. 1997. *Invertebr. Neurosci.* 3: 15–25.

[49]Kristan, W. B., Jr., Lockery, S. R., and Lewis, J. E. 1995. *J. Neurobiol.* 27: 380–389.

[50]Lewis, J. E., and Kristan, W. B., Jr. 1998. *Nature* 391: 76–79.

[51]Ayali, A., and Harris-Warrick, R. M. 1999. *J. Neurosci.* 19: 6712–6722.

[52]Bartos, M., et al. 1999. *J. Neurosci.* 19: 6650–6660.

[53]Elson, R. C., et al. 1999. *J. Neurophysiol.* 82: 115–122.

[54]Nicholls, J. G., and Wallace, B. G. 1978. *J. Physiol.* 281: 157–170.

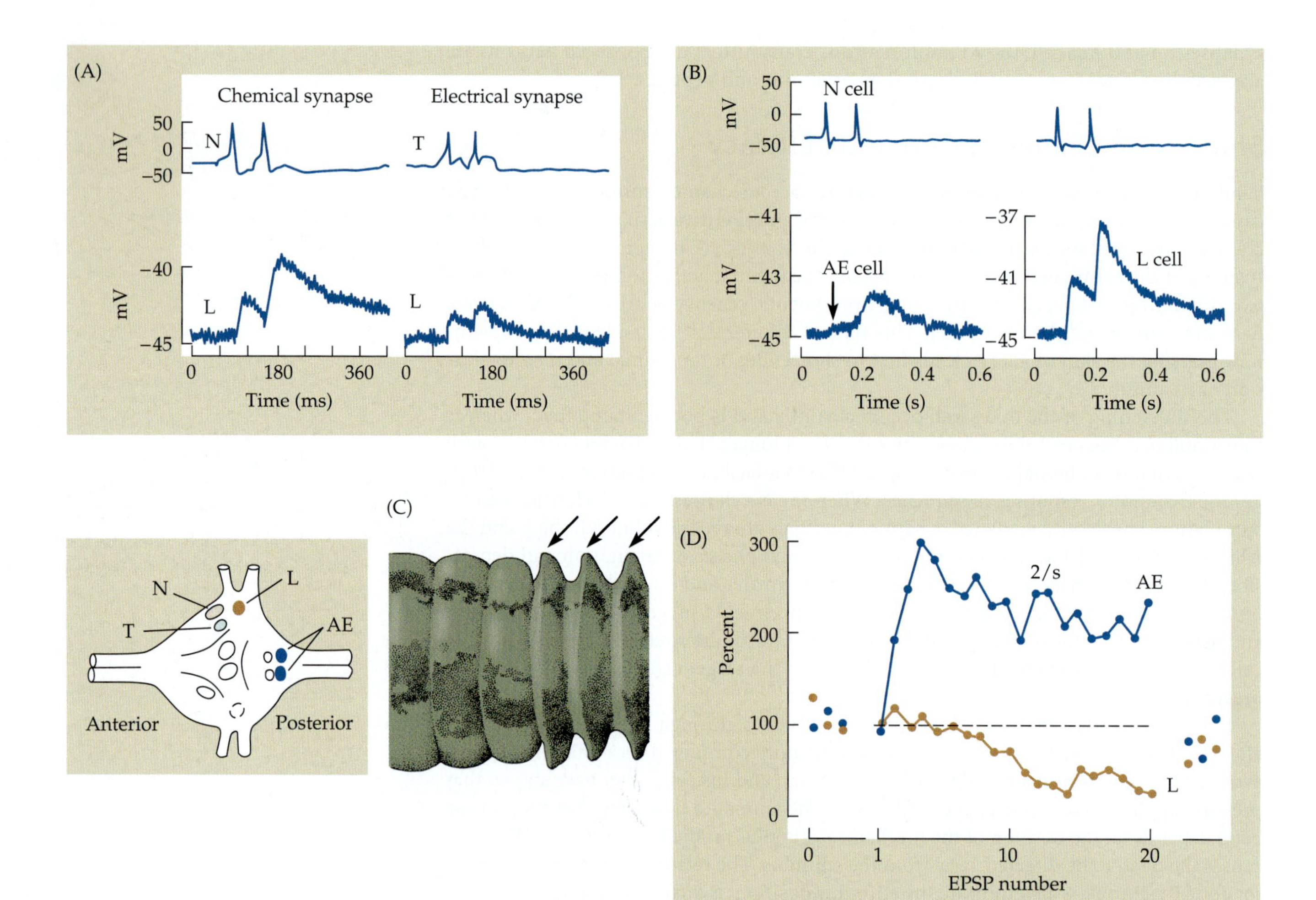

FIGURE 15.6 Facilitation and Depression at synapses between leech sensory and motor neurons. (A) Comparison of chemical and electrical transmission in a ganglion. An N cell is stimulated twice in succession, and at the same time its impulses are recorded (upper traces). At the chemical synapse between N and L cells, facilitation occurs, so a second impulse leads to a larger synaptic potential (bottom left). In contrast, two potentials evoked by T cell impulses in an L cell cause unchanged postsynaptic potentials with double stimulation. This is typical of electrical synapses. (B) Characteristics of transmitter release at different synapses made by a presynaptic neuron. An N cell is stimulated, and responses are recorded in L and AE cells. Facilitation is greater at the N–AE synapse. (The small first synaptic potential in the AE cell is marked by an arrow.) (C) Impulses in an AE cell produce annular erection (arrows). (D) When the N cell is stimulated at a rate of 2 per second, the synaptic potentials in the AE cell are facilitated to more than double their original size, whereas those in the L cell decrease in amplitude. The abscissa indicates the number of the synaptic potential recorded in the train. The ordinate gives the proportional height of the synaptic potentials compared with the average value before the train (100%). (A after Nicholls and Purves, 1972; B–D after Muller and Nicholls, 1974.)

an instance of presynaptic inhibition that is mediated by hyperpolarization of the terminals, and in which the mechanism has been demonstrated directly by quantal analysis. The presynaptic inhibition is timed to ensure that while the heart vessel on one side of the animal is contracted, that on the other side is relaxed. Similar effects of presynaptic membrane potential on tonic release can be analyzed in detail at synapses that form between pairs of neurons in culture.[55]

[55]Fernandez-de-Miguel, F., and Drapeau, P. 1995. *J. Neurobiol.* 27: 367–379.

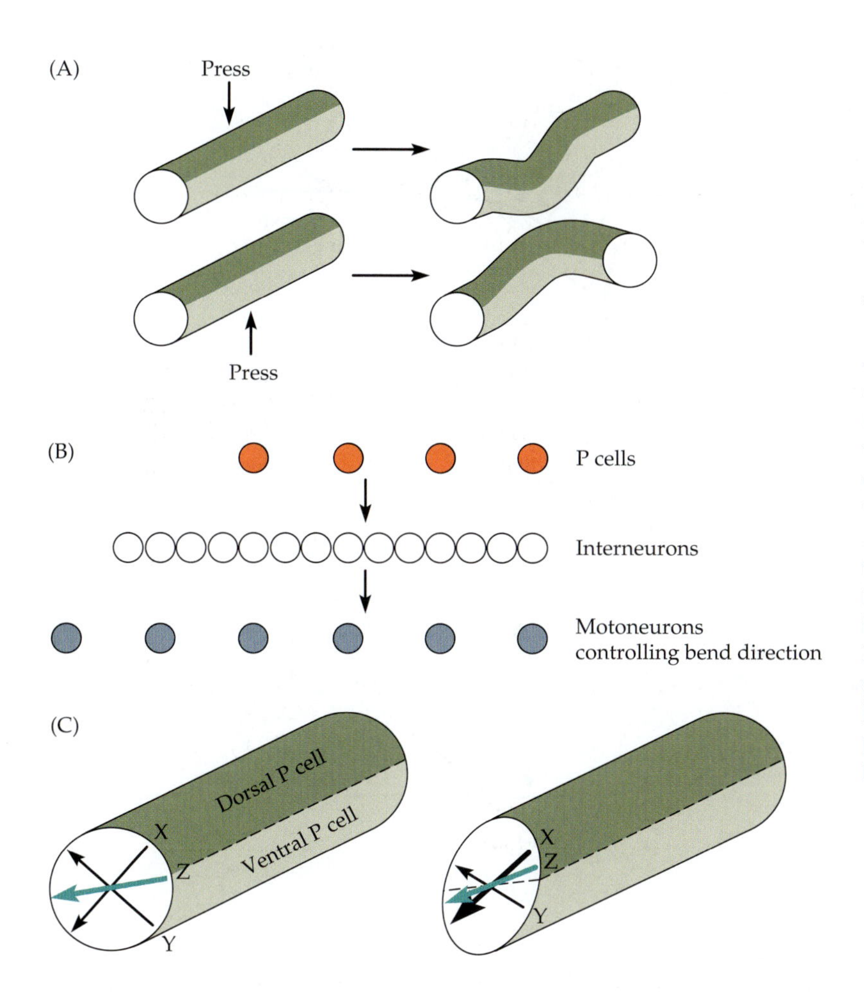

FIGURE 15.7 Integration by Interneurons used in leech bending reflex. When the surface of a leech is pressed on one part, a reflex bend occurs in the body so as to move that region away from the stimulus. Touch (T) and pressure (P) sensory neurons convey signals to an array of interneurons. These in turn make excitatory and inhibitory connections on motoneurons. (A) When the P cell innervating dorsal skin is activated by pressure (arrow), the longitudinal muscles immediately underneath the skin contract, while those on the opposite side of the body relax through inhibition. Similarly, when ventral skin is pressed, the ventral longitudinal muscle contracts, while dorsal muscle on the opposite side of the animal relaxes. (B) Scheme for connections of the four P cells that connect to the interneurons that in turn excite or inhibit motoneurons used for bending. (C) When both P cells are activated, by pressing simultaneously at X and at Y, or by pressing at Z, the interneurons integrate the information and give rise to a vector producing an appropriate bend. The direction of the bend can be accurately predicted from the firing frequencies of the two P cells. (After Lewis and Kristan, 1998.)

Repetitive Firing and Conduction Block

Failure of impulse conduction represents another well-defined mechanism for altering the synaptic action of one cell upon its postsynaptic targets. In the central nervous system of the leech and the cockroach, and in crustacean motor axons,[56] repeated trains of action potentials occurring at natural frequencies cause conduction to fail at specific axonal branch points. In T, P, and N sensory neurons of the leech, the mechanism depends on the hyperpolarization caused by the electrogenic sodium pump and by long-lasting changes in a calcium-activated potassium conductance. For example, repeatedly stroking or pressing the skin of the leech causes trains of impulses and a maintained hyperpolarization of P sensory cells. As a result, propagation of impulses becomes blocked at branch points within the neuropil where the geometry for impulse conduction is unfavorable.[31] These are sites at which a small-diameter axon feeds into a larger one (Chapter 7). At the same time other branches of the same neuron continue to transmit.[57]

Conduction block represents a nonsynaptic mechanism that temporarily disconnects a P cell or a T cell from one defined set of its postsynaptic targets. When some, but not all, of the presynaptic fibers connected to a cell fail to conduct, transmitter release and the efficacy of transmission are reduced. By making lesions with a laser at specific selected branch points, Muller and his colleagues have assessed the contribution made by discrete branches of the sensory P cell to its synaptic action on motor cells.[58] An example of conduction block in a P cell is shown in Figure 15.8. The synaptic potentials recorded in the postsynaptic AE motor cell are abolished when impulses become blocked at the point marked by "X" (Figure 15.8B). At the same time, synaptic potentials still arise in the L motoneuron, because it receives inputs from other branches of the P cell in which conduction continued. Repetitive activity at natural frequencies in effect reduces the area of skin from which responses in the AE cell (in this example) are evoked. Hence, it causes temporary shrinkage of the receptive field.[57]

[56]Grossman, Y., Parnas, I., and Spira, M. E. 1979. *J. Physiol.* 295: 307–322.

[57]Gu, X. 1991. *J. Physiol.* 441: 755–778.

[58]Gu, X. N., Muller, K. J., and Young, S. R. 1991. *J. Physiol.* 441: 733–754.

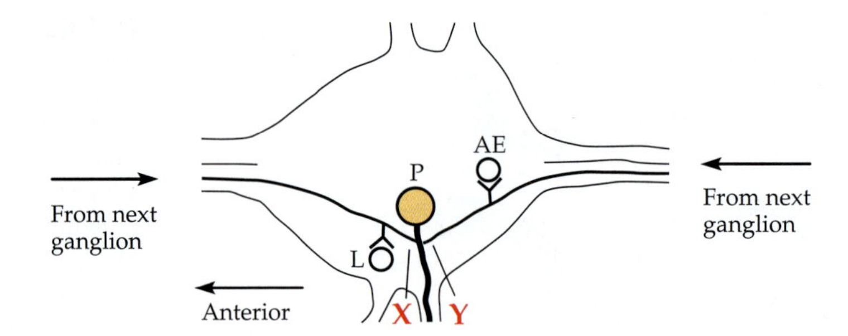

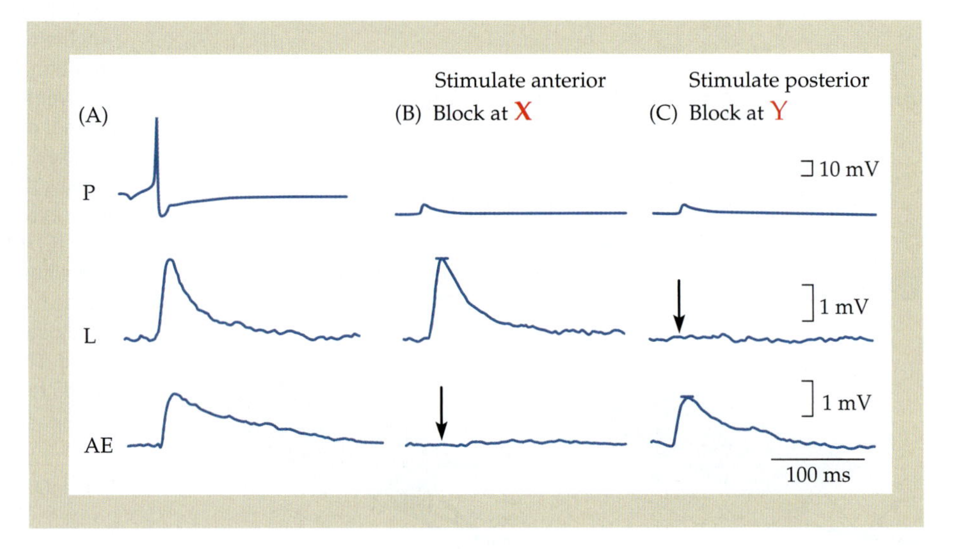

FIGURE 15.8 Effects of Conduction Block on synaptic transmission and integration. The medially situated P cell (innervating dorsal skin and similar to that shown in Figure 15.4) sends presynaptic axons to motor cells that cause the longitudinal muscles to contract (L cell) and the annuli to erect (AE cell). (A) The synaptic potentials in the L cell and the AE cell were evoked by stimulating the P cell intracellularly. When the impulses originate in the neighboring segment by pressure, they travel along the small axons in the connective, successfully invade all the branches of the P cell, and produce synaptic potentials in the AE and the L cell. (B) When the skin in the next most anterior segment is pressed repeatedly, the P cell becomes hyperpolarized and impulse conduction fails at the point labeled X. When this occurs, the L cell still receives its input and produces a synaptic potential, but the AE cell has been temporarily disconnected (arrow). (C) Conversely, when the posterior field is stimulated, impulses initially invade all the branches in the P cell, as in A. After repetitive stimulation, however, they become blocked at Y, where the small axon encounters the larger axon. Once again cells become disconnected (arrow). This time, however, it is the L cell that fails to show a synaptic potential, while the AE cell continues to do so. (After Gu, 1991.)

An unexpected event can occur when conduction becomes blocked. The action potential that has failed to invade the larger axon sets up a local potential in it. This depolarization of the large axon beyond the point of failure lasts longer than the action potential in the incoming fiber. It can thereby initiate a second action potential in that axon traveling away from the site of block to the synaptic terminals of the sensory neuron. There it initiates transmitter release a second time, with minimal delay after the first impulse.[59]

Conduction block was described first for dorsal root axons in the mammalian spinal cord by Barron and Matthews in 1935,[60] was largely forgotten, and only much later was analyzed in invertebrates. It is now apparent that conduction block is a key feature of integrative mechanisms in dendrites in mammalian CNS.[61]

Higher Levels of Integration

One aim of the studies on an animal like the leech is to analyze how complex behavioral acts are built up from simple, elementary reflexes. The leech has been particularly valuable for tracing the circuits and for identifying one by one the individual cells that act in concert to produce and maintain swimming. This complex movement has been examined by Stent, Kristan, Friesen, and their colleagues. The head and tail "brains" are not essential for generating the swimming rhythm, which can occur in a few isolated segments or even a single segment of the animal. As in other invertebrates (such as the cockroach, the locust, and the cricket), in which central motor programs involving a small number of individual cells have been shown to control complex patterns of movement, the basic rhythm in the leech is established by inhibitory and excitatory synaptic interactions within the CNS. Peripheral receptors serve to trigger, enhance, depress, or halt leech swimming. Similarly, mammalian CNS preparations isolated from the body generate the respiratory rhythm in vitro (Chapter 22).

Again, as in vertebrates (Chapter 22), an interesting modulatory role is played by biogenic amines in regulating motor activity.[62,63] Quiescent, sluggish(!), nonswimming

[59]Baccus, S. A. 1998. *Proc. Natl. Acad. Sci. USA* 95: 8345–8350.

[60]Barron, D. H., and Matthews, B. H. C. 1935. *J. Physiol.* 85: 73–103.

[61]Tsubokawa, H., and Ross, W. N. 1997. *J. Neurosci.* 17: 5782–5791.

[62]Glover, J. C., and Kramer, A. P. 1982. *Science* 216: 317–319.

[63]Brodfuehrer, P. D., et al. 1995. *J. Neurobiol.* 27: 403–418.

leeches have lower blood levels of 5-HT (serotonin) than do active leeches. Moreover, stimulation of identified cells (known as Retzius cells) that secrete 5-HT promotes an increase in its concentration in the blood and promotes swimming in the animal as well. The level of 5-HT can be depleted in embryos by means of a specific chemical (5,6-dehydroxytryptamine) that selectively destroys the 5-HT neurons in the developing ganglia. After development, the adult leeches do not swim spontaneously, but they will do so if immersed in a weak solution of 5-HT.

The S Interneuron and Sensitization

The way in which an identified neuron can play a coordinating role in the behavior of an animal has been studied extensively in insects and crustaceans, in which an identified "command" neuron can initiate or orchestrate a behavioral response of the animal. In the leech, a complex function of this sort is well illustrated by a single, unpaired interneuron in each ganglion called the S cell (Figures 15.2 and 15.9).[64] The S neuron receives excitatory inputs from touch and pressure sensory cells and in turn excites the L motoneuron, which causes shortening (as already discussed). Each S cell is electrically connected by a large, rapidly conducting axon to the axon of its homologue in the neighboring ganglion. That these junctions occur in the midregion of the connective (see Figure 15.9) is revealed by injecting one S cell with horseradish peroxidase, which cannot cross gap junctions.

[64]Muller, K. J., and Carbonetto, S. 1979. *J. Comp. Neurol.* 185: 485–516.

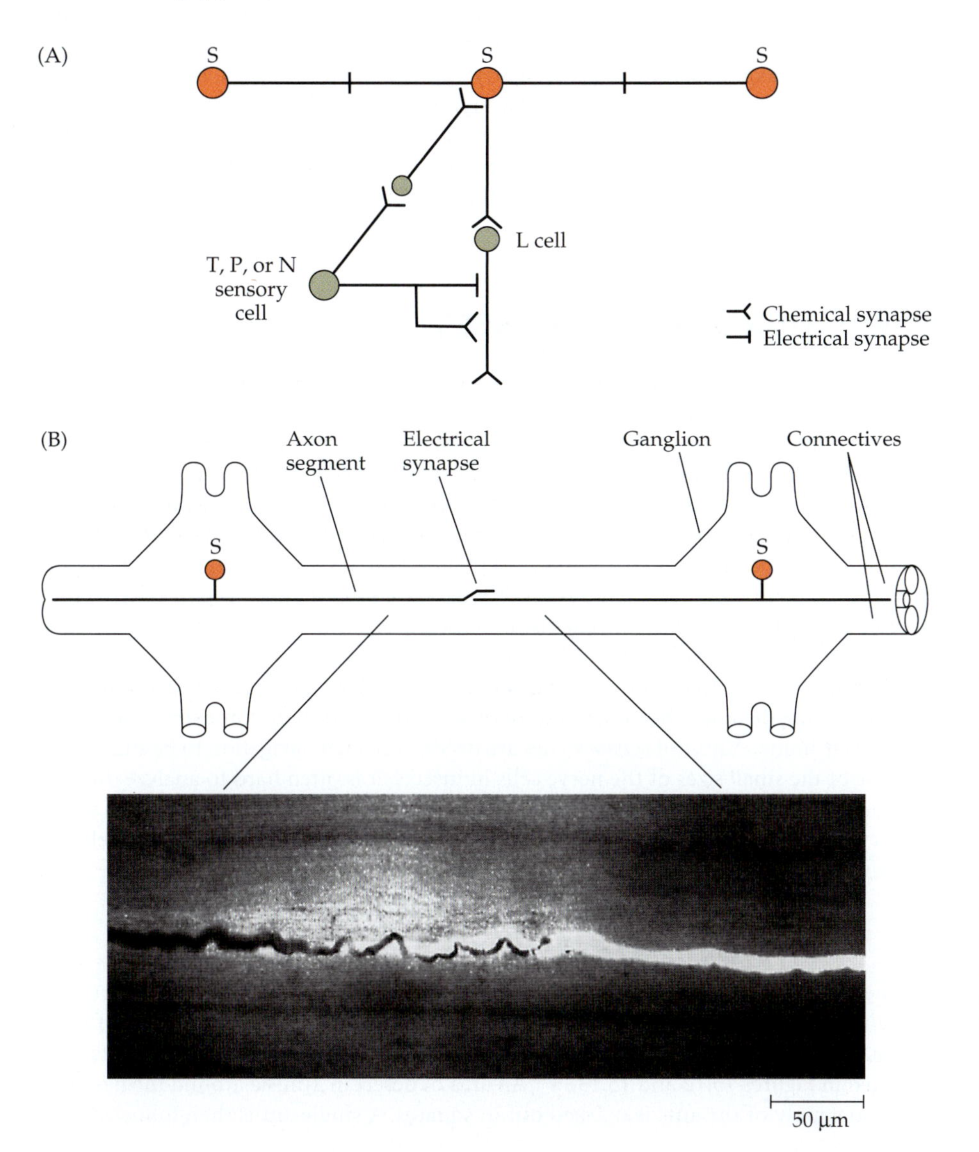

FIGURE 15.9 Connections of S Cells that conduct rapidly and make electrical synapses in the midregion of the connective. (A) Electrical and chemical connections of sensory T, P, or N cells, and S cells onto the L motoneuron. Each ganglion contains only one S cell, the axons of which make electrical synapses on its homologues. Through the S cell chain along the animal, sensory stimuli cause rapid shortening. (B) S cells are connected to their homologues in neighboring ganglia by electrical synapses. The photomicrograph shows an S cell that was filled with a mixture of Lucifer yellow and horseradish peroxidase (HRP). HRP cannot cross electrical junctions, whereas Lucifer yellow can. Accordingly, the neighboring S cell axon became labeled with Lucifer yellow but not with HRP. S cell axons regenerate and re-form their connections with very high specificity after injury. The micrograph, which resembles that of a normal animal, was in fact taken from a preparation in which the axon of one S cell had been severed and had re-formed its connections after regeneration. (After Mason and Muller, 1996; micrograph kindly provided by K. Muller.)

The chain of S cells is essential for adaptive types of behavior. If repeated touch stimuli are applied to a segment of the leech's body wall, a reflex shortening is produced. The response becomes progressively weaker with each repeated touch—a process known as **habituation** (Figure 15.10).[65] After a stronger stimulus that activates P (pressure) and N (nociceptive) cells, as well as T (touch) cells, has been applied to a different region of the body wall, the shortening response to touch once again becomes evident. This recovery process is known as **dishabituation**. Similarly, if a strong stimulus is applied without previous habituation, **sensitization** occurs; that is, the strength of the response to touch becomes greater than normal. The firing patterns of S cells do not change during habituation. But S cell activity is increased during both sensitization and dishabituation (see Figure 15.10). In technically difficult experiments, the S cell axon was cut or an S cell was killed in its entirety by injection of pronase. Killing the S cell did not interfere with shortening or habituation as such, but it did abolish dishabituation and sensitization.

In a second series of experiments, the axons of S cells were lesioned and then allowed to regenerate. A remarkable property of the S cell is that after its axon is severed, it grows back to re-form its electrical connections with the target S axon with extraordinary precision.[66] As expected, breaking the train of transmission along the S cells throughout the length of the animal abolished sensitization. Some weeks later, as shown in Figure 15.10, when S cell axons had regenerated and re-formed their connections, sensitization once again reappeared.[67]

These experiments represent a clear demonstration of the way in which a single cell can play a part in a highly complex response such as sensitization. The actual mechanism of sensitization in the leech has been shown to involve serotonin.[68] (A peculiar series of coincidences make the name "S" particularly appropriate for this cell. It was originally called S because its impulse constituted the largest "spike" in the connective. When it was found that S cells were very tightly coupled, they were thought to be a "syncytium." Now they are "S" because of sensitization and serotonin!) In *Aplysia*, sensitization (for which serotonin is similarly responsible) has been studied in great detail at the cellular and molecular levels.[69]

NAVIGATION BY ANTS AND BEES

An essential technique for understanding the workings of the nervous system is the quantitative analysis of behavior. By analyzing behavior, one can reveal fundamental principles. For example, key concepts that explain color vision and dark adaptation in humans were first established through psychophysical experiments. From the responses of subjects exposed to different wavelengths or intensities of lights, it was possible to deduce mechanisms of perception. Only much later could cellular mechanisms be demonstrated in photoreceptors and visual pathways (Chapters 19 and 20). Similarly, in invertebrates, observation of animals under natural conditions has led to valuable insights into the role of receptors and of integrative mechanisms of the CNS.

The extraordinary performance of invertebrate nervous systems can be appreciated by considering complex navigation by ants or bees.[70,71] A desert ant can wander and a bee can fly for long distances from the nest, search for food and then, somehow, unerringly find its way home. A host of sensory cues are needed for such navigation to be successful. Because of the small sizes of the nerve cells in insects, it is often hard to analyze directly membrane properties and synaptic transmission. And yet, as the following sections show, sensory mechanisms in these animals can be inferred and then analyzed at the cellular level through insights derived from behavioral experiments.

The Desert Ant's Pathway Home

Wehner and his colleagues have made experiments to analyze how the desert ant, *Cataglyphis bicolor* (Figure 15.11), is able to wander for long distances in search of food and then return toward the nest in a straight line. The principle of their experiments is illustrated in Figures 15.12 and 15.13.[72,73] An area of desert in Tunisia around the nest and the food supply of the ants is marked out in squares. A single ant is then followed as it

[65] Sahley, C. L., et al. 1994. *J. Neurosci.* 14: 6715–6721.

[66] Elliott, E. J., and Muller, K. J. 1983. *J. Neurosci.* 3: 1994–2006.

[67] Modney, B. K., Sahley, C. L., and Muller, K. J. 1997. *J. Neurosci.* 17: 6478–6482.

[68] Ehrlich, J. S., et al. 1992. *J. Neurobiol.* 23: 270–279.

[69] Antonov, I., Kandel, E. R., and Hawkins, R. D. 1999. *J. Neurosci.* 19: 10438–10450.

[70] Wehner, R., and Menzel, R. 1990. *Annu. Rev. Neurosci.* 13: 403–414.

[71] Wehner, R. 1994. *Fortschr. Zool.* 31: 11–53.

[72] Wehner, R. 1997. In *Orientation and Communication in Arthropods.* Birkhauser, Basel, Switzerland, pp. 145–185.

[73] Collett, M., Collett, T. S., and Wehner, R. 1999. *Curr. Biol.* 9: 1031–1034.

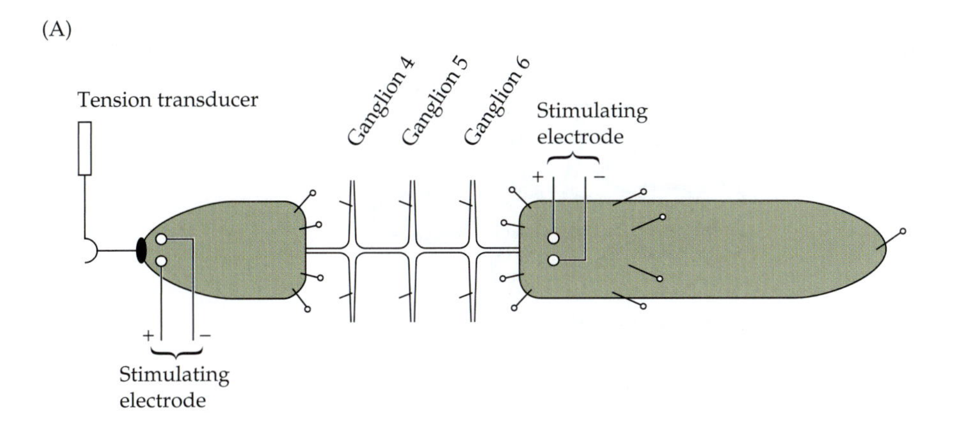

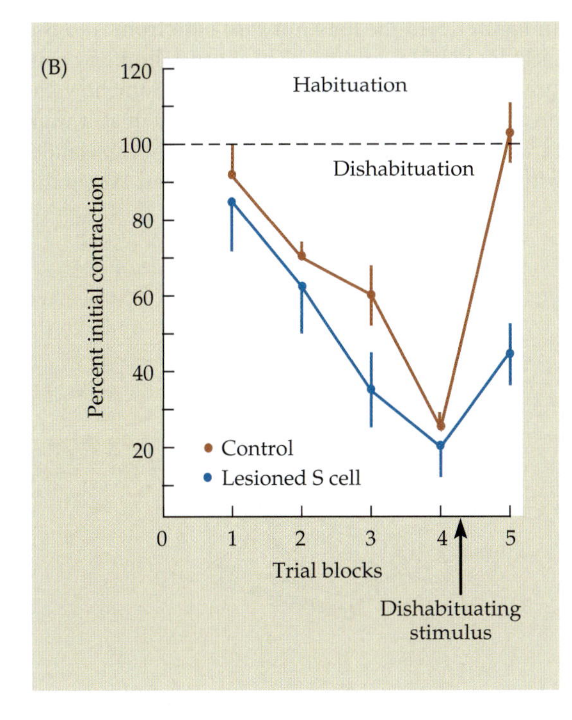

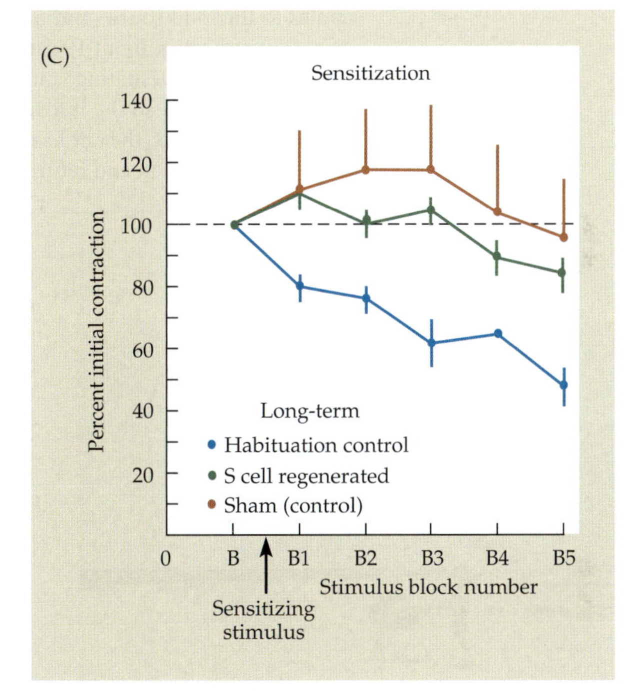

FIGURE 15.10 Habituation, Dishabituation, and Sensitization of leech reflexes, and the role of the S cell. (A) Isolated ganglia connected to the anterior and posterior parts of the body. Stimuli are applied to anterior or posterior skin by electrodes or by mechanical stimuli that activate T, P, or N cells. Intracellular recordings are made from S cells, sensory cells, and motoneurons, and muscle contractions are measured with a tension transducer. Weak repetitive stimuli applied to the tail end of the animal give rise to habituation. A stronger stimulus gives rise to sensitization and to dishabituation. (B) Responses to weak electrical shocks to posterior skin. In repeated trials, the responses became smaller (habituation, red circles). After the fourth trial, a dishabituating strong stimulus was given that produced a bigger response. After elimination of the S cell by killing it in one segment, only habituation occurred (blue circles). (C) When a strong sensitizing stimulus was given (sensitization, red circles), the responses became stronger than normal instead of habituating. After elimination of the S cell by killing it in one segment or by cutting its axon acutely, sensitization could no longer be elicited (blue circles). Repeated stimuli gave rise only to habituation. After an S cell axon had been severed and then regenerated in the manner shown in Figure 15.9, a strong stimulus once again gave rise to clear sensitization (green circles) and dishabituation (not shown). This experiment demonstrates that a single cell is essential for these complex responses. (After Sahley et al., 1994, and Modney, Sahley, and Muller, 1997.)

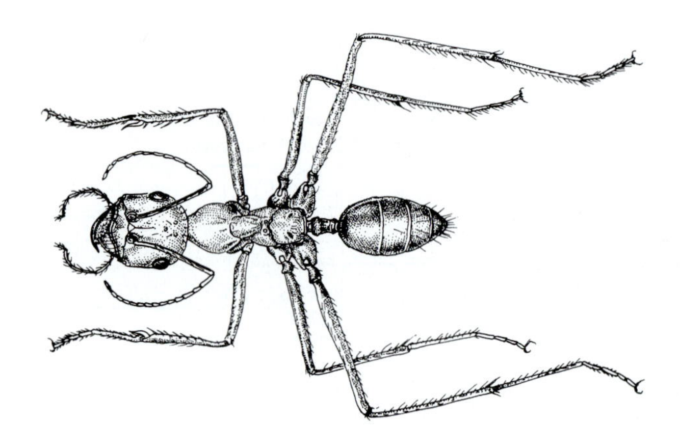

FIGURE 15.11 The Desert Ant, ***Cataglyphis bicolor*****,** which navigates successfully by means of detection of ultraviolet light. It is able to return home in a direct path after having searched for food. (After Wehner, 1994b.)

walks to the food source and back. In Figure 15.13 the long outward path from A to B is tortuous and takes about 19 min. (Each dot marks 1 minute.) The journey home by contrast is direct, unhesitating, and accurate, and takes only about 6 minutes. Somehow, the ant has integrated all the information about its movements over about 250 m as it wanders out. Somehow, the ant has kept track of the angles and distances. An ant can wander for more than 100 m and return to within 1 m of its nest, an error of less than 1% (in this

(A)

(B)

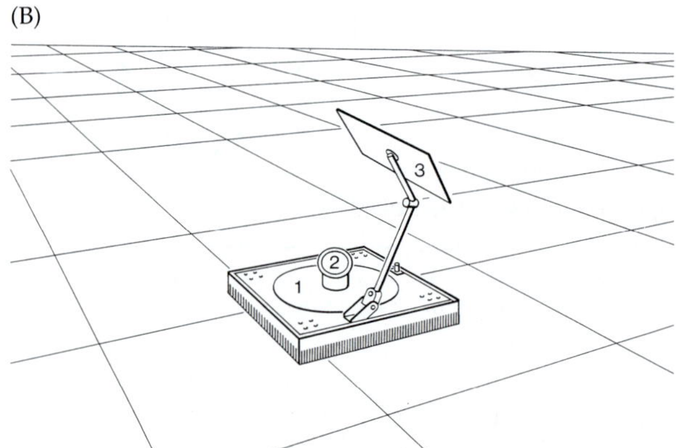

FIGURE 15.12 To Measure the Movements of an Ant it is placed on a grid marked out on the desert floor (A). A desert ant is tracked in the field by manipulating its skylight vision experimentally and simultaneously recording its walking trajectory using a rolling optical laboratory for following ants as they migrate along the desert floor and for controlling the portion of sky that they can observe (B). The trolley is moved by the experimenter in such a way as to keep the ant always centered within the optical setup. The horizontal aperture (1) can be fitted with filters that cut out all light apart from ultraviolet or that allow polarization in one direction. A small aperture (2) sits atop a circular tube, and the screen (3) can be used to prevent the ant from seeing sunlight directly. Since the trolley has a frame, the little ant cannot see the skyline or markers on the ground. It is also shielded from the wind. The white lines are 1 m apart and are painted on the desert floor to enable the observers to track the ant's progress accurately. (After Wehner, 1994b; photograph kindly provided by R. Wehner.)

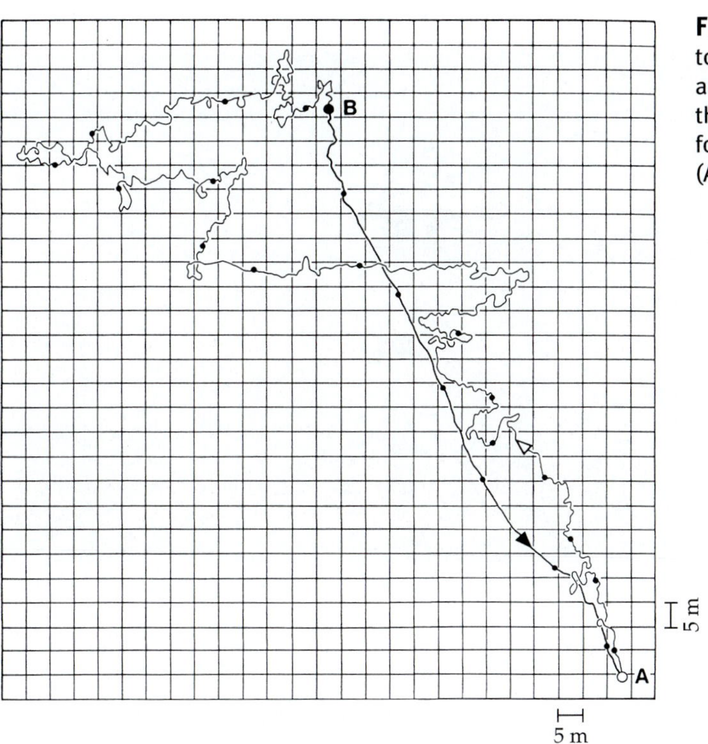

FIGURE 15.13 Pathway Taken by an Ant from its nest (A) to the source of food (B). Each dot on the trace toward food and back represents 1 minute of time. The ant progresses to the food in a tortuous path over 592 m. But it heads straight for home (140 m) with no deviation and amazing accuracy. (After Wehner, 1994a.)

respect the ant is doing far better than at least one author of this book, who does not have the mathematical skills for making computations of this sort).

What cues does the ant receive from the environment to do this? A first guess might be pheromones or chemical signals. These, however, are not used and would in any case be inoperative at the extremely high (up to 45°C) desert temperatures.[74] Spatial cues and useful landmarks do not abound in the desert. Although objects along the pathway and at the nest can indeed provide information (as we shall see shortly), they are not the principal compass that the ant makes use of. This was shown by experiments in which the ant made the round trip while able to view only a portion of the sky.[71] To eliminate the sun and all other cues, the experimenter walked along with the ant, pushing a trolley (see Figure 15.12) that kept the ant centered under an aperture to the sky. In the aperture could be placed filters to select the direction, the wavelength, or the angle of polarization of the light seen by the ant through the aperture to the sky. In the absence of the sun, landmarks, and odors, the ant seeing only sky still headed straight for home, as the experimenter trundled along, screening it with the little wagon.

The Use of Polarized Light as a Compass

Polarization of electromagnetic radiation refers to the situation in which the electrical vector of an electromagnetic wave becomes restricted to a single plane. That light from the sun becomes polarized as it traverses the atmosphere is a familiar observation: A properly oriented polarizing filter makes the sky dark (by blocking polarized light). Light reflected from the uneven surface of clouds, however, is no longer polarized. Hence, the brightness of the clouds is much less affected by such a filter. When the sun is vertically overhead, the pattern of polarized light is very simple. All polarization orientations would be in the horizontal plane and equally represented, so the pattern could not be used as a compass. If, however, the sun shines at any angle other than vertical, an asymmetric array of polarization orientations occurs, as shown in Figure 15.14. This pattern could in principle be used as a map for navigation. Although human eyes of course cannot detect polarized light, those of the desert ant can, as can those of bees, wasps, and crustaceans.[75] The pattern of polarized light defines the position of the sun, whether or not it can be seen directly.

[74]Wehner, R., Marsh, A. C., and Wehner, S. 1992. *Nature* 586–587.

[75]Muller, K. J. 1973. *J. Physiol.* 232: 573–595.

(A)

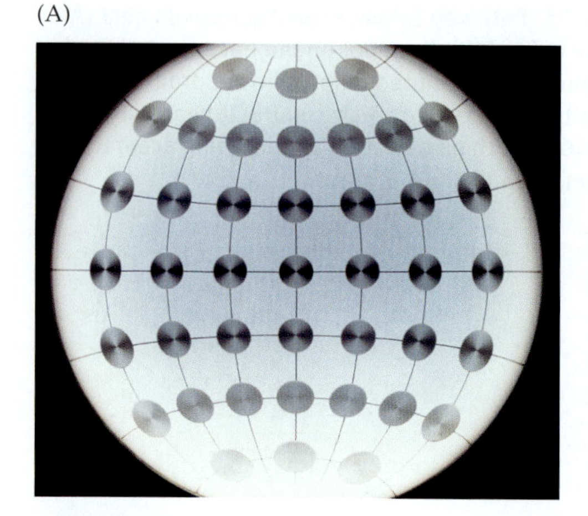

(B)

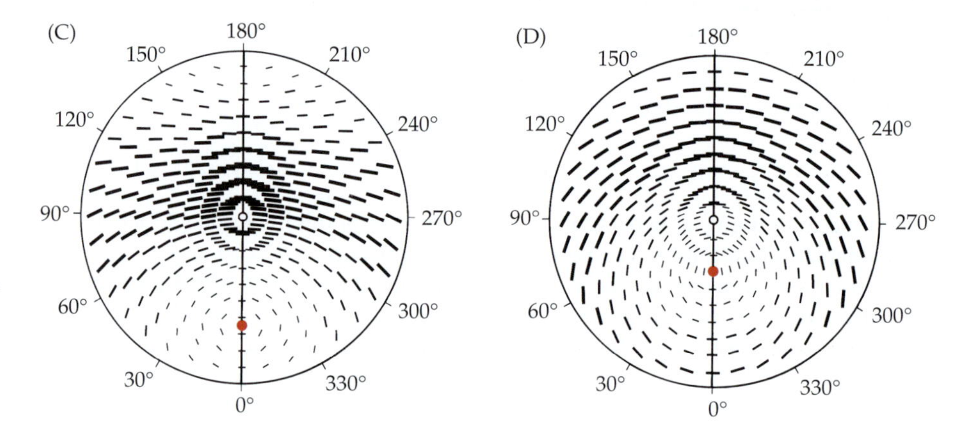

FIGURE 15.14 Patterns of Polarized Light and Their Detection by an array of polarizing filters. (A, B) The arrangement of polarizing filters mounted in a dome so as to detect patterns of polarized light in the sky. For a person standing at the equator at noon with the sun shining directly overhead, the patterns would be completely symmetrical. (C, D) The patterns produced by light from the sun at two different positions, indicated by the red dots. The ant need only observe a small part of the heavens in order to compute the position of the sun and thereby navigate successfully. (A and B after Wehner, 1994a; C and D after Wehner, 1997.)

Polarized Light Detection by the Ant's Eye

The multifaceted, complex eye of an ant consists of an array of units, the ommatidia, each supplied with its own nerve tracts. Within each ommatidium are eight photoreceptors. Each ommatidium sees the world from a different perspective.[76] In the desert ant, groups of these ommatidia are sensitive to polarized light of a particular orientation. They are located in the dorsal rim area of the compound eye and respond preferentially to light in the ultraviolet range. Figures 15.15 and 15.16 show the distribution of these detectors. Their sensitivity to polarization arises from the precise orientation of the membranes containing the photopigment rhodopsin that absorbs light. The rhodopsin molecules are arranged in parallel, in uniform register in the microvilli of a photoreceptor cell as shown in Figure 15.16.[77] Since rhodopsin absorbs light optimally along the long axis of the molecule, one particular plane of polarized light will be most effective in generating electrical signals in that photoreceptor. Moreover, the orientations of the rhodopsin molecules in four different photoreceptors of an ommatidium in the dorsal rim are aligned precisely at 90° to one another, an arrangement seen only in those photoreceptors concerned with polarized light. This orthogonal arrangement of four receptors in one ommatidium seems designed to allow a determination of the angle of polarization. (A single photoreceptor on its own could not differentiate between differences of intensity, wavelength, and polarization.) Very similar arrangements of polarized light receptors have been found in crustacean eyes.

The evidence that polarized light is essential for the ant's navigation has been provided by the following experiments. First, if the eye is covered by a contact lens leaving only the dor-

[76]Muller, K. J. 1973. *J. Physiol.* 232: 573–595.

[77]Zollikofer, C., Wehner, R., and Fukushi, T. 1995. *J. Exp. Biol.* 198: 1637–1646.

(A)

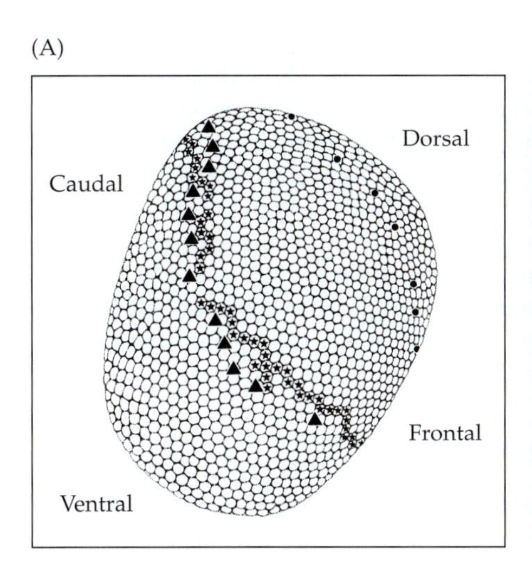

(B)

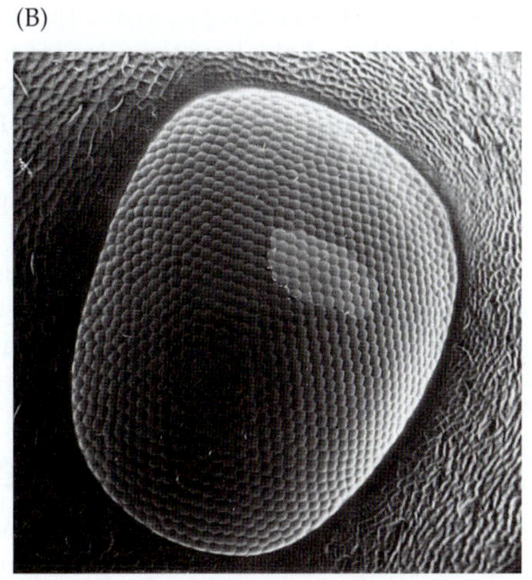

FIGURE 15.15 Polarized Light Detectors in the Eye of the Desert Ant. (A) The arrangement of photoreceptors. The receptors for polarized light lie in the region dorsal to the line marked by triangles. This is the region of the eye that the ant uses to navigate. (B) Scanning electron micrograph of the eye of the ant. A contact lens was placed over part of this eye, blurring the outlines of the photoreceptors. (After Wehner, 1994a.)

sal anterior quadrant exposed, the ant can still take a direct way home. Second, if the dorsal rim of the eye is blocked, pathfinding becomes disturbed. Third, if the pattern of polarization reaching the ant's eye is shifted by appropriate filters (placed on the wagon) the return course becomes deviated by a precise and calculable extent.

For navigation to be successful, information arriving at the eye must be correlated with a celestial map in which the position of the sun determines the orientations of polarized light. As the ant walks away from the nest, constant reference is made to the complex but regular pattern in the sky by the ommatidia. This provides the nervous system with information about distance traversed and angles. The compound eye is hemispherical, and this allows for a faithful spatial representation of orientations like those shown in Figure 15.14. Moreover, when the orientation preferences of the ommatidia are analyzed, one finds a coherent arrangement that is suited to a representation in neural terms of the polarized light patterns of the sky. As a result, the amount of match or mismatch of the pattern of polarized light can be used to determine compass direction in a predictable manner (predictable for ants and for ant researchers).

Strategies for Finding the Nest

A problem arises because the sun is not stationary. Hence, the ant has to compensate for the shift that occurs during the day. That it can do so has been shown in the experi-

(A)

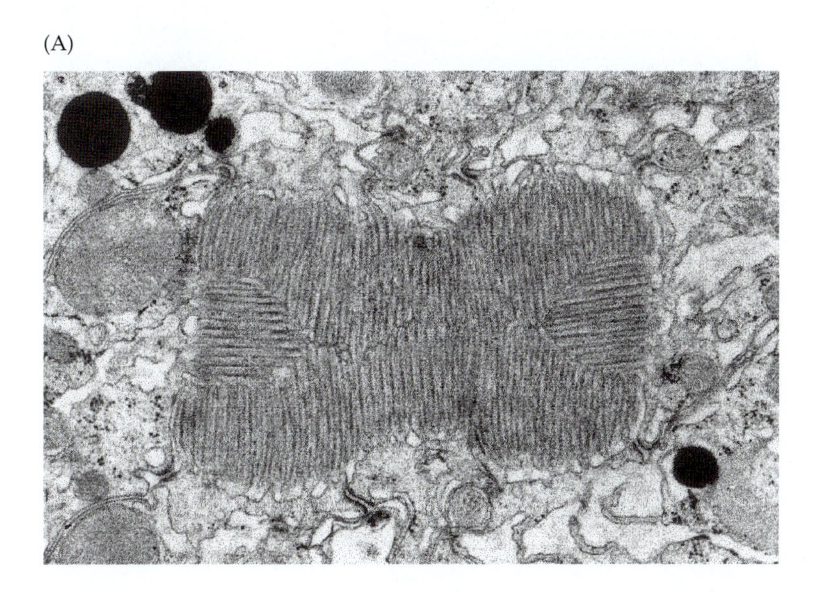

(B)

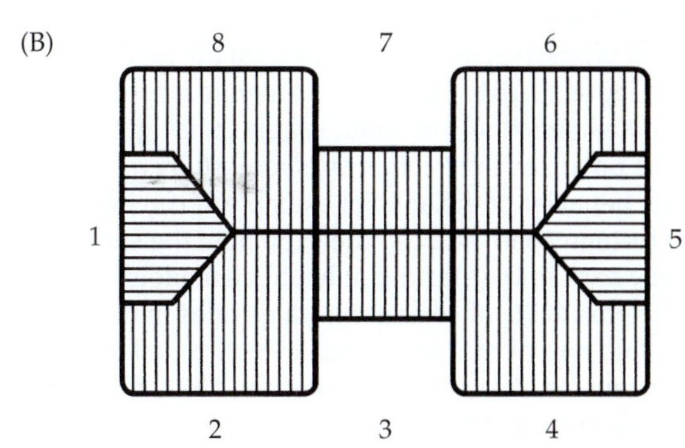

FIGURE 15.16 Arrangement of Photoreceptors Responsive to Polarized Light. (A) Electron micrograph of the dorsal region of the eye of an ant. The microvilli in the ommatidia are orthogonally arranged in a remarkably precise manner. (B) Diagram of the arrangement of microvilli. The numbers 1 and 5 respond preferentially to polarized light at a direction at right angles to those preferred by the other receptors. Elsewhere in the eye the microvilli are not necessarily at right angles in this way. (Electron micrograph kindly provided by Rudiiger Wehner; B after Wehner, 1996.)

ments in which an ant is placed at a spot removed from the nest at one hour, kept there, and then released at a later time. An ant that has become familiar with the sun's rate of movement for at least one day is able to correct its trajectory in an appropriate manner, as though it had memorized or known of the patterns of polarized light at different times of day. In addition to the polarized light compass, the sun itself and external objects situated along the path can be used to aid navigation.[78] Distinctive features of the terrain and objects are of principal importance in the last part of the return to the nest, which constitutes a tiny hole in the desert. If the homing vector has led to an error, such that the ant has not arrived precisely at its nest, a new strategy is introduced.[79] A series of exploratory loops is made, increasingly larger but always returning to the starting point. This represents an optimal strategy for reconnoitering without getting lost.

Neural Mechanisms for Navigation

One satisfactory aspect of these studies is the detailed information now available about the initial sensory mechanisms. Since each ommatidium of the dorsal rim contains two sets of photoreceptors sensitive to planes of polarized light at right angles to one another, and since each ommatidium views the sky at a different angle, the array of ommatidia provides information to the brain about the spatial distribution of polarized light vectors. Moreover, one can compute how the system behaves when challenged. In behavioral studies, objects are placed in the path so that detours must be made and corrected for. Another challenge for the ants is to displace them at different times of day or with different distortions of the polarized light input. At the same time, detailed recordings like those possible in the leech cannot yet be made about integrative steps performed by neurons within the ant brain. Hence, the link between sensory input and motor performance in the ant remains unknown. Nevertheless, by a computational approach in which the known properties of neurons are used, one can produce models or even robots that mimic accurately the navigational behavior of the desert ant using cues provided by polarized light (Figure 15.17).

Since the ant neurons are so small, recordings have been made by Labhart and his colleagues from interneurons in crickets that receive inputs from receptors for polarized light.[80] As in ants and crustaceans, the microvilli of two receptors are arranged orthogonally: They project to the interneuron, which provides information about the vector.

[78]Collett, M., et al. 1998. *Nature* 394: 269–272.

[79]Muller, M., and Wehner, R. 1994. *J. Comp. Physiol. [A]* 175: 525–530.

[80]Labhart, T. 1988. *Nature* 331: 435–437.

FIGURE 15.17 Mobile Robot Known as Sahabot devised by Wehner and his associates. This robot is equipped with six polarized light sensors arranged in pairs. Each pair forms a polarization opponent unit with properties resembling those of the neuron shown in Figure 15.18A. Each can be tuned to a specific evoked direction. The robot is able to navigate with this polarized light compass, successfully recreating the behavior of the ant. For example, it can be driven in a tortuous path and can then compute the shortest way back to the starting point. Such models illustrate the possibility of computing trajectories in this way but do not, of course, provide evidence that this is the system the ant uses. (After Wehner, 1997; photograph kindly provided by R. Wehner.)

Electrical recordings from such cells are shown in Figure 15.18. Their responses are just what would be predicted from behavioral studies with polarized light.

Polarized Light and Twisted Photoreceptors in Bees

The presence of receptors for polarized light in the eye of a bee could in principle be a mixed blessing. On the one hand the precise arrays of microvilli allow the heavens to be scanned for the orientation of polarized light. But for discrimination of colors, polarization can cause difficulties. As they fly, bees need to identify flowers reliably by their colors. Leaves and petals, however, vary in the way they reflect light in a manner that depends on how waxy the surfaces are. Leaves that are shiny reflect polarized light more than those that are matte. Consequently, the angle at which a leaf or a petal is illuminated and viewed will affect the amount and the direction of polarized light that is reflected.

The photopigments necessary for color vision in the bee are contained in microvilli (like those of the ant) of specific receptors sensitive to green, blue, or ultraviolet light. In the bee eye, as in the ant, the photopigments are arranged in a precise, parallel series of rhabdomeres. With variable uncalibrated contributions by polarized light, the signals regarding color could become ambiguous because the appreciation of color depends not only on wavelength but on the relative absorption by different classes of color receptors. As Wehner and Bernard[81] have written, "This means that for the bee the hue of a given part of a plant would change, whenever an approaching bee changed its direction of flight and thus, its direction of view—a completely unwanted phenomenon. For example, when zigzagging over a meadow with all its differently inclined surfaces of leaves, the bee would experience pointillistic fireworks of false colors that would make it difficult to impossible to detect the real colors of the flowers." To avoid this problem (which does not of course affect us, since we are not sensitive to the orientation of polarized light), the retina of the bee contains "twisted" receptors. By light and electron microscopy it was found that the photoreceptors were twisted along their long axis, producing a progressive change in the orientation of their microvilli (Figure 15.19). The twist, which is about 1°/μm length, means that the microvilli are no longer in a parallel

[81]Wehner, R., and Bernard, G. D. 1993. *Proc. Natl. Acad. Sci. USA* 90: 4132–4135.

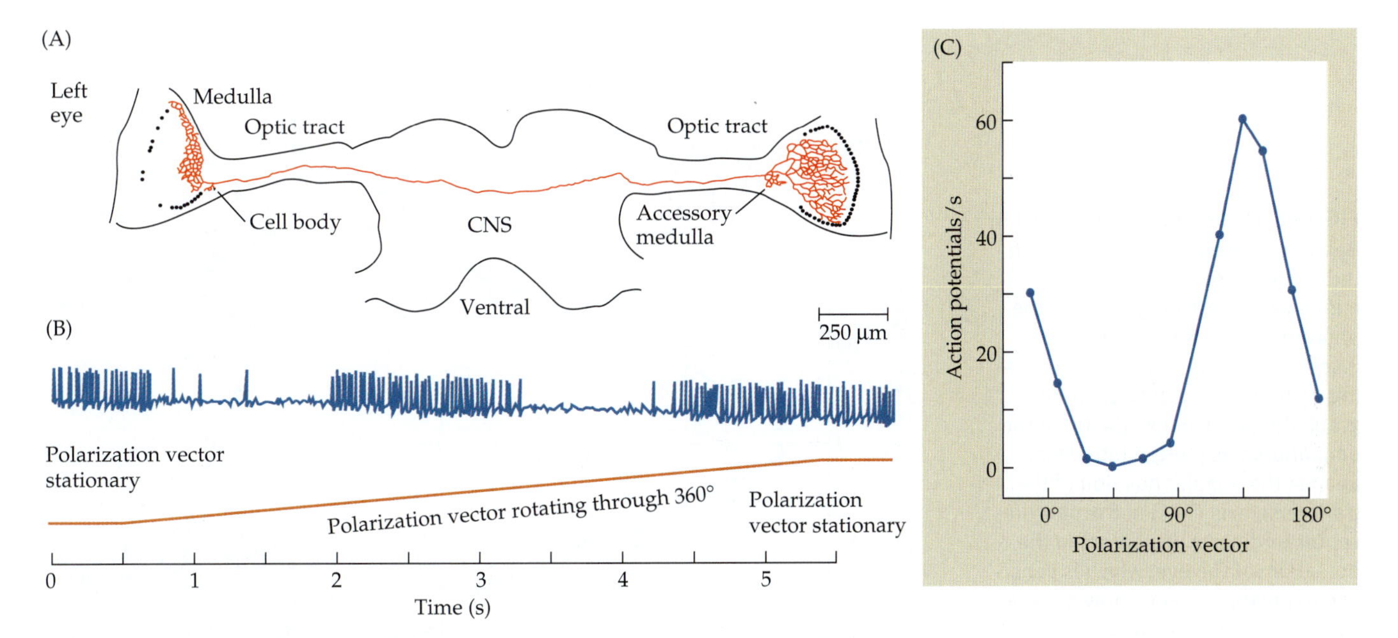

FIGURE 15.18 Electrical Responses of Polarization-Sensitive Interneurons in cricket CNS. These interneurons are large enough to be impaled by microelectrodes. (A) Reconstruction of the interneuron stained by intracellular application of neurobiotin. This neuron receives its input from the left eye. (B) Responses of a neuron of this type to polarized light when the vector is rotated through 360°. (C) Graphical representation of responses plotted against polarization vector. (After Labhart, 1988.)

array throughout the depth of the retina. Thus, the receptor no longer responds selectively to polarized light. For many years the photoreceptor twist had seemed like an anatomical quirk of no significance.

As in the ant, the dorsal rim of the bee's eye still contains untwisted receptors. These do respond to the plane of polarized light and can therefore be used for navigation and orientation. As Land has said, "One has to be impressed by the simplicity and versatility of nature's optical engineering."[82]

Use of Magnetic Fields by Bees for Navigation

In addition to visual cues and polarized light,[83] bees can use a magnetic compass to orient themselves while searching for a target. This was shown in experiments by Collett and his colleagues in which they trained bees to collect sugar from a small bottle cap on a board.[84] For orientation, a black cylinder was placed in a constant compass direction at a fixed distance from the bottle cap. The cylinder and the sucrose were moved to different places on the board between trials. Periodically the bottle cap was removed, leaving only the cylinder. The exploration by a trained bee was monitored by video recording as it hunted for the sucrose and then returned home. Figure 15.20 shows the trajectory of a bee as it approached and later left the cylinder (black circle) and the food source (cross). What is clear is that the bee turned so as to face south before landing, and again faced south shortly after taking off. In this way it viewed the visual cue and the attractant (sugar) each time from a constant direction. Observation of the sky alone was not a sufficient explanation for this behavior. Bees faced south in the rain, under a completely overcast sky, or when the sky compass was eliminated. From such observations one can conclude that somehow the animal can distinguish south from north, east, or west.

That the bees are sensitive to magnetic fields was shown by training bees under tarpaulin with imposed magnetic fields that shifted the magnetic north. The bees oriented themselves again to the "south," but this direction was the south of the imposed magnetic field. How changes in imposed magnetic fields produce changes in behavior and pattern

[82]Land, M. F. 1993. *Nature* 363: 581–582.

[83]Lehrer, M., and Collett, T. S. 1994. *J. Comp. Physiol. [A]* 175: 171–177.

[84]Collett, T. S., and Baron, J. 1994. *Nature* 368: 137–140.

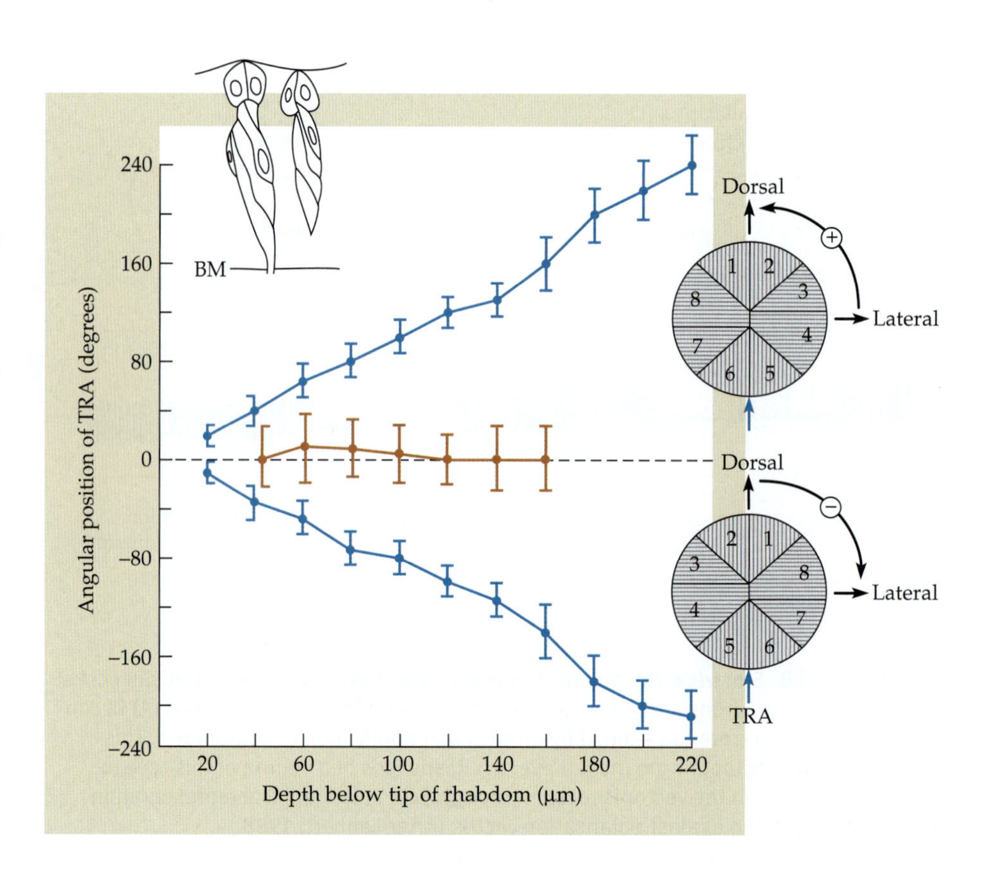

FIGURE 15.19 Anatomical Twist in Bee Photoreceptors through which the influence of polarized light is minimized. The arrangement of the photoreceptors is shown (in the inset) with twists evident at increasing depth from the surface. In the graph the depth below the tip of the ommatidium is plotted (abscissa) against the angular position of the transverse axis (TRA) of the photoreceptor (ordinate). Receptors in the dorsal rim of the eye, which is sensitive to polarized light, show no twist (red circles). As a result of the twist, a photoreceptor outside the dorsal rim does not respond selectively to polarized light of a particular orientation. BM = basement membrane of compound eye. (After Wehner and Bernard, 1993.)

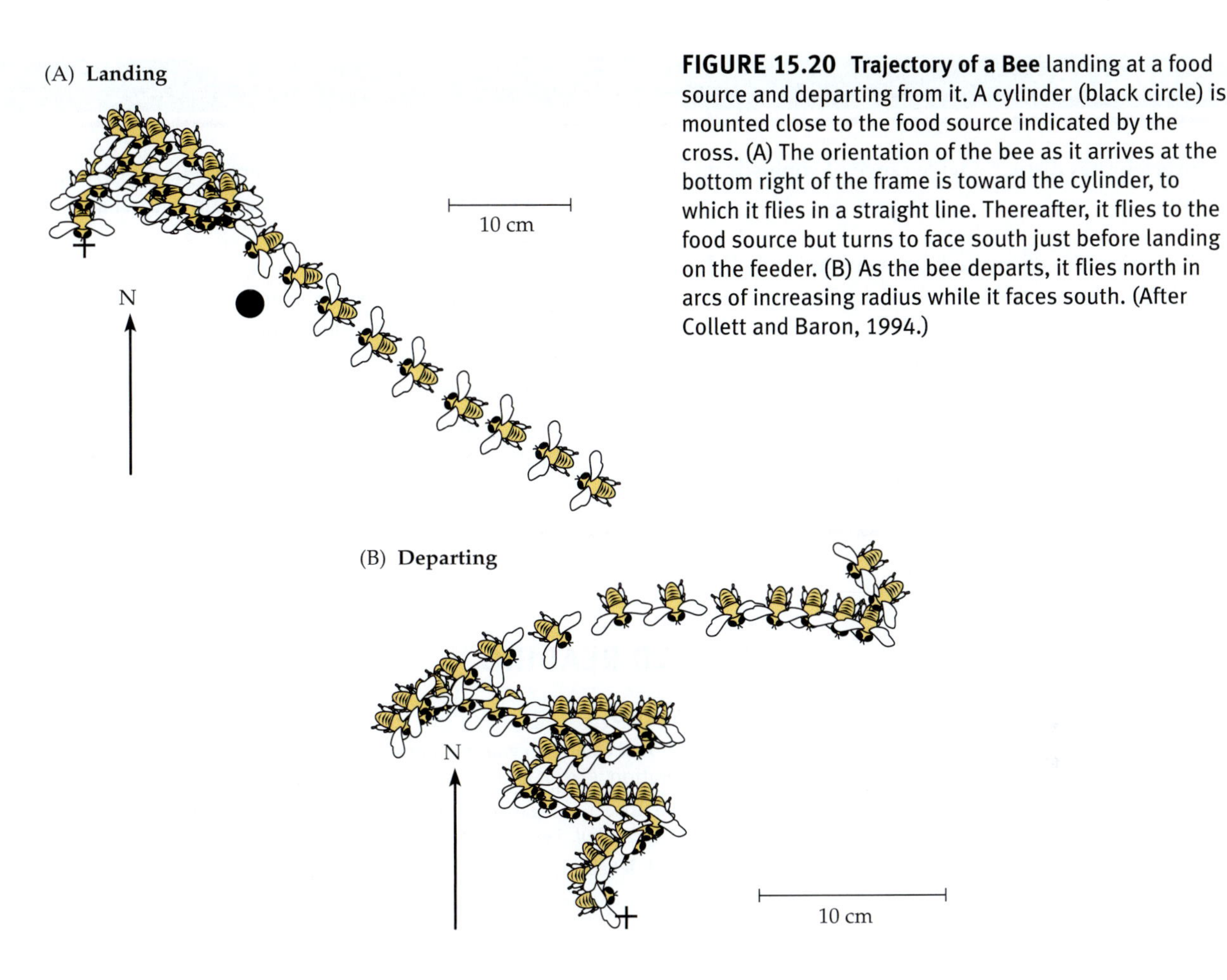

FIGURE 15.20 Trajectory of a Bee landing at a food source and departing from it. A cylinder (black circle) is mounted close to the food source indicated by the cross. (A) The orientation of the bee as it arrives at the bottom right of the frame is toward the cylinder, to which it flies in a straight line. Thereafter, it flies to the food source but turns to face south just before landing on the feeder. (B) As the bee departs, it flies north in arcs of increasing radius while it faces south. (After Collett and Baron, 1994.)

recognition is not known. These are sensory mechanisms not represented in our nervous system but present in animals such as birds, turtles, and certain invertebrates.

WHY SHOULD ONE WORK ON INVERTEBRATE NERVOUS SYSTEMS?

Throughout this book, and from the examples described in this chapter, it is evident that invertebrate nervous systems have been essential for approaching problem after problem relating to the biophysics, cell biology, and development of nerve cells. Particularly striking is the conservation of fundamental mechanisms in species after species through evolution. Often experiments on an invertebrate have produced the very insight needed for starting the investigation of similar problems in a mammal. For example, a great impetus for the development of the slice technique for mammalian brain came from the work on invertebrate ganglia in which identified neurons could be seen directly in the microscope as they were impaled. Hartline's work on the horseshoe crab eye provided the key stimulus for Kuffler's experiments on the cat retina.[85,86] At the same time it would be futile to hope to understand how the visual cortex of a monkey carries out its functions from studies made on an invertebrate.

What then is the "use" of studying, say, navigation by ants or bees? First, one can guess that even though we do not see polarized light or sense magnetic fields, the underlying principles for analyzing sensory information and translating it to an effective output will be used again in higher nervous systems, in one way or another. Second, the work on invertebrates illustrates a fundamental attitude toward biology: Neurobiology is not restricted to the study of "the" brain (i.e., ours). Rather, there is an inherent fascination in trying to understand how the tiny, finite brain of a leech, an ant, or a bee can do wonderful and sophisticated computations that are essential for its survival.

[85]Hartline, H. K. 1940. *Am. J. Physiol.* 130: 690–699.

[86]Kuffler, S. W. 1953. *J. Neurophysiol.* 16: 37–68.

Summary

- Invertebrates display a wide range of sophisticated behaviors.
- The properties of invertebrate neurons and glial cells resemble those of vertebrates.
- Invertebrate nervous systems consist of hundreds or thousands of neurons.
- Each type of invertebrate can offer advantages for investigation of a specific type of problem.
- Properties of individual nerve cells and synapses can be used to explain the behavior of the animal and its modifications.
- Quantitative measurements of behavior shed light on fundamental principles in neurobiology.
- Not all work on invertebrate central nervous systems is necessarily directed toward understanding mechanisms in human brain. It is fascinating in its own right.

SUGGESTED READING

General Reviews

Calabrese, R. L. 1998. Cellular, synaptic, network, and modulatory mechanisms involved in rhythm generation. *Curr. Opin. Neurobiol.* 8: 710–717.

Collett, T. S. 1996. Insect navigation en-route to the goal—Multiple strategies for the use of landmarks. *J. Exp. Biol.* 199: 227–235.

Neurobiology and development of the leech (Special Issue). 1995. *J. Neurobiol.* 27(Pt.3).

Wehner, R. 1994. The polarization-vision project: Championing organismic biology. *Fortschr. Zool.* 31: 11–53.

Original Papers

Antonov, I., Kandel, E. R., and Hawkins, R. D. 1999. The contribution of facilitation of monosynaptic PSPs to dishabituation and sensitization of the *Aplysia* siphon withdrawal reflex. *J. Neurosci.* 19: 10438–10450.

Baccus, S. A. 1998. Synaptic facilitation by reflected action potentials: Enhancement of transmission when nerve impulses reverse direction at axon branch points. *Proc. Natl. Acad. Sci. USA* 95: 8345–8350.

Bruns, D., and Jahn, R. 1995. Real-time measurement of transmitter release from single synaptic vesicles. *Nature* 377: 62–65.

Collett, T. S., and Baron, J. 1994. Biological compasses and the coordinate frame of landmark memories in honeybees. *Nature* 368: 137–140.

Gelperin, A. 1999. Oscillatory dynamics and information processing in olfactory systems. *J. Exp. Biol.* 18: 1855–1864.

Gu, X. 1991. Effect of conduction block at axon bifurcations on synaptic transmission to different postsynaptic neurones in the leech. *J. Physiol.* 441: 755–778.

Lewis, J. E., and Kristan, W. B., Jr. 1998. Quantitative analysis of a directed behavior in the medicinal leech: Implications for organizing motor output. *J. Neurosci.* 18: 1571–1582.

Modney, B. K., Sahley, C. L., and Muller, K. J. 1997. Regeneration of a central synapse restores nonassociative learning. *J. Neurosci.* 17: 6478–6482.

Muller, M., and Wehner, R. 1994. The hidden spiral: Systematic search and path integration in desert ants, *Cataglyphis fortis*. *J. Comp. Physiol. [A]* 175: 525–530.

Wang, H., and Macagno, E. R. 1997. The establishment of peripheral sensory arbors in the leech: In vivo time-lapse studies reveal a highly dynamic process. *J. Neurosci.* 17: 2408–2419.

Wehner, R., and Bernard, G. D. 1993. Photoreceptor twist: A solution to the false-color problem. *Proc. Natl. Acad. Sci. USA* 90: 4132–4135.

16 AUTONOMIC NERVOUS SYSTEM

THE AUTONOMIC NERVOUS SYSTEM controls essential functions of the vertebrate body. Thus, neurons of the autonomic nervous system supply smooth muscles in the eye, lung, gut, blood vessels, bladder, genitalia, and uterus. They regulate glandular secretion, blood pressure, heart rate, cardiac output, and body temperature, as well as food and water intake. In contrast to speedy conduction and muscle contractions required for limb movements, these "housekeeping" or "vegetative" functions are slower, last longer, and are often less focused.

Four distinct groupings of neurons make up the autonomic nervous system. The **sympathetic** division consists of neurons, the axons of which leave the spinal cord through ventral roots from thoracic and lumbar segments. They form synapses on nerve cells in sympathetic ganglia situated alongside and at a distance from the spinal cord, and on chromaffin cells in the adrenal medulla. Sympathetic postganglionic axons are unmyelinated and extend over long distances to target areas. The **parasympathetic** division consists of axons leaving through certain cranial and sacral nerves. They form synapses in ganglia situated within the target organs. Parasympathetic postganglionic axons are in general shorter than those of the sympathetic nervous system. A third, highly complex division consists of millions of nerve cells in the intestinal wall, the **enteric** nervous system. The fourth division comprises neurons in the spinal cord, hypothalamus, and brainstem. Within the CNS, boundaries between the autonomic and somatic nervous systems are not sharply defined.

Synaptic transmission in the autonomic nervous system is extraordinary in its diversity, including all the known transmitters. Principles of transmission and integration that were revealed at autonomic synapses include the chemical nature of synaptic transmission, reuptake of transmitter, autoreceptors on presynaptic terminals, co-release of more than one transmitter at a single terminal, and the role of second messengers. Transmitters used at autonomic ganglia include acetylcholine acting on both nicotinic and muscarinic receptors, peptides, and dopamine. Postganglionic parasympathetic nerve terminals release acetylcholine as the primary transmitter that acts on muscarinic receptors in the target organs. Postganglionic sympathetic neurons release norepinephrine, epinephrine, acetylcholine, purines, or peptides as primary transmitters. Sympathetic and parasympathetic neurons co-release ATP and peptides. Whereas much is known about the regulation of activity in smooth muscle and gland cells, less information is available about integrative mechanisms within the CNS that regulate autonomic functions.

The periodic 24-hour cycle of activity, known as circadian rhythm, influences many autonomic functions. Experiments in which recordings were made from specific neurons in the hypothalamus have revealed one of the cellular mechanisms that generate the rhythm. Slow increases in intracellular chloride concentration in daytime cause inhibition by γ-aminobutyric acid (GABA) to be converted to excitation. Thereby firing is increased during the day and decreased at night.

The name "autonomic" implies an independent system that runs on its own. In part this is true. The autonomic nervous system controls blood vessels, the heart, glands and smooth muscle throughout the gut, bronchi, bladder, and spleen, without our having to think about it. By a simple act of will one cannot increase the diameter of the pupil, or the blood flow through one's little finger. It is possible of course to cheat to some extent by the use of tricks: Thus, the generation of emotion by deliberately thinking of an exam, a dental appointment, or a film star can increase the heart rate.

In practice, the performance of the autonomic nervous system is closely linked to voluntary movements. Exercise results in appropriate diversion of blood to the muscles and in stimulation of sweat glands; the action of standing up from a recumbent position requires circulatory adjustments so as to maintain blood flow to the brain. Ingestion of a meal reroutes blood to the stomach and intestines. Through the turning on or off of activity in a widespread group of target cells, the autonomic nervous system deals with the housekeeping maintenance work of the body. The priorities are made by the brain, which sets in motion digestion, reproduction, micturition, defecation, or focusing in dim light, through mechanisms that are not decided by our conscious will. Of key concern for mankind are disorders of the autonomic nervous system that lead to conditions such as asthma, constipation, diarrhea, ulcers, hypertension, heart disease, stroke, and retention of urine (or lack thereof).

Recent experiments and classical work on the autonomic nervous system represent such an extensive and varied field that a comprehensive review is impossible in this chapter. Indeed, entire textbooks[1–3] and specialized journals[4,5] are devoted to important functions of the autonomic nervous system. A large amount of information is available about mechanisms that control the enteric nervous system and bladder, the diameter of the pupil, secretion by glands and by chromaffin cells of the adrenal medulla, and regulation by the CNS of respiration, temperature, body weight, appetite, and reproduction.[6–11] In this chapter, as in others, the main emphasis is on a few selected examples that illustrate cellular, molecular, and integrative mechanisms.

It will be shown that although much is now known about the autonomic nervous system, many open questions remain, particularly about integrative mechanisms within the CNS. There the distinction between autonomic and somatic systems has no hard-and-fast boundaries. It is convenient to begin with a brief description of the principal features of the peripheral autonomic nervous system.

FUNCTIONS UNDER INVOLUNTARY CONTROL

Sympathetic and Parasympathetic Nervous Systems

The principal anatomical features are shown in Figure 16.1. Virtually all the organs of the body are supplied by autonomic neurons. Even skeletal muscle fibers, which receive no direct innervation, are dependent on the autonomic nervous system; their blood supply is regulated according to need. Sympathetic preganglionic neurons are situated in the intermediolateral horn of the spinal cord of segments T1 to L3. Their myelinated axons pass through ventral roots to form synapses in ganglia situated alongside the vertebral column and more peripherally (Figure 16.1A). From these ganglia run unmyelinated axons to the tissues. By contrast, the parasympathetic outflow is restricted to cranial nerves III, VII, IX, and X and sacral roots S2, S3, and S4 (Figure 16.1B). The parasympathetic ganglia are located close to or in the tissues themselves. Hence, the parasympathetic myelinated preganglionic axon is long, whereas the unmyelinated postganglionic axon is short.

The actions of the two systems are often but not always antagonistic (Table 16.1). For example, excitation of sympathetic neurons leads to dilatation of the pupil, increased heart rate, and decreased gut motility. Parasympathetic excitation produces opposite effects: pupillary constriction, slowed heart rate, and increased gut motility. On the other hand, glandular secretion can be increased by activation of either system. Both systems can cause smooth muscles to contract or to relax, depending on the transmitter that is released and the types of receptors that are present on the muscle.

[1]Gabella, G. 1976. *The Structure of the Autonomic Nervous System*. Chapman and Hall, London.

[2]Appenzeller, O. 1990. *The Autonomic Nervous System: An Introduction to Basic and Clinical Concepts*. Elsevier, New York.

[3]Broadley, K. J. 1996. *Autonomic Pharmacology*. Taylor & Francis, London.

[4]*J. Auton. Nerv. Syst.*

[5]*J. Auton. Pharmacol.*

[6]deGroat, W. C., et al. 1996. *Prog. Brain Res.* 107: 97–111.

[7]Janig, W., and McLachlan, E. M. 1987. *Physiol. Rev.* 67: 1332–1404.

[8]Gamlin, P. D., et al. 1998. *Vision Res.* 38: 3353–3358.

[9]Taraskevich, P. S., and Douglas, W. W. 1984. *Fed. Proc.* 43: 2373–2378.

[10]Refinetti, R., and Menaker, M. 1992. *Physiol. Behav.* 51: 613–637.

[11]Inui, A. 1999. *Trends Neurosci.* 22: 62–67.

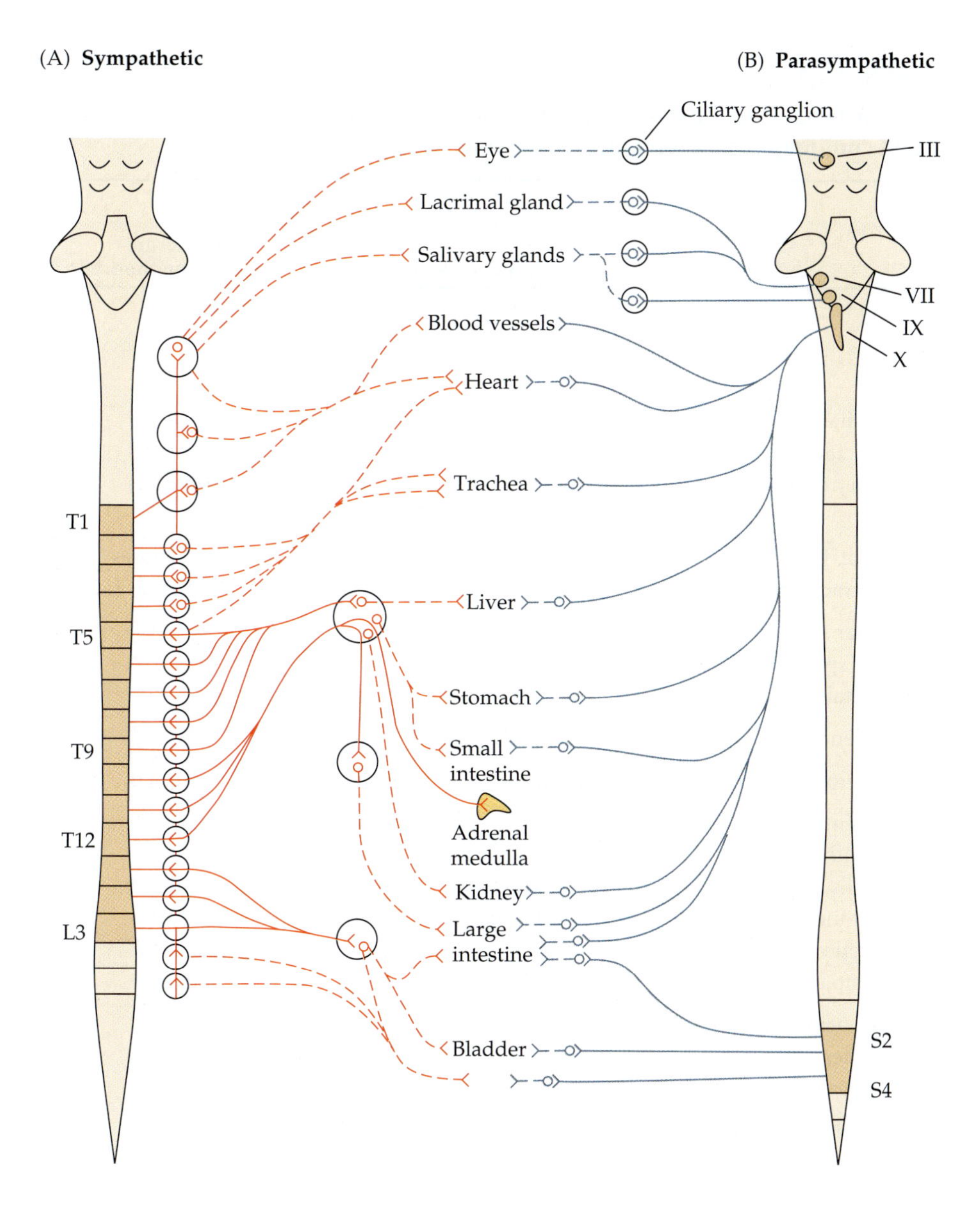

FIGURE 16.1 The Autonomic Nervous System and the target organs that it innervates. (A) Principal features of the sympathetic nervous system, including the paravertebral ganglia, peripheral ganglia, and adrenal medulla. (B) The more restricted output from the parasympathetic nervous system and the targets that it innervates. Solid lines are myelinated preganglionic axons; dashed lines are unmyelinated postganglionic axons.

A major difference between the two autonomic systems is that the sympathetic nervous system tends to be thrown into action as a whole, with widespread generalized consequences for the body. It is typically activated under conditions of fright, fight, and flight, and during intense exercise. The symptoms are familiar: dilated pupils, dry mouth, pounding heart, sweating, and enhanced emotions. The systemic effects of sympathetic neuronal activity are enhanced further by chromaffin cells in the adrenal medulla.[12] These cells are modified ganglionic neurons. They receive cholinergic input from preganglionic axons and secrete epinephrine, norepinephrine, peptides, and ATP as hormones into the bloodstream. Epinephrine in the blood reinforces and extends sympathetic activity: It can reach and bind to receptors in smooth muscle of bronchi, far from sympathetic nerve endings; epinephrine also binds to receptors in blood vessels that are insensitive to norepinephrine (see Box 16.1). Unlike norepinephrine, epinephrine can produce vasodilatation as well as contraction of blood vessels.

By contrast, the parasympathetic nervous system is more focused in its activity. It is surely a considerable advantage that the pupil can constrict in bright light and the lens of the eye accommodate for viewing nearby objects selectively, without concomitant and ill-timed arousal of bladder contractions or even more embarrassing parasympathetic effects.

[12]Douglas, W. W. 1966. *Pharmacol. Rev.* 18: 471–480.

Table 16.1
Characteristic actions of adrenergic sympathetic and cholinergic parasympathetic nervous systems

	Effect of		
	Adrenergic sympathetic		Cholinergic parasympathetic
Organ	**Action**[a]	**Receptor**[b]	**Action**
Eye			
Iris			
Radial muscle	Contracts	α_1	—
Circular muscle	—	—	Contracts
Ciliary muscle	(Relaxes)	β	Contracts
Heart			
Sinoatrial node	Accelerates	β_1	Decelerates
Contractility	Increases	β_1	Decreases (atria)
Vascular smooth muscle			
Skin, splanchnic vessels	Contracts	α	—
Skeletal muscle vessels	Relaxes	β_2	—
Nerve endings	Inhibits release	α_2	—
Bronchiolar smooth muscle	Relaxes	β_2	Contracts
Gastrointestinal tract			
Smooth muscle			
Walls	Relaxes	α_1, β_2	Contracts
Sphincters	Contracts	α_1	Relaxes
Secretion	—	—	Increases
Myenteric plexus	Inhibits	α	Activates
Genitourinary smooth muscle			
Bladder wall	Relaxes	β_2	Contracts
Sphincter	Contracts	α_1	Relaxes
Metabolic functions			
Liver	Gluconeogenesis	α/β_2	—
	Glycogenolysis	α/β_2	—

[a]Accounts of the actions of the autonomic nervous system on target organs listed in this table that are not dealt with in this chapter are given in reviews and textbooks of physiology and pharmacology (see references in text).

[b]Not all the adrenergic receptors or effector cells are included; purinergic, peptidergic, and cholinergic mechanisms are dealt with in the text. Whereas epinephrine acts on all the adrenergic receptors, norepinephrine is effective on α_1, α_2, and β_1 receptors, but not β_2. The various types of adrenergic and muscarinic receptors are characterized by the specific agonists and antagonists that bind to them, and by their molecular structures.

Synaptic Transmission in Autonomic Ganglia

Certain mechanisms of transmission in the autonomic nervous system have already been described in earlier chapters (10, 12, and 13). These include the co-release of multiple transmitters from nerve endings, modulatory actions of autonomic transmitters, the properties of receptors that use second messengers, and the effects of acetylcholine and epinephrine on cardiac muscle. The way in which such mechanisms interact to influence signaling is well illustrated by experiments made on synaptic transmission in autonomic ganglia.

Autonomic ganglia constitute relay stations, the functional significance of which is not immediately obvious.[13] At first glance the mechanism of direct transmission is strikingly

[13]McLachlan, E. M. (ed.). 1995. *Autonomic Ganglia*. Gordon and Breach, London.

similar to that at the skeletal neuromuscular junction. Each presynaptic impulse releases acetylcholine, which acts on nicotinic receptors in the postsynaptic cell to open channels and produce a fast depolarization (Chapter 9).[14,15] As at the nerve–muscle junction, a single presynaptic action potential is followed by one in the postsynaptic cell (Figure 16.2).

Transmission at the synapse between the preganglionic axon and the ganglion cell is, however, far more elaborate than would at first appear. Thus, a very different picture emerges with repetitive stimulation of the presynaptic axon at frequencies comparable to those occurring normally in the animal. Under these conditions, the ganglion is not simply a throughway, but is a site of complex interactions. With trains of impulses, and prolonged depolarizing and hyperpolarizing, long-latency synaptic potentials arise in the ganglion cell. They summate to produce a steady, subthreshold depolarization maintained for seconds, minutes, or even hours. During the depolarization a single presynaptic action potential can now give rise to multiple postsynaptic impulses. Both the fast and the slow synaptic potentials are evoked by the release of acetylcholine from the presynaptic nerve terminals. As before, the fast, direct synaptic potential results from activation of nicotinic acetylcholine receptors. The slow potential is due to activation of muscarinic ACh receptors that are coupled to G proteins (Chapter 10).

Kuffler and his colleagues found that a second transmitter also contributes to slow synaptic potentials.[16,17] Certain presynaptic axons release a decapeptide that resembles luteinizing hormone–releasing hormone (LHRH). (LHRH is also known as GnRH, gonadotropin-releasing hormone; see Figure 16.9.) Hence, in autonomic ganglia, neuronal firing and excitability are controlled by both ACh and LHRH, secreted by presynaptic neurons. Complex interactions between transmitters and receptors are also observed in chromaffin cells of the adrenal medulla. These modified ganglion cells are stimulated by preganglionic axons of the splanchnic nerve that release ACh and ATP.[18]

[14]Ullian, E. M., McIntosh, J. M., and Sargent, P. B. 1997. *J. Neurosci.* 17: 7210–7219.

[15]Sargent, P. B. 1993. *Annu. Rev. Neurosci.* 16: 403–443.

[16]Kuffler, S. W. 1980. *J. Exp. Biol.* 89: 257–286.

[17]Jan, Y. N., Jan, L. Y., and Kuffler, S. W. 1980. *Proc. Natl. Acad. Sci. USA* 77: 5008–5012.

[18]Holman, M. E., et al. 1994. *J. Physiol.* 478: 115–124.

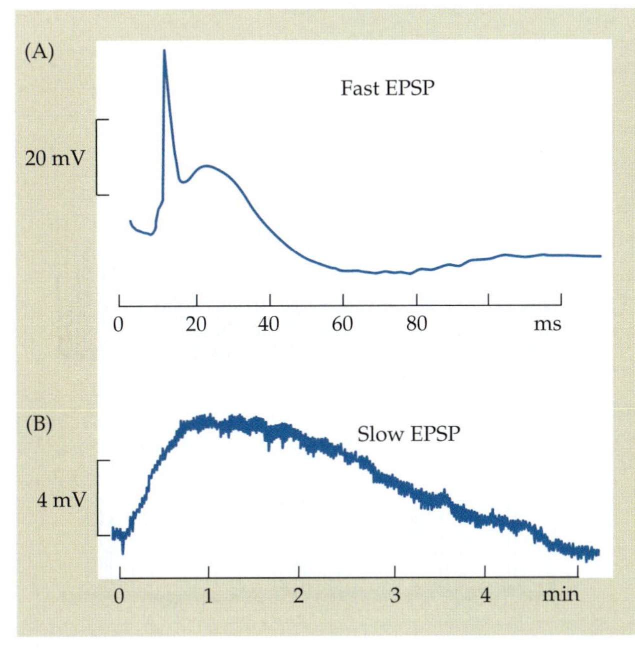

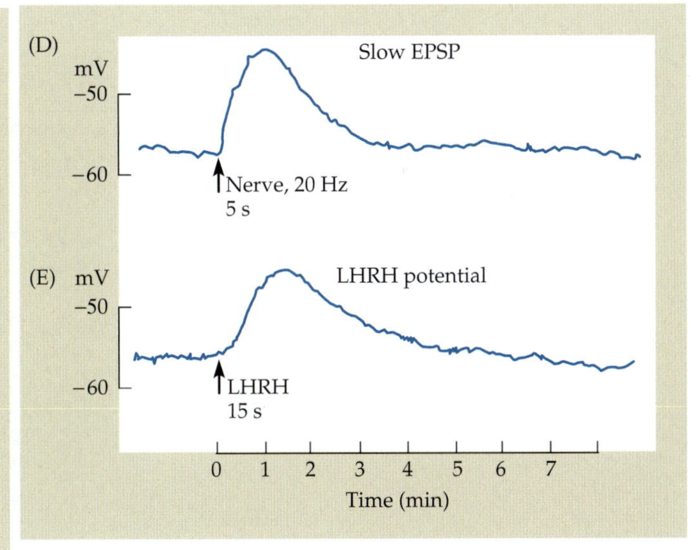

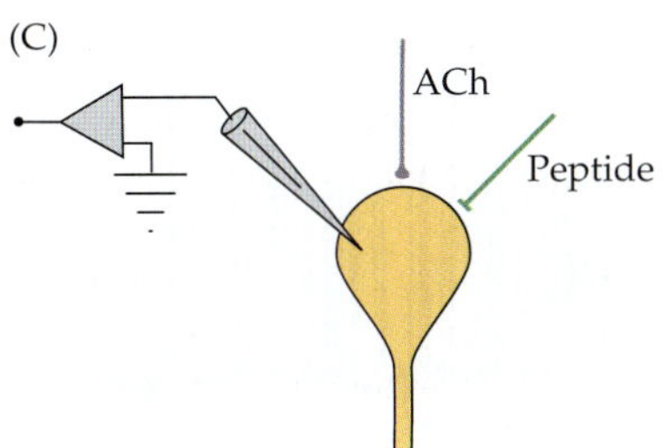

FIGURE 16.2 Fast and Slow Synaptic Potentials in Sympathetic Ganglion Cells in a bullfrog. (A) A single stimulus to the preganglionic inputs evokes a large, fast EPSP and an action potential. (B) Trains of stimuli (10/s for 5 s) are required to elicit slow synaptic potentials. Note that the timescale is in minutes instead of milliseconds. (C) Cells in ganglia receive cholinergic and peptidergic inputs, which can be stimulated selectively. (D) Slow excitatory potential evoked by stimulating peptidergic input (20/s for 5 s). As in A, the depolarization lasts several minutes. (E) Application of LHRH to the same neuron by pressure from a micropipette. The peptide mimics the action of the naturally released transmitter. (After Kuffler, 1980.)

Unlikely as it might seem, this description of ganglionic transmission has been oversimplified. Integration in sympathetic and parasympathetic ganglia is somehow modulated by interneurons known as SIF cells (small intensely fluorescent cells) containing catecholamines, and also by presynaptic endings that release vasoactive intestinal peptide (VIP) and enkephalins.[19,20]

M-Currents in Autonomic Ganglia

What is the mechanism responsible for the slow depolarizations produced by ACh and LHRH? This question was resolved by Brown, Adams, and their colleagues, who identified an unusual potassium current carried by "M-channels" (so called because they are influenced by muscarinic ACh receptors).[21–23] M-channels open frequently at rest, making a substantial contribution to the resting potassium conductance, and their probability of opening increases with depolarization. Activation of muscarinic receptors causes the channels to *close* (Figures 16.3 and 16.4). As a consequence, the resting influx of sodium ions is no longer in balance with potassium efflux and the cell depolarizes.

M-channels have a dramatic effect on the synaptic responses to ACh.[24] A single presynaptic action potential leads to activation of nicotinic receptors, producing an EPSP that is very brief and elicits at most a single postsynaptic action potential. The EPSP is short because synaptic depolarization results in M-channel activation, thus increasing the rate of repolarization (see Figure 16.2). When muscarinic receptors are activated as well, the resulting decrease in potassium conductance has two effects: First, the cell is depolarized. Second, EPSPs produced by activation of nicotinic receptors are greatly prolonged. As a result, each directly mediated EPSP remains above threshold for many milliseconds and evokes a train of impulses (see Figure 16.4). Peptide-evoked slow potentials are also mediated by closure of M-channels, and they have a similar effect on synaptic activity.

[19] Prud'homme, M. J., et al. 1999. *Brain Res.* 821: 141–149.

[20] Kiyama, H., et al. 1993. *Brain Res. Mol. Brain Res.* 19: 345–348.

[21] Adams, P. R., and Brown, D. A. 1980. *Br. J. Pharmacol.* 68: 353–355.

[22] Adams, P. R., Brown, D. A., and Constanti, A. 1982. *J. Physiol.* 332: 223–262.

[23] Brown, D. 1988. *Trends Neurosci.* 11: 294–299.

[24] Jones, S. W., and Adams, P. R. 1987. In *Neuromodulation: The Biochemical Control of Neuronal Excitability.* Oxford University Press, New York, pp. 159–186.

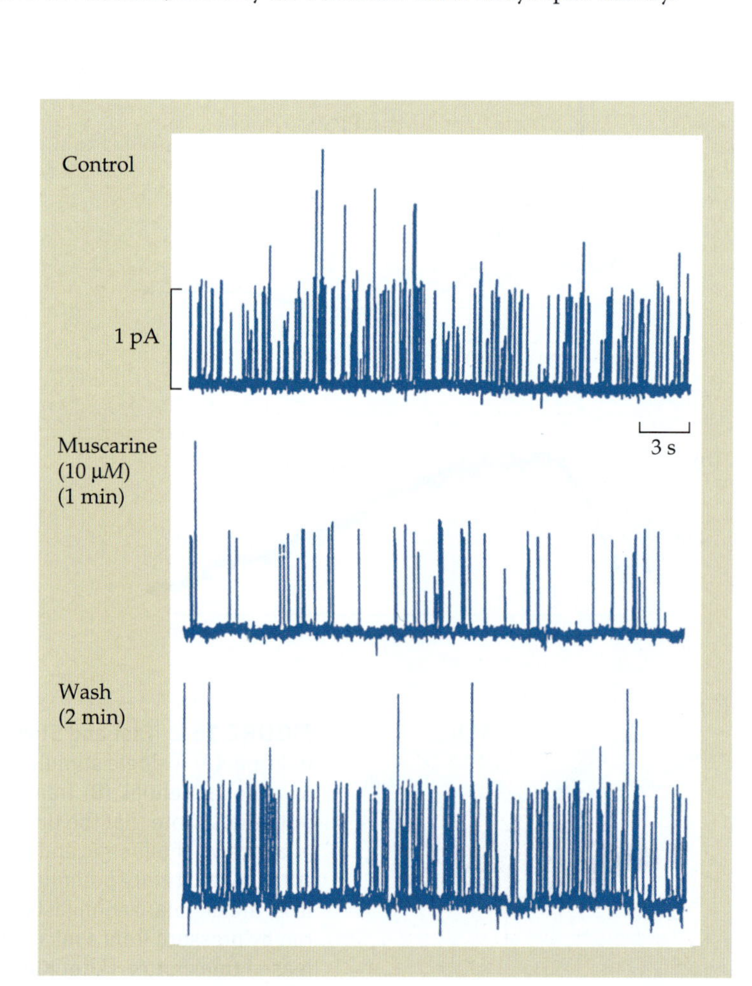

FIGURE 16.3 Properties of M-Channels, which are closed by application of muscarine. The records were made by cell-attached patch recording from a transformed neuroblastoma cell in culture that expressed M-receptors and M-channels. When muscarine was applied, the probability of channel opening was drastically decreased. (After Brown et al. 1993.)

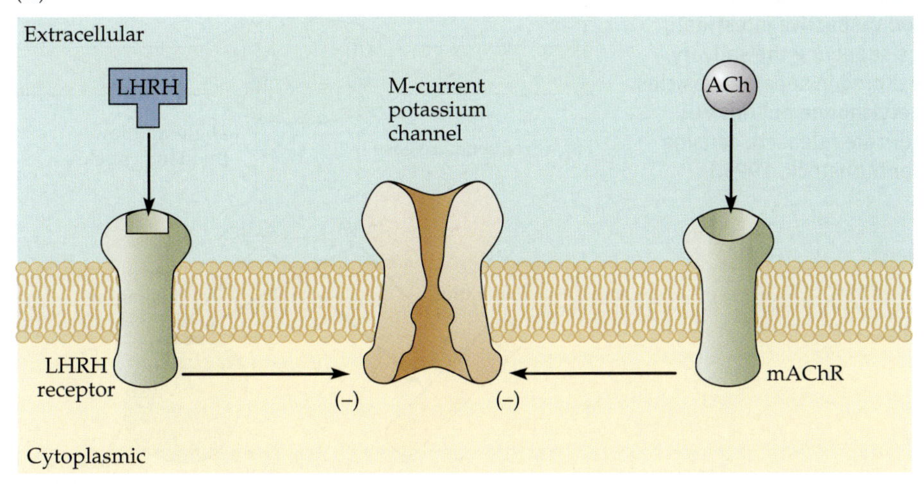

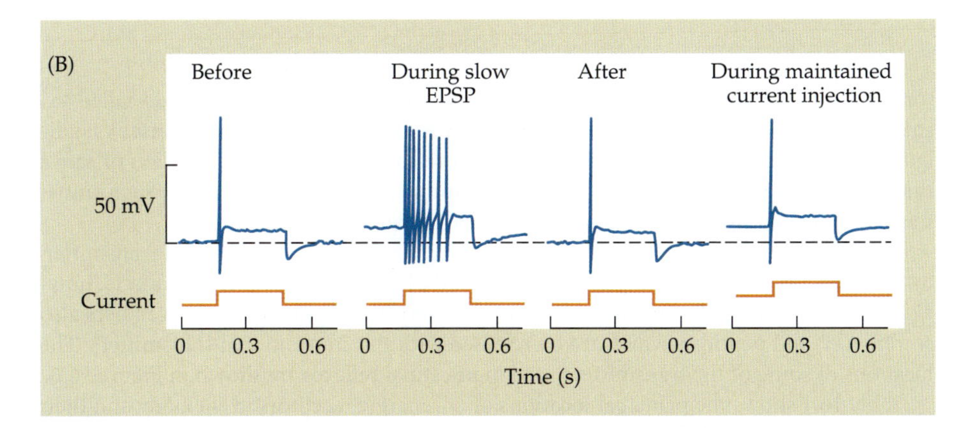

FIGURE 16.4 Inhibition of Potassium Currents in Sympathetic Ganglion Cells modulates responses to presynaptic stimulation. (A) Binding of ACh to muscarinic receptors (mAChR) and binding of LHRH to its receptor both inhibit M-current potassium channels. (B) The effect of the decrease in the M-current during the slow synaptic potential is to increase the excitability of the ganglion cell. Depolarizing current pulses applied through the microelectrode (lower traces) before and after a slow synaptic potential produce a single action potential. During the slow potential, the same current pulse elicits a burst of action potentials. Depolarizing the ganglion cell (to the same extent as occurs during closure of M-channels) by injecting a maintained current has no such effect on the responsiveness of the cell. (After Jones and Adams, 1987.)

After their discovery in autonomic ganglia, M-channels were found in neurons in the spinal cord, hippocampus, and cerebral cortex.[25] McKinnon, Brown, and their colleagues have cloned genes for subunits that contribute to the structure of M potassium channels.[26,27] Although it has been shown that increased intracellular calcium is involved in the second messenger system leading to M-channel closure,[28] the complete mechanism has not been elucidated.

M-channels have a major effect on firing patterns in the autonomic nervous system.[29] In cells with large M-currents, such as those that cause dilatation of the pupil, presynaptic inputs do not fire tonically and the output is roughly one to one. The responses of the ganglion cells are therefore phasic, firing solely when commands are given. By contrast, cells in lumbar ganglia that cause vasoconstriction receive a continuous bombardment from presynaptic inputs. This inhibits their M-currents through the muscarinic effect of acetylcholine. Accordingly, they fire tonically at varying frequencies, depending on the input, and produce greater or reduced tonic vasoconstriction. These results fit with the requirement of discontinuous, episodic dilatation of the pupil on demand and maintained control of blood vessel diameter. Tonic and phasic discharges have additional effects; they determine perhaps which types of transmitter are to be released by the terminals of a ganglion cell onto its targets.

[25]Marrion, N. V. 1997. *Annu. Rev. Physiol.* 59: 483–504.

[26]Wang, H. S., et al. 1998. *Science* 282: 1890–1893.

[27]Selyanko, A. A., et al. 1999. *J. Neurosci.* 19: 7742–7756.

[28]Selyanko, A. A., and Brown, D. A. 1996. *Neuron* 16: 151–162.

[29]Janig, W., and McLachlan, E. M. 1992. *Trends Neurosci.* 15: 475–481.

SYNAPTIC TRANSMISSION BY POSTGANGLIONIC AXONS

Although acetylcholine is the principal transmitter used by postganglionic parasympathetic axons, they co-release peptides (Figure 16.5). For example, ACh released by parasympathetic axons causes salivary glands to secrete by acting on muscarinic recep-

FIGURE 16.5 Cotransmission by Parasympathetic Postganglionic Neurons that release both acetylcholine and vasoactive intestinal peptide (VIP). Parasympathetic nerve fibers supplying the salivary gland secrete both transmitters, which are stored in separate vesicles. Stimulation at low frequencies releases acetylcholine but not VIP, while at higher frequencies both transmitters are released, causing vasodilatation and secretion of saliva. (After Burnstock, 1995.)

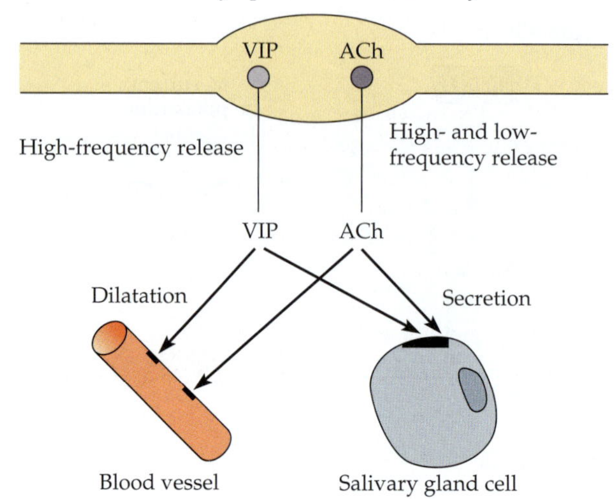

Geoff Burnstock

tors. With high-frequency stimulation, the same axons also liberate a peptide called vasoactive intestinal peptide (VIP). VIP, originally found in gut and brain, causes vasodilatation, increased intracellular calcium concentration, and increased secretion of saliva that is not blocked by atropine, an antagonist for muscarinic receptors.[30,31] Such transmission is also known as nonadrenergic, noncholinergic, or NANC.[32]

For sympathetic postganglionic neurons, norepinephrine is the principal transmitter. Sympathetic axons innervating sweat glands and blood vessels in skeletal muscle, however, do secrete acetylcholine instead of norepinephrine.[33] Sympathetic nerve fibers also secrete ATP and peptides, which are co-released with the conventional transmitters. The locations of some of the transmitters used in intestinal reflexes are shown in Figure 16.6.

Table 16.1 shows the principal locations of adrenergic receptors in the body and their mechanisms of action. With the advent of molecular biology it became clear that all the α_1, α_2, β_1, and β_2 adrenergic receptors, like muscarinic receptors activated by ACh, are similar in their amino acid sequences. All these receptors contain seven transmembrane segments, are coupled to G proteins, and use second messengers (Chapters 10 and 13; Box 16.1).

[30]Ekstrom, J., Asztely, A., and Tobin, G. 1998. *Eur. J. Morphol.* 36 Suppl.: 208–212.

[31]Zhang, Z., Evans, R. L., and Culp, D. J. 1998. *Eur. J. Morphol.* 36 Suppl.: 219–221.

[32]Burnstock, G. 1993. *Drug Dev. Res.* 28: 195–206.

[33]Guidry, G., and Landis, S. C. 1998. *Dev. Biol.* 199: 175–184.

[34]Burnstock, G. 1995. *J. Physiol. Pharmacol.* 46: 365–384.

[35]Burnstock, G., and Holman, M. E. 1961. *J. Physiol.* 155: 115–133.

Purinergic Transmission

In a remarkable series of experiments, Burnstock and his colleagues demonstrated the existence of a major class of sympathetic transmitters, the purines ATP and adenosine.[34] Certain sympathetic nerve fibers secrete ATP from their terminals, either as the principal transmitter or together with norepinephrine or acetylcholine. The experiments showing that ATP is a sympathetic transmitter were originally designed for a quite different purpose. Burnstock and Holman[35] recorded intracellularly from smooth muscle fibers of the

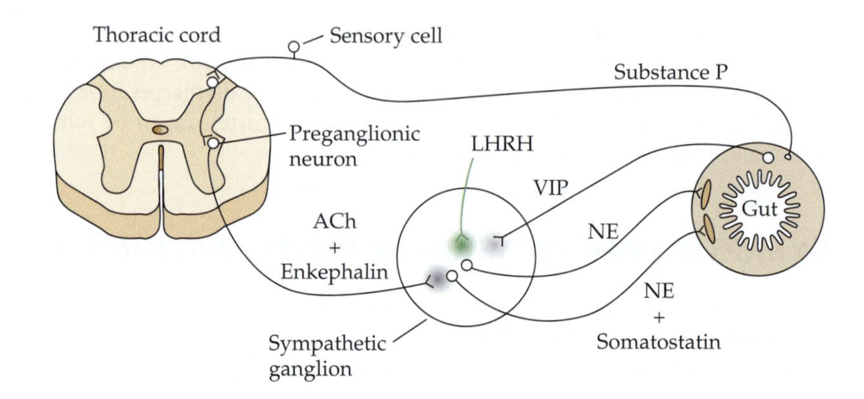

FIGURE 16.6 Localization of Transmitters and neuropeptides in neurons of the sympathetic nervous system. All known transmitters are found in the intestine. NE = norepinephrine. (After Hökfelt et al., 1980.)

reproductive system to investigate sympathetic synaptic transmission, which at the time was supposedly mediated by norepinephrine. Their recordings of spontaneous miniature potentials constituted an important finding, since hitherto quantal release had been observed only at the skeletal neuromuscular junction. Later, however, Burnstock and his colleagues[36] showed that norepinephrine is not the sole transmitter, and that the sympathetic neurons co-release a purine (ATP) and a peptide (neuropeptide Y). The miniature potentials in the smooth muscle are in fact due to ATP rather than norepinephrine acting directly on ion channel receptors.

Two main families of receptors for purines have been identified, sequenced, and cloned (Chapter 3). A family of eight or more P1 receptors are situated in pre- and postsynaptic structures in the periphery and in brain. They are activated preferentially by adenosine and are G protein coupled. Although ATP is the transmitter liberated from presynaptic endings, it is rapidly broken down by enzymes to adenosine, which is the natural agonist for the P1 receptors. ATP itself acts directly on a family of P2 receptors[37] in the central and the peripheral nervous systems, where it regulates endocrine secretion and smooth muscle contractility.

Sensory Inputs to the Autonomic Nervous System

This description of the autonomic nervous system has failed to mention essential components: sensory inputs and reflex regulation. Indeed in textbooks of physiology and pharmacology, the autonomic nervous system is often considered as though it functioned as a purely motor system for smooth muscle, cardiac muscle, and glands. One reason for neglect of autonomic regulation is the paucity of our knowledge. Mechanisms of smooth

[36]Kasakov, L., et al. 1988. *J. Auton. Nerv. Syst.* 22: 75–82.

[37]Soto, F., Garcia-Guzman, M., and Stühmer, W. 1997. *J. Membr. Biol.* 160: 91–100.

Box 16.1 The Path to Understanding Sympathetic Mechanisms

The development of concepts about mechanisms of synaptic transmission in the autonomic nervous system spanned many years. The original work of Henry Dale in the 1930s and 1940s showed that excitatory and inhibitory transmission in target organs is extremely complicated. One source of confusion arose from the idea that epinephrine might be the transmitter molecule liberated by sympathetic nerves (as originally suggested by Elliott in 1904). It was not until von Euler[38] discovered that norepinephrine, not epinephrine, is the principal transmitter released by sympathetic nerves that the distinction between hormonal and transmitter actions could be accounted for.

Epinephrine released from the adrenal medulla reaches receptors in cells that are not innervated by sympathetic axons—for example, on smooth muscle fibers of bronchioles in the lung. It can also act on receptors that are insensitive to norepinephrine released from sympathetic nerve fibers. A major advance was made by Ahlquist,[39,40] who devised the scheme for classifying α and β adrenergic receptors by comparing specific agonists and antagonists. This was essential for explaining the varied excitatory and inhibitory sympathetic actions on blood pressure, smooth muscle, the gut, bronchi, and glands. Also essential was the discovery that some sympathetic axons release acetylcholine or purines.

An understanding of the transmitters and receptors used by the autonomic nervous system has allowed new drugs to be developed for treatment of diseases. One example is provided by the control of bronchioles in the lung. Bronchoconstriction occurs in asthma or severe anaphylactic reactions of the immune system. To initiate relaxation of smooth muscle in the bronchi, β_2 receptors are activated by giving epinephrine or a more specific agonist (such as salbutamol, a β_2 agonist used for the treatment of asthma), but not norepinephrine, which has no effect on β_2 receptors. Bronchodilatation allows the patient to breathe again. Another example is provided by the beta-blockers that are widely used and highly effective in the treatment of high blood pressure, coronary artery disease, and glaucoma. As their name implies, these drugs act by blocking the actions of norepinephrine and epinephrine on β receptors in the heart, in the smooth muscle of blood vessels, in the kidney, and in the eye.[41]

[38]von Euler, U. S. 1956. *Noradrenaline.* Charles Thomas, Springfield, IL.

[39]Ahlquist, R. P. 1948. *Am. J. Physiol.* 153: 586–600.

[40]Black, J. W. 1976. *Postgrad. Med. J.* 52: 11–13.

[41]Black, J. W., and Prichard, B. N. 1973. *Br. Med. Bull.* 29: 163–167.

muscle contraction, such as blood vessel constriction and dilatation, the forward propulsion of material through the gut, and bladder emptying seem relatively straightforward compared to, say, playing tennis. And yet the integration that is required is far from simple. On the afferent side, sensory receptors in the eye, the lungs, the blood vessels, the viscera, and other target tissues provide information about the organs concerned (see Figure 16.6). These include nociceptive receptors capable of producing intense pain.

A deceptively simple and well-studied reflex is the response of the circulation to changes in body position. With the human body lying flat, the brain is supplied by blood without differences in pressure between the legs and the head. The assumption of a vertical stance causes a drop in blood pressure above the level of the heart, as blood accumulates in the gut and the legs. In the absence of autonomic regulation, loss of consciousness results from standing up, owing to diminished blood flow through the brain. The receptors that signal the need for a change in the pattern of circulation are situated in a large artery in the neck, the carotid artery. The endings are stretch receptors embedded in a swelling of the arterial wall, known as the carotid sinus. Distension of the wall causes increased firing, as shown in Figure 16.7.

The sensory axons run to the brainstem and terminate in a well-defined nucleus (the nucleus of the solitary tract). These neurons in turn project to the brainstem reticular formation, and those in turn to the autonomic preganglionic neurons. In the horizontal position, a high rate of sensory firing gives rise to inhibition of cardiovascular sympathetic outputs (see Figure 16.7). Blood pressure, heart rate, and cardiac output are depressed, and blood vessels in skin and gut are dilated, when one is lying down. With assumption of vertical posture, the pressure in the artery falls and the rate of firing of the carotid sinus axons is reduced dramatically, removing the central inhibition. The resulting release of sympathetic activity causes blood vessels in the skin and gut to constrict, and cardiac output and heart rate to rise. The increase in pressure maintains blood flow through the brain. The ancient recordings in Figure 16.7C, made by Anrep and Starling using a pointer scratching the surface of a rotating smoked drum, still provide good illustrations of these effects.[42] (It was the same Starling who, with Bayliss in 1902, first coined the word "hormone" and proposed the concept of hormonal action while they worked on the control of secretion in the gut.)

The description of the reflex presented in Figure 16.7 appears simple. But the central sympathetic and parasympathetic integrative mechanisms for rerouting blood where it is urgently needed remain a black box.[43] This is also the case for other autonomic reflexes for which the sensory and motor limbs are known—for example, enteric, excretory, and respiratory reflexes.[44–46]

The Enteric Nervous System

Local regulatory reflexes in the gut (see Figure 16.6) are extremely complex and are brought about by vast numbers of neurons. The enteric nervous system contains more than 10 million nerve cells arranged in the wall of the intestine as sensory neurons, interneurons, and motor neurons. Every known transmitter is represented there (and many of them were first discovered in the gut). To analyze the intrinsic circuits is difficult because of the profuse local reflexes and numbers of connections.[47] Functional analysis has been a major challenge even in simpler systems, such as the viscera of the lobster. When Selverston and his colleagues[48] began to study the stomatogastric ganglion, with its complement of only 30 neurons, it seemed that it could perhaps be worked out completely. Yet, although great progress has been made by recording electrically from identified neurons, and although principles of general significance for neurobiology have been discovered, a complete understanding is still not at hand. What appeared at first to be a simple circuit for regulating gut functions turned out to be plastic and modifiable rather than static and hardwired.

Regulation of Autonomic Functions by the Hypothalamus

An essential aspect of control of the autonomic nervous system is provided by hormones. The secretion of hormones by glands (such as the thyroid, the ovary, the adrenal cortex)

[42]Starling, E. H. 1941. *Starling's Principles of Human Physiology*. Churchill, London.

[43]Sato, T., et al. 1999. *Am. J. Physiol.* 276: H2251–H2261.

[44]Raybould, H. E., et al. 1991. *Adv. Exp. Med. Biol.* 298: 109–127.

[45]Morgan, C., deGroat, W. C., and Nadelhaft, I. 1986. *J. Comp. Neurol.* 243: 23–40.

[46]Coleridge, H. M., and Coleridge, J. C. 1994. *Respir. Physiol.* 98: 1–13.

[47]Obaid, A. L., et al. 1999. *J. Neurosci.* 19: 3073–3093.

[48]Selverston, A., et al. 1998. *Ann. N.Y. Acad. Sci.* 860: 35–50.

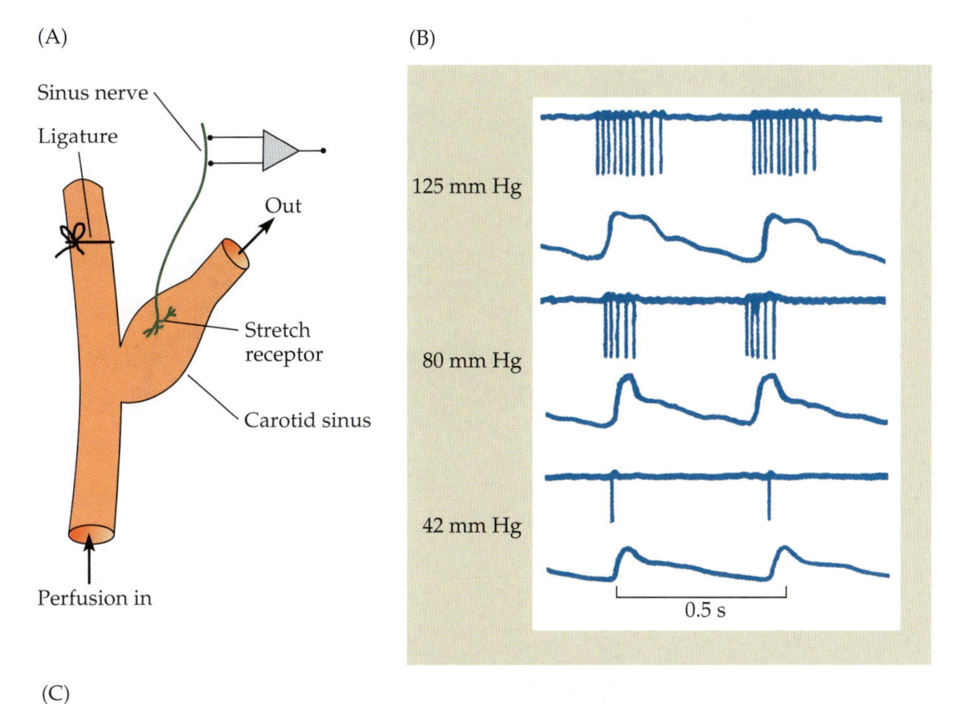

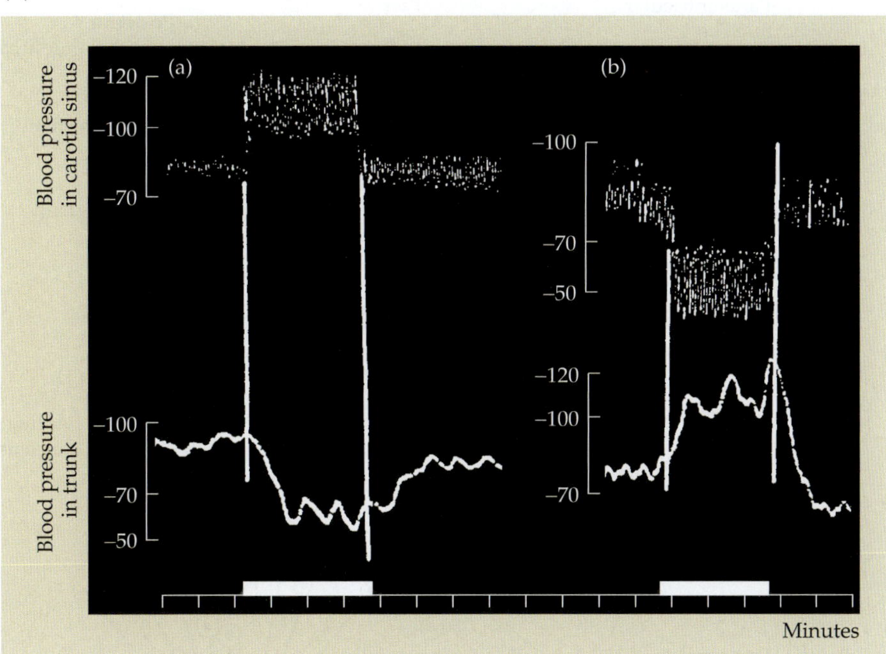

FIGURE 16.7 Firing of Carotid Sinus Stretch Receptors in response to raised blood pressure. (A) Experimental arrangement for recording from sensory nerve fibers in the carotid sinus while it is distended by the circulation or, as in the diagram, artificially perfused. (B) Relationship between blood pressure (lower trace) and the firing of a single afferent fiber from the carotid sinus at different levels of mean arterial pressure (from top down: 125, 80, and 45 mm of mercury, measured with a manometer). (C) A classic record made in 1924. The head of this animal was supplied with blood from a different animal so that blood pressure in the head arteries could be controlled separately by the experimenters. (a) Increased pressure in the head caused a fall in systemic blood pressure in the trunk of the animal. (b) Decreased pressure in the head caused an increase in systemic pressure. Such records were made before electrical recordings were possible; experimenters determined blood pressure using a mercury manometer and registered the movements with a fine pointer on a smoked drum. (B redrawn from Neil, 1954; C after Starling, 1941.)

is regulated by releasing factors secreted in the CNS (discussed in the sections that follow). The hormones in turn act back on the CNS to regulate the secretion of releasing factors, creating a feedback loop.

The hypothalamus (Figures 16.8 and 16.9) is a brain area that controls integrative autonomic functions, including body temperature, appetite, water intake, defecation, micturition, heart rate, arterial pressure, sexual activity, lactation, and, on a slower timescale, growth.[49] The precision of these homeostatic mechanisms makes it possible for all of us to keep our body temperature at about 37°C, our blood pressure at about 120/80 mm Hg, our heart rate at 70 beats/min, and our intake and output of water at 1.5 l/day, and for food to be propelled inexorably along the alimentary tract with appropriate secretions for

[49]Brooks, C. M. 1988. *Brain Res. Bull.* 20: 657–667.

FIGURE 16.8 Position of the Hypothalamus and the Pituitary Gland in the human brain. (A) Sagittal section of brain, with the area shown in B outlined. (B) Nuclei of the hypothalamus and adjacent structures.

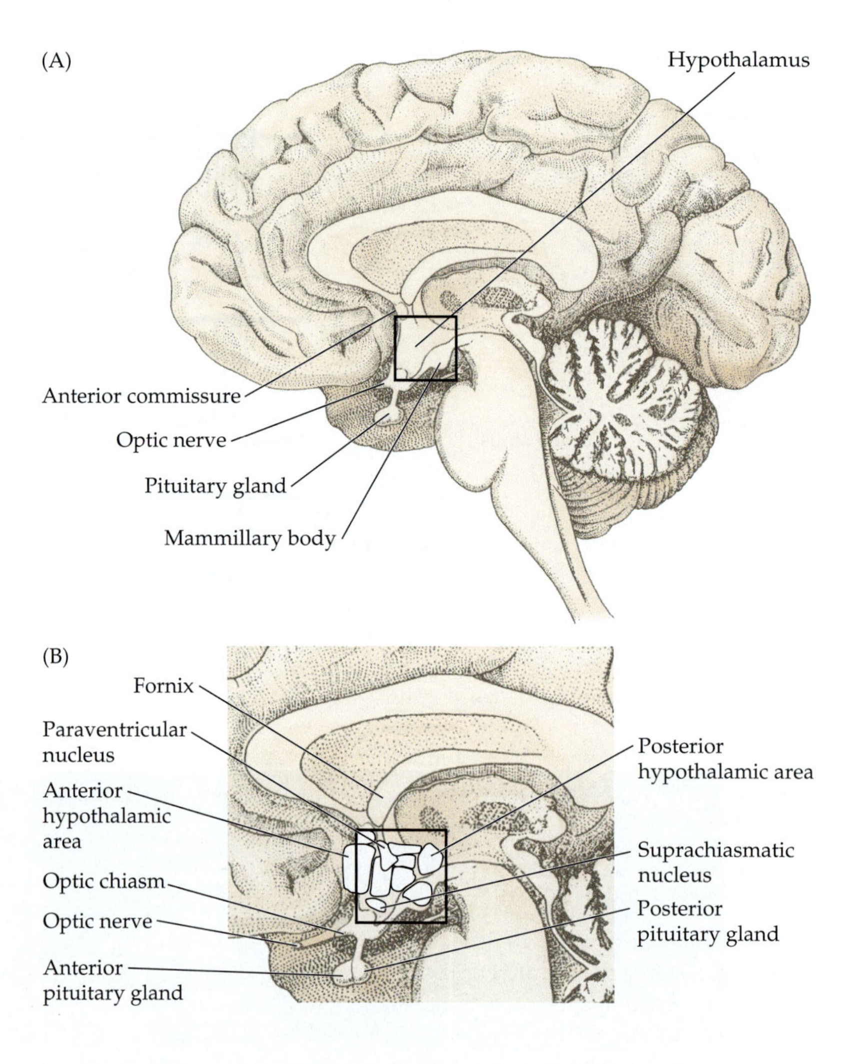

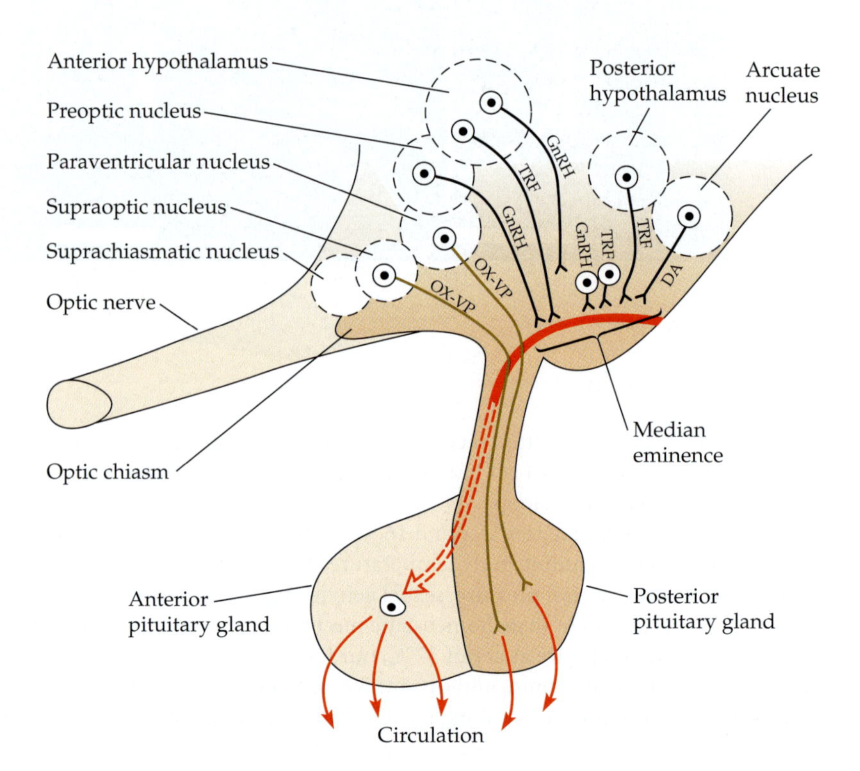

FIGURE 16.9 Connections of Hypothalamic Neurons with the neurohypophysis (posterior pituitary gland) and adenohypophysis (anterior pituitary gland). Axons run directly to the neurohypophysis. There the terminals secrete hormones into the circulation. By contrast, releasing hormones released by neurons in the hypothalamus reach the adenohypophysis in high concentration through a dedicated group of "portal" vessels (red dashed lines). There they activate secretory cells, which liberate hormones into the circulation. DA = dopamine, GnRH = gonadotropin-releasing hormone, TRF = thyroid hormone–releasing factor, OX-VP = oxytocin–vasopressin.

digestion and absorption at every level. The hypothalamus is also a brain area in which emotions are coupled to autonomic responses: The thought of food leads to secretion of saliva, the anticipation of exercise gives rise to increased sympathetic activity, and so on.

Other delicately controlled functions are the extraordinarily precise and regular rhythms generated by the hypothalamus. Slow rhythms include those that control endocrine secretion. For example, sexual and reproductive functions oscillate with periods of weeks that depend on secretion of peptide hormones by hypothalamic cells. These act on the anterior pituitary gland to stimulate secretion of other hormones into the bloodstream.

Hypothalamic Neurons That Release Hormones

A well-studied example of hormonal release is provided by neurons within the hypothalamus that secrete gonadotropin-releasing hormone (GnRH, which is the same as LHRH).[50] A prime action of these neurons is to secrete GnRH into a "portal system" of blood vessels that flow directly from the hypothalamus to the anterior pituitary gland (see Figure 16.9). Neurally released GnRH thereby acts selectively on a gland that is not directly innervated, enabling the central nervous system to control hormonal secretion. Thereafter the releasing hormone is diluted in the major vessels of the circulation and cannot, for example, influence synaptic transmission in autonomic ganglia. In the anterior pituitary gland (adenohypophysis), GnRH stimulates specific cells to secrete gonadotropin, a hormone that is essential for sexual and reproductive rhythms and functions.

This brief, oversimplified account cannot do justice to the beautiful experiments of G. W. Harris, who first demonstrated that the local release of a releasing hormone from the hypothalamus could provide an essential control mechanism.[51,52] His description of delivery of a chemical message through a system of blood vessels by highly localized transport was a revolutionary concept.

Distribution and Numbers of GnRH Cells

GnRH cells are dispersed throughout the hypothalamus, with no clearly defined nucleus or aggregate. Only those GnRH cells close to the anterior pituitary gland (in the median eminence) are the ones mentioned in the previous section that promote its gonadotropin secretion (see Figure 16.9). The release of the releasing hormone itself is influenced by hormones, such as those secreted by the ovary that feed back into the brain, and by synaptic inputs mediated by a variety of transmitters, including norepinephrine, dopamine, histamine, glutamate, and GABA.[53,54]

One extraordinary feature of the GnRH cells is their small number: 1300 in rats and 800 in mice.[55] Rats and mice (and we) would become extinct without these few, scattered cells in the brain. A second remarkable feature is their development (Chapter 23). During embryonic days 10 to 15 in rats, the precursor cells first appear in a region known as the olfactory placode. This is the region destined to be the future olfactory mucosa. After dividing, the cells migrate along axons of the olfactory nerve and end up in the hypothalamus.[55,56] The pathways and molecular mechanisms of GnRH cell migration have been studied in embryos, in newborn opossums, and in culture systems.[57] Since all the cells can be reliably marked by specific antibodies to GnRH, they can be counted quantitatively at the site of origin, and counts can be made as they migrate. Other types of neurons migrate along the same axonal pathway as the GnRH cells. Before reaching the hypothalamus, however, they branch off along other axons to reach distinctively different destinations.[58]

Figure 16.9 shows that in addition to the GnRH cells in the hypothalamus, there exist specific populations of neurons that secrete other hormones required for autonomic functions. Metabolism, thyroid function, absorption of salts by the kidney, and growth all depend on releasing hormones that are secreted into the portal system and that act on the anterior pituitary gland.

Specific hypothalamic neurons in the supraoptic and paraventricular nuclei (see Figure 16.9) innervate the posterior pituitary gland directly. Their endings release antidiuretic hormone (ADH, also known as vasopressin) and oxytocin into the blood.[59–61] Hence, the control of water absorption by the kidney and the contractions of the uterus depend directly on the firing of hypothalamic neurons.

[50]Dellovade, T., et al. 1998. *Gen. Comp. Endocrinol.* 112: 276–282.

[51]Harris, G. W., and Naftolin, F. 1970. *Br. Med. Bull.* 26: 3–9.

[52]Harris, G. W., and Ruf, K. B. 1970. *J. Physiol.* 208: 243–250.

[53]Rissman, E. F. 1996. *Biol. Reprod.* 54: 413–419.

[54]Segovia, C. T., et al. 1996. *Life Sci.* 58: 1453–1459.

[55]Wray, S., Grant, P., and Gainer, H. 1989. *Proc. Natl. Acad. Sci. USA* 86: 8132–8136.

[56]Tarozzo, G., et al. 1998. *Ann. N.Y. Acad. Sci.* 839: 196–200.

[57]Fueshko, S., and Wray, S. 1994. *Dev. Biol.* 166: 331–348.

[58]Tarozzo, G., et al. 1995. *Proc. R. Soc. Lond. B* 262: 95–101.

[59]Cunningham, E. T., Jr., and Sawchenko, P. E. 1991. *Trends Neurosci.* 14: 406–411.

[60]Gainer, H., et al. 1986. *Neuroendocrinology* 43: 557–563.

[61]Obaid, A. L., and Salzberg, B. M. 1996. *J. Gen. Physiol.* 107: 353–368.

Circadian Rhythms

Of particular importance in the life of an animal are the circadian rhythms that control the day–night sleep–wakefulness cycle. In the absence of all external cues, 24-hour rhythmical cycles are maintained by an internal clock for prolonged periods, weeks or months, in invertebrates as well as vertebrates,[62–64] and even in explants or neurons in culture.[65–67] The internal timing mechanism can be altered (or "entrained") by providing regularly spaced light and dark stimuli.[68] Autonomic functions are strongly influenced by biological clocks that act on the pineal gland and the secretion of melatonin.[69,70]

Information about cellular and molecular mechanisms that allow neurons to produce regular night and day cycles has been obtained in both invertebrates and vertebrates. For example, there is an aggregate of secretory nerve cells in the visual pathway of crustaceans known as the eyestalk. In this structure it has been possible to maintain rhythms of metabolic activity, secretion, and impulse firing while the isolated organ is kept in culture.[71] Electrical recordings have been made from pacemaker cells, the peptides that they release have been characterized, and their mechanisms of action have been analyzed. Moreover, in culture the activity of the pacemaker neurons can be entrained by the imposition of alternating light and dark cycles. Examples of circadian rhythms produced by cells in culture are shown in Figures 16.10 and 16.11.

In mammals, a key structure in the hypothalamus for generating the rhythm of the internal clock is the suprachiasmatic nucleus (SCN). An important input to this nucleus is from the eye.[72] After destruction of the suprachiasmatic nucleus in rats, light and dark entrainment of endogenous rhythms becomes lost. Locomotor activity, drinking, and sleep–waking cycles, as well as rhythms of hormone secretion, become disrupted.[73,74] If fetal hypothalamic tissue containing the SCN is transplanted to a host previously rendered arrhythmic by a complete lesion of the SCN, then rhythmicity is restored with a free running period corresponding to the donor genotype.

In neurons of the suprachiasmatic nucleus, the frequency of spontaneous action potentials increases during the day and decreases at night, as shown in Figure 16.12. By what mechanism is the rhythm produced? This problem has been investigated in slices of rat SCN in culture.[75] GABA has been shown to be a major transmitter used by neurons in this

[62]Pohl, C. R., and Knobil, E. 1982. *Annu. Rev. Physiol.* 44: 583–593.

[63]Moore, R. Y. 1997. *Annu. Rev. Med.* 48: 253–266.

[64]Arechiga, H. 1993. *Curr. Opin. Neurobiol.* 3: 1005–1010.

[65]Tosini, G., and Menaker, M. 1996. *Science* 272: 419–421.

[66]Martinez de la Escalera, G., Choi, A. L. H., and Weiner, R. L. 1992. *Proc. Natl. Acad. Sci. USA* 89: 1852–1855.

[67]Herzog, E. D., Takahashi, J. S., and Block, G. I. D. 1998. *Nature Neurosci.* 1: 708–713.

[68]Boivin, D. B., et al. 1996. *Nature* 379: 540–542.

[69]Borjigin, J., Li, X., and Snyder, S. H. 1999. *Annu. Rev. Pharmacol. Toxicol.* 39: 53–65.

[70]Cassone, V. M. 1998. *Chronobiol. Int.* 15: 457–473.

[71]Arechiga, H., et al. 1993. *Chronobiol. Int.* 10: 1–19.

[72]Kornhauser, J. M., Mayo, K. E., and Takahashi, J. S. 1996. *Behav. Genet.* 26: 221–240.

[73]Ralph, M. R., et al. 1990. *Science* 247: 975–978.

[74]Kaufman, C. M., and Menaker, M. J. 1993. *Neural Transplant Plast.* 4: 257–265.

[75]Wagner, S., et al. 1997. *Nature* 387: 598–603.

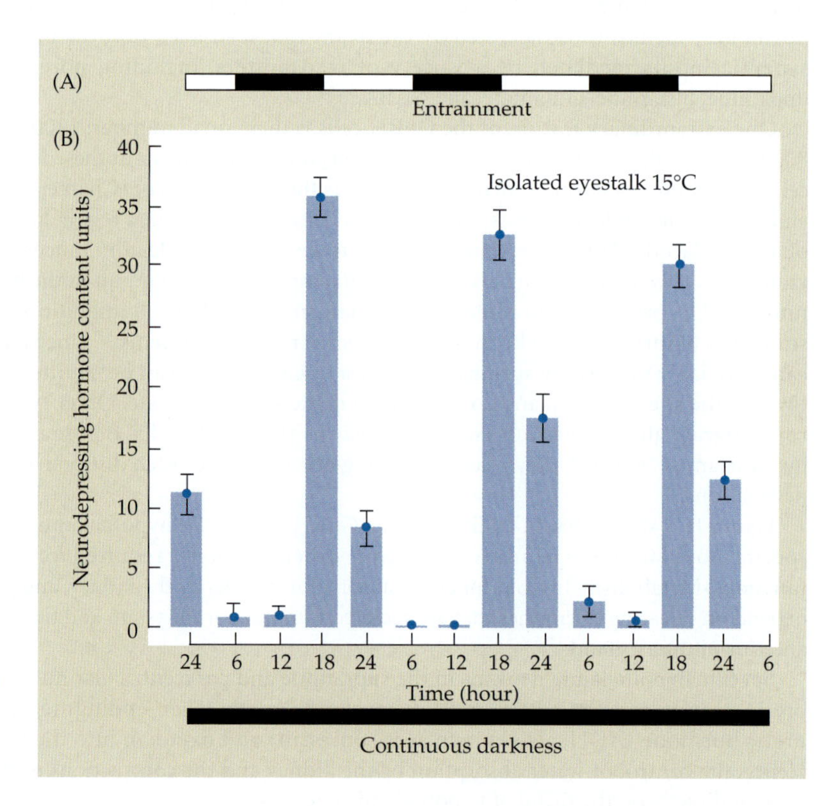

FIGURE 16.10 Circadian Rhythm of Crayfish Eyestalks in Isolation. (A) The isolated eyestalk was kept in organ culture and entrained to light–dark cycles (indicated by the alternating light and dark bars above). (B) Later, at the time of the experiment when the hormone content was measured, the eyestalk was deprived of all stimuli and kept in darkness (indicated by the black bar below). The day–night rhythm clearly persisted. (After Arechiga et al., 1993.)

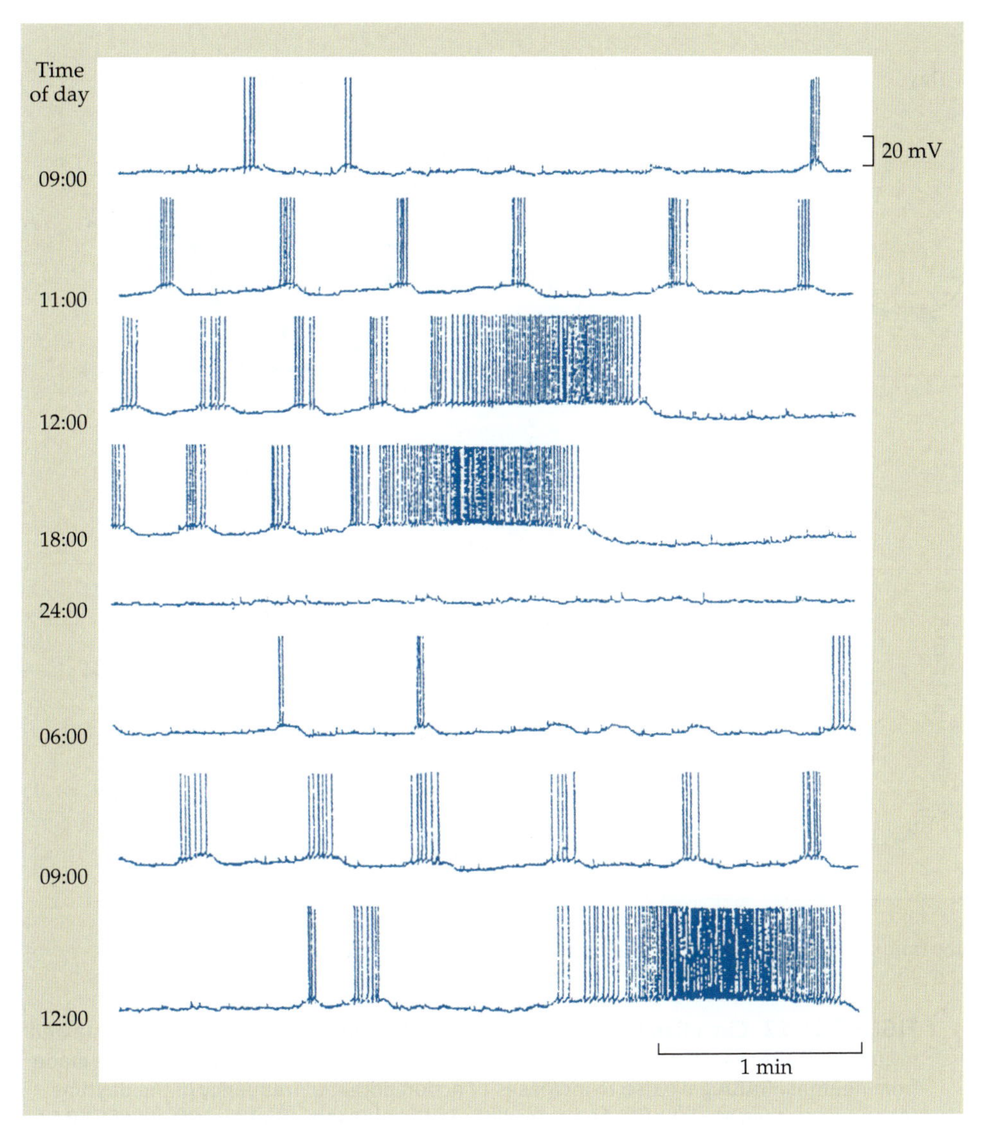

FIGURE 16.11 Circadian Firing of Neurons of crayfish eyestalk maintained in culture. During the 24-hour cycle the action potential activity recorded intracellularly from a single neuron undergoes cyclical changes. Regular bursts occur at 10.00 A.M., followed by irregular firing between 12:00 noon and 6:00 P.M., and silence at midnight; next morning the regular bursts begin again. (From records kindly provided by H. Arechiga and U. Garcia.)

nucleus.[76] Yarom and his colleagues[73] have shown that suprachiasmatic neurons in a slice respond to GABA with depolarization and increased rates of firing during the day (Figure 16.12A). The same concentration of GABA applied at night causes hyperpolarization and a decrease in rate (Figure 16.12B). Hence, as in developing central nervous system,[77] GABA can be an excitatory or an inhibitory transmitter. The type of response depends on the level of internal chloride. As described in Chapter 5, when the internal chloride concentration is low, the chloride equilibrium potential (E_{Cl}) is more negative than the resting potential: Opening of channels by GABA allows chloride ions to enter and the membrane to hyperpolarize. With raised internal chloride concentrations, E_{Cl} shifts to a value positive with respect to the resting membrane potential. As a result, GABA causes chloride ions to move out of the cell and gives rise to a depolarization (see also Gribkoff et al. for alternative schemes for rhythm generation).[78] It is not known whether changes in internal chloride concentrations are solely responsible for producing the rhythm in neurons of the suprachiasmatic nucleus, or what factors are responsible for the concentration changes.

Common proteins that are associated with periodicity throughout the animal kingdom have been revealed by genetic techniques. Genes and proteins that control circadian rhythms have been identified and cloned in *Drosophila*.[79] In many species, one such gene known as *per* has been found to be present in pacemaker regions such as the suprachiasmatic nucleus. In flies, deletion of the *per* gene abolishes circadian rhythm. Reintroduction

[76]Strecker, G. J., Wuarin, J. P., and Dudek, F. E. 1997. *J. Neurophysiol.* 78: 2217–2220.

[77]Cherubini, E., Gaiarsa, J. L., and Ben-Ari, Y. 1991. *Trends Neurosci.* 14: 515–519.

[78]Gribkoff, V. K., et al. 1999. *J. Biol. Rhythms* 14: 126–130.

[79]Hall, J. C. 1995. *Trends Neurosci.* 18: 230–240.

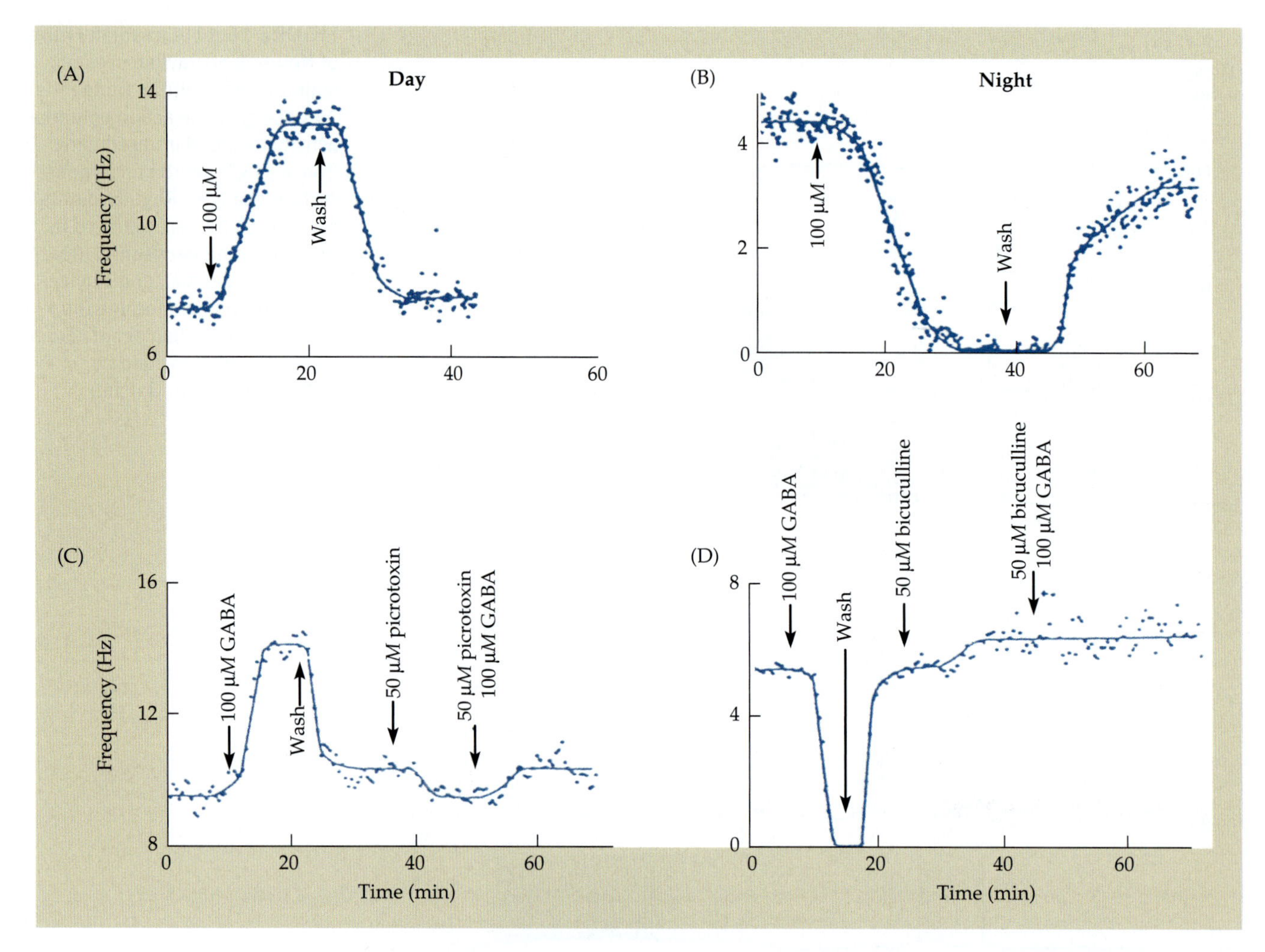

FIGURE 16.12 Circadian Rhythm of Slice of Rat Suprachiasmatic Nucleus maintained in culture. GABA was applied at different times while extracellular recordings were made from neurons. GABA gave rise to increases of action potential frequency in the daytime (A) and decreases at night (B). The recordings in C and D show that the effects of GABA were blocked by GABA antagonists (bicuculline and picrotoxin). The change from excitation to inhibition can be accounted for in terms of changed intracellular chloride concentrations, which were assessed by whole-cell patch recordings (not shown). The conductance change produced by GABA remains unchanged during the day–night cycle. The mechanisms that rhythmically change intracellular chloride concentration are not yet known. (After Wagner et al., 1997.)

of the *per* gene reestablishes the rhythm.[80,81] Although no link has been found between regulatory proteins and intracellular chloride concentration, it is gratifying that one can now begin to explain circadian rhythms in terms of genes and ion concentrations in well-defined groups of neurons.

[80]Zehring, W. A., et al. 1984. *Cell* 39: 369–376.

[81]Ewer, J., Rosbash, M., and Hall, J. C. 1998. *Nature* 333: 82–84.

SUMMARY

- The autonomic nervous system regulates essential functions of all internal organs and is itself regulated by hormonal and sensory feedback.

- Parasympathetic effects are focused, compared to the widespread effects of sympathetic activation.

- ACh is the principal transmitter used for transmission in autonomic ganglia, at parasympathetic nerve endings, and at certain sympathetic nerve endings.

- Norepinephrine is the principal transmitter for most sympathetic endings. Other transmitters include acetylcholine, peptides, and ATP.

- A single molecule—for example, LHRH, which is also known as GnRH—can act as a transmitter at synapses and a hormone within the brain.

- Analysis of effects mediated by the autonomic nervous system is complex, owing to the variety of receptors and the large numbers of peptide and nonpeptide transmitters.

- Epinephrine released as a hormone into the circulation from the adrenal medulla reaches receptors in target cells that are not affected by transmitter released from nerve endings.

- The hypothalamus is the region of the brain that controls the overall activities of the autonomic nervous system and also regulates the secretion of hormones.

- The hypothalamus itself is influenced by higher centers of the central nervous system and by hormones.

SUGGESTED READING

General Reviews

Arechiga, H. 1993. Circadian rhythms. *Curr. Opin. Neurobiol.* 3: 1005–1010.

Brown, D. 1988. M-currents: An update. *Trends Neurosci.* 11: 294–299.

Burnstock, G. 1993. Physiological and pathological roles of purines: An update. *Drug Dev. Res.* 28: 195–206.

Cooper, J. R., Bloom, F. E., and Roth, R. H. 1996. *The Biochemical Basis of Pharmacology*. Oxford University Press, New York.

Hall, J. C. 1995. Tripping along the trail to the molecular mechanisms of biological clocks. *Trends Neurosci.* 18: 230–240.

Janig, W., and McLachlan, E. M. 1992. Characteristics of function-specific pathways in the sympathetic nervous system. *Trends Neurosci.* 15: 475–481.

Marrion, N. V. 1997. Control of M-current. *Annu. Rev. Physiol.* 59: 483–504.

Moore, R. Y. 1997. Circadian rhythms: Basic neurobiology and clinical applications. *Annu. Rev. Med.* 48: 253–266.

Rang, H. P., Dale, M. M., and Ritter, J. M. 1999. *Pharmacology*, 4th Ed. Churchill Livingstone, Edinburgh, Scotland, Chapters 6–10.

Tosini, G., and Menaker, M. 1996. Circadian-rhythms in cultured mammalian retina. *Science* 272: 419–421.

Original Papers

Ewer, J., Rosbash, M., and Hall, J. C. 1998. An inducible promoter fused to the period gene in *Drosophila* conditionally rescues adult per-mutant arrhythmicity. *Nature* 333: 82–84.

Holman, M. E., Coleman, H. A., Tonta, M. A., and Parkington, H. C. 1994. Synaptic transmission from splanchnic nerves to the adrenal medulla of guinea pigs. *J. Physiol.* 478: 115–124.

Kaufman, C. M., and Menaker, M. J. 1993. Effect of transplanting suprachiasmatic nuclei from donors of different ages into completely SCN lesioned hamsters. *Neural Transplant Plast.* 4: 257–265.

Kuffler, S. W. 1980. Slow synaptic responses in autonomic ganglia and the pursuit of a peptidergic transmitter. *J. Exp. Biol.* 89: 257–286.

Ralph, M. R., Foster, R., Davis, F. C., and Menaker, M. 1990. Transplanted suprachiasmatic nucleus determines circadian period. *Science* 247: 975–978.

Selyanko, A. A., and Brown, D. A. 1996. Intracellular calcium directly inhibits potassium M channels in excised membrane patches from rat sympathetic neurons. *Neuron* 16: 151–162.

Soto, F., Garcia-Guzman, M., and Stühmer, W. 1997. Cloned ligand-gated channels activated by extracellular ATP (P2X receptors). *J. Membr. Biol.* 160: 91–100.

Ullian, E. M., McIntosh, J. M., and Sargent, P. B. 1997. Rapid synaptic transmission in the avian ciliary ganglion is mediated by two distinct classes of nicotinic receptors. *J. Neurosci.* 17: 7210–7219.

Wagner, S., Castel, M., Gainer, H., and Yarom, Y. 1997. GABA in the mammalian suprachiasmatic nucleus and its role in diurnal rhythmicity. *Nature* 387: 598–603.

Wang, H. S., Pan, Z., Shi, W., Brown, B. S., Wymore, R. S., Cohen, I. S., Dixon, J. E., and McKinnon, D. 1998. KCNQ2 and KCNW potassium channel subunits: Molecular correlates of the M-channel. *Science* 282: 1890–1893.

Wray, S., Grant, P., and Gainer, H. 1989. Evidence that cells expressing luteinizing hormone-releasing hormone mRNA in the mouse are derived from progenitor cells in the olfactory placode. *Proc. Natl. Acad. Sci. USA* 86: 8132–8136.

17 Transduction of Mechanical and Chemical Stimuli

STIMULUS INTENSITY AND TIMING ARE ENCODED in receptor potentials arising in the receptive endings of sensory cells. Receptor potentials can be depolarizing or hyperpolarizing; they increase in amplitude with increased stimulus intensity and saturate at higher stimulus levels. During sustained stimulation the receptor potential adapts to a lower level. Adaptation can occur quickly or slowly. It arises from mechanical, electrical, or biochemical processes in different cell types. Receptors that adapt slowly encode stimulus duration. Rapidly adapting receptors are specialized to detect changes in the stimulus.

Transduction of mechanical stimuli occurs in a wide variety of sensory cells situated in skin, muscles, joints, and internal organs. Sensory hair cells of the inner ear provide an example of the mechanisms by which deformation produces electrical signals. Thus, movement of the head or sounds arriving in the ear deflect hair bundles and cause nonselective cation channels to open, resulting in depolarization. The channels close again during sustained deflection due to calcium-dependent processes of adaptation.

Olfactory receptors consist of ciliated receptor cells in the nose; the mechanism of transduction is very different from that of mechanotransduction. Olfactants bind to a large family of G protein–binding receptors. A subsequent rise in cyclic adenosine monophosphate (cAMP) causes cation channels to open, and the resulting depolarization generates action potentials. Similarly, in taste buds, certain taste stimuli (amino acids, sugar, and bitter compounds) are transduced by G protein–coupled membrane receptors. Again, a rise in cAMP causes cation channels to open, and action potentials are initiated. Salts and acids (sour) can act directly on ion channels in the taste bud receptor cell. Transduction in receptors specific for pain and temperature sensation involves direct action on cation channels in the sensitive endings as well as activation of metabotropic receptors. In addition, cells in damaged tissue release substances that sensitize pain fibers.

We know the physical world through our senses. We reach out to touch nearby objects, or receive signals transmitted from afar. Sensory receptors are the gateways through which these signals pass. Right at the outset the receptors set the stage for all the analyses of sensory events that are subsequently made by the central nervous system. They define the limits of sensitivity and determine the range of stimuli that can be detected and acted upon. With rare exceptions, each type of receptor is specialized to respond preferentially to only one type of stimulus energy, called the **adequate stimulus**: Rods and cones in the eye respond to light (Chapter 19), nerve endings in the skin to touch, pressure, or vibration, receptors in the tongue to chemical tastants. The stimulus, whatever its modality, is always converted (or **transduced**) to an electrical signal, the **receptor potential**. In general, the strength and duration of a stimulus are coded in electrical signals; recognition by the CNS of the modality of the stimulus and its position depend on the nature of the sensory ending and its anatomical location. Thus, a temperature receptor in the foot has its own pathway into the nervous system, quite distinct from that of a vibration receptor in the hand, but in both axons the signals are action potential trains of variable frequency and duration.

For sensory signals there is a great deal of amplification at the receptor level, so that very small external stimuli provide a trigger to release stored charges that appear as electrical potentials. For example, odors from only a few molecules of specific odorant substances (pheromones) are able to act as sex attractants for moths. Similarly, a few quanta of light trapped by receptors in the retina are sufficient to produce a visual sensation. This extreme sensitivity extends to the inner ear, where mechanical displacements of 10^{-10} m are detectable.[1] Equally remarkable are electroreceptors in some fish that can detect electrical fields of a few nanovolts per centimeter.[2,3] This is smaller in magnitude than the field that would be produced if two wires connected to either pole of an ordinary flashlight battery could be dipped into the Atlantic Ocean—one at Bordeaux, the other at New York!

Sensory receptors have a well-defined range of stimuli to which they respond. For example, our auditory hair cells can respond to sound only within a bandwidth of about 20 to 20,000 Hz. The response by receptors in our retina to electromagnetic radiation is restricted similarly to wavelengths between about 400 and 750 nm. Shorter (near-ultraviolet) and longer (near-infrared) wavelengths go undetected. Restrictions of this kind are not usually because of unavoidable physical limitations. Instead each system is tuned to the particular needs of the organism: Whales and bats can hear much higher frequencies; snakes can detect infrared, and bees ultraviolet, radiation. For dogs and pigs the sense of smell is more refined than in humans.

What mechanisms provide such great sensitivity and selectivity to sensory receptor cells? In this chapter we focus specifically on the transduction of mechanical and chemical sensory stimuli. To describe **mechanotransduction** we have chosen stretch receptors of muscle and mechanoreceptive hair cells of the inner ear. Mechanisms of **chemotransduction** are illustrated by olfactory and gustatory receptors. We conclude with a discussion of **nociception** that underlies pain sensation and combines the transduction of chemical and mechanical stimuli. (Phototransduction by retinal rods and cones is described separately in Chapter 19.)

STIMULUS CODING BY MECHANORECEPTORS

Short and Long Receptors

The receptor potential generated by the transduction process reflects the intensity and duration of the original stimulus. In some receptors, such as rods and cones in the retina that do not have long axons, receptor potentials spread passively from the sensory region to the synaptic region of the cell (Figure 17.1A). Such receptors are known as **short receptors**. The passage of information from the receptor end to the synaptic end of the cell does not require the intervention of action potentials. In some cells, passive spread of the receptor potential can reach a surprisingly distant point. For example, in some crustacean[4] and leech[5] mechanoreceptors, and in photoreceptors in the barnacle eye,[6] the receptor potential spreads passively over a distance of several millimeters. In such cells the membrane resistance, and hence the length constant for spread of passive depolarization, is unusually

[1]Bialek, W. 1987. *Annu. Rev. Biophys. Biophys. Chem.* 16: 455–478.

[2]Kalmidjn, A. J. 1982. *Science* 218: 916–918.

[3]Heiligenberg, W. 1989. *J. Exp. Biol.* 146: 255–275.

[4]Roberts, A., and Bush, B. M. H. 1971. *J. Exp. Biol.* 54: 515–524.

[5]Blackshaw, S. E., and Thompson, S. W. 1988. *J. Physiol.* 396: 121–137.

[6]Hudspeth, A. J., Poo, M. M., and Stuart, A. E. 1977. *J. Physiol.* 272: 25–43.

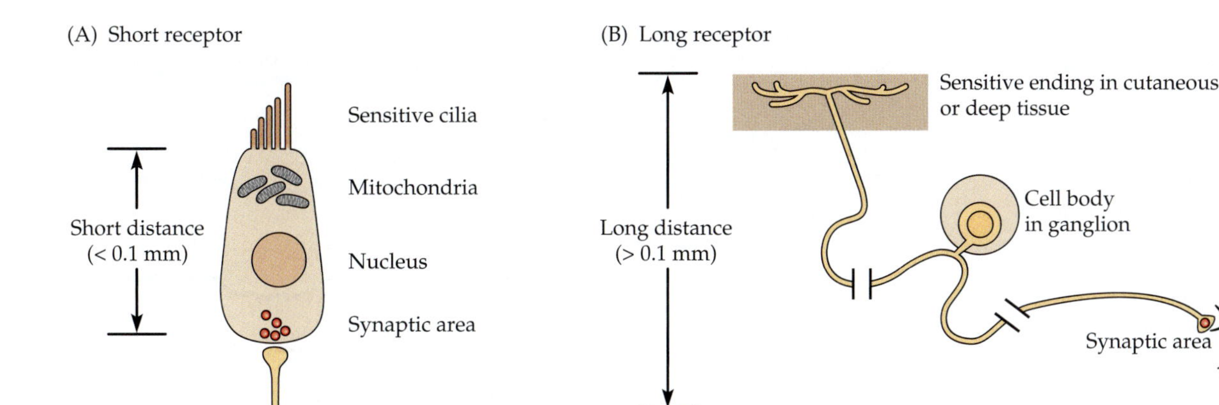

FIGURE 17.1 Short and Long Sensory Receptor Cells. (A) Short receptors—such as retinal rods and cones, and mechanosensory hair cells of the inner ear—are less than 100 μm in length. Thus, receptor potentials generated within the sensitive ending spread effectively throughout the cell, altering transmitter release at synaptic areas. (B) Long receptors, such as muscle spindle afferents and cutaneous mechanoreceptors, employ action potentials to conduct their signals to a distant second-order neuron. The amplitude of the receptor potential is encoded in the frequency of action potentials it generates.

high. Whereas receptor potentials are usually depolarizing, certain short receptors respond to their adequate stimulus with a hyperpolarizing potential change. This occurs, for example, in photoreceptors of the vertebrate retina (Chapter 19) and in cochlear hair cells that have both hyperpolarizing and depolarizing responses. Whatever the polarity of the receptor potential, short receptors release neurotransmitter tonically from their synaptic regions; depolarization increases or hyperpolarization decreases the rate of release.

In **long receptors** (Figure 17.1B), such as those in skin or muscle, information from single receptors must be sent over a much longer distance to reach the second-order sensory cell (e.g., from the big toe to the spinal cord). In order to accomplish this, the receptor performs a second transformation process: Receptor potentials give rise to trains of action potentials whose duration and frequency code information about the duration and intensity of the original stimulus. These then carry the information to the synaptic terminals of the cell.

The **frequency code** for stimulus intensity is established through the interaction of the maintained receptor current from sensory terminals and the conductance changes associated with the action potential. At the end of each action potential, the increased potassium conductance that occurs on the recovery phase drives the membrane in the hyperpolarizing direction, toward E_K (the potassium equilibrium potential). This increase in potassium conductance is transient, while the sustained transduction current depolarizes the membrane once more to the firing level. The stronger the receptor current, the sooner the firing level is reached again, and the higher the impulse frequency. Similar considerations also apply to all neurons in which synaptic input, rather than a receptor potential, sums to alter the frequency of action potentials.

Encoding Stimulus Parameters by Stretch Receptors

The way in which sensory receptors generate electrical signals was studied early on by Adrian and Zotterman[7] using extracellular recording from sensory nerve fibers arising in vertebrate muscle stretch receptors. The first demonstration of the link between sensory stimuli and electrical signals in a mechanoreceptor was made by Katz,[8] who recorded receptor potentials and showed that stretch caused a depolarization of the **sensory ending**. When the receptor potential was observed in isolation by blocking the nerve discharge with procaine, a local anesthetic, its amplitude could be seen to increase in a graded fashion with muscle stretch.

Figure 17.2 shows the relationship between receptor potential amplitude and stretch. The function begins with a slope of roughly 0.1 mV per millimeter of stretch, but flattens out at higher levels. The sensitivity of the sensory ending, in millivolts per millimeter of stretch, decreases as the stimulus grows. Many sensory receptors take advantage of this nonlinear relationship to provide amplitude coding over a wide range of stimulus intensities. In such receptors the response amplitude continues to increase, but in proportion to the logarithm of stimulus intensity. This is of great utility in receptors such as hair cells and photoreceptors that respond to stimuli whose amplitudes vary by several orders of magnitude.

These relationships between stimulus intensity and sensitivity correspond to those first described in 1846 by Weber. He measured the ability of subjects to discriminate

E. D. Adrian

[7]Adrian, E. D., and Zotterman, Y. 1926. *J. Physiol.* 61: 151–171.

[8]Katz, B. 1950. *J. Physiol.* 111: 261–282.

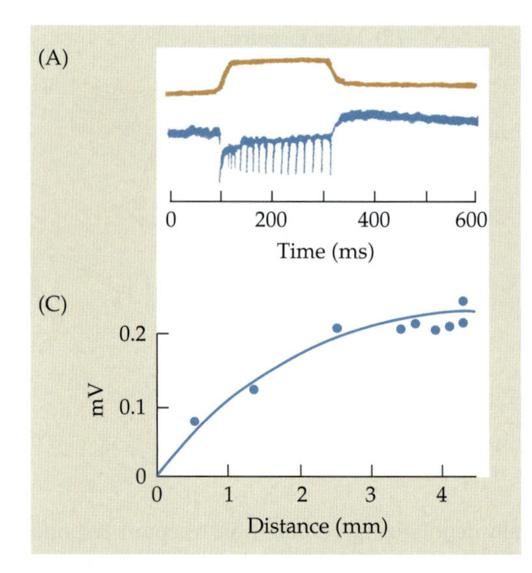

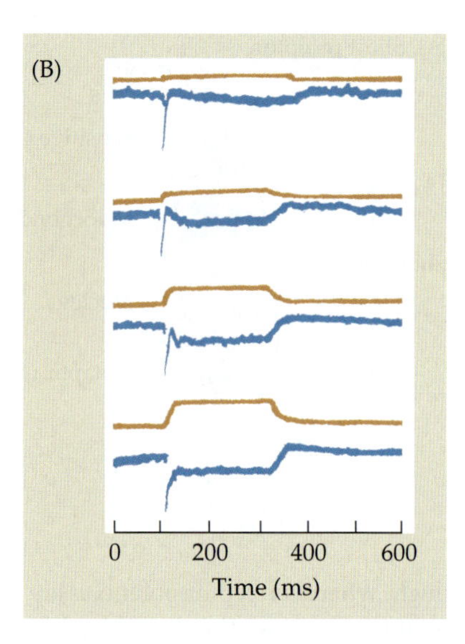

FIGURE 17.2 Receptor Potentials Recorded Extracellularly from a sensory nerve fiber supplying a muscle spindle. The recording electrode is placed as close as possible to the receptor. Downward deflection of the voltage record (lower traces) indicates receptor depolarization. (A) Stretching the muscle (upper trace) produces a receptor potential, upon which is superimposed a series of action potentials (lower trace). (B) Four stretches of increasing magnitude applied to the muscle after procaine has been added to the bathing solution. Action potentials (except for the first) are abolished by procaine, but the receptor potentials remain. (C) Plot of receptor potential amplitude against increase in muscle length. (After Katz, 1950.)

weights held in their two hands and showed that it varied in proportion to the size of the weights. That is, the subjects were just able to detect a 3 g difference between weights of about 100 g each, whereas kilogram weights had to differ by 30 g (in each case the detectable difference was about 3% of each object's weight). This relationship was formalized by Fechner, who pointed out that this result implied a logarithmic relationship between stimulus and response. The Weber–Fechner law[9] is one formulation of the nonlinear relationship between stimulus strength and sensation. Although the exact form of the relationship depends on stimulus modality, this process applies to many aspects of perception and behavior. (For example, we might pinch pennies when pricing pencils, but happily spend hundreds of dollars more for a classier computer.)

The Crayfish Stretch Receptor

Stimulus coding was analyzed in detail in crayfish stretch receptors by Eyzaguirre and Kuffler.[10] This preparation is particularly useful because the cell body of the stretch receptor lies in isolation—not in a ganglion, but on its own in the periphery, where it can be seen in live preparations (Figure 17.3A). It is large enough for penetration by intracellular microelectrodes. The cell inserts its dendrites into a fine muscle strand nearby and sends an axon centrally to a segmental ganglion (Figure 17.3B). In addition, the receptor receives inhibitory innervation from the ganglion; the muscle fibers into which it inserts receive excitatory and inhibitory innervation. Thus, receptor sensitivity is regulated by the CNS.

There are two types of crustacean stretch receptors with distinct structural and physiological characteristics, and their dendrites are embedded in different types of muscle. One responds well at the beginning of a stretch, but its response quickly wanes. This decrease in response to a steady stimulus is called **adaptation**. In contrast to the **rapidly adapting** receptor, the second type is **slowly adapting**; that is, its response is well maintained during prolonged stretch. The responses of a rapidly adapting and a slowly adapting stretch receptor are shown in Figure 17.4. In the slowly adapting receptor (Figure 17.4A), mild stretch of the muscle produces a depolarizing receptor potential of about 5 mV, lasting for the duration of the stretch. A larger stretch produces a larger potential that depolarizes the cell to above threshold and produces a train of action potentials that propagate centrally along the axon. A similar stretch of the muscle produces only transient responses in the rapidly adapting receptor (Figure 17.4B).

Muscle Spindles

Stretch receptors in mammalian skeletal muscles show mechanisms of action similar to those in crustaceans. Such stretch receptors were called **muscle spindles** by early anatomists because of their resemblance to the spindles used by weavers. Details of the

[9]Boring, E. G. 1942. *Sensation and Perception in the History of Experimental Psychology*. Appleton-Century, New York.

[10]Eyzaguirre, C., and Kuffler, S. W. 1955. *J. Gen. Physiol.* 39: 87–119.

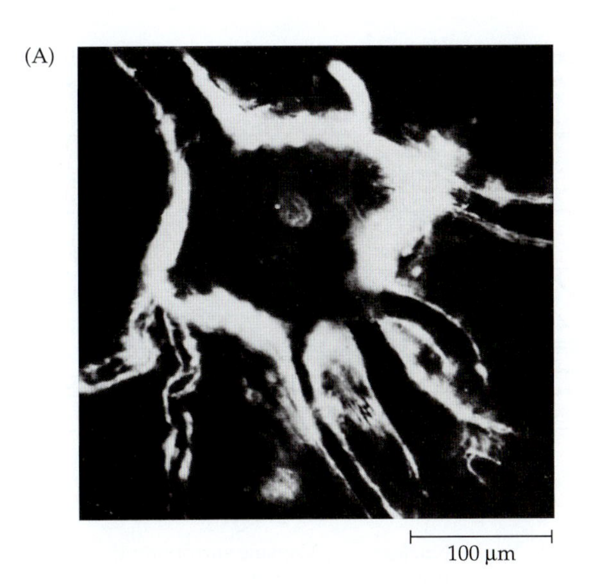

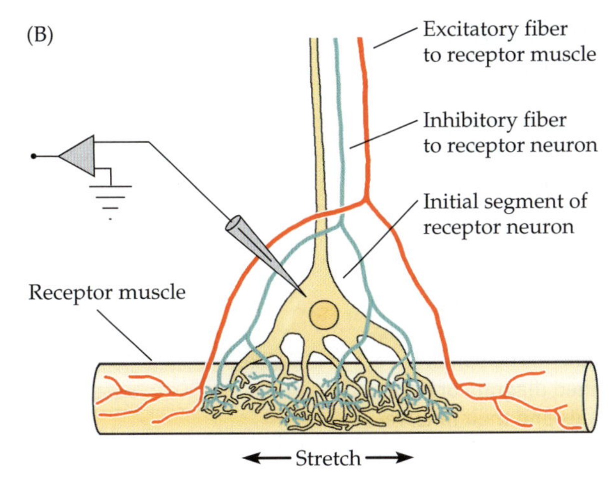

FIGURE 17.3 Crustacean Stretch Receptor. (A) Living receptor neuron viewed with dark-field illumination. Distal portions of six dendrites insert into the receptor muscle, which is not visible. (B) Relation between stretch receptor neuron and muscle, indicating the method of intracellular recording. The excitatory fiber to the muscle produces contraction; the inhibitory fiber innervates the neuron. Two additional inhibitory fibers are not shown. (After Eyzaguirre and Kuffler, 1955.)

way in which muscle spindles respond were worked out by B. H. C. Matthews in the early 1930s.[11–13] For many years his experiments provided one of the best attempts to describe comprehensively a sensory end organ and its control. Matthews was able to record impulses in single nerve fibers from individual spindles in frogs and cats with an oscilloscope he designed for the purpose (no mean feat in 1930).

Muscle fibers within the spindle (**intrafusal fibers**) differ from the main muscle mass (**extrafusal fibers**) in numerous ways (reviewed by Hunt[14]), including the molecular structure of myosin they contain.[15] They are called intrafusal fibers after the Latin word for "spindle," *fusus*. Figure 17.5 illustrates schematically the sensory apparatus of spindles in leg muscles of the cat. The spindle consists of a capsule containing 8 to 10 intrafusal

[11]Matthews, B. H. C. 1931. *J. Physiol.* 71: 64–110.

[12]Matthews, B. H. C. 1931. *J. Physiol.* 72: 153–174.

[13]Matthews, B. H. C. 1933. *J. Physiol.* 78: 1–53.

[14]Hunt, C. C. 1990. *Physiol. Rev.* 70: 643–663.

[15]Walro, J. M., and Kucera, J. 1999. *Trends Neurosci.* 22: 180–184.

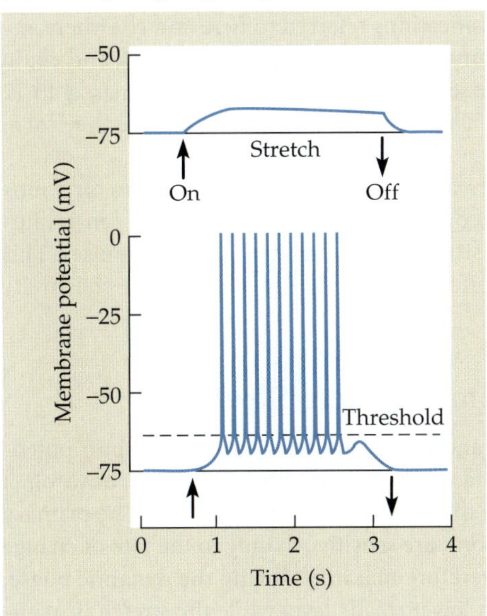

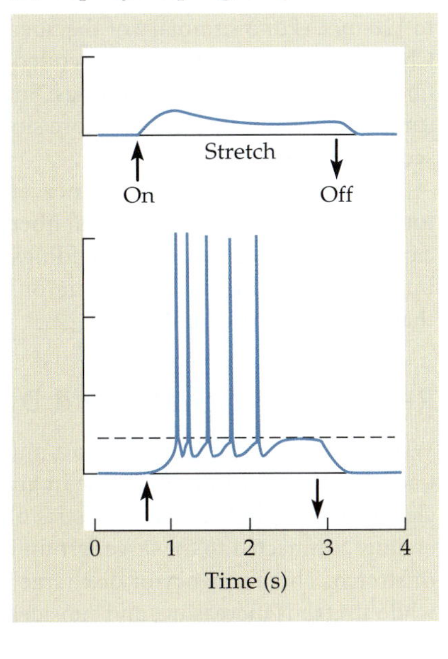

FIGURE 17.4 Responses of Stretch Receptor Neurons to increases in muscle length, recorded intracellularly as indicated in Figure 17.3B. (A) In a slowly adapting receptor, a weak stretch for about 2 s produces a subthreshold receptor potential that persists throughout the stretch (upper record). With a stronger stretch, a larger receptor potential sets up a series of action potentials (lower record). (B) In a rapidly adapting receptor, the receptor potential is not maintained (upper record), and during the large stretch the action potential frequency declines (lower record). (After Eyzaguirre and Kuffler, 1955.)

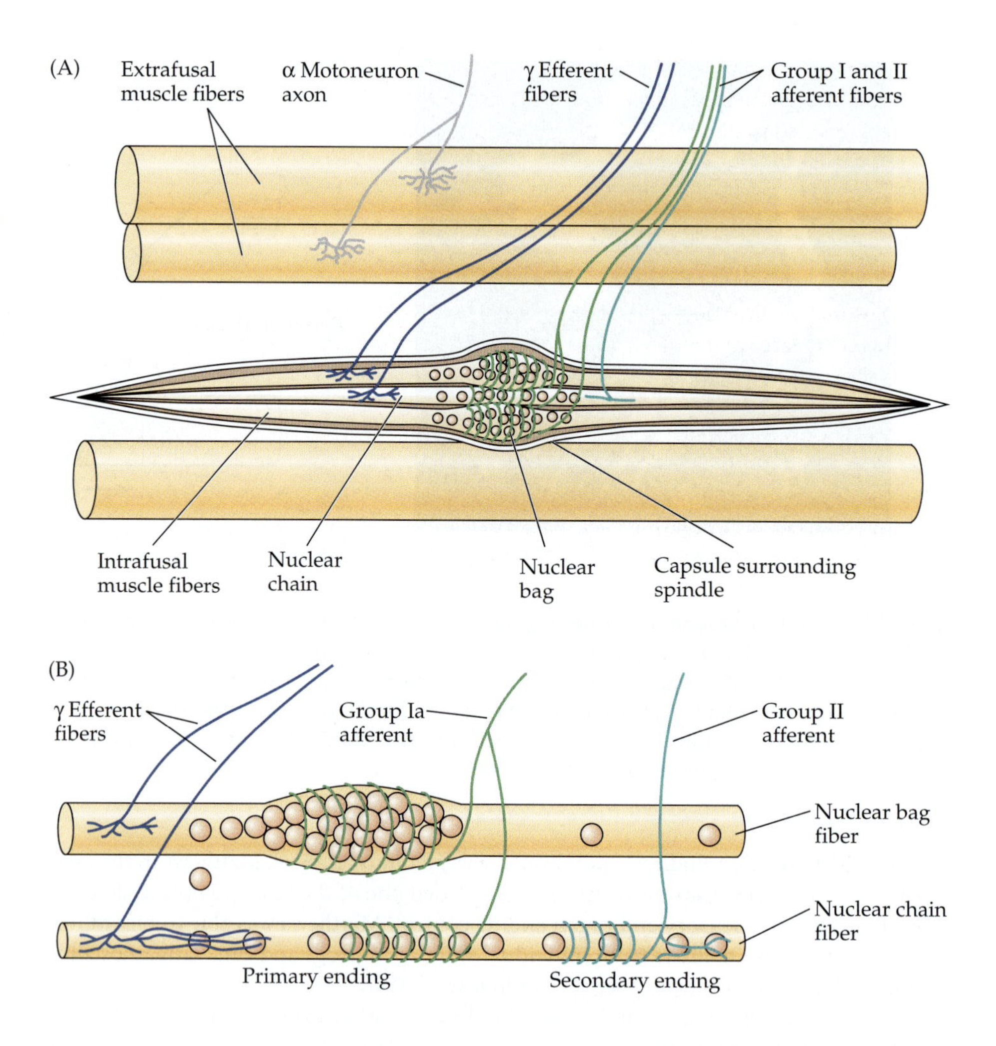

FIGURE 17.5 Mammalian Muscle Spindle. (A) Scheme of mammalian muscle spindle innervation. The spindle, composed of small intrafusal fibers, is embedded in the bulk of the muscle, which is made up of large muscle fibers supplied by α motoneurons. γ Motor (fusimotor) fibers supply the intrafusal muscle fibers, and group I and group II afferent fibers carry sensory signals from the muscle spindle to the spinal cord. (B) Simplified diagram of intrafusal muscle types and their innervation. (B after Matthews, 1964.)

fibers. In the central, or equatorial, region there is in each fiber a large aggregation of nuclei. Their arrangement provides the basis for the classification of intrafusal fibers as bag or chain fibers, depending on whether the nuclei are grouped together centrally or are arranged linearly.

Two types of sensory neurons innervate each muscle spindle. The larger nerve fibers, group Ia afferents, have diameters of 12 to 20 μm and conduct impulses at velocities up to 120 m/s. (For a summary of the fiber classifications referred to here and elsewhere, see Chapter 7.) Their terminals are coiled around the central parts of both bag and chain fibers to form the **primary endings**. Smaller sensory nerves (Group II fibers) are 4 to 12 μm in diameter and conduct more slowly. They contact chain fibers, where they form **secondary endings**.

The muscle spindle also is innervated by motoneurons (**fusimotor fibers**, or γ motoneurons). They cause intrafusal fibers to contract and thereby stretch the central nuclear region where the sensory endings are situated, causing them to fire impulses. This interaction provides a mechanism for the efferent control of muscle spindle sensitivity that will be described in Chapter 22.

Responses to Static and Dynamic Muscle Stretch

When a rapid stretch is applied to a muscle and to the spindles within it, receptor potentials and bursts of impulses arise in group Ia and II sensory fibers. There is, however, a clear difference in the characteristics of the discharges in the two endings. The primary endings, connected to the larger group Ia axons, are sensitive mainly to the rate of change of stretch. The frequency of discharge is therefore maximal during the dynamic phase, while stretch is increasing, and subsides to a lower steady level while the stretch is maintained. The secondary endings, connected to the smaller group II fibers, are relatively unaffected by the rate of stretch but are sensitive to the level of static tension.[16] This behavior is illustrated in Figure 17.6. The group Ia (dynamic) and group II (static) afferents are analogous to the rapidly adapting and slowly adapting receptors in the crayfish muscle and in other sensory systems.

Mechanisms of Adaptation in Mechanoreceptors

In mammalian muscle spindles, the viscoelastic properties of intrafusal fibers allow a gradual decrease in deformation of the sensory terminals.[17] A variety of processes have been shown to contribute to adaptation of crustacean stretch receptors.[16,18–20]

In the slowly adapting stretch receptor, trains of impulses lead to an increase in internal sodium concentration and activation of the sodium pump. The net outward transport of positive charges by the pump reduces the amplitude of the receptor potential and hence the discharge frequency. Yet another factor contributing to adaptation is an increase in potassium conductance. For example, during a train of impulses in crayfish stretch receptors, calcium entry through voltage-activated channels causes the opening of calcium-activated potassium channels. The effect of this increase in potassium conductance is to "short out" the receptor potential, reducing its amplitude and the frequency of the sensory impulses.

The rapidly adapting crayfish stretch receptor shows prompt adaptation of its firing rate even when a steady depolarizing current is applied experimentally. During imposed stretch, calcium influx through the transduction channels activates nearby calcium-dependent potassium channels to hyperpolarize the cell.[21]

Adaptation in the Pacinian Corpuscle

The Pacinian corpuscle is a rapidly adapting cutaneous mechanoreceptor.[22] Its nerve terminal is enclosed in an onionlike capsule. Pressure applied slowly to the capsule produces no response at all; more rapidly applied pressure produces only one or two action potentials. However, the receptors are exquisitely sensitive to vibration up to frequencies of 1000/s. Although they are found generally in subcutaneous tissue, they are particularly common around footpads and claws of mammals, and in the interosseus membranes bridging the bones of the leg and forearm, where they act as sensitive detectors of ground vibration.[23] A similar structure, the Herbst corpuscle, is found in the legs, bills, and cutaneous tissue of birds (and in the tongues of woodpeckers!). Speculation about their physiological function includes detection by the duck's bill of aquatic vibrations due to small prey, and in soaring birds, detection of vibration of flight feathers due to improper aerodynamic trim.[24]

The mechanism of adaptation in the Pacinian corpuscle was studied in detail by Werner Loewenstein and his associates, who showed that it was due, in part, to the dynamic properties of the capsule.[25] When a mechanical pulse was applied to an isolated, intact corpuscle, a brief receptor potential appeared at the onset and withdrawal of the

[16]Matthews, P. B. C. 1981. *J. Physiol.* 320: 1–30.

[17]Fukami, Y., and Hunt, C. C. 1977. *J. Neurophysiol.* 40: 1121–1131.

[18]Nakajima, S., and Takahashi, K. 1966. *J. Physiol.* 187: 105–127.

[19]Nakajima, S., and Onodera, K. 1969. *J. Physiol.* 200: 187–204.

[20]Sokolove, P. G., and Cooke, I. M. 1971. *J. Gen. Physiol.* 57: 125–163.

[21]Erxleben, C. F. 1993. *Neuroreport* 4: 616–618.

[22]Bell, J., Bolanowski, S., and Holmes, M. H. 1994. *Prog. Neurobiol.* 42: 79–128.

[23]Quilliam, T. A., and Armstrong, J. 1963. *Endeavour* 22: 55–60.

[24]McIntyre, A. K. 1980. *Trends Neurosci.* 3: 202–205.

[25]Loewenstein, W. R., and Mendelson, M. 1965. *J. Physiol.* 177: 377–397.

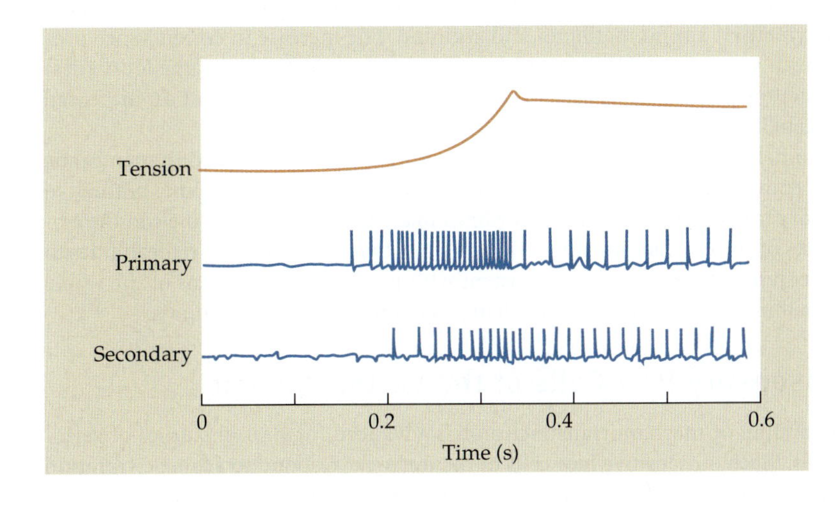

FIGURE 17.6 Differences in Muscle Spindle Responses. Recordings of action potentials from single primary (group Ia) and secondary (group II) sensory afferent fibers originating in a cat muscle spindle. The primary fiber greatly increases its discharge rate as tension develops during the stretch; during the maintained phase of the stretch, it quickly adapts to a lower rate. The secondary fiber increases its firing rate more slowly as tension develops and maintains its discharge during the steady stretch. (After Jansen and Matthews, 1962.)

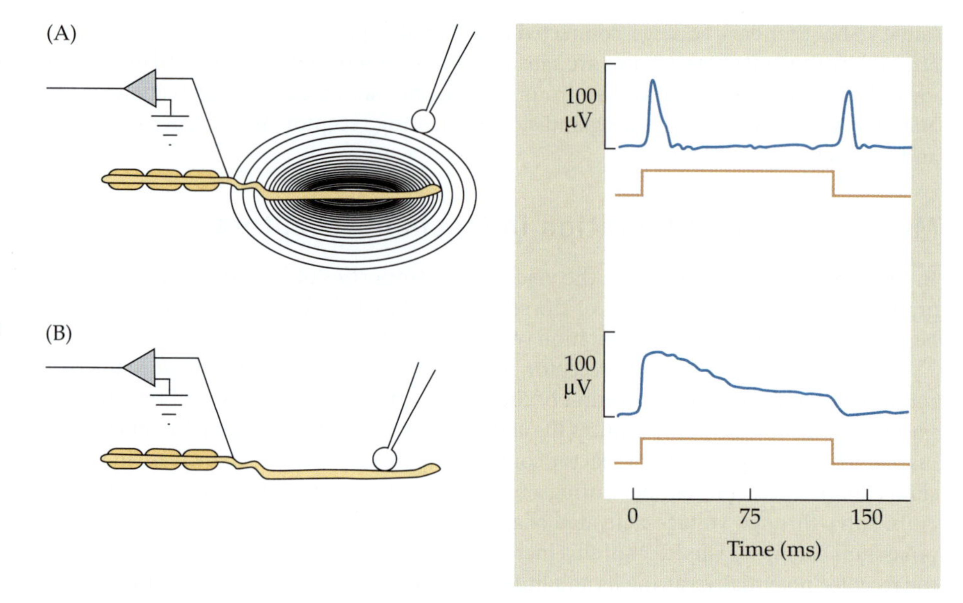

FIGURE 17.7 Adaptation in a Pacinian Corpuscle. (A) A pressure step applied to the body of the corpuscle (lower trace) produces a rapidly adapting receptor potential (upper trace), as a result of a transient wave of deformation that travels through the capsule to the nerve terminal. A similar response occurs on removal of the pulse. (B) After removal of the capsule layers, pressure applied to the nerve terminal produces a receptor potential that lasts for the duration of the pulse. (After Loewenstein and Mendelson, 1965.)

pulse (Figure 17.7A). The responses to sustained pressure are transient because compression of the sensitive ending is relieved by redistribution of fluid within the capsule. After the capsule was stripped carefully from the nerve ending, the receptor potential decayed only slowly during the pulse (Figure 17.7B). Nonetheless, even when the receptor potential was prolonged, there was still only a brief burst of action potentials in the afferent axon (not shown); that is, the properties of the axon itself are matched to those of the intact receptor.

TRANSDUCTION OF MECHANICAL STIMULI

It is apparent that there must be stretch-sensitive ion channels localized to mechanosensory nerve terminals, the sites of receptor potential generation. Furthermore, those transduction channels must differ from the channels that support the action potential, since responses to stretch continue in the presence of a local anesthetic that blocks conduction. Mechanosensitive ion channels are found in a wide variety of cell types and organs, including endothelial cells of blood vessels, baroreceptors in the carotid sinus, touch and pressure receptors in the skin, muscle stretch receptors, and mechanosensitive hair cells of the inner ear.[26–28]

By voltage clamp it has been shown that the current underlying the receptor potential in crayfish stretch receptors is associated with an increase in permeability to sodium and potassium,[29,30] as well as to divalent cations[31] and to larger organic cations such as tris (tris [hydroxymethyl] amino methane) and arginine. The increase in conductance produced by stretch is unaffected by tetrodotoxin[32] but can be altered by some local anesthetics.[33] Receptor potentials in vertebrate muscle spindles also depend on increased cation permeability.[34]

Single channels activated by membrane distortion were first observed in membrane patches from embryonic chick muscle cells[35] and other cell membranes having nothing to do with sensory transduction.[36] Patch recordings made from similar channels in the primary dendrites of the crayfish stretch receptor[37] show that their relative permeabilities to sodium, potassium, and calcium are consistent with previous observations of the whole cell. How membrane deformation causes channels to open is largely unknown.

Mechanosensory Hair Cells of the Vertebrate Ear

Our understanding of mechanotransduction has advanced furthest in studies of vertebrate hair cells. Mechanosensitive **hair cells** of the inner ear respond to acoustic vibration or head motion that causes fluid movements within the chambers of the inner ear. The

[26]Takahashi, M., et al. 1997. *J. Vasc. Res.* 34: 212–219.
[27]Ingber, D. E. 1997. *Annu. Rev. Physiol.* 59: 575–599.
[28]Garcia-Anoveros, J., and Corey, D. P. 1997. *Annu. Rev. Neurosci.* 20: 567–594.
[29]Brown, H. M., Ottoson, D., and Rydqvist, B. 1978. *J. Physiol.* 284: 155–179.
[30]Rydqvist, B., and Purali, N. 1993. *J. Physiol.* 469: 193–211.
[31]Edwards, C., et al. 1981. *Neuroscience* 6: 1455–1460.
[32]Nakajima, S., and Onodera, K. 1969. *J. Physiol.* 200: 161–185.
[33]Lin, J. H., and Rydqvist, B. 1999. *Acta Physiol. Scand.* 166: 65–74.
[34]Hunt, C. C., Wilkerson, R. S., and Fukami, Y. 1978. *J. Gen. Physiol.* 71: 683–698.
[35]Guharay, R., and Sachs, F. 1984. *J. Physiol.* 352: 685–701.
[36]Sachs, F. 1988. *Crit. Rev. Biomed. Eng.* 16: 141–169.
[37]Erxleben, C. 1989. *J. Gen. Physiol.* 94: 1071–1083.

exact form of the fluid movement depends on the shape and composition of the particular end organ involved. We discuss in Chapter 18 the frequency-specific vibration pattern of the membranes within the cochlear spiral, and the different roles of inner and outer hair cells in audition. For now it will suffice to point out that hair cells in the **cochlea** are stimulated by fluid movements within the acoustic frequency range—in humans, from 20 to 20,000 Hz (Box 17.1). The vestibular end organs of the inner ear are constructed quite differently and respond to the much lower frequencies generated during head movement. Mass loading of the saccule and utricle by an overlying **otolithic membrane** makes these epithelial sheets sensitive to linear acceleration. The hair cells in the semicircular canals are stimulated by angular acceleration during head rotation. Whatever the form of the fluid movement, in each case it causes deflection of a bundle of modified microvilli, or stereocilia, that project from the apical surface of the hair cell. Bundle deflection results directly in the opening of mechanosensitive ion channels.

Structure of Hair Cell Receptors

Hair cells and surrounding supporting cells form epithelial sheets that separate dissimilar fluid spaces of the inner ear. The basolateral membranes of hair cells are bathed by perilymph, similar in composition to ordinary extracellular fluid containing high sodium and low potassium (Figure 17.8). The apical, hair-bearing surface of the hair cell faces en-

FIGURE 17.8 The Mechanosensory Hair Cell. (A) Schematic drawing highlighting the functional specializations of the hair cell. A bundle of specialized microvilli called stereocilia projects from the cuticular plate into the endolymphatic space. In some hair cells a true cilium, the kinocilium, is found at one side of the hair bundle. Below the nucleus, hair cells form synapses with afferent and efferent neurons. Synaptic vesicles surround a dense body opposite the ending of an afferent neuron. Efferent neurons projecting from the brainstem form cholinergic synapses. Inside the hair cell a synaptic cistern lies in close apposition to the plasma membrane underlying the efferent contact. (B) Transmission electron micrograph of a hair cell from chick inner ear. The hair bundle was bent over during fixation. In this type of hair cell the cuticular surface is expanded. Synaptic contacts are made on the basal pole of the cell. (Micrograph kindly provided by R. Michaels.)

(A)

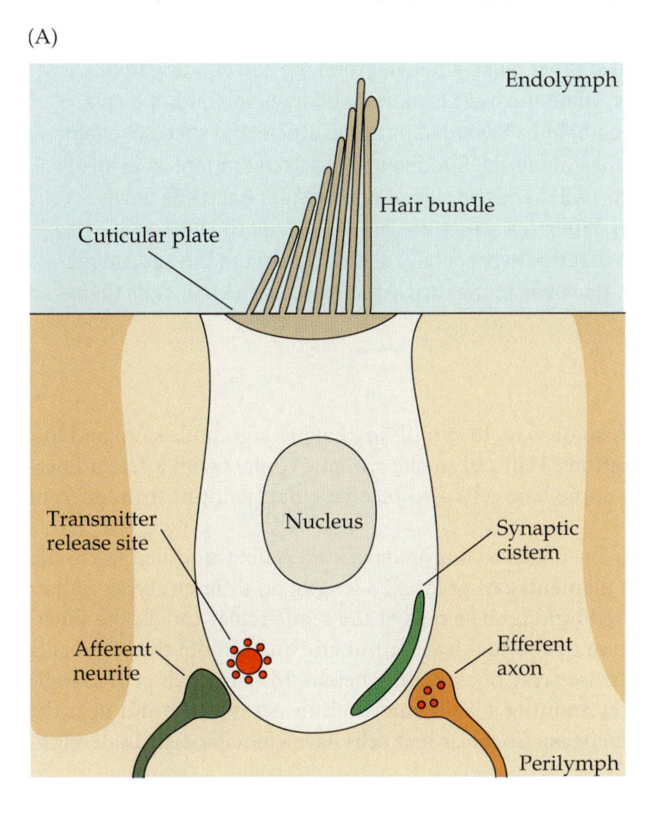

(B)

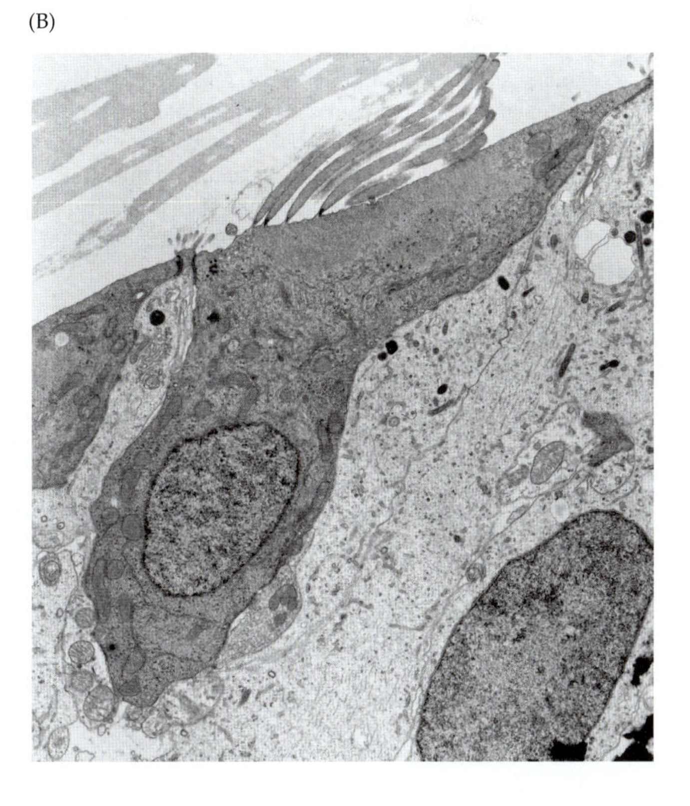

Box 17.1 Sensory Epithelia of the Inner Ear

The adequate stimulus for mechanosensory hair cells is determined by the structure of the end organ in which they reside. Cochlear hair cells respond to acoustic stimulation at frequencies ranging from 20 to 20,000 cycles/s (Hz) in humans. Cochlear frequency selectivity in mammals is determined largely by the mechanical properties of the **basilar membrane** on which the hair cells rest. This structure varies in stiffness along the coiled cochlear duct, so hair cells at different positions are subject to different amplitude motion depending on the frequency of acoustic stimulation. In contrast, the hair cells within vestibular epithelia signal head motion and position at much lower frequencies, and we consider here the features of these end organs that confer their particular sensitivities.

Changes of the orientation of the head in space are detected by the **vestibular apparatus**, an organ of extreme importance in maintaining balance, as well as orientation in relation to the visual world. When we walk, run, or swim, information is sent continuously from the vestibular apparatus to the CNS. This vestibular information, combined with sensory input from muscle receptors in the neck that tell us how the body is oriented in relation to the head, is essential for the maintenance of balance and position. Although we can make some quite rapid head movements, vestibular input usually occurs at a few cycles per second or less.

Shown in part A of the figure are the five components of the vestibular system, as seen from the front. The three **semicircular canals** are filled with fluid that, because of its inertia, flows within them when the head is rotated. The canals are oriented in three planes more or less at right angles to one another. The anterior canal, A, is in a vertical plane angled forward to about 45° off the midline; the posterior canal, P, is angled backward the same amount. (The relations of the planes to the head are shown in the inset.) The horizontal canal, H, is in a plane that is nearly horizontal. The **utricle** (green) and **saccule** (red) are responsible for detection of linear acceleration.

The mechanism responsible for detection of rotation is illustrated in part B, which shows a cross section through the fluid-filled **ampulla** of the horizontal canal. When the head is rotated clockwise, the fluid, because of its inertia, moves in the reverse direction within the rotating canal (arrow). The fluid pushes on the gelatinous mass called the **cupula**, in the base of which are embedded the **stereocilia** of hair cells. When the stereocilia are deflected, ion channels open and the hair cells are depolarized. If the rotation is continued (as when one is sitting on a rotating stool), the fluid will catch up with the body rotation and then cease to move inside the canal. When the rotation is stopped, however, transient fluid movement in the reverse direction will occur, bending stereocilia in the opposite direction and causing hair cell hyperpolarization.

It can be seen that the system detects only changes in velocity of rotation (i.e., rotary acceleration). In the same way the anterior and posterior semicircular canals signal rotations of the head in any vertical plane. Furthermore, the horizontal canals in the two ears provide signals of opposite sign during head rotation, as do contralateral pairs of anterior and posterior canals.

Hair cells in the utricle and saccule respond to linear acceleration by a different mechanism. The utricle is oriented in a horizontal plane, and the saccule is oriented vertically (part C). As in the ampullae, hair cell cilia are embedded in an overlying gelatinous material. However, in the utricle and saccule this material also contains crystals of calcium carbonate called **otoconia** ("ear dust") or **otoliths** ("ear stones") that make it heavier than the surrounding fluid. Thus, when the head is moved suddenly forward, the cilia will be bent backward in both the utricle and saccule, again because of inertia. Sudden upward acceleration, as in an elevator, will bend the cilia of the saccular hair cells downward. When the structures are moved, all the hair cells associated with them have their cilia bent in the same direction. However, in the utricle and saccule the hair cells them-

dolymph, a solution similar in some ways to cytoplasm, having high potassium and low sodium and calcium concentrations. Hair cells make synaptic contact with afferent fibers on their basolateral surfaces. Some hair cells also receive synaptic input from efferent neurons from the brainstem.

There are anywhere from a few dozen to hundreds of stereocilia (modified microvilli containing polymerized actin filaments), of graduated length, on different types of hair cells. The longest stereocilia are found on hair cells of the semicircular canals, the shortest in the high-frequency region of the cochlea. Within any one bundle the stereocilia occur in an organ pipe or staircase array of ascending height. In many hair cells a single true cilium, the kinocilium (containing a 9 + 2 microtubule array), is found near the middle of the tallest row of stereocilia. Cochlear hair cells have kinocilia early in develop-

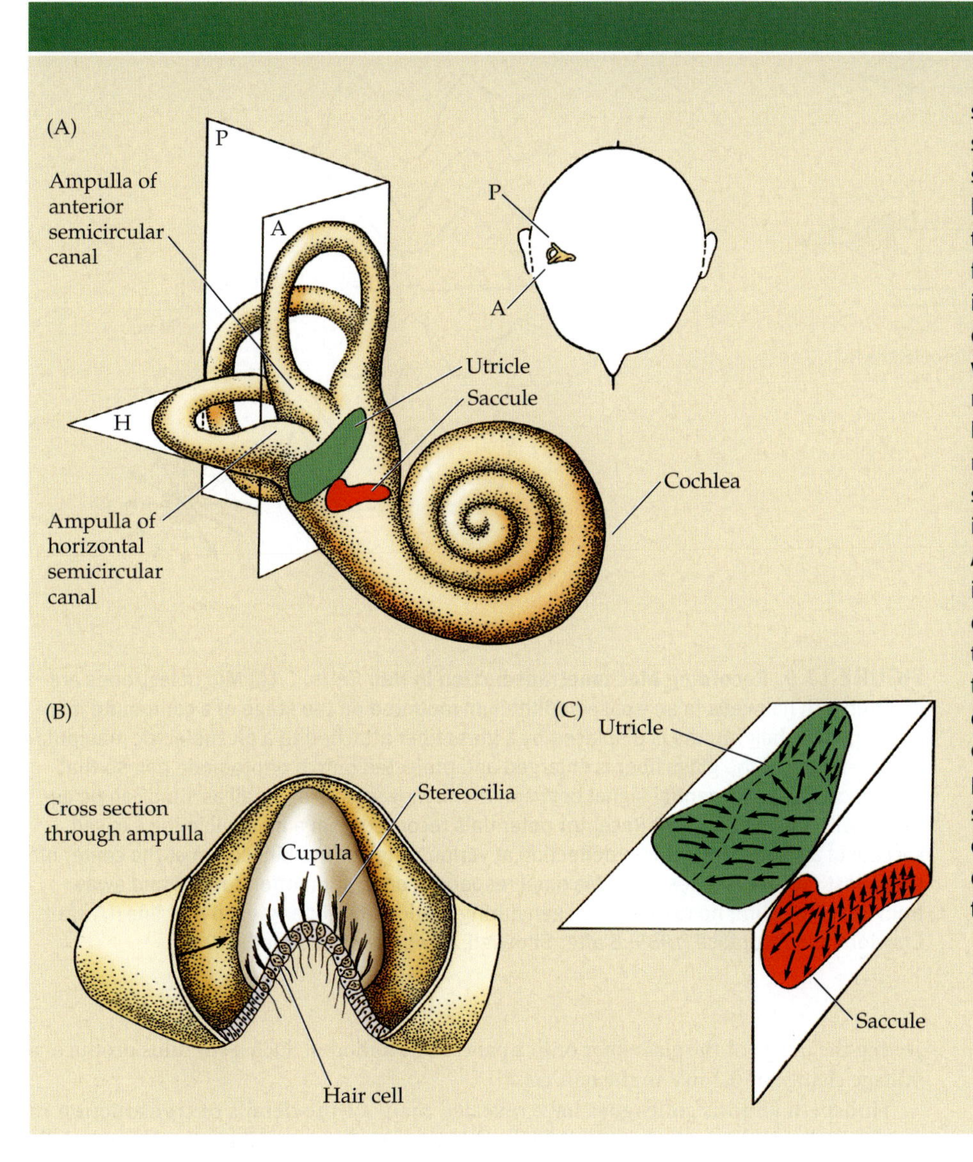

selves have different orientations, so a movement that depolarizes some will have no effect on, or will hyperpolarize, others. In this way the direction of movement is defined. Hair cell orientation varies in a systematic manner, as indicated by the arrows in part C, which show the direction of movement that causes depolarization of hair cells in each region. As with rotary movement, the system responds only to change in velocity—that is, to acceleration. Although the acceleration of gravity exerts a static force on hair cells of the saccule, this is not detected because the hair cells themselves adapt quickly. As a consequence, in the absence of other clues we have no internal perception of up or down, and scuba divers, for example, must observe the direction of movement of rising bubbles for such information.

ment, but these disappear later. The stereocilia narrow to insert into a cuticular plate. During bundle deflection the stereocilia behave like rigid rods and bend at this insertion point.[38] A variety of lateral linkages cause the assembly of stereocilia to move as a unified hair bundle.

Transduction by Hair Bundle Deflection

It has been known for a number of years that electrical responses in hair cells are produced by deformation of the hair bundle;[39,40] however, direct experimental confirmation required the development of sensitive techniques for producing and measuring very small movements while recording from hair cells. Auditory and vestibular epithelia from cold-blooded vertebrates such as turtles and frogs have proved particularly advantageous for these experiments. Procedures used by Crawford and Fettiplace to stimulate hair cells in the turtle's basilar papilla (the auditory epithelium analogous to the organ of Corti in mammals) are shown in Figure 17.9A. The basilar papilla was dissected from the inner ear and mounted in saline on the stage of a compound microscope. A microelectrode records the hair cell's membrane potential while a glass fiber attached to a piezoelectric manipulator pushes the hair bundle. Movements as small as 1 nm can be detected by pro-

[38] Flock, A., Flock, B., and Murray, E. 1977. *Acta Otolaryngol. (Stockh.)* 83: 85–91.

[39] Loewenstein, O., and Wersall, J. 1959. *Nature* 184: 1807–1808.

[40] Flock, A. 1965. *Cold Spring Harb. Symp. Quant. Biol.* 30: 133–145.

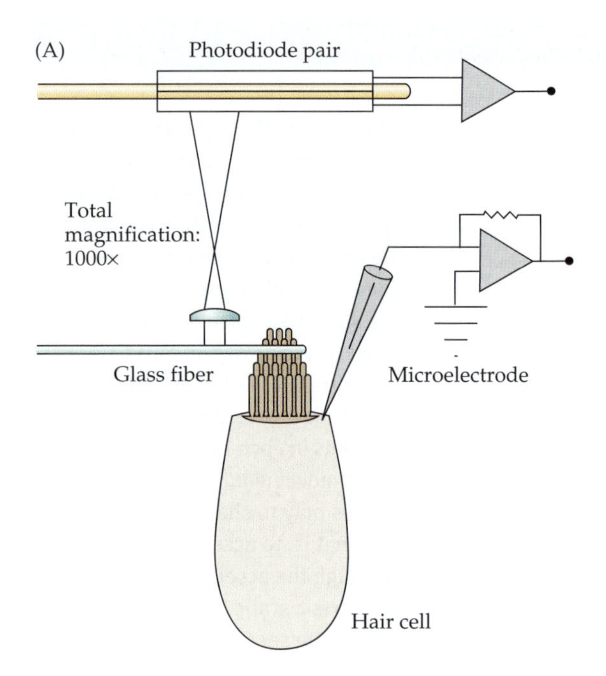

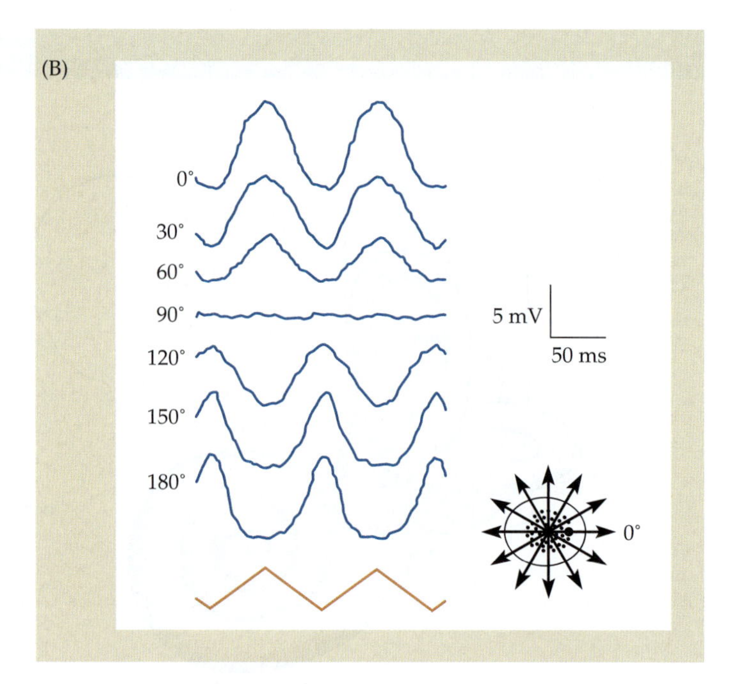

FIGURE 17.9 Recording Mechanotransduction in Hair Cells. (A) Microelectrodes are inserted into hair cells in an excised epithelium mounted on the stage of a compound microscope. The hair bundle is displaced by a glass fiber attached to a piezoelectric manipulator. The image of the glass fiber is enlarged and projected onto a photodiode pair so that motion causes a differential signal between them. Movements as small as 1 nm can be detected with this method. (B) Receptor potentials recorded from a hair cell in the excised saccule of a frog during bundle deflection at various angles. The kinocilium at the center of the tallest row of hairs lies at 0°. Maximal responses occur for motion toward and away from the kinocilium; no response is seen during motion at right angles to that line. (A after Crawford and Fettiplace, 1985; B after Shotwell, Jacobs, and Hudspeth, 1981.)

jecting the image of the glass fiber onto a pair of photodiodes. Such a stimulus produces a voltage change of 0.2 mV in the hair cell.[41]

Hudspeth and his colleagues have revealed many of the details of transduction in vestibular hair cells of the frog.[42–44] In one series of experiments they demonstrated directly the functional orientation of the hair bundle by varying the direction of stimulation with a piezoelectric manipulator. Deflections toward the kinocilium depolarized the cell, while movement away resulted in hyperpolarization. Bundle deflection perpendicular to that axis caused no change in membrane current.[45] The results of such an experiment are seen in Figure 17.9B, where the magnitude of the voltage change generated in the hair cell varies with the angle of bundle deflection.

Tip Links and Gating Springs

What structural feature of the hair bundle might underlie the directional selectivity of transduction? Pickles and colleagues used the scanning electron microscope to describe a unique class of extracellular linkages connecting the top of one hair with the side of the adjacent taller hair.[46] These **tip links** (Figure 17.10A) were observed only along the axis of mechanical stimulation (i.e., oriented up and down the staircase). The position of the tip links suggested that they might be involved in mechanotransduction, and treatments that break the tip link abolish transduction.[47,48] Indeed, extracellular recordings indicated that the channels activated by mechanical stimuli are located near the top of the hair bundle.[49–51]

Quantitative measures of transduction and the identification of tip links are combined in the **gating spring** hypothesis of mechanotransduction in hair cells. Deflection of the hair bundle in the "positive" direction (toward the taller hairs) separates the tips and stretches the gating spring, thus pulling open the transduction channel's gate (Figure 17.10B). When

[41]Crawford, A. C., and Fettiplace, R. 1985. *J. Physiol.* 364: 359–379.
[42]Hudspeth, A. J., and Corey, D. P. 1977. *Proc. Natl. Acad. Sci. USA* 74: 2407–2411.
[43]Hudspeth, A. J., and Jacobs, R. 1979. *Proc. Natl. Acad. Sci. USA* 76: 1506–1509.
[44]Corey, D. P., and Hudspeth, A. J. 1979. *Nature* 281: 675–677.
[45]Shotwell, S. L., Jacobs, R., and Hudspeth, A. J. 1981. *Annu. N.Y. Acad. Sci.* 374: 1–10.
[46]Pickles, J. O., Comis, S. D., and Osborne, M. P. 1984. *Hear. Res.* 15: 103–112.
[47]Crawford, A. C., Evans, M. G., and Fettiplace, R. 1991. *J. Physiol.* 434: 369–398.
[48]Assad, J. A., Shepherd, G. M. G., and Corey, D. P. 1991. *Neuron* 7: 985–994.
[49]Hudspeth, A. J. 1982. *J. Neurosci.* 2: 1–10.
[50]Jaramillo, F., and Hudspeth, A. J. 1991. *Neuron* 7: 409–420.
[51]Lumpkin, E. A., and Hudspeth, A. J. 1995. *Proc. Natl. Acad. Sci. USA* 92: 10297–10301.

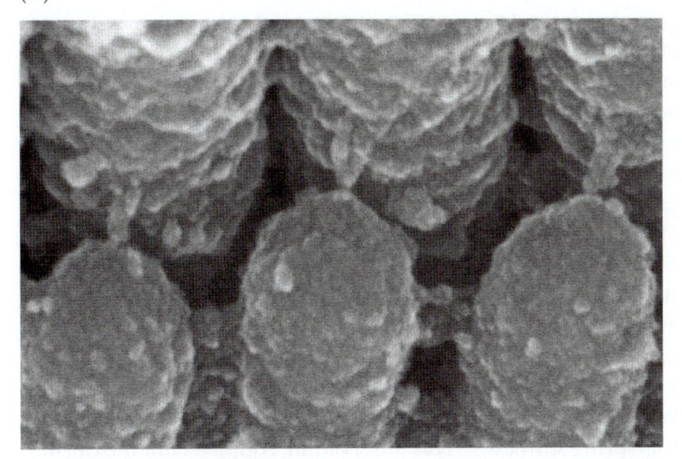

(B)

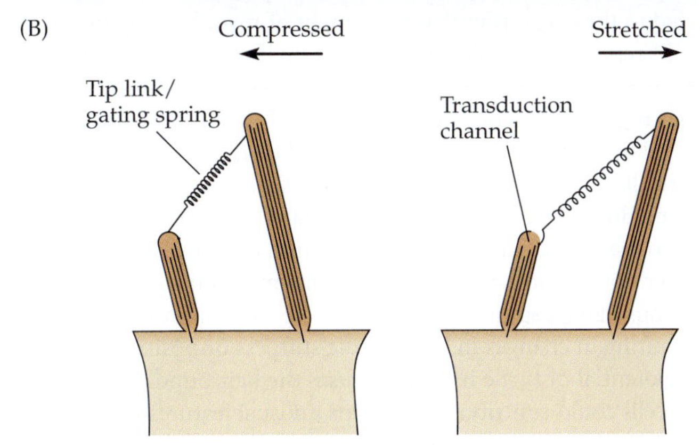

FIGURE 17.10 Tip Links on Hair Cell Stereocilia. (A) Scanning electron micrograph showing extracellular fibers that run from the tips of shorter stereocilia to the sides of adjacent taller stereocilia. In some cases these appear to bifurcate. (B) Tip links are positioned so that deflection of the hair bundle in the excitatory direction stretches the tip link and pulls open the transduction channel (right), while the opposite motion compresses the tip link (left). The orientation of the tip link is appropriate to serve as a gating spring that pulls directly on the gate of the mechanical transduction channel. (Micrograph kindly provided by D. Furness.)

the bundle is pushed away from the tallest hairs, the spring is compressed and the channels close. Although such a scheme might seem somewhat fanciful, a direct physical connection between bundle mechanics and channel gating is required by the great speed at which transduction occurs in hair cells, with time constants of opening of about 40 μs.[52,53] Further, the energetics and mechanics of transduction are consistent with this model. For example, it is possible to measure a decrease in bundle stiffness as the transduction channels open, as though this molecular motion relieves some tension on the gating spring.[54]

Transduction Channels in Hair Cells

What types of channels are opened in the tips of hair cells? These appear to be nonselective cation channels that have considerable calcium permeability. The single-channel conductance is about 100 pS.[53,55,56] From this and measurements of total transducer current it is possible to calculate that each hair cell has only about 100 transduction channels. This corresponds to only a few channels per stereocilium!

The very small numbers of channels in each hair cell makes biochemical and molecular biological investigation especially challenging. Hair cell transduction channels have no intrinsic voltage dependence, nor are they ligand gated in any traditional sense. It is unlikely, therefore, that strong homologies exist between these and other classes of gated ion channels whose genes have been described. However, a growing number of mechanosensitive ion channels are being cloned from bacteria, yeast, nematodes, and flies,[27,28] so candidate genes are likely to be identified in the near future.

Adaptation of Hair Cells

Hair cells are enormously sensitive, with threshold responses to hair bundle motions of less than 10^{-9} m. It seems likely, then, that some type of adaptive process exists to restore sensitivity in the face of a "background" stimulus. For example, vestibular hair cells in the

[52]Corey, D. P., and Hudspeth, A. J. 1983. *J. Neurosci.* 3: 962–976.

[53]Crawford, A. C., Evans, M. G., and Fettiplace, R. 1989. *J. Physiol.* 419: 405–434.

[54]Howard, J., and Hudspeth, A. J. 1988. *Neuron* 1: 189–199.

[55]Ohmori, H. 1985. *J. Physiol.* 359: 189–217.

[56]Denk, W., et al. 1995. *Neuron* 15: 1311–1321.

saccule and utricle must remain sensitive to subtle head movements while continuously being subjected to gravitational force acting on the overlying otolithic membrane (see Box 17.1). Indeed, it can be shown by direct measurement that during prolonged displacement, hair cells do adapt.

Figure 17.11 shows the result of an experiment in which a steady deflection of 0.65 μm was imposed on the hair bundle of a turtle hair cell whose membrane potential was set to different levels in voltage clamp. At –72 mV, the inward transduction current first rose to a peak of 100 pA, then decayed with a time constant of approximately 10 ms. This adaptive change depends on the influx of calcium ions through the open transduction channels. When the membrane potential was changed to +60 mV, the now outward transduction current (the reversal potential is 0 mV) was well maintained and adaptation was abolished. At +60 mV, the driving force on calcium is small and little calcium enters the cell. In other experiments, adaptation also was eliminated by the removal of calcium from the extracellular solution, or by intracellular injection of strong calcium buffers.[57]

In some hair cells, adaptation is accompanied by a change in bundle stiffness that suggests a "resetting" of gating-spring stiffness.[58] This observation, coupled with the calcium dependence of adaptation, led to the suggestion that movement of myosin along the actin core of the stereocilium might reposition the site at which the tip link is attached, altering bundle stiffness.[59] Myosins have been cloned from the inner ear,[60] and specific antibodies have been used to show that myosin Iβ (a nonmuscle form) is located near the tips of stereocilia in frog hair cells.[61] This relatively slow, myosin-based adaptation may be most prominent in vestibular hair cells. A more rapid form of adaptation in auditory hair cells arises from calcium ions acting directly on the transduction channel, causing it to close.[62]

The tight coupling between mechanical input and transduction channel gating has important consequences for inner ear function. Such tight coupling implies that feedback will occur during transduction. Thus, the receptor potential alters the calcium influx (through changes in driving force), resulting in changes in hair bundle stiffness or position. Indeed, alteration of the membrane potential of turtle hair cells causes the hair bundle to move.[41] The active mechanics of hair cells could contribute to various unusual features of audition, including sounds (called otoacoustic emissions) that can be elicited from the ears of most species, including humans.[63] The ability to elicit emissions from the ear has provided a way for audiologists to assess hair cell function directly, even in infants or comatose patients.[64]

[57]Ricci, A. J., Wu, Y-C., and Fettiplace, R. 1998. *J. Neurosci.* 18: 8261–8277.

[58]Hacohen, N., et al. 1989. *J. Neurosci.* 9: 3988–3997.

[59]Hudspeth, A. J., and Gillespie, P. G. 1994. *Neuron* 12: 1–9.

[60]Solc, C. K., et al. 1994. *Auditory Neurosci.* 1: 63–75.

[61]Gillespie, P. G., Wagner, M. C., and Hudspeth, J. A. 1993. *Neuron* 11: 581–594.

[62]Ricci, A. J., and Fettiplace, R. 1997. *J. Physiol.* 501: 111–124.

[63]Kemp, D. T. 1978. *J. Acoust. Soc. Am.* 64: 1386–1391.

[64]Lonsbury-Martin, B. L., et al. 1995. *Otolaryngol. Head Neck Surg.* 112: 50–63.

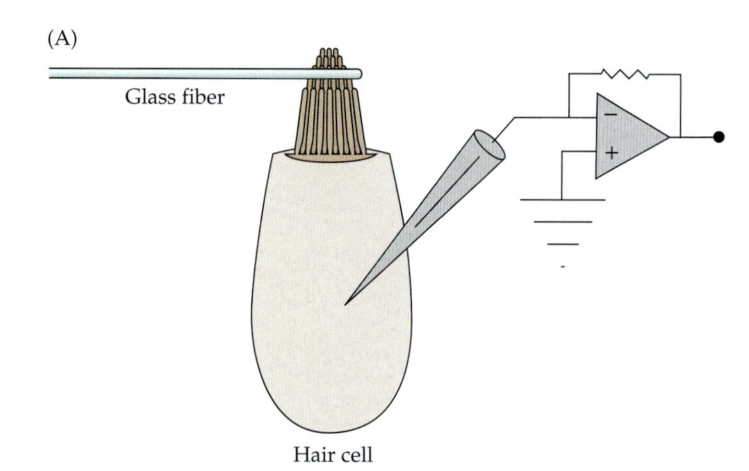

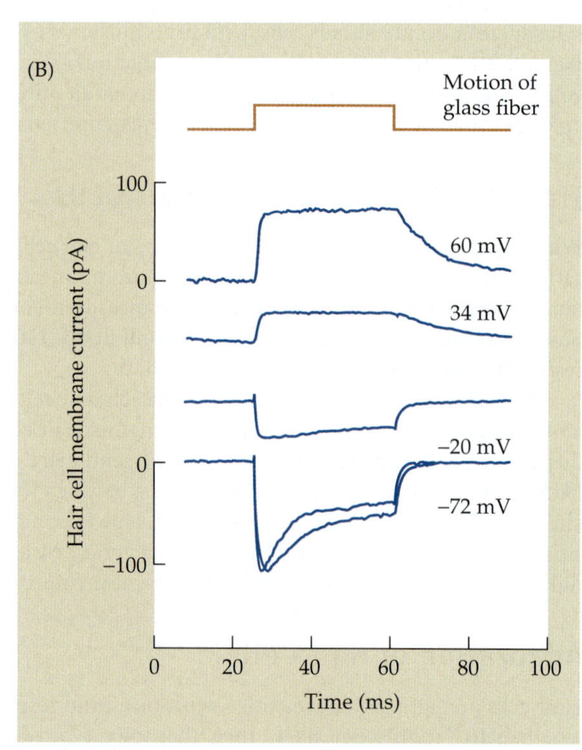

FIGURE 17.11 Adaptation of Transducer Current in Hair Cells. (A) A sustained deflection of the hair bundle was imposed with a piezoelectric manipulator, and transducer current was recorded by voltage clamp. (B) Membrane currents recorded from a hair cell during mechanotransduction. At negative membrane potentials the current was inward and decayed during the displacement. At positive membrane potentials the current was reversed in polarity and showed no decay; calcium influx is also reduced, so calcium-dependent adaptation is eliminated. (After Crawford, Evans, and Fettiplace, 1989.)

OLFACTION

Mechanotransduction in the ear attains high sensitivity by tightly coupling stimulus energy to the hair cell's membrane potential. In contrast, great sensitivity is obtained in olfaction and vision, and in some forms of taste, by chemical amplification—that is, second messenger pathways in which enzymatic cascades produce large numbers of intermediate products, thereby increasing by a thousandfold the effect of one activated receptor molecule.

Olfaction (smell) is poorly developed in humans compared to dogs or pigs or butterflies. But at the same time, considerable effort (and advertising time) goes into what might be considered human olfactory behavior (consider the numbers of soaps, deodorants, and perfumes that are aimed at securing a socially acceptable personal bouquet). Olfactory signals are essential to human survival, stimulating feeding, reproduction, and mother–infant bonding.[65,66] In addition, olfactory stimuli have a powerful ability to evoke emotions and to elicit long-stored memories (perhaps an evolutionary legacy from the use of olfactory cues for homing, as in other species[67]). This haunting effect of olfactory memory was described by Marcel Proust in *Remembrance of Things Past*:

> When, from a long-distant past nothing remains.
> After the people are dead;
> after the things are broken and scattered.
> Still, alone, more fragile,
> but with more vitality,
> more insubstantial,
> but more persistent, more faithful,
> the smell and taste of things remain poised forever,
> like ghosts, ready to remind us . . .

Detection and discrimination of the unique blend of individual odors linked to those memories begins with a family of molecular receptors in the olfactory receptor neurons.

[65]Stern, K., and McClintock, M. K. 1998. *Nature* 392: 177–179.

[66]Kelly, D. R. 1996. *Chem. Biol.* 8: 595–602.

[67]Bingman, V. P., and Benvenuti, S. 1996. *J. Exp. Zool.* 276: 186–192.

[68]Pevsner, J., et al. 1988. *Science* 241: 336–339.

[69] Farbman, A. I. 1994. *Semin. Cell. Biol.* 5:3–10.

[70]Adrian, E. D. 1953. *Acta Physiol. Scand.* 29: 5–14.

[71]Ottoson, D. 1956. *Acta Physiol. Scand.* 35(Suppl. 122): 1–83.

[72]Maue, R. A., and Dionne, V. E. 1987. *J. Gen. Physiol.* 90: 95–125.

Olfactory Receptors

Odors are detected in a patch of about 100,000 olfactory receptor neurons whose axons project through a thin portion of the frontal skull (the cribriform plate) to the olfactory bulb (Figure 17.12). The long cilia of the olfactory receptors extend into the nasal cavity, where they lie in a layer of mucus, approximately 50 μm thick in humans, that is entirely replaced every 10 min. The mucous layer protects the sensory epithelium, washing out potentially toxic airborne compounds, and all odorants must dissolve through this to reach the sensory cilia. An odorant-binding protein helps to concentrate hydrophobic odorants in this aqueous layer.[68] Olfactory receptors are unusual in that they are continuously replaced throughout the animal's lifetime. Each receptor lives for a month or two, and new receptors arise from a layer of basal cells in the olfactory epithelium.[69]

The Olfactory Response

Early measurements of olfactory responses were made by Adrian[70] and Ottoson.[71] Since then, evidence has accumulated that odorant molecules interact with receptors in the ciliary membrane to produce an increase in conductance resulting in depolarization. Action potentials then travel along the olfactory receptor axon into the central nervous system. Patch clamp techniques have been used to record odorant-induced currents from isolated olfactory cells,[72] and to record the precise time course and localization of the odor response.

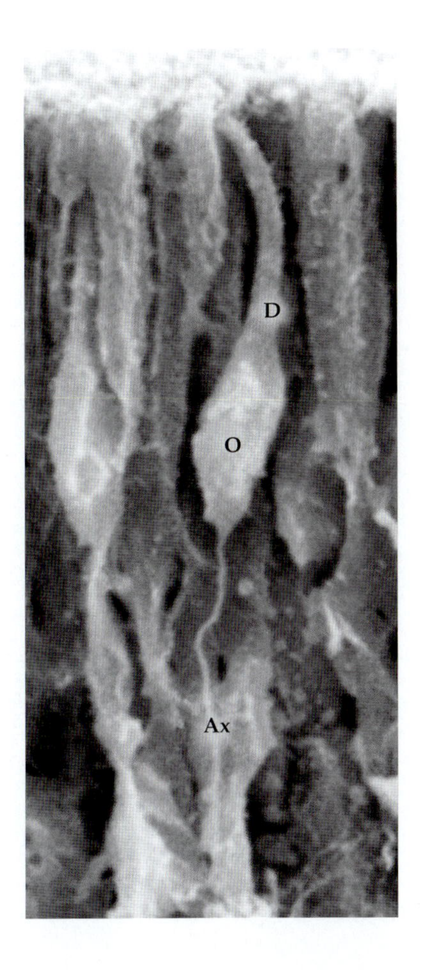

FIGURE 17.12 Olfactory Receptor Neurons. Scanning electron micrograph of the olfactory epithelium of a hamster. Each receptor neuron (O) has a long dendrite (D) that extends to the surface and an axon (Ax) that projects from the epithelium to the olfactory bulb. The long sensory cilia at the tip of the dendrite form a dense mat in this preparation and are not individually resolved. (Micrograph kindly provided by R. Costanzo.)

An example of such an experiment on a cell isolated from the olfactory mucosa of the salamander[73] is shown in Figure 17.13. The membrane potential of the cell is held at –65 mV, and a solution containing a mixture of odorant molecules (approximately 0.1 m*M*) in 100 m*M* KCl is applied from a second pipette by a brief (35 ms) pressure pulse—first to the soma, then to the distal portion of the dendrite and the cilia. Pipette solution applied to the soma produces a rapid inward current, due to the local increase in potassium concentration. The time course of the potassium response provides a measure of the speed of application and subsequent dissipation of the solution by diffusion into the surrounding bath. A second smaller and slower inward current appears when the odorants reach the apical dendrite.

Solution applied to the apical dendrite and cilia produces only a small potassium response; presumably there are fewer potassium channels in that portion of the cell. However, the odorant itself produces a large inward current that outlasts the time of application by several seconds. The experiment clearly indicates that the region of sensitivity to the odorants is the distal dendrite and cilia, and the prolonged time course of the dendritic response is consistent with the idea that the conductance change is produced by a second messenger system.

Cyclic Nucleotide–Gated Channels in Olfactory Receptors

The depolarization produced by olfactants arises from the opening of nonselective cation channels[74] and by additional current flow through calcium-activated chloride channels.[75,76] Nonselective cation channels in olfactory receptors are opened by intracellular cAMP (Figure 17.14), and these channels are closely related to the cation channels opened by cGMP (cyclic guanosine monophosphate) in rod photoreceptors.[77] Like the rod photoreceptor channel (Chapter 19), the olfactory receptor channel is neither activated nor inactivated by changes in membrane potential. These channels are permeable to Na^+, K^+, and Ca^{2+} ions. Calcium entry may contribute to processes of adaptation. Only a few open cation channels are required to initiate action potentials because of the very high input resistance in olfactory receptors, suggesting that even single odorant molecules can be detected.[78]

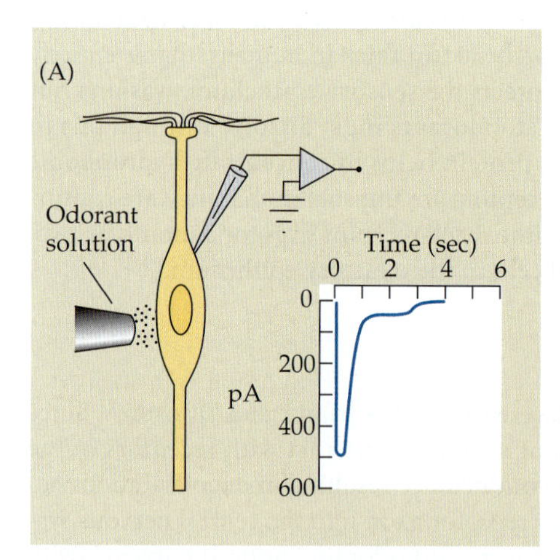

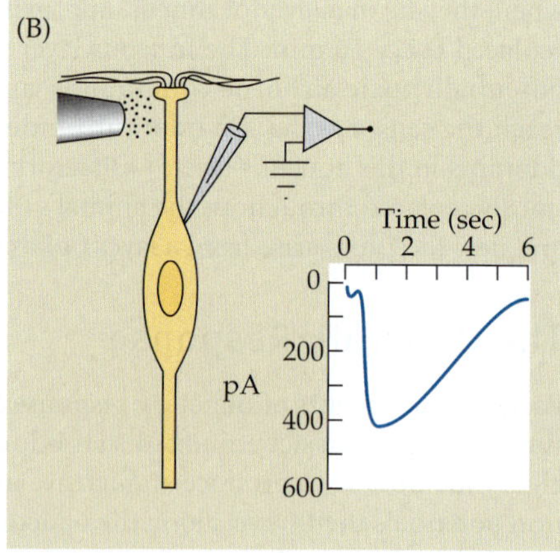

FIGURE 17.13 Responses of Isolated Olfactory Cells from the salamander. A patch clamp electrode is used to record whole-cell current. A solution containing 0.1 m*M* odorant mixture in 100 m*M* KCl is applied to the cell by a brief (35 ms) pressure pulse. (A) When solution is applied to the cell body, there is a rapid transient inward current due to the increased KCl concentration, followed by a smaller, slower current as the odorant reaches the apical end of the dendrite. The time course of the fast inward current is indicative of the time course of application and dissipation of the electrode solution. (B) When the solution is applied to the dendrite, there is only a small rapid current due to the KCl, but a large current, due to the odorant, lasts for several seconds after the electrode solution has washed away. (After Firestein, Shepherd, and Werblin, 1990.)

[73]Firestein, S., Shepherd, G. M., and Werblin, F. 1990. *J. Physiol.* 430: 135–158.

[74]Nakamura, T., and Gold, G. H. 1987. *Nature* 532: 442–444.

[75]Kleene, S. J., and Gesteland, R. C. 1991. *J. Neurosci.* 11: 3624–3629.

[76]Kurahashi, T., and Yau, K. Y. 1993. *Nature* 363: 71–74.

[77]Dhallan, R. S., et al. 1990. *Nature* 347: 184–187.

[78]Menini, A., Picco, C., and Firestein, S. 1995. *Nature* 373: 435–437.

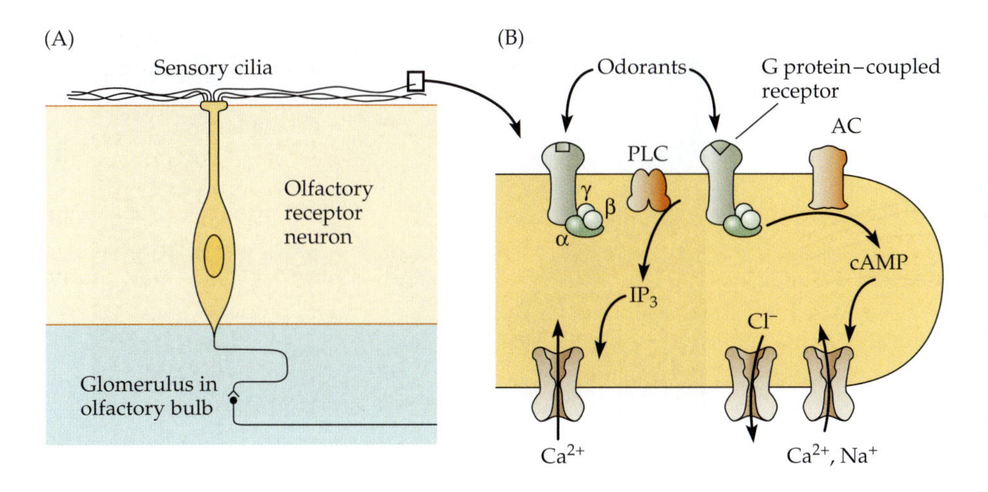

FIGURE 17.14 Transduction in Olfactory Cilia. (A) The molecular receptors for olfactants are found in sensory cilia that project into the mucous layer of the olfactory epithelium. Depolarizing receptor potentials in these long receptors give rise to action potentials that propagate along the olfactory receptor neuron's axon into the central nervous system. (B) Odorant molecules bind to specific G protein–coupled receptors in the plasma membrane of the olfactory cilia. This frees the α subunit to activate adenylyl cyclase (AC) and raise the concentration of cAMP, which causes nonselective cation channels to open, depolarizing the membrane. Calcium-gated chloride current can enhance this effect. Other pathways may involve the activation of phospholipase C (PLC) and the consequent rise in IP_3 to act directly on plasma membrane calcium channels.

Coupling the Receptor to Ion Channels

How is odorant binding coupled to the gating of cAMP-dependent cation channels? It is thought that molecular receptors coupled to G proteins first bind the odorant. The activated G protein releases the α subunit, which stimulates adenylyl cyclase to produce cAMP (see Figure 17.14). An extensive family of candidate odorant receptor genes has been identified,[79] and these encode seven-transmembrane, G protein–coupled proteins related in structure to metabotropic neurotransmitter receptors (Chapter 10). The greatest variability between encoded proteins occurs in the sequence of three of the transmembrane domains that form the ligand-binding pocket. A G protein specifically expressed in olfactory epithelia (G_{olf}) has been identified.[80] Release of $G_{olf\alpha}$ by odorant binding leads to increased activity in an olfactory adenylyl cyclase, which exists at high levels in olfactory cilia.[81,82]

The efficacy of this transduction pathway is illustrated by the rapid appearance of cAMP. Breer and colleagues[83] used a stop-flow apparatus to show that a 10-fold increase in cAMP concentration occurred within 50 ms of odorant application in a preparation of isolated olfactory cilia. There is also evidence that olfactory neurons use G protein activation of phospholipase C and production of IP_3 (inositol trisphosphate) in transduction.[84] In this case IP_3 may act directly to open calcium channels in the plasma membrane.[85] IP_3 appears to be especially important for invertebrate olfaction.[86] The role of IP_3 in mammalian olfaction is less clear, as suggested by the finding that transgenic mice lacking cAMP-gated channels lost all ability to discriminate odors.[87]

Odorant Specificity

Humans can discriminate a very large number of odors, and the existence of hundreds, or possibly thousands, of olfactory receptor molecules seems to provide a substrate for this capability. A remaining difficulty is the lack of odorant specificity of individual olfactory receptor neurons, each of which recognize a spectrum of odors rather than being highly selective.[88] One approach to further understanding of this issue has been to determine the expression pattern of cloned receptor molecules among olfactory receptor neurons using in situ hybridization. Each particular odorant receptor was found in a restricted area of the olfactory epithelium.[89,90] Different families of receptor genes appear to be expressed in zones extending along the length of the epithelium (Figure 17.15), with any given gene expressed in only a small number of olfactory receptor neurons.

In addition to the main olfactory epithelium, mammals possess a vomeronasal organ that is involved in the detection of pheromones that stimulate mating and other behaviors. Vomeronasal receptor neurons project to an accessory olfactory bulb, which in turn projects to the limbic system. The neurons express two additional families of olfactant receptors. These are also G protein–coupled receptors, but they are distinct from those encoded in the main olfactory epithelium. Each vomeronasal neuron may express just one type of molecular receptor, and the pattern of expression differs between male and female rats.[91]

[79]Buck, L., and Axel R. 1991. *Cell* 65: 175–187.

[80]Jones, D. T., and Reed, R. R. 1989. *Science* 244: 790–796.

[81]Pace, U., et al. 1985. *Nature* 316: 255–258.

[82]Bakalyar, H. A., and Reed, R. R. 1990. *Science* 250: 1403–1406.

[83]Breer, H., Boekhoff, I., and Tareilus, E. 1990. *Nature* 345: 65–68.

[84]Boekhoeff, I., et al. 1990. *EMBO J.* 9: 2453–2458.

[85]Restrepo, D., et al. 1990. *Science* 249: 1166.

[86]Ache, B. W., and Zhainazarov, A. 1995. *Curr. Opin. Neurobiol.* 5: 461–466.

[87]Brunet, L. J., Gold, G. H., and Ngai, J. 1996. *Neuron* 17: 681–693.

[88]Sicard, G., and Holley, A. 1984. *Brain Res.* 292: 283–296.

[89]Ressler, K. J., Sullivan, S. L., and Buck, L. B. 1993. *Cell* 73: 597–609.

[90]Vassar, R., Ngai, J., and Axel, R. 1993. *Cell* 74: 309–318.

[91]Herrada, G., and Dulac, C. 1997. *Cell* 90: 763–773.

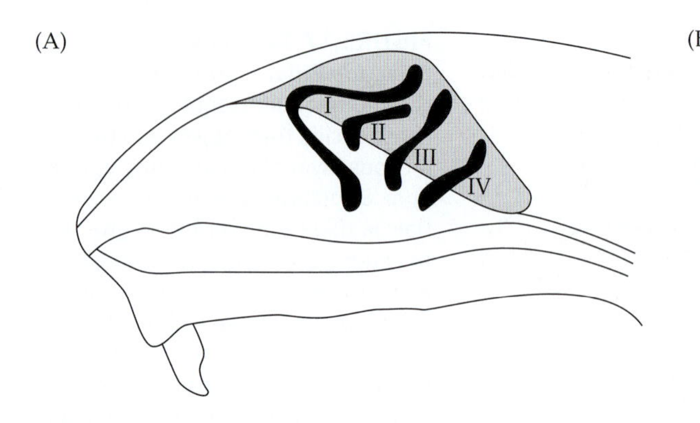

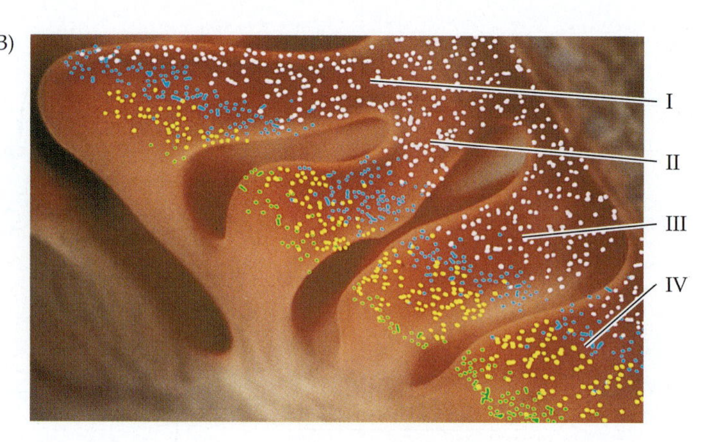

FIGURE 17.15 Expression of Specific Odorant Receptor Genes by Subsets of Olfactory Receptor Neurons. (A) The olfactory epithelium of a rat lies on a series of convolutions of the nasal cavity called turbinates—labeled I to IV. (B) Bands of olfactory receptor neurons label positively for olfactory receptor mRNAs. Probes for different mRNAs label nonoverlapping populations of neurons (depicted as green, yellow, blue, and white dots). (After Vassar, Ngai, and Axel, 1993; B kindly provided by R. Vassar.)

MECHANISMS OF TASTE (GUSTATION)

Discussions of taste and smell are often combined because both senses are activated by chemical stimuli arriving from the outside world. Indeed, some tastants (taste stimuli) act on G protein–coupled receptors in ways quite similar to those discussed for olfaction. However, other tastants, principally salts and acids, act directly on membrane conductances, and taste receptor cells differ anatomically from olfactory receptor neurons.

Taste Receptor Cells

Taste receptors are ciliated neuroepithelial cells found in taste buds on the tongue surface (Figure 17.16). Like olfactory receptors, taste cells are regenerated throughout life. Unlike olfactory receptors, taste cells do not have axons, but form chemical synapses with afferent neurites in the taste bud. Microvilli project from the apical pole of the taste cell into the open pore of the taste bud, where they come into contact with tastants dissolved in saliva on the tongue's surface.

Taste stimuli are usually subdivided into five categories: salt, sour, bitter, sweet, and umami—this last being a Japanese word for the taste of monosodium glutamate (MSG), or more generally, amino acid taste. This variety of tastant molecules, from simple ions to

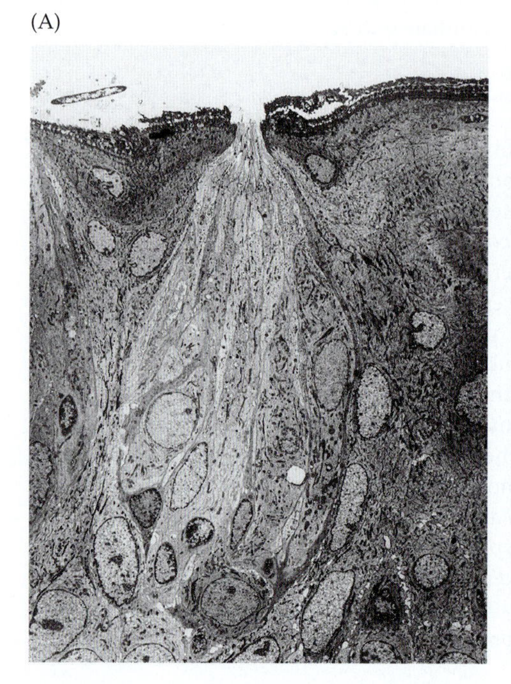

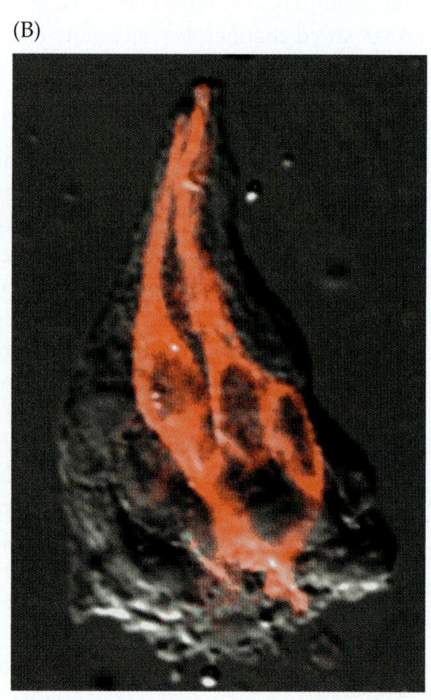

FIGURE 17.16 Taste Receptor Cells Are Found in Taste Buds within the lingual epithelium. (A) Transmission electron micrograph of a taste bud in rat tongue. Individual taste receptor cells have microvilli that project into the taste pore to sample the saliva. (B) An individual taste bud dissected from rat tongue. The taste receptor cells (red) are labeled with an antibody to gustducin, a G protein involved in taste transduction. (A kindly provided by R. Yang and J. Kinnamon; B kindly provided by I. Wanner and S. D. Roper.)

complex carbohydrates and proteins, is coupled with a wide variety of transduction mechanisms. These fall into two broad categories: direct action of the tastant on ion channels, and mediation of tastant reception through second messenger pathways, often involving G proteins.[92]

Salt Taste and Sour Taste

There is good agreement that the taste of salt is mediated by direct flux of sodium (or other monovalent cations) through channels in the apical membrane of the taste cell that are open at rest (Figure 17.17A).[93] Sodium is present at higher concentration in salty foods (>100 m*M*) than in saliva, and it simply diffuses down its electrochemical gradient into taste cells. The resulting depolarization leads to transmitter release from chemical synapses made by the taste cell onto afferent neurites. Taste cell sodium channels are not voltage gated, and they are similar to epithelial sodium channels found in frog skin and kidney. These sodium channels are blocked by the diuretic compound amiloride and consist of three homologous subunits.[94] The α subunit has been detected in lingual epithelium.[95]

Sour taste is caused by the high concentration of protons in acidic foods, which may enter taste cells through the amiloride-blockable channels (Figure 17.17B).[96] Another mechanism results from the blockade of K^+ channels by protons, leading to depolarization.[97] Frog taste cells have cation channels that are opened by protons,[98] also leading to depolarization. In addition to acting at the taste cell's cilia, salts and protons may percolate through the taste pore (a paracellular pathway) to act on the same or other ion channels (including some that are amiloride-insensitive) in the basolateral membranes of the cell.[99] This illustrates what seems to be a general principle of gustation: Several parallel transduction pathways can exist for any one class of tastants.

Sweet Taste and Bitter Taste

Sweet substances (sugars) and bitter substances (often plant alkaloids; the prototype is quinine) tend to be large molecules that are bound by macromolecular receptors with high degrees of specificity. With few exceptions, sweet and bitter tastants activate second messenger pathways through interaction with G protein–coupled receptors (Figure 17.17C and D). Only a few of these macromolecular taste receptors have been identified (see the

[92]Kinnamon, S. C., and Margolskee, R. F. 1996. *Curr. Opin. Neurobiol.* 6: 506–513.

[93]Avenet, P., and Lindemann, B. 1991. *J. Membr. Biol.* 124: 33–41.

[94]Canessa, C. M., et al. 1994. *Nature* 367: 463–467.

[95]Li, X-J., Blackshaw, S., and Snyder, S. H. 1994. *Proc. Natl. Acad. Sci. USA* 91: 1814–1818.

[96]Gilbertson, T. A., Roper, S. D., and Kinnamon, S. C. 1993. *Neuron* 10: 931–942.

[97]Kinnamon, S. C., Dionne, V. E., and Beam, K. G. 1988. *Proc. Natl. Acad. Sci. USA* 85: 7023–7027.

[98]Okada, Y., Miyamoto, T., and Sato, T. 1994. *J. Exp. Biol.* 187: 19–32.

[99]Lindemann, B. 1996. *Physiol. Rev.* 76: 718–766.

FIGURE 17.17 Mechanisms of Taste Transduction. Tastant molecules range from protons (acids) to simple salts to complex organic compounds. This wide range of chemical stimuli is transduced by a multiplicity of mechanisms. Salts (A) and acids (B) can permeate ion channels in the sensitive ending or block normally open potassium channels. (C) Some bitter compounds block potassium channels to cause depolarization. (D) Sugars and amino acids (umami) interact with G protein–coupled receptors to initiate second messenger cascades. All these mechanisms lead eventually to depolarization, voltage-gated calcium influx, and increased release of transmitter onto associated afferent dendrites.

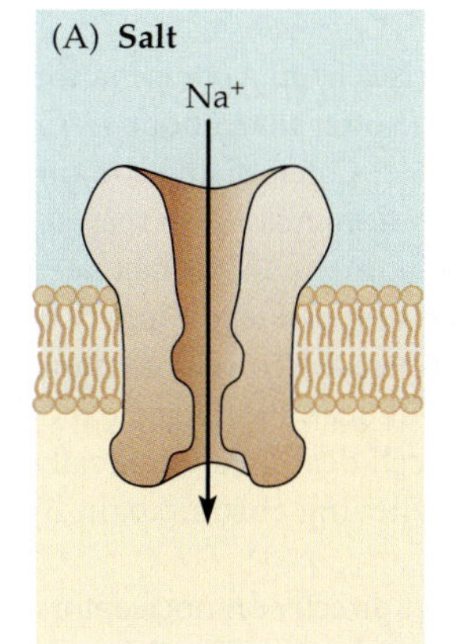

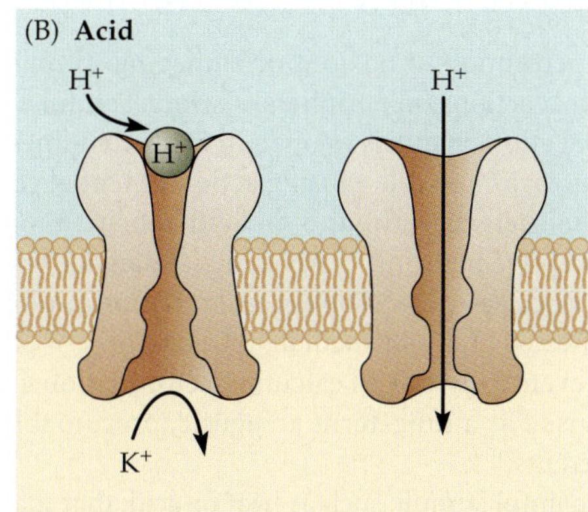

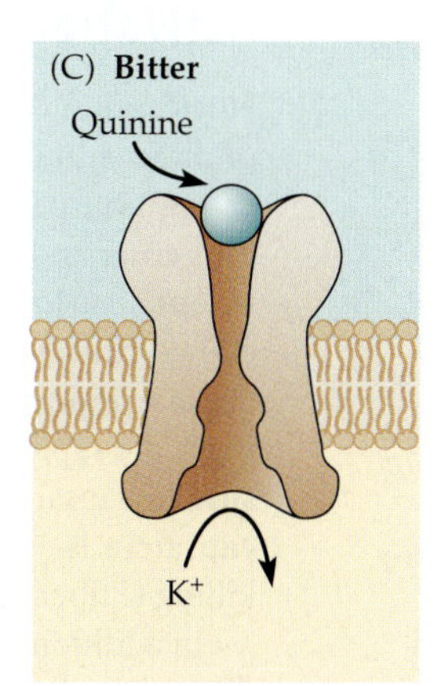

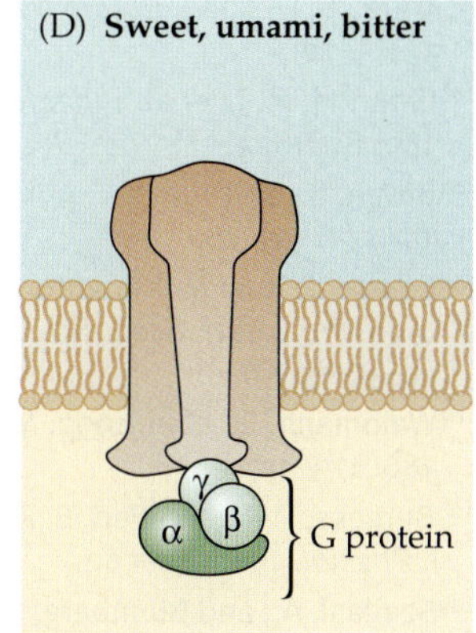

next section), but a taste cell–specific G protein, gustducin, has been cloned and is homologous to the photoreceptor G protein transducin.[100] A variety of downstream effectors have been identified in taste cells and include cAMP phosphodiesterase, phospholipase C and the other components of the pathways leading to cyclic nucleotides and IP_3.[92] The G protein hypothesis for bitter and sweet taste is supported by the observation that a transgenic mouse lacking the gustducin gene could taste salt and sour, but not bitter or sweet.[101]

Molecular Receptors for Glutamate and Chili

Free glutamate occurs in many foods, including meats, cheeses, and some vegetables, and is used as a flavor enhancer in the form of monosodium glutamate (MSG). Glutamate taste is subserved by a G protein–coupled metabotropic glutamate receptor that is expressed specifically in taste buds, but not in surrounding nonsensory lingual epithelium.[102,103] Conditioned taste aversion studies confirmed that MSG and a specific agonist of the mGluR4 receptor, L-2-amino-4-phosphonobutyrate (L-AP4), elicit similar tastes in rats. Such studies take advantage of the strong association between food taste and subsequent nausea to determine whether given substances taste alike. Rats that were nauseated (by intraperitoneal LiCl injection) subsequent to MSG taste then avoided the specific mGluR4 agonist L-AP4 (and MSG) but not specific agonists of other glutamate receptors.

Thus, a molecular receptor for glutamate like those that serve at synapses elsewhere is a specific taste receptor in the tongue. The "hot" taste of chili peppers provides another example of multiuse molecular receptors. Hot chilis are not sensed by taste cells per se, but rather by pain fibers in the tongue that are activated by the compound capsaicin. A capsaicin receptor has been cloned and shown to be a calcium-selective cation channel.[104] The "hot" receptor (labeled VR1 because it was the first member of the vanilloid receptor family) is present in small-caliber (C fiber) dorsal root ganglion cells that signal pain. Thus, nature has provided chili peppers with a chemical targeted to this receptor, possibly to discourage herbivores by activating pain fibers—a not entirely successful strategy in the case of humans with a preference for spicy foods.

TRANSDUCTION OF NOCICEPTIVE AND THERMAL STIMULI

At a skin temperature of about 33°C we are usually unaware of any temperature sensation. Raising or lowering skin temperature above this neutral point produces a sensation of warming or cooling. There are two kinds of temperature receptors in the skin—one signaling warmth, the other cold. You can readily demonstrate this on the back of your hand: If a probe at room temperature (such as a dull pencil point) is pressed against the skin at various places, it will occasionally produce a punctate sensation of cold. Outside such spots, only touch is felt. A warm metal probe can be used to find other points, spatially distinct from cold spots, where sensations of warmth are felt. Such spots are fewer and require extensive searching. Skin temperature afferents are distinct from those signaling painful extremes of temperature.

Activation and Sensitization of Nociceptors

Nociception (the perception of noxious or damaging stimuli) arises from a combination of direct and indirect actions on peripheral sensors. Painful heat (hotter than about 43°C) causes nonspecific cation channels to open in C fiber endings.[105,106] Calcium and sodium ions enter and depolarize the cell, causing action potential generation. Acids also may act to open cation channels directly, and an acid-sensitive ion channel (ASIC) has been cloned from nociceptive neurons.[107] Mechanical damage to the skin can lead directly to depolarization of nociceptors.[108] The capsaicin receptor, VR1, opens when cells containing it are rapidly heated and may mediate the sensation of painful heat. Prolonged exposure to capsaicin eventually causes calcium accumulation and cell death. Paradoxically, capsaicin is also used as a long-term analgesic, presumably relieving chronic pain by killing C fiber afferents.[109]

In addition to painful stimuli, such as heat or acid, that may act directly on nociceptors, damaged cells release chemical activators such as adenosine triphosphate (ATP). One sub-

[100]McLaughlin, S. K., McKinnon, P. J., and Margolskee, R. F. 1992. *Nature* 357: 563–569.
[101]Wong, G. T., Gannon, K. S., and Margolskee, R. F. 1996. *Nature* 381: 796–800.
[102]Chaudhari, N., et al. 1996. *J. Neurosci.* 16: 3817–3826.
[103]Chaudhari, N., Landin, A. M., and Roper, S. D. 2000. *Nature Neurosci.* 3: 113–119.
[104]Caterina, M. J., et al. 1997. *Nature* 389: 816–824.
[105]Bevan, S., and Yeats, J. 1991. *J. Physiol.* 433: 145–161.
[106]Cesare, P., and McNaughton, P. A. 1996. *Proc. Natl. Acad. Sci. USA* 93: 15435–15439.
[107]Waldmann, R., et al. 1997. *Nature* 386: 173–177.
[108]Burgess, P. R., and Perl, E. R. 1967. *J. Physiol.* 190: 541–562.
[109]Szallasi, A., and Blumberg, P. M. 1996. *Pain* 68: 195–208.

unit of ATP receptor ($P2X_3$) occurs specifically in C fiber somata in dorsal root ganglia and may combine with other subunits to mediate the slowly desensitizing excitation of nociceptors by ATP.[110–112] Cellular damage also leads to the release of cytoplasmic proteases, which then cleave serum proteins. In this manner the nine–amino acid peptide bradykinin is produced from kininogen, a ubiquitous inactive precursor. Bradykinin is a potent activator of C fiber endings, leading to the production of inward current and generation of action potentials. The bradykinin receptor may act by increasing the second messenger IP_3.[113]

Bradykinin and other chemicals in damaged skin also act to increase the sensitivity of nociceptive endings. Heat-activated cation current is larger and occurs at lower temperatures as a result of the activation of protein kinase C by bradykinin.[30] Other inflammatory mediators include prostaglandins, serotonin, histamine, and substance P. Prostaglandin E2 and serotonin raise cyclic AMP to increase the amplitude and voltage sensitivity of voltage-gated sodium current in nociceptors.[114] Tissue damage also gives rise to increased expression of α-adrenergic receptors in dorsal root ganglion neurons—another mechanism of increased excitability.[115] Activated pain fibers release substance P not only from their synapses within the spinal cord (Chapter 14), but also from their terminals in the skin. In the periphery, substance P may increase the excitability of C fibers by blocking K^+ channels.[116] The process of sensitization is accompanied by local vasodilatation and edema. The affected area becomes "hyperalgesic," having a reduced threshold for pain.

[110]Chen, C. C., et al. 1995. *Nature* 377: 428–431.

[111]Lewis, C., et al. 1995. *Nature* 377: 432–435.

[112]Cook, S. P., et al. 1997. *Nature* 387: 505–508.

[113]Burgess, G. M., et al. 1989. *J. Neurosci.* 9: 3314–3325.

[114]Gold, M. S., et al. 1996. *Proc. Natl. Acad. Sci. USA* 93: 1108–1112.

[115]Perl, E. R. 1999. *Proc. Natl. Acad. Sci. USA* 96: 7664–7667.

[116]Adams, P. R., Brown, D. A., and Jones, S. W. 1983. *Br. J. Pharmacol.* 79: 330–333.

Summary

- Each type of sensory receptor responds preferentially to one type of stimulus energy, the adequate stimulus.

- Short and long receptors differ morphologically and functionally. Short receptors encode stimulus intensity directly in the amplitude of the receptor potential. Long receptors take the additional step of converting the receptor potential amplitude into a frequency code of action potential firing.

- The response of many receptors varies with the log of the stimulus intensity. This enables some receptor types to have a wide dynamic range.

- Most sensory receptors adapt somewhat during prolonged stimuli. Adaptation arises from both mechanical and electrical factors. In some receptors very rapid adaptation makes them "tuned" to rapidly varying stimuli, such as vibration.

- Mechanosensory hair cells of the inner ear couple movement directly to the gating of ion channels by physical connection. The tip link that connects adjacent stereocilia is stretched by deflection of the hair bundle and so pulls open an ion channel.

- Calcium entry through the nonselective mechanotransduction channel of hair cells leads to adaptation and closure of the channel.

- Olfactory neurons employ G protein–coupled membrane receptors that lead to the opening of cAMP-gated cation channels in the plasma membrane.

- Each member of the large family of olfactory receptor proteins is expressed in a small number of olfactory receptors. All neurons expressing a particular receptor protein project to a single glomerulus in the olfactory bulb.

- Amino acids, sugars, and bitter compounds bind to G protein–coupled receptors in taste sensory cells.

- Salt and protons (sour) act directly on ion channels to generate receptor potentials in taste cells.

- Pain and temperature sensations are mediated by a variety of chemical messengers. Direct mechanical damage or excessive heat initiate action potentials in pain fibers. Compounds released from damaged tissue, such as bradykinin, sensitize nociceptive endings.

SUGGESTED READING

General Reviews

Bell, J., Bolanowski, S., and Holmes, M. H. 1994. The structure and function of Pacinian corpuscles: A review. *Prog. Neurobiol.* 42: 79–128.

Bialek, W. 1987. Physical limits to sensation and perception. *Annu. Rev. Biophys. Biophys. Chem.* 16: 455–478.

Garcia-Anoveros, J., and Corey, D. P. 1997. The molecules of mechanosensation. *Annu. Rev. Neurosci.* 20: 567–594.

Hudspeth, A. J., and Gillespie, P. G. 1994. Pulling springs to tune transduction: Adaptation by hair cells. *Neuron* 12: 1–9.

Hunt, C. C. 1990. Mammalian muscle spindle: Peripheral mechanisms. *Physiol. Rev.* 70: 643–663.

Kinnamon, S. C., and Margolskee, R. F. 1996. Mechanisms of taste transduction. *Curr. Opin. Neurobiol.* 6: 506–513.

Lindemann, B. 1996. Taste reception. *Physiol. Rev.* 76: 718–766.

Mombaerts, P. 1999. Molecular biology of odorant receptors in vertebrates. *Annu. Rev. Neurosci.* 22: 487–509.

Perl, E. R. 1999. Causalgia, pathological pain, and adrenergic receptors. *Proc. Natl. Acad. Sci. USA* 96: 7664–7667.

Original Papers

Assad, J. A., Shepherd, G. M. G., and Corey, D. P. 1991. Tip-link integrity and mechanical transduction in vertebrate hair cells. *Neuron* 7: 985–994.

Buck, L., and Axel, R. 1991. A novel multigene family may encode odorant receptors: A molecular basis for odor recognition. *Cell* 65: 175–187.

Caterina, M. J., Schumacher, M. A., Tominaga M., Rosen T. A., Levine J. D., and Julius D. 1997. The capsaicin receptor: A heat-activated ion channel in the pain pathway. *Nature* 389: 816–824.

Cesare, P., and McNaughton, P. A. 1996. A novel heat-activated current in nociceptive neurones and its sensitization by bradykinin. *Proc. Natl. Acad. Sci. USA* 93: 15435–15439.

Chaudhari, N., Landin, A. M., and Roper, S. D. 2000. A metabotropic glutamate receptor variant functions as a taste receptor. *Nature Neurosci.* 3: 113–119.

Cook, S. P., Vulchanova, L., Hargreaves, K. M., Elde, R., and McCleskey, E. W. 1997. Distinct ATP receptors on pain-sensing and stretch-sensing neurons. *Nature* 387: 505–508.

Crawford, A. C., and Fettiplace, R. 1985. The mechanical properties of ciliary bundles of turtle cochlear hair cells. *J. Physiol.* 364: 359–379.

Crawford, A. C., Evans, M. G., and Fettiplace, R. 1991. The actions of calcium on the mechanoelectrical transducer current of turtle hair cells. *J. Physiol.* 434: 369–398.

Dhallan, R. S., Yau, K. W., Schrader, K. A., and Reed, R. R. 1990. Primary structure and functional expression of a cyclic nucleotide activated channel from olfactory neurons. *Nature* 347: 184–187.

Eyzaguirre, C., and Kuffler, S. W. 1955. Processes of excitation in the dendrites and soma of single isolated sensory nerve cells of the lobster and crayfish. *J. Gen. Physiol.* 39: 87–119.

Gillespie, P. G., Wagner, M. C., and Hudspeth, A. J. 1993. Identification of a 120 kd hair-bundle myosin located near stereociliary tips. *Neuron* 11: 581–594.

Hudspeth, A. J., and Corey, D. P. 1977. Sensitivity, polarity and conductance change in the response of vertebrate hair cells to controlled mechanical stimuli. *Proc. Natl. Acad. Sci. USA* 74: 2407–2411.

Loewenstein, W. R., and Mendelson, M. 1965. Components of adaptation in a Pacinian corpuscle. *J. Physiol.* 177: 377–397.

Nakamura, T., and Gold, G. H. 1987. A cyclic nucleotide-gated conductance in olfactory receptor cilia. *Nature* 532: 442–444.

Ressler, K. J., Sullivan, S. L., and Buck, L. B. 1993. A zonal organization of odorant receptor gene expression in the olfactory epithelium. *Cell* 73: 597–609.

Ricci A. J., Wu, Y-C., and Fettiplace R. 1998. The endogenous calcium buffer and the time course of transducer adaptation in the auditory hair cells. *J. Neurosci.* 18: 8261–8277.

Shotwell, S. L., Jacobs, R., and Hudspeth A. J. 1981. Directional sensitivity of individual vertebrate hair cells to controlled deflection of their hair bundles. *Annu. N.Y. Acad. Sci.* 374: 1–10.

Vassar, R., Ngai, J., and Axel R. 1993. Spatial segregation of odorant receptor expression in the mammalian olfactory epithelium. *Cell* 74: 309–318.

18 Processing of Somatosensory and Auditory Signals

IN THIS CHAPTER WE EXAMINE THE WAYS IN WHICH two sensory systems, somatosensory and auditory, generate signals and analyze information about the body and about sound. Both systems depend in the first instance on sensory receptors that respond to mechanical stimuli, arising from direct deformation by touch or limb motion, or from displacements by sound waves. The central nervous system is provided with information about the location of a touch on the body surface through signals arising from mechanoreceptor neurons with specific modalities in skin and subcutaneous tissues. The receptive field of a somatosensory neuron in the CNS is defined as that area of the periphery in which the adequate stimulus alters activity. A receptive field may be very small, as on the fingertips, or large, as on the middle of the back. The concept of receptive field for auditory sensory neurons is more abstract. The primary sensory afferents encode the frequency composition of sound with extraordinary sensitivity and selectivity, but they do not specify its spatial location. Rather, the map of auditory space is derived by central neurons from analysis of the timing and intensity of inputs to the two ears.

These differences in the two systems are reflected in the pathways and processing of signals in the CNS. Somatosensory information is passed from peripheral afferents to second-order neurons in dorsal column nuclei to the ventrobasal complex of the thalamus with little obvious modification. Thus, neurons in the primary somatosensory cortex receiving thalamic input have response characteristics that correspond closely to those of the sensory cells that innervate the skin or joints directly. The somatosensory cortex is somatotopically mapped. Vertical columns of cells within the primary somatosensory cortex have similar receptive field locations and response modalities. Secondary (and association) somatosensory areas are also somatotopically mapped, but they contain neurons whose stimulus requirements are more complex than those of cells in primary somatosensory cortex, suggesting a hierarchical arrangement for feature extraction. The "meaning" of a stimulus in the somatosensory system is fundamentally one of position on the body surface.

The meaning of sound is derived from an analysis of its spectral (frequency) and temporal characteristics. The frequency selectivity of auditory hair cells arises from mechanical and electrical tuning. The hair cell epithelium is tonotopically arranged. Afferent fibers innervate hair cells selectively and so have tuning curves, meaning that they respond best to a characteristic frequency of sound. Efferent feedback onto cochlear hair cells reduces their sensitivity and frequency selectivity. Afferent fibers from the cochlea synapse within brainstem nuclei. Second-order neurons project to the superior olivary complex, or into pathways that ascend through the inferior colliculus to the medial geniculate nucleus of the thalamus. Neurons in primary auditory cortex receive input from both ears and encode features of sound that are more complex than those detected in the periphery. Sound localization arises from neural computations involving comparisons between inputs to the two ears. Accordingly, the central auditory pathway includes a complex set of synaptic relays and feedback connections at which binaural comparisons are made or other aspects of timing and frequency composition are determined.

Our knowledge of the world depends on the transduction of environmental energy into neuronal signals. Once electrical signals have been produced, how do they become endowed with meaning? Four organizing principles are fundamental to sensory processing. The first is the **preservation of nearest-neighbor relationships** in the organization of the nervous system, from the receptive surface through to the cortex. The resulting somatotopy (tonotopy in the auditory system) underlies synaptic processes such as lateral inhibition, whereby excitatory field centers are more sharply defined. The second principle is based on the fact that the nervous system pays particular attention to signals that vary in time. Thus **temporal**, or **frequency**, **tuning** is fundamental to sensory analysis. The third principle is that of **parallel processing** of different functional aspects of a stimulus. Fourth is the concept of **hierarchical**, or **serial**, processing, whereby higher levels of the sensory pathway combine inputs from lower levels to derive new, more complex sensory constructs. In this chapter the somatosensory and auditory systems are examined. Similar rules of neural organization can be found in both systems. At the same time, the differing functional requirements of somatic sensation and audition provide illustrative examples of the diversity of sensory mechanisms.

THE SOMATOSENSORY SYSTEM: TACTILE RECOGNITION

When you brush your fingertips across your clothing, or touch them with a pencil tip, you are activating touch receptors that inform you of the texture of an object or the location of a point stimulus. If you firmly grip the handle of a hammer or a tennis racket, you activate receptors deeper in the skin that respond to pressure or stretching forces due to that contact. It is evident that quite distinct sensations arise in these different situations. What receptor types provide us with this range of sensations, and how are they organized? How do we distinguish rough from smooth, an egg from a hand grenade? Discriminative touch, vibration sense, and proprioception (limb position) are signaled by mechanoreceptors that project through the dorsal columns of the spinal cord and into the lemniscal portion of the somatosensory system, so called because the receptors pass to the thalamus through a well-defined structure, the **medial lemniscus** (see Figure 18.3A). In this section we describe the processing of cutaneous mechanoreception by this system. The roles of muscle stretch receptors and joint receptors that serve proprioception are discussed in the context of motor control (Chapter 22).

Organization of Receptors for Fine Touch

The hairless surfaces of the palm and fingers are among the most sensitive areas of the body, being innervated by approximately 17,000 cutaneous receptors.[1] Their afferent fibers have been studied in monkeys and humans using a technique called **microneurography**:[2] Fine tungsten electrodes are inserted into a nerve and positioned to record impulses from an individual afferent axon during mechanical stimulation of the skin surface. The area of skin in which touch elicits activity from that fiber defines its **receptive field**. Figure 18.1 shows the receptive fields of one type of skin afferent fiber. An area roughly 3 mm in diameter has a very low threshold for indentation. The boundaries of this receptive field are marked by a steep rise in stimulus threshold (Figure 18.1B); much deeper indentation outside the area is required to activate firing. Within the receptive field are local hot spots of especially high sensitivity (low threshold). These may correspond to the location of individual receptive endings.

The tactile afferents of glabrous (hairless) skin can be characterized as rapidly adapting or slowly adapting, with large or small receptive fields.[3] Rapidly adapting receptors with small receptive fields end in **Meissner's corpuscles** in superficial layers of skin; slowly adapting afferents with small receptive fields end in **Merkel's disks** (Figure 18.2). **Pacinian corpuscles** are vibration-sensitive, rapidly adapting receptors with large receptive fields and are found in deeper layers of skin. Afferents ending in **Ruffini's capsules** are slowly adapting with large receptive fields.

Slowly and rapidly adapting mechanoreceptors with small receptive fields occur at the highest density in the fingertips (100/cm^2), falling off sharply even in the middle pha-

[1]Johansson, R. S., and Vallbo, A. B. 1979. *J. Physiol.* 297: 405–422.

[2]Vallbo, A. B., and Hagbarth, K.-E. 1968. *Exp. Neurol.* 21: 270–289.

[3]Iggo, A. 1974. In *The Peripheral Nervous System*. Plenum, New York, pp. 347–404.

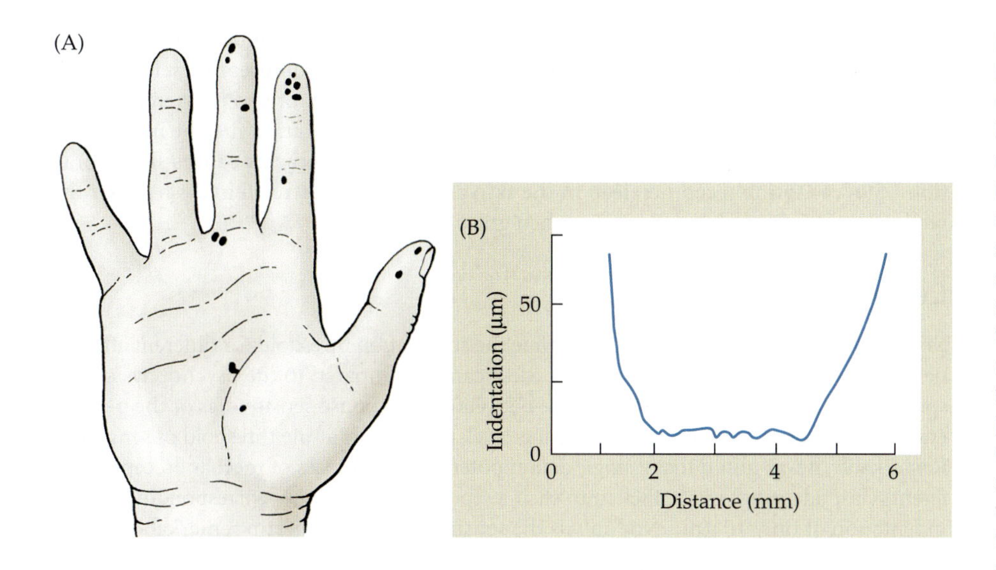

FIGURE 18.1 Receptive Fields of Rapidly Adapting Touch Receptors (probably Meissner's corpuscles) in a human hand. Fields were mapped by recording discharges from single fibers in a sensory nerve, using tungsten electrodes inserted through the skin. (A) Receptive fields consist of small, well-defined areas on the fingers and palm. Each dot represents the receptive field of one fiber. (B) Indentation (in micrometers) required to produce a response in one unit is plotted against location within a receptive field. The region of maximum sensitivity is about 3 mm in radius, within which indentations of only a few micrometers are sufficient to produce a response. Points of maximum sensitivity within the region presumably correspond to the position of individual endings of branches of the same afferent fiber. (After Johansson and Vallbo, 1983.)

langes of the finger.[1] This differential innervation pattern is reflected in the expanded cortical map of the distal digits (as we will see shortly) and underlies the preferential use of the fingers for tactile exploration. Only the lips and tongue are more richly innervated.

Stimulus Coding

How do the response properties of single touch afferent neurons relate to perception? Do specific types of receptors encode qualities such as rough or smooth, or are these sensations derived by the convergence of several touch modalities? Humans describe vibration of the skin at frequencies below 40 Hz as flutter, and stimulation frequencies of 80 to 300 Hz as hum or buzz. The frequency dependence of these sensations correlates with the tuning of groups of rapidly adapting receptors. Meissner's corpuscle afferents respond best to stimulation at 30 Hz; Pacinian corpuscle afferents respond best to skin vibration at frequencies of 250 Hz.[4] Thus, there is a suggestive correlation between the response properties of identified receptor types and psychophysically defined stimuli.

However, passive vibration of the skin has limited relevance to the cognitive process of **active touch**, whereby the hands and fingers move rapidly over an object to assess surface

[4]LaMotte, R. H., and Mountcastle, V. B. 1975. *J. Neurophysiol.* 38: 539–559.

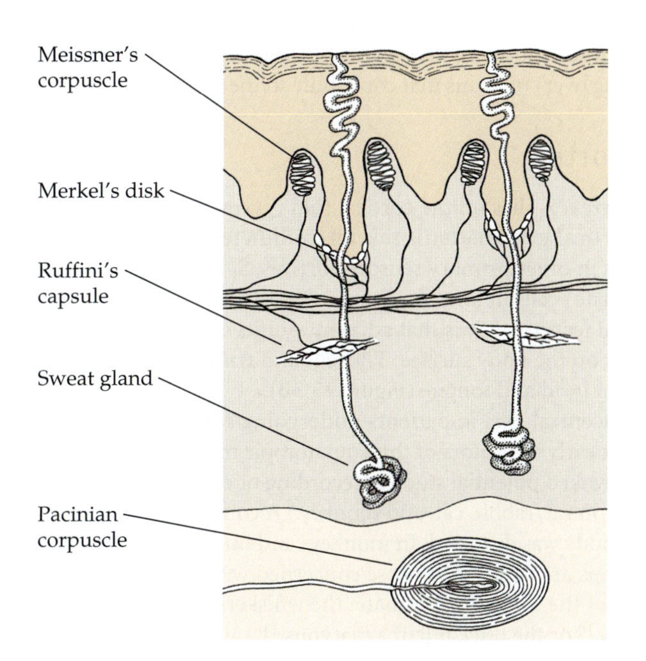

FIGURE 18.2 Mechanoreceptors in Skin. Rapidly adapting Meissner's corpuscles and slowly adapting Merkel's disks respond to skin indentation within small, well-defined fields. Ruffini's capsules are slowly adapting and have wider fields with poorly defined borders. Pacinian corpuscles are very rapidly adapting and sensitive to vibration over a wide area of skin. Not shown are rapidly adapting hair follicle receptors that respond to hair movement, and free nerve endings that mediate temperature, itch, and pain sensations. (After Johansson and Vallbo, 1983.)

texture and shape. Afferent recordings in monkeys trained to discriminate surface shapes (embossed letters) show that slowly adapting receptors with small receptive fields play an important role.[5] This class of receptors also contributes importantly to the perception of roughness.[6] Rapidly adapting receptors that are sensitive to vibration may be involved in the assessment of grip, possibly because slipping of grasped objects causes skin vibration.[7] Slow adapting receptors deep in the skin are sensitive to stretch in a particular direction, another stimulus that occurs when grasping objects.

Central Pathways

Microneurography can be used to determine the stimulation threshold for different afferent neuronal types, and in a human subject these can be compared to the psychophysical response. There is a strong correspondence between the response sensitivities of the peripheral afferents and the perception of stimulus level, particularly at the threshold of sensation. Remarkably, it was found that a single action potential in one class of mechanoreceptive afferents (fast-adapting) was sufficient to elicit a reportable sensation—corresponding to skin indentation of only 10 μm.[1] Not only is the sensitivity of the fingertip remarkable, but so too is transmission through to the brain. The flow of information from periphery to consciousness occurs with extremely high fidelity. Only a few synaptic contacts exist along this path to the cortex, and each operates with a high safety factor, so that transmission can be one for one. It is clearly beneficial that stimuli acting directly on the body surface have quick access to consciousness (if it's close enough to touch, you'd better pay attention).

Figure 18.3A shows the main somatosensory pathways for carrying sensations of touch, pressure, and vibration to the brain (see also Appendix C). The peripheral axons of afferent neurons connect to cell bodies located in the dorsal root ganglia. Central processes of afferents from skin, deep tissue, muscle, and joints enter the dorsal roots, give off axon collaterals to make synapses on spinal neurons, and ascend in the dorsal columns to end on second-order cells in the cuneate and gracile nuclei (which are known as the **dorsal column nuclei**). The axons of second-order cells in the dorsal column nuclei cross the midline and ascend in the medial lemniscus to end on cells in the ventroposterior lateral (VPL) nucleus of the thalamus. The pathways from the head and neck are anatomically distinct, but functionally analogous to the spinal pathway, ending in the ventroposterior medial (VPM) nucleus of the thalamus. The third-order cells in VPL and VPM thalamus project to the postcentral gyrus of the cerebral cortex, the primary somatosensory area.

Only three synapses lie between excitation of the primary afferent ending in the skin and activation of a cortical neuron. Nearest-neighbor relationships are preserved at each level in the pathway, resulting in a somatotopic projection up to the level of cortex. The left side of the body is mapped onto the right somatosensory cortex, a contralateral organization arising from the decussation (crossing over) of axons that contribute to the medial lemniscus.

The Somatosensory Cortex

The **primary somatosensory cortex**, S_1, lies on the cortical fold (gyrus) immediately behind the central sulcus. The cortical gray matter is several millimeters thick and composed of six distinctive layers. As in other primary sensory cortices, S_1 is distinguished by the high density of granule (stellate) cells in layer 4, the main destination of thalamic inputs. A significant organizational feature of S_1 is that adjacent points on the cortical surface represent adjacent locations on the body surface. The legs and trunk are found most medially, followed by hands, then head and tongue (Figure 18.3B).

Stimulation of pre- and postcentral gyri in patients undergoing surgical removal of epileptic foci provided important early indicators of this somatotopic map.[8] More detailed cortical maps were drawn from evoked potential studies (recording of cortical signals during skin stimulation) performed on rat, rabbit, cat, and monkey.[9] A consistent observation was that the cortical map of the body was distorted. In monkeys and humans, regions of S_1 mapping to hands, fingers, and lips are larger than those concerned with trunk or legs. In other animals, different regions of the body predominate: the whiskers of a rat or mouse (see Figure 18.7), a raccoon's paw,[10] or the duck bill of a platypus.[11]

[5]Johnson, K. O., Hsiao, S. S., and Twombly, I. A. 1995. In *The Cognitive Sciences*. MIT Press, Cambridge, MA, pp. 253–268.

[6]Hsiao, S. S., Johnson, K. O., and Twombly, I. A. 1993. *Acta Psychol. (Amst.)* 84: 53–67.

[7]Brisben, A. J., Hsiao, S. S., and Johnson, K. O. 1999. *J. Neurophysiol.* 81: 1548–1558.

[8]Penfield, W., and Rasmussen, T. 1950. *The Cerebral Cortex of Man: A Clinical Study of Localization of Function*. Macmillan, New York.

[9]Woolsey, C. N. 1958. In *The Biological and Biochemical Bases of Behavior*. University of Wisconsin Press, Madison, pp. 63–82.

[10]Rasmusson, D. D. 1982. *J. Comp. Neurol.* 205: 313–326.

[11]Krubitzer, L., et al. 1995. *J. Comp. Neurol.* 351: 261–306.

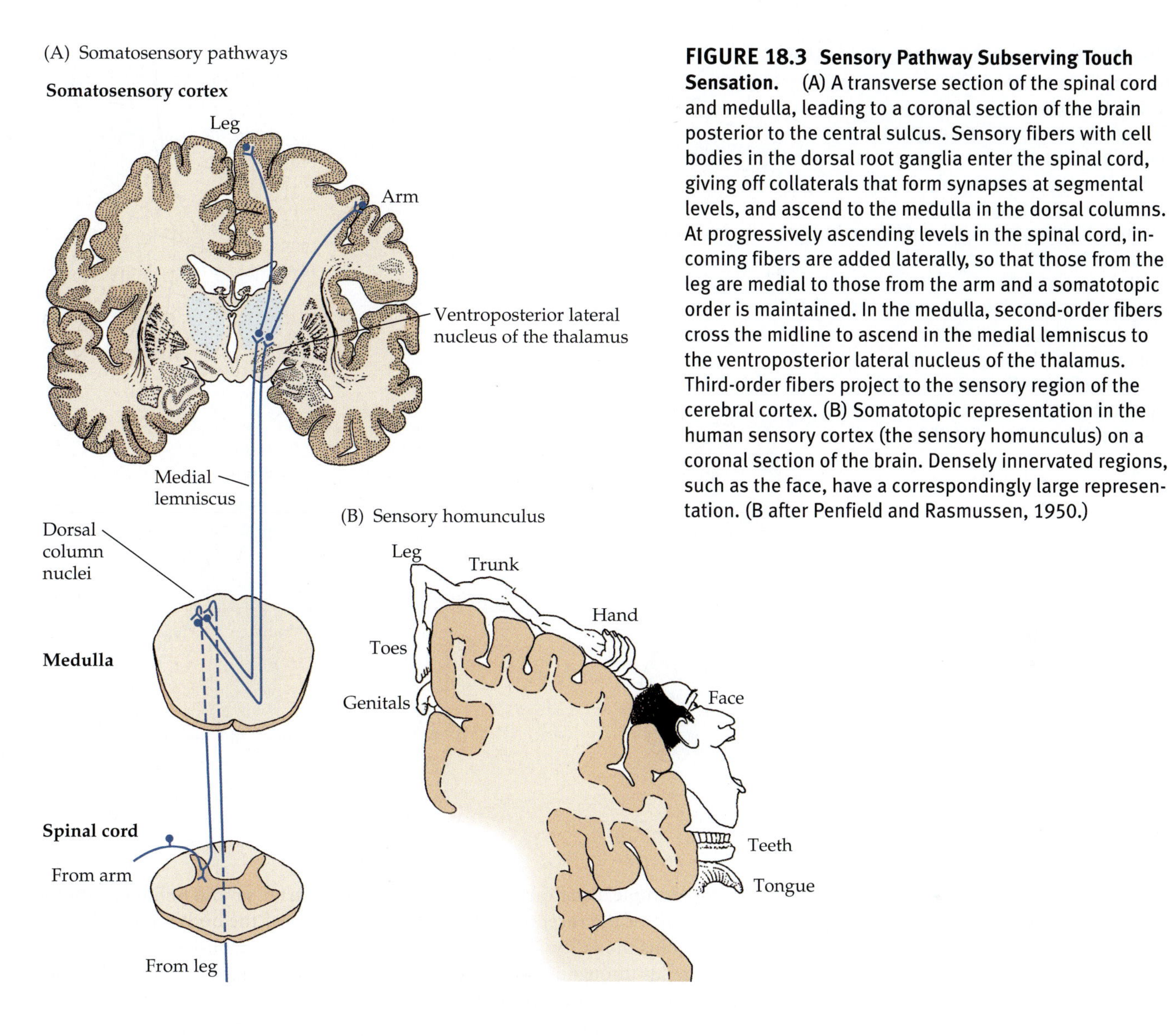

FIGURE 18.3 Sensory Pathway Subserving Touch Sensation. (A) A transverse section of the spinal cord and medulla, leading to a coronal section of the brain posterior to the central sulcus. Sensory fibers with cell bodies in the dorsal root ganglia enter the spinal cord, giving off collaterals that form synapses at segmental levels, and ascend to the medulla in the dorsal columns. At progressively ascending levels in the spinal cord, incoming fibers are added laterally, so that those from the leg are medial to those from the arm and a somatotopic order is maintained. In the medulla, second-order fibers cross the midline to ascend in the medial lemniscus to the ventroposterior lateral nucleus of the thalamus. Third-order fibers project to the sensory region of the cerebral cortex. (B) Somatotopic representation in the human sensory cortex (the sensory homunculus) on a coronal section of the brain. Densely innervated regions, such as the face, have a correspondingly large representation. (B after Penfield and Rasmussen, 1950.)

The distortion of the somatotopic map arises from the differential innervation of the periphery by fine-touch receptors in the fingertips, lips, and tongue. At the same time, it has been shown that the cortical representations are not fixed immutably but can be altered by lesions that modify the sensory input to cortex, or even by differential stimulation of the periphery. For example, the cortical representation of the left hand of stringed-instrument (e.g., violin) players was found to be larger than that of control individuals.[12] Observations like these imply that thalamic inputs to cortex are plastic, with synaptic fields whose size or efficacy can change as a result of activity.[13]

Response Properties of Cortical Neurons

How is information processed in the somatosensory cortex? We previously described the properties of cutaneous mechanoreceptors by recording their action potentials during stimulation of the skin. Similar experiments have been done to determine how cortical neurons respond to skin stimulation. Metal or glass microelectrodes advanced into the brain with micromanipulators were used to record the action potentials of single cortical neurons, so-called single-unit recording. In the 1950s, Powell and Mountcastle made electrode penetrations in S_1 perpendicular to the cortical surface.[14,15] They found that cells encountered along one vertical track had overlapping receptive fields on the body surface (Figure 18.4). Moreover, each cell preferred the same type of stimulation—skin

[12]Elbert, T., et al. 1995. *Science* 270: 305–307.

[13]Xerri, C., et al. 1999. *Cerebral Cortex* 9: 264–276.

[14]Mountcastle, V. B. 1957. *J. Neurophysiol.* 20: 408–434.

[15]Powell, T. P. S., and Mountcastle, V. B. 1959. *Bull. Johns Hopkins Hosp.* 105: 133–162.

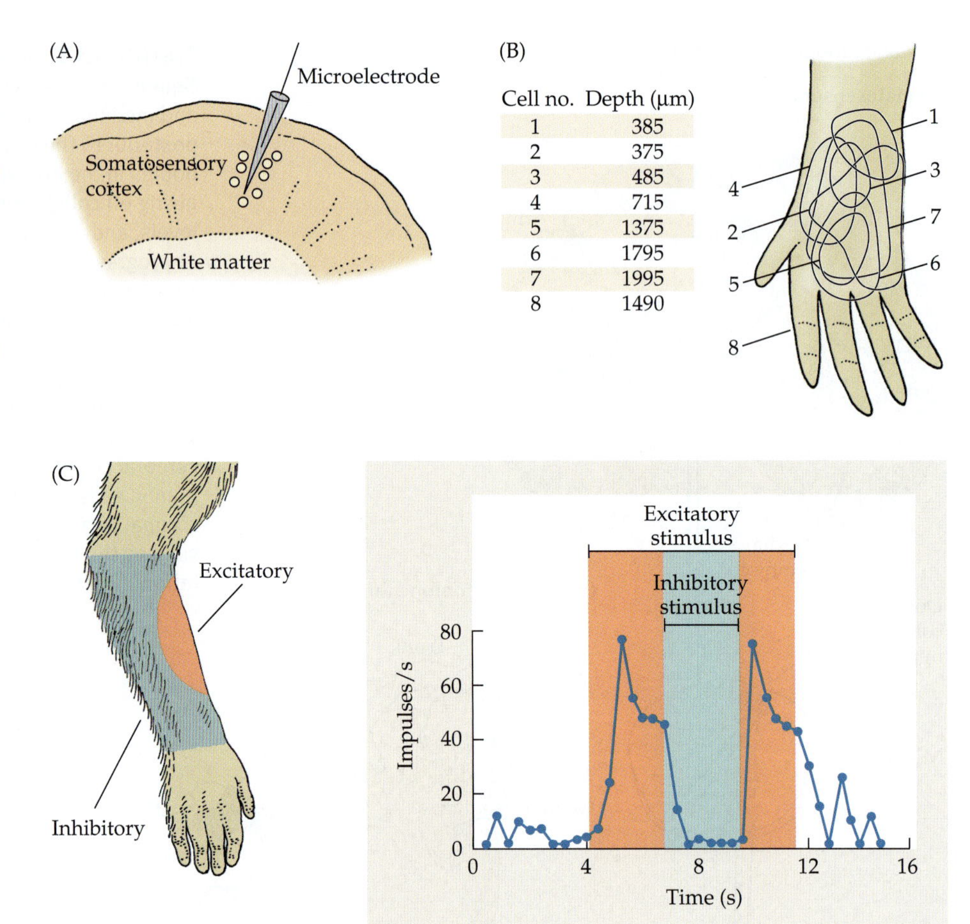

FIGURE 18.4 Receptive Fields of Neurons in monkey somatosensory cortex. (A) An electrode is inserted through the cortex at right angles to the surface. (B) Each cell encountered by the electrode successively deeper within the cortex responds to touch over roughly the same region of the hand. (C) Receptive field of a cell in another region of the cortex. The excitatory region on the ventral surface of the forearm is surrounded by an inhibitory region that suppresses the response of the cell. (B after Powell and Mountcastle, 1959; C after Mountcastle and Powell, 1959.)

vibration, for example. Penetrations through an area of cortex several millimeters away resulted in a shift of the receptive field to another region of the body, and perhaps a change of stimulus modality. Thus, Powell and Mountcastle established an important principle that holds throughout the cortex: Cortical cells are arranged in columns according to receptive field location (somatotopy) and stimulus modality.[16]

Surround Inhibition

The receptive fields of cortical neurons differ in important ways from those of primary afferent neurons. The receptive field of a primary touch afferent neuron is simply that area of skin where stimulation causes excitation. The receptive field of a cortical neuron, however, is more complex. Thus, stimulation of appropriate areas of skin gives rise to excitation or inhibition of a cortical neuron. Often a central area of skin in which touch produces excitation is surrounded by inhibitory regions (see Figure 18.4C). This **inhibitory surround** arises from synaptic interactions along the somatosensory pathway, presumably by inhibitory interneurons that project from neighboring cells—a process of **lateral inhibition**.

Lateral inhibition is an important mechanism for improving tactile resolution. When two closely spaced probes are applied to the skin (Figure 18.5), afferent endings immediately under each probe may be maximally activated, but afferent fibers innervating the intervening skin will also be excited, as a result of the spreading indentation of elastic skin. If each afferent fiber drives inhibitory interneurons proportionately in the dorsal column nuclei, then maximally activated afferents will win out over their less activated neighbors. As a consequence, the firing of fibers whose activity is derived from the intervening skin is suppressed. This improves the ability of the cortex to discriminate between one large probe and two smaller probes.

[16]Mountcastle, V. B. 1997. *Brain* 120: 701–722.

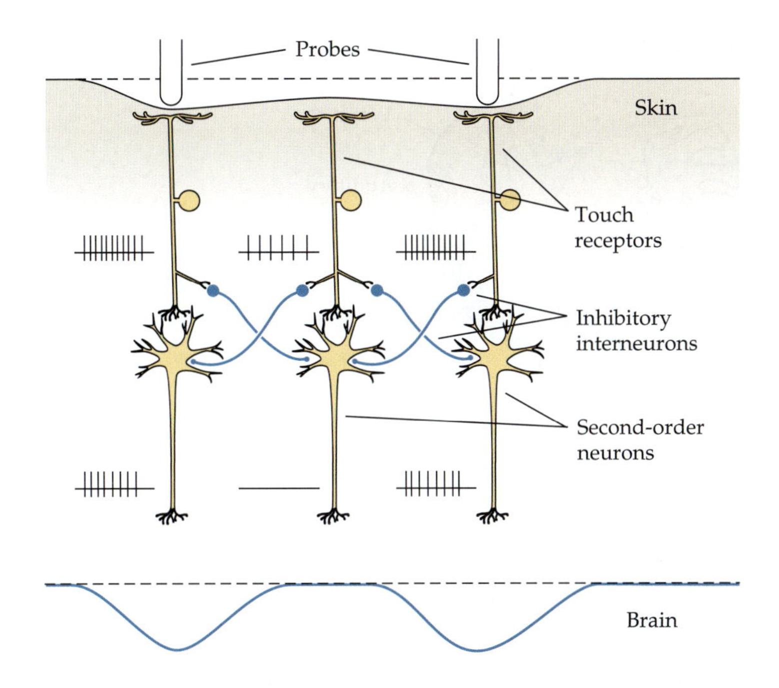

FIGURE 18.5 Lateral Inhibition. Two probes deflect the skin and activate three touch receptors, one of which lies between the points of contact. All three receptors excite second-order neurons, as well as inhibitory interneurons in a central nucleus. The inhibitory interneurons project to neighboring pathways and fire in proportion to the input they receive. Thus, the second-order neurons have more highly differentiated patterns of activity than do the primary afferents. The bottom blue line labeled "brain" represents the stimulus profile as it might be represented in cortex after being refined by lateral inhibition. These effects are exaggerated here and would normally occur at several stages en route to cortex.

Lateral inhibition can be seen as a means to improve the detection of edges as well. A stimulus that falls across both the excitatory center and the inhibitory surround will be less effective than a stimulus that stops (has an edge) within the excitatory center. Again, this illustrates the emphasis that the nervous system places on detecting change, in this case a spatial edge.

Parallel Processing of Sensory Modalities

The postcentral gyrus can be subdivided into several regions on the basis of differences in cellular organization and sensory modality (Figure 18.6). Area 3a is immediately adjacent to primary motor cortex, and neurons there respond primarily to muscle stretch receptors and other deep receptors.[17] In contrast, area 3b receives inputs from rapidly and slowly adapting cutaneous receptors, especially those with small receptive fields. The specific localization of receptor type within subregions of S_1 illustrates the concept of parallel processing within each sensory system. Each area of the body is represented in the cutaneous zone (3b), as well as in the deep receptor zone (3a). Thus, the somatotopic cortical map actually consists of multiple representations of the body plan, one for each sensory modality.

The precise anatomical and functional organization of somatosensory cortex is highlighted by the projection of sensory innervation from whiskers of the mouse (Figure 18.7). As in humans, the face area of the mouse cortex is disproportionately large. Histological studies by Woolsey and Van der Loos[18] have shown that this area of the mouse somatosensory cortex contains characteristic groups of nerve cells clustered in the form of cylinders that lie perpendicularly

[17]Tanji, J., and Wise, S. P. 1981. *J. Neurophys.* 45: 467–481.

[18]Woolsey, T. A., and Van der Loos, H. 1970. *Brain Res.* 17: 205–242.

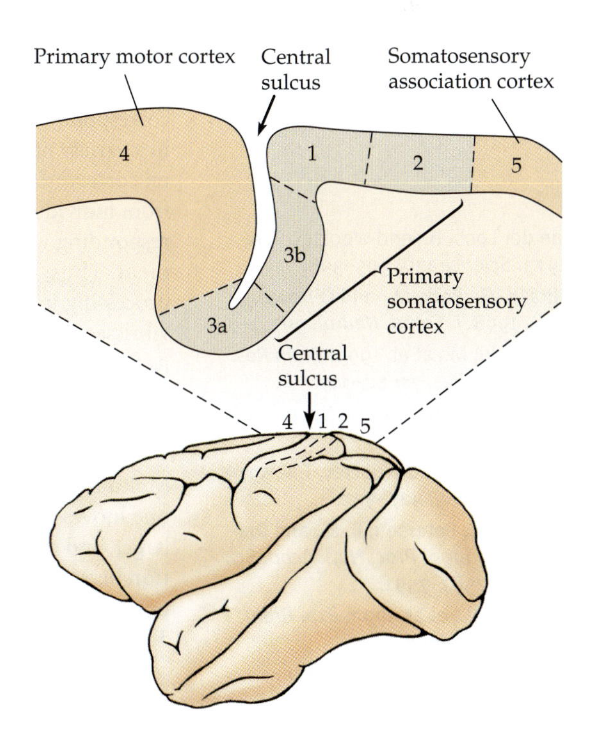

FIGURE 18.6 Somatosensory Cortex. The subdivisions of primary somatosensory cortex (S_1) are shown here on a sagittal section through the postcentral gyrus of the monkey cortex. S_1 is subdivided into areas 3a, 3b, 1, and 2, proceeding posteriorly from the central sulcus. Area 5 is part of the somatosensory association cortex. Area 4 lying anterior to the central sulcus is primary motor cortex. (See Box 18.1 for numeration.)

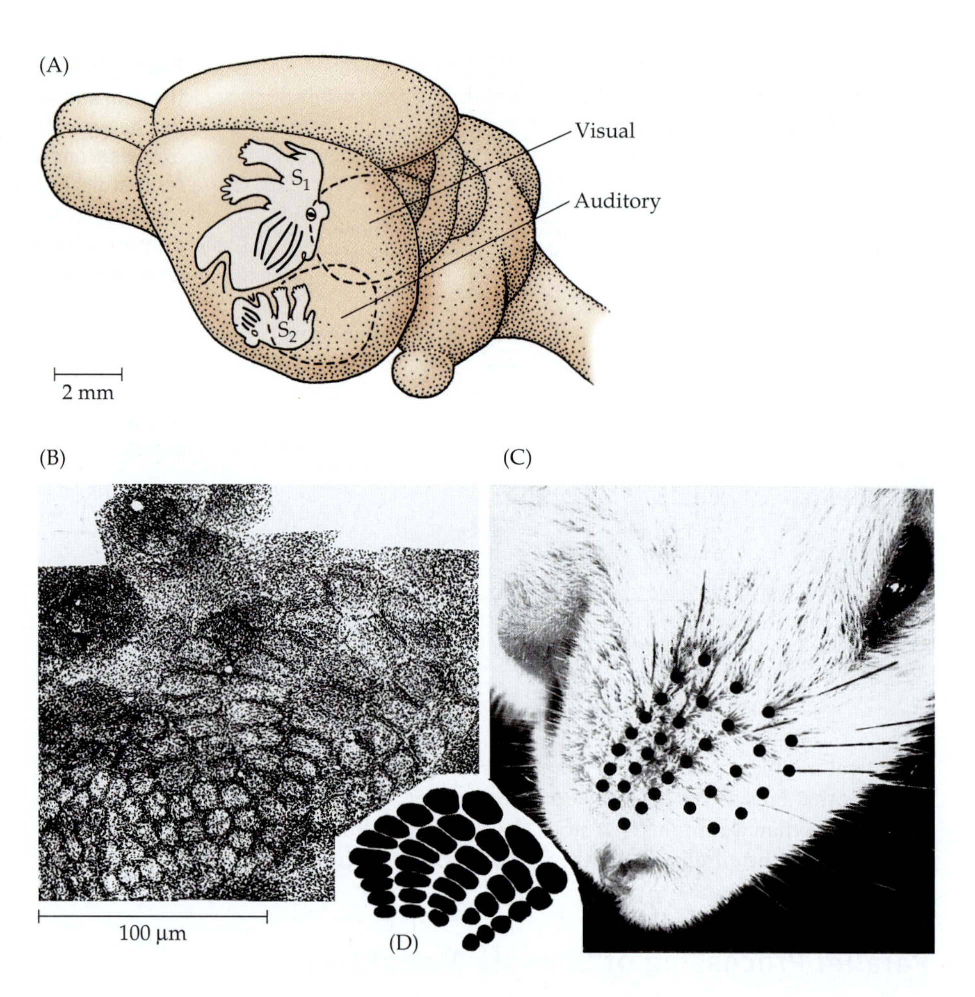

FIGURE 18.7 Barrels in Mouse Somatosensory Cortex corresponding to vibrissae. (A) Diagram of a mouse brain showing the somatotopic representation on the sensory cortex (the "musculus," which corresponds to the homunculus in Figure 18.3B), particularly the face and vibrissae. S_2 is the secondary somatosensory cortex. (B) Horizontal section through the cortex in the area representing the vibrissae, showing barrels in cross section. (C) Close-up of whiskers, the origin of each marked by a black dot. (D) Schematic diagram of barrel arrangement from B, showing correspondence with vibrissae. (After Woolsey and Van der Loos, 1970.)

within the cortex. These assemblies have been termed **barrels** for their shapes, as determined from serial reconstruction. Each barrel is 100 to 400 μm in diameter and is composed of a ring of cells surrounding a core containing fewer cells.

The array of barrels in the mouse is consistently organized into five rows. Remarkably, this pattern corresponds exactly to the rows of whiskers, or vibrissae, on the mouse's muzzle—one barrel for one whisker. If a whisker is removed early in postnatal life, the corresponding barrel disappears from the cortex.[19] Deafferentation at later stages results in a variety of changes, including altered expression of neurotransmitter receptors in barrel cortex.[20,21] The functional role of the barrels was confirmed by recording electrically from individual cells within barrel cortex.[22,23] Each cell responds to movement of the corresponding whisker—some firing selectively for a particular direction of whisker movement. Thus, an impressive fraction of the mouse somatosensory cortex is devoted to processing information acquired by these muzzle mechanoreceptors. The mouse uses its vibrissae as sensitive antennae, waving them back and forth as it walks along to detect objects on either side of its path.[24]

Secondary and Associated Somatosensory Cortices

Multiple representations of the body occur in the primary somatosensory cortex, S1. As one moves away from the central sulcus, from area 3b to 1 to 2, receptive fields become larger, and center–surround organization, direction selectivity, and submodality integration become prominent features of the neuronal response.[25,26] Area 1 lies immediately posterior to area 3b and receives activating input from it. Area 2 receives input from both 3a and 3b.

In addition to the primary somatosensory cortex, multiple representations of the body occur in other, secondary areas. In general, these are zones or strips of cortex that

[19]Van der Loos, H., and Woolsey, T. A. 1973. *Science* 179: 395–398.

[20]Bina, K. G., Park, M., and O'Dowd, D. K. 1998. *J. Comp. Neurol.* 397: 1–9.

[21]Gierdalski, M., et al. 1999. *Brain Res. Mol. Brain Res.* 71: 111–119.

[22]Welker, C. 1976. *J. Comp. Neurol.* 166: 173–190.

[23]Simons, D. J., and Woolsey, T. A. 1979. *Brain Res.* 165: 327–332.

[24]Harris, J. A., Petersen, R. S., and Diamond, M. 1999. *Proc. Natl. Acad. Sci. USA* 96: 7587–7591.

[25]Iwamura, U., et al. 1993. *Exp. Brain Res.* 92: 360–368.

[26]Kaas, J. H. 1993. *Ann. Anat.* 175: 509–518.

lie posterior and ventrolateral to S_1. In primates, the **secondary somatosensory cortex**, S_2, lies on the anterior wall of the lateral sulcus in a position ventral to S_1 (Figure 18.8). S_2 is also somatotopically organized, and it receives input from the thalamus, as well as from S_1. If primate S_1 is removed surgically, the activity of cells in S_2 is much reduced.[27] If only areas 3b and 2 (where superficial, or cutaneous, stimulation is processed) are taken from S_1, then the number of cells in S_2 that respond to cutaneous stimulation is specifically reduced.[28] The results of these manipulations suggest that somatosensory information flows from S_1 to S_2. In other species a direct thalamic path to S_2 appears to be more important.[29]

Information flow from S_1 to S_2 also may be hierarchical, in the sense that information is processed in S_1, and S_2 then deals with derived features of somatosensation. For instance, when a monkey's fingertip touches a rotating drum, many neurons in S_1 increase firing in proportion to increasing texture of the drum surface, as though they were driven directly by afferent input. In S_2, however, many neurons signal a change of texture independent of the magnitude of change.[30] In other words, neurons in S_2 have derived a concept, the difference in surface quality, from more literal information in S_1. Other observations also suggest that S_2 deals with complex aspects of sensory coding. For example, many cells in S_2 of primates have bilateral receptive fields,[31] responding to stimulation of similar areas on both sides of the body.

Brodmann's areas 5 and 7 make up the somatosensory association cortex (Box 18.1). Here neurons respond to skin stimulation or limb movement, but the stimulus requirements are complex and different from the response profiles of primary sensory afferents. Furthermore, neurons here also may respond to visual stimulation. Although the specific functions of the accessory areas of somatosensory cortex are not understood, it is clear that there is a trend to more complex stimulus requirements as one proceeds away from the central sulcus.

Pain and Temperature Pathways

Information about noxious stimuli and temperature is conveyed to higher centers by specific receptors and along pathways largely distinct from those used for position sense, touch, or pressure. The senses of pain and temperature are served by small-diameter, myelinated and unmyelinated afferents with free nerve endings in the skin and other tissues. The endings are not morphologically specialized, but can be divided into different subtypes responding best to touch, heating or cooling, or damaging and noxious stimuli.[32] Nociceptive and temperature-sensitive afferent axons fall into two size classes. Aδ axons are 1 to 4 μm in diameter and conduct at 6 to 25 m/s. Unmyelinated C axons, 0.1 to 1 μm in diameter, conduct at 0.5 to 2 m/s (Chapter 7).

The two fiber types mediate different pain sensations. In humans a brief, intense stimulus delivered to a distal limb gives rise first to a sharp and relatively brief pricking sensation (first pain), followed by a dull, prolonged burning sensation (second pain). Electrophysiological experiments have shown that first pain is related to activation of small myelinated fibers and second pain to C fiber activation.[33–36]

[27] Pons, T. P., Garraghty, P. E., and Mishkin, M. 1987. *Science* 237: 417–420.

[28] Pons, T. P., Garraghty, P. E., and Mishkin, M. 1992. *J. Neurophysiol.* 68: 518–527.

[29] Turman, A. B., et al. 1992. *J. Neurophysiol.* 67: 411–429.

[30] Jiang, W., Tremblay, F., and Chapman, C. E. 1997. *J. Neurophysiol.* 77: 1656–1662.

[31] Whitsel, B. L., Petrucelli, L. M., and Werner, G. 1969. *J. Neurophysiol.* 32: 170–183.

[32] Perl, E. R. 1996. *Prog. Brain Res.* 113: 21–37.

[33] Collins, W. F., Jr., Nelson, F. E., and Randt, C. T. 1960. *Arch. Neurol.* 3: 381–385.

[34] Konietzny, F., et al. 1981. *Exp. Brain Res.* 42: 219–222.

[35] Torebjork, H. E., and Hallin, R. G. 1973. *Exp. Brain Res.* 16: 321–332.

[36] Ochoa, J., and Torebjork, E. 1989. *J. Physiol.* 415: 583–599.

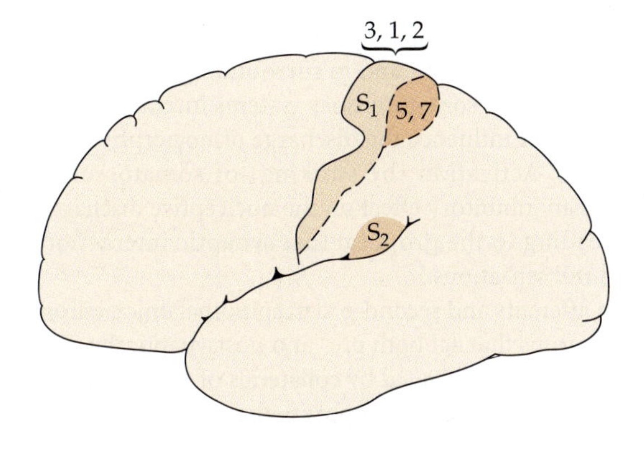

FIGURE 18.8 Somatosensory Cortices. The primary somatosensory cortex, S_1, lies on the postcentral gyrus immediately behind the central sulcus. Somatosensory association cortices (Brodmann's areas 5 and 7) lie posterior to S_1. The secondary somatosensory cortex, S_2, lies below S_1, tucked into the lateral sulcus.

Box 18.1 Brodmann's Areas

Cross sections of cortex reveal that the size of constituent cells, the thickness of cellular layers and fiber tracts, and other features all vary regionally. These anatomical distinctions were mapped into 52 "cytoarchitectonic" areas by Brodmann,[37] many of which correspond to functionally defined regions and are commonly referred to today. For example, area 4 corresponds to primary motor cortex; areas 3, 1, and 2 correspond to primary somatosensory cortex; areas 5 and 7 correspond to somatosensory association cortex; areas 41 and 42 are auditory cortex; and areas 17 and 18 are visual. A reduced form of Brodmann's map is shown here for both lateral and medial views of the brain, illustrating some of the primary and secondary cortical areas.

[37]von Economo, G., and Koskinas, G. N. 1925. *Die Cytoarchitecktonik der Hirnrinde des erwachsenen Menschen*. Julius Springer, Heidelberg.

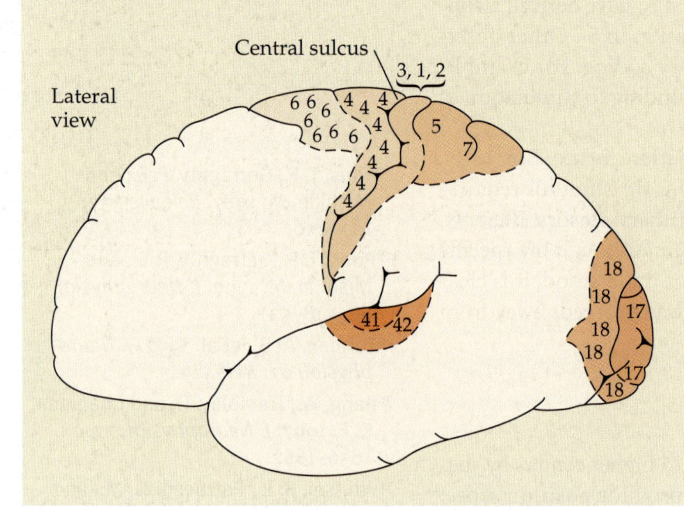

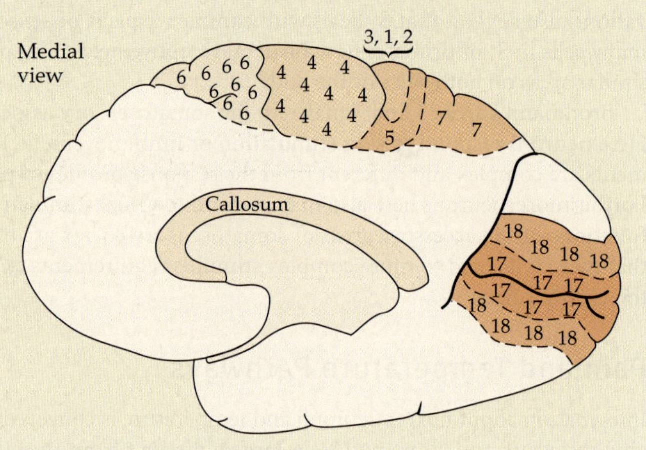

Central Pathways for Pain

Nociceptive and temperature-sensitive afferent axons form synapses on second-order cells within the dorsal horn of the spinal cord. Fibers from second-order cells cross the midline and ascend through two major pathways: the lateral spinothalamic tract and the ventral spinothalamic tract (Figure 18.9). The ascending fibers end in the ventrobasal and medial nuclei of the thalamus—hence its name as the **spinothalamic tract.** Cells in the ventrobasal and medial nuclei of the thalamus project to somatosensory cortex and widely throughout the brain. Unlike higher-order neurons in the dorsal column (lemniscal) pathway, those in the thalamus and cortex receiving nociceptive inputs have large, ill-defined receptive fields, often covering wide areas of the body bilaterally.[38]

Although afferent nociceptors have their own through-lines to the cortex, it is clear that no noxious stimulus can fail to activate receptors of the dorsal column (lemniscal) pathway responding to other stimulus modalities. For example, a painful pinch at one point on the skin will activate touch receptors both at that point and in surrounding skin. Indeed, numerous experiments have shown that the two somatosensory systems interact. In particular, light touch or stroking of the skin can influence the discharge of nociceptive neurons within the central nervous system. Activation (by stroking) of somatosensory afferents with nearby receptive fields has an inhibitory effect on the nociceptive discharges of cortical and thalamic neurons,[39] leading to the proposal that synaptic interactions within the spinal cord might "gate" painful sensations.[40]

Synaptic transmission between pain afferents and second-order, spinothalamic neurons in the dorsal horn is inhibited by interneurons that act both pre- and postsynaptically (Figure 18.10). These inhibitory interneurons can be activated by collaterals of large-diameter (lemniscal) afferents. Thus, touch receptors in skin can suppress the sensation of pain

[38]Willis, W. D., and Westlund, K. N. 1997. *J. Clin. Neurophysiol.* 14: 2–31.

[39]Poggio, G. F., and Mountcastle, V. B. 1960. *Bull. Johns Hopkins Hosp.* 106: 266–316.

[40]Melzack, R., and Wall, P. D. 1965. *Science* 150: 971–979.

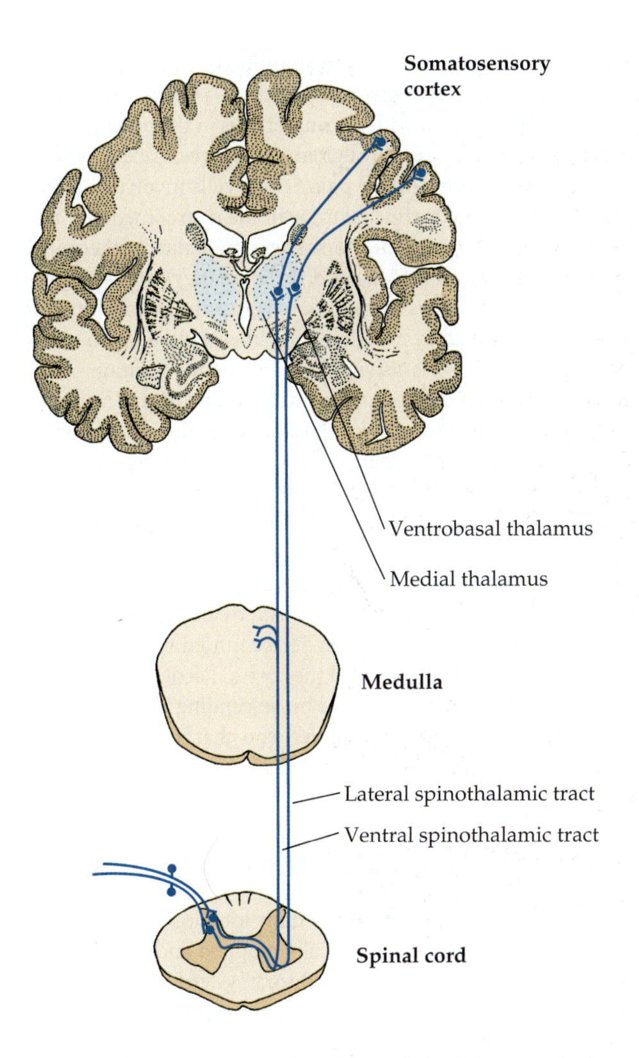

FIGURE 18.9 Sensory Pathways Mediating Pain and Temperature, shown on cross sections of the spinal cord and medulla and a coronal section of the brain. Small myelinated and unmyelinated fibers enter the spinal cord to form synapses with second-order neurons. The second-order axons cross the midline to ascend in the ventral and lateral spinothalamic tracts to end in the medial and ventrobasal nuclei of the thalamus. Third-order cells then project to the cerebral cortex. Cells in the ventral spinothalamic tract give off collateral branches in the medulla, and some terminate at this level (the spinoreticular tract; not shown).

through this interaction. The subject of modulation of pain sensation received considerable attention with the discovery of opiate receptors in the central nervous system and with the identification of naturally occurring opiatelike peptides in the specific neurons concerned with nociceptive function[1] (see the discussion of peptide neurotransmitters in Chapter 14).[41]

As in other sensory systems, descending influences from higher centers can dramatically modify the flow of sensory information about noxious stimuli that reaches conscious-

[41]Hughes, J. (ed.). 1983. *Br. Med. Bull.* 39: 1–106.

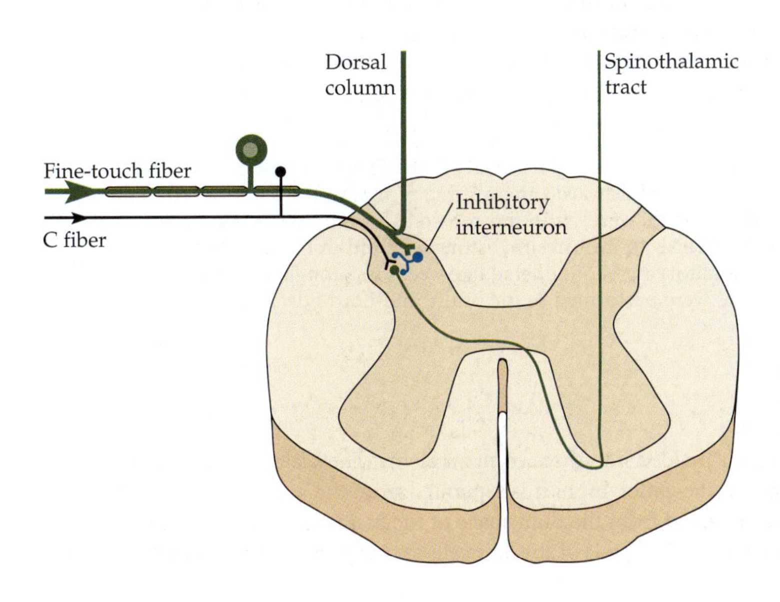

FIGURE 18.10 Somatosensory Inhibition of Pain Transmission. Synaptic transmission from pain afferents to second-order neurons in the spinal cord is reduced by inhibitory interneurons. The inhibitory interneurons can be activated by collaterals of large-diameter sensory fibers (such as those that mediate fine touch or vibration sense).

ness.[42,43] A discussion of the specific pathways subserving these influences can be found in an informative monograph on the neurophysiological and clinical aspects of pain by Fields.[44] The known pathway starts in the midbrain with a group of cells in the periaqueductal gray matter. In some cases, stimulation of this area through implanted electrodes has been found to produce a selective reduction of severe clinical pain.[45] The cells involved are believed to be enkephalinergic and project to serotonergic neurons in the rostral medulla. The medullary neurons then send descending fibers along the dorsolateral funiculus of the spinal cord to end in the dorsal horns, particularly in laminae I and II. There they form synapses with interneurons and terminals of afferent fibers to modulate transmission in the pain pathway. The descending fibers are accompanied by noradrenergic fibers belonging to a second pathway originating in dorsolateral pons that is also involved in pain modulation.

Pain itself remains an elusive and difficult concept, beyond the scope of this book. In sharp contrast to the analysis of visual, auditory, or somatosensory systems, a discussion of pain, with its high emotional content, of necessity deals with subjective matters, feelings akin to "anguish" and "suffering" that cannot at present be expressed in the language of neurobiology (just as "seeing a sunset" or "feeling warm" cannot be considered in those terms). Nevertheless, certain correlates between neural activity and pain are apparent. Thus, the common experience of a sharp, well-localized stab of pain, followed by a dull, gradually swelling, poorly localized burning ache can be attributed to the activity of the Aδ and C fibers. Similarly, the analgesic effect of stroking the skin can be accounted for by synaptic interactions at the level of the spinal cord, the thalamus, and the cortex. Particularly appealing is the idea that the level of pain perceived may be reduced by descending influences that involve the action of morphinelike peptides, enabling a soldier who charges into battle to ignore injuries that under normal circumstances would be excruciatingly painful.

THE AUDITORY SYSTEM: ENCODING SOUND FREQUENCY

The somatosensory system analyzes stimuli on the basis of their locations on the body map. By an analogous process, visual stimuli are retinotopically mapped and analyzed for their location in the visual (spatial) world. However, the auditory system employs a significantly different method of analysis. The auditory periphery, the **cochlea**, maps sound stimuli according to their frequency content rather than spatial location, and successive stages of the auditory system employ this tonotopic analysis to determine the meaning of speech and other complex sounds. How is sound frequency encoded by cochlear hair cells, and how are these signals processed at higher levels of the auditory system? How is the spatial location of a sound determined?

In terrestrial vertebrates, sound waves in the air enter the outer ear, strike the **tympanic membrane** (eardrum), and, by a series of mechanical couplings in the middle ear, are converted to fluid waves in the cochlea. The fluid waves, in turn, cause vibration of the **basilar membrane**, on which sit sensory **hair cells** in the **organ of Corti**. This process has been summarized poetically by Aldous Huxley:[46]

> Pongileoni's blowing and the scraping of the anonymous fiddlers had shaken the air in the great hall, had set the glass of the windows looking on to it vibrating; and this in turn had shaken the air in Lord Edward's apartment on the further side. The shaking air rattled Lord Edward's membrana tympani; the interlocked malleus, incus and stirrup bones were set in motion so as to agitate the membrane of the oval window and raise an infinitesimal storm of fluid in the labyrinth. The hairy endings of the auditory nerve shuddered like weeds in a rough sea; a vast number of obscure miracles were performed in the brain, and Lord Edward ecstatically whispered "Bach"!

The Cochlea

The cochlear duct is divided into three compartments: The **scala media** contains a high-potassium solution, the endolymph. It is separated from the overlying **scala vestibuli** by Reissner's membrane and from the fluid space of the **scala tympani** by intercellular tight junctions between the apical ends of the hair cells and their surrounding supporting cells

[42]Basbaum, A. I., and Fields, H. L. 1979. *J. Comp. Neurol.* 187: 513–522.

[43]Fields, H. L., and Besson, J-M. (eds.). Pain Modulation Issue. 1988. *Prog. Brain Res.* 77.

[44]Fields, H. L. 1987. *Pain*. McGraw-Hill, New York.

[45]Fields, H. L., and Basbaum, A. I. 1978. *Annu. Rev. Physiol.* 40: 217–248.

[46]Huxley, A. 1928. *Point Counter Point*. Harper Collins, New York, Chapter 3.

(see Figure 18.11B). The scala tympani and scala vestibuli contain perilymph, similar in composition to cerebrospinal fluid; the scala media contains endolymph, whose ionic composition is like that of cytoplasm: high potassium, low sodium, and with calcium ions held to micromolar concentration.[47,48] This unusual extracellular fluid is established by the pumping actions of cells in the **stria vascularis**, a secretory epithelium that lines the lateral wall of the scala media.

There are two distinct groups of hair cells in the mammalian cochlea: **inner hair cells** and **outer hair cells**. These differ by position (inner hair cells are closer to the central axis of the cochlear coil) and innervation pattern. Inner hair cells receive more than 90% of the afferent contacts to the cochlea;[49–51] outer hair cells are the postsynaptic targets of the efferent nerve supply (see Figure 18.11C).[52] This differential innervation pattern raises interesting questions concerning the functional roles of these two cell types that will be discussed later in this chapter. The **tectorial membrane** overlies the hair bundles of both inner and outer hair cells, and differential motion of the tectorial and basilar membranes causes lateral shear to gate the mechanotransduction channels (as described in Chapter 17).

Frequency Selectivity: Mechanical Tuning

Auditory perception is not simply a matter of detecting a sound stimulus. The information content of sound is a function of its frequency composition. Thus, the auditory system depends on the ability of its mechanosensory hair cells to be frequency-selective. In the mammalian cochlea, frequency tuning depends on the mechanical properties of the specialized structure in which the sensory cells reside, in a manner analogous to the mechanical adaptation provided by the multilamellate capsule of the Pacinian corpuscle. The width and thickness of the basilar membrane vary systematically along the length of the cochlear duct (Figure 18.11A). At the base of the cochlea (nearest the oval window), the basilar membrane is narrow and rigid; at the opposite end (the cochlear apex), it is wide and flex-

[47]Bosher, S. K., and Warrren, R. L. 1978. *Nature* 273: 377–378.

[48]Anniko, M., and Wroblewski, R. 1986. *Hear. Res.* 22: 279–293.

[49]Spoendlin, H. 1972. *Acta Otolaryngol.* 73: 235–248.

[50]Kiang, N. Y., et al. 1982. *Science* 217: 175–177.

[51]Brown, M. C. 1987. *J. Comp. Neurol.* 260: 591–604.

[52]Warr, W. B. 1975. *J. Comp. Neurol.* 161: 159–182.

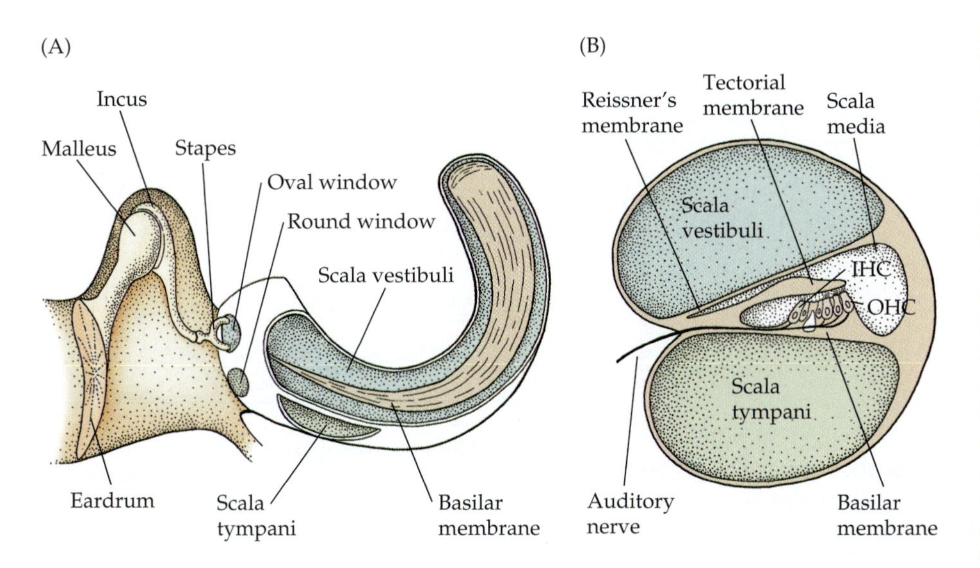

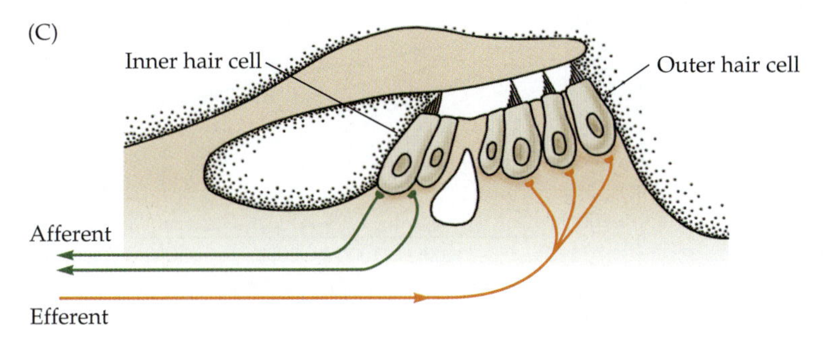

FIGURE 18.11 Structure of the Cochlea. (A) The middle ear and cochlea, showing the eardrum and its bony connections to the oval window. The cochlea is represented as unwound and cut open to show the major internal compartments (scala vestibuli and scala tympani), and overlying structures have been removed to reveal the shape of the basilar membrane, which is narrow near the cochlear base and broad at the apex. (B) Cross section of the cochlea, showing the scala media (which contains endolymph, a high-potassium solution) and the structural relations among the basilar membrane, inner hair cells (IHC) and outer (OHC) hair cells, and tectorial membrane. Hair cells form synapses on the terminals of auditory nerve fibers that have their cell bodies in the spiral ganglion. (C) Innervation of cochlear hair cells. Approximately 95% of afferent fibers are postsynaptic to IHCs. As many as 20 afferent fibers contact a single IHC. OHCs have few afferent contacts but instead are the postsynaptic targets of cholinergic efferent neurons that project from the olivary complex in the brainstem.

ible. Because of this varying stiffness, incoming sound waves cause fluid motions that vibrate the basilar membrane at different positions depending on sound frequency.

This feature of cochlear function was described by von Bekesy,[53] who used stroboscopic illumination of reflective particles scattered on the cochlea to visualize the pattern of vibration. What he observed was that high-frequency sounds cause maximal vibration of the thicker, stiffer basal end of the membrane, and low-frequency sounds cause maximal vibration of the more flexible apical membrane. Von Bekesy referred to this as a traveling-wave pattern of vibration.

As a consequence of mechanical tuning of the basilar membrane, hair cells and their postsynaptic afferent fibers at the cochlear base are preferentially stimulated by high-frequency sound, whereas those at the cochlear apex respond best to low-frequency sound (Figure 18.12). The frequency selectivity of cochlear afferents can be determined by recording action potentials during presentation of pure tones whose frequency and intensity is systematically varied. The resulting **tuning curve** is V-shaped, with a best or characteristic frequency defined as the pure tone to which the fiber is most sensitive. The characteristic frequency of each fiber is determined by where along the cochlear duct it contacts a hair cell.

Sound frequency is mapped along an epithelial array of receptors, producing a cochleotopic, or **tonotopic, map** for frequency. Therefore, the auditory system can use principles of neuronal processing—such as lateral inhibition—based on tonotopy to interpret and process information encoded in sound frequency in the same way that somatotopic maps underlie the processing of somatosensory signals.

Efferent Inhibition of the Cochlea

Like muscle spindles, auditory receptors are subject to efferent regulation. Neurons of the superior olivary complex in the mammalian brainstem project to ipsilateral and contralateral cochleas.[54] Activation of this pathway causes the release of acetylcholine at efferent synapses onto hair cells and suppresses the response to sound of cochlear afferent fibers.[55,56] Efferent feedback reduces cochlear sensitivity in the presence of back-

[53]von Bekesy, G. 1960. *Experiments in Hearing*. McGraw-Hill, New York.

[54]Rasmussen, G. L. 1946. *J. Comp. Neurol.* 84: 141–220.

[55]Galambos, R. 1956. *J. Neurophysiol.* 19: 424–437.

[56]Jasser, A., and Guth, P. S. 1973 *J. Neurochem.* 20: 45–53.

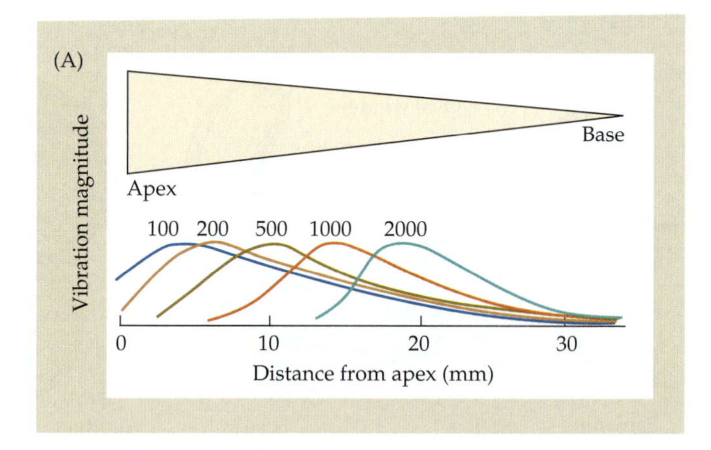

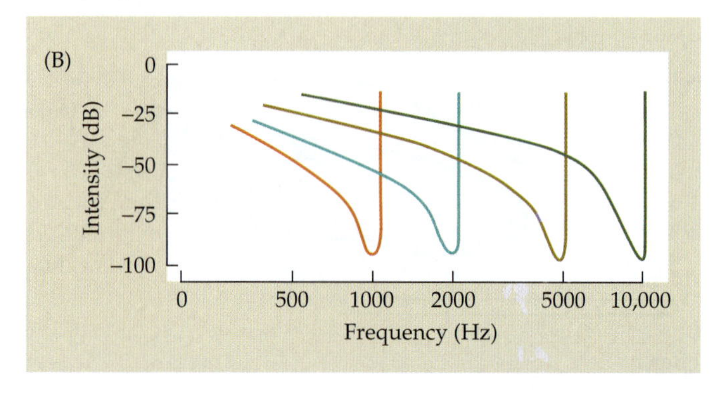

FIGURE 18.12 Cochlear Tuning. (A) The location of maximum displacement of the basilar membrane in the cochlea by sound waves depends on frequency. The curves represent relative displacement at the indicated frequencies (100–2000 Hz). At the low frequencies, maximum displacement is near the wider (more flexible) apical membrane; higher frequencies produce maximal displacement near the narrower (stiffer) base. (B) Typical tuning curves of four individual eighth-nerve fibers innervating different locations on the cochlea. Sound intensity in decibels (dB) needed to produce discharges in a fiber is plotted against frequency of the auditory stimulus. The best frequencies for the fibers (i.e., the frequencies requiring the least stimulus intensity) are 1000, 2000, 5000, and 10,000 Hz. (A after von Bekesy, 1960; B after Katsuki, 1961.)

ground noise and minimizes saturation effects.[57] This is analogous to the resetting of muscle spindle sensitivity by the gamma efferents, maintaining the flow of information during limb movements over a wide range of positions (Chapter 22). Efferent fibers are activated by sound, and they innervate restricted portions of the cochlea; as a result, noise suppression is frequency-specific.[58–61] In addition to restoring dynamic range, inhibition broadens the tuning curve of afferent fibers. Finally, efferent feedback can protect the cochlea from loud sound damage.[62] Indeed, the strength of efferent feedback is inversely correlated with the degree of acoustic injury from loud sound.[63]

Activation of the efferent pathway to the turtle ear causes large hyperpolarizing inhibitory postsynaptic potentials (IPSPs) in hair cells.[64,65] The effect of inhibition on the response of a hair cell to sound is shown in Figure 18.13A. The cell was stimulated with pure tones at three frequencies, one that corresponded to the cell's best or characteristic frequency, one at a higher frequency, and one at a lower frequency. The intensities of the tones were adjusted so that each evoked an oscillating receptor potential of the same amplitude. A short train of shocks to the efferent axons hyperpolarized the cell and severely attenuated its response to a tone at 220 Hz (the characteristic frequency). At lower and higher frequencies of acoustic stimulation, activation of the efferents still resulted in hyperpolarization, but the low-frequency oscillations were actually enhanced in amplitude, whereas those at high frequencies were unchanged. This differential effect of inhibition results in broadening of the hair cell's frequency response (Figure 18.13B).

The mechanism of cholinergic inhibition has been studied in chick hair cells.[66,67] The ACh receptors are ligand-gated cation channels through which sodium and calcium enter

[57]Winslow, R. L., and Sachs, M. B. 1987. *J. Neurophysiol.* 57: 1002–1021.

[58]Liberman, M. C. 1988. *J. Neurophysiol.* 60: 1779–1798.

[59]Robertson, D., Anderson, C-J., and Cole, K. S. 1987. *Hear. Res.* 25: 69–76.

[60]Liberman, M. C., and Brown, M. C. 1986. *Hear. Res.* 24: 17–36.

[61]Wiederhold, M. L., and Kiang, N. Y. S. 1970. *J. Acoust. Soc. Am.* 48: 950–965.

[62]Rajan, R. 1995. *J. Neurophysiol.* 74: 598–615.

[63]Maison, S. F. and Liberman, M. C. 2000. *J. Neurosci.* 20: 4701–4707.

[64]Art, J. J., Fettiplace, R., and Fuchs, P. A. 1984. *J. Physiol.* 356: 525–550.

[65]Art, J. J., et al. 1985. *J. Physiol.* 360: 397–421.

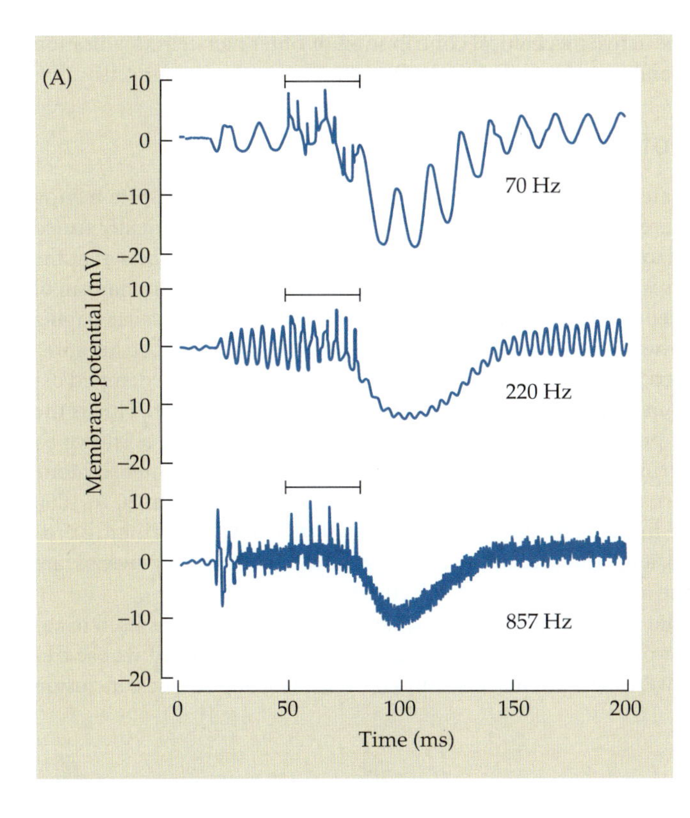

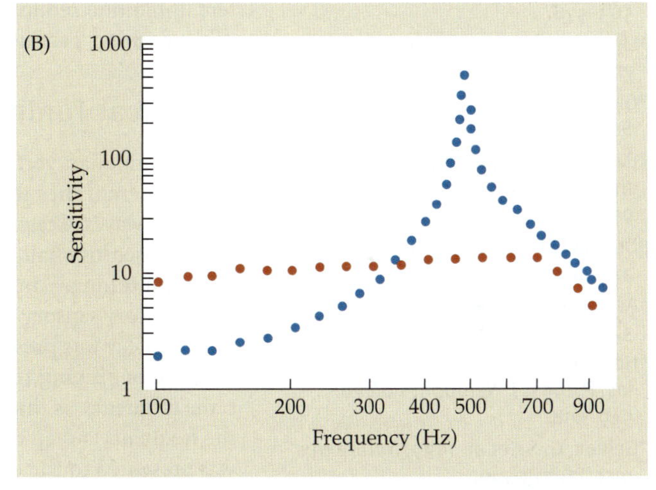

FIGURE 18.13 Effect of Efferent Stimulation on cochlear hair cell responses to acoustic stimuli. (A) Center record: The oscillatory response of a cell to an acoustic stimulus of 220 Hz (its resonant frequency) is inhibited by a brief train of efferent stimuli (indicated by the bar), and the cell is hyperpolarized. Upper record: The response to an acoustic stimulus at 70 Hz. The stimulus intensity was adjusted so that the oscillatory response was similar in magnitude to that at 220 Hz. Efferent stimulation produces a hyperpolarization, but an increase rather than a decrease in the oscillatory response. Lower record: The oscillatory response to an 857 Hz stimulus (again adjusted to produce a response similar to that at the resonant frequency) is unchanged by efferent stimulation. (B) Sensitivity of another cell (in millivolts per unit of sound pressure) as a function of frequency in the absence (blue) and presence (red) of efferent stimulation. Efferent inhibition reduces the response at the resonant frequency and increases the sensitivity at low frequencies, thereby degrading tuning. (A after Art et al., 1985; B after Fettiplace, 1987.)

the cell, leading to the activation of calcium-dependent potassium channels and membrane hyperpolarization (Chapter 10). A similar cholinergic inhibitory mechanism is found in mammalian hair cells.[68–70] The hair cell's response to ACh is antagonized by α-bungarotoxin, and there is good evidence that an unusual member of the nicotinic receptor family of genes, α9, is a ligand-binding subunit of the hair cell ACh receptor.[71–73]

Electromotility of Mammalian Cochlear Hair Cells

Efferent neurons that inhibit the cochlear sound response synapse onto outer hair cells, but not onto inner hair cells that generate the neural output of the mammalian cochlea. How is the inhibitory effect communicated from the outer to the inner hair cell?

An important property of outer hair cells in the mammalian cochlea is **electromotility**. Outer hair cells isolated from the mammalian cochlea shorten in response to depolarization and lengthen during hyperpolarization of their membranes (Figure 18.14).[74,75] These movements are not generated by actin–myosin but result from direct effects of voltage on charged "motor" proteins in the basolateral membrane.[76–78] Whatever the mechanism, electromotility of outer hair cells adds to the vibration of the cochlear partition during stimulation by sound, and thus increases the deflection of the stereocilia of the inner hair cell.[79]

Outer hair cells enhance cochlear tuning by adding mechanical energy to movement of the basilar membrane (Figure 18.15). This explains how efferent synapses made on outer hair cells suppress the sound response of inner hair cells. Since outer hair cell motility is driven by changes in membrane potential, synaptic inhibition acts to suppress electromotility in a manner analogous to inhibitory effects elsewhere in the nervous system. Inhibition reduces the active mechanical contribution of outer hair cells to vibration of the cochlear partition, and so reduces indirectly the excitation of inner hair cells.

Electrical Tuning of Hair Cells

Nonmammalian vertebrate ears achieve acoustic frequency selectivity despite having basilar membranes that are not elongated and so do not support mechanically tuned traveling waves. Crawford and Fettiplace undertook studies in the turtle to show that the mechanosensory hair cells were intrinsically frequency-selective through a mechanism of electrical tuning.[80] Intracellular recordings from a hair cell in the turtle's basilar papilla (the auditory sensory epithelium) are shown in Figure 18.16. When a brief acoustic stimulus (a click) was presented, the hair cell's membrane potential underwent a damped oscillation or ringing response that occurred at a frequency of about 350 Hz. This is the same frequency as that of pure tones to which the hair cell is most sensitive, as shown by the frequency sweep experiment shown in Figure 18.16C. Here a constant intensity tone was presented to the external ear and its frequency was gradually swept from 20 to 1000 Hz. The voltage response in the hair cell peaked near 350 Hz. The voltage ringing produced by an acoustic transient, and the tuning curve produced by frequency sweeps, are equivalent measures of the hair cell's frequency selectivity.

Figure 18.16B shows the important result that when a microelectrode was used to inject a rectangular current pulse into the cell, voltage ringing was produced whose frequency and rate of decay were identical to those caused by acoustic input. The conclusion

[66]Fuchs, P. A., and Murrow, B. W. 1992. *J. Neurosci.* 12: 800–809.
[67]Martin, A. R., and Fuchs, P. A. 1992. *Proc. R. Soc. Lond. B* 250: 71–76.
[68]Housley, G., and Ashmore, J. 1991. *Proc. R. Soc. Lond. B* 244: 161–167.
[69]Blanchet, C., et al. 1996. *J. Neurosci.* 16: 2574–2584.
[70]Evans, M. 1996. *J. Physiol.* 491: 563–578.
[71]Elgoyhen, A. B., et al. 1994. *Cell* 79: 705–715.
[72]Glowatzki, E., et al. 1995. *Proc. R. Soc. Lond. B* 262: 141–147.
[73]Park, H-J., Niedzielski, A. S., and Wenthold, R. J. 1997. *Hear. Res.* 112: 95–105.
[74]Brownell, W. E., et al. 1985. *Science* 227: 194–196.
[75]Ashmore, J. F. 1987. *J. Physiol.* 388: 323–347.
[76]Hallworth, R., Evans, B. N., and Dallos, P. 1993. *J. Neurophysiol.* 70: 549–558.
[77]Geleoc, G. S., et al. 1999. *Nature Neurosci.* 2: 713–719.
[78]Zheng, J., et al. 2000. Nature 405: 149–55.
[79]Patuzzi, R. 1996. In *The Cochlea.* Springer, New York, pp. 186–257.
[80]Crawford, A. C., and Fettiplace, R. 1981. *J. Physiol.* 312: 377–412.

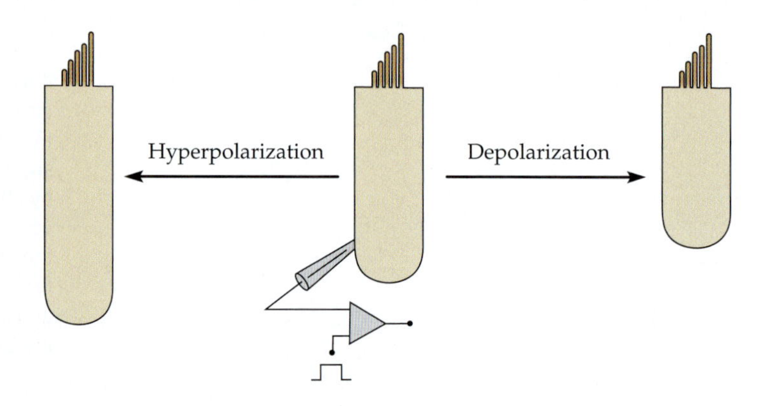

FIGURE 18.14 Electromotility of Outer Hair Cells. Whole-cell patch pipettes are used to clamp the membrane potential of outer hair cells isolated from the mammalian cochlea. Depolarization causes the cells to shorten; hyperpolarization makes them longer. Such length changes can be as large as 30 nm/mV.

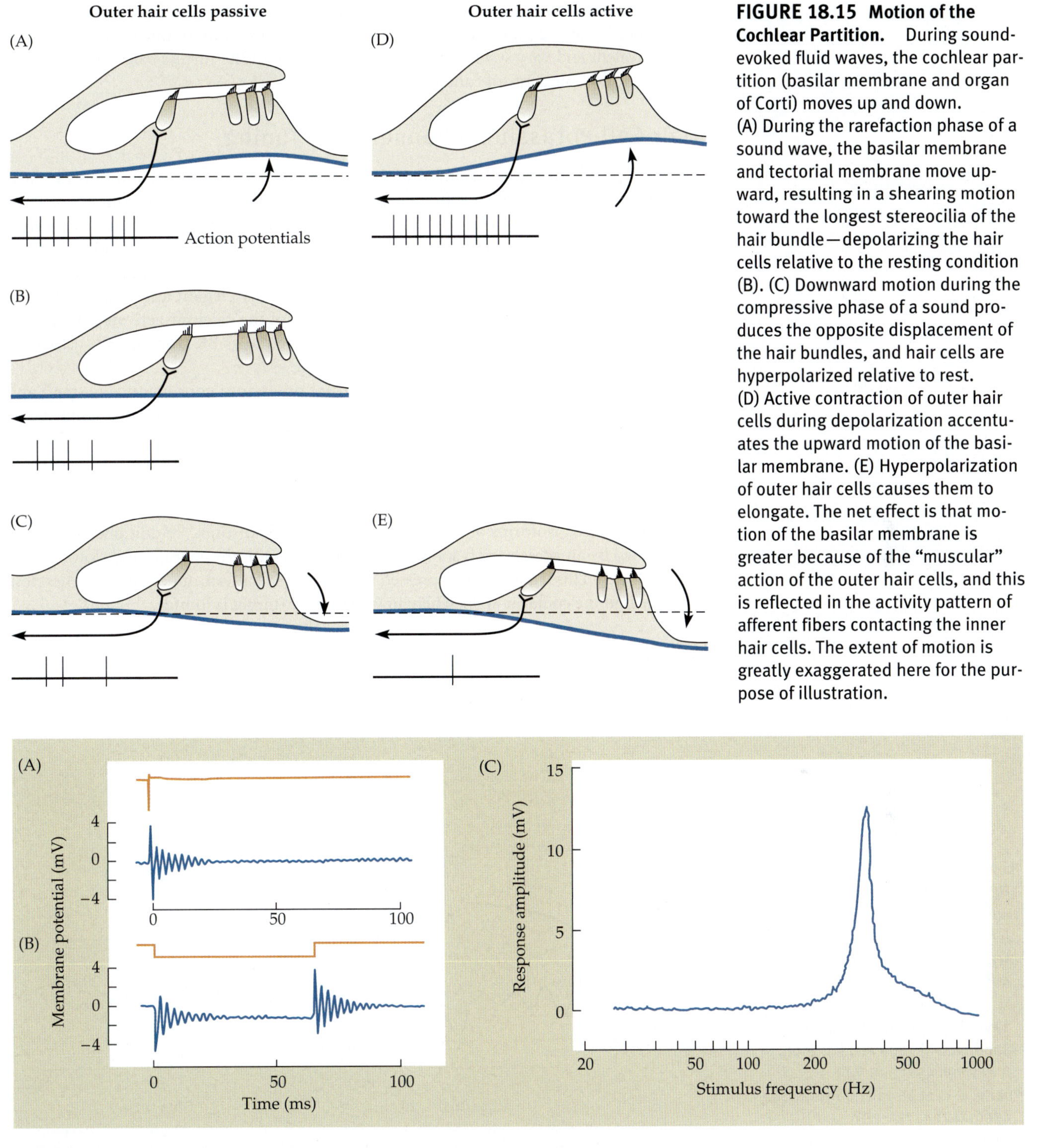

FIGURE 18.15 Motion of the Cochlear Partition. During sound-evoked fluid waves, the cochlear partition (basilar membrane and organ of Corti) moves up and down. (A) During the rarefaction phase of a sound wave, the basilar membrane and tectorial membrane move upward, resulting in a shearing motion toward the longest stereocilia of the hair bundle—depolarizing the hair cells relative to the resting condition (B). (C) Downward motion during the compressive phase of a sound produces the opposite displacement of the hair bundles, and hair cells are hyperpolarized relative to rest. (D) Active contraction of outer hair cells during depolarization accentuates the upward motion of the basilar membrane. (E) Hyperpolarization of outer hair cells causes them to elongate. The net effect is that motion of the basilar membrane is greater because of the "muscular" action of the outer hair cells, and this is reflected in the activity pattern of afferent fibers contacting the inner hair cells. The extent of motion is greatly exaggerated here for the purpose of illustration.

FIGURE 18.16 Hair Cell Tuning in the turtle cochlea. (A) The effect of an acoustic click (indicated by the upper trace) on the membrane potential of a hair cell (lower trace, relative to the resting membrane potential of −50 mV), recorded with an intracellular microelectrode. The click produces a damped oscillation in membrane potential at a frequency of about 350 Hz, with an initial peak-to-peak amplitude of about 8 mV. (B) A hyperpolarizing current pulse (upper trace) applied to the same cell produces similar oscillations at both the onset and termination of the pulse, indicating that the frequency of oscillation is an intrinsic electrical property of the hair cell. (C) When a hair cell with oscillatory responses such as those seen in A and B is stimulated with pure tones ranging from 25 to 1000 Hz, the peak-to-peak amplitude of the receptor potential has a sharp maximum at about 350 Hz. (After Fettiplace, 1987.)

from these experiments is that the hair cell's frequency selectivity depends on the electrical properties of the membrane. Electrical and acoustic tuning frequencies of hair cells are equivalent and vary systematically along the length of the turtle's basilar papilla, producing a tonotopic array of tuned detectors.

Hair Cell Potassium Channels and Tuning

What properties of the cell membrane provide this electrical tuning, and how do these properties vary to determine different tuning frequencies? Recordings from hair cells in the frog's saccule (a vestibular sensory organ) showed that interactions between voltage-gated calcium channels and calcium-activated potassium channels can produce voltage ringing.[81,82] Studies of hair cells isolated from turtle ear[83,84] showed that the characteristic frequency of each cell is determined in a remarkably elegant and straightforward way—namely, by the density and kinetic properties of the calcium-activated potassium channels in each cell (Figure 18.17). These are referred to as BK (for "big K^+") channels because of their large single-channel conductance.

In cells tuned to lower frequencies, the total potassium conductance is small and slower to activate, and thus gives rise to relatively slow voltage oscillations. In higher-frequency cells, the potassium conductance is larger and more rapidly activating. In the lowest-frequency hair cells a still slower, purely voltage-gated potassium channel supports the oscillation.[85] The BK channels in hair cells of chicks[86] and turtles[87] are encoded by a gene whose mRNA is subject to alternative splicing of its composite exons,[88,89] and some differentially spliced isoforms of the channel are kinetically distinct.[90] Additional variability is provided by an accesssory β subunit that combines with the channel and slows its gating kinetics.[91] This β subunit is expressed in a tonotopic pattern, appearing at highest levels in cells at the cochlear apex, where low-frequency tuning is found.

The Auditory Pathway

The main auditory pathways are illustrated schematically in Figure 18.18. Auditory fibers of the eighth nerve travel centrally and send branches to both the dorsal and ventral

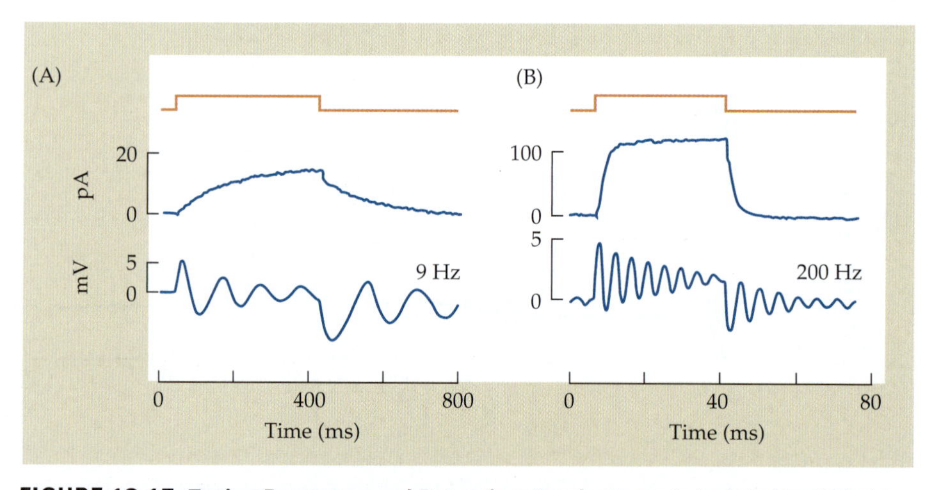

FIGURE 18.17 Tuning Frequency and Potassium Conductance in isolated turtle hair cells measured by whole-cell patch clamp recording. (A) The middle record shows outward current, carried mainly by potassium, produced by a depolarizing voltage command (duration indicated in the top record). Current rises slowly to a maximum of 15 pA, with a time constant of about 200 ms. A small current step of the same duration produces oscillatory voltage responses at the beginning and end of the pulse (lower record) with a resonant frequency of 9 Hz. (B) In another cell, a similar depolarizing pulse produces a much larger, rapidly rising outward current (middle record; note the changes in current and timescales), indicating a greater density of potassium channels with faster kinetics. The oscillatory response to a small current pulse reveals a concomitant increase in tuning frequency to 200 Hz (lower record). (After Fettiplace, 1987.)

[81]Lewis, R. S., and Hudspeth, A. J. 1983. *Nature* 304: 538–541.

[82]Hudspeth, A. J., and Lewis, R. S. 1988. *J. Physiol.* 400: 237–274.

[83]Art, J. J., and Fettiplace, R. 1987. *J. Physiol.* 385: 207–242.

[84]Art, J. J., Fettiplace, R., and Wu, Y-C. 1995. *J. Gen. Physiol.* 105: 49–72.

[85]Goodman, M., and Art, J. J. 1996. *J. Physiol.* 497: 395–412.

[86]Jiang, G-J., et al. 1997. *Proc. R. Soc. Lond. B* 264: 731–737.

[87]Jones, E. M. C., Laus, C., and Fettiplace, R. 1998. *Proc. R. Soc. Lond. B* 265: 685–692.

[88]Navaratnam, D. S., et al. 1997. *Neuron* 19: 1077–1085.

[89]Rosenblatt, K. P., et al. 1997. *Neuron* 19: 1061–1075.

[90]Jones, E. M. C., Gray-Keller, M., and Fettiplace, R. 1999. *J. Physiol.* 518: 653–665.

[91]Ramanathan, K., et al. 1999. *Science* 283: 215–217.

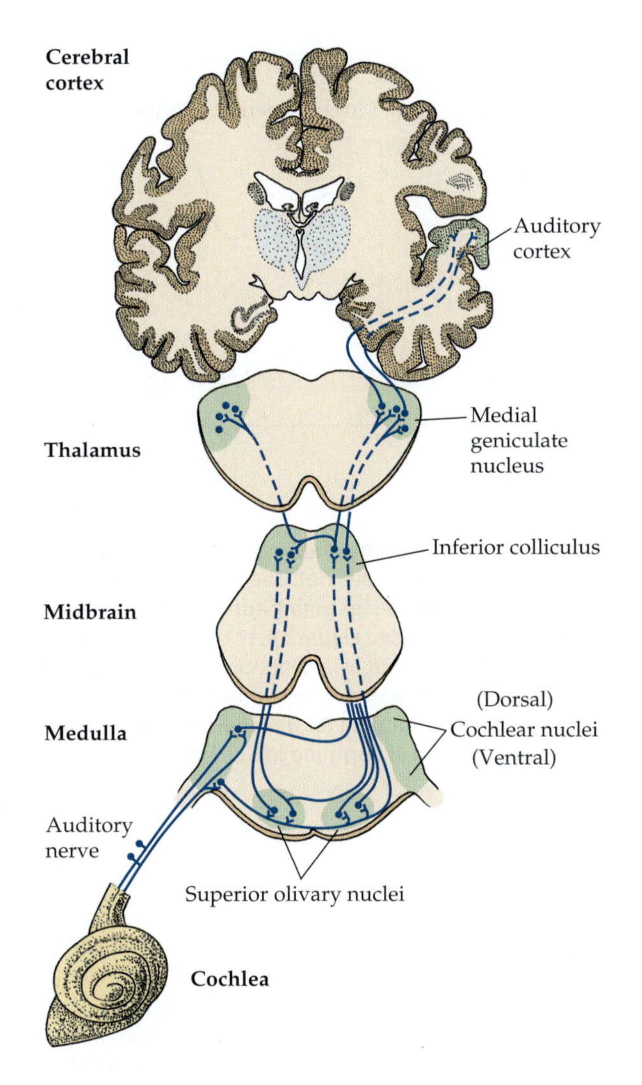

FIGURE 18.18 Central Auditory Pathways shown schematically on transverse sections of the medulla, midbrain, and thalamus, and a coronal section of the cerebral cortex. Auditory nerve fibers end in the dorsal and ventral cochlear nuclei. Second-order fibers ascend to the contralateral inferior colliculus; those from the ventral cochlear nucleus also supply collaterals bilaterally to the superior olivary nuclei. Further bilateral interaction occurs at the level of the inferior colliculus. Fibers then ascend to the medial geniculate nucleus of the thalamus and hence to the auditory cortex. (After Berne and Levy, 1988.)

cochlear nuclei.[92] Second-order axons ascend in the contralateral lateral lemniscus to innervate cells in the inferior colliculus (the nucleus of the lateral lemniscus is a synaptic way station for some of these fibers). Neurons in the ventral cochlear nucleus also provide collateral branches to both the ipsilateral and contralateral superior olivary nuclei. Third-order cells in the olivary nuclei, in turn, send ascending fibers to the inferior colliculus. The pathway continues through the medial geniculate nucleus of the thalamus to the auditory region of the transverse surface of the temporal lobe of the cerebral cortex. Each level in the auditory pathway is tonotopically mapped.

As one ascends higher in the auditory system, pure tones become less and less important as stimuli for individual cells. For instance, cells in the dorsal cochlear nucleus and inferior colliculus are inhibited by tones off their characteristic frequency.[93,94] Cells in the auditory cortex respond to combinations of tones (e.g., sweeping upward or downward in frequency), to binaural but not monaural sounds, and to other complex auditory stimuli.[95–97] When cells are found that do respond to pure tones, their frequency selectivity often is enhanced over that of the primary afferent fibers. Thus, the tuning curve (threshold versus frequency) of the cortical neuron is sharper than that of the primary afferent[98] (Figure 18.19A). This sharpening of frequency response corresponds to a sharpening of spatial localization on the basilar membrane and is accomplished by lateral inhibition, just as in the somatosensory system. Thus, the receptive field of this cortical neuron is an excitatory strip on the basilar membrane flanked on either side by inhibitory bands. This translates to a narrow frequency range that excites the cell, flanked by higher and lower frequencies that are inhibitory (see Figure 18.19B). Inhibitory sidebands are generated in part by descending feedback from the cortex itself that shapes the response properties of cells in lower levels.[99]

[92]Fekete, D. M., et al. 1984. *J. Comp. Neurol.* 229: 432–450.

[93]Young, E. D. 1984. In *Hearing Science*. College-Hill Press, San Diego, CA, pp. 423–460.

[94]Ramachandran, R., Davis, K. A., and May, B. J. 1999. *J. Neurophysiol.* 82: 152–163.

[95]Shamma, S. A., et al. 1993. *J. Neurophysiol.* 69: 367–383.

[96]Wang, X., et al. 1995. *J. Neurophysiol.* 74: 2685–2706.

[97]Rauschecker, J. P. 1998. *Curr. Opin. Neurobiol.* 8: 516–521.

[98]Arthur, R. M., Pfeiffer, R. R., and Suga, N. 1971. *J. Physiol.* 212: 593–609.

[99]Suga, N., Zhang, Y., and Yan, J. 1997. *J. Neurophysiol.* 77: 2098–2114.

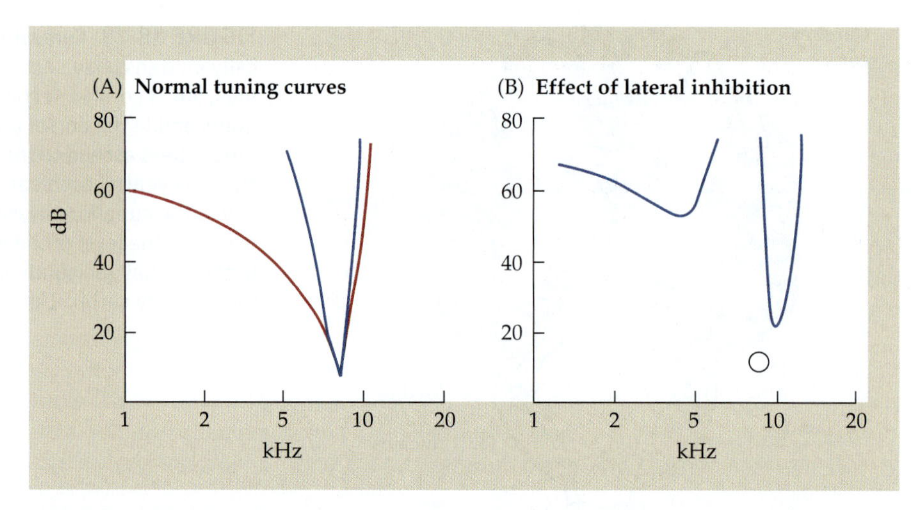

FIGURE 18.19 Lateral Inhibition of a Neuron in the Auditory Cortex. (A) Tuning curve for a neuron in the auditory cortex of a cat (blue line). The best frequency is about 7 kHz. The red line shows, for comparison, the broader tuning curve of a mammalian auditory nerve fiber with the same best frequency (see Figure 18.12). (B) Lateral inhibition of the same cortical neuron. A response is produced by a continuous tone at the best frequency (open circle); an additional tone is then added, which inhibits the best-frequency response. Curves indicate the sound pressure required for the additional tone to reduce the response by 20%, thereby defining the area of surround inhibition. (After Arthur, Pfeiffer, and Suga, 1971.)

Auditory Cortex

Auditory input processed through both dorsal and ventral cochlear nuclei ascends to the auditory cortex. The **primary auditory cortex** (A_1) is located on the superior bank of the temporal lobe, corresponding to Brodmann's areas 41 and 42 (see the figure in Box 18.1). In cats, A_1 is exposed on the lateral surface of the brain, so most combined anatomical–physiological studies have been performed in this species.[100] Microelectrode studies have shown that A_1 also has columnar organization, with cells along a vertical track all having the same best frequency.[101–103] In fact, A_1 is organized into isofrequency bands or slabs of cortex running orthogonal to the tonotopic axis.

The auditory cortex in monkeys contains three complete cochleotopic maps, with parallel projections from the medial geniculate nucleus to all these areas. The most posterior of the three corresponds to primary auditory cortex on the basis of histological features.[104] Surrounding this central core are secondary auditory areas that interconnect with primary cortex, but also with subdivisions of the medial geniculate nucleus. Thus, both serial and parallel processing take place in the auditory cortex.[105] Electroencephalographic mapping and functional magnetic resonance studies provide evidence for similar organization of auditory cortex in humans,[106,107] and this area of cortex is activated even during silent lipreading![108]

By analogy to other sensory cortices, one would expect the cochleotopic map of A_1 to be subdivided into different functional zones. If one moves an electrode along an isofrequency slab, most cells are either excited by sound in either ear (EE), or excited by one ear and inhibited by the other (EI) (Figure 18.20B). These binaural sensitivities vary systematically and are thought to constitute strips running at right angles to the isofrequency contours.[109] Some evidence also exists for systematic variations of intensity and bandwidth coding in A_1.[110–112] The EE–EI organization may be derived from binaural interactions that first arise in the olivary nuclei (see the next section) and has led to the suggestion that A_1 also is mapped for auditory space, although at present this is not proven.

The processing of auditory signals is complex.[113] Behaviorally important sounds must be extracted from a rich and variable acoustic environment.[114] Not only the frequency content but also the sequence in time of incoming sounds must be analyzed in some way[115] (e.g., playing a tape recording of human speech backward produces gibberish). In

[100]Woolsey, C. N. 1972. In *Physiology of the Auditory System*. National Educational Consultants, Baltimore, MD, pp. 271–282.

[101]Merzenich, M. M., and Brugge, J. F. 1973. *Brain Res.* 50: 275–296.

[102]Merzenich, M. M., Knight, P. L., and Roth, G. L. 1975. *J. Neurophysiol.* 38: 231–249.

[103]Reale, R. A., and Imig, T. J. 1980. *J. Comp. Neurol.* 192: 265–291.

[104]Kaas, J. H., Hackett, T. A., and Tramo, M. J. 1999. *Curr. Opin. Neurobiol.* 9: 164–170.

[105]Rauschecker, J. P., et al. 1997. *J. Comp. Neurol.* 382: 89–103.

[106]Romani, G. L., Williamson, S. J., and Kaufman, L. 1982. *Science* 216: 1339–1340.

[107]Pantev, C., et al. 1995. *Electroencephalogr. Clin. Neurophysiol.* 94: 26–40.

[108]Calvert, G. A., et al. 1997. *Science* 276: 593–596.

[109]Middlebrooks, J. C., Dykes, R. W., and Merzenich, M. M. 1980. *Brain Res.* 181: 31–48.

[110]Sutter, M. L., and Schreiner, C. E. 1995. *J. Neurophysiol.* 73: 190–204.

[111]Hiel, P., Rajan, R., and Irvine, D. 1994. *Hear. Res.* 76: 188–202.

[112]Clarey, J. C., Barone, P., and Imig, T. J. 1994. *J. Neurophysiol.* 72: 2383–2405.

[113]Sachs, M. B. 1984. *Annu. Rev. Physiol.* 46: 261–273.

[114]Nelken, I., Rotman, Y., and Bar Yosef, O. 1999. *Nature* 397: 154–157.

[115]Griffiths, T. D., et al. 1998. *Nature Neurosci.* 1: 422–427.

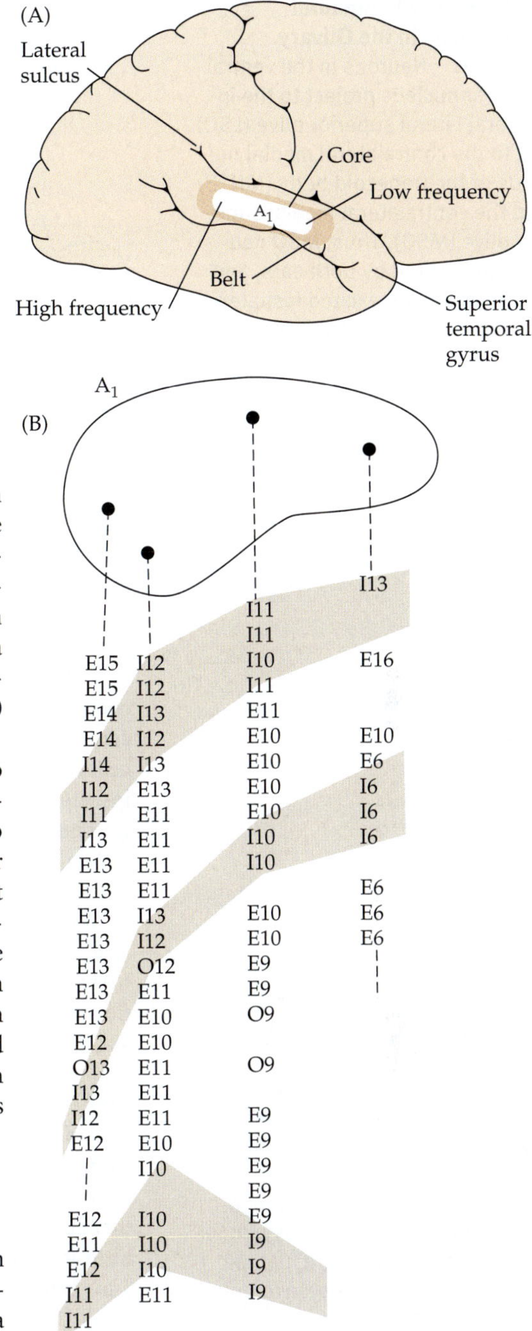

FIGURE 18.20 Auditory Cortex. (A) Afferents from the medial geniculate nucleus project to a region of the superior temporal gyrus of the monkey brain, shown here as though the lateral sulcus were spread open. A central core region (A_1) is tonotopically mapped, with low frequencies found anteriorly and high frequencies found posteriorly. Additional tonotopic areas are found in a surrounding belt region. (B) Successive tangential electrode penetrations were made along isofrequency regions of A_1 in a cat, and the effects of binaural versus monaural sound were examined. For each cell encountered, the effect of contralateral sound (E = excitatory; I = inhibitory; O = no effect) and best frequency are listed. As expected, the best frequency was relatively constant in any penetration. However, binaural stimulation enhanced (E), inhibited (I), or left unchanged (O) the monaural response, and these effects occurred in clusters. (B modified from Middlebrooks, Dykes, and Merzenich, 1980.)

addition, such processing in any animal species must deal not only with analysis of environmental sound, but also with vocalization within the species. In humans the basic elements of speech, called **phonemes**, are common to all languages and are the sounds first babbled by babies before particular ones are selected to be combined into words.[116] The basic sounds can be analyzed as combinations of frequency–time relations—for example, a continuous component at 1000 Hz accompanied by a second, frequency-modulated component starting at 5000 Hz and descending rapidly to 500 Hz. The components are called **formants**.

By analogy with the visual system, which contains cells that respond to slits, corners, edges, and other geometrical forms (Chapter 20), we might expect to find higher-order cells in the human auditory cortex that respond to particular formants, or perhaps phonemes. This principle does hold for other animals. For example, some cells in the auditory cortex of the mustached bat respond to specific combinations of constant-frequency and frequency-modulated tones that are equivalent to the bat's own vocalizations.[117] At the same time, we might expect to search in vain for cells responsive to human sound elements in the auditory systems of bats or other mammals. Such higher-order cells *are* found, however, in mynah birds that have been trained to talk.[118] Cortical mechanisms of language coding may be discovered in species like the marmoset, which uses a limited and stereotypic series of calls ("words") for communication.[96]

Sound Localization

The impressive sensitivity and frequency selectivity of the auditory system were present in animals long before human language evolved. Rather, auditory system refinements arose to improve the organism's ability to locate a sound in space. The advantages of doing so are obvious; long-range signals emitted as sound waves can reveal predator or prey in the absence of visual or other cues. However, in contrast to the visual or somatosensory systems, the auditory neuroepithelium cannot be used for coding location, since it is dedicated to tonotopic mapping. Instead, sound location is computed from binaural comparisons of timing and intensity that occur within the central auditory system. The auditory pathway, therefore, is correspondingly complex, involving numerous subcortical synaptic relays and extensive crossing over at nearly every level.

The **dorsal cochlear nucleus** is largely for monaural frequency analysis[119] and provides a relatively direct, tonotopically organized projection onto contralateral A_1. In contrast, second-order neurons in the **ventral cochlear nucleus** project both ipsilaterally and contralaterally to the superior olivary complex in the brainstem. Most neurons in the **medial superior olive** (MSO) are excited by stimulation of either ear (and so correspond to EE neurons; see Figure 18.21) but respond best when a tone is presented to the two ears with a

[116]De Boysson-Bardies, B., et al. 1989. *J. Child Lang.* 16: 1–17.

[117]Tsuzuki, K., and Suga, N. 1988. *J. Neurophysiol.* 60: 1908–1923.

[118]Langer, G., Bronke, D., and Scheich, H. 1981. *Exp. Brain Res.* 43: 11–24.

[119]Young, E. D., et al. 1992. *Philos. Trans. R. Soc. Lond. B* 336: 407–413.

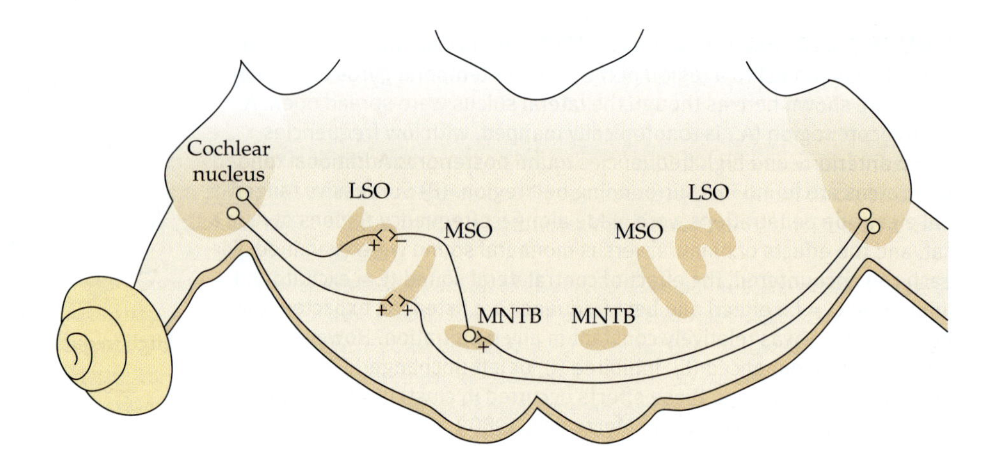

FIGURE 18.21 Binaural Connections in the Olivary Complex. Neurons in the ventral cochlear nucleus project to the ipsilateral lateral superior olive (LSO) but to the contralateral medial nucleus of the trapezoid body (MNTB) and the contralateral medial superior olive (MSO). Thus, MSO neurons are excited by both ears, and LSO neurons are excited ipsilaterally but inhibited contralaterally by way of the intervening inhibitory interneuron in the MNTB.

characteristic delay, corresponding to arrival first at one ear, then the other. The velocity of sound in air is 340 m/s, so the maximal time difference imparted by the human head (about 18 cm in diameter) is 0.5 ms for a sound arriving along the axis of the two ears, and successively more frontal locations impart smaller time differences. In addition to differences in arrival time, continuous sound sources give rise to phase differences at the two ears.

Cells in the **lateral superior olive** (**LSO**) receive excitation from the ipsilateral ventral cochlear nucleus (Figure 18.21). Cells in the contralateral ventral cochlear nucleus project across the midline to synapse in the **medial nucleus of the trapezoid body** (**MNTB**). The cells of the MNTB inhibit neurons in the LSO. Thus, neurons in the LSO are excited by ipsilateral but inhibited by contralateral sound (EI). Such an interaction would be useful for detecting differences in the intensity of sound at the two ears. As much as a 10-fold difference in intensity is found at high frequencies, for which the head serves as an effective sound shadow.

Both phase and intensity differences vary as a function of frequency. For the human head, phase differences are significant below 2 kHz, whereas intensity differences become more prominent at higher frequencies. Psychophysical studies have shown that localization is accomplished by combining differences between the two ears in time of arrival and/or intensity of the incoming sound.[120,121] Thus, if clicks are presented through earphones with different delays, the sound is localized toward the ear in which the click arrives first. If the clicks are made simultaneous, but of different intensities, localization is to the side with the loudest click. In fact, localization cues depend precisely on frequency composition. The head and external ear filter out specific frequency components depending on where in space a sound originates.[122] These **spectral notches** provide critical signals that are used to localize sound.[123,124] Humans can detect interaural time differences of as little as 5 μs, remarkable resolution considering that action potentials are approximately 1 ms in duration—and emphasizing the importance of precise timing to auditory function.

[120]Blauert, J. 1982. *Scand. Audiol. Suppl.* 15: 7–26.

[121]Buell, T. N., and Hafter, E. R. 1988. *J. Acoust. Soc. Am.* 84: 2063–2066.

[122]Rice, J. J., et al. 1992. *Hear. Res.* 58: 132–152.

[123]May, B. J., and Huang, A. Y. 1996. *J. Acoust. Soc. Am.* 100: 1059–1069.

[124]Huang, A. Y., and May, B. J. 1996. *J. Acoust. Soc. Am.* 100: 1070–1080.

Summary

- Somatosensory receptors in the skin respond to a variety of stimuli, such as touch, pressure, vibration, and change in temperature. The receptors differ in their rates of adaptation and in their receptive field size.

- Adaptation of sensory responses depends on the mechanical properties of accessory structures, and on changes in excitability. Rapidly adapting receptors are "tuned" to higher-frequency stimulation.

- The central somatosensory pathway is somatotopically mapped. The somatotopic map of cortex is subdivided into functional areas, reflecting the stimulus specificity of peripheral receptors.

SUMMARY (CONTINUED)

- The receptive fields of neurons in somatosensory cortex show center–surround organization.
- Mechanical tuning of the basilar membrane confers frequency selectivity onto the responses of auditory afferents in mammals. Voltage-driven motility of outer hair cells contributes to acoustic sensitivity of the mammalian cochlea.
- Electrical tuning provides frequency selectivity in hair cells of lower vertebrates. Voltage-gated calcium channels and calcium-activated potassium channels interact to enhance the voltage response at a frequency that corresponds to the characteristic acoustic frequency for each hair cell.
- Cochlear hair cells are subject to efferent inhibition by cholinergic brainstem neurons. Efferent inhibition alters the sensitivity and tuning of cochlear afferents.
- The central auditory pathway, including cortex, is tonotopically mapped. Response properties of cells in auditory cortex are complex, showing binaural interactions and dependence on temporal combinations of tones.
- Binaural comparisons of sound intensity and timing are used to compute the locations of sounds in space. These computations are made with synaptic connections in nuclei of the superior olive.

SUGGESTED READING

General Reviews

Fettiplace, R., and Fuchs, P. A. 1999. Mechanisms of hair cell tuning. *Annu. Rev. Physiol.* 61: 809–834.

Kaas, J. H., Hackett, T. A., and Tramo, M. J. 1999. Auditory processing in primate cerebral cortex. *Curr. Opin. Neurobiol.* 9: 164–170.

Mountcastle, V. B. 1995. The parietal system and some higher brain functions. *Cerebral Cortex* 5: 377–390.

Mountcastle, V. B. 1997. The columnar organization of the neocortex. *Brain* 120: 701–722.

Patuzzi, R. 1996. Cochlear micromechanics and macromechanics. In P. Dallos, A. N. Popper, and R. R. Fay (eds.), *The Cochlea*. Springer, New York, pp. 186–257.

Perl, E. R. 1996. Cutaneous polymodal receptors: Characteristics and plasticity. *Prog. Brain Res.* 113: 21–37.

von Bekesy, G. 1960. *Experiments in Hearing*. McGraw-Hill, New York.

Willis, W. D., and Westlund, K. N. 1997. Neuroanatomy of the pain system and of the pathways that modulate pain. *J. Clin. Neurophysiol.* 14: 2–31.

Original Papers

Art, J. J., and Fettiplace, R. 1987. Variation of membrane properties in hair cells isolated from the turtle cochlea. *J. Physiol.* 385: 207–242.

Art, J. J., Fettiplace, R., and Fuchs, P. A. 1984. Synaptic hyperpolarisation and inhibition of turtle cochlea hair cells. *J. Physiol.* 356: 525–550.

Ashmore, J. F. 1987. A fast motile response in guinea-pig outer hair cells: The cellular basis of the cochlear amplifier. *J. Physiol.* 388: 323–347.

Crawford, A. C., and Fettiplace, R. 1981. An electrical tuning mechanism in turtle cochlear hair cells. *J. Physiol.* 312: 377–412.

Harris, J. A., Petersen, R. S., and Diamond, M. 1999. *Proc. Natl. Acad. Sci. USA* 96: 7587–7591.

Hudspeth, A. J., and Lewis, R. S. 1988. Kinetic analysis of voltage- and ion-dependent conductances in saccular hair cells of the bull-frog, *Rana catesbeiana*. *J. Physiol.* 400: 237–274.

Johansson, R. S., and Vallbo, A. B. 1979. Detection of tactile stimuli. Thresholds of afferent units related to psychophysical thresholds in the human hand. *J. Physiol.* 297: 405–422.

Middlebrooks, J. C., Dykes, R. W., and Merzenich, M. M. 1980. Binaural response-specific bands in primary auditory cortex (AI) of the cat: Topographical organization orthogonal to isofrequency contours. *Brain Res.* 181: 31–48.

Mountcastle, V. B. 1957. Modality and topographic properties of single neurons of cat's somatic sensory cortex. *J. Neurophysiol.* 20: 408–434.

Nelken, I.; Rotman, Y., and Bar Yosef, O. 1999. Responses of auditory cortex neurons to structural features of natural sounds. *Nature* 397: 154–157.

Pons, T. P., Garraghty, P. E., and Mishkin, M. 1987. Physiological evidence for serial processing in somatosensory cortex. *Science* 237: 417–420.

Winslow, R. L., and Sachs, M. B. 1987. Effect of electrical stimulation of the crossed olivocochlear bundle on auditory nerve response to tones in noise. *J. Neurophysiol.* 57: 1002–1021.

19 TRANSDUCTION AND SIGNALING IN THE RETINA

THE WAY IN WHICH NEURONAL SIGNALS ARE EVOKED by light to produce our perception of scenes with objects and background, movement, shade, and color begins in the retina. Responses to light start at receptors known as rods and cones that contain visual pigments. Rods are highly sensitive and can be activated by a single quantum of light. Color and daylight vision depend on cones. Absorption of light by the visual pigment of a photoreceptor activates a G protein, leading to a cascade of biochemical reactions. As a result, nucleotide-gated cation channels in the membrane close, causing the photoreceptor to become hyperpolarized. Light thereby reduces ongoing transmitter release onto postsynaptic bipolar and horizontal cells. Signals from photoreceptors finally reach ganglion cells, whose axons enter the optic nerve and constitute the sole output from the eye.

The connections between receptors and ganglion cells involve bipolar, horizontal, and amacrine cells. Like rods and cones, bipolar and horizontal cells produce graded local potentials, not action potentials. Signaling by individual neurons in the retina and at successive levels of the visual system is best analyzed in terms of receptive fields, which are the building blocks for perception. **Receptive field** of a neuron in the visual system refers to the restricted area of the retinal surface that, upon illumination, enhances or inhibits the signaling of that cell. The receptive field of a retinal ganglion cell is a small circular area on the retina. Action potentials are evoked in "on" ganglion cells by small spots of light shone onto the center of the field surrounded by darkness, or in "off" cells by small dark spots surrounded by light. Two groups of ganglion cells are functionally important. Known as parvocellular (P) and magnocellular (M), they are distinguished by their sizes, positions, connections, and physiological responses. Smaller P ganglion cells exhibit fine spatial discrimination and color sensitivity. Larger M ganglion cells respond better to moving stimuli and to small changes in contrast. These distinctive properties of M and P divisions are maintained through successive relays in the brain up to consciousness.

This chapter sets the stage for the analysis of structure and function in the cerebral cortex discussed in Chapters 20 and 21. The performance of nerve cells in the retina is described more fully here than in Chapter 1, which served to illustrate principles of signaling and organization. Recent years have produced an overwhelming body of work on psychophysics, color vision, dark adaptation, retinal pigments, transduction, transmitters, and the organization of the retina. Each of these topics can form the basis of a self-contained monograph (see the "Suggested Reading" section at the end of the chapter). The same applies to comparative aspects of the visual system in invertebrates and in lower vertebrates, as well as in mammals. Since a comprehensive account is not possible within the scope of this book, we have selected experiments that provide a continuous thread, extending from the properties of cells in the retina to mechanisms that underlie perception.

THE EYE

The eye acts as a self-contained outpost of the brain. It collects information, analyzes it, and hands it on for further processing by higher centers through a well-defined pathway, the optic nerve. The initial step in visual processing is the formation on each retina of a sharp upside-down image of the outside world. Essential for clear vision are (1) correct focus of the image by adjustment of the thickness of the lens (accommodation), (2) regulation by the pupil diameter of light entering the eye, and (3) convergence of the two eyes to ensure that matching images fall on corresponding points of both retinas. Our vision depends critically on the region of the visual field that is being inspected. We can read small print at the center of gaze, but not in the peripheral field of vision. This loss of acuity arises from the way in which visual information is processed by the retina; it is not the result of blurred images or optical distortion outside the central region. We describe first the principal anatomical features of the visual pathway and then the stepwise transformation of signals through the retina as light is trapped by visual pigments to generate electrical currents.

Anatomical Pathways in the Visual System

The pathways from the eye to the cerebral cortex are illustrated in Figure 19.1, which depicts some of the major landmarks of the visual system. The optic nerve fibers arise from ganglion cells in the retina and end on layers of cells in a relay station of the thalamus, the lateral geniculate nucleus. Geniculate axons in turn project through the optic radiation to the cerebral cortex (Chapter 20). From there the progression becomes ever more complex, with no end point in sight.

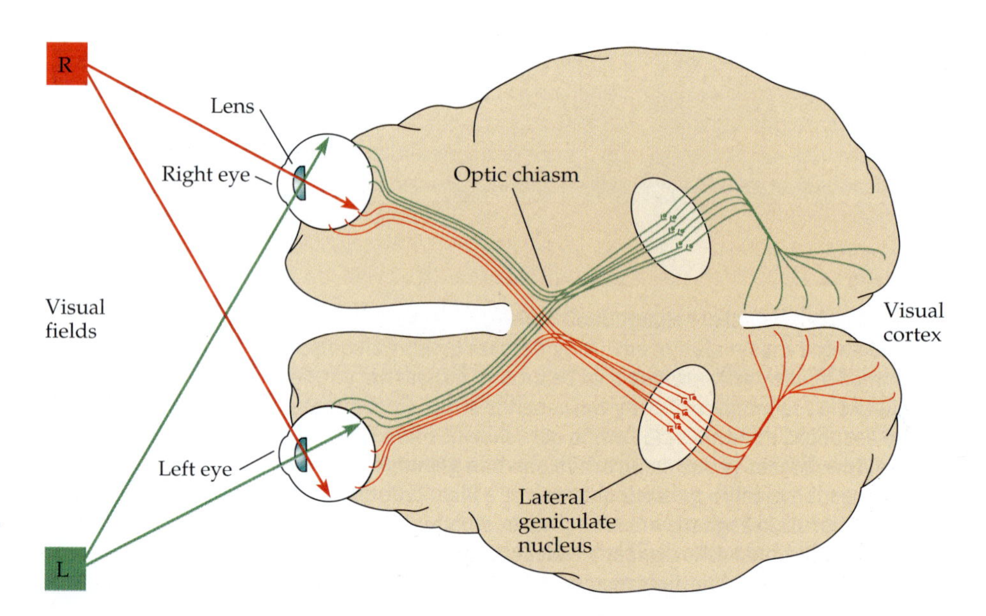

FIGURE 19.1 Visual Pathways. The right side of each retina, shown in green, projects to the right lateral geniculate nucleus. Thus, the right visual cortex receives information exclusively from the left half of the visual field.

Figure 19.1 shows how the output from each retina divides in two at the optic chiasm to supply the lateral geniculate nucleus and visual cortex on each side of the animal. The right side of each retina projects to the right cerebral hemisphere. Because of optical reversal by the lens, this means that the right side of each retina receives the image of the visual world on the left side of the animal. Each cerebral hemisphere, therefore, "sees" the visual field on the opposite side. Accordingly, people with damage to the right side of the cortex caused by trauma or disease become blind in the left visual field, and vice versa. Other pathways that branch off to the midbrain are not described here. In higher vertebrates they are concerned primarily with regulating eye movements and pupillary responses.

Convergence and Divergence of Connections

By examining the cellular anatomy of the various structures in the visual pathway, one can exclude the possibility that information is handed on unchanged from level to level. The neurons converge and diverge extensively at every stage; that is, each cell receives and makes connections with a number of other cells. For example, the human eye contains over 100 million primary receptors, the rods and cones, but only about 1 million optic nerve fibers are sent by ganglion cells along the optic nerve into the brain. In the monkey and the cat the same principle holds: a step-down in neuronal numbers from receptors to ganglion cells. Therefore, within the eye there occurs a funneling of information. An individual neuron that receives inputs from numerous incoming nerve terminals cannot reflect separately the signals of any one of them. Instead, converging impulses of different origin are integrated at each stage into an entirely new message that takes account of all the inputs. After leaving the retina, each ganglion cell axon supplies many geniculate cells.

THE RETINA

Layering of Cells in the Retina

What makes the retina so especially inviting for physiological research are the neat layering and stereotyped morphology of the relatively few cell types—there are only five main classes.[1,2] The arrangement and typical positions of various cells are illustrated in Figure 19.2, which shows a cross section of a human retina, similar to that shown in Fig-

[1]Boycott, B. B., and Dowling, J. E. 1969. *Philos. Trans. R. Soc. Lond. B* 255: 109–184.

[2]Sterling, P. 1997. Retina. In *Synaptic Organization of the Brain*. Oxford University Press, New York, Chapter 6.

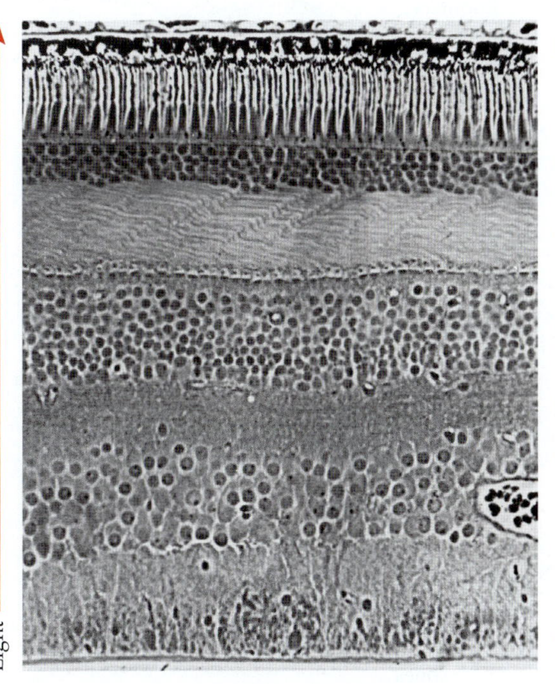

FIGURE 19.2 Section through a Human Retina showing the five principal cell types arranged in layers. Light enters the retina and reaches the rods and cones, where it is absorbed, initiating signals in the outer segments. Synaptic connections are made in the outer and inner plexiform layers. (From Boycott and Dowling, 1969.)

ure 1.2. On the deep surface, farthest from the incoming light, lie the **rods and cones**, which are concerned with night and daytime vision, respectively. They are connected to the **bipolar cells**, which in turn connect to the **ganglion cells**, which give rise to the optic nerve fibers.

Apart from this through-line, there are other cells that make predominantly lateral, or side-to-side, connections. These are the **horizontal cells** and the **amacrine cells**. Only amacrine and ganglion cells give propagated action potentials. Photoreceptors, horizontal cells, and bipolar cells give only local graded signals. Within each of these major classes, there are subgroups exhibiting important differences in structure and function. Müller cells, the properties of which are described in Chapter 8, constitute the glial cells of the retina.

Rods and Cones

Photoreceptors set the stage for vision and define the limits for how the outside world can be perceived. Certain snakes have specialized receptors to detect infrared radiation; ants and bees can use the polarized properties of skylight to navigate (Chapter 15). Our photoreceptors detect neither of these signals. Cats, lacking appropriate receptors, are color-blind—as we are at night (when all cats are gray). The sensitivity of our rods in darkness is such that a single quantum of light can give rise to a measurable signal, and only seven or so rods need to be activated by single quanta for a conscious sensation.[3] Yet with different sets of cone photoreceptors, we can detect subtle tints and differences in contrast or color on a bright day, when the light intensity is 100 million times greater.

Arrangement and Morphology of Photoreceptors

The rods and cones constitute a densely packed array of photodetectors in the layer of retina adjacent to the pigment epithelium (Figure 19.3), farthest from the cornea and the incoming light. With the exception of one small area, the fovea, light must traverse dense layers of cells and fibers before reaching the outer segments of receptors, where photons are absorbed. As Helmholtz wrote in 1867,[4]

[3]Hecht, S., Shlaer, S., and Pirenne, M. H. 1942. *J. Gen. Physiol.* 25: 819–840.

[4]Helmholtz, H. von. 1962/1927. *Helmholtz's Treatise on Physiological Optics*. Dover, New York.

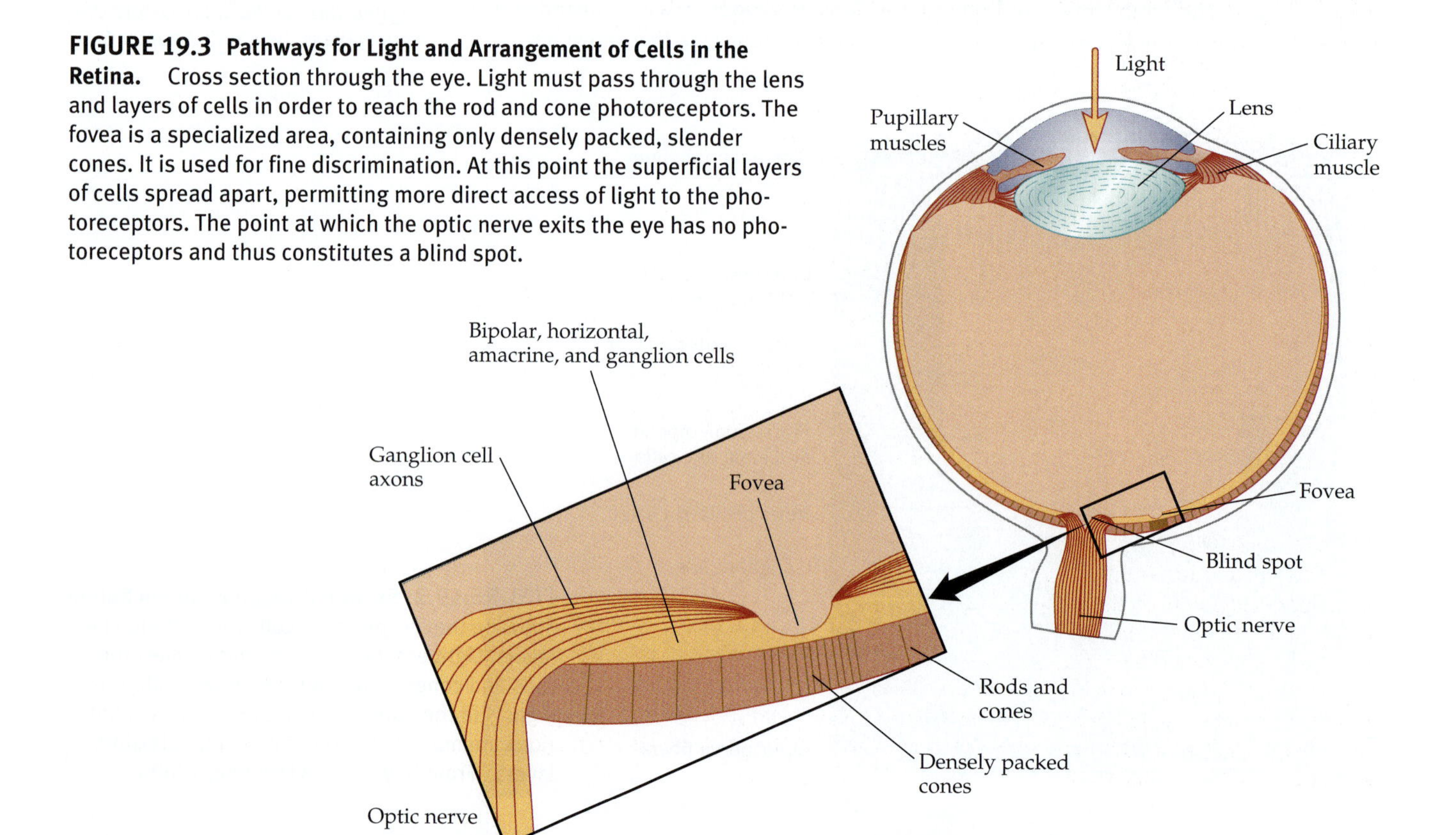

FIGURE 19.3 Pathways for Light and Arrangement of Cells in the Retina. Cross section through the eye. Light must pass through the lens and layers of cells in order to reach the rod and cone photoreceptors. The fovea is a specialized area, containing only densely packed, slender cones. It is used for fine discrimination. At this point the superficial layers of cells spread apart, permitting more direct access of light to the photoreceptors. The point at which the optic nerve exits the eye has no photoreceptors and thus constitutes a blind spot.

> There is in the retina a remarkable spot which is placed near its center . . . and which . . . is called the fovea or pit . . . [It] is of great importance for vision since it is the spot where the most exact discrimination is made. The cones are here packed most closely together and receive light which has not been impeded by other semi-transparent parts of the retina. We may assume that a single nervous . . . connection . . . runs from each of these cones through the trunk of the optic nerve to the brain . . . and there produces its special impression so that the excitation of each individual cone will produce a distinct and separate effect upon the sense.

It is extraordinary that Helmholtz could write this paragraph before the word "synapse" or even the cell doctrine existed.

Counts made at the fovea reveal that cones are closely packed, with a density of 200,000/mm^2, and that rods are excluded. Moreover the cones at the fovea are more slender than those in peripheral parts of the retina.[2] Since the fovea contains no rods, it constitutes a blind spot at night. A different blind spot corresponds to the region of the retina from which the optic nerve fibers leave the eye; at this spot there are no photoreceptors.

Figure 19.4 shows three principal features of photoreceptor structure: (1) an outer segment within which light is absorbed by visual pigments; (2) an inner segment containing the nucleus, ion pumps, transporters, ribosomes, mitochondria, and endoplasmic reticulum; and (3) the synaptic terminal, which releases glutamate onto second-order cells and which also receives synaptic inputs. The release site of the synaptic ending is highly characteristic with one or more "ribbon" structures, along which are aligned the vesicles containing transmitter (see Figure 19.17).[2]

Hermann von Helmholtz (1821–1894), together with one of his drawings and the frontispiece of his book on vision. Helmholtz made equally important and original contributions to the study of medicine, hearing, neurophysiology, and thermodynamics. It seems refreshing to read his prose today. (Photomontage courtesy of Dr. Rolf Boch.)

Electrical Responses of Vertebrate Photoreceptors to Light

As shown in Chapter 17, sensory receptors typically respond to appropriate stimuli by graded local depolarizations that initiate action potentials. Although most invertebrate

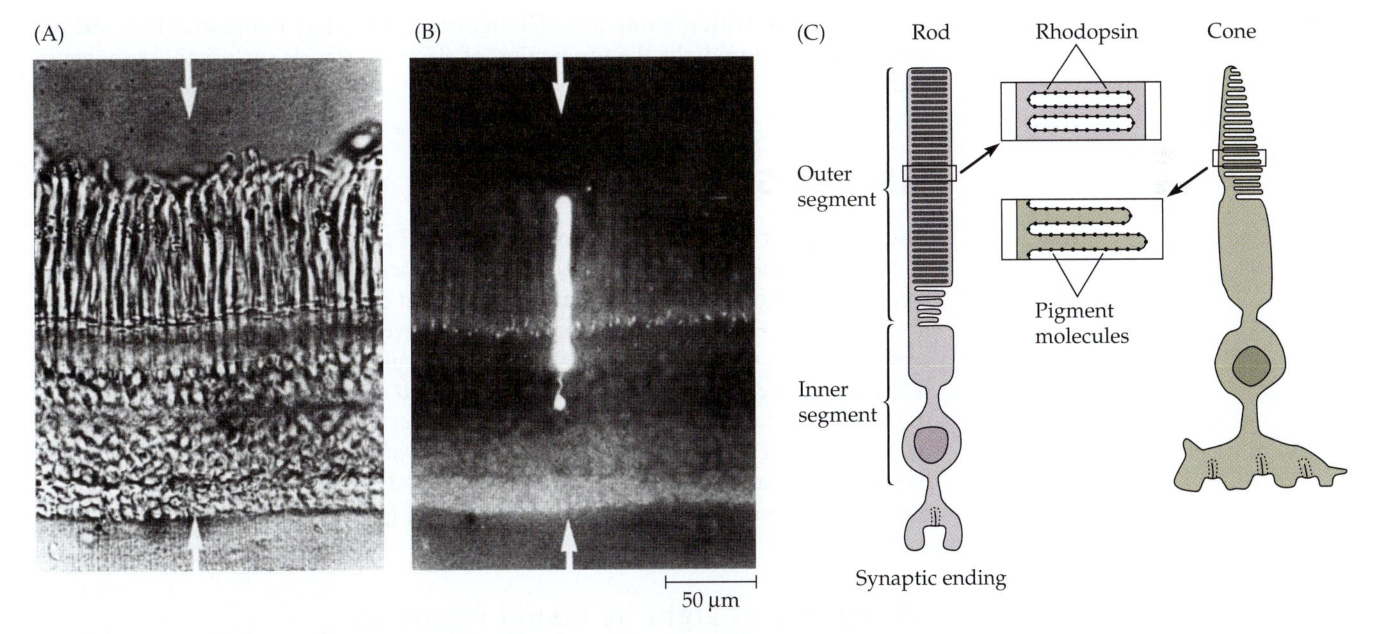

FIGURE 19.4 Photoreceptors in Retina. (A and B) Rod in toad retina injected with a fluorescent dye, Lucifer yellow, as seen in visible (A) and ultraviolet (B) light. Arrows mark identical points on the retina. (C) Diagram of a rod and a cone. In the rod, the pigment rhodopsin (black dots) is embedded in membranes arranged in the form of disks, not continuous with the outer membrane of the cell. In the cone, the pigment molecules are on infolded membranes that are continuous with the surface membrane. The outer segment is connected to the inner segment by a narrow stalk. The synaptic endings continually release transmitter in the dark. (A and B kindly provided by the late B. Nunn, unpublished; C after Baylor 1987.)

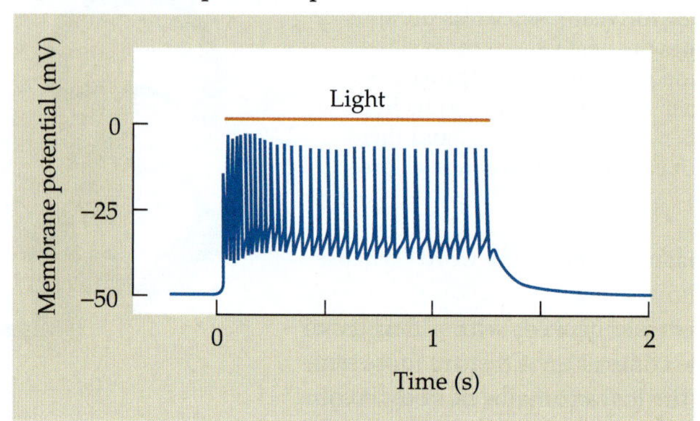

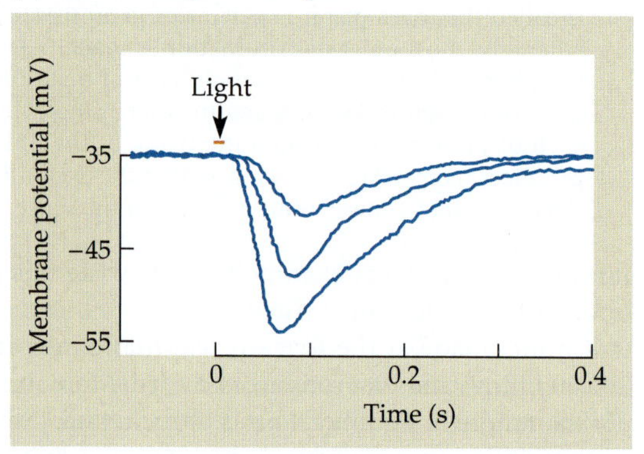

FIGURE 19.5 Responses of Photoreceptors. (A) Photoreceptors of an invertebrate (a horseshoe crab) respond to light with a depolarization that gives rise to impulses. This is the usual type of response elicited from sensory receptors activated by various stimuli, such as touch, pressure, or stretch (Chapter 17). (B) Photoreceptors of a vertebrate (a turtle) respond with a hyperpolarization that is graded according to the intensity of the flash. (A after Fuortes and Poggio, 1963; B after Baylor, Fuortes, and O'Bryan, 1971.)

photoreceptors behave in this way (Figure 19.5A),[5] the responses of most vertebrate photoreceptors to light are very different. Figure 19.5B shows the responses of a turtle rod recorded with an intracellular microelectrode.[6] In the dark (at rest), the photoreceptor is depolarized by a continuous inward current flowing into the outer segment. Light leads to a hyperpolarization by turning off the ongoing inward current. The following paragraphs deal with the way in which light is absorbed by photoreceptors and the mechanisms that underlie the production of electrical signals such as those shown in Figure 19.5.

VISUAL PIGMENTS

Visual pigments are concentrated in membranes of the outer segments. Each rod contains approximately 10^8 pigment molecules. They are aggregated on several hundred discrete disks (approximately 750 in a monkey rod) that do not make contact with the outer membrane (see Figure 19.4C). By contrast, cones have pigment-laden infoldings that are continuous with the cell membrane. Pigment molecules make up about 80% of the total disk protein. The visual pigment is so closely packed on outer segment membranes that the distance between two visual pigment molecules in a rod is less than 10 nm.[7] This dense packing of sensitive molecules in serial layers of membranes traversed by light enhances the probability that a photon will be trapped on its way through the outer segment. The following question arises: How are signals generated when light is absorbed by visual pigments?

Absorption of Light by Visual Pigments

The events that occur when light is absorbed by the rod pigment rhodopsin have been studied by psychophysical, biochemical, physiological, and molecular techniques. Visual pigment molecules consist of two moieties: a protein known as opsin, and a chromophore, 11-*cis* vitamin A aldehyde, known as **retinal** (Figure 19.6). (A chromophore is a chemical group producing color in a compound.) The absorption characteristics of the visual pigments have been measured quantitatively by spectrophotometry.[8,9] When different wavelengths of light are shone through a solution of the rod visual pigment rhodopsin, blue-green light at a wavelength of about 500 nm is absorbed most effectively. The absorption spectrum is similar when small spots of light of different wavelengths are

[5]Fuortes, M. G. F., and Poggio, G. F. 1963. *J. Gen. Physiol.* 46: 435–452.

[6]Baylor, D. A., Fuortes, M. G. F., and O'Bryan, P. M. 1971. *J. Physiol.* 214: 265–294.

[7]Dowling, J. E. 1987. *The Retina: An Approachable Part of the Brain.* Harvard University Press, Cambridge, MA.

[8]Brown, P. K., and Wald, G. 1963. *Nature* 200: 37–43.

[9]Marks, W. B., Dobelle, W. H., and MacNichol, E. F. 1964. *Science* 143: 1181–1183.

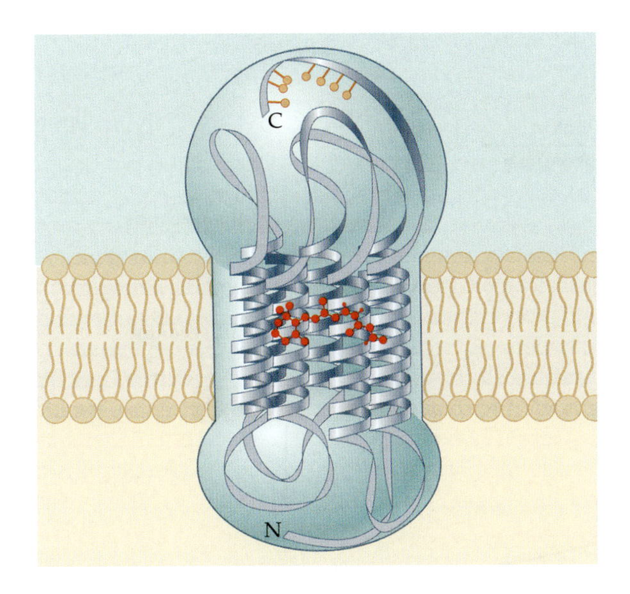

FIGURE 19.6 Structure of Vertebrate Rhodopsin in the Membrane. The helix is partly opened to show the position of retinal (red). C = carboxy terminus; N = amino terminus. (After Stryer and Bourne, 1986.)

shone onto single rods under the microscope. An elegant correspondence has been demonstrated between the absorption characteristics of rhodopsin and our perception in dim light. Quantitative psychophysical measurements made in human subjects show that blue-greenish lights at about 500 nm are optimal for perception of a dim light in the dark. In daylight, when rods are inactive and cones are used, we are more sensitive to red light, corresponding to the absorption spectra of cones (which will be discussed shortly).[7]

Once a photon has been absorbed by rhodopsin, retinal undergoes photoisomerization and changes from the 11-*cis* to an all-*trans* configuration. This transition is extremely rapid; it takes place in approximately 10^{-12} seconds. The protein then undergoes a series of transformational changes through various intermediates.[10] One conformation of the protein, metarhodopsin II, is of crucial importance for transduction (discussed a little later in the chapter). Figure 19.7 shows the sequence of changes occurring in bleaching and in regeneration of active rhodopsin. Metarhodopsin II appears after about 1 ms. Regeneration of pigments after bleaching is slow, taking many minutes; it entails transfer of retinal between the photoreceptors and the pigment epithelium (see Figure 19.2).[11]

Wilhelm Kühne (1837–1900). Under the portrait, on the left, is shown the view from his room that was presented to the retina of a rabbit, causing bleaching and the clearly discernible image of the window arrangement (shown on the right). Kühne isolated visual purple for the first time. (Photomontage courtesy of Dr. Rolf Boch.)

Structure of Rhodopsin

At the molecular level, the protein opsin consists of 348 amino acid residues with seven hydrophobic regions, each containing 20 to 25 amino acids, corresponding to seven transmembrane helices.[12] The amino terminus is located in the extracellular space (i.e., within the disk in rods) and the carboxy terminus in the cytoplasm. Retinal is attached to opsin through a lysine residue in the seventh membrane-spanning segment. Opsin is a member of the family of seven-transmembrane-domain proteins that includes metabotropic transmitter receptors, such as adrenergic and muscarinic receptors (see Figure 19.9). Like rhodopsin, those receptors exert their effects through second messengers by activating G proteins (Chapters 2 and 10). The stability of rhodopsin in the dark is extraordinary. Baylor has calculated that spontaneous thermal isomerization of a single rhodopsin molecule occurs about once every 3000 years, or 10^{23} times more slowly than photoisomerization.[13]

Cones and Color Vision

Extraordinary insights and experiments by Young and Helmholtz in the nineteenth century defined crucial questions for color vision and at the same time provided clear, unequivocal explanations. Their conclusion that there must be three types of sensory photoreceptors for color has stood the test of time and been confirmed at the molecular level. To set the stage we quote again from Helmholtz, who compares the perception of light and sound, color and tone. One envies the clarity, force, and timeless beauty of his

[10] Matthews, R. G., et al. 1963. *J. Gen. Physiol.* 47: 215–240.

[11] Pepperberg, D. R., et al. 1993. *Mol. Neurobiol.* 7: 61–85.

[12] Nathans, J., and Hogness, D. S. 1984. *Proc. Natl. Acad. Sci. USA* 81: 4851–4855.

[13] Baylor, D. A. 1987. *Invest. Ophthalmol. Vis. Sci.* 28: 34–49.

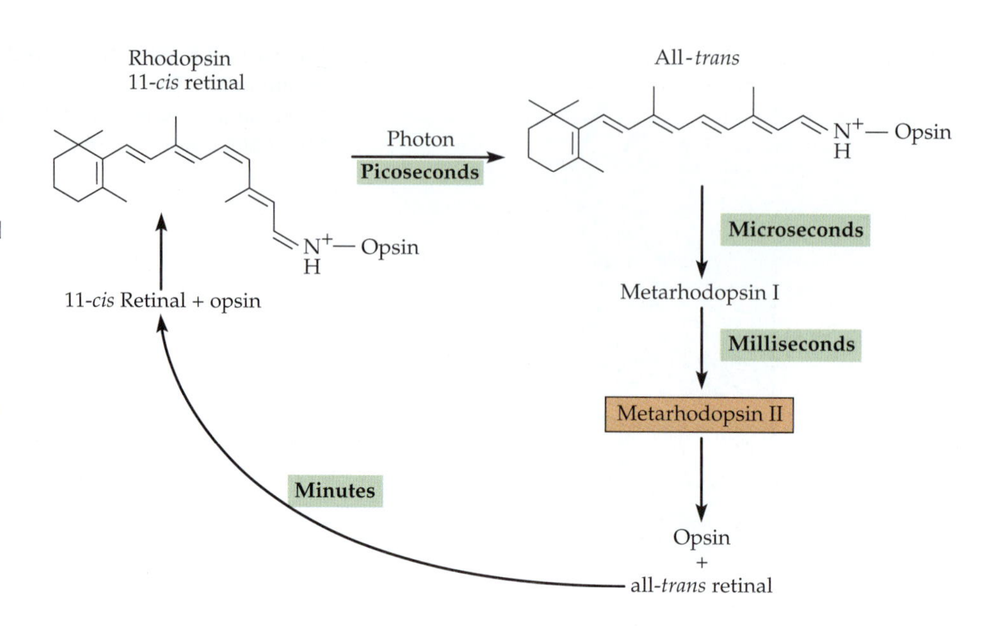

FIGURE 19.7 Bleaching of Rhodopsin by Light. In the dark, 11-*cis* retinal is bound to the protein opsin. Capture of a photon causes isomerization of the 11-*cis* retinal to all-*trans* retinal. The opsin–all-*trans* molecule in turn is rapidly converted to metarhodopsin II, which is dissociated to opsin and all-*trans* retinal. The regeneration of rhodopsin depends on interactions between photoreceptors and cells of the retinal pigment epithelium. Metarhodopsin II is the trigger that sets in motion activation of the second messenger system. (After Dowling, 1987.)

Denis Baylor, 1991

thinking, especially in view of the confusing, vitalistic concepts that were current in the nineteenth century:

> All differences of hue depend upon combinations in different proportions of the three primary colors . . . red, green and violet. . . . Just as the difference of sensation of light and warmth depends . . . upon whether the rays of the sun fall upon nerves of sight or nerves of feeling, so it is supposed in Young's hypothesis that the difference of sensation of colors depends simply upon whether one or the other kind of nervous fibers are more strongly affected. When all three kinds are equally excited, the result is the sensation of white light. . . . If we allow two different colored lights to fall at the same time upon a white screen . . . we see only a single compound, more or less different from the two original ones. We shall better understand the remarkable fact that we are able to refer all the varieties in the composition of external light to mixtures of three colors if we compare the eye with the ear. . . . In the case of sound . . . we recognize the long waves as low notes, the short as high-pitched, and the ear may receive at once many waves of sound, that is to say many notes. But here these do not melt into compound notes in the same way that colors . . . melt into compound colors. The eye cannot tell the difference if we substitute orange for red and yellow; but if we hear the notes C and E sounded at the same time, we cannot put D instead of them . . . if the ear perceived musical tones as the eye colors, every accord might be completely represented by combining only three constant notes, one very low, one very high and one intermediate, simply changing the relative strength of these three primary notes to produce all possible musical effects. . . . But we find a continuous transition of colors into one another through numberless intermediate gradations . . . the way in which (colors) appear . . . depends chiefly upon the constitution of our nervous system. . . . It must be confessed that both in man and in quadrupeds we have at present no anatomical basis for this theory of colors.

These farsighted, accurate predictions were validated by quite different sets of observations. By spectrophotometry, Wald, Brown, MacNichol, Dartnall, and their colleagues[8,9,14] showed the existence of three types of cones with different pigments in human retina (Figure 19.8A). Second, Baylor and his colleagues recorded currents from monkey and human cones. The results are shown in Figure 19.8B.[15] Three populations of cones were found with distinct but overlapping sensitivities in the blue, green, or red part of the spectrum. The wavelengths of light optimal for initiating electrical signals coincided precisely with the absorbance peaks of the visual pigments demonstrated by spectrophotometry, as well as by psychophysical measurements of spectral sensitivity (Figure 19.8C). Finally, the genes for the red, green, and blue cone opsin pigments, as well as the gene for rhodopsin, were cloned and sequenced by Nathans.[16–18]

What accounts for the ability of different visual pigment molecules to trap specific wavelengths of light preferentially? It turns out that rhodopsin, the rod visual pigment, and all three cone visual pigments contain the same chromophore, 11-*cis* retinal. How-

[14]Dartnall, H. J. A., Bowmaker, J. K., and Molino, J. D. 1983. *Proc. R. Soc. Lond. B* 220: 115–130.

[15]Schnapf, J. L., et al. 1988. *Vis. Neurosci.* 1: 255–261.

[16]Nathans, J. 1987. *Annu. Rev. Neurosci.* 10: 163–194.

[17]Nathans, J. 1989. *Sci. Am.* 260(2): 42–49.

[18]Nathans, J. 1999. *Neuron* 24: 299–312.

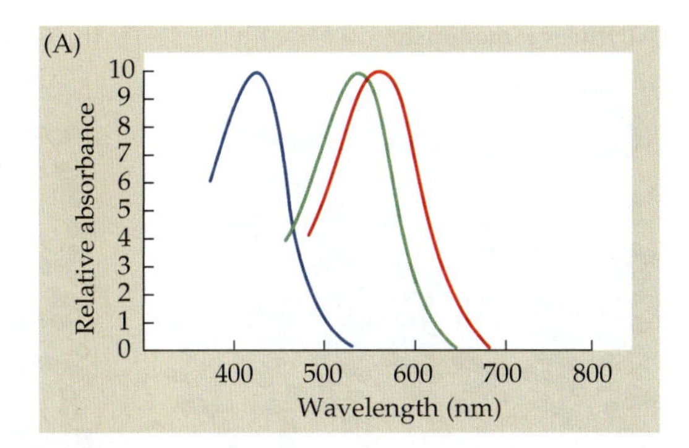

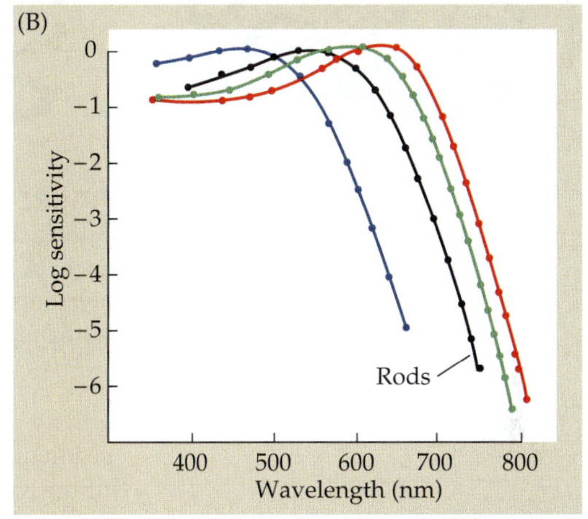

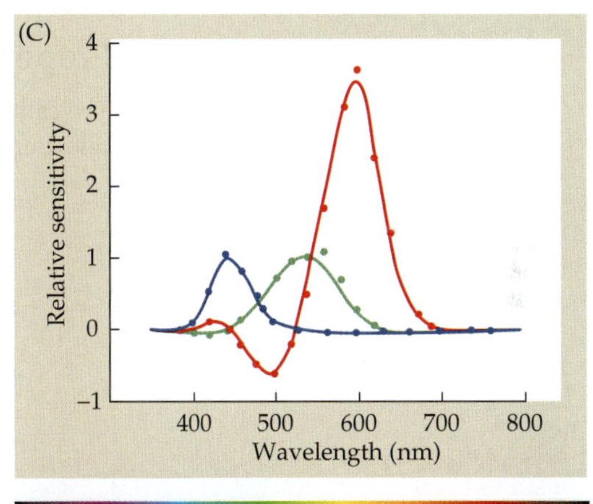

FIGURE 19.8 Spectral Sensitivity of Photoreceptors of human subjects and of visual pigments. (A) Spectral sensitivity curves of the three colored visual pigments showing absorbance peaks at wavelengths corresponding to blue, green, and red. (B) Spectral sensitivities of blue-, green-, and red-sensitive cones (as colored) and rods (black) from macaque monkeys. The responses were recorded by suction electrodes, averaged and normalized. The curve through the rod spectrum was obtained from visual pigments in human subjects. (C) Comparison of spectral sensitivity of monkey cones with those obtained by human color matching. The continuous curves represent color-matching experiments in which the sensitivity at various wavelengths was determined in human subjects. The dots show results predicted from electrical measurements made by recording currents from single cones, after correcting for absorption in the lens and by pigments on the path to the outer segment. The correspondence between results obtained on single cells and by color matching is extraordinarily good. (A after Schnapf and Baylor, 1987; B after Baylor, 1987; C after Dowling, 1987.)

ever, the amino acid sequences of the various pigment proteins differ from one another (Figure 19.9). Differences in just a few amino acids account for the differences in spectral sensitivity.

Color Blindness

Although a single type of photoreceptor cannot on its own provide information about color, three types of cones like those in Figure 19.9 can. In principle, two types of cones with different pigments might be sufficient to recognize colors, but many different mixtures of wavelengths would then appear identical. This is the situation in certain color-blind people, in whom Nathans has shown that a genetic defect results in an absence of one of the pigments. From our present perspective one can only marvel that explanations at the molecular level so beautifully confirm the brilliant but rigorous speculations of Young and Helmholtz. Their idea that major attributes of color vision and color blindness are to be found within the receptors themselves has now been confirmed by direct physiological measurements and corresponding differences in genes and protein structure.[18]

TRANSDUCTION BY PHOTORECEPTORS

How does photoisomerization of rhodopsin give rise to a change in membrane potential? For many years it was clear that some sort of internal transmitter was required for the generation of electrical signals in rods and cones. One reason is that information about the capture of photons in a rod outer segment must somehow be conveyed from rhodopsin, in the disk, across the cytoplasm to the outer membrane. A second reason is the enormous amplification of the response. Baylor and his colleagues,[19] working on turtle photoreceptors, showed that decreases in membrane conductance and measurable electrical signals were produced when a single photon was absorbed and activated one pigment molecule out of about 10^8.

The sequence of events through which activated photopigment molecules change membrane potential has been elucidated by patch clamp recordings from rod and cone outer segments and by molecular techniques. The scheme for transduction from light to electrical signals is shown in Figure 19.10.

In darkness, a continuous "dark" current flows into the outer segment of rods and cones.[20] As a result they have membrane potentials of approximately −40 mV, far from the potassium equilibrium potential, E_K (−80 mV). The inward current in the dark is carried mainly by sodium, moving down its electrochemical gradient through cation channels in the outer

[19]Baylor, D. A., and Fuortes, M. G. F. 1970. *J. Physiol.* 207: 77–92.

[20]Baylor, D. A., Lamb, T. D., and Yau, K. W. 1979. *J. Physiol.* 288: 589–611.

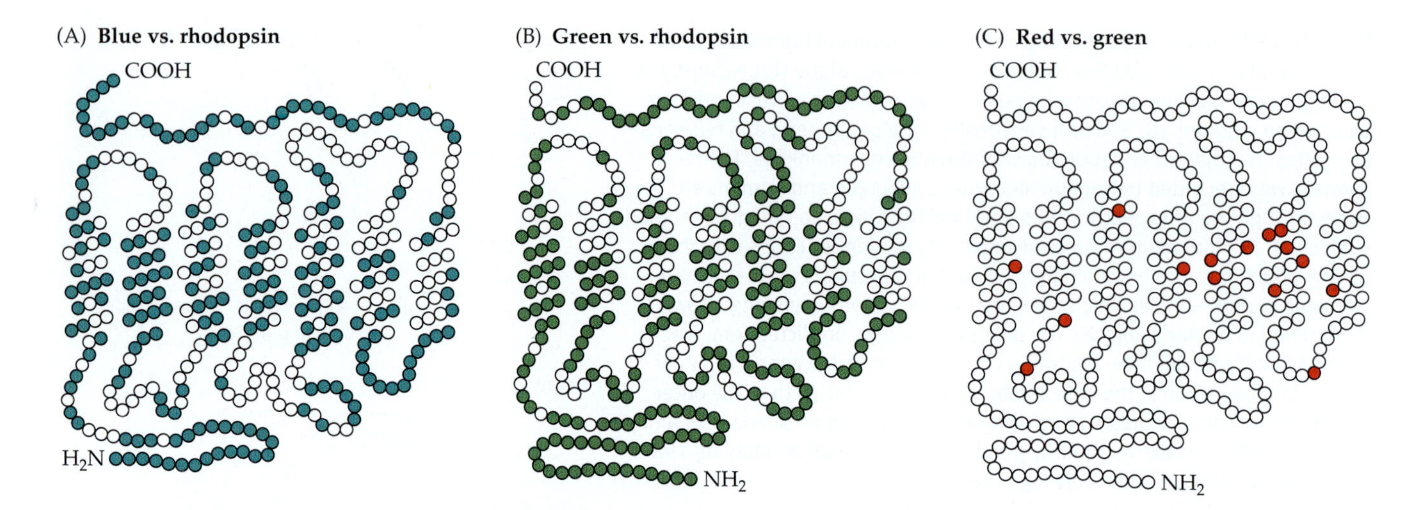

FIGURE 19.9 Comparisons of Amino Acid Sequences of red, green, and blue pigments with each other and with rhodopsin. Each colored dot represents an amino acid difference. (A and B) Blue and green pigments compared with rhodopsin. (C) Green and red pigments compared. The sequences of red and green pigments are highly similar. (From Nathans, 1989.)

segment. Hyperpolarization of the photoreceptor by light is brought about by closure of the channels, allowing the membrane potential to move toward E_K.

Properties of the Photoreceptor Channels

The cation channels in the outer segment, under normal physiological conditions, have calcium/sodium/potassium permeability ratios of 12.5:1.0:0.7, and a single-channel con-

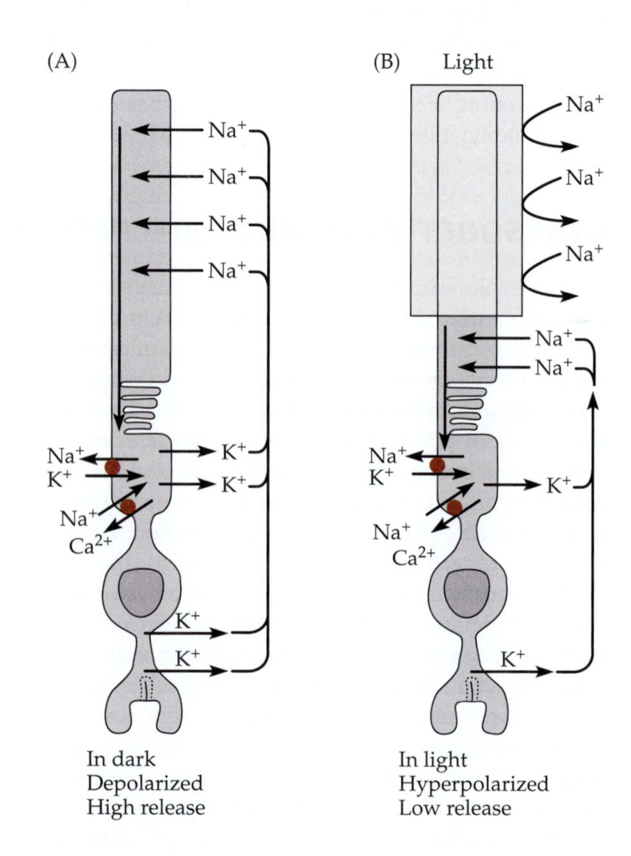

FIGURE 19.10 Dark Current in a Rod. (A) In darkness, sodium ions flow through cation channels of the rod outer segment, causing a depolarization; calcium ions also enter through the cation channels. The current loop is completed through the neck of the rod with the outward movement of potassium through the inner segment membrane. (B) When the outer segment is illuminated, the channels close because of a decrease in intracellular cyclic GMP, and the rod then becomes hyperpolarized. This hyperpolarization reduces transmitter release. Sodium, potassium, and calcium concentrations of the rod are maintained by pumps and exchangers in the inner segment (red circles); calcium exchangers are also present in the outer segment (see Box 19.1). (After Baylor, 1987.)

ductance of about 0.1 pS.[21] Because the sodium concentration is much higher than that of calcium, about 85% of the inward current is carried by sodium. The driving force for potassium movement is, of course, outward. As calcium ions move through the channels, they are tightly bound to sites within the pore and thus interfere with the passage of other cations. Because of this property, removal of calcium from the solution around the cell allows sodium and potassium to move much more freely through the channel, increasing its conductance to about 25 pS.

Fesenko, Yau, Baylor, Stryer, and their colleagues[22–24] have shown that cyclic GMP acts as the internal transmitter from disk to surface membrane and fulfills the requirements of appropriate kinetics and great amplification. As shown in Figure 19.11, a high cytoplasmic concentration of cyclic GMP keeps the cation channels in an open state. When the cGMP concentration of the fluid facing the inside of the membrane is reduced, channel openings become rare events. Thus, the membrane potential of the photoreceptors is a reflection of the cytoplasmic cGMP concentration: The higher the concentration, the more the cell is depolarized. The cGMP concentration in turn is inversely related to the intensity of ambient light. Increasing light intensity reduces cGMP concentration and reduces the fraction of open channels. In the absence of cGMP, almost all the channels are closed and the resistance of the outer segment membrane approaches that of a channel-free lipid bilayer.

Molecular Structure of Cyclic GMP–Gated Channels

Complementary DNAs for rod outer segment channels have been isolated and the amino acid sequences determined for channel subunits from human, bovine, mouse, and chicken retinas. There is a pronounced sequence similarity between cDNAs of outer segment channel subunits and the subunits of other cyclic nucleotide–gated channels—for example, those found in the olfactory system.[25,26] Their membrane regions share structural similarities with other cation-selective channels, particularly in the S4 region and the region of the pore (Chapter 3). The photoreceptor channels are tetramers made up of at least two different subunit proteins, α and β, with apparent molecular sizes of 63 and 240 kD, respectively. The intracellular nucleotide-binding site is near the carboxy terminus of the α and β subunits.[27] Expression of the subunits in oocytes produces cation

[21]Yau, K. W., and Chen, T. Y. 1995. In *Handbook of Receptors and Channels: Ligand- and Voltage-Gated Ion Channels.* CRC Press, Boca Raton, FL, pp. 307–335.

[22]Fesenko, E. E., Kolesnikov, S. S., and Lyubarsky, A. L. 1985. *Nature* 313: 310–313.

[23]Yau, K. W., and Nakatani, K. 1985. *Nature* 317: 252–255.

[24]Stryer, L., and Bourne, H. R. 1986. *Annu. Rev. Cell Biol.* 2: 391–419.

[25]Torre, V., et al. 1995. *J. Neurosci.* 15: 7757–7768.

[26]Kaupp, U. B. 1995. *Curr. Opin. Neurobiol.* 5: 434–442.

[27]Finn, J. T., Grunwald, M. E., and Yau, K-W. 1996. *Annu. Rev. Physiol.* 58: 395–426.

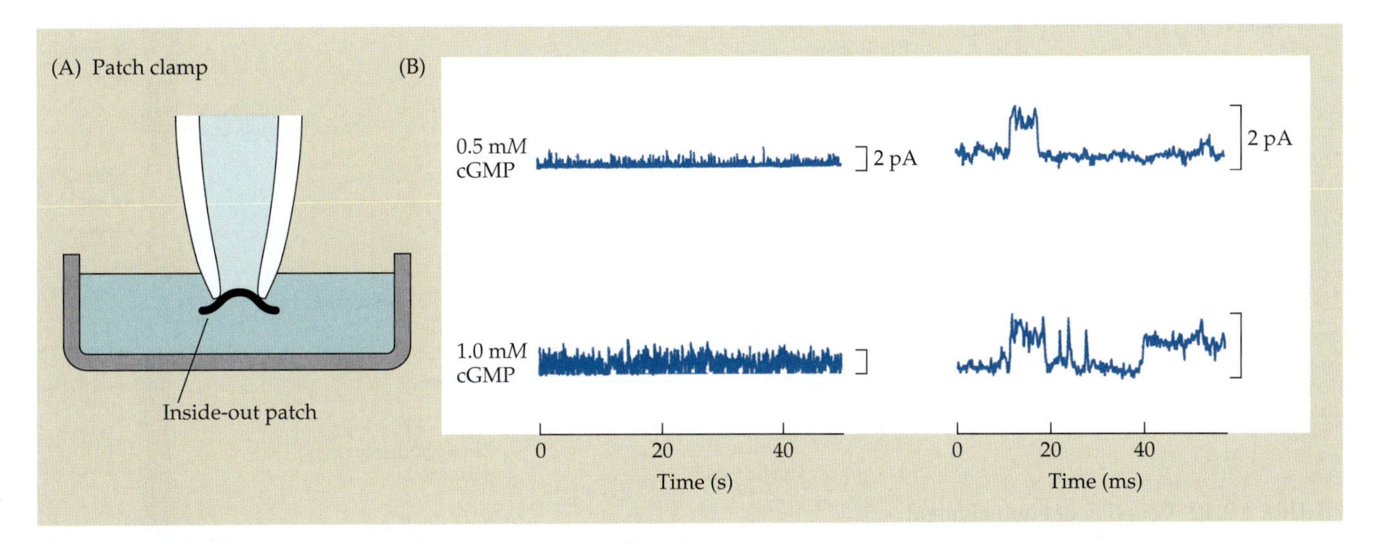

FIGURE 19.11 Role of Cyclic GMP in Opening Sodium Channels in rod outer segment membranes. Single-channel recordings made from inside-out patches bathed in various concentrations of cyclic GMP. Channel opening causes deflections in the upward direction. The frequency of channel opening is extremely low in control recordings. Addition of cyclic GMP causes single-channel openings, the frequency of which increases with increased concentration. (After Baylor, 1987.)

channels with properties similar to those found in rod outer segments: They are activated by cGMP and have the expected permeability ratios and conductances.[28]

The Cyclic GMP Cascade

The sequence of events leading to reduction in cGMP concentration and consequent closing of the cation channels is shown in Figure 19.12. The decrease in internal cyclic GMP concentration triggered by light is brought about by metarhodopsin II, an intermediary in the bleaching process (see Figure 19.7). Metarhodopsin II acts on the G protein **transducin**, which consists of three polypeptide chains, α, β, and γ.[29,30]

The interaction of metarhodopsin II and transducin causes GDP bound to the α subunit to be exchanged for GTP (Chapter 10). This activates the α subunit, which separates from the β and γ subunits, and in turn activates a membrane-bound phosphodiesterase, the enzyme that hydrolyzes cGMP. The cGMP concentration falls, fewer sodium channels open, and the rods hyperpolarize. The cascade is terminated by phosphorylation of the carboxy-terminal region of the active metarhodopsin II (Box 19.1). The key role of cyclic GMP in controlling the channels during the response is supported by biochemical experiments. Illumination of photoreceptors can cause a 20% decrease in the internal cGMP concentration.[13]

Vertebrate Photoreceptors with Depolarizing Responses to Light

There is an interesting exception to the account that we have given for transduction in vertebrate photoreceptors. Lizards have a third eye on the top of the head. Within it there are small cones that resemble those seen in the lateral eyes. These photoreceptors are, however, remarkable in that they depolarize in response to illumination.[31,32] The nucleotide channels are similar to those in other vertebrate photoreceptors, and so is the cGMP cascade, except in one respect: Activation of the photoreceptor and the G protein causes an increase in the concentration of cGMP. As a result, the outer segment channels

[28]Bucossi, G., Nizzari, M., and Torre, V. 1997. *Biophys. J.* 72: 1165–1181.
[29]Stryer, L. 1987. *Sci. Am.* 257(1): 42–50.
[30]Stryer, L. 1991–1992. *Harvey Lect.* 87: 129–143.
[31]Finn, J. T., Solessio, E. C., and Yau, K. W. 1997. *Nature* 385: 815–819.
[32]Finn, J. T., et al. 1998. *Vision Res.* 38: 1353–1357.

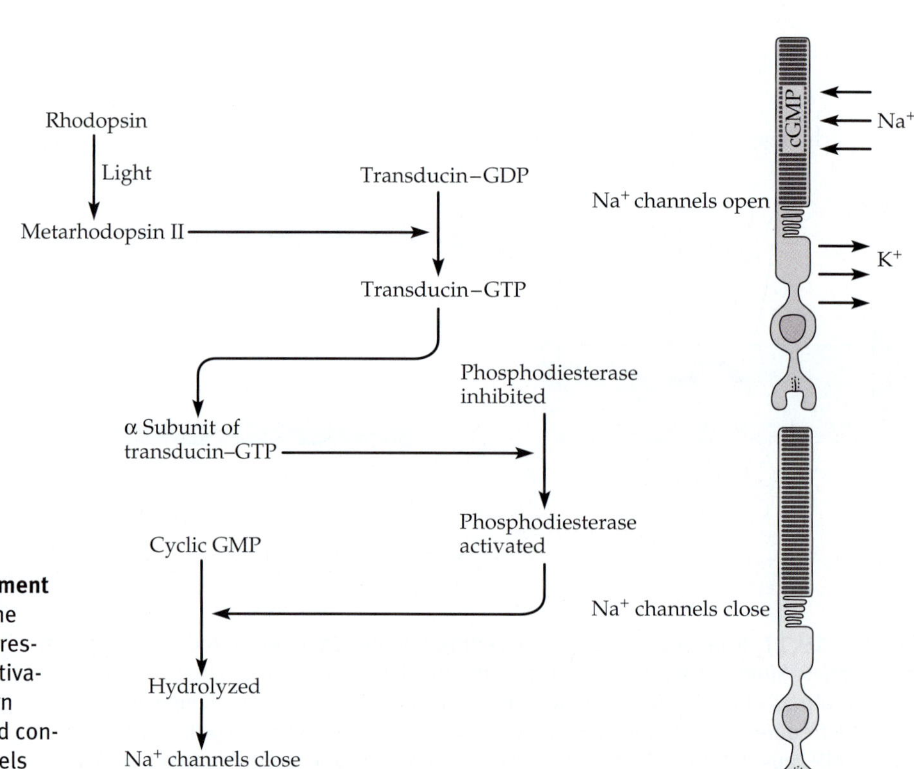

FIGURE 19.12 Coupling of Photopigment Activation to G Protein Activation. The G protein transducin binds GTP in the presence of metarhodopsin II, leading to activation of phosphodiesterase, which in turn hydrolyzes cyclic GMP. With the reduced concentration of cyclic GMP, sodium channels close. (After Baylor, 1987.)

Box 19.1 Adaptation of Photoreceptors

It is essential to ensure that responses to light can occur at different background levels of illumination. If bright ambient light were able to close all the nucleotide-gated channels, the receptor would be unable to register any further increase in intensity. Moreover, very different levels of sensitivity are required to respond to low light intensities, when only a few quanta may be absorbed, as compared with bright light, involving the absorption of millions of quanta. One major mechanism is the use of separate sets of receptors for vision in dim light and bright light. Another is adaptation in individual photoreceptors.[33] Through adaptation, photoreceptors continue to respond over a wide range of illumination.

Calcium is one factor of key importance for adaptation of photoreceptors.[34,35] In the dark the nucleotide-gated channels are open and calcium ions continually flow into the photoreceptor. Calcium is extruded by ion pumps and exchangers in the outer segment (Chapter 4, and see Figure 19.10).[36] Under conditions of steady illumination, the channels close and calcium entry is reduced. Because calcium is still actively extruded (see Figure 19.10), the intracellular calcium concentration falls. The reduced intracellular calcium concentration opposes closure of the nucleotide-gated channels by several mechanisms.

One action of calcium is to reduce the affinity of channels to cGMP through a calcium-binding protein, calmodulin. Lowered calcium concentrations therefore increase the affinity of the channels for cGMP, and thereby potentiate channel opening and current flow during illumination. Second, lowered intracellular calcium favors intracellular cGMP accumulation: It increases the activity of guanylate cyclase (i.e., promotes cGMP synthesis) and inhibits the activation of phosphodiesterase (i.e., slows cGMP hydrolysis). Third, lowered intracellular calcium causes phosphorylation of metarhodopsin II and thereby speeds up its inactivation. The rate of termination of the catalytic activity of metarhodopsin II is of importance for transduction, since while it is active the cascade continues to generate a signal. The phosphorylation of activated rhodopsin is mediated by recoverin,[37] a calcium-binding molecule that is also involved in the inhibition of phosphodiesterase.

[33]Fain, G. L., Matthews, H. R., and Cornwall, M. C. 1996. *Trends Neurosci.* 19: 502–507.

[34]Baylor, D. 1996. *Proc. Natl. Acad. Sci. USA* 93: 540–565.

[35]Koutalos, Y., and Yau, K. W. 1996. *Trends Neurosci.* 19: 73–81.

[36]Morgans, C. W., et al. 1998. *J. Neurosci.* 18: 2467–2474.

[37]Baylor, D. A., and Burns, M. E. 1998. *Eye* 12: 521–525.

[38]Chen, J., et al. 1995. *Science* 267: 374–377.

Molecular mechanisms for adaptation were analyzed by Baylor and his colleagues, who measured the rate of adaptation of photoreceptor responses in normal and transgenic mice in which 15 amino acids were deleted from the C terminus of rhodopsin, the presumptive site for phosphorylation of activated rhodopsin.[38] In a normal rod (part B of the figure), a flash of light produces an outward current that adapts (i.e., decreases) as expected. Part C of the figure shows the response of a rod in a transgenic mouse to a flash: The response, instead of declining, lasts about 20 times longer than normal. Thus, the 15 amino acids that had been deleted constitute the region of the molecule required for recovery from the effects of a flash of light.

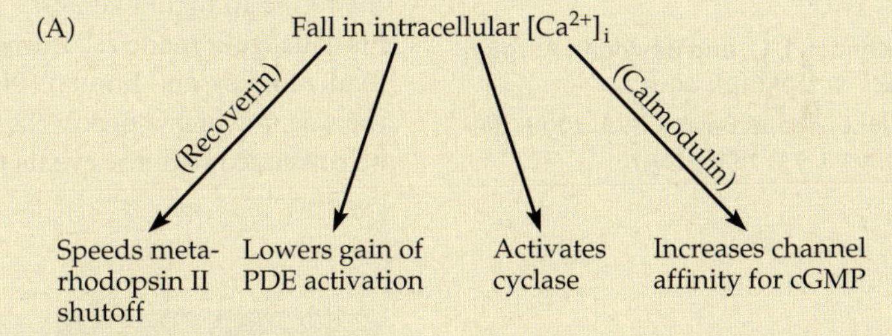

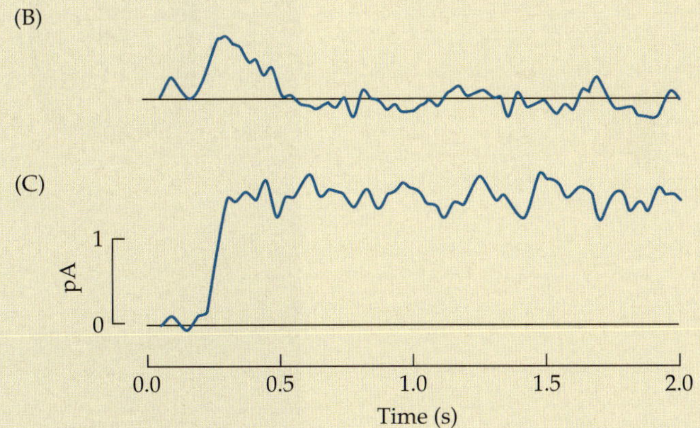

Mechanisms of Adaptation. (A) Diagram showing effects of a fall in the intracellular calcium concentration on mechanisms that influence adaptation photoreceptors to steady light. (B) Responses to flashes delivered at time zero were recorded from a rod by suction electrodes. The response elicited by activation of a normal rhodopsin molecule declined as expected. (C) The trace was recorded from a rod in a transgenic mouse in which rhodopsin molecules had been truncated by the deletion of 15 amino acids from the C terminus. The duration of the response was prolonged through failure of the altered rhodopsin molecule to become shut off after the flash. (After Baylor, 1996.)

open and cations flow into the cell—a "light" current. This is brought about through inhibition of the activity of phosphodiesterase, which is high in the dark in this photoreceptor. In summary, the steps in the lizard third eye photoreceptor are:

light → [cGMP]↑ → cation channels in outer segment open → depolarization

Amplification through the Cyclic GMP Cascade

Two steps of the cyclic GMP cascade provide the great amplification that accounts for the exquisite sensitivity of rods to light. First, a single molecule of active metarhodopsin II catalyzes the exchange of many molecules of GDP for GTP and thereby liberates hundreds of G protein α subunits.[34] Second, each α subunit activates a molecule of phosphodiesterase in the disk that can hydrolyze a large number of cytoplasmic cGMP molecules and thereby close many channels.

Responses to Single Quanta of Light

The finding that single quanta of light can give rise to a conscious sensation raises a number of tantalizing questions. How large is the unitary response, how is it distinguished from ongoing noise, and how is such minimal information transferred faithfully through the retina to higher centers? To measure unitary responses to single quanta, Baylor and his colleagues made recordings of currents generated by individual rods in retinas from toad, monkey, and human (Figure 19.13).[39] These experiments provide a rare example of the way in which a process as complex as seeing the dimmest possible flashes of light can be correlated with the events that occur in single molecules.[40]

[39]Schnapf, J. L., and Baylor, D. A. 1987. *Sci. Am.* 256(4): 40–47.

[40]Rieke, F., and Baylor, D. A. 1998. *Biophys. J.* 75: 1836–1857.

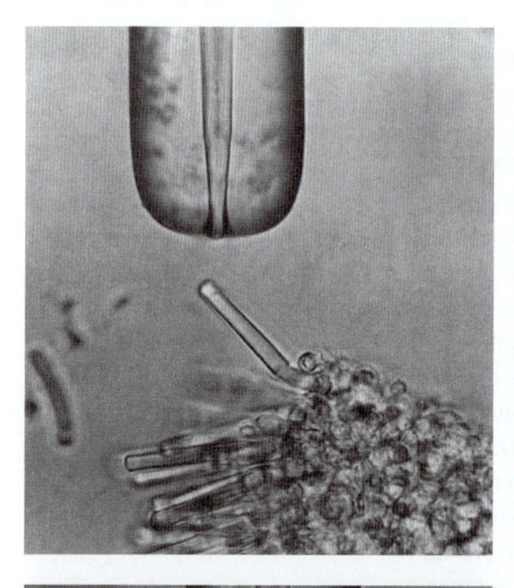

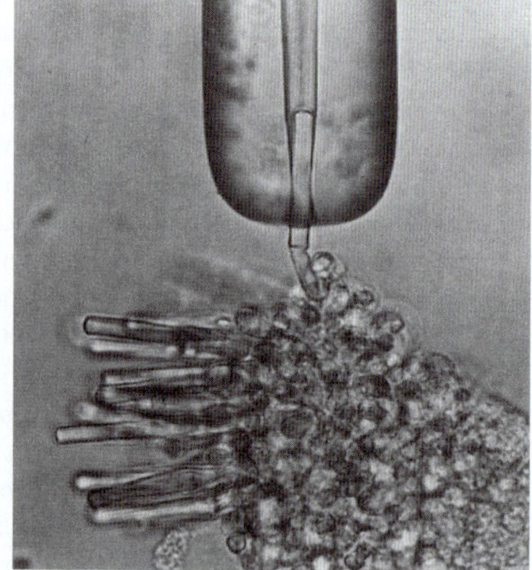

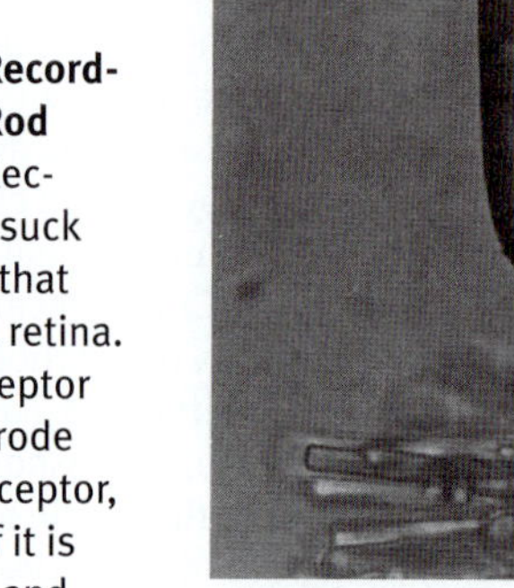

FIGURE 19.13 Method for Recording Membrane Currents of a Rod Outer Segment. A suction electrode with a fine tip is used to suck up the outer segment of a rod that protrudes from a piece of toad retina. Slits of light illuminate the receptor with precision. Since the electrode fits tightly around the photoreceptor, current flowing into it or out of it is recorded. (From Baylor, Lamb, and Yau, 1979.)

The procedure was to isolate a piece of retina from the animal or from a cadaver and maintain it in darkness. To measure currents, the outer segment of the rod was sucked into a fine pipette (see Figure 19.13). As expected, in darkness a current flowed continuously into the outer segment. Flashes of light closed channels in the outer segment, causing a decrease in the dark current. Figure 19.14A shows responses to very dim light flashes, corresponding to one or two quanta of light on the outer segment. The currents are small and quantal in nature. That is, sometimes a dim flash evokes a unitary response, sometimes a doublet, and sometimes nothing at all.

In monkey rods the current reduction caused by a single photon is about 0.5 pA. This corresponds to the closure of approximately 300 channels, about 3 to 5% of the rod channels that are open in the dark, and is the result of the large amplification through the cyclic GMP cascade. Moreover, because of the extreme stability of visual pigments mentioned earlier, random isomerizations and spurious channel closings are rare events. This

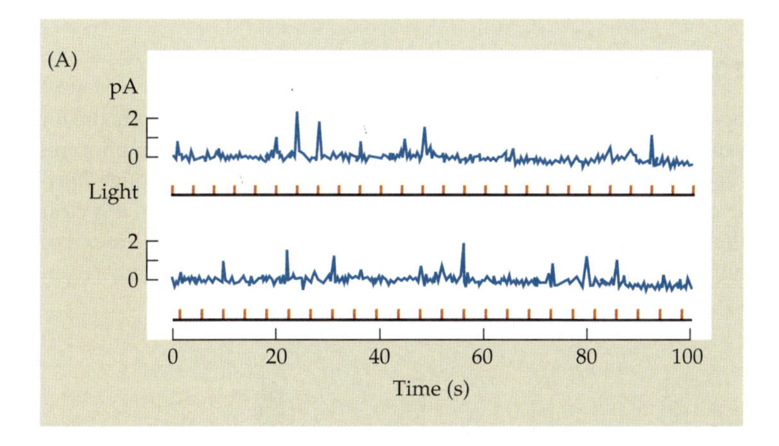

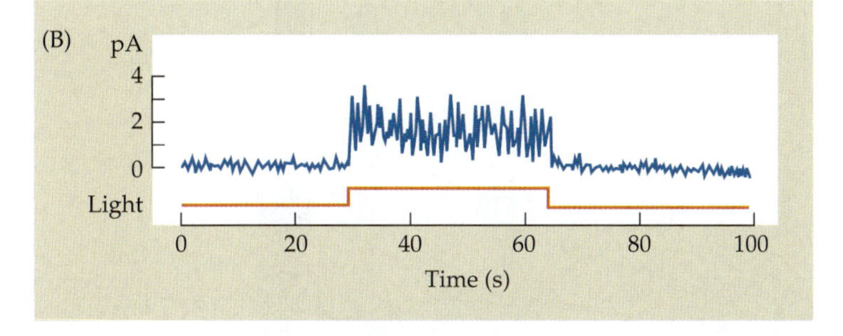

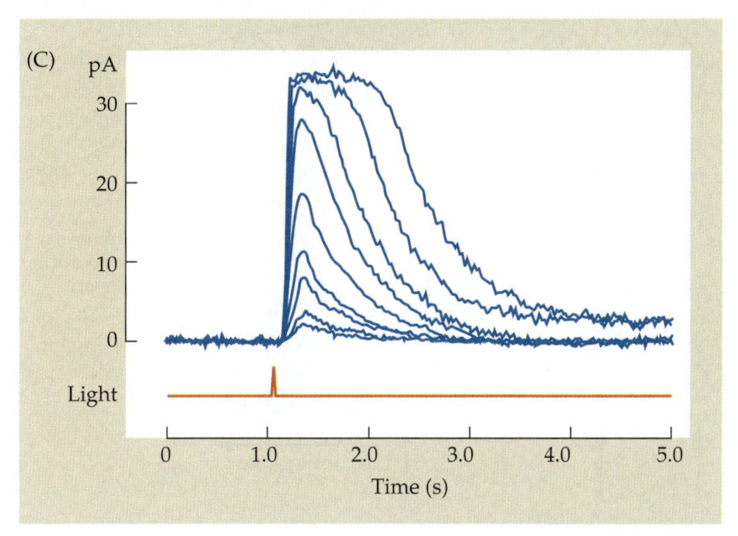

FIGURE 19.14 Recordings Made by Suction Electrode from Monkey Rod Outer Segment. (A) Responses to dim flashes (applied as indicated in the red traces labeled "Light") are shown in the two current traces. The currents fluctuate in a quantal manner. Smaller deflections are the currents generated by single photons interacting with visual pigments. Often photoisomerizations failed to occur. (B) Steady, more intense illumination (bottom trace) gives rise to a burst of signals. (C) Records from a rod in a monkey retina with flashes of increasing intensity. These currents are the counterpart of voltage traces shown in Figure 19.5B. (From Baylor, Nunn, and Schnapf, 1984.)

allows the effects of single light quanta to stand out against an extremely quiet background. It has been shown that electrical coupling through gap junctions between photoreceptors provides an additional smoothing effect that reduces the background noise and improves the signal-to-noise ratio for responses of rods to single quanta.[36]

TRANSMISSION FROM PHOTORECEPTORS TO BIPOLAR CELLS

Numerous questions arise about the transmission of signals in the retina. How are bipolar cells influenced by rods and cones? And how are horizontal and amacrine cells involved in signaling? The analysis of signal processing by these neurons in the outer and inner plexiform layers (Figure 19.15) has required a combination of techniques, including microelectrode recordings, dye injection, morphological studies, cellular neurochemistry, and identification of transmitters and receptors.

Bipolar, Horizontal, and Amacrine cells

From the morphological descriptions of Ramón y Cajal (Chapter 1) a general scheme for the wiring diagram of the retina emerges: a through-line of transmission from photoreceptors to ganglion cells by way of bipolar cells, with interactions mediated through horizontal and amacrine cells.[42] From the pattern of connections in primate retina, it is clear that the output of the eye must be the result of highly complex integrative processes. For example, the horizontal cell shown in Figure 19.16D receives synapses from many recep-

[41]Baylor, D. A., and Hodgkin, A. L. 1973. *J. Physiol.* 234: 163–198.

[42]Boycott, B., and Wässle, H. 1999. *Invest. Ophthalmol. Vis. Sci.* 40: 1313–1327.

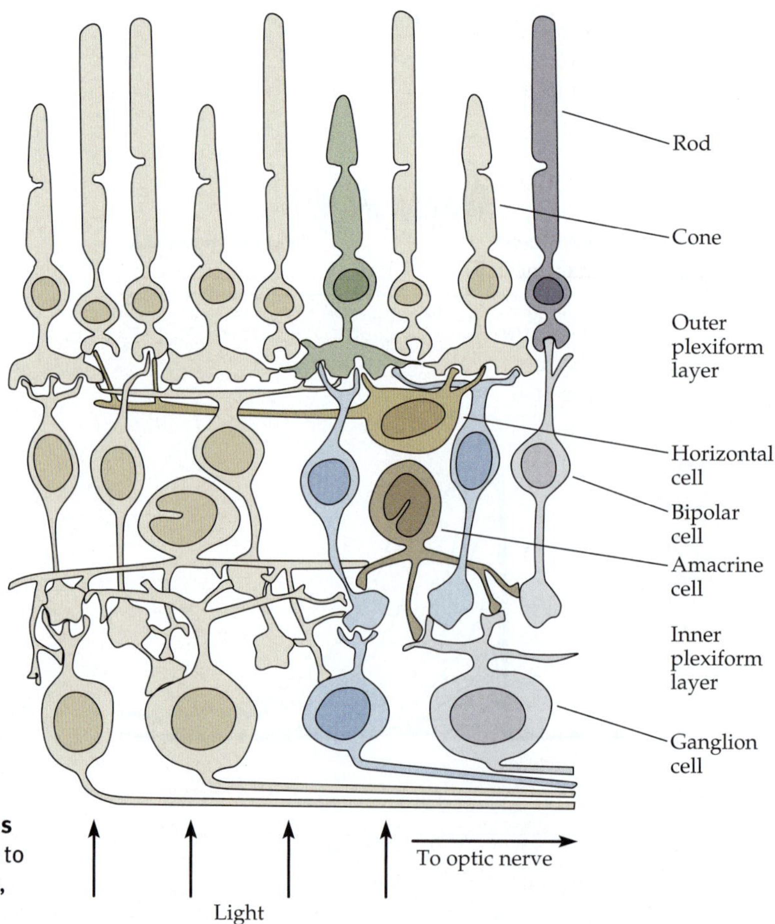

FIGURE 19.15 Principal Cell Types and Connections of Primate Retina to illustrate rod and cone pathways to ganglion cells. (After Dowling and Boycott, 1966; Daw, Jensen, and Brunken, 1990.)

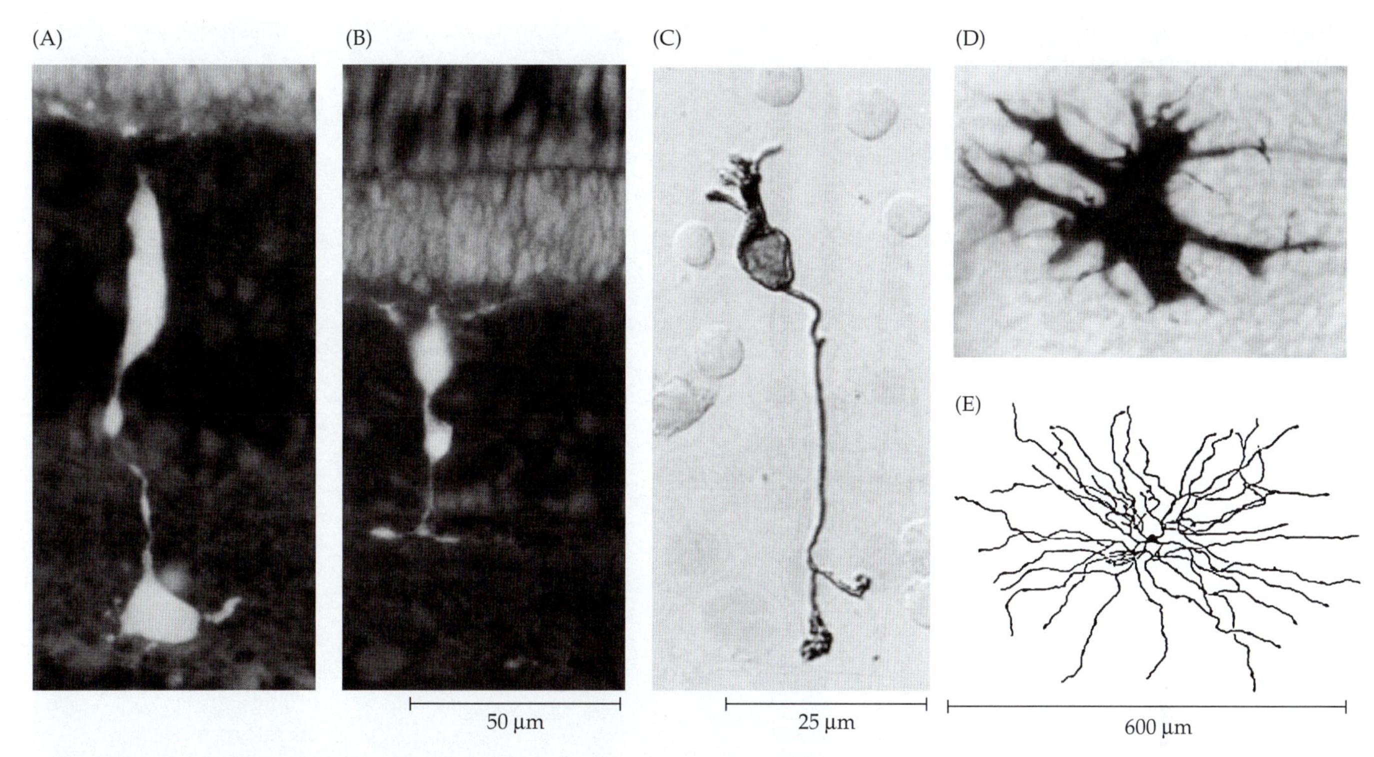

FIGURE 19.16 Bipolar, Horizontal, and Amacrine Cells. (A) A depolarizing "on"-center bipolar cell of goldfish injected with fluorescent dye. (B) A hyperpolarizing "off"-center goldfish bipolar cell. (C) A bipolar cell isolated from rat retina stained for protein kinase C. (D) A horizontal cell in dogfish retina injected with horseradish peroxidase. (E) Indolamine-accumulating amacrine cell from rabbit retina injected with Lucifer yellow. (A, B, and D kindly provided by A. Kaneko, unpublished; C from Yamashita and Wässle, 1991; E from Masland, 1988.)

tors and in turn feeds back onto them. Horizontal cells also end on bipolar cells. Similarly, certain amacrine cells (Figure 19.16E), which receive inputs from bipolar cells, send synapses back to the bipolar cells as well as to ganglion cells. One can conclude that horizontal cells and amacrine cells transmit and modify signals traveling through the retina. An additional source of complexity is that each of the major classes of neurons shown in Figures 19.15 and 19.16 has numerous morphological and pharmacological subtypes.[2] By physiological, biochemical, and anatomical criteria, several major classes of bipolar cells, more than two types of horizontal cell, and at least 20 types of amacrine cells have been described.[43–45]

Transmitters in the Retina

Virtually all the known transmitters have been found in the retina.[46–49] Glutamate is the transmitter liberated by photoreceptors and bipolar cells. Horizontal cells secrete GABA. Some amacrine cells secrete dopamine, others indolamines, others acetylcholine. The distributions and functional significance of the transmitters, receptors, receptor subunits, and transporters have been explored in detail by immunohistochemistry and by in situ hybridization. Peptide transmitters found in the eye,[50] such as VIP (vasoactive intestinal peptide), can play a subtle trophic role in development and have been linked to the production of myopia (nearsightedness).[51]

The continued release of quanta of glutamate from photoreceptors in the dark is achieved by specialized endings of the type known as **ribbon synapses**.[2,52,53] An example is shown in Figure 19.17. The cone terminal contains vesicles that are docked, ready for release, along a long, flat organelle, the ribbon. The postsynaptic elements are terminals of bipolar and horizontal cells that invaginate the photoreceptor ending.

[43]Strettoi, E., and Masland, R. H. 1996. *Proc. Natl. Acad. Sci. USA* 93: 14906–14911.

[44]Kolb, H. 1997. *Eye* 11: 904–923.

[45]MacNeil, M. A., et al. 1999. *J. Comp. Neurol.* 413: 305–326.

[46]Wässle, H., et al. 1998. *Vision Res.* 38: 1411–1430.

[47]Brandstatter, J. H., Koulen, P., and Wässle, H. 1988. *Vision Res.* 38: 1385–1397.

[48]Pourcho, R. G. 1996. *Curr. Eye Res.* 15: 797–803.

[49]Qian, H., et al. 1997. *J. Neurophysiol.* 78: 2402–2412.

[50]Herbst, H., and Their, P. 1996. *Exp. Brain Res.* 111: 345–355.

[51]Raviola, E., and Wiesel, T. N. 1990. *Ciba Found. Symp.* 155: 22–38.

[52]Vardi, N., et al. 1998. *Vision Res.* 38: 1359–1369.

[53]von Gersdorff, H., et al. 1996. *Neuron* 16: 1221–1227.

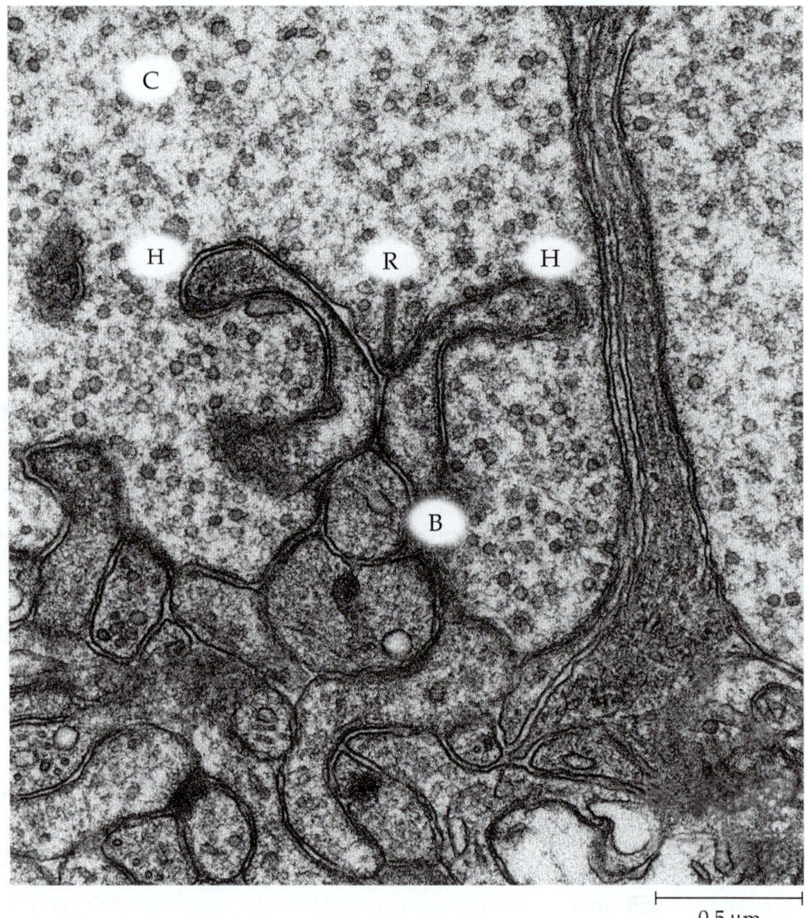

FIGURE 19.17 Ribbon Synapses Made by a Photoreceptor Terminal on bipolar and horizontal cell endings. Presynaptic vesicles in a cone terminal (C) are aligned along the ribbon (R). This type of synapse is adapted for maintained release of quanta of glutamate onto bipolar (B) and horizontal (H) cells in darkness. The horizontal cell, which releases GABA, feeds back onto the receptor terminal. (Micrograph kindly provided by P. Sterling.)

The Concept of Receptive Fields

The technique of illuminating selected areas of the retina introduced the important concept of the **receptive field**, a concept that provided a key for understanding the significance of the signals not only in the retina but at successive stages in the cortex. The term "receptive field" was coined originally by Sherrington in relation to reflex actions (see also Chapter 18) and was introduced to the visual system by Hartline.[54] The receptive field of a neuron in the visual system can be defined as *the area of the retina from which the activity of a neuron can be influenced by light* (see also Chapter 18). For example, a record of the activity of one particular neuron in the optic nerve or cortex of a cat shows that the rate of firing increases or decreases only if the illumination is changed over a defined area of retina (see Figures 19.18 and 19.21). This area is its receptive field. By definition, illumination outside a receptive field produces no effect on firing. The area itself can be subdivided into distinct regions, some of which increase activity and others of which suppress it. This description of receptive fields also applies to neurons such as bipolar or horizontal cells in which graded local potentials are produced by light falling on the retina (discussed later in the chapter). It will be shown that diffuse flashes of light are of little or no use for assessing function in the visual system.

Responses of Bipolar Cells

Each bipolar cell receives its direct input either from rods or from cones. Rod bipolar cells are typically supplied by 15 to 45 receptors. One type of cone bipolar, the midget bipolar, receives its input from a single cone.[2] As one might expect, midget bipolar cells are found in the through-line from the fovea, where acuity is highest. They end on specialized ganglion cells. Other bipolar cells are supplied by a convergent input from 5 to 20 adjacent

[54]Hartline, H. K. 1940. *Am. J. Physiol.* 130: 690–699.

cones. Bipolar cells and horizontal cells respond to illumination with graded depolarizations or hyperpolarizations.

The responses and receptive fields of bipolar cells depend on two mechanisms: First, the continuous release of glutamate from photoreceptors in the dark keeps some bipolar cells depolarized and others hyperpolarized, depending on whether the cells have excitatory or inhibitory glutamate receptors. Second, light causes photoreceptors to be hyperpolarized, thereby causing a reduction in glutamate release. Accordingly, decreased tonic release from illuminated photoreceptors will reduce excitation of bipolar cells with excitatory receptors, giving rise to hyperpolarization.[55] These are called H (hyperpolarizing) bipolar cells (Figure 19.18).

Conversely, decreased tonic release from illuminated photoreceptors will give rise to depolarization of bipolar cells with inhibitory receptors. These are D (depolarizing) bipolar cells. D bipolar cells constitute one of the few cell types in which glutamate has been shown to have an inhibitory action (Chapter 14). Kaneko and his colleagues have shown that the inhibition is mediated by metabotropic glutamate receptors in the bipolar cell that act through G proteins and second messengers to produce a decrease in conductance;[56,57] hence, in the dark, as a result of receptor activation, cyclic nucleotide–gated channels in the D bipolar cell are closed (the opposite of what one observes in photoreceptors). After illumination, glutamate release is reduced, cation channels open, and the bipolar cell becomes depolarized.[58]

Receptive Field Organization of Bipolar Cells

The receptive field of a hyperpolarizing (H) bipolar cell is shown in Figure 19.18. A small spot of light shone onto the central part of the field causes a sustained hyperpolarization. Illumination by an annulus, leaving the center dark, causes depolarization. Thus, the central area driven directly by photoreceptors is enveloped by an antagonistic **surround**. The H bipolar cell of Figure 19.18 can be described as having an "off"-center receptive field, since it becomes depolarized when the spot of light goes off.

D bipolar cells have similarly shaped concentric fields, except that illumination of the center causes depolarization and illumination of the surround causes hyperpolarization. Because it is depolarized when the light goes on, the D bipolar cell has an "on"-center re-

[55]Kaneko, A., and Hashimoto, H. 1969. *Vision Res.* 9: 37–55.

[56]Kikkawa, S., et al. 1993. *Biochem. Biophys. Res. Commun.* 195: 374–379.

[57]Kaneko, A., et al. 1994. *Biomed. Res.* 15(Suppl. 1): 41–45.

[58]Nakanishi, S., et al. 1998. *Brain Res. Rev.* 26: 230–235.

(A) **Central illumination**

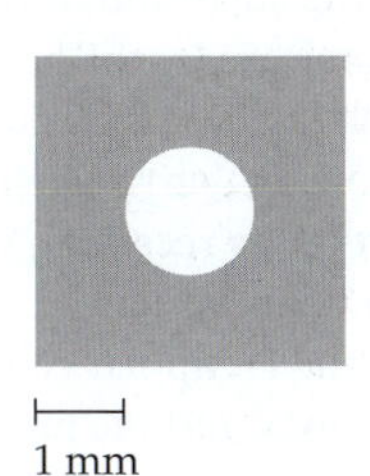

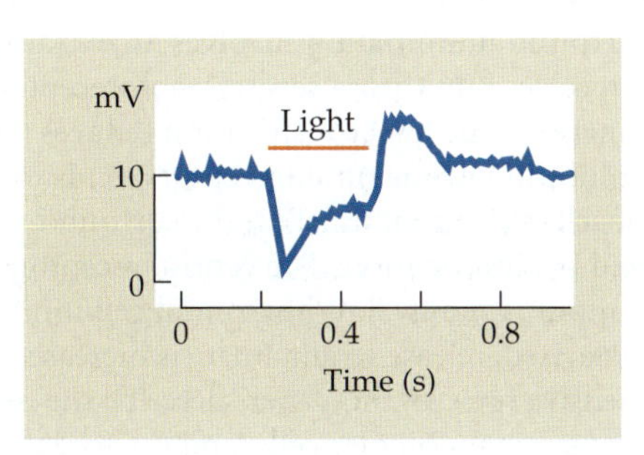

(B) **Annular illumination**

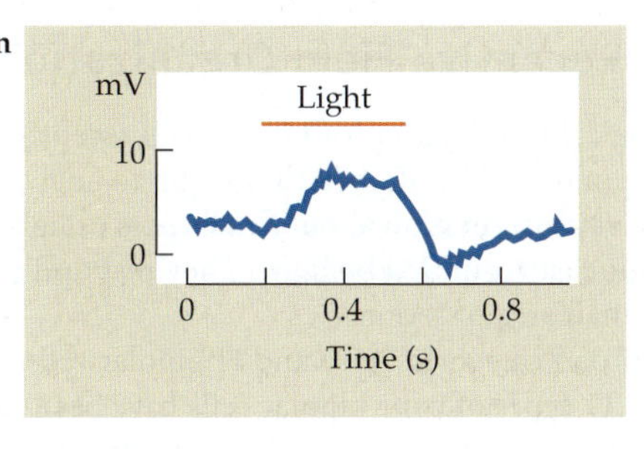

FIGURE 19.18 Receptive Field Organization of a Hyperpolarizing (H) Bipolar Cell. Records made from the bipolar cell in the goldfish retina show a hyperpolarization (A) in response to illumination of the center of the receptive field. Annular illumination causes the cell to respond with a depolarization (B). Diffuse light would have little effect on the cell. For a depolarizing (D) bipolar cell, illumination of the center would produce depolarization, while the annulus would produce hyperpolarization. (After Kaneko, 1970.)

ceptive field. The terminology of "on" and "off" responses will be used extensively to describe receptive field properties at successive levels of the visual system. An important principle is that a single photoreceptor can contribute to the receptive field centers of both "on" and "off" bipolar cells and to the surrounds of others.

Horizontal Cells and Surround Inhibition

The responses of D and H bipolar cells to surround illumination are mediated by horizontal cells. Each horizontal cell receives inputs from a large number of photoreceptors. Horizontal cells, like H bipolar cells, respond to illumination of photoreceptors by *hyperpolarization* (once again, because glutamate release from photoreceptor terminals decreases with illumination). Another feature of horizontal cells is that they are electrically coupled to each other.[59] Lucifer yellow dye injected into one horizontal cell spreads readily to others through gap junctions. Any one horizontal cell is therefore influenced by light shone on a large area of retina because of current flow from its neighbors.

Horizontal cells make inhibitory synaptic connections; they release GABA back onto the photoreceptors and bipolar cells.[60–62] Thus, depolarization of photoreceptors in the dark is antagonized by inhibitory input from horizontal cells. Receptor illumination results in horizontal cell hyperpolarization and a reduction in GABA. Hence, hyperpolarization of photoreceptors by diffuse light is countered by a reduction in the GABA inhibition coming from horizontal cells. It has been suggested that nitric oxide, which is synthesized by photoreceptors and horizontal cells, also contributes to the inhibition of glutamate release by photoreceptors.[63] In summary, there is negative feedback onto the photoreceptors through the horizontal cells:

illumination→ photoreceptor hyperpolarization→ horizontal cell hyperpolarization→ photoreceptor depolarization

The connections involved in the "off"-center, "on"-surround responses of an H bipolar cell are shown schematically in Figure 19.19. For simplicity, the center is represented by a single photoreceptor and the surround by a few neighboring receptors connected to a single horizontal cell. The response to illumination of the central photoreceptor is straightforward (Figure 19.19A). Photoreceptor activation results in hyperpolarization and therefore a reduction in glutamate release. As a result, the bipolar cell is hyperpolarized. The horizontal cell receives a hyperpolarizing input as well, but it is from only one photoreceptor and the effect is small, as is the negative feedback onto the central photoreceptor.

The response to surround illumination involves an extra step (Figure 19.19B). The horizontal cell, which receives input from several photoreceptors in the surround, is hyperpolarized by illumination. The hyperpolarization reduces GABA release by the horizontal cell. Reduced inhibition of the photoreceptors tends to produce depolarization. The depolarizing feedback effect is minimal on the surround receptors, which are being strongly hyperpolarized by illumination. The central receptor, however, is receiving no illumination; its only input is removal of horizontal cell inhibition. Consequently, the central receptor is depolarized, release of glutamate is increased, and the H bipolar cell is depolarized. Comprehensive reviews and papers describe the morphology and the properties of photoreceptor terminals, bipolar cells, and the feedback synapses of horizontal cells onto bipolar cells.[2,7,46]

Significance of Receptive Field Organization of Bipolar Cells

What are the physiological implications of bipolar cell receptive fields? D and H bipolar cells do not simply respond to light. Rather, they begin to analyze information about patterns. Their signals convey information about small spots of light surrounded by darkness or about small dark spots surrounded by light. They respond to contrasting patterns of light and dark over a small area of retina.

In addition to the broad categories of D and H bipolar cells mentioned in the list that follows, approximately 11 types of cone bipolar cells have been distinguished by morphological and immunohistochemical criteria.[42] Here we summarize three principal types.

[59]Kaneko, A. 1971. *J. Physiol.* 213: 95–105.

[60]Kaneko, A., and Tachibana, M. 1986. *J. Physiol.* 373: 443–461.

[61]Schwartz, E. A. 1987. *Science* 238: 350–355.

[62]Yang, X. L., Gao, F., and Wu, S. M. 1999. *Vis. Neurosci.* 16: 967–979.

[63]Savchenko, A., Barnes, S., and Kramer, R. H. 1997. *Nature* 390: 694–698.

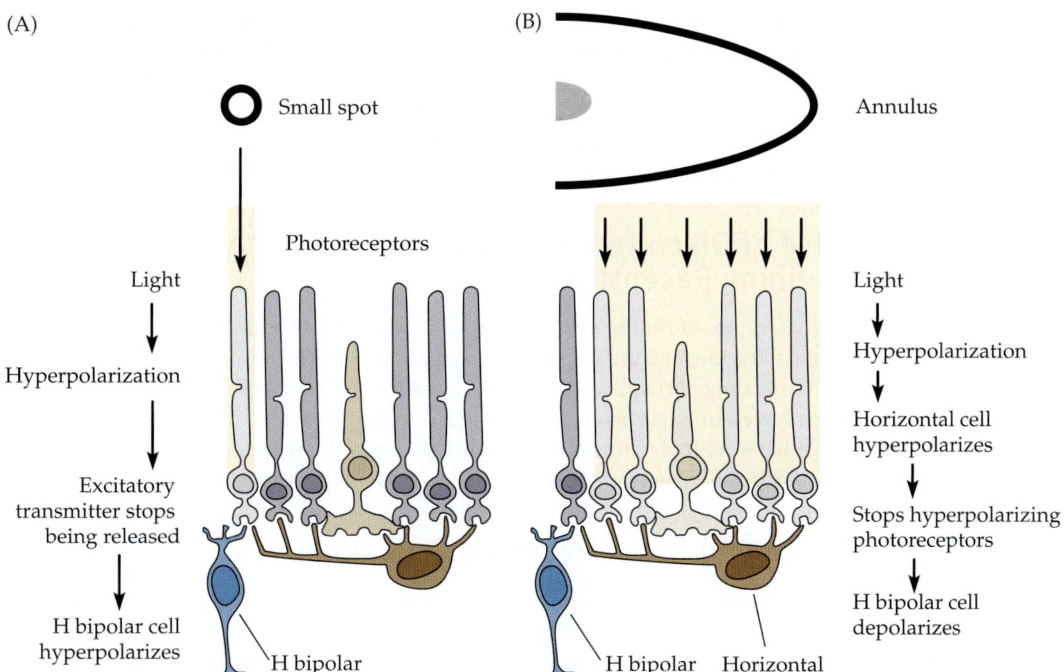

FIGURE 19.19 Connections of Photoreceptors, Bipolar Cells, and Horizontal Cells. The figure illustrates connections required to elicit responses in bipolar cells. (A) Light falling on a single photoreceptor causes it to become hyperpolarized. As a result, glutamate stops being released and the H bipolar cell, as in Figure 19.18, becomes hyperpolarized through loss of excitation. (B) Light falling on the surrounding area in the form of an annulus again prevents glutamate from being released by photoreceptors. As a result, the horizontal cell becomes hyperpolarized; this hyperpolarization prevents the horizontal cell from releasing its inhibitory transmitter, GABA, onto the photoreceptor. The photoreceptor that is connected to the H bipolar cell therefore becomes depolarized (through removal of inhibition). It once again releases glutamate and depolarizes the bipolar cell. With diffuse light, the depolarizing and hyperpolarizing effects cancel each other out. Thus, horizontal cells play an essential part in the construction of the receptive field properties of bipolar cells.

1. D and H cone bipolar cells respond best to small light or dark spots.
2. The centers of D and H midget bipolar cells are supplied by single cones.
3. Rod bipolar cells are D, or "on"-center, and respond best to small bright spots.

RECEPTIVE FIELDS OF GANGLION CELLS

The Output of the Retina

Many years before the electrical responses of photoreceptors or bipolar cells could be measured, important information was obtained by recording from ganglion cells. Thus, the first analysis of signaling in the retina was made at the output stage, the end result of synaptic interactions. An advantage of this approach was that ganglion cells give all-or-nothing action potentials: Records could be made with extracellular electrodes at a time before intracellular microelectrodes had been perfected or dye injections developed. Second, going straight to the output was a simplification and shortcut. It was in ganglion cells that concentric "on" and "off" receptive fields were first described. This made it possible to understand the responses seen later in bipolar and horizontal cells.

It was Stephen Kuffler who pioneered the experimental analysis of the mammalian visual system by concentrating on receptive field organization and the meaning of signals in the cat.[64] The end results of synaptic interactions rather than the synaptic mechanisms themselves were the objective. Hubel[65] has succinctly put the achievement in perspective:

> What is especially interesting to me is the unexpectedness of the results, as reflected in the failure of anyone before Kuffler to guess that something like center-surround receptive fields could exist or that the optic nerve would virtually ignore anything so boring as diffuse light levels.

The principal new approach was not so much a matter of recording technique; rather it consisted of formulating the following question: What is the best way to stimulate individual ganglion cells? This question led to the use of discrete circumscribed spots for stimulation of selected areas of the retina, instead of diffuse uniform illumination. Such procedures had been foreshadowed by pioneering work on the eye of a simple inverte-

[64]Kuffler, S. W. 1953. *J. Neurophysiol.* 16: 37–68.

[65]Hubel, D. H. 1988. *Eye, Brain and Vision.* Scientific American Library, New York.

brate, the horseshoe crab *Limulus*,[54] and on the retina of the frog.[66,67] Kuffler's initial choice of the cat was a lucky one; in the rabbit, for example, the situation would have been more complicated. Rabbit ganglion cells have elaborate receptive fields that respond to such complex features as edges or to movement in a particular direction.[68] Equally complex are lower vertebrates, such as frogs. A general law seems to emerge: The dumber the animal, the smarter its retina (D. A. Baylor, personal communication).

The Use of Discrete Visual Stimuli in Intact Animals for Defining Receptive Fields

An essential feature of Kuffler's early experiments was the use of the intact undissected eye, the normal refraction channels of which served as pathways for stimulation.[64] A convenient way of illuminating particular portions of the retina is to anesthetize the animal and place it facing a screen or a television monitor at a distance for which its eyes are properly refracted. When one then shines patterns of light onto the screen or displays computer-generated television images, these will be well focused on the retinal surface (Figure 19.20).

Ganglion Cell Receptive Field Organization

When one records from a particular ganglion cell, the first task is to find the location of its receptive field. Characteristically, most ganglion cells and neurons throughout the visual system show discharges at rest even in the absence of illumination. Appropriate stimuli do not necessarily initiate activity but may modulate the resting discharge; responses of ganglion cells can consist of either an increase or a decrease of frequency.

Figure 19.21, adapted from a paper by Kuffler, shows that for a ganglion cell a small spot of light, 0.2 mm in diameter, shone onto a part of the receptive field is far more effective than diffuse illumination in producing excitation. Furthermore, the same spot of light can have opposite effects, depending on the exact position of the stimulus within the receptive field. For example, in one area the spot of light excites a ganglion cell for the duration of illumination. Such an "on" response can be converted into an inhibitory "off" response by simply shifting the spot by 1 mm or less across the retinal surface. As is the case for bipolar cells, which were studied many years later, there are two basic receptive field types: "on"-center and "off"-center ganglion cells. The receptive fields of both types are roughly concentric, with the ganglion cell soma in the geometrical central region of its field.

In an "on"-center receptive field, light produces the most vigorous response if it completely fills the center, whereas for most effective inhibition of firing, the light must cover the entire ring-shaped surround (annular illumination in Figure 19.21). When the inhibitory annular light is turned off, the ganglion cell gives an exuberant "off" discharge. An "off"-center field has a converse organization, with inhibition arising in the circular center. For either cell, the spotlike center and its surround are antagonistic; therefore, if both center and surround are illuminated simultaneously, they tend to cancel each other's contribution.

[66]Barlow, H. B. 1953. *J. Physiol.* 119: 69–88.

[67]Maturana, H. R., et al. 1960. *J. Gen. Physiol.* 43: 129–175.

[68]Barlow, H. B., Hill, R. M., and Levick, W. R. 1964. *J. Physiol.* 173: 377–407.

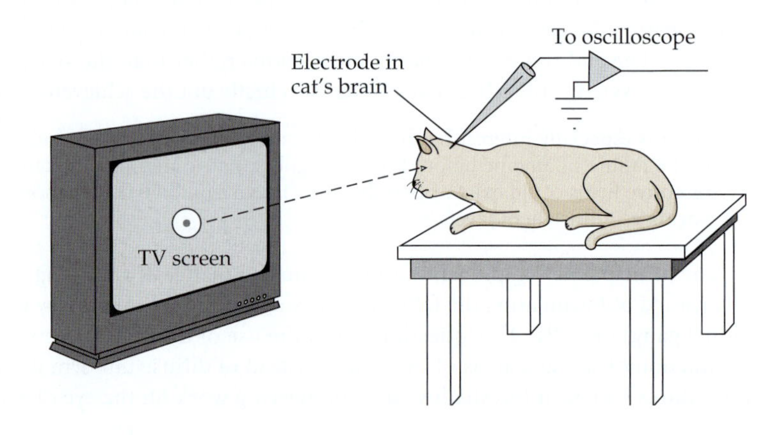

FIGURE 19.20 Stimulation of Retina with Patterns of Light. The eyes of an anesthetized, light-adapted cat or monkey focus on a movie or a television screen with various patterns of light generated by a computer or shone by a projector. An electrode records the responses from a single cell in the visual pathway. Light or shadow falling onto a restricted area of the screen may increase or decrease the frequency of signals given by the neuron. One can delineate the receptive field of the cell by determining the areas on the screen from which the neuron's firing is influenced. In his original experiments, Kuffler shone light directly into the eye by means of a specially constructed ophthalmoscope.

Sizes of Receptive Fields

Neighboring ganglion cells collect information from very similar, but not quite identical, areas of the retina. Even a small (0.1 mm) spot of light on the retina covers the receptive fields of many ganglion cells. Some are inhibited, others excited. This characteristic organization, with neighboring groups of receptors projecting onto neighboring ganglion cells in the retina, is retained at all levels in visual pathways. Throughout the visual system, the systematic analysis of the positions of cells and their receptive fields demonstrates the general principle that *neurons processing related information are clustered together*. In sensory systems this means that the central neurons dealing with a particular area of the surface can communicate with each other over short distances. This appears to be an economical arrangement, as it saves long lines of communication and simplifies the making of connections (Chapter 17).

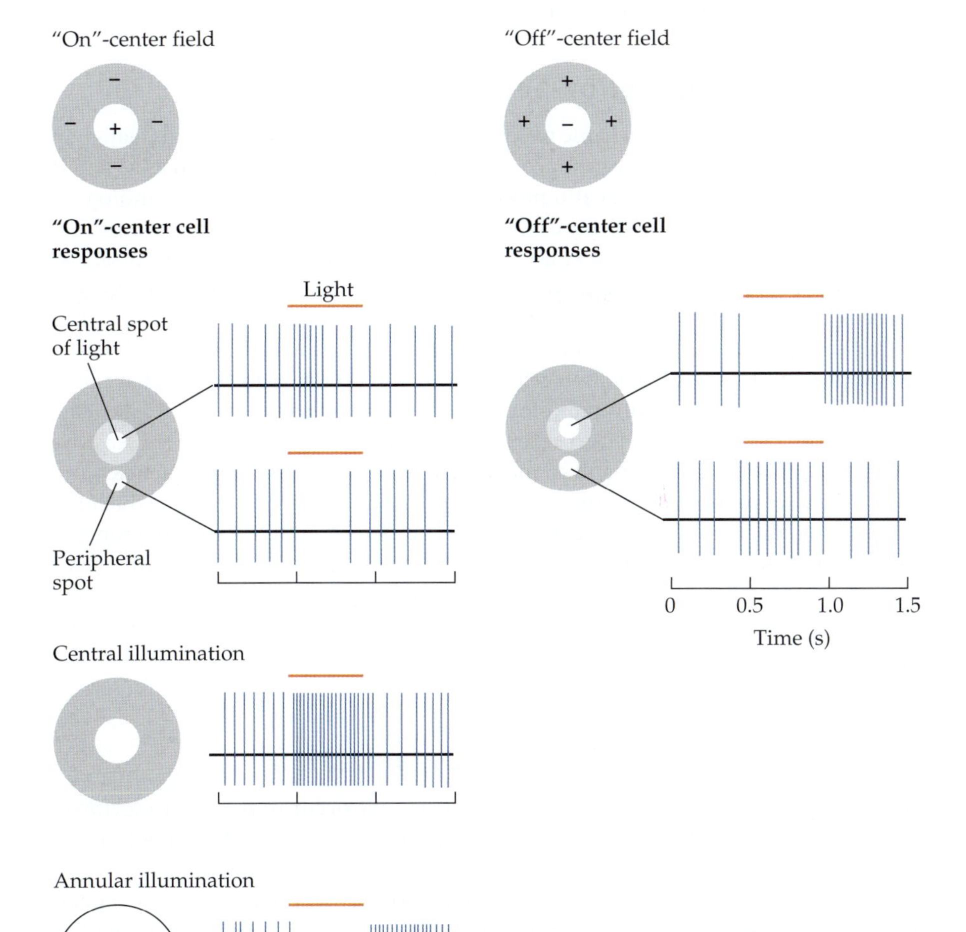

FIGURE 19.21 Receptive Fields of Ganglion Cells in the retinas of cats and monkeys are grouped into two main classes: "on"-center and "off"-center fields. "On"-center cells respond best to a spot of light shone onto the central part of the receptive field. Illumination (indicated by the red bar above records) of the surrounding area with a spot or a ring of light reduces or suppresses the discharges and causes responses when the light is turned off. Illumination of the entire receptive field elicits weak discharges because center and surround antagonize each other's effects, as with bipolar cells. "Off"-center cells slow down or stop signaling when the central area of their field is illuminated and accelerate when the light is turned off. Light shone onto the surround of an "off"-center receptive field causes excitation of the neuron. (After Kuffler, 1953.)

The size of the receptive field of a ganglion cell depends on its location in the retina. The receptive fields of cells situated in the central areas of the retina have much smaller centers than those at the periphery; receptive fields are smallest in the fovea, where the acuity of vision is highest.[69,70] The central "on" or "off" region of such a "midget" ganglion cell's receptive field can be supplied by a single cone and is accordingly only about 2.5 μm in diameter, subtending 0.5 *minutes* of arc—smaller than the period at the end of this sentence. Note that receptive fields can be described either as dimensions on the retina or as degrees of arc subtended by the stimulus. In our eyes, 1 mm on the retina corresponds to about 4°. For reference, the image of the moon has a diameter of ⅛ mm on our retina, corresponding to 0.5°, or 30 minutes of arc.

There are similar gradations of receptive field size in relation to fine resolution in the visual and somatosensory systems. A higher-order sensory neuron in the brain responding to a fine touch applied to the skin of the fingertip has a receptive field that is very small compared with that of a neuron having a field on the skin of the upper arm (Chapter 18). To discern the form of an object, we use our fingertips and fovea, not the less discriminating regions on the receptor surfaces with poorer resolution.

Classification of Ganglion Cells

Superimposed on the general scheme of "on"- or "off"-center receptive fields, ganglion cells in the monkey retina can be grouped into two main categories denoted as M and P. The criteria are both anatomical and physiological. The "M" and "P" terminology is based on the anatomical projections of these neurons to the lateral geniculate nucleus and from there to the cortex (Chapter 20). P ganglion cells project to the four dorsal layers of smaller cells in the lateral geniculate nucleus (the parvocellular division), M ganglion cells to the larger cells in two ventral layers (the magnocellular division). The characteristics of neurons in the M and P pathways are maintained at successive levels in the visual system. In brief, P ganglion cells have small receptive field centers, have high spatial resolution, and are sensitive to color. P cells provide information about fine detail at high contrast.[71] M cells have larger receptive fields than P cells and are more sensitive to small differences in contrast and to movement; they fire at higher frequencies and conduct impulses more rapidly along their larger-diameter axons. In the cat, which has no color vision, the classification of ganglion cells is different, with X, Y, and W groups.[72] X and Y are in some respects parallel to P and M in their properties, but there are major differences and the two classifications are not interchangeable.[73] The magno-/parvo-classification in the visual system provides a convenient and useful framework for studying pathways and properties.

Synaptic Inputs to Ganglion Cells Responsible for Receptive Field Organization

Inputs to ganglion cells from bipolar and amacrine cells occur in the inner plexiform layer (see Figure 19.15). A full description of the intricate connections from photoreceptors to ganglion cells is given in a lucid and comprehensive review by Sterling.[2]

As expected, depolarizing "on"-center and hyperpolarizing "off"-center cone bipolar cells make chemical synapses with corresponding "on"- and "off"-center ganglion cells. These synapses are excitatory; consequently, the change in membrane potential of a bipolar cell causes the ganglion cell to which it is connected to change its membrane potential in the same direction. Transmission from rods to ganglion cells is not so simple. One of the pathways is by way of rod bipolar cells that in turn connect to a special type of amacrine cell (known as the A2 amacrine cell).[74–76] The connections are so precise that both rods and cones in the same part of the retina supply the same ganglion cell appropriately, but by way of different interposed cells.

An elegant demonstration of the efficacy of the through-line from a single photoreceptor to a ganglion cell was provided by Baylor and Fettiplace.[77] They changed the membrane potential of a single photoreceptor by passing current through an intracellular microelectrode. At the same time they found a ganglion cell to which the photorecep-

[69]Kier, C. K., Buchsbaum, G., and Sterling, P. 1995. *J. Neurosci.* 15: 7673–7683.

[70]Croner, L. J., and Kaplan, E. 1995. *Vision Res.* 35: 7–24.

[71]Kaplan, E., and Shapley, R. M. 1986. *Proc. Natl. Acad. Sci. USA* 83: 2755–2757.

[72]Enroth-Cugell, C., and Robson, J. G. 1966. *J. Physiol.* 187: 517–552.

[73]Benardete, E. A., Kaplan, E., and Knight, B. W. 1992. *Vis. Neurosci.* 8: 483–486.

[74]Daw, N. W., Jensen, R. J., and Brunken, W. J. 1990. *Trends Neurosci.* 13: 110–115.

[75]Dacheux, R. F., and Raviola, E. 1986. *J. Neurosci.* 6: 331–345.

[76]Soucy, E., et al. 1998. *Neuron* 21: 481–493.

[77]Baylor, D. A., and Fettiplace, R. 1977. *J. Physiol.* 271: 391–424.

tor was connected by way of bipolar, horizontal, and amacrine cells. Hyperpolarization of the single receptor, a cone sensitive to red light, initiated firing in the ganglion cell. This experiment showed unequivocally that the hyperpolarization or depolarization of one receptor is sufficient to provide information to other cells in the brain about light or darkness in the outside world.

What Information Do Ganglion Cells Convey?

The most striking feature of ganglion cell signals is that they tell a different story from that of primary sensory receptors. They do not convey information about absolute levels of illumination, because they behave in a similar fashion at different background levels of light. They ignore much of the information of the photoreceptors, which work more like a photographic plate or a light meter. Rather, they measure differences within their receptive fields by comparing the degree of illumination between the center and the surround. Apparently they are designed to notice simultaneous contrast and ignore gradual changes in overall illumination. They are exquisitely tuned to detect such contrast as the edge of an image crossing the opposing regions of a receptive field.

Experiments made in the salamander retina by Baylor, Meister, and their colleagues suggest that temporal aspects of firing by ganglion cells can also contribute to spatial resolution.[78,79] Throughout the previous discussion, the spike trains from individual neurons have been treated as separate lines from which the analysis of visual input is made by the brain. Synchrony of firing by two cells may also be an additional variable.[80] Thus, when recordings were made simultaneously from ganglion cells with closely adjacent fields, a high degree of synchrony was found with some visual stimuli but not others. Synchronization of the action potentials in pairs of ganglion cells depended critically on the size and the position of the spot of light. For example, stimulation of two "off"-center ganglion cells with overlapping field centers by two small spots in appropriate sites produced activity in both cells as expected. The requirement for both cells to fire in synchrony was, however, a spot of light that was smaller than the receptive field of either ganglion cell and that straddled the border of the two fields.

In principle, analysis of the degree of synchrony could be used by higher centers to obtain information about light falling on the retina that could not be deduced from looking at the firing of the two ganglion cells separately. It is known that synchronicity of synaptic inputs can have pronounced effects on integration, but it is not yet known whether the CNS uses this variable to extract information from the firing of ganglion cells.

It is appropriate to close this chapter with a quotation from Sherrington, written long before receptive fields were mapped for single cells. Unlike that of Helmholtz, Sherrington's somewhat opaque style often makes it difficult to read his profoundly original papers and books. Yet the following paragraph reveals his poetic insight into the physiology of vision:[81]

> The chief wonder of all we have not touched on yet. Wonder of wonders, though familiar even to boredom. So much with us that we forget it all the time. The eye sends, as we saw, into the cell-and-fibre forest of the brain throughout the waking day continual rhythmic streams of tiny, individual evanescent, electrical potentials. This throbbing streaming crowd of electrified shifting points in the spongework of the brain bears no obvious semblance in space-pattern, and even in temporal relation resembles but a little remotely the tiny two-dimensional upside-down picture of the outside world which the eye-ball paints on the beginnings of its nerve-fibres to electrical storm. And the electrical storm so set up is one which affects a whole population of brain-cells. Electrical charges having in themselves not the faintest elements of the visual—having, for instance, nothing of "distance," "right-side-upness," no "vertical," nor "horizontal," nor "color," nor "brightness," nor "shadow," nor "roundness," nor "squareness," nor "contour," nor "transparency," nor "opacity," nor "near," nor "far," nor visual anything—yet conjure up all these. A shower of little electrical leaks conjures up for me, when I look, the landscape; the castle on the height, or when I look at him, my friend's face and how distant he is from me they tell me. Taking their word for it, I go forward and my other senses confirm that he is there.

[78]Meister, M., Lagnado, L., and Baylor, D. A. 1995. *Science* 270: 1207–1210.

[79]Meister, M., and Berry, M. J., II 1999. *Neuron* 22: 435–450.

[80]Maffei, L., and Galli-Resta, L. 1990. *Proc. Natl. Acad. Sci. USA* 87: 2861–2964.

[81]Sherrington, C. S. 1951. *Man on His Nature*. Cambridge University Press, Cambridge.

SUMMARY

- Rod and cone photoreceptors respond to illumination in dim and bright light.
- The visual pigments are densely packed on rod and cone membranes.
- Transduction occurs in a series of steps involving a G protein and cyclic GMP.
- In darkness, photoreceptors are depolarized and continuously release glutamate.
- Light causes the closing of nucleotide-gated cation channels, hyperpolarization, and reduction of glutamate release.
- Two main classes of bipolar cells respond to glutamate released by photoreceptors.
- H bipolar cells are depolarized in the dark and hyperpolarized by light.
- D bipolar cells are hyperpolarized in the dark and depolarized by light.
- Receptive field refers to the area of the visual field or retina, illumination of which influences the signals of a cell in the visual system.
- Photoreceptors, horizontal cells, and bipolar cells do not produce action potentials.
- Ganglion cells and amacrine cells give action potentials.
- Bipolar cells and ganglion cells have concentric receptive fields with "on" or "off" centers and antagonistic surrounds.
- Ganglion cells respond poorly to diffuse light.
- Large ganglion cells, known as magnocellular, or M, cells have large receptive fields and respond well to movement.
- Smaller ganglion cells, known as parvocellular, or P, cells, have smaller receptive fields and respond to color and fine detail.

SUGGESTED READING

General Reviews

Baylor, D. 1996. How photons start vision. *Proc. Natl. Acad. Sci. USA* 93: 540–565.

Boycott, B., and Wässle, H. 1999. Parallel processing in the mammalian retina: The Proctor Lecture. *Invest. Ophthalmol. Vis. Sci.* 40: 1313–1327.

Dowling, J. E. 1987. *The Retina: An Approachable Part of the Brain.* Harvard University Press, Cambridge, MA.

Finn, J. T., Grunwald, M. E., and Yau, K-W. 1996. Cyclic nucleotide-gated ion channels: An extended family with diverse functions. *Annu. Rev. Physiol.* 58: 395–426.

Nakanishi, S., Nakajima, Y., Masu, M., Ueda, Y., Nakahara, K., Watanabe, D., Yamaguchi, S., Kawabata, S., and Okada, M. 1998. Glutamate receptors: Brain function and signal transduction. *Brain Res. Rev.* 26: 230–235.

Nathans, J. 1989. The genes for color vision. *Sci. Am.* 260(2): 42–49.

Sterling, P. 1997. Retina. In G. M. Shepherd (ed.), *Synaptic Organization of the Brain.* Oxford University Press, New York, Chapter 6.

Stryer, L. 1991–1992. Molecular mechanism of visual excitation. *Harvey Lect.* 87: 129–143.

Original Papers

Baylor, D. A., Lamb, T. D., and Yau, K. W. 1979. The membrane current of single rod outer segments. *J. Physiol.* 288: 589–611.

Boycott, B. B., and Dowling, J. E. 1969. Organization of primate retina: Light microscopy. *Philos. Trans. R. Soc. Lond. B* 255: 109–184.

Chen, J., Makino, C. L., Peachey, N. S., Baylor, D. A., and Simon, M. I. 1995. Mechanisms of rhodopsin inactivation in vivo as revealed by a COOH-terminal truncation mutant. *Science* 267: 374–377.

Croner, L. J., and Kaplan, E. 1995. Receptive fields of P and M ganglion cells across the primate retina. *Vision Res.* 35: 7–24.

Finn, J. T., Xiong, W. H., Solessio, E. C., and Yau, K. W. 1998. A cGMP-gated cation channel and phototransduction in depolarizing photoreceptors of the lizard parietal eye. *Vision Res.* 38: 1353–1357.

Kaneko, A. 1970. Physiological and morphological identification of horizontal, bipolar and amacrine cells in goldfish retina. *J. Physiol.* 207: 623–633.

Kaneko, A., Delavilla, P., Kurahashi, T., and Sasaki, T. 1994. Role of L-glutamate for formation of on-responses and off-responses in the retina. *Biomed. Res.* 15(Suppl. 1): 41–45.

Kuffler, S. W. 1953. Discharge patterns and functional organization of the mammalian retina. *J. Neurophysiol.* 16: 37–68.

Meister, M., Lagnado, L., and Baylor, D. A. 1995. Concerted signaling by retinal ganglion cells. *Science* 270: 1207–1210.

Schnapf, J. L., Kraft, T. W., Nunn, B. J., and Baylor, D. A. 1988. Spectral sensitivity of primate photoreceptors. *Vis. Neurosci.* 1: 255–261.

20 Signaling in the Lateral Geniculate Nucleus and the Primary Visual Cortex

RETINAL GANGLION CELLS PROJECT TO THE LATERAL GENICULATE NUCLEUS, where they form a retinotopic map. The six layers of the mammalian lateral geniculate nucleus are each innervated by one eye or the other, and they receive input from distinct subtypes of retinal ganglion cells, resulting in magnocellular, parvocellular, or koniocellular geniculate layers. Lateral geniculate neurons have center–surround receptive fields like those of retinal ganglion cells.

Lateral geniculate neurons project to and form a retinotopic map in primary visual cortex, V_1, also called area 17 or striate cortex. The receptive fields of cortical cells, rather than having a center–surround organization, consist of lines or edges, representing an additional step in visual analysis. The six layers of V_1 have specific organizational properties: afferent fibers from the geniculate end primarily in layer 4 (and some in layer 6); cells in layers 2, 3, and 5 receive cortical input. Cells in layers 5 and 6 project to subcortical areas, and cells in 2 and 3 project to other cortical areas. Each vertical stack of cortical cells functions as a module, operating on input from one location in visual space and forwarding the processed information to secondary visual areas. This columnar organization of visual cortex is evident in the constancy of receptive field location throughout the depth of the cortex, and in the segregation of the projection of the two eyes into ocular dominance columns.

Two classes of neurons in V_1 have been defined by their response properties. Receptive fields of simple cells are elongated, with adjacent “on” and “off” areas. Thus, the optimal stimulus for a simple cell is a specifically oriented light or dark bar. A complex cell also responds to oriented bars, but the bar can fall in any region of the receptive field. End inhibition of simple or complex cells gives rise to still more detailed stimulus requirements, such as a line of specific length, or a corner in the receptive field.

The receptive fields of simple cells result from the convergent input of a number of geniculate afferents whose adjoining field centers define the receptive area. The fields of complex cells depend on input from simple cells and other cortical cells. The progression of receptive field properties from retina to lateral geniculate, to cortical simple and then complex cells suggests a hierarchical flow of information whereby the neural constructs from one level are combined to produce still more abstract concepts at the next. Throughout the pathways, the emphasis is on contrast and the detection of edges, rather than on diffuse illumination. Thus, the complex cells of the visual cortex can “see” the lines that define the edges of a box, but they care little about the absolute level of light inside that box.

A clear, continuous thread extending from signaling to perception is provided by the pioneering experiments begun by Kuffler[1] in the retina and followed into the visual cortex by Hubel and Wiesel. Hubel has given a vivid description of the early experiments on visual cortex in Stephen Kuffler's laboratory at Johns Hopkins University in the 1950s.[2] Since then, our understanding of the physiology and anatomy of the cerebral cortex has blossomed through the experiments of Hubel and Wiesel and through the large body of work for which they supplied the starting point and inspiration. Our aim is to give a brief, narrative description of signaling and cortical architecture in relation to perception, based on the classic work of Hubel and Wiesel and on later experiments made by them and their colleagues and by many others. In this chapter we delineate the functional architecture of the lateral geniculate nucleus and visual cortex and how they support the first steps in the analysis of the visual scene: the construction of lines and shapes from the center–surround output of the retina.

Proceeding from the retina and lateral geniculate nucleus to the cerebral cortex raises questions that go beyond simple matters of technique. It has long been acknowledged that understanding the workings of any part of the nervous system requires knowledge of the cellular properties of its neurons: how they conduct and carry information, and how they transmit that information from one cell to the next at synapses. Yet, the monitoring of activity in single neurons might seem an unprofitable way to study higher functions in which large numbers of cells take part. The argument usually has taken (and still does at times) the following form: The brain contains some 10^{10} or more cells. Even the simplest task or event engages hundreds of thousands of nerve cells in various parts of the nervous system. What chance do physiologists have of gaining insight into complex actions within the brain when they sample only one or a few of these units, a hopelessly small fraction of the total number?

On closer scrutiny, the logic of the argument about basic difficulties introduced by large numbers and complex higher functions is not so impeccable as it seems. As so frequently happens, a simplifying principle turns up, opening a new and clarifying view. What simplifies the situation in the visual cortex is that the major cell types are laid out in an apparently well-ordered manner as repeating units. This repetitive neural fabric is interwoven with the **retinotopic map** of visual cortex. That is, adjacent points in the retina project to adjacent points on the cortical surface. Thus, the visual cortex is designed to bring an identical set of neural analyzers to bear on each tiny segment of the visual field. In addition, techniques that label functionally related cellular assemblies have begun to reveal larger-scale patterns of cortical organization. Indeed, the cortical architecture provides the structural basis for cortical function, and novel anatomical techniques have inspired new analytical insights. Thus, before we describe the functional connectivity of visual neurons, it is useful to summarize briefly the general structure of the central visual pathways, beginning with that of the lateral geniculate nucleus.

THE LATERAL GENICULATE NUCLEUS

The optic nerve fibers running from each eye terminate on cells of the right and left **lateral geniculate nucleus (LGN)** (Figure 20.1), a distinctively layered structure ("geniculate" means "bent like a knee").[3] In the LGN of the cat, there are three obvious, well-defined layers of cells (A, A_1, C), one of which (A_1) has a complex structure that has been further subdivided. In monkeys and other primates, including humans, the LGN has six layers of cells. The cells in the deeper layers, 1 and 2, are larger than those in layers 3, 4, 5, and 6, giving rise to the terms **magnocellular** (M, large cell) and **parvocellular** (P, small cell) layers. This classification corresponds to that of the large (M) and small (P) retinal ganglion cells that project to the lateral geniculate nucleus. Between each of the M and P layers lies a zone of very small cells: the interlaminar, or **koniocellular** (K), layers. K cells are functionally and neurochemically[4] distinct from M and P cells and provide a third channel to visual cortex.[5]

In both cat and monkey, each layer of the LGN is supplied by one eye or the other. In the monkey, layers 6, 4, and 1 are supplied by the contralateral eye, layers 5, 3, and 2 by the ipsilateral eye. The segregation of endings from each eye into separate layers has been

[1]Kuffler, S. W. 1953. *J. Neurophysiol.* 16: 37–68.

[2]Hubel, D. H. 1982. *Nature* 299: 515–524.

[3]Guillery, R. W. 1970. *J. Comp. Neurol.* 138: 339–368.

[4]Hendry, S. H. C., and Yoshioka, T. 1994. *Science* 264: 575–577.

[5]Casagrande, V. A. 1994. *Trends Neurosci.* 17: 305–310.

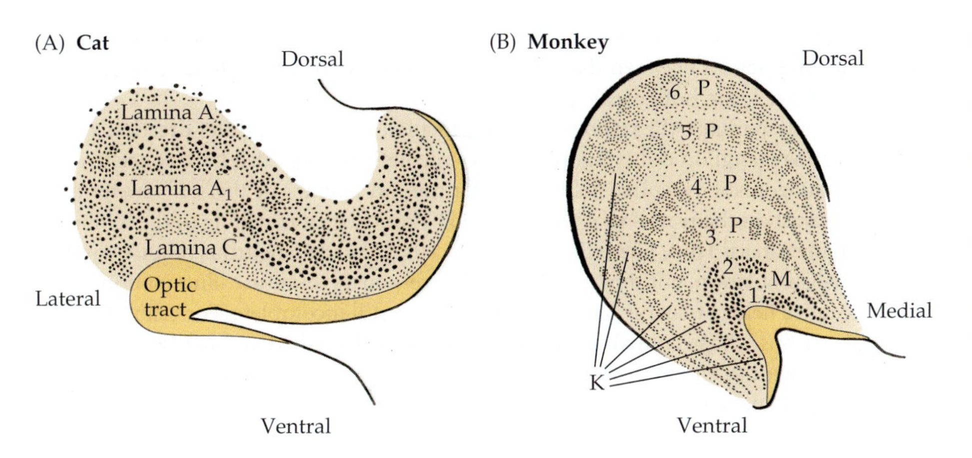

FIGURE 20.1 Lateral Geniculate Nucleus. (A) In cat LGN, there are three layers of cells: A, A_1, and C. (B) Monkey LGN has six major layers designated parvocellular, or P (3, 4, 5, 6) and magnocellular, or M (1, 2), separated by the koniocellular (K) layers. In both animals, each layer is supplied by only one eye and contains cells with specialized response properties. (A from Szentágothai, 1973; B after Hendry and Calkins, 1998.)

shown by electrical recording and by a variety of anatomical techiques.[6–8] Particularly striking is the arborization of a single optic nerve fiber that has been injected with the enzyme horseradish peroxidase (Figure 20.2). The terminals are all confined to the layers supplied by that eye, with no spillover across the border. Because of the orderly and systematic separation of fibers at the chiasm, the receptive fields of the cells in the LGN are all situated in the visual field on the opposite side of the animal (Chapter 19).

Visual Field Maps in the Lateral Geniculate Nucleus

An important topographical feature is the highly ordered arrangement of receptive fields within each of the geniculate layers. Neighboring regions of the retina make connections with neighboring geniculate cells, so the receptive fields of adjacent neurons overlap over most of their area.[4,5] The *area centralis* in the cat (the region of the cat retina with small receptive field centers) and the fovea in the monkey project onto the greater portion of each geniculate layer, and a similar distribution has been found in humans by the use of functional magnetic resonance imaging.[9] There are relatively few cells devoted to the peripheral retina. This extensive representation of the fovea reflects the high density of foveal receptors necessary for high-acuity vision.[10] Although there are probably equal numbers of optic nerve fibers and geniculate cells, each geniculate cell receives convergent input from several optic nerve fibers. Each optic nerve fiber in turn diverges to make synapses with several geniculate neurons.

Not only is each layer topographically ordered, but also cells in different layers are in retinotopic register with each other. Thus, if a penetration is made perpendicularly through the LGN, records are obtained from successive cells, driven by first one eye and then the other, as the microelectrode passes from one layer to the next. The positions of the receptive fields remain in corresponding positions on the two retinas, representing

[6]Hubel, D. H., and Wiesel, T. N. 1972. *J. Comp. Neurol.* 146: 421–450.

[7]Hubel, D. H., and Wiesel, T. N. 1961. *J. Physiol.* 155: 385–398.

[8]Bowling, D. B., and Michael, C. R. 1980. *Nature* 286: 899–902.

[9]Chen, W., et al. 1999. *Proc. Natl. Acad. Sci. USA* 96: 2430–2434.

[10]Azzopardi, P., Jones, K. E., and Cowey, A. 1999. *Vision Res.* 39: 2179–2189.

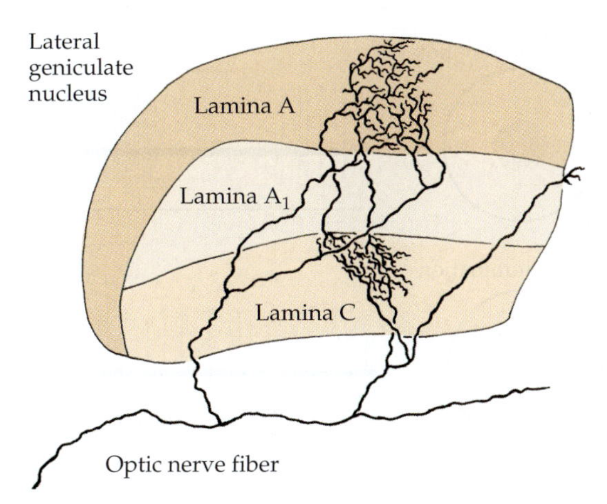

FIGURE 20.2 Termination of an Optic Nerve Fiber in cat LGN. A single "on"-center axon from the contralateral eye was injected with horseradish peroxidase. Branches end in layers A and C, but not in A_1. (After Bowling and Michael, 1980.)

the same region in the visual field.[4,5] For cells in the LGN, no extensive mixing of information or interaction between the eyes occurs, and binocularly excited cells (neurons with receptive fields in both eyes) are rare.

Surprisingly, perhaps, the responses from geniculate cells do not differ drastically from those of retinal ganglion cells (Figure 20.3). Geniculate neurons also have concentrically arranged antagonistic receptive fields, of either the "off"-center or the "on"-center type, but the contrast mechanism is more finely tuned by more equal matching of the inhibitory and excitatory areas. Thus, like retinal ganglion cells, geniculate neurons require contrast for optimal stimulation, but they give still weaker responses to diffuse illumination. Studies of receptive fields of lateral geniculate neurons are still incomplete. For example, there are interneurons whose contributions have not been established and pathways that descend from cortex to end in the nucleus.[11] Cortical feedback is necessary for synchronized firing by geniculate neurons.[12–14]

Functional Layers of the Lateral Geniculate Nucleus

Why is more than one layer of the LGN devoted to each eye? It is now becoming known that different functional properties are represented in the different layers. For example, cells in the four dorsal parvocellular layers of the monkey LGN, like P ganglion cells, respond to lights of different colors with fine discrimination.[15] In contrast, layers 1 and 2 (the magnocellular layers) contain M-like cells that give brisk responses and are not color selective, whereas cells in the K layers receive input from "blue-on" retinal ganglion cells[16] and may play a special role in color vision.[17] In the cat, X and Y fibers (see the section "Classification of Ganglion Cells" in Chapter 19) end in different sublayers of A, C, and A_1, and specific inactivation of A but not C layers degrades the accuracy of eye movements.[18] "On"- and "off"-center cells are also segregated into different layers of the LGN of mink and ferret and, to some extent, monkey.[19–21] In summary, the lateral geniculate nucleus is a way station in which ganglion cell axons are sorted so that nearby cells receive inputs from the

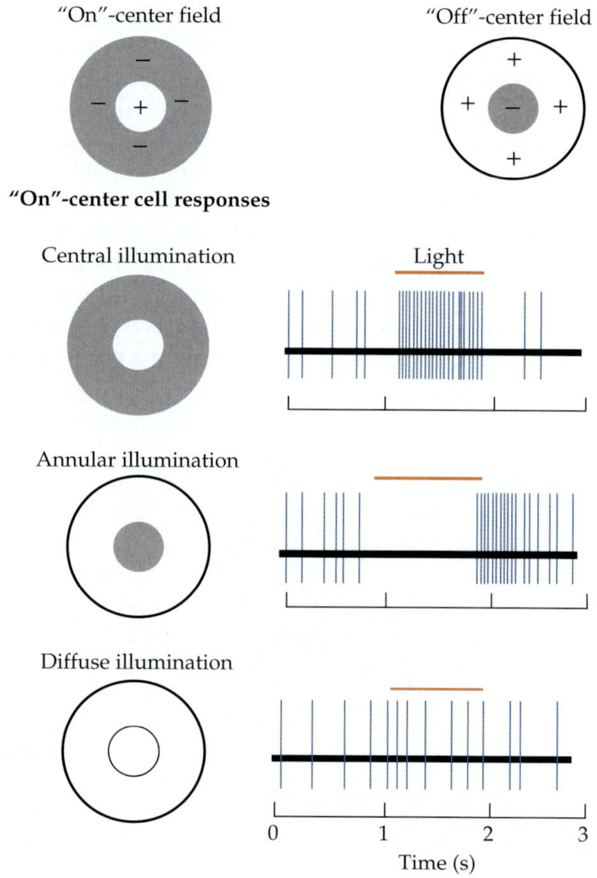

[11]Gilbert, C. D. 1983. *Annu. Rev. Neurosci.* 6: 217–247.

[12]Sillito, A. M., et al. 1994. *Nature* 369: 479–482.

[13]Funke, K., and Worgotter, F. 1997. *Prog. Neurobiol.* 53: 67–119.

[14]Weliky, M., and Katz, L. C. 1999. *Science* 285: 599–604.

[15]White, A. J., et al. 1998. *J. Neurophysiol.* 80: 2063–2076.

[16]Dacey, D. M., and Lee, B. B. 1994. *Nature* 367: 731–735.

[17]Martin, P. R., et al. 1997. *Eur. J. Neurosci.* 9: 1536–1541.

[18]Take, A. K., and Malpeli, J. G. 1998. *J. Neurophysiol.* 80: 2206–2209.

[19]LeVay, S., and McConnell, S. K. 1982. *Nature* 300: 350–351.

[20]Stryker, M. P., and Zahs, K. R. 1983. *J. Neurosci.* 10: 1943–1951.

[21]Schiller, P. H., and Malpeli, J. G. 1978. *J. Neurophysiol.* 41: 788–797.

"On"-center field

"Off"-center field

"On"-center cell responses

Central illumination

Light

Annular illumination

Diffuse illumination

0 1 2 3

Time (s)

FIGURE 20.3 Receptive Fields of Lateral Geniculate Nucleus Cells. The concentric receptive fields of cells in the LGN resemble those of ganglion cells in the retina, consisting of "on"-center and "off"-center types. The responses illustrated are from an "on"-center cell in cat LGN. The bar above each record indicates illumination. The central and surround areas antagonize each other's effects, so diffuse illumination of the entire receptive field gives only weak responses (bottom record), less pronounced than in retinal ganglion cells (compare Figure 19.21). (After Hubel and Wiesel, 1961.)

same region of the visual field, and neurons processing like information are clustered. The anatomical basis for parallel processing of information in the visual system is evident here.

CYTOARCHITECTURE OF THE VISUAL CORTEX

Visual information passes to the cortex from the lateral geniculate nucleus through the optic radiation. In the monkey, the optic radiation ends on a folded plate of cells about 2 mm thick (Figure 20.4). This region of the brain—known as primary visual cortex, visual area 1, or V_1—is also called the striate cortex or area 17, older terms that were based on anatomical criteria developed at the beginning of the twentieth century. V_1 lies posteriorly in the occipital lobe and can be recognized in cross section by its characteristic appearance. Incoming bundles of fibers in this area form a clear stripe that can be seen by the naked eye (hence the name "striate") (Figure 20.4B). Adjacent extrastriate regions of cortex are also concerned with vision. The area that immediately surrounds V_1 is called V_2 (or area 18) and receives inputs from V_1 (see Figure 20.4C). The exact boundaries of so-called extrastriate visual cortices (V_2–V_5) cannot be defined by simple inspection of the brain, but a number of criteria exist.[22–26] For example, in V_2 the striate appearance is lost, large cells are found superficially, and coarse, obliquely running myelinated fibers are seen in the deeper layers. It will be shown in Chapter 21 that the different visual areas perform different types of analysis.

Each area contains its own representation of the visual field projected in an orderly, retinotopic manner. Projection maps were made before the era of single-cell analysis by shining light onto small parts of the retina and recording with gross electrodes. These maps, and those made more recently using brain imaging techniques such as positron emission tomography and functional magnetic resonance imaging, demonstrated that much more cortical area is devoted to representation of the fovea than to representation of the rest of the retina.[27–30] This is as expected, since form vision is served principally by the higher densities of photoreceptors in foveal and parafoveal areas and is analogous to the enlarged areas de-

[22]Hubel, D. H., and Wiesel, T. N. 1965. *J. Neurophysiol.* 28: 229–289.
[23]Shipp, S., and Zeki, S. 1985. *Nature* 315: 322–325.
[24]DeYoe, E. A., et al. 1990. *Vis. Neurosci.* 5: 67–81.
[25]Maunsell, J. H., and Newsome, W. T. 1987. *Annu. Rev. Neurosci.* 10: 363–401.
[26]Kaas, J. H. 1996. *Prog. Brain Res.* 112: 213–221.
[27]Talbot, S. A., and Marshall, W. H. 1941. *Am J. Ophthalmol.* 24: 1255–1264.
[28]Daniel, P. M., and Whitteridge, D. 1961. *J. Physiol.* 159: 203–221.
[29]Fox, P. T., et al. 1987. *J. Neurosci.* 7: 913–922.
[30]Engel, S. A., Glover, G. H., and Wandell, B. A. 1997. *Cerebral Cortex* 7: 181–192.

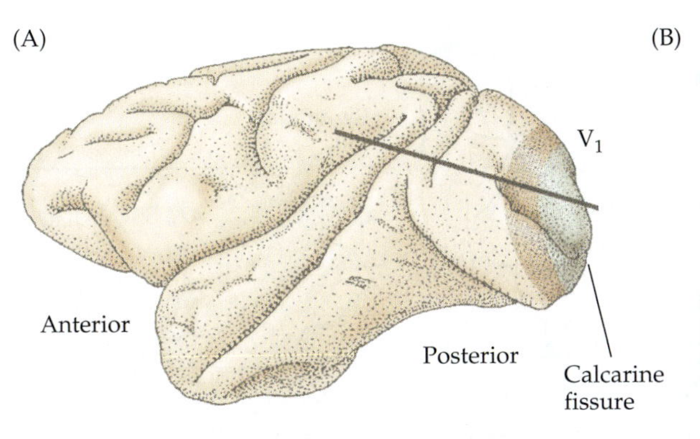

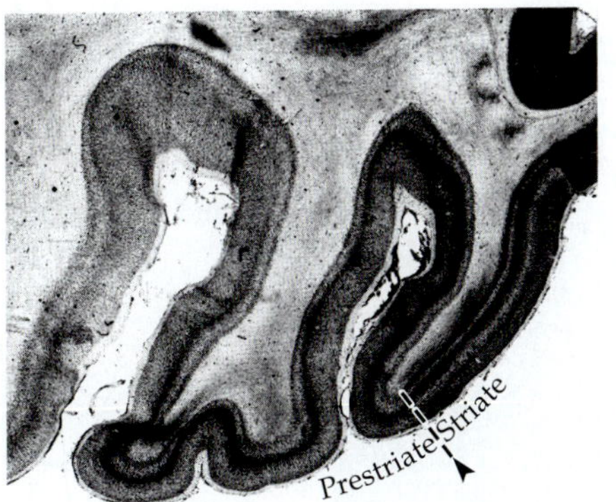

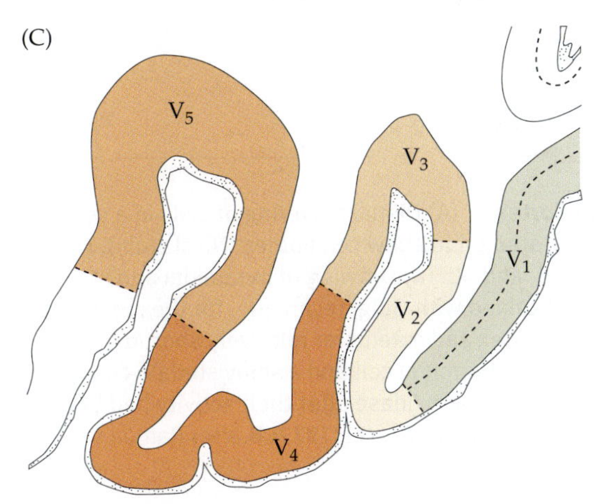

FIGURE 20.4 Relation of Primary Visual Cortex (V_1) to V_2, V_3, V_4, and V_5 in the monkey. (A) The cortex and plane of section passing through V_1 and V_2. The boundary between V_1 and V_2 is unambiguous. (B, C) A section through the occipital cortex. In B, the boundary between V_1 (striate) and V_2 (prestriate) occurs at the dotted line, where the striped appearance is lost. The boundaries between V_2, V_3, V_4, and V_5 are revealed by a combination of physiological and anatomical studies. (After Zeki, 1990; micrograph kindly provided by S. Zeki and M. Rayan.)

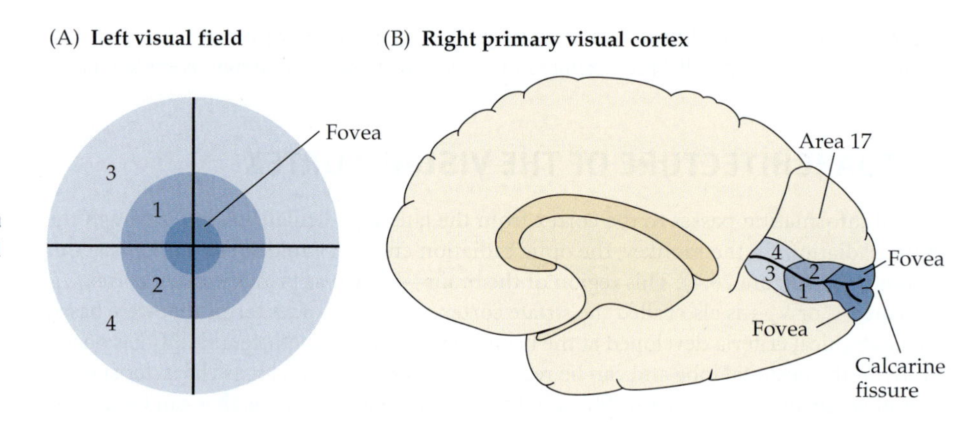

FIGURE 20.5 Visual Field Map of the Cortex. (A) The visual field is divided into a central zone (the fovea), and eight hemiquadrants. The left visual field maps onto the right primary visual cortex. (B) In humans, the primary visual cortex (Area 17) lies almost entirely on the medial surface of the occipital lobe. The foveal region is found most posteriorly, with the peripheral zones mapping anteriorly. The fovea claims a disproportionate share of the primary visual cortex.

voted to hand and face in the primary somatosensory cortex (Chapter 18). The retinal fovea is mapped onto the occipital pole of the cerebral cortex; the map of the retinal periphery extends anteriorly along the medial surface of the occipital lobe (Figure 20.5).[31] Because of image reversal by the lens, the upper visual field appears on the lower retina and is projected to V_1 below the calcarine fissure; lower visual fields map above the calcarine fissure.

In sections of the cortex, neurons can be classified according to their shapes. The two principal groups of neurons are stellate cells and pyramidal cells. Examples of these cells are shown in Figure 20.6B. The main differences are in the lengths of the axons and the shapes of the cell bodies. The axons of pyramidal cells are longer, descend into white matter, and leave the cortex; those of stellate cells tend to terminate locally. The two groups of cells exhibit other variations, such as the presence or absence of spines on the dendrites, that bear on their functional properties.[32] There are other fancifully named neurons (double bouquet cells, chandelier cells, basket cells, and crescent cells), as well as neu-

[31]Van Essen, D. C., and Drury, H. A. 1997. *J. Neurosci.* 17: 7079–7102.

[32]Lund, J. S. 1988. *Annu. Rev. Neurosci.* 11: 253–288.

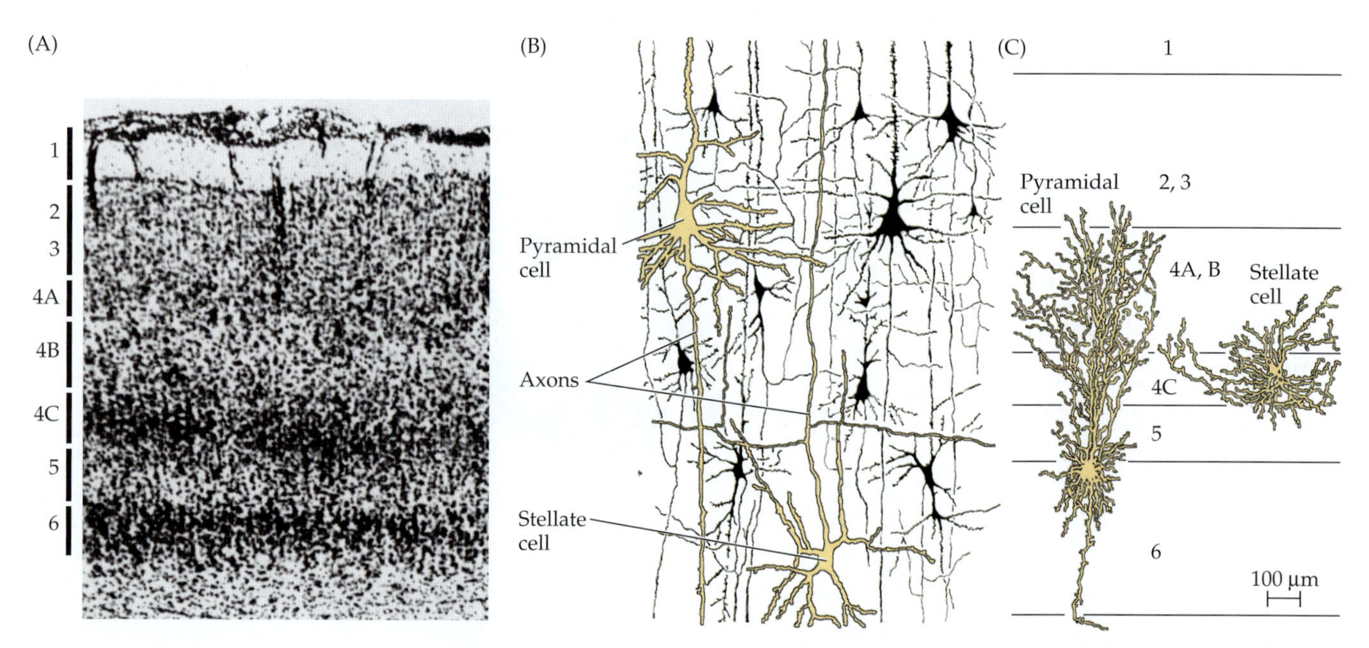

FIGURE 20.6 Architecture of Visual Cortex. (A) Distinct layering of cells in a section of striate cortex of the macaque monkey, stained to show cell bodies (Nissl stain). Fibers arriving from the LGN end in layers 4A, 4B, and 4C. (B) Drawing of pyramidal and stellate cells in cat visual cortex. The processes, stained with Golgi technique, for the most part run radially through the thickness of the cortex and extend for relatively short distances laterally. (C) Drawing from photographs of a pyramidal cell and a spiny stellate cell in cat cortex that had been injected with horseradish peroxidase after their activity had been recorded. Both were simple cells. (A from Hubel and Wiesel, 1972; B after Ramón y Cajal 1955; C from Gilbert and Wiesel, 1979.)

roglial cells. Characteristically, the processes of cells run for the most part in a radial direction, up and down through the thickness of the cortex (at right angles to the surface). In contrast, many (but not all) of their lateral processes are short. Connections between primary and higher-order visual cortices are made by axons that run in bundles through the white matter underlying the cellular layers.

Inputs, Outputs, and Layering of Cortex

A general feature of the mammalian cortex is that the cells are arranged in six layers within the gray matter (Figure 20.6A). The layers vary in appearance depending on the density of cell packing and the thickness of each area of cortex. The inputs are shown in the left-hand side of Figure 20.7A. Incoming geniculate fibers end for the most part in layer 4, with some contacts also made in layer 6. Superficial layers receive inputs from the pulvinar, another region of the thalamus. Numerous cortical cells, especially those in layer 2 and the upper portions of layers 3 and 5, receive inputs from neurons within the cortex. The predominant geniculate input to layer 4 further divides among identified sublayers, the details and functional significance of which will be addressed in Chapter 21.

The outputs from layers 6, 5, 4, 3, and 2 are shown in the right-hand side of Figure 20.7A.[11,33] A single cell that sends efferent signals out of the cortex can also mediate intracortical connections from one layer to another. For example, axons of a cell in layer 6, in

[33]Callaway, E. M. 1998. *Annu. Rev. Neurosci.* 21: 47–74.

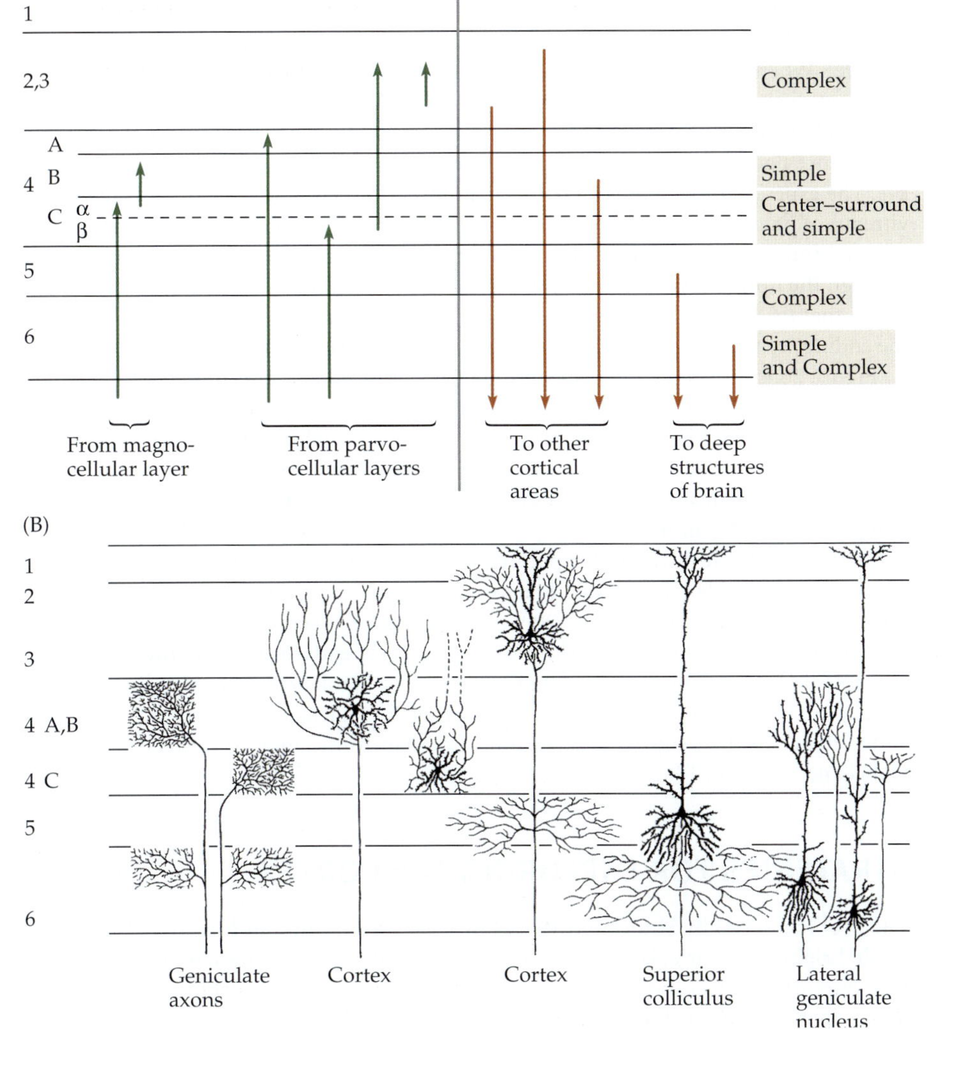

FIGURE 20.7 Connections of Visual Cortex. (A) The layers with their various inputs, outputs, and types of cells. Note that inputs from the LGN end mainly in layer 4. Those arising from magnocellular layers end principally in 4Cα and 4B, whereas those from parvocellular layers end in 4A and 4Cβ. Simple cells are found mainly in layers 4 and 6, complex cells in layers 2, 3, 5, and 6. Cells in layers 2,3, and 4B send axons to other cortical areas; cells in layers 5 and 6 send axons to the superior colliculus and the lateral geniculate nucleus. (B) Principal arborizations of geniculate axons and cortical neurons in the cat. In addition to these vertical connections, many cells have long horizontal connections running within a layer to distant regions of the cortex. (A after Hubel, 1988; B after Gilbert and Wiesel, 1981.)

addition to supplying the geniculate, can end in one of several other cortical layers, depending on response type.[34] From this anatomy a general pattern emerges: Information from the retina is transmitted to cells (mainly in layer 4) by geniculate axons, handed on from neuron to neuron through the thickness of the cortex, and then sent out to other regions of the brain by fibers looping down through white matter. Thus, the radial, or vertical, organization of cortex suggests that *columns* of neurons serve as computational units, processing some feature of the visual scene and passing it on to other cortical regions.

Segregation of Geniculate Inputs in Layer 4

The termination of geniculate afferents in layer 4 of primary visual cortex confers further organization onto this layer that can be observed both physiologically and anatomically. The first such property to be demonstrated was the segregation of inputs derived from the two eyes. In adult cats and monkeys, the cells in one layer of the LGN receiving their inputs from one eye project to aggregates of target cells in layer 4C that are separate from those supplied by the other eye. The aggregates are grouped as alternating stripes or bands of cortical cells that are supplied exclusively by one eye or the other. In deeper and more superficial cortical layers, cells are driven by both eyes, although one eye usually dominates. Hubel and Wiesel made the original demonstration of eye segregation and ocular dominance in the primary visual cortex with electrical recording techniques. They used the term **ocular dominance columns** to describe their observations, following the concept of cortical columns introduced by Mountcastle for somatosensory cortex (Chapter 18).[35]

A variety of experimental procedures have been developed for demonstrating the alternating groups of cells in layer 4 supplied by the left or right eye. One of the first was to make a small lesion in just one layer of the LGN (recall that each layer is supplied by just one eye). When this was done, degenerating terminals subsequently appeared in layer 4 in a characteristic pattern of alternating patches.[6] These correspond to areas driven by the eye supplying the lesioned area of the LGN. Later a striking demonstration of the ocular dominance pattern was provided by the transport of radioactive amino acids from one eye. The experiment is to inject a tritiated amino acid such as proline or leucine into the vitreous of one eye, from which it is taken up by nerve cell bodies of the retina and incorporated into protein. In time, the labeled protein is transported from ganglion cells through the optic nerve fibers to their terminals within the LGN. An extraordinary feature is that the label is also transferred from neuron to neuron across synapses.[36–38] Thus, the endings of the geniculate fibers in visual cortex are also labeled.

Figure 20.8 shows the disposition in layer 4 of radioactive terminals of geniculate axons supplied by the injected eye (radioactivity causes deposition of silver grains in overlaid photographic emulsion, which then appear as white areas in dark-field photographs). The labeled patches interdigitate with unlabeled areas supplied by the noninjected eye. The center-to-center distance between ocular dominance patches for one eye is approximately 1.0 mm.

At the cellular level a similar pattern has been revealed in layer 4 by injection of horseradish peroxidase into individual axons of LGN neurons as they approach the cortex.[39,40] The axon shown in Figure 20.9 is that of an "off"-center geniculate afferent that gives transient responses to dark, moving spots. It ends in two distinct clusters of processes in layer 4. The processes are separated by a blank area corresponding in size to the territory supplied by the other eye. Such morphological studies have borne out and added depth to the original description of ocular dominance columns presented by Hubel and Wiesel in 1962.

STRATEGIES FOR EXPLORING THE CORTEX

The problem faced by Hubel and Wiesel in 1958 was to find out how signals denoting small, bright, dark, or colored spots in the retina could be transmuted into signals that conveyed information about the shape, size, color, movement, and depth of objects. Techniques that are routine now—such as optical recording, horseradish peroxidase injection,

[34]Hirsch, J. A., et al. 1998. *J. Neurosci.* 18: 8086–8094.

[35]Mountcastle, V. B. 1957. *J. Neurophysiol.* 20: 408–434.

[36]Specht, S., and Grafstein, B. 1973. *Exp. Neurol.* 41: 705–722.

[37]LeVay, S., Hubel, D. H., and Wiesel, T. N. 1975. *J. Comp. Neurol.* 159: 559–576.

[38]LeVay, S., et al. 1985. *J. Neurosci.* 5: 486–501.

[39]Gilbert, C. D., and Wiesel, T. N. 1979. *Nature* 280: 120–125.

[40]Blasdel, G. G., and Lund, J. S. 1983. *J. Neurosci.* 3: 1389–1413.

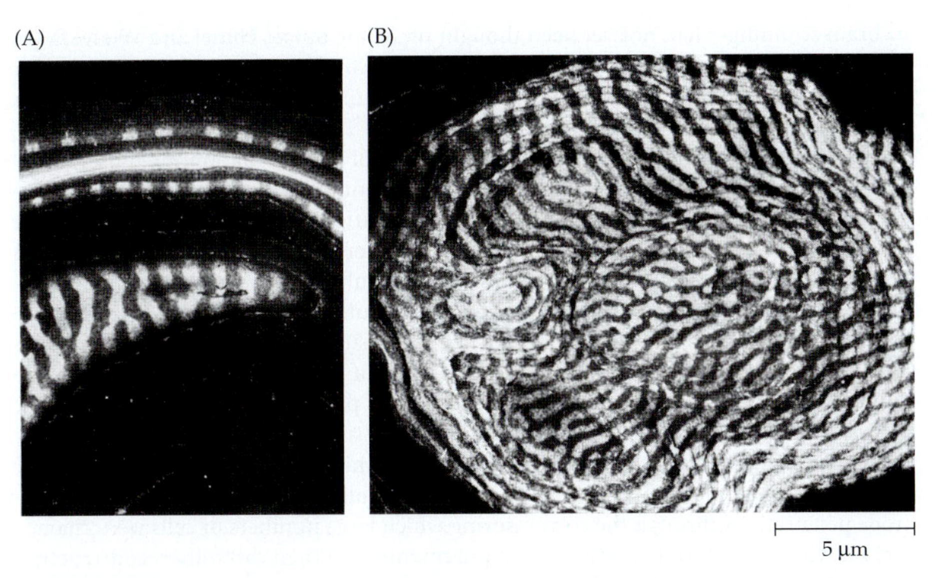

FIGURE 20.8 Ocular Dominance Columns in monkey cortex demonstrated by injection of radioactive proline into one eye. Autoradiographs photographed with dark-field illumination in which the silver grains appear white. (A) At the top of the picture the section passes through layer 4 of the visual cortex at right angles to the surface, displaying columns cut perpendicularly. In the center, layer 4 has been cut horizontally, showing that the columns consist of longer slabs. (B) Reconstruction made from numerous horizontal sections of layer 4C in another monkey in which the ipsilateral eye had been injected. (No single horizontal section can encompass more than a part of layer 4 because of the curvature of the cortex.) In both A and B, the ocular dominance columns appear as stripes of equal width supplied by one eye or the other. (Autoradiographs kindly provided by S. LeVay.)

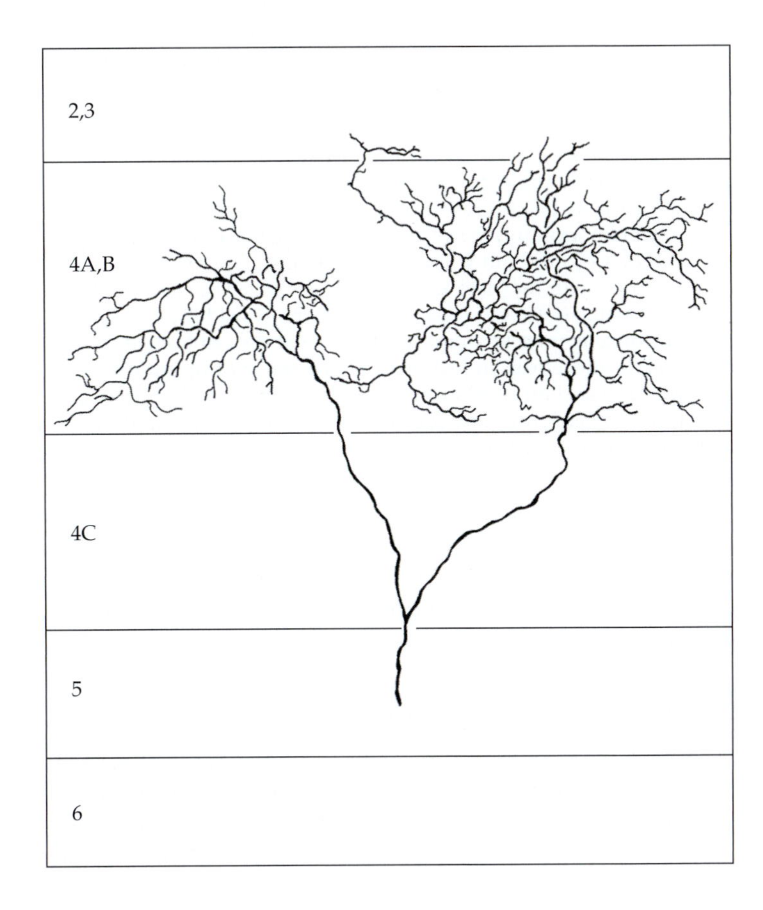

FIGURE 20.9 "Off"-Center Axon from LGN terminating in layer 4 of cat visual cortex. The axon was injected with horseradish peroxidase through a microelectrode. The terminals are grouped in two clusters separated by a vacant zone supplied by the other eye. (After Gilbert and Wiesel, 1979.)

or brain scanning—had not yet been thought of. At the outset Hubel and Wiesel faced completely unanswered questions, which they tackled by assuming that visual centers in the cortex would perform their processing according to principles similar to those in the retina, but at a more advanced level.

One crucial strategy in their analysis was the use of stimuli that mimic those occurring under natural conditions. For example, edges, contours, and simple patterns presented to the eye revealed features of the organization that could never have been detected by using bright flashes without form. Another key to the success of Hubel and Wiesel's approach lay in asking not simply what stimulus evokes a response in a particular neuron, but what is the *most effective* stimulus. Pursuit of this question through the various stages of the visual system has elicited many surprising and remarkable results. Early papers demonstrated that the receptive fields of simple and complex cells in the primary visual cortex constitute the initial stages of pattern recognition. In addition, analysis of receptive fields clearly revealed the helpful, simplifying principle that neurons located along radial tracks from cortical surface to white matter perform similar tasks, thereby defining functional columns. This was accomplished by making long microelectrode penetrations through the cortex during which large numbers of cells were characterized with respect to their stimulus requirements, and then carefully reconstructing the electrode tracks histologically.

Cortical Receptive Fields

Responses of cortical neurons, like those of the retinal ganglion and geniculate cells, tend to occur on a background of maintained activity. A consistent observation is that discharges of cortical neurons are not significantly influenced by diffuse illumination of the retina. Almost complete insensitivity to diffuse light is a furthering of the process already noted in the retina and the lateral geniculate nucleus; it results from an equally matched antagonistic action of the inhibitory and excitatory regions in the receptive fields of cortical cells. The neuronal firing rate is altered only when certain demands about the position and form of the stimulus on the retina are met. The receptive fields of most cortical neurons have configurations that differ from those of retinal or geniculate cells, so spots of light often have little or no effect. In his Nobel address, Hubel described the experiment in which Wiesel and he first recognized this essential property:[2]

> Our first real discovery came about as a surprise. For three or four hours, we got absolutely nowhere. Then gradually we began to elicit some vague and inconsistent responses by stimulating somewhere in the midperiphery of the retina. We were inserting the glass slide with its black spot into the slot of the ophthalmoscope when suddenly, over the audio monitor, the cell went off like a machine gun. After some fussing and fiddling, we found out what was happening. The response had nothing to do with the black dot. As the glass slide was inserted, its edge was casting onto the retina a faint but sharp shadow, a straight dark line on a light background. That was what the cell wanted, and it wanted it, moreover, in just one narrow range of orientations. This was unheard of. It is hard now to think back and realize just how free we were from any idea of what cortical cells might be doing in an animal's daily life.

David H. Hubel (left) and Torsten N. Wiesel during an experiment, about 1969. The cat, not shown, also faces the screen.

By following a progression of clues, Hubel and Wiesel worked out the appropriate light stimuli for various cortical cells; initially they classified the receptive fields as **simple** or **complex**. Each of these categories includes a number of subgroups and important variables that bear on perceptual mechanisms.

Responses of Simple Cells

Most simple cells are found in layers 4 and 6 and deep in layer 3. All these layers receive direct input from the LGN (although layer 4C is the most favored destination, as described earlier). The receptive fields of simple cells can be mapped with stationary spots of light, and they exhibit several variations.[41,42] One type of simple cell has a receptive field that consists of an extended narrow central portion flanked by two antagonistic areas. The center may be either excitatory or inhibitory. Figure 20.10 shows a receptive field of a simple cell in the striate cortex mapped out with spots of light that excited only weakly in the center (because the spots covered only a small fraction of the central "on" area).

The requirements of such a simple cell are exacting, as illustrated in Figure 20.10. For optimal activation, it needs a bar of light that is not more than a certain width, that entirely fills the central area, and that is oriented at a certain angle. Illumination of the surrounding areas suppresses any ongoing activity, or reduces the efficacy of a simultaneous center excitation. As predicted by mapping with spots of light, a vertically oriented bar is the most effective stimulus. Even small deviations from that pattern result in a diminished response. Different cells have receptive fields requiring a wide range of different orientations and positions. A new population of simple cells is therefore activated by rotating the stimulus or by shifting its position in the visual field. The distribution of inhibitory–excitatory flanks in various simple cell receptive fields may not be symmetrical, or the field may consist of two longitudinal regions facing each other—one excitatory, the other inhibitory.

Figure 20.11 shows examples of four such receptive fields, all with a common axis of orientation but with differences in the distribution of areas within the field. For the receptive field in Figure 20.11A, a narrow slit of light oriented 1 o'clock to 7 o'clock (assuming the visual field corresponds to a clock face with 12 o'clock high) elicits the best response. A dark bar in the same place with light flanks suppresses ongoing spontaneous activity. Cells with the field shapes shown in Figure 20.11B and C fire best with a dark bar in the central area. For the field shown in Figure 20.11D, an edge with light on the left and darkness on the right is the most effective "on" response, whereas reversing the dark and light areas is best for "off" discharges. In simple cells, the optimal width of the narrow light or dark bar is comparable to the diameters of the "on"- or "off"-center regions in the doughnut-shaped receptive fields of ganglion or lateral geniculate cells. Thus, cortical cells that have fields derived from the fovea are best excited by narrower bars than those that excite cells with fields in retinal periphery, corresponding to the smaller receptive fields of foveal ganglion cells.

Another type of simple cell has been found. Once again the orientation and position of the stimulus are critical, and the field is made of antagonistic "on" and "off" areas. But in addition, the length of the bar or edge is important: Stretching the bar beyond an optimal length reduces its effectiveness as a stimulus.[43] It is as though there is an additional "off" area that exists at the top or bottom end of the fields shown in Figure 20.11 that tends to suppress

[41]Hubel, D. H., and Wiesel, T. N. 1959. *J. Physiol.* 148: 574–591.

[42]Hubel, D. H., and Wiesel, T. N. 1968. *J. Physiol.* 195: 215–243.

[43]Gilbert, C. D. 1977. *J. Physiol.* 268: 391–421.

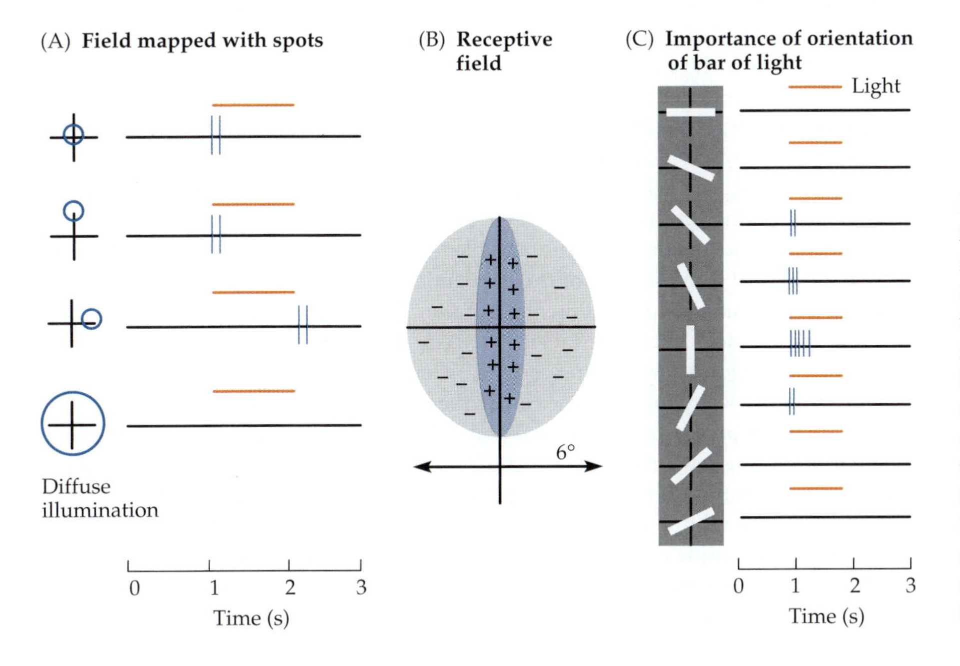

FIGURE 20.10 Responses of a Simple Cell in cat striate cortex to spots of light (A) and bars (C). The receptive field (B) has a narrow central "on" area (+) flanked by symmetrical antagonistic "off" areas (–). The best stimulus for this cell is a vertically oriented light bar in the center of its receptive field (fifth record from the top in C). Other orientations are less effective or ineffective. Diffuse light does not stimulate. The bar above each record in A and C indicates the duration of stimulation. (After Hubel and Wiesel, 1959.)

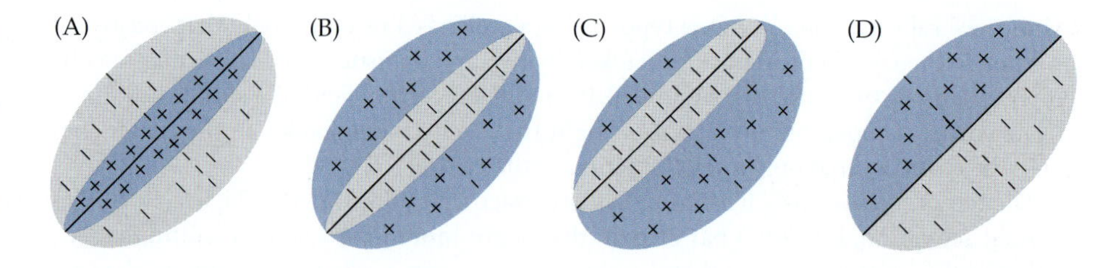

FIGURE 20.11 Receptive Fields of Simple Cells in cat striate cortex. In practice, all possible orientations are observed for each type of field. The optimal stimuli are a narrow slit or bar of light in the center for A; a dark bar for B and C; and an edge with dark on the right for D. Considerable asymmetry can be present, as in C. (After Hubel and Wiesel, 1962.)

firing when illuminated (**end inhibition** or **end stopping**). Hence, for such simple cells the best stimulus is an appropriately oriented bar or edge that stops in a particular place.

The common properties of all simple cells are (1) that they respond best to a properly oriented stimulus positioned so as not to encroach on antagonistic zones and (2) that stationary slits or spots can be used to define "on" and "off" areas. Another constant and remarkable feature is that in spite of all the different proportions of inhibitory and excitatory areas, the two contributions match exactly and cancel each other's effectiveness, so diffuse illumination of the entire receptive field produces a feeble response at best. The "off" areas in cortical fields are not always able to initiate impulses in response to dark bars. Frequently (particularly in end inhibition and in the more elaborate fields to be described shortly), illumination of the "off" area can be detected only as a reduction in the discharge evoked from the "on" area. Moving edges or bars of the appropriate orientation are highly effective at initiating impulses. Once again there is a specialization for detecting differences, but the spotlike contrast representation of ganglion cells has been transformed and extended into a line or an edge. Resolution has not been lost, but has been incorporated into a more complex pattern.

Synthesis of the Simple Receptive Field

A scheme of organization was proposed early on by Hubel and Wiesel to explain the origin of cortical receptive fields. This scheme had the advantage of using known mechanisms to explain how a nerve cell can respond so selectively to a visual pattern—such as the oriented lines that excite simple cells. They proposed that in the cortex, simple cells behave as if they were built up of large numbers of geniculate fields. This idea is illustrated in Figure 20.12A, where the fields of geniculate neurons connected to a cortical cell are lined up in such a way that a properly oriented bar of light through their centers would excite them all strongly. If the bar were widened or displaced slightly to either side, it would fall on the inhibitory surround of each cell and reduce or stop the excitatory output. Convergence of these geniculate neurons could produce a cortical cell whose optimal stimulus would be just such an oriented bar of light.

Connections such as these were postulated by Hubel and Wiesel as the simplest that could account for orientation selectivity. That is, the pattern of geniculate innervation itself determines the response characteristics of cortical neurons. These cortical receptive fields are then refined by intrinsic interneurons. Ferster has provided experimental evidence to show how the receptive fields of simple cells are synthesized in the cat cortex by making intracellular recordings from neurons in layer 4 of the cat visual cortex that had simple receptive fields.[44] In such recordings, it is possible to observe the synaptic potentials that are activated by visual stimuli. Many of these synaptic events are presumed to be due to transmitter release from geniculate afferents, and they sum to greater amplitudes at preferred orientations, as would be expected if an arrangement such as that of Figure 20.12A existed.

A still more rigorous test was performed by local cooling of the cortex so that all polysynaptic activity was suppressed.[45] Under these conditions, only direct monosynaptic geniculate input persists (albeit slowed and reduced in amplitude) (Figure 20.12B), but orientation tuning remains (Figure 20.12C). Although the pattern of geniculate input is sufficient for orientation tuning of simple cells in visual cortex, additional refinement is provided by both inhibitory[46,47] and excitatory intracortical connections.[48] Intracellular recordings from simple cells show that illumination of surrounding "off" areas produces inhibitory synaptic potentials.[49,50] These serve to sharpen the orientation selectivity and to maintain tuning as visual contrast varies.[51]

[44]Ferster, D. 1988. *J. Neurosci.* 8: 1172–1180.

[45]Ferster, D., Chung, S., and Wheat, H. 1996. *Nature* 380: 249–252.

[46]Sillito, A. M. 1979. *J. Physiol.* 289: 33–53.

[47]Sato, H., et al. 1996. *J. Physiol.* 494: 757–771.

[48]Stratford, K. J., et al. 1996. *Nature* 382: 258–261.

[49]Pei, X., et al. 1994. *J. Neurosci.* 14: 7130–7140.

[50]Hirsch, J. A., et al. 1998. *J. Neurosci.* 18: 9517–9528.

[51]Sompolinsky, H., and Shapley, R. 1997. *Curr. Opin. Neurobiol.* 7: 514–522.

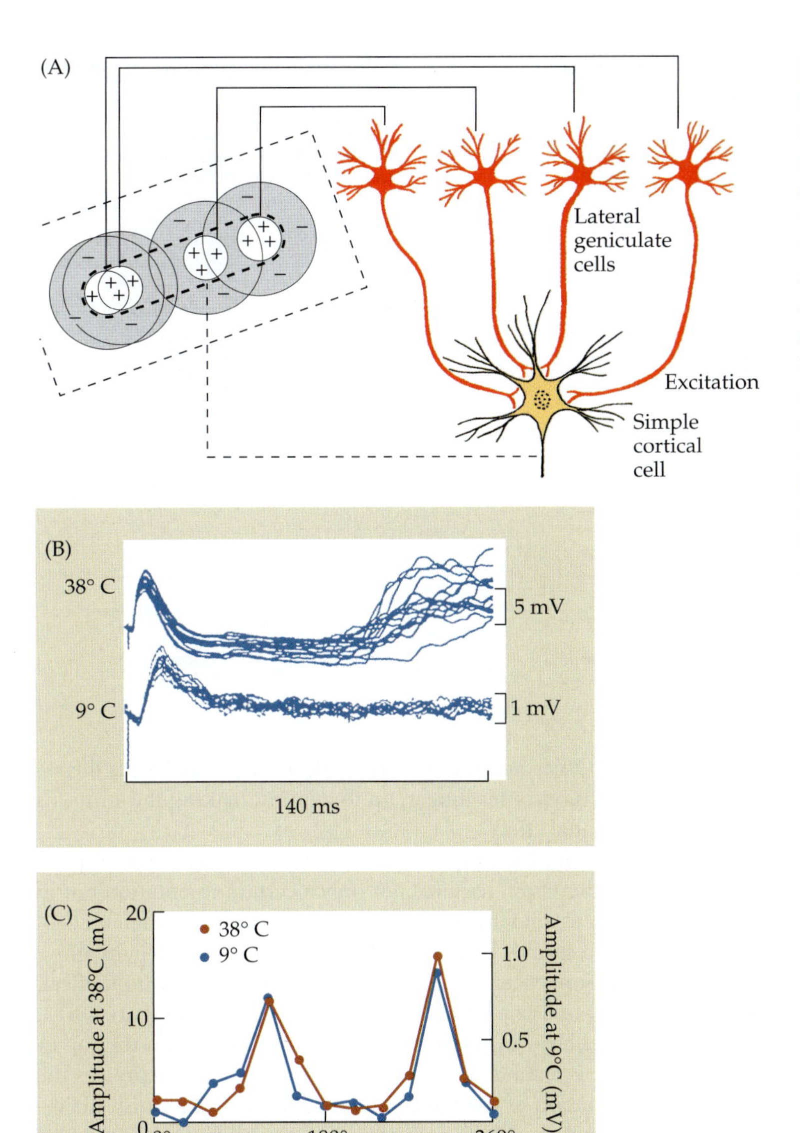

FIGURE 20.12 Synthesis of Simple Receptive Fields. Hypothesis devised by Hubel and Wiesel to explain the synthesis of simple-cell receptive fields. (A) The elongated receptive field of a simple cell is produced by the convergence of inputs from many geniculate neurons (only four are shown) whose concentric receptive fields are aligned on the retina. (B) Effect of cortical cooling on the response of simple cells. Electrical stimulation of the LGN causes short latency (monosynaptic) synaptic potentials at 38°C in a simple cell. These become smaller and slower at 9°C. Note that cooling eliminates longer-latency (polysynaptic) signals. (C) Orientation tuning to visual stimuli is comparable at both temperatures, consistent with the hypothesis that geniculate input specifies the simple-cell receptive field and does not require intracortical feedback. (A after Hubel and Wiesel, 1962; B and C after Ferster, Chung, and Wheat, 1996.)

Responses of Complex Cells

In recordings made from individual neurons in the visual cortex, one finds, in addition to simple cells, other neurons that behave quite differently. These complex cells, which are abundant in layers 2, 3, and 5, have two important properties in common with simple cells: Illumination of the entire field is ineffective, and they require specific field axis orientation of a dark–light boundary.[52] The demand, however, for precise positioning of the stimulus, observed in simple cells, is relaxed in complex cells. In addition, there are no longer distinct "on" and "off" areas that can be mapped with small spots of light. As long as a properly oriented stimulus falls within the boundary of the receptive field, most complex cells will respond, as in the examples illustrated in Figure 20.13. The meaning of the signals arising from complex cells, therefore, differs significantly from that of simple cells. The simple cell localizes an oriented bar of light to a particular position within the receptive field; the complex cell signals the abstract concept of *orientation without strict reference to position.*

Two main classes of complex cells can be distinguished; both respond best to moving edges or slits of fixed width and precise orientation. One type of cell gives the responses shown in Figure 20.13. Such cells respond to an oriented edge anywhere within a certain area. In the example, the vertical edge causes nearly equivalent responses at any of four lo-

[52]Hubel, D. H., and Wiesel, T. N. 1962. *J. Physiol.* 160: 106–154.

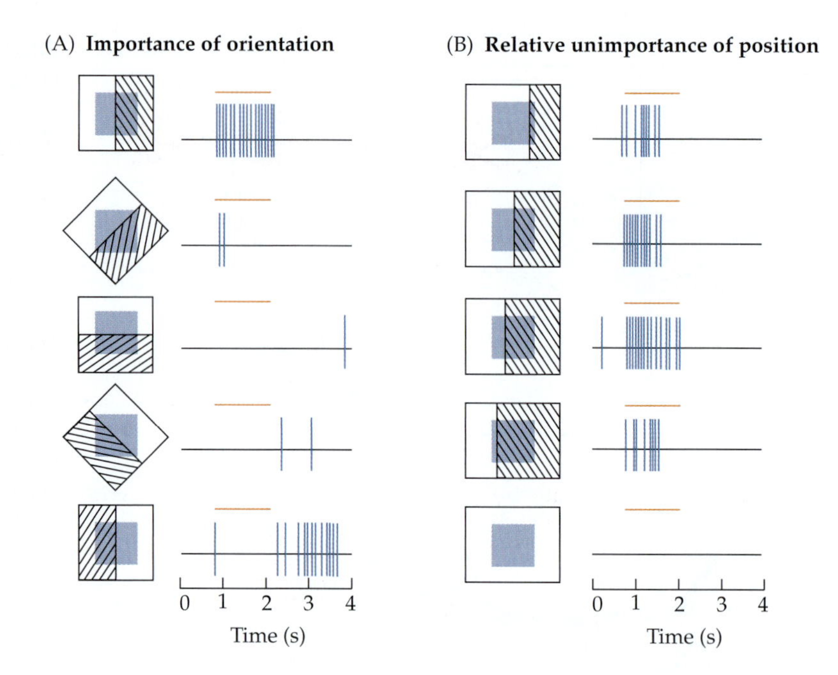

FIGURE 20.13 Responses of a Complex Cell in cat striate cortex. The cell responds best to a vertical edge located within its receptive field (the blue square). (A) With light on the left and dark (hatching) on the right (first record), there is an "on" response. With light on the right (fifth record), there is an "off" response. Orientation other than vertical is less effective. (B) The position of the border within the field is not important. Illumination of the entire receptive field (bottom record) gives no response. (After Hubel and Wiesel, 1962.)

cations, as seen in Figure 20.13B. Other orientations are ineffective. For such cells the response improves as the edge or slit becomes longer, up to a point; making the stimulus even longer gives rise to no additional effect.

Other complex cells, like end-stopped simple cells, require slits or edges that stop.[3,53] The best stimulus for such cells, therefore, requires not only a certain orientation but in addition some discontinuity, such as a line that stops, an angle, or a corner. Figure 20.14 shows the responses of a cell that responds best to an edge angled at 45°, with light above and dark below. Diffuse illumination, other axis orientations, or spots are without effect. At first glance, one might classify this as a complex cell similar to that in Figure 20.13. The difference, however, becomes clear in the fourth and fifth records from the top in Figure 20.14, when the dark edge is extended; this elongation to the right depresses the response. However, *diffuse* illumination of the right-hand field does not diminish the response (last record). It is therefore not an "off" area as such. One description of the best stimulus for this cell is a corner. Moreover, the stimulus must move in one direction. Such directional sensitivity is a feature commonly found in complex cells. Other, still more demanding complex cells can be found, particularly in extrastriate cortex.[3]

Synthesis of the Complex Receptive Field

In the same way that the receptive field of a simple cortical cell can be built up by the convergence of geniculate afferents, so too could the receptive field of a complex cortical cell be synthesized by combining those of simple cells. Figure 20.15 presents a hypothetical complex cell that is excited by a vertical edge stimulus that falls anywhere within the area of the receptive field. This is so because wherever the edge falls, one of the simple fields is traversed at its vertical inhibitory–excitatory boundary. The other simple fields do not respond because both of their components are illuminated or darkened uniformly. Diffuse illumination of the entire field covers all component fields equally, and therefore none fires.

One can postulate that only one or a few of the simple cells need fire at any one position of the stimulus to evoke a near maximal response in a complex cell. Consistent with this hypothesis, intracellular recording from complex cells reveals few monosynaptic contacts from the LGN but rather a preponderance of long-latency, disynaptic inputs, presumably arising from cortical simple cells.[46] An end-stopped complex cell is illustrated in Figure 20.15B. Here two complex cells with opposing synaptic effects combine to produce the end-stopped complex cell that could detect a corner.

[53]Palmer, L. A., and Rosenquist, A. C. 1974. *Brain Res.* 67: 27–42.

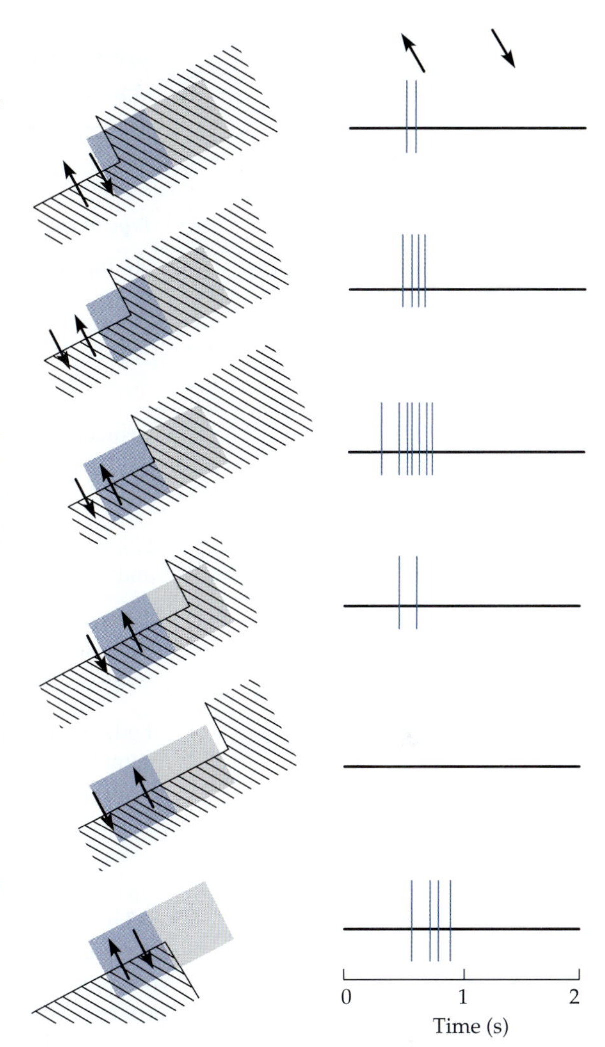

FIGURE 20.14 End Inhibition of a Complex Cell in V_2 (area 18) of the cat cortex. The best stimulus for this cell (third record from the top) is a moving (arrows), oriented edge (a corner) that does not encroach on the antagonistic right-hand portion of the receptive field. The records also show the selective sensitivity of the cell to upward movement. (After Hubel and Wiesel, 1965.)

Receptive Fields: Units for Form Perception

Together these results lend support to the idea of hierarchical organization whereby increasing complexity in receptive field organization is produced by convergence of appropriate inputs in an orderly manner. This does not mean that each receptive field of succeeding complexity is generated solely by combining inputs derived from the immediately preceding level. For instance, complex cells can receive inputs from LGN cells.[13] Also, both feedback and horizontal connections are present throughout the cortex.[54] As mentioned previously, cortical input serves to sharpen the orientation tuning of simple cells. Nonetheless, the original working hypothesis proposed by Hubel and Wiesel in 1962 continues to provide a clear, elegant, and reasonable conceptual framework upon which to design new experimental tests.

Table 20.1 summarizes some of the characteristics of receptive fields at successive levels of the visual system. Each eye conveys to the brain information collected from regions of various sizes on the retinal surface. The emphasis is not on diffuse illumination or the absolute amount of energy absorbed by photoreceptors. Rather the visual system extracts information about contrast by comparing the level of activity in cells with adjoining receptive fields. At each higher level, such neural computations define ever more elaborate spatial features.

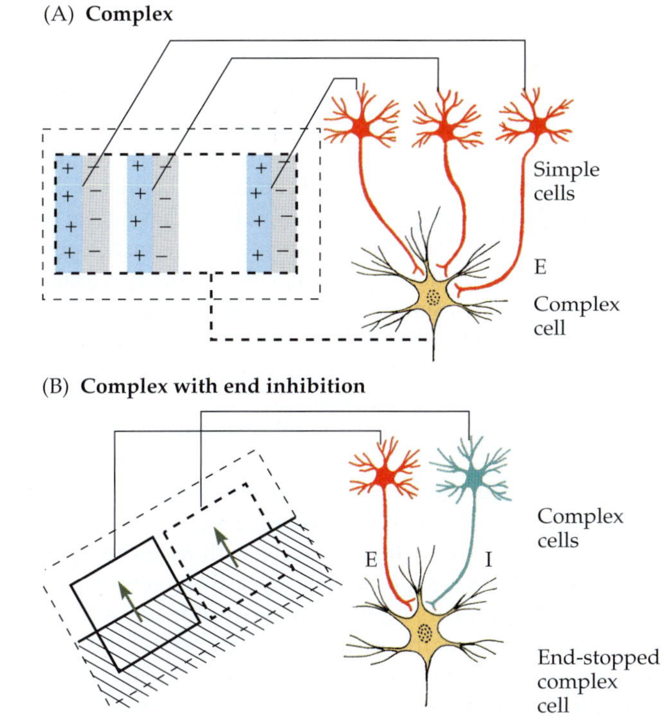

[54]Lamme, V. A. F., Super, H., and Spekreijse, H. 1998. *Curr. Opin. Neurobiol.* 8: 529–535.

FIGURE 20.15 Synthesis of Complex Receptive Fields. (A) Convergent input from simple cells responding best to a vertically oriented edge at slightly different positions could bring about the behavior of a complex cell that responds well to a vertically oriented edge situated anywhere within its field. (B) Each of two complex cells responds best to an obliquely oriented edge. But one cell is excitatory and the other is inhibitory to the end-stopped complex cell. Hence, an edge that covers both fields, as in the sketch, is ineffective, whereas a corner restricted to the left field would excite. E = excitation; I = inhibition. (After Hubel and Wiesel, 1962, 1965.)

Table 20.1
Characteristics of receptive fields at successive levels of the visual system

Type of cell	Shape of field	What is best stimulus?	How good is diffuse light as a stimulus?
Photoreceptor	⊕	Light	Good
Ganglion	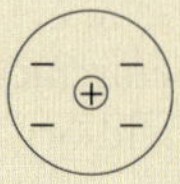	Small spot or narrow bar over center	Moderate
Geniculate	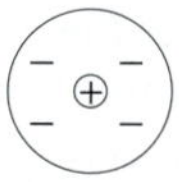	Small spot or narrow bar over center	Poor
Simple (layers 4 and 6 only)		Narrow bar or edge (some end-inhibited)	Ineffective
Complex (outside layer 4)	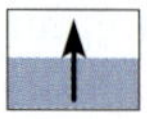	Bar or edge	Ineffective
End-inhibited complex (outside layer 4)	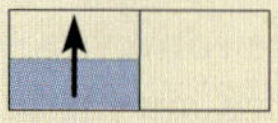	Line or edge that stops; corner or angle	Ineffective

This process can be appreciated by considering the types of signals generated by a square patch of light as shown in Figure 20.16. "On"-center retinal ganglion cells within the square will increase their discharge (at least initially), while"off"-center cells are suppressed. The best-stimulated ganglion cells, however, are those subjected to the maximum contrast—that is, those having centers lying immediately adjacent to the boundary between the light and dark areas and consequently having less activation of their inhibitory regions. Neurons in the LGN behave similarly. Cortical cells having receptive fields lying either completely within the square or outside it send no signals because diffuse illumination is not an effective stimulus. Only those simple cells with receptive fields oriented to coincide with the horizontal or vertical boundaries of the square will be stimulated.

Similar considerations apply to the stimulation of complex cells, which also require properly oriented bars or edges. End-inhibited complex cells detect the corner of the square or a line that stops. There is an important difference, however, which is related to the fact that the eyes continually make small saccadic eye movements. These are not perceived as motion but are essential to prevent photoreceptor adaptation as the eyes fixate. Each "microsaccade" causes a new population of simple cells with exactly the same orientation but slightly different receptive field location to be thrown into action. For those complex cells that "see" the square, however, a boundary of appropriate orientation can be anywhere within the field. Thus, many of the same complex cells will continue to fire during eye movements, as long as the displacement is small and the pattern does not pass outside the receptive field of the cell.

If the preceding considerations are valid, the surprising conclusion is that the primary visual cortex receives little information about the absolute level of uniform illumination within the square. Signals arrive only from the cells with receptive fields situated close to the border. This hypothesis is supported by an easily replicated psychophysical experiment. A square that appears light when surrounded by a black border can be made to appear dark merely by increasing the brightness of the surround. In other words, we perceive the difference or contrast at the boundary, and it is by that standard that the brightness in the uniformly illuminated central area is

Is orientation of stimulus important?	Are there distinct "on" and "off" areas within receptor fields?	Are cells driven by both eyes?	Can cells respond selectively to movement in one direction?
No	No	No	No
No	Yes	No	No
No	Yes	No	No
Yes	Yes	Yes (except in layer 4)	Some can
Yes	No	Yes	Some can
Yes	No	Yes	Some can

judged. This is not to say that general luminance is entirely ignored by the nervous system. For example, the pupil of the eye varies with ambient light intensity over a wide range. Pupillary size is adjusted by a feedback mechanism, the incoming loop of which leaves the eye through the optic nerve.

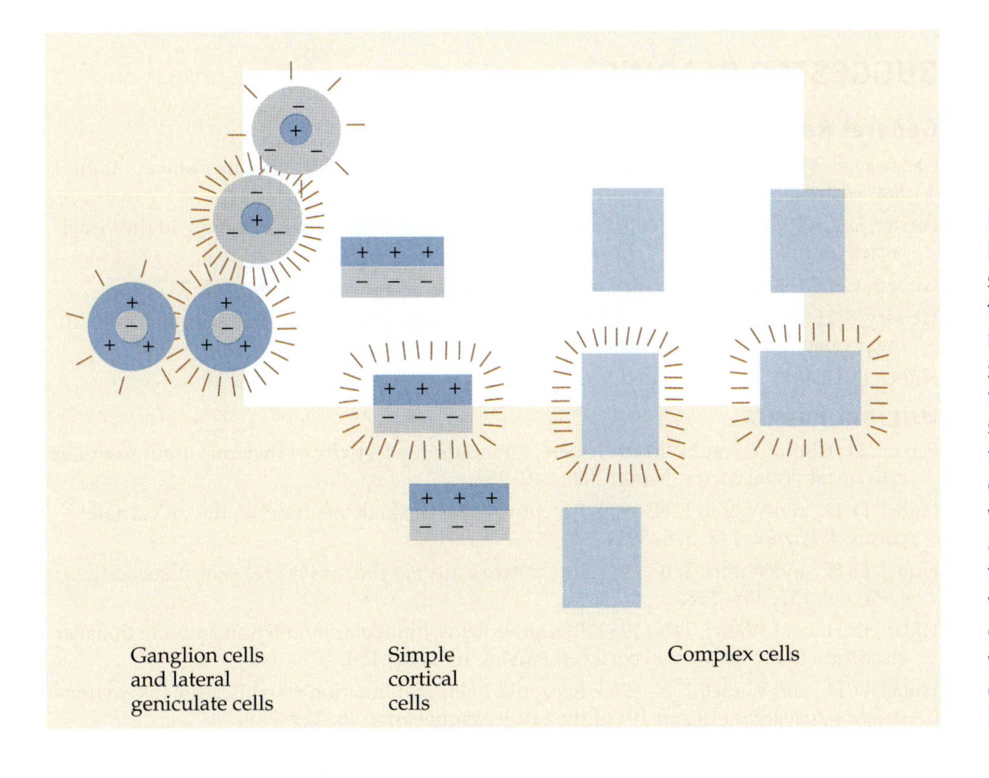

FIGURE 20.16 Responses of Neurons to a Pattern. When a square patch of light is presented to the retina, signals arise predominately from ganglion cells and lateral geniculate cells whose receptive fields lie close to the border of the square, not those subjected to uniform light or darkness. Simple and complex cells having receptive fields with the correct position (situated along the border or at a corner) and the correct orientation preference will also fire; those not on the border or with an inappropriate orientation will remain silent. Activity level is indicated by the number of radiating lines around each field.

The work of Hubel and Wiesel and many others since has made clear that the first general step in visual analysis is to construct lines or edges from the center–surround, spotlike, receptive fields of the retina. In V_1, the visual system begins to derive form from the retinal map. Finding these connections has given us our first glimpse into how the brain "computes." But these first steps to detect a line, or even a corner, remain a long way from complete visual recognition during which shape, color, size, and motion are all combined so that we can recognize a car, a cow, or the face of a friend. In Chapter 21 we will describe the parallel paths through cortex that may mediate these various submodalities of vision.

Summary

■ The lateral geniculate nucleus of the thalamus receives input from retinal ganglion cells. Inputs from the two eyes segregate to different layers that are in retinotopic register.

■ The lateral geniculate layers are functionally distinct, comprising magnocellular, parvocellular, or koniocellular response types.

■ The six layers of primary visual cortex serve as input and output stages of cortical processing.

■ Geniculate afferents from the two eyes are segregated in layer 4C of striate cortex, establishing ocular dominance columns that can be detected physiologically and anatomically.

■ Magnocellular and parvocellular layers of the LGN project specifically to different sublayers of cortical layer 4C.

■ The receptive field of an LGN cell has a concentric center–surround organization resembling that of a retinal ganglion cell and responds poorly to uniform illumination.

■ Simple cells of striate cortex respond to oriented light or dark bars. Their receptive fields can be mapped with spots of light as though composed of adjoining lateral geniculate center–surround receptive fields.

■ Complex cells of striate cortex also respond to oriented bars or edges. However, their receptive fields cannot be mapped with spots of light, but instead result from the convergence of multiple simple cells with adjoining receptive areas.

■ End inhibition results when an additional suppressive zone specifies the optimal length for a simple or complex cell.

SUGGESTED READING

General Reviews

Callaway, E. M. 1998. Local circuits in primary visual cortex of the macaque monkey. *Annu. Rev. Neurosci.* 21: 47–74.

Ferster, D., and Miller, K. D. 2000. Neural mechanisms of orientation selectivity in the visual cortex. *Annu. Rev. Neurosci.* 23: 441–471.

Gilbert, C. D. 1983. Microcircuitry of the visual cortex. *Annu. Rev. Neurosci.* 6: 217–247.

Hendry, S. H. C., and Calkins, D. J. 1998. Neuronal chemistry and functional organization in the primate visual system. *Trends Neurosci.* 21: 344–349.

Hubel, D. H. 1988. *Eye, Brain and Vision*. Scientific American Library. New York.

Original Papers

Ferster, D., Chung, S., and Wheat, H. 1996. Orientation selectivity of thalamic input to simple cells of cat visual cortex. *Nature* 380: 249–252.

Hubel, D. H., and Wiesel, T. N. 1959. Receptive fields of single neurones in the cat's striate cortex. *J. Physiol.* 148: 574–591.

Hubel, D. H., and Wiesel, T. N. 1961. Integrative action in the cat's lateral geniculate body. *J. Physiol.* 155: 385–398.

Hubel, D. H., and Wiesel, T. N. 1962. Receptive fields, binocular interaction and functional architecture in the cat's visual cortex. *J. Physiol.* 160: 106–154.

Hubel, D. H., and Wiesel, T. N. 1965. Receptive fields and functional architecture in two nonstriate visual areas (18 and 19) of the cat. *J. Neurophysiol.* 28: 229–289.

Hubel, D. H., and Wiesel, T. N. 1968. Receptive fields and functional architecture of monkey striate cortex. *J. Physiol.* 195: 215–243.

Hubel, D. H., and Wiesel, T. N. 1972. Laminar and columnar distribution of geniculo-cortical fibers in the macaque monkey. *J. Comp. Neurol.* 146: 421–450.

Kuffler, S. W. 1953. Discharge patterns and functional organization of the mammalian retina. *J. Neurophysiol.* 16: 37–68.

Van Essen, D. C., and Drury, H. A. 1997. Structural and functional analyses of human cerebral cortex using a surface-based atlas. *J. Neurosci.* 17: 7079–7102.

21 FUNCTIONAL ARCHITECTURE OF THE VISUAL CORTEX

THE VISUAL CORTEX IS ORGANIZED INTO VERTICAL CLUSTERS OF CELLS with similar functional attributes. Neurons that are preferentially driven by the right or the left eye are grouped in ocular dominance columns. Orientation columns consist of neurons whose line or edge preferences are at similar angles. The ocular dominance and orientation columns were first discovered by recording electrical activity from series of cortical cells as electrodes traversed the cortical thickness. Ocular dominance and orientation columns also can be visualized by biochemical and optical techniques that reveal activated regions in the cortex of a living animal.

The axons of magnocellular (M) and parvocellular (P) neurons of the lateral geniculate nucleus project to different subdivisions of layer 4 of primary visual cortex. From here M and P channels distribute differentially to "blobs" and "stripes" revealed by cytochrome oxidase labeling in primary and secondary visual cortex, respectively. Neurons in the M pathway are concerned with the detection of moving stimuli and are sensitive to differences in contrast and depth. Neurons in the P pathway deal with fine detail and color.

During visual perception, features such as color and motion are analyzed separately. This is illustrated by the fact that lesions in discrete regions of the brain result in selective loss of such features, rather than an overall reduction in quality of visual images. Lesions of the parietal cortex in an area known as MT (or V_5) lead to loss of motion detection and impairment of depth perception. In the occipitotemporal lobe, lesions of area V_4 result in loss of ability to recognize color.

One remarkable development is the use of noninvasive functional magnetic resonance imaging to detect cellular activity in the brains of animals, including humans. Although individual cortical ocular dominance and orientation columns are below the current limit of resolution of this technique, regions of the visual cortex specialized for specific tasks, such as detection of motion or recognition of faces, have been localized.

In Chapter 20 we discussed the flow of information through primary visual cortex by describing the effect of a visual stimulus on successively higher-order cortical cells. This approach has provided an understanding of the cellular mechanisms for the analysis of *form* at each point on the visual field. Our task now is to examine how other aspects of vision, such as *color* and *motion*, are encoded in the cortex, and how these properties might be combined in the perception of a visual image.

We have described the point-to-point representation of the retina in the cortex of visual area 1 (V_1) and the manner in which inputs from the two eyes are sorted into ocular dominance columns. Within this retinotopic map are other functional groupings, for example, columns of cells that all respond to a specific line orientation. We begin this chapter by examining the relationship between ocular dominance and orientation columns. We next examine the evidence that motion and color are analyzed in parallel channels through visual cortex. Finally, we consider examples of higher levels of processing in visual areas beyond the primary visual cortex.

OCULAR DOMINANCE SLABS AND ORIENTATION COLUMNS

Hubel and Wiesel's early experiments showed cortical cells with similar properties to be aggregated together in a vertical, columnar organization.[1,2] In any one penetration through the cortex as the electrode moved from the surface through layer after layer to white matter, all cells were found to have the same field axis orientation, ocular dominance, and position in the visual field. The columns for eye preference have already been mentioned. As discussed in Chapter 20, the inputs from the two eyes are segregated in

[1]Hubel, D. H., and Wiesel, T. N. 1963. *J. Physiol.* 165: 559–568.

[2]Hubel, D. H., and Wiesel, T. N. 1974. *J. Comp. Neurol.* 158: 267–294.

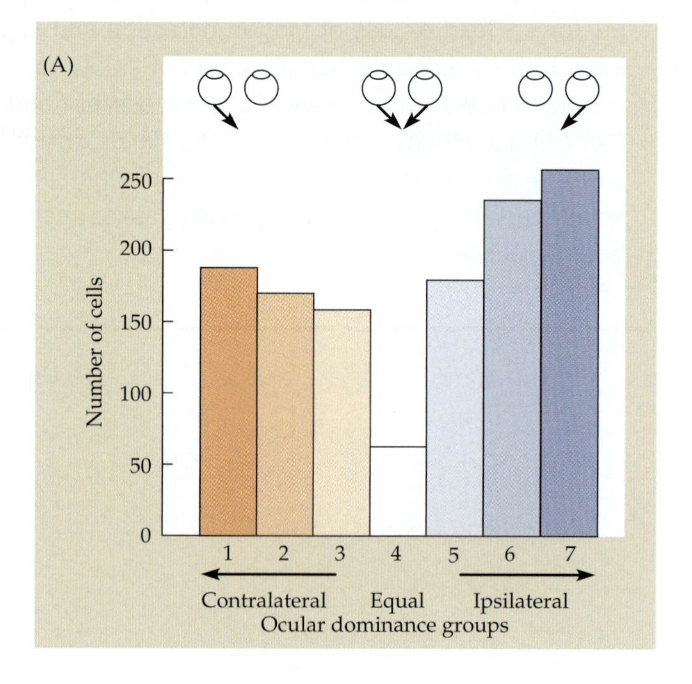

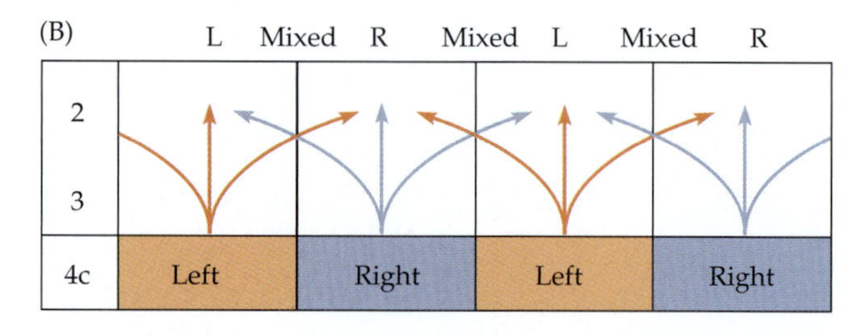

FIGURE 21.1 Physiological Demonstration of Ocular Dominance Columns. (A) Eye preference of 1116 cells in V_1 of 28 rhesus monkeys. Most cells (groups 2 through 6) are driven by both eyes. (B) Diagram to show how inputs from two eyes arriving in layer 4 of the cortex are combined in more superficial layers through horizontal or oblique connections to create cells with binocular fields. (After Hubel and Wiesel, 1968; Hubel, 1988.)

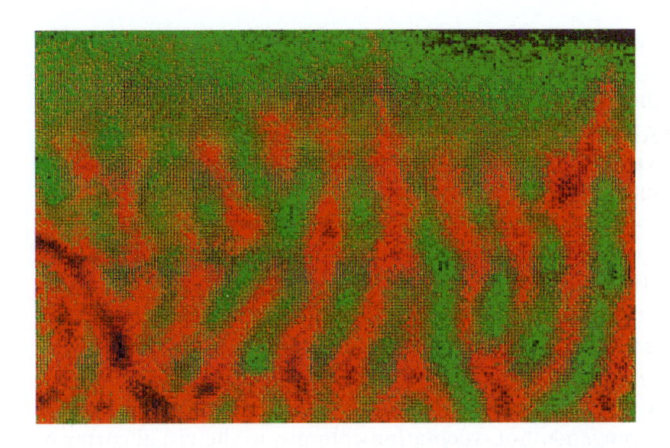

FIGURE 21.2 Display of Ocular Dominance Columns by Optical Imaging. A sensitive camera detects changes in light reflected from the monkey cortex following activity induced in just one eye. The intensity changes are color-coded so that active areas are red. The pattern of red stripes corresponds to ocular dominance columns revealed by anatomical labeling methods (see Figure 20.8). (After T'so et al., 1990.)

layer 4, where cortical neurons are driven monocularly. In any given column extending above and below layer 4, all the cortical neurons, although driven by both eyes, have the same eye preference. Figure 21.1 illustrates the variation in ocular dominance shown by neurons in the striate cortex in the monkey. The cells (total of 1116) are subdivided into seven groups. Groups 1 and 7 are driven exclusively by one of the two eyes and are found in layer 4 of the cortex. In groups 2, 3, 5, and 6 the effect of one eye is stronger than that of the other, and the cells in the middle group (4) are equally influenced. It is clear from the histogram that the majority of cells respond preferentially to one or the other eye.

As described in Chapter 20, ocular dominance columns form an interdigitating pattern on the cortex. Eye preference patterns in cortex can be visualized directly by using activity-dependent optical signals that arise from changes in tissue reflectivity or changes in fluorescence of extrinsically applied dyes.[3–5] In this way a large area of cortex can be examined for activity while visual stimuli are presented to the animal. Figure 21.2 shows ocular dominance columns detected by this kind of experiment. The striped pattern is like that obtained by injection of radioactive tracers into one eye (see Figure 20.8 in Chapter 20). Applying a visual stimulus to only one eye revealed stripes of cells driven by the activated eye, interspersed with less active cells. The projection of these surface stripes through the depth of cortex describes ocular dominance "slabs" subdividing the retinotopic map. In this way, the information in both eyes regarding a stimulus in one position in the visual field can be integrated by neighboring cells in the visual cortex (see Figure 21.1B).

Orientation Columns

What other functional groupings occur among cells of V_1? In Chapter 20 we described the orientation preferences of simple and complex cells, so we might ask if this feature is systematically organized across the cortex. A sample experiment is shown in Figure 21.3.

[3]Grinvald, A., et al. 1986. *Nature* 324: 361–364.

[4]Ts'o, D. Y., et al. 1990. *Science* 249: 417–420.

[5]Blasdel, G. G. 1989. *Annu. Rev. Physiol.* 51: 561–581.

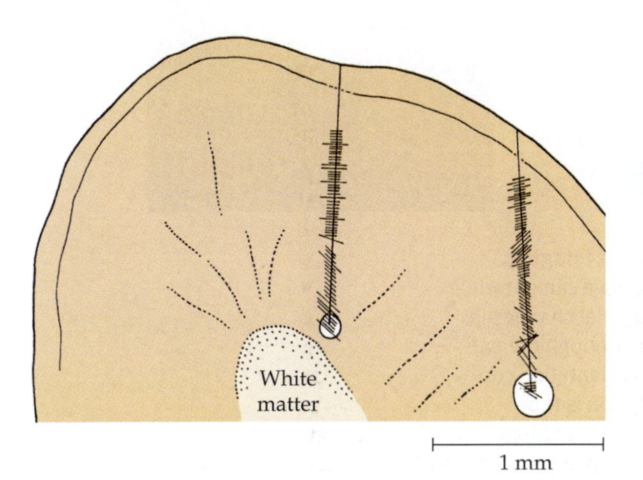

FIGURE 21.3 Axis Orientation of Receptive Fields of Neurons encountered as an electrode traverses the cortex of a cat. Cell after cell tends to have the same axis orientation, indicated by the angle of the bar to the electrode track. The penetration to the right is more oblique; consequently, the track crosses several columns and the axis orientations change frequently. The position of each cell is determined by making lesions repeatedly and at the end of the penetration (circle), and reconstructing the electrode track in serial sections of the brain. Such experiments have established that cat and monkey cells with similar axis orientation are stacked in columns running at right angles to the cortical surface. (After Hubel and Wiesel, 1962.)

A microelectrode is inserted perpendicular to the surface of the cortex in area V_1 of the cat. Each bar indicates the location of one cell and its preferred receptive field orientation in the progression through the cortical depth. Small lesions are made at significant points along the electrode track by passing current through the electrode. From these and an end point (indicated by the circle at the end of each electrode track), the position of each recorded cell was reconstructed. In the left-hand track the first 38 cells were optimally driven by bars or edges at about 90° to the vertical, at one position in the visual field. After penetrating about 0.6 mm, the axis of the receptive field orientation changed to about 45°. In a second track (to the right), more tangential to the cortical surface, successive cells have slightly different receptive field positions and field axis orientations. With this oblique penetration the field axes changed in a regular manner as though moving through a series of columns with different axis orientations. The orientation columns receive their input from cells with largely overlapping receptive fields on the retinal surface.

Information about the arrangement of orientation columns in the visual cortex of monkeys and cats was first obtained by making tangential rather than vertical electrode penetrations through the cortex.[2] Each 50 μm advance of the electrode horizontally along the cortex was accompanied by a change in field axis orientation of about 10°, sometimes in a regular sequence through 180°. The field axis orientation columns were narrower than those for ocular dominance—20 to 50 μm wide, compared with 250 to 500 μm. The first anatomical demonstration of orientation columns based on relative activity was by Sokoloff, using uptake of 2-deoxyglucose by active cells.[6] The principle is that active cells take up but cannot metabolize radioactive 2-deoxyglucose. As a result, metabolically active cells become radioactive and their distribution can then be seen by autoradiography. In monkeys and cats whose eyes had been exposed to horizontal or vertical stripes, bands of radioactivity corresponding to horizontal or vertical orientation columns were seen in the cortex.

The organization of orientation columns also has been studied by optical imaging techniques in living animals. An example is the experiment by Bonhoeffer and Grinvald[7] shown in Figure 21.4. The presentation of visual stimuli in a variety of orientations produced activity in different cortical areas. The response to each orientation is represented by a different color. What is striking is the arrangement of the orientation columns with respect to one another. At first the organization appears disorderly. However, careful inspection reveals the presence of **pinwheel centers**, focal points at which all the orientations come together. From there the cells responsive to a particular orientation radiate in an extraordinarily regular manner. Some pinwheels are systematically organized with

[6]Sokoloff, L. 1977. *J. Neurochem.* 29: 13–26.

[7]Bonhoeffer, T., and Grinvald, A. 1991. *Nature* 353: 429–431.

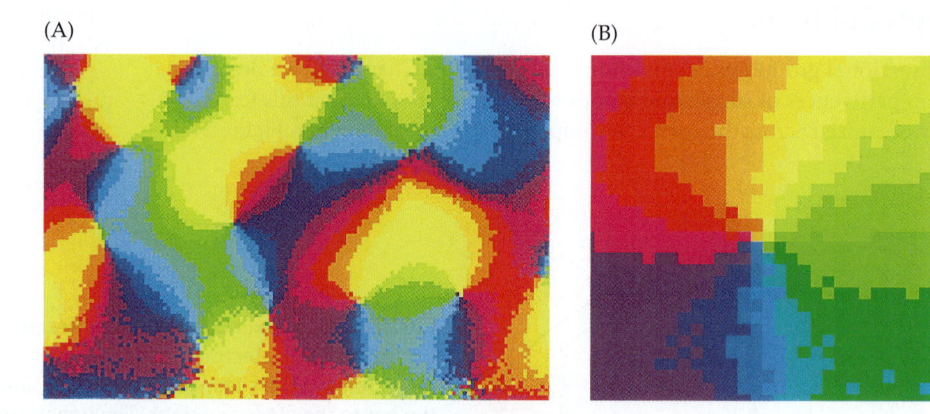

FIGURE 21.4 Detection of Orientation Columns (Pinwheels) by Optical Imaging. The activity-dependent reflectance of visual cortex was recorded by a sensitive camera while an eye was stimulated with oriented bars. (A) Each orientation caused maximal changes in different regions (an orientation "contour," encoded by a different color). Although the pattern seems at first disorderly, close inspection reveals centers at which all orientation contours come together in a pinwheel, as shown in (B). Note that each orientation is represented only once and that the sequence is beautifully precise. Such pinwheel centers occur at regular distances from each other. (After Bonhoeffer and Grinvald, 1991.)

clockwise progression, others counterclockwise. Thus, orientation is represented in a radial rather than a linear fashion. Each orientation appears only once in the cycle,[8] vertical and horizontal contours claim slightly more cortical area.[9] One or two such centers, evenly spaced, occur in each square millimeter of cortex. This kind of pattern was proposed earlier on theoretical grounds.[10]

The Relation between Ocular Dominance and Orientation Columns

Optical imaging methods also have been used to reveal the relationship between orientation and ocular dominance columns. The pattern of cortical activity is imaged first for eye-specific stimulation, then again with a series of oriented bars.[11,12] The results of one such experiment are shown in Figure 21.5A. Each orientation contour is indicated by a separate colored line (an iso-orientation contour), and ocular dominance zones (or columns) are shown as clear or shaded. The orientation pinwheels are clearly seen as the convergence of iso-orientation contours, and a set of contour lines between pinwheels typically crosses an ocular dominance boundary. That is, most orientation domains are split into ipsilateral and contralateral halves, and thus serve the two eyes for that region of visual space. In fact, each pinwheel center tends to occur near the center of an ocular dominance patch,[13,14] and iso-orientation contours tend to cross ocular dominance boundaries at right angles (Figure 21.5B).[15]

[8]Swindale, N. V., Matsubara, J. A., and Cynader, M. S. 1987. *J. Neurosci.* 7: 1414–1427.

[9]Coppola, D. M., et al. 1998. *Proc. Natl. Acad. Sci. USA* 95: 2621–2623.

[10]Linsker, R. 1989. *Proc. Natl. Acad. Sci. USA* 83: 8779–8783.

[11]Hubener, M., et al. 1997. *J. Neurosci.* 17: 9270–9284.

[12]Blasdel, G. G., Obermayer, K., and Kiorpes, L. 1995. *Vis. Neurosci.* 12: 589–603.

[13]Crair, M. C., et al. 1997. *J. Neurophysiol.* 77: 3381–3385.

[14]Löwel, S., et al. 1998. *Eur. J. Neurosci.* 10: 2629–2643.

[15]Obermayer, K., and Blasdel, G. G. 1993. *J. Neurosci.* 13: 4114–4129.

(A)

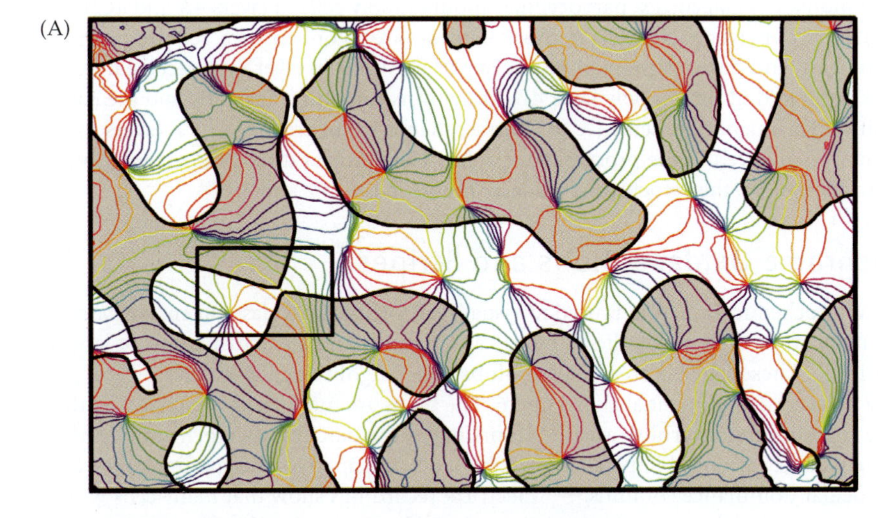

(B)

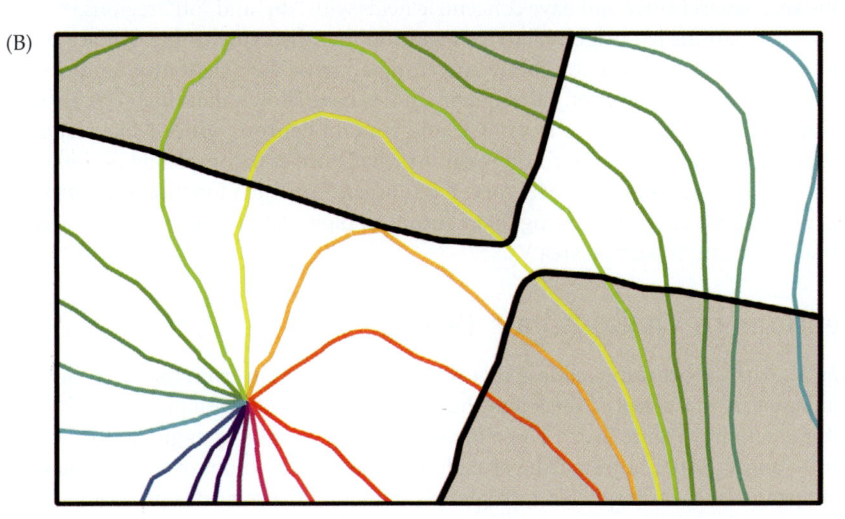

FIGURE 21.5 Orientation Pinwheels and Ocular Dominance Columns. (A) Activity-dependent reflectance was imaged in a region of V_1. Ocular dominance stripes and orientation contours were determined as in Figures 21.2 and 21.4; then the resulting maps were overlaid. Gray areas denote the contralateral eye. Each orientation contour is represented by a colored line, and pinwheel centers tend to occur near the centers of ocular dominance columns. This map extends approximately 7 mm across the cortical surface. (B) An enlargement of the boxed area in A. Note that orientation contours tend to cross ocular dominance boundaries at right angles. This is consistent with the idea that each orientation domain (the set of all orientation contours) is shared by the two eyes for one region of the retinal map. (From Hubener et al., 1997.)

PARALLEL PROCESSING OF FORM, MOTION, AND COLOR

Ocular dominance and orientation columns provide just two functional arrangements that apply to neurons in visual cortex. Direction of movement,[16] spatial frequency (essentially a reflection of receptive field size),[17] and image disparity (an important determinant in depth perception)[18] also appear in columnar arrangements in the visual cortex. The question then arises as to how all necessary aspects of image analysis could be carried out for each point on the retinotopically mapped cortex. The intermingling of functional columns shown in Figure 21.5 provides part of the answer. Indeed, long before optical imaging was used to observe such relationships, Hubel and Wiesel[19] proposed a conceptual scheme they termed a **hypercolumn**, in which all possible orientations could be represented for corresponding regions of the visual field for both eyes. An adjacent hypercolumn would analyze information in the same way for an adjacent but overlapping part of the visual field and so on, until the entire retina was mapped across the cortex. The challenge remains to incorporate additional features of visual analysis into this scheme.

Magnocellular, Parvocellular, and Koniocellular "Channels"

In Chapter 20 we saw that the lateral geniculate nucleus is divided into parvocellular (P), magnocellular (M), and koniocellular (K) layers. Each of these receives inputs from particular classes of retinal ganglion cells and projects to specific layers of V_1 (see Figure 21.7). Neurons in the P and M layers of the LGN project to different subdivisions of layer 4 in V_1.[20] Layer 4 contains sublayers A, B, and C, with this last subdivided into 4Cα and 4Cβ. In monkey visual cortex, parvocellular inputs supply cells in layers 4A and 4Cβ and the upper part of layer 6. Neurons in this P pathway are color-sensitive, require relatively high contrast, and have a sustained response pattern. Cells in layer 4Cα receive magnocellular input and in turn supply layer 4B. M neurons respond transiently and are color-*in*sensitive but much more sensitive to contrast than are P neurons.[21] The K layers in the LGN may be involved in the processing of color. They project directly to an anatomical specialization of V_1, the cytochrome oxidase "blobs" in layers 2 and 3.[22,23]

Cytochrome Oxidase Blobs and Stripes

When a histochemical reaction is used to visualize cytochrome oxidase, an enzyme associated with areas of high metabolic activity,[24] an intermittent or patchy pattern of label is found in V_1. These labeled areas, called **blobs**, are roughly circular collections of cells mainly in layers 2 and 3, but also in 5 and 6. The blobs are more richly vascularized than are the surrounding **interblob** areas.[25] In primates, the patches of cytochrome oxidase stain are precisely arranged in parallel rows about 0.5 mm apart, corresponding to the centers of ocular dominance columns.[26,27] Electrical recordings show that many cells within the blobs are color-sensitive and have concentric fields with "on" and "off" regions.[28]

These observations led originally to a suggestion that the cytochrome oxidase blobs represent a separate pathway for color, intermingled with the orientation and ocular dominance columns.[29] However, subsequent studies have shown that the blobs also receive inputs from M sublayers of layer 4C[30] and contain neurons with M-like response properties.[31] Thus, the blob–interblob regions provide as yet incompletely understood sorting and recombination of the M, P, and K pathways. That such functional assortment does occur is suggested by the finding that neurons in blob and interblob areas project to specific subdivisions of cortical area V_2.

Projections to Visual Area 2 (V_2)

Staining V_2 with cytochrome oxidase reveals a pattern different from that seen in V_1.[32] The stain appears in a series of thick and thin stripes, alternating with paler areas having less enzyme activity (Figure 21.6). These parallel stripes run at right angles to the border between V_1 and V_2. After horseradish peroxidase is injected into blobs of V_1, it is taken up by axon terminals and transported retrogradely, revealing that neurons providing input

16Weliky, M., Bosking, W. H., and Fitzpatrick, D. 1996. *Nature* 379: 725–728.
17Tootell, R. B., Silverman, M. S., and De Valois, R. 1981. *Science* 214: 813–815.
18LeVay, S., and Voight, T. 1988. *Vis. Neurosci.* 1: 395–414.
19Hubel, D. H., and Wiesel, T. N. 1972. *J. Comp. Neurol.* 146: 421–450.
20Fitzpatrick, D., Lund, J. S., and Blasdel, G. G. 1985. *J. Neurosci.* 5: 3329–3349.
21Sclar, G., Maunsell, J. H. R., and Lennie, P. 1990. *Vision Res.* 30: 1–10.
22Lachica, E. A., and Casagrande, V. A. 1992. *J. Comp. Neurol.* 319: 141–158.
23Komatsu, H. 1998. *Curr. Opin. Neurobiol.* 8: 503–508.
24Wong-Riley, M. 1989. *Trends Neurosci.* 12: 94–101.
25Zheng, D., LaMantia, A. S., and Purves, D. 1991. *J. Neurosci.* 11: 2622–2629.
26Livingstone, M. S., and Hubel, D. H. 1984. *J. Neurosci.* 4: 309–356.
27Hendrickson, A. E. 1985. *Trends Neurosci.* 8: 406–410.
28Ts'o, D. Y., and Gilbert, C. D. 1988. *J. Neurosci.* 8: 1712–1727.
29Livingstone, M. S., and Hubel, D. 1988. *Science* 240: 740–749.
30Yabuta, N. H., and Callaway, E. M. 1998. *J. Neurosci.* 18: 9489–9499.
31Merigan, W. H., and Maunsell, J. H. R. 1993. *Annu. Rev. Neurosci.* 16: 369–402.
32Olavarria, J. F., and Van Essen, D. C. 1997. *Cerebral Cortex* 7: 395–404.

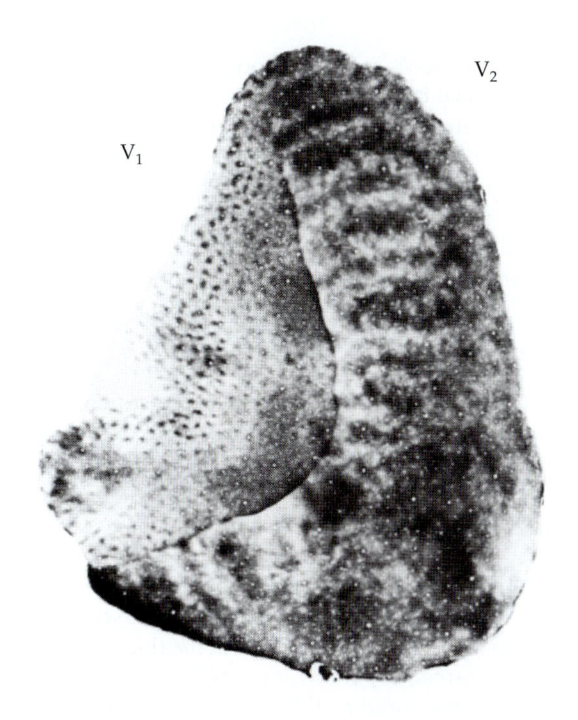

FIGURE 21.6 Blobs in V_1 and Stripes in V_2 of monkey visual cortex stained by cytochrome oxidase. The blobs are arranged in a polka-dot pattern. A clear boundary can be discerned between V_1 and V_2. At this line the blobs change to stripes, thick and thin, running at right angles to the border. (From Livingstone and Hubel, 1988.)

to the blobs are located in the thin stripes in V_2. The connections are reciprocal: Injections into thin stripes label blobs in V_1.[33] In contrast, the interblob regions project to the pale stripes. The thick stripes receive primarily magnocellular information from layers 4B and 4Cα. Remarkably, this functional subdivision can even be distinguished at a molecular level; the monoclonal antibody Cat-301 preferentially labels magnocellular pathways throughout the monkey visual cortex.[34]

Association Areas of Visual Cortex

Considerable effort has been expended to understand the processing of M and P channels in visual cortex, motivated by the functional and anatomical distinction of retinal ganglion cells and geniculate neurons (Figure 21.7). Additional motivation comes from psychophysical measurements, brain imaging, and studies of patients with specific lesions[35]

[33]Livingstone, M. S., and Hubel, D. H. 1987. *J. Neurosci.* 7: 3371–3377.

[34]DeYoe, E. A., et al. 1990. *Vis. Neurosci.* 5: 67–81.

[35]Grusser, O. J., and Landis, T. 1991. *Visual Agnosias*. Vol. 12 of *Vision and Visual Dysfunction*. CRC, Boca Raton, FL.

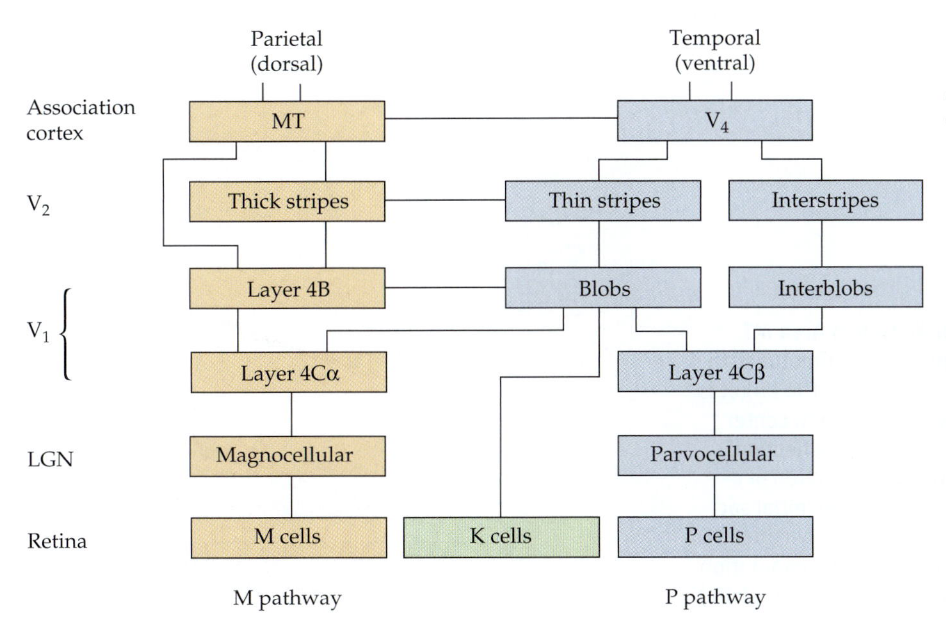

FIGURE 21.7 Schematic Organization of M, P, and K Channels to Visual Cortex. Functionally distinct layers of the LGN project to different layers in V_1. K layers project to blobs in layers 2 and 3. The M and P layers of 4C may interact preferentially with blob and interblob regions in layers 2 and 3. Blobs project preferentially to thin stripes in V_2. Thin stripes project to V_4. Thick stripes in V_2 receive input from layer 4B in V_1 and project to association area MT (V_5). M channels appear to project to dorsal (parietal) visual cortex, where movement is analyzed; P channels project preferentially to area V_4, where color vision is processed. (After Merigan and Maunsell, 1993.)

suggesting that different aspects of visual analysis are carried out in physically separate brain regions, the association areas of visual cortex. A dorsal, or parietal, pathway (so called because the relevant areas of cortex are in parietal cortex, dorsal to the primary visual cortex) is thought to be important for assessing motion and the spatial relationships of form—properties similar to the M channels already described. Lesions in the dorsal, parietal path result in neglect of a portion of the visual field, and disruption of visuomotor orientation. Lesions in the temporal cortex (ventral to V_1) diminish visual identification of objects, their colors, and fine details[36]—reflecting the properties of the P channels.

Motion Detection and Area MT

For most of us, an offhand definition of vision would consist largely of a description of object recognition, involving foveal analysis of fine detail—of the sort that occurs while reading these words. However, an equally important function of vision is the analysis of motion, although this remains largely automatic and unconscious (unless one has the misfortune to suffer inappropriate movement of the visual field because of a vestibular disorder).

As already described, motion is analyzed by the magnocellular–parietal component of the visual pathway (see Figure 21.7). Magnocellular-pathway neurons are sensitive to moving stimuli, and this trait is maintained through V_1 and V_2 (via the thick stripes) to the middle temporal association cortex (area MT, or V_5). Area MT is retinotopically mapped.[37] Neurons there are selective for the speed and direction of a moving stimulus[38,39] and are clustered together into columns with similar preferred directions.[40,41] When small regions of MT were chemically lesioned with a neurotoxin, a monkey's ability to detect a moving pattern of dots in a corresponding region of the visual field was impaired, while thresholds for contrast were unaffected.[42]

[36]Ungerleider, L. G., and Mishkin, M. 1982. In *The Analysis of Visual Behavior.* MIT Press, Cambridge, MA, pp. 549–586.

[37]Maunsell, J. H., and Newsome, W. T. 1987. *Annu. Rev. Neurosci.* 10: 363–401.

[38]Zeki, S. M. 1974. *J. Physiol.* 236: 549–573.

[39]Maunsell, J. H. R., and Van Essen, D. C. 1983. *J. Neurophysiol.* 49: 1127–1147.

[40]Albright, T. D. 1984. *J. Neurophysiol.* 52: 1106–1130.

[41]Malonek, D., Tootell, R. B. H., and Grinvald, A. 1994. *Proc. R. Soc. Lond. B* 258: 109–119.

[42]Newsome, W. T., and Pare, E. B. 1988. *J. Neurosci.* 8: 2201–2211.

[43]Dursteler, R. M., Wurtz, R. H., and Newsome, W. T. 1987. *J. Neurophysiol.* 57: 1262–1287.

Area MT and Visual Tracking

Area MT is involved in visual tracking. This was shown by experiments[43] in which a monkey was trained to track a moving target with its eyes (Figure 21.8). The normal pattern of eye movements is seen in the upper record in Figure 21.8, in which the moving target (trajectory begins at time 0) was acquired by a rapid saccade (the downward deflection occurring 200 ms later) and then retained on the fovea by an accurate tracking, or smooth pursuit, process. After a small injection of neurotoxin (ibotenic acid) into the foveal region of MT, the monkey's ability to track the moving target was markedly impaired. In particular, after the initial saccade the subsequent tracking velocity was much slower than the target velocity. That underestimate also applies to the initial saccade made by the lesioned animal: It positioned the eye as though the estimated velocity were slower

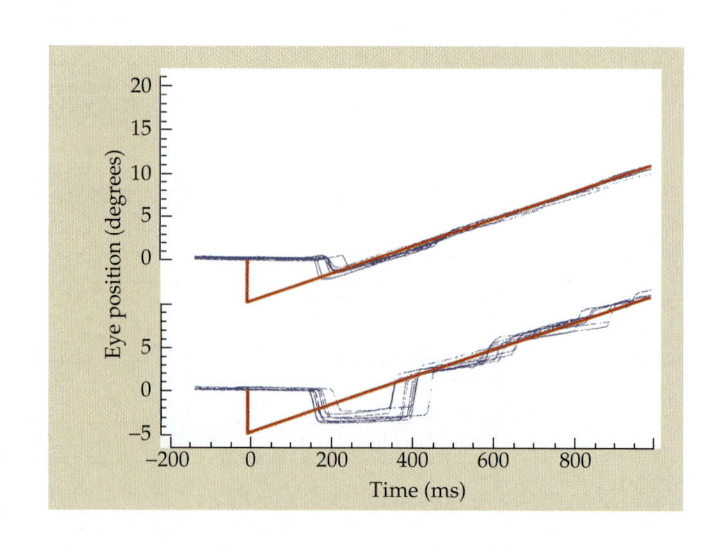

FIGURE 21.8 Organization of Visual Motion in Area MT. A monkey was trained to track a moving target (stimulus path shown by red line), and its eye position relative to the target is shown in the upper record. After an initial saccade to center the target on the fovea (the rapid downward eye deflection), the eye closely followed the target path. After injection of a neurotoxin in area MT (lower record), however, the initial saccade was too large and overshot the target, and subsequent tracking was slower than required, as though the computation of target speed were faulty. (From Newsome and Wurtz, 1988.)

than the actual velocity. Somehow the neurotoxin lesion perturbed the ability of area MT to compute an accurate estimate of target velocity.

How is motion computed in area MT? Cells are clustered into columns of similar preferred directions across the retinotopic map. Thus, the movement of a target across the retina ought to activate those columns aligned with the direction of movement. But moving visual targets will not activate only one such column, being more likely to exhibit complex patterns of motion that activate many sets of directionally tuned neurons to varying degrees. Thus, it will require some form of neural computation to derive an averaged movement vector. The neural arithmetic performed by multiple direction columns was studied by Newsome and his colleagues using electrical microstimulation to alter eye movements in trained monkeys.[44] A microelectrode recorded the preferred direction of a column of cells, and then was used to pass small amounts of current to activate that same column during a target-tracking eye movement (Figure 21.9). Then the tracking eye movement was compared with and without electrical stimulation of MT to ask how components of the visual motion map add. Eye position tracks target position closely in the control condition (Figure 21.9A). When electrical stimulation activated an MT column whose directional preference was different from that of the moving target, the resulting eye movement lay somewhere between the two (Figure 21.9B). The conclusion is that the movement map is used to compute the vectorial average of the activated direction columns. An appealing feature of these experiments is that the monkey's behavior (eye movements) indicates the type of analysis made by higher centers in the cortex. Similar vector averaging has long been known to occur during the generation of saccadic eye movements.[45]

Color Vision

Parvocellular neurons of the visual pathway carry information regarding fine details of form and color. At the level of the cones we have seen a clear correlation between neuronal signals and the wavelength of light falling on the retina (Chapter 19): Red, green,

[44]Groh, J. M., Born, R. T., and Newsome, W. T. 1997. *J. Neurosci.* 17: 4312–4330.

[45]Robinson, D. A., and Fuchs, A. F. 1969. *J. Neurophysiol.* 32: 637–648.

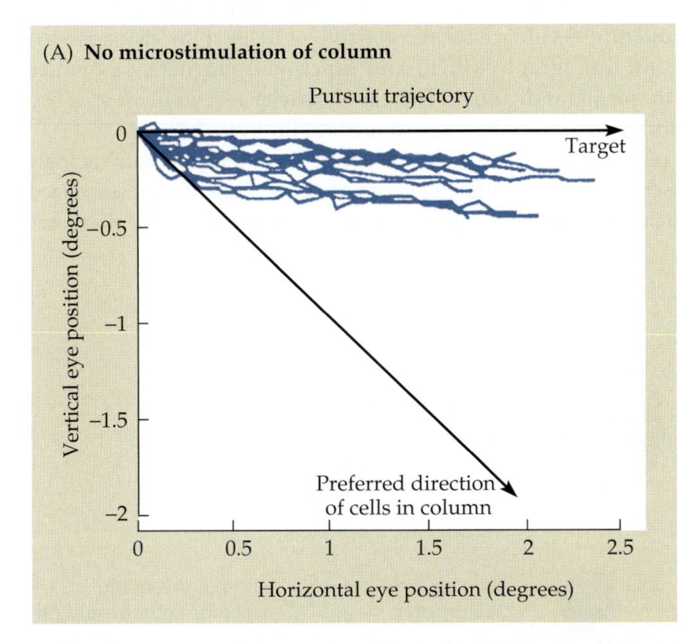

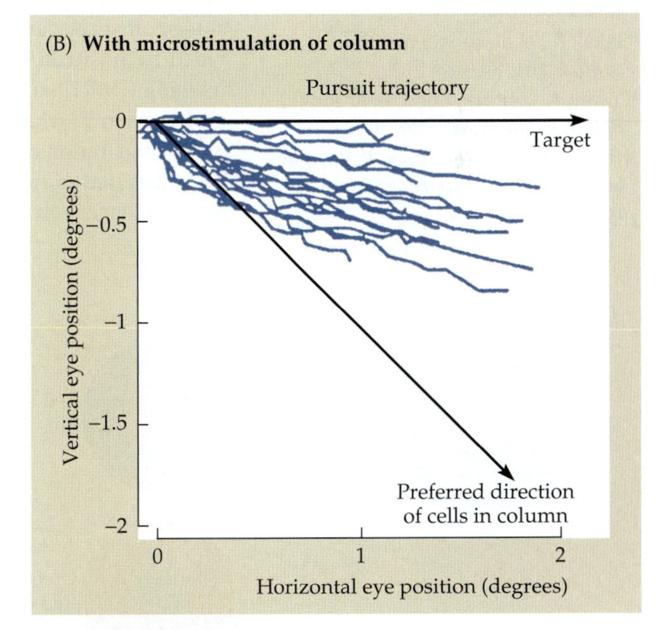

FIGURE 21.9 The Direction of Eye Movements Can Be Altered by Electrical Stimulation in Area MT. (A) Eye movements were recorded in response to a moving visual target. Earlier, an electrode had been inserted into area MT and the preferred direction of cells in that location was noted. This preferred direction differed from that of the moving target. (B) When this location in area MT was stimulated electrically, the resulting eye movements were biased in the preferred direction of cells in the stimulated region. These results suggest that visual motion is computed as the vector sum of several preferred directions in area MT. (After Groh, Born, and Newsome, 1997.)

and blue cones preferentially absorb light in long-, medium-, and short-wavelength ranges. In principle, therefore, by comparing the activity of each type of cone, the nervous system could compute wavelength. Is this how color perception is achieved by the visual system?

The convergence of inputs from cones begins with color-coded horizontal cells and continues with ganglion cells and lateral geniculate cells of the parvocellular division. The properties of such color-coded ganglion and geniculate cells are shown in Figure 21.10. No major transformation of receptive field properties has been found between optic nerve fibers and geniculate cells.[46] The receptive field of the geniculate cell in Figure 21.10A is concentric, with a red "on" center and a green "off" surround. A small red spot illuminating the center causes a brisk discharge; a larger green light over the surround inhibits it. Such a cell responds best to red light, on a neutral or blue, but not on a green, background. It has a conventional center–surround response to spots of white light. Other cells show blue–yellow antagonistic areas (yellow from a mixed red and green cone contribution). Representative types of center–surround organizations that have been observed in the parvocellular layers of the monkey geniculate nucleus are summarized in Figure 21.10B.

Red–green and blue–yellow neurons are examples of **color-opponent cells** that analyze wavelength by comparing their cone inputs in just the way that Young or Helmholtz would have enjoyed finding out about (Chapter 19). A red, green, blue, yellow, black, or white ball on a billiard table evokes, in an array of color-opponent cells, unambiguous signals that are then presented to the brain.

Pathways to Color Vision

A picture of the subsequent steps in cortical analysis of color and color perception has been provided by experiments of Zeki, Hubel, Daw, Land, and their colleagues. It has been mentioned already that the pathways for color tend to be segregated from those for analysis of other properties, such as depth, movement, contrast, and form. Zeki has demonstrated parvocellular pathways that lead from V_1 through V_2 to V_4, in which color-coded cells are abundant.[47] Evidence for the key role of V_4 in color vision is provided by positron emission tomography (PET) and functional magnetic resonance imaging (fMRI) studies in normal individuals. Increased activity is seen in an area homologous to V_4 when colored patterns of light are shone into the eyes.[48–50]

The separation of the perception of color from the perception of form is convincingly demonstrated in rare cases of patients suffering a loss of color vision due to localized brain damage (cerebral **achromatopsia**).[51,52] For example, an unusual adult patient has

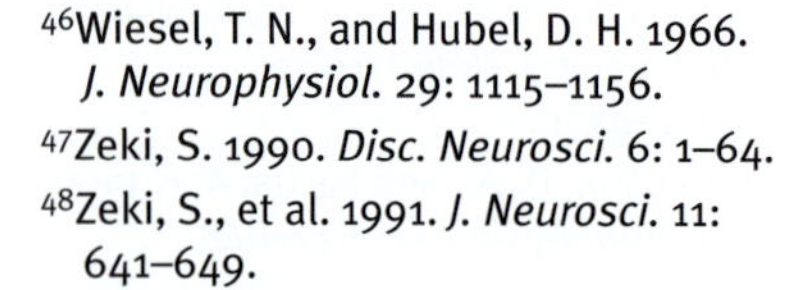
[46]Wiesel, T. N., and Hubel, D. H. 1966. *J. Neurophysiol.* 29: 1115–1156.

[47]Zeki, S. 1990. *Disc. Neurosci.* 6: 1–64.

[48]Zeki, S., et al. 1991. *J. Neurosci.* 11: 641–649.

[49]Takechi, H., et al. 1997. *Neurosci. Lett.* 230: 17–20.

[50]Sakai, K., et al. 1995. *Proc. R. Soc. Lond. B* 261: 89–98.

[51]Cowey, A., and Heywood, C. A. 1995. *Behav. Brain Res.* 71: 89–100.

[52]Heywood, C. A., Kentridge, R. W., and Cowey, A. 1998. *Exp. Brain Res.* 123: 145–153.

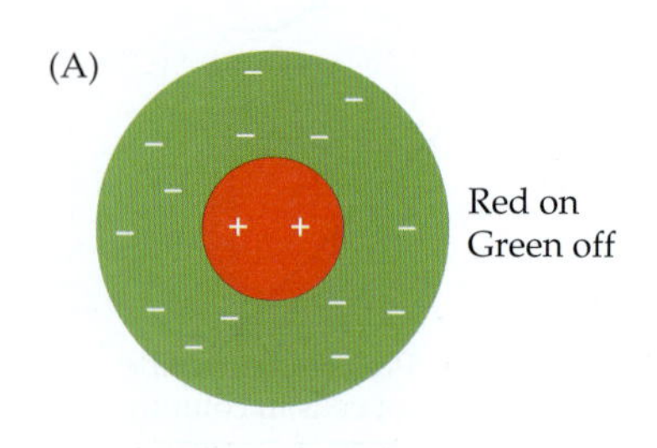

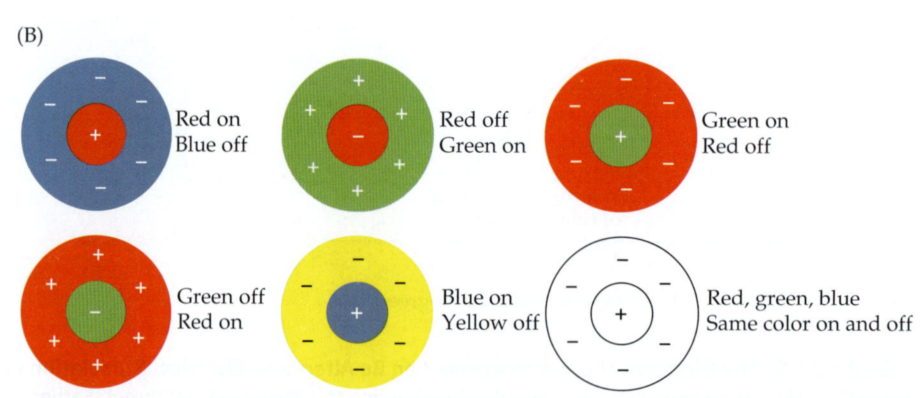

FIGURE 21.10 Receptive Field Organization of Cells in parvocellular layers of the monkey LGN, responding to color. (A) Receptive field organization with a red "on" center and a green "off" surround. A cell of this sort would fire best in response to a small or large red spot on a neutral background. A red spot on a green background would be relatively ineffective. (B) Example of receptive field organizations with various spectral sensitivities. (After Wiesel and Hubel, 1966; Hubel, 1988.)

been described who suffered bilateral lesions to cortex anterior to V_1, probably corresponding to V_4.[53] Before the lesion he had normal color vision, but afterward he became unable to see colors. He knew from prior experience that a strawberry was red and a banana yellow, but after the lesion the world appeared like a black-and-white movie. Other functions, such as memory and form vision, were little affected and he was able to continue his job as a customs inspector(!). When a familiar object was pointed out, he could describe what color it ought to have, but he could not match it with an appropriate colored marker. His deficit was not in speech or in object recognition, but in color perception per se.

Psychophysical tests in normal subjects confirm the separation of color pathways right up to perception.[54] Full descriptions are provided in papers[55] and reviews by Zeki,[40] Hubel,[56] and Livingstone.[57] For example, it is hard, if not impossible, to discern patterns or forms in a scene unless the magnocellular pathways are activated by the contrast that is normally present as different degrees of brightness and shadow. The parvocellular system, with its emphasis on color and high spatial resolution, has only limited ability on its own to provide information about the form of objects. Hence, a colored picture of a complex structure with all its components reflecting equal quantities of light appears formless. This is because the magnocellular pathway is not activated. Similarly, our sensations of depth and movement tend to be lost when black-and-white contrast is insufficient to bring in magnocellular pathways. An impressive demonstration is to move a pattern of green and red stripes across a computer monitor. The intensity of each color can then be adjusted to be equiluminant (i.e., each red or green stripe emitting the same effective amount of light as its neighbor, although at a different wavelength). One still sees the colored stripes, but they seem to stop moving.

Color Constancy

A major problem in our understanding of color vision has been to understand how the cortex decides on the color of an object in the visual scene. Our brains program these calculations so successfully that we do not intuitively realize that there is indeed a problem. Surely blue illustrations in this book appear blue because they reflect light at short wavelength. From all that has been said so far, one might imagine that the colors we see are determined directly and simply by the wavelength of light. That this is not so was apparent to Helmholtz (again!).[58] He pointed out that an apple seen in daylight, in sunset, or by candlelight appears red. Yet the light reflected from its surface contains far more red light at sunset and far more yellow in candlelight. Somehow the brain has assigned a red color to the apple that does not shift under very different conditions; it "discounts the illuminant."

A familiar example is the tint of two correctly exposed photographs taken on the same film in daylight and in the artificial room light provided by electrical lamp bulbs. The daylight colors seem realistic, whereas those taken indoors appear too yellow. Yet we are not at all aware of this yellowness when in an artificially lit room. (This phenomenon until recently was a commonplace observation; the flash present on nearly every camera nowadays emits wavelengths of light balanced to resemble daylight.) The biological advantages of color constancy are clear: Green berries should not turn red at sunset; pink lips should not turn yellow by candlelight.

Spectacular demonstrations of color constancy designed by Land[59,60] have acted as a powerful stimulus for neurobiological research into color vision. What his demonstrations showed was that the way we see the color of an object depends critically on the light reflected from the entire scene, not just from the object itself. We cannot assign a color—yellow, green, blue, or white—to an area just by determining the wavelength of light reflected from it. We need also to know the composition of light reflected from surrounding areas. This uncomfortable conclusion, referred to as the Land phenomenon, seems counterintuitive. As with black and white, however, the brain assigns the sensation of color by comparing light falling on different parts of the retina, instead of measuring absolute brightness or wavelengths at one spot on the retina. Rather, it is as though overall comparisons were made in the cortex of contrasts at boundaries, within three separate pictures of a complex scene viewed through short-, medium-, and long-wavelength filters (Box 21.1).

[53]Pearlman, A. L., Birch, J., and Meadows, J. C. 1979. *Ann. Neurol.* 5: 253–261.

[54]Leonards, U., and Singer, W. 1997. *Vision Res.* 37: 1129–1140.

[55]Barbur, J. L., Harlow, A. J., and Plant, G. T. 1994. *Proc. R. Soc. Lond. B* 258: 327–334.

[56]Hubel, D. H. 1988. *Eye, Brain and Vision*. Scientific American Library, New York.

[57]Livingstone, M. S., and Hubel, D. H. 1987. *J. Neurosci.* 7: 3416–3468.

[58]Helmholtz, H. von. 1962/1927. *Helmholtz's Treatise on Physiological Optics*. Dover, New York.

[59]Land, E. H. 1986. *Vision Res.* 26: 7–21.

[60]Land, E. H. 1986. *Proc. Natl. Acad. Sci. USA* 83: 3078–3080.

Box 21.1 Color Constancy

Part A of the figure shows a color image, taken from one of Land's extraordinary demonstrations, in which a complex pattern of about 100 rectangles and squares of colored paper is illuminated by lights shining from three separate projectors.[61] The picture is abstract, nonrepresentational, and reminiscent of a painting by Mondrian. The slide projectors contain no slides. The brightness of each projector is controlled by a rheostat, and the wavelength by an interference filter. As expected, when the three projectors shine lights of equal intensity at long, medium, and short wavelengths, corresponding to the red, green, and blue, respectively, of our cone system, we see the colors of the figure.

The first surprising result occurs when one changes the relative intensities of red, green, and blue by increasing the current through the red projector while reducing it through the green and even more through the blue. The adjustments are made so that the overall light reflected back has the same intensity as before. Although the wavelength composition of the light has changed, the colors of the squares do not change when the pattern is viewed as a whole. The yellow, green, red, blue, and white squares retain their colors even when the red or the green or the blue component of the light is increased or decreased over a wide range.

Land further describes the following experiment: A photometer equally sensitive to all wavelengths is pointed at a yellow panel. With the green and blue projectors turned off, the red projector is adjusted until a reading of 1 (in arbitrary units) is obtained. It is then turned off and the procedure is repeated with the blue and green projectors. From this yellow panel, then, equal amounts of red, green, and blue light are reflected when all three projectors are turned on. It looks yellow when we view the entire picture under this illumination. Now exactly the same procedure is followed while illuminating a green panel, adjusting the red, green, and blue lights so that the same intensity at each wavelength is reflected back to the photometer. By definition, the light coming back to our eyes from the green panel has exactly the same wavelength composition as that previously reflected from the yellow. What color will the green panel appear to have? The intuitive answer is yellow. Yet in the context of the entire picture, the green panel still appears green. If, however, all that can be seen is the small patch of uniform color, surrounded by darkness, its color constancy is lost (see part B of the figure where the formerly green panel is shown as yellow to illustrate the point.). Only in the context of the whole pattern is color constancy preserved. This remarkable effect does not depend on judgment, adaptation, or afterimages. From these and other experiments Land concluded,

> If you contemplate the Mondrian while making these sweeping changes in relative illumination on the separate wavebands, you cannot possibly understand why there are not concomitant changes in color—as long as you think in terms of color-mixing at each point. If, however, you can imagine the Mondrian as being a composite of three independent images, one carried by long waves, one carried by middle waves, and one carried by short waves, I can then introduce the proposition that, first, each of these images is unaltered by a change in the flux by which it is carried and second, that it is the comparing of these three complete images, rather than the merging of their fluxes, which produces the array of colors.

The phenomenon of color constancy tells us that color vision depends on the analysis of contrast. Land's "retinex theory" ("retinex" is "retina" and "cortex" combined) provides a framework for computing the color seen at a particular part of the retina on the basis of the relative intensities of three wavelengths and their spatial interactions.

(A)

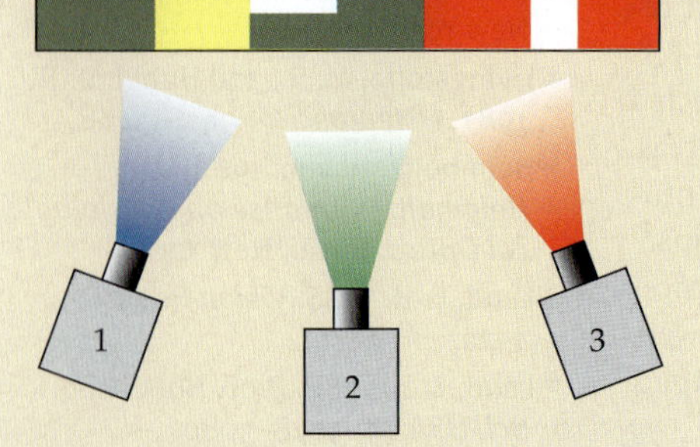

(B)

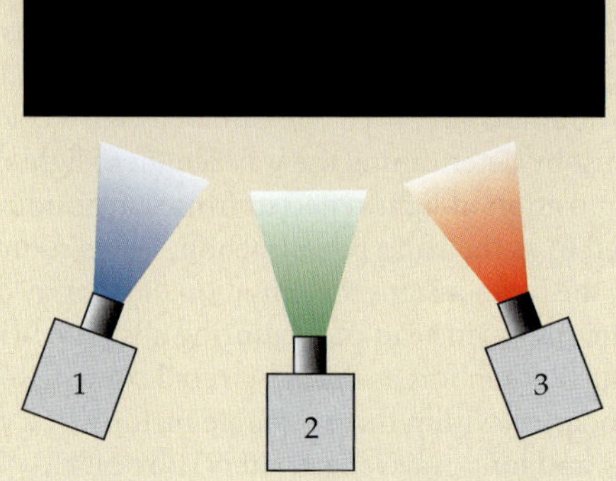

[61] Land, E. H. 1983. *Proc. Natl. Acad. Sci. USA* 80: 5163–5169.

It has not been possible to provide a comprehensive and satisfactory explanation of the Land phenomenon in terms of the receptive field properties of color-coded cells in V_1, V_2, and V_4. However, one type of cell, known as **double opponent**, has properties that could participate in color constancy. Originally described in the goldfish retina by Daw,[62,63] double-opponent cells have been found in primate cortex, but not in the lateral geniculate nucleus or retina.[28,64–66] They therefore appear to represent a higher stage of color processing. In brief, such cells have roughly concentric center–surround receptive fields with red–green or blue–yellow antagonism (Figure 21.11). But unlike color-opponent cells in the LGN, each color produces antagonistic effects in both center and surround of the double-opponent cell. Thus, shining red light in the center evokes an "on" discharge, and red in the surround evokes an "off." Green in the surround evokes an "on" discharge, in the center an "off."

Suppose one evokes a discharge with a small red spot on the center of such a cell's receptive field, in the presence of white ambient light. If one now increases the proportion of red in the ambient light, the discharge will change only a little: Increased excitation of the central area by red will be antagonized by increased inhibition from the periphery by red. Indeed, the balance of red, green, and blue cone inputs to center and surround regions varies from cell to cell in V_1 such that a continuum of hue preferences is observed.[67,68] This contrasts with the color preferences of LGN neurons, which coincide largely with the primary colors. Presumably, long horizontal connections from blob to blob (see the next section) play a role in the spatial interactions required to explain the Land phenomenon.

THE INTEGRATION OF VISUAL INFORMATION

Horizontal Connections within Primary Visual Cortex

A scheme for visual processing such as that shown in Figure 21.7 represents a working model that serves to organize one's thinking. However, the separation of magnocellular and parvocellular pathways—for detection of contrast, motion, and depth on the one hand and color and form on the other—is by no means absolute. Interactions can be found even in V_1, where magnocellular input can be found in blobs and interblobs.[20,30,69,70] Further, although the identification of V_1 and V_2 is generally agreed on, additional areas of association visual cortex do not have precisely defined boundaries. Receptive field properties of cells in these regions may vary and representation of visual fields may appear disorderly.

[62]Daw, N. W. 1984. *Trends Neurosci.* 7: 330–335.

[63]Daw, N. W. 1968. *J. Physiol.* 197: 567–592.

[64]Hubel, D. H., and Livingstone, M. S. 1987. *J. Neurosci.* 7: 3378–3415.

[65]Hubel, D. H., and Livingstone, M. S. 1990. *J. Neurosci.* 10: 2223–2237.

[66]Tootell, R. B., et al. 1988. *J. Neurosci.* 8: 1569–1593.

[67]Lennie, P., Krauskopf, J., and Sclar, G. 1990. *J. Neurosci.* 10: 649–669.

[68]Yoshioka, T., Dow, B. M., and Vautin, R. G. 1996. *Behav. Brain Res.* 76: 51–70.

[69]Lachica, E. A., Beck, P. D., and Casagrande, V. A. 1992. *Proc. Natl. Acad. Sci. USA* 89: 3566–3570.

[70]Sawatari, A., and Callaway, E. M. 1996. *Nature* 380: 442–446.

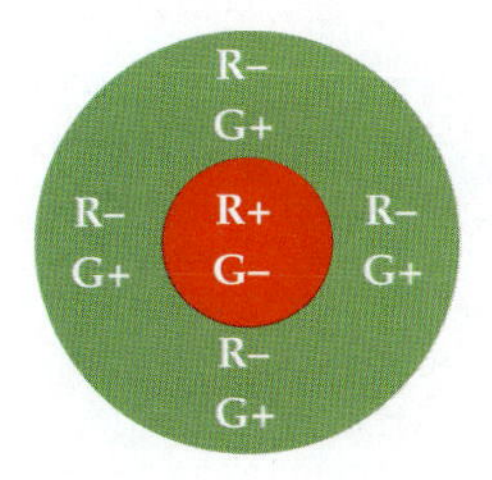

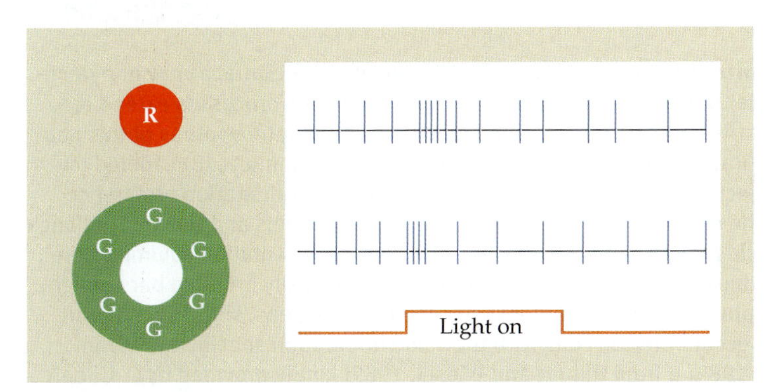

FIGURE 21.11 Receptive Field of a Double-Opponent Cell, the properties of which could help to explain color constancy. A cell of this type responds best to a small red spot shining on a green background (upper record), or to green annular illumination (lower record). (After Daw, 1984.)

In V_1 itself, a variety of connections have been described that suggest organizational principles more complex than those we have considered thus far. Whereas classical staining techniques, such as Golgi impregnation, revealed a preponderance of neuronal processes that run perpendicularly from layer to layer, intracellular injections of single cells have demonstrated that cortical neurons also have long horizontal processes that extend laterally from column to column (Figure 21.12A).[71–78] Connections such as these contribute to the synthesis of elongated receptive fields of simple cells in layer 6 of V_1: The receptive fields of layer 5 cells are combined and added end to end on the layer 6 simple cell by way of long horizontal axons. Many simple and complex cells are found with long horizontal projections that extend more than 8 mm, corresponding to several hypercolumns. An individual neuron can therefore integrate information over an area of retina several times larger than the receptive field as measured by conventional techniques.[79]

Of particular interest is that connections are made only between columns that have similar orientation specificity. Evidence for such specific interconnections has been obtained by two additional methods. First, when labeled microbeads are injected into one column, they are transported to a distant hypercolumn with the same orientation preference (Figure 21.12B). Second, by cross-correlation of the firing patterns of neurons with the same orientation preference in two widely separated columns, one can show that they are interconnected.[30,80,81] Moreover, after a lesion has been made in the retina, cortical cells that have been deprived of input can show responses to distant stimuli that would be outside their "normal" receptive fields.[28]

[71]Gilbert, C. D., and Wiesel, T. N. 1979. *Nature* 280: 120–125.

[72]Gilbert, C. D., and Wiesel, T. N. 1983. *J. Neurosci.* 3: 1116–1133.

[73]Gilbert, C. D., and Wiesel, T. N. 1989. *J. Neurosci.* 9: 2432–2442.

[74]Katz, L. C., Gilbert, C. D., and Wiesel, T. N. 1989. *J. Neurosci.* 9: 1389–1399.

[75]Callaway, E. M., and Wiser, A. K. 1996. *Vis. Neurosci.* 13: 907–922.

[76]Bosking, W. H., et al. 1997. *J. Neurosci.* 17: 2112–2127.

[77]Kisvardy, Z. F., et al. 1997. *Cerebral Cortex* 7: 605–618.

[78]Schmidt, K. E., et al. 1997. *J. Neurosci.* 17: 5480–5492.

[79]Ts'o, D. Y., Gilbert, C. D., and Wiesel, T. N. 1986. *J. Neurosci.* 6: 1160–1170.

[80]Gray, C. M., et al. 1989. *Nature* 338: 334–337.

[81]Schwarz, C., and Bolz, J. 1991. *J. Neurosci.* 11: 2995–3007.

Receptive Fields from Both Eyes Converging on Cortical Neurons

When we look at an object with one or both eyes, we see only one image, even though the size and the position of the object's projection are slightly different on the two retinas. In-

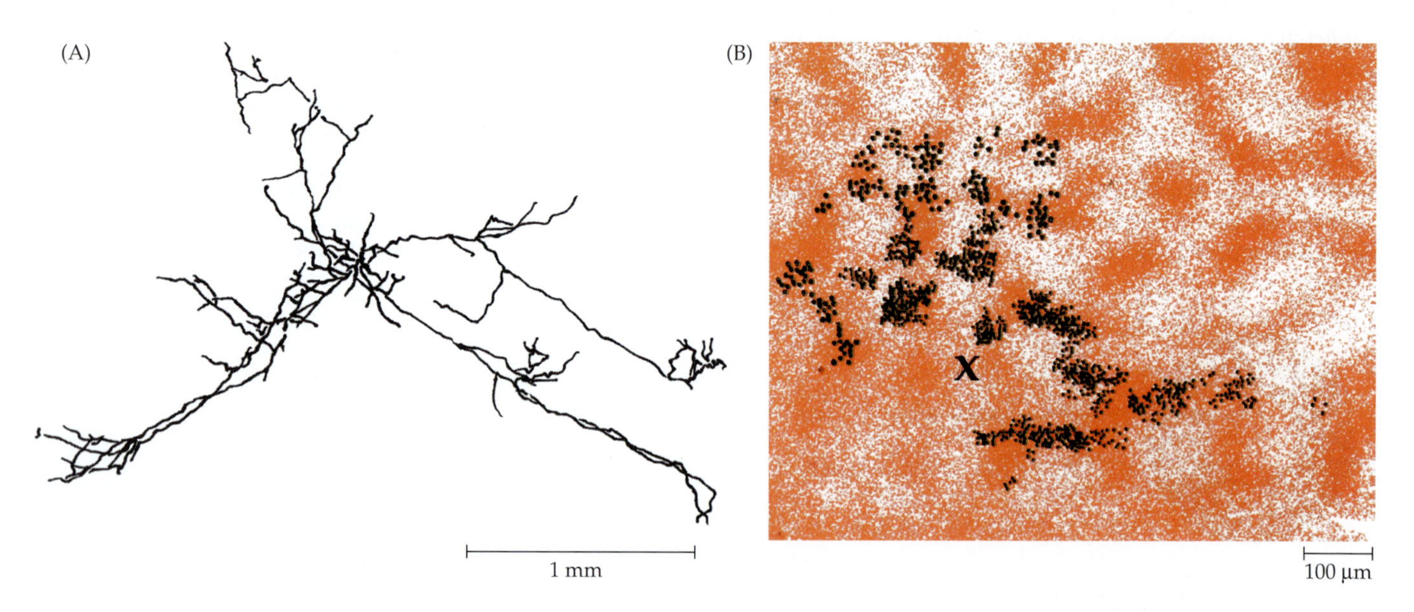

FIGURE 21.12 Horizontal Connections in Visual Cortex. (A) Surface view of pyramidal cell in cat V_1, after labeling with horseradish peroxidase. The processes extend for nearly 3 mm across the cortical surface. Fine branches and synaptic boutons of this neuron occurred in several discrete clusters separated by 800 μm or more. (B) Labeled microspheres were injected into a region where cells had vertical orientation preference (large black "X"). The microspheres are taken up by axon terminals and transported back to the somata of cells projecting to the injection site. Vertical orientation columns were also labeled using deoxyglucose during stimulation with vertically oriented bars of light. The labeled microspheres are found in deoxyglucose-labeled areas, showing that horizontal connections occur between cells with the same orientation specificity. (A from Gilbert and Wiesel, 1983; B from Gilbert and Wiesel, 1989; kindly provided by C. Gilbert.)

terestingly, well over 100 years ago Johannes Müller suggested that individual nerve fibers from the two eyes might fuse or become connected to the same cells in the brain. Thereby, he almost exactly anticipated Hubel and Wiesel's results.[82,83] They found that about 80% of all cortical neurons in the visual areas of the cat can be driven from both eyes. Since the neurons in the various layers in the LGN are predominantly innervated from one eye or the other, the first opportunity for significant interaction between the eyes must occur in the cortex. As mentioned earlier, the separation is maintained in layer 4 of V_1, where each simple cell is driven by only one eye, the other being without effect. Mixing between the two eyes occurs in the subsequent relay stations—that is, in layers deeper toward the white matter and in layers closer to the cortical surface.

Examination of the receptive fields of a binocularly driven cell shows that (1) they are usually in exactly equivalent positions in the visual field of the two eyes, (2) their preferred orientation is the same, and (3) the corresponding areas in the receptive fields add to each other's effect. An example of synergistic action between the two eyes is shown for a simple cell in Figure 21.13. Shining light onto an "on" region in the left eye sums with illumination onto the "on" area of the right eye. Simultaneous illumination of antagonis-

[82]Hubel, D. H., and Wiesel, T. N. 1962. *J. Physiol.* 160: 106–154.

[83]Hubel, D. H., and Wiesel, T. N. 1959. *J. Physiol.* 148: 574–591.

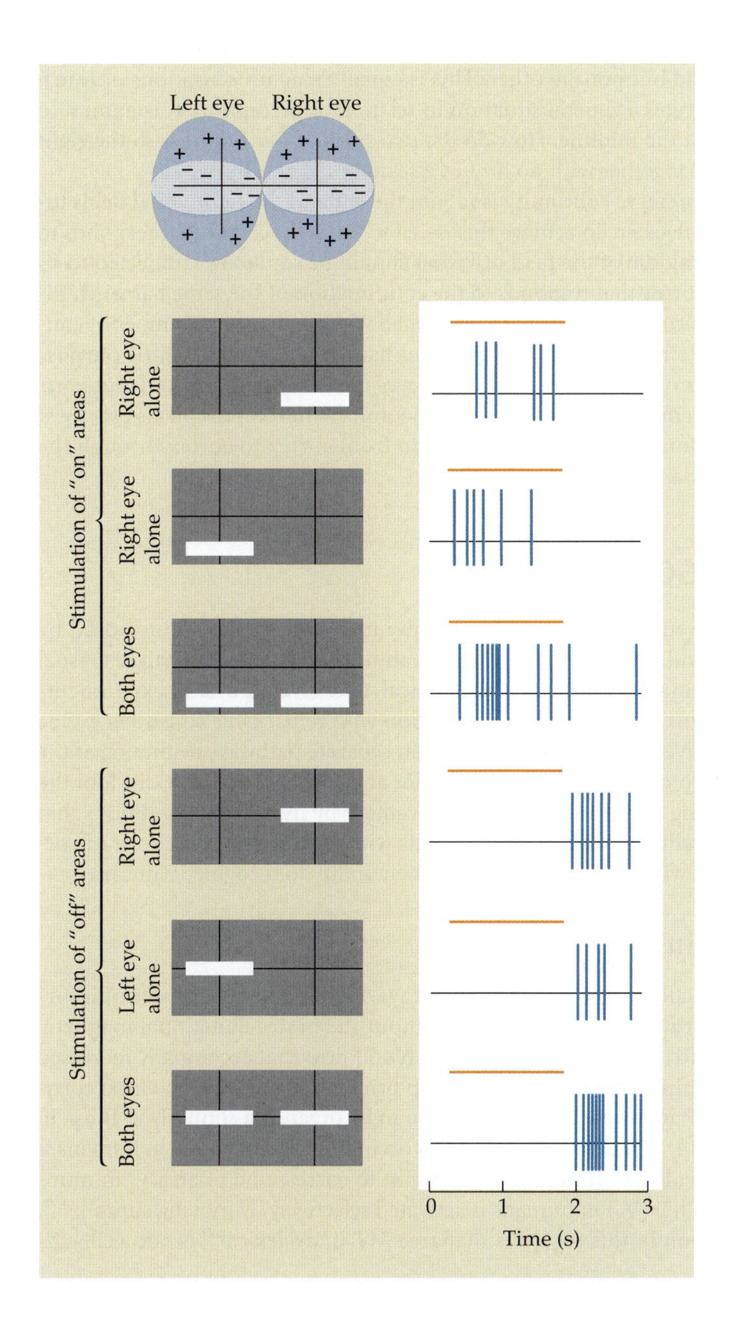

FIGURE 21.13 Binocular Activation of a simple cortical neuron that has identical receptive fields in both eyes. Simultaneous illumination of corresponding "on" areas (+) of right and left receptive fields is more effective than stimulation of one alone (upper three records). In the same way, stimulation of "off" areas (–) in the two eyes reinforces each other's "off" discharges (lower three records). In contrast, cells used for depth perception have receptive fields in the two eyes in disparate regions of the visual field. Such cells require that the bar is farther from or closer to the eye than the plane of focus. (After Hubel and Wiesel, 1959.)

tic areas in the two eyes reduces ongoing responses and increases "off" discharges. Such cells signal a unified image in the two eyes.

For depth perception there exists another binocular specialization of receptive fields.[84] An object out of the plane of focus casts images on disparate parts of the two retinas. Neurons with properties that fit them for depth perception have been found in primary and association visual cortex. For such cells the best stimulus is an appropriately oriented bar in front of the plane of focus (for certain cells), or beyond it (for others).[64,85,86] Impulses fail to be evoked by presenting the bar only to one eye or to the other, or to both in the plane of focus. It is the disparity of the position on the two retinas that these cells require. Disparity coding by complex cells in the primary visual cortex may drive vergence eye movements used to bring objects into focus.[87] Depth perception as such may arise in higher cortical areas. For example, clusters of neurons with similar binocular disparity preferences are found in visual association cortex V_5 (area MT).[88] The depth perception of a trained monkey was predictably altered when such a cluster was electrically stimulated.

Connections for Combining Right and Left Visual Fields

A separate problem concerns the way in which the two cortices are knit together to produce a single image of the body and the world. Each hemicortex is wired to perceive one half of the external world but not the other. This is equally true for sensations of touch and position and constitutes a general situation in relation to perception. It is natural to wonder what happens at the midline. How do the two sides of our brain mesh the right world and the left world together with no hint of a seam or discontinuity?

The obvious way to preserve continuity is to join the left and the right visual fields together at the midline in register. To achieve this, a cell in the right cortex that responds to a horizontal bar in the middle of the field of vision should be somehow connected to its counterpart in the left cortex that responds to the continuation of the same bar. Such interactions would allow a complete picture to be formed with a minimum number of connections between the two hemispheres. On the other hand, there would be little purpose in linking fields seen out of the corners of the two eyes that look on quite different parts of the world. Highly specific connections between neurons with receptive fields exactly at the midline have been found experimentally to run from cortex to cortex through the corpus callosum.[89] They are described in Box 21.2.

WHERE DO WE GO FROM HERE?

It has become possible to approach experimentally many of the questions posed by Helmholtz, Hering,[90] and, in our own time, Land about how the brain analyzes visual scenes falling on the retina. Anatomical, physiological, and psychophysical experiments have each revealed designs in the cortex for perception with remarkable consistency. One important principle derived from such studies is that separate pathways beginning at the retina extend through to consciousness (see Schiller and Lee, 1991[91]). You can fool the systems used for assessing depth and detecting movement by using light conditions that bring in only the parvocellular channels, and patients with lesions to specific areas of cortex can lose color vision with little impairment of their ability to recognize patterns.

Functional Imaging

Noninvasive imaging technologies provide novel ways of studying the transfer of information in the visual pathways, and indeed throughout the brain. Functional magnetic resonance imaging (fMRI) detects local changes in blood flow that accompany increased neural activity.[92] This technique can be used to map primary and associated visual cortical areas in humans,[93–96] and these have been shown to be organized similarly to those in the monkey (Figure 21.14). Thus, an area of ventral occipitotemporal cortex in humans is activated selectively by colored stimuli and is thought to correspond to area V_4 in monkeys.[50,97] Similarly, area MT (V_5) in humans is activated selectively by moving stimuli.[98,99] An intriguing observation in this regard is that area MT in dyslexics[100] is not well acti-

[84]Barlow, H. B., Blakemore, C., and Pettigrew, J. D. 1967. *J. Physiol.* 193: 327–342.
[85]Fischer, B., and Poggio, G. F. 1979. *Proc. R. Soc. Lond. B* 204: 409–414.
[86]Ferster, D. 1981. *J. Physiol.* 311: 623–655.
[87]Masson, G. S., Busettini, C., and Miles, F. A. 1997. *Nature* 389: 283–286.
[88]DeAngelis, G. C., Cumming, B. G., and Newsome, W. T. 1998. *Nature* 394: 677–680.
[89]Hubel, D. H., and Wiesel, T. N. 1967. *J. Neurophysiol.* 30: 1561–1573.
[90]Hering, E. 1986. *Outline of a Theory of the Light Sense.* Harvard University Press, Cambridge, MA.
[91]Schiller, P. H., and Lee, K. 1991. *Science* 251: 1251–1253.
[92]Fox, P. T., and Raichle, M. E. 1986. *Proc. Natl. Acad. Sci. USA* 83: 1140–1144.
[93]Sereno, M. I., et al. 1995. *Science* 268: 889–893.
[94]DeYoe, E. A., et al. 1996. *Proc. Natl. Acad. Sci. USA* 93: 2382–2386.
[95]Tootell, R. B., et al. 1996. *Trends Neurosci.* 19: 481–489.
[96]Courtney, S. M., and Ungerleider, L. G. 1997. *Curr. Opin. Neurobiol.* 7: 554–561.
[97]Martin, A., et al. 1995. *Science* 270: 102–105.
[98]Watson, J. D., et al. 1993. *Cerebral Cortex* 3: 79–94.
[99]Tootell, R. B., et al. 1995. *J. Neurosci.* 15: 3215–3230.
[100]Eden, G. F., et al. 1996. *Nature* 382: 66–69.

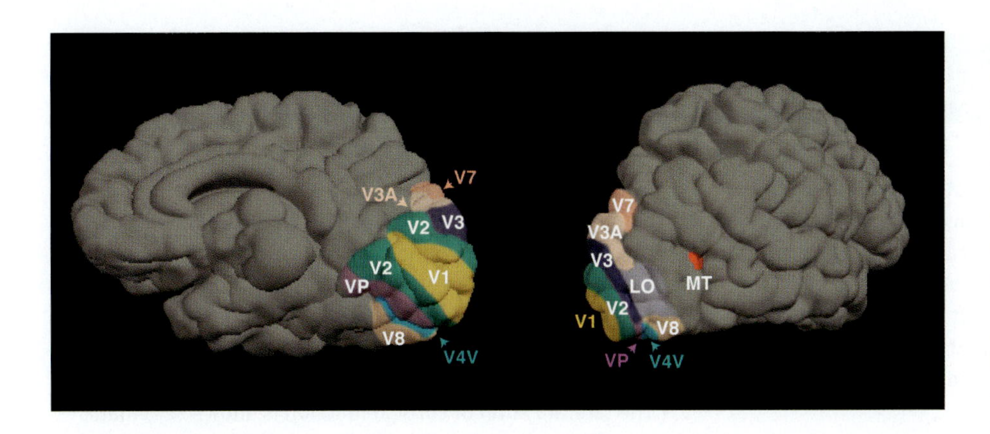

FIGURE 21.14 Functional Maps of Human Visual Cortex can be constructed by presenting visual stimuli to subjects undergoing fMRI. Image content is varied to provide selective stimulation of primary and association visual cortices. Functionally defined zones are color-coded in a medial (left) and lateral (right) view of the right hemisphere of a human brain. The primary visual cortex (V_1) is located most posteriorly and immediately surrounds the calcarine fissure, while successively "higher" association areas are found more anteriorly and at greater distance from the calcarine fissure. Motion sensitive areas (V_3) and area MT are found more dorsally, color- and form-processing areas (V_4, LO) tend to be more ventrally located. VP = ventral posterior; LO = lateral occipital; MT = middle temporal. (Image kindly provided by N. Hadjikhani and R. Tootell.)

vated by moving stimuli, supporting the notion that this perceptual deficit may be associated with a dysfunction of the magnocellular visual pathway.[101]

Faces and Letters

An extension of the hypothesis of hierarchical organization predicts that cells should be discovered that bring together larger and larger sets of the information that appears in the field of vision. Indeed, in higher visual areas, neurons that specifically respond to faces have been detected by microelectrode recording.[102] Functional imaging has reinforced the hypothesis that a specific locus within the occipitotemporal cortex (the fusiform gyrus) is activated selectively during the viewing of faces rather than other objects.[103] Seen in the upper coronal section scan in Figure 21.15, the area outlined in green in the cortex was activated during viewing of a human face, whereas nonface objects (e.g., a spoon) activated more posterior regions bilaterally. Face recognition may be lateralized in human cortex, as is the language center. In right-handed subjects the right fusiform gyrus was activated preferentially or exclusively by face stimuli. In two left-handed subjects, the left fusiform gyrus was preferentially activated.

[101]Livingstone, M. S., et al. 1991. *Proc. Natl. Acad. Sci. USA* 88: 7943–7947.

[102]Damasio, A. R., Tranel, D., and Damasio, H. 1990. *Annu. Rev. Neurosci.* 13: 89–109.

[103]Kanwisher, N., McDermott, J., and Chun, M. M. 1997. *J. Neurosci.* 17: 4302–4311.

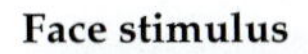

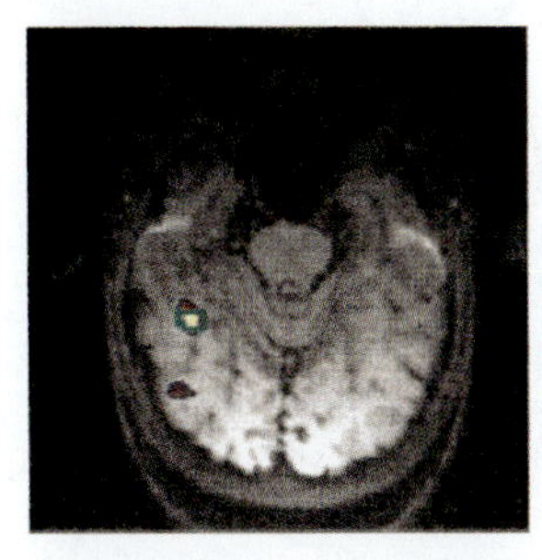

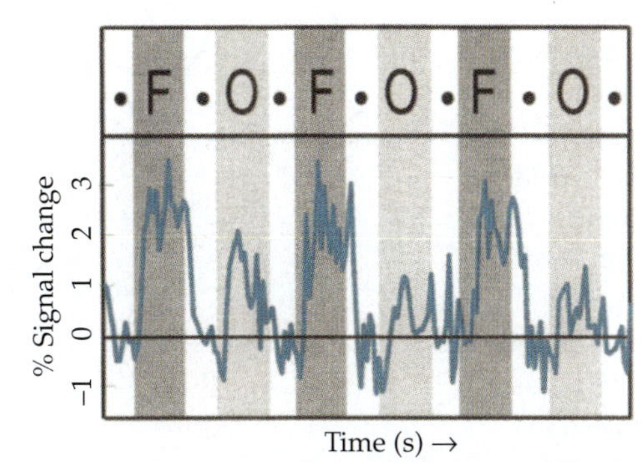

Object stimulus

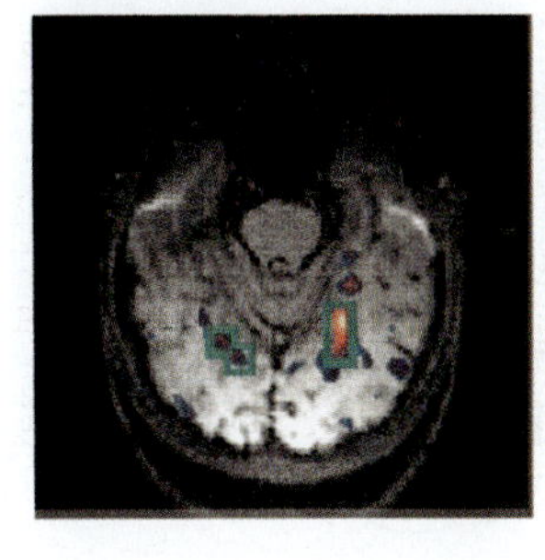

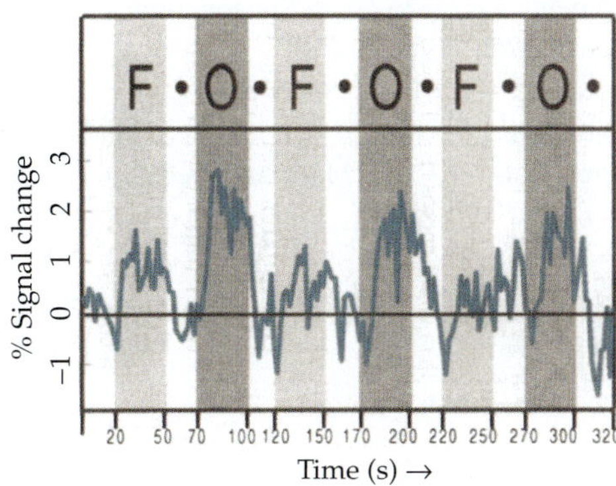

FIGURE 21.15 Face-Specific Regions of Cortex, as revealed by functional MRI. During fMRI scanning, a subject was shown either faces or nonface objects (e.g., a spoon). A region in the right fusiform gyrus (left side of fMRI) was identified in which pictures of faces, more than objects, produced significant and consistent increases in activity (top). Nonface objects produced bilateral stimulation of further posterior regions (bottom). The average signal percent change in the outlined areas is shown to the right of each fMRI. Vertical shading indicates the presentation of faces (F) or nonface objects (O). (From Kanwisher, McDermott, and Chun, 1997.)

Box 21.2 Corpus Callosum

The general question of transfer of information between the hemispheres has been studied in humans and in monkeys by Sperry, Myers, Gazzaniga, and their colleagues.[104–106] Concentrating on the coordinating role of the **corpus callosum**, a bundle of fibers that runs between the two hemispheres, they have shown that the fibers are involved in the transfer of information and learning from one hemisphere to the other.

To cite one example, a normal person can name an object, such as a coin or a key, when it is placed in either hand (stereognosis). After section of the corpus callosum (for treatment of severe epilepsy), however, a right-handed person can name the object only when it is placed in the right hand; the information from the right hand crosses before reaching the cortex and projects to the left hemisphere. It is in the *left* hemisphere that the main area responsible for language lies. What happens when the object is placed in the left hand, which projects to the right hemisphere? Information still reaches consciousness; there is, however, no way in which the concept "key" can be *verbally* expressed because the language center in the left cortex cannot be reached with transfer across the corpus callosum. Thus, a right-handed person without an intact corpus callosum can recognize a key with the left hand, in the sense of being able to open a lock, but cannot say the word "key" in naming the object. Other experiments on the corpus callosum provide surprising insight into higher function. When deprived of these cross-connections, the two hemispheres can lead virtually separate existences.

The fusing, or knitting together, of the two fields of vision is mediated by fibers in the corpus callosum. Certain cells have receptive fields that straddle the midline and receive information about both sides of the visual world. These cells lie at the boundary of V_1 and V_2, and they combine inputs from both hemispheres by way of the corpus callosum. Interestingly, these cells lie in layer 3, which contains neurons known to send their axons to other regions of cortex. The organization and orientation preference of two receptive fields that are brought together in this way are similar. Projections from nerve cells in one hemisphere to the other have been shown anatomically by injecting horseradish peroxidase into the cortex at the boundary of V_1 and V_2.[107] Enzyme taken up by terminals is transported to neuronal cell bodies situated at exactly corresponding sites in opposite hemispheres. Cutting or cooling (which blocks conduction) the corpus callosum causes the receptive field to shrink and become confined to just one side of the midline (the usual arrangement for cortical cells). Furthermore, recordings from single fibers in the callosum show that they have receptive fields close to the midline, not in the periphery.

The role of callosal fibers is clearly demonstrated in an experiment of Berlucchi and Rizzolatti.[108] They made a longitudinal cut through the optic chiasm, thereby severing all direct connections to the cortex from the contralateral eye. Yet, provided the callosum was intact, some cells in the cortex with fields close to the midline could still respond to appropriate visual stimulation of the contralateral eye.

The lateralization of cortical function is shown schematically in the figure below. The left hand and left visual field project to the right hemisphere. The right hand and right visual field project to the left hemisphere, in which specialized language function resides.

[104]Sperry, R. W. 1970. *Proc. Res. Assoc. Nerv. Ment. Dis.* 48: 123–138.
[105]Gazzaniga, M. S. 2000. *Brain* 123: 1293–1326.
[106]Gazzaniga, M. S. 1989. *Science* 245: 947–952.
[107]Shatz, C. J. 1977. *J. Comp. Neurol.* 173: 497–518.
[108]Berlucchi, G., and Rizzolatti, G. 1968. *Science* 159: 308–310.

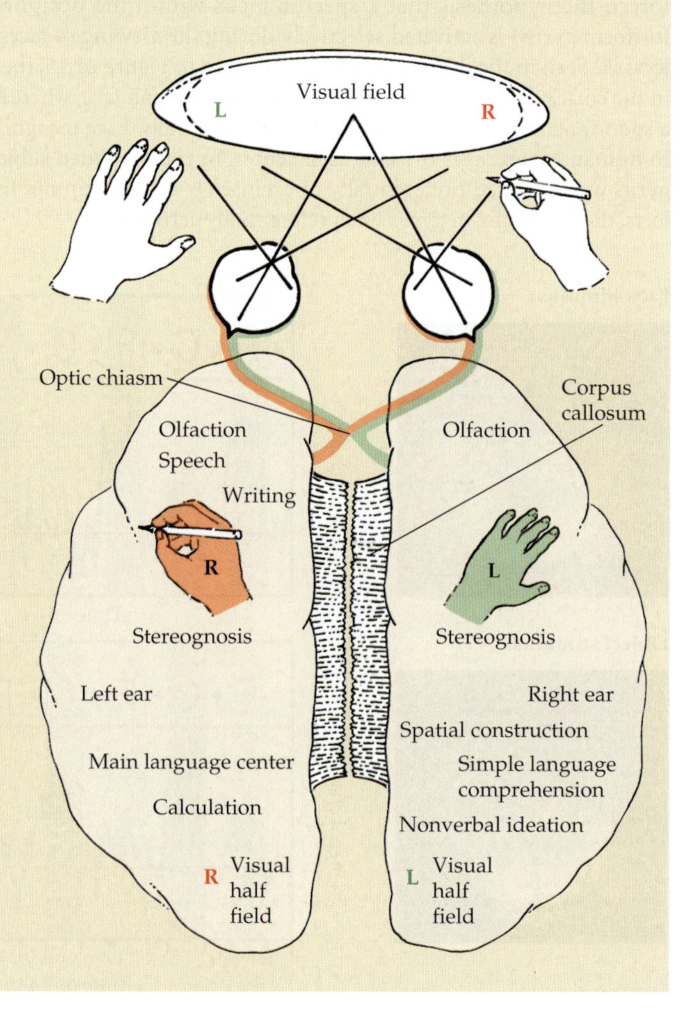

The specialized localization of face recognition is supported by a well-defined clinical condition in which face recognition, but not other visual processing, is specifically damaged, called **prosopagnosia.**[109,110] In one such case, a man with above-average intelligence and good memory could not recognize faces, including that of his own wife.[111] Indeed he reported, "At the club I saw someone strange staring at me, and asked the steward who it was. You'll laugh at me. I'd been looking at myself in a mirror."

The failure of visual recognition can extend to other categories, such as a bird-watcher who no longer could distinguish between species, or a racetrack gambler who lost the ability to identify individual racehorses. On more than one occasion prosopagnosic farmers have complained of losing their ability to identify the individual cows in their herds. Neurological and neuropathological studies have shown that prosopagnosia is associated with lesions in the right, and sometimes bilateral, occipitotemporal cortex.[112] Other kinds of visual stimuli also produce specific patterns of activation in occipitotemporal cortex. For instance, sequences of printed letters (letter strings) cause preferential activation of the inferior occipital sulcus in the left cortex.[113] Correspondingly, brain lesions that specifically interrupt the ability to identify letters ("pure" alexia) involve the left occipitotemporal cortex.[114]

Are specialized areas of visual cortex innate, or are they acquired by experience? Although one might imagine that "face neurons" could arise ontogenetically, a similar basis for the neural coding of letter strings seems unlikely. Rather the cortex must be tuned to important stimuli during the lifetime of the organism. Indeed, the face area of cortex also was activated preferentially when an expert bird-watcher was shown pictures of birds.[115] Do specialized regions of visual cortex develop as a result of repeated practice? The resolution and reproducibility of functional imaging techniques suggests that we might soon be able to observe such changes directly, as has been seen in motor cortex during skill practice (Chapter 22). The development and plasticity of neocortex is discussed further in Chapter 25.

[109]Farah, M. J. 1990. *Visual Agnosia.* MIT Press, Cambridge, MA.

[110]De Renzi, E. 1997. Prosopagnosia. In *Behavioral Neurology and Neuropsychology*. McGraw-Hill, New York, pp. 245–255.

[111]Pallis, C. A. 1955. *J. Neurol. Neurosurg. Psychiatry* 18: 218–224.

[112]Meadows, J. C. 1974. *J. Neurol. Neurosurg. Psychiatry* 37: 489–501.

[113]Puce, A., et al. 1996. *J. Neurosci.* 16: 5205–5215.

[114]Damasio, A. R., and Damasio, H. 1983. *Neurology* 33: 1573–1583.

[115]Gauthier, I., et al. 2000. *Nature Neurosci.* 3: 191–197.

Summary

- Neurons in primary visual cortex are organized according to eye preference and orientation selectivity.
- The layout of ocular dominance slabs and orientation pinwheels can be visualized by imaging activity-dependent optical signals from the brain surface. Iso-orientation contours tend to intersect ocular dominance domains at right angles, and each orientation domain is shared between two ocular dominance columns.
- Magnocellular, parvocellular, and koniocellular pathways form parallel channels from retina to visual cortex. Magnocellular neurons are sensitive to motion and low levels of contrast. Parvocellular neurons signal spatial detail and color. Koniocellular neurons carry color information directly to cytochrome oxidase blobs.
- Cytochrome oxidase blobs occur in the center of each ocular dominance column and represent a site of signal synthesis in V_1.
- Cytochrome oxidase–positive stripes and intervening cortex in V_2 are specifically interconnected with blobs in V_1.
- Visual motion is encoded by neurons in V_5 (area MT) in parietal cortex.
- Area V_4 in occipitotemporal cortex contains a preponderance of color-coded neurons.
- Double-opponent color neurons in visual cortex have properties that could contribute to the perceptual phenomenon of color constancy.
- Integration of receptive fields in cortex is provided by long horizontal axons that interconnect columns of cells with similar response properties.
- Most cortical neurons receive input from corresponding points in the visual field of the two eyes, but some neurons respond to stimuli that fall at different points in the two retinas. Such binocular disparities mediate stereoscopic depth perception in area MT.
- Functional magnetic resonance imaging (fMRI) enables primary and secondary visual areas, as well as still more highly specialized regions, to be mapped in human cortex.

SUGGESTED READING

General Reviews

Callaway, E. M. 1998. Local circuits in primary visual cortex of the macaque monkey. *Annu. Rev. Neurosci.* 21: 47–74.

Casagrande, V. A. 1994. A third parallel visual pathway to primate area V1. *Trends Neurosci.* 17: 305–310.

Courtney, S. M., and Ungerleider, L. G. 1997. What fMRI has taught us about human vision. *Curr. Opin. Neurobiol.* 7: 554–561.

Hubel, D. H. 1988. *Eye, Brain and Vision*. Scientific American Library, New York.

Hubel, D. H., and Wiesel, T. N. 1977. Functional architecture of macaque monkey visual cortex (Ferrier Lecture). *Proc. R. Soc. Lond. B* 198: 1–59.

Komatsu, H. 1998. Mechanisms of central color vision. *Curr. Opin. Neurobiol.* 8: 503–508.

Merigan, W. H., and Maunsell, J. H. R. 1993. How parallel are the primate visual pathways? *Annu. Rev. Neurosci.* 16: 369–402.

Newsome, W. T., and Wurtz, R. H. 1988. Probing visual cortical function with discrete chemical lesions. *Trends Neurosci.* 11: 394–400.

Tootell, R. B., Dale, A. M., Sereno, M. I., and Malach, R. 1996. New images from human visual cortex. *Trends Neurosci.* 19: 481–489.

Zeki, S. 1990. Colour vision and functional specialisation in the visual cortex. *Disc. Neurosci.* 6: 1–64.

Original Papers

Gilbert, C. D., and Wiesel, T. N. 1989. Columnar specificity of intrinsic horizontal and corticocortical connections in cat visual cortetx. *J. Neurosci.* 9: 2432–2442.

Groh, J. M., Born, R. T., and Newsome, W. T. 1997. How is a sensory map read out? Effects of microstimulation in visual area MT on saccades and smooth pursuit eye movements. *J. Neurosci.* 17: 4312–4330.

Hubel, D. H., and Wiesel, T. N. 1962. Receptive fields, binocular interaction and functional architecture in the cat's visual cortex. *J. Physiol.* 160: 106–154.

Hubener, M., Shoham, D., Grinvald, A., and Bonhoeffer, T. 1997. Spatial relationships among three columnar systems in cat area 17. *J. Neurosci.* 17: 9270–9284.

Kanwisher, N., McDermott, J., and Chun, M. M. 1997. The fusiform face area: A module in human extrastriate cortex specialized for face perception. *J. Neurosci.* 17: 4302–4311.

Land, E. H. 1983. Recent advances in retinex theory and some implications for cortical computations: Color vision and the natural image. *Proc. Natl. Acad. Sci. USA* 80: 5163–5169.

Livingstone, M. S., and Hubel, D. H. 1987. Connections between layer 4B of area 17 and the thick cytochrome oxidase stripes of area 18 in the squirrel monkey. *J. Neurosci.* 7: 3371–3377.

Zeki, S., Watson, J. D., Lueck, C. J., Friston, K. J., Kennard, C., and Frackowiack, R. S. 1991. A direct demonstration of functional specialization in human visual cortex. *J. Neurosci.* 11: 641–649.

22 Cellular Mechanisms of Motor Control

AS IN SENSORY SYSTEMS, THE NEURAL ORGANIZATION OF MOTOR CONTROL is hierarchical, with smaller, simpler elements integrated into more complex patterns at higher levels of the nervous system. Sensory input and higher motor commands eventually influence the "final common path," the spinal motoneuron that supplies skeletal muscle. One input to the motoneuron is from muscle spindle stretch (group Ia afferents) receptors. Each Ia afferent fiber diverges to contact all the motoneurons that innervate its muscle of origin. The effect of a single Ia impulse on a motoneuron is weak, but the total synaptic input from this source is strengthened by temporal and spatial summation.

Spinal interneurons mediate reflexes that result in coordinated movements of one or more limbs, showing that complex programs of motor control are present in the spinal cord. Sensory input is important for modulating motor programs in the face of a varying external environment. The spinal motor apparatus has a medial component that controls the musculature of the trunk and proximal limbs, especially in the context of automatic movements such as postural control, walking, and breathing. Lateral motor pathways control the distal limbs and play an important role in more complex voluntary movements.

The motor cortex is somatotopically mapped. Cortical neurons fire in relation to the force of contraction of a muscle or small group of muscles mediating movement about a particular joint. The direction of movement is encoded in the directional selectivity of cortical neurons. Sensorimotor integration is carried out in accessory motor areas such as premotor cortex and parietal association cortex.

The cerebellum serves as an accessory sensorimotor integration area. Cerebellar deficits involve a loss of coordination and balance, but little change in strength or sensation. The basal ganglia form a feedback circuit with the cortex. Diseases that affect the basal ganglia give rise to spontaneous, disruptive motor outputs or a reduction of voluntary movement.

For a leech or a fish to swim, for an owl or a cat to catch a mouse, or for a bear or a child to ride a bicycle, the muscles of the body must be brought into play in rapid succession, contracting and relaxing in harmony. Where in the brain of a higher animal the decisions are made, or how voluntary actions begin, are complex questions with no complete solution in sight. At the same time, the importance of studying the underlying cellular mechanisms is evident. A challenge arises, however, from the intricacy of motor control. More than in sensory systems, our understanding of motor events decreases rapidly as we move from the periphery into the central nervous system. This is a consequence of the fact that there are many pathways converging onto motoneurons from higher centers, and even more onto the higher-order cells. Thus, as we explore in the reverse direction, moving from the periphery into the web of motor activity, we must consider an ever increasing number of divergent pathways or pick a specific path that, by necessity, can provide only limited information.

As in sensory systems, principles of organization have emerged to simplify the task. The first is that motor control is arranged hierarchically. Increasingly complex motor tasks are organized in successively higher centers (Figure 22.1). Thus, at the simplest level, sensory neurons synapse with motoneurons within the spinal cord to mediate simple reflexes, without involvement of higher centers being required. Next, networks of interneurons in spinal cord and brainstem make up central pattern generators to coordinate the interplay of multiple motor groups (such as during locomotion or respiration). Over these reside neurons in motor cortex, cerebellum, and basal ganglia that oversee the activity of the lower levels and create novel motor patterns.

A second important principle is that, with few exceptions, motor output is continually updated and adjusted by feedback. Sensory information is used at every level of the motor axis to help shape and inform movement. Both negative and positive feedback loops through the basal ganglia and cerebellum are essential to the timing and coordination of cortical motor programs.

Finally, parallel streams of motor commands can be delineated, corresponding to the regulation of axial and distal musculature. This is an anatomical distinction of clear functional significance. The musculature of body trunk and proximal limbs serves postural control and rhythmic activities such as locomotion and breathing. Distal musculature, especially that of the hands and fingers in primates (and some other animals, such as rac-

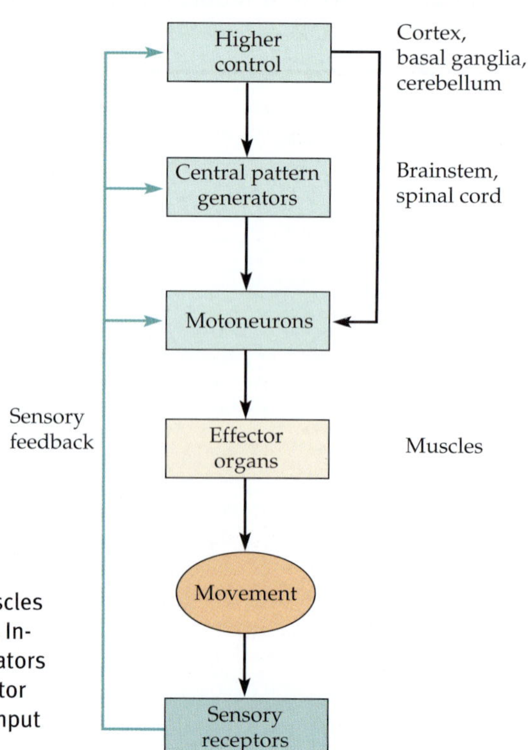

FIGURE 22.1 A General Scheme for Motor System Organization. Limb muscles are controlled by motoneurons and interneurons of the spinal motor apparatus. Interneurons within the spinal cord and brainstem make up central pattern generators that direct the motor apparatus. Motor output is planned and refined by the motor cortex, basal ganglia, and cerebellum. At every level of motor control, sensory input serves to initiate, inform, and modulate output.

coons), is employed in complex learned behaviors. Different brainstem nuclei and descending cortical pathways serve the medial and lateral divisions.

The cellular mechanisms underlying spinal reflexes are relatively well understood, and we begin with a discussion of the properties of spinal motoneurons, and the synaptic organization underlying reflexes. The analysis is then extended by considering how specific groups of neurons within the spinal cord and brainstem produce rhythmic coordinated movements such as locomotion and respiration. Finally, descending influences from motor cortex, cerebellum, and basal ganglia are considered.

THE MOTOR UNIT

Sherrington called the spinal motoneuron the **final common path** because all the neural influences that have to do with movement or posture converge upon it. The major motoneurons of the spinal cord are called α motoneurons (smaller motoneurons that regulate the sensitivity of muscle spindles, called γ motoneurons, will be discussed later). A single α motoneuron innervates a group of muscle fibers, and together the motoneuron and its target fibers form a functional element known as a **motor unit**. The number of muscle fibers in a motor unit ranges from a few—for example, in muscles used to extend or flex the fingers—to several thousand in the large proximal muscles of the limbs.

C. S. Sherrington with one of his pupils (J. C. Eccles) in the mid-1930s.

When a motor neuron discharges, all the muscle fibers to which it is connected contract. The motor unit constitutes the elementary component of normal movement. The smoothness and precision of our movements are brought about by varying the number and timing of motor units brought into play.[1] The actions of individual motor units are not apparent when the whole muscle contracts because the individual contributions are asynchronous and are smoothed out by the elastic properties of the muscles. For example, the 25,000 muscle fibers in the cat soleus are supplied by 100 α motoneurons. Single contractions of the whole muscle can therefore be graded in 100 steps by **recruitment** of motor units. Sustained contractions, involving repetitive activation of motor units, can be tuned even more finely. This is because the contribution of each motor unit is itself graded by the rate at which its motoneuron fires—that is, by the frequency of contraction of its muscle fibers.

Synaptic Inputs to Motoneurons

The recruitment and fine control of motoneurons to produce coordinated movements requires that multiple influences should play upon them in the appropriate sequence and with appropriate balance. It is therefore not surprising that the average motoneuron receives many thousands of synaptic inputs[2] (see Figure 1.13A in Chapter 1) that relay instructions from higher centers and provide information from various sensory modalities in the periphery. Inputs to the cell produce excitatory and inhibitory postsynaptic potentials (EPSPs and IPSPs), and presynaptic inhibition selectively regulates the efficacy of the incoming signals. Each time the motoneuron is sufficiently depolarized, an impulse originates in a particular region of the cell, the **axon hillock**.

Through extensive work, pioneered by Lloyd[3] and by Eccles and his colleagues,[1] much is now known about the mechanisms of synaptic transmission onto this cell and the interaction of excitatory and inhibitory synapses. One important excitatory input arises from the muscle spindles (Chapter 17): The group Ia afferent fibers make monosynaptic excitatory connections on motoneurons. By painstakingly recording from all the motor neurons supplying a particular muscle (its **motor pool**), Mendell and Henneman[4] have shown that each Ia afferent fiber from a muscle sends an input to as many as 300 motoneurons, virtually all those supplying that muscle. In some large muscles, small groups of axons branch off from the motor nerve to innervate fibers in defined subcompartments of the muscle. In this situation, arborizations of group Ia afferents from receptors in a subcompartment make contact preferentially with motoneurons supplying that same compartment.[5]

The arborization pattern of an individual Ia afferent can be seen directly using intracellular injection of the enzyme horseradish peroxidase (HRP). The labeled Ia afferent

[1]Adrian, E. D. 1959. *The Mechanism of Nervous Action*. University of Pennsylvania Press, Philadelphia.

[2]Brannstrom, T. 1993. *J. Comp. Neurol.* 330: 439–454.

[3]Lloyd, D. P. C. 1943. *J. Neurophysiol.* 6: 317–326.

[4]Mendell, L. M., and Henneman, E. 1971. *J. Neurophysiol.* 34: 171–187.

[5]Lucas, S. M., and Binder, M. D. 1984. *J. Neurophysiol.* 51: 50–63.

axon branches extensively along the rostrocaudal axis of the spinal cord to contact members of the motor pool (Figure 22.2). Close examination of HRP-filled afferent fibers allows their contacts with individual motoneurons to be mapped (Figure 22.2B). The convergence of Ia afferents onto motoneurons is anatomically precise, each afferent making two to five contacts on the dendritic tree.[6] All the Ia contacts onto a given motoneuron tend to occur within the same general region of the dendritic tree, although not necessarily at the same electrical distance from the soma.[7] It is remarkable that a single axon collateral provides all these contacts, while other branches of the same axon pass by to supply other motoneurons.

Unitary Synaptic Potentials in Motoneurons

An impulse in a single Ia afferent fiber gives rise to only a very small monosynaptic excitatory potential in a motoneuron, corresponding on average to the release of one or two quanta of transmitter in total from the four to seven synaptic contacts (i.e., not every bouton releases a quantum with every impulse). Quantitative measurements of release were first made by Kuno,[8] who dissected small sensory nerve bundles in dorsal roots and recorded the potentials produced by stimulation of single Ia afferent fibers.

Another way of assessing the effects of individual inputs is to average the potentials evoked in motoneurons by a single muscle receptor.[9,10] The scheme is illustrated in Figure 22.3. Incoming sensory impulses from a single Ia fiber, recorded from a dorsal root filament, are used to trigger an averaging device that sums the potentials recorded from the motoneuron after each incoming event. During the summation process, random changes in potential cancel each other and become lost. On the other hand, the small EPSPs that follow the sensory impulse with a consistent latency become progressively reinforced. Because afferents from muscle spindles discharge typically at the rate of 50 to 400/s, it is easy to obtain the average of many hundreds of responses.

The EPSP evoked by a single Ia afferent has an averaged amplitude, measured at the cell body, of about 200 μV. Single potentials of this magnitude can be expected to have little influence on the firing pattern of a motoneuron, but they can sum during brief bursts of activity and add to other inputs to have a meaningful effect (Figure 22.4). When a single Ia afferent fires at high frequency, the EPSPs overlap, resulting in a buildup of depolarization called **temporal summation** (Figure 22.4A). In addition, sufficient stretch of a muscle like the soleus could result in activation of all 50 of its Ia stretch receptors. Since each Ia afferent diverges to all the members of the motor pool, each motoneuron in turn receives the convergent input from 50 Ia afferents. Thus, excitation arises from the **spatial summation** of all the inputs contacting the motoneuron's dendritic tree (Figure 22.4B).

[6]Brown, A. G., and Fyffe, R. E. W. 1982. *J. Physiol.* 313: 121–140.

[7]Burke, R. E., and Glenn, L. L. 1996. *J. Comp. Neurol.* 372: 465–485.

[8]Kuno, M. 1971. *Physiol. Rev.* 51: 647–678.

[9]Kirkwood, P. A., and Sears, T. A. 1982. *J. Physiol.* 322: 287–314.

[10]Honig, M. C., Collins, W. F., and Mendell, L. 1983. *J. Neurophysiol.* 49: 886–901.

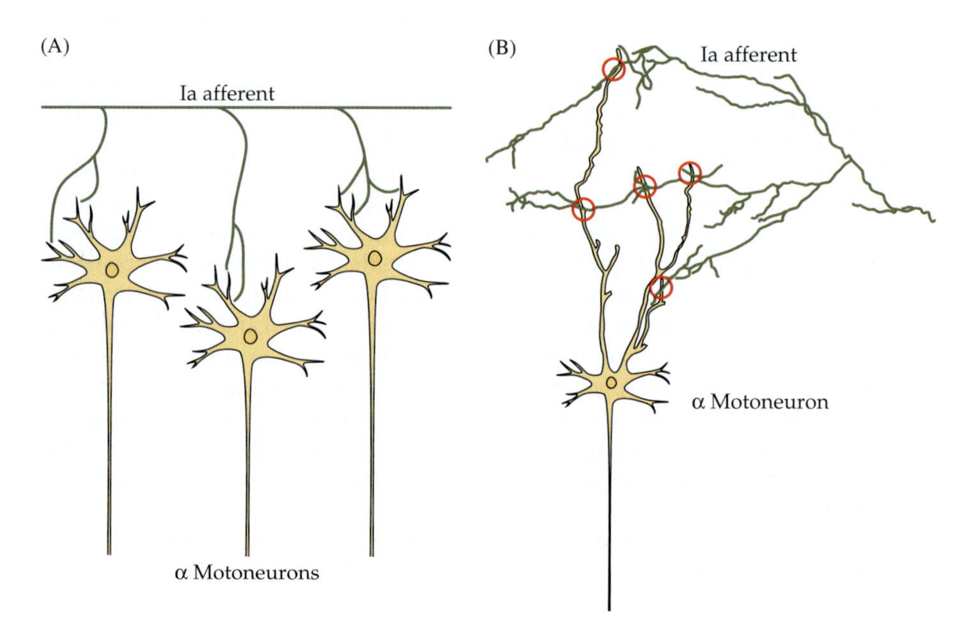

FIGURE 22.2 The Contacts between Stretch Receptor Afferents and Spinal Motoneurons. (A) A single muscle spindle fiber (Ia) afferent sends branches to several motoneurons. (B) More detailed view shows the afferent fiber passing over multiple dendritic branches, indicating possible points of synaptic contact (red circles). Innervation patterns of this kind can be seen experimentally by labeling afferent fibers and motoneurons with histological markers such as horseradish peroxidase (HRP). (After Burke and Glenn, 1996.)

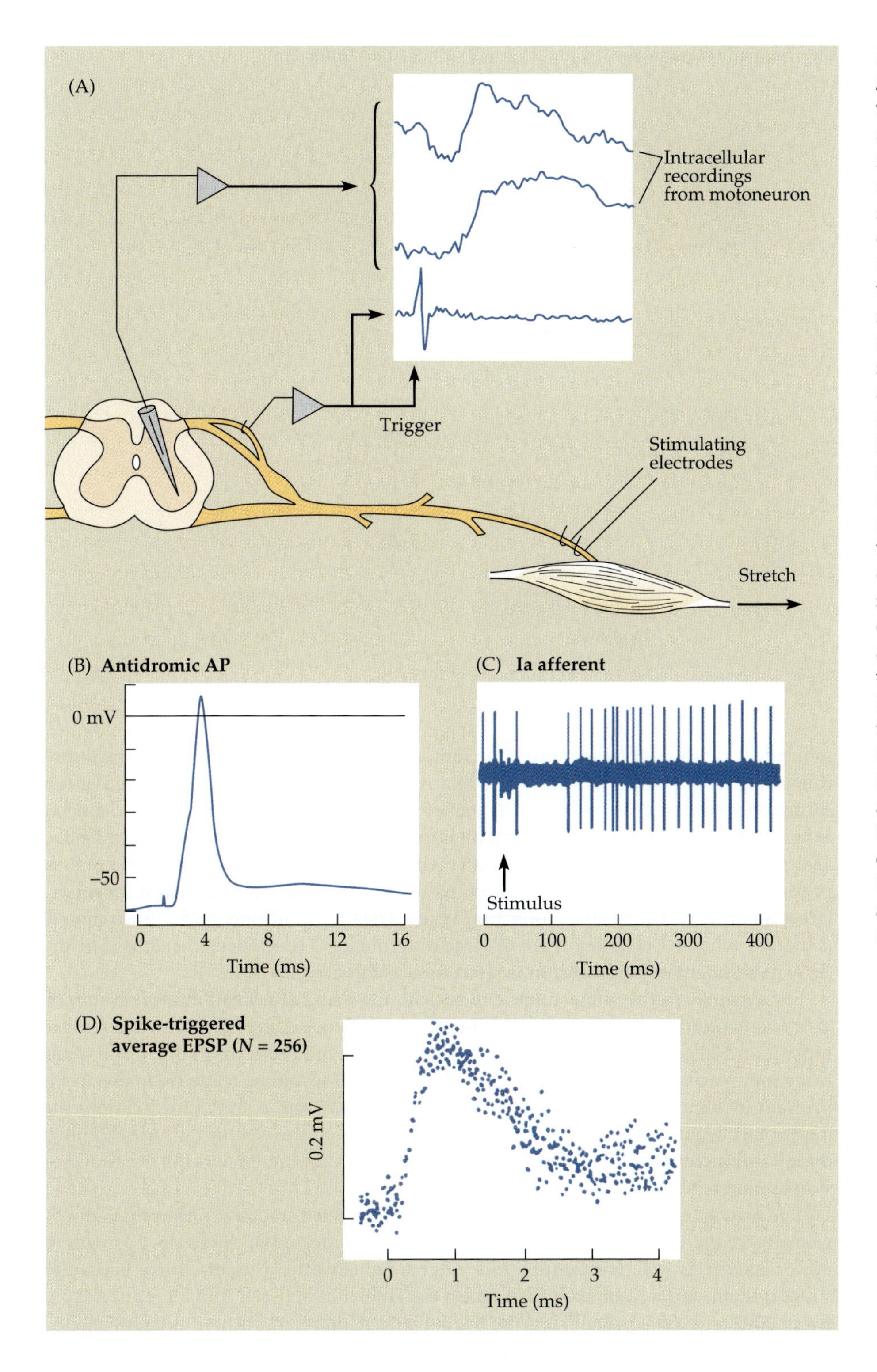

FIGURE 22.3 Spike-Triggered Averaging of Excitatory Postsynaptic Potentials. (A) Scheme for averaging: The impulse from a single stretch receptor recorded in a dorsal root filament is used to trigger the sweep of an oscilloscope. An intracellular microelectrode records the membrane potential of a spinal motoneuron. The oscilloscope records show the incoming sensory impulse (lowest trace) and two successive sweeps of excitatory potentials that are difficult to quantify because of unrelated fluctuations in the baseline. (B) Stimulation of the muscle nerve evokes an antidromic action potential (AP) in the intracellular record, identifying the cell as a motoneuron. (C) Extracellular recording of a single-fiber discharge in the dorsal root filament. Discharge is produced when the muscle is stretched and pauses when a muscle contraction is evoked by stimulation. This identifies the fiber as a Ia afferent from a muscle spindle (Chapter 17). (D) Averaged EPSPs recorded as in (A). When a large number of sweeps are averaged, potentials not time-locked to the sensory impulse cancel out, leaving only the average evoked EPSP. This record shows the digitized average of 256 such events. (After Hilaire, Nicholls, and Sears, 1983.)

Integration of a multitude of excitatory and inhibitory synaptic inputs determines whether or not a motoneuron will reach threshold for the action potential.

The Size Principle and Graded Contractions

How is the activation of motoneurons coordinated to produce smoothly graded movements? As already described, the force of contraction can be increased by recruiting additional motoneurons, and by increasing their rate of firing. However, even further

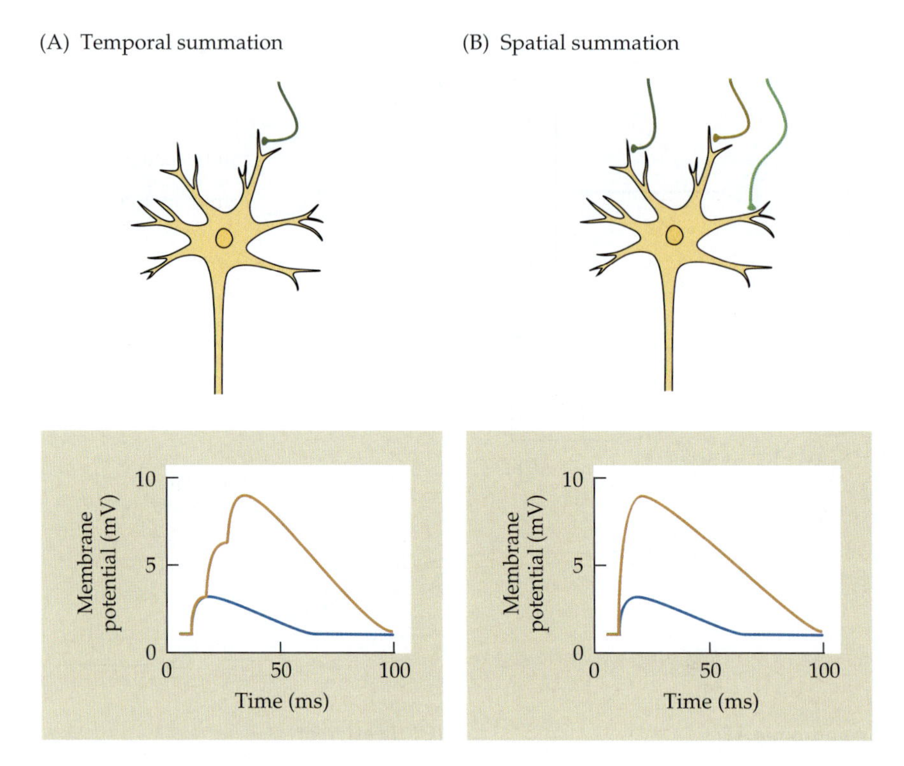

FIGURE 22.4 Temporal and Spatial Summation. (A) An action potential in a single Ia afferent produces a synaptic potential in a motoneuron that is only a fraction of a millivolt. When the afferent fires at a higher frequency, the successive EPSPs ride on the falling phase of the previous one, so they build up to a larger depolarization—temporal summation. (B) A muscle, such as the soleus in a cat, may have as many as 50 muscle spindles, and so an equivalent number of Ia afferent fibers. These all diverge to contact the majority of motoneurons in the motor pool. Thus, 50 Ia afferents converge onto each motoneuron. A strong stretch of the muscle is likely to activate all Ia afferents, so their individual EPSPs add to depolarize the motoneuron by spatial summation.

refinement is provided by recruiting the motoneurons according to their size. Small motoneurons with small axons innervate relatively few muscle fibers, so their activation causes only modest changes in muscle tension. Large motoneurons with large-diameter axons contact many muscle fibers, and an impulse from such a motoneuron gives rise to a large increase in muscle tension. When a contraction occurs, small motor units fire first, producing small increments in tension. As the strength of the contraction increases, larger units are recruited, each contributing progressively more tension.[11] A nice control is thereby achieved, enabling small or large movements to be graded efficiently. The orderly recruitment of motoneurons is referred to as the **size principle**.

For example, in the soleus muscle of the cat, the firing of a small motoneuron may give rise to an increase in tension of about 5 g, whereas a larger unit may contribute more than 100 g, and the maximum contraction brought about by all the motor units firing may reach over 3.5 kg. Plainly, a small motor unit would be relatively ineffective if brought in when the contraction is near its maximum, and a large unit firing in the lower range would perturb fine movements. The fact that the motor units are recruited in order of increasing size means that each additional unit increments the existing tension by about 5%.

The principle that motor unit recruitment adds a fixed fraction, rather than an absolute force increment, to existing muscle tension is reflected in the sensory aspects of motor activity as well. For example, we judge weights by the muscular force needed to support them, and we can easily distinguish the difference between 2 and 3 g, but not between 2002 and 2003 g. Again, it is the *relative* change that is important, as enunciated in the Weber–Fechner relationship (Chapter 17).

How do the cellular properties of motoneurons help to explain the size principle? Suppose that all the motoneurons innervating a muscle received the same synaptic input. The voltage change produced by the synaptic current in each motoneuron depends on input resistance, which is a function of cell size (Figure 22.5). As shown in Chapter 7, the input resistance varies inversely with the cell radius. Thus, any given input will produce a larger voltage change in smaller motoneurons, making them more likely to reach threshold than the large motoneurons. As the strength of sensory input increases, larger and larger motoneurons will be brought to threshold. The size principle will hold for any synaptic input that is uniformly distributed among motoneurons.[12–14]

[11]Henneman, E., Somjen, G., and Carpenter, D. O. 1965. *J. Neurophysiol.* 28: 560–580.

[12]Bawa, P., and Lemon, R. N. 1993. *J. Physiol.* 471: 445–464.

[13]Rossi, A., Zalaffi, A., and Decchi, B. 1996. *Brain Res.* 714: 76–86.

[14]Schmied, A., et al. 1997. *Exp. Brain Res.* 113: 214–229.

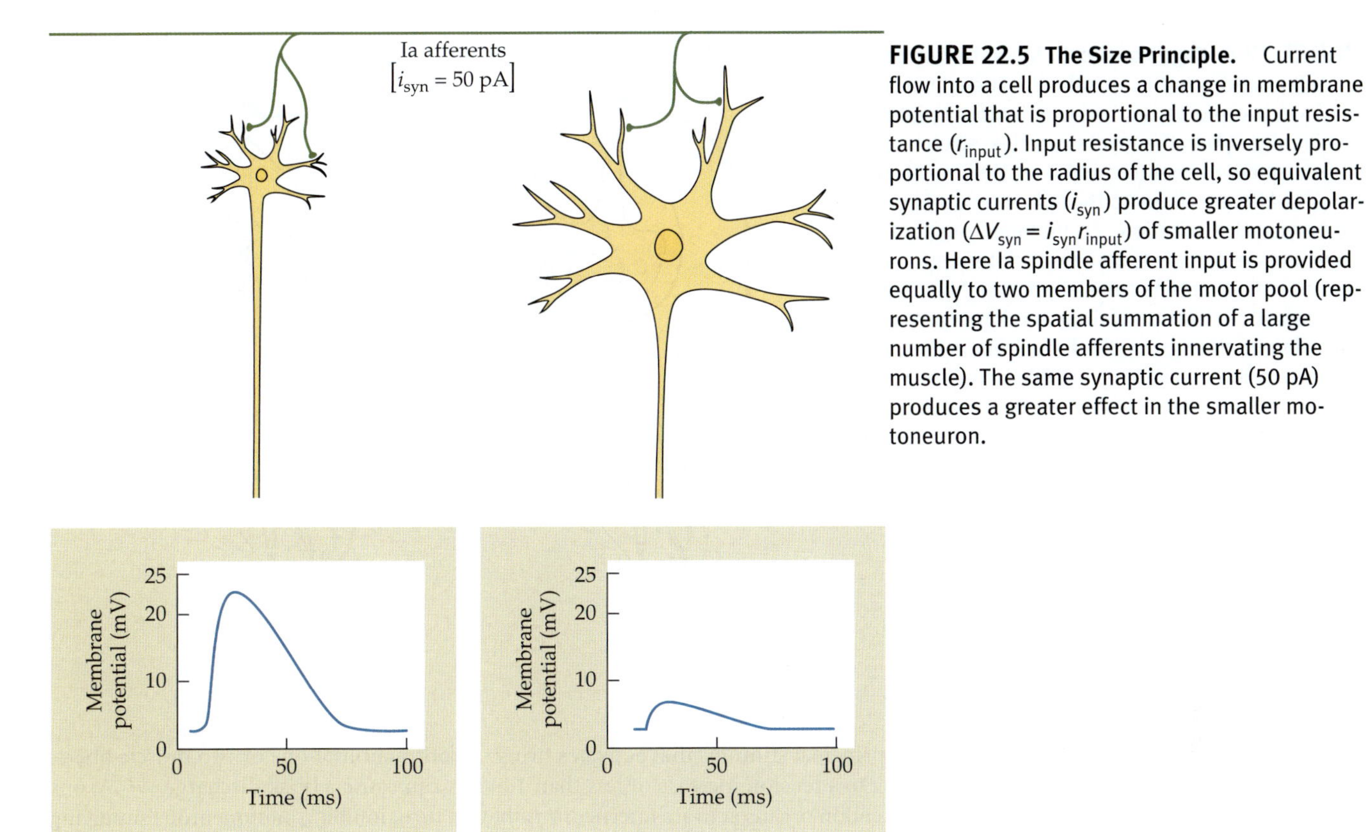

FIGURE 22.5 The Size Principle. Current flow into a cell produces a change in membrane potential that is proportional to the input resistance (r_{input}). Input resistance is inversely proportional to the radius of the cell, so equivalent synaptic currents (i_{syn}) produce greater depolarization ($\Delta V_{syn} = i_{syn} r_{input}$) of smaller motoneurons. Here Ia spindle afferent input is provided equally to two members of the motor pool (representing the spatial summation of a large number of spindle afferents innervating the muscle). The same synaptic current (50 pA) produces a greater effect in the smaller motoneuron.

SPINAL REFLEXES

Reciprocal Innervation

Limb movements are produced by the coordinated contraction of groups of muscles that work together, referred to as **agonists**. At the same time, opposing muscles, called **antagonists** are made to relax. **Extensor** muscles open, or extend, the joints and oppose the force of gravity; **flexor** muscles close, or flex, the joints and pull the limbs toward the body. When a **myotatic reflex** is activated by muscle stretch—for example, by tapping the patellar tendon to produce a knee jerk—the primary sensory endings in the muscle spindles of the extensors are deformed and initiate impulses in group Ia afferent fibers going to the spinal cord. These impulses produce monosynaptic excitation of the α motoneurons going back to the muscle that has been stretched, resulting in a reflex contraction. Extensor contraction is accompanied by simultaneous inhibition of the α motoneurons that innervate antagonistic flexor muscles. This occurs because Ia afferents also activate spinal interneurons that inhibit the antagonist α motoneurons (Figure 22.6A). The principle of one group of muscles being excited while its antagonists are inhibited was first described by Sherrington, who called it **reciprocal innervation**.

For the sake of simplicity, a number of pathways are omitted from Figure 22.6A. For example, discharges from the smaller group II afferents reinforce the reflex largely by way of interneurons, but also monosynaptically.[15] Such intraspinal connections have been worked out in detail by Lundberg, Jankowska, and their colleagues.[16–18]

Inhibitory interneurons are also activated by Golgi tendon organs whose sensitive endings are encapsulated near the tendon–muscle junctions (Figure 22.6B). Their afferent fibers are designated type Ib, to distinguish them from the primary spindle afferent fibers. These stretch receptors are in series with contracting skeletal muscles. They can be made to discharge impulses by passive stretch, but muscle contraction, to which they are more sensitive,

[15]Kirkwood, P. A., and Sears, T. A. 1974. *Nature* 252: 243–244.

[16]Lundberg, A. 1979. *Prog. Brain Res.* 50: 11–28.

[17]Jankowska, E., and Riddell, J. S. 1995. *J. Physiol.* 483: 461–471.

[18]Gladden, M. H., Jankowska, E., and Czarkowska-Bauch, J. 1998. *J. Physiol.* 512: 507–520.

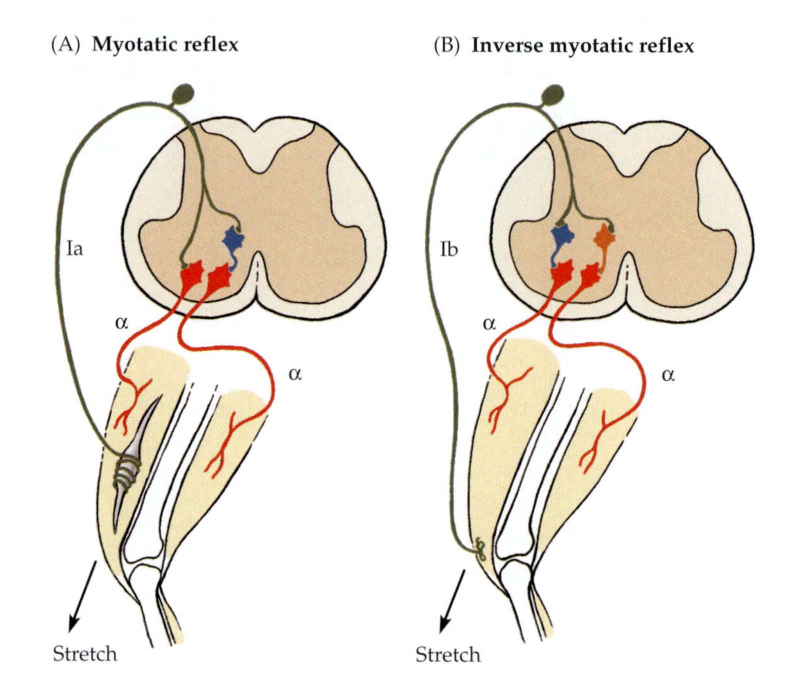

FIGURE 22.6 Organization of Synaptic Connections for reflex actions in the spinal cord. The spinal cord is shown in transverse section, with inhibitory interneurons in blue. (A) In the myotatic reflex, stretch of the muscle spindle generates impulses that travel along group Ia afferent fibers to the spinal cord and produce monosynaptic excitation of α motoneurons to that same muscle. Impulses also excite interneurons that, in turn, inhibit motoneurons supplying the antagonist muscles. (B) Activation of Golgi tendon organs produces impulses in group Ib afferent fibers that, through interneuronal connections, provide inhibition to motoneurons supplying the same muscle and excitation to antagonist motoneurons. (This is sometimes called the inverse myotatic reflex.)

is the principal stimulus that activates firing. A contraction of one or two muscle fibers, leading to a tension increase of less than 100 mg, can cause a brisk discharge.[19,20] Axons from tendon organs activate interneurons that, in turn, inhibit α motoneurons supplying their muscle of origin (see Figure 22.6B).[21] The information that they provide about muscle tension helps to shape the motor pattern[22] and is relayed to higher centers.

[19]Crago, P. E., Houk, J. C., and Rymer, W. Z. 1982. *J. Neurophysiol.* 47: 1069–1083.

[20]Fukami, T. 1982. *J. Neurophysiol.* 47: 810–846.

[21]Matthews, P. B. C. 1972. *Mammalian Muscle Receptors and Their Central Action*. Edward Arnold, London.

[22]Gossard, J. P., et al. 1994. *Exp. Brain Res.* 98: 213–228.

Sensory Information from Muscle Receptors

Golgi tendon organs and muscle spindle afferents produce opposite effects on the motoneurons to their muscle of origin. Muscle spindle activity excites the α motoneuron, and the ten-

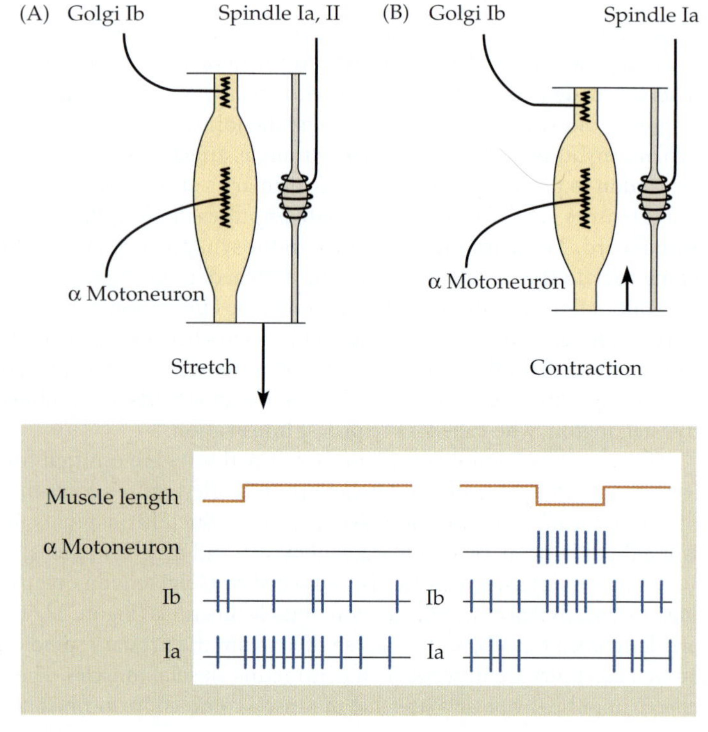

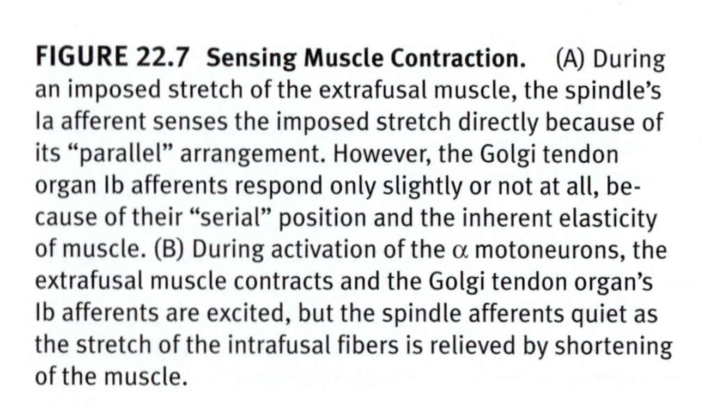

FIGURE 22.7 Sensing Muscle Contraction. (A) During an imposed stretch of the extrafusal muscle, the spindle's Ia afferent senses the imposed stretch directly because of its "parallel" arrangement. However, the Golgi tendon organ Ib afferents respond only slightly or not at all, because of their "serial" position and the inherent elasticity of muscle. (B) During activation of the α motoneurons, the extrafusal muscle contracts and the Golgi tendon organ's Ib afferents are excited, but the spindle afferents quiet as the stretch of the intrafusal fibers is relieved by shortening of the muscle.

don organ inhibits (through an interneuron). The differential arrangements of these two receptor types provides information about limb position and force that has broader significance beyond their reflex actions (Figure 22.7). The tendon organ is in series with the working muscle and is activated principally by contraction; it responds poorly to extrinsic stretch. In contrast, the muscle spindle is functionally in parallel and is turned off by muscle contraction, but responds well when the muscle is moved passively. Thus, the muscle spindle could in principle serve as an indicator of muscle length, whereas the activity of tendon organs relates directly to muscular force. Information from both types of receptors is relayed to higher centers, and the motor system monitors both to tune performance optimally.

Efferent Control of Muscle Spindles

Muscle spindle responses are complicated by the fact that the spindles themselves have their own contractile elements, the intrafusal muscle fibers (Chapter 17). These differ in a number of interesting ways from the extrafusal fibers that make up the working muscle.[23] They contract in response to excitation by **fusimotor fibers** (Figure 22.8A) of the small γ motoneurons in the spinal cord, 2–8 μm in diameter. Fusimotor fibers were first described by Eccles and Sherrington[24] and were studied in more detail by Leksell.[25]

The role of fusimotor fibers was firmly established in a series of technically difficult and definitive experiments by Kuffler, Hunt, and Quilliam.[26] The procedure was to record in the dorsal root the activity in an individual afferent fiber coming from a spindle in an anesthetized cat while stimulating a fusimotor fiber in the ventral root going to the same spindle. Fusimotor stimulation produced an increase in sensory activity, but no increase

[23]Walro, J. M., and Kucera, J. 1999. *Trends Neurosci.* 22: 180–184.

[24]Eccles, J. C., and Sherrington, C. S. 1930. *Proc. R. Soc. Lond. B* 106: 326–357.

[25]Leksell, L. 1945. *Acta Physiol. Scand.* 10(Suppl. 31): 1–84.

[26]Kuffler, S. W., Hunt, C. C., and Quilliam, J. P. 1951. *J. Neurophysiol.* 14: 29–51.

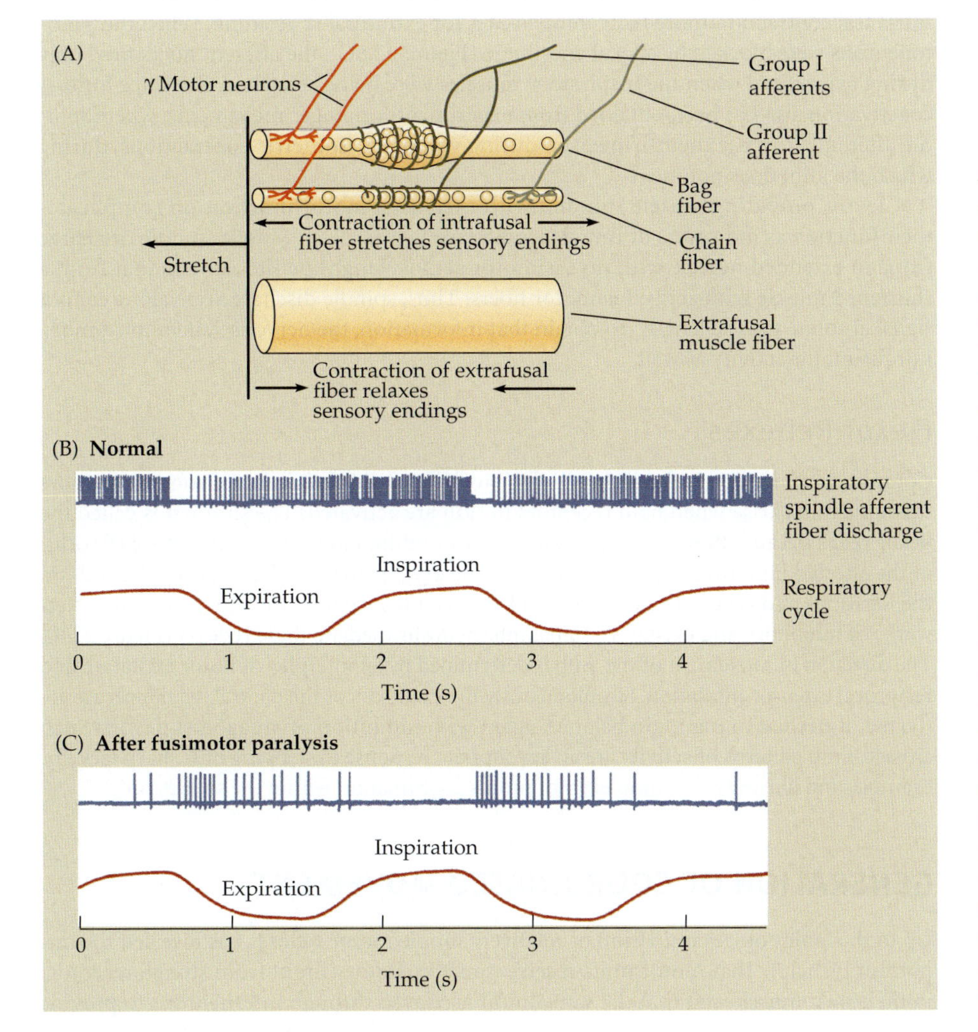

FIGURE 22.8 Efferent Regulation of Muscle Spindles. (A) Intrafusal muscle fibers of mammalian spindles receive innervation by γ efferent fibers that, when activated, cause the intrafusal fibers to contract. This contraction distorts the sensitive endings of the spindle afferents, causing them to fire. The mass of extrafusal muscle fibers is drawn separately; when it contracts, stretch on the intrafusal fibers is reduced. (B) Extracellular recording from a muscle spindle in an inspiratory muscle during the respiratory cycle (lower trace). Normally the sensory discharge from inspiratory muscle spindles is highest during inspiration, even though the muscles are shortening. This is due to simultaneous activation of γ efferents to the spindle. (C) After the fusimotor fibers are blocked selectively by procaine, the spindles behave passively, with increased discharges during expiration, when the muscles are stretched, and cessation of activity on inspiration, when the muscles are shortened. (After Critchlow and von Euler, 1963.)

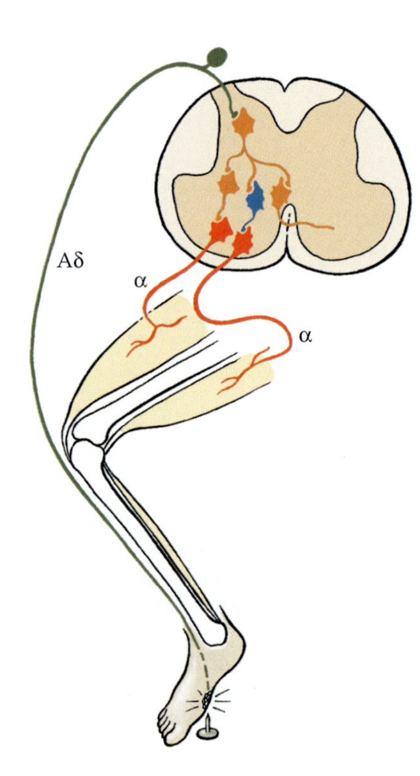

FIGURE 22.9 The Flexor Reflex is a limb-withdrawal reflex, produced in this example by stepping on a tack. Excitation of Aδ pain fibers results in elevation of the thigh (synaptic connections not shown) and flexing of the knee joint by polysynaptic excitation of flexor motoneurons and inhibition of extensors (blue interneurons are inhibitory). Also not shown are contralateral connections that subserve extension of the opposite leg for support.

in muscle tension. Trains of impulses in fusimotor neurons accelerated the sensory discharge if the muscle was stretched, or initiated a sensory discharge in a shortened muscle. The sensory discharge occurs because fusimotor activity causes the intrafusal muscle fibers to contract, thus stretching the sensory endings of group I and II stretch receptors.

The role of the γ motor system in regulating muscle spindle discharge is as follows: Without the γ system the spindle receptors would be unloaded during muscle shortening and, being slack, would be insensitive to further changes in muscle length. When α motoneurons are activated to produce shortening of extrafusal muscle fibers, however, γ motoneurons are also activated to cause contraction of the intrafusal fibers as well. In this way the fusimotor system maintains the tension on the muscle spindle receptors over the full range of limb position, so that their sensitivity is undiminished. The fusimotor innervation of muscle spindles can be thought of as a "gain control" system, continually adjusting sensitivity to maintain dynamic range.

Evidence for such coactivation of α and γ motoneurons was obtained by Sears, von Euler, and others,[27–29] who recorded discharges in spindle afferents to inspiratory muscles during respiration. Figure 22.8B shows that afferent discharges in a spindle from an inspiratory muscle are, in fact, highest during inspiration, when the muscle is contracted and short. This can be explained if α and γ motoneurons are activated together, so that intrafusal contraction more than compensates for extrafusal shortening. When the γ motoneurons were blocked by a local anesthetic (Figure 22.8C), the afferent fibers fired only during expiration, when the inspiratory muscles were being stretched. Evidence for **α–γ coactivation** also has been obtained from experiments on finger movements, where spindle afferents increase their firing even during voluntary isometric contractions, during which the joint does not move.[30]

Clearly, providing muscle spindles with their own motor innervation complicates their function as indicators of muscle length (see Figure 22.8). Muscle spindle discharge from an extended muscle with no fusimotor activity might be the same as that from a shortened muscle with active fusimotor fibers. Thus, spindle discharge rate alone cannot signal absolute muscle length. To obtain that information, the nervous system must monitor fusimotor activity as well.

Flexor Reflexes

Complex combinations of muscle activity, involving multiple joints and sometimes more than one limb, arise when pain receptors in skin are activated. The simplest is called the **flexor reflex** because flexor activation and extensor inhibition leads to withdrawal (flexion) of the affected limb from a painful stimulus. The exact pattern of the response depends on the locations and types of nociceptive and skin tactile receptors that are activated.[31] Networks of inhibitory and excitatory spinal interneurons mediate these reflexes (Figure 22.9). The direction of movement of the limb is determined by an interplay of flexor excitation and reciprocal extensor inhibition. Simultaneously the extensors of the contralateral limb are activated, if needed to maintain balance, again with inhibition of antagonist flexors. Such crossed-limb patterns of activity are an appropriate response to stepping on a sharp tack, for example, and illustrate the complexity of motor coordination within the spinal cord.

GENERATION OF COORDINATED MOVEMENT

Up to this point our examination of relatively simple motor reflexes has revealed the important principle that motor units receive direct excitatory input from stretch receptors in their own muscle and that the same input is relayed through interneurons to provide

[27]Sears, T. A. 1964. *J. Physiol.* 174: 295–315.

[28]Critchlow, V., and von Euler, C. 1963. *J. Physiol.* 168: 820–847.

[29]Greer, J. J., and Stein, R. B. 1990. *J. Physiol.* 422: 245–264.

[30]Vallbo, A. B. 1990. *J. Neurophysiol.* 63: 1307–1313.

[31]Sherrington, C. S. 1910. *J. Physiol.* 40: 28–121.

inhibitory input to antagonist muscles. For repetitive, rhythmic movements such as walking or breathing it seems logical to ask whether the alternating activation of flexor and extensor motor units and alternating movement of the legs make use of similar principles of reciprocal innervation. This is in fact the case, since all such movements involve reciprocal contraction and relaxation of antagonist muscles. However, there is an important distinction: Even though similar (or perhaps even identical) interneuronal circuits might be used for reflex activity on the one hand, and rhythmic activity on the other, sensory feedback is not necessary for movement. Rhythmic patterns of motor activity can occur in the isolated central nervous system of both invertebrates and vertebrates.[32]

Central Pattern Generators

If rhythmic patterns of motor activity are not derived from sensory feedback, then what intrinsic cellular mechanisms generate this output? Two mechanisms contribute to the formation of neuronal pattern generators: (1) rhythmic alterations in excitability within a single neuron, called a **pacemaker cell**, and (2) synaptic interactions between members of a neuronal network. It is likely that these mechanisms work together to generate various patterns of rhythmic activity in the nervous system.

Rhythmic firing in pacemaker cells results from oscillations in membrane potential. Such oscillations can be driven by a number of ionic mechanisms but often involve feedback between calcium entry and calcium-activated potassium channels.[33] Potassium channels are important generally in determining the timing and pattern of excitation in neurons. For example, cell-specific expression of potassium channels is found in pattern-generating neurons of the lobster.[34] Pacemaker neurons have been found in both invertebrates and vertebrates (see Chapter 15) (including some responsible for circadian rhythms in the retina of the snail, with a period of 24 hours![35]). Studies on spinal cord preparations in vitro have revealed rhythm-generating interneurons having the properties necessary for producing locomotor patterns.[36–39]

It is clear that pacemaker cells could provide rhythmic excitation directly to other cells—for example, to motoneurons. Providing excitation that alternates between two different groups of neurons, such as flexors and extensors, requires more circuitry. One way of achieving this is shown in Figure 22.10, in which two identical pacemaker cells (1 and 2) drive the extensor and flexor motoneurons, respectively. They also have inhibitory connections (via interneurons) on one another. Because of this mutual inhibition, neither fires while the other is active, so that their rhythmic activity is kept out of phase. Thus, inhibitory synaptic interactions ensure reciprocity in the overall pattern of activity. Central pattern generation in many cases involves a combination of pacemaker and network mechanisms.[40,41]

[32]Kiehn, O., et al. (eds.). 1998. *Neuronal Mechanisms for Generating Locomotor Activity*. New York Academy of Sciences, New York.

[33]Smith, S. J., and Thompson, S. H. 1987. *J. Physiol.* 382: 425–448.

[34]Baro, D. J., et al. 1997. *J. Neurosci.* 17: 6597–6610.

[35]Michel, S., et al. 1993. *Science* 259: 239–241.

[36]Cazalets, J. R., Borde, M., and Clarac, F. 1996. *J. Neurosci.* 16: 298–306.

[37]Kremer, E., and Lev-Tov, A. 1997. *J. Neurophysiol.* 77: 1155–1170.

[38]Bracci, E., Beato, M., and Nistri, A. 1998. *J. Neurophysiol.* 79: 2643–2652.

[39]Ballerini, L., et al. 1999. *J. Physiol.* 517: 459–475.

[40]Getting, P. A. 1989. *Annu. Rev. Neurosci.* 12: 185–204.

[41]Arshavsky, Y. I., Deliagina, T. G., and Orlovsky, G. N. 1997. *Curr. Opin. Neurobiol.* 7: 781–789.

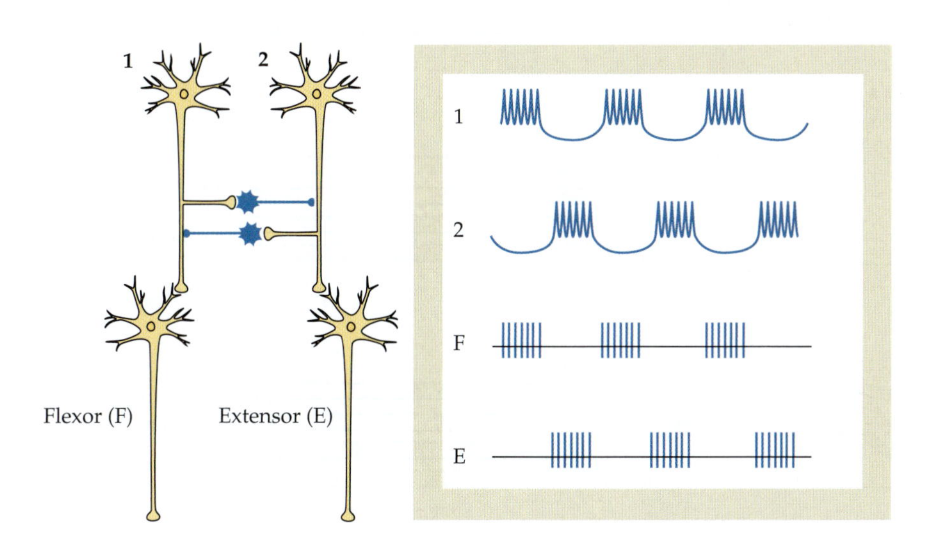

FIGURE 22.10 A Hypothetical Pattern Generator consisting of two pacemaker cells (1 and 2) that excite flexor and extensor motoneuron pools, respectively. The pacemaker cells discharge with a spontaneous rhythm and are coupled by inhibitory interneurons (blue) that ensure that their discharge is out of phase. When pacemaker 1 fires, the flexor motoneuron is activated and pacemaker 2 is suppressed. When pacemaker 1 falls silent, pacemaker 2 can fire, maintaining the silence of pacemaker 1 and activating the extensor motoneuron. The upper two traces represent intracellular records from pacemakers 1 and 2; the lower traces represent extracellular recordings from flexor (F) and extensor (E) motoneurons.

Locomotion

During locomotion each leg executes an elementary stepping movement that consists of two phases:[42] (1) a swing phase, during which the leg, having been extended to the rear, is flexed, raised off the ground, swung forward, and extended again to contact the ground, and (2) a stance phase, during which the leg is in contact with the ground, moving backward in relation to the direction taken by the body. The sequence of movements among the limbs is also stereotyped.

In Figure 22.11, for a walking cat, the left hindlimb is lifted off the ground first, then the left forelimb, the right hindlimb, and the right forelimb. As the cat moves faster, two types of changes occur: (1) The pattern generator for each limb executes a shorter stance phase, consistent with a more forceful push off the ground. The swing phase duration remains relatively constant. (2) Increased speed can also involve a change of "gait," meaning the pattern of coordination among the limbs. The "trot pattern generator" directs the lifting of two legs at once; all four legs might leave the ground together in the fastest gallop. The stereotypy of these patterns suggests that there are **motor programs** for gaits resident within the central nervous system that can be played out upon command. There is substantial evidence that intrinsic motor programs do exist, are genetically predetermined, and appear "spontaneously" during development, independent of experience.[43,44]

As early as 1911, Graham Brown[45] showed that the elementary circuits required for walking movements in cats appeared to reside within the spinal cord. The raising and placing of two hindfeet in alternation could be achieved in a cat after its thoracic spinal cord had been transected (the cat was suspended above the treadmill with a sling). Moreover, other experiments have shown that after certain drugs are given, such as DOPA (a precursor to biogenic amines; Chapter 9), the hindlegs of a spinally transected cat will keep pace as treadmill speed is accelerated.[46]

[42]Pearson, K. 1976. *Sci. Am.* 235(6): 72–86.

[43]Bekoff, A. 1992. *J. Neurobiol.* 23: 1486–1505.

[44]Robinson, S. R., and Smotherman, W. P. 1992. *J. Neurobiol.* 23: 1574–1600.

[45]Brown, T. G. 1911. *Proc. R. Soc. Lond. B* 84: 308–319.

[46]Grillner, S. 1975. *Physiol. Rev.* 55: 247–304.

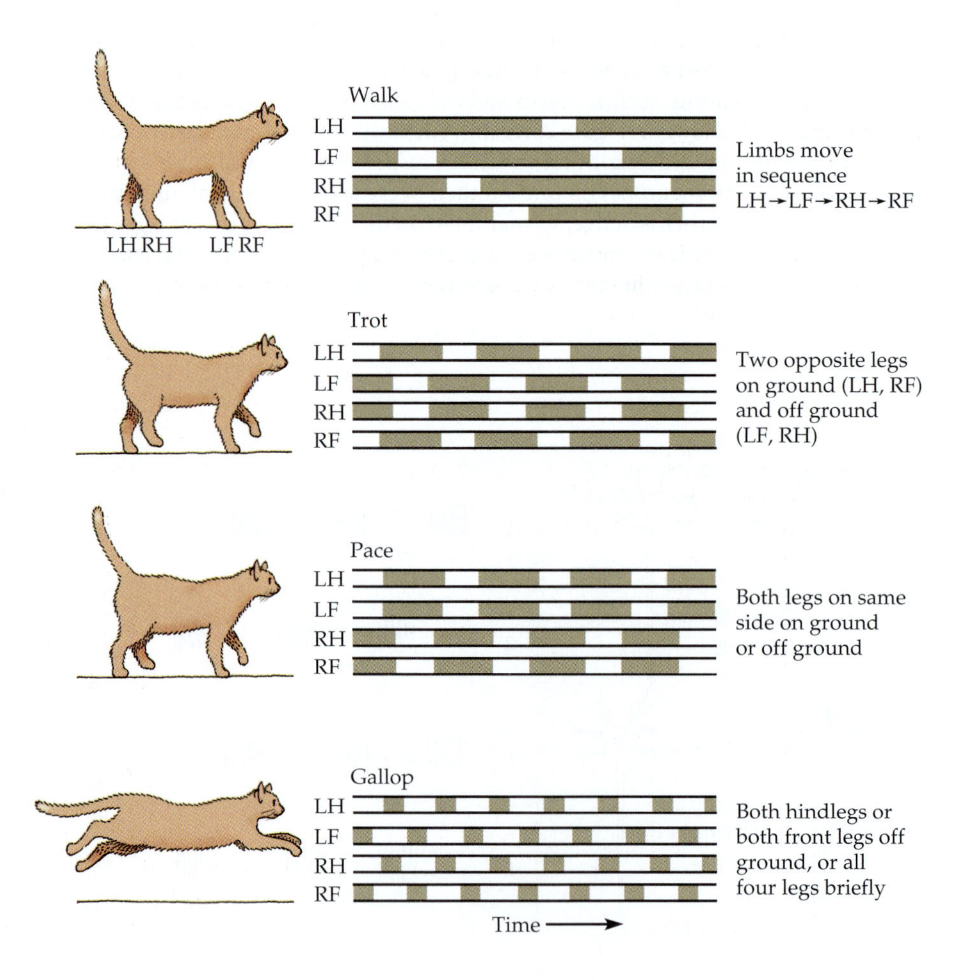

FIGURE 22.11 The Stepping Pattern of a Cat during walk, trot, pace, and gallop, four different gaits of locomotion. The white bars show the time that a foot is off the ground (the swing phase, during which flexor motoneurons are active); the grey bars show the time a foot is on the ground (the stance phase, during which extensor motoneurons are active). During walking, the legs are moved in sequence, first on one side, then on the other. During a trot, a different pattern of interlimb coordination is used: Diagonally opposite limbs are raised together. In a pace, the rhythm changes again, the limbs on the same side being raised together. Faster still is the gallop, in which the hindlimbs and then the front limbs leave the ground. LF = left foreleg, LH = left hindleg, RF = right foreleg, RH = right hindleg. (After Pearson, 1976.)

Perhaps most importantly from a clinical point of view, treatment with noradrenergic agonists and treadmill training have been shown to enhance the ability of cats to execute load-bearing walking movements with their hindlegs after a thoracic spinal transection,[47,48] suggesting that the expression of these intrinsic motor programs can be enhanced with practice. Treadmill training may also induce spinal learning in paraplegic humans.[49] In these studies the limbs move in response to contact with a moving treadmill, and continued practice improves performance. Thus, the spinal pattern generators can be activated by sensory input (the treadmill moves the feet) and strengthened over time by repeated exposure to that input.

The Interaction of Sensory Feedback and Central Motor Programs

As already noted, sensory feedback is not essential for the generation of rhythmic motor activity. Thus, after the nervous system of cockroaches has been disconnected from all sensory input, motoneurons continue to generate impulses in patterns that would normally result in walking.[50] Similarly, in isolated spinal cord segments of the lamprey,[51,52] turtle,[53,54] and rat,[39] maintained rhythmic discharges of motoneurons can be elicited that are appropriate for swimming or walking in the intact animal.

Many movements, rhythmic or not, can be completed reasonably well in the absence of sensory feedback, provided there are no external perturbations. For example, during birdsong the movements of the muscles follow each other in a rapid, orderly sequence without sufficient time for a feedback loop to be completed: The next instructions are sent out from the central nervous system before the first movement or sound can be analyzed by the central nervous system. Both lower primates and humans can perform relatively fine, previously learned movements with deafferented limbs.[55] In the absence of sensory feedback, however, the maintenance of fine control tends to deteriorate as the task progresses, possibly because of the accumulation of small errors. In humans, for example, handwriting becomes illegible. In addition, deafferentation severely impairs the learning of new movements, as does the removal of somatosensory cortex. After unilateral ablation of areas 1, 2, and 3 (primary somatosensory cortex), learning an unfamiliar task (retrieval of small pieces of food by lifting them across a gap) was very much prolonged on the side contralateral to the lesion.[56] A similar lesion made after the task had been learned did not degrade performance.

Thus, the presence of autonomous internal pattern generators does not mean that during behavior of the intact animal, feedback from the periphery is ignored entirely. For example, if a small twig touches the dorsum of the foot of a walking cat during the swing phase, the foot will be lifted elegantly over the twig. It has been shown that Ib afferents from Golgi tendon organs provide important information about extensor loading during locomotion and thereby influence the strength and duration of the stance phase of the normal locomotor cycle.[22,57,58] The γ efferent control of muscle spindles described earlier provides still another example of an avenue for sensory input to modulate or shape the motor pattern. α–γ Coactivation provides a set point for extrafusal muscle length. If an external load prevents the muscle from moving the limb the expected amount, additional excitation will be provided by spindle afferents. The role of sensory feedback is to modulate ongoing motor programs in accord with the organism's needs and in response to challenges imposed by the external world. The interaction of sensory input and motor pattern generators is illustrated nicely in a simple, rhythmic behavior—respiration.

Respiration

From the moment the umbilicus is severed to our last day of life, a reliable flow of oxygen is delivered to the blood and carbon dioxide is carried away by the ceaseless pumping of the lungs. Two antagonistic sets of muscles are responsible for drawing air into the lungs and expelling it. During inspiration, the rib cage is raised by the external intercostal muscles, and the diaphragm contracts. As a result, the volume of the chest is increased, the lungs expand, and air enters. Expiration is achieved by relaxation of the diaphragm and contraction of the internal intercostal muscles. Other muscles of the thorax and abdomen

[47]Chau, C., Barbeau, H., and Rossignol, S. 1998. *J. Neurophysiol.* 79: 2941–2963.

[48]De Leon, R. D., et al. 1998. *J. Neurophysiol.* 79: 1329–1340.

[49]Wernig, A., et al. 1995. *Eur. J. Neurosci.* 7: 823–829.

[50]Pearson, K. G., and Iles, J. F. 1970. *J. Exp. Biol.* 52: 139–165.

[51]Grillner, S., Wallen, P., and Brodin, L. 1991. *Annu. Rev. Neurosci.* 14: 169–199.

[52]Grillner, S., Parker, D., and el Manira, A. 1998. *Ann. N.Y. Acad. Sci.* 860: 1–18.

[53]Guertin, P. A., and Hounsgaard, J. 1998. *Neurosci. Lett.* 245: 5–8.

[54]Stein, P. S., McCullough, M. L., and Currie, S. N. 1998. *Ann. N.Y. Acad. Sci.* 860: 142–154.

[55]Marsden, C. D., Rothwell, J. C., and Day, B. L. 1984. *Trends Neurosci.* 7: 253–257.

[56]Sakamoto, T., Arissan, K., and Asanuma, H. 1989. *Brain Res.* 503: 258–264.

[57]Pearson, K. G., Ramirez, J. M., and Jiang, W. 1992. *Exp. Brain Res.* 90: 557–566.

[58]Pearson, K. G., Misiaszek, J. E., and Fouad, K. 1998. *Ann. N.Y. Acad. Sci.* 860: 203–215.

also contribute to a variable extent, depending on the posture of the animal and the rate and depth of respiration.[59] An example of the respiratory rhythm is shown in Figure 22.12. Activity of each muscle can be registered by strain gauges or by recording its electrical activity with wire electrodes embedded in the body of the muscle—electromyography (EMG). Figure 22.12 shows that inspiratory and expiratory muscle contractions are accompanied by bursts of potentials, indicating motor unit discharges; it is apparent that the two sets of muscles contract out of phase.

During respiration, the stretch reflex contributes to the inspiratory and expiratory movements, maintaining the excitability of the motoneurons as the internal and external intercostal muscles are stretched alternately. Each Ia afferent fiber firing at approximately 100/s contributes excitatory synaptic potentials to the homonymous motor neurons (i.e., motor neurons supplying the same muscle) and, through interneurons, inhibition to the motor neurons of antagonist muscles. Throughout the cycle, the afferent discharge from muscle spindles is maintained at a high frequency, even when the muscles are actively contracting, owing to activity of the fusimotor (γ efferent) fibers[27,28] (see Figure 22.8). Fusimotor activity presets muscle spindles for the length change expected during that half cycle. If an unexpected obstruction were to prevent the expected movement, the shortened muscle spindle would be "stretched," and would fire and increase excitation to the motoneurons.

Figure 22.13 shows the effect on muscle fiber discharge activity produced by lengthening or shortening an inspiratory muscle (the levator costae) by pulling on its tendon. With each inspiration of the animal, the electromyogram shows a burst of spikes. The activity is enhanced when the muscle is lengthened (Figure 22.13A), but the basic pattern is little altered. Although individual spindle afferent EPSPs are small, their combined effect is powerful, since in this preparation the motoneurons cease to drive the muscle when stretch is removed (Figure 22.13B).

[59]Da Silva, K. M. C., et al. 1977. *J. Physiol.* 266: 499–521.

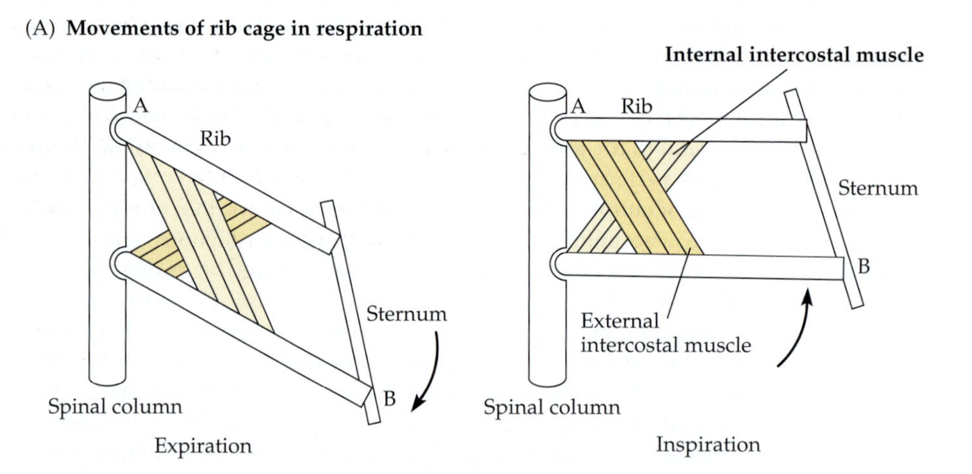

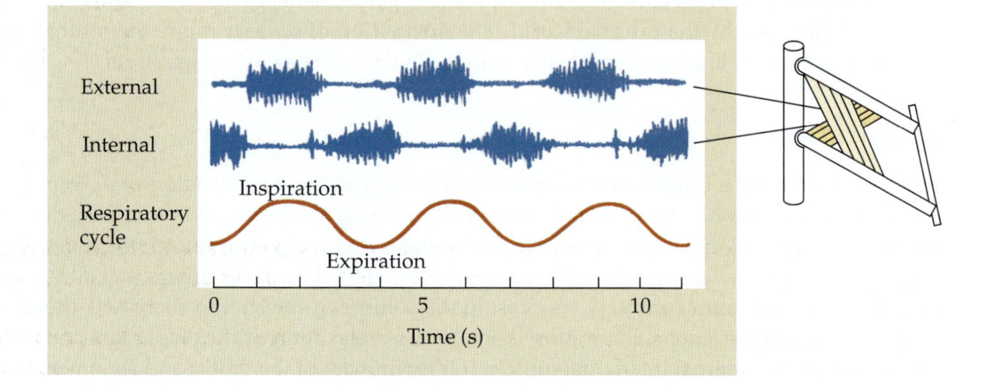

FIGURE 22.12 Movements of Rib Cage and Respiratory Muscles during expiration and inspiration. (A) Actions of the internal intercostal muscles (depressing the ribs during expiration) and external intercostal muscles (raising the ribs during inspiration). (B) Activity of respiratory muscles in the cat, recorded with needle electrodes. Discharges of the external and internal intercostal muscles are out of phase.

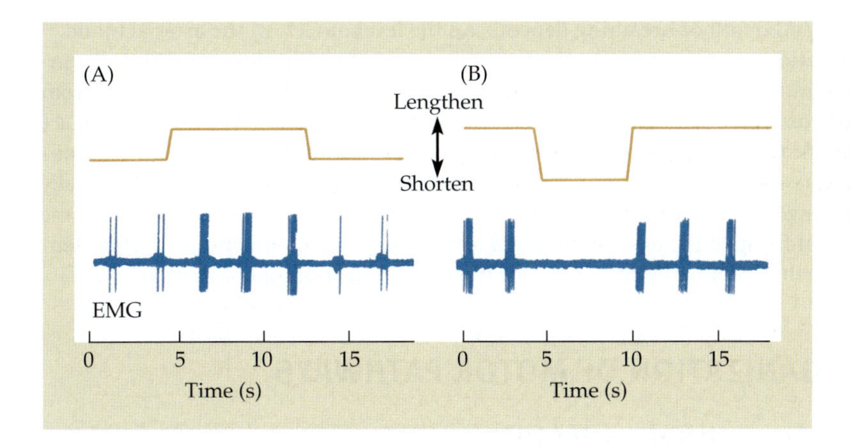

FIGURE 22.13 Stretch Reflex of an Inspiratory Muscle. During each inspiration the electromyogram (EMG) from a small muscle, the levator costae, shows bursts of action potentials, and the muscle contracts. (A) Lengthening of the muscle by pulling on its tendon increases the burst activity by increasing the reflex drive on the motor unit. (B) Shortening of the muscle, which removes the reflex drive, reduces the EMG activity. (From Hilaire, Nicholls, and Sears, 1983.)

How is the respiratory rhythm generated? A transection at the level of the medulla abolishes breathing in cats.[59] Thus, the pattern generator does not lie in the spinal cord. Within the pons and medulla, however, there are pools of neurons that fire during inspiration or expiration and that produce excitation and inhibition of the appropriate respiratory motor neurons. For example, during inspiration the motoneurons supplying external intercostal (inspiratory) muscles are depolarized by a barrage of EPSPs arising from neurons in higher centers of the medulla or pons, causing bursts of action potentials. The inspiratory phase is terminated by a burst of inhibitory potentials to the inspiratory neurons from other central neurons in closely adjacent regions that are associated with expiration.[60,61]

An example of such a rhythmically active brainstem neuron is shown in Figure 22.14A, recorded from the isolated central nervous system of a neonatal opossum.[62] The upper trace in Figure 22.14B is an extracellular record from a single neuron in the medulla, showing short bursts of impulses with a period of about 2 s. In the lower trace, recordings from a thoracic ventral root show corresponding rhythmic discharges of motoneurons supplying respiratory muscles, occurring at the same frequency and with a slight delay. The properties and interconnections of the medullary neurons that generate the respiratory rhythm are not fully understood. Endogenously bursting pacemaker neurons have been found in neonatal rats within a region of the ventral medulla called the pre-Botzinger complex and may generate the respiratory rhythm.[63–65] Reciprocal inhibition between interneurons is also involved.

[60]Sears, T. A. 1964. *J. Physiol.* 175: 404–424.

[61]Hilaire, G. G., Nicholls, J. G., and Sears, T. A. 1983. *J. Physiol.* 342: 527–548.

[62]Nicholls, J. G., et al. 1990. *J. Exp. Biol.* 152: 1–15.

[63]Smith, J. C., et al. 1991. *Science* 254: 726–729.

[64]Rekling, J. C., and Feldman, J. L. 1998. *Annu. Rev. Physiol.* 60: 385–405.

[65]Onimaru, H. 1995. *Neurosci. Res.* 21: 183–190.

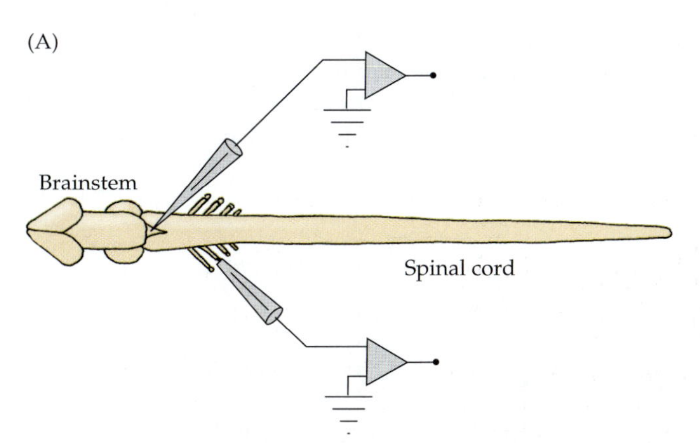

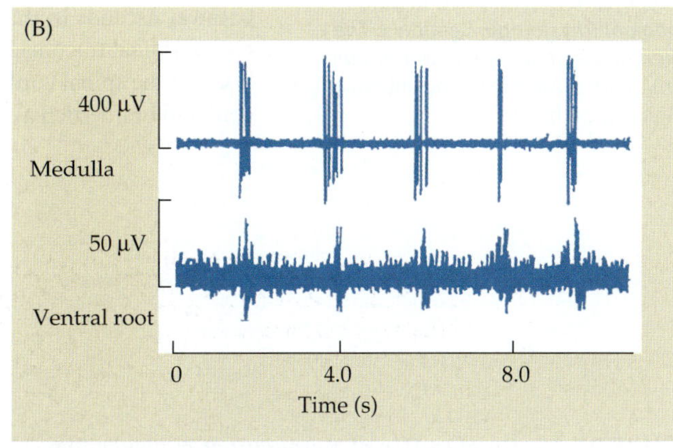

FIGURE 22.14 Respiratory Rhythm Recorded from Brainstem Neurons in an isolated central nervous system from a neonatal opossum (A). (B)Two brainstem cells discharge regularly with an interburst interval of about 2 s. Simultaneous recording from a thoracic ventral root shows the corresponding discharge of respiratory motoneurons. (Kindly provided by D. J. Zou and J. G. Nicholls.)

The rate and depth of breathing depends on the level of CO_2 in the arterial blood.[66] Under conditions of reduced CO_2, the rate and depth of respiration are reduced; conversely, respiration is increased by raised levels of CO_2. This effect depends on inputs from chemoreceptors in the carotid arteries and aorta, and from neurons in the medulla that are sensitive to the CO_2 levels.[67] The firing patterns of individual medullary neurons that control expiratory and inspiratory motoneurons have been shown to be influenced critically by CO_2: Changes in the steady level of CO_2 are translated into pronounced changes in the frequencies of firing of the interneurons and, therefore, of the motoneurons.[68] Thus, sensory input to the brainstem is important for modulating the respiratory rhythm.

THE ORGANIZATION OF MOTOR PATHWAYS

Organization of Spinal Motoneurons

Figure 22.15 shows the orderly arrangement of motoneurons within the cross section of the cervical spinal cord. Extensor motoneurons tend to be ventral to flexor motoneurons. Motoneurons supplying trunk and proximal muscles are medial and ventral. They are concerned mainly with sustained activities such as stance and postural adjustments. Lateral motoneurons innervate distal muscles that tend to be concerned more directly with phasic activities such as manipulation.[69] The activities of medial motoneurons are coordinated by long spinal interneurons that extend over many spinal segments. Short spinal interneurons are restricted to cervical or lumbar enlargements of the spinal cord and coordinate lateral motoneuron pools.

Supraspinal Control of Motoneurons

The major pathways descending onto motoneurons arise in the cerebral cortex and the brainstem (Figure 22.16; see also Appendix C). In accordance with their sites of termination within the ventral horn of the spinal gray matter, these pathways can be classified into two groups: lateral and medial.[70] The two major lateral pathways are the **lateral corticospinal tract**, originating in the cerebral cortex, and the **rubrospinal tract**, originating in the red nucleus of the midbrain. The medial pathways include the **ventral corticospinal tract**, the lateral and medial **vestibulospinal tracts**, the pontine and medullary **reticulospinal tracts**, and the **tectospinal tract**.

For the reader not familiar with the anatomy of the central nervous system, these terms may seem bewildering. Fortunately, some generalizations can be made. A pathway is named according to its origin first, then its termination; in addition, if two or more pathways run in the spinal cord, their individual locations are specified. Thus, the lateral spinothalamic pathway ascends in the lateral portion of the spinal cord to the thalamus; the medullary reticulospinal tract originates in the reticular formation of the medulla and ends at various levels in the spinal cord. Terms indicating direction with respect to main axes of the spinal cord and brain, such as "rostral," "anterior," and "ventral," are defined in Appendix C.

[66]Bainton, C. R., Kirkwood, P. A., and Sears, T. A. 1978. *J. Physiol.* 280: 249–272.

[67]Bruce, E. N., and Cherniak, N. S. 1987. *J. Appl. Physiol.* 62: 389–402.

[68]Eugenin, J., and Nicholls, J. G. 1997. *J. Physiol.* 501: 425–437.

[69]Crosby, E. C., Humphrey, T., and Lauer, E. W. 1966. *Correlative Anatomy of the Nervous System.* Macmillan, New York.

[70]Kuypers, H. G. J. M. 1981. In *Handbook of Physiology*, Section 1: *The Nervous System*, Vol. 2: *Motor Control*, Part 2. American Physiological Society, Bethesda, MD.

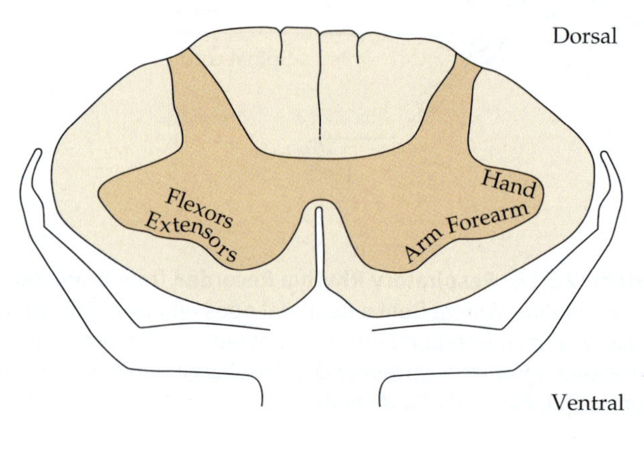

FIGURE 22.15 Organization of Motoneurons supplying the upper extremities, shown in a transverse section of the spinal cord in the cervical region. Muscles of the shoulder and arm are represented most medially, those of the hand most laterally. Extensor motoneurons are located nearest the margin of the gray matter; flexor motoneurons are more central.

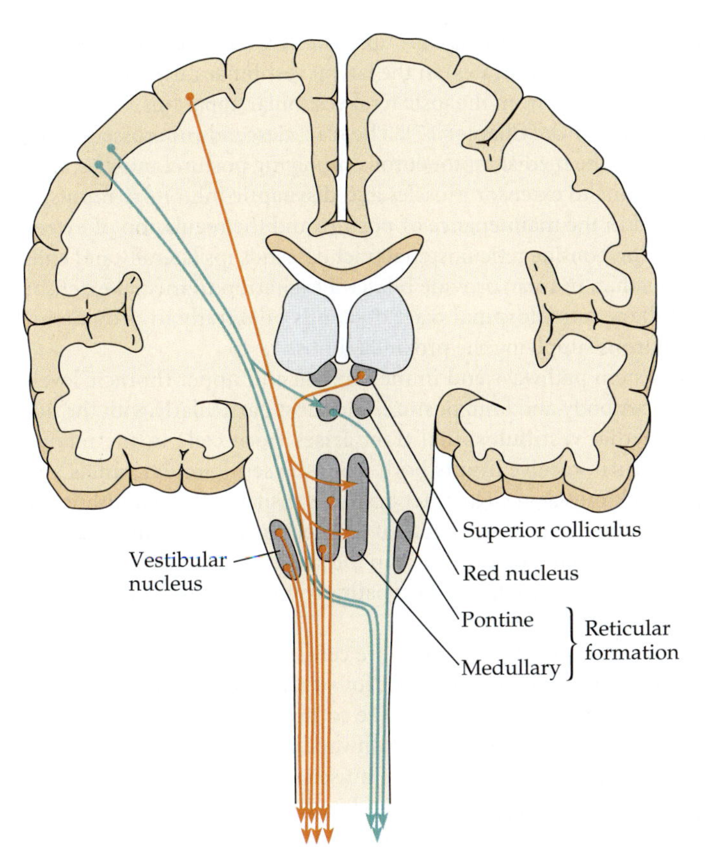

FIGURE 22.16 Major Motor Pathways in the vertebrate central nervous system supplying lateral motoneurons (blue) and medial motoneurons (orange), shown schematically on a coronal section of the cerebral hemispheres, continuing to a longitudinal section of the brainstem and spinal cord. Cells in the primary motor area of the cerebral cortex send axons to the contralateral spinal cord to form the lateral corticospinal tract, with collateral connections to the red nucleus. Axons from cells in the red nucleus cross the midline and descend in the rubrospinal tract. These tracts supply monosynaptic and polysynaptic innervation largely to lateral motoneurons (i.e., those supplying distal musculature). Some cortical fibers descend without crossing to form the ventral corticospinal tract, supplying collaterals to brainstem nuclei. The axial musculature is supplied predominantly by the motor regions of the brainstem through the reticulospinal tracts, originating in the pontine and medullary reticular formations; the vestibulospinal tracts, originating in the vestibular nuclei; and the tectospinal tract, originating in the superior colliculus.

Lateral Motor Pathways

The lateral corticospinal tract originates in the motor and premotor areas of the cerebral cortex in front of the central sulcus (Brodmann's areas 4 and 6; see Appendix C), as well as from a small strip of the postcentral region (area 3) of the cerebral cortex (see Figure 22.17). Fibers pass downward through the internal capsule and cerebral peduncles to the medullary pyramids, after which most cross the midline (decussate) and continue their descent laterally in the spinal cord. Fibers in the lateral corticospinal tracts terminate most prominently on interneurons and motoneurons in the lateral gray matter that control the distal muscles concerned with fine manipulation.[71] An important feature of the tract is that many of the fibers have terminal branches that end directly on motoneurons controlling muscles of the digits.[72,73] In humans and other primates, interruption of the lateral corticospinal tract results primarily in a loss of the ability to move the fingers independently and a consequent deficit in the ability to perform fine, precise tactile movements.[74,75]

The rubrospinal tracts originate in the red nuclei (see Figure 22.16) and cross the midline before descending in the spinal cord to end on interneurons and occasional motoneurons[76] associated with the lateral motor system. Cells in the red nucleus are arranged somatotopically and receive excitatory inputs from the motor cortex and from the cerebellum. The precise functional role of the rubrospinal tract is unclear. It is thought to duplicate many of the functions of the corticospinal tract, and therefore constitutes a parallel pathway from the cortex. In primates, lesions of the tract have little obvious effect, but after interruption of both the rubrospinal and corticospinal tracts, coordinated positioning of the hands and feet is severely impaired.[77,78]

Medial Motor Pathways

Except for a small component from the ventral corticospinal tract, the medial pathways originate primarily in the brainstem (see Figure 22.16) and distribute their inputs to me-

[71]Biber, M. P., Kneisley, L. W., and LaVail, J. H. 1978. *Exp. Neurol.* 59: 492–508.

[72]Cheney, D. P., and Fetz, E. E. 1980. *J. Neurophysiol.* 44: 773–791.

[73]Rouiller, E. M., et al. 1996. *Eur. J. Neurosci.* 8: 1055–1059.

[74]Lawrence, D. G., and Kuypers, H. G. J. M. 1968. *Brain* 91: 1–14.

[75]Hepp-Reymond, M. C. 1988. In *Comparative Primate Biology*, Vol. 4. Liss, New York, pp. 501–624.

[76]Fujito, Y., and Aoki, M. 1995. *Exp. Brain Res.* 105: 181–190.

[77]Kennedy, P. R. 1990. *Trends Neurosci.* 13: 474–479.

[78]Pettersson, L. G., et al. 1997. *Neurosci. Res.* 29: 241–256.

dial motoneurons innervating proximal muscle groups. The cells of origin of the lateral vestibulospinal tract lie (as the name indicates) in the lateral vestibular nucleus. Each lateral vestibular nucleus receives input from the ipsilateral vestibular apparatus, in particular from the utricles of the labyrinth (Chapter 17). The tract descends uncrossed in the spinal cord to provide input to the medial motoneurons supplying postural muscles, with monosynaptic excitatory inputs to extensor muscles and disynaptic inhibitory inputs to flexors. The tract is involved in the maintenance of posture and the regulation of extensor (i.e., antigravity) tone. The pontine reticulospinal tract descends ipsilaterally and ends on segmental interneurons that, in turn, provide bilateral excitation to medial extensor motoneurons. The medullary reticulospinal tract descends bilaterally to provide inhibitory inputs to motoneurons supplying the proximal limbs.

Two other medial brainstem pathways end in the cervical and upper thoracic levels and are concerned with upper body and limb posture, and most particularly with the position of the head. The medial vestibulospinal tract arises from cells in the medial vestibular nucleus that, in turn, receive inputs both from the semicircular canals and from stretch receptors in neck muscles.[79] The tract descends ipsilaterally to midthoracic levels and is concerned with postural adjustments of the neck and upper limbs during angular acceleration. The tectospinal tract originates in the superior colliculus and decussates before descending to upper cervical levels. This pathway mediates orientation of the head and eyes to visual and auditory targets.

In summary, descending pathways arising from the cerebral cortex and the red nucleus supply lateral motoneurons and are important for organized movement of small groups of muscles, especially the distal musculature. The corticospinal tract is particularly important for fine movements of the digits. Motor pathways descending from the brainstem, in contrast, are directed toward large groups of muscles, especially proximal musculature, concerned with the regulation of position and posture of the lower and upper body and the head. These pathways have prominent inputs from the vestibular apparatus.

[79]Kaspar, J., Schor, R. H., and Wilson, V. J. 1988. *J. Neurophysiol.* 60: 1765–1768.

[80]Jones, E. G., and Wise, S. P. 1977. *J. Comp. Neurol.* 175: 391–438.

[81]Porter, L. L., and Sakamoto, K. 1988. *J. Comp. Neurol.* 271: 387–396.

[82]Kaneko, T., Caria, M. A., and Asanuma, H. 1994. *J. Comp. Neurol.* 345: 172–184.

MOTOR CORTEX AND THE EXECUTION OF VOLUNTARY MOVEMENT

The corticospinal tract originates in primary and secondary motor cortices located on the precentral gyrus, and from primary and secondary somatosensory cortices on the postcentral gyrus (Figures 22.17 and 22.18).[80] Cells from which the pathway originates are arranged in an orderly manner to form a somatotopic pattern in primary motor cortex—M_1 (Figure 22.17). This can be demonstrated by electrical stimulation of small regions of cortex to activate muscle. In addition, somatosensory cortex projects somatotopically onto columnar arrays of neurons in M_1.[81,82]

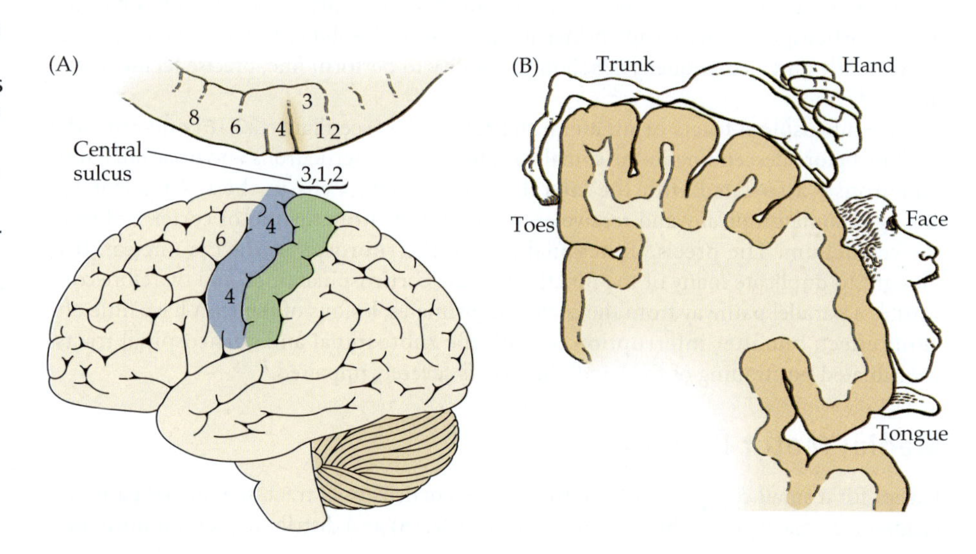

FIGURE 22.17 Motor Representation on the Cerebral Cortex. (A) Lateral view of the surface of the cerebral cortex. Motor activity is associated with activation of cells in area 4 of the cerebral cortex, including the cells of origin of the corticospinal tract. This is the primary motor area M_1. The motor system also includes area 6 (premotor area), extending onto the medial surface of the hemisphere. (B) Coronal section through the cerebral hemisphere anterior to the central sulcus. The musculature of the human body is represented in an orderly, but distorted, fashion, with the leg and foot on the medial surface of the hemisphere, and the head most lateral. The very large area devoted to the hand is indicative of the number of neurons involved in control of manipulations by the digits.

Box 22.1 Extracellular Recording of Motor Activity

The effects of synaptic activation on motoneurons can be determined without intracellular recording. Simple plate electrodes on the skin overlying a muscle can record the compound action potentials produced during contraction of individual motor units. Recall that each motor unit (a collection of muscle fibers) is activated by a single motoneuron. Thus, using a window discriminator to select one particular extracellular compound action potential by its unique amplitude allows one to follow the activity of a single motoneuron in the spinal cord. Such recordings can be combined with electrical stimulation of afferent nerves, or with activation of descending motor pathways, by electrical or magnetic trans-cranial stimulation.[85] Thus, by combining extracellular motor unit recording with trans-cranial stimulation, synaptic responses of a single motoneuron to activation of a particular motor pathway can be observed in an awake human subject.

[85]Rothwell, J. C. 1997. *J. Neurosci. Methods* 74: 113–122.

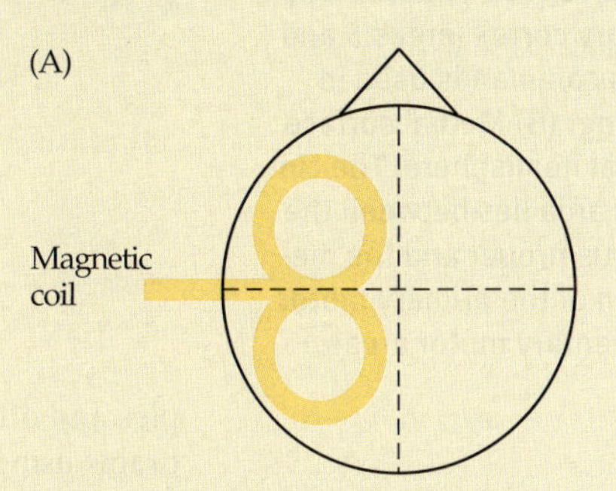

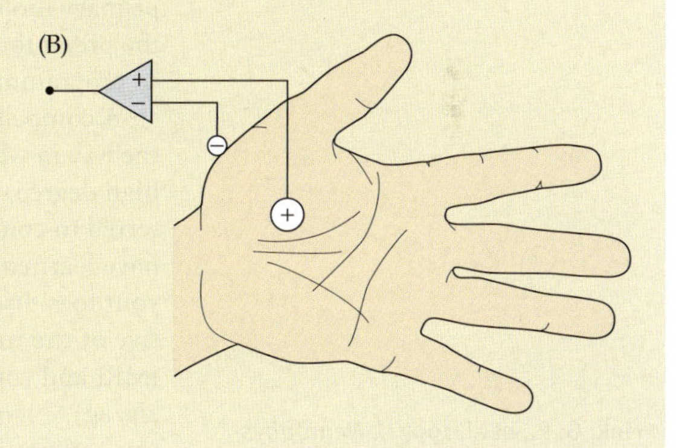

(A) Trans-cranial magnetic stimulation. An electromagnet is used to generate strong magnetic fields on the surface of the skull. These induce electrical fields and excite a subset of neurons within the cortex. The procedure is used with conscious subjects and is not painful; thus, it is possible to activate cortex experimentally in awake humans. (B) Surface electrodes placed over a muscle in the hand record extracellular currents that flow when single motor units are activated. These extracellular compound action potentials vary in amplitude with the size of the motor unit. By using a window discriminator to select a particular size of action potential, it is possible to study the activity of a single motor unit (and so motoneuron) in an awake human.

The somatomotor map was first demonstrated by Fritsch and Hitzig in 1870 by stimulation of the cerebral cortex of animals.[83] The somatotopic representation in humans was first mapped by Penfield and his colleagues during neurosurgery.[84] Localized stimulation of the cortical surface with brief electrical shocks produced movements of restricted regions of the body. As in the somatosensory cortex, the mapping in humans is distorted, with enlarged representation of the face and hands compared to the trunk. Noninvasive recording techniques such as functional magnetic resonance imaging (fMRI) provide similar views of the motor cortex map.[86] Functional MRI and trans-cranial magnetic electrical stimulation (Box 22.1), as well as other techniques, have been used to show that the map of motor cortex is plastic and can be altered following peripheral lesions.[87] Indeed, alterations of the cortical map can even be demonstrated as a consequence of practice to acquire a novel skill.[88–90] It has been suggested that synaptic rearrangements within M_1 constitute one substrate of motor learning.[91]

Association Motor Cortex

The secondary, or association, motor cortex consists of the premotor cortex (Brodmann's area 6), which lies anterior and somewhat lateral to M_1, and the supplemental motor area, also anterior to M_1 (Figure 22.18). Both these areas are somatotopically organized[92,93] (though not as clearly as M_1) and receive input from sensory association cortex (posterior parietal areas 5 and 7). Premotor cortex is strongly influenced by the cerebellum, and the supplementary motor area is connected with the basal ganglia. Movements elicited by electrical stimulation in premotor and supplementary motor cortex are com-

[83]Fritsch, G., and Hitzig, E. 1870. *Arch. Anat. Physiol. Wiss. Med.* 37: 300–332.

[84]Penfield, W., and Rasmussen, T. 1950. *The Cerebral Cortex of Man: A Clinical Study of Localization of Function.* Macmillan, New York.

[86]Rao, S. M., et al. 1995. *Neurology* 45: 919–924.

[87]Sanes, J. N., Suner, S., and Donoghue, J. P. 1990. *Exp. Brain Res.* 79: 479–491.

[88]Karni, A., et al. 1995. *Nature* 377: 155–158.

[89]Nudo, R. J., et al. 1996. *J. Neurosci.* 16: 785–807.

[90]Classen, J., et al. 1998. *J. Neurophysiol.* 79: 1117–1123.

[91]Asanuma, H., and Pavlides, C. 1997. *Neuroreport* 8: 1–4.

[92]Mitz, A. R., and Wise, S. P. 1987. *J. Neurosci.* 7: 1010–1021.

[93]Preuss, T. M., Stepniewska, I., and Kaas, J. H. 1996. *J. Comp. Neurol.* 371: 649–676.

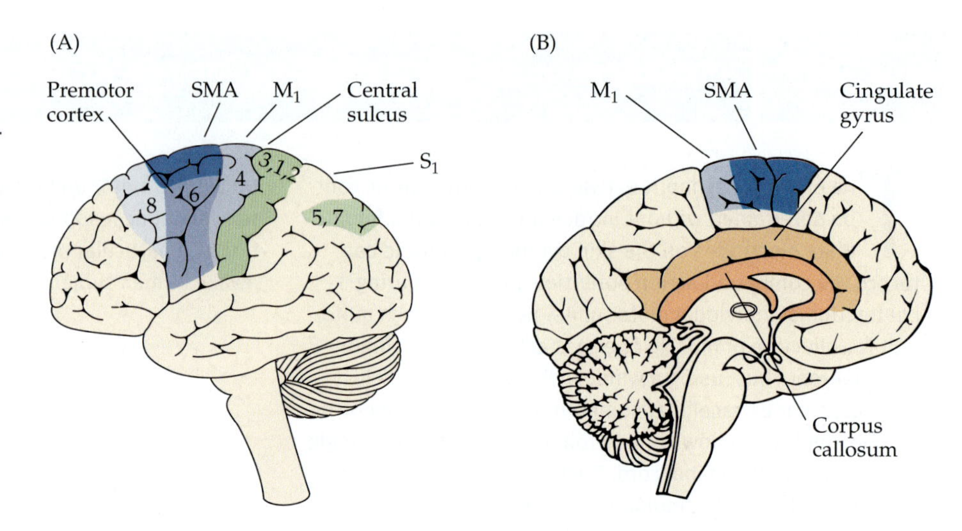

FIGURE 22.18 The Association Motor Cortices. (A) The primary and association motor cortices lie anterior to the central sulcus in Brodmann's areas 4 (primary motor cortex) and 6 (premotor cortex and supplemental motor area, or SMA). Frontal eye fields are found in area 8. Primary somatic cortex, S_1 (areas 3, 1, and 2), and especially association somatosensory cortex (areas 5 and 7), generates commands used in motor planning. (B) Medial surface of the cerebral hemisphere. The cingulate motor area lies between the cingulate gyrus proper and the medial extension of the primary motor and supplementary motor areas.

plex and often bilateral. Motor-related activity is observed in both these areas of human brains using positron emission tomography.[94] Both areas project somatotopically to the primary motor cortex. Together with sensory areas such as the posterior parietal cortex, the premotor and supplemental motor cortex are thought to participate in the planning, or programming, of motor acts.[95–97]

A compelling example of programming by premotor cortex is provided by studies of the way in which one's signature is written. This process is normally accomplished with a high degree of skill by the dominant hand. However, that motor program can be transferred to completely different muscle groups. Thus, Adrian[98] pointed out that once you have learned to write your name, you can do it at once by holding the pencil between your toes! In the motor system must be lodged a scheme for writing one's name irrespective of the muscles to be used. Signature writing by finger or toe was examined using fMRI and compared to patterns of cortical activation for a generic movement pattern (a zigzag).[99] Dorsal and lateral premotor cortex in the *hand* area of the somatotopic map was activated whether the signature was made by finger or toe (Figure 22.19). Thus, the learned motor pattern is stored in association motor cortex of the trained limb but can be accessed to direct movement of a different limb.

An additional area with distinctive motor functions is a region of the cingulate cortex situated under the supplemental motor area.[100] The cingulate cortex lies above the corpus callosum (see Figure 22.18) and is part of the limbic system mediating behavioral motivation.[101] Cellular activity there is correlated with reward-based motor selection; particular motor patterns are initiated when a reward can be expected to follow the appropriate movement.[102]

The Activity of Cortical Neurons

How is the activity of neurons in motor cortex related to the initiation and performance of a movement? Do individual neurons in M_1 direct the activity of a single muscle, the strength of contraction of specific muscle groups, the magnitude of displacement around a joint, or the direction of movement? These kinds of questions were first asked by Evarts,[103,104] who recorded the activity of pyramidal tract cells in the motor cortex during the performance of trained wrist movements by awake monkeys (Figure 22.20). By loading the wrist to oppose either flexion or extension, Evarts could dissociate the force required for a movement from its direction. It was found that the cortical cells tended to be associated with either extension or flexion, and their discharges were related to the force required to execute the movement.[105] This behavior of the cortical cells was not unlike the behavior of the spinal motoneurons to which they projected. Subsequent experiments found that this particular kind of behavior is characteristic of corticospinal cells that end directly on spinal motoneurons.[72] Other classes of cells exhibit more complex behavior, depending on the imposed load or starting position of the limb.[106]

[94]Fink, G. R., et al. 1997. *J. Neurophysiol.* 77: 2164–2174.

[95]Georgopoulos, A. P. 1991. *Annu. Rev. Neurosci.* 14: 361–377.

[96]Donoghue, J. P., and Sanes, J. N. 1994. *J. Clin. Neurophysiol.* 11: 382–396.

[97]Rizzolatti, G., Luppino, G., and Matelli, M. 1998. *Electroencephalogr. Clin. Neurophysiol.* 106: 283–296.

[98]Adrian, E. D. 1946. *The Physical Background of Perception.* Clarendon, Oxford, England.

[99]Rijntjes, M., et al. 1999. *J. Neurosci.* 19: 8043–8048.

[100]Picard, N., and Strick, P. L. 1996. *Cerebral Cortex* 6: 342–353.

[101]Devinsky, O., Morrell, M. J., and Vogt, B. A. 1995. *Brain* 118: 279–306.

[102]Shima, K., and Tanji, J. 1998. *Science* 282: 1335–1338.

[103]Evarts, E. V. 1965. *J. Neurophysiol.* 28: 216–228.

[104]Evarts, E. V. 1966. *J. Neurophysiol.* 29: 1011–1027.

[105]Evarts, E. V. 1968. *J. Neurophysiol.* 31: 14–27.

[106]Humphrey, D. R., and Reed, D. J. 1983. *Adv. Neurol.* 39: 347–372.

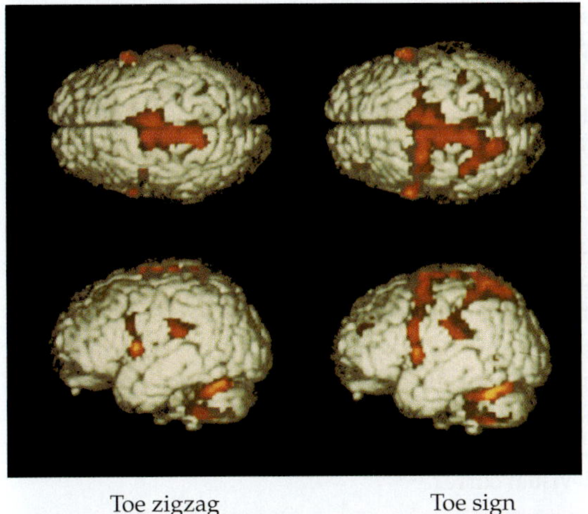

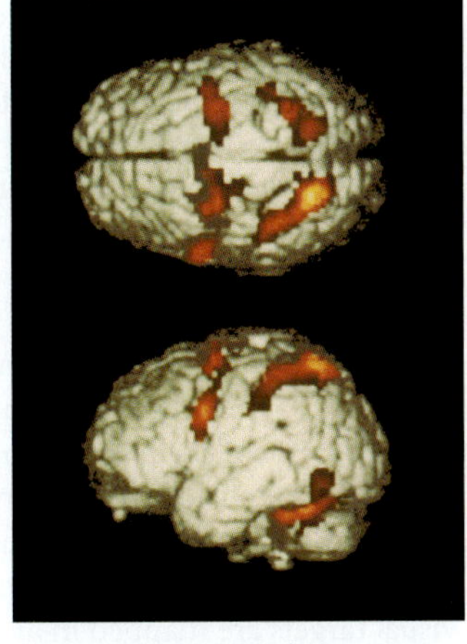

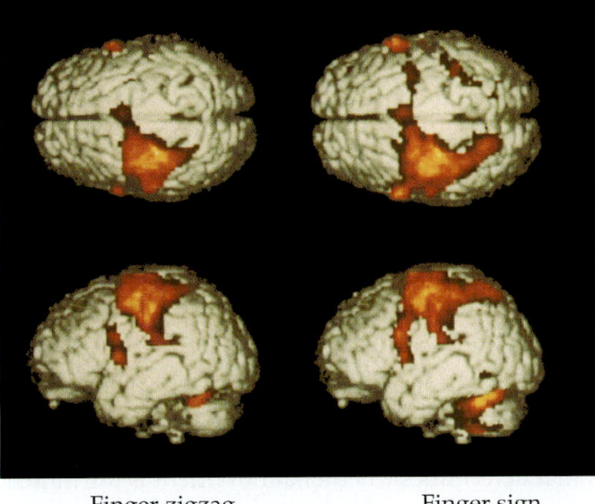

FIGURE 22.19 Functional MRI of Human Motor Cortex activated during signature writing. Each cortical image is color-coded to indicate areas in which activity is higher than at rest. The upper four images show activity when the toe was used to write a zigzag (left pair) or the subject's name (right pair). The bottom four images are arranged similarly for writing with the finger. Subtractive analysis was used to construct the two images on the right, which show areas that were activated during writing of the signature, whether finger or toe was used. The premotor cortex in the region of the hand representation was active, as well as the somatosensory association cortex and a region of the lateral cerebellum. (From Rijntjes et al., 1999.)

Cortical Cell Activity Related to Direction of Arm Movements

Reaching out to grasp a desired object requires an elaborate series of neural computations. Visual processing identifies the object and its spatial location. The position of the target must then be compared with that of the hand, and a trajectory computed to join them. Finally, that spatial trajectory must be converted into coordinated contractions of muscles that will move the hand to that location in space. Which of these processes take place in primary motor cortex? Experiments with visually guided whole-arm movements indicated that some neurons in the arm area of the primary motor cortex discharged at maximum rates when the movement was in a particular direction.[107] Preferred directions were not absolute; the discharges of such neurons fell off as the angle of reach was altered. Further, the preferred direction varied with the initial position or posture of the limb.[108,109]

[107]Schwartz, A. B., Kettner, R. E., and Georgopoulos, A. P. 1988. *J. Neurosci.* 8: 2913–2927.

[108]Caminiti, R., Johnson, P. B., and Urbano, A. 1990. *J. Neurosci.* 10: 2039–2058.

[109]Scott, S. H., and Kalaska, J. F. 1995. *J. Neurophysiol.* 73: 2563–2567.

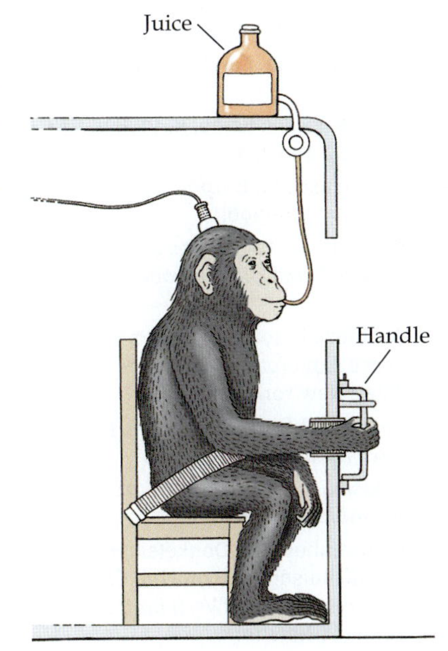

FIGURE 22.20 Experimental Arrangement for Recording Cellular Activity related to wrist movement. A monkey, previously trained to move a handle to a designated position, is seated in the chair with its forearm placed in a cuff. The monkey deflects a handle to the left or right between stops, by flexion or extension of the wrist. A system of weights, or a torque motor (not shown), is used to load the handle to oppose either flexion or extension. For visually guided movements, the handle position is indicated on a display screen. When the monkey places the handle in the designated position, it receives a reward of fruit juice. Single-unit activity is recorded with a microelectrode positioned in an appropriate area of the brain, by means of a microdrive fixed to the skull.

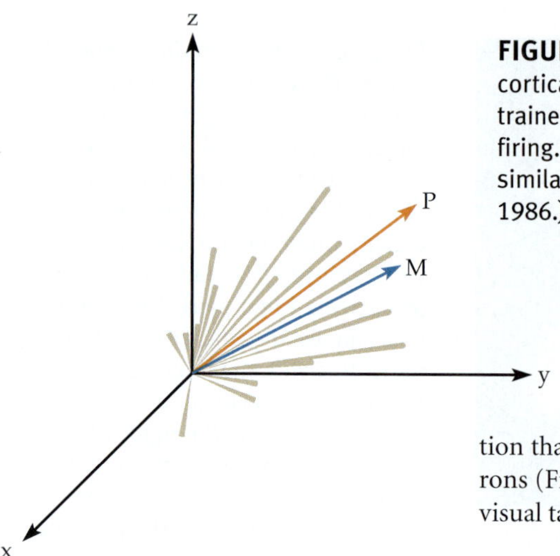

FIGURE 22.21 Encoding Movement in the Motor Cortex. The preferred direction of cortical neurons is shown in three dimensions. The activity of each neuron during a trained limb movement is plotted as a scalar whose length is proportional to its rate of firing. The vector sum for that population of neurons (P) is an arrow whose direction is similar to that of the movement (M). (After Georgopoulos, Schwartz, and Kettner, 1986.)

Neurons with similar preferred directions are clustered together in M_1 and are interconnected with excitatory synapses. Neurons with opposing preferred directions tend to inhibit one another.[110] Georgopoulos and his colleagues[111–113] have suggested that movement trajectory is determined by the activity within an ensemble of neurons. The outcome encodes a direction that is equivalent to the vector sum of the preferred directions of all the active neurons (Figure 22.21). Such vectorial mapping is reminiscent of the way in which moving visual targets are mapped in area MT visual cortex.[114]

Planning a Movement

How do neurons in M_1 locate a target in space? Shen and Alexander[115] trained monkeys to make movements that were aligned either with a visual target, or with a location rotated 90° away from the visual target. In this way they were able to differentiate activity that was location-dependent as opposed to movement-dependent. Some neurons in M_1 fired only during the movement of the limb, as described earlier. However, many neurons fired initially with respect to the location of the target, but eventually according to the resulting movement. This finding suggests that a transformation from spatial location to muscle activity takes place in M_1. Mountcastle, Romo, and colleagues[116,117] used a testing paradigm that required a monkey to categorize the frequency of a tactile stimulus before making a trained movement. They found neurons in M_1 whose activity reflected the category choice ("what kind of stimulus is it?"), independent of the following movement—a neural correlate of decision making. Thus, M_1 not only directs muscular contraction, but also contributes to the neural analysis that determines whether a movement is warranted.

Clearly, in order to move a limb to a point in space, the motor system must be able to plot coordinates on a neural map of space. Such spatial maps are not restricted to motor, or even premotor, cortex. Association areas of parietal cortex bring together many sensory modalities to integrate body, visual, and auditory space.[118,119] Lesions in parietal cortex result in a variety of so-called neglect syndromes in which regions of the body or portions of visual space are ignored.[120] These syndromes can be quite discrete, affecting only one side of the body, or involving only intimate space (within reaching distance) or more distant space.[121] These specific deficits suggest that space is represented in different forms in anatomically distinct regions of the brain.

Motor control is widely represented, involving M_1, premotor cortex, supplementary motor areas, and even somatosensory cortex.[122] As is true for the sensory systems, this multiplicity arises from the parallel (as well as serial) arrangement of motor control. It also reflects the difficulty of segregating the coordination and planning of motor acts from the sensory input that necessarily drives and modulates them. Indeed, the computational demands for motor planning are so great that a second computing center evolved in higher vertebrates where motor performance and sensory feedback are finely tuned—the cerebellum.

THE CEREBELLUM

The cerebellum is an outgrowth of the pons and consists of a three-layered cortex overlying deep nuclei. Its anatomical features are summarized in Appendix C. Comparison of cerebellar structure among vertebrates suggests that it arose during evolution as an elaboration of vestibular nuclei,[123] and direct connections with primary vestibular afferents and other vestibular nuclei remain in all species. However, the role of the cerebellum in brain function extends far beyond the mediation of balance and posture suggested by these vestibular connections. The cerebellum also has widespread influence over the neo-

[110]Georgopoulos, A. P., Taira, M., and Lukashin, A. 1993. *Science* 260: 47–52.

[111]Georgopoulos, A. P., Schwartz, A. B. and Kettner R. E. 1986. *Science* 233: 1416–1419.

[112]Georgopolous, A. P., Kettner, R. E., and Schwartz, A. B. 1988. *J. Neurosci.* 8: 2928–2937.

[113]Georgopoulos, A. P. 1994. *Neuron* 13: 257–268.

[114]Groh, J. H., Born, R. T., and Newsome, W. T. 1997. *J. Neurosci.* 17: 4312–4330.

[115]Shen, L., and Alexander, G. E. 1997. *J. Neurophysiol.* 77: 1171–1194.

[116]Mountcastle, V. B., Atluri, P. P., and Romo, R. 1992. *Cerebral Cortex* 2: 277–294.

[117]Salinas, E., and Romo, R. 1998. *J. Neurosci.* 18: 499–511.

[118]Mountcastle, V. B. 1995. *Cerebral Cortex* 5: 377–390.

[119]Rizzolatti, G., Fogassi, L., and Gallese, V. 1997. *Curr. Opin. Neurobiol.* 7: 562–567.

[120]De Renzi, E. 1982. *Disorders of Space Exploration and Cognition.* Wiley, New York.

[121]Halligan, P. W., and Marshall, J. C. 1991. *Nature* 350: 498–500.

[122]Kalaska, J. F., et al. 1997. *Curr. Opin. Neurobiol.* 7: 849–859.

[123]Nieuwenhuys, R., Donkelaar, H. J., and Nicholson, C. 1998. *The Central Nervous System of Vertebrates.* Springer, New York.

cortex and is implicated in various forms of plasticity[124] and some types of cognitive deficits.[125] Nonetheless, our best understanding of the cerebellum concerns its role in motor coordination and timing. Cerebellar lesions are characterized by a deficit of coordination, with little loss of strength or sensation. The influence of the cerebellum on motor control is mediated by its extensive interconnections with the premotor cortex, as well as the spinal motor apparatus, via brainstem motor nuclei.

Connections of the Cerebellum

Output from the cerebellar cortex is carried by the axons of **Purkinje cells** that make inhibitory synapses onto cells of the deep cerebellar nuclei or vestibular nuclei. The projections are organized in an orderly fashion (Figure 22.22): Those from the flocculus and nodulus (vestibulocerebellum) project directly to the vestibular nuclei.[126,127] The remainder project to the deep nuclei in mediolateral progression. As we move laterally across the cerebellar cortex from the midline, Purkinje cells project first to the fastigial nucleus, then to the interposed nucleus, and finally to the dentate.

Details of the projections from the cerebellar nuclei are shown in Figure 22.22. An important feature is that the dentate and interposed nuclei send their outputs to the motor cortex via the ventrolateral nucleus of the thalamus and thus exert their primary influence on the lateral motor system. The interposed nucleus also projects to the red nucleus. The fastigial nucleus, on the other hand, projects to the vestibular nucleus and the reticular formation, thereby influencing the vestibulospinal and reticulospinal tracts—that is, the medial motor system. Somatotopic order is maintained in each cortical region and carried on through each nuclear projection.

Inputs to the cerebellum (see Figure 22.22) also are segregated. The lateral hemispheres receive input from a wide area of cerebral cortex (via relay nuclei in the pons) and from the red nucleus (via the inferior olive). The flocculonodular lobe receives inputs from the vestibular nucleus. The medial zone of the cerebellar cortex receives proprioceptive and cutaneous input from all levels of the spinal cord. For this reason Sherrington referred to the cerebellum as "the head ganglion of the proprioceptive system." The sensory inputs form multiple somatotopic representations on the cerebellar cortex, overlying the motor representations in the same regions. In summary, the cerebellum receives proprio-

[124]Thach, W. T. 1998. *Neurobiol. Learn. Mem.* 70: 177–188.

[125]Schmahmann, J. D., and Sherman, J. C. 1998. *Brain* 121: 561–579.

[126]Wylie, D. R., et al. 1994. *J. Comp. Neurol.* 349: 448–463.

[127]Tan, J., Epema, A. H., and Voogd, J. 1995. *J. Comp. Neurol.* 356: 51–71.

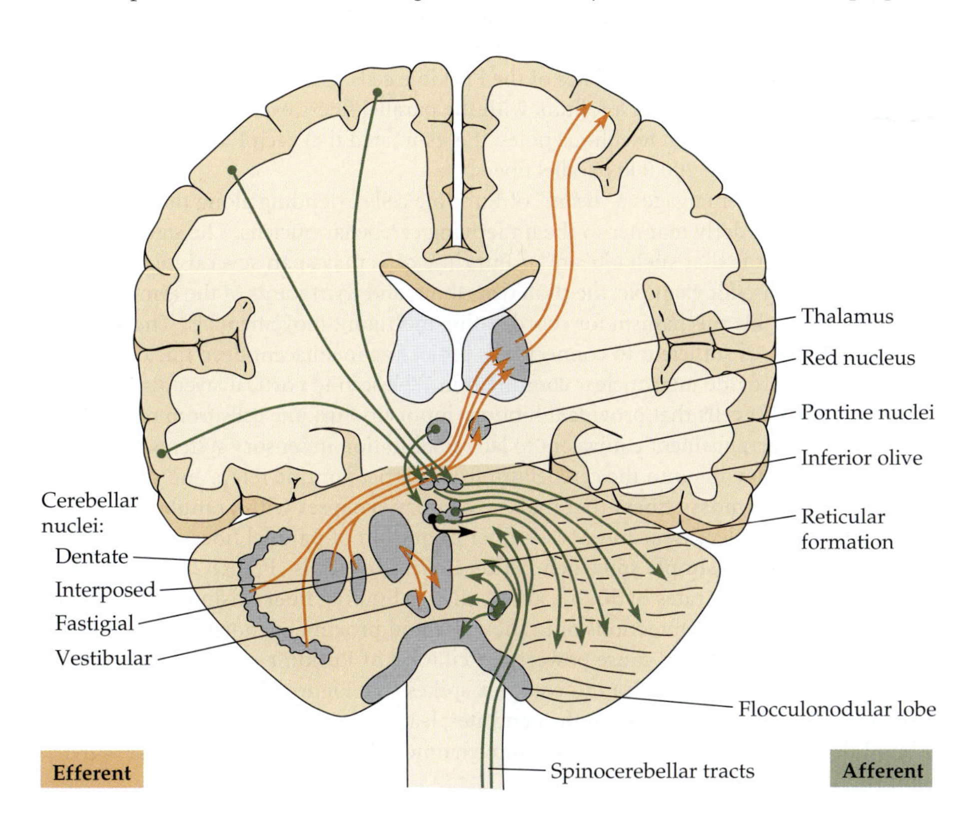

FIGURE 22.22 Efferent and Afferent Pathways in the Cerebellum. Cerebral hemispheres, brainstem, and spinal cord, together with a view of the superior surface of the cerebellum (right) and the underlying cerebellar nuclei (left). Outputs from the cerebellum (left side, orange) are through the dentate, interposed, and fastigial nuclei. Fibers from the dentate nucleus supply the contralateral motor cortex through the ventrolateral and parts of the ventroposterolateral nuclei of the thalamus. The interposed nuclei project to the contralateral red nucleus. Both these pathways, therefore, are associated with the lateral motor system. The fastigial nucleus projects to the vestibular nucleus and the pontine and medullary reticular formations, contributing to the medial motor system. Inputs (right side, green) to the lateral hemispheres of the cerebellum are from wide areas of the cerebral cortex, through the pontine nuclei. Afferent input from the red nucleus is relayed through the inferior olive. More medially, the cerebellum receives extensive input from the spinocerebellar tracts. The flocculonodular lobe is supplied by the vestibular nucleus.

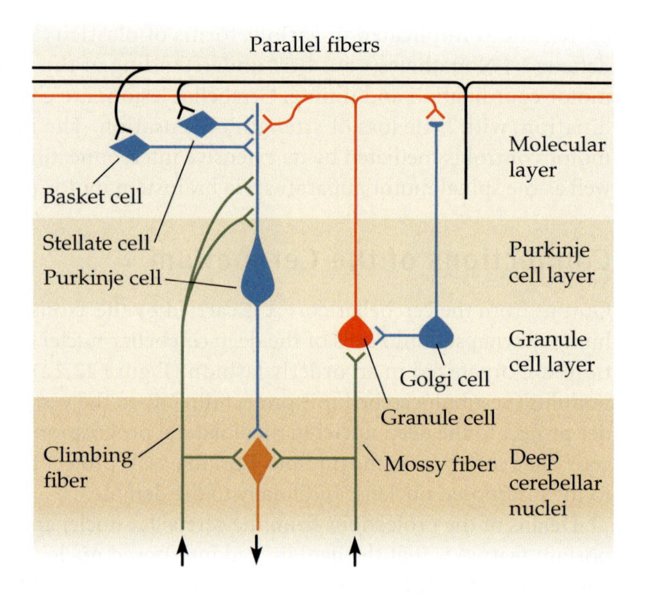

FIGURE 22.23 Synaptic Organization of the Cerebellum. The cerebellar cortex is subdivided into the granule cell, Purkinje cell, and molecular layers. Red and orange cells are excitatory, blue cells inhibitory. The sole output from the cortical layer to cells of the deep cerebellar nuclei is by inhibitory axons from Purkinje cells. The axons of deep nuclear neurons form the output paths of the cerebellum. Mossy-fiber inputs excite granule cells, whose axons ascend to the molecular layer to form a parallel fiber network. Parallel fibers form excitatory synapses on Purkinje cells, stellate cells, basket cells, and dendrites of Golgi cells. Climbing fibers form excitatory synapses on Purkinje cells. Both climbing fibers and mossy fibers make excitatory connections with cells in the deep cerebellar nuclei.

ceptive, vestibular, and other sensory input, as well as a massive projection from motor and association cortex, presumably reflecting actual or intended movement.

Cytoarchitecture of the Cerebellar Cortex

The cerebellar cortex is composed of three layers (Figure 22.23).[128,129] The innermost layer is packed with 10^{10} to 10^{11} **granule cells**—approximating the sum of all other cells in the nervous system! They send axons to the outermost (molecular) layer to form a system of **parallel fibers**, each extending several millimeters along the folium. Also in the granule cell layer are **Golgi cells**, which make inhibitory synapses onto granule cells.

The second cortical layer is occupied by Purkinje cells, whose axons constitute the sole output from the cerebellar cortex. The Purkinje cell dendrites extend into the outer molecular layer of the cortex, with their planar arborizations oriented at right angles to the streams of parallel fibers. The parallel fibers make excitatory synaptic contacts onto spiny processes of the distal dendrites of the Purkinje cells. One can imagine the Purkinje cells as stacked in a row along a folium, with the parallel fibers extending through them, rather like the wires laid on telephone poles. It is estimated that each Purkinje cell receives inputs from more than 200,000 parallel fibers.

Each parallel fiber engages a "beam" of Purkinje cells extending along the folium and projecting in an orderly manner to the underlying cerebellar nucleus. The significance of this arrangement is that such a beam of Purkinje cells may span several joints in a somatotopic region—for example, the shoulder, elbow, and wrist joints of the arm—thereby providing a possible mechanism for coordinating multijoint movements.[129] The length of the parallel fibers is sufficient to connect cells projecting to adjacent deep nuclei, possibly functioning to provide internuclear coordination. The second cortical layer also contains **stellate** and **basket cells** that provide inhibitory inputs to Purkinje cells from remote parallel fibers—an arrangement equivalent to lateral inhibition in sensory systems.

Information flowing into the cerebellum from corticopontine relays and sensory systems is carried by **mossy fibers** that make excitatory synapses with granule cells, Golgi cells, and deep-nuclear neurons. Mossy-fiber excitation of parallel fibers (the axons of granule cells) causes **simple spike** generation in Purkinje cells (Figure 22.24).[130] These occur continuously at rates of 50 to 150/s. A single **climbing fiber** arising in the inferior olive makes extensive connections onto the soma and proximal dendrite of 1 to 10 Purkinje cells. Climbing fibers cause powerful excitation of Purkinje cells,[131,132] producing large plateau potentials that lead to **complex spikes** (see Figure 22.24). This activity involves calcium action potentials in the dendrites, leading to a large calcium influx,[133] and is modulated by voltage-sensitive potassium channels.[134] During complex spike activity, simple spikes are suppressed.[135,136]

[128]Llinás, R. R. 1975. *Sci. Am.* 232(1): 56–71.

[129]Ito, M. 1984. *The Cerebellum and Neural Control.* Raven, New York.

[130]Martinez, F. E., Crill, W. E., and Kennedy, T. T. 1971. *J. Neurophysiol.* 34: 348–356.

[131]Ito, M., and Simpson, J. I. 1971. *Brain Res.* 31: 215–219.

[132]Konnerth, A., Llano, I., and Armstrong, C. M. 1990. *Proc. Natl. Acad. Sci. USA* 87: 2662–2665.

[133]Miyakawa, H., et al. 1992. *J. Neurophysiol.* 68: 1178–1189.

[134]Midtgaard, J., Lasser-Ross, N., and Ross, W. N. 1993. *J. Neurophysiol.* 70: 2455–2469.

[135]Welsh, J. P., and Llinás, R. R. 1997. *Prog. Brain Res.* 114: 449–461.

[136]Lang, E. J., et al. 1999. *J. Neurosci.* 19: 2728–2739.

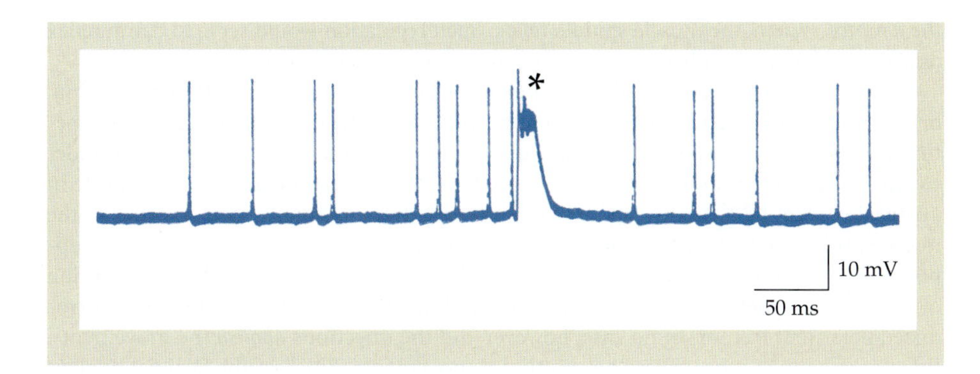

FIGURE 22.24 Intracellular Recording of Simple and Complex Spikes in a cerebellar Purkinje cell from an anesthetized cat. Complex spikes (star) due to climbing-fiber input have pronounced "plateau potentials" and occur at much lower frequencies than do simple spikes caused by mossy fiber–granule cell–parallel fiber input. (After Martinez, Crill, and Kennedy, 1971.)

Cellular Activity in Cerebellar Nuclei

Thach and his colleagues[137,138] examined the relations between trained movements and cellular activity in the deep cerebellar nuclei. Monkeys were trained to perform a sequence of flexion and extension movements with flexor and extensor loading. Varying the load served to dissociate muscle activity from joint position and direction of movement. Cells in the cerebellar nuclei discharged in one of three distinct patterns: some in relation to work against a load (i.e., muscle activity), some in relation to joint position, and some in relation to direction of intended movement.

During trained movements, cells in the dentate nucleus were active prior to those in primary motor cortex, followed by activity in the interposed nuclei and muscular contraction. This sequence is consistent with the idea that information about planned movements is relayed from the association motor cortex to the lateral lobe of the cerebellum, where it is processed and sent back to the primary motor cortex through the dentate nucleus. Signals from the motor cortex are then relayed to the appropriate spinal motoneurons, as well as through the interposed nuclei, where ongoing performance is refined.

The idea that the dentate nucleus plays a role in the initiation of planned movements is supported by other experiments in which cooling of the nucleus slowed the onset of both volitional movement and related cellular activity in the motor cortex.[139] Purkinje cells in the intermediate zone of the cerebellum that project to the interposed nucleus receive proprioceptive input from muscles participating in a movement, and can modulate the output of the interposed nucleus accordingly. During ramp movements, the discharge patterns of cells in the nucleus were found to mimic those of Ia afferent fibers under these conditions. The cerebellar cells discharge vigorously at the beginning of both flexion and extension (Figure 22.25). Further, tremors during ramp movements were reflected in the discharges of both Ia afferents and interposed-nucleus cells. It has been suggested from these and other experiments that one function of the interposed nucleus is to monitor and control, through

[137]Thach, W. T. 1978. *J. Neurophysiol.* 41: 654–676.

[138]Schreiber, M. H., and Thach, W. T., Jr. 1985. *J. Neurophysiol.* 54: 1228–1270.

[139]Meyer-Lohmann, J., Hore, J., and Brooks, V. B. 1977. *J. Neurophysiol.* 40: 1038–1050.

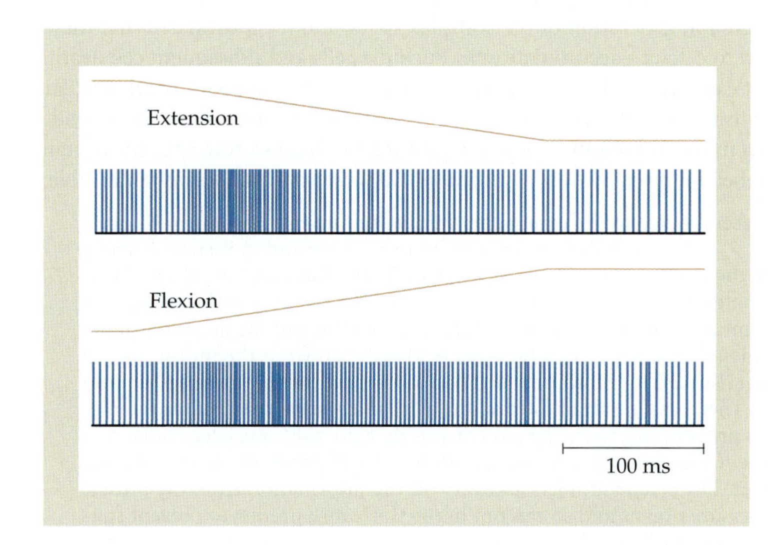

FIGURE 22.25 Discharge Pattern of a Cell in the Interposed Nucleus of the cerebellum during a visually guided ramp movement. The monkey was trained to track a cursor on a visual display with the handle position, by flexing or extending the wrist (see Figure 22.20). The cell is bidirectional, increasing its firing rate on both flexion and extension. Discharge is phasic, rapid near the beginning of the movement and declining gradually in frequency as movement is completed. This neuron provides information on timing, helping to shape the dynamics of wrist movement. (After Thach, 1978.)

the γ motor system, the muscle spindle reflex. Such regulation would serve to damp reflex oscillations that might otherwise occur during slow changes in length.

In summary, it appears that the dentate nucleus may be concerned with the initiation, organization, and execution of planned movements, and the interposed nuclei with fine-tuning of reflexes and movements, particularly with respect to keeping oscillations under control. The afferent and efferent connections of the fastigial nucleus suggest a role in postural control.[140] Experiments in which synaptic transmission in one or other of the nuclei was blocked either temporarily or permanently by local chemical injections support this view. In the dentate nuclei, such injections introduced a delay in the execution of trained movements around a single joint, and block in the interposed nucleus introduced persistent tremor. In no case, however, did the injections *abolish* the movements. Effects were much more severe on multijoint movements. Block of the fastigial nucleus impaired sitting, standing, and walking; block of the interposed nucleus produced severe tremor during reaching; and block of the dentate was followed by overreaching targets and incoordination of compound finger movements.

Deficits in Patients with Cerebellar Damage

The functional organization of the cerebellum is revealed by the patterns of deficits found in patients. For example, a medulloblastoma that arises on the midline of the fourth ventricle leads first to deficits in balance because of pressure on the vestibulocerebellum.[141] In general, cerebellar dysfunction is characterized by a failure of movement coordination. Locomotion is still possible, but the gait is unsteady and the legs are widespread to compensate for poor balance. Problems arise especially when rapidly alternating motions are attempted, and this is related to difficulties of initiation and termination of movement. A classic neurological test is to ask a patient to move his finger rapidly between two points in space (typically from his own nose to the physician's finger). Cerebellar lesions result in a wobbly trajectory and "past pointing" as the target is overshot. An "intention tremor" appears during movement, but is absent at rest. These problems are attributed to a lack of the accurate timing and balance between opposing muscle groups required to complete movements precisely.

Two types of motor diseases in patients (**cerebellar ataxias**)—episodic ataxia and spinocerebellar ataxia (SCA)—have been associated with inherited genetic mutations,[142] often leading to Purkinje cell degeneration. As in Huntington's disease (see the next section), some of the mutations associated with SCA involve so-called triplet repeat expansion of the nucleotides (CAG) coding for the amino acid glutamine. Many of the affected genes are of unknown function, but one (*SCA6*) involves a CAG expansion in the pore-forming region of the calcium channel subunit α_{1A}, which is expressed in cerebellar Purkinje cells.[143,144] Other mutations of the same gene have been shown to be associated with type 2 episodic ataxia, as well as with familial migraine.[145] Type 1 episodic ataxia is associated with mutation of a voltage-gated potassium channel.[146] Both channel types are thought to be involved in generation of the complex spike, and it is postulated that these mutations may lead to calcium accumulation in Purkinje cells and subsequent cell death.

Although recent experiments have provided valuable information about the molecular and cellular mechanisms of cerebellum function, and about the specific roles of the individual cerebellar nuclei in control of movement and posture, the lucid summary of the overall function of the cerebellum given by Adrian more than 50 years ago[98] still seems remarkable:

> In spite of its resemblance to the cerebrum, the cerebellum has nothing to do with mental activity. . . . The cerebellum has the more immediate and quite unconscious task of keeping the body balanced whatever the limbs are doing and of insuring that the limbs do whatever is required of them. Its actions show what complex things can be done by the mechanism of the nervous system in carrying out the decisions of the mind. If I decide to raise my arm, a message is dispatched from the motor area of one cerebral hemisphere to the spinal cord and a duplicate of that message goes to the cerebellum. There, as a result of interactions with other sensory impulses, supplementary orders are sent out to the spinal cord so that the right muscles come in at the exact moment when they are needed, both to raise my arm and keep my body from falling over. The cerebellum has access to all the information from the muscle spindles and pressure organs and so can put in the staff work needed to prevent traffic jams and bad coordination. If it is injured the timing breaks down, muscles come

[140] Thach, W. T., Goodkin, H. G., and Keating, J. G. 1992. *Annu. Rev. Neurosci.* 15: 403–442.

[141] Holmes, G. 1939. *Brain* 62: 1–30.

[142] Klockgether, T., and Evert, B. 1998. *Trends Neurosci.* 21: 413–418.

[143] Zhuchenko, O., et al. 1997. *Nature Genet.* 15: 62–69.

[144] Ludwig, A., Flockerzi, V., and Hofmann, F. 1997. *J. Neurosci.* 17: 1339–1349.

[145] Ophoff, R. A., et al. 1996. *Cell* 87: 543–552.

[146] Browne, D. L., et al. 1994. *Nature Genet.* 8: 136–140.

in too early or too late and with the wrong force. The staff work needs to be elaborate, particularly when the body has to be balanced on two legs and uses its arms for all manner of movement, but it is done by the machinery of the nervous system after the mind has given its orders. The cerebellum has nothing to do with formulating the general plan of the campaign. Its removal would not affect what we feel or think, apart from the fact that we should be aware that our limbs were not under full control and so should have to plan our activities accordingly.

THE BASAL GANGLIA

Beneath the outer cortical layers of the cerebral hemispheres are groups of neurons collected into nuclear masses known collectively as the basal ganglia. These consist of the **caudate nucleus** and the **putamen** (known together as the **neostriatum**), and the external and internal divisions of the **globus pallidus** (Figure 22.26). Two midbrain structures—the **substantia nigra** and the **subthalamic nucleus**—have afferent and efferent connections with the basal ganglia and are part of the system. Dopaminergic neurons in the substantia nigra (Chapter 14) project to the striatum (the nigrostriatal pathway). The basal ganglia receive widespread inputs from the cerebral cortex, particularly from the precentral gyrus. Their major outputs are directed to the ventrolateral and ventroanterior nuclei of the thalamus (overlapping with regions receiving input from the cerebellum) and hence back to the cortex.[147] The basal ganglia provide essential modulation of motor output through this complex feedback circuitry.[148,149]

Functional Circuitry of the Basal Ganglia

The caudate and putamen together function as the input stage of the basal ganglia, receiving glutamatergic excitation from cortical neurons (Figure 22.27). The putamen receives its input from the sensorimotor strip surrounding the central sulcus, so its activity relates most directly to motor activity. The caudate is innervated by frontal cortex and is involved in higher-order cognitive processing. This parallel arrangement underlies the role of the basal ganglia in cognition and affect, as well as motor processing.[150,151]

[147]Hoover, J. E., and Strick, P. L. 1999. *J. Neurosci.* 19: 1446–1463.

[148]Mink, J. W., and Thach, W. T. 1993. *Curr. Opin. Neurobiol.* 3: 950–957.

[149]Graybiel, A. M., et al. 1994. *Science* 265: 1826–1831.

[150]Alexander, G. E., DeLong, M. R., and Strick, P. L. 1986. *Annu. Rev. Neurosci.* 9: 357–381.

[151]Brown, L. L., Schneider, J. S., and Lidsky, T. I. 1997. *Curr. Opin. Neurobiol.* 7: 157–163.

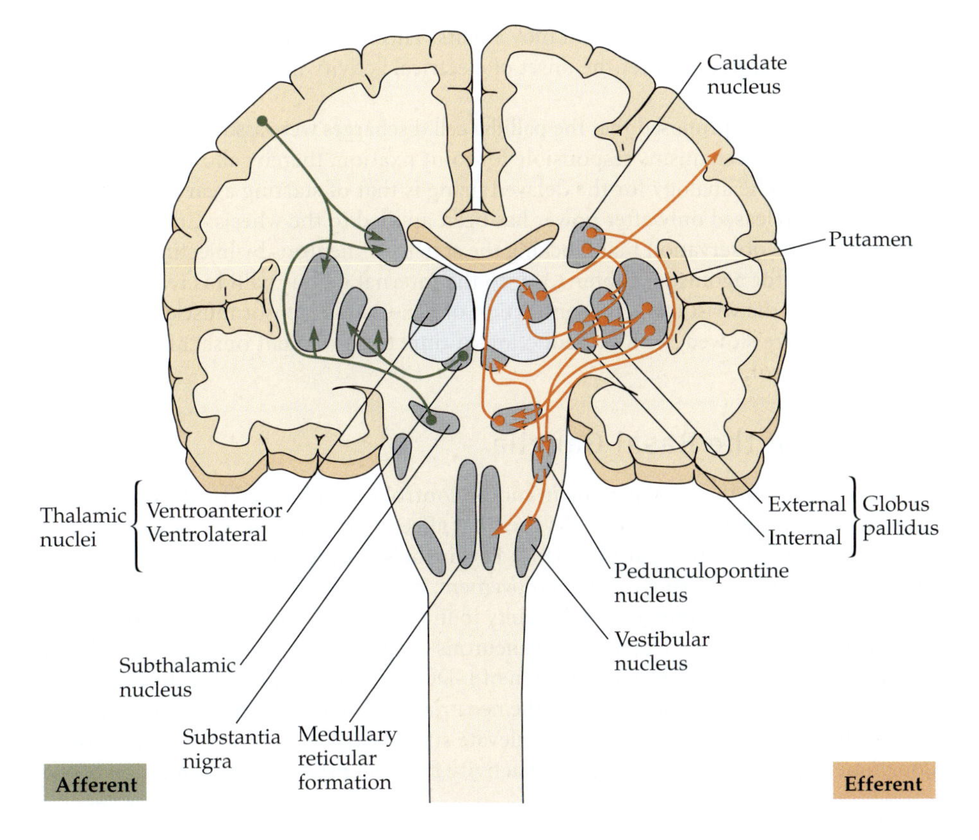

FIGURE 22.26 The Basal Ganglia. Coronal section through the cerebral hemispheres, continuing to a longitudinal section of the brainstem and spinal cord. Basal ganglia include the caudate nucleus, putamen, and globus pallidus (external and internal divisions). Two additional nuclei, the substantia nigra and the subthalamic nucleus, interconnect extensively with the basal ganglia and are sometimes included with them. The predominant source of input to the basal ganglia is the cortex. Outputs from the basal ganglia go to the ventroanterior and ventrolateral nuclei of the thalamus, which in turn project to cortex, completing a cortical feedback circuit. Additional output pathways project to the vestibular nucleus and medullary reticular formation through the pedunculopontine nucleus.

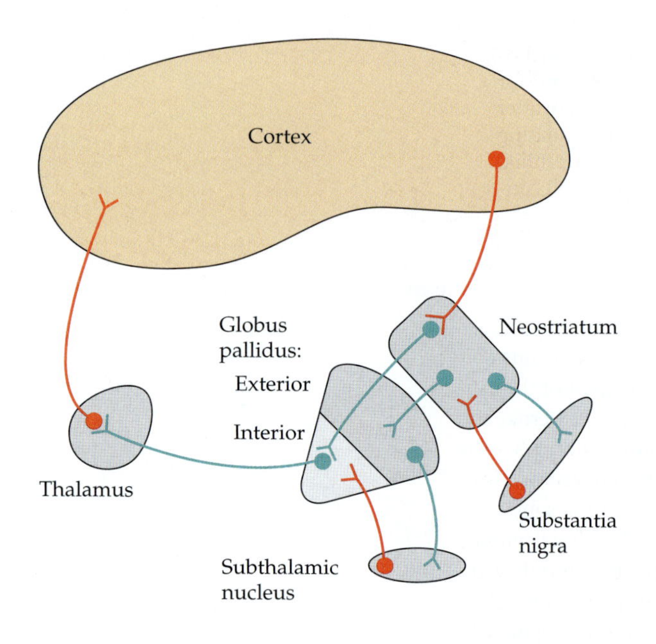

FIGURE 22.27 Functional Circuits of the Basal Ganglia. Glutamatergic neurons in cortex excite GABAergic cells of the neostriatum (caudate and putamen). Striatal neurons project to the external globus pallidus (the indirect pathway) and internal globus pallidus (the direct pathway) to inhibit GABAergic neurons in those nuclei. GABAergic neurons of the internal globus pallidus project to and inhibit the thalamus. Dopaminergic neurons of the substantia nigra produce net excitation of the striatum. Glutamatergic neurons of the subthalamic nucleus are inhibited by the projection from the external globus pallidus and excite GABAergic neurons of the internal globus pallidus. Excitatory neurons are shown in red, inhibitory neurons in blue.

GABAergic neurons of the caudate and putamen project to the globus pallidus and inhibit its activity. Neurons of the globus pallidus are also inhibitory, releasing GABA onto thalamic neurons in the ventroanterior and ventrolateral nuclei. Neurons in the globus pallidus fire continuously at approximately 50 action potentials per second, thereby continuously inhibiting the flow of excitation from thalamus to cortex. Thus, increased cortical activity excites the caudate nucleus and putamen, inhibiting the globus pallidus, and resulting in *disinhibition* of the thalamus. The basal ganglia inhibit the thalamocortical pathway until instructed otherwise.

Cellular Activity in Basal Ganglia

Mink and Thach[152,153] have examined the behavior of cells in the globus pallidus of monkeys during a variety of visually guided movements. Movements consisted of wrist flexion or extension against a fixed load, often involving a single muscle group. Monkeys were required to keep a light spot in register with a moving cursor on a screen. Cell discharge rates were relatively unaffected by initial wrist position, movement velocity, or load, but they increased or decreased in relation to movement. Cells in globus pallidus changed their activity during sudden movements. This occurred well after activity in the dentate nucleus and often after the onset of electrical activity in the muscles driving the movement.

Mink and Thach proposed that the pallidal cell discharges were associated with the release of holding mechanisms responsible for joint fixation, thereby allowing the movement to occur. The analogy for the delayed firing is that of starting a car on a hill: The hand brake is released only after power has been applied to the wheels. Consistent with this idea was the observation that blocking the activity of neurons by injecting muscimol (a GABA receptor agonist; Chapter 13) into the internal globus pallidus resulted in increased tone in the wrist due to cocontraction of flexor and extensor muscles.[154] Trained movements were slowed, with no reduction in time to movement onset after displacement of the cursor.

Diseases of the Basal Ganglia

The importance of the basal ganglia in motor control is emphasized by the devastating consequences of neurodegenerative diseases that affect their function.[155,156] James Parkinson described the "shaking palsy" in 1817; **Parkinson's disease** is characterized by a resting tremor that is lost during intended movement, by increased tone due to simultaneous activation of antagonist muscles, by difficulty in initiating movements, and by slowness of movement once begun.[157] Dopaminergic neurons of the substantia nigra degenerate, and dopamine replacement (through ingestion of L-DOPA) is the standard therapy.

Nigral neurons release dopamine in the neostriatum, inhibiting some neurons and exciting others, and the overall effect is to elevate striatal activity.[158] Thus, with the loss of dopamine in Parkinson's disease, striatal activity falls and produces less inhibition of the

[152]Mink, J. W., and Thach, W. T. 1991. *J. Neurophysiol.* 65: 273–300.

[153]Mink, J. W., and Thach, W. T. 1991. *J. Neurophysiol.* 65: 301–329.

[154]Mink, J. W., and Thach, W. T. 1991. *J. Neurophysiol.* 65: 330–351.

[155]Albin, R. L., Young, A. B., and Penney, J. B. 1995. *Trends Neurosci.* 18: 63–64.

[156]Wichmann, T., and DeLong, M. R. 1996. *Curr. Opin. Neurobiol.* 6: 751–758.

[157]Olanow, C. W., and Tatton, W. G. 1999. *Annu. Rev. Neurosci.* 22: 123–144.

[158]Gerfen, C. R. 1995. *Clin. Neuropharmacol.* 18: 162–177.

globus pallidus. The increase in activity in the globus pallidus reduces firing of cells in the thalamus, which, in turn, decreases excitatory input to the motor cortex. The result is the hypokinesis that is the dominant deficit of this disease.

Another well-known basal ganglia disorder is **Huntington's disease**, whose hallmark is one of *hyper*kinesis—in particular, the generation of spontaneous, disruptive movements that gives this disease its other name: Huntington's chorea (from the Greek word for "dance"). In this disease, striatal neurons projecting to and inhibiting external globus pallidus neurons degenerate. These pallidal neurons project to and normally inhibit the subthalamic nucleus, which in turn excites the output neurons of the globus pallidus. With the loss of striatal disinhibition, the subthalamic nucleus no longer excites the globus pallidus (see Figure 22.27), so inhibition of the thalamus by the globus pallidus is reduced, allowing the inappropriate release of thalamic excitation to motor cortex—hyperkinesia.[159] This interpretation is supported by the similar effects of a condition called **hemiballismus**, which occurs when the subthalamic nucleus is damaged by an infarct of the posterior cerebral artery. After this damage, flailing movements of the contralateral arm and leg occur, again due to reduced inhibition by the pallidal output neurons.

Huntington's disease is genetically determined.[160,161] Linkage analysis of affected families led to cloning of the mutated gene. There is an expansion of a CAG triplet repeat (CAG codes for the amino acid glutamine). The CAG triplet repeats 17 to 34 times in genes of the normal population, but 40 to 121 times in Huntington's patients.[162] A greater number of repeats is correlated with earlier onset of the disease (usually in late middle age), and expansion of the repeated sequence is seen in successive generations. The Huntington's protein is large, greater than 3000 amino acids in length; its function remains unknown.[163]

[159]Albin, R. L. 1995. *Parkinsonism Rel. Disord.* 1: 3–11.

[160]Reddy, P. H.,Williams, M., and Tagle, D. A. 1999. *Trends Neurosci.* 22: 248–255.

[161]Paulson, H. L., and Fischbeck, K. H. 1996. *Annu. Rev. Neurosci.* 19: 79–107.

[162]Huntington's Disease Collaborative Research Group. 1993. *Cell* 72: 971–983.

[163]Gusella, J. F., and MacDonald, M. E. 1998. *Curr. Opin. Neurobiol.* 8: 425–430.

Summary

- A motor unit consists of a single α motor neuron and the muscle fibers it innervates.
- Muscle spindle afferents diverge to make weak synaptic contacts onto all the motoneurons innervating the muscle of origin. Spatial and temporal summation of spindle afferents brings motoneurons to threshold.
- Muscular contraction begins with small motor units and progresses to large (the size principle of motor recruitment) because small motoneurons are more easily excited than large motoneurons by a given synaptic input.
- The stretch reflex excites agonist muscles, and it inhibits antagonists through inhibitory interneurons.
- Golgi tendon organs are sensitive to muscle tension and inhibit motoneurons through interneurons.
- The muscle spindle's sensitivity to stretch is modulated by activation of γ motoneurons (fusimotor) that cause the intrafusal muscle fibers to contract. α–γ Coactivation continuously adjusts the spindle to maintain its sensitivity during programmed movements.
- Flexor reflexes and crossed extensor reflexes initiated by painful stimuli demonstrate the elements of interlimb coordination essential for locomotion.
- Medial and lateral pools of spinal motoneurons innervate the muscles of the trunk and distal limbs, respectively. These are under the control of distinct motor centers in the brainstem and cortex.
- Pacemaker neurons and synaptic interactions combine to form central pattern generators that drive intrinsic motor programs such as locomotion and respiration.
- Central motor programs are triggered and continuously modulated by sensory feedback.
- Respiration provides an example of a motor program arising from a brainstem pattern generator that is modulated by the level of CO_2 in the blood, and by stretch receptors in respiratory muscle.
- The primary motor cortex M_1 lies anterior to the central sulcus and is somatotopically mapped. Corticospinal neurons (projecting to the spinal cord) fire in proportion to muscle force.
- Many neurons in M_1 are direction-selective. Their activity also can depend on the starting position of the limb or the type of triggering stimulus.

SUMMARY (CONTINUED)

■ The cerebellum helps to plan and execute motor commands through feedback with the cortex, and by descending commands through the red nucleus and brain stem nuclei. Lesions of the cerebellum disrupt coordination, with no loss of sensation or strength.

■ The basal ganglia provide negative feedback to the cerebral cortex and act to restrict the available motor outputs. The consequences of basal ganglia disease reflect the complex pattern of feedback loops that underlie their function.

SUGGESTED READING

General Reviews

Arshavsky, Y. I., Deliagina, T. G., and Orlovsky, G. N. 1997. Pattern generation. *Curr. Opin. Neurobiol.* 7: 781–789.

Asanuma, H. 1989. *The Motor Cortex*. Raven, New York.

Georgopoulos, A. P. 1994. New concepts in generation of movement. *Neuron* 13: 257–268.

Getting, P. A. 1989. Emerging principles governing the operation of neural networks. *Annu. Rev. Neurosci.* 12: 185–204.

Kiehn, O., et al. (eds.) 1998. *Neuronal Mechanisms for Generating Locomotor Activity* (Annals of the New York Academy of Sciences, Vol. 860). New York Academy of Sciences, New York. [22]

Klockgether, T., and Evert, B. 1998. Genes involved in hereditary ataxias. *Trends Neurosci.* 21: 413–418.

Paulson, H. L., and Fischbeck, K. H. 1996. Trinucleotide repeats in neurogenetic disorders. *Annu. Rev. Neurosci.* 19: 79–107.

Penfield, W., and Rasmussen, T. 1950. *The Cerebral Cortex of Man: A Clinical Study of Localization of Function*. Macmillan, New York.

Rekling, J. C., and Feldman, J. L. 1998. PreBotzinger complex and pacemaker neurons: Hypothesized site and kernel for respiratory rhythm generation. *Ann. Rev. Physiol.* 60: 385–405.

Thach, W. T., Goodkin, H. G., and Keating, J. G. 1992. The cerebellum and the adaptive coordination of movement. *Annu. Rev. Neurosci.* 15: 403–442.

Wichmann, T., and DeLong, M. R. 1996. Functional and pathophysiological models of the basal ganglia. *Curr. Opin. Neurobiol.* 6: 751–758.

Original Papers

Henneman, E., Somjen, G., and Carpenter, D. O. 1965. Functional significance of cell size in spinal motoneurons. *J. Neurophysiol.* 28: 560–580.

Karni, A., Meyer, G., Jezzard, P., Adams, M. M., Turner, R., and Ungerleider, L. G. 1995. Functional MRI evidence for adult motor cortex plasticity during motor skill learning. *Nature* 377: 155–158.

Lang, E. J., Sugihara, I., Welsh, J. P., and Llinás, R. R. 1999. Patterns of spontaneous Purkinje cell complex spike activity in the awake rat. *J. Neurosci.* 19: 2728–2739.

Meyer-Lohmann, J., Hore, J., and Brooks, V. B. 1977. Cerebellar participation in generation of prompt arm movements. *J. Neurophysiol.* 40: 1038–1050.

Mink, J. W. and Thach, W. T. 1991a. Basal ganglia motor control. I. Nonexclusive relation of pallidal discharge to five movement modes. *J. Neurophysiol.* 65: 273–300.

Salinas, E. and Romo, R. 1998. Conversion of sensory signals into motor commands in primary motor cortex. *J. Neurosci.* 18: 499–511.

Sherrington, C. S. 1910. Flexor-reflex of the limb, crossed extension reflex, and reflex stepping and standing (cat and dog). *J. Physiol.* 40: 28–121.

Thach, W. T. 1978. Correlation of neural discharge with pattern and force of muscular activity, joint position, and direction of next intended movement in motor cortex and cerebellum. *J. Neurophysiol.* 41: 654–676.

PART 4 Development of the Nervous System

23 Development of the Nervous System

NERVE CELLS ACQUIRE THEIR IDENTITIES AND ESTABLISH ORDERLY and precise synaptic connections during development in response to genetic and environmental influences. These include cell lineage, inductive and trophic interactions between cells, cues that guide cell migration and axon outgrowth, specific cell–cell recognition, and activity-dependent refinement of connections.

The development of the vertebrate nervous system begins with the formation of the neural plate in the dorsal ectoderm. The neural plate then curls to give rise to the neural tube and the neural crest. Neurons and glial cells of the central nervous system are produced by division of precursor cells in the ventricular zone of the neural tube. Postmitotic neurons migrate away from the ventricular surface to form the gray matter of the adult nervous system. Within each region of the developing nervous system the fates of cells become progressively restricted according to anteroposterior, dorsoventral, and local patterns. In the last few years explanations at the molecular level have become available for many aspects of development, such as anteroposterior and dorsoventral patterning, that previously could be described only phenomenologically, with no idea of the underlying mechanisms. For example, the expression of a series of homeobox genes along the anteroposterior axis establishes the identity of segments in the hindbrain; subsequently the dorsoventral pattern of differentiation is determined, in part, by a gradient of a protein known as Sonic hedgehog.

Neural crest cells form the peripheral nervous system. The phenotype adopted by a neural crest cell is determined by signals from cells in its environment. Thus, a neural crest cell transplanted early in development assumes the fate appropriate to its new location.

To establish synaptic contacts with their targets, neurons extend axons tipped with growth cones that explore the environment. Two classes of molecules have been identified as important substrates for growth cone movements: cell adhesion molecules of the immunoglobulin superfamily and extracellular matrix adhesion molecules. Growth cone navigation is controlled by long- and short-range attractive and repulsive cues. Chemoattractants guide axons either to their ultimate synaptic partners or to an intermediate target, such as a guidepost cell. Chemorepellents prevent axons from entering inappropriate territories. Axonal projections made during development are often more extensive than those seen in the adult, and are trimmed to the adult pattern by trophic and activity-dependent mechanisms.

Functional synaptic contacts are formed rapidly, but at first they lack specializations characteristic of adult junctions. Over the course of several weeks synapses mature to their adult form.

A common feature of vertebrate central nervous system development is an initial overproduction of neurons followed by a period of cell death. Neuronal death is regulated by competition for trophic substances. Nerve growth factor is one member of a family of proteins, called neurotrophins, each of which sustains particular neuronal populations.

The orderliness of the connections made by nerve cells with one another and with tissues in the periphery is a prerequisite for normal nervous system function. To create this precise neural architecture during development, the correct number and type of neurons must be generated, assume their appropriate positions, and make synapses on the proper target cells. For example, for the stretch reflex to work, the Ia afferent sensory neuron in a dorsal root ganglion must send one axon to end on the appropriate region of a muscle spindle, and another axon centrally to make synapses exclusively on those motoneurons that innervate the muscle that contains the spindle. Other branches of the central axon end on spinal interneurons or run in the dorsal columns to innervate cells in the dorsal column nuclei. In addition, the number of sensory neurons and motor neurons must be matched to the size of the muscle and the number of spindles it contains.

A variety of questions arise when one considers such an example. How do cells acquire their identities as neurons or glial cells? What cues guide neurons to their correct positions? What cellular mechanisms enable a neuron to extend an axon to a particular target, selected from a myriad of choices, and form a synapse? The answers to such questions influence thinking about the genetic blueprint for wiring a brain containing 10^{10} to 10^{12} cells with a much smaller number of genes, 10^5 or so. Moreover, a certain degree of flexibility must be maintained during critical periods in development, and even in the adult, allowing synapses to be formed, modified, or removed, often simply as a result of altered patterns of activity (Chapter 25).

The scope of all the problems relating to development, synapse formation, neural specificity, and changes in efficacy is too great for a comprehensive review. Many aspects are covered in detail elsewhere.[1,2] In this chapter we provide a brief account of vertebrate neuroembryology and describe selected experimental approaches to questions of neural development. The topics are discussed in developmental sequence, beginning with the induction of neuroectoderm and early neural morphogenesis. Then the regional specification of neural tissue and the factors that determine the identity of individual neurons and glial cells are described. Finally, the mechanisms of axon outgrowth, target innervation, and synapse formation are considered, together with the role of growth factors and competitive interactions in shaping the final form of the nervous system.

Terminology

A major problem of terminology arises as studies of development enter the molecular realm. In recent years more and more molecules have been identified and their mechanisms of action elucidated. They include genes and proteins important for cell survival, growth, and differentiation, for axon extension and navigation, and for synapse formation and modification. To make matters worse, many of these proteins and genes are given peculiar names based on the history of their discovery and the quirks of the investigators in the particular laboratory where they were characterized. As a result, the reader (who has already had to deal with alienating terms such as "delayed rectification" in our discussion of signaling) now has to face name after name, such as "Sonic hedgehog" and "Ephrin," as well as an endless host of acronyms such as "N-CAM," "BDNF" and "Elf-2," none of which provide any direct indication of function to a reader outside the field. We have tried to keep the number of proteins and genes described to a minimum and, wherever feasible, have identified them by their full names.

Genetic Approaches for Understanding Development

Three major advances in recent years have led to a rapid increase in our ability to explain in terms of the underlying molecular mechanisms phenomena that in the past could only be described. First is the development of new molecular biological techniques to monitor and manipulate gene expression. Second is the discovery that mechanisms and molecules that mediate neural development are remarkably similar throughout the animal kingdom. For example, as mentioned in Chapter 1, the gene that directs the formation of the eye in a developing chick, mouse, or human is similar to that controlling eye formation in *Drosophila*.[3] Thus, genes that play a role in the development of *Drosophila*, yeast, or *C. el-*

[1]Gilbert, S. F. 2000. *Developmental Biology*, 6th Ed. Sinauer, Sunderland, MA.

[2]Zigmond, M. J., et al. (eds.). 1999. *Fundamental Neuroscience*. Academic Press, New York.

[3]Quiring, R., et al. 1994. *Science* 265: 785–789.

egans often have vertebrate homologues that are also important in development. Third is the emergence of the zebrafish, first introduced to neurobiology by Streisinger,[4] as a particularly favorable vertebrate preparation for developmental studies.[5] The zebrafish embryo is transparent, allowing direct observation of individual cells throughout embryogenesis, and development is rapid. Most important, techniques have been developed by which new mutations can be induced, identified, and maintained in zebrafish, paving the way for the discovery of important vertebrate genes that might not have homologues in simpler invertebrate species.[6]

EARLY NEURAL MORPHOGENESIS

Early in vertebrate embryogenesis, the region of the gastrula that will give rise to the nervous system is a simple sheet of ectoderm (Figure 23.1). The cells in the sheet are under the influence of growth factors (including two proteins of the bone morphogenic protein family, BMP-2 and BMP-4) that suppress neural differentiation and promote the formation of epidermal tissue.[7] Next, diffusible signaling proteins that block the action of these growth factors are released from a particular "organizer" region of the gastrula, called the **Spemann organizer** (in amphibian eggs) or **Hensen's node** (in chick and mammalian embryos). These proteins (known as follistatin, noggin, and chordin) allow a signaling cascade to proceed in cells near the organizer that promotes neural differentiation, forming the **neural plate.** The neural plate is a sheet of elongated neuroectodermal cells from which the nervous system will form.

Over most of its length the edges of the neural plate thicken and move upward, forming **neural folds,** which fuse at the midline to give rise to a hollow **neural tube** (Figure 23.2). The process by which the neural plate is formed and converted into a neural tube is called **neurulation.** Some of the cells at the lips of the neural fold come to lie between the neural tube and the overlying ectoderm. These cells form the **neural crest.** Neural crest cells migrate away from the neural tube and give rise to a wide variety of peripheral tissues, including neurons and satellite cells of the sensory, sympathetic, and parasympathetic nervous systems, cells of the adrenal medulla, pigmented cells of the epidermis, and bones and connective tissue in the head.

As development proceeds, the anterior (cephalic or rostral) portion of the neural tube undergoes a series of swellings, constrictions, and flexures that form anatomically defined regions of the brain (Figure 23.3). The caudal portion of the neural tube retains a relatively simple tubular structure and forms the spinal cord.

[4]Streisinger, G., et al. 1981. *Nature* 291: 293–296.

[5]Detrich, H. W., III, Westerfield, M., and Zon, L. I. 1999. *Methods Cell Biol.* 59: 3–10.

[6]Wylie, C. (ed.). Zebrafish Issue. 1996. *Development* 123: 1–481.

[7]Sasai, Y. 1998. *Neuron* 21: 455–458.

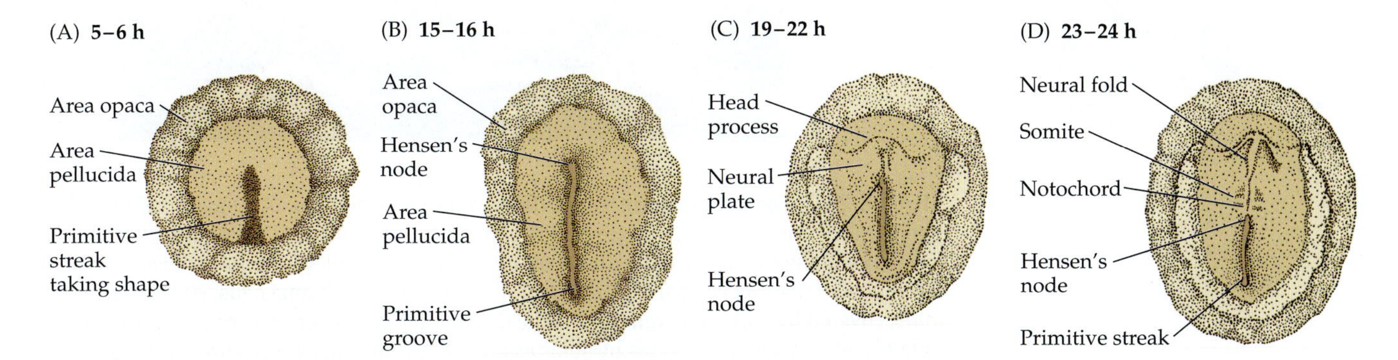

FIGURE 23.1 Early Morphogenesis in the vertebrate embryo. Dorsal views of the first day in the development of a chick embryo. (A) 5–6 h: formation and elongation of the primitive streak. (B) 15–16 h: formation of the primitive groove and Hensen's node. (C) 19–22 h: formation of the head process and neural plate. (D) 23–24 h: formation of the neural fold, notochord, and mesodermal somites. (After Gilbert, 2000.)

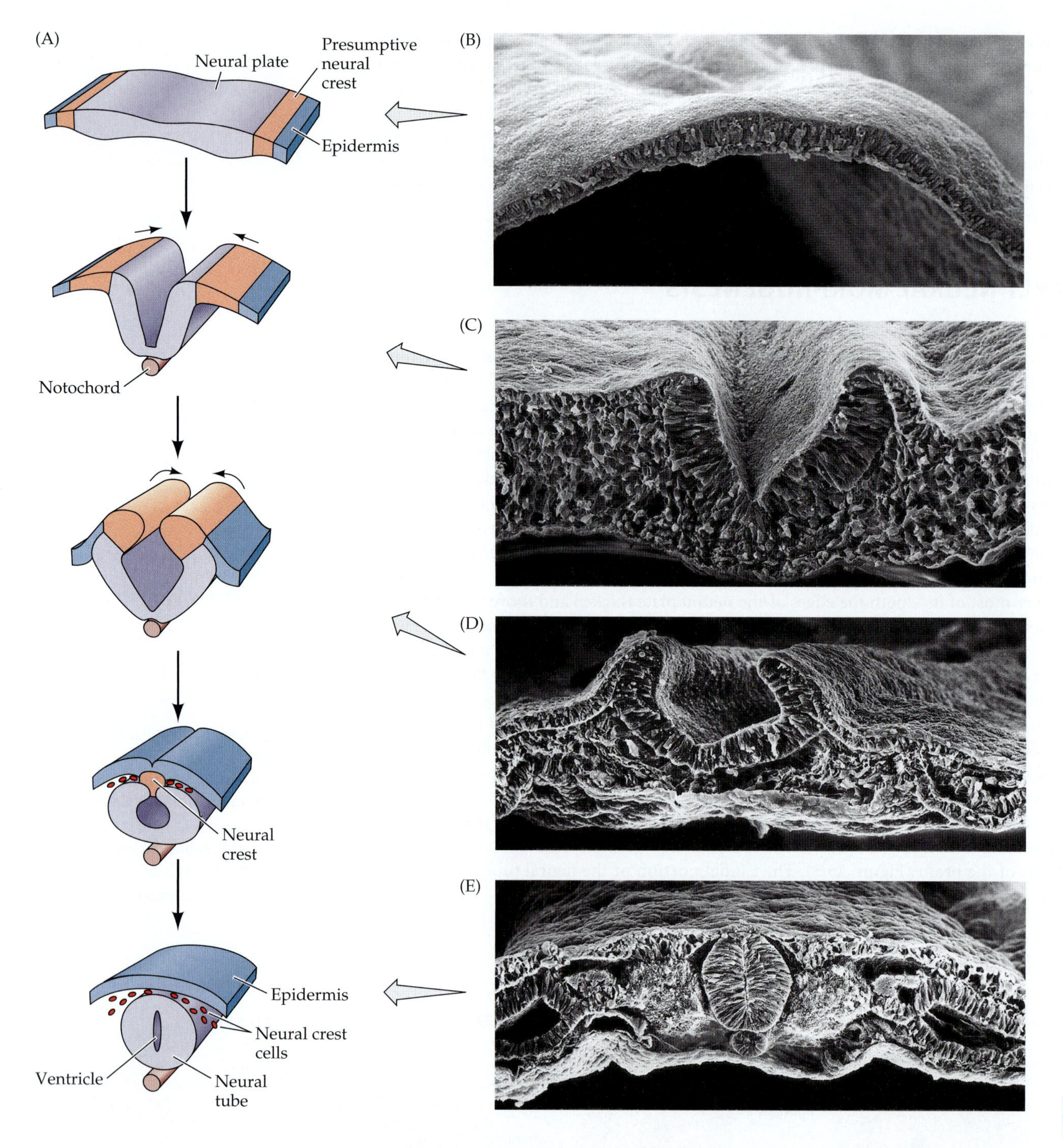

FIGURE 23.2 Formation of the Neural Tube in the chick embryo. (A) Diagram of neurulation. (B–E) Scanning electron micrographs of neural tube formation. (B) Neural plate, formed by elongated cells in the dorsal region of the ectoderm. (C) Neural groove, formed by elongated neuroepithelial cells and surrounded by mesenchymal cells. (D) Neural folds, covered by flattened epidermal cells. (E) Neural tube, covered by presumptive epidermis and flanked on the sides by somites and on the bottom by the notochord. (After Gilbert, 2000; photographs kindly provided by K. W. Tosney.)

Production of Neuronal and Glial Cell Precursors

The wall of the neural tube is initially composed of a single, rapidly dividing layer of cells; each cell extends from the luminal, or **ventricular**, edge to the external, or **pial**, surface (Figure 23.4A). As each cell progresses through the cell cycle, its nucleus migrates back

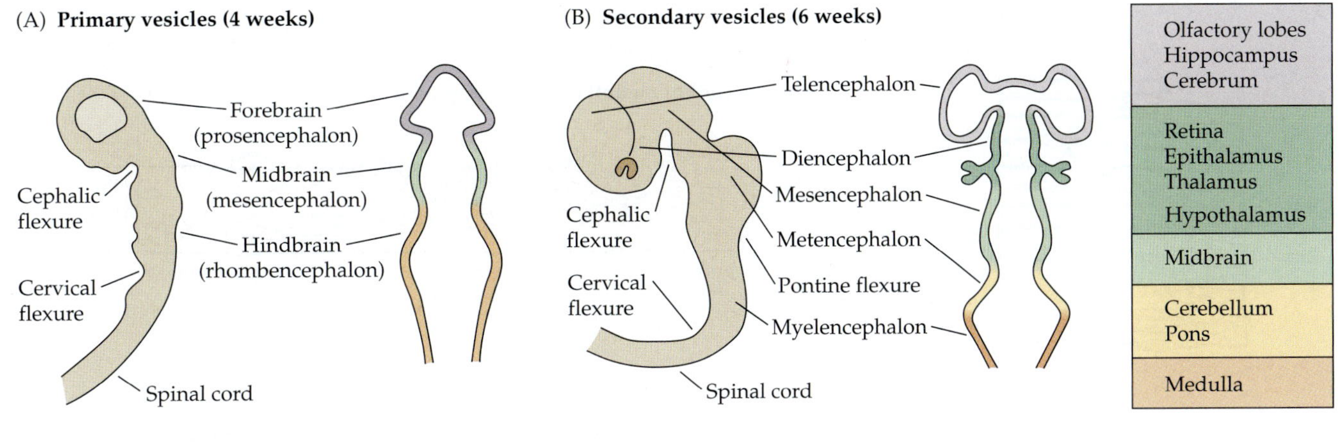

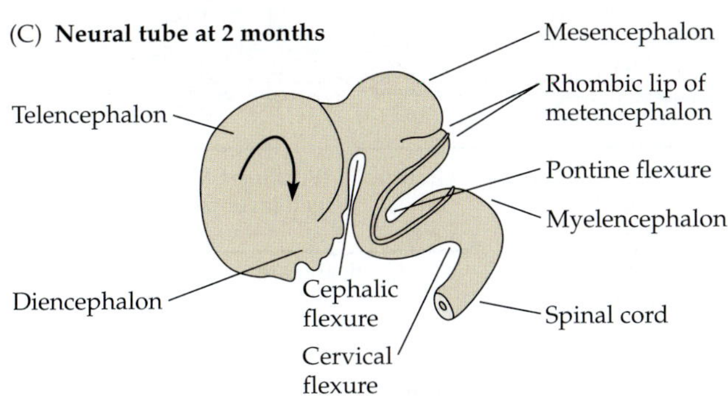

FIGURE 23.3 Early Human Brain Development. Lateral views of the developing brain and corresponding schematic horizontal sections through the vesicles. (A) At 4 weeks of development, the human CNS consists of three primary vesicles. (B) At 6 weeks of development, five secondary vesicles can be distinguished. (C) By 2 months, a series of flexures, constrictions, and swellings form the various regions of the brain. Further development is dominated by rapid growth of the telencephalon in a "C" shape (arrow). (After Nolte, 1988.)

and forth between the ventricular and pial surfaces. DNA synthesis occurs while the nucleus is near the pial surface; during cell division (**cytokinesis**) the nucleus lies near the ventricular surface and the pial connection is temporarily lost. Following cell division, one or both of the daughter cells may lose contact with the ventricular surface and migrate away. This is the point at which they become neurons or glial cells. Once they migrate away from the ventricular zone, most cells destined to become neurons are postmitotic (they will never divide again). Glial cell precursors, on the other hand, can divide even after they have reached their final locations.

As more and more postmitotic cells are produced, the neural tube thickens and assumes a three-layered configuration: an innermost **ventricular zone** (where proliferation continues), an intermediate **mantle zone** containing the cell bodies of the migrating neurons, and a superficial **marginal zone** composed of the elongating axons of the underlying neurons (Figure 23.4B). This three-layered structure persists in the spinal cord and medulla (Figure 23.4C). In other regions, such as the cerebrum and cerebellum, some neurons migrate into the marginal zone to form a **cortical plate**, which matures into the adult cortex.

Migration of Neurons in the CNS

In many regions of the developing brain, such as the cerebrum and cerebellum, the migration of neurons is dependent on **radial glial cells** (Chapter 8). These cells maintain their contacts with both the ventricular and the pial surfaces of the neural tube. As the walls of the neural tube thicken because of the continued division of cells in the ventricular layer and the accumulation of neurons in the mantle zone and cortical plate, the radial glial cells become extremely elongated. From detailed light- and electron-microscopic studies of the development of the cerebrum and cerebellum, Rakic and his colleagues[8] showed that neurons move along this scaffolding of radial glial cells to reach their appropriate positions in the cortex (see Figure 8.10). Observations in mutant mice[9] and on cells maintained in culture[10] have confirmed this pattern of migration. The proteins mediating such neuronal migration are beginning to be identified. They include a neural glycoprotein known as astrotactin[11] and isoforms of the integrin family of receptors for extracellular matrix adhesion molecules (discussed later in the chapter).[12]

[8]Rakic, P. 1981. *Trends Neurosci.* 4: 184–187.

[9]Goldowitz, D. 1989. *Neuron* 2: 1565–1575.

[10]Hatten, M. E., Liem, R. K. H., and Mason, C. A. 1986. *J. Neurosci.* 6: 2676–2683.

[11]Zheng, C., Heintz, N., and Hatten, M. E. 1996. *Science* 272: 417–419.

[12]Anton, E. S., Kreidberg, J. A., and Rakic, P. 1999. *Neuron* 22: 277–289.

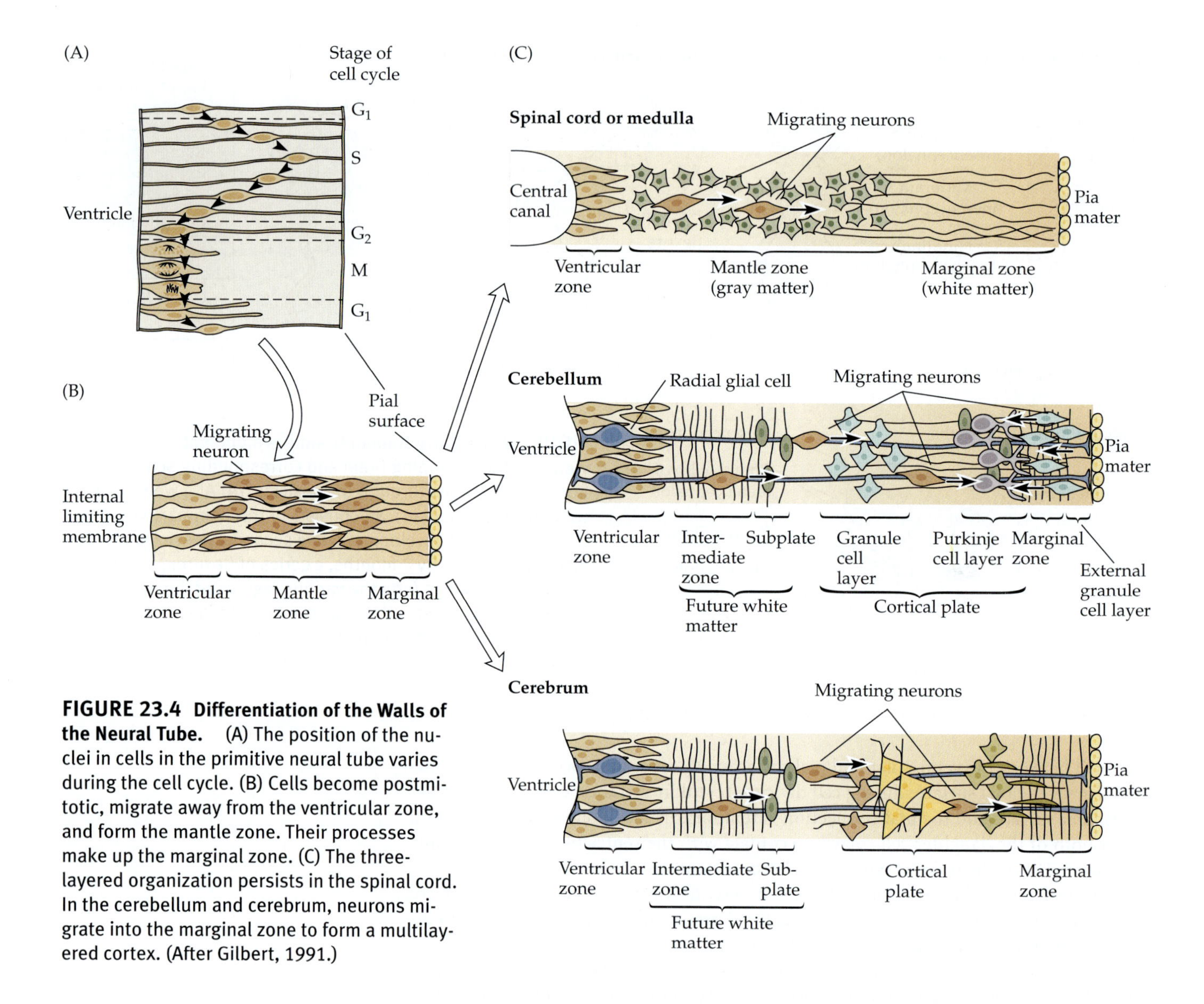

FIGURE 23.4 Differentiation of the Walls of the Neural Tube. (A) The position of the nuclei in cells in the primitive neural tube varies during the cell cycle. (B) Cells become postmitotic, migrate away from the ventricular zone, and form the mantle zone. Their processes make up the marginal zone. (C) The three-layered organization persists in the spinal cord. In the cerebellum and cerebrum, neurons migrate into the marginal zone to form a multilayered cortex. (After Gilbert, 1991.)

Neurons can also migrate through regions of the central nervous system in which there are no radial glial cells. One remarkable population of neurons expressing gonadotropin-releasing hormone (GnRH) migrates from the periphery *into* the central nervous system (see also Chapter 16). The GnRH cells travel about 2 mm, moving from the olfactory pit, an ectodermal derivative (**placode**) that gives rise to the nasal epithelium, into the hypothalamus along a previously established axon tract.[13]

Extracellular Matrix Adhesion Proteins and Neural Crest Cell Migration

In the peripheral nervous system, neural crest cells migrate along pathways that lack both axon tracts and organized glial structures. Such migrations are guided by attractive and repulsive interactions with cell surface and extracellular matrix components. Two extracellular matrix adhesion proteins, laminin and fibronectin, are concentrated along neural crest cell migratory pathways in the embryo (Figure 23.5).[14] Agents that inhibit the interaction between integrin receptors and extracellular matrix components block the movement of neural crest cells on surfaces coated with extracellular matrix adhesion molecules in vitro and the migration of cranial neural crest cells in vivo. Another extracellular matrix protein, F-spondin, is expressed in regions adjacent to the neural crest cell migratory route. It inhibits the movement of crest cells, thereby confining them to their proper pathway.[15]

[13]Wray, S., Grant, P., and Gainer, H. 1989. *Proc. Natl. Acad. Sci. USA* 86: 8132–8136.

[14]Bronner-Fraser, M. 1985. *J. Cell Biol.* 101: 610–617.

[15]Debby-Brafman, A., et al. 1999. *Neuron* 22: 475–488.

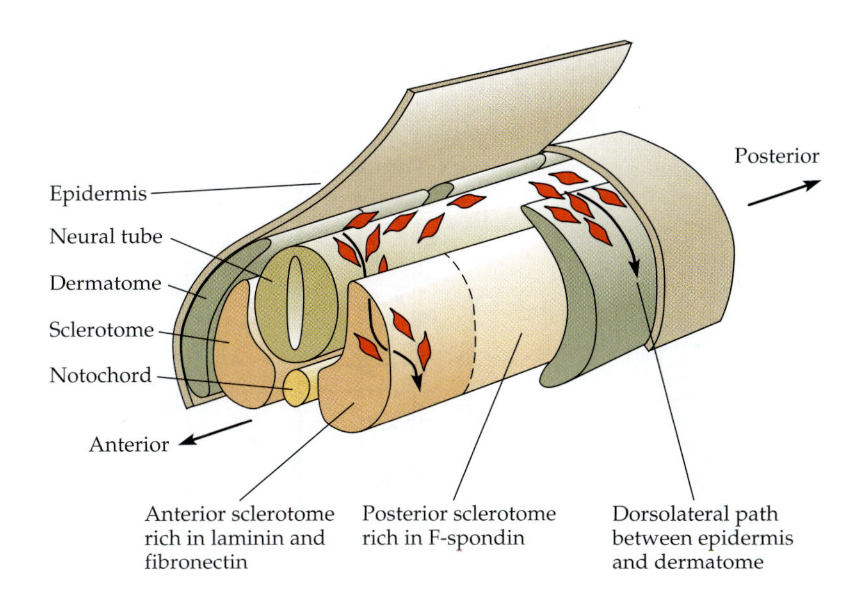

FIGURE 23.5 Migration of Neural Crest Cells in the trunk region of the chick embryo. Crest cells migrate ventrally through the anterior sclerotome, a region rich in laminin and fibronectin, and avoid the posterior sclerotome, a region rich in F-spondin. These cells become neurons of the dorsal root and autonomic ganglia, Schwann cells, and adrenal chromaffin cells. Later in development, crest cells migrate dorsolaterally beneath the epidermis to form pigment-producing melanocytes.

REGIONAL SPECIFICATION OF NEURAL TISSUE

In the adult, different regions of the nervous system have conspicuously different phenotypes, according to the function each performs. Thus, the cerebellum differs from the cerebral cortex, which differs from the retina. How are such differences established during development? As is true for other cells, the phenotype of a neuron is determined by its pattern of gene expression. This is controlled by transcription factors, proteins that bind to regulatory regions of one or more genes and influence the rate at which they are transcribed. Development is characterized by the sequential and hierarchical expression of transcription factors. Each influences the expression of the next and restricts the ultimate phenotype.

Investigation of the regional specification of neural tissue in the vertebrate brain was aided immensely by the finding that homologues of genes specifying position and position-dependent differentiation in the embryo of the fruit fly, *Drosophila*, are conserved in vertebrates and often serve similar functions. Many of these genes encode transcription factors.

Homeotic Genes and Segmentation

A remarkable example of such conservation is found in the development of the vertebrate hindbrain. Unlike the rest of the vertebrate brain, the embryonic hindbrain (**rhombencephalon**) has a conspicuously segmented structure. Each segment exhibits the same general pattern of neuronal differentiation, but from segment to segment the pattern is modified in specific ways (Figure 23.6). Several genes have been identified whose pattern of expression at early stages in development correlates with the segmental boundaries of the hindbrain (Figure 23.7).[16] These genes fall into two categories: (1) Certain genes in the first category play a role in establishing the overall architecture of repeating segmental units. Some genes in this group encode transcription factors (*kreisler*, *Krox-20*); others encode receptor tyrosine kinases (*Sek-1* through *Sek-4*) or their ligands (*Elf-2*). (A receptor tyrosine kinase is a transmembrane protein whose intracellular tyrosine kinase domain is activated when a ligand binds to its extracellular domain.) (2) The second category contains genes that determine the fate of each segment. These genes make up the highly conserved *Hox* family.

Hox genes were first characterized in *Drosophila*, where they were shown to act as **homeotic regulator genes.**[17] Homeotic genes are master genes that coordinate the expression of many other genes during development to create a particular structure. For example, mutations of *Hox* homeotic genes in *Drosophila* cause one body part to be replaced with another; thus a leg can develop where an antenna should be. Homeotic genes contain a conserved stretch of DNA, dubbed the **homeobox**. The homeobox encodes a sequence of 60 amino acids that recognize and bind to specific DNA sequences in a series of

[16]Lumsden, A., and Krumlauf, R. 1996. *Science* 274: 1109–1115.

[17]Graba, Y., Aragnol, D., and Pradel, J. 1997. *BioEssays* 19: 379–388.

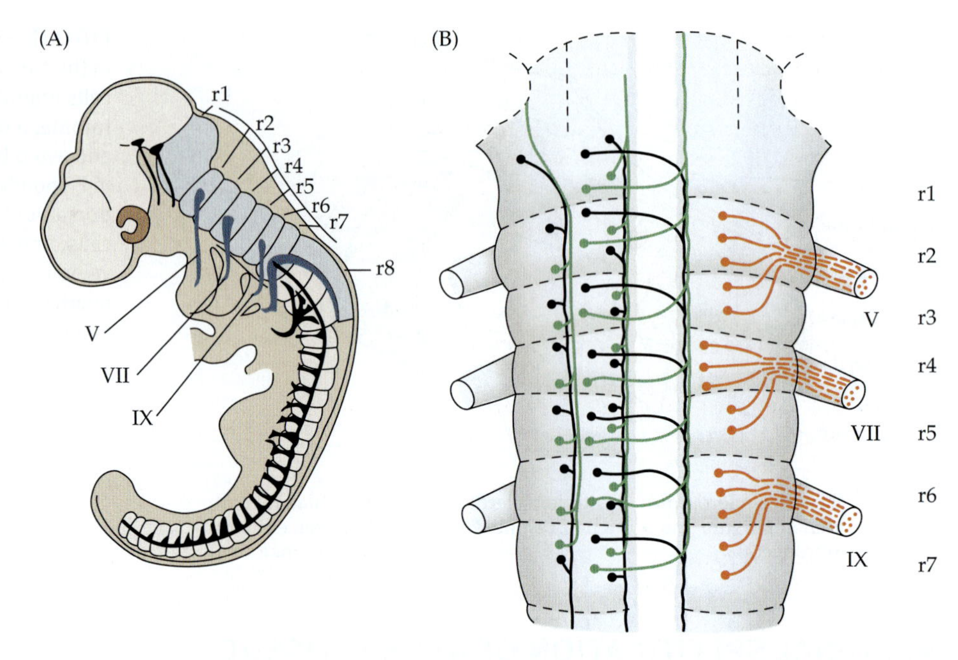

FIGURE 23.6 The Vertebrate Hindbrain Develops as a Conspicuously Segmented Structure. (A) Diagram of a 3-day chick embryo, illustrating the segmental arrangement of rhombomeres (r1–r8) in the hindbrain. (B) Pattern of cell organization in rhombomeres r1 to r7 of the 3-day chick embryonic hindbrain. Reticular neurons (left side) and branchiomotor neurons (right) occur in a segmentally repeating pattern. Motor neurons send their axons into cranial nerves V, VII, and IX. (After Keynes and Lumsden, 1990, and Lumsden and Krumlauf, 1996.)

subordinate genes. Each homeotic gene thereby coordinates the expression of a number of genes that, together, determine the structure of one region of the embryo.

The segmental pattern of *Hox* gene expression observed in the chick and rodent hindbrain suggests that *Hox* genes could likewise be acting as homeotic master genes in vertebrate development, creating structures appropriate to particular anteroposterior positions in the embryonic hindbrain. Evidence from transplantation, gene knockout, and ectopic expression studies are consistent with this idea.[18,19] Additional support is provided by the finding that mutations of *Hox* and other homeobox-containing genes in humans give rise to malformations of the corresponding CNS regions.[20,21]

The next obvious question is, What determines the pattern of *Hox* gene expression? The answer is, at least in part, a gradient of **retinoic acid.**[22] Retinoic acid is produced by

[18]Capecchi, M. R. 1997. *Cold Spring Harb. Symp. Quant. Biol.* 62: 273–281.

[19]Morrison, A. D. 1998. *BioEssays* 20: 794–797.

[20]Boncinelli, E., Mallamaci, A., and Broccoli, V. 1998. *Adv. Genet.* 38: 1–29.

[21]Walsh, C. A. 1999. *Neuron* 23: 19–29.

[22]Gould, A., Itasaki, N., and Krumlauf, R. 1998. *Neuron* 21: 39–51.

FIGURE 23.7 Segmental Expression of Genes in Rhombomeres r1 to r8 of the vertebrate hindbrain. Grey bars indicate rhombomeres in which each gene is expressed; black bars indicate a high level of expression. Early transcription factors, Eph family receptor tyrosine kinases, and Eph ligands establish the segmental pattern of rhombomeres. The *Hox* homeobox genes determine the fate of cells within each rhombomere in a segmentally specific way. Data from chick and mouse. (After Lumsden and Krumlauf, 1996.)

Hensen's node, the Spemann organizer for avian and mammalian embryos (see Figure 23.1). Not only does retinoic acid activate the transcription of *Hox* genes, but also there are systematic differences in the sensitivity of different *Hox* genes to retinoic acid. Thus, diffusion of retinoic acid from Hensen's node establishes a gradient that contributes to the orderly anteroposterior sequence of *Hox* gene expression in the hindbrain.

Notochord and Floor Plate

The characteristics of the vertebrate nervous system vary along a dorsoventral as well as an anteroposterior axis. For example, a band of specialized glial cells called the **floor plate** lies along the ventral midline of the spinal cord. Adjacent, more laterally situated **basal** regions of the neural tube give rise to motoneurons, more dorsal **alar** regions give rise to interneurons, and the most dorsal region forms the neural crest. Characteristic features of the ventral spinal cord, such as the differentiation of floor plate cells and motoneurons, are induced by a signal from the notochord.[23] Thus, if an additional notochord is transplanted into an embryo adjacent to the neural tube, a second floor plate and population of motoneurons form (Figure 23.8). Alternatively, if the notochord is removed from an embryo, floor plate cells and motoneurons fail to form.

The notochord signal that induces the formation of floor plate cells and motoneurons is the product of the *Sonic hedgehog* gene.[24] Sonic hedgehog protein is synthesized by notochord cells (and later by floor plate cells), concentrated on their surfaces, and released to diffuse to neighboring cells. The high levels of Sonic hedgehog protein on the surface of the notochord induce cells of the neural tube to become floor plate cells. Lower levels of Sonic hedgehog induce the expression of a homeobox gene (*Nkx-2.2*) that causes cells to develop into visceral motor neurons.[25] Still lower levels of Sonic hedgehog protein in-

[23] Yamada, T., et al. 1991. *Cell* 64: 635–647.

[24] Roelink, H., et al. 1995. *Cell* 81: 445–455.

[25] Brisco, J., et al. 1999. *Nature* 398: 622–627.

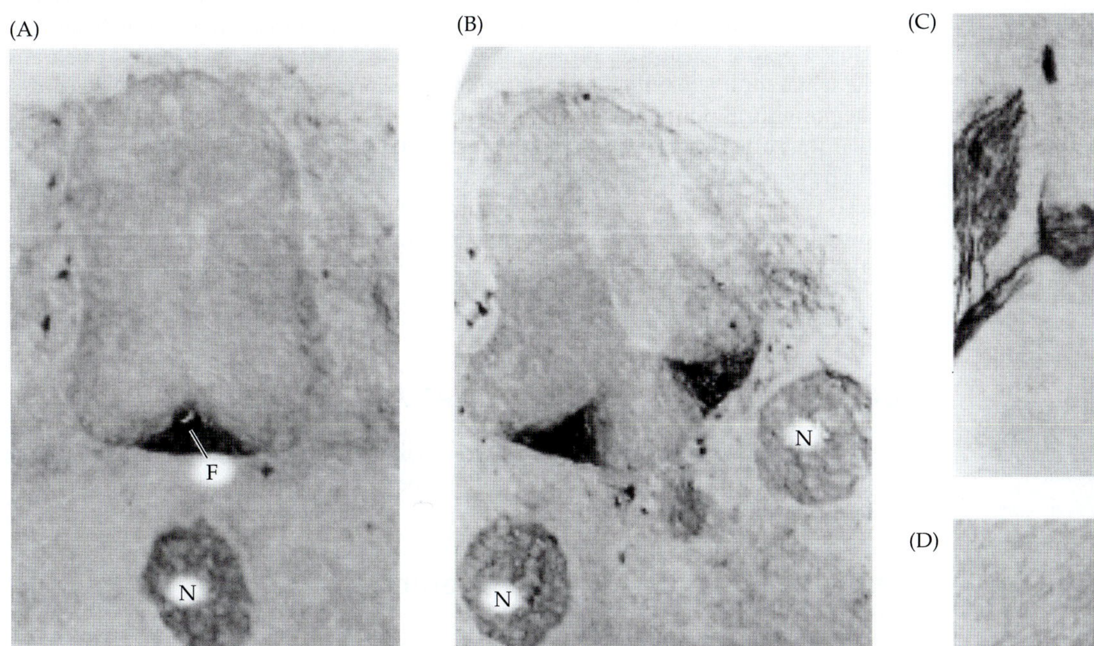

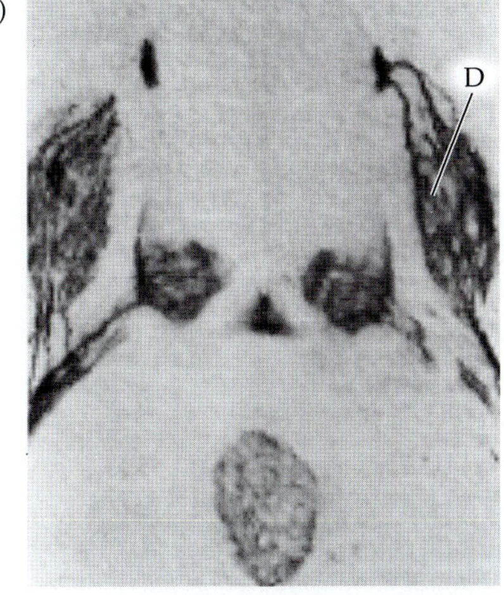

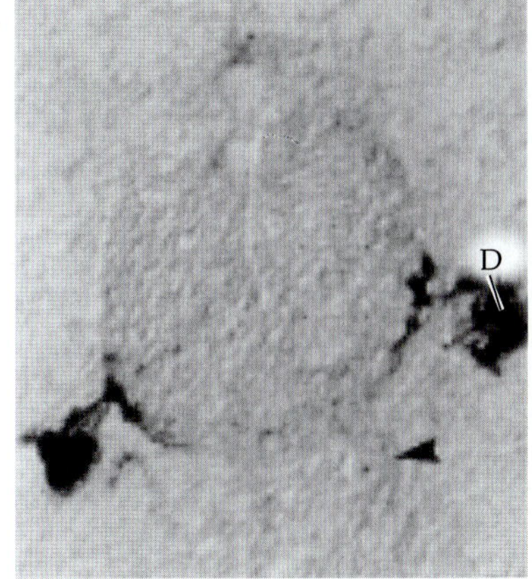

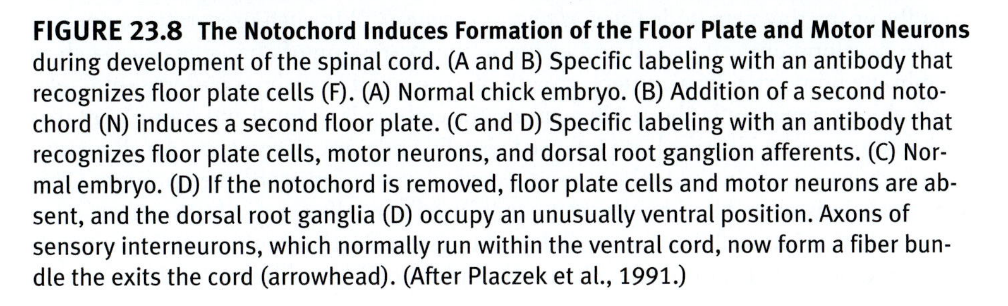

FIGURE 23.8 The Notochord Induces Formation of the Floor Plate and Motor Neurons during development of the spinal cord. (A and B) Specific labeling with an antibody that recognizes floor plate cells (F). (A) Normal chick embryo. (B) Addition of a second notochord (N) induces a second floor plate. (C and D) Specific labeling with an antibody that recognizes floor plate cells, motor neurons, and dorsal root ganglion afferents. (C) Normal embryo. (D) If the notochord is removed, floor plate cells and motor neurons are absent, and the dorsal root ganglia (D) occupy an unusually ventral position. Axons of sensory interneurons, which normally run within the ventral cord, now form a fiber bundle the exits the cord (arrowhead). (After Placzek et al., 1991.)

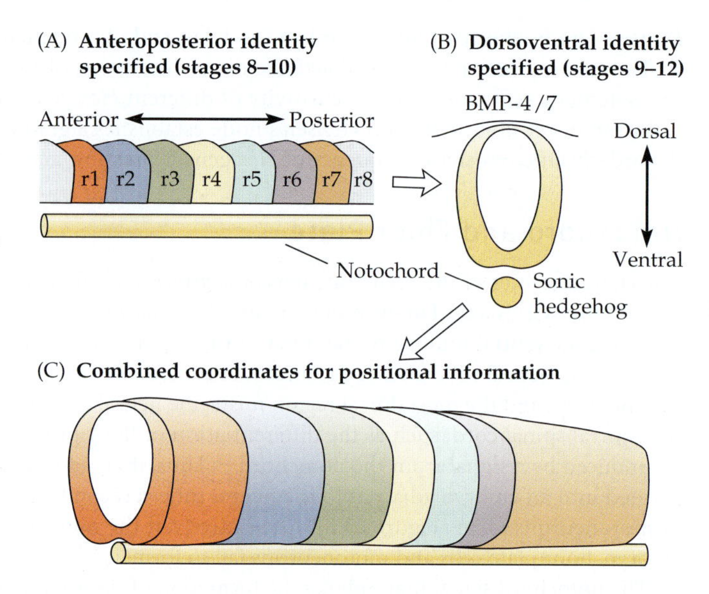

FIGURE 23.9 A Coordinate System of Positional Information in the Vertebrate Hindbrain is established in two steps. (A) First, anteroposterior position is encoded—for example, by *Hox* gene expression. (B) Subsequently, dorsoventral position is encoded by gradients of midline signals such as Sonic hedgehog and BMP-4/7. (C) The resulting two-dimensional coordinate system of positional information restricts the repertoire of cell fates available to pluripotent precursor cells. (After Simon, Hornbruch, and Lumsden, 1995.)

duce *Pax-6*, which inhibits expression of *Nkx-2.2* and allows cells to develop along the default pathway to become somatic motor neurons.

General Scheme for Regional Specification

Anteroposterior and dorsoventral gradients of transcription factors determine regional identity throughout the CNS.[26,27] The same factor can induce quite different characteristics, depending on where in the embryo it is expressed. For example, Sonic hedgehog specifies a ventral phenotype all along the anteroposterior axis, inducing motor neurons in the spinal cord, serotonergic neurons in the anterior hindbrain, dopaminergic neurons in the posterior hindbrain, and oculomotor neurons in the anterior midbrain.[28,29] In a similar way, other transcription factors specify a dorsal phenotype (BMP-4 and BMP-7).[30] The general rule appears to be that the responses available to a pluripotent precursor cell within a particular region of the developing nervous system are first restricted according to its anteroposterior position—for example, by *Hox* gene expression (Figure 23.9).[31] Cell phenotype is then further restricted according to dorsoventral position by midline signals such as *Sonic hedgehog*.

DETERMINATION OF NEURONAL AND GLIAL CELL IDENTITY

In vertebrates, inductive interactions between cells are an important determinant of cell fate. In simpler organisms, the fate of a cell may be determined autonomously as a consequence of cell lineage.

Cell Lineage and Inductive Interactions in Simple Nervous Systems

Cell lineage is most readily followed in invertebrates like the leech, the grasshopper, the fruit fly, or the tiny nematode *Caenorhabditis elegans*.[32–34] In such preparations one can follow development cell by cell and examine the expression of characteristics such as membrane properties, transmitters, growth of axons, and branching patterns. In *C. elegans*, which contains only about 300 neurons, the embryo is so small and transparent that each neuron can be identified and followed visually under the microscope. An alternative approach is to mark individual cells and see what types of progeny they produce. This type of analysis, introduced by Weisblat, Stent, and their colleagues in leech embryos, involves injecting intracellular tracers, such as fluorescent dextran or the enzyme horseradish peroxidase (HRP), into individual cells and following cell lineage either in living

[26]Rubenstein, J. L., and Beachy, P. A. 1998. *Curr. Opin. Neurobiol.* 8: 18–26.
[27]Rubenstein, J. L., et al. 1998. *Annu. Rev. Neurosci.* 21: 445–477.
[28]Ericson, J., et al. 1995. *Cell* 81: 747–756.
[29]Ye, W., et al. 1998. *Cell* 93: 755–766.
[30]Lee, K. J., and Jessell, T. M. 1999. *Annu. Rev. Neurosci.* 22: 261–294.
[31]Simon, H., Hornbruch, A., and Lumsden, A. 1995. *Curr. Biol.* 5: 205–214.
[32]Shankland, M. 1995. *J. Neurobiol.* 27: 294–309.
[33]Doe, C. Q., and Skeath, J. B. 1996. *Curr. Opin. Neurobiol.* 6: 18–24.
[34]Sengupta, P., and Bargmann, C. I. 1996. *Dev. Genet.* 18: 73–80.

embryos or after staining the embryos to visualize cells containing the enzyme.[35] Comparable experiments can be done by injecting cDNA constructs encoding fluorescent protein reporter genes, or by creating transgenic animals that express such proteins.[36]

Experiments of this kind reveal that the pattern of cell lineage in simple invertebrates is remarkably reproducible. Therefore, laser beams can be used to kill individual, identified cells and so determine how the fate of remaining cells is altered. In many cases the surviving cells ignore the loss of their neighbor; their fate is determined by an autonomous, cell lineage–dependent mechanism. In such cells gene expression is determined by inherited cytoplasmic or nuclear factors that act independently of intercellular signals.[37] In other cases, however, the fate of surviving cells is modified.[38] Thus, even in animals with a rigorously stereotyped pattern of cell lineage, cell fate may be determined by inductive interactions.

Inductive Interactions in Development of *Drosophila* Eyes

The patterned development of the compound eyes of *Drosophila* affords another system in which direct observation can be used to identify individual cells and follow their fate.[39] In addition, *Drosophila* genetics offers an especially powerful approach to testing the effects of lineage and inductive interactions between cells on neuronal differentiation. Mutants can be isolated that lack a particular cell type, or in which the pattern of differentiation is perturbed in subtle ways, and the effects on the fate of surviving cells determined.

Such techniques have been used in experiments by Benzer, Ready, Rubin, Tomlinson, Zipursky, and their colleagues to examine the differentiation of neurons and supporting cells in the eye. The *Drosophila* eye consists of a crystalline array of repeating units, the ommatidia (Figure 23.10A), each of which contains eight photoreceptors (R1–R8). The first cell to begin to differentiate in each ommatidium is one of the photoreceptors, R8. R8 cells appear at random in the neuroepithelium. Once an R8 cell begins to differentiate, it inhibits any neighboring cell from becoming an R8 cell. R2 and R5 are specified next, then R3 and R4, R1 and R6, and finally R7 (Figure 23.10B). Two mutants were found whose eyes developed normally except that no R7 cell appeared (see Figure 23.10B). They were called *sevenless* (*sev*$^-$) and *bride-of-sevenless* (*boss*$^-$) to denote the loss of R7. Detailed analysis of these mutants provided one of the first examples of the molecular mechanisms by which inductive interactions determine cell fate.[40–42]

The *sevenless* gene encodes a receptor (called Sevenless, or Sev), for which the product of the *bride-of-sevenless* gene (called Boss) is a ligand. During development R8 induces the formation of R7. This occurs when Boss, which is expressed on the surface of the R8 cell, binds to Sev, which is found in the membrane of the presumptive R7 cell (Figure 23.10C). The Boss–Sev interaction activates the intracellular tyrosine kinase domain of Sev, initiating a signaling cascade in R7 that leads to its differentiation. The signaling pathway is complex, involving the sequential activation of a series of protein kinases (enzymes that phosphorylate proteins) that leads to inhibition of negative regulators as well as activation of positive regulators of gene expression in R7. Many signals that induce changes in gene expression do so by interacting with receptor tyrosine kinases to trigger such intracellular signaling cascades.

Cell Lineage in the Mammalian CNS

It is technically more difficult to study cell lineage in the mammalian central nervous system because individual cells can rarely be identified and injected with tracers. Approaches that have proven useful include mapping the fate of genetically marked cells in embryonic and adult chimeras,[43] and infecting cells in the CNS of developing animals with specially engineered viruses (Figure 23.11).[44–46] Such viruses are constructed so that they will become permanently incorporated into the chromosomes of the host cell, be replicated during cell division, and thus be passed on to the descendants of that cell. In this way the signal carried by the virus is not diluted during successive cell divisions. The presence of the virus can be detected at any stage by visualizing the protein it encodes. Provided the number of cells infected is low, it can be concluded that a cluster of stained cells found later in development is a clone, the progeny of a single infected parent cell.

[35] Stent, G. S., et al. 1992. *Int. Rev. Neurobiol.* 33: 109–193.

[36] Long, Q., et al. 1997. *Development* 124: 4105–4111.

[37] Nelson, B. H., and Weisblat, D. A. 1992. *Development* 115: 103–115.

[38] Isaken, D. E., Liu, N. J., and Weisblat, D. A. 1999. *Development* 126: 3381–3390.

[39] Ready, D. 1989. *Trends Neurosci.* 12: 102–110.

[40] Harris, W. A., and Hartenstein, V. 1999. *Fundamental Neuroscience.* Academic Press, New York, pp. 481–517.

[41] Zipursky, S.L., and Rubin, G.M. 1994. *Annu. Rev. Neurosci.* 17: 373-397.

[42] Dickson, B .J. 1998. *Curr. Biol.* 8: R90–R92.

[43] Rossant, J. 1985. *Phil. Trans. R. Soc. Lond. B* 312: 91–100.

[44] Turner, D. L., and Cepko, C. L. 1987. *Nature* 328: 131–136.

[45] Luskin, M. B., Pearlman, A. L., and Sanes, J. R. 1988. *Neuron* 1: 635–647.

[46] Moriyoshi, K., et al. 1996. *Neuron* 16: 255–260.

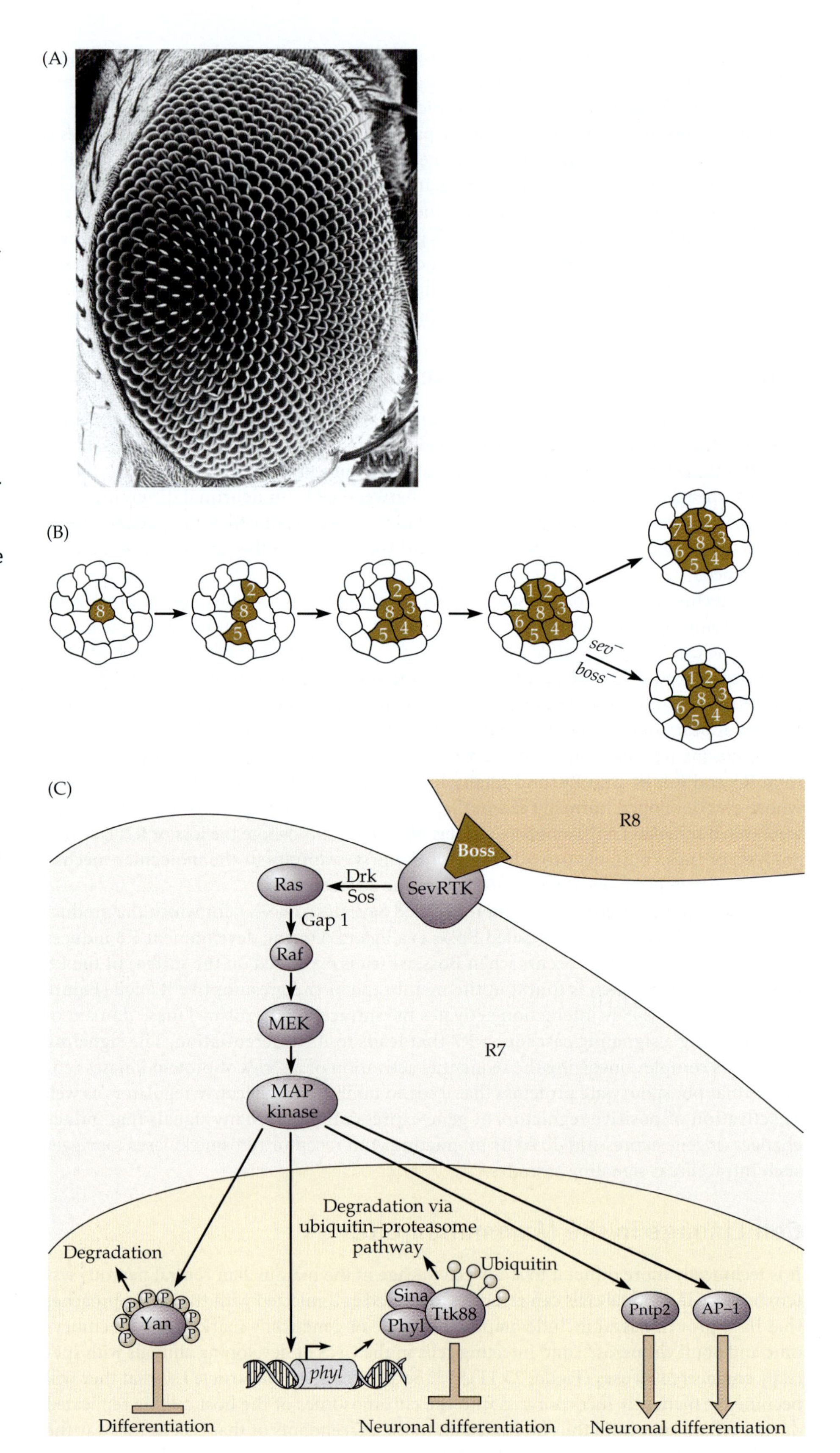

FIGURE 23.10 Inductive Interactions Regulate Development of photoreceptor cells in *Drosophila*. (A) Scanning electron micrograph of a compound eye in *Drosophila*. Each facet is an ommatidium. (B) Normal progression of differentiation of the eight photoreceptors in each ommatidium. *Sevenless* (*sev*⁻) and *bride of sevenless* (*boss*⁻) mutations each prevent differentiation of R7. (C) Signaling pathway regulating the differentiation of R7. The product of the *boss* gene, an integral membrane protein expressed in R8 (Boss), activates the product of the *sev* gene, a receptor tyrosine kinase (SevRTK). The Sev kinase triggers an intracellular signaling cascade that activates MAP kinase, which has several targets. MAP kinase phosphorylates the protein Yan, causing it to be degraded. Yan would otherwise block differentiation. MAP kinase also causes expression of the protein Phyl, which together with a second protein, Sina, causes the transcription factor Ttk88 to be degraded by the ubiquitin–proteasome pathway. Ttk88 prevents neuronal differentiation. MAP kinase also activates Pntp2 and AP-1, two transcription factors that promote neuronal differentiation. (After Dickson, 1998; micrograph kindly provided by D. F. Ready.)

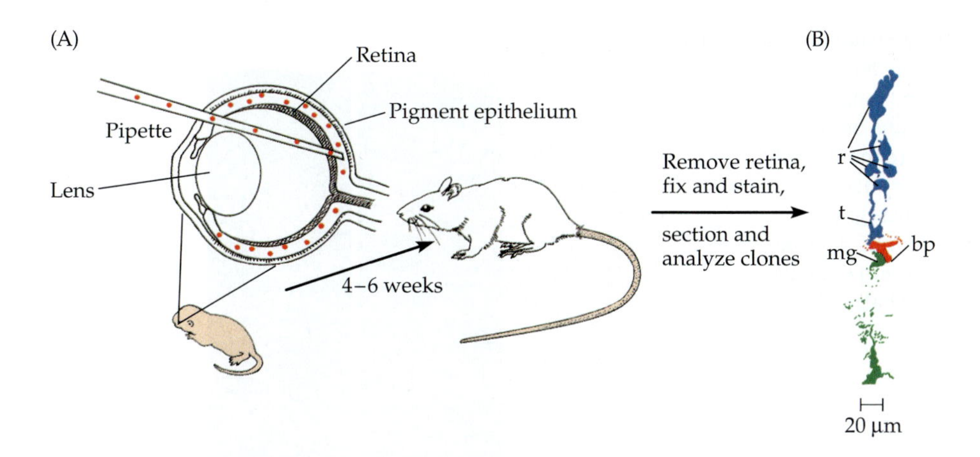

FIGURE 23.11 Clonally Related Cells Are Labeled by Injecting Retroviral Markers into the rat retina. (A) A retrovirus encoding β-galactosidase is injected into the eye between the retina and the pigment epithelium early in development, infecting a few retinal precursor cells. (B) Staining the adult retina with a histochemical reaction for β-galactosidase reveals a cluster of labeled progeny of a single precursor cell. Camera lucida drawing of a clone that includes five rods (r), one bipolar cell (bp) juxtaposed to the terminal of a rod cell (t), and a Müller glial cell (mg). (After Turner and Cepko, 1987.)

For example, when such a virus was injected into the eye of a newborn rat and the retina was examined after the animal had reached adulthood, clusters of stained cells were seen that frequently included both a glial cell and several types of neurons.[44] Thus, a single progenitor cell in the retina at the time of birth must have divided to give rise to both neurons and glial cells.

Unlike the leech or *Drosophila*, the rodent retina shows no evidence for specific lineages that give rise to particular types of neurons. Rather the intrinsic competence of the progenitor cells to respond to external cues and the cues present in the environment both change with time, generating a succession of different cell types.[47] When similar experiments were made in cerebral cortex, on the other hand, clones containing both neurons and glial cells were rare, suggesting that at the time of viral infection separate populations of progenitors for neurons and glial cells were present in the cortical ventricular zone.[45] Moreover, clones tended to contain exclusively pyramidal cells or nonpyramidal cells, suggesting that these lineages diverge early in neurogenesis.[48]

In summary, in the nervous systems of simple organisms the lineage history of a cell limits its developmental potential. In the central nervous systems of more complex organisms, inductive interactions between cells are of overriding importance for determining cell fate. In a well-known vivid analogy, Sidney Brenner has characterized cell fate as determined by the European plan or the American plan: In the European plan, who you are as a neuron is determined by your ancestry; in the American plan, by your neighbors.

The Relationship between Neuronal Birthday and Cell Fate

Does the time at which a presumptive neuron stops dividing and migrates away from the ventricular zone influence its fate? This question can be addressed by marking neurons that become postmitotic, or are "born," at a particular time. In a technique pioneered by Angevine and Sidman,[49] a pulse of [^{3}H]thymidine is given by intrauterine or intravenous injection on a particular day during development. The label is taken up and incorporated into the DNA of cells undergoing cell division at that time; unincorporated thymidine rapidly disappears from the blood. Postmitotic cells will not be labeled. In cells that continue to divide after the pulse (glial cells and neural precursor cells that remain in the ventricular zone), the label will be diluted during subsequent rounds of DNA synthesis and cell division. However, if a cell divides during the pulse of [^{3}H]thymidine and one or both of its progeny stop dividing, migrate away from the ventricular zone, and differentiate into neurons, then those neurons will remain heavily labeled. Thus, the fate of neurons born on a particular day can be visualized by exposing an embryo to a pulse of [^{3}H]thymidine on that day, allowing development to continue, and then analyzing the nervous system by autoradiography to detect labeled cells.

Use of this technique has revealed that in the mammalian cerebrum there is a systematic relationship between the time a neuron is born and its final position in the cerebral cortex; development proceeds in an inside-out fashion (Figure 23.12).[50,51] Neurons of the deepest cortical layers are born first. Neurons in more superficial layers are born later and migrate through the cells of the deeper layers to assume their final positions within the

[47]Cepko, C. L., et al. 1996. *Proc. Natl. Acad. Sci. USA* 93: 589–595.

[48]McConnell, S. K. 1995. *Neuron* 15: 761–768.

[49]Angevine, J. B., and Sidman, R. L. 1961. *Nature* 192: 766–768.

[50]Rakic, P. 1974. *Science* 183: 425–427.

[51]Luskin, M. B., and Shatz, C. J. 1985. *J. Comp. Neurol.* 242: 611–631.

FIGURE 23.12 Neurogenesis of the Primary Visual Cortex of the cat. (A) Autoradiographs of sections of the adult visual cortex of animals injected with [^{3}H]thymidine on embryonic day 33 (E33) or 56 (E56). Bright-field micrographs of the same sections stained with cresyl violet indicate that heavily labeled cells are located in layer 6 after the E33 injection, and layers 2 and 3 after injection on E56. (B) Histograms showing the distribution of cells labeled on various days between E30 and E56 illustrate the inside-out pattern of neurogenesis in the visual cortex. (After Luskin and Shatz, 1985; micrograph kindly provided by M. B. Luskin.)

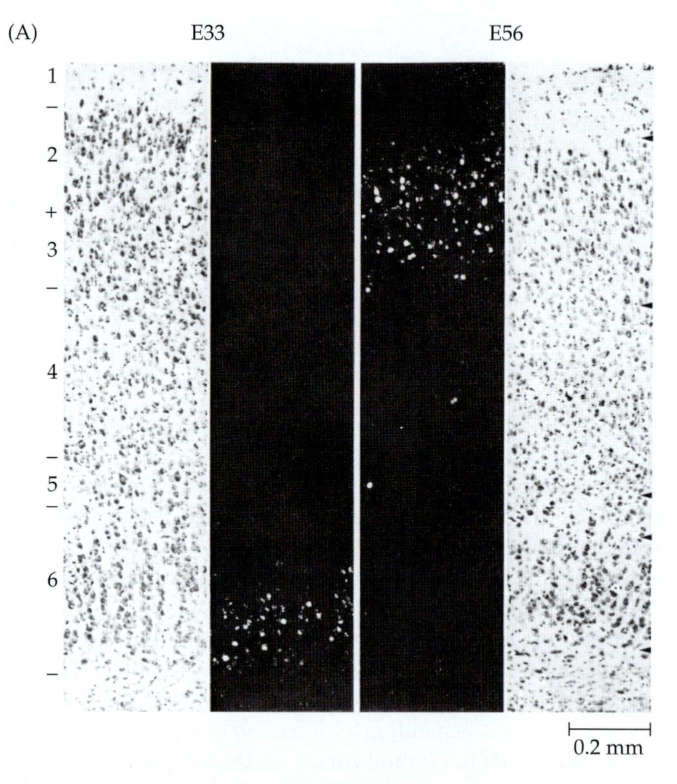

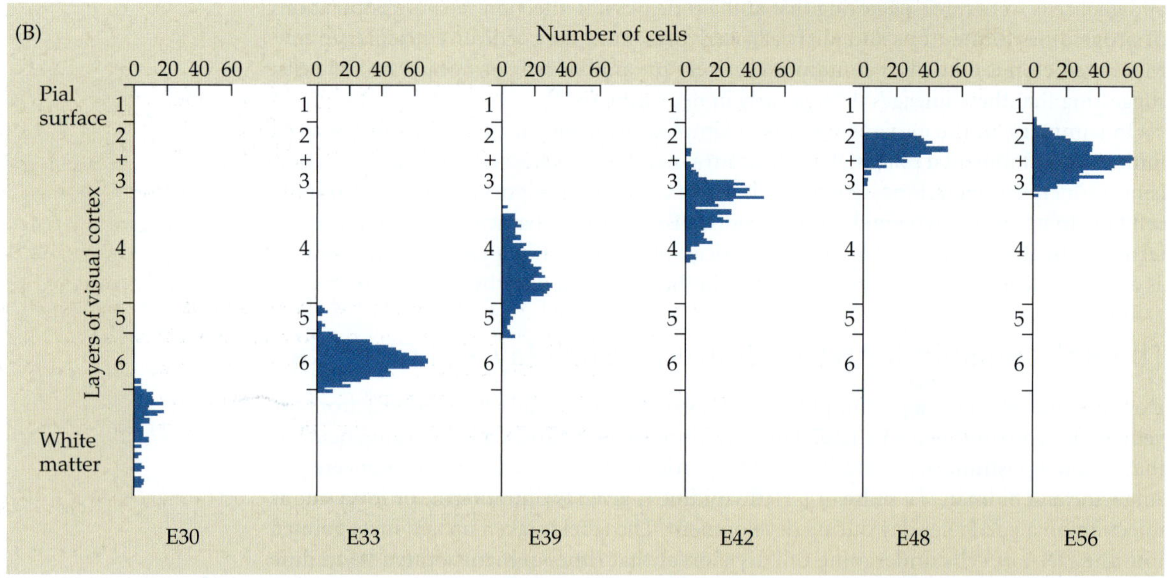

cortex. A similar correspondence between time of birth and the ultimate position occupied by a neuron is found throughout the CNS, although not all regions develop in the inside-out manner of the cerebral cortex.

Do neurons simply migrate outward until they reach the surface of the developing cerebral cortex, or can a migrating neuron recognize a particular laminar destination? Experiments in developing ferrets have demonstrated that cortical neurons become committed to occupy a particular position (Figure 23.13).[48,52] Cells, including embryonic precursors destined to give rise to neurons of layers 5 and 6, were taken from the ventricular zone of young embryos and transplanted into the ventricular zone of older hosts, amid cells destined for the more superficial layers 2 and 3. Precursor cells that were transplanted while in the early S phase of the cell cycle (when DNA synthesis takes place) were respecified: They gave rise to neurons that migrated to layers 2 and 3. However, if the

[52]McConnell, S. K. 1988. *J. Neurosci.* 8: 945–974.

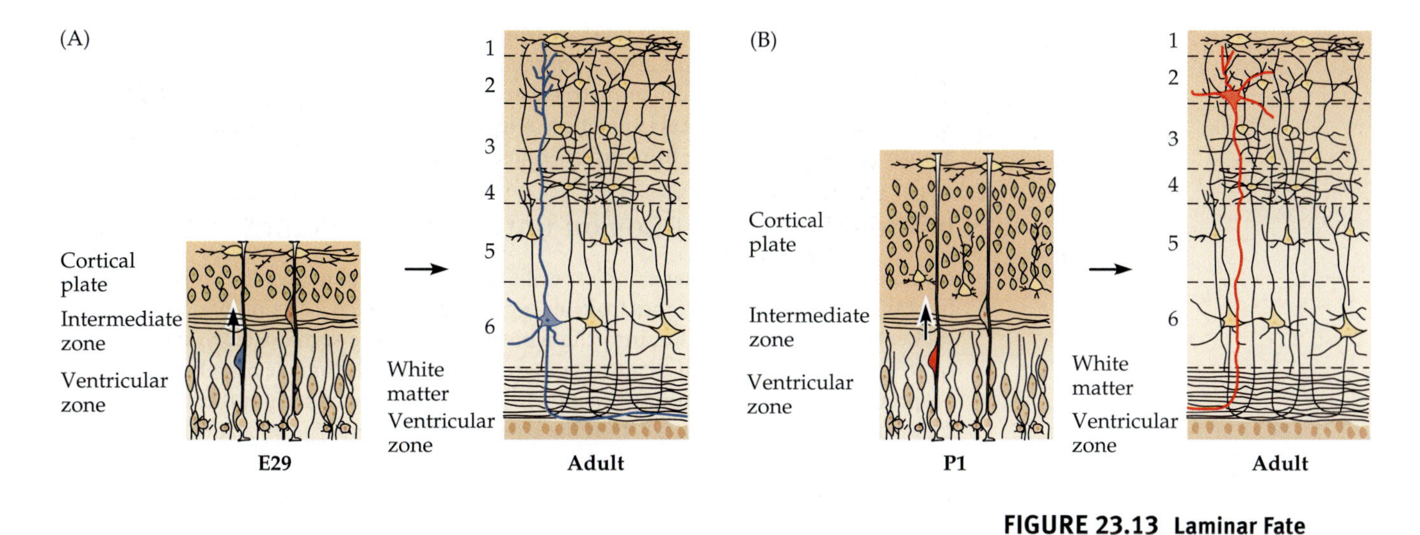

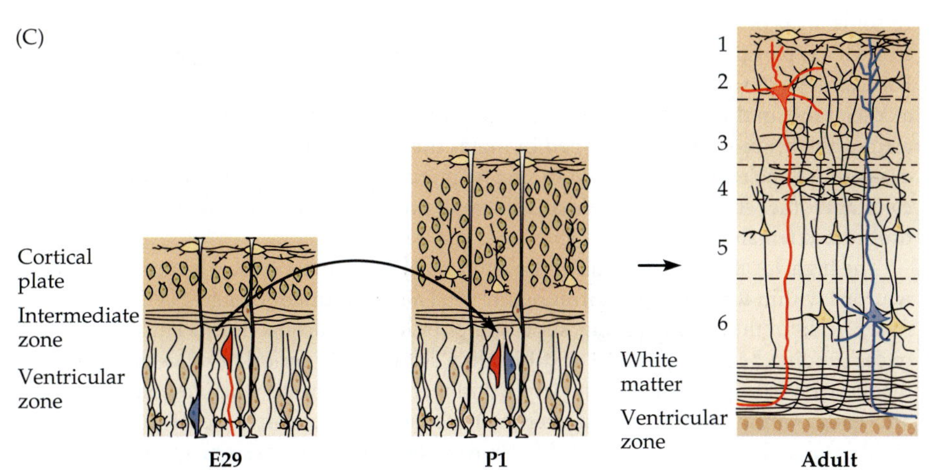

FIGURE 23.13 Laminar Fate Determination in the Cerebral Cortex of the ferret. (A) Neurons born on day E29 migrate to form layer 6 of the cortical plate in the adult. (B) Neurons born on postnatal day 1 (P1) migrate to form layer 2/3 of the adult cortical plate. (C) Cells transplanted from the ventricular zone of E29 embryos to the ventricular zone of P1 neonates adopted one of two fates. Cells transplanted while in the late stages of their final cell cycle or postmitotic neurons that had not yet migrated from the ventricular zone (blue) maintained their original identity and migrated to layer 6. Cells transplanted while in the S phase of the cell cycle (red) were respecified and migrated to layer 2/3.

transplanted cells were in the late stages of their final cell cycle or were postmitotic neurons that had not yet migrated from the ventricular zone, then they maintained their original identity. They migrated only as far as layer 6, then stopped and formed connections appropriate for their birthdays. Thus, the laminar fate of a cortical neuron is determined while it is still in the ventricular zone, just prior to the cell's final mitotic division.

Genetic Abnormalities of Cortical Layers in *Reeler* Mice

Another example of how a neuron's fate can be determined before it reaches its final position is found in the mutant mouse *reeler* (so called because of the uneven gait it displays). In the developing cortex of *reeler* mice, neurons fail to migrate past one another.[53] Therefore, their relative positions in the adult are inverted: Neurons born at early times end up in the most superficial layers, whereas neurons born later end up in deeper layers. In spite of their aberrant positions, the misplaced neurons acquire the morphological appearance and make connections appropriate to their time of birth. Thus, the morphology of cortical neurons and the nature of their synaptic interactions are determined at the time the neuron is born and can be expressed independently of position. The *reeler* phenotype results from more subtle consequences of the mutation in the cortex and elsewhere in the brain.

The product of the *reeler* gene is a large extracellular matrix glycoprotein called reelin.[54] It is not expressed in the cortical cells that fail to migrate properly, but rather in Cajal–Retzius cells in the marginal zone of the cortex.[55] Disruption of other genes in cortical cells can mimic *reeler*. For example, mutation of a cytoplasmic phosphotyrosine-containing protein called Disabled-1 produces a *reeler* phenotype,[56] as does genetic knockout of the very

[53] Caviness, V. S., Jr. 1982. *Dev. Brain Res.* 4: 293–302.

[54] D'Arcangelo, G., et al. 1997. *J. Neurosci.* 17: 23–31.

[55] Meyer, G., et al. 1998. *J. Comp. Neurol.* 397: 493–518.

[56] Howell, B. W., et al. 1997. *Nature* 389: 733–737.

low-density lipoprotein receptor together with the apolipoprotein E receptor 2.[57] This suggests that these three proteins are part of the intracellular machinery that enables cortical cells to recognize the extracellular matrix–associated reelin signal and use it to establish the appropriate pattern of cortical lamination. A number of other genetic malformations of the human cerebral cortex arise from disorders of neuronal migration.[21]

Influence of Local Cues on Cortical Architecture

An example of a feature that can be specified after a cortical neuron has migrated away from the ventricular zone comes from an experiment in which a section of developing visual cortex was transplanted into the whisker area of the somatosensory cortex in a rat.[58,59] Neurons in the transplant rearranged to form the distinctive barrel structure appropriate to their new location (Chapter 18), a phenotype not seen in the visual cortex. Thus, extrinsic influences can determine aspects of the phenotype of cortical neurons.

Hormonal Control of Development

In some areas of the central nervous system the fate of neurons is under hormonal control. This is particularly conspicuous in regions subserving sexually dimorphic behaviors. For example, in songbirds such as the canary, the high vocal center (HVC) nucleus plays a crucial role in the acquisition and retention of song—a uniquely male behavior.[60] This area of the brain is more developed in males than in females (Figure 23.14). However, even adult females can be induced to sing by injections of testosterone, and the HVC nucleus and other structures associated with song production become enlarged in such androgenized females.

The high vocal center nucleus in the adult male and female canary is unusual because neurons within it turn over continuously.[61,62] The recruitment of new HVC neurons in males peaks in the fall and spring, shortly after peaks in neuronal death. This is just when males modify their song for the next breeding season. The period of neuronal death coincides with a drop in testosterone levels, and recruitment peaks when testosterone levels are rising. Administration of testosterone to females induces an increase in the recruitment of new HVC neurons. New HVC neurons, like their predecessors, receive appropriate synaptic input and project axons to the proper targets. These remarkable observations indicate not only that new neurons arise in the adult brain, but also that they can be assimilated into its complex architecture so as to provide the substrate for the remodeling of a behavior as intricate as birdsong.

Neural Stem Cells

The HVC of songbirds is not the only example of new neurons being added to the adult nervous system. Neurons are also continuously added to the hippocampus and olfactory bulb in adult mammals.[63,64] Where do the cells come from? **Neural stem cells** that have the capacity for self-renewal have been isolated from the walls of the ventricles and from

[57]Trommsdorff, M., et al. 1999. *Cell* 97: 689–701.

[58]Schlaggar, B. L., and O'Leary, D. D. M. 1991. *Science* 252: 1556–1560.

[59]O'Leary, D. D. M., et al. 1995. *Ciba Found. Symp.* 193: 214–230.

[60]Nottebohm, F. 1989. *Sci. Am.* 260(2): 74–79.

[61]Kirn, J. R., and Nottebohm, F. 1993. *J. Neurosci.* 13: 1654–1663.

[62]Rasika, S., Alvarez-Buylla, A., and Nottebohm, F. 1999. *Neuron* 22: 53–62.

[63]Altman, J., and Das, G. D. 1965. *J. Comp. Neurol.* 124: 319–335.

[64]Kornack, D. R., and Rakic, P. 1999. *Proc. Natl. Acad. Sci. USA* 96: 5768–5773.

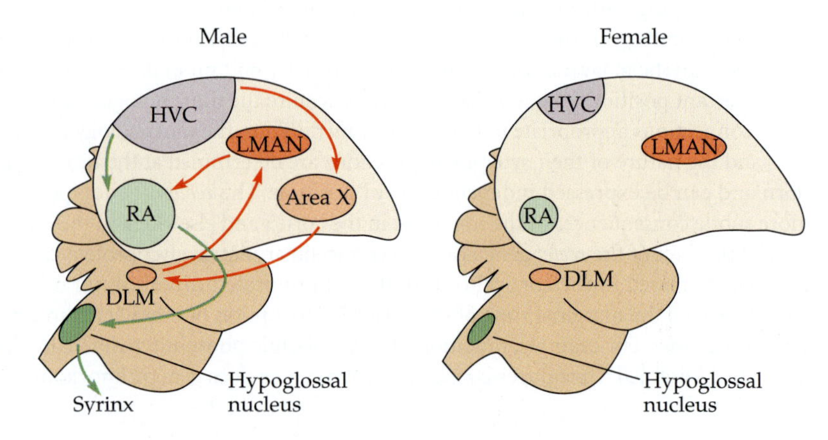

FIGURE 23.14 Sexual Dimorphism in the Avian Brain. Schematic diagram of the major brain areas and pathways involved in production of song in songbirds. The higher vocal center (HVC), robust nucleus of the archistriatum (RA), and hypoglossal nucleus form the posterior, vocal motor pathway (green). The HVC, area X, medial dorsolateral nucleus of the thalamus (DLM), and lateral magnocellular nucleus of the anterior neostriatum (LMAN) form the anterior pathway (red). HVC, hypoglossal nucleus, and RA are significantly larger in male birds; area X has not been observed in brains of female finches.

the hippocampus of the adult brain and propagated in vitro.[65] Such cells can differentiate into neurons, oligodendrocytes, and astrocytes. Neurons that are added to the olfactory bulb in vivo arise from slowly dividing stem cells in the innermost, ependymal layer of the walls of the lateral ventricles (the vestige of the original ventricular zone).[66] One of the progeny of each division enters the subependymal zone to become a progenitor cell. The progenitor cell divides rapidly, forming immature neurons that migrate rostrally into the bulb, where they differentiate as interneurons and integrate into the existing circuitry.[67] When a lesion is made in the central nervous system, the progeny of the stem cells become astrocytes rather than neurons, migrate to the lesion, and participate in scar formation at the site of the injury. Thus, the fate of stem cell progeny in the adult can be influenced by environmental signals.

Neural stem cells therefore represent a population of cells that can multiply in culture and be induced to differentiate into either glial cells or neurons. This raises the possibility that such stem cells could be harvested to generate neurons or glial cells for use in the treatment of nervous system diseases in which cell loss or dysfunction occurs (Chapter 24).[68]

[65]Reynolds, B. A., and Weiss, S. 1996. *Dev. Biol.* 175: 1–13.

[66]Johansson, C. B., et al. 1999. *Cell* 96: 25–34.

[67]Luskin, M. 1993. *Neuron* 11: 173–189.

[68]Yandava, B. G., Billinghurst, L. L., and Snyder, E. Y. 1999. *Proc. Natl. Acad. Sci. USA* 96: 7029–7034.

[69]Weston, J. 1970. *Adv. Morphogenesis* 8: 41–114.

[70]Dupin, E., Ziller, C., and Le Douarin, N. M. 1998. *Curr. Top. Dev. Biol.* 36: 1–35.

Control of Neuronal Phenotype in the Peripheral Nervous System

Are the mechanisms that determine cell fate in the vertebrate peripheral nervous system similar to those that operate in the central nervous system? For example, what roles do cell lineage, neuronal birthday, or local cues play in determining the fate of peripheral neurons and glial cells? Are the progeny of a particular ancestor destined to form cells with specified properties, such as being autonomic rather than sensory cells, or using ACh as the transmitter rather than norepinephrine?

Such questions have been studied in chick and quail nervous systems by Le Douarin, Weston, and others.[69,70] In vertebrate embryos, neural crest cells at different positions along the neuraxis give rise to different cell types of the peripheral nervous system (Figure 23.15A). To investigate whether the phenotype of cells derived from the neural crest was fixed early in development or could be altered by moving the cells to a new position along the neuraxis, Le Douarin transplanted cells from one region of the neural crest to a different region of a host embryo and then followed the fates of the transplanted cells. In these experiments, donor cells were taken from quail embryos and were implanted into a host chick embryo so that the transplanted cells could be recognized on the basis of unambiguous cytological differences between quail and chick cells (Figure 23.15B). After transplantation, quail cells assumed the fate appropriate to their new position. For example, cells removed from a region that would normally become the adrenal gland could instead innervate the gut.

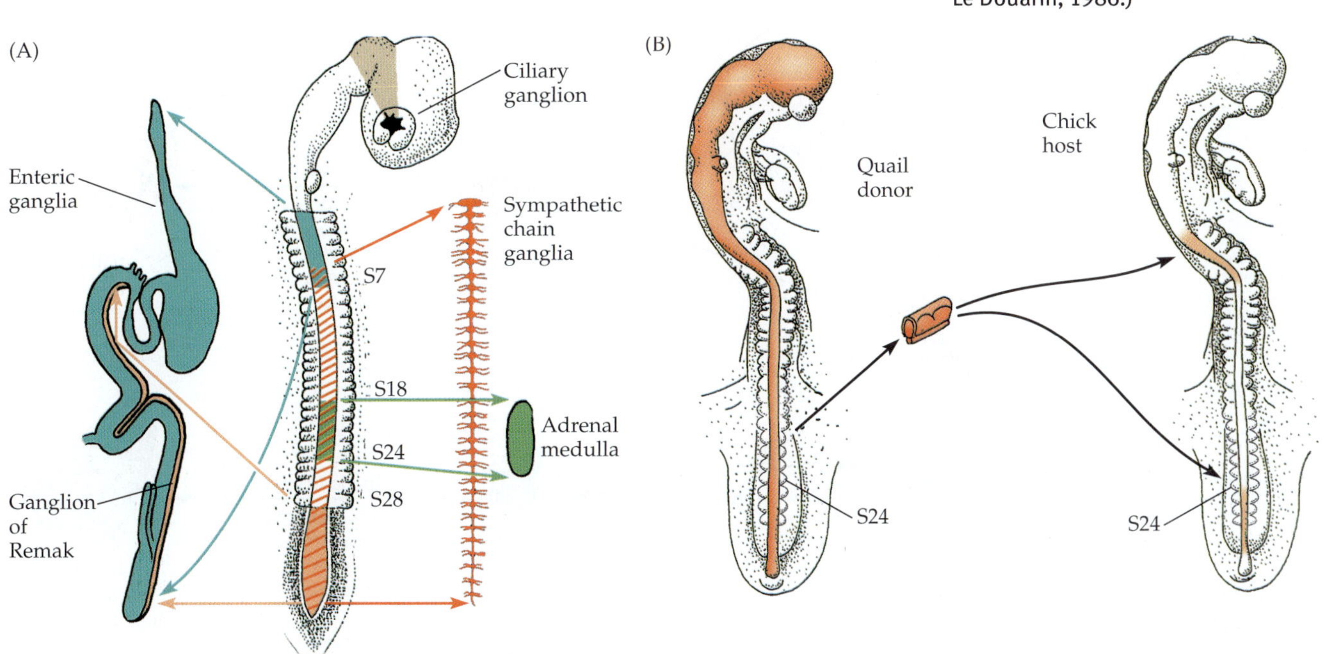

FIGURE 23.15 The Fate of a Neural Crest Cell is determined by environmental cues. (A) Neural crest cells give rise to a variety of peripheral ganglia. The ciliary ganglion is formed by cells from the mesencephalic neural crest. The ganglion of Remak and the enteric ganglia of the gut are formed by cells from the vagal (somites 1–7) and lumbosacral (caudal to S28) regions of the neural crest. The ganglia of the sympathetic chain are derived from all regions of the neural crest caudal to S5. The adrenal medulla is populated by crest cells from S18–S24. (B) If crest cells from S18–S24, which are destined to form the adrenal medulla, are transplanted from a quail donor to the vagal or lumbosacral region of a host chick embryo, they will adopt the fate appropriate to their new location and populate the ganglion of Remak and the enteric ganglia of the gut. (After Le Douarin, 1986.)

A related question is whether the phenotype of neural crest cells is determined before the cells migrate away from the neural tube or after they reach their position in the periphery. The answer is provided by studies of neural crest cells isolated from one such peripheral destination, the adrenal gland primordium of embryonic mammals.[71] In culture, these cells give rise to either chromaffin cells, which are found in the adult adrenal medulla, or adrenergic sympathetic neurons. Glucocorticoid hormone (synthesized by cells of the adrenal cortex) promotes expression of chromaffin cell–specific enzymes (Figure 23.16). By contrast, two proteins present at sites where sympathetic ganglia form, called basic fibroblast growth factor (bFGF) and nerve growth factor (NGF), cause progenitor cells in culture to differentiate as sympathetic neurons. Thus, the phenotype of a neural crest cell can be determined after it has migrated away from the neural tube, by factors in the periphery.

Transmitter Choice in the Peripheral Nervous System

When neural crest cells are transplanted at a sufficiently early stage, their fate can be altered so that they make an entirely different transmitter, such as acetylcholine instead of norepinephrine. In some cases such a change occurs during the normal course of development. For example, sympathetic neurons that innervate sweat glands initially synthesize norepinephrine, but during the second and third weeks of postnatal development they are induced to synthesize acetylcholine by factors associated with their target.[72] At later stages, crest cells become committed and lose the ability to differentiate in a manner determined by their environment.

Sympathetic ganglion cells in culture have been used to explore the mechanism of this neurotransmitter switch. When neurons are dissociated from the superior cervical ganglia of newborn rats and grown in culture in the absence of other cell types, all of them

[71]Anderson, D. J. 1993. *Annu. Rev. Neurosci.* 16: 129–158.

[72]Francis, N. J., and Landis, S. C. 1999. *Annu. Rev. Neurosci.* 22: 541–566.

FIGURE 23.16 Cell Fate Determination in the neural crest sympathoadrenal (SA) lineage. Pluripotent neural crest cells give rise to bipotent SA progenitor cells. The fate of SA progenitor cells depends on environmental signals. SA progenitors that migrate to the region of the adrenal medulla are exposed to glucocorticoids, which inhibit neural differentiation and induce chromaffin cell differentiation. Alternatively, SA progenitor cells that migrate to the region of the developing sympathetic chain are exposed to basic fibroblast growth factor (bFGF), which induces the expression of nerve growth factor (NGF) receptors in the SA progenitors. Subsequent exposure to NGF renders the cells insensitive to glucocorticoid-induced differentiation and induces them to differentiate as sympathetic neurons.

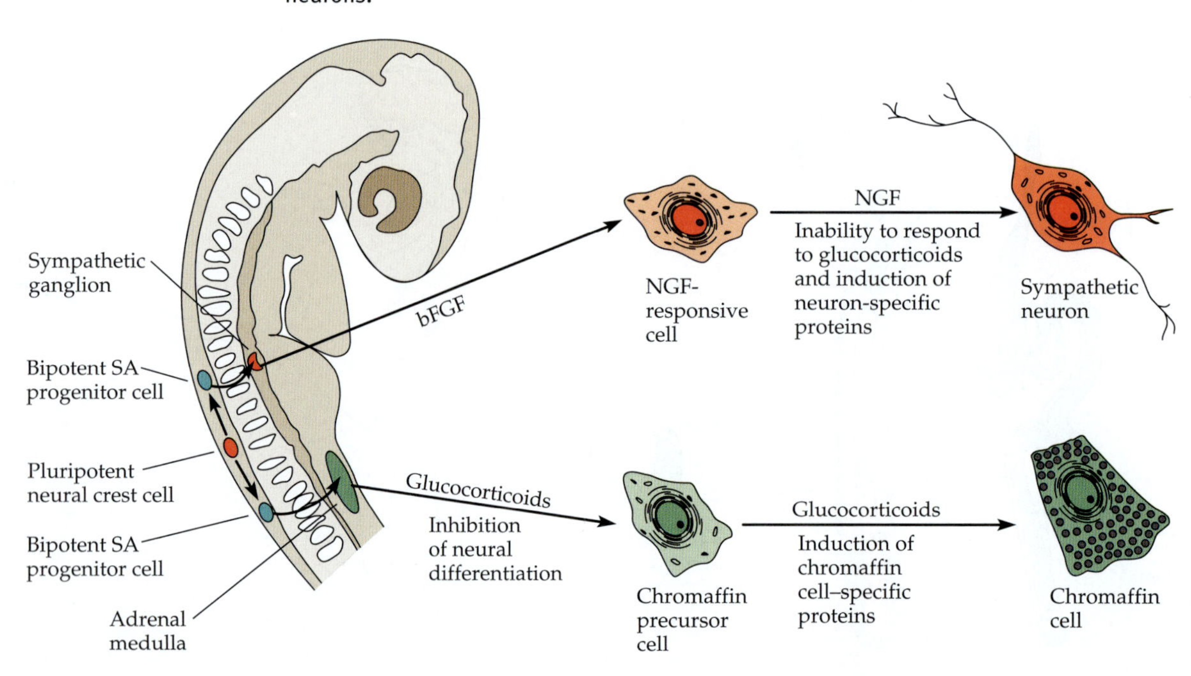

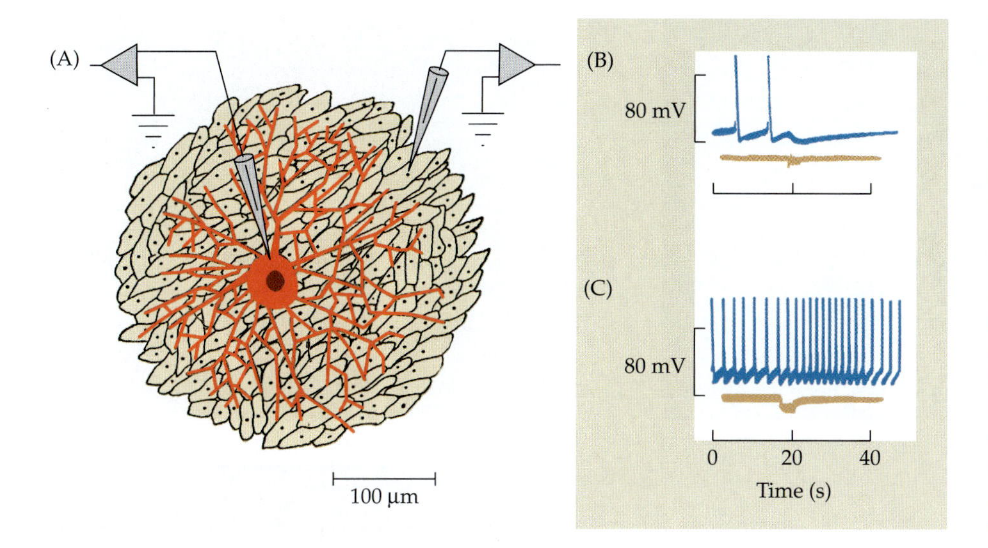

FIGURE 23.17 Single Neurons from Sympathetic Ganglia Can Release Both Acetylcholine and Norepinephrine at synapses on heart cells in culture. (A) A microculture containing a single sympathetic neuron grown on an island of cardiac muscle cells. (B) A brief train of impulses in the neuron (10 Hz, deflection of lower trace) produced inhibition of spontaneous myocyte activity due to release of ACh (upper trace). (C) Addition of atropine (10^{-7} *M*) blocked the inhibitory cholinergic response, revealing an excitatory effect, which is due to the release of norepinephrine. (After Furshpan et al., 1976.)

contain tyrosine hydroxylase and synthesize and store catecholamines.[73] However, if the neurons are grown in the presence of certain nonneuronal cells, such as heart muscle cells or sweat glands, the neurons slowly stop synthesizing catecholamines and begin synthesizing choline acetyltransferase and acetylcholine instead.[74] To establish unequivocally that this change was occurring in individual cells, single neurons were cultured on microislands of heart cells (Figure 23.17).[75] The neurons rapidly extended neurites and established synaptic contact with the heart cells. Initially these synapses were purely adrenergic, but over the course of several days cells began to release both ACh and norepinephrine. Eventually transmission became purely cholinergic.

A factor that induces cholinergic differentiation of sympathetic neurons was purified from heart-conditioned medium and cloned.[76] It turned out to be leukemia inhibitory factor (LIF), a protein that had been characterized previously on the basis of its ability to induce differentiation of cells in the immune system. Two other closely related cytokines, ciliary neurotrophic factor (CNTF) and cardiotrophin-1, were found to have a similar effect on cultured neurons.[72] All three activate the same receptor complex (called the LIFRβ-gp130 receptor), and blockade of this receptor inhibits induction of cholinergic properties in sympathetic neurons cultured with sweat glands. However, the sweat gland innervation of mutant mice lacking both LIF and CNTF is normal, and results of other experiments suggest that cardiotrophin-1 is not the factor that mediates the switch in cells innervating sweat glands in vivo. Although the triple knockout has not been described, it would appear that the sweat gland–derived factor is another, as yet unidentified, ligand for the LIFRβ-gp130 receptor.

AXON OUTGROWTH

Growth Cones, Axon Elongation, and the Role of Actin

The tips of growing axons expand to form **growth cones**. Ramón y Cajal was the first to recognize the growth cone as the region of the axon responsible for navigation and elongation toward a target (Figure 23.18). Growth cones extend and retract broad membranous sheets called lamellipodia, and slender, spikelike protrusions termed filopodia, for distances of tens of micrometers, as if sampling the substrate in every direction.[77] Filopodia adhere to the substrate and pull the growth cone in their direction.

Actin plays a key role in growth cone motility (Figure 23.19).[78–80] Both lamellipodia and filopodia are rich in filamentous actin, and agents that inhibit actin polymerization, such as the fungal toxin cytochalasin B, immobilize growth cones. The protrusion and retraction of lamellipodia and filopodia, as well as the forward movement of the body of the growth cone, appear to be mediated by two processes: (1) polymerization and disas-

[73]Mains, R. E., and Patterson, P. H. 1973. *J. Cell Biol.* 59: 329–345.

[74]Patterson, P. H., and Chun, L. L. Y. 1977. *Dev. Biol.* 56: 263–280.

[75]Furshpan, E. J., et al. 1976. *Proc. Natl. Acad. Sci. USA* 73: 4225–4229.

[76]Yamamori, T., et al. 1989. *Science* 246: 1412–1416.

[77]Smith, S. J. 1988. *Science* 242: 708–715.

[78]Letourneau, P. C. 1996. *Perspect. Dev. Neurobiol.* 4: 111–123.

[79]Lin, C-H., Thompson, C. A., and Forscher, P. 1994. *Curr. Opin. Neurobiol.* 4: 640–647.

[80]Suter, D. M., et al. 1998. *J. Cell Biol.* 141: 227–240.

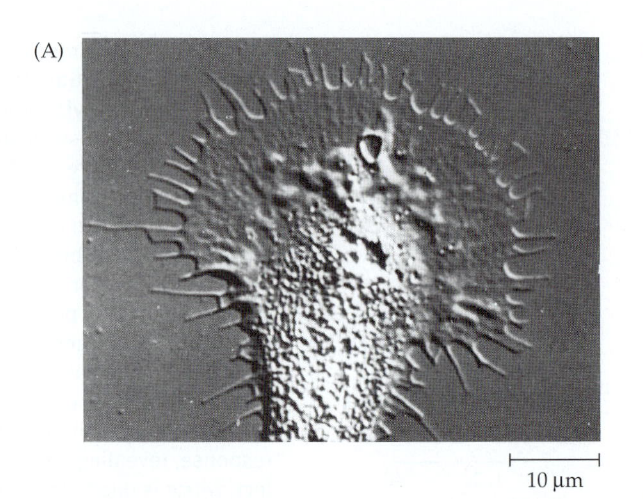

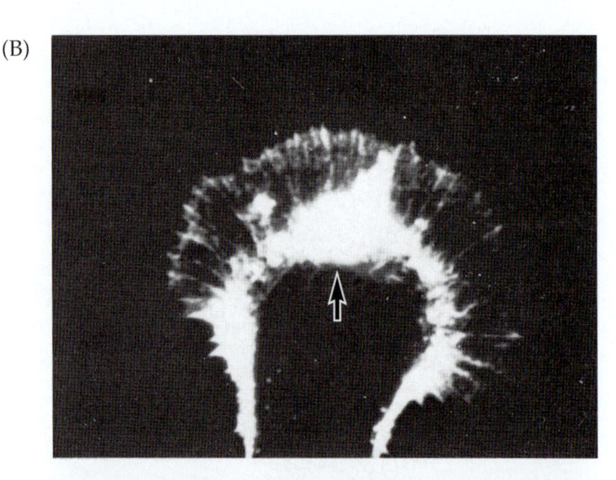

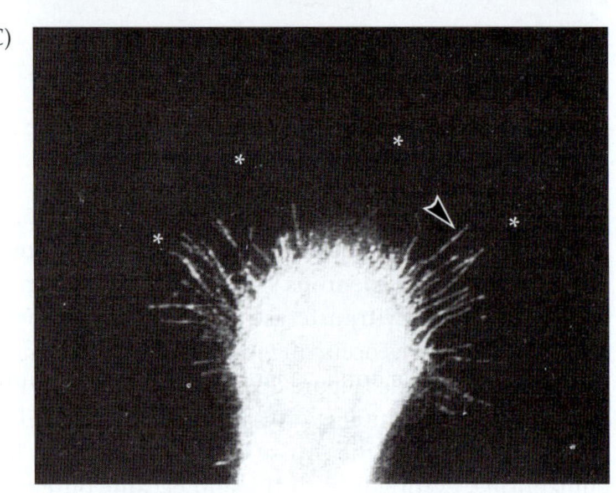

FIGURE 23.18 The Morphology of Growth Cones. (A) Growth cone observed by differential interference contrast microscopy. (B) Fluorescence micrograph showing the distribution of filamentous actin visualized with rhodamine-conjugated phalloidin. Actin filaments align with filopodia, or microspikes, in the periphery of the growth cone; randomly oriented filaments are often concentrated near the central domain (arrow). (C) Microtubule distribution visualized with antitubulin antibodies and fluorescein-conjugated secondary antibodies. Microtubules are concentrated in the axon. Most terminate in the central domain of the growth cone; some (arrowhead) extend toward the growth cone margin (asterisks). (After Forscher and Smith, 1988; micrographs kindly provided by S. J. Smith.)

sembly of actin filaments and (2) myosin-mediated translocation of actin filaments away from the leading edge of the growth cone. Both processes can harness the energy of ATP hydrolysis to generate force and are regulated by actin-binding proteins. Calcium, protein kinases, and other intracellular second messengers modulate the activity of actin-binding proteins. For example, stalling and retraction of growth cones, two common events in axon growth, are associated with calcium influx and an increased frequency of transient elevations in intracellular calcium concentration.[81,82]

[81]Grumbacher-Reinert, S., and Nicholls, J. 1992. *J. Exp. Biol.* 167: 1–14.

[82]Gomez, T. M., and Spitzer, N. C. 1999. *Nature* 397: 350–355.

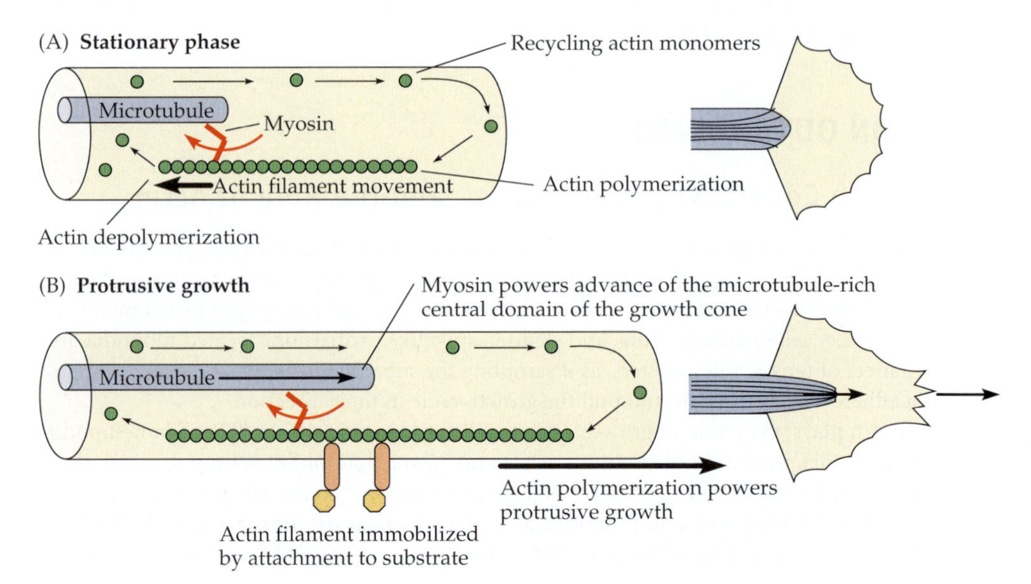

FIGURE 23.19 Model for Actin-Based Motility of Growth Cones. (A) Cross section (left) and top view (right) of a growth cone in stationary phase. Microtubule-attached myosin powers rearward movement of actin filaments, while filaments are undergoing continuous polymerization at the leading edge of the growth cone and depolymerization centrally. (B) Similar views of a growth cone during protrusive growth. Actin filaments are immobilized by attachment to the substrate. Actin polymerization now causes protrusion of the growth cone, while myosin cycling moves microtubules forward, advancing the central domain of the growth cone. (After Suter et al., 1998.)

Cell and Extracellular Matrix Adhesion Molecules and Axon Outgrowth

Cell adhesion molecules mediate axon outgrowth by providing a favorable environment for growth cone extension. **Cell adhesion molecules** are transmembrane or membrane-associated glycoproteins characterized by structural motifs in their extracellular portions that are homologous to immunoglobulin constant-region domains and fibronectin type III domains (Figure 23.20). Members of this immunoglobulin (Ig) superfamily include neural cell adhesion molecule (N-CAM), neuroglial CAM (NgCAM), TAG-1, MAG, and DCC.[83] These molecules mediate cell–cell adhesion, either through homophilic binding to their counterparts on other cells (e.g., N-CAM to N-CAM) or through heterophilic binding involving a distinct Ig superfamily member (e.g., NrCAM to TAG-1). An additional ubiquitous cell adhesion molecule is N-cadherin (see Figure 23.20), which mediates homophilic, calcium-dependent cell adhesion.[84]

In culture, expression of N-CAM and N-cadherin in cells promotes their aggregation, the extension of axons on cellular but not extracellular matrix substrates, and the binding together of growing axons into fascicles. Stimulation of axonal growth by cell adhesion molecules is not simply a matter of making the substrate "sticky"; it is mediated by activation of receptor tyrosine kinases such as the fibroblast growth factor (FGF) receptor.[83] The FGF receptor triggers an intracellular tyrosine phosphorylation signaling cascade that promotes axon elongation. Protein tyrosine phosphatases, enzymes that remove phosphate residues from tyrosine, regulate such responses.

Extracellular matrix adhesion molecules, including laminin, fibronectin, tenascin (J1, cytotactin), and perlecan, also provide favorable substrates for neurite extension.[85,86] These large extracellular glycoproteins have two or more identical or similar subunits held together by disulfide bonds (Figure 23.21). Each subunit is characterized by repeated structural motifs. Extracellular matrix proteins interact with cells via a family of receptors known as integrins. Many isoforms of the α and β subunits of integrin have been identified; each αβ combination produces a receptor with distinctive binding properties. Integrins provide

[83]Walsh, F. S., and Doherty, P. 1997. *Annu. Rev. Cell Dev. Biol.* 13: 425–456.

[84]Takeichi, M. 1995. *Curr. Opin. Cell Biol.* 7: 619–627.

[85]Lander, A. D. 1989. *Trends Neurosci.* 12: 189–195.

[86]Reichardt, L. F., and Tomaselli, K. J. 1991. *Annu. Rev. Neurosci.* 14: 531–570.

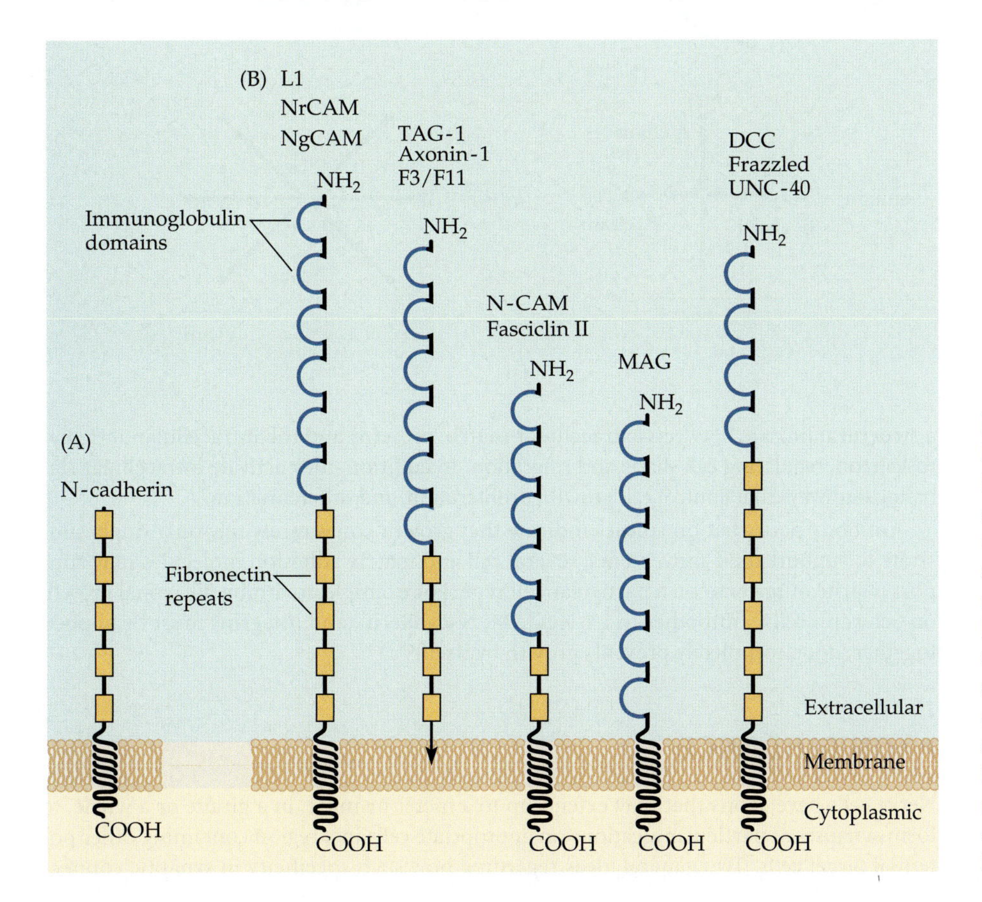

FIGURE 23.20 Two Classes of Neural Cell Adhesion Molecules. (A) N-cadherin mediates homophilic, calcium-dependent cell adhesion. (B) Members of the immunoglobulin superfamily are characterized by multiple repeats of disulfide-linked loops homologous to domains first characterized in the constant region of immunoglobulin molecules. Many of these cell adhesion molecules also contain multiple domains similar to the type III repeat of fibronectin (rectangles). Multiple names denote homologous proteins in different species.

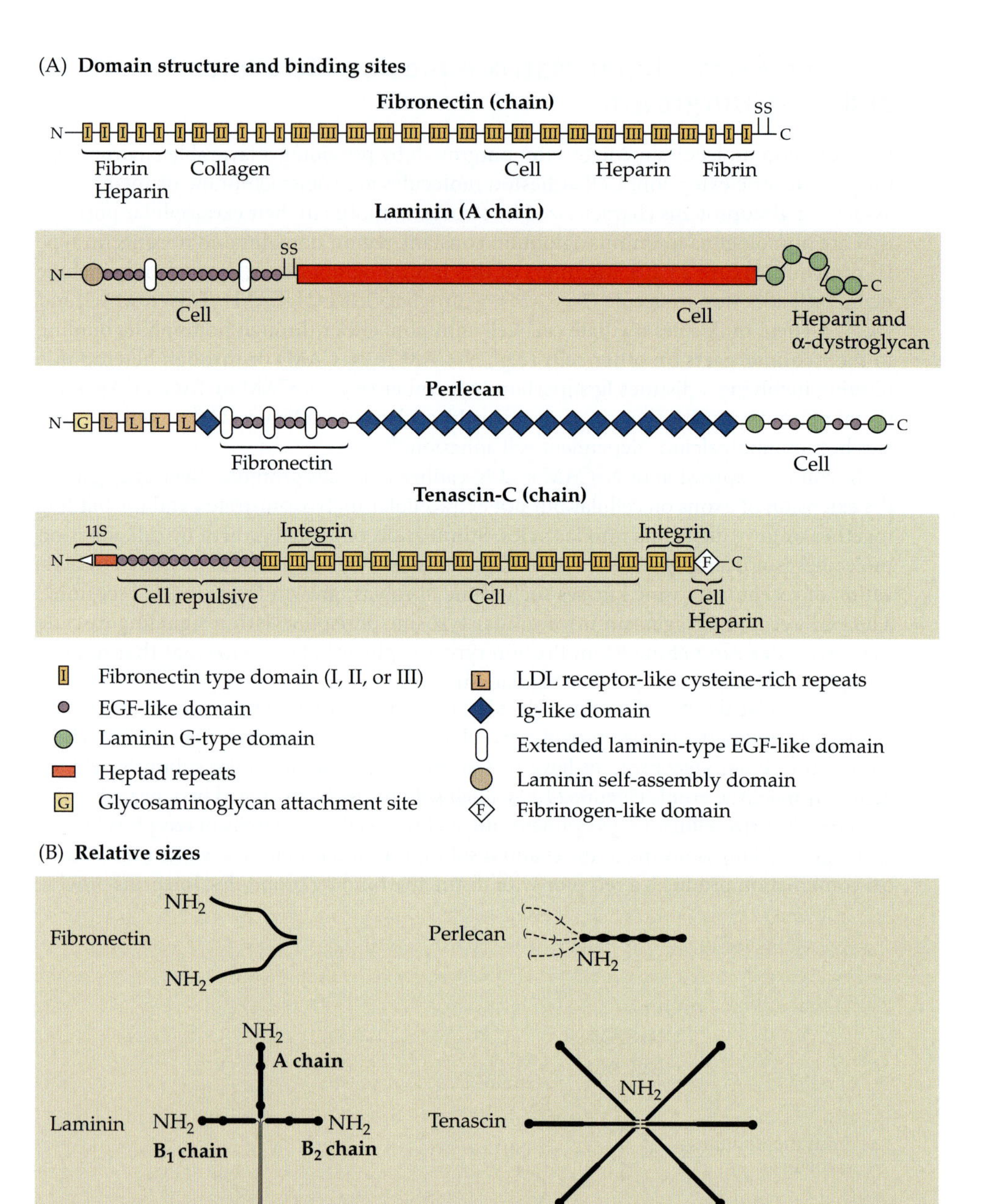

FIGURE 23.21 Extracellular Matrix Proteins that mediate cell–substrate adhesion and repulsion. (A) Schematic representations of the domain structure of the chains that comprise fibronectin, laminin (A chain), perlecan, and tenascin-C, and the sites at which cells, collagen, heparin, fibronectin, integrin, fibrin, and α-dystroglycan bind. S indicates positions of Cys residues involved in interchain disulfide bridging. Tenascin-C contains both binding and repulsive domains; the predominate effect of the whole protein is repulsion of cells and growth cones. (B) Molecules drawn approximately to scale. All but perlecan are oligomeric proteins; the chains are linked by disulfide bonds. Dashed lines indicate the 90- to 170-nm-long heparan sulfate chains attached to the N-terminal domain of perlecan. (After Engel, J., 1992.)

a structural linkage between extracellular matrix proteins and the intracellular actin cytoskeleton, regulating cell shape and migration. In addition, they activate intracellular signaling pathways that control cell growth, proliferation, and differentiation.[87]

Antibody perturbation studies indicate that growth cones rarely rely on a single substrate to support their movements; several cell and matrix adhesion molecules may support neurite outgrowth on a particular cell type. For example, to inhibit all axonal growth on Schwann cells, antibodies to L1/NgCAM, N-cadherin, and integrins must be applied together; none completely prevents growth by itself.[88,89]

[87]Giancotti, F. G., and Ruoslahti, E. 1999. *Science* 285: 1028–1032.

[88]Bixby, J. L., Lilien, J., and Reichardt, L. F. 1988. *J. Cell Biol.* 107: 353–361.

[89]Seilheimer, B., and Schachner, M. 1988. *J. Cell Biol.* 107: 341–351.

AXON GUIDANCE

Nerve cells have axons that may extend up to a meter or more, in a giraffe or a whale, to form synapses at particular locations on appropriate cells in a region containing other potential target cells. Two general ideas regarding how such specificity in synaptic connections is established during development were proposed during the first quarter of the twentieth century. One held that neurons and their targets were prespecified in such a way that only appropriate synaptic connections were established.[90] The other idea was that initial connections were established more or less randomly and were then sorted out by target-induced specification of neurons, elimination of incorrect synapses, or death of inappropriately connected cells.[91] A wealth of experimental evidence now supports the idea that the growth of axons and the formation of synaptic connections are selective; an axon is guided accurately to its target by cues in its environment.[92]

Target-Dependent and Target-Independent Navigation

How do extracellular cues guide growth cones? Ramón y Cajal originally proposed a **chemoattractant model** of axon guidance, whereby the growth cone navigates along a gradient of some molecule released by its target. This can account for directed axon outgrowth when the distance from the nerve cell body to its target is very short. For example, Lumsden and Davies have studied the growth of axons from the trigeminal ganglion in the head of the mouse into the adjacent epithelial tissue, a distance of less than 1 mm.[93] (These axons ultimately give rise to the sensory innervation of the whiskers [Chapter 18].) If the developing trigeminal ganglion is placed in culture near explants from several peripheral tissues, neurites grow from the ganglion toward their appropriate target, ignoring other tissues. Explants of target epithelium have this effect on axon outgrowth only if they are taken from embryos at the time innervation normally occurs.

In contrast, the ability of spinal motor axons to grow to the appropriate region in the limb does not depend on the presence of their target muscles. This was shown by removing the somites, which give rise to limb musculature, early in development.[94] Motor axons extended normally from the spinal cord, grew into the limb, and formed the appropriate pattern of muscle nerves, all in the absence of muscle. Thus, the factors that guide motor axons to their correct destinations in the limb are not supplied by the muscles that the axons ultimately innervate.

Navigation via Guidepost Cells

When the distance from a neuron to its target is more than several hundred microns, the pathway is marked with intermediate targets. For example, growth cones arising from sensory cells in the limbs of developing grasshoppers make abrupt changes in direction as they extend toward the CNS (Figure 23.22).[95] The turns occur when the growth cones contact so-called **guidepost cells**.[96,97] Such behavior suggests that interactions with guidepost cells, which are often immature neurons, are responsible for redirecting the growth cones. That this is so can be demonstrated by removing the guidepost cells by laser ablation before the growth cone arrives, in which case the appropriate change in trajectory is not made.

Synaptic Interactions with Guidepost Cells

In some cases axons make transient synaptic contacts with guidepost cells during development. In the developing hippocampus, for example, axons from the entorhinal cortex first make synapses with a transient population of neurons, the Cajal–Retzius cells.[98,99] Later, as granule cells appear and mature, entorhinal axons abandon the Cajal–Retzius

[90]Langley, J. N. 1895. *J. Physiol.* 18: 280–284.

[91]Weiss, P. 1936. *Biol. Rev.* 11: 494–531.

[92]Mueller, B. K. 1999. *Annu. Rev. Neurosci.* 22: 351–388.

[93]Lumsden, A. G. S., and Davies, A. M. 1986. *Nature* 323: 538–539.

[94]Phelan, K. A., and Hollyday, M. 1990. *J. Neurosci.* 10: 2699–2716.

[95]Bentley, D., and Caudy, M. 1983. *Cold Spring Harb. Symp. Quant. Biol.* 48: 573–585.

[96]Palka, J., Whitlock, K. E., and Murray, M. A. 1992. *Curr. Opin. Neurobiol.* 2: 48–54.

[97]Sato, Y., et al. 1998. *J. Neurosci.* 18: 7800–7810.

[98]Deller, T., Drakew, A., and Frotscher, M. 1999. *Exp. Neurol.* 156: 239–253.

[99]Del Río, J.A., et al. 1997. *Nature* 385: 70–74.

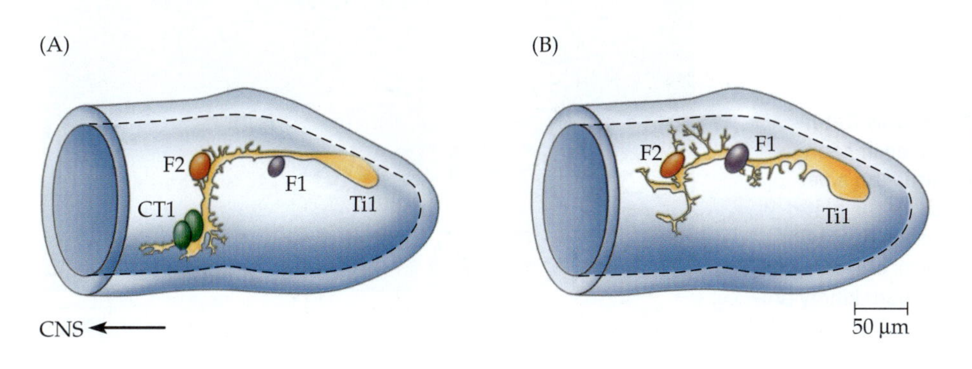

FIGURE 23.22 Growth Cones of Peripheral Neurons Rely on Guidepost Cells to navigate through the limb of the grasshopper. (A) In normal embryos, the axon of the Ti1 neuron encounters a series of guidepost cells on its route to the central nervous system: F1, F2, and two CT1 cells. (B) If the CT1 cells are killed early in development, the Ti1 neuron forms several axonal branches at the site of cell F2, with growth cones extending in abnormal directions. (After Bentley and Caudy, 1983.)

cells and contact granule cell dendrites. The Cajal–Retzius cells then disappear. In another example, axons from neurons in the lateral geniculate nucleus of the mammalian visual system reach the developing cortical plate before their synaptic targets, the pyramidal cells of layer 4, have been born. The geniculate axons form synaptic connections with the subplate neurons,[100] which are produced early in embryogenesis. The subplate neurons lie beneath the developing cortical plate and are destined to disappear shortly after birth.[101] After a few weeks, when the layer 4 pyramidal cells have reached their position in the cortex, geniculate axons abandon their connections with the subplate neurons and invade the cortex to establish the adult pattern of innervation. If the subplate neurons are eliminated early in development by local application of neurotoxins, lateral geniculate axons grow past the developing visual cortex and fail to form synaptic contacts with their targets.[102]

Mechanisms of Axon Guidance

The molecules that guide growth cones act in four basic ways, as **long- or short-range attractants** or **repellents.** Table 23.1 lists many of the guidance molecules and receptors that we discuss in this chapter.

Some short-range axon guidance cues are provided by contact of growth cones with the cell surface or extracellular matrix adhesion proteins that were described earlier in this chapter as promoters of axon outgrowth.[103] Adhesion molecules and their receptors also play essential roles in bundling of axons together (called fasciculation). Interfering with the activity of specific adhesion molecules or their receptors by gene knockout or application of antibodies disrupts the pattern of axon outgrowth and target innervation in animals and in culture.[104–106]

Another effect of extracellular matrix molecules on growing axons is illustrated by experiments on identified cells isolated from the leech central nervous system and grown in culture (Figure 23.23).[107] Substrates that contain tenascin or laminin not only support rapid extension of neurites from leech neurons, but also influence the pattern of neurite outgrowth and the distribution of calcium channels on the cells. Different neurons respond to particular extracellular matrix molecules in individual ways, thus providing an economical scheme by which a few molecules can exert diverse effects.

On the other hand, studies of growth cone behavior in cell culture indicate that cell and extracellular matrix adhesion molecules do not provide long-range guidance cues. For example, growth cones do not extend preferentially up or down gradients of cell and matrix adhesion molecules.[108] Nor do growth cones navigate using differences in the strength of adhesion of different surfaces; adhesive molecules are either permissive for growth or not.[109] Long-range guidance involves growth cone navigation along a gradient of a diffusible factor.

[100]Shatz, C. J., and Luskin, M. B. 1986. *J. Neurosci.* 6: 3655–3658.

[101]Luskin, M. B. and Shatz, C. J. 1985. *J. Neurosci.* 5: 1062–1075.

[102]Ghosh, A., et al. 1990. *Nature* 347: 179–181.

[103]Goodman, C. S. 1996. *Annu. Rev. Neurosci.* 19: 341–377.

[104]Brittis, P.A., et al. 1996. *Mol. Cell Neurosci.* 8: 120–128.

[105]McFarlane, S., et al. 1996. *Neuron* 17: 245–254.

[106]Hoang, B., and Chiba, A. 1998. *J. Neurosci.* 18: 7847–7855.

[107]Masuda-Nakagawa, L. M. and Nicholls, J. G. 1991. *Phil. Trans. R. Soc. Lond. B* 331: 323–335.

[108]McKenna, M. P. and Raper, J. A. 1988. *Dev. Biol.* 130: 232–236.

[109]Isbister, C. M. and O'Connor, T. P. 1999. *J. Neurosci.* 19: 2589–2600.

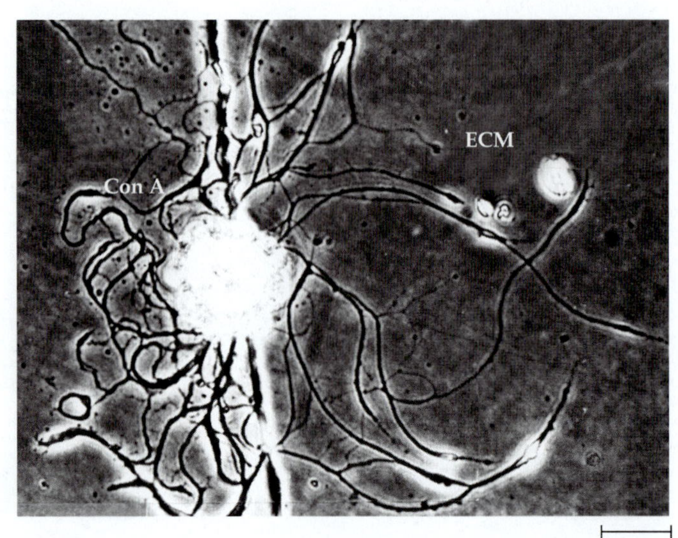

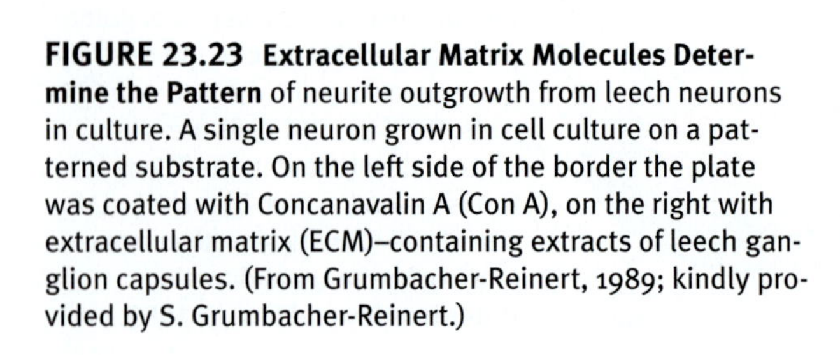

FIGURE 23.23 Extracellular Matrix Molecules Determine the Pattern of neurite outgrowth from leech neurons in culture. A single neuron grown in cell culture on a patterned substrate. On the left side of the border the plate was coated with Concanavalin A (Con A), on the right with extracellular matrix (ECM)–containing extracts of leech ganglion capsules. (From Grumbacher-Reinert, 1989; kindly provided by S. Grumbacher-Reinert.)

Table 23.1
Examples of short- and long-range chemoattraction and chemorepulsion

Ligand	Receptor/ligand	Attraction/repulsion
Short-range (contact mediated)		
TAG-1	NrCAM	Attraction
Slit	Robo	Repulsion
N-CAM	N-CAM	Attraction
ECM adhesion proteins	Integrins	Attraction
Ephrins	Eph receptors	Repulsion
Long-range (diffusible ligand)		
Netrin	DCC	Attraction
Semaphorins	Neuropilins	Repulsion
Slit	?	Repulsion
Netrin	?	Repulsion

Growth Cone Navigation in the Spinal Cord

A good example of the variety of mechanisms and molecules that growth cones employ to navigate to their targets is provided by axons of commissural interneurons in the vertebrate spinal cord. Early in development commissural interneurons, which lie in the dorsal part of the spinal cord, extend axons that grow ventrally, cross the midline, and then run longitudinally along the spinal cord toward their synaptic targets (Figure 23.24).[110]

Axons of commissural interneurons are initially attracted to the ventral midline by the protein netrin-1, a diffusible long-range chemoattractant released by specialized floor plate cells that lie along the ventral midline of the cord (Figure 23.24A).[111] Netrin-1 interacts with a receptor expressed by commissural neuron axons called DCC (which was mentioned earlier because it also acts as a cell adhesion molecule; see Figure 23.20).[112] The existence of a diffusible factor produced by the floor plate that attracted commissural axons was first demonstrated by culturing pieces of dorsal spinal cord either alone or together with pieces of floor plate (Figure 23.25).[110] Axons of commissural neurons specifically grew toward the floor plate, even when the explants were separated by as much as several hundred microns. This distance is too great to be spanned by a filopodium from a growth cone, implicating a diffusible factor. The factor, netrin-1, is a member of a family of secreted proteins characterized by domains homologous to the amino-terminal do-

[110]Tessier-Lavigne, M., et al. 1988. *Nature* 336: 775–778.
[111]Kennedy, T. E., et al. 1994. *Cell* 78: 425–435.
[112]Fazeli, A., et al. 1997. *Nature* 386: 796–804.

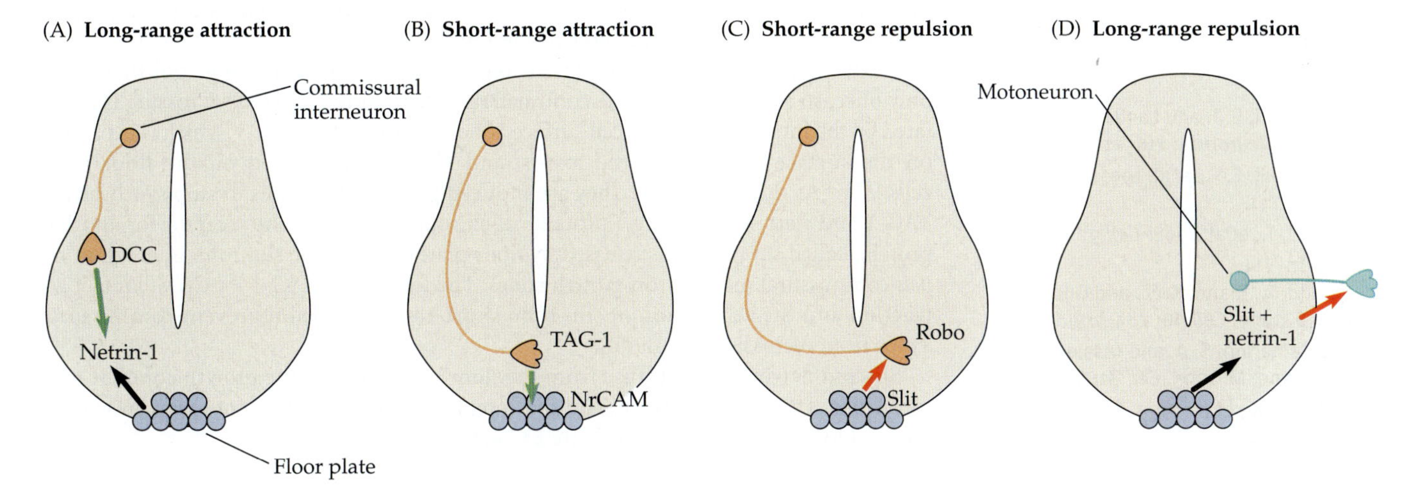

FIGURE 23.24 Long- and Short-Range Chemoattraction and Chemorepulsion guide developing axons in the vertebrate spinal cord. (A) Netrin-1, acting as a long-range chemoattractant, is released by cells of the floor plate and binds to its receptor (DCC) on commissural neurons, attracting their growth cones. (B) TAG-1 on commissural axon growth cones binds to NrCAM on floor plate cells. This short-range chemoattraction facilitates extension of commissural axon growth cones across the floor plate. (C) Robo on commissural axon growth cones binds to slit on floor plate cells. This short-range chemorepulsion prevents the axons from recrossing the floor plate. (D) Slit and netrin-1, released from the floor plate, interact with receptors on growth cones of motor neurons and repel them. This long-range chemorepulsion helps direct the growth of motor axons away from the cord.

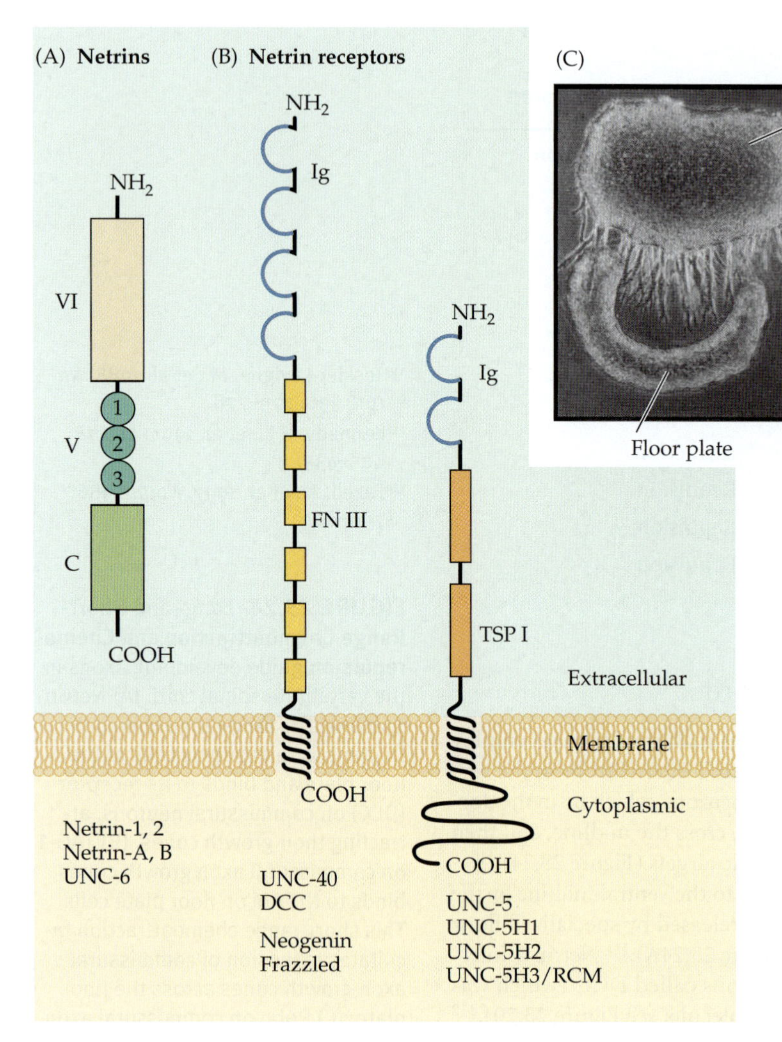

FIGURE 23.25 Netrins and Netrin Receptors function in long-range attraction and repulsion. (A) The amino-terminal portion of the secreted protein netrin is made up of domains VI and V, which are homologous to domains at the amino terminus of the γ chain of laminin. Domain V contains three EGF-like repeats. The carboxy-terminal domain C has no homology to laminin. (B) Netrin receptors have an extracellular domain, a single membrane-spanning region, and an intracellular domain. The extracellular domain of the DCC family of netrin receptors has four immunoglobulin domains (Ig) and six fibronectin III repeats (FN III). DCC and its homologues also function as neural cell adhesion molecules. The UNC-5 family of netrin receptors has two immunoglobulin and two thrombospondin type I domains (TSP I) extracellularly, and a large cytoplasmic region. (C) Micrographs of pieces of dorsal spinal cord from embryonic rats (top in each panel) cultured with a piece of floor plate tissue (left), an aggregate of COS cells secreting recombinant netrin-1 (middle), or control COS cells (right). The floor plate and netrin-1 both elicit the profuse and directed outgrowth of bundles of commissural axons from the dorsal spinal cord tissue. (After Tessier-Lavigne et al., 1988 and Kennedy et al., 1994; micrographs kindly provided by M. Tessier-Lavigne).

mains of the γ chain of laminin-1 (Figure 23.25A).[111, 92] Homologues of netrin are also involved in axon guidance in *Drosophila* and in *C. elegans*.

Next, the commissural axon growth cones cross the ventral midline, but they do so only once, so as to remain on the contralateral side (Figure 23.24B,C). Crossing is facilitated by the interaction of two cell surface adhesive molecules: TAG-1, which is expressed on the surface of commissural axons, and NrCAM, expressed on the floor plate cells.[113,114] As the axons cross, they are induced by floor plate cells to stop synthesizing TAG-1 and start synthesizing a protein called robo.[115] Robo is the receptor for another protein, called slit , which is released by floor plate cells.[116] The slit–robo interaction *repels* commissural interneuron growth cones. The loss of the TAG-1–NrCAM contact attraction and the acquisition of slit–robo short-range repulsion prevent commissural axons from recrossing the midline.

Slit and netrin-1 also diffuse from the floor plate to repel the growth cones of motoneurons. This is a long-range repulsive interaction that directs motor axons away from the cord toward the periphery (Figure 23.24D).[117,118]

[113]Stoeckli, E. T. and Landmesser, L. T. 1995. *Neuron* 14: 1165–1179.

[114]Stoeckli, E. T., et al. 1997. *Neuron* 18: 209–221.

[115]Kidd, T., et al. 1998. *Cell* 92: 205–215.

[116]Kidd, T., Bland, K. S., and Goodman, C. S. 1999. *Cell* 96: 785–794.

[117]Colamarino, S. A. and Tessier-Lavigne, M. 1995. *Cell* 81: 621–629.

[118]Brose, K., et al. 1999. *Cell* 96: 795–806.

Semaphorin Family of Chemorepellents

Additional proteins that act as chemorepellents are the **semaphorins**, a large family of secreted and transmembrane proteins (Figure 23.26A).[119] Originally identified and implicated in axon guidance in grasshoppers, semaphorins were recognized as chemorepellents when a vertebrate homologue, collapsin-1, was purified and shown to cause retraction or

[119]Kolodkin, A. L. 1998. *Prog. Brain Res.* 117: 115–132.

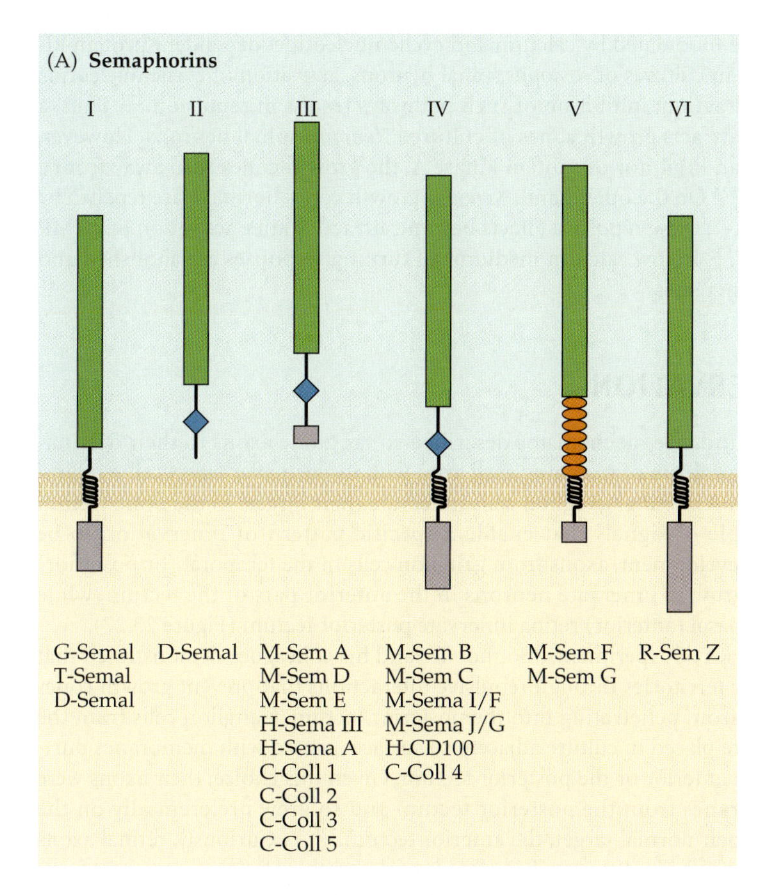

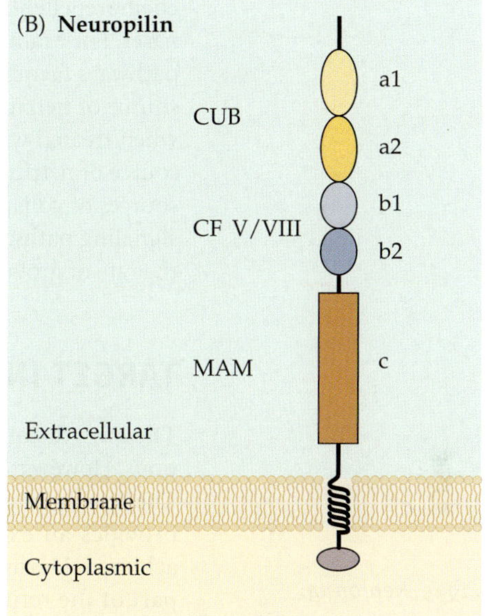

(C)

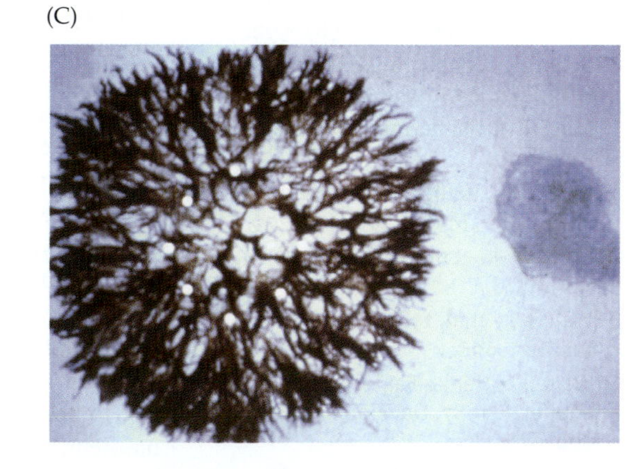

(D)

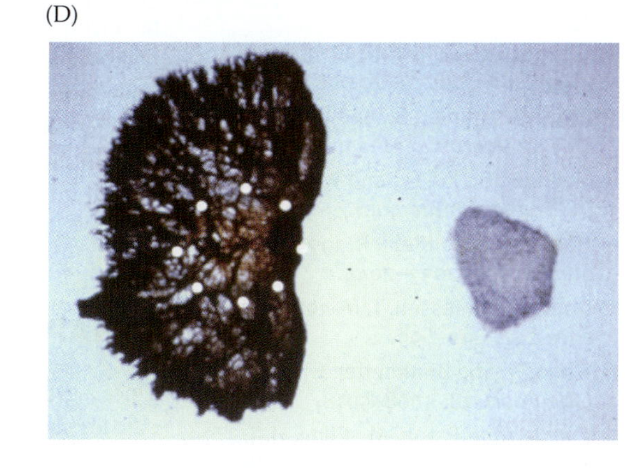

FIGURE 23.26 Semaphorins and Neuropilin mediate short- and long-range chemoattraction and chemorepulsion. The sema domain (green bars) characterizes all six classes of semaphorins, which include secreted (II and III) and membrane-bound (I, IV–VI) forms. Extracellular regions also contain immunoglobulin-like domains (blue diamonds), thrombospondin type I repeats (orange ovals), and domains rich in basic amino acids (purple bars). Cytoplasmic domains are short and highly variable. (B) Neuropilin is a semaphorin receptor. In the extracellular region, domains a1 and a2 (also called CUB motifs) are similar to domains in complement factors C1a and C1s, bone morphogenetic protein 1, and several metalloproteinases. The b1 and b2 domains resemble domains in coagulation factors V and VIII. The c region contains the MAM domain, a motif found in the tyrosine phophatase Mu, A5/neuropilin, and Meprin, a metalloendopeptidase. The short cytoplasmic tail is highly conserved across species. (C and D) Semaphorin III (Sema3A) is a long-range chemorepellent for sensory axons. Embryonic rat dorsal root ganglia (left in each panel) were cultured for 48 h adjacent to aggregates of COS cells. White dots outline the ganglion. NGF was included in the culture medium to elicit outgrowth of small-diameter sensory fibers. (C) Control COS cells. Axons grow out in a radial fashion. (D) COS cells secreting recombinant Sema3A. Sema3A repels axons. (After Messersmith et al., 1995; micrographs kindly provided by M. Tessier-Lavigne.)

collapse of growth cones in cell culture and to mediate long-range repulsion of axons in explant cultures.[120] Neuropilins are highly conserved receptors for the semaphorins (Figure 23.26B).[121]

Modulation of Response to Chemorepellents and Chemoattractants

The signal transduction pathways that mediate responses to chemoattractants and chemorepellents are modulated by calcium and cyclic nucleotide–dependent protein kinases. For example, in cultures of *Xenopus* spinal neurons, activation of cyclic nucleotide pathways favors attraction; inhibition of such pathways results in repulsion.[122] Thus, a source of netrin-1 attracts growth cones of cultured *Xenopus* spinal neurons. However, when treated with an inhibitor of protein kinase A, the growth cones turn away from a source of netrin.[123,124] On the other hand, *Xenopus* growth cones normally are repelled by sources of collapsin-1; these repulsive effects become attractive after activation of cGMP signaling pathways.[125] In low calcium medium, all turning responses are abolished and the rate of growth increases.

TARGET INNERVATION

The growth cone guidance mechanisms described so far guide axons to their destinations. However, the problem of matching each axon with its particular target cell remains. The way in which chick retinal ganglion cells innervate their targets in the optic tectum provides an example of signals that enable a specific pattern of innervation to be achieved. During development, axons from ganglion cells in the temporal (or posterior) part of the retina grow to innervate neurons in the anterior part of the tectum, while those arising from nasal (anterior) retina innervate posterior tectum (Figure 23.27).

In an elegant series of experiments, Bonhoeffer and his colleagues demonstrated that axons sort out their territories through repulsive interactions that prevent growth cones of temporal axons from penetrating into the posterior tectum. Ganglion cells from the temporal retina were placed in culture adjacent to surfaces coated with membranes purified from either the anterior or the posterior tectum. Given this choice, their axons were repelled by membranes from the posterior tectum and so grew preferentially on the membranes from their normal target, the anterior tectum.[126,127] Curiously, retinal axons elongated rapidly on either substrate when not presented with a choice.

The molecules responsible for this repulsive interaction are members of a family of receptor tyrosine kinases (known as Eph kinases) and their ligands (called ephrins).[128,129] Ephrin-A2 and ephrin-5 are expressed in the tectum during the time retinotectal connections are being formed, and their concentration increases in a graded manner from anterior to posterior. The Eph A3 receptor is expressed on retinal axons in a corresponding nasotemporal gradient. When incorporated into lipid vesicles and added to the medium bathing growing temporal retinal axons, ephrins A2 and A5 cause the growth cones to detach from the substrate and retract.[130] Ephrins and the Eph family of receptor tyrosine kinases act throughout the developing nervous system to influence axon pathfinding, cell migration, and cell intermingling, typically by a similar repellent mechanism.[131,132]

A single anteroposterior gradient is not sufficient to enable retinal axons to reach their appropriate destinations in the tectum.[92,133] Later refinement of the anteroposterior projection involves both position-dependent chemical cues[134] and activity-dependent mechanisms[135] (Chapter 25).

SYNAPSE FORMATION

Once a growth cone has reached its target, it must establish synaptic contact, often at a specific location on the cell. To analyze the mechanisms by which such precise connections are formed within the CNS constitutes a major problem. A favorable preparation for studying the formation of synaptic connections is the vertebrate skeletal neuromuscular junction.

[120]Luo, Y., et al. 1995. *Neuron* 14: 1131–1140.

[121]Fujisawa, H., and Kitsukawa, T. 1998. *Curr. Opin. Neurobiol.* 8: 587–592.

[122]Song, H.-J., and Poo, M. M. 1999. *Curr. Opin. Neurobiol.* 9: 355–363.

[123]Ming, G.-L., et al. 1997. *Neuron* 19: 1225–1235.

[124]Song, H.-J., Ming, G.-L., and Poo, M.-M. 1997. *Nature* 388: 275–279.

[125]Song, H.-J., et al. 1998. *Science* 281: 1515–1518.

[126]Walter, J., et al. 1987. *Development* 101: 685–696.

[127]Walter, J., Henke-Fahle, S., and Bonhoeffer, F. 1987. *Development* 101: 909–913.

[128]Drescher, U., et al. 1995. *Cell* 82: 359–370.

[129]O'Leary, D. D. and Wilkinson, D.G. 1999. *Curr. Opin. Neurobiol.* 9: 65–73.

[130]Cox, E. C., Muller, B., and Bonhoeffer, F. 1990. *Neuron* 4: 31–47.

[131]Mellitzer, G., Xu, Q., and Wilkinson, D. G. 1999. *Nature* 400: 77–81.

[132]Holder, N., and Klein, R. 1999. *Development* 126: 2033–2044.

[133]Dodd, J. and Jessell, T. M. 1988. *Science* 242: 692–699.

[134]Ichijo, H. and Bonhoeffer, F. 1998. *J. Neurosci.* 18: 5008–5018.

[135]Constantine-Paton, M., Cline, H. T., and Debski, E. 1990. *Annu. Rev. Neurosci.* 13: 129–154.

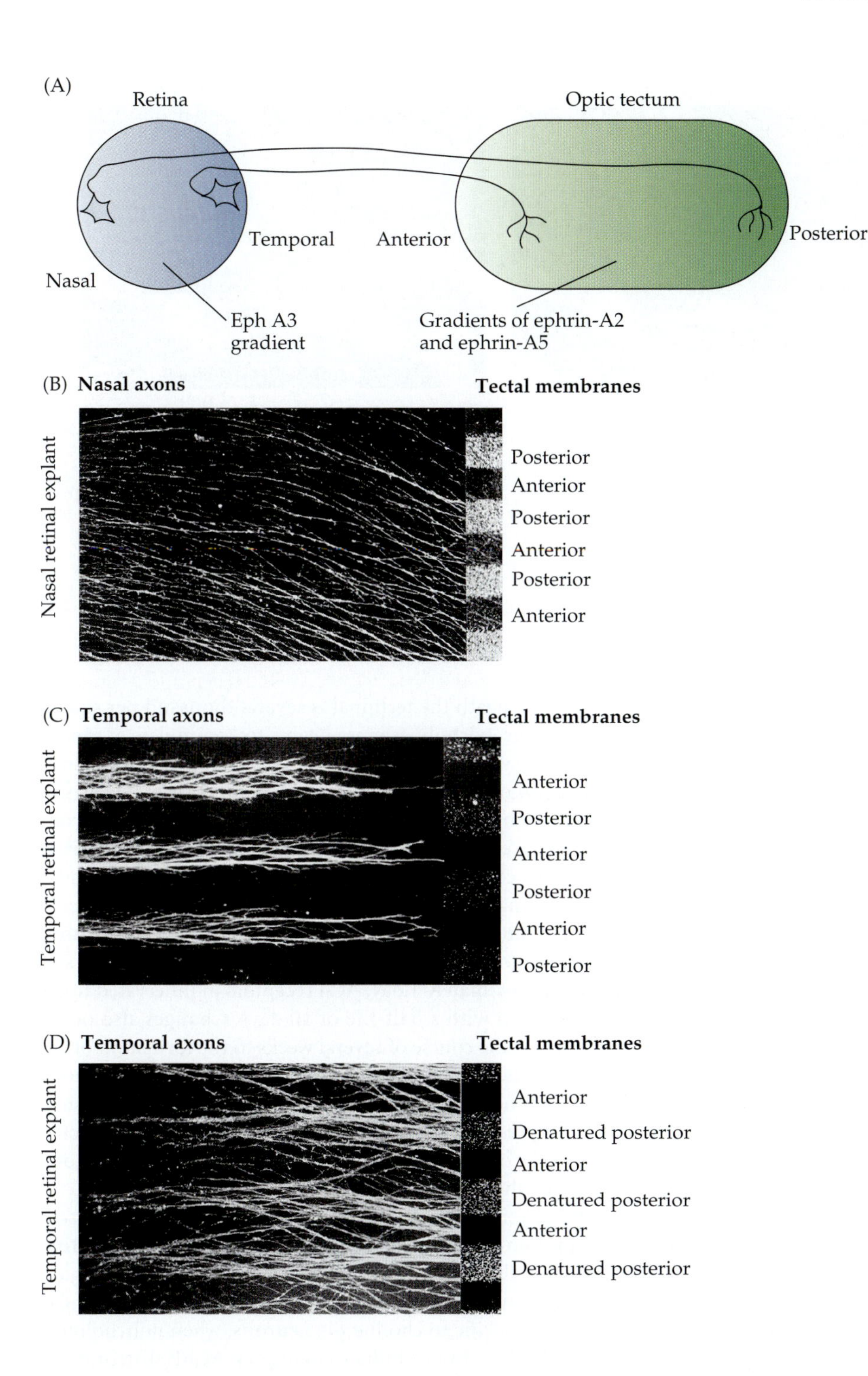

FIGURE 23.27 The Role of Repulsive Interactions in Innervation of the optic tectum in the chick. (A) Ganglion cells in the nasal retina innervate neurons in the posterior tectum; ganglion cells in the temporal retina innervate neurons in the anterior tectum. There is a nasotemporal gradient of the Eph A3 receptor tyrosine kinase in retinal ganglion cells and anteroposterior gradients of the Eph receptor ligands ephrin-A2 and ephrin-A5 in the tectum. Axons of temporal ganglion cells are prevented from entering the posterior tectum by the repulsive interaction of Eph receptors and ligands. (B) In cell culture, axons from neurons in the nasal retina grow equally well on lanes coated with membranes isolated from anterior or posterior tectum. (C) Axons from neurons in the temporal retina prefer to grow on anterior membranes. (D) Axons from temporal retina grow equally well on intact anterior membranes and denatured posterior membranes, indicating that normally they are repelled by heat-sensitive components of the posterior membranes. (B and D from Walter, Henke-Fahle, and Bonhoeffer, 1987; C from Walter et al., 1987; micrographs kindly provided by F. Bonhoeffer.)

Accumulation of ACh Receptors

Studies by Fischbach, Cohen, Changeux, Salpeter, Steinbach, Poo, Kidokoro, and others[136,137] have shown that early in development ACh receptors are distributed diffusely over the surface of uninnervated myotubes at a density of a few hundred receptors per square micrometer. As the growth cone of a motor axon approaches a myotube, depolarizing potentials arise due to the release of ACh from the growth cone (Figure 23.28).[138] Upon contact, the rate of spontaneous release of quanta of ACh rapidly increases, as does the size of the synaptic potential evoked by stimulating the axon. Thus, within minutes a functional synaptic connection is established.

The first synaptic specialization to form is an accumulation of ACh receptors (AChRs) beneath the axon terminal.[137] This begins within hours of the initial contact. Within a

[136]Salpeter, M. M. (ed.). 1987. *The Vertebrate Neuromuscular Junction.* Alan R. Liss, New York.

[137]Sanes, J. R., and Lichtman, J. W. 1999. *Annu. Rev. Neurosci.* 22: 389–442.

[138]Evers, J., et al. 1989. *J. Neurosci.* 9: 1523–1539.

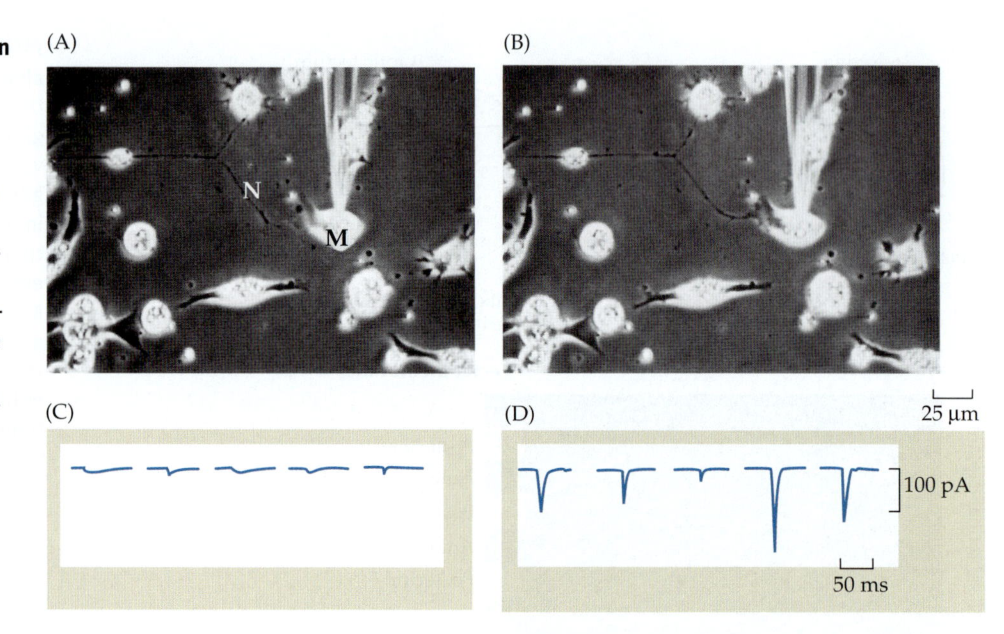

FIGURE 23.28 Rapid Formation of Functional Synaptic Connections between motor axons and muscle cells. (A, B) Phase-contrast photographs of a growing neurite (N) and a spindle-shaped myocyte (M) in a *Xenopus* neuron–muscle cell culture at the beginning (A) and end (B) of electrical recording. (C, D) Whole-cell patch clamp records from the myocyte. Spontaneous synaptic currents can be recorded within 1 min of contact (C) and have increased in strength several-fold by 18 min (D). (From Evers et al., 1989.)

day or two the density of receptors beneath the terminal is several thousand per square micrometer. At about the same time acetylcholinesterase begins to accumulate at synaptic sites, and wisps of basal lamina are visible within the synaptic cleft. Further differentiation of the neuromuscular junction occurs gradually during the next several weeks of development. In many species the γ subunit of the AChR is replaced by an ε subunit, converting the embryonic form of the receptor to the adult form (Chapter 3). The distribution of the receptor also changes: The concentration beneath the axon terminal continues to increase to adult levels of approximately 10^4 receptors/μm^2, and the density of receptors in nonsynaptic portions of the muscle fiber decreases to less than 10 receptors/μm^2. The metabolic stability of ACh receptors also changes. Before innervation, receptors in the membrane have a half-life of approximately 1 day; ACh receptors in innervated fibers are remarkably stable, being degraded with a half-life of 10 days. Changes also occur within the axon terminal, leading over the course of several weeks to the formation of active zones. These and many other studies indicate that the formation of a synapse is not a single all-or-none event. Although functional synaptic transmission may be established very rapidly, the differentiation of pre- and postsynaptic specializations is a protracted process, occurring over several weeks of development, and it relies on the exchange of a variety of molecular signals between the nerve terminal and muscle fiber.

Detailed morphological and physiological experiments demonstrate that growth cones contact muscle cells at random positions on their surface, ignoring preexisting AChR clusters and rapidly inducing the formation of new receptor aggregates (Figure 23.29).[139,140] Thus, axon terminals must release a signal that induces AChR accumulation in the muscle cell. The signal is specific to cholinergic neurons; when noncholinergic neurons grow over muscle cells, they do not induce changes in AChR distribution. The signal is not ACh itself, however; ACh receptors accumulate beneath axon terminals in cultures grown in the presence of curare or α-bungarotoxin, which blocks the interaction of ACh with its receptor. Experiments originally aimed at identifying signals controlling regeneration of the neuromuscular junction (Chapter 24) have identified a protein, called agrin, that is released by motor nerve terminals and triggers the accumulation of ACh receptors, cholinesterase, and other components of the postsynaptic apparatus at synaptic sites.[141]

Agrin-Induced Synaptic Differentiation

Agrin occurs in several isoforms that arise from a single gene by alternative splicing.[142] Motoneurons, muscle cells, and Schwann cells all express agrin, but only motoneurons

[139] Anderson, M. J., and Cohen, M. W. 1977. *J. Physiol.* 268: 757–773.

[140] Frank, E., and Fischbach, G. D. 1979. *J. Cell Biol.* 83: 143–158.

[141] McMahan, U. J., and Wallace, B. G. 1989. *Dev. Neurosci.* 11: 227–247.

[142] Bowe, M. A., and Fallon, J. R. 1995. *Annu. Rev. Neurosci.* 18: 443–462.

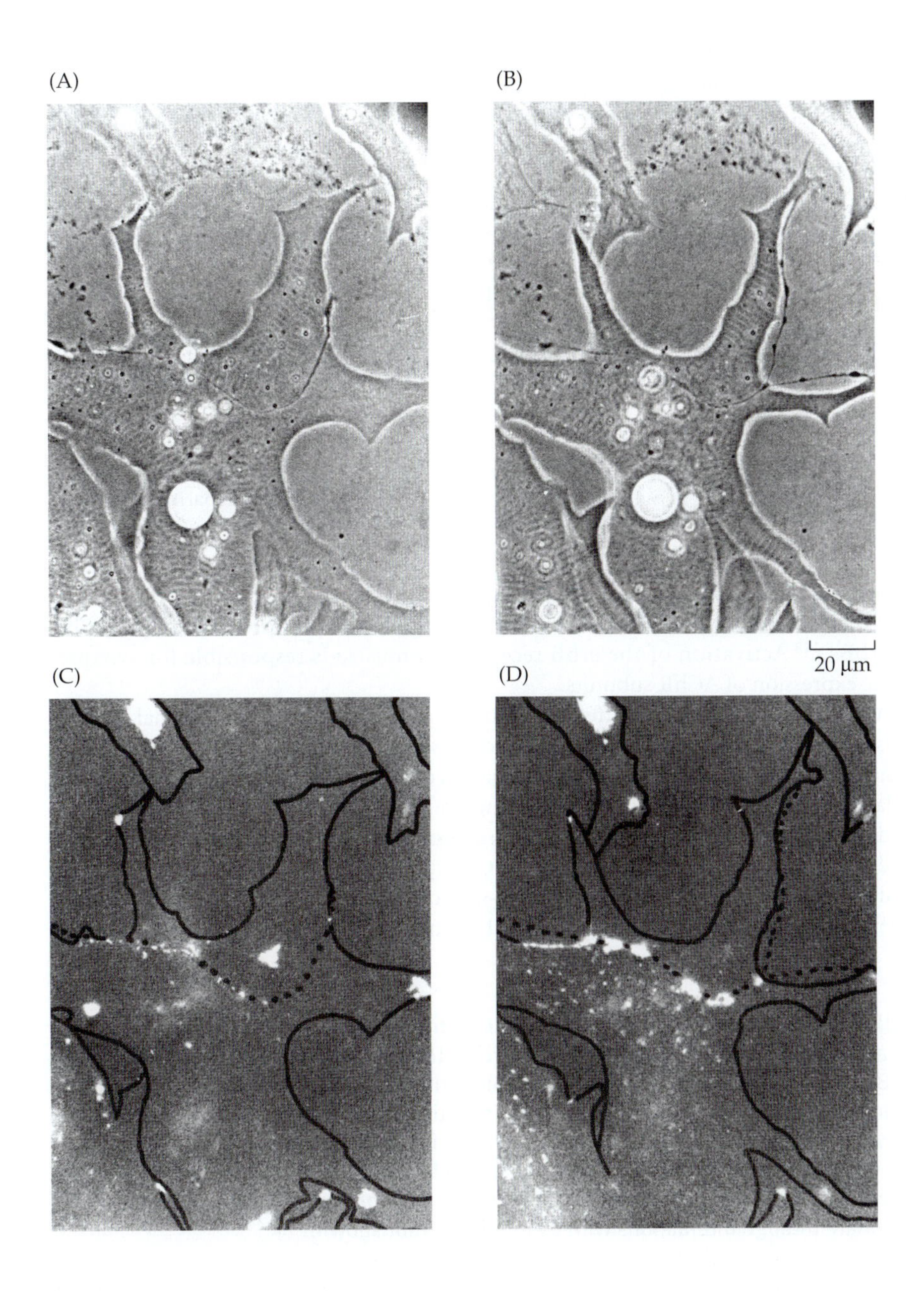

FIGURE 23.29 Axons Induce Aggregation of Acetylcholine Receptors at sites of contact with muscle cells. Phase-contrast (A,B) and fluorescence (C, D) micrographs of a *Xenopus* neuron–muscle cell culture. Acetylcholine receptors are labeled with rhodamine α-bungarotoxin. (A, C) Before and immediately after contact there are spontaneous clusters of ACh receptors on the myocyte. (B, D) After 24 h, the spontaneous AChR patches have disappeared, and new patches have been induced to form beneath the axon. (From Anderson and Cohen, 1977; micrographs kindly provided by M. Cohen.)

express the isoform that is potent in inducing postsynaptic differentiation. Agrin is a large heparan sulfate proteoglycan, with domains that interact with laminin, heparin-binding proteins, α-dystroglycan, heparin, and integrins (Figure 23.30).[143] The ability to induce the formation of postsynaptic specializations resides in the most C-terminal domain.

The essential role of agrin in the formation of the neuromuscular junction is evident in mice in which agrin expression is prevented by homologous recombination.[144,145] In such agrin gene knockouts, myofibers appear normally and axons grow into the developing muscles, but neuromuscular junctions fail to form. A similar phenotype is seen in mice in which the muscle-specific receptor tyrosine kinase MuSK is knocked out.[146] Considerable evidence suggests that MuSK forms part of the agrin receptor and that agrin-induced autophosphorylation of MuSK triggers an intracellular signaling cascade that recruits components to the postsynaptic apparatus (Figure 23.31). One essential component is rapsyn, a protein thought to mediate interactions among ACh receptors, MuSK, α- and β-dystroglycan, and members of the Src family of cytoplasmic tyrosine kinases.[147] Thus, at neuromuscular junctions in rapsyn-deficient mutant mice, MuSK accumulates, selective expression of AChR genes occurs in synaptic nuclei, and some presynaptic specializations form, but no accumulation of ACh receptors occurs.

[143]Denzer, A. J., et al. 1998. *EMBO J.* 17: 335–343.
[144]Gautam, M., et al. 1996. *Cell* 85: 525–535.
[145]Burgess, R. W., et al. 1999. *Neuron* 23: 33–44.
[146]DeChiara, T. M., et al. 1996. *Cell* 85: 501–512.
[147]Gautam, M., et al. 1995. *Nature* 377: 232–236.

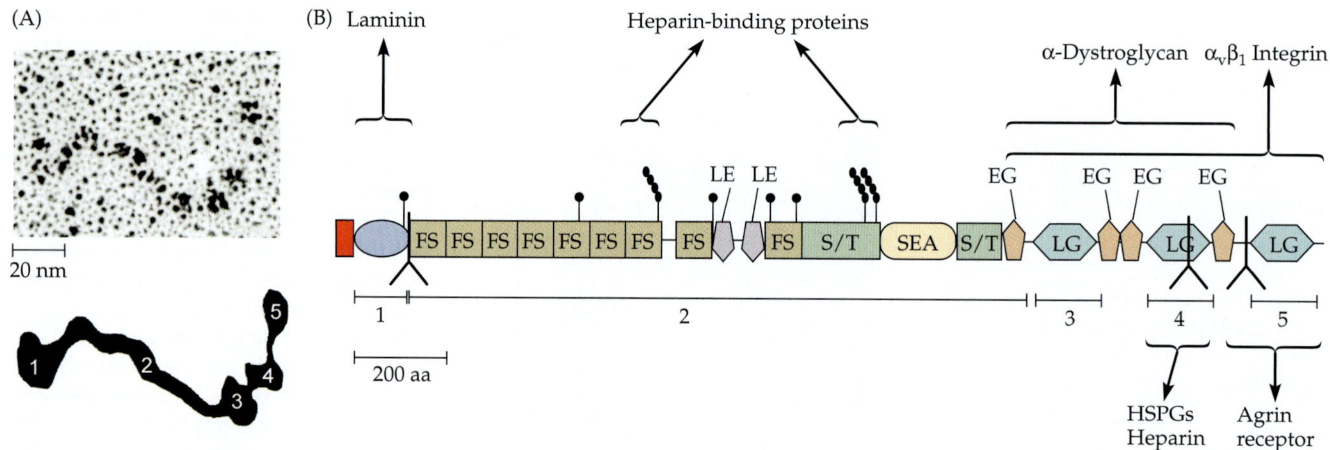

FIGURE 23.30 Agrin Is a Large Heparan Sulfate Proteoglycan (400 to 600 kDa) with domains that interact with laminin, heparan sulfate proteoglycans (HSPGs), heparin, α-dystroglycan, integrin, heparin-binding proteins, and the agrin receptor that causes AChR aggregation. (A) Electron micrograph of agrin after rotary shadowing. (B) Schematic diagram of the structural and binding domains of chick agrin. EG = epidermal growth factor–like domain; FS = follistatin-like domain; LE = laminin EGF–like domain; LG = laminin G–like domain; SEA = motif found in sea urchin sperm protein, enterokinase, and agrin; S/T = serine and/or threonine-rich domain. Binding regions are indicated, as are globular (1, 3–5) and extended (2) regions of the molecule that can be recognized in part A. (After Denzer et al., 1998; micrograph kindly provided by M. Rueg.)

Among the proteins that accumulate in agrin-induced specializations is ARIA, a member of the neuregulin protein family, and the neuregulin receptor proteins erbB2, erbB3, and erbB4.[148] Activation of the erbB receptors in muscle is responsible for synapse-specific expression of AChR subunits.

Much less is known about differentiation of the presynaptic nerve terminal. Experiments by McMahan demonstrated that molecules stably associated with the synaptic basal lamina in adult muscle could induce the formation of active zones in regenerating axons.[141] The lack of presynaptic specializations in agrin- and MuSK-deficient mutant mice suggests that during development presynaptic differentiation is triggered by retrograde signals released by muscle cells in response to agrin.[145] One such retrograde, basal lamina–associated signal is laminin-β2; it accumulates in agrin-induced postsynaptic specializations,[149] and presynaptic differentiation is clearly abnormal in mutant mice deficient in laminin-β2 .[150]

FIGURE 23.31 Interaction of Agrin with MuSK triggers differentiation of postsynaptic specializations in muscle cells at which ACh receptors, rapsyn, and dystroglycan accumulate. Binding of agrin to MuSK requires an unidentified coreceptor (MASC) and results in tyrosine autophosphorylation of MuSK and activation of intracellular tyrosine kinases Src and Fyn. Activated MuSK recruits rapsyn, via an unidentified transmembrane protein, RATL. Rapsyn, in turn, recruits ACh receptors, which become phosphorylated on tyrosine residues of the β subunit, and dystroglycan. Many additional postsynaptic components accumulate through interactions with dystroglycan (not shown).

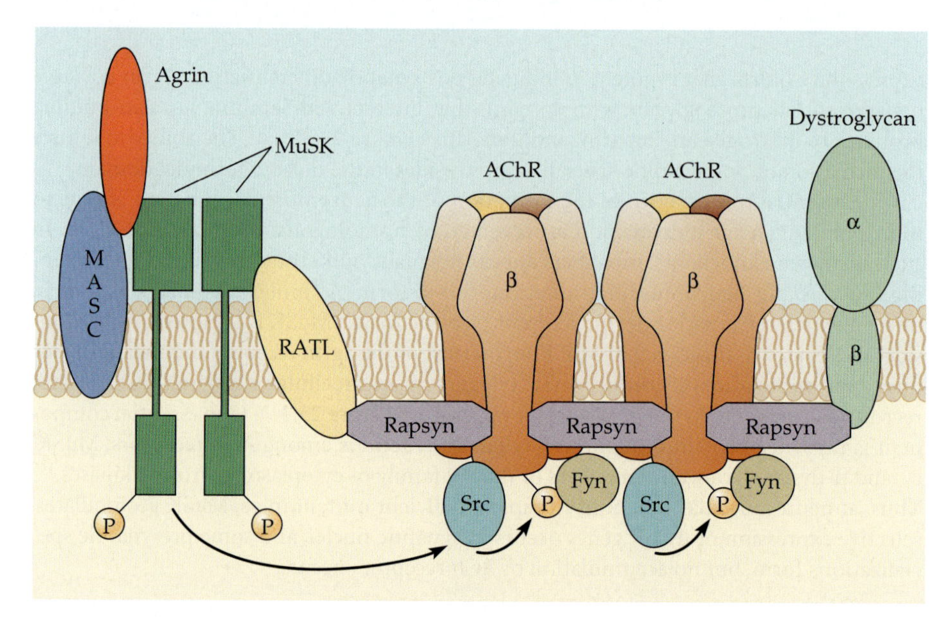

[148]Fischbach, G. D., and Rosen, K. M. 1997. *Annu. Rev. Neurosci.* 20: 429–458.

[149]Cohen, I., et al. 1997. *Mol. Cell. Neurosci.* 9: 237–253.

[150]Noakes, P. G., et al. 1995. *Nature* 374: 258–262.

Formation of CNS Synapses

What happens during synapse formation in the central nervous system, where presynaptic cells releasing GABA, glutamate, ACh, and 5-HT might all converge on an individual postsynaptic neuron? Does each terminal induce the accumulation of the appropriate receptors, do terminals seek out preexisting clusters of corresponding receptors, or do postsynaptic specializations contain a mix of receptors for different transmitters? Studies by Banker, Sheng, Craig, and their colleagues in cultures of hippocampal neurons demonstrate that receptors for glutamate are concentrated beneath glutamatergic nerve terminals on dendritic spines, while receptors for GABA are concentrated beneath GABAergic axon terminals on dendritic shafts (Figure 23.32).[151] Cells developing in culture form receptor clusters spontaneously, but they are ignored by axon terminals. Instead, axon terminals induce de novo the formation of a postsynaptic scaffold of cytoskeletal proteins to which appropriate receptors later attach.[152]

Mammalian retina offers another preparation where receptor localization and synapse formation can be studied. In the adult, $GABA_A$ receptors, $GABA_C$ receptors, and glycine receptors are all clustered in postsynaptic membranes, but at distinct synaptic sites.[153] Moreover, in individual ganglion cells expressing several isoforms of the $GABA_A$ receptor α subunit, different isoforms are found at different synapses.[154] Thus, GABA and glycine receptor subunits are selectively assembled into multimeric receptors, which, in turn, are

[151]Craig, A. M., et al. 1994. *Proc. Natl. Acad. Sci. USA* 91: 12373–12377.

[152]Rao, A., et al. 1998. *J. Neurosci.* 18: 1217–1229.

[153]Koulen, P. et al. 1998. *Euro. J. Neurosci.* 10: 115–127.

[154]Wassle, H., et al. 1998. *Vision Res.* 38: 1411–1430.

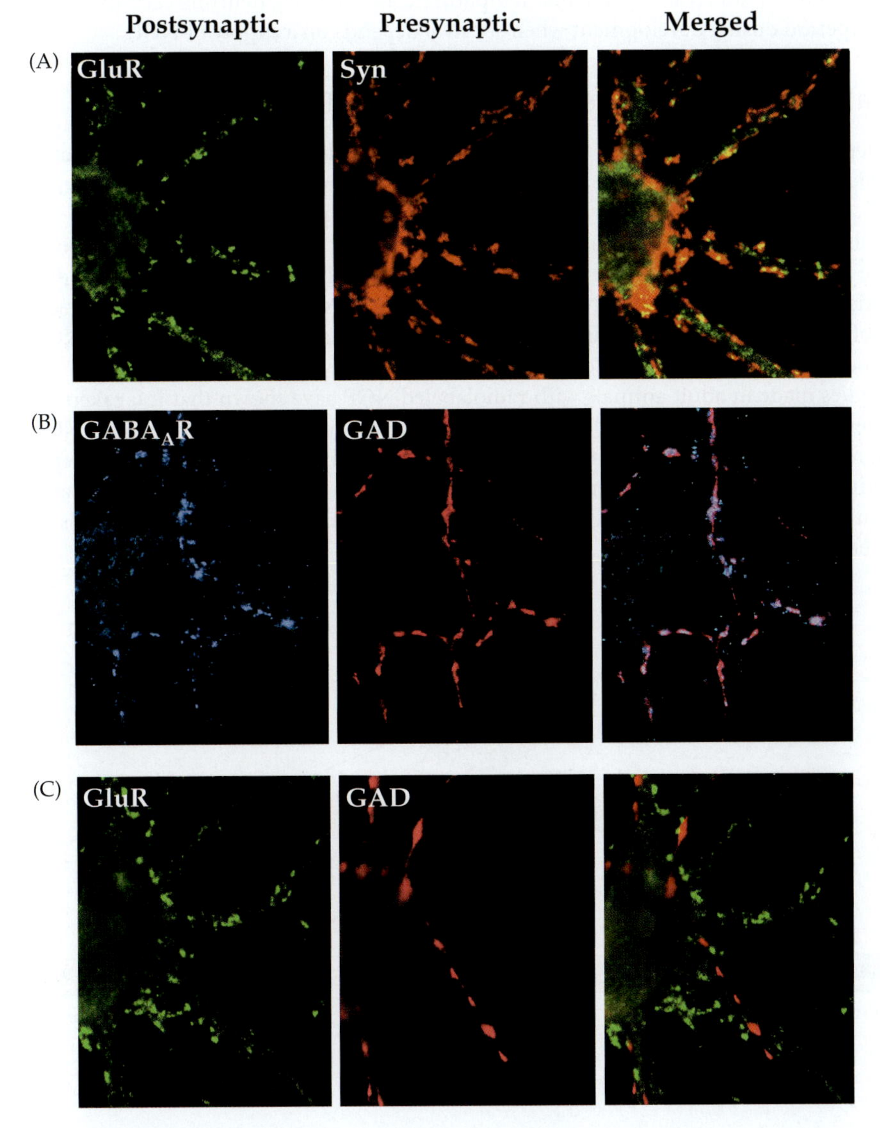

FIGURE 23.32 Selective Accumulation of Postsynaptic Receptors at synapses between rat hippocampal neurons grown in cell culture. Fluorescence micrographs of pyramidal cells, which receive two types of synaptic input: excitatory synapses, at which glutamate is the transmitter, and inhibitory synapses, at which GABA is the transmitter. In the first column, postsynaptic receptors are visualized, in the second, presynaptic terminals, and in the third, the images are merged. (A) Glutamate AMPA receptors are labeled with antibodies to the GluR1 subunit (GluR), all synaptic terminals are labeled with antibodies to synaptophysin (Syn). Glutamate receptors cluster beneath a subpopulation of nerve terminals (yellow regions in merged image). (B) $GABA_A$ receptors are labeled with antibodies to the β2/3 subunit ($GABA_AR$), nerve terminals releasing GABA are labeled with antibodies to glutamic acid decarboxylase (GAD). $GABA_A$ receptors cluster opposite GABAergic terminals (purple regions in merged view). (C) Glutamate AMPA receptors are labeled with antibodies to the GluR1 subunit (GluR), nerve terminals releasing GABA are labeled with antibodies to glutamic acid decarboxylase (GAD). Glutamate receptor clusters do not colocalize with GABAergic nerve terminals. Thus, glutamate receptors cluster specifically beneath terminals releasing glutamate, and $GABA_A$ receptors cluster selectively beneath terminals releasing GABA. (Micrographs kindly provided by A. M. Craig.)

Rita Levi-Montalcini

differentially targeted to synapses. During development, retinal neurons initially stain diffusely for glycine and GABA receptors; aggregates of receptors appear as morphologically identifiable synapses form.[153,155]

GROWTH FACTORS AND SURVIVAL OF NEURONS

Nerve Growth Factor

Developing neurons rely on a supply of specific proteins, called growth factors, for survival. The first growth factor to be identified was nerve growth factor (NGF). In pioneering experiments, Levi-Montalcini, Cohen, and their colleagues first demonstrated that NGF stimulated the outgrowth of neurites from sensory and sympathetic neurons (Box 23.1). They went on to show that NGF was also required for neuronal survival. Blocking NGF action in newborn mice with antibodies led to the death of sympathetic neurons. The parasympathetic nervous system was not affected, and the dorsal root ganglia were only slightly smaller than normal. In subsequent experiments it was shown that at an earlier, fetal stage in development, dorsal root ganglion sensory neurons likewise required NGF for survival; they did not survive when exposed to NGF antibodies (by immunizing the mother).[156] In adults, antibodies to NGF were much less effective on either cell population. Null mutation of NGF and its high-affinity receptor (discussed later in the chapter) in transgenic mice has confirmed that both sympathetic and many sensory neurons require NGF for survival.[157,158] Thus, sympathetic and sensory neurons each display a critical period during development when survival depends on a supply of NGF.

[155]Sassoe-Pognetto, M., and Wassle, H. 1997. *J. Comp. Neurol.* 381: 158–174.
[156]Dolkart-Gorin, P. and Johnson, E. M. 1979. *Proc. Natl. Acad. Sci. USA* 76: 5382–5386.
[157]Crowley, C., et al. 1994. *Cell* 76:1001–1011.
[158]Smeyne, R. J., et al. 1994. *Nature* 368: 246–249.
[159]Campenot, R. B. 1977. *Proc. Natl. Acad. Sci. USA* 74: 4516–4519.
[160]Hendry, I. A., et al. 1974. *Brain Res.* 68: 103–121.
[161]Black, I. B. 1978. *Annu. Rev. Neurosci.* 1: 183–214.

Uptake and Retrograde Transport of NGF

The dependence on NGF for cell survival suggests an action on the cell soma. When embryonic sympathetic neurons are grown in three-compartment culture chambers (Figure 23.33), the central compartment in which the neurons are placed must contain NGF for the cells to survive.[159] However, after neurites have reached the side compartments, NGF can be removed from the central compartment and the cells will remain alive, provided the side compartments contain NGF (Figure 23.33B). This suggests that the trophic effects of NGF on developing neurons can be mediated by retrograde transport of NGF from nerve terminals to the cell soma.

Studies made in adult animals with radiolabeled NGF have shown that it is taken up into nerve terminals and actively transported back to the soma.[160] In adult neurons, NGF is not required for survival, but it regulates, among other things, the synthesis of norepinephrine by inducing two enzymes required for its synthesis: tyrosine hydroxylase and dopamine β-hydroxylase.[161] If NGF transport in adult neurons is impaired, the levels of these enzymes fall.

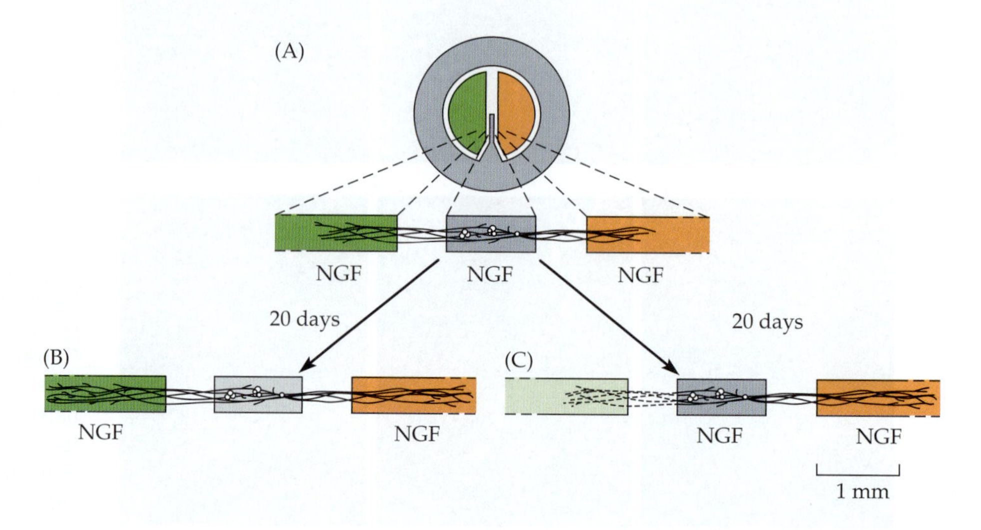

FIGURE 23.33 Nerve Growth Factor and the Survival of Axon Branches from sympathetic ganglion cells grown in cell culture. (A) Neurons dissociated from neonatal sympathetic ganglia plated in the central compartment send neurites under a Teflon divider and into the adjacent compartments; all compartments contain NGF. (B) After initial outgrowth has occurred, removal of NGF from the central compartment for 20 days has no effect; neurons in the central compartment are maintained by NGF transported retrogradely from their terminals in the side compartments. (C) After initial outgrowth has occurred, removal of NGF from the left compartment causes the neurites entering it to degenerate, while those in the compartment containing NGF remain. (After Campenot, 1982.)

Box 23.1 Discovery of Nerve Growth Factor

A pivotal series of experiments that opened the door to investigation of the molecular basis of neuronal development was provided by the work of Levi-Montalcini, Cohen, and their colleagues.[162] Their studies, initially focused on the control of neurite outgrowth, led to the discovery of a family of proteins essential for the survival of particular classes of neurons. These studies provided the framework for approaching many of the problems raised in this chapter; the course of the investigations also illustrates the manner in which research can progress in the hands of extraordinarily perceptive investigators. The search for the growth factor is a remarkable sequence of coincidences, false but profitable leads, extraordinary and apparently fortunate choices, all leading to important insights into the mechanisms of neuronal survival and differentiation.

Transplanting an extra leg onto the back of a tadpole, a lizard, or a newt causes the outgrowth of nerve fibers from the central nervous system.[163] To follow up on the idea that there must be substances in transplanted limbs capable of attracting nerve fibers, it was reasonable to test the effect of rapidly growing tissues on the growth of neurons. The initial experiments were made by implanting onto chick embryos a connective tissue tumor (sarcoma) obtained from mice. On the side where the sarcoma had been implanted there was a profuse outgrowth of sensory and sympathetic nerve fibers from the embryo into the tumor. To show that the effect was caused by a humoral factor, sarcomas were grafted onto the chorioallantoic membrane, a tissue that surrounds the embryo. There was no direct contact between the embryo and the tumor, but once again the chick's dorsal root ganglia and sympathetic neurons grew profusely.[164] Next, it was shown that sarcoma cells produce a similar dramatic effect on chick ganglia maintained in culture, providing a simple and reliable bioassay (Figure A).

The active factor in the sarcoma initially appeared to be a nucleoprotein. To see if nucleic acids were essential components of the growth-promoting factor, tumors were incubated with a snake venom whose action would hydrolyze the nucleic acids and thereby render the tumor fraction inactive. With venom present, however, the growth, far from being inactivated, was further increased. In fact, the control experiment of adding snake venom without the sarcoma extract revealed, surprisingly, that the venom itself was a far richer source of growth factor than was the sarcoma.[165] This, in turn, gave rise to the speculation that, since venom is secreted by the salivary gland, possibly salivary glands from other animals might also contain a similar factor.

The animal selected was the mouse.[166] It was fortunate that adult male mice were chosen because the salivary glands of female or immature mice contain far less growth factor, as is also true for glands of other animals that have since been tried. Extracts of salivary glands of adult male mice were found to be potent in causing the growth of neurites from sensory ganglion explanted into culture (see Figure A). The substance extracted from snake venom and salivary glands of mice was called nerve growth factor (NGF).

When antiserum to NGF was injected into newborn mice, their sympathetic nervous system failed to develop (see Figure B).[167] The animals lived normally but responded poorly to stress conditions. Thus, NGF was found not only to enhance the outgrowth of neurites from sympathetic neurons, but to be essential for their survival as well. The functional role of NGF activity in salivary glands in the animal is still not clear, especially since removal of these glands from young animals has only minor effects on nerve growth.

[162] Levi-Montalcini, R. 1982. *Annu. Rev. Neurosci.* 5:341–362.
[163] Hamburger, V. 1939. *Physiol. Zool.* 12:268–284.
[164] See Levi-Montalcini, R. and Angeletti, P. U. 1968. *Physiol. Rev.* 48:534–569 for references to earlier work.
[165] Cohen, S. 1959. *J. Biol. Chem.* 234:1129–1137.
[166] Cohen, S. 1960. *Proc. Natl. Acad. Sci. USA* 46:302–311.
[167] Levi-Montalcini, R. and Cohen, S. 1960. *Ann. N.Y. Acad. Sci.* 85:324–341.

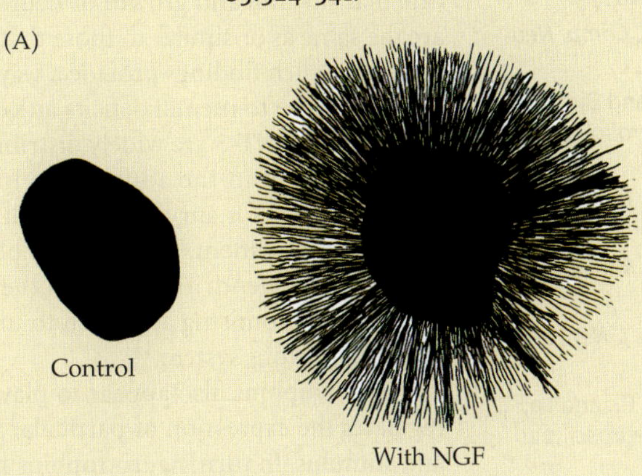

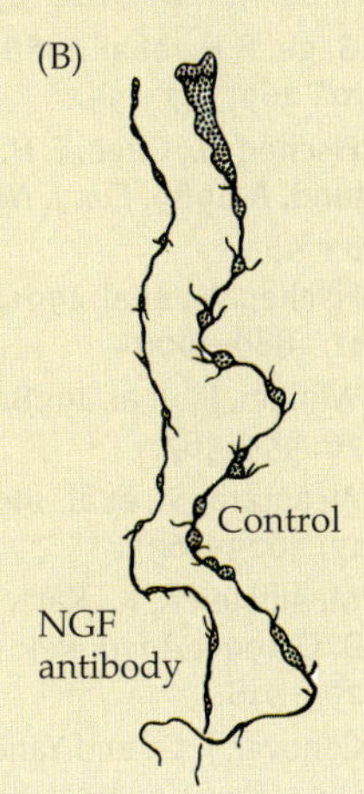

Effect of nerve growth factor on neurons in a sensory ganglion from a 7-day-old chick embryo. (A) Sensory ganglion maintained in culture for 24 h in control medium (left) or medium supplemented with nerve growth factor (right). Nerve growth factor induces prolific outgrowth of axons. (B) Thoracic sympathetic chain of ganglia from a control animal (mouse) is on the right. To the left is a ganglion chain, much smaller in size, from a mouse injected 5 days after birth with antiserum to nerve growth factor. (A after Levi-Montalcini, 1964; B after Levi-Montalcini and Cohen, 1960.)

The Neurotrophin Family of Growth Factors

These observations set the stage for a molecular analysis of the mechanism of action of nerve growth factor by a number of groups, including those of Levi-Montalcini, Shooter, Thoenen, and Barde.[168] Their studies explored such questions as the characteristics of NGF, the receptors on the membrane that interact with NGF, and the subsequent metabolic events. In salivary glands NGF is present as a complex made up of three types of subunits: α, β, and γ. It is the β subunit that is responsible for promoting nerve growth and survival; it consists of two identical peptide chains, each containing 118 amino acids and 3 disulfide bridges.

Sensory neurons not only innervate targets in the periphery, but extend axons into the central nervous system as well. It is natural to ask whether a factor similar to NGF exists in the central nervous system. Indeed, a protein was identified in extracts of the CNS, called brain-derived neurotrophic factor (BDNF), that promoted the survival of dorsal root ganglion neurons in culture and rescued them in vivo if administered to embryos during the period of natural neuronal death.[169] Purification and characterization of BDNF revealed that it has a high degree of homology to NGF, indicating that NGF and BDNF are members of a family of growth factors, which have been given the name **neurotrophins**. Molecular genetic experiments have identified additional neurotrophins: NT-3, NT-4/5, and NT-6.[170] All consist of dimers of a small basic peptide, held together by disulfide linkages between the conserved cysteine residues. The dimer appears as a symmetrical protein with the variable, basic regions that determine receptor specificity exposed on the surface (Figure 23.34).[171]

Early in development, before sensory neurons innervate their peripheral targets, neural crest cells and sensory neurons in dorsal root and trigeminal ganglia require BDNF or NT-3 for proliferation, differentiation, and survival.[172] At such early stages, neurotrophins appear to be provided by the neurons themselves and by the mesenchymal tissues through which the sensory axons grow.[173] Later, after their axons reach their targets, sensory neurons begin to express NGF receptors and become dependent on target-derived NGF for survival.[174]

Neurotrophins in the Central Nervous System

Of particular interest is the finding of a population of NGF-sensitive cells within the CNS.[175] These cholinergic neurons are located in the basal forebrain and innervate several structures, including the hippocampus, a region of the CNS thought to be involved in learning and memory (Chapters 12 and 14). If the axons of these neurons are cut in the adult rat, the cells die. If, however, NGF is infused into the CNS, then these neurons survive axotomy. The number of these cells that stain with markers for cholinergic function declines with age, as does the ability of rats to learn a maze or other spatial memory tasks.[176] If NGF is infused into aged rats, the number of cells that can be stained increases and the rat's performance in spatial memory tasks improves.[177] These observations indicate that survival and growth of neurons within the CNS are likely to rely on factors that are the same as or similar to those that have been identified for peripheral neurons. At the same time such findings provide a way of thinking, in molecular terms, about defects that might give rise to mental deficits and how they might be ameliorated.[178]

BDNF and NT-3 are widely distributed in the central nervous system, both during development and in the adult. Cells in the cortex and hippocampus appear to require BDNF or NT-3 for survival, although changes in cell survival within the CNS have been difficult to document.[179] BDNF has been shown to influence the growth and complexity of axonal and dendritic arbors in the developing CNS.[180] Analysis of knockout mice is providing a promising approach to understanding the roles of neurotrophins in the developing nervous system.[181]

Neurotrophins also appear to play a role in the adult brain.[180] Physiological activity regulates the expression of particular neurotrophins in ways that vary with brain region and stimulus. In turn, neurotrophins have been shown to influence physiological activity, potentiating synaptic transmission and increasing or decreasing neuronal excitability by regulating ion channel expression.

[168]Greene, L. A., and Shooter, E. M. 1980. *Annu. Rev. Neurosci.* 3: 353–402.

[169]Barde, Y-A. 1989. *Neuron* 2: 1525–1534.

[170]Lewin, G. R., and Barde, Y-A. 1996. *Annu. Rev. Neurosci.* 19: 289–317.

[171]McDonald, N. Q., and Chao, M. V. 1995. *J. Biol. Chem.* 270: 19669–19672.

[172]Farinas, I., et al. 1996. *Neuron* 17: 1068–1078.

[173]Kokaia, Z., et al. 1993. *Proc. Natl. Acad. Sci. USA* 90: 6711–6715.

[174]Davies, A. M., and Lumsden, A. 1990. *Annu. Rev. Neurosci.* 13: 61–73.

[175]Gage, F. H., et al. 1988. *J. Comp. Neurol.* 269: 147–155.

[176]Fischer, W., Gage, F. H., and Bjorklund, A. 1989. *Eur. J. Neurosci.* 1: 34–45.

[177]Fischer, W., et al. 1991. *J. Neurosci.* 11: 1889–1906.

[178]Winkler, J., et al. 1998. *J. Mol. Med.* 76: 555–567.

[179]Alcantara, S., et al. 1997. *J. Neurosci.* 17: 3623–3633.

[180]McAllister, A. K., Katz, L. C., and Lo, D. C. 1999. *Annu. Rev. Neurosci.* 22: 295–318.

[181]Conover, J. C., and Yancopoulos, G. D. 1997. *Rev. Neurosci.* 8: 13–27.

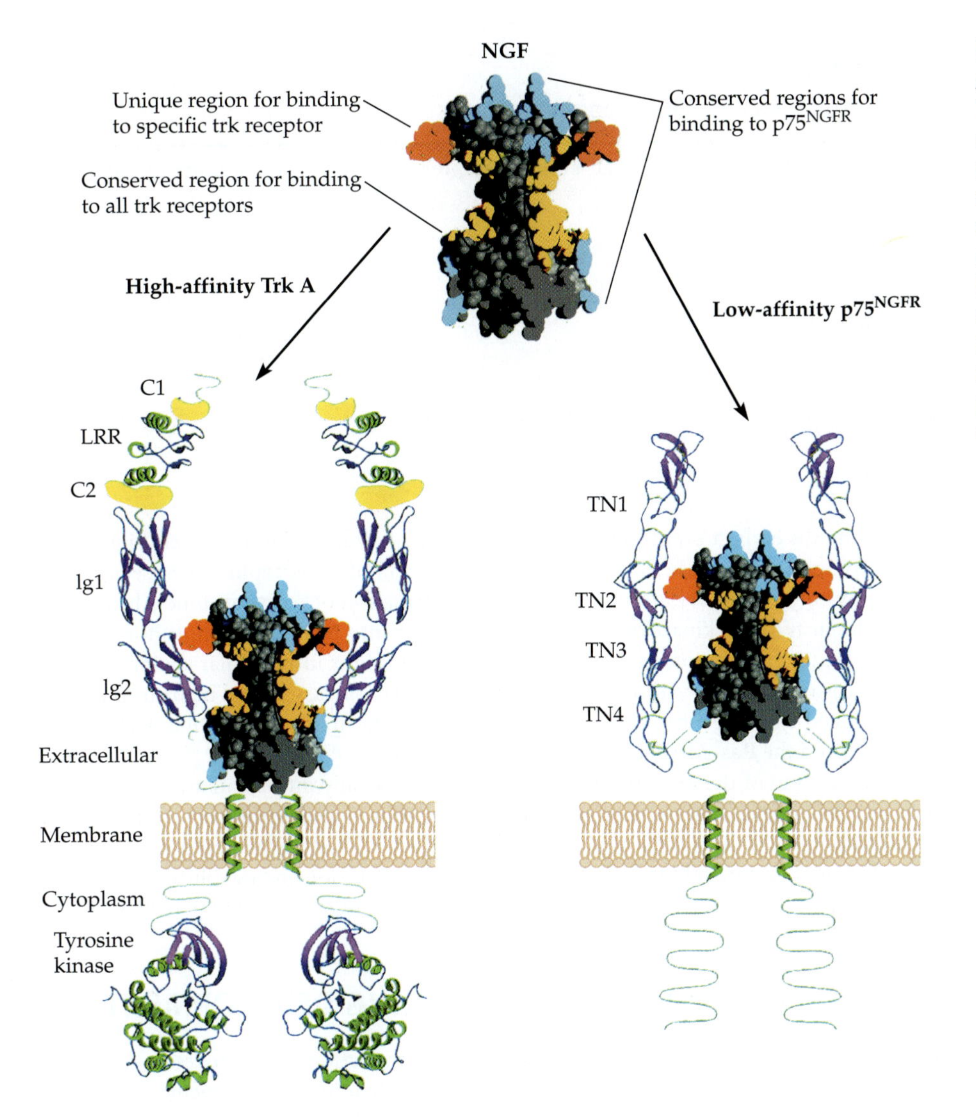

FIGURE 23.34 High- and Low-Affinity Neurotrophin Receptors. The neurotrophin dimer (NGF is shown) binds to two high-affinity Trk receptors through conserved regions of the neurotrophins that interact with all Trk receptors (orange) and unique regions specific for a particular Trk receptor (red). Regions that are conserved in all neurotrophins (blue) mediate binding of the neurotrophin dimer to two low-affinity $p75^{NGFR}$ molecules. (After Wiesmann et al., 1999.)

Neurotrophin Receptors

The neurotrophins interact with two types of receptors on the surface of their target neurons (see Figure 23.34).[182,183] All neurotrophins bind with relatively equal and low affinity ($K_d = 10^{-9}\ M$) to a membrane receptor originally referred to as the low-affinity, fast NGF receptor or $p75^{NGFR}$; it might more appropriately be called the low-affinity neurotrophin receptor, $p75^{LNTR}$ (Table 23.2). This receptor is found on both neurons and nonneuronal cells. There are also high-affinity ($K_d = 10^{-11}\ M$) receptors for the neurotrophins. Results of bioassays indicate that the effects of neurotrophins on cell survival and neurite outgrowth are mediated by binding to the high-affinity receptors.

Although the high-affinity receptor for NGF is normally found only on neurons, it was originally identified in human colon carcinoma cells as part of the product of the *trk* oncogene (oncogenes are genes involved in mediating cell transformation). The counterpart of the *trk* oncogene found in normal cells encodes a protein of 140 kD that is referred to as $p140^{prototrk}$ or simply Trk. The structure of the Trk protein, predicted from its deduced amino acid sequence, consists of an extracellular domain containing the neurotrophin-binding site, a short transmembrane segment, and an intracellular domain encoding a tyrosine kinase (see Figure 23.34). There are at least three members of the *trk* family of proto-oncogenes, each acting as the high-affinity receptor for one or more neurotrophins: TrkA is the receptor for NGF and NT-6, TrkB appears to be a receptor for BDNF and NT-4/5, and TrkC is the receptor for NT-3 (see Table 23.2).

[182]Bothwell, M. 1995. *Annu. Rev. Neurosci.* 18: 223–253.

[183]Frade, J. M., and Barde, Y. A. 1998. *BioEssays* 20: 137–145.

Table 23.2
Neurotrophins and neurotrophin receptors

Neurotrophin	High-affinity receptor	Low-affinity receptor
NGF	TrkA	$p75^{LNTR a}$
BDNF	TrkB	$p75^{LNTR}$
NT-3	TrkC[b]	$p75^{LNTR}$
NT-4/5	TrkB	$p75^{LNTR}$
NT-6	TrkA	$p75^{LNTR}$

[a]Originally termed $p75^{NGFR}$.
[b]NT-3 also binds TrkB and TrkA, but with lower affinity.

Among the earliest events detected after binding of neurotrophins to their high-affinity receptors is an increase in tyrosine phosphorylation of the receptors themselves. This occurs by autophosphorylation; ligand-induced formation of receptor dimers brings the intracellular tyrosine kinase domain of each receptor into position to phosphorylate the other receptor. Trk receptor phosphorylation activates four intracellular signaling pathways: tyrosine phosphorylation of a protein called SNT (*suc*-associated neurotrophic factor–induced tyrosine-phosphorylated target), phospholipase C, phosphatidylinositol 3-kinase, and the Ras–MAP kinase cascade (Figure 23.35).[184]

The function of the low-affinity receptor, which lacks any intracellular domain, is not known. On some cells it may interact with the high-affinity receptor during the binding of neurotrophins. On other cells, especially those that lack high-affinity receptors, it may trigger cell death[185] or provide a mechanism for restricting diffusion and establishing high local concentrations of neurotrophins, as during peripheral nerve regeneration (Chapter 24).

COMPETITIVE INTERACTIONS DURING DEVELOPMENT

After axons reach their targets and make synaptic connections, two processes refine the pattern of innervation. One is a period of cell death, when many of the neurons that had established synaptic connections die. The second is a reduction in the number of axons and synapses, accompanied by a reorganization of surviving connections, to achieve the adult pattern. Each process involves competition for a limited supply of growth factors.

Neuronal Cell Death

A conspicuous feature of the development of the nervous system is that many neurons are born to die. In invertebrates, extensive neuronal death accompanies the sweeping changes that occur during metamorphosis and is under hormonal regulation.[186] However, in both the developing invertebrate and vertebrate nervous systems, cell death also occurs in the absence of such gross morphological changes.[187]

Experiments by Hamburger and Levi-Montalcini first documented such programmed neuronal death in vertebrate embryos and demonstrated that its extent can be influenced by manipulating the size of the target tissue.[160,188] They showed, for example, that in the developing limb, at about the same time as synaptic connections were first being formed on myofibers, 40 to 70% of the motoneurons that had sent axons into the limb died. Implantation of a supernumerary limb reduced the fraction of motoneurons that died, while removal of the limb bud exacerbated the death of motoneurons, suggesting that motoneurons were competing for some trophic substance supplied by their target tissue.

Indeed, a remarkable variety of neurotrophic proteins, many muscle derived, have been identified that can sustain motoneurons: BDNF, NT-3 and NT-4/5, IGF (insulin-like growth factor), CNTF (ciliary neurotrophic factor), GDNF (glial-derived neurotrophic factor), CDF (cholinergic differentiation factor, also called LIF [leukemia inhibitory factor]), and IGF1

[184]Segal, R. A., and Greenberg, M. E. 1996. *Annu. Rev. Neurosci.* 19: 463–489.

[185]Frade, J. M., Rodriguez-Tebar, A., and Barde, Y-A. 1996. *Nature* 383: 166–168.

[186]Truman, J. W., Thorn, R. S., and Robinow, S. 1992. *J. Neurobiol.* 23: 1295–1311.

[187]Oppenheim, R. W. 1991. *Annu. Rev. Neurosci.* 14: 453–501.

[188]Hollyday, M., and Hamburger, V. 1976. *J. Comp. Neurol.* 170: 311–320.

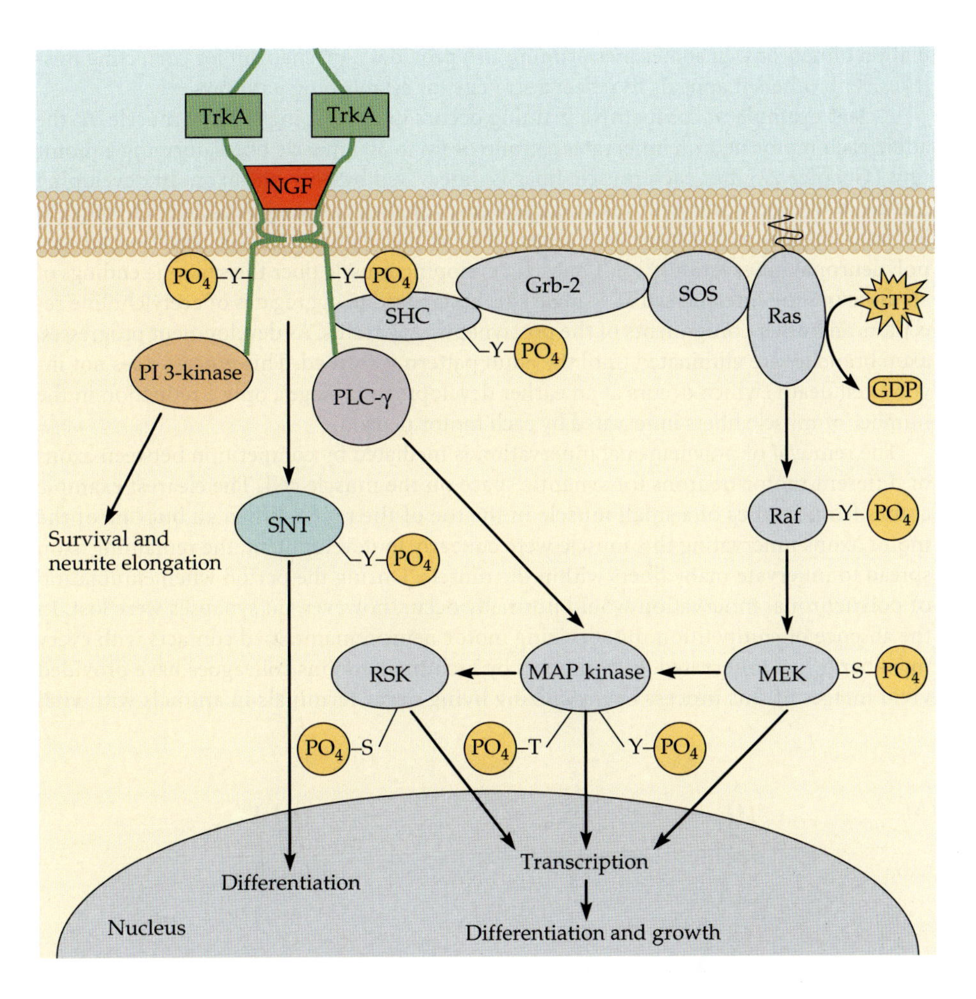

FIGURE 23.35 Binding of the Nerve Growth Factor Dimer to two TrkA receptors brings each TrkA receptor's intracellular protein tyrosine kinase domain into position to phosphorylate tyrosine residues on the other TrkA receptor. This triggers four intracellular signaling cascades, which lead to growth and differentiation. Activation of phosphatidylinositol 3-kinase (PI 3-kinase) promotes survival and neurite elongation. Tyrosine phosphorylation of SNT promotes differentiation. Activation of phospholipase C-γ (PLC-γ) stimulates MAP kinase, which induces gene expression, differentiation, and growth, both directly and by phosphorylating RSK. MAP kinase is also stimulated through a signaling cascade that involves the coupling proteins SHC, Grb-2, SOS, and Ras, and the kinases Raf and MEK. $-S-PO_4$ = serine phosphorylation; $-T-PO_4$ = threonine phosphorylation; $-Y-PO_4$ = tyrosine phosphorylation.

(insulin-like growth factor 1).[189–192] When injected into embryos, these proteins rescue motoneurons that otherwise would have died. However, several lines of evidence, including analysis of mutant mice lacking one or more of these proteins or their receptors, suggest that none is required for motoneuron survival during development. The best candidate for such a motoneuron survival factor is an as yet unidentified ligand of the CNTF receptor.[193]

An overproduction of neurons followed by a period of cell death is a common pattern throughout the developing vertebrate nervous system. Some of the neurons that die may not have made any synapses, or may have innervated an inappropriate target. In such cases cell death contributes to the specificity of innervation.[194] Most of the cells that die, however, appear to have reached and innervated their correct targets. Thus, cell death is primarily a mechanism by which the size of the neuronal input is matched to the size of the target.

A surprising finding was that inhibitors of mRNA or protein synthesis prevented the death of neurons deprived of their appropriate neurotrophin.[195] Results of this and later experiments indicated that neuronal cell death typically occurs by apoptosis. **Apoptosis** is a process by which cells activate intrinsic "suicide" machinery that mediates an orderly breakdown of DNA and proteins within the cell, a process that usually requires synthesis of proteolytic enzymes or their activators.[196,197]

Pruning and the Removal of Polyneuronal Innervation

Once the population of neurons innervating a particular target has been restricted through cell death, the surviving neurons compete with one another for synaptic territory. This competition typically results in the loss of some of the terminal branches and synapses made initially, a process referred to as **pruning**.[198] Pruning provides a mechanism for ensuring appropriate and complete innervation of a target by a particular pop-

[189]Sendtner, M., Holtmann, B., and Hughes, R. A. 1996. *Neurochem. Res.* 21: 831–841.

[190]Oppenheim, R. W., et al. 1995. *Nature* 373: 344–346.

[191]Caroni, P., and Grandes, P. 1990. *J. Cell Biol.* 110: 1307–1317.

[192]McManaman, J. L., Haverkamp, L. J., and Oppenheim, R. W. 1991. *Adv. Neurol.* 56: 81–88.

[193]DeChiara, T. M., et al. 1995. *Cell* 83: 313–322.

[194]O'Leary, D. D. M., Fawcett, J. W., and Cowan, W. M. 1986. *J. Neurosci.* 6: 3692–3705.

[195]Martin, D. P., et al. 1988. *J. Cell Biol.* 106: 829–844.

[196]Friedlander, R. M., and Yuan, J. 1998. *Cell Death Differ.* 5: 823–831.

[197]Bergmann, A., Agapite, J., and Steller, H. 1998. *Oncogene* 17: 3215–3223.

[198]O'Leary, D. D. 1992. *Curr. Opin. Neurobiol.* 2: 70–77.

ulation of neurons. In some cases pruning also provides a mechanism for correcting mistakes;[199] in others it appears to reflect a strategy for establishing pathways.[200]

A clear example of competitive pruning occurs in developing skeletal muscle. In the adult, each motor neuron innervates a group of up to 300 muscle fibers, forming a motor unit (Chapter 22), but each muscle fiber is innervated by only one axon. In developing muscle, however, motor neurons branch extensively such that each muscle fiber comes to be innervated by axons from several motor neurons (Figure 23.36), a situation termed **polyneuronal innervation.**[201,202] On each developing muscle fiber the synaptic endings of all the axons are interspersed at a single site, juxtaposed to aggregates of acetylcholine receptors and other components of the postsynaptic apparatus. As development progresses, axon branches are eliminated until the adult pattern is formed. This process does not involve cell death (which occurs at an earlier developmental stage), only a reduction in the number of muscle fibers innervated by each motor neuron.

The removal of polyneuronal innervation is mediated by competition between axons of different motor neurons for synaptic space on the muscle cell. The clearest example comes from studies of a small muscle in the toe of the rat.[203] When all but one of the motor axons innervating this muscle were cut early in development, the remaining axon spread to innervate many fibers within the muscle. During the period when elimination of polyneuronal innervation would normally occur, however, no synapses were lost. In the absence of competition the surviving motor neuron maintained contacts with every myofiber it had innervated. Experiments by Lichtman and his colleagues have provided vivid images of this process by visualizing living nerve terminals in animals with vital

[199]Nakamura, H., and O'Leary, D. D. M. 1989. *J. Neurosci.* 9: 3776–3795.

[200]O'Leary, D. D. M., and Terashima, T. 1988. *Neuron* 1: 901–910.

[201]Redfern, P. A. 1970. *J. Physiol.* 209: 701–709.

[202]Brown, M. C., Jansen, J. K. S., and Van Essen, D. 1976. *J. Physiol.* 261: 387–422.

[203]Betz, W. J., Caldwell, J. H., and Ribchester, R. R. 1980. *J. Physiol.* 303: 265–279.

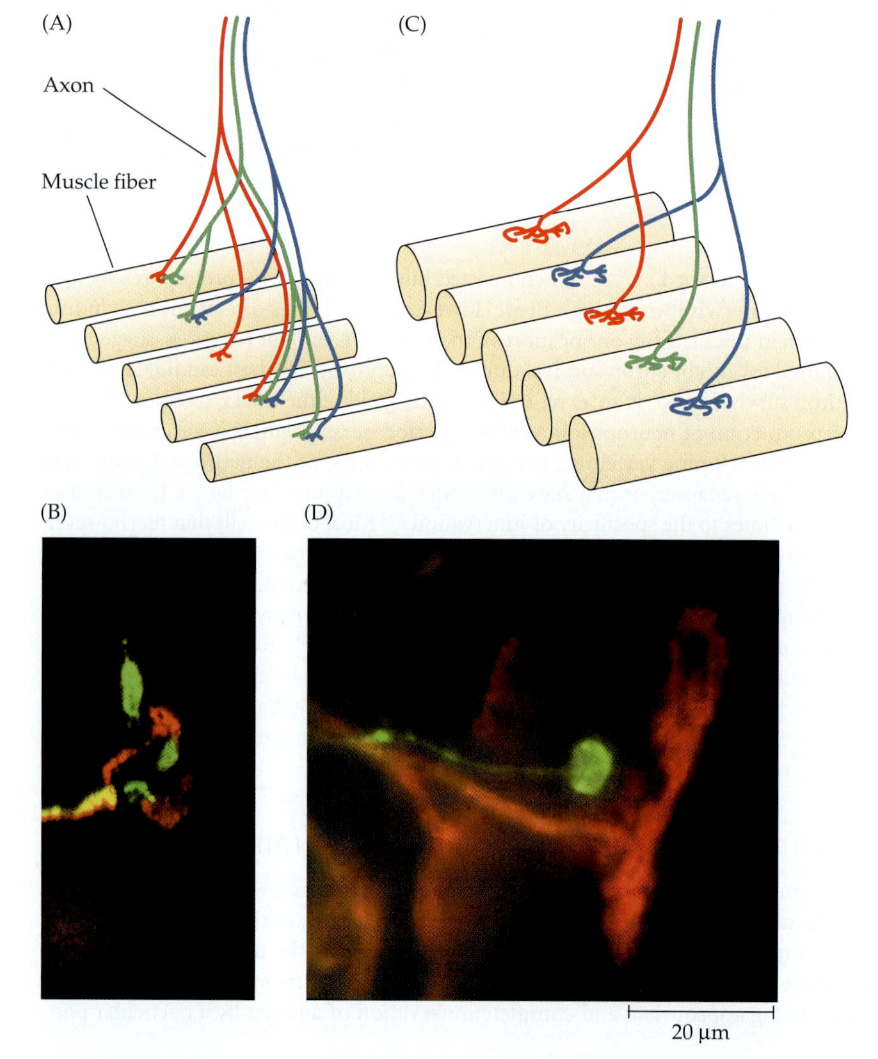

FIGURE 23.36 Polyneuronal Innervation and Its Elimination at the vertebrate skeletal neuromuscular junction. (A) During embryonic development, motor axons branch to innervate many muscle fibers, and each muscle fiber is innervated by several motor axons (polyneuronal innervation). (B) Fluorescence micrograph of a neuromuscular junction of an E18 mouse showing the distribution of terminals of two axons, each labeled with a lipophilic probe (diI in red, diA in green). During the period of polyneuronal innervation, the terminal arbors of all motor axons innervating a particular muscle fiber interdigitate at a single synaptic site. (C) After birth, polyneuronal innervation is eliminated as axon branches retract, leaving each muscle fiber innervated by a single motor axon. (D) Fluorescence micrograph of a mouse neuromuscular junction during the period of removal of polyneuronal innervation. Two axons innervating the junction were labeled as in B. All terminals of one axon (green) have been eliminated, and the axon is being withdrawn. (Micrographs kindly provided by J. W. Lichtman).

dyes and observing changes in synaptic structure during synapse elimination.[204,205] Similar retraction of multiple inputs has been shown to occur in the autonomic ganglia of neonatal rats and guinea pigs.[206] Each ganglion cell is initially supplied by multiple inputs—about five—but by about 5 weeks after birth only one usually remains.

Neuronal Activity and Synapse Elimination

Physiological experiments indicate that activity plays a role in synapse elimination, influencing both the rate and outcome of the competition between axon terminals. Stimulating the nerve to a muscle via implanted wire electrodes increases the rate of synapse elimination.[207,208] If activity is reduced, by applying tetrodotoxin in a cuff around the nerve to block action potentials or by inhibiting synaptic transmission, synapse elimination is slowed.[209,210] Muscles that receive input from axons that run in two different nerves allow the interesting experiment of blocking impulses in one nerve and not the other.[211,212] In such cases inactive axons are clearly at a competitive disadvantage: Axons in the blocked nerve innervate smaller-than-normal motor units; those in the active nerve innervate more fibers than usual. Domination by the active nerve is not complete, however, suggesting that other factors are involved.

Activity-dependent competition extends to the level of branches of a single motor axon.[213] If a small region of an adult junction is inactivated by focal application of α-bungarotoxin, the inactive region of the junction is eliminated (Figure 23.37). If the entire junction is silenced, no elimination occurs. The molecular basis of this competition and the mechanism by which activity contributes to synapse elimination are not known.

Similar competition for synaptic targets occurs during the development of CNS pathways.[214] One example is the formation of the ocular dominance columns in visual cortex (Chapter 25), in which axons from the lateral geniculate nucleus conveying information from the two eyes initially overlap extensively in layer 4 of the cortex and then sort out

[204]Balice-Gordon, R. J., and Lichtman, J. W. 1993. *J. Neurosci.* 13: 834–855.

[205]Gan, W-B., and Lichtman, J. W. 1998. *Science* 282: 1508–1511.

[206]Purves, D., and Lichtman, J. W. 1983. *Annu. Rev. Physiol.* 45: 553–565.

[207]O'Brien, R. A. D., Ostberg, A. J. C., and Vrbova, G. 1978. *J. Physiol.* 282: 571–582.

[208]Thompson, W. 1983. *Nature* 302: 614–616.

[209]Thompson, W., Kuffler, D. P., and Jansen, J. K. S. 1979. *Neuroscience* 4: 271–281.

[210]Brown, M. C., Hopkins, W. G., and Keynes, R. J. 1982. *J. Physiol.* 329: 439–450.

[211]Ribchester, R. R., and Taxt, T. 1983. *J. Physiol.* 344: 89–111.

[212]Ribchester, R. R., and Taxt, T. 1984. *J. Physiol.* 347: 497–511.

[213]Balice-Gordon, R. J., and Lichtman, J. W. 1994. *Nature* 372: 519–524.

[214]Katz, L. C., and Shatz, C. J. 1996. *Science* 274: 1133–1138.

FIGURE 23.37 Activity-Dependent Competition between Branches of a single motor axon. (A, B) Fluorescence micrographs of a neuromuscular junction in the mouse sternomastoid muscle showing the presynaptic terminal (A), stained with 4-Di-2-Asp (which labels mitochondria) and the postsynaptic membrane (B), stained with a low dose of rhodamine-conjugated α-bungarotoxin (which labels ACh receptors). (C) ACh receptors in the lower part of the junction are blocked by a saturating dose of unlabeled α-bungarotoxin, which results in a blockade of neuromuscular transmission selectively in this portion of the junction. (D, E) After 31 days, the same junction is labeled as in A and B and reexamined. Both the axon terminal (D) and the postsynaptic ACh receptors (E) have been removed from the blocked region. Thus, focal blockade of transmission resulted in focal synapse elimination. (After Balice-Gordon and Lichtman, 1994; micrographs kindly provided by J. W. Lichtman.)

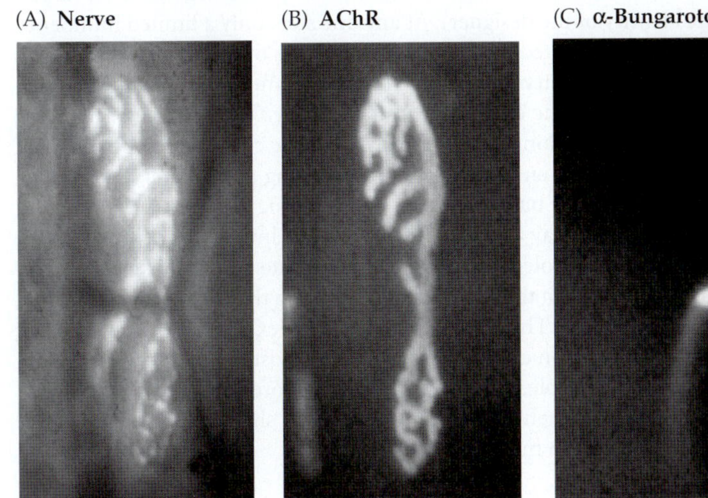

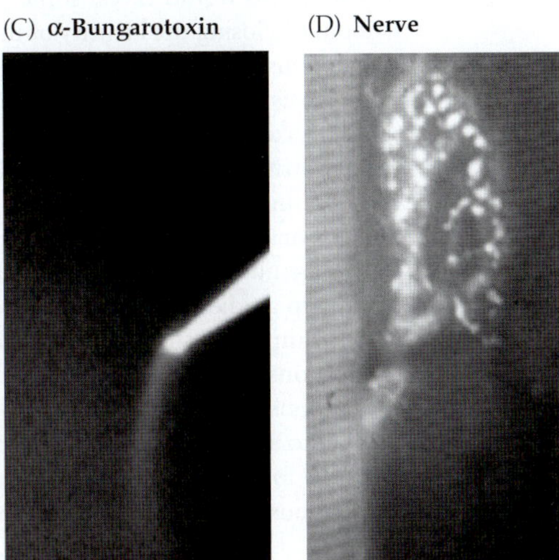

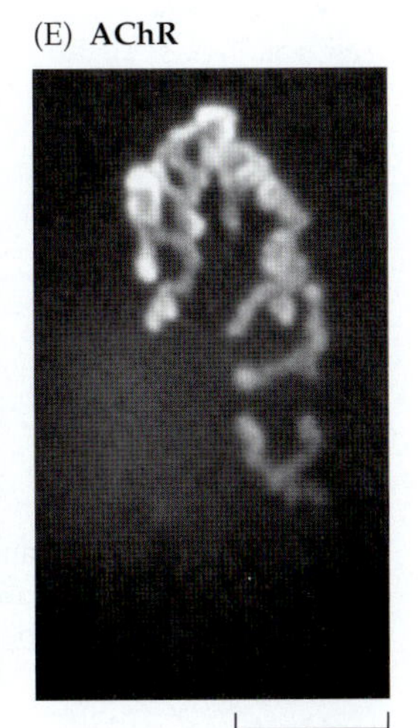

into left-eye and right-eye columns. Here the pattern of activity in the terminals from the two eyes plays a decisive role in determining the outcome of the competition.

Neurotrophins and Pruning

Pruning may occur through competition for a limited amount of target-derived trophic factor. For example, in developing visual cortex, adding excess BDNF blocks the formation of ocular dominance columns by preventing the loss of branches of lateral geniculate axons (Chapter 25).[215] Experiments in which sympathetic neurons were grown in three-compartment culture chambers illustrate that the supply of a neurotrophin can determine the survival or death of individual branches of a neuron (see Figure 23.33).[216] In this experiment, the medium in all three compartments initially contained NGF. Cells were added to the central compartment and extended axons into each of two lateral compartments. NGF was then removed from one of the lateral compartments. Axons in the central compartment and those projecting into the compartment containing NGF survived, but the axons that had extended into the compartment from which NGF was removed degenerated (see Figure 23.33C). Thus, within the target tissue, competition between axon terminals for a limited supply of neurotrophin may shape the terminal arbors of axons by sustaining some branches while others degenerate.

GENERAL CONSIDERATIONS OF NEURAL SPECIFICITY

Considerable progress has been made in understanding how nerve cells find their targets and establish synaptic contacts. However, when one contemplates the innumerable connections that must be formed when wiring up the nervous system, the problem of specificity seems daunting. A commonplace analogy may be encouraging. Let us assume that we are ignorant about the workings and design of the postal system. A chapter from a book on the nervous system, without its illustrations, is posted in Trieste, Italy, and addressed to Sunderland, Massachusetts, where it arrives a few days later. How does it get there? The writer knows only the closest letter box and is unaware even of the post office in his district. The postal worker who empties the letter box knows the post office; there the clerk who handles the mail may not know where Sunderland is but does know how to direct the package to the airport, and so on, to the right country, city, street, building, and eventually the correct person. If this were not enough, the illustrations that complement the chapter are posted by separate mail from Denver and Baltimore to the same destination, where they arrive almost simultaneously with the chapter from Trieste. All the while mail is moving through the same letter boxes and post offices in different directions to different final destinations.

The comforting feature of this analogy is that the problem seems altogether baffling at first sight. Yet, one can solve the postal puzzle by following the mail step by step to its destination. This would reveal some of the logic and design of postal organization (albeit without disclosing the identity of the designer). At any one step, only a limited number of instructions are followed and a limited number of mechanisms operate.

Some aspects of neural specificity may not be too different. A retinal ganglion cell sends its axon toward the back of the eye, where it makes a turn to enter the optic nerve together with fibers from other regions of the retina. The optic chiasm presents the next choice point, where the decision to enter the optic tract leading to one lateral geniculate nucleus or the other may be made based on local chemical signals. Within the lateral geniculate nucleus, retinal axons may arrange themselves and innervate their targets according to gradients of repulsive molecules. Axons of geniculate neurons likewise follow a fairly simple path to their targets in the cortex, stopping along the way to form transient connections with subplate neurons. Thus, the seemingly complex task of forming specific connections between retinal ganglion cells and neurons in the visual cortex can be broken down into a series of relatively simple, independent events. Moreover, when the connections are formed during development the distances are very short; pathways elongate tremendously as the nervous system matures.

[215]Cabelli, R. J., Hohn, A., and Shatz, C. J. 1995. *Science* 267: 1662–1666.

[216]Campenot, R. B. 1982. *Dev. Biol.* 93: 13–21.

Summary

- During vertebrate embryogenesis, proteins diffuse from the Spemann organizer to induce formation of the neural plate, the edges of which fold upward to form the neural tube.

- Cells divide rapidly in the wall of the neural tube. Postmitotic neurons and glial progenitor cells migrate away from the ventricular surface of the neural tube to form the CNS.

- Neuronal migration occurs along radial glial cells and pathways marked by cell surface and extracellular matrix components.

- The ultimate identity of a cell is determined by cell lineage and inductive interactions.

- Homeotic genes are master genes that control and coordinate the expression of groups of other genes, thereby determining the formation of body parts.

- In the vertebrate CNS, the fates of developing neurons are first restricted according to anteroposterior position and then by dorsoventral position. The *Hox* family of homeotic genes determines anteroposterior identity in the hindbrain. A protein called Sonic hedgehog, produced by the notochord, induces a ventral fate in cells along the length of the neural tube.

- Signals that influence cell differentiation often act through receptor tyrosine kinases, activating complex intracellular signaling cascades that induce changes in gene expression. An example is the regulation of photoreceptor differentiation in *Drosophila.*

- In mammalian cerebral cortex, development proceeds in an inside-out fashion. Neurons of the deepest cortical layers are born first.

- Neural crest cells arise from the edge of the neural fold and migrate away from the neural tube to form the peripheral nervous system, pigment cells, and bones of the head.

- Division of neural stem cells in the central nervous system of adult birds and mammals continually produces new neurons.

- The tip of a growing axon expands to form a growth cone.

- Cell surface and extracellular matrix adhesion molecules guide growth cones by short-range attractive and repulsive mechanisms.

- Netrins act as long-range chemoattractants for many types of axons, semaphorins as long-range chemorepellents.

- The ephrins and the Eph family of receptor tyrosine kinases influence pathfinding, cell migration, and cell intermingling by a chemorepellent mechanism.

- When a motoneuron growth cone contacts a muscle cell, functional synaptic transmission is established within minutes.

- The release of agrin from presynaptic terminals induces the formation of postsynaptic specializations in skeletal muscle.

- Neurons rely on trophic factors for survival and differentiation.

- Programmed neuronal death is a common feature of neural development.

- Synaptic connections, once established, are pruned to ensure appropriate and complete innervation of the target. Pruning occurs through activity-dependent competition among axon terminals for target-derived trophic substances.

SUGGESTED READING

General Reviews

Capecchi, M. R. 1997. *Hox* genes and mammalian development. *Cold Spring Harb. Symp. Quant. Biol.* 62: 273–281.

Dupin, E., Ziller, C., and Le Douarin, N. M. 1998. The avian embryo as a model in developmental studies: Chimeras and in vitro clonal analysis. *Curr. Top. Dev. Biol.* 36: 1–35.

Frade, J. M., and Barde, Y. A. 1998. Nerve growth factor: Two receptors, multiple functions. *BioEssays* 20: 137–145.

Francis, N. J., and Landis, S. C. 1999. Cellular and molecular determinants of sympathetic neuron development. *Annu. Rev. Neurosci.* 22: 541–566.

Gilbert, S. F. 2000. *Developmental Biology*, 6th Ed. Sinauer Associates, Sunderland, MA.

Levi-Montalcini, R. 1982. Developmental neurobiology and the natural history of nerve growth factor. *Annu. Rev. Neurosci.* 5: 341–362.

Lumsden, A., and Krumlauf, R. 1996. Patterning the vertebrate neuroaxis. *Science* 274: 1109–1115.

McAllister, A. K., Katz, L. C., and Lo, D. C. 1999. Neurotrophins and synaptic plasticity. *Annu. Rev. Neurosci.* 22: 295–318.

McConnell, S. K. 1995. Constructing the cerebral cortex: Neurogenesis and fate determination. *Neuron* 15: 761–768.

Mueller, B. K. 1999. Growth cone guidance: First steps towards a deeper understanding. *Annu. Rev. Neurosci.* 22: 351–388.

O'Leary, D. D., and Wilkinson, D. G. 1999. Eph receptors and ephrins in neural development. *Curr. Opin. Neurobiol.* 9: 65–73.

Oppenheim, R. W. 1991. Cell death during development of the nervous system. *Annu. Rev. Neurosci.* 14: 453–501.

Sanes, J. R., and Lichtman, J. W. 1999. Development of the vertebrate neuromuscular junction. *Annu. Rev. Neurosci.* 22: 389–442.

Song, H-J., and Poo, M-M. 1999. Signal transduction underlying growth cone guidance by diffusible factors. *Curr. Opin. Neurobiol.* 9: 355–363.

Walsh, F. S., and Doherty, P. 1997. Neural cell adhesion molecules of the immunoglobulin superfamily: Role in axon growth and guidance. *Annu. Rev. Cell Dev. Biol.* 13: 425–456.

Zigmond, M. J., Bloom, F. E., Landis, S. C., Roberts, J. L., and Squire, L. R. (eds.). 1999. *Fundamental Neuroscience*. Academic Press, New York.

Original Papers

Anderson, M. J., and Cohen, M. W. 1977. Nerve-induced and spontaneous redistribution of acetylcholine receptors on cultured muscle cells. *J. Physiol.* 268: 757–773.

Balice-Gordon, R. J., and Lichtman, J. W. 1993. In vivo observations of pre- and postsynaptic changes during the transition from multiple to single innervation at developing neuromuscular junctions. *J. Neurosci.* 13: 834–855.

Briscoe, J., Sussel, L., Serup, P., Hartigan-O'Connor, D., Jessell, T. M., Rubenstein, J. L. R., and Ericson, J. 1999. Homeobox gene Nkx2.2 and specification of neuronal identity by graded Sonic hedgehog signaling. *Nature* 398: 622–627.

Brown, M. C., Jansen, J. K. S., and Van Essen, D. 1976. Polyneuronal innervation of skeletal muscle in new-born rats and its elimination during maturation. *J. Physiol.* 261: 387–422.

Campenot, R. B. 1982. Development of sympathetic neurons in compartmentalized cultures. II. Local control of neurite survival by nerve growth factor. *Dev. Biol.* 93: 13–21.

Cox, E. C., Muller, B., and Bonhoeffer, F. 1990. Axonal guidance in the chick visual system: Posterior tectal membranes induce collapse of growth cones from the temporal retina. *Neuron* 4: 31–47.

Craig, A. M., Blackstone, C. D., Huganir, R. L., and Banker, G. 1994. Selective clustering of glutamate and γ-aminobutyric acid receptors opposite terminals releasing the corresponding transmitters. *Proc. Natl. Acad. Sci. USA* 91: 12373–12377.

Ghosh, A., Antonini, A., McConnell, S. K., and Shatz, C. J. 1990. Requirement for subplate neurons in the formation of thalamocortical connections. *Nature* 347: 179–181.

Gomez, T. M., and Spitzer, N. C. 1999. In vivo regulation of axon extension and pathfinding by growth-cone calcium transients. *Nature* 397: 350–355.

Hollyday, M., and Hamburger, V. 1976. Reduction of the naturally occurring motor neuron loss by enlargement of the periphery. *J. Comp. Neurol.* 170: 311–320.

Kennedy, T. E., Serafini, T., de la Torre, J. R., and Tessier-Lavigne, M. 1994. Netrins are diffusible chemotropic factors for commissural axons in the embryonic spinal cord. *Cell* 78: 425–435.

Lumsden, A. G. S., and Davies, A. M. 1986. Chemotropic effect of specific target epithelium in the developing mammalian nervous system. *Nature* 323: 538–539.

Luskin, M. 1993. Restricted proliferation and migration of postnatally generated neurons derived from the forebrain subventricular zone. *Neuron* 11: 173–189.

Luskin, M. B., Pearlman, A. L., and Sanes, J. R. 1988. Cell lineage in the cerebral cortex of the mouse studied in vivo and in vitro with a recombinant retrovirus. *Neuron* 1: 635–647.

Schlaggar, B. L., and O'Leary, D. D. M. 1991. Potential of visual cortex to develop an array of functional units unique to somatosensory cortex. *Science* 252: 1556–1560.

Song, H-J., Ming, G-L., and Poo, M-M. 1997. cAMP-induced switching in turning direction of nerve growth cones. *Nature* 388: 275–279.

Turner, D. L., and Cepko, C. L. 1987. A common progenitor for neurons and glia persists in rat retina late in development. *Nature* 328: 131–136.

Walter, J., Henke-Fahle, S., and Bonhoeffer, F. 1987. Avoidance of posterior tectal membranes by temporal retinal axons. *Development* 101: 909–913.

Yamada, T., Placzek, M., Tanaka, H., Dodd, J., and Jessell, T. M. 1991. Control of cell pattern in the developing nervous system: Polarizing activity of the floor plate and notochord. *Cell* 64: 635–647.

24 Denervation and Regeneration of Synaptic Connections

WHEN AN AXON IN THE VERTEBRATE NERVOUS SYSTEM IS SEVERED, the distal portion degenerates and changes occur in the damaged neuron, its targets, and cells presynaptic to it. The changes result from the interruption of axonal transport of trophic substances that control neuronal differentiation and survival and from alterations in the pattern of electrical activity.

After denervation, vertebrate skeletal muscle fibers become more sensitive to acetylcholine by expressing acetylcholine receptors over their entire surface. Direct electrical stimulation of denervated supersensitive muscles causes the region sensitive to acetylcholine to shrink back to the original end plate. Activity also influences the rate at which acetylcholine receptors are degraded and replaced. The effects of activity are mediated by calcium influx and activation of intracellular second messengers. Unlike innervated muscle fibers, denervated muscles accept innervation anywhere along their length. Denervated muscle fibers not only are amenable to innervation, but also induce undamaged nerve terminals to sprout new branches. Similarly, neurons that have been deprived of their innervation become supersensitive to transmitters and cause nearby nerve terminals to sprout.

The ability of damaged axons to regrow and innervate appropriate targets varies widely among species. In invertebrates and in lower vertebrates such as frogs and newts, severed axons regenerate and reconnect precisely with their original synaptic partners, as do axons in fetal or neonatal higher vertebrates, including mammals. Axons in the peripheral nervous system of adult higher vertebrates also regrow after they are damaged. If a peripheral nerve is crushed, regenerating axons are guided back to their peripheral targets by their original endoneurial tubes and sheaths of Schwann cell basal lamina, restoring normal function. If a peripheral nerve is cut, reinnervation of targets is often incomplete and imprecise. Agrin and other factors associated with the synaptic portion of the muscle fiber's basal lamina sheath trigger differentiation of pre- and postsynaptic specializations in regenerating nerve and muscle cells.

In the adult mammalian central nervous system, regeneration typically fails altogether. However, adult mammalian CNS neurons can sprout new axons and form new synapses over short distances, and they can extend axons for long distances through peripheral nerve grafts or undamaged regions of the CNS to reestablish appropriate synaptic connections. In addition, embryonic neurons and neural stem cells grafted into the adult CNS differentiate, extend axons, and can become integrated appropriately into the existing synaptic circuitry. Such transplantation techniques hold promise for the amelioration of functional deficits resulting from CNS lesions and neurodegenerative diseases.

In many species the nervous system has a remarkable ability to reestablish with a high degree of specificity synaptic connections that have been disrupted by trauma. The regenerative powers of neurons in the CNS were first demonstrated by Matthey, who in the 1920s sectioned the optic nerve of a newt and found that vision was restored within a few weeks.[1] Beginning in the 1940s, Sperry, Stone, and their colleagues took advantage of this regenerative capacity to explore how specific connections form within the nervous system. Their experiments on regenerating retinotectal connections in frogs and fish provided support for the idea that neurons selectively innervate their targets during regeneration, rather than making connections at random that are subsequently reorganized.[2] Later, detailed studies in leeches, crickets, and crayfish by Edwards, Hoy, Macagno, Muller, Nicholls, Palka, and others[3] demonstrated that, after being severed, axons of individual identified neurons in invertebrates can find and reconnect precisely with their original synaptic partners, ignoring a multitude of other potential targets. In contrast, regeneration of severed connections in the adult mammalian nervous system is typically incomplete or absent altogether.

In this chapter we first describe the changes that occur in a neuron and its surrounding glial cells after its axon is cut, and the effects of denervation on postsynaptic targets of the lesioned axon. Then we consider the ability of neurons to sprout new axons, reestablish synaptic connections with their targets, and restore lost function. Finally, we discuss the ability to replace neurons lost to injury or disease.

CHANGES IN AXOTOMIZED NEURONS AND THE SURROUNDING GLIAL CELLS

Wallerian Degeneration

If a sensory or motor axon in a vertebrate peripheral nerve is severed, a characteristic sequence of changes occurs (Figure 24.1).[4] The distal portion of the axon degenerates, as does a short length of the proximal portion. The Schwann cells that had formed the myelin sheath of the distal segment of the nerve dedifferentiate, proliferate, and, together with invading macrophages, phagocytize the axonal and myelin remnants. This response is called **Wallerian degeneration**, after the nineteenth-century anatomist Augustus Waller, who first described it. The cell body and its nucleus swell, the nucleus moves from its typ-

[1]Matthey, R. 1925. *C. R. Soc. Biol.* 93: 904–906.

[2]Purves, D., and Lichtman, J. W. 1985. *Principles of Neural Development.* Sinauer, Sunderland, MA.

[3]von Bernhardi, R., and Muller, K. J. 1995. *J. Neurobiol.* 27: 353–366.

[4]Grafstein, B. 1983. In *Nerve, Organ, and Tissue Regeneration: Research Perspectives.* Academic Press, New York, pp. 37–50.

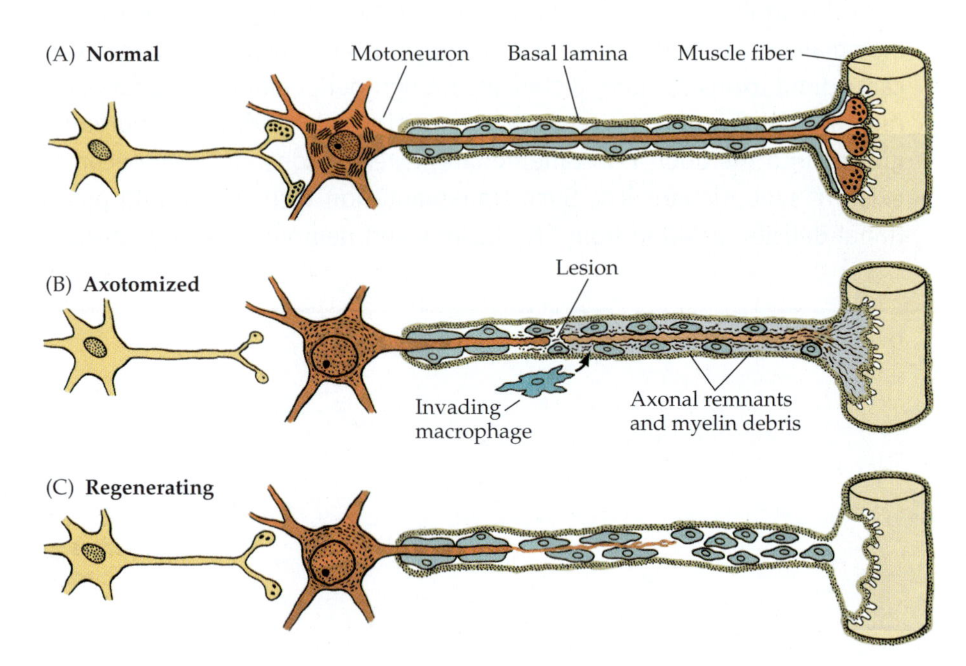

FIGURE 24.1 Degenerative Changes After Axotomy. (A) A typical motoneuron in an adult vertebrate. (B) After axotomy, the nerve terminal, the distal segment of the axon, and a short length of the proximal segment of the axon degenerate. Schwann cells dedifferentiate, proliferate, and, together with invading microglial cells and macrophages, phagocytize the axonal and myelin remnants. The axotomized neuron undergoes chromatolysis, presynaptic terminals retract, and degenerative changes may occur in pre- and postsynaptic cells. (C) The axon regenerates along the column of Schwann cells within the endoneurial tube and sheath of basal lamina that had surrounded the original axon.

ical position in the center of the cell soma to an eccentric location, and the ordered arrays of endoplasmic reticulum, called **Nissl substance**, disperse. Since the Nissl substance stains prominently with commonly used basic dyes, its dispersal following axotomy causes a decrease in staining that is referred to as **chromatolysis**.

Within a few hours, new axonal sprouts emerge from near the tip of the proximal stump and begin regenerating. If the neuron successfully reestablishes contact with a target, the cell body regains its original appearance. Chromatolysis also occurs after axons are severed in the central nervous system.

The response of adult neurons that fail to reestablish contact with their targets varies.[5,6] In mammals, retinal ganglion cells rapidly die if their axons in the optic nerve are severed; most cells in the ventrobasal thalamus survive cortical lesions but atrophy. After lesion of their axons in a peripheral nerve, many motor and dorsal root ganglion sensory neurons die; those that remain atrophy. Most axotomized autonomic ganglion cells survive but become less sensitive to acetylcholine and shrink in size.

Retrograde Trans-Synaptic Effects of Axotomy

Axotomy can also cause changes in the neurons that provide synaptic inputs to a damaged cell. For example, after axotomy of an autonomic ganglion cell in a chick, a rat, or a guinea pig, synaptic inputs onto the ganglion cell become less effective.[7–9] This is due in part to a decrease in the sensitivity of the axotomized cell to the neurotransmitter acetylcholine. In addition, retrograde trans-synaptic effects cause many of the presynaptic terminals to retract from the axotomized cell and the remaining terminals to release fewer quanta of transmitter (Figure 24.2).[10] Thus, damage to a neuron alters its ability to "hold on" to its presynaptic input. Rotshenker has shown an additional, retrograde trans-synaptic effect in motoneurons in the frog and the mouse.[11] When a motor nerve is cut on one side of the animal, axon terminals of the intact, undamaged motor neurons innervating the corresponding muscle on the other side sprout new branches and form additional synapses. A signal spreads from the axotomized neurons, crosses the spinal cord, and in-

[5]Lieberman, A. R. 1971. *Int. Rev. Neurobiol.* 14: 49–124.

[6]Muller, H. W., and Stoll, G. 1998. *Curr. Opin. Neurol.* 11: 557–562.

[7]Purves, D. 1975. *J. Physiol.* 252: 429–463.

[8]Brenner, H. R., and Johnson, E. W. 1976. *J. Physiol.* 260: 143–158.

[9]Brenner, H. R., and Martin, A. R. 1976. *J. Physiol.* 260: 159–175.

[10]Matthews, M. R., and Nelson, V. H. 1975. *J. Physiol.* 245: 91–135.

[11]Rotshenker, S. 1988. *Trends Neurosci.* 11: 363–366.

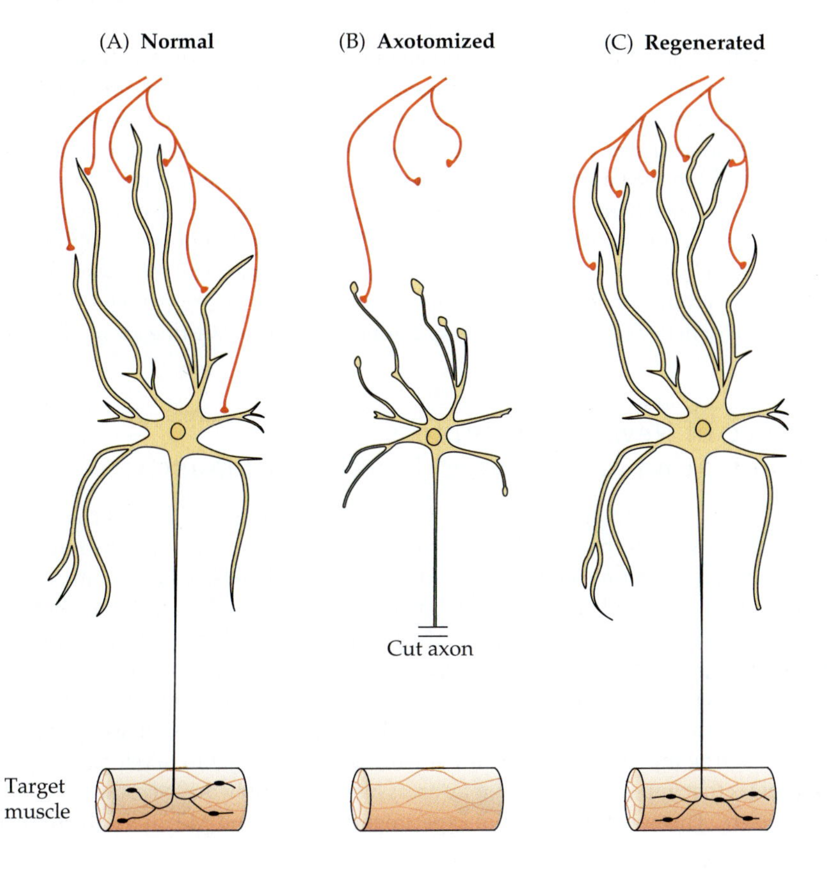

FIGURE 24.2 Axotomized Autonomic Ganglion Cells Atrophy and Lose Presynaptic Inputs. (A) Normal neuron. (B) Within a few days after axotomy, neurons atrophy and many dendrites show large varicosities. Many presynaptic terminals retract from dendrites; those that remain release less transmitter. (C) If the postganglionic axon regenerates and reinnervates its peripheral target, the cell and synaptic inputs recover. (After Purves, 1975.)

fluences undamaged motoneurons on the other side of the animal. Motoneurons innervating other muscles are not affected.

Trophic Substances and the Effects of Axotomy

Certain effects of axotomy—chromatolysis, atrophy, and cell death—result from the loss of trophic substances produced by the target tissue and transported retrogradely along the axon to the cell body.[12] Clear examples come from studies of the effects of nerve growth factor (NGF) on sensory and sympathetic neurons, discussed in Chapter 23. Thus, in guinea pig autonomic ganglia the effects of axotomy are mimicked by injecting antibodies to NGF subcutaneously for several days or by blocking retrograde transport in postganglionic nerves. Conversely, the effects of axotomy are largely prevented by application of NGF to the ganglion.[13] In a similar way, retrograde trans-synaptic atrophy may result from a decrease in the production of trophic factors by damaged neurons.

EFFECTS OF DENERVATION ON POSTSYNAPTIC CELLS

Neuromuscular synapses have provided useful preparations for studying mechanisms of synaptic transmission that are important at both peripheral and central synapses. Similarly, the properties of denervated muscles clearly demonstrate consequences of removal of synaptic inputs that are also seen in neurons in the CNS.[14]

The Denervated Muscle Membrane

Toward the end of the nineteenth century it was found that denervated skeletal muscles show certain clear-cut changes, such as spontaneous, asynchronous contractions called **fibrillation**. Fibrillation is initiated by the muscle membrane itself, not by ACh,[15] although most of the spontaneous action potentials producing fibrillation originate in the region of the former end plate.[16] The onset of fibrillation may be as early as 2 to 5 days after denervation in rats, guinea pigs, or rabbits, or well over a week in monkeys and humans.

Before or at the start of fibrillation, mammalian muscle fibers become supersensitive to a variety of chemicals. This means that the concentration of a substance required to excite a muscle is reduced by a factor of several hundred to a thousand. For example, a denervated mammalian skeletal muscle is about 1000 times more sensitive to ACh, applied either directly in the bathing fluid or injected into an artery supplying the muscle, than is a normally innervated muscle.[17] The action potential in denervated muscles also changes, becoming more resistant to tetrodotoxin, the puffer fish poison that blocks sodium channels (Chapter 3). This change is due to the reappearance of tetrodotoxin-resistant sodium channels that are the prevailing form in immature muscle.[18] Other changes occur in denervated muscle, such as a gradual atrophy, or wasting away, of muscle fibers.[19–21]

Appearance of New ACh Receptors after Denervation or Prolonged Inactivity of Muscle

Supersensitivity to acetylcholine is explained by an increase in the number and change in the distribution of ACh receptors in denervated muscles. This has been demonstrated by applying ACh to small regions of the muscle surface by ionophoretic release from an extracellular micropipette while recording the membrane potential. In a normally innervated frog, snake, or mammalian muscle, only the end-plate region—where the nerve fiber makes a synapse—is sensitive to ACh; the rest of the muscle membrane has a very low sensitivity. After denervation, the area sensitive to ACh increases until the surface of the muscle is almost uniformly sensitive to ACh (Figure 24.3).[22] In mammals this takes about a week; in frog muscle the changes are smaller and take longer to develop.[23]

The receptors that appear in extrasynaptic areas have not simply drifted away from the original end plate. This was first shown in experiments by Katz and Miledi in which frog

[12]Lindsay, R. M. 1996. *Ciba Found. Symp.* 196: 39–48.

[13]Nja, A., and Purves, D. 1978. *J. Physiol.* 277: 55–75.

[14]Cannon, W. B., and Rosenbluth, A. 1949. *The Supersensitivity of Denervated Structures: Law of Denervation.* Macmillan, New York.

[15]Purves, D., and Sakmann, B. 1974. *J. Physiol.* 239: 125–153.

[16]Belmar, J., and Eyzaguirre, C. 1966. *J. Neurophysiol.* 29: 425–441.

[17]Brown, G. L. 1937. *J. Physiol.* 89: 438–461.

[18]Kallen, R. G., et al. 1990. *Neuron* 4: 233–242.

[19]Guth, L. 1968. *Physiol. Rev.* 48: 645–687.

[20]Spector, S. A. 1985. *J. Neurosci.* 5: 2189–2196.

[21]Attaix, D., et al. 1994. *Reprod. Nutr. Dev.* 34: 583–597.

[22]Axelsson, J., and Thesleff, S. 1959. *J. Physiol.* 147: 178–193.

[23]Miledi, R. 1960. *J. Physiol.* 151: 1–23.

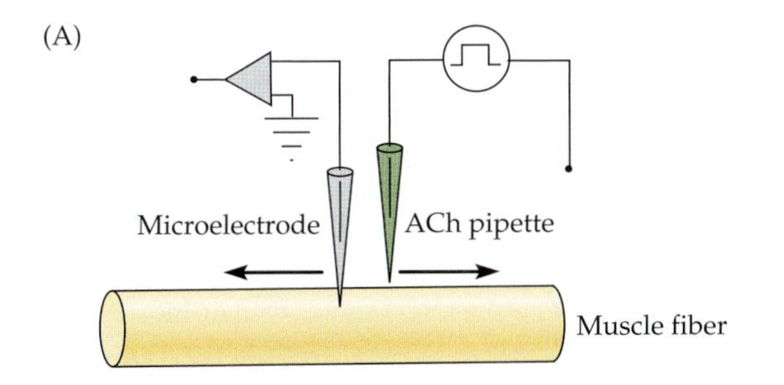

FIGURE 24.3 New ACh Receptors Appear after Denervation in cat muscle. (A) Pulses of ACh are applied from an ACh-filled pipette at different positions along the surface of a muscle fiber, while the membrane potential is recorded with an intracellular microelectrode. (B) In a muscle fiber with intact innervation, a response is seen only in the vicinity of the end plate. (C) After 14 days of denervation, a muscle fiber responds to ACh along its entire length. (After Axelsson and Thesleff, 1959.)

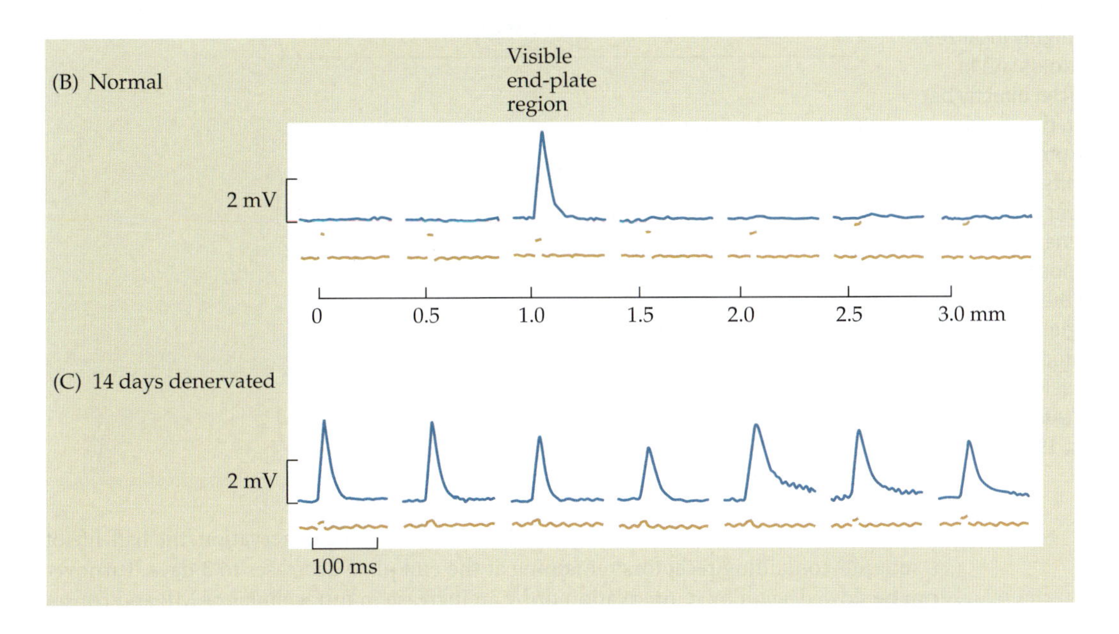

muscles were cut in two; nucleated fragments that were physically separated from the original end plate survived and developed increased sensitivity to ACh.[24] Thus, new ACh receptors are synthesized in extrajunctional regions of denervated muscles.

Synthesis and Degradation of Receptors in Denervated Muscle

A valuable technique for studying the distribution and turnover of ACh receptors is to label them with radioactive α-bungarotoxin, which binds to ACh receptors strongly and with high specificity. Bathing normal and denervated muscles in toxin and measuring binding at end-plate and end-plate-free areas confirmed that the number and distribution of binding sites is changed after denervation.[25,26] In normal muscle there are 10^4 binding sites/μm^2 in the postsynaptic membrane, compared with fewer than $10/\mu m^2$ in end-plate-free areas. After denervation, receptor sites in the extrasynaptic regions increase to about $10^3/\mu m^2$, with little change in density in the synaptic region.

The increase in the number of receptors in denervated muscle is attributable to enhanced receptor synthesis.[25,27] Thus, the rate of appearance of new ACh receptors increases markedly after denervation, and substances that block protein synthesis (such as actinomycin and puromycin) prevent the increase in extrasynaptic receptor density in muscles maintained in organ culture. Northern blot analysis and in situ hybridization demonstrate that in normal muscle, only those few nuclei located immediately beneath the end plate are synthesizing ACh receptor subunit mRNAs; in contrast ACh receptor genes are transcribed by nuclei all along the length of denervated muscle fibers (Figure 24.4).[28–30]

Denervation also affects the subunit composition and rate of degradation of ACh receptors. In mature muscle, junctional and extrajunctional ACh receptors contain an

[24] Katz, B., and Miledi, R. 1964. *J. Physiol.* 170: 389–396.

[25] Fambrough, D. M. 1979. *Physiol. Rev.* 59: 165–227.

[26] Salpeter, M. M., and Loring, R. H. 1985. *Prog. Neurobiol.* 25: 297–325.

[27] Scheutze, S. M., and Role, L. M. 1987. *Annu. Rev. Neurosci.* 10: 403–457.

[28] Merlie, J. P., and Sanes, J. R. 1985. *Nature* 317: 66–68.

[29] Bursztajn, S., Berman, S. A., and Gilbert, W. 1989. *Proc. Natl. Acad. Sci. USA* 86: 2928–2932.

[30] Fontaine, B., and Changeux, J.-P. 1989. *J. Cell Biol.* 108: 1025–1037.

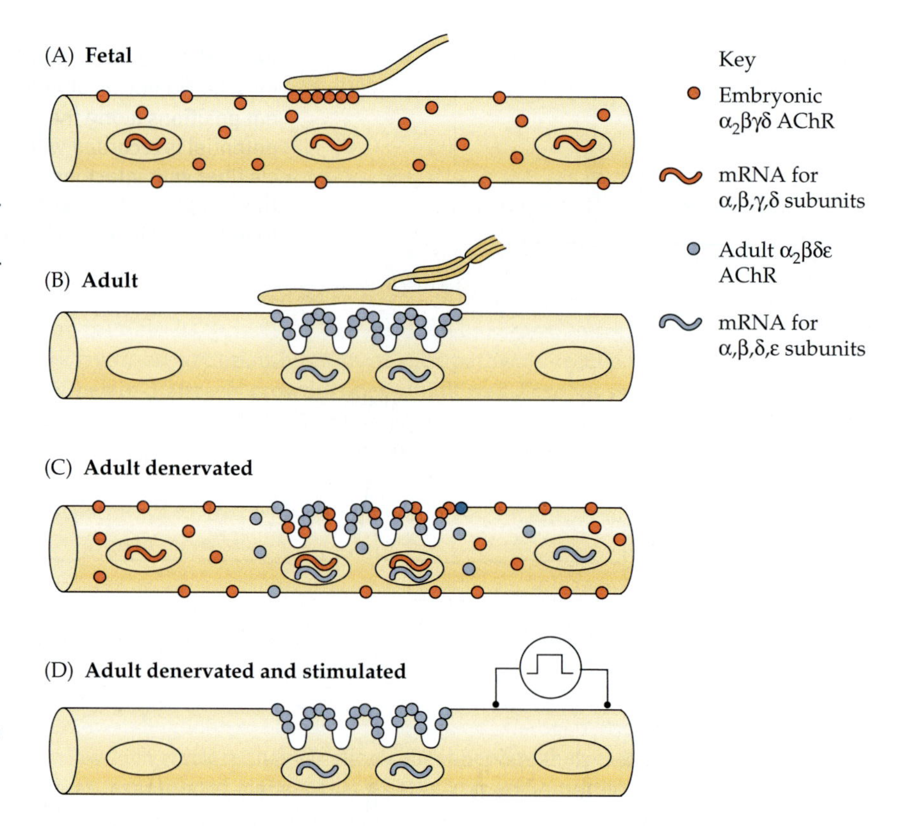

FIGURE 24.4 Synthesis and Distribution of ACh Receptors in rat muscle. (A) In fetal muscles, mRNAs for the α, β, γ, and δ subunits of the ACh receptor (AChR) are expressed in nuclei all along the length of the myofiber. The embryonic $\alpha_2\beta\gamma\delta$ form of the receptor is found over the entire surface of the myofiber and accumulates at the site of innervation. (B) In adult muscles, mRNAs for the α, β, δ, and ε subunits are expressed only in nuclei directly beneath the end plate. The adult $\alpha_2\beta\delta\varepsilon$ form of the receptor is highly localized to the crests of the junctional folds. (C) In denervated adult muscles, nuclei directly beneath the end plate express α, β, γ, δ, and ε subunits; all other nuclei reexpress the fetal pattern of α, β, γ, and δ subunits. Embryonic ACh receptors are found all over the surface of the myofiber (producing denervation supersensitivity), including the postsynaptic membrane; the adult form of the receptor is restricted to the end-plate region. (D) If denervated muscles are stimulated directly, the pattern of ACh receptor expression resembles that in innervated myofibers. (After Witzemann, Brenner, and Sakmann, 1991.)

ε subunit and have a half-life of about 10 days.[31,32] Following denervation, the half-life of ε subunit–containing receptors remaining at the end plate decreases to 3 days. Turnover can be slowed again by reinnervation or by an increase in intracellular cAMP and consequent activation of protein kinase A.[33]

New receptors synthesized in denervated muscle (whether synaptic or extrasynaptic) resemble those in embryonic muscle. They contain a γ subunit and turn over with a half-life of 1 day;[34,35] this rate can be slowed somewhat by exogenous ATP acting through purinergic receptors.[36]

Role of Muscle Inactivity in Denervation Supersensitivity

How does section of a nerve lead to the appearance of new receptors—through inactivity of the muscle or through some other mechanism? Lømo and Rosenthal[37] investigated this problem by blocking conduction in rat nerves by application of a local anesthetic or diphtheria toxin. The substances were applied by means of a cuff to a short length of the nerve some distance from the muscle. With this technique, the muscles became inactive because motor impulses failed to conduct past the cuff. Test stimulation of the nerve distal to the block produced a twitch of the muscle as usual, and miniature end-plate potentials still occurred normally, showing that synaptic transmission was intact. And yet after 7 days of nerve block, the muscle had become supersensitive (Figure 24.5). Other experiments have shown that new extrajunctional receptors appear when neuromuscular transmission is blocked by long-term application of curare or α-bungarotoxin to a muscle.[38] These results show that denervation supersensitivity is produced by the loss of synaptic activation of the muscle.[39]

The importance of muscle activity itself as a factor in controlling supersensitivity was confirmed in experiments in which supersensitive denervated muscles in the rat were stimulated directly through permanently implanted electrodes. Repetitive direct stimulation of muscles over several days caused the sensitive area to become restricted, so that once again only the synaptic region was sensitive to ACh (Figures 24.4D and 24.6).[37] The frequency of spontaneous activity in fibrillating muscle is too low to reverse the effects of denervation on the distribution of ACh receptors.[25,40]

[31]Salpeter, M. M., and Marchaterre, M. 1992. *J. Neurosci.* 12: 35–38.

[32]Sala, C., et al. 1997. *J. Neurosci.* 17: 8937–8944.

[33]Xu, R., and Salpeter, M. M. 1995. *J. Cell. Physiol.* 165: 30–39.

[34]Mishina, M., et al. 1986. *Nature* 321: 406–411.

[35]Shyng, S.-L., and Salpeter, M. M. 1990. *J. Neurosci.* 10: 3905–3915.

[36]O'Malley, J., Moore, C. T., and Salpeter, M. M. 1997. *J. Cell Biol.* 138: 159–165.

[37]Lømo, T., and Rosenthal, J. 1972. *J. Physiol.* 221: 493–513.

[38]Berg, D. K., and Hall, Z. W. 1975. *J. Physiol.* 244: 659–676.

[39]Witzemann, V., Brenner, H-R., and Sakmann, B. 1991. *J. Cell Biol.* 114: 125–141.

[40]Purves, D., and Sakmann, B. 1974. *J. Physiol.* 237: 157–182.

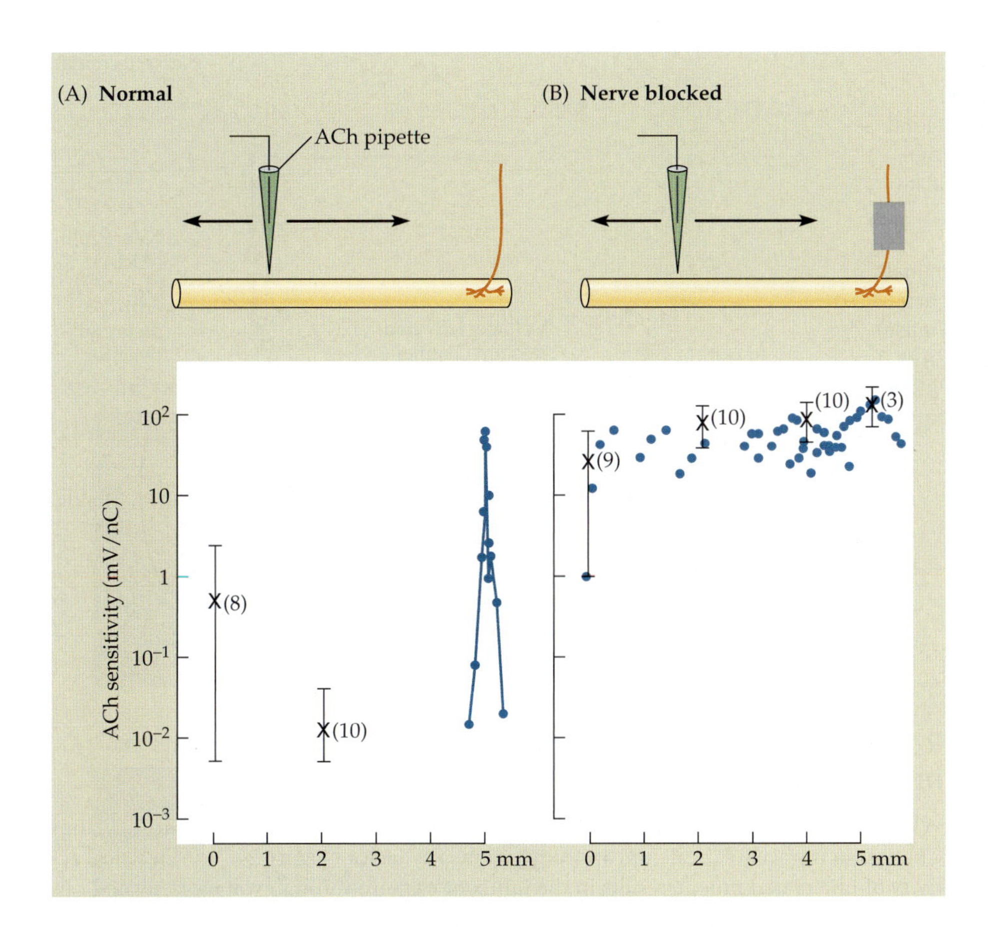

FIGURE 24.5 New ACh Receptors appear after block of nerve conduction in rat muscle. (A) In the normal muscle, ACh sensitivity is restricted to the end-plate region (near the 5 mm position). (B) After the nerve to the muscle was blocked for 7 days by a local anesthetic, the ACh sensitivity is distributed over the entire muscle fiber surface. Sensitivity is expressed numerically in millivolts of depolarization per nanocoulomb of charge ejected from the pipette (Chapter 9). The crosses and bars represent the mean and range of sensitivities of a number (in parentheses) of adjacent muscle fibers. (From Lømo and Rosenthal, 1972.)

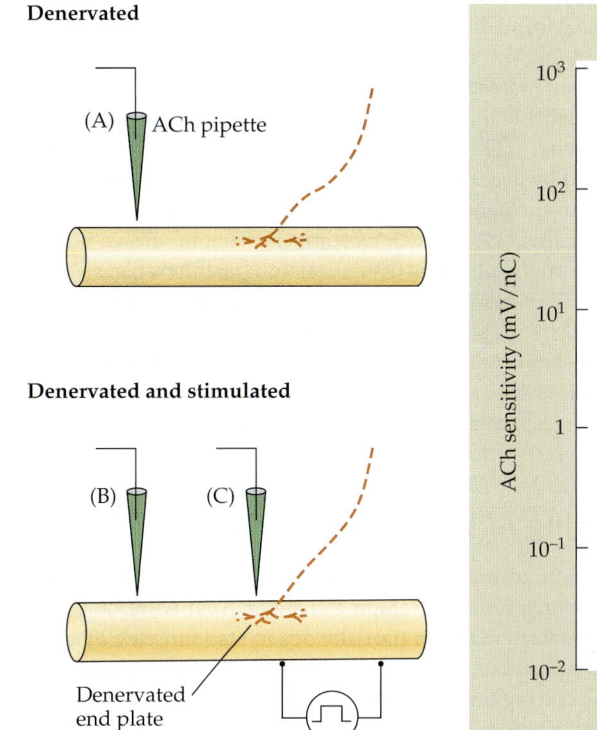

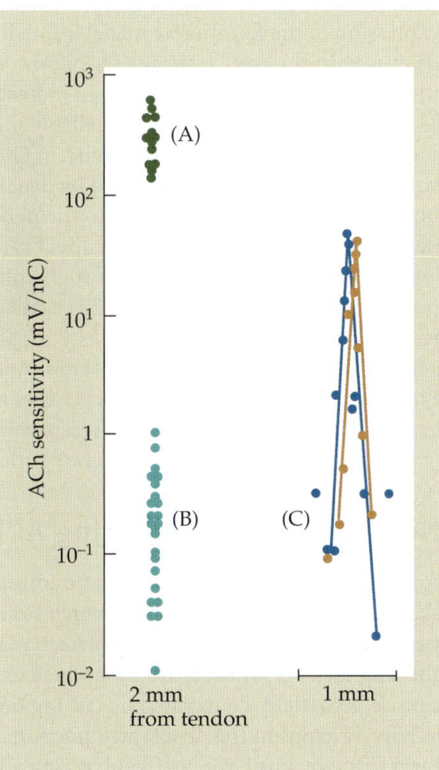

FIGURE 24.6 Reversal of Supersensitivity in a denervated rat muscle by direct stimulation of the muscle fibers. (A) Increased sensitivity in the extrasynaptic portion of a muscle fiber after 14 days of denervation. (B) Sensitivity in the extrasynaptic region of a muscle that had been denervated for 7 days without stimulation, and then stimulated intermittently for another 7 days. This treatment reversed the denervation supersensitivity. (C) ACh sensitivity in two stimulated fibers of the same muscle near their denervated end-plate regions. The high sensitivity is confined to this region in the stimulated muscle. (After Lømo and Rosenthal, 1972.)

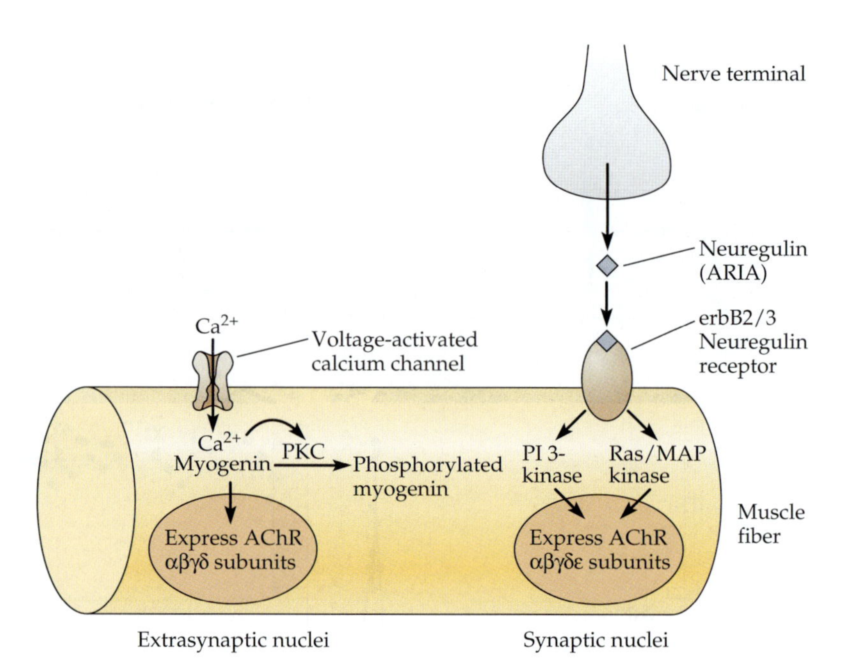

FIGURE 24.7 Control of ACh Receptor Synthesis by Calcium and Neural Factors. In extrasynaptic regions of a vertebrate skeletal muscle fiber, influx of calcium through voltage-activated calcium channels activates protein kinase C (PKC), which phosphorylates and inactivates myogenin. At the synapse, the neuregulin ARIA is released from nerve terminals and interacts with erbB2/3 receptors. This activates phosphatidylinositol 3-kinase (PI 3-kinase) and Ras/mitogen-activated protein (Ras/MAP) kinase pathways, leading to expression of AChR α, β, γ, δ, and ε subunits. Other neural signals suppress expression of the γ subunit.

Role of Calcium in Development of Supersensitivity in Denervated Muscle

What is it about the lack of muscle activity that causes supersensitivity to develop? Changes in intracellular calcium appear to be the key factor (Figure 24.7).[41] Electrical activity of innervated muscle results in the influx of calcium through voltage-activated calcium channels. Increased intracellular calcium activates protein kinase C, which in turn phosphorylates and inhibits myogenin. Myogenin is a transcription factor that induces the expression of AChR subunit genes, as well as regulating many other aspects of muscle differentiation. Thus, in innervated muscle, calcium influx inhibits AChR gene expression, keeping overall AChR levels low. (Additional signals that specifically induce AChR expression in the few muscle nuclei immediately beneath the postsynaptic membrane are discussed in the next section.) In inactive muscle, calcium influx is reduced, which removes the inhibition and leads to an increase in AChR expression.

The changes in ACh receptor half-life that occur in denervated muscle are also a consequence of reduced muscle activity.[42] The rate of receptor degradation increases to a similar extent in muscles paralyzed by denervation and in those made inactive by continuous application of tetrodotoxin to the nerve. Conversely, direct electrical stimulation of denervated muscle restores the turnover rate of ACh receptors at synaptic sites to normal levels. Again influx of calcium through voltage-activated calcium channels plays an important role.[43] The reduction in degradation rate produced by muscle activity is mimicked by treating inactive muscles with the calcium ionophore A23187. Conversely, activity-dependent ACh receptor stabilization is prevented by calcium channel blockers. Elevation of intracellular cAMP also slows receptor degradation in inactive muscles, suggesting that calcium influx causes receptor stabilization through activation of adenylyl cyclase and changes in protein phosphorylation.[44,45]

[41]Duclert, A., and Changeux, J. P. 1995. *Physiol. Rev.* 75: 339–368.

[42]Fumagalli, G., et al. 1990. *Neuron* 4: 563–569.

[43]Rotzler, S., Schramek, H., and Brenner, H. R. 1991. *Nature* 349: 337–339.

[44]Shyng, S.-L., Xu, R., and Salpeter, M. M. 1991. *Neuron* 6: 469–475.

[45]Caroni, P., et al. 1993. *J. Neurosci.* 13: 1315–1325.

Neural Factors Regulating ACh Receptor Synthesis

Activity is not the only factor that maintains the normal complement of receptors in skeletal muscles. Experiments in which slowly developing changes occur without activity per se playing an obvious role have been made on partially denervated muscles. Fibers in the frog sartorius muscle are innervated at more than one site along their length. If the muscle is partially denervated by cutting intramuscular branches of the nerve, supersensitivity develops in the denervated portions of the muscle fibers. Yet these fibers have not been inactive, but have kept contracting all along.[23] In situ hybridization studies of the distribution of mRNAs encoding ACh receptor subunits in normal, denervated, and toxin-paralyzed rat muscles suggest that at least two neural factors regulate ACh receptor expression independently of muscle activity: One induces the expression of the adult ε subunit in nuclei at the end plate, and the other suppresses expression of the γ subunit and to a lesser extent mRNAs for the other subunits.[39]

One neural factor that regulates ACh receptor expression is ARIA (*a*cetylcholine *r*eceptor-*i*nducing *a*ctivity), which was originally isolated from chick brain.[41,46] ARIA is synthesized and secreted by motoneurons and induces the expression of all five AChR subunits, most notably the ε subunit (see Figure 24.7).[47] ARIA is a member of a family of proteins called neuregulins. Receptors for the neuregulins are members of a family of receptor tyrosine kinases related to the type I epidermal growth factor receptor; several of the receptor subunits (erbB2, erbB3, and erbB4) accumulate at the neuromuscular junction. Stimulation of the erbB receptors in muscle activates two intracellular signaling cascades, the Ras/mitogen-activated protein (Ras/MAP) kinase and phosphatidylinositol 3-kinase (PI 3-kinase) pathways, and thereby causes enhanced AChR gene expression.[48–50]

Distribution of Receptors in Nerve Cells after Denervation

The effect of denervation on the distribution of transmitter receptors in neurons has been studied in autonomic ganglion cells in frogs and chicks. In the living frog heart, parasympathetic neurons can be seen in the transparent interatrial septum, greatly facilitating application of ACh to discrete spots on the cell surface. In addition, the cells have no dendrites; synapses are made on the cell soma (Figure 24.8A). Like skeletal muscle fibers, these neurons are sensitive to the transmitter ACh at selected spots on their surfaces, immediately under the presynaptic terminals.[51] When the distribution of ACh receptors was assessed by immunofluorescence microscopy, ganglion cells were found to

[46]Lemke, G. 1996. *Mol. Cell. Neurosci.* 7: 247–262.

[47]Si, J., and Mei, L. 1999. *Brain Res. Mol. Brain Res.* 67: 18–27.

[48]Tansey, M. G., Chu, G. C., and Merlie, J. P. 1996. *J. Cell Biol.* 134: 465–476.

[49]Si, J., Luo, Z., and Mei, L. 1996. *J. Biol. Chem.* 271: 19752–19759.

[50]Altiok, N., Altiok, S., and Changeux, J. P. 1997. *EMBO J.* 16: 717–725.

[51]Harris, A. J., Kuffler, S. W., and Dennis, M. L. 1971. *Proc. R. Soc. Lond. B* 177: 541–553.

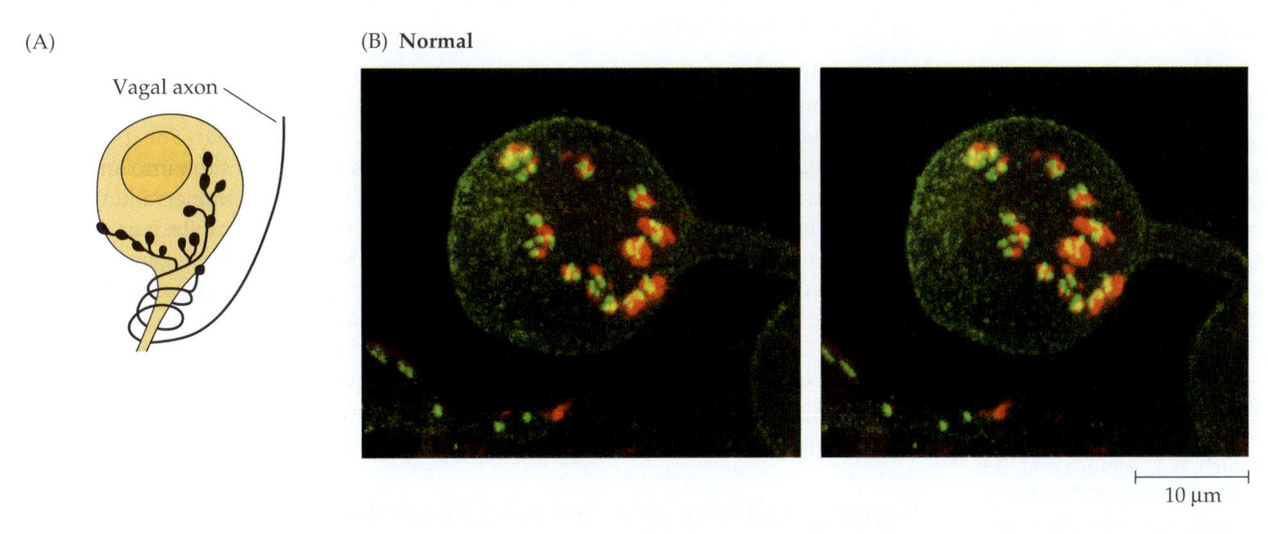

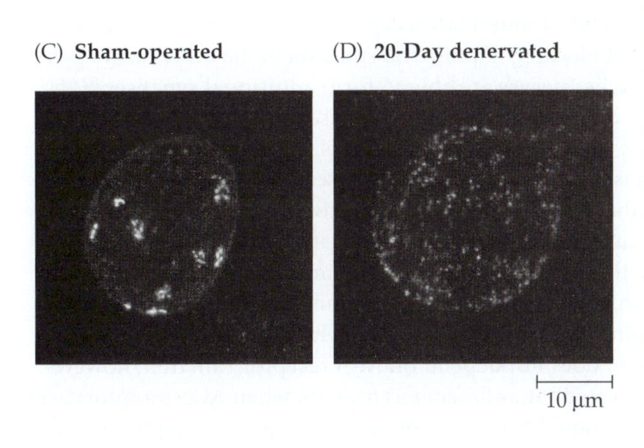

FIGURE 24.8 Development of Supersensitivity in parasympathetic nerve cells in frog heart after denervation. (A) Parasympathetic ganglion cells are innervated by axons in the vagus nerve, which form terminal boutons scattered over the cell surface. (B) Stereo-pair immunofluorescence micrographs of a ganglion cell in a normal animal, labeled with antibodies to AChRs (green) and to synaptic vesicles (red). Large, dense clusters of AChRs are located at synaptic sites; more than a hundred small extrasynaptic clusters are spread over the rest of the cell surface. (C, D) Images of sham-operated (C) and 20-day denervated (D) ganglion cells labeled with antibodies to AChRs. Denervation causes a decrease in the number of synaptic clusters and a marked increase in small extrasynaptic clusters, producing supersensitivity. (From Wilson Horch and Sargent, 1996; micrographs kindly provided by P. B. Sargent.)

have approximately 30 large, dense AChR clusters located at synaptic sites, and more than 100 small extrasynaptic clusters spread over the cell surface (Figure 24.8B).[52] Approximately 20% of the receptors were at extrasynaptic sites.[53]

To study the effects of denervation, the two vagus nerves to the heart were cut and the frog was left to recover.[54,55] Synaptic transmission between vagal nerve terminals and ganglion cells failed rapidly, starting on the second day after denervation. At the same time, the area of the neuronal surface membrane sensitive to ACh increased. By 4 to 5 days, ACh caused a membrane depolarization when applied anywhere on the cell surface. In other respects the cells were normal. It was not the number of ACh receptors that changed; rather it was the distribution (Figure 24.8C and D).[52,56] Denervation reduced the number of synaptic clusters by 90% and caused a two- to threefold increase in the number of small extrasynaptic clusters spread over the cell surface. Denervation also caused a decrease in the level of acetylcholinesterase.[57] Thus, the sensitivity to ACh was increased. If the original nerve was allowed to grow back, the sensitive area became restricted once more to synaptic sites.[55]

In other chick and frog ganglia, denervation has little or no effect on the number or distribution of surface ACh receptors.[58–61] For example, there is no difference in the sensitivity of neurons in denervated frog sympathetic ganglia to *ionophoretically* applied ACh. If the sensitivity to ACh is measured by recording the gross extracellular response to *bath-applied* ACh, however, an 18-fold increase in sensitivity is seen. This is due to the loss of acetylcholinesterase activity associated with preterminal and terminal portions of the preganglionic axons, which normally prevents ACh from reaching cells buried within the ganglion. Thus, the effects of denervation on neurotransmitter receptor distribution are somewhat different in nerve and muscle cells, and they vary among neuronal cell types.

Susceptibility of Normal and Denervated Muscles to New Innervation

In the adult mammal and frog, an innervated muscle fiber will not accept innervation by an additional nerve.[62] Thus, if a cut motor nerve is placed on an innervated muscle, it will not form additional new end plates on the muscle fibers. In contrast, nerve fibers do grow out and reinnervate denervated or injured muscle fibers. Unlike the situation during development, in which growth cones contact muscle fibers at random, reinnervation usually occurs at the site of the original end plate. Regenerating axons appear to be guided to the original synaptic sites by the endoneurial tubes of the former axons (see Figure 24.1) and by processes extended by the Schwann cells that had capped the former axon terminal (Chapter 8).[63] However, if a cut nerve is placed far enough away from the original end plate, or if an axon is in some way prevented from reaching it, then an entirely new end plate can be formed. This means that nerve fibers can grow out to an adult muscle fiber and form synapses in a region that had never been innervated, with the induction of both pre- and postsynaptic specializations.

Supersensitivity and Synapse Formation

What conditions enable a denervated muscle to accept a nerve? After rat muscles are made supersensitive as a result of blocking impulse transmission in the nerve or by application of botulinum toxin, foreign nerves are able to form additional synapses.[62,64,65] After release of the block, each of the two nerves can give rise to synaptic potentials and evoke contractions. Conversely, when a denervated muscle is stimulated directly, its ability to accept extra innervation is lost together with its supersensitivity.

Is it a normal prerequisite that the muscle be supersensitive for innervation to occur? As described in Chapter 23, muscle fibers in fetal and neonatal rats are sensitive to ACh along their length at the time they become innervated.[66] Similarly, myofibers grown in cell culture are sensitive to ACh and amenable to innervation throughout their length.[67] Both initial innervation and reinnervation therefore occur when the muscle fibers are supersensitive. Synapse formation does not depend on ACh receptor function, however. Reinnervation occurs in denervated rat and *Xenopus* muscles when ACh receptors are blocked by α-bungarotoxin or curare.[68,69] One factor determining the susceptibility of a

[52]Wilson Horch, H. L., and Sargent, P. B. 1996. *J. Neurosci.* 16: 1720–1729.
[53]Sargent, P. B., and Pang, D. Z. 1989. *J. Neurosci.* 9: 1062–1072.
[54]Kuffler, S. W., Dennis, M. J., and Harris, A. J. 1971. *Proc. R. Soc. Lond. B* 177: 555–563.
[55]Dennis, M. J., and Sargent, P. B. 1979. *J. Physiol.* 289: 263–275.
[56]Sargent, P. B., et al. 1991. *J. Neurosci.* 11:3610–3623.
[57]Streichert, L. C., and Sargent, P. B. 1992. *J. Physiol.* 445: 249–260.
[58]McEachern, A. E., Jacob, M. H., and Berg, D. K. 1989. *J. Neurosci.* 9: 3899–3907.
[59]Loring, R. H., and Zigmond, R. E. 1987. *J. Neurosci.* 7: 2153–2162.
[60]Dunn, P. M., and Marshall, L. M. 1985. *J. Physiol.* 363: 211–225.
[61]Wilson Horch, H. L., and Sargent, P. B. 1995. *J. Neurosci.* 15: 7778–7795.
[62]Jansen, J. K. S., et al. 1973. *Science* 181: 559–561.
[63]Son, Y. J., and Thompson, W. J. 1995. *Neuron* 14: 125–132.
[64]Thesleff, S. 1960. *J. Physiol.* 151: 598–607.
[65]Fex, S., et al. 1966. *J. Physiol.* 184: 872–882.
[66]Diamond, J., and Miledi, R. 1962. *J. Physiol.* 162: 393–408.
[67]Frank, E., and Fischbach, G. D. 1979. *J. Cell Biol.* 83: 143–158.
[68]Van Essen, D., and Jansen, J. K. 1974. *Acta Physiol. Scand.* 91: 571–573.
[69]Cohen, M. W. 1972. *Brain Res.* 41: 457–463.

muscle to innervation is MuSK, the receptor tyrosine kinase through which agrin triggers differentiation of the postsynaptic apparatus (Chapter 23).[70]

Denervation-Induced Axonal Sprouting

Not only are denervated muscles amenable to innervation, but they actively induce undamaged nerves to sprout new terminal branches. For example, if a muscle is partially denervated, the remaining axon terminals will sprout and innervate the denervated fibers (Figure 24.9).[71] As with regulation of ACh receptor synthesis and degradation, muscle inactivity triggers this process. Sprouting and hyperinnervation occur if muscle activity is prevented by blocking action potential propagation in the nerve with a cuff impregnated with tetrodotoxin,[72] or if neuromuscular transmission is blocked with botulinum toxin or α-bungarotoxin.[73,74] The terminal Schwann cell plays an important role in the regulation of axon terminal sprouting (Chapter 8).[75]

The molecular mechanisms that induce sprouting of axon terminals have yet to be worked out in detail; however, the signals are quite selective. In leech skin, for example, killing a particular sensory or motor neuron by injecting it with pronase induces axon sprouting into the denervated territory, but only of axons of cells that have the same sensory or motor modality.[76–78] Molecular biological techniques can be applied to single leech neurons to identify genes that are specifically upregulated under such conditions.[79]

[70]Meier, T., and Wallace, B. G. 1998. *BioEssays* 20: 819–829.

[71]Brown, M. C., Holland, R. L., and Hopkins, W. G. 1981. *Annu. Rev Neurosci.* 4: 17–42.

[72]Brown, M. C., and Ironton, R. 1977. *Nature* 265: 459–461.

[73]Duchen, L. W., and Strich, S. J. 1968. *Q. J. Exp. Physiol.* 53: 84–89.

[74]Holland, R. L., and Brown, M. C. 1980. *Science* 207: 649–651.

[75]Son, Y-J., and Thompson, W. J. 1995. *Neuron* 14: 133–141.

[76]Bowling, D., Nicholls, J. G., and Parnas, I. 1978. *J. Physiol.* 282: 169–180.

[77]Blackshaw, S. E., Nicholls, J. G., and Parnas, I. 1982. *J. Physiol.* 326: 261–268.

[78]Parnas, I. 1987. *J. Exp. Biol.* 132: 231–247.

[79]Korneev, S., et al. 1997. *Inv. Neurosci.* 3: 185–192.

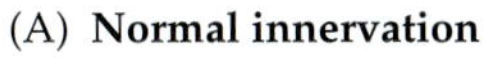

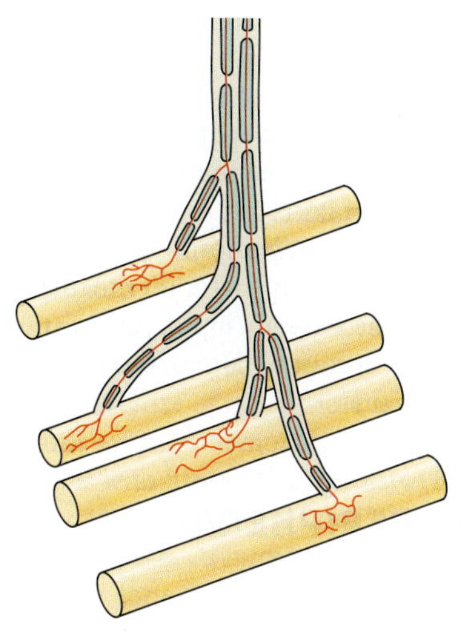

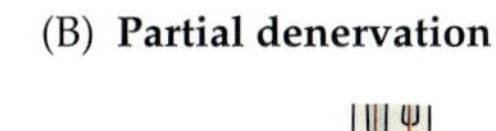

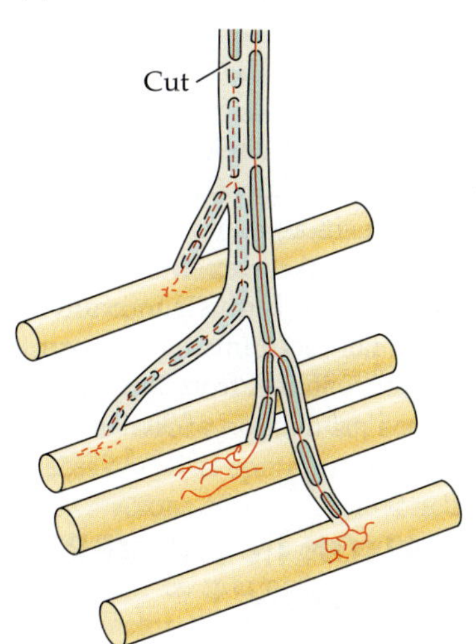

(C) **Sprouting**

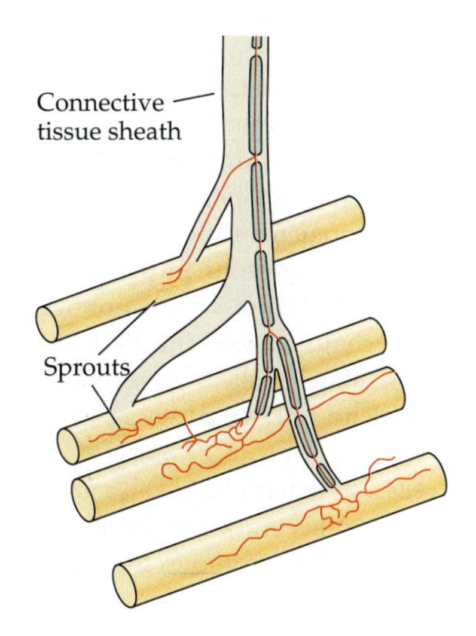

(D) **Reinnervation**

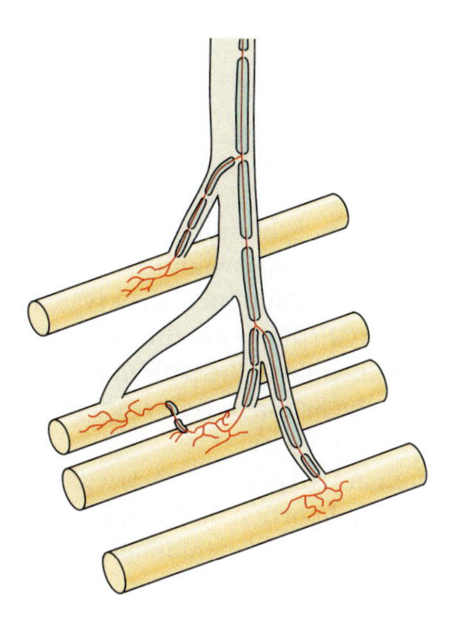

FIGURE 24.9 Nerve Terminals Sprout in Response to Partial Denervation of a mammalian skeletal muscle. (A) Normal pattern of innervation. (B) Some fibers are denervated by cutting a few of the axons innervating the muscle. (C) Axons sprout from the terminals and from nodes along the preterminal axons of undamaged motoneurons to innervate the denervated fibers. (D) After 1 or 2 months, sprouts that have contacted vacant end plates are retained, while other sprouts disappear. (After Brown, Holland, and Hopkins, 1981.)

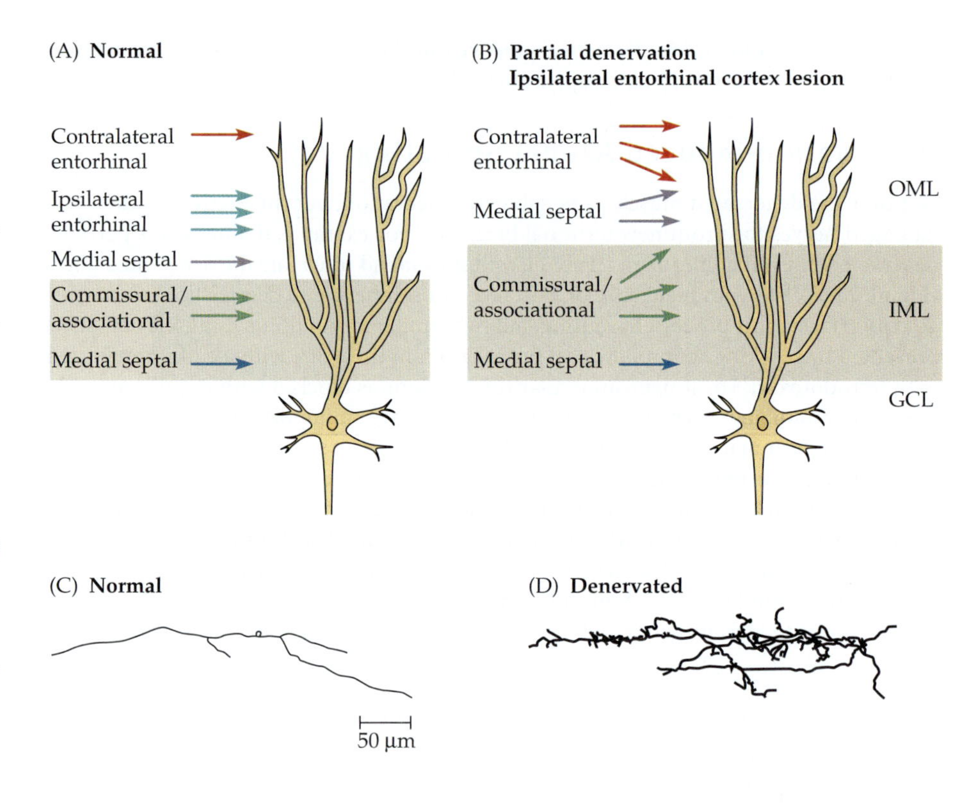

FIGURE 24.10 Sprouting of Axons in the Mammalian CNS. (A) A typical granule cell in the dentate gyrus receives dense synaptic input from the ipsilateral entorhinal cortex, and sparse input from the contralateral cortex and medial septum on its dendrites in the outer molecular layer (OML). Dendrites in the inner molecular layer (IML) receive medial septal and commissural/associational inputs. GCL = granule cell layer. (B) Following ablation of the ipsilateral entorhinal cortex, axons from the contralateral entorhinal cortex sprout extensively and replace the ipsilateral entorhinal input in the OML. There is also some sprouting of axons from the medial septum in the OML. Commissural/associational fibers in the IML expand their termination zone. (C,D) Examples of terminal arbors of axons from the contralateral entorhinal cortex in the OML of the dentate gyrus. (C) Normal. (D) Two months after lesion of the ipsilateral entorhinal cortex. (After Deller and Frotscher, 1997.)

Similarly, when cells in the dentate gyrus of the rat hippocampus are partially denervated by ablating the ipsilateral entorhinal cortex, the three remaining inputs sprout and form new synapses, in a precise and orderly manner (Figure 24.10).[80] Axons from the contralateral entorhinal cortex, which normally provide only sparse input, sprout most extensively and replace, both structurally and functionally, the ipsilateral entorhinal input, which normally accounts for approximately 80% of the synaptic input. Among the fibers responding are inputs from the medial septum; their sprouting appears to be triggered by increased production of NGF at the site of the lesion.

REGENERATION IN THE VERTEBRATE PERIPHERAL NERVOUS SYSTEM

Regrowth of Severed Axons

In the peripheral nervous system Schwann cells provide an environment especially conducive to axon regeneration. The growth-promoting activity of Schwann cells is due to secretion of numerous trophic factors, surface expression of cell adhesion molecules and integrins, and production of extracellular matrix components such as laminin.[81] For example, experiments in which the sciatic nerve is lesioned have shown that, as the peripheral portion of the axon degenerates, proliferating Schwann cells synthesize high levels of two neurotrophic factors: BDNF (brain-derived neurotrophic factor) and NGF (Figure 24.11).[82,83] Thus, Schwann cells may temporarily supply regenerating motor, sensory, and sympathetic axons with BDNF and NGF as they grow back to their peripheral targets. It is interesting that such "denervated" Schwann cells also express large numbers of low-affinity NGF/BDNF receptors on their surface, perhaps to hold the NGF and BDNF they produce along the path that regenerating axons should take.[84] As regeneration progresses, the Schwann cells cease production of NGF and BDNF and once again ensheathe the axons.

Apolipoprotein E (ApoE), synthesized by Schwann cells and macrophages, also accumulates in the distal portion of damaged peripheral nerves and becomes associated with the Schwann cell basement membrane (see Figure 24.11).[85,86] ApoE promotes the health

[80]Deller, T., and Frotscher, M. 1997. *Prog. Neurobiol.* 53: 687–727.

[81]Guénard, V., Xu, X. M., and Bunge, M. B. 1993. *Sem. Neurosci.* 5: 401–411.

[82]Heumann, R., et al. 1987. *J. Cell Biol.* 104: 1623–1631.

[83]Meyer, M., et al. 1992. *J. Cell Biol.* 119: 45–54.

[84]Johnson, E. M., Jr., Taniuchi, M., and DiStefano, P. S. 1988. *Trends Neurosci.* 11: 299–304.

[85]Skene, J. H. P., and Shooter, E. M. 1983. *Proc. Natl. Acad. Sci. USA* 80: 4169–4173.

[86]Fullerton, S. M., Strittmatter, W. J., and Matthew, W. D. 1998. *Exp. Neurol.* 153: 156–163.

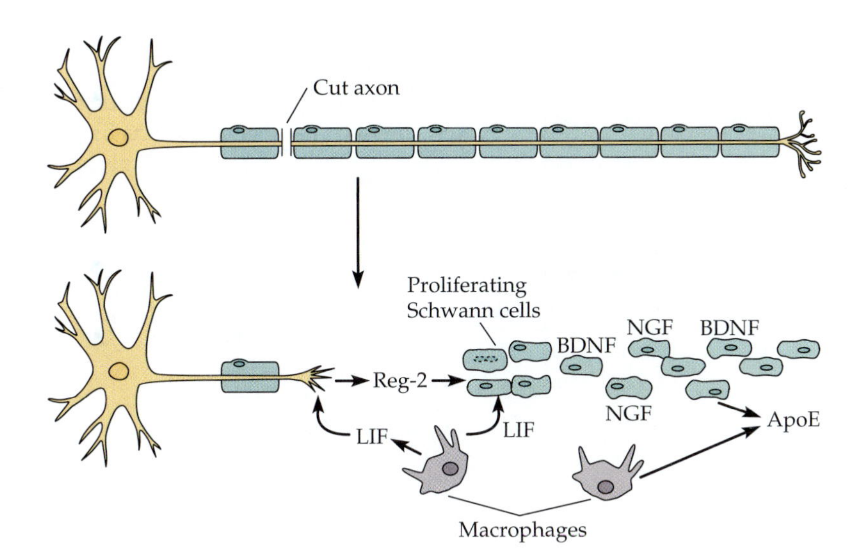

FIGURE 24.11 Schwann Cells Promote Axon Regrowth in the vertebrate peripheral nervous system. After axotomy, the distal portion of the axon and the myelin degenerate and are phagocytized. Schwann cell proliferation is stimulated by two cytokines: leukemia inhibitory factor (LIF) from macrophages and Reg-2 from axon terminals. Expression of Reg-2 is enhanced by LIF. Proliferating Schwann cells synthesize two neurotrophic factors, BDNF and NGF, which are held on the cell surface by low-affinity BDNF/NGF receptors and help sustain regenerating axons and guide them to their targets. Schwann cells and macrophages also synthesize apolipoprotein E (ApoE), which may help promote neuron survival and axon regrowth.

and survival of neurons by virtue of its protective effects against oxidative damage, and it promotes neurite outgrowth and adhesion. However, regeneration and remyelination of peripheral axons is not changed in ApoE knockout mice.[87] Mice lacking ApoE do have abnormal and reduced numbers of unmyelinated axons, and a corresponding reduction in sensitivity to noxious thermal stimuli.[86]

When a peripheral nerve is lesioned, factors that promote Schwann cell proliferation increase at the site of injury. They include two cytokines: leukemia inhibitory factor (LIF)[88] and Reg-2, which is a potent mitogen for Schwann cells (see Figure 24.11).[89] Reg-2 is specifically expressed in developing and regenerating motor and sensory neurons, and its expression is enhanced by LIF. Antibodies to Reg-2 slow regeneration after a sciatic nerve crush.

Specificity of Reinnervation

For complete recovery of function, regenerating axons must reestablish connections with their original targets. Classic experiments of Langley, now confirmed by analysis of single cells, demonstrated that regenerating mammalian preganglionic autonomic axons tended to reinnervate the appropriate postganglionic neurons.[2] This trend results in part from positional cues that bias synapse formation between neurons and target cells on the basis of anteroposterior position. These cues are shared by the sympathetic and motor systems. Thus, if an anteriorly located intercostal muscle is transplanted into a site in the neck and reinnervated by preganglionic autonomic axons in the cervical sympathetic trunk, most of the regenerated inputs are made by axons of neurons located in more anterior portions of the spinal cord (Figure 24.12).[90] Conversely, muscles transplanted from a more posterior region tend to be innervated by more posteriorly derived axons. A similar trend is seen after transplantation of anterior or posterior sympathetic ganglia.[91] In young rats, comparable positional bias has been shown in the reinnervation of fibers within muscles that span several segments.[92]

Neuromuscular connections are restored accurately following transection of motor nerves in neonatal rats, larval frogs, and adult newts.[93] One mechanism for reestablishing connections selectively is competition between axons; in salamander muscles that have been innervated by inappropriate axons, the foreign synapses are eliminated after the normal nerve reestablishes its connection.[94] In adult mammals, regenerating sensory, motor, and postganglionic autonomic axons show little ability to navigate back selectively to their original targets, and foreign nerves can be as effective as the original ones in innervating muscle fibers.[93] Indeed, a foreign nerve can even displace the original motor axons in intact adult rat muscles.[95] Selective regeneration can occur in adult mammals if a peripheral nerve is crushed rather than cut, so as to leave the endoneurial tubes and Schwann cell basal lamina that surround the axons intact (see Figure 24.1).[93] Under such conditions, axons regenerate within their parent tubes and are guided back to their origi-

[87] Goodrum, J. F., et al. 1995. *J. Neurochem.* 64: 408–416.

[88] Banner, L. R., and Patterson, P. H. 1994. *Proc. Natl Acad. Sci. USA* 91: 7109–7113.

[89] Livesey, F. J., et al. 1997. *Nature* 390: 614–618.

[90] Wigston, D. J., and Sanes, J. R. 1985. *J. Neurosci.* 5: 1208–1221.

[91] Purves, D., Thompson, W., and Yip, J. W. 1981. *J. Physiol.* 313: 49–63.

[92] Laskowski, M. B., and Sanes, J. R. 1988. *J. Neurosci.* 8: 3094–3099.

[93] Fawcett, J. W., and Keynes, R. J. 1990. *Annu. Rev. Neurosci.* 13: 43–60.

[94] Dennis, M. J., and Yip, J. W. 1978. *J. Physiol.* 274: 299–310.

[95] Bixby, J. L., and Van Essen, D. C. 1979. *Nature* 282: 726–728.

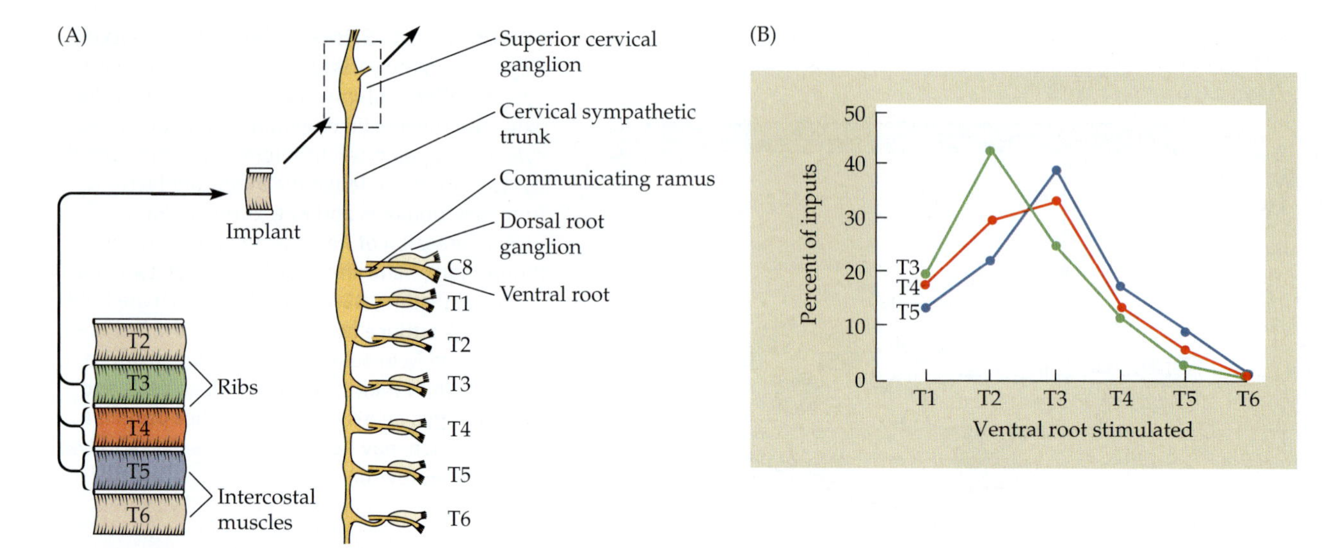

FIGURE 24.12 Selective Reinnervation of Muscles Based on Segmental Origin. (A) An intercostal muscle from thoracic segment T3, T4, or T5 was transplanted to the neck of an adult rat and reinnervated by preganglionic autonomic axons in the cervical sympathetic trunk, after removal of the superior cervical ganglion. The segmental origin of the inputs was determined by recording from the muscle and stimulating individual ventral roots that supply preganglionic axons to the trunk. (B) Distribution of inputs to transplanted T3, T4, and T5 muscles. More anterior muscles tend to be innervated by more anterior preganglionic axons. (After Wigston and Sanes, 1985.)

nal targets. If the endoneurial tubes are disrupted, as when a nerve is cut, then regenerating axons enter tubes in the distal portion of the nerve at random and frequently are guided to and make synapses with inappropriate targets.

Properties of Nerve and Muscle after Formation of Aberrant Contacts

Observations concerning the consequences of inappropriate contacts date back to 1904, when Langley and Anderson made the remarkable observation that muscles of the cat could become innervated by cholinergic preganglionic sympathetic fibers,[96] which normally make synapses on nerve cells in ganglia (Chapter 16). Similar synapses have been shown to be formed by autonomic nerves on frog and rat skeletal muscle (as in Figure 24.12).[90,97] In such experiments, many of the properties of the nerve and muscle were found to remain unchanged, despite the abnormal innervation.

In other experiments the properties of muscles have been shown to become markedly changed with foreign innervation. Slow skeletal muscle fibers in the frog are quite distinctive: They are diffusely innervated, have a characteristic fine structure, and do not give regenerative impulses or twitches.[98] After denervation, slow fibers can become reinnervated by nerves that normally innervate twitch muscles at discrete end plates. Under these conditions, slow fibers become able to give conducted action potentials and twitches.[99] Eccles, Buller, Close, and their colleagues cut and interchanged the nerves to rapidly and more slowly contracting skeletal muscles in kittens and rats. Both these types of mammalian muscle fibers produce propagating action potentials and are called **slow-twitch** and **fast-twitch** fibers. After the muscles were reinnervated by the inappropriate nerves, the slow-twitch muscles became faster and the fast-twitch ones slower.[100] A major factor in the transformation is the pattern of impulses in the nerve and the resulting muscle contractions; the motoneurons innervating slow- and fast-twitch muscle fibers tend to fire at different frequencies.[101]

ROLE OF BASAL LAMINA AT REGENERATING NEUROMUSCULAR SYNAPSES

A structure that plays a key role in the regeneration of neuromuscular synapses is the synaptic basal lamina, which lies between the nerve terminal and the muscle membrane. The synaptic basal lamina constitutes a densely staining extracellular matrix made up of proteoglycans and glycoproteins. As shown in Figure 24.13A, basal lamina surrounds the muscle, the nerve terminal, and the Schwann cell, and dips into the folds in the postsynaptic membrane.

McMahan and his colleagues have made a series of systematic and elegant studies of the effects of the synaptic basal lamina on the differentiation of nerve and muscle.[102–105]

[96] Langley, J. N., and Anderson, H. K. 1904. *J. Physiol.* 31: 365–391.
[97] Grinnell, A. D., and Rheuben, M. B. 1979. *J. Physiol.* 289: 219–240.
[98] Kuffler, S. W., and Vaughan Williams, E. M. 1953. *J. Physiol.* 121: 289–317.
[99] Miledi, R., Stefani, E., and Steinbach, A. B. 1971. *J. Physiol.* 217: 737–754.
[100] Close, R. I. 1972. *Physiol. Rev.* 52: 129–197.
[101] Salmons, S., and Sreter, F. A. 1975. *J. Anat.* 120: 412–415.
[102] Sanes, J. R., Marshall, L. M., and McMahan, U. J. 1978. *J. Cell Biol.* 78: 176–198.
[103] Burden, S. J., Sargent, P. B., and McMahan, U. J. 1979. *J. Cell Biol.* 82: 412–425.
[104] McMahan, U. J., and Slater, C. R. 1984. *J. Cell Biol.* 98: 1453–1473.
[105] Anglister, L., and McMahan, U. J. 1985. *J. Cell Biol.* 101: 735–743.

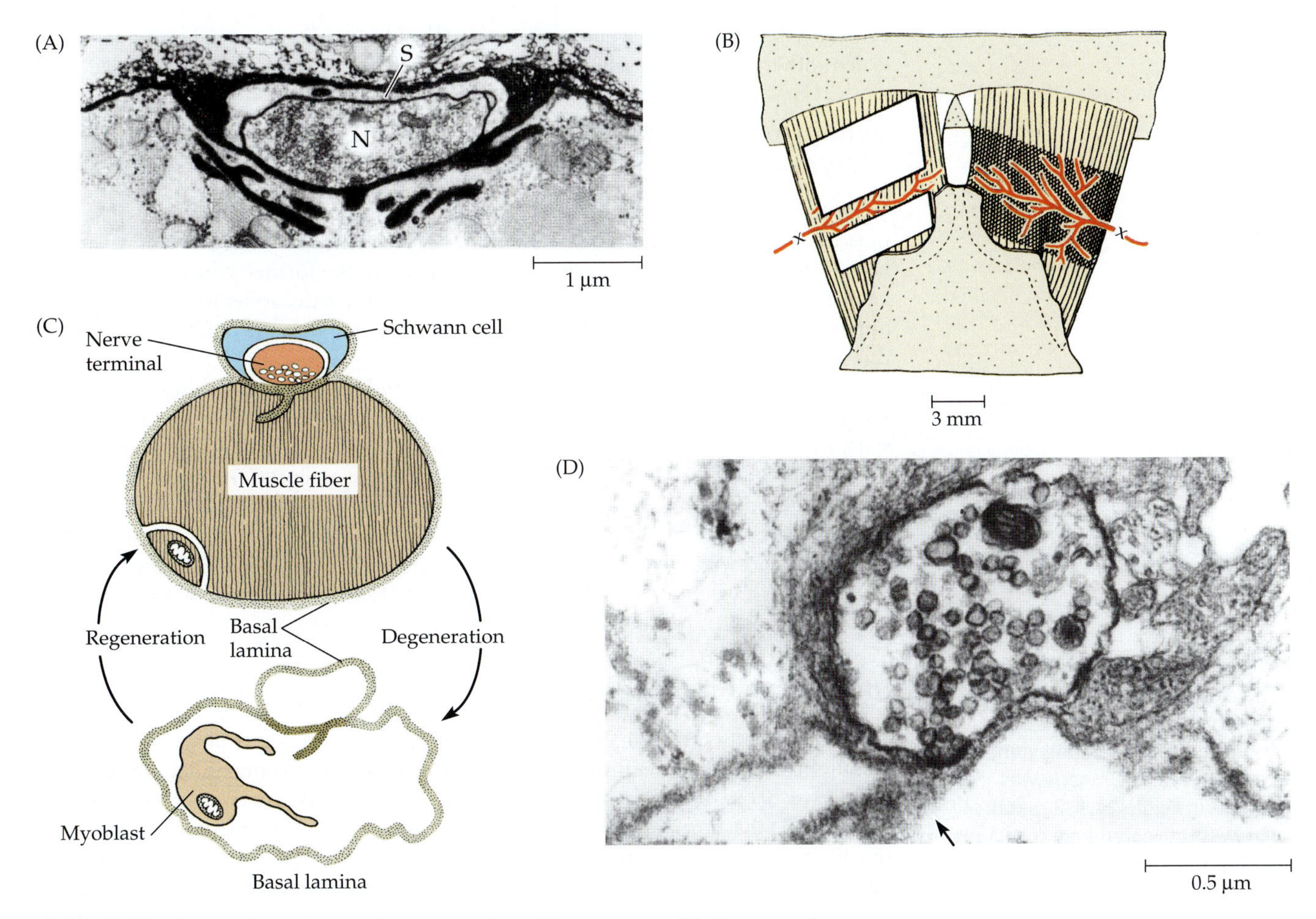

FIGURE 24.13 Basal Lamina and Regeneration of Synapses. (A) Electron micrograph of a normal neuromuscular synapse in the frog, stained with ruthenium red to show the basal lamina that dips into the postsynaptic folds and surrounds the Schwann cell (S) and nerve terminal (N). (B) Diagram of the cutaneous pectoris muscle, showing the region frozen (right) or cut away (left) to damage muscle fibers. (C) Freezing causes all cellular elements of the neuromuscular junction to degenerate and be phagocytized, leaving only the basal lamina sheath of the muscle fiber and Schwann cell intact. New neuromuscular junctions are restored by regenerating axons and muscle fibers. (D) Nerve and muscle were damaged, and regeneration of muscle fibers was prevented by X-irradiation. In the absence of muscle fibers, axons regenerated; contacted original synaptic sites, marked by the tongue of basal lamina that had extended into the junctional fold (arrow); and formed active zones. (After McMahan, Edgington, and Kuffler, 1980; micrographs kindly provided by U. J. McMahan.)

The key to their analysis was to use an easily accessible, very thin muscle in the frog—the cutaneous pectoris—in which the position of the end plates is highly ordered and easily seen in a living muscle. As a first step, cells in the region of innervation were killed by cutting the nerve and muscle fibers or by repeated application of a brass bar cooled in liquid nitrogen (Figure 24.13B). Within days the portion of the muscle fibers in the damaged region, together with the nerve terminals, degenerated and were phagocytized, but the basal lamina sheaths remained intact (Figure 24.13C). The location of the original neuromuscular junctions could be recognized by the distinctive morphology of the basal lamina sheaths of the muscle and Schwann cell at the junctional sites, and because cholinesterase remained concentrated in the basal lamina of the synaptic cleft and folds for weeks following the operation.

Two weeks after damage to the muscles, new myofibers had formed within the basal lamina sheaths and were contacted by regenerated axon terminals, which evoked muscle twitches when the nerves were stimulated. Nearly all of the regenerated synapses were located precisely at original synaptic sites, as marked by cholinesterase. Thus, signals associated with the synaptic basal lamina specified where regenerating synapses formed.

Synaptic Basal Lamina and Formation of Synaptic Specializations

To investigate further the nature of the signals associated with the synaptic basal lamina, muscles were damaged and the nerve was crushed, but muscle fiber regeneration was prevented by X-irradiation. Regenerating axons grew to the former synaptic sites on the basal lamina—as marked by cholinesterase—and formed active zones for release precisely opposite portions of the basal lamina that had projected into the junctional folds—all this without a postsynaptic target (Figure 24.13D).

In a parallel series of experiments, McMahan and his colleagues demonstrated that synaptic basal lamina in the adult also contains factors that trigger differentiation of postsynaptic specializations in regenerating myofibers. Muscles were damaged as described, but reinnervation was prevented by removing a long segment of the nerve. When new muscle fibers regenerated within the basal lamina sheaths, they formed junctional folds and aggregates of ACh receptors and acetylcholinesterase precisely at the point where they came in contact with the original synaptic basal lamina (Figure 24.14). Thus, signals stably associated with synaptic basal lamina can trigger the formation of synaptic specializations in both regenerating myofibers and nerve terminals.

Identification of Agrin

In order to identify the signal in synaptic basal lamina that triggers postsynaptic differentiation, McMahan and his colleagues prepared basal lamina–containing extracts from the electric organ of the marine ray *Torpedo californica*, a tissue derived embryologically from muscle that receives very dense cholinergic innervation. When added to myofibers in culture, these extracts mimicked the effects of synaptic basal lamina on regenerating muscle fibers; that is, they induced the formation of specializations at which ACh recep-

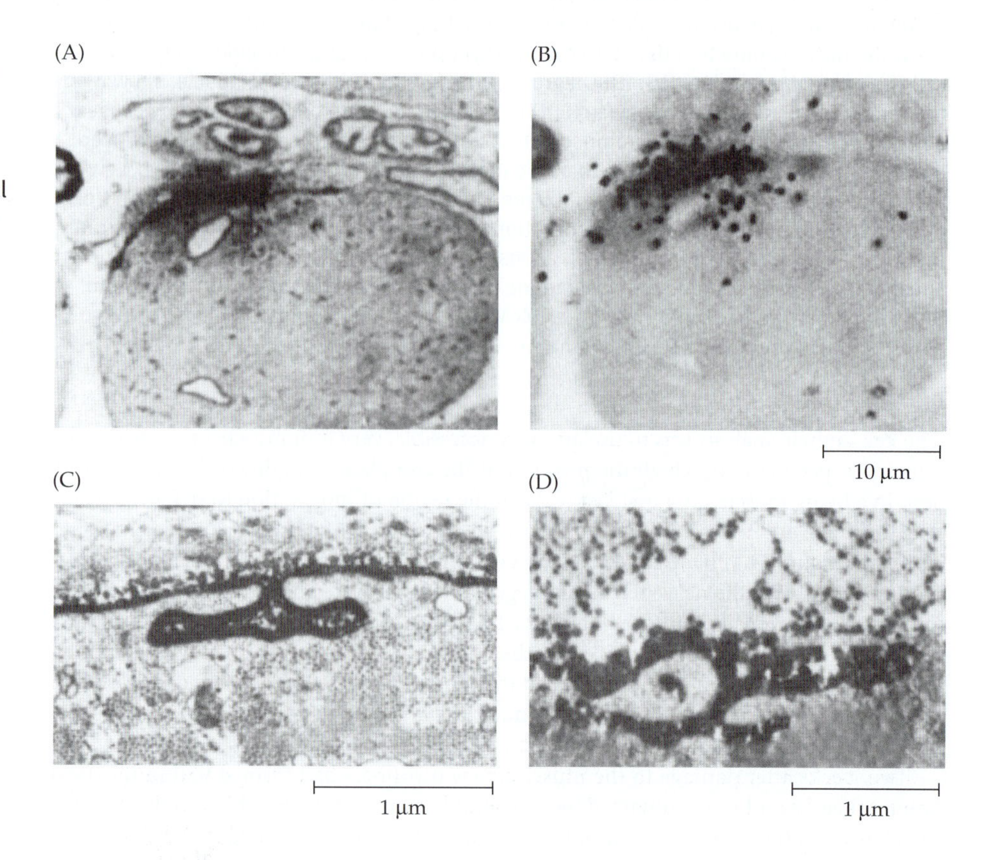

FIGURE 24.14 Accumulation of ACh Receptors and Acetylcholinesterase at Original Synaptic Sites on muscle fibers regenerating in the absence of nerve. The muscle was frozen as in Figure 24.13B, but the nerve was prevented from regenerating. New muscle fibers formed within the basal lamina sheaths. (A and B) Light-microscope autoradiography of a regenerated muscle stained for cholinesterase to mark the original synaptic site (in focus in part A) and incubated with radioactive α-bungarotoxin to label ACh receptors (silver grains in focus in part B). (C) Electron micrograph of the original synaptic site in a regenerated muscle labeled with HRP–α-bungarotoxin. The distribution of ACh receptors is indicated by the dense HRP reaction product, which lines the muscle fiber surface and the junctional folds. (D) Electron micrograph of the original synaptic site in a regenerated muscle stained for cholinesterase. The original cholinesterase was permanently inactivated at the time the muscle was frozen. Thus, the dense reaction product is due to cholinesterase synthesized and accumulated at the original synaptic site by the regenerating muscle fiber. (A and B after McMahan, Edgington, and Kuffler, 1980; C after McMahan and Slater, 1984; D after Anglister and McMahan, 1985; micrographs kindly provided by U. J. McMahan.)

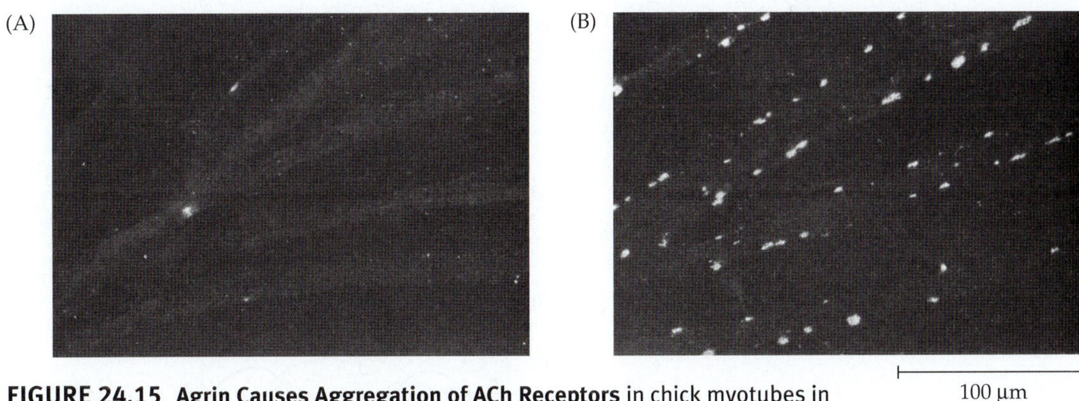

FIGURE 24.15 Agrin Causes Aggregation of ACh Receptors in chick myotubes in culture. Fluorescence micrographs of myotubes labeled with rhodamine-conjugated α-bungarotoxin to mark ACh receptors. (A) Receptors are distributed over the surface of control myotubes at low density. (B) Overnight incubation with agrin causes the formation of patches at which ACh receptors accumulate, together with other components of the postsynaptic apparatus. (After McMahan and Wallace, 1989.)

tors accumulated, together with several other components of the postsynaptic apparatus (Figure 24.15).[106]

The active component in the extracts, called **agrin**, has been purified and characterized, and cDNAs encoding it cloned from chick, rat, and ray.[107] Results of in situ hybridization and immunohistochemical studies demonstrate that agrin is synthesized by motor neurons, transported down their axons, and released to induce differentiation of the postsynaptic apparatus at developing neuromuscular junctions (Chapter 23).[108] Agrin then becomes incorporated into the synaptic basal lamina, where it helps maintain the postsynaptic apparatus in the adult and trigger its differentiation during regeneration.

As described in Chapter 23, a muscle-specific receptor tyrosine kinase called MuSK forms part of the agrin receptor. Activation of the protein tyrosine kinase domain in MuSK triggers an intracellular tyrosine phosphorylation cascade that is an early event in ACh receptor aggregation.

REGENERATION IN THE MAMMALIAN CNS

The adult mammalian central nervous system has limited capacity for regeneration. Transection of major axon tracts is not followed by axon regrowth and restitution of function. As described earlier, however, after lesions in the central nervous system, undamaged axons can sprout and form new synapses with considerable specificity. Moreover, it has become apparent through the work of Aguayo, Bunge, Kawaguchi, Nicholls, Raisman, Schwab, and their colleagues that even after major tracts in the central nervous system are severed, axons can, under suitable circumstances, regrow for distances of several centimeters and form synapses with appropriate targets.[109–111]

Role of Glial Cells in CNS Regeneration

Of prime importance for limiting axonal regeneration in the central nervous system is the immediate environment provided by CNS glial cells (Chapter 8). Clues to an inhibitory role of CNS glial cells are provided by several experiments. First, although axons severed in the CNS typically do not regrow, motor neurons, whose cell bodies lie within the spinal cord, can regenerate severed peripheral axons (Figure 24.16). Likewise, axons of sensory neurons regrow to their targets in the periphery, but fail to regenerate when severed within the CNS. Indeed, after a dorsal root is cut, sensory axons regenerate toward the spinal cord but stop growing when they reach the astrocytic processes that delimit the surface of the central nervous system. Moreover, axons in the periphery will not enter an

[106]McMahan, U. J., and Wallace, B. G. 1989. *Dev. Neurosci.* 11: 227–247.

[107]Bowe, M. A., and Fallon, J. R. 1995. *Annu. Rev. Neurosci.* 18: 443–462.

[108]McMahan, U. J. 1990. *Cold Spring Harb. Symp. Quant. Biol.* 50: 407–418.

[109]Bray, G. M., et al. 1991. *Ann. N.Y. Acad. Sci.* 633: 214–228.

[110]Nicholls, J., and Saunders, N. 1996. *Trends Neurosci.* 19: 229–234.

[111]Schwab, M. E., and Bartholdi, D. 1996. *Physiol. Rev.* 76: 319–370.

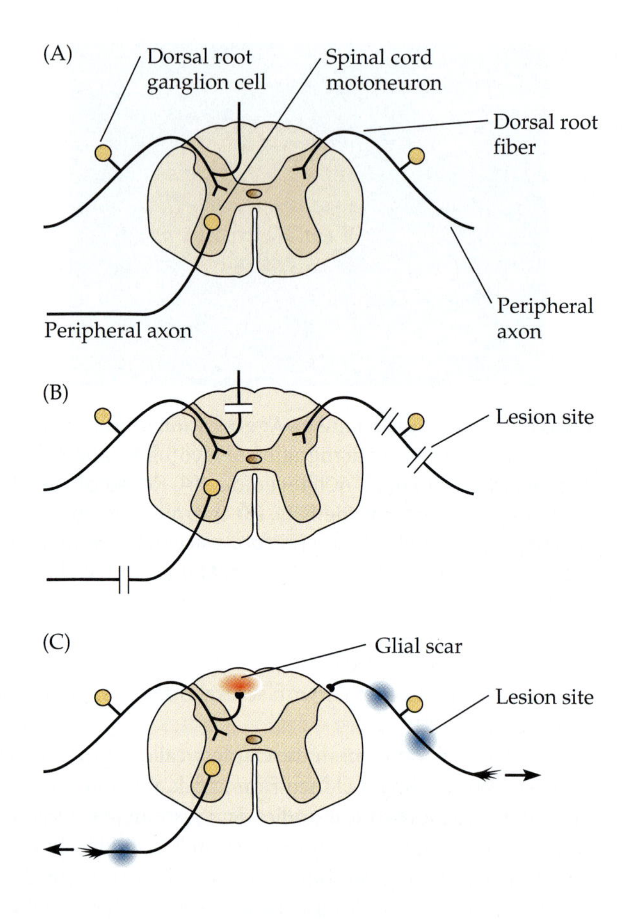

FIGURE 24.16 Axons of Sensory and Motor Neurons Regenerate in the Periphery but Not in the CNS. (A) Motoneurons, dorsal root ganglion sensory neurons, and their axonal processes in the mammalian nervous system. (B) Sites of axon lesions. (C) Extent of regeneration. Axons of dorsal root ganglion neurons and motoneurons regenerate through lesion sites in peripheral nerves and dorsal roots (blue). However, regenerating dorsal root fibers stop when they reach the astrocytic processes that delimit the surface of the spinal cord. Axons of dorsal root ganglion sensory neurons also do not regenerate through glial scars that form at lesion sites in the CNS (red).

optic nerve graft, which consists of CNS glial cells.[112] Such findings suggest that CNS glial cells actively inhibit growth.

On the other hand, when dorsal root ganglion neurons are injected into CNS white matter tracts in such a way as to minimize trauma, they frequently extend axons for long distances in the white matter, invade gray matter, and form terminal arbors.[113] Thus, when there is no trauma-induced glial reaction, regeneration of axons by adult neurons is not prevented by contact with CNS glial cells.

When tracts in the CNS are lesioned, astrocytes, microglial cells, meningeal cells, and oligodendrocyte precursor cells accumulate at the site of the lesion to form a glial scar.[114] These cells produce a variety of molecules that have been shown to inhibit axon growth, including free radicals, nitric oxide, arachidonic acid derivatives, and a variety of proteoglycans. For example, Schwab and his colleagues found that oligodendrocytes from the mature central nervous system have proteins on their surface, designated NI-35 and NI-250, that induce long-lasting collapse of growth cones and inhibit neurite outgrowth in vitro.[111,115] A monoclonal antibody was raised that neutralized this activity. In the presence of the antibody, axons could regenerate across a spinal cord lesion and partial locomotor function could be restored, although the extent of regeneration under such conditions was still meager.[111,116,117] Antibody application also promoted sprouting of intact fibers and formation of new synaptic contacts, which may account in part for the recovery of function.[118] It has been suggested that local, antibody-induced inflammatory responses contribute to axon sprouting and regrowth.[119,120]

Schwann Cell Bridges and Regeneration

Schwann cells produce a favorable environment for the growth of axons of CNS neurons. For example, when segments of peripheral nerves are grafted between the cut ends of the spinal cord in a mouse or rat, fibers grow across and fill the gap.[121] (The graft is composed of Schwann cells and connective tissue; the peripheral axons degenerate.) Similarly, cultures of Schwann cells implanted into the spinal cord promote growth.[122] This effect

[112]Aguayo, A. J., et al. 1978. *Neurosci. Lett.* 9: 97–104.
[113]Davies, S. J. A., et al. 1997. *Nature* 390: 680–683.
[114]Fawcett, J. W., and Asher, P. A. 1999. *Brain Res. Bull.* 49: 377–391.
[115]Schwab, M. E., and Caroni, P. 1988. *J. Neurosci.* 8: 2381–2393.
[116]Schnell, L., and Schwab, M. E. 1990. *Nature* 343: 269–272.
[117]Bregman, B. S., et al. 1995. *Nature* 378: 498–501.
[118]Z'Graggen, W. J., et al. 1998. *J. Neurosci.* 18: 4744–4757.
[119]Rabchevsky, A. G., and Streit, W. J. 1997. *J. Neurosci. Res.* 47: 34–48.
[120]Fitch, M. T., and Silver, J. 1999. In *CNS Regeneration: Basic Science and Clinical Advances.* Academic Press, San Diego, pp. 55–88.
[121]Richardson, P. M., McGuinness, U. M., and Aguayo, A. J. 1980. *Nature* 284: 264–265.
[122]Xu, X. M., et al. 1997. *J. Neurocytol.* 26: 1–16.

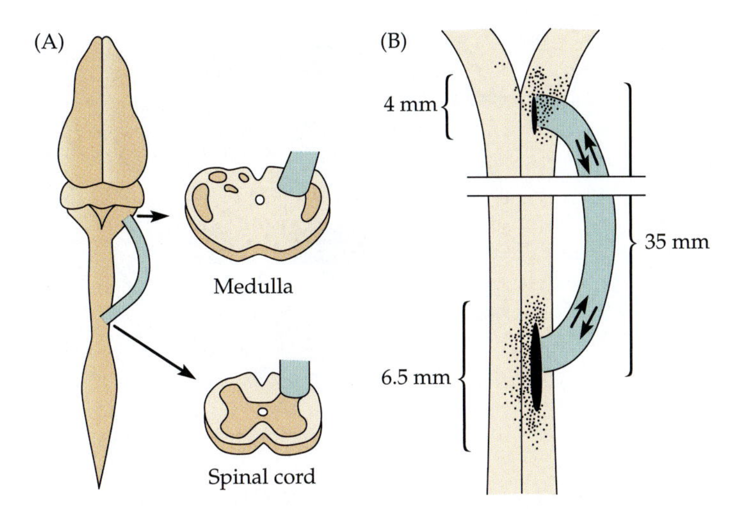

FIGURE 24.17 Bridges between Medulla and Spinal Cord enable CNS neurons to grow for prolonged distances. The grafted bridge consists of a segment of adult rat sciatic nerve in which axons have degenerated, leaving Schwann cells. These act as a conduit along which central axons can grow. (A) Sites of insertion of the graft. (B) Neurons are labeled by cutting the graft and applying HRP to the cut ends. Positions of 1472 neuronal cell bodies were labeled by retrograde transport of HRP in seven grafted rats. Most of the cells sending axons into the graft are situated close to its points of insertion. (After David and Aguayo, 1981.)

can be enhanced by genetically engineering the Schwann cells to produce supranormal amounts of neurotrophic factors.[123] Injecting suspensions of ensheathing glial cells into the stumps of a transected spinal cord or at the site of an electrolytic lesion in the corticospinal tract likewise enhances regeneration of axons.[124,125] Ensheathing glial cells are found only in the olfactory system, where new neurons are born and extend axons into the CNS throughout adulthood.

A dramatic effect is observed by the use of bridges of the type shown in Figure 24.17.[126] One end of a segment of sciatic nerve is implanted into the spinal cord, the other into a higher region of the nervous system (upper spinal cord, medulla, or thalamus). Bridges have even been made to extend from cortex to another part of the CNS or to muscle. After several weeks or months, the graft resembles a normal nerve trunk filled with myelinated and unmyelinated axons. These neurons fire impulses and are electrically excited or inhibited by stimuli applied above or below the sites of implantation. By cutting the bridge and dipping the cut ends into horseradish peroxidase or other markers, the cells of origin become labeled and their distribution can be mapped (Figure 24.17B). Such experiments show that axons in the bridge, which have grown over distances of several centimeters, arise from neurons whose cell bodies lie within the central nervous system. Usually only those neurons with somata not more than a few millimeters from the bridge send axons into it. Similarly, axons leaving the bridge to enter the central nervous system grow only a short distance before terminating.

Not all CNS neurons extend axons into permissive environments. For example, if the axons of cerebellar Purkinje cells are severed in the adult, the cells survive indefinitely, but no axonal regrowth occurs, even if pieces of embryonic cerebellum are grafted adjacent to the severed axons.[127] Axons of other cerebellar cells readily innervate such grafts. Thus, regeneration depends both on growth-permissive or growth-promoting conditions and on the intrinsic properties of the neuron. The inability of Purkinje cells to regrow severed axons is correlated with their failure to up-regulate proteins involved in axon growth in response to axotomy.

Formation of Synapses by Axons Regenerating in the Mammalian CNS

Can axons regenerating in the CNS of mammals locate their correct targets and make functional synapses? Experiments on regenerating retinal ganglion cell axons indicate that the answer is yes.[109] If the optic nerve is cut and a peripheral nerve bridge is inserted between the eye and the superior colliculus, retinal ganglion cell axons that grow through the bridge can extend into the target, arborize, and form synapses (Figure 24.18). The regenerated synapses are formed on the correct regions of their target cells, have a normal structure when visualized by electron microscopy, and are functional, in that the postsynaptic cells can be driven by illumination of the eye.

[123]Menei, P., et al. 1998. *Eur. J. Neurosci.* 10: 607–621.

[124]Ramón-Cueto, A., et al. 1998. *J. Neurosci.* 18: 3803–3815.

[125]Li, Y., Field, P. M., and Raisman, G. 1998. *J. Neurosci.* 18: 10514–10524.

[126]David, S., and Aguayo, A. J. 1981. *Science* 214: 931–933.

[127]Zagrebelsky, M., et al. 1998. *J. Neurosci.* 18: 7912–7929.

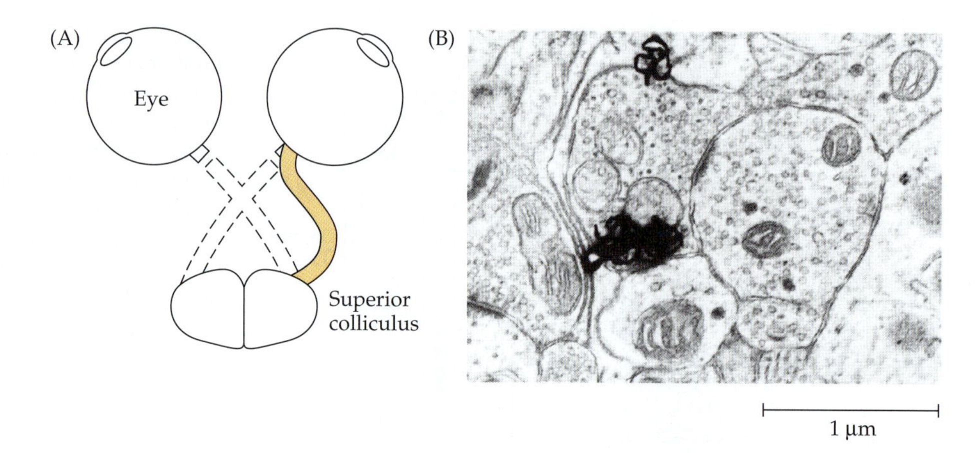

FIGURE 24.18 Reconnection of the Retina and Superior Colliculus through a peripheral nerve graft in an adult rat. (A) The optic nerves were severed, and one was replaced by a 3 to 4 cm segment of the peroneal nerve (yellow). Regeneration was tested by injecting anterograde tracers into the eye or recording responses of superior colliculus neurons to light flashed onto the retina. (B) Electron microscope autoradiogram of a regenerated retinal ganglion cell axon terminal in the superior colliculus. [^{3}H]-labeled amino acids were injected into the eye 2 days before the brain was fixed and sectioned; silver grains exposed by radiolabeled proteins transported from the injected eye identify ganglion cell axon terminals. The regenerated terminal resembles those seen in control animals; it is filled with round synaptic vesicles and forms asymmetric synapses. (After Vidal-Sanz, Bray, and Aguayo, 1991; micrograph kindly provided by A. J. Aguayo.)

Regeneration in the Immature Mammalian CNS

Compared to the adult, the immature mammalian CNS provides a more favorable environment for regeneration.[110] For example, if the spinal cord of a neonatal opossum is crushed or cut, axons grow across the lesion and conduction through the damaged region is restored within a few days, even when the spinal cord is removed from the animal and maintained in culture.[128] Similar results have been obtained in embryonic rat or mouse spinal cord in culture. Even after complete transection of the spinal cord in a newborn opossum, prolonged survival leads to substantial and precise regeneration and excellent functional recovery. For example, sensory axons reestablish direct synaptic connections onto motor neurons,[129] and the animal can walk, swim, and climb in a coordinated manner.[130]

The spinal cord of a 9-day-old opossum will regenerate well, while that of a 12-day-old animal will not. A striking feature of the opossum spinal cord at 9 days of age is the absence of myelin and the small number of glial cells it contains. Indeed, the end of the critical period during which successful regeneration can occur corresponds with the appearance of oligodendrocytes, myelin sheaths, and the neurite growth-inhibiting proteins NI-35 and NI-250.[131] Likewise, in embryonic chicks, neurons in the brainstem will regenerate their spinal axons if the spinal cord is transected prior to the onset of myelination.[132] At later stages regeneration fails, except when myelination is delayed or disrupted. Such findings offer the possibility of exploring in detail the molecular mechanisms that facilitate and inhibit regeneration in the central nervous system.

Neuronal Transplants

Among the most devastating of human diseases are those resulting from the spontaneous degeneration of CNS neurons, such as Parkinson's disease, Alzheimer's disease, and Huntington's disease. In the adult, most nerve cells are postmitotic; at present, no physiological mechanisms are known for replacing neurons that have been lost. One approach to cell replacement, studied by Björklund, Gage, Dunnett, Lund, Sotelo, Vrbova, Lindvall, and their colleagues, has been to transplant embryonic nerve cells into the adult brain.[133] Unlike neurons from the adult central nervous system, which die following transplantation, cells taken from fetal or neonatal animals survive and grow after being inserted into the gray matter of the adult central nervous system (Figure 24.19). There they differentiate, extend axons, and release transmitters.

An example of this is provided by experiments in which neurons were transplanted into the basal ganglia of rats after destruction of dopamine-containing neurons in the substantia nigra, a loss that mimics in some ways the deficits caused by Parkinson's disease in humans.[134] In normal animals, the dopaminergic neurons in the substantia nigra (a region in the midbrain) innervate cells in the basal ganglia (a region involved in programming movements; Chapters 14 and 22 and Appendix C). If a lesion of this dopamine pathway is made on one side of a rat, a disorder of movement results; the animal turns toward the side of the lesion in response to stress or certain drug treatments. This asymmetry of movement disappears after dopamine-containing neurons from the substantia

[128]Treherne, J. M., et al. 1992. *Proc. Natl. Acad. Sci. USA* 89: 431–434.

[129]Lepre, M., Fernandéz, J., and Nicholls, J. G. 1998. *Eur. J. Neurosci.* 10: 2500–2510.

[130]Saunders, N. R., et al. 1998. *J. Neurosci.* 18: 339–355.

[131]Varga, Z. M., et al. 1995. *Eur. J. Neurosci.* 7: 2119–2129.

[132]Keirstead, H. S., et al. 1995. *J. Neurosci.* 15: 6963–6974.

[133]Björklund, A. 1991. *Trends Neurosci.* 14: 319–322.

[134]Ridley, R. M., and Baker, H. F. 1991. *Trends Neurosci.* 14: 366–370.

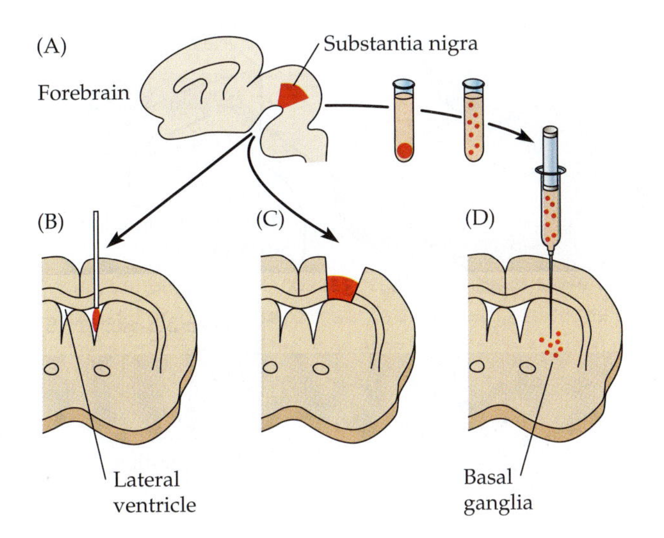

FIGURE 24.19 Procedures for Transplanting Embryonic Tissue into adult rat brain. Tissue rich in cells containing dopamine is dissected from the substantia nigra (A) and is injected into the lateral ventricle (B) or grafted into a cavity in the cortex overlying the basal ganglia (C). Alternatively, a suspension of dissociated substantia nigra cells can be injected directly into the basal ganglia (D). Such embryonic cells survive, sprout, and secrete transmitter. (After Dunnett, Björklund, and Stenevi, 1983.)

nigra of immature animals are transplanted into the basal ganglia on the lesioned side.[135] Ultrastructural studies have shown that the transplanted neurons extend axons into the surrounding region and form synapses with host neurons.

The degree of functional recovery following grafts depends on the extent to which synaptic connections are reestablished. Remarkably, appropriate integration of neurons into the complex synaptic circuitry of the brain occurs normally in the adult,[136] as well as after grafting embryonic tissue into lesioned adult cortex, hippocampus, and striatum.[133] For instance, fetal retinas transplanted into neonatal rat brains are capable of making specific functional connections, thereby restoring appropriate visual reflexes.[137] Grafts of embryonic entorhinal cortex into adult rats with entorhinal cortex lesions can reinnervate deafferented zones in the hippocampus, form synaptic connections, and partially ameliorate deficits in spatial memory.[138,139]

A remarkable example is the anatomical and functional integration of transplanted embryonic cerebellar Purkinje cells in the adult *pcd* (*P*urkinje *c*ell *d*egeneration) mouse, a mutant whose cerebellar Purkinje cells degenerate shortly after birth (Figure 24.20).[140] Sotelo and his colleagues grafted either dissociated cells or solid pieces of the cerebellar primordium into the cerebellum of the adult mutant mouse. Donor Purkinje cells migrated out of the graft to the positions originally occupied by the degenerated Purkinje cells. They did so along the host Bergmann radial glial cells, which were induced by the graft to reexpress proteins involved in guiding Purkinje cells.[141] Within 2 weeks many transplanted cells formed dendritic arbors that resembled those of normal Purkinje cells, climbing fibers formed synapses—first on the cell body, then on the proximal dendrites—and parallel fibers innervated the distal dendrites. Characteristic synaptic potentials were recorded following stimulation of the climbing-fiber and mossy-fiber inputs. However, the implanted cells rarely succeeded in establishing synaptic connections with their normal targets in the deep cerebellar nuclei of the host, innervating instead nearby donor deep nuclear neurons that survived in the remnant of the graft. Nevertheless, such experiments demonstrate that transplanted cells can become incorporated into the synaptic circuitry of an adult host to a remarkable extent.

It is clear that many neurons in the mammalian CNS retain in the adult an impressive ability to regenerate axons and dendrites and to reestablish appropriate synaptic connections. The prominent failure of regeneration that is so evident after most CNS lesions results from inhibition of these innate regenerative capacities, by glial cell–derived factors and changes in receptors and trophic molecules associated with growth. Identification of ways to suppress endogenous growth-inhibiting factors is an active area of research, as is the development of neural stem cell lines that offer the potential of providing a readily available source of glial cells and neurons whose properties can be manipulated by genetic engineering (Chapter 23).[142] Such advances, combined with improved transplantation techniques, provide hope for the amelioration of functional deficits resulting from CNS lesions and neurodegenerative diseases.

[135]Björklund, A., et al. 1980. *Brain Res.* 199: 307–333.

[136]Luskin, M. 1993. *Neuron* 11: 173–189.

[137]Radel, J. D., Kustra, D. J., and Lund, R. D. 1995. *Neuroscience* 68: 893–907.

[138]Zhou, W., Raisman, G., and Zhou, C. 1998. *Brain Res.* 788: 202–206.

[139]Zhou, W., et al. 1998. *Brain Res.* 792: 97–104.

[140]Sotelo, C., and Alvarado-Mallart, R. M. 1991. *Trends Neurosci.* 14: 350–355.

[141]Sotelo, C., et al. 1994. *J. Neurosci.* 14: 124–133.

[142]Martinez-Serrano, A., and Björklund, A. 1997. *Trends Neurosci.* 20: 530–538.

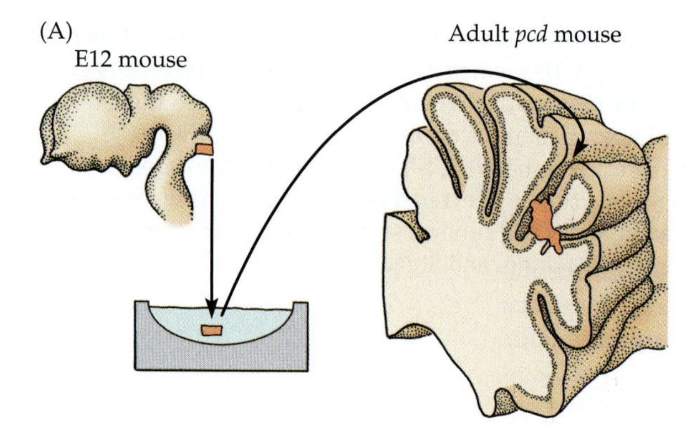

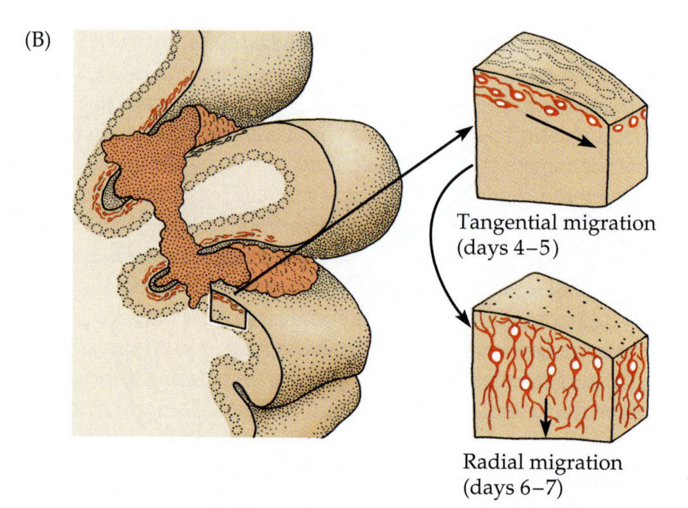

FIGURE 24.20 Reconstruction of Cerebellar Circuits by transplantation of embryonic cerebellar tissue (shown in red) into an adult *pcd* mouse, a mutant in which Purkinje cells degenerate shortly after birth. (A) Solid pieces of cerebellar primordium from a 12-day embryo (E12) were injected into the cerebellum of a 2- to 4-month-old *pcd* mouse. (B) By four to five days after transplantation, Purkinje cells have migrated out of the graft tangentially along the cerebellar surface. During days six and seven after transplantation, Purkinje cells migrate radially inward along Bergmann glial cells, penetrating the host molecular layer. (C) Donor Purkinje cells that lie within 600 μm of the host deep cerebellar nuclei (DCN) extend axons into the DCN and make synaptic contacts on their specific targets. Donor Purkinje cells farther from the host DCN make contact with donor DCN cells in the graft remnant. (After Sotelo and Alvarado-Mallart, 1991.)

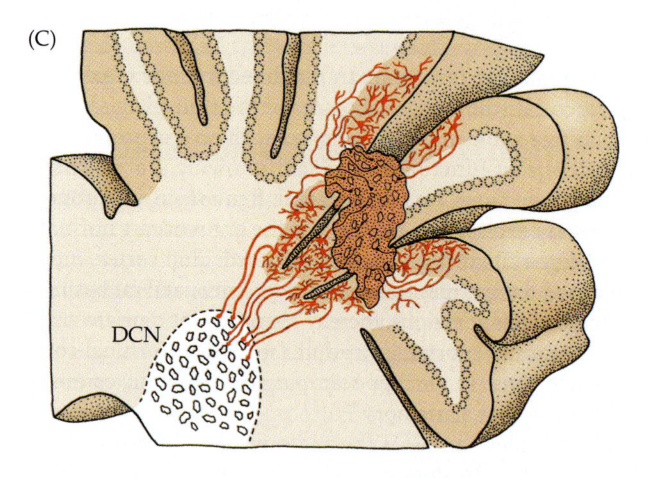

Summary

- When an axon is severed in the vertebrate nervous system, the distal portion degenerates. The axotomized cell either undergoes chromatolysis or dies.
- Many of the presynaptic terminals innervating an axotomized neuron retract; those that remain release fewer than normal quanta of transmitter.
- In denervated skeletal muscle fibers, new ACh receptors are synthesized and inserted in extrasynaptic regions, making the muscle supersensitive to ACh. Denervated neurons also become supersensitive to the transmitters released by damaged presynaptic axons.
- Muscle activity is an important factor determining receptor number and distribution. Muscle activity also influences the rate at which ACh receptors are degraded and replaced.
- In the adult mammal or frog, an innervated muscle will not accept innervation by an additional nerve. In contrast, nerve fibers will form new synapses on denervated or injured muscle fibers.
- Partially denervated muscles and neurons cause nearby undamaged nerves to sprout new branches and form new synapses.
- In the peripheral nervous system, Schwann cells provide an environment especially conducive to axon regrowth.
- The synaptic portion of the basal lamina sheath that surrounds muscle fibers has associated with it factors, such as agrin, that induce synaptic specializations in regenerating axon terminals and muscle fibers. Agrin is a proteoglycan synthesized by motor neurons and released from their axon terminals. It becomes associated with the synaptic basal lamina and induces the formation of postsynaptic specializations.

SUMMARY (CONTINUED)

- The adult mammalian central nervous system has limited capacity for regeneration.

- Schwann cells, in the form of a peripheral nerve graft or injected as a cell suspension at the site of a lesion, produce a favorable environment for regrowth of axons of mammalian CNS neurons.

- Regeneration occurs in the CNS of immature mammals.

- Neurons from fetal or neonatal animals, as well as neurons and glial cells derived from neural stem cell lines, survive and grow when transplanted into the adult mammalian CNS. Transplanted cells can become incorporated into the existing synaptic circuitry and partially restore normal function.

SUGGESTED READING

General Reviews

Björklund, A. 1991. Neural transplantation—An experimental tool with clinical possibilities. *Trends Neurosci.* 14: 319–322.

Fitch, M. T., and Silver, J. 1999. Beyond the glial scar: Cellular and molecular mechanisms by which glial cells contribute to CNS regenerative failure. In M. H. Tuszynski and J. H. Kordower (eds.), *CNS Regeneration: Basic Science and Clinical Advances.* Academic Press, San Diego, CA, pp. 55–88.

Martinez-Serrano, A., and Björklund, A. 1997. Immortalized neural progenitor cells for CNS gene transfer and repair. *Trends Neurosci.* 20: 530–538.

Muller, H. W., and Stoll, G. 1998. Nerve injury and regeneration: Basic insights and therapeutic interventions. *Curr. Opin. Neurol.* 11: 557–562.

Nicholls, J., and Saunders, N. 1996. Regeneration of immature mammalian spinal cord after injury. *Trends Neurosci.* 19: 229–234.

Sanes, J. R., and Lichtman, J. W. 1999. Development of the vertebrate neuromuscular junction. *Annu. Rev. Neurosci.* 22: 389–442.

Schwab, M. E., and Bartholdi, D. 1996. Degeneration and regeneration of axons in the lesioned spinal cord. *Physiol. Rev.* 76: 319–370.

von Bernhardi, R., and Muller, K. J. 1995. Repair of the central nervous system: Lessons from lesions in leeches. *J. Neurobiol.* 27: 353–366.

Original Papers

Björklund, A., Dunnett, S. B., Stenevi, U., Lewis, N. E., and Iversen, S. D. 1980. Reinnervation of the denervated striatum by substantia nigra transplants: Functional consequences as revealed by pharmacological and sensorimotor testing. *Brain Res.* 199: 307–333.

Blackshaw, S. E., Nicholls, J. G., and Parnas, I. 1982. Expanded receptive fields of cutaneous mechanoreceptor cells after single neurone deletion in leech central nervous system. *J. Physiol.* 326: 261–268.

Brown, M. C., and Ironton, R. 1977. Motor neurone sprouting induced by prolonged tetrodotoxin block of nerve action potentials. *Nature* 265: 459–461.

Burden, S. J., Sargent, P. B., and McMahan, U. J. 1979. Acetylcholine receptors in regenerating muscle accumulate at original synaptic sites in the absence of the nerve. *J. Cell Biol.* 82: 412–425.

David, S., and Aguayo, A. J. 1981. Axonal elongation into peripheral nervous system "bridges" after central nervous system injury in adult rats. *Science* 214: 931–933.

Davies, S. J. A., Fitch, M. T., Memberg, S. P., Hall, A. K., Raisman, G., and Silver, J. 1997. Regeneration of adult axons in white matter tracts of the central nervous system. *Nature* 390: 680–683.

Lømo, T., and Rosenthal, J. 1972. Control of ACh sensitivity by muscle activity in the rat. *J. Physiol.* 221: 493–513.

Miledi, R. 1960. The acetylcholine sensitivity of frog muscle fibers after complete or partial denervation. *J. Physiol.* 151: 1–23.

Mishina, M., Takai, T., Imoto, K., Noda, M., Takahashi, T., Numa, S., Methfessel, C., and Sakmann, B. 1986. Molecular distinction between fetal and adult forms of muscle acetylcholine receptor. *Nature* 321: 406–411.

Sanes, J. R., Marshall, L. M., and McMahan, U. J. 1978. Reinnervation of muscle fiber basal lamina after removal of myofibers. *J. Cell Biol.* 78: 176–198.

Saunders, N. R., Kitchener, P., Knott, G. W., Nicholls, J. G., Potter, A. and Smith, T. J. 1998. Development of walking, swimming and neuronal connections after complete spinal cord transection in the neonatal opossum, *Monodelphis domestica*. *J. Neurosci.* 18: 339–355.

Schwab, M. E., and Caroni, P. 1988. Oligodendrocytes and CNS myelin are nonpermissive substrates for neurite growth and fibroblast spreading in vitro. *J. Neurosci.* 8: 2381–2393.

Son, Y-J., and Thompson, W. J. 1995. Nerve sprouting in muscle is induced and guided by processes extended by Schwann cells. *Neuron* 14: 133–141.

Wilson Horch, H. L., and Sargent, P. B. 1996. Effects of denervation on acetylcholine receptor clusters on frog cardiac ganglion neurons as revealed by quantitative laser scanning confocal microscopy. *J. Neurosci.* 16: 1720–1729.

25 Critical Periods in Visual and Auditory Systems

THE EFFECTS OF USE AND DISUSE ON THE ESTABLISHMENT of connections have been analyzed in the visual systems of newly born kittens and monkeys. At birth, the receptive fields of neurons in the retina, lateral geniculate nucleus, and visual cortex resemble those of adults, except in layer 4 of the visual cortex. At birth, cortical cells in layer 4 are driven by both eyes. During the first 6 weeks the adult pattern is established so that cells in layer 4 respond to only one eye, while cells in other layers continue to be driven binocularly. Closure of the lids of one eye during the first 3 months of life leads to blindness in that eye and loss of its ability to drive cortical cells. Cortical columns supplied by the deprived eye shrink while those supplied by the normal, undeprived eye expand.

Lid closure in adult animals has no effect on columnar architecture or responses. During the critical period, changes produced by sensory deprivation are reversed by opening the sutured eye and closing the undeprived eye. Additional evidence for competition between the two eyes is provided by experiments made with both eyes closed in early life. When neither eye has an advantage, normal columnar structure develops; however, each cell in the cortex is driven by only one eye. Similarly, when strabismus is produced by cutting extraocular muscles in immature monkeys, few cortical neurons are driven by both eyes, although each eye receives its normal input. The role of activity in competition is shown by experiments in which blockade of impulses in both optic nerves with tetrodotoxin prevents segregation of ocular dominance columns. Spontaneous impulse activity and neurotrophins can contribute to the formation of appropriate connections.

In immature owls, development of the auditory system exhibits critical periods. Prisms placed over the eyes result in displaced receptive field positions. A mismatch thereby occurs between maps of space in the tectum corresponding to visual and auditory inputs. During the first months of life this discrepancy becomes corrected by remapping of auditory fields in the tectum. After a critical period such shifts no longer occur. In owls brought up in an enriched environment with enhanced sensory experience, the critical period during which maps can be brought into register is prolonged. Sensory deprivation experiments are significant for considering the development of higher brain functions.

We have emphasized repeatedly the specificity of the wiring that is necessary for the nervous system to function properly. It is also clear that development of connections between neurons continues after birth for various periods in different animals. For example, kittens are born with their eyes closed. If the lids are opened and light is shone into an eye, the pupil constricts, although the animal had not previously been exposed to light and appears to be completely blind.[1] By 10 days, the kitten shows evidence of vision and thereafter begins to recognize objects and patterns. When kittens are brought up in darkness instead of in their normal environment, the pupillary reflex continues to function, but they remain blind. It is as though there were a hierarchy of susceptibility, with "hard" and "soft" wiring in different parts of the brain.

Changes in the performance of the visual system during development raise a number of questions. What are the relative contributions of genetic factors and experience? To what extent are the neuronal circuits required for vision already present and ready to work at birth? What effect on their development does light falling into the eyes have? Is a kitten brought up in darkness blind because the connections fail to form, or because some of the connections that had originally been there have withered away? These concepts can be characterized by the phrase "nature and nurture." Over the years the pendulum has tended to swing as new findings have become available, often as though the phrase were "nature *or* nurture" (or worse still, "nature versus nurture"). This chapter shows that both are (as one would expect) essential.

The visual system offers great advantages for directly approaching questions relating to development because the relay stations are accessible and the amount of light and natural stimulation can readily be altered. Within the visual system we again emphasize experiments made in monkeys and cats. For our purposes, it is convenient to start by focusing on work that follows logically from the material presented in Chapters 20 and 21. Studies of the immature visual system set the stage for analyzing the effects of deprivation in other sensory circuits, notably the auditory system in the barn owl, and for considering cellular mechanisms that could play a part in the modification of connections in early life.

THE VISUAL SYSTEM IN NEWLY BORN MONKEYS AND KITTENS

A good deal is known about the organization of the connections that underlie visual perception in the adult cat and monkey. A simple cell in the cortex selectively "recognizes" one well-defined type of visual stimulus, such as the movement of a narrow bar of light oriented vertically, in a particular region of the visual field of either eye. Such responses are possible because of the precise and orderly connections made to cortical cells from the retina and lateral geniculate nucleus (Chapters 20 and 21). It is natural to wonder whether cells and connections of this type are already present in the newborn animal, or whether they develop as a result of visual experience, in which case visual stimuli in early life would somehow enable a random set of preexisting connections to be re-formed or modified for such a specific task.

To study visually naive animals, monkeys were taken immediately after birth, or after delivery by cesarean section, care being taken to avoid exposure to light. To prevent form vision until animals were old enough to be studied, the lids were sutured or the cornea was covered by a translucent occluder, which blurs images but allows light to pass. Similarly, visually naive kittens and ferrets were examined during the first weeks after birth.[2–5]

Receptive Fields and Response Properties of Cortical Cells in Newborn Animals

A newborn monkey appears visually alert and is able to fixate. The responses of cortical neurons in many ways resemble those of the adult animal. For example, recordings made from individual cells in the primary visual cortex (V_1) show that the cells are not driven by diffuse illumination. As in a mature animal, they fire best when light or dark bars with a particular orientation are shone onto a particular region of the retina.[2] In recordings made in animals lacking prior visual experience, the range of orientations cannot be dis-

[1]Riesen, A. H., and Aarons, L. 1959. *J. Comp. Physiol. Psychol.* 52: 142–149.

[2]Wiesel, T. N., and Hubel, D. H. 1974. *J. Comp. Neurol.* 158: 307–318.

[3]Hubel, D. H., and Wiesel, T. N. 1963. *J. Neurophysiol.* 26: 994–1002.

[4]Crair, M. C., Gillespie, D. C., and Stryker, M. P. 1998. *Science* 279: 566–570.

[5]Chapman, B., Stryker, M. P., and Bonhoeffer, T. 1996. *J. Neurosci.* 16: 6443–6453.

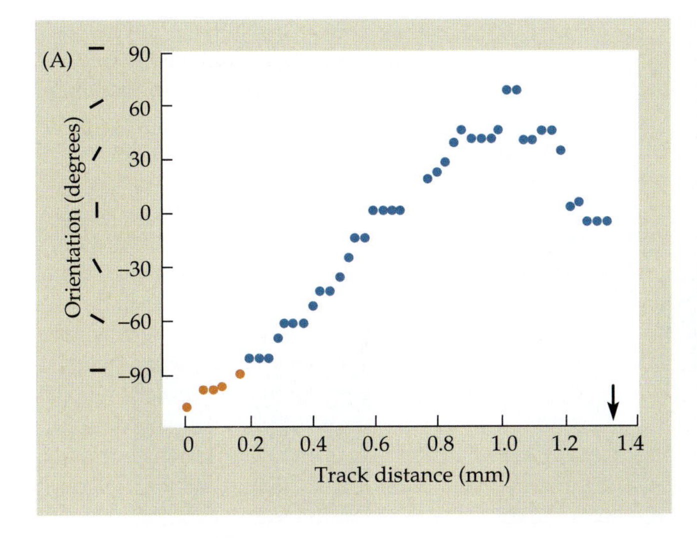

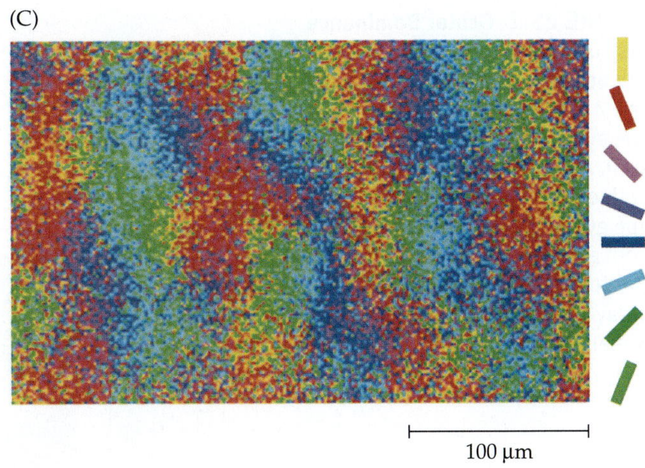

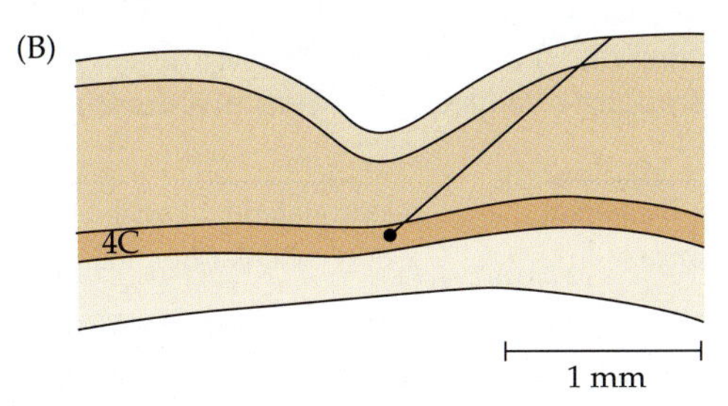

FIGURE 25.1 Orientation Columns in the absence of visual experience. (A) Axis orientation of receptive fields encountered by an electrode during an oblique penetration through the cortex of a 17-day-old baby monkey whose eyes had been sutured closed on the second day after birth. The receptive field orientation changes progressively as columns are traversed, indicating that normal orientation columns are present in the visually naive animal. Red dots are from the ipsilateral eye, blue dots from the contralateral eye. (B) The black dot marks the lesion made at the end of the electrode track in layer 4. (C) Orientation columns displayed by imaging in a 14-day-old kitten with lids sutured at birth. Colored bars (right) represent the orientation of the stimulus. Note that pinwheels are already present. (A and B from Wiesel and Hubel, 1974; C from Crair, Gillespie, and Stryker, 1998; micrograph kindly provided by M. C. Crair and M. P. Stryker.)

tinguished from that in adults. The receptive fields are also organized into antagonistic "on" and "off" areas that are driven by both eyes. Moreover, with oblique penetrations the preferred orientation changes in a regular sequence as the electrode moves through the cortex[2] (Figure 25.1). Figure 25.1 also shows that orientation preference maps with characteristic pinwheels (Chapter 21) are already evident, with all orientations represented equally at the time of birth.[4]

Ocular Dominance Columns in Newborn Monkeys and Kittens

At birth, most cells in all layers of the primary visual cortex are already driven by both eyes—some better by one eye, some by the other, and some equally well by both. Figure 25.2 shows the distribution of responses of neurons in all cortical layers according to eye preference in immature and adult monkeys.[2]

The degree of dominance can be conveniently expressed in a histogram by grouping neurons into seven categories according to the discharge frequency with which they responded to stimulation of one or the other eye (Chapter 21). The majority of cells respond to appropriate illumination of either eye. Cells in groups 1 and 7 in Figure 25.2 were driven only by visual stimuli applied to one eye, while those in groups 2 through 6 responded to both eyes.

The histograms of Figure 25.2A and B appear similar, with a range of eye preferences. However, they do not reveal a striking and important difference between newborn and adult monkeys in the properties of cells in layer 4: In that layer, cells in the newborn monkey are driven by both eyes; after 6 weeks of development, cells in layer 4 are driven by one eye only.[6] Outside of layer 4, cortical cells in newborn monkeys appear similar in their responses to those in adults, except that in some cells discharges are less vigorous or absent.

[6]Wiesel, T. N. 1982. *Nature* 299: 583–591.

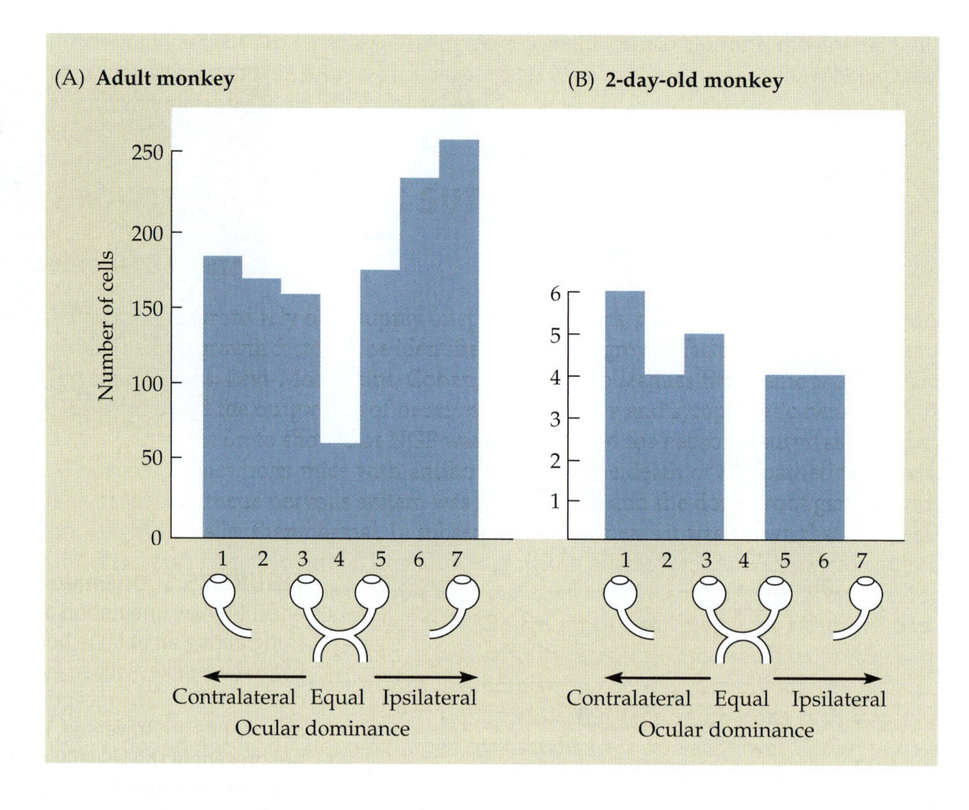

FIGURE 25.2 Ocular Dominance Distribution in the visual cortex of newborn monkey. Cells in groups 1 and 7 of the histograms are driven by one eye only (ipsilateral or contralateral). All other cells have input from both eyes. In groups 2, 3, 5, and 6, one eye predominates. In group 4, both eyes have equal influence. (A) Normal adult monkey. (B) Normal 2-day-old monkey. (After Wiesel and Hubel, 1974.)

Development of Ocular Dominance Columns

In layer 4, LeVay, Wiesel, and Hubel[7] (also Rakic[12,13]; see the next section) found a striking and important anatomical difference between adult and newborn. An explanation for cells in layer 4 of the newly born monkey being driven by both eyes is that the arborizations of geniculate fibers ending in layer 4 overlap extensively (Figures 25.3 and 25.4).[8] The individual axons reaching layer 4 from the geniculate nucleus spread over a wider area than in the adult. As a result, the territories supplied by each eye overlap. This is different from the pat-

[7]LeVay, S., Wiesel, T. N., and Hubel, D. H. 1980. *J. Comp. Neurol.* 191: 1–51.

[8]LeVay, S., Stryker, M. P., and Shatz, C. J. 1978. *J. Comp. Neurol.* 179: 223–244.

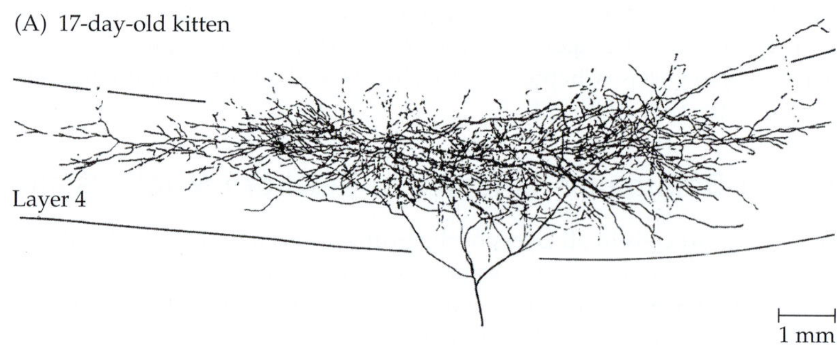

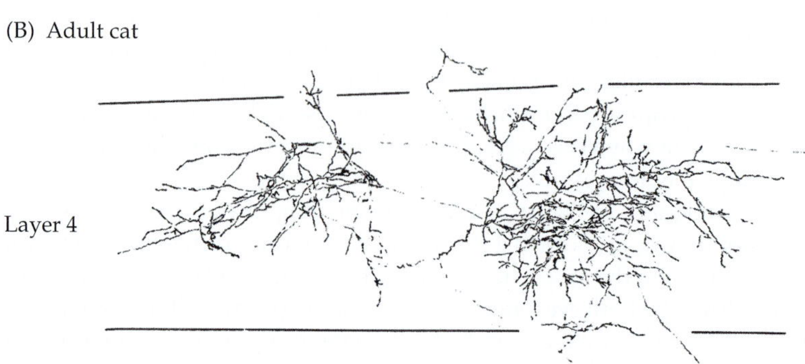

FIGURE 25.3 Age Dependence of Branching Patterns of Axons from Lateral Geniculate Nucleus ending in layer 4, labeled by injection with horseradish peroxidase. (A) Axon of a 17-day-old kitten. The axon spreads over a large uninterrupted territory in layer 4 of the visual cortex. (B) In the adult cat, the geniculate axon ends in two discrete tufts, interrupted by unlabeled fibers coming from the other eye. (After Wiesel, 1982.)

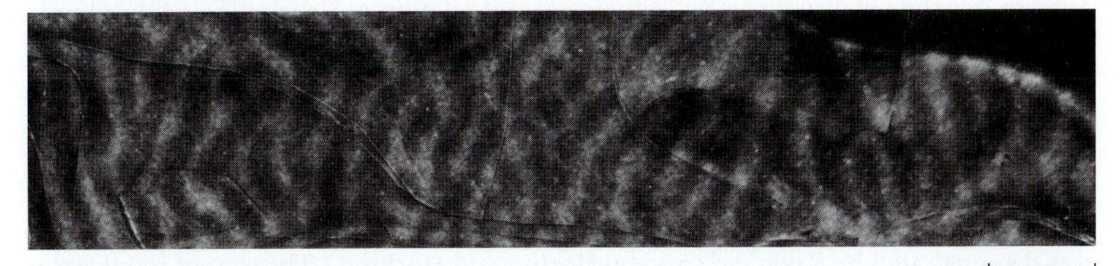

FIGURE 25.4 Ocular Dominance Columns in Layer 4 of the visual cortex in a visually naive monkey and in an older animal that had been exposed to light. (A) The monkey was delivered 8 days prematurely by cesarean section. Infrared night vision goggles were used in complete darkness for the delivery and for the injection of radioactive proline into the right eye 1 day later. The animal was maintained in complete darkness for 7 more days. The autoradiograph shows layer 4 of the ipsilateral primary visual cortex. Ocular dominance columns can be discerned. (B) Similar section through layer 4 in a 16-day-old monkey that had been born naturally and raised under normal lighting conditions. The right eye was injected with radioactive proline at 9 days. The boundaries of the ocular dominance columns are better defined. (After Horton and Hocking, 1996a; photographs kindly provided by J. Horton.)

tern in the adult, in which the ocular dominance columns corresponding to the inputs from the two eyes are precisely defined with sharp boundaries. Figure 25.4 shows that ocular dominance columns can be resolved in a monkey at birth, even though development is not yet complete. The borders in the visually naive animal appear fuzzier and less well defined.[9]

During the first 6 weeks of the animal's life, the axons of geniculate cells in layer 4 retract to form smaller arborizations as though pruned (see Figure 25.3). In this way, restricted separate domains of cortex in layer 4 become established, each supplied exclusively by one eye or the other (Figure 25.5). Comparable changes during development occur at the preceding stage in the visual pathway, in the lateral geniculate nucleus.[10,11] As optic nerve fibers from the two eyes grow into the nucleus, their arborizations overlap extensively before they separate into distinct layers. The postnatal development of ocular dominance columns and geniculate layers proceeds in animals reared in total darkness. These morphological observations can explain and provide a quantitative measure of changes in ocular dominance that occur during development.

[9]Horton, J. C., and Hocking, D. R. 1996. *J. Neurosci.* 16: 1791–1807.

[10]Rakic, P. 1977. *Philos. Trans. R. Soc. Lond. B* 278: 245–260.

[11]Shatz, C. J. 1996. *Proc. Natl. Acad. Sci. USA* 93: 602–608.

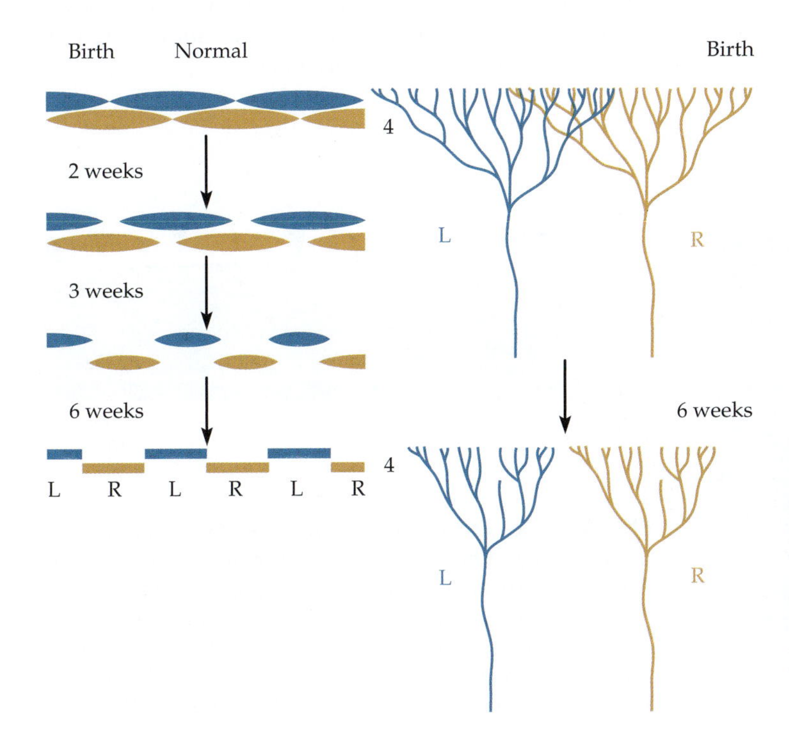

FIGURE 25.5 Retraction of Lateral Geniculate Nucleus Axons ending in layer 4 of the cortex during the first 6 weeks of life. The figure shows the overlap of inputs from the right (R) and left (L) eyes present at birth and the subsequent segregation into separate clusters corresponding to ocular dominance columns. The overlap at birth is greater in kittens than in monkeys. (After Hubel and Wiesel, 1977.)

The “blobs” revealed by cytochrome oxidase staining are already defined anatomically at birth in primary visual cortex, in their normal relation to developing ocular dominance columns.[9] The stripes in visual area 2 (area 18) are also evident in newborn monkeys that have had no visual experience (Figure 25.6).

Development of Cortical Architecture in Utero

The extent to which ocular dominance columns and blobs start to develop before birth was demonstrated by Rakic[7,12,13] and by Horton and Hocking.[9] The eyes of monkeys at different stages in utero were injected with radioactive amino acids (Chapter 20), so as to label cells in the lateral geniculate nucleus and their axons in layer 4 of the cortex. Alternatively, columns and blobs were revealed by staining for cytochrome oxidase (Chapter 21). At early stages in utero, in the absence of any visual input, the overlap of the territories supplied by the two eyes is virtually complete. Several days before birth the ocular dominance columns, blobs, and thick and thin stripes of visual area 2 are recognizable and in their normal relation to each other.

In kittens, ocular dominance columns in the visual cortex become well defined only at about 30 days of age.[8] Ferrets and marsupials are born at even earlier stages and thus have a greater degree of postnatal development of the visual circuitry.

Genetic Factors in the Development of Visual Circuits

The principal point to be made here comes as no great surprise: Certain features of the basic wiring are already established before the animal has had a chance to see anything. Other features become fully developed only in the first few weeks of life. The retraction of geniculate axons in layer 4 is reminiscent of the events occurring during the development of nerve–muscle synapses in neonatal rats: At birth, each motor end plate is supplied by numerous motoneurons, but in a few weeks, most of the axons retract, leaving each muscle fiber supplied by just one motoneuron[14] (Chapter 23).

As a result of genetic defects, interesting and important systematic abnormalities arise in the visual system. Examples are provided by color blindness and by *reeler* mice, in which cortical layering is abnormal.[15,16] A third example is a defect of the albino gene in Siamese cats and in albino animals; this not only determines the color of the animal through the synthesis of melanin, but also causes it to be cross-eyed, as a result of systematic errors in connections in the visual system.[17–19] In this chapter we focus rather on abnormal sensory experience in early life and the way in which it can drastically affect the anatomy and physiology of the brain for the rest of an animal’s life.

[12]Rakic, P. 1986. *Trends Neurosci.* 9: 11–15.
[13]Kuljis, R. O., and Rakic, P. 1990. *Proc. Natl. Acad. Sci. USA* 87: 5303–5306.
[14]Bernstein, M., and Lichtman, J. W. 1999. *Curr. Opin. Neurobiol.* 9: 364–370.
[15]Caviness, V. S., Jr. 1982. *Dev. Brain Res.* 4: 293–302.
[16]Rakic, P., and Caviness, V. S., Jr. 1995. *Neuron* 14: 1101–1104.
[17]Guillery, R. W. 1974. *Sci. Am.* 230(5): 44–54.
[18]Hubel, D. H., and Wiesel, T. N. 1971. *J. Physiol.* 218: 33–62.
[19]Kliot, M., and Shatz, C. J. 1985. *J. Neurosci.* 5: 2641–2653.

(A)

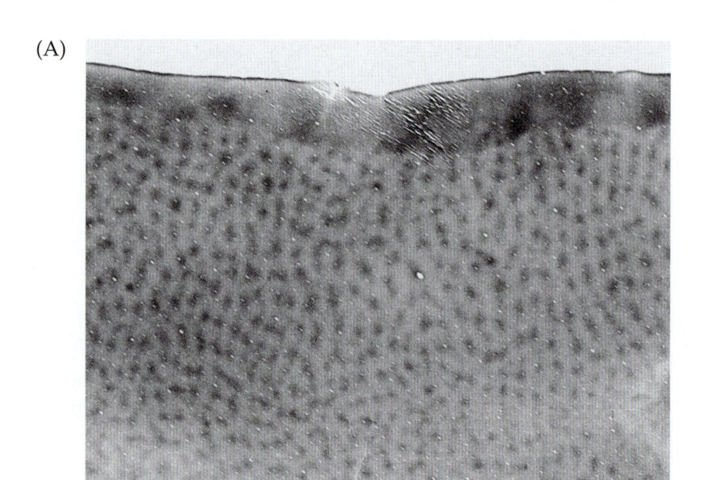

(B)

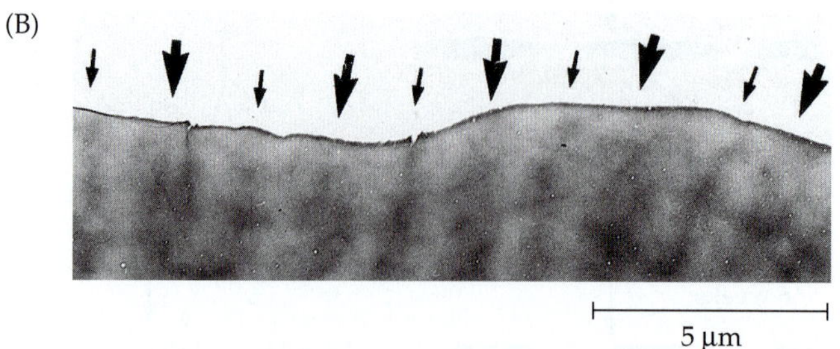

FIGURE 25.6 Architecture of the Visual Cortex of Newborn Monkey without visual experience (as in Figure 25.4A), delivered by cesarean section 8 days prematurely and kept in total darkness for 7 days. (A) Cytochrome oxidase staining shows blobs in area 17. (B) Thick and thin stripes in visual area 2, labeled with thick and thin arrows. (After Horton and Hocking, 1996a; photographs kindly provided by J. Horton.)

EFFECTS OF ABNORMAL EXPERIENCE IN EARLY LIFE

This section describes three types of experiments, for the most part made in the first instance by Hubel and Wiesel, in which animals were deprived of normal visual stimuli.[6,20] They studied the effects on the physiological responses of nerve cells and the structure of the visual system after (1) closing the lids of one or both eyes; (2) preventing form vision, but not access of light to the eye; and (3) leaving light and form vision intact, but producing an artificial strabismus (squint) in one eye. These procedures consistently cause remarkable abnormalities in the function and anatomy of the cortex. A feature of Hubel and Wiesel's work is that the results are so reproducible, clear, and dramatic.

Blindness after Lid Closure

When the lids of one eye were sutured during the first 2 weeks of life, monkeys and kittens still developed normally and used their unoperated eye. At the end of 1 to 3 months, however, when the operated eye was opened and the normal one closed, it was clear that the animals were practically blind in the operated eyes. For example, kittens would bump into objects and fall off tables.[6,21] There was no gross evidence of a physiological defect within such eyes; pupillary reflexes appeared normal, and so did the electroretinogram, which is an index of the average electrical activity of the eye. Records made from retinal ganglion cells in deprived animals showed no changes in their responses, and their receptive fields appeared normal.

Responses of Cortical Cells after Monocular Deprivation

Although responses of cells in the lateral geniculate nucleus appeared relatively unchanged after monocular deprivation,[22] there were major changes in the responses of cortical cells.[6,7,23] When electrical recordings were made in the visual cortex, very few cells could be driven by the eye that had been closed. The majority of those that did respond had abnormal receptive fields. Responses of cells driven by the undeprived eye were normal. Figure 25.7 shows ocular dominance histograms obtained from the cells examined in monkeys and kittens raised with closure of one eye during the first weeks of life.

[20]Hubel, D. H. 1988. *Eye, Brain and Vision.* Scientific American Library, New York.

[21]Wiesel, T. N., and Hubel, D. H. 1963. *J. Neurophysiol.* 26: 1003–1017.

[22]Wiesel, T. N., and Hubel, D. H. 1963. *J. Neurophysiol.* 26: 978–993.

[23]Wiesel, T. N., and Hubel, D. H. 1965. *J. Neurophysiol.* 28: 1029–1040.

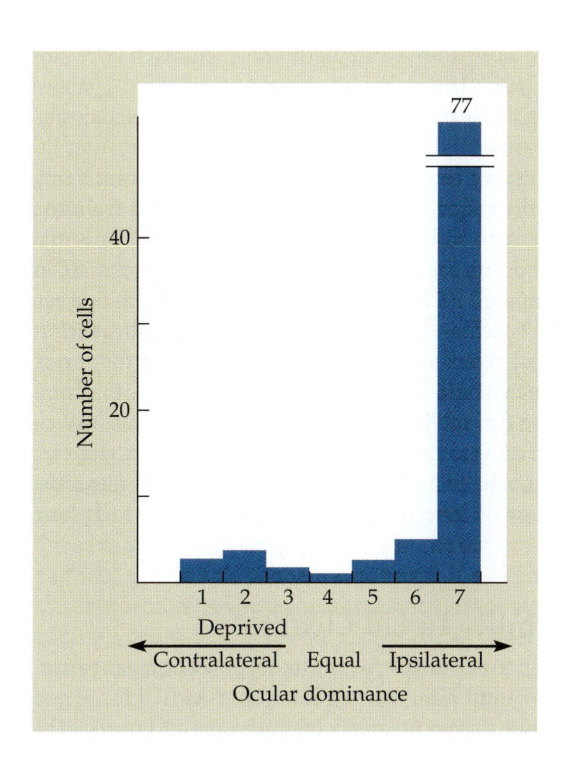

FIGURE 25.7 Damage Produced by Closure of One Eye. Ocular dominance distribution in a monkey whose right eye was closed from 21 to 30 days of age. In spite of subsequent 4 years of binocular vision, most cortical neurons were unresponsive to stimulation of the right eye. (From LeVay, Wiesel, and Hubel, 1980.)

Relative Importance of Diffuse Light and Form for Maintaining Normal Responses

The results described so far indicate that if one eye is not used normally in the first weeks of life, its power wanes and it ceases to be effective in the visual cortex. These far-reaching changes are produced by the relatively minor procedure of closing the lids, without cutting any nerves. What is the important condition for maintaining and developing proper visual responses? Is diffuse light adequate?

Lid closure reduces the level of light that reaches the retina but does not exclude it. One might suspect, therefore, that diffuse light alone would not keep an eye functioning normally. That form vision, rather than the presence of light, is the important stimulus required to prevent abnormal development of cortical connections was tested in a series of experiments made in newborn kittens. A plastic occluder (like a Ping-Pong ball) was placed over the cornea instead of closing the eyelids; the occluder prevented form vision but admitted light. All these cats were blind in the deprived eye.[21] Furthermore, cortical cells were no longer driven by the deprived eye. Neither retinal nor geniculate responses were noticeably changed under such conditions.

Morphological Changes in the Lateral Geniculate Nucleus after Visual Deprivation

Cells in the lateral geniculate nucleus of cat and of monkey are arranged in layers, each supplied predominantly by one or the other eye (Chapter 20). In the same animals that showed marked abnormalities in the cortex after lid closure, the geniculate cells seemed at first to be normal. Nevertheless, it was shown that marked changes in morphology occurred after lid closure of one eye: The cells were noticeably smaller than in the layers supplied by the normal eye.[22] The reduction in size depended on the duration of lid closure. Surprisingly, cells in the lateral geniculate nucleus showed obvious morphological changes but little significant physiological deficit. Several lines of evidence suggest that the size of cells in the lateral geniculate nucleus may reflect the extent of their arborization in the cortex.[24,25]

Morphological Changes in the Cortex after Visual Deprivation

The morphological consequences of eye closure are particularly conspicuous in layer 4 of the visual cortex V_1, where geniculate fibers terminate in an orderly manner.[7,26] Changes in ocular dominance columns following lid closure in monkeys have been revealed by autoradiography of the cortex after injection of radioactive materials into one eye. After lid closure there is a marked reduction in the width of ocular dominance columns receiving projections from the occluded eye. At the same time, the columns with inputs from the normal eye show a corresponding increase in width compared with that seen in a normal adult monkey. The shrinkage of ocular dominance columns is evident in Figure 25.8, in which the normal columns can be compared with columns in animals in which one eye had been closed at 2 weeks and left closed for 18 months. The changes indicate that geniculate axons activated by the normal eye retained or captured territory in the cortex lost by their weaker, visually deprived neighbors. These results are consistent with physiological observations made by recording from cells in layer 4. Almost all cells were driven only by the eye that had not been deprived. Certain features of the cortex are less vulnerable to deprivation than layer 4. Thus, enucleation of one or both eyes does not affect the distributions of blobs in visual area 1, or the alternating striped pattern of cytochrome oxidase staining in visual area 2.

[24]Guillery, R. W., and Stelzner, D. J. 1970. *J. Comp. Neurol.* 139: 413–422.

[25]Humphrey, A. L., et al. 1985. *J. Comp. Neurol.* 233: 159–189.

[26]Hubel, D. H., Wiesel, T. N., and LeVay, S. 1977. *Philos. Trans. R. Soc. Lond. B* 278: 377–409.

Critical Period of Susceptibility to Lid Closure

When the lids of one eye are closed in an adult cat or monkey, no abnormal consequences are seen.[6,7] For example, in an adult animal, even if an eye is closed for over a year the cells in the cortex continue to be driven normally by both eyes and display the

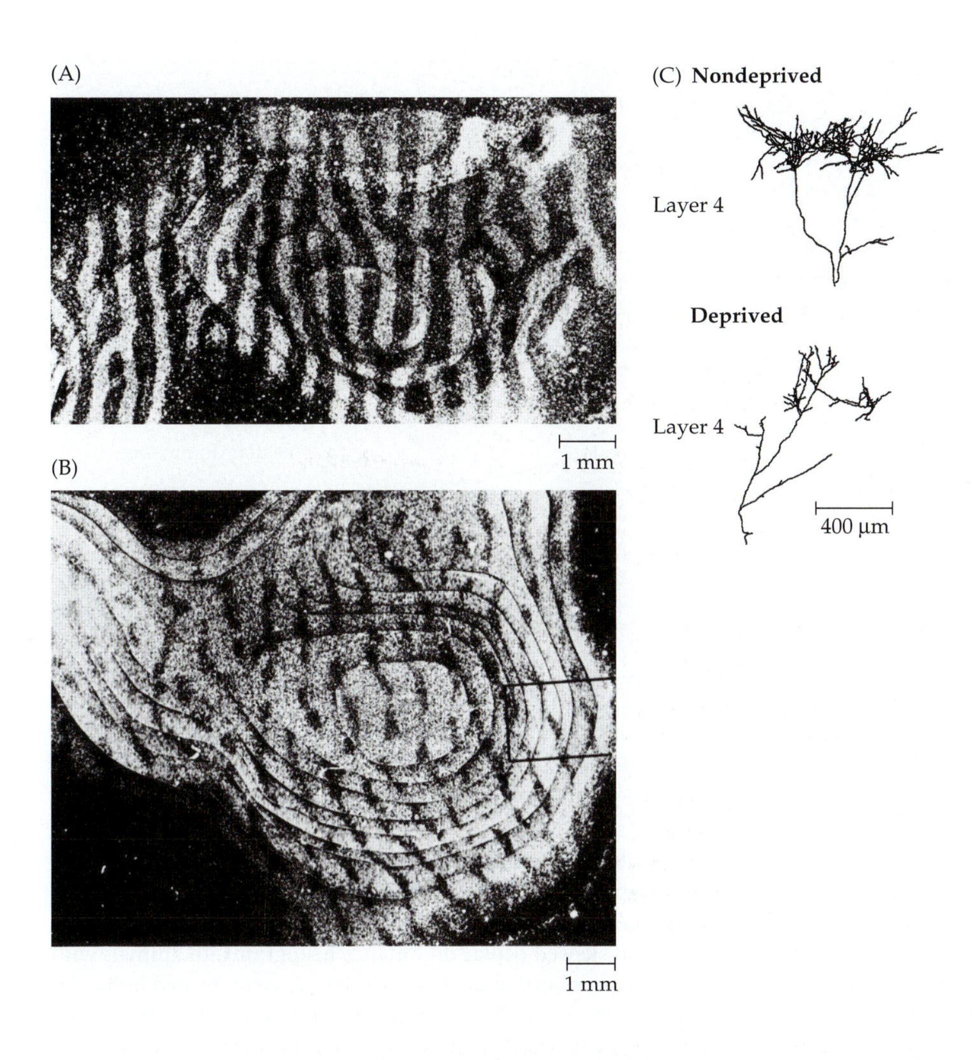

FIGURE 25.8 Ocular Dominance Columns in Layer 4 after Closure of One Eye. (A) Normal adult monkey. The right eye had been injected with a radioactive proline–fucose mixture 10 days previously. Layer 4 displays alternating light and dark stripes of equal width. Radioactively labeled geniculate axons in layer 4 of the right hemisphere appear as fine white granules forming columns. Intervening dark bands correspond to the other eye. This image was made as a photomontage reconstruction from parallel sections of layer 4 with autoradiography. (B) Reconstruction of layer 4 in 18-month-old monkey whose right eye had been closed at the age of 2 weeks. Radioactive material was injected into the normal left eye. White grains demonstrate columns in layer 4 from the nondeprived eye, which are larger than normal. Columns supplied by the eye that had been closed (black) are narrower than normal. (C) Cortical arborization of labeled geniculate axons ending in layer 4 of a kitten in which one eye had been closed for 33 days. The terminal arborization of the geniculate axon from the deprived eye shows a dramatic reduction of branches compared to that from the nondeprived eye. (A and B from Hubel, Wiesel, and LeVay, 1977; C after Antonini and Stryker, 1993b.)

normal ocular dominance histogram. Moreover, after one eye has been completely removed in an adult monkey, the structure of layer 4 remains normal when observed with autoradiography or other staining methods, even though there is atrophy in the lateral geniculate nucleus. This finding indicates a remarkable resistance to change of layer 4 in the adult animal when compared with the changes seen in the immature animal.

In monkeys the greatest sensitivity to lid closure is during the first 6 weeks of life.[6,7,26] At any time before 6 weeks, with a peak at 1 week of age,[27] substantial changes in eye preference and columnar architecture develop if one eye is closed for a few days. During the subsequent months (up to about 12 to 16 months), several weeks of closure are required to produce obvious changes in ocular dominance histograms or the width of columns in layer 4. At later times, changes do not develop even after surgical removal of one eye.

The period of greatest susceptibility to lid closure in kittens has been narrowed down to the fourth and fifth weeks after birth.[28,29] During the first 3 weeks or so of life, eye closure has little effect. This is not surprising, since the kittens' eyes are normally closed for the first 10 days. But abruptly, during weeks 4 and 5, sensitivity increases. Closure at that age for as little as 3 to 4 days leads to a sharp decline in the number of cells that can be driven by the deprived eye. An experiment in which littermates are compared is shown in Figure 25.9. In this example, 6- and 8-day closures starting at the age of 23 and 30 days (Figure 25.9A and B) caused about as great an effect as 3 months of monocular deprivation from birth. The susceptibility to lid closure declines after the critical period has passed and eventually disappears by about 3 months of age (Figure 25.9C). The critical period can, however, be prolonged by rearing kittens in the dark.[30,31] In the absence of visual experience, the susceptibility to monocular closure can still be demonstrated at 6 months of age. However, there is evidence that even a brief exposure of the kitten to light for a few hours may be sufficient to prevent such extension of the critical period.

[27]Horton, J. C., and Hocking, D. R. 1997. *J. Neurosci.* 17: 3684–3709.

[28]Hubel, D. H., and Wiesel, T. N. 1970. *J. Physiol.* 206: 419–436.

[29]Malach, R., Ebert, R., and Van Sluyters, R. C. 1984. *J. Neurophysiol.* 51: 538–551.

[30]Cynader, M., and Mitchell, D. E. 1980. *J. Neurophysiol.* 43: 1026–1040.

[31]Daw, N. W., et al. 1995. *Ciba Found. Symp.* 193: 258–276.

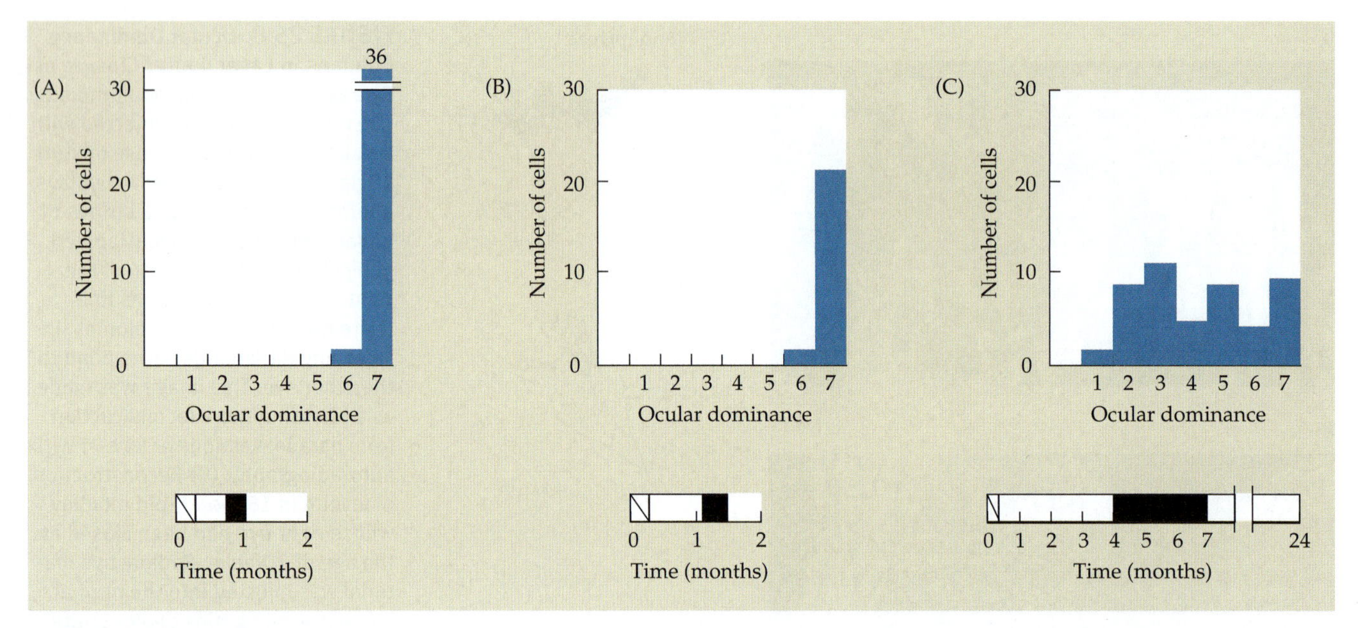

FIGURE 25.9 Critical Period in Kittens. Histograms showing eye preference in the visual cortex of kittens that were littermates, in which the right eye was closed at different ages. The period during which the eye was closed is indicated under the histograms. (A) Eyelids sutured for 6 days at 23 days of age. (B) Eyelids sutured for 9 days at 30 days of age. (C) The right eye was open the first 4 months, then closed for 3 months, and then kept open until 2 years of age, when the recordings were made. (After Hubel and Wiesel, 1970.)

Recovery during the Critical Period

To what extent is recovery possible after lid closure during the critical period? Even if the deprived eye in a cat or monkey is subsequently opened for months or years, the damage remains permanent, with little or no recovery: The animal continues to be blind in that eye, with shrunken columns and skewed ocular dominance histograms. In animals with monocular closure, experiments have been made in which the lids were opened in the deprived eye and closed over the normal eye. This procedure, termed **reverse suture**, leads to a dramatic recovery of vision, provided it is carried out during the critical period.[7,32,33] Monkeys and kittens not only begin to see again with the initially deprived eye, but they become blind in the other eye. Accompanying these changes, the ocular dominance histograms switch; that is, the newly opened eye drives most cells, while the eye that had been opened for the first weeks (now closed) cannot. Moreover, the anatomical pattern in layer 4 revealed by autoradiography shows a corresponding change: The shrunken regions supplied by the initially closed eye expand at the expense of the other eye. Figure 25.10 shows ocular dominance histograms and autoradiographs of the cortex in a monkey in which the right eye had been closed at 2 days and left closed for 3 weeks. By this time the deprived eye would no longer be able to drive cortical cells and the columns supplied by it would have retracted. The right eye was then opened and the left eye closed for the next 9 months. Nearly all neurons responded only to the initially deprived right eye, and the areas of cortex supplied by it had expanded.

The conclusions from these experiments are that (1) during the critical period in a normal animal, geniculate fibers supplying layer 4 of the cortex retract so that each eye supplies areas of comparable extent; (2) lid closure of one eye during the critical period leads to unequal retraction; and (3) reverse suture during the critical period produces *sprouting* of the geniculate axons so that an eye can recapture the cells it had lost (see Figure 25.10).[34] If postponed until adulthood, reverse suture is without effect. For example, in a monkey in which the reverse suture was performed at 1 year of age, the labeled columns for the initially deprived eye remained shrunken.

The concept of a well-defined, hard-and-fast critical period may be an oversimplification. Experiments on reverse suture in monkeys suggest that different layers of the striate cortex may develop at different rates; the critical period may be over for one layer while an adjacent layer is still capable of being modified in structure and function (see Figures 25.10C and 25.11).

[32]Blakemore, C., and Van Sluyters, R. C. 1974. *J. Physiol.* 237: 195–216.

[33]Kim, D. S., and Bonhoeffer, T. 1994. *Nature* 370: 370–372.

[34]Antonini, A., et al. 1998. *J. Neurosci.* 18: 9896–9909.

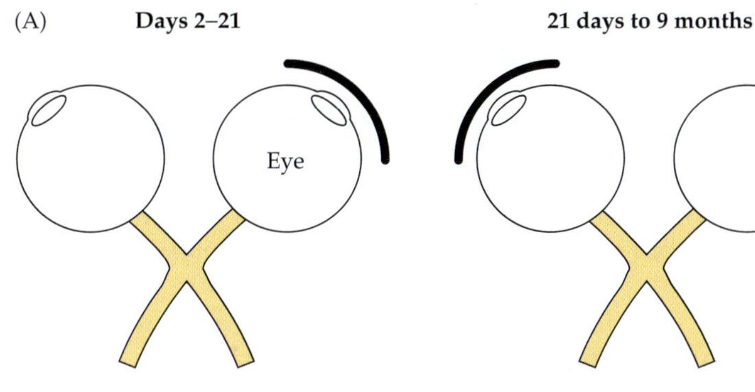

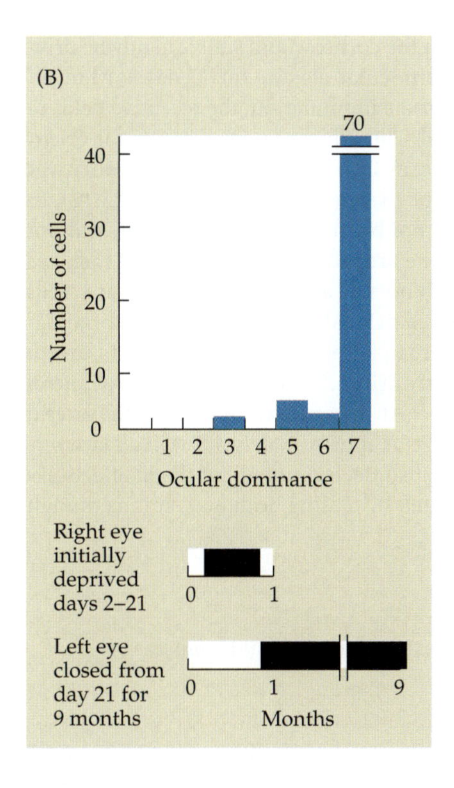

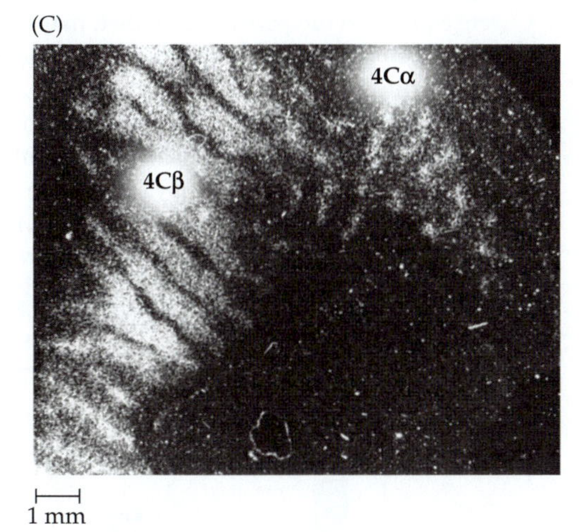

FIGURE 25.10 Effects of Reverse Suture on Ocular Dominance histogram and columnar organization in a monkey. (A) The procedure was as follows: The right eye was closed from days 2 to 21 after birth, after which it was opened while the left eye was closed from day 21 for 9 months. (B) The ocular dominance histogram shows that almost all cells were driven exclusively by the right eye, which had been initially deprived. Virtually no cortical cells were driven by the left eye. Had both eyes been kept open at 21 days, the histogram would be reversed. Accordingly, fibers driven by the right eye recaptured cortical cells they had previously lost. (C) Tangential section of cortex passing through layer 4Cβ and 4Cα. The bands labeled by the right eye are expanded in layer 4Cβ even though it had been deprived of light for 19 days. During those first days, the columns supplied by the right eye had shrunk before expanding. Recovery did not occur equally well in other layers, such as 4Cα. (After LeVay, Wiesel, and Hubel, 1980.)

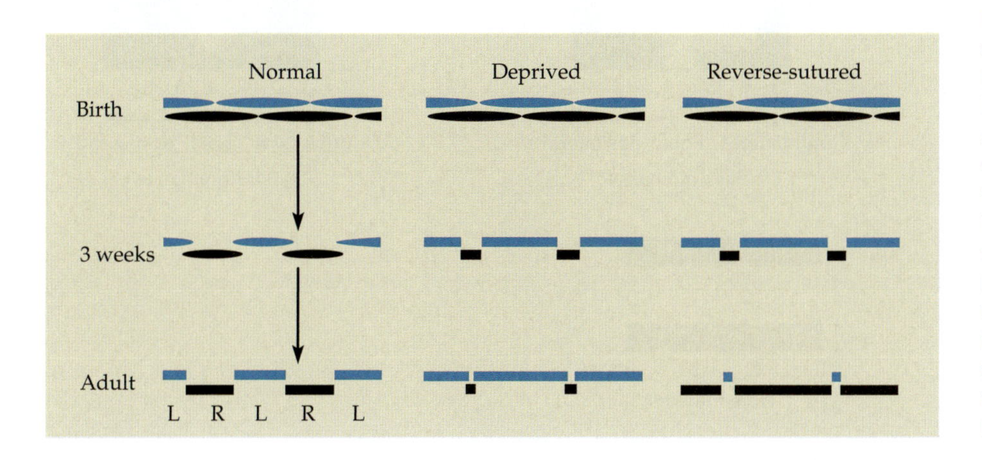

FIGURE 25.11 Summary of Effects of Eye Closure. In a normal monkey, ocular dominance columns have become well defined in layer 4 of the cortex by 6 weeks. Lid closure causes excessive retraction of geniculate fibers supplied by the deprived eye. Those supplied by the open eye retract less than usual, so their columns in layer 4 of the cortex are larger than normal in the adult. After reverse suture during the critical period, the initially deprived eye can recapture the territory it had lost in layer 4. (After Hubel and Wiesel, 1977.)

REQUIREMENTS FOR MAINTENANCE OF FUNCTIONING CONNECTIONS IN THE VISUAL SYSTEM

At this stage one might be tempted to conclude that loss of activity in the visual pathways is the main factor that tends to disrupt normal responses of cortical neurons. After all, cortical cells are driven not by diffuse illumination but by shapes and forms. The following discussion shows that there must be additional causes of a far more subtle nature. In particular, interactions must occur between the effects of impulse activity from the two eyes.

Binocular Lid Closure and the Role of Competition

The first clue that loss of visually evoked activity cannot on its own account for the changed performance of neurons is shown by the following experiments. Both eyes were closed in monkeys, newborn or delivered prematurely by cesarean section.[2,6] From the preceding discussion one might guess that cells in the cortex would subsequently be driven by neither eye. Surprisingly, however, after binocular closure for 17 days or longer, most cortical cells could still be driven by appropriate illumination; the receptive fields of simple and complex cells appeared largely normal. The columnar organization for orientation was similar to that in controls (see Figure 25.1). The principal abnormality was that a substantial fraction of the cells could not be driven binocularly (Figure 25.12). In addition, some spontaneously active cells could not be driven at all, and others did not require specifically oriented stimuli. However, the areas of cortex supplied by each eye were equal, and the pattern resembled that seen in normal adult monkeys: In layer 4, cells were driven by one eye only, and columns were well defined when marked by autoradiography or by cytochrome oxidase. Binocular closure in kittens led to similar effects, except that more cortical cells continued to be binocularly driven.[23] The arborizations of lateral geniculate axons in layer 4 were not shrunken.[35] At the same time, cells in the lateral geniculate body showed atrophy (a decrease in size of approximately 40%) in all layers.

[35]Antonini, A., and Stryker, M. P. 1998. *Vis. Neurosci.* 15: 401–409.

The conclusion from these experiments is that some, but not all, of the ill effects expected from closing one eye are reduced or averted by closing both eyes. It is as though

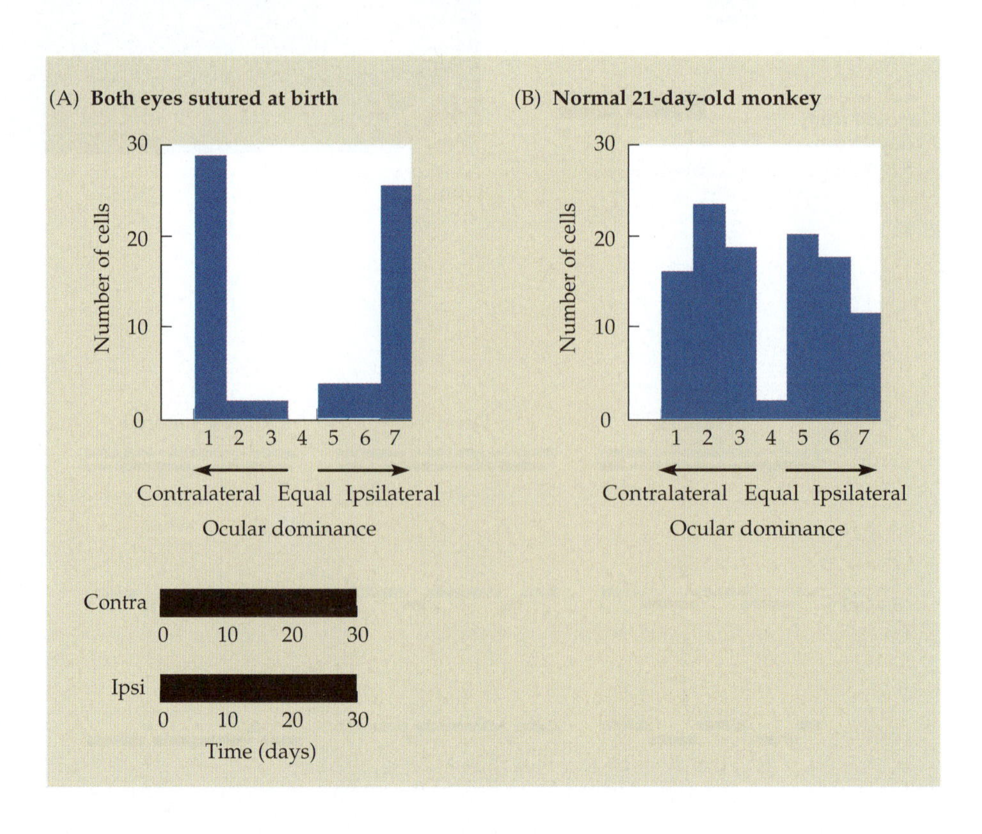

FIGURE 25.12 Ocular Dominance after Closure of Both Eyes at Birth. (A) The monkey studied here was delivered by cesarean section, and the lids of both eyes were immediately sutured. Recordings were made at 30 days of age. Each of the two deprived eyes could still drive cells in the visual cortex. Receptive fields appeared normal, except that few cells were driven by both eyes. Closure time is indicated at the bottom of the histogram. (B) Ocular dominance histogram from a normal 21-day-old monkey. (After Wiesel and Hubel, 1974.)

inputs from the two eyes are in competition for representation in cortical cells, and with one eye closed the contest becomes unequal.

Effects of Strabismus (Squint)

The abnormal effects described in the preceding discussion were produced by suturing eyelids or by using translucent diffusers, implicating loss of form vision. Following the clue that cross-eyed children (i.e., those with strabismus, or "squint") or wall-eyed children can become blind in one eye, Hubel and Wiesel produced artificial strabismus in kittens and newborn monkeys by cutting an eye muscle.[6,36] The optical axis of that eye was thereby deflected from normal. Under such conditions, illumination and pattern stimulation for each eye remained unchanged.

The results at first seemed disappointing because after several months vision in both eyes of the operated animals appeared normal, and Hubel and Wiesel were about to abandon a laborious set of experiments (personal communication). Nevertheless, they recorded from cortical cells and consistently obtained the following results. Individual cortical cells had normal receptive fields and responded briskly to precisely oriented stimuli. But almost every cell responded only to one eye; some were driven only by the ipsilateral eye and others only by the contralateral, but almost none were driven by both. The cells were, as usual, grouped in columns with respect to eye preference and field axis orientation.[6,37] As expected, no atrophy occurred in the lateral geniculate nucleus, and the columnar architecture of layer 4 was unchanged. The almost complete lack of binocular representation in cortical cells is shown in a histogram from a monkey with artificial strabismus (Figure 25.13). The critical period for displaced images to produce changes is comparable to that for monocular deprivation.[38]

These experiments provide an example in which all the usual parameters of light are normal—the amount of illumination and form and pattern stimuli. The only change consists of a failure of the images to fall on corresponding regions of the two retinas. The factor that seems important for the loss of binocular convergence of geniculate projections to the cortex is lack of congruity of input from the two eyes. It is as though the homologous receptive fields in both eyes must be in register with one another, so that excitation will be simultaneous. The following experiments further support this idea.

During the first 3 months of its life, the eyes of a kitten were occluded with an opaque plastic cover that was switched on alternate days from one eye to the other, so that the two eyes received the same total experience, but at different times.[36] The result was the same as in the cross-eyed experiments: Cells were driven predominantly by either one eye or the other, but not by both. The maintenance of normal binocularity, therefore, depends not only on the amount of impulse traffic but also on the appropriate spatial and temporal overlap of activity in the different incoming fibers.

[36]Hubel, D. H., and Wiesel, T. N. 1965. *J. Neurophysiol.* 28: 1041–1059.

[37]Löwel, S., and Singer, W. 1992. *Science* 255: 209–212.

[38]Baker, F. H., Grigg, P., and van Noorden, G. K. 1974. *Brain Res.* 66: 185–208.

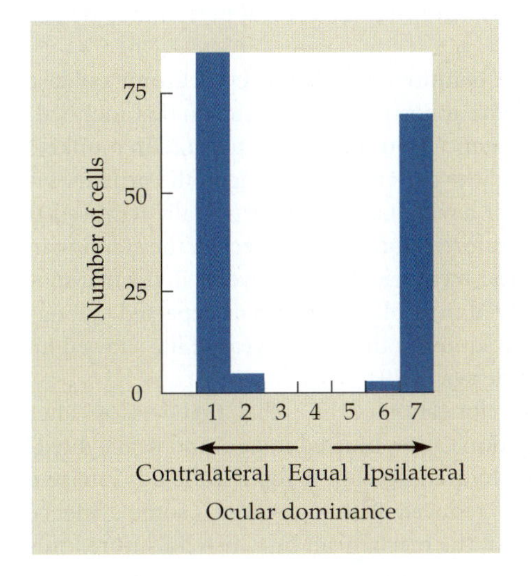

FIGURE 25.13 Effect of Strabismus on Monkey Ocular Dominance. The eyes in this 3-year-old monkey had faced in different directions since one eye muscle had been cut at 3 weeks of age. Cells are driven by one eye or the other, but not both. The ocular dominance columns in layer 4 appear normal in such animals. (After Hubel and Wiesel, unpublished, in Wiesel, 1982.)

Changes in Orientation Preference

A logical extension is to ask whether the orientation preference of cortical cells can be changed by raising animals in an environment in which they see only one orientation. An experimental approach that involved competition as well as deprivation was used by Carlson, Hubel, and Wiesel.[39] The lids of one eye were sutured in a newborn monkey. The animal was kept in darkness except when it placed its head in a holder. Then, with the head held vertically, it would see vertical stripes with the unsutured eye. Since the monkey received orange juice each time it placed its head in the holder correctly, it performed this maneuver frequently. Thus, during the critical period, one eye received no visual input while the other saw only vertical stripes. After 57 hours of experience between 12 and 54 days after birth, normal levels of cortical activity were found, with cells of all orientations arranged as usual in columns. As expected, the open eye tended to dominate.

When tests were made for orientation preference, the results shown in Figure 25.14 were obtained. Both eyes could drive cells well when horizontal lines were the stimulus. However, the left eye (the open eye) was considerably more effective for vertical stripes. The probable explanation for this result is that neither eye saw horizontal bars or edges during the critical period. Hence, the stimulation for horizontality is analogous to binocular closure; that is, the competition is equal. For the vertical input, however, the open eye had an enriched experience and "captured" cells in vertical orientation columns that had previously been supplied by the deprived eye. Similar results have been obtained by Bonhoeffer and his colleagues in kittens reared in a striped environment.[40]

Critical Periods in Development of Human Visual System and Clinical Implications

The susceptibility of kittens and monkeys during early life is reminiscent of clinical observations made in humans. It has long been known that removal of a clouded or opaque lens (or "cataract") can lead to a restoration of vision, even though the patient has been blind for many years. In contrast, a cataract that develops in a newborn or premature baby would in the past often have led to blindness. Before Hubel and Wiesel's experiments, cataracts were removed from children at a late stage, when they were considered to be ready for the operation. The result was that blindness was permanent, without the possibility of recovery.[41] Nowadays cataracts in newborn babies are removed by surgery as early as possible in the critical period, with excellent prospects for full vision.[42,43]

A second clinical application is provided by a familiar clinical procedure used in the past for the treatment of cross-eyed or wall-eyed children. Routinely, the good eye would be patched for prolonged periods in order to encourage the weaker eye to be used. There is evidence that damage in acuity of the patched eye may result, depending on the child's age at the time and the duration of the patching.[44,45] Such prolonged patching is no longer routinely practiced. Clinical observations suggest that the greatest sensitivity occurs in babies during the first year but that the critical period may persist for several years.

It is of interest that the ocular dominance columns revealed by cytochrome oxidase staining of layer 4 of the visual cortex in patients at postmortem has shown that monocular deprivation causes changes in ocular dominance columns similar to those in monkeys or kittens. Horton and Hocking[46] have studied the patterns of staining in the primary visual cortex of children that had grown up after having had one eye surgically removed at 1 week of age because of a tumor. In the postmortem brains, as expected, the staining in layer 4C was uniform instead of displaying clear territories for each eye, as in normal subjects or patients who had had one eye removed in adult life. Again as expected, a long-term strabismus in a patient who developed a squint in the second year of life showed no changes in column width at postmortem at the age of 79.[47]

A third example of clinical relevance is the elongation of the eyeball that develops after lid closure in neonatal monkeys. Such elongation causes blurred images and nearsightedness (myopia).[48] Similarly, it is known that children develop myopia if the lids interfere with vision or if the transparency of the eye is reduced. Although there is some evidence that activity and transmitters could play a part, the mechanism by which lid suture leads to elongation of the eye is not known.

[39]Carlson, M., Hubel, D. H., and Wiesel, T. N. 1986. *Brain Res.* 390: 71–81.

[40]Sengpiel, F., Stawinski, P., and Bonhoeffer, T. 1999. *Nature Neurosci.* 2: 727–732.

[41]Francois, J. 1979. *Ophthalmology* 86: 1586–1598.

[42]Birch, E. E., et al. 1993. *Invest. Ophthalmol. Vis. Sci.* 34: 3687–3699.

[43]Hamill, M. B., and Koch, D. D. 1999. *Curr. Opin. Ophthalmol.* 10: 4–9.

[44]Daw, N. W. 1998. *Arch. Ophthalmol.* 116: 502–505.

[45]Van Noorden, G. K. 1990. *Binocular Vision and Ocular Motility.* Mosby, St. Louis, MO.

[46]Horton, J. C., and Hocking, D. R. 1998. *Vis. Neurosci.* 15: 289–303.

[47]Horton, J. C., and Hocking, D. R. 1996. *Vis. Neurosci.* 13: 787–795.

[48]Stone, R. A., et al. 1988. *Proc. Natl. Acad. Sci. USA* 85: 257–260.

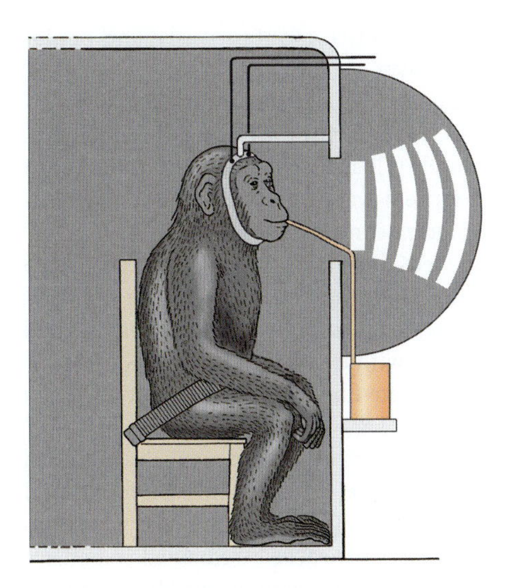

FIGURE 25.14 Orientation Preferences of Cortical Cells in a monkey with altered visual experience. The monkey was kept in a dark room. At 12 days the right eye was closed. Whenever the monkey placed its head in a holder, it received orange juice. At that time it also saw vertical stripes with its left eye. (The head holder ensured that the head was not tilted.) For a total of 57 hours of exposure between 12 and 54 days, one eye saw only vertical lines, the other nothing. (A) When horizontally oriented light stimuli was shone onto the screen, cortical cells driven by the left eye or the right eye responded equally well. In this histogram no deprivation is apparent for horizontal orientation, except for a lack of binocular cells. (B) With vertically oriented stimuli, the left eye, which had been kept open, was much more effective in driving cortical cells. The histogram resembles that seen after monocular deprivation. The results suggest that competition was equal for horizontal stimuli that neither eye had ever seen, and unequal for vertical stimuli (favored by the left, open eye). (After Carlson, Hubel, and Wiesel, 1986.)

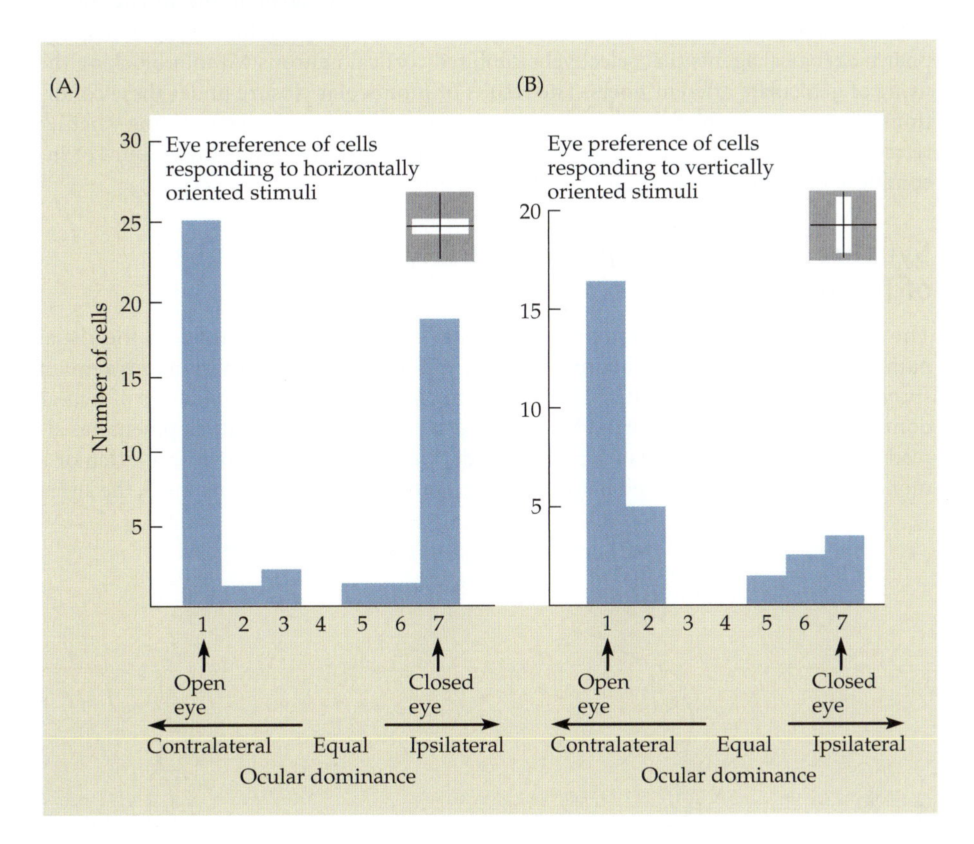

CELLULAR AND MOLECULAR MECHANISMS OF DEPRIVATION CHANGES

Effects of Impulse Activity on Structure

Two separate but related problems are raised by the experiments on visual deprivation in early life. First, what are the effects of impulse activity on the outgrowth and retraction patterns of neurons? Second, how do coherent and incoherent activity in two pathways determine how they compete for territory and define their fields?

For neurons in the visual system, the role of the action potentials themselves in shaping the architecture has been shown in experiments made in kittens. When the lids are closed or the animal is brought up in complete darkness, impulse traffic in the visual pathways does not stop entirely. Neurons continue to fire spontaneously, and ocular dominance columns develop as segregated areas in layer 4.

Experiments by Stryker, Shatz, and their colleagues[49,50] have shown that this presumably equal low level of activity from the two eyes is important for normal development. Tetrodotoxin (TTX), which blocks action potentials (Chapter 6), was injected into both eyes of newly born kittens. Several days later, after removal of the toxin, the visual pathways from retina through the geniculate to the cortex conducted once again. One interesting result was that in the lateral geniculate nucleus, inputs from the two eyes failed to segregate into separate layers. Moreover, cells in layer 4 of the visual cortex were still driven by both eyes as in the newborn animal, and the ocular dominance columns revealed by autoradiography resembled the neonatal pattern, with extensive overlap and no clear boundaries. Thus, in the absence of *all* firing, geniculate fibers failed to retract normally in layer 4 of the cortex (Figures 25.15 and 25.16).

The effects of lid closure on ocular dominance columns can be dramatically modified by blocking activity in the cortex with TTX. For example, TTX was infused for several days into the visual cortex of a kitten during the critical period, while one eye was deprived of form and light.[51] After the TTX was removed, cortical cells remained responsive to stimuli in both eyes, even though one had been deprived. Again, in the absence of activity, retraction failed to occur. Instead of using TTX, Stryker and his colleagues applied pharmacological agents that selectively inhibited cortical neurons without blocking the firing of geniculate afferent fibers. The results of monocular closure under these conditions suggested that the firing of postsynaptic cells plays a part in determining whether retraction occurs after monocular deprivation.[52] Thus, it is not simply the amount of incoming activity that is important.

[49]Stryker, M. P., and Harris, W. A. 1986. *J. Neurosci.* 6: 2117–2133.

[50]Sretavan, D. W., Shatz, C. J., and Stryker, M. P. 1988. *Nature* 336: 468–471.

[51]Reiter, H. O., Waitzman, D. M., and Stryker, M. P. 1986. *Exp. Brain Res.* 65: 182–188.

[52]Hata, Y., Tsumoto, T., and Stryker, M. P. 1999. *Neuron* 22: 375–381.

Synchronized Spontaneous Activity in the Absence of Inputs during Development

The experiments with tetrodotoxin suggest that action potential activity in the visual pathway is necessary for retraction of axons. Without ongoing activity, axons remain spread across layers in the lateral geniculate nucleus and across boundaries of ocular dominance columns in the cortex. Yet, as we have seen, much of the development has already proceeded by the time of birth. In the darkness of the womb, before a kitten or a monkey has seen anything, and before photoreceptors have become functional, the layers

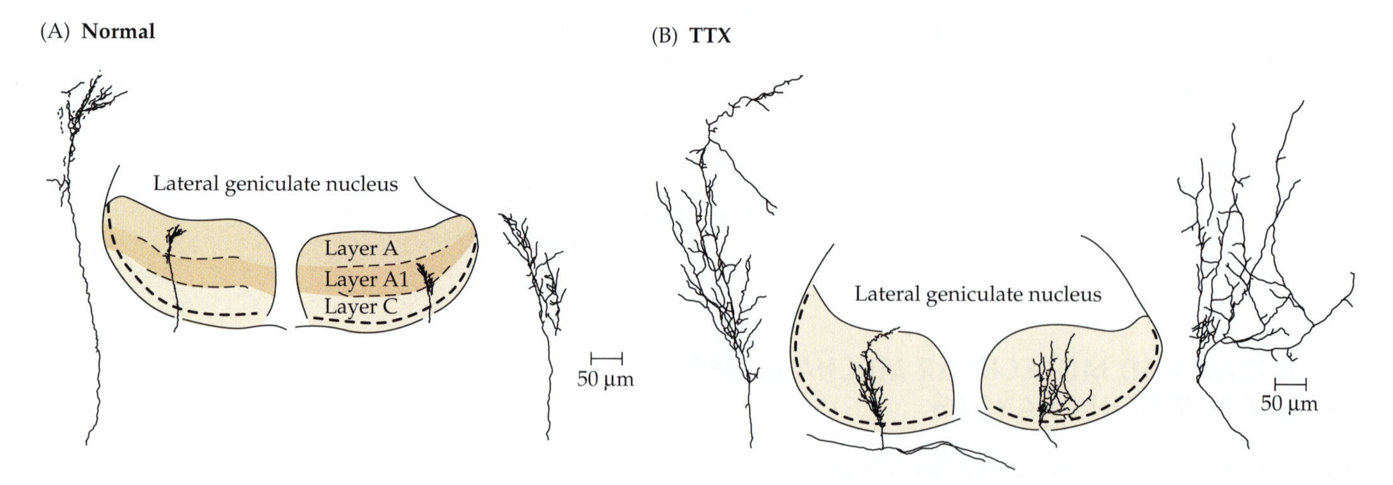

FIGURE 25.15 Effect of Abolition of Electrical Activity by Tetrodotoxin on arborization of optic nerve fibers terminating in the lateral geniculate nucleus. (A) In a normal kitten the terminals of optic nerve fibers labeled with horseradish peroxidase are restricted to the single layer where they end. (B) After application of tetrodotoxin for 16 days during embryonic life, labeled axons show much larger arborizations that are not restricted to individual layers. (After Sretavan, Shatz, and Stryker, 1988.)

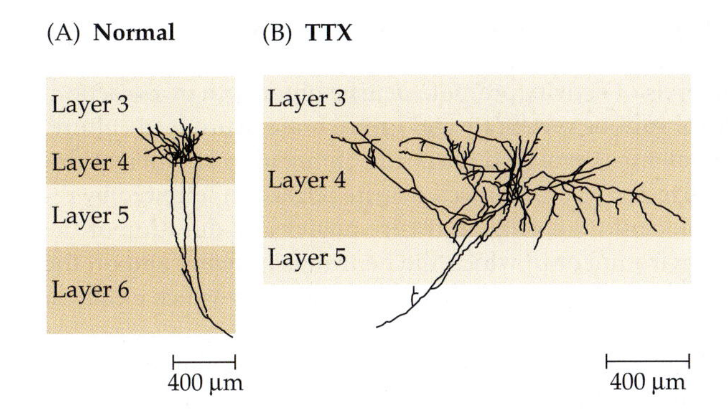

FIGURE 25.16 Increased Arborization of Lateral Geniculate Fibers ending in layer 4 of visual cortex after application of tetrodotoxin to both eyes. (A) Normal arborization of a labeled geniculate axon in layer 4 (30-day-old animal). (B) Labeled geniculate axon in a kitten in which tetrodotoxin had been applied to the eyes for 12 days (29-day-old animal). The axons of this neuron cover a much larger area of cortex. (After Antonini and Stryker, 1993a.)

of the lateral geniculate nucleus and the cortical columns are recognizable. Does this mean that early development proceeds without action potential activity, or is there intrinsic impulse activity in the system that guides development?

Maffei and his colleagues demonstrated that synchronous bursts of action potential traffic propagate along the optic nerve in utero.[53] There has been considerable speculation about how such firing patterns might be generated and what role they might play in the formation of coherent fields of innervation.

That there are periodic, synchronized waves of firing of neighboring ganglion cells was shown by Meister, Baylor, and their colleagues in retinas isolated from immature ferrets and fetal kittens.[54] The retina was placed in a chamber over an array of 61 electrodes. In the record from each electrode it was possible to identify discharges from up to four different ganglion cells. Recordings of this type revealed an ordered pattern of activity sweeping across the retina from ganglion cell to ganglion cell. An example is shown in Figure 25.17. Small dots represent the electrode positions, large blue dots the location of action potential discharges; the size of the dot indicates discharge frequency. Successive frames were taken at 0.5 s intervals. The wave of activity spread across the retina over a period of about 3 s. Typically such waves recurred repeatedly, separated by silent periods on the order of 2 s in duration. Wong[55] has reported a similar wavelike spread of transient changes in intracellular calcium concentration and suggested that they may play a role in synchronization of the electrical activity. There is evidence that cholinergic neurons, starburst amacrine cells, and electrical coupling play a part in the generation of the coordinated firing of ganglion cells in the immature retina.[56,57]

It is still not known, however, how synchronized activity in one eye leads to segregation of its optic nerve fibers in geniculate layers and to the prevention of overlap between layers.

[53]Maffei, L., and Galli-Resta, L. 1990. *Proc. Natl. Acad. Sci. USA* 87: 2861–2964.

[54]Meister, M., et al. 1991. *Science* 252: 939–943.

[55]Wong, R. O. 1999. *Annu. Rev. Neurosci.* 22: 29–47.

[56]Zhou, Z. J. 1998. *J. Neurosci.* 18: 4155–4165.

[57]Brivanlou, I. H., Warland, D. K., and Meister, M. 1998. *Neuron* 20: 527–539.

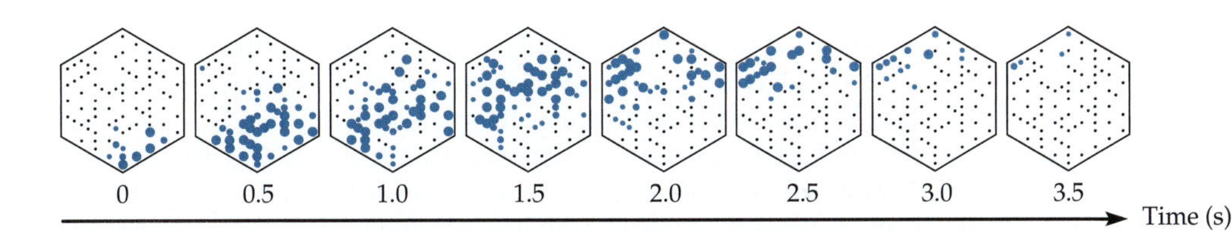

FIGURE 25.17 Wave of Impulse Activity Spreading across Isolated Retina of a neonatal ferret. The isolated retina was placed on recording electrodes, embedded in a regular array in the dish. The position of each of 82 retinal neurons is represented by a small black spot. Electrically active neurons are marked by larger blue spots, the sizes of which are proportional to the firing rates. Each frame represents the activity averaged over successive 0.5 s intervals. During the time represented by the eight frames (3.5 s), action potentials begin with one small group of cells and spread slowly across the retina. A new wave begins shortly after, and then another, each spreading in a different direction. At this stage of development, photoreceptors in the ferret are not responsive to light. (After Meister et al., 1991.)

Cellular Mechanisms for Plasticity of Connections

To test directly how different levels of activity promote neurite outgrowth or retraction, experiments have been made on various vertebrate and invertebrate neurons in culture. For example, trains of action potentials evoked by electrical stimulation at different frequencies have been shown to give rise to pronounced neurite retraction followed by regrowth. Moreover, this effect depends not only on the frequency and duration of the trains but on the molecular environment in which the neuron is growing and on the stage of outgrowth. Chapter 23 provides a review of the mechanisms by which electrical activity influences neurite outgrowth.

The experiments with strabismus show that, for connections to be maintained, neurons must somehow detect whether the firing from two inputs is in or out of phase. Recordings made from cortical neurons show that they can in fact respond with marked amplification to two synaptic inputs arising in precise synchrony.[58] How this could preferentially allow one set of endings to remain while others retract is not known.

Trophic Molecules and the Maintenance of Connections

An area of intense research is to assess whether competition for trophic molecules secreted by postsynaptic target cells plays a role in determining which of the inputs maintain connections (Chapter 23). Maffei and his colleagues first provided experimental evidence that neurotrophins such as NGF and BDNF (nerve growth factor and brain-derived neurotrophic factor) can prevent the effects of monocular lid closure in the developing rat visual system. Their results suggest that geniculate fibers compete for growth factors.[59,60]

The supposition is that axons that fail to receive enough of a trophic molecule from the target lose their connections and withdraw, as in other systems (Chapter 23). Accordingly, an excessive or abundant concentration of NGF was provided to the rat cortex by direct infusion or by secretion from explanted Schwann cells. Under these conditions monocular deprivation no longer gave rise to column shrinkage, and cortical cells were still driven by both eyes. The introduction of anti-NGF antibody blocked these effects. Moreover, in developing visual system, NGF antibodies caused shrinkage of cells and prolongation of the critical period, as though a normal trophic action of NGF were being antagonized.[61]

Although information is becoming available about local differences in the molecular environment and about the inherent properties of neurons themselves, it is still not possible to provide a full and comprehensive explanation for describing critical periods and plasticity at the molecular level. Thus, we cannot yet explain why connections in the visual cortex should be so much more vulnerable than those in the retina or the spinal cord.

Segregation of Inputs without Competition

In the experiments described so far, an underlying principle has been that the two eyes compete for connections and territory in the lateral geniculate nucleus and in layer 4 of the primary visual cortex, starting off with roughly equal opportunity. Rakic and his colleagues have used a different approach to study how neighboring groups of cells with defined properties sort out their terminals and their targets as they develop, *without* competition.[62]

As we have already described, the magnocellular (M) and parvocellular (P) systems occupy distinct layers in the lateral geniculate nucleus and in visual cortex. By staining individual M and P axons as they grow during development into the lateral geniculate nucleus, Rakic and his colleagues have shown that axons from the outset arrive in the correct M and P layers, where they form characteristic, nonoverlapping arborizations. M fibers end only in geniculate layers 1 and 2, while P fibers end only in layers 3, 4, 5, and 6 (Figure 25.18), with no spillover.

Hence, when the two eyes develop their connections, one can suppose that competition plays a role for inputs that provide similar information about the visual world. By contrast, the M and P systems carry quite different types of information. Their connections (like those involved in the formation of blobs and or the stripes in visual area 2)[46] develop according to principles in which competition is not of the essence. A further ex-

[58] Larkum, M. E., Zhu, J. J., and Sakmann, B. 1999. *Nature* 398: 338–341.

[59] Berardi, N., and Maffei, L. 1999. *J. Neurobiol.* 41: 119–126.

[60] Pizzorusso, T., et al. 1994. *Proc. Natl. Acad. Sci. USA* 91: 2572–2576.

[61] Capsoni, S., et al. 1999. *Neuroscience* 88: 393–403.

[62] Meissirel, C., et al. 1997. *Proc. Natl. Acad. Sci. USA* 94: 5900–5905.

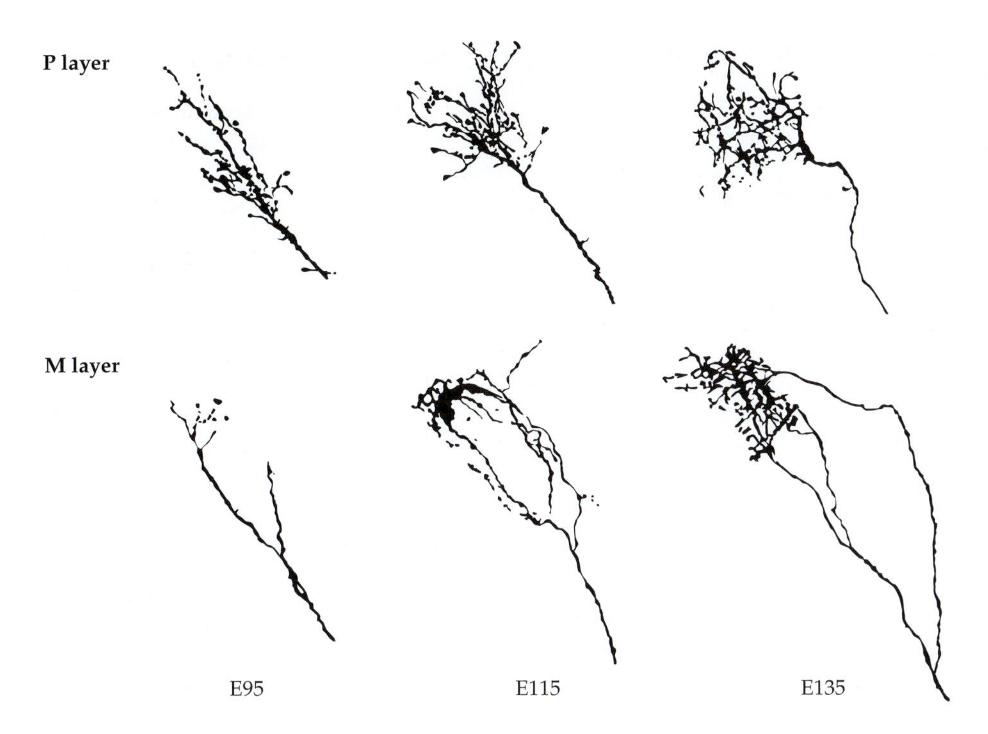

FIGURE 25.18 Absence of Competition for Territory by Magno- and Parvocellular Retinogeniculate Axons. Labeled optic nerve fibers ending in M and P layers of monkey lateral geniculate nucleus continue to grow during embryonic days 95 to 135 (E95 to E135). From the outset, axons are confined to their appropriate layers without showing spread into adjacent layers of retraction during development. Thus, M and P fibers show no competition for space. (After Meissirel, et al., 1997.)

ample of connections that form without signs of competition is the development of orientation maps in the visual cortex of the kitten.[63] (Note that findings of this sort reinforce one's belief that not *everything* that happens can be best accounted for by competition in the nervous system, or even in the marketplace.)

CRITICAL PERIODS IN THE AUDITORY SYSTEM

The effects produced by altered visual inputs in kittens and immature monkeys have a number of important implications for our understanding of the nervous system. A remarkable example of how adaptable the auditory system is during its critical period is provided by the experiments of Knudsen and his colleagues on barn owls.[64] Early auditory experience shapes the auditory spatial tuning of neurons in the barn owl's optic tectum in a frequency-dependent manner.[65] The following account shows how altered visual input in early life influences the representation of the auditory system in the owl's brain.

Eric Knudsen

Auditory and Visual Experience in Newborn Barn Owls

An owl turns its head to the exact spot at which a sound arises (important for catching squeaking mice). Horizontal location is achieved by measuring the interaural time difference—that is, the delay between sound waves impinging on first one ear and then the other. (The owl can also use the intensity of sound in the two ears to measure vertical position; asymmetrically distributed groups of feathers on the face deflect sound from above into one ear, from below to the other.) Another index of the position and path taken by the mouse is provided by the eyes. Figure 25.19A shows that in normal adult owls the neural maps for visual and auditory space are precisely aligned in one layer of the optic tectum (which corresponds to the superior colliculus in mammals). Such maps were produced by measuring responses evoked in individual neurons in the tectum by sound stimuli from different locations and by light presented in different parts of the visual field.

Baby owls were then raised with visual fields displaced by 23° to the left or the right by prisms placed over their eyes (Figure 25.20A).[66,67] This shifted the image of the visual field on the retina and hence on the tectum, so that the visual and auditory maps were no longer in alignment (Figure 25.20B,C). Over the next 6 to 8 weeks, the auditory space map became displaced until it once again came to fit in exact register with the new visual

[63]Godecke, I., and Bonhoeffer, T. 1996. *Nature* 379: 251–254.

[64]Knudsen, E. I. 1999. *J. Comp. Physiol. [A]* 185: 305–321.

[65]Gold, J. I., and Knudsen, E. I. 1999. *J. Neurophysiol.* 82: 2197–2209.

[66]Knudsen, E. I., and Knudsen, P. F. 1990. *J. Neurosci.* 10: 222–232.

[67]Feldman, D. E., and Knudsen, E. I. 1997. *J. Neurosci.* 17: 6820–6837.

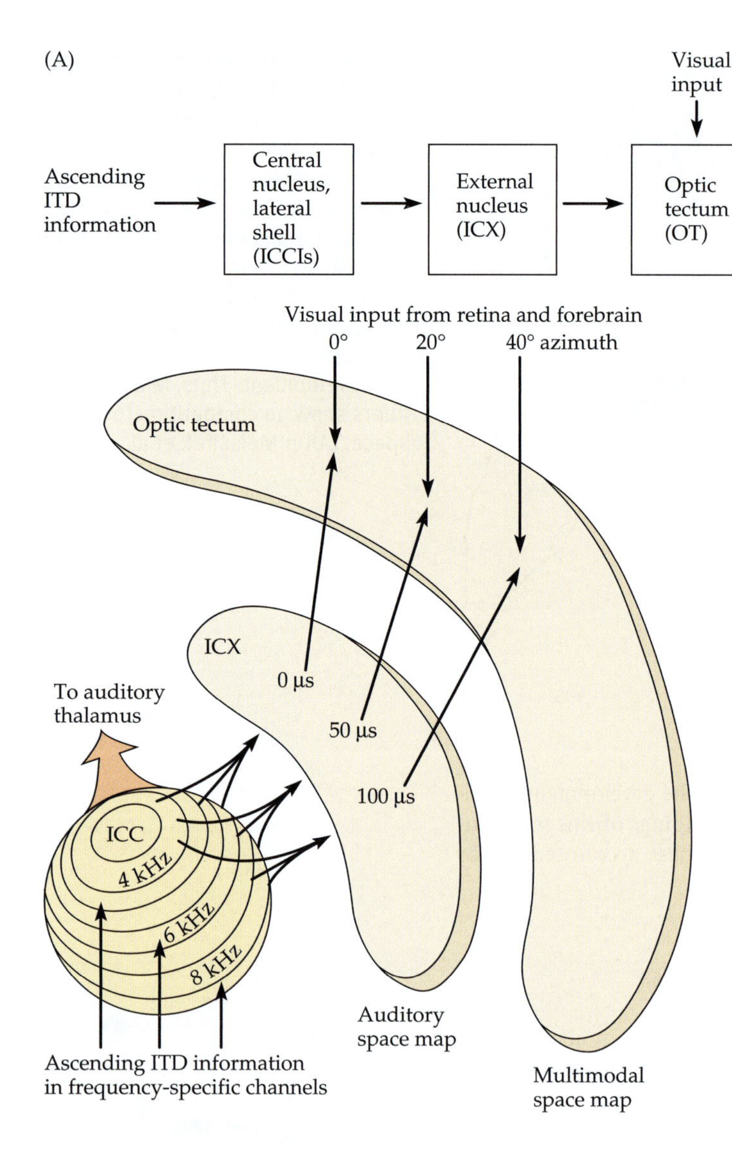

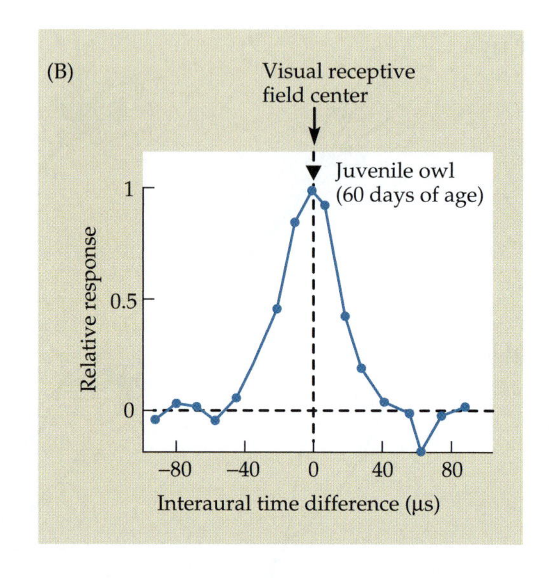

FIGURE 25.19 Superimposition of Auditory and Visual Space Maps in the optic tectum of barn owl. (A) Ascending auditory pathway to the optic tectum. Auditory neurons in the internal (ICC) and external (ICX) nuclei of the inferior colliculus are tonotopically arranged. They project to the optic tectum, maintaining an orderly sequence. The auditory space map depends on the time difference of the arrival of sounds in the two ears. Auditory and visual space maps are in register with each other. Thus, neurons that one records from the position marked "0 µs" respond to visual and auditory stimuli that are directly in front of the bird. ITD = interaural time difference. (B) Graph of responses to interaural time differences plotted in a juvenile owl 60 days of age. The time difference between the two sounds indicates the position to the left or to the right. The neuron that responds to the time value of a 0 µs interval in interaural time difference fires best when the stimulus is directly in front of the animal, corresponding to a receptive field position at the center of the visual field. Sounds coming from the left or the right of the owl reach the ears with a delay, activating neurons with different response curves, peaks being displaced from 0 µs and in register with optical stimuli. (A after Knudsen, 1999a; B after Feldman and Knudsen, 1997.)

map. Plasticity in the critical period therefore enabled fine-tuning of brain function to be carried out in response to experience and knowledge of the world. As a result, the owl became able to orient its eyes toward sounds in spite of the prisms.

At later stages the prisms were removed from the owls' eyes. Now the visual and auditory maps were again out of register. Provided that the owls were less than 200 days of age when the prisms were removed, the auditory map shifted a second time and returned to its original alignment, in register with the visual map (Figure 25.20, but see also Figure 25.21A).

Effects of Enriched Sensory Experience in Early Life

Since the flexibility of connections and structure in early life make the brain vulnerable to sensory deprivation, it is natural to wonder whether enrichment and a fuller early life during the critical period could enhance cortical function. Such tests are in practice difficult to devise: First, newly born animals need to be with their mothers for much of the time. Second, it is hard to know for sure what would constitute an extra rich and pleasant set of stimuli for, say, baby birds, rats, mice, or monkeys in the wild.

(A)

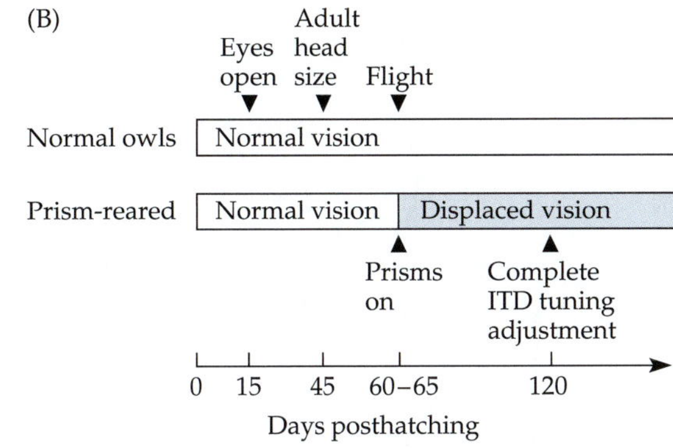

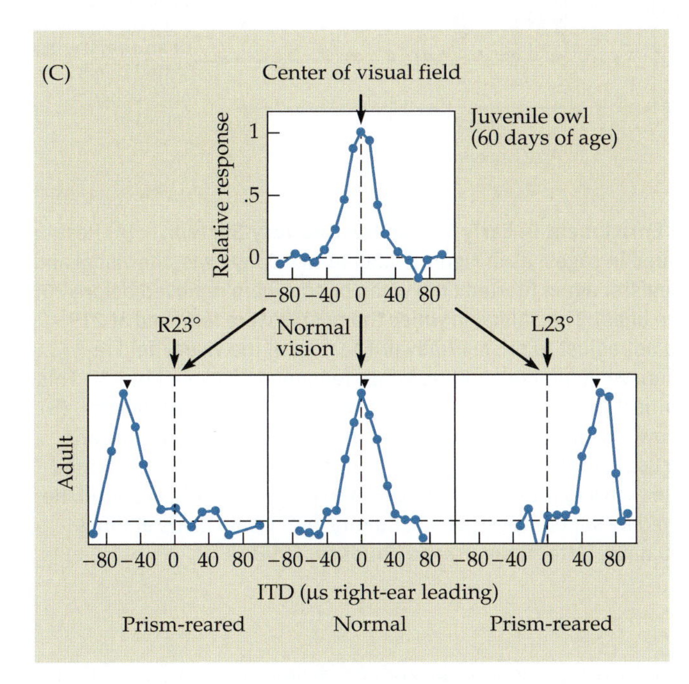

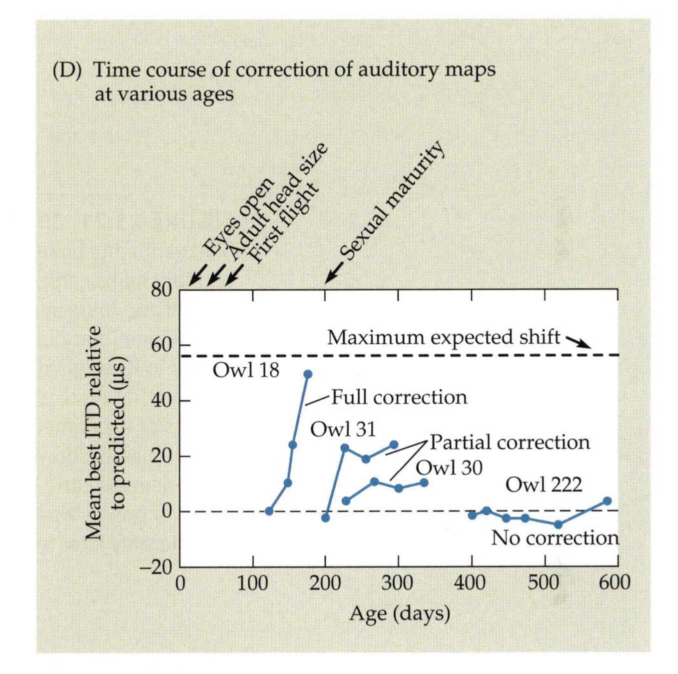

FIGURE 25.20 Shift of Auditory Receptive Fields after application of prisms during the critical period. (A) Baby owl with glasses consisting of prisms that offset its visual field by 23° to the right (or to the left depending on the spectacles). (B) Stages at which prisms were placed on owl eyes. (C) Tuning curves showing responses to sounds with different interaural time intervals. In the normal juvenile owl at 60 days, the response of the neuron resembles that of Figure 25.19: An interaural time difference (ITD) value of 0 µs corresponds to the center of the visual field. The tuning curves were shifted after rearing with prisms. Visual receptive fields were displaced to the left or the right by 23°. Now the best responses of the auditory neurons occurred at ITDs that corresponded to the displaced visual fields. The auditory and visual maps were again in register. (D) Time course of shift in auditory tuning curves in 3 immature owls. The adult owl (owl 222) showed no correction. (A kindly provided by E. Knudsen; B after Feldman and Knudsen, 1997; C and D after Brainard and Knudsen, 1998.)

An interesting and unexpected effect of enrichment in early life was found by Brainard and Knudsen[68] in barn owls with the visual field displacements already mentioned. In a second set of experiments, they repeated their earlier procedures, with one important difference in the way the baby owls were brought up. Instead of the baby owl being confined on its own in a cage (as in the earlier experiments), it spent the first weeks of life in an aviary where it lived with other owls and was able to fly around. In these owls with enriched experience, they confirmed that, as before, the effect of a visual field displacement in early life

[68]Brainard, M. S., and Knudsen, E. I. 1998. *J. Neurosci.* 18: 3929–3942.

(A) **Owls in individual cages; not enriched. Critical period for recovery at about 200 days.**

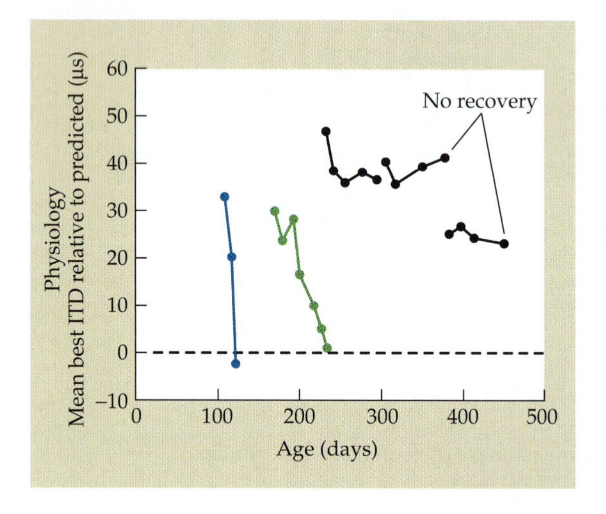

(B) **"Enriched" owls in aviary. Absence of critical period for recovery after prism removal.**

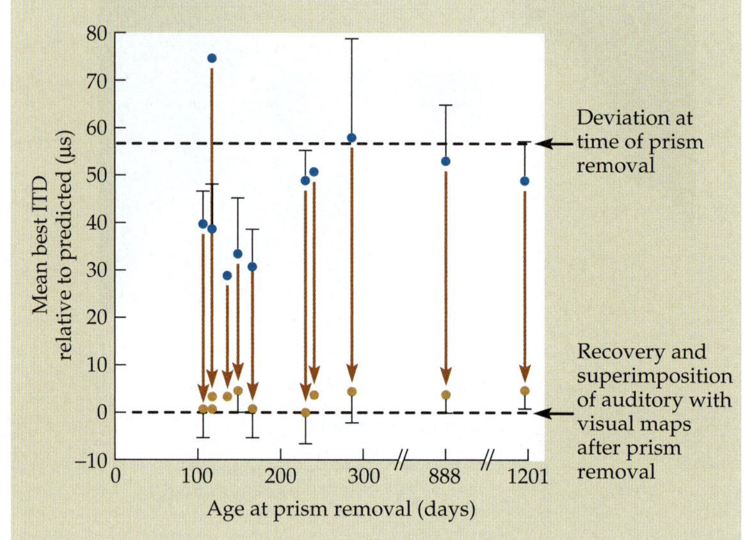

FIGURE 25.21 Effect of Enrichment in Early Life on the Auditory System. (A) Results of experiments on owls reared in cages, as in Figure 25.20. In two baby owls, the visual and auditory maps came into register again (dotted line) when prisms were removed before 200 days of age (blue and green lines). The animal in which the prisms were removed at 219 days of age (black line) did not adjust to prism removal. The map of the visual field returned to its original position, while that of auditory space remained displaced by 23°. This showed that there was a critical period in these owls for correction after prism removal. (B) Results of experiments on owls with enriched experience in early life reared in an aviary. These animals showed the usual shift in auditory maps and response curves when reared with prisms in early life. The recovery after prism removal did not show a critical period. Removal of prisms was followed by superimposition of auditory with visual maps at times considerably later than 200 days. (After Brainard and Knudsen, 1998.)

was corrected in the tectum by realignment of the auditory field onto the new position of the visual field. When the prisms were removed in the young animals up to 200 days, as before, the auditory field moved back to its original mapping in the cortex, thereby allowing auditory and visual responses to be matched. What was novel and different in these later results is shown in Figure 25.21B. Long after the 200 days, at times when the owl had matured and become adult, the correction of the auditory field still occurred when the prisms were removed, as though there were no critical period. Similarly, the critical period in early life during which the application of prisms caused remapping in the tectum was extended to 200 days instead of 70 days. Thus, the richer environment lengthened the period of adaptability in the auditory system. In other tests it was shown that owls that had once learned to adapt to prisms in early life could reestablish such connections as adults, unlike animals that had no such experience beforehand.[69]

Enrichment in early life can also result in morphological changes in neurons. In the visual cortex of rats and mice that had been exposed to frequent, daily, mild and nonaversive sensory stimuli in early life, dendritic arborizations were more profuse than those in controls.[70]

[69]Knudsen, E. I. 1998. *Science* 279: 1531–1533.

[70]Venable, N., et al. 1989. *Brain Res. Dev. Brain Res.* 49: 140–144.

[71]Lorenz, K. 1970. *Studies in Animal and Human Behavior*, Vols. 1 and 2. Harvard University Press, Cambridge, MA.

CRITICAL PERIODS FOR HIGHER FUNCTIONS

There exists a wealth of literature reporting other complex behavioral processes in a variety of animals that show periods of susceptibility. Imprinting is one example. Lorenz[71] has shown that a duck will follow any moving object, provided that it is presented during the first day after hatching, and will act throughout life as if it were the real mother. Similarly, the melodic patterns produced by birds, such as zebra finches, depend on the songs that the baby birds hear in early life during a critical period.[72] Birds of a particular species will remember and reproduce appropriate (but not inappropriate) melodies that they hear in early life.

In higher animals—for example, in dogs—behavioral studies indicate that if they are handled by humans during a critical period of 4 to 8 weeks after birth, they are far more tractable and tame than animals that have been isolated from human contact.[73] The critical period in an animal's development seems to correspond to a time during which a significant sharpening of senses or faculties occurs.

What Is the Biological Advantage of Critical Periods?

Why should plasticity in early life be so pronounced? The developing brain must not only form itself but must be able to represent the outside world, the body, and its movements.[74] The eye, for example, must grow to be the right size for distant objects to be in focus on the retina through the relaxed lens. And the two eyes, separated by different distances in different newborn babies, must act together. As if this were not enough, gross changes occur in limb length, skull diameter, and therefore body image in the first months and years of life.

At all times the maps in the brain for different functions must be in register, as in the owl experiments of Knudsen. In adult life, too, there is evidence for plasticity.[75–78] At a functional level we can rapidly and reversibly adapt to prisms over our eyes that displace the visual fields, even as adults. Blind spots produced in the visual fields by lesions in the eye can be filled in.[79] In adult monkeys, changes in the representation of the body occur in the somatosensory cortex after lesions or deafferentation.

It is tempting to speculate about the effects of deprivation on higher functions in humans. One can imagine, as Hubel[20] has said,

> Perhaps the most exciting possibility for the future is the extension of this type of work to other systems besides sensory. Experimental psychologists and psychiatrists both emphasize the importance of early experience on subsequent behavior patterns—could it be that deprivation of social contacts or the existence of other abnormal emotional situations early in life may lead to a deterioration or distortion of connections in some yet unexplored parts of the brain?

To find a physiological basis for such behavioral questions seems a distant, but not impossible, goal.

[72]Arnold, A. P., and Schlinger, B. A. 1993. *Brain Behav. Evol.* 42: 231–241.

[73]Fuller, J. L. 1967. *Science* 158: 1645–1652.

[74]Singer, W. 1995. *Science* 270: 758–764.

[75]Polley, D. B., Chen-Bee, C. H., and Frostig, R. D. 1999. *Neuron* 24: 623–637.

[76]Buonomano, D. V., and Merzenich, M. M. 1998. *Annu. Rev. Neurosci.* 21: 149–186.

[77]Welker, E., et al. 1992. *J. Neurosci.* 12: 153–170.

[78]Sterr, A., et al. 1998. *J. Neurosci.* 18: 4417–4423.

[79]Gilbert, C. D. 1998. *Physiol. Rev.* 78: 467–485.

SUMMARY

- Receptive fields and cortical architecture in newly born monkeys and kittens resemble those of adults in many respects.
- In layer 4 of the cortex, however, geniculate axons overlap and cells are driven by both eyes instead of one during the first 6 weeks of life.
- A critical period of about 3 months exists after birth during which closure of the lids of one eye causes changes in structure and function.
- Closure of the lids of one eye leads to blindness in that eye.
- Cortical cells are no longer driven by the deprived eye, and its ocular dominance columns shrink.
- After the critical period, closure of lids or enucleation of an eye does not change cortical architecture.
- Binocular lid closure and induction of squint during the critical period do not cause changes in ocular dominance columns but do prevent binocular responses.
- Such results suggest that the two eyes compete for cells in the visual cortex.
- In newly born owls, the processing of auditory inputs is modified by sensory input during a critical period.
- Enrichment of experience in early life increases the duration of the critical period.

SUGGESTED READING

General Reviews

Hubel, D. H. 1988. *Eye, Brain and Vision*. Scientific American Library, New York.

Hubel, D. H., and Wiesel, T. N. 1977. Ferrier Lecture. Functional architecture of macaque monkey visual cortex. *Proc. R. Soc. Lond. B* 198: 1–59.

Knudsen, E. I. 1999. Early experience and critical periods. In S. C. Landis, M. J. Zigmond, L. R. Squire, J. L. Roberts, and F. E. Bloom (eds.), *Fundamental Neuroscience*. Academic Press, New York, pp. 637–654.

Singer, W. 1995. Development and plasticity of cortical processing architectures. *Science* 270: 758–764.

Wiesel, T. N. 1982. The postnatal development of the visual cortex and the influence of environment. *Nature* 299: 583–591.

Wong, R. O. 1999. Retinal waves and visual system development. *Annu. Rev. Neurosci.* 22: 29–47.

Original Papers

Antonini, A., Gillespie, D. C., Crair, M. C., and Stryker, M. P. 1998. Morphology of single geniculocortical afferents and functional recovery of the visual cortex after reverse monocular deprivation in the kitten. *J. Neurosci.* 18: 9896–9909.

Brainard, M. S., and Knudsen, E. I. 1998. Sensitive periods for visual calibration of the auditory space map in the barn owl optic tectum. *J. Neurosci.* 18: 3929–3942.

Carlson, M., Hubel, D. H., and Wiesel, T. N. 1986. Effects of monocular exposure to oriented lines on monkey striate cortex. *Brain Res.* 390: 71–81.

Horton, J. C., and Hocking, D. R. 1996. An adult-like pattern of ocular dominance columns in striate cortex of newborn monkeys prior to visual experience. *J. Neurosci.* 16: 1791–1807.

Hubel, D. H., and Wiesel, T. N. 1965. Binocular interaction in striate cortex of kittens reared with artificial squint. *J. Neurophysiol.* 28: 1041–1059.

Hubel, D. H., Wiesel, T. N., and LeVay, S. 1977. Plasticity of ocular dominance columns in monkey striate cortex. *Philos. Trans. R. Soc. Lond. B* 278: 377–409.

Knudsen, E. I., and Knudsen, P. F. 1990. Sensitive and critical periods for visual calibration of sound localization by barn owls. *J. Neurosci.* 10: 222–232.

LeVay, S., Wiesel, T. N., and Hubel, D. H. 1980. The development of ocular dominance columns in normal and visually deprived monkeys. *J. Comp. Neurol.* 191: 1–51.

Meissirel, C., Wikler, K. C., Chalupa, L. M., and Rakic, P. 1997. Early divergence of magnocellular and parvocellular functional subsystems in the embryonic primate visual system. *Proc. Natl. Acad. Sci. USA* 94: 5900–5905.

Meister, M., Wong, R. O., Baylor, D. A., and Shatz, C. J. 1991. Synchronous bursts of action potentials in ganglion cells of the developing mammalian retina. *Science* 252: 939–943.

Wiesel, T. N., and Hubel, D. H. 1963. Single-cell responses in striate cortex of kittens deprived of vision in one eye. *J. Neurophysiol.* 26: 1003–1017.

PART 5

Conclusion

26 Open Questions

With each new edition of *From Neuron to Brain*, our understanding of how nerve cells produce electrical signals, how they communicate with one another, how they act in concert, and how they become connected during development has become deeper. In the last few years major advances have been obtained through novel molecular biological and imaging techniques. There seems, however, to be no way to guess what new techniques will become available in the future, and what new questions will arise as a consequence. Thus, at the time of the first edition of this book in 1976, few could conceive of the use of site-directed mutagenesis to study gating currents or the optical demonstration of functional columns in the living brain. Many problems that now seem unapproachable will require techniques and advances not yet imagined.

What can one predict today about novel concepts that might be incorporated into the next edition of this book? One reasonable guess is that more intensive collaboration among basic scientists working at the cellular and molecular levels, cognitive neuroscientists, and neurologists will be important for understanding integrative and higher brain functions relating to perception, movement, and memory. One can hope also that an increase in fundamental knowledge of the nervous system will lead to the prevention and alleviation of diseases of the nervous system that arise from unknown causes and that cannot yet be treated effectively.

Open questions about the nervous system and the brain are very different from those in subjects such as physics, chemistry, or even biology in general. It is not only the reader of a book such as this who can point to the important deficiencies in our knowledge and understanding. A layperson outside science is aware that we do not know the mechanisms of higher functions such as consciousness, learning, sleep, the production of coordinated movements, or even how one initiates the bending of a finger as an act of will. The same person, even if highly sophisticated and well educated, would probably have more difficulty in pointing out what things one still needs to know about relativity, particle physics, chemical reactions, or genetics. It is this wealth of obvious, unsolved, and important human questions that makes neuroscience so appealing today.

To illustrate one everyday example of our present ignorance of how the brain performs its functions, consider a sport such as tennis. An expert player—say, Martina Hingis—sees her opponent hit the ball. She can compute rapidly where it will land and how high it will bounce. The ball may be traveling at 100 km/h, but she can rush to the right spot, with her arm extended so as to hit the ball in the center of the racket, with exactly the right force to send it exactly onto the line in the other court (exploiting the remembered weakness of her opponent's backhand). We could just as well have picked as examples the way in which a pelican dives for a fish, a frog catches a fly with its tongue, or a bee drinks from a particular flower. In each of these examples objects must be recognized against a rich background, and highly coordinated movements must be planned, initiated, regulated, and brought to fulfillment. And somehow the necessary neuronal connections must have been formed.

In the following paragraphs we consider selected unsolved problems in neuroscience that might become approachable in the future, particularly in relation to the topics emphasized in this book.

Cellular and Molecular Studies of Neuronal Functions

So much new information has become available with such rapidity in the past few years about channels, receptors, transmitters, transporters, second messengers, and long-term changes at synapses that open questions posed today might already have solutions by the time the book goes to print.

Still not worked out are the intimate structural changes that mediate the opening, closing, and inactivation of channels. An obvious unresolved problem concerns the functional roles played by so many subtypes of ion channels, transmitter receptors, and transporters. Another major problem, still in its early stages, concerns the way in which molecules are transported to precise regions of neurons, sodium channels to nodes, receptors to dendritic spines, and synaptic vesicles to active zones in presynaptic terminals. The formation of postsynaptic specializations at the neuromuscular junction constitutes an example in which the signal triggering the localization of key molecules has been identified (see the discussion of agrin in Chapter 24), but the detailed mechanism remains to be determined.

In the pursuit of problems of high interest, such as long-term potentiation and long-term depression, an enormous number of detailed experiments have been made to unravel the underlying mechanisms. One fundamental question concerns the role of retrograde signals from postsynaptic neurons to presynaptic terminals. Another is whether long-term potentiation and depression are essential for the laying down and retrieval of memory.

Functional Importance of Intercellular Transfer of Materials

Numerous experiments have demonstrated trans-synaptic transfer of amino acids or proteins from neuron to neuron—for example, from the retina through the lateral geniculate nucleus to the visual cortex. That such transfer occurs is certain, yet we do not have crucial information about the mechanisms for transfer or the functional significance. Intercellular transfer of small molecules also occurs between cells linked by gap junctions (Chapter 7). It has been suggested that intercellular transfer represents a mechanism for controlling growth and development. A related question concerns the role of glial cells in relation to neuronal signaling (Chapter 8), particularly in relation to quantitative measures of exchange between glial cells and neurons and the importance of such exchange for function.

Development and Regeneration

In spite of remarkable progress, it is still not known how neurons select their precise targets. One can now approach in molecular terms problems such as directed neurite outgrowth toward targets, termination of growth, and the refinement of connections by selective pruning and cell death. At the same time, we can only wonder how the extraordinary precision of connections is achieved, as, for example, by terminals of muscle spindle afferent fibers on motoneurons in the spinal cord. With several thousand neurons in each cubic millimeter of tissue, how are the appropriate motor cells selected and innervated at the appropriate sites? And through what mechanisms does the same sensory cell form synapses with quite different release characteristics on specific neurons in the medulla? As for the failure of regeneration after injury to the mammalian central nervous system, the reasons are still not known, in spite of considerable advances in our understanding of molecular mechanisms that promote and inhibit neurite outgrowth.

Another major question concerns the development that proceeds in early life under the influence of experience. In particular, we know next to nothing about the effects of critical periods on the maturation of higher functions, including emotional states and personality.

Genetic Approaches to Understanding the Nervous System

It is hard to predict the consequences of the revolution in genetic techniques for understanding brain function. The present use of transgenic animals in which identified genes have been altered or deleted provides a powerful tool, but one still hampered by difficulties of interpretation owing to redundancy of function and unexpected side effects. With the completion of the Human Genome Project, candidate genes and molecules that are altered in disease and development will be known. To analyze this large array of information and separate the important from the incidental constitutes an immense task. The scope of the problem is illustrated by the study of inherited diseases, such as Huntington's disease, in which the altered gene can be identified by linkage analysis of the affected families.[1] Yet, although the altered sequences of the Huntington's disease gene were identified long ago, the function of the protein remains unknown. Similarly, mutations in genes coding for voltage-gated calcium channels are associated with familial hemiplegic migraine and cerebellar ataxia.[2] But again there is no clear link in terms of mechanisms. A long-term hope is that genetic therapies will be developed for such conditions. Possibly the way to devise treatments will become clearer with better means of delivery, as well as better temporal and spatial control of expression. It is perhaps worth commenting that this paragraph deals with techniques and information in search of questions.

Sensory and Motor Integration

A serious deficiency in our knowledge concerns the enormous numbers of neurons with no obvious function, particularly unmyelinated fibers that greatly outnumber myelinated fibers. One example from this book is how the large variety of amacrine cell types (more than 20) contributes to processing in the retina. Another is the role of group II afferents from muscle spindles in spinal cord function.

Mechanisms for the initiation and control of coordinated movements represent problems that have seen progress but still remain open. Thanks to noninvasive techniques for imaging and stimulation, one can for the first time obtain detailed images of brain activity. Yet more than 50 years ago, in remarkably prescient comments, Adrian pointed out that once you have learned to write your name, you can do it at once by holding the pencil between your toes.[3] For our ability to transfer such programs from one effector system to another we have no explanation.

Similarly for sensory systems, the neural mechanisms for integrating the entire picture of, say, a bulldog or an artichoke, let alone the outside world, remain beyond our reach at present. In discussions of this type at this stage, the dreaded homunculus makes his ap-

[1]Haque, N. S., Borghesani, P., and Isacson, O. 1997. *Mol. Med. Today* 3: 175–183.

[2]Burgess, D. L., and Noebels, J. L. 1999. *Ann. N.Y. Acad. Sci.* 868: 199–212.

[3]Adrian, E. D. 1946. *The Physical Background of Perception*. Clarendon, Oxford, England.

pearance—the cell or little man in the brain who actually sees what we see. To ridicule this concept is both fashionable and a sure sign of sophistication. Nevertheless, the homunculus does have a useful function: He represents and continually reminds us of our ignorance about higher cortical function. As soon as answers are found, he will die a natural death like phlogiston. We have no way as yet to replace him by a computer.

In addition to these obvious gaps in our knowledge, the mechanisms for the precise control of body temperature, blood pressure, and intestinal functions remain black boxes. Interactions of the brain with the immune system represent another major field of active research that is still at an early stage, with many open questions.

Mathematical modeling and computational neuroscience represent fields that depend critically on measurements made in channels, membranes, individual neurons, synapses, and networks. As yet, successes comparable to those of the Hodgkin–Huxley equations in describing comprehensively the permeability changes responsible for the action potential have not been realized in other areas. One principal reason is the incompleteness of the facts available for modeling complex processes, such as synaptic plasticity and integration. For example, how fully could one have hoped to model cortical circuits before the discovery of NMDA receptors or conduction block, and how many more such mechanisms await discovery?

Rhythmicity

Neuronal rhythms considered in this book include respiration and circadian rhythms, as well as the periodicity of firing by neurons in the cerebellum, hippocampus, thalamus, and spinal cord. Except in a few examples, such as the stomatogastric ganglion of the lobster and the swimming of the leech, we have no detailed information about the mechanisms that underlie the genesis or the regularity of firing patterns. Moreover, it is not at all clear what functions are played by current oscillations in well-known phenomena such as the alpha and delta waves of the electroencephalogram.

Input from Clinical Neurology to Studies of the Brain

For many years neurology was not only inseparable from neurobiology but provided the only method for studying higher functions in relation to brain structure. A triumph of the early neurologists was to use nature's own experiments to describe functions of various brain areas from careful correlation of symptoms with lesions. Their achievements are all the more remarkable because the use of lesions to assess function is fraught with pitfalls. With the newer techniques now available, such as magnetic resonance imaging (MRI) and positron emission tomography (PET), the neurologist is now for the first time in position to locate and observe lesions directly, to follow their progress in the living brain and to make inferences about higher cortical functions.[4] Improvements in spatial and temporal resolution seem around the corner. This could allow one to follow in real time the sequence of neuronal events that give rise to a decision, to perception, or to the laying down of a memory.

The dramatic story of Phineas Gage emphasizes the advantages and pitfalls of lesions and deficits as a means of analyzing brain function.[5] In 1848, at the age of 25, Phineas Gage suffered a massive lesion to the brain while working as a construction foreman on a railway in Vermont. As he pushed on a tamping iron to place a charge of gunpowder into a rock, the gunpowder exploded and blew the iron rod clear through the front of his skull. Gage lost consciousness only briefly and could soon sit up and speak. What astounded the doctor was that he recovered rapidly and was able to lead a relatively normal life for more than 12 years. Gage's personality, however, underwent a major change. From being a well-liked, quiet, sober, industrious, and careful worker, he changed into a loud-mouthed, boastful, impatient, and restless braggart. At a time when nothing was known of sensory, motor, visual, or auditory cortex, the neurological investigation showed that the prefrontal area was associated with the very highest functions of human conduct and personality. But Gage's accident was unlike most neurological cases. The site of the lesion in the frontal cortex and its extent were obvious in the living patient.

Other examples of neurological observations made in the nineteenth century that defined specific brain areas involved in higher functions were those of Broca and Wernicke,

[4]Adams, R. D., Victor, M., and Ropper, A. H. 1997. *Principles of Neurology*, 6th Ed. McGraw-Hill, New York.

[5]Harlow, J. M. 1868. *Publ. Mass. Med. Soc.* 2: 328–334.

FIGURE 26.1 Drawing of a Cat (right) made by a patient who had a large lesion of the right parietal lobe. Only the right side of the drawing was copied. All details on the left were overlooked. Such deficits are commonly seen following lesions of this type. (After Driver and Halligan; drawing kindly provided by J. Driver, 1991.)

who correlated defects in speech with the areas of cortex that had been damaged by vascular accidents or tumors. Even when precise areas are not known, clinicians and neuropsychologists can reliably define and separate processes such as long- and short-term memory.

Highly counterintuitive and difficult to comprehend at first sight are effects of lesions to the parietal lobe on one side, usually the right. Patients with such lesions may no longer recognize that there are two sides to the body and to the outside world. The left side of the body ceases to exist, so a patient will not recognize his left hand as his own. When such patients with a lesion of the right parietal lobe are asked to draw a daisy, all the petals are on the right, as are all the spokes of a bicycle wheel.[6] A drawing of a cat made by a right-handed, 61-year-old patient with a parietal lesion is shown in Figure 26.1. It is important to emphasize that these are true neurological defects, not hysterical reactions of the patient. Such clinical observations show that our inner world, which seems so complete, so unitary, and so perfect, is composed of elementary components welded together to form a continuum.

As information becomes available from brain scans and from sophisticated tests of language and performance, one can expect the exploration of higher functions to depend ever more on input from cognitive neuroscience and neurology.

Input from Basic Neuroscience to Neurology

There clearly exists a two-way street between basic and applied neuroscience. Molecular biological and genetic techniques are already beginning to play a role in diagnoses of such conditions as retinoblastoma and Huntington's disease, and the possibility of treatment with genetically engineered cells is now being intensively investigated in muscular dystrophy and Parkinson's disease. Sophisticated electrophysiological techniques are used by neurosurgeons for recording from individual neurons, for implanting electrodes (as for control of the bladder), for noninvasive stimulation, and for devising prostheses to replace lost functions.

One example from experiments described in this book can illustrate how research in basic neuroscience can help to provide new treatments for serious conditions. As a result of the work of Hubel and Wiesel on sensory deprivation in newly born kittens and monkeys, it became evident that a newborn baby that suffers from a cataract should have it removed as soon as possible. Such procedures have prevented countless cases of blindness. This was not an outcome that the investigators had in mind at the time they were doing their initial experiments on receptive fields in the visual cortex.

For most diseases of the nervous system that afflict mankind (e.g., Alzheimer's disease or amyotrophic lateral sclerosis), we have little or no knowledge of the root cause and no effective treatment. It might be argued that it would be better to invest the money used

[6]Driver, J., and Halligan, P. W. 1991. *Cogn. Neuropsychol.* 8: 475–496.

for basic neuroscience in applied science or neurology. Surely it would be better to find cures for the diseases directly rather than trying to find out how the nervous system works? In instances in which applied research has been emphasized over basic biological research the results have been disappointing, to say the least. For example, the Soviet Union established and supported massive institutes for applied research in physiology and pharmacology, each with hundreds of research workers—the type of research in which the investigator followed a scientific problem for its interest and its beauty was considered "bourgeois." Yet, during the existence of the Soviet Union, not one new drug was developed there that came into routine clinical use.

The Rate of Progress

Although books on the brain and consciousness are appearing at an alarming rate, it is surely a disservice to the field to pretend that answers to exceedingly difficult questions are just around the corner. For example, if one considers the circuitry necessary for playing tennis, it seems somewhat rash to have prophesied in 1996 (in an editorial in *Science*) that "the main principles of neural development will have been discovered by the end of this century."[7] There is a natural tendency for scientists and journalists to be optimistic and to offer hope that solutions for difficult problems are imminent. Thus, the time required for the cure for spinal cord injuries has been stated on occasion to be 7 or 10 years (more than 7 years ago!). Although such pronouncements might well encourage neuroscientists interested in the field, they can have a devastating effect on patients if they are not fulfilled in the time specified, as unfortunately is often the case.[8]

Conclusions

When one is faced with the fantastic range of animal behavior, from navigation by an ant to the reading of a textbook by a student, it is clear that understanding how the nervous system works is a fascinating, open-ended task of first-order interest in its own right.

An obvious inference from history is that approaches to the treatment of diseases often arise unexpectedly from experiments devoted to quite different questions. Further, an increase of natural knowledge on its own is a worthy objective, for without it, logical approaches to prevention and treatment of neurological problems can be only partially realized. In this context, it is almost impossible to define the "relevance" of any particular project at the time it is undertaken. Indeed, when asked about the "significance" of a research plan, complete honesty usually requires a very simple answer: "Don't know"!

Quite apart from the treatment of disease, the dividends that can accrue for society as a result of understanding the development and the functions of the nervous system are beyond today's imagination.

[7]Raff, M. 1996. *Science* 274: 1063.

[8]Krauthammer, C. 2000. *Time* February 14: 76.

SUGGESTED READING

Adams, R. D., Victor, M., and Ropper, A. H. 1997. *Principles of Neurology*, 6th Ed. McGraw-Hill, New York.

Burgess, P. W., and Shallice, T. 1996. Bizarre responses, rule detection and frontal lobe lesions. *Cortex* 32: 241–259.

Crick, F. 1999. The impact of molecular biology on neuroscience. *Philos. Trans. R. Soc. Lond. B* 354: 2021–2025.

Goldman-Rakic, P. S. 1999. The physiological approach: Functional architecture of working memory and disordered cognition in schizophrenia. *Biol. Psychiatry* 46: 650–661.

Henson, R., Shallice, T., and Dolan, R. 2000. Neuroimaging evidence for dissociable forms of repetition priming. *Science* 287: 1269–1272.

Jeannerod, M., and Frak, V. 1999. Mental imaging of motor activity in humans. *Curr. Opin. Neurobiol.* 9: 735–739.

Mountcastle, V. 1998. Brain science at the century's ebb. *Daedalus* 127: 1–36.

Zeki, S. 1999. Splendours and miseries of the brain. *Philos. Trans. R. Soc. Lond. B* 354: 2053–2065.

APPENDIX A
CURRENT FLOW IN ELECTRICAL CIRCUITS

A FEW BASIC CONCEPTS ARE REQUIRED to understand the electrical circuits used in this presentation. For our purposes it is sufficient to describe the properties of circuit elements and explain how they work when connected together in ways that correspond to the circuits described for nerves. The difficulties sometimes encountered on first reading accounts of electrical circuits often stem from the apparently abstract nature of the forces and movements involved. It is reassuring, therefore, to realize that many of the original pioneers in the field must have been faced with similar problems, since the terms devised in the last century are mainly related to the movement of fluids. Thus, the words "current," "flow," "potential," "resistance," and "capacitance" apply equally well to both electricity and hydraulics. The analogy between the two systems is illustrated by the fact that complex problems in hydraulics may be solved by using solutions to equivalent electrical circuits.

The analogy between a simple electrical circuit and its hydraulic equivalent is illustrated in Figure 1. The first point to be made is that a source of energy is required to keep the current flowing. In the hydraulic circuit, it is a pump; in the electrical circuit, a battery. The second point is that neither water nor electrical charge is created or lost within such a system. Thus, the flow rate of water is the same at points a, b, and c in the hydraulic circuit, since no water is added or removed between them. Similarly the electrical current in the equivalent circuit is the same at the three corresponding points. In both circuits, there are a number of *resistances* to current flow. In the hydraulic circuit, such resistance is offered by narrow tubes; similarly, thinner wires offer greater resistance to electrical current flow.

Terms and Units Describing Electrical Currents

The unit used to express rate of flow is to some extent a matter of choice; one can measure flow of water through a pipe in cubic feet per minute, for example, although in some other situation milliliters per hour might be more suitable. Electrical current flow is conventionally measured in **coulombs/sec** or **amperes** (abbreviated A). One coulomb is equal to the charge carried by 6.24×10^{18} electrons. In electrical circuits and equations, current is usually designated by I or i. As with flow of water, flow of current is a vector quantity, which is just a way of saying that it has a specified direction. The direction of flow is often indicated by arrows, as in Figure 1, current always being assumed to flow from the positive to the negative pole of a battery.

What do *positive* and *negative* mean with regard to current flow? Here the hydraulic analogy does not help. It is useful instead to consider the effects of passing current through a chemical solution. For example, suppose two copper wires are dipped into a solution of copper sulfate and connected to the positive and negative poles of a battery. Copper ions in solution are repelled from the positive wire, move through the solution, and are deposited from the solution onto the negative wire. In short, positive ions move in the direction conventionally designated for current: from positive to negative in the circuit. At the same time, sulfate ions move in the opposite direction and are deposited onto the positive wire. The direction specified for current, then, is the direction in which positive charges move in the circuit; negative charges move in the reverse direction.

To explain the energy source for current flow and the meaning of electrical **potential**, the hydraulic analogy is again useful. The flow of fluid depicted in Figure 1 depends on a pressure difference, the direction of flow being from high to low. No net movement occurs between two parts of the circuit at the same pressure. The overall pressure in the cir-

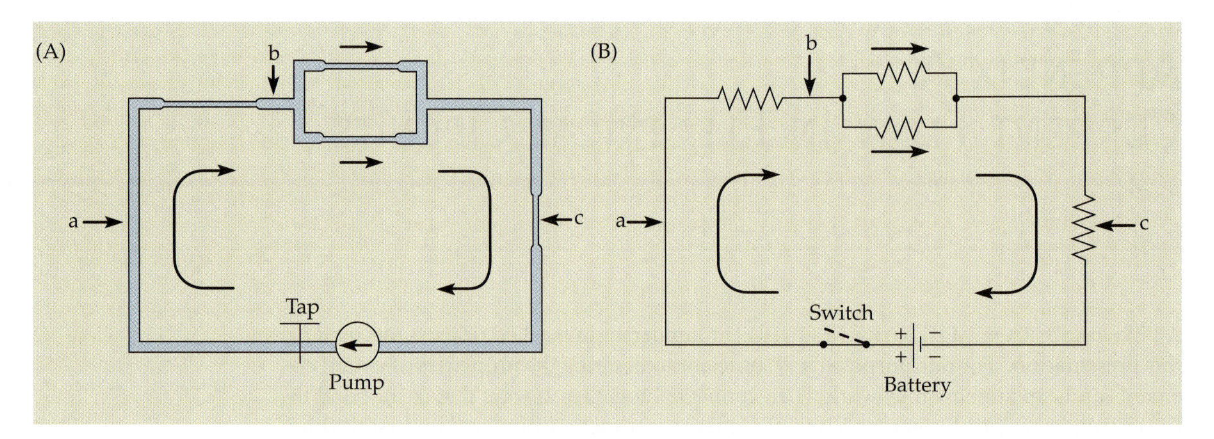

FIGURE 1 Hydraulic and Electrical Circuits. (A,B) Corresponding circuits for the flow of water and of electrical current. A battery is analogous to a pump that operates at constant pressure, the switch to a tap in the hydraulic line, and resistors to constrictions in the tubes.

cuit is supplied by expenditure of energy in driving the pump. In the electrical circuit shown here, the electrical "pressure" or **potential** is provided by a **battery** in which chemical energy is stored. Hydraulic pressure is measured in gm/cm^2; electrical potential is measured in **volts.**

Symbols used in electrical circuit diagrams and arrangements of circuit elements in series and in parallel are illustrated in Figure 2. As the names imply, a **voltmeter** measures electrical potential and is equivalent to a pressure gauge in hydraulics; an **ammeter** measures current flowing in a circuit and is equivalent to a flowmeter.

Ohm's Law and Electrical Resistance

In hydraulic systems, at least under ideal circumstances, the amount of current flowing through the system increases with pressure. The factor that determines the relation between pressure and flow rate is an inherent characteristic of the pipes, their **resistance.** Small-diameter, long pipes have greater resistances than large-diameter, short ones. Similarly, current flow in electrical circuits depends on the resistance in the circuit. Again, small, long wires have larger resistances than large, short ones. If current is being passed through an ionic solution, the resistance of the solution will increase as the solution is made more dilute. This is because there are fewer ions available to carry the current. In conductors such as wires, the relation between current and potential difference is de-

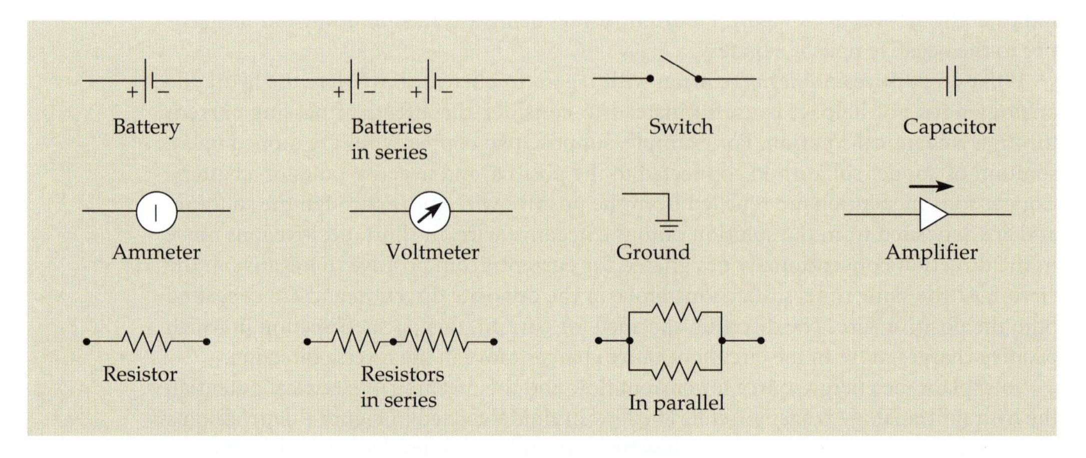

FIGURE 2 Symbols in Electrical Circuit Diagrams

scribed by Ohm's law, formulated by Ohm in the 1820s. The law says that the amount of current (I) flowing in a conductor is related to the potential difference (V) applied to it, $I = V/R$. The constant R is the resistance of the wire. If I is in amperes and V is in volts, then R is in units of **ohms** (Ω). The reciprocal of resistance is called conductance, and is a measure of the ease with which current flows through a conductor. Conductance is indicated by $g = 1/R$; the units of conductance are **siemens** (S). Thus, Ohm's law may also be written $I = gV$.

Use of Ohm's Law in Understanding Circuits

Ohm's law holds whenever the graph of current against potential is a straight line. In any circuit or part of a circuit for which this is true, any one of the three variables in the equation may be calculated if the other two are known. For example,

1. We can pass a known current through a nerve membrane, measure the change in potential, and then calculate the membrane resistance ($R = V/I$).
2. If we measure the potential difference produced by an unknown current and know the membrane resistance, we can calculate the applied current ($I = V/R$).
3. If we pass a known current through the membrane and know its resistance, then we can calculate the change in potential ($V = IR$).

Two additional simple, but important, rules (Kirchoff's laws) should be mentioned:

1. The algebraic sum of all the battery voltages is equal to the algebraic sum of all the IR voltage drops in a loop. An example of this is shown in Figure 3B: $V = IR_1 + IR_2$ (this is a statement of the conservation of energy).
2. The algebraic sum of all the currents flowing toward any junction is zero. For example, at point a in Figure 4, $I_{total} + I_{R1} + I_{R3} = 0$, which means that I_{total} (arriving) $= -I_{R1} - I_{R3}$ (leaving) (this is merely a statement that charge is neither created nor destroyed anywhere in the circuit).

We can now examine in more detail the circuits of Figures 3 and 4, which are needed to construct a model of the membrane. Figure 3A shows a battery (V) of 10 V connected to a resistance (R) of 10 Ω. The switch S can be opened or closed, thereby interrupting or establishing current flow. The voltage applied to R is 10 V; therefore the current measured by the ammeter, I, is, by Ohm's law, 1.0 A. In Figure 3B, the resistor is replaced by two resistors, R_1 and R_2, **in series**. By the first of Kirchoff's laws, the current flowing into point b must be equal to that leaving. Therefore, the same current, I, must flow through both the resistors. By the second of Kirchoff's laws, then, $IR_1 + IR_2 = V$ (10 V). It follows that the current, $I = V/(R_1 + R_2) = 0.5$ A. The voltage at b, then, is 5 V positive to that at c and a is 5 V positive to b. Note that because there is only one path for the current, the total resistance, R_{total}, seen by the battery is simply the sum of the two resistors; that is,

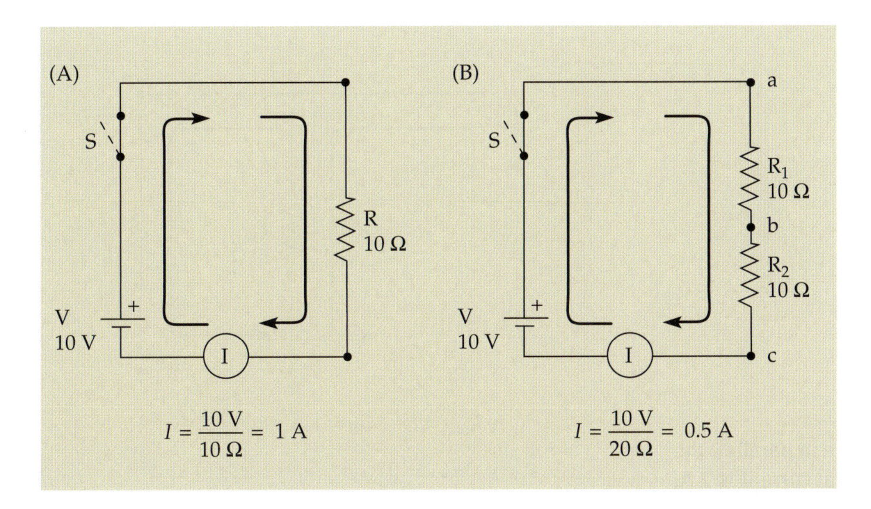

FIGURE 3 Ohm's Law applied to simple circuits. (A) Current I = (10 V)/(10 Ω) = 1 A. (B) Current = (10 V)/(20 Ω) = 0.5 A, and the voltage across each resistor is 5 V.

$$R_{total} = R_1 + R_2$$

What happens if, as shown in Figure 4, we add a second resistor, also of 10 Ω, **in parallel** rather than in series? In the circuit, the two resistors R_1 and R_3 provide two separate pathways for current. Both have a voltage V (10 V) across them, so the respective currents will be

$$I_{R_1} = V/R_1 = 1\text{ A}$$
$$I_{R_3} = V/R_3 = 1\text{ A}$$

Therefore, to satisfy the first of Kirchoff's laws, there must be 2 A arriving at point a and 2 A leaving point b. The ammeter, then, will read 2 A. Now the combined resistance of R_1 and R_3 is $R_{total} = V/I = (10\text{ V})/(2\text{ A}) = 5\ \Omega$, or half that of the individual resistors. This makes sense if one thinks of the hydraulic analogy: Two pipes in parallel will offer less resistance to flow than one pipe alone. In the parallel electrical circuit the *conductances* add: $g_{total} = g_1 + g_3$, or $1/R_{total} = 1/R_1 + 1/R_3$.

If we now generalize to any number (n) of resistors, resistances in series simply add:

$$R_{total} = R_1 + R_2 + R_3 + \ldots + R_n$$

and in parallel their reciprocals add:

$$1/R_{total} = 1/R_1 + 1/R_2 + 1/R_3 + \ldots + 1/R_n$$

Applying Circuit Analysis to a Membrane Model

Figure 5A shows a circuit similar to that used to represent nerve membranes. Notice that the two batteries drive current around the circuit in the same direction and that the resistors R_1 and R_2 are in series. What is the potential difference between points b and d (which represent the outside and inside of the membrane)? The total potential across the two resistors between a and c is 150 mV, a being positive to c. Therefore, the current flowing between a and c through the resistors is 150 mV/100,000Ω = 1.5 μA. When 1.5 μA flows across 10,000 Ω, as between a and b, a potential drop of 15 mV is produced, a being positive with respect to b. The potential difference between the inside and the outside is therefore 100 mV – 15 mV = 85 mV. We can obtain the same result by considering the voltage drop across R_2 (1.5 μA × 90,000 Ω = 135 mV) and adding it to V_2 (135 mV – 50 mV = 85 mV). This *must* be so, as the potential between b and d must have a unique value.

In Figure 5B, R_1 and R_2 have been exchanged. As the total resistance in the circuit is the same, the current must be the same, 1.5 μA. Now the potential drop across R_2, between a and b, is 90,000 Ω × 1.5 μA = 135 mV, a being positive to b. Now the potential across the membrane is 100 mV – 135 mV = –35 mV, **outside negative**; the same result can, of course, be obtained from the current through R_1. This simple circuit illustrates an important point about membrane physiology: *The potential across a membrane can change as a result of resistance changes while the batteries remain unchanged.* A general expression for the membrane potential in the circuit shown in Figure 5A can be derived simply, as follows:

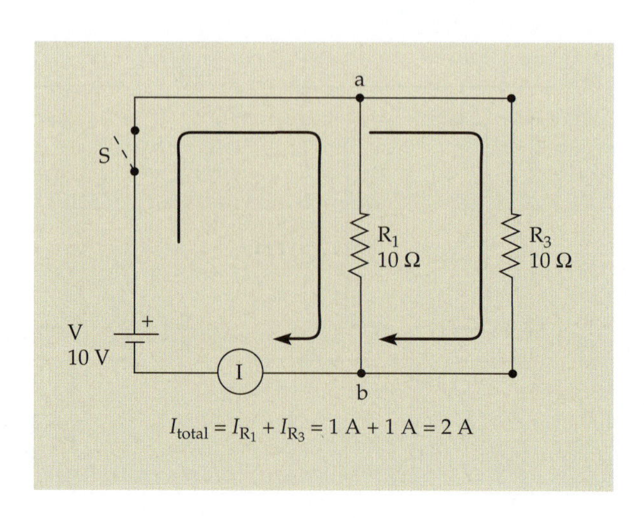

FIGURE 4 Parallel Resistors. When R_1 and R_3 are in parallel, the voltage drop across each resistor is 10 V and the total current is 2 A.

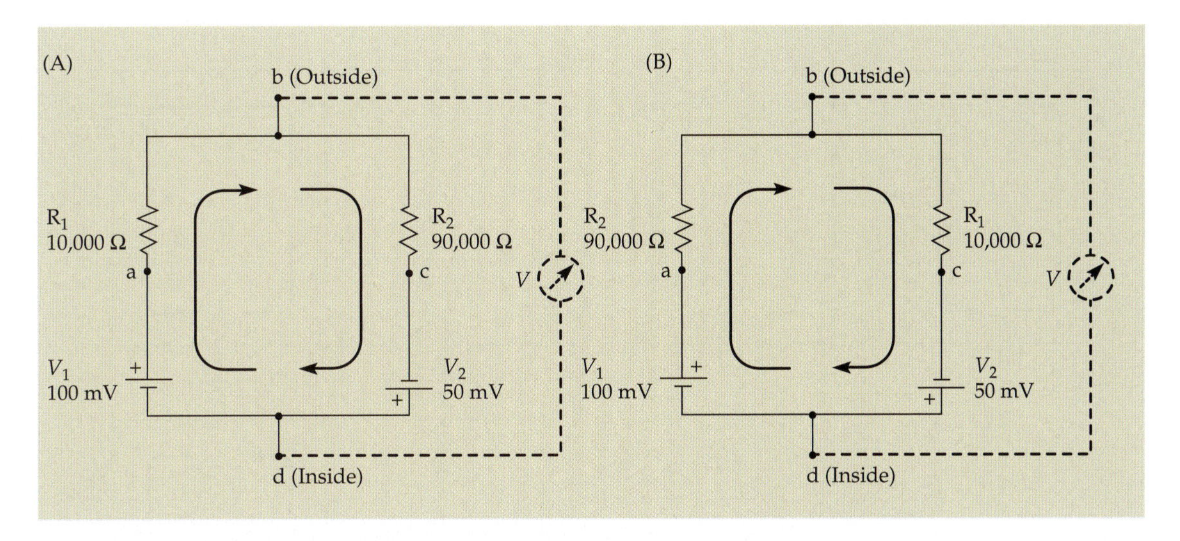

FIGURE 5 Analogue Circuits for Nerve Membranes. In A and B the resistors R_1 and R_2 are reversed; otherwise the circuits are the same. The batteries V_1 and V_2 are in series. In (A), point b (the "outside" of the membrane) is positive with respect to d (the "inside") by 85 mV; in (B) it is negative by 35 mV. These circuits illustrate how changes in resistance can give rise to membrane potential changes even though the batteries (which represent ionic equilibrium potentials) remain constant.

$$V_m = V_1 - IR_1$$

As $I = (V_1 + V_2)/(R_1 + R_2)$:

$$V_m = V_1 - \frac{(V_1 + V_2)R_1}{R_1 + R_2}$$

On rearranging:

$$V_m = \frac{V_1R_2/R_1 - V_2}{1 + R_2/R_1}$$

Electrical Capacitance and Time Constant

In the circuits described in Figures 3 and 4, closing or opening the switch produces instantaneous and simultaneous changes in current and potential. Capacitors introduce a time element into the consideration of current flow. They accumulate and store electrical charge, and, when they are present in a circuit, current and voltage changes are no longer simultaneous. A capacitor consists of two conducting plates (usually of metal) separated by an insulator (air, mica, oil, or plastic). When voltage is applied between the plates (Figure 6A), there is an instantaneous displacement of charge from one plate to the other through the external circuit. Once the capacitor is fully charged, however, there is no further current, as none can flow across the insulator. The **capacitance** (C) of a capacitor is defined by how much charge (q) it can store for each volt applied to it:

$$C = q/V$$

The units of capacitance are coulombs/volt or **farads** (F). The larger the plates of a capacitor and the closer together they are, the greater its capacitance. A one-farad capacitor is very large; capacitances in common use are in the range of microfarads (μF) or smaller.

When the switch in Figure 6A is closed, then, there is an instantaneous charge separation at the plates. The amount of charge stored in the capacitor is proportional to its capacitance and to the magnitude of the applied voltage (V_0). When the switch is opened, as in B, the charge on the capacitor remains, as does the voltage (V) between the plates. (One can sometimes get a surprising shock from electronic apparatus after it has been turned off because some of the capacitors in the circuits may remain charged.) The capacitor can

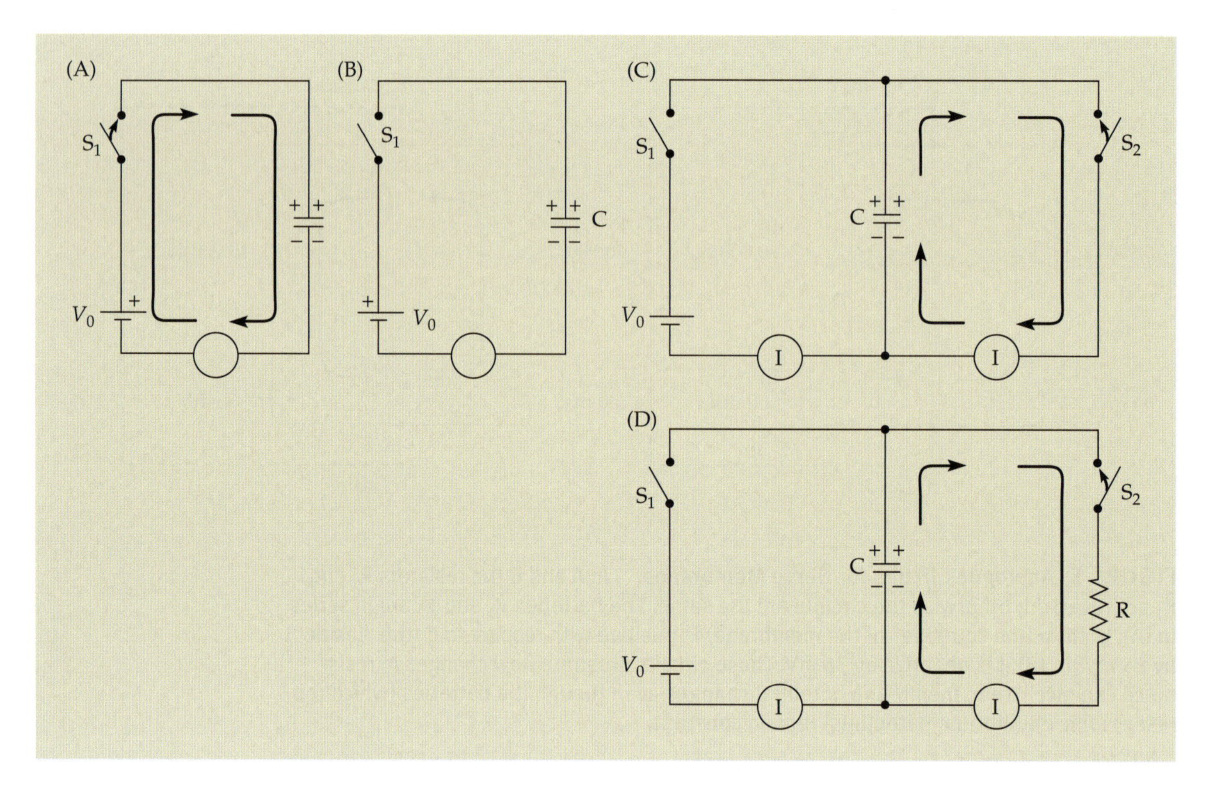

FIGURE 6 Capacitors in electrical circuits. A, B, and C are idealized circuits having no resistance. When S_1 is closed in (A), the capacitor is charged instantaneously to voltage V_0. If S_1 is then opened (B), the potential remains on the capacitor. Closing switch S_2 (C) discharges the capacitor instantaneously. In (D) the capacitor is discharged through resistor R. The maximum discharge current is $I = V_0/R$.

be discharged by shorting it with a second switch, as in Figure 6C. Again, the current flow is instantaneous, returning the charge and the voltage on the capacitor to zero. If, instead, the capacitor is discharged through a resistor (R, Figure 6D), the discharge is no longer instantaneous. This is because the resistor limits the current flow. If the voltage on the capacitor is V, then by Ohm's law the maximum current is $I = V/R$. With no resistor in the circuit, the current becomes infinitely large and the capacitor is discharged in an infinitesimal time period; if R is very large, the discharge process takes a very long time. The rate of discharge at any given time, dq/dt, is the current flowing at that particular time. In other words, $dq/dt = -V/R$ (negative because the charge is decreasing with time), where V initially is equal to the battery voltage and decreases as the capacitor is discharged. As $q = CV$, $dq/dt = CdV/dt$, and we can then write $CdV/dt = -V/R$, or

$$dV/dt = -V/RC$$

The equation says that the rate of loss of voltage from the capacitor is proportional to the voltage remaining. Thus, as the voltage decreases, the rate of discharge decreases. The constant of proportionality, $1/RC$, is the **rate constant** for the process: RC is its **time constant**. This kind of process arises over and over again in nature. For example, the rate at which water drains from a bathtub decreases as the depth, and hence the pressure at the drain, decreases. In this kind of situation, the discharge process is described by an exponential function:

$$V = V_0\, e^{-t/\tau}$$

where V_0 is the initial charge on the capacitor and the time constant $\tau = RC$. Similarly, when the capacitor is charged through a resistor, as in Figure 7, the charging process takes a finite time. The voltage between the plates increases with time until the battery voltage is reached and no further current flows. The charging process is now a rising exponential, with a time constant $\tau = RC$:

$$V = V_0(1 - e^{-t/\tau})$$

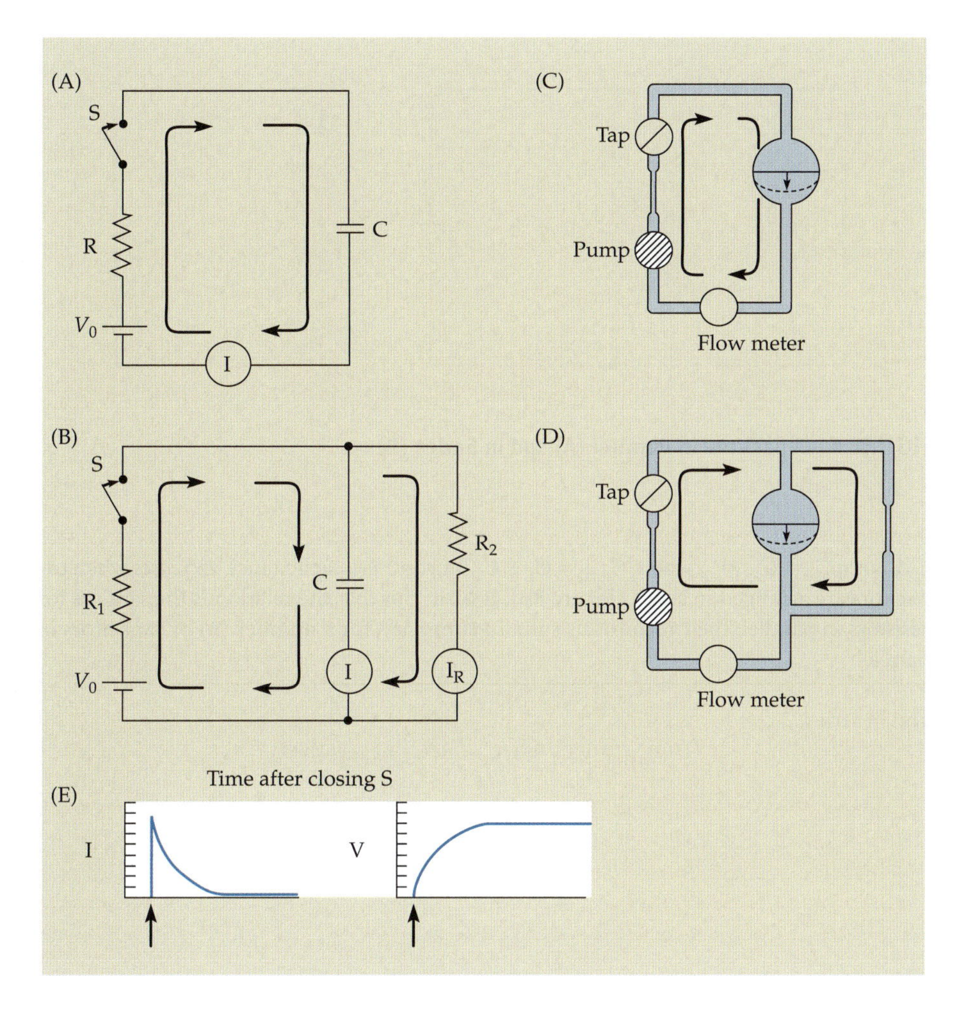

FIGURE 7 Charging a Capacitor. In (A) the capacitor is charged at a rate limited by the resistor, the initial rate being $I = V_0/R$. In (B) the charging rate depends on both resistors in the circuit. In (E) the capacitative current and the voltage across the capacitor are shown as functions of time. The voltage reaches its final value only when the capacitor is fully charged, i.e., when no more current flows into the capacitor. (C) and (D) are hydraulic analogues of the circuits in A and B.

These examples illustrate another property of a capacitor. Current flows into and out of the capacitor only when the potential is changing:

$$I_c = dQ/dt = CdV/dt$$

When the voltage across the capacitor is steady ($dV/dt = 0$), the capacitative current, I_c, is zero. In other words, the capacitance has an "infinite resistance" for a steady potential difference and a "low resistance" for a rapidly changing potential. Figure 7B shows a circuit in which current flows through a resistor and capacitor in parallel and Figure 7E the time courses of the capacitative current and voltage.

The properties of a capacitor in a circuit can be illustrated by the slightly more elaborate hydraulic analogy shown in Figure 7C. The capacitor is represented by an elastic diaphragm that forms a partition in a fluid-filled chamber. When the tap is opened, fluid is pumped from one side of the chamber to the other. The pressure generated by the pump causes the diaphragm to bulge. Fluid continues to flow until, because of its elasticity, the diaphragm provides an equal and opposite pressure; then there is no more fluid flow and the chamber is fully charged. If a tube is placed alongside, as in Figure 7D, some fluid flows through the tube and some is used to expand the diaphragm. The rate of expansion depends on the resistance of the tube, and on the capacity of the chamber. If the tube is of high resistance, then for a given flow the pressure difference between its two ends will be relatively large. In that case, the distention of the diaphragm will be large and take a relatively long time to achieve. Similarly, if the capacity of the chamber is larger, more fluid is diverted during the filling (or "charging") process and a longer time is required to reach a steady state. Thus, the characteristic time constant of the system is determined by the product of resistance and capacitance.

When capacitors are arranged in parallel, as in Figure 8A, the total capacitance is increased. The total charge stored is the sum of the charges stored in each: $q_1 + q_2 = C_1V_0 +$

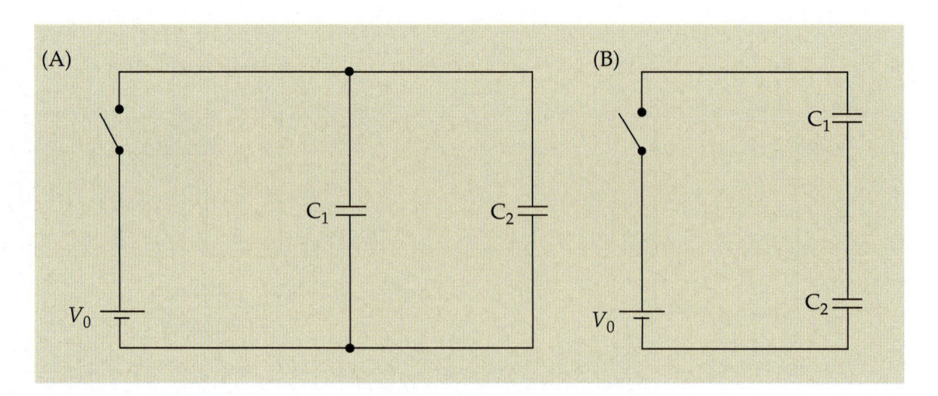

FIGURE 8 Capacitors in Parallel (A) and in Series (B)

C_2V_0 or $q_{total} = C_{total}V_0$, where $C_{total} = C_1 + C_2$. In contrast, capacitance *decreases* when capacitors are arranged in series (Figure 8B). It turns out that the relation is the same as for resistors in parallel: their reciprocals sum. In summary, for a number (n) of capacitors in parallel:

$$C_{total} = C_1 + C_2 + C_3 + \ldots + C_n$$

and in series,

$$1/C_{total} = 1/C_1 + 1/C_2 + 1/C_3 + \ldots + 1/C_n$$

Appendix B
Metabolic Pathways for the Synthesis and Inactivation of Low-Molecular-Weight Transmitters

THE FIGURES ON THE FOLLOWING PAGES summarize the predominant metabolic pathways for the low-molecular-weight transmitters acetylcholine, GABA, glutamate, dopamine, norepinephrine, epinephrine, 5-HT, and histamine. Glycine, purines, NO, and CO are not included; there appear to be no special neuronal pathways for their synthesis or degradation. For each metabolic step, the portion of the molecule being modified is highlighted in color. Further information can be found in several comprehensive texts:

Gilman, A. G., Rall, T., Nies, A. S., and Taylor, P. (Eds.). 1990. *Goodman and Gilman's The Pharmacological Basis of Therapeutics*, 8th Ed. Pergamon Press, New York.

Siegel, G. J., Agranoff, B. W., Albers, R. W., Fisher, S. K., and Uhler, M. D. (Eds.). 1999. *Basic Neurochemistry*, 6th Ed. Lippincott-Raven, New York.

Stryer, L. 1995. *Biochemistry*, 4th Ed., W. H. Freeman, New York.

Acteylcholine (ACh)

Synthesis

$$\underset{\text{Acetyl-CoA}}{H_3C-\overset{O}{\overset{\|}{C}}-S-CoA} + \underset{\text{Choline}}{HO-CH_2-CH_2-\overset{+}{N}-(CH_3)_3} \xrightleftharpoons{\text{Choline acetyltransferase}} \underset{\text{CoA}}{HS-CoA} + \underset{\text{Acetylcholine}}{H_3C-\overset{O}{\overset{\|}{C}}-O-CH_2-CH_2-\overset{+}{N}-(CH_3)_3}$$

Degradation

$$\underset{\text{Acetylcholine}}{H_3C-\overset{O}{\overset{\|}{C}}-O-CH_2-CH_2-\overset{+}{N}-(CH_3)_3} + H_2O \xrightleftharpoons{\text{Acetylcholinesterase}} \underset{\text{Acetate}}{H_3C-C(=O)O^-} + \underset{\text{Choline}}{HO-CH_2-CH_2-\overset{+}{N}-(CH_3)_3} + H^+$$

γ-Aminobutyric acid (GABA)

Synthesis

$$\underset{\text{Glutamate}}{{}^+H_3N-\overset{H}{\underset{COO^-}{C}}-CH_2-CH_2-COO^-} \xrightarrow[\;CO_2\;]{\text{Glutamic acid decarboxylase}} \underset{\gamma\text{-Aminobutyric acid (GABA)}}{{}^+H_3N-CH_2-CH_2-CH_2-COO^-}$$

Degradation

GLUTAMATE

Synthesis

Glutamine + H_2O —(Glutaminase)→ Glutamate + NH_4^+

$^{+}H_3N-CH(COO^-)-CH_2-CH_2-C(=O)NH_2$ (Glutamine) + H_2O → $^{+}H_3N-CH(COO^-)-CH_2-CH_2-C(=O)O^-$ (Glutamate) + NH_4^+

Degradation

Glutamate + NH_4^+ —(Glutamine synthetase; ATP → ADP + P_i)→ Glutamine + H^+

$^{+}H_3N-CH(COO^-)-CH_2-CH_2-C(=O)O^-$ (Glutamate) + NH_4^+ → $^{+}H_3N-CH(COO^-)-CH_2-CH_2-C(=O)NH_2$ (Glutamine) + H^+

CATECHOLAMINES: DOPAMINE

Synthesis

Tyrosine —(Tyrosine hydroxylase; O_2 + Tetrahydrobiopterin → H_2O + Dihydrobiopterin)→ 3,4-Dihydroxyphenylalanine (DOPA) —(Aromatic L-amino acid decarboxylase; → CO_2)→ Dopamine

Degradation

Dopamine —(COMT)→ 3-Methoxytyramine

Dopamine —(MAO)→ 3,4-Dihydroxy-β-phenylacetaldehyde

3,4-Dihydroxy-β-phenylacetaldehyde —(AR)→ 3,4-Dihydroxy-β-phenylethanol

3,4-Dihydroxy-β-phenylethanol —(COMT)→ 3-Methoxy-4-hydroxy-β-phenylethanol

3-Methoxytyramine —(MAO)→ 3-Methoxy-4-hydroxy-β-phenylacetaldehyde

3-Methoxy-4-hydroxy-β-phenylacetaldehyde —(AR)→ 3-Methoxy-4-hydroxy-β-phenylethanol

3,4-Dihydroxy-β-phenylacetaldehyde —(ADH)→ 3,4-Dihydroxyphenylacetic acid (DOPAC)

3,4-Dihydroxyphenylacetic acid (DOPAC) —(COMT)→ 3-Methoxy-4-hydroxyphenylacetic acid (HVA)

3-Methoxy-4-hydroxy-β-phenylacetaldehyde —(ADH)→ 3-Methoxy-4-hydroxyphenylacetic acid (HVA)

CATECHOLAMINES: NOREPINEPHRINE AND EPINEPHRINE

Synthesis

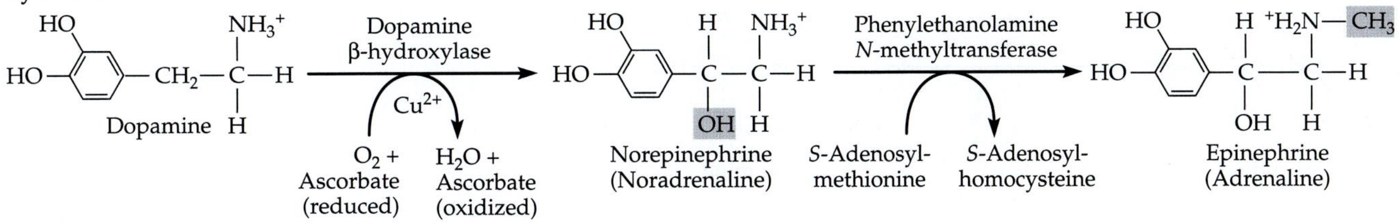

Dopamine → (Dopamine β-hydroxylase, Cu^{2+}; O_2 + Ascorbate (reduced) → H_2O + Ascorbate (oxidized)) → Norepinephrine (Noradrenaline) → (Phenylethanolamine N-methyltransferase; S-Adenosyl-methionine → S-Adenosyl-homocysteine) → Epinephrine (Adrenaline)

Degradation

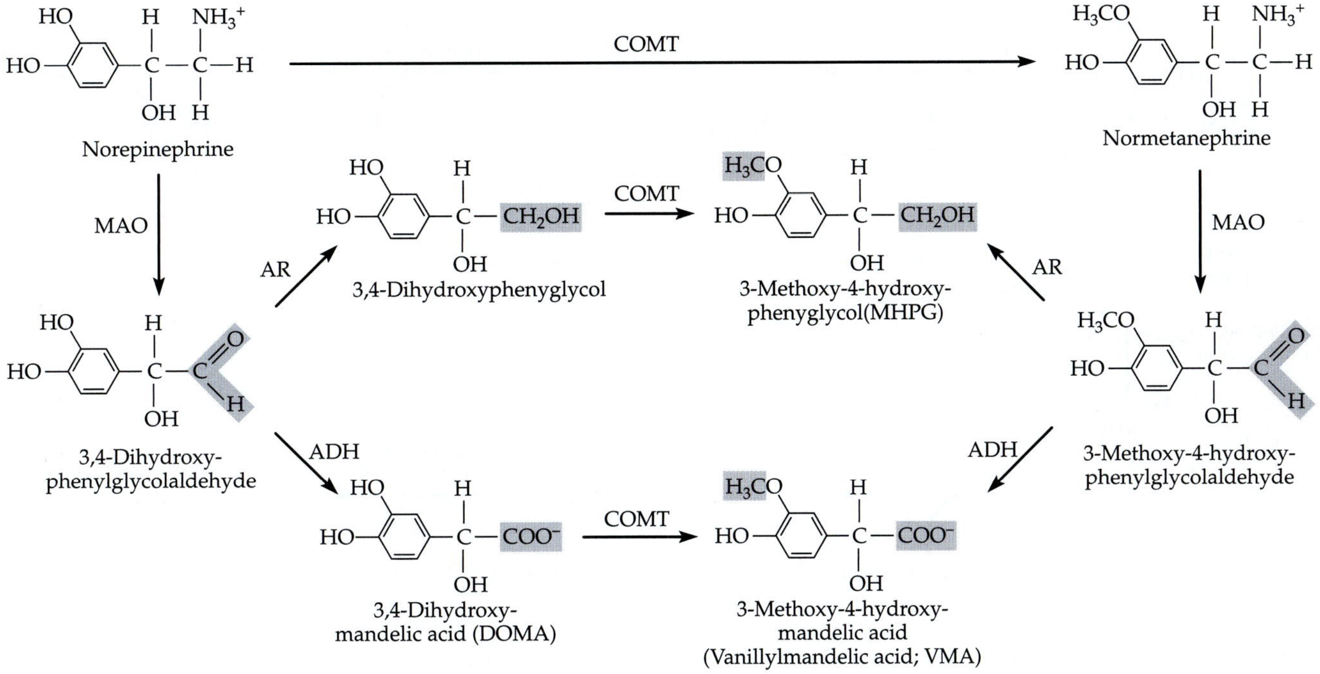

Norepinephrine → (COMT) → Normetanephrine

Norepinephrine → (MAO) → 3,4-Dihydroxy-phenylglycolaldehyde

3,4-Dihydroxy-phenylglycolaldehyde → (AR) → 3,4-Dihydroxyphenyglycol → (COMT) → 3-Methoxy-4-hydroxy-phenyglycol(MHPG)

3,4-Dihydroxy-phenylglycolaldehyde → (ADH) → 3,4-Dihydroxy-mandelic acid (DOMA) → (COMT) → 3-Methoxy-4-hydroxy-mandelic acid (Vanillylmandelic acid; VMA)

Normetanephrine → (MAO) → 3-Methoxy-4-hydroxy-phenylglycolaldehyde

3-Methoxy-4-hydroxy-phenylglycolaldehyde → (AR) → 3-Methoxy-4-hydroxy-phenyglycol(MHPG)

3-Methoxy-4-hydroxy-phenylglycolaldehyde → (ADH) → 3-Methoxy-4-hydroxy-mandelic acid (Vanillylmandelic acid; VMA)

5-HYDROXYTRYPTAMINE (5-HT; SEROTONIN)

Synthesis

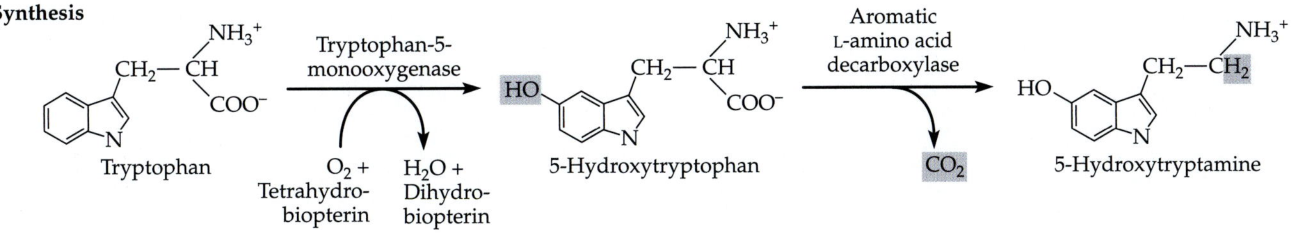

Tryptophan → (Tryptophan-5-monooxygenase; O_2 + Tetrahydro-biopterin → H_2O + Dihydro-biopterin) → 5-Hydroxytryptophan → (Aromatic L-amino acid decarboxylase; → CO_2) → 5-Hydroxytryptamine

Degradation

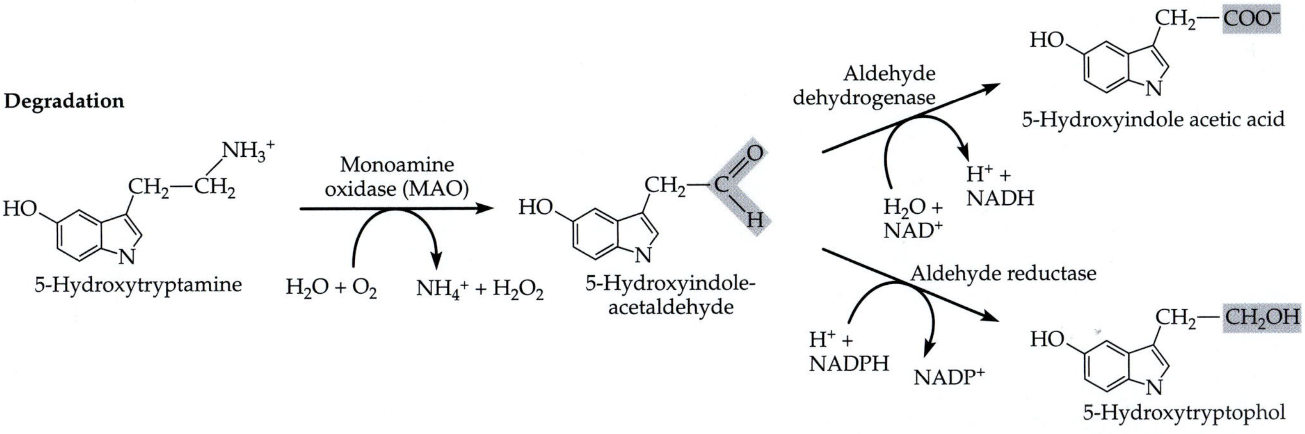

5-Hydroxytryptamine → (Monoamine oxidase (MAO); $H_2O + O_2$ → $NH_4^+ + H_2O_2$) → 5-Hydroxyindole-acetaldehyde

5-Hydroxyindole-acetaldehyde → (Aldehyde dehydrogenase; H_2O + NAD^+ → H^+ + NADH) → 5-Hydroxyindole acetic acid

5-Hydroxyindole-acetaldehyde → (Aldehyde reductase; H^+ + NADPH → $NADP^+$) → 5-Hydroxytryptophol

Histamine

Synthesis

NH_3^+

$CH_2—C—COO^-$

H

HN N

L-Histidine

Histidine decarboxylase

CO_2

NH_3^+

$CH_2—C—H$

H

HN N

Histamine

Degradation

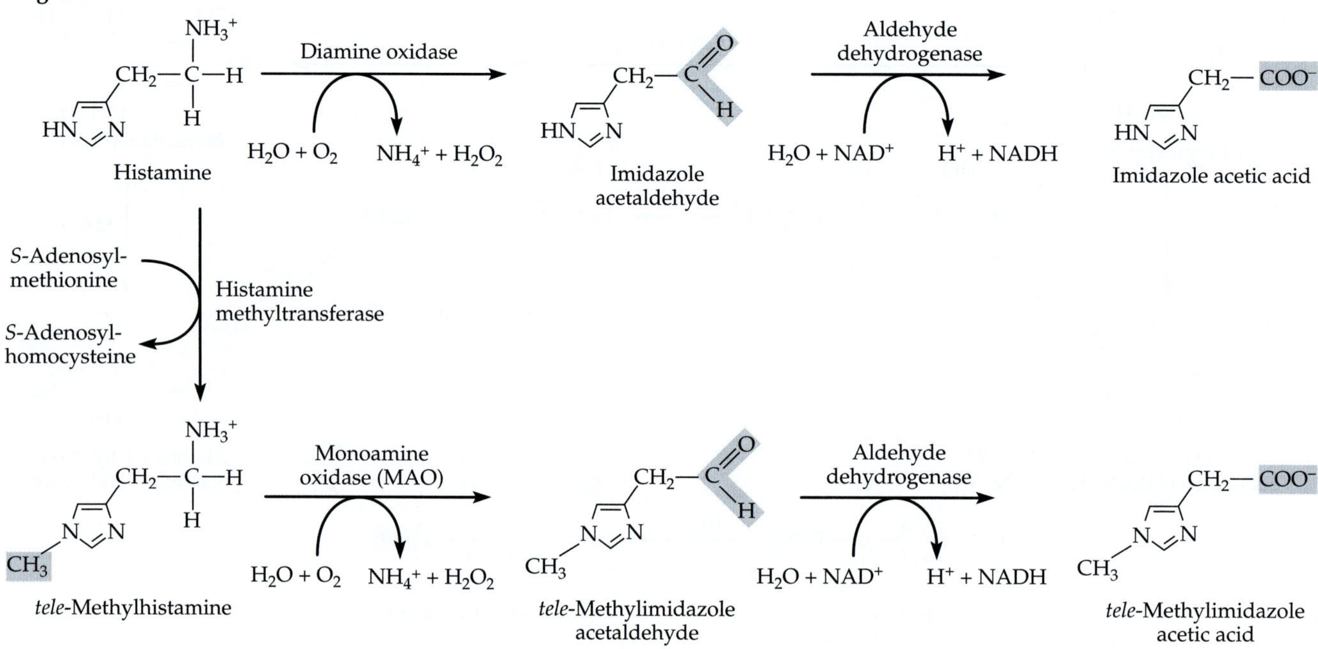

DEGRADATION OF BIOGENIC AMINES

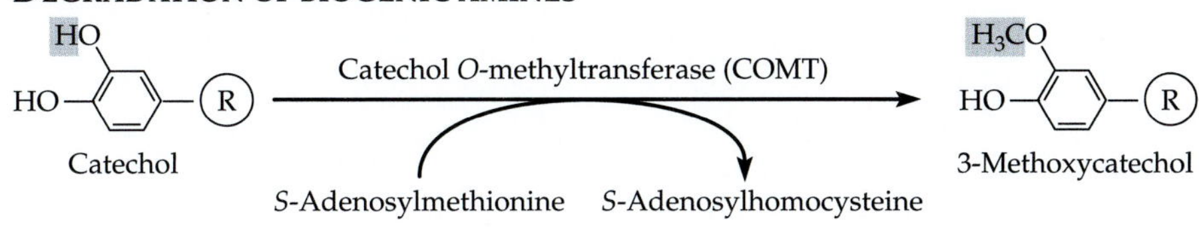

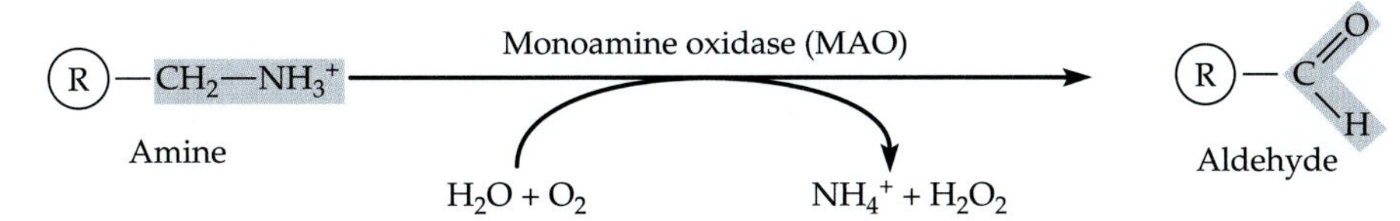

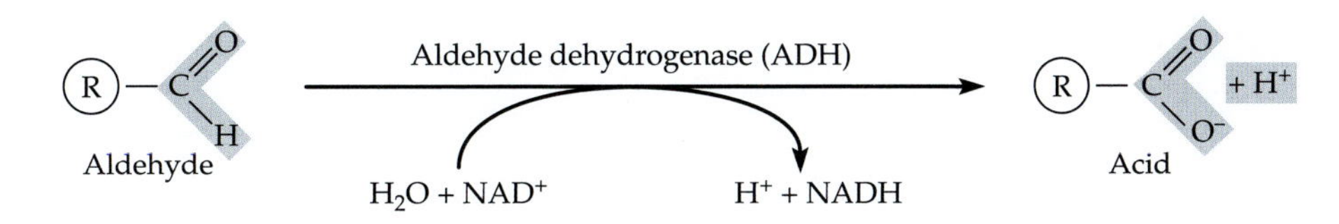

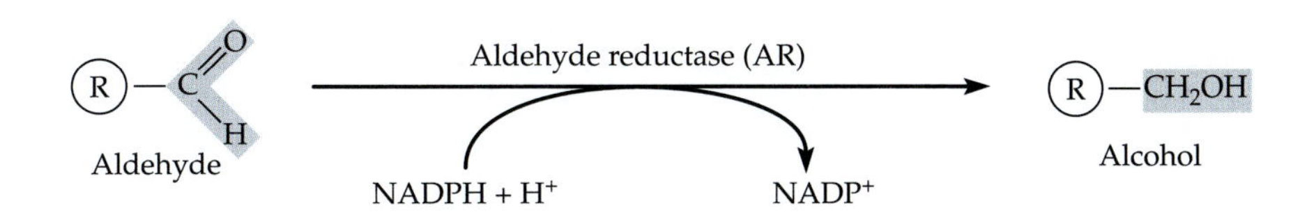

APPENDIX C
STRUCTURES AND PATHWAYS OF THE BRAIN

THE FOLLOWING FIGURES SHOW THE BRAIN viewed from different aspects and cut in different sections. The aim is to provide a visual equivalent of a glossary relating to material in the text, rather than to present a full atlas. Consequently, only key landmarks and structures are illustrated. Further anatomical information can be found in a number of comprehensive texts:

Carpenter, M. B. 1991. *Core Text of Neuroanatomy*, 4th Ed. Williams and Wilkins, Baltimore.

Martin, J. H. 1996. *Neuroanatomy: Text and Atlas*, 2nd Ed. Appleton and Lange, Stamford, CT.

Nolte, J. 1998. *The Human Brain: An Introduction to Its Functional Anatomy*, 4th Ed., Mosby-Year Book, St. Louis.

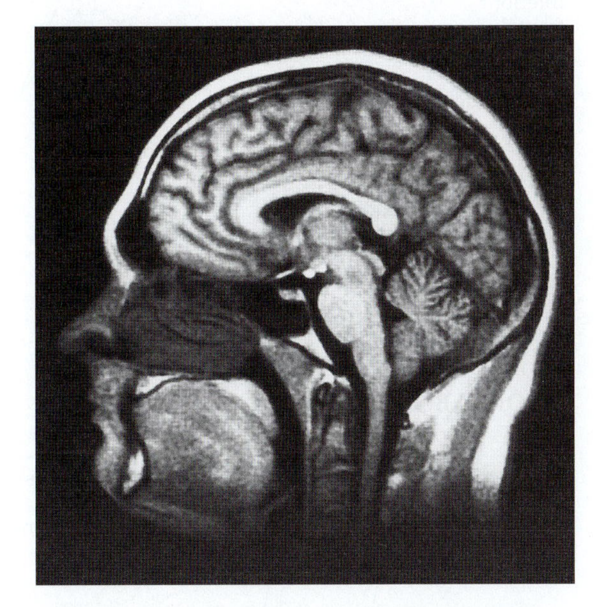

Magnetic resonance image of a living human brain (sagittal section). Copyright 1984 by the General Electric Company. Reproduced with permission.

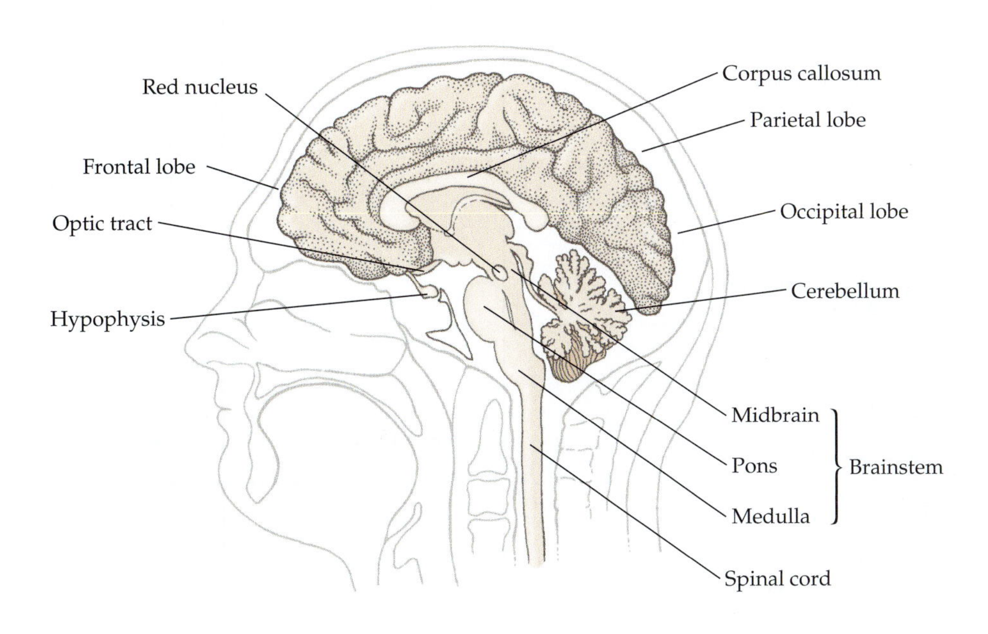

SIDE VIEW

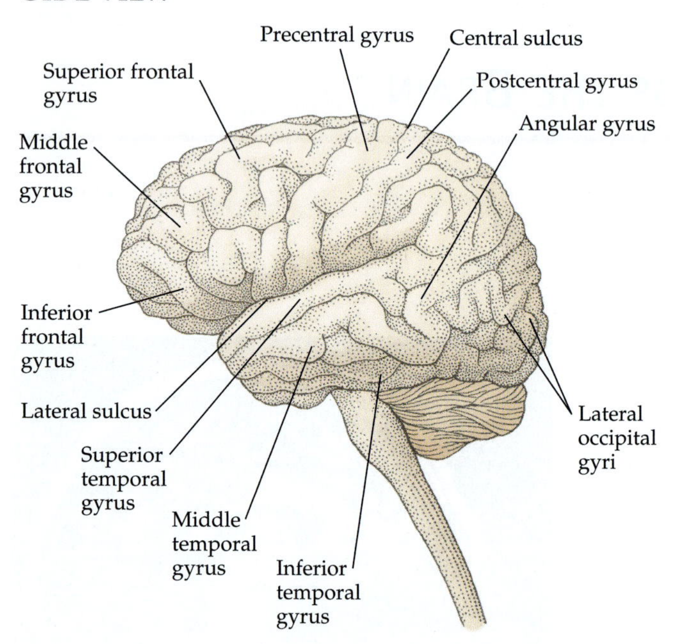

DIRECTIONAL TERMS

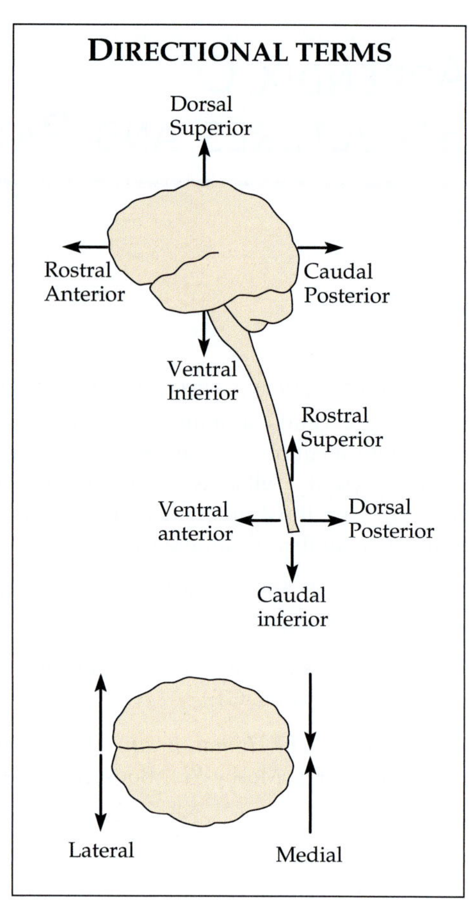

FROM ABOVE

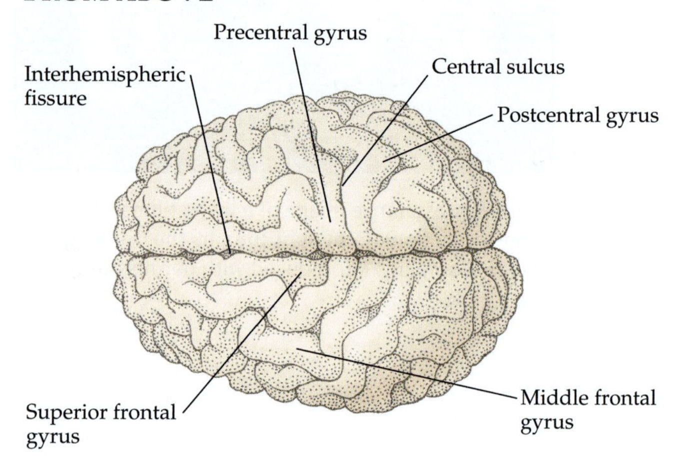

PLANES OF SECTION

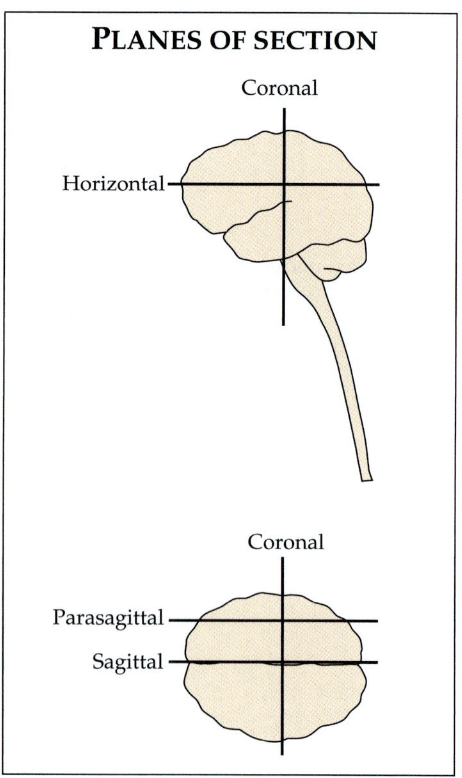

FROM BELOW

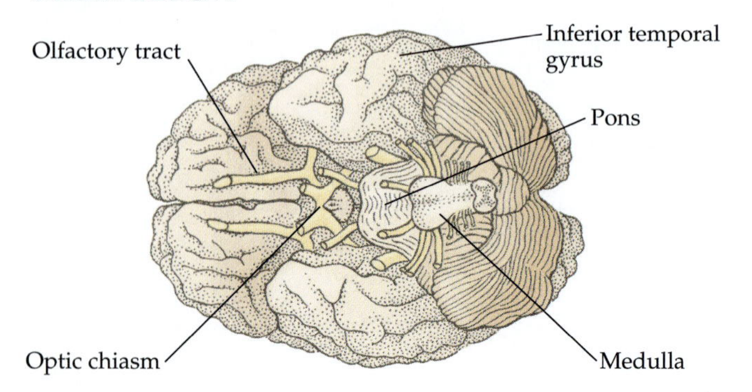

NUMBERED ANATOMICAL AREAS OF THE CEREBRAL CORTEX (BRODMANN'S AREAS)

LOCALIZATION OF MOTOR AND SENSORY FUNCTIONS

LATERAL VIEW

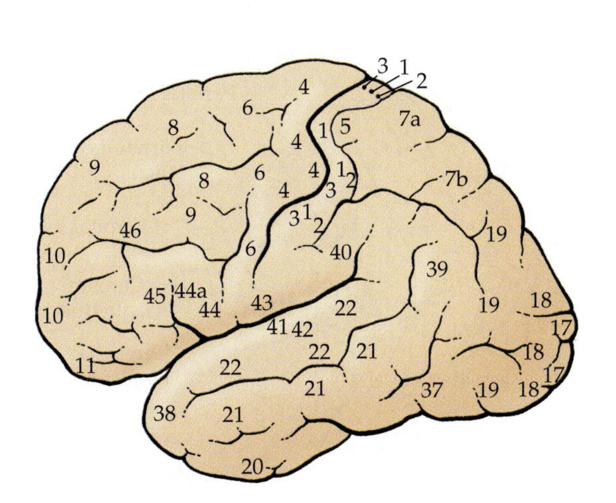

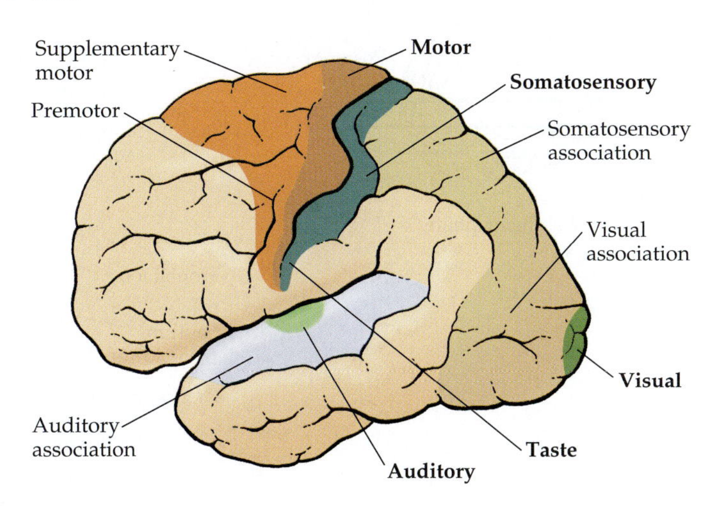

SAGITTAL VIEW

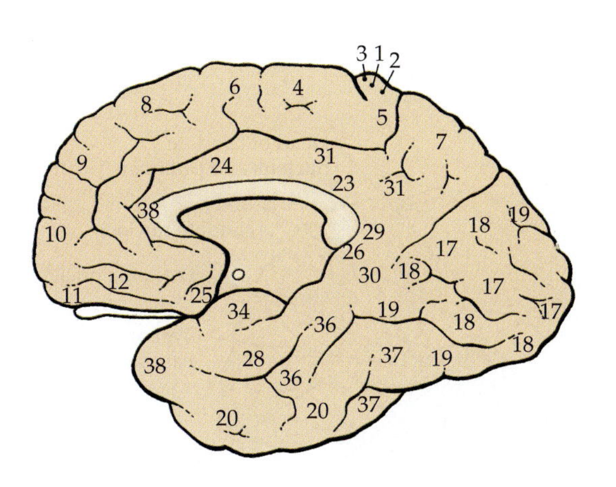

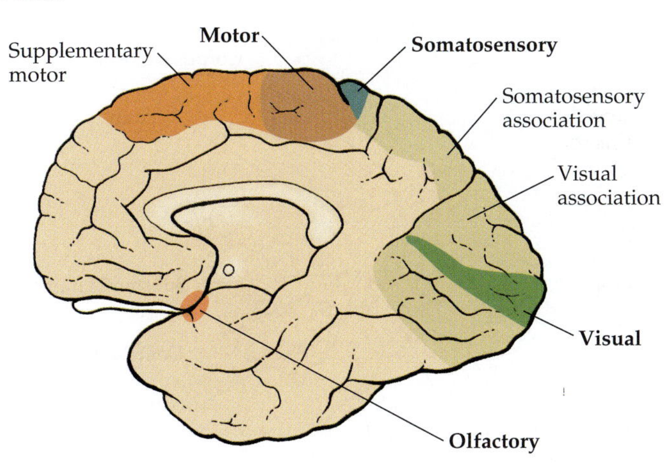

SAGGITAL SECTIONS

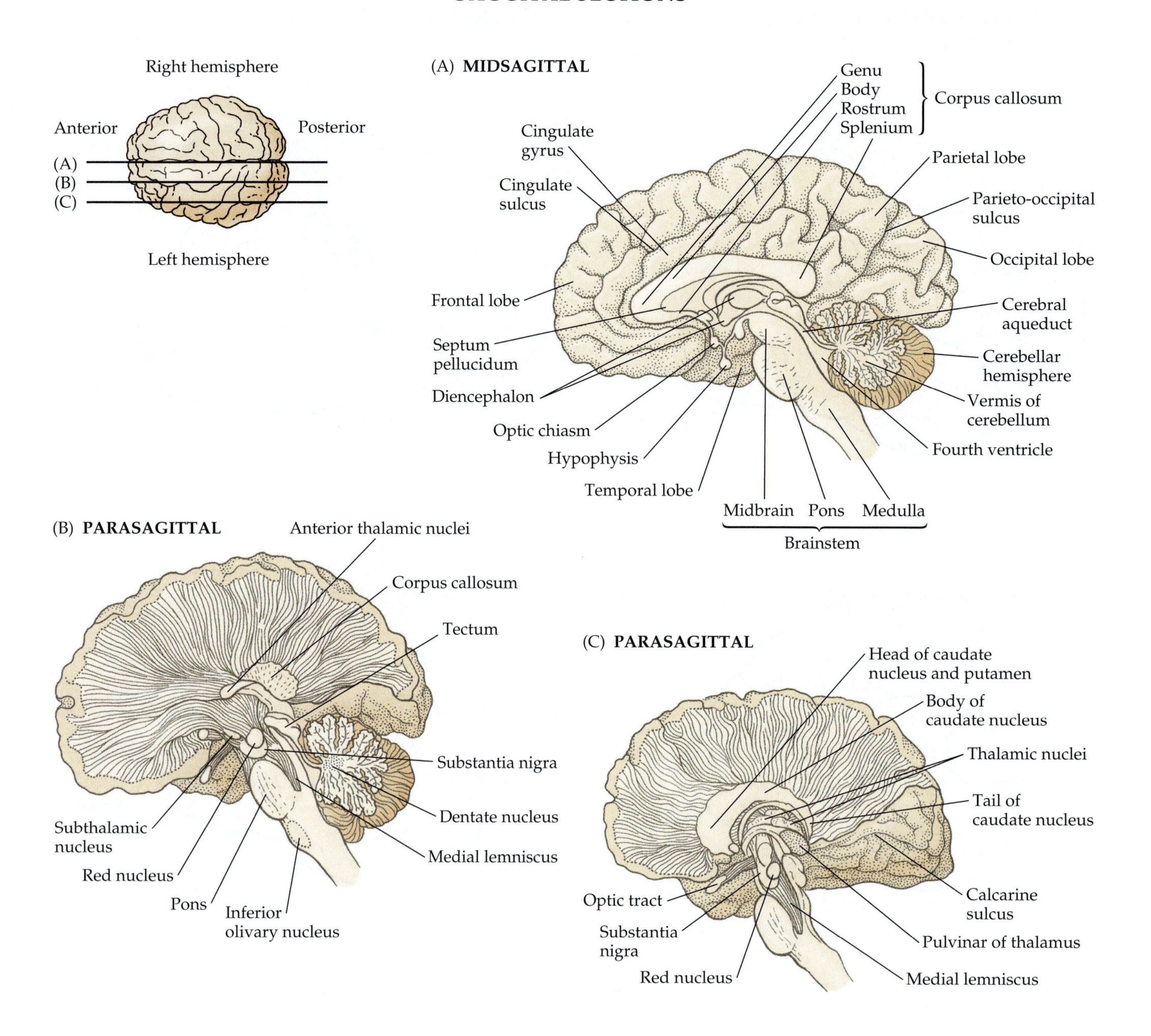

HORIZONTAL SECTIONS

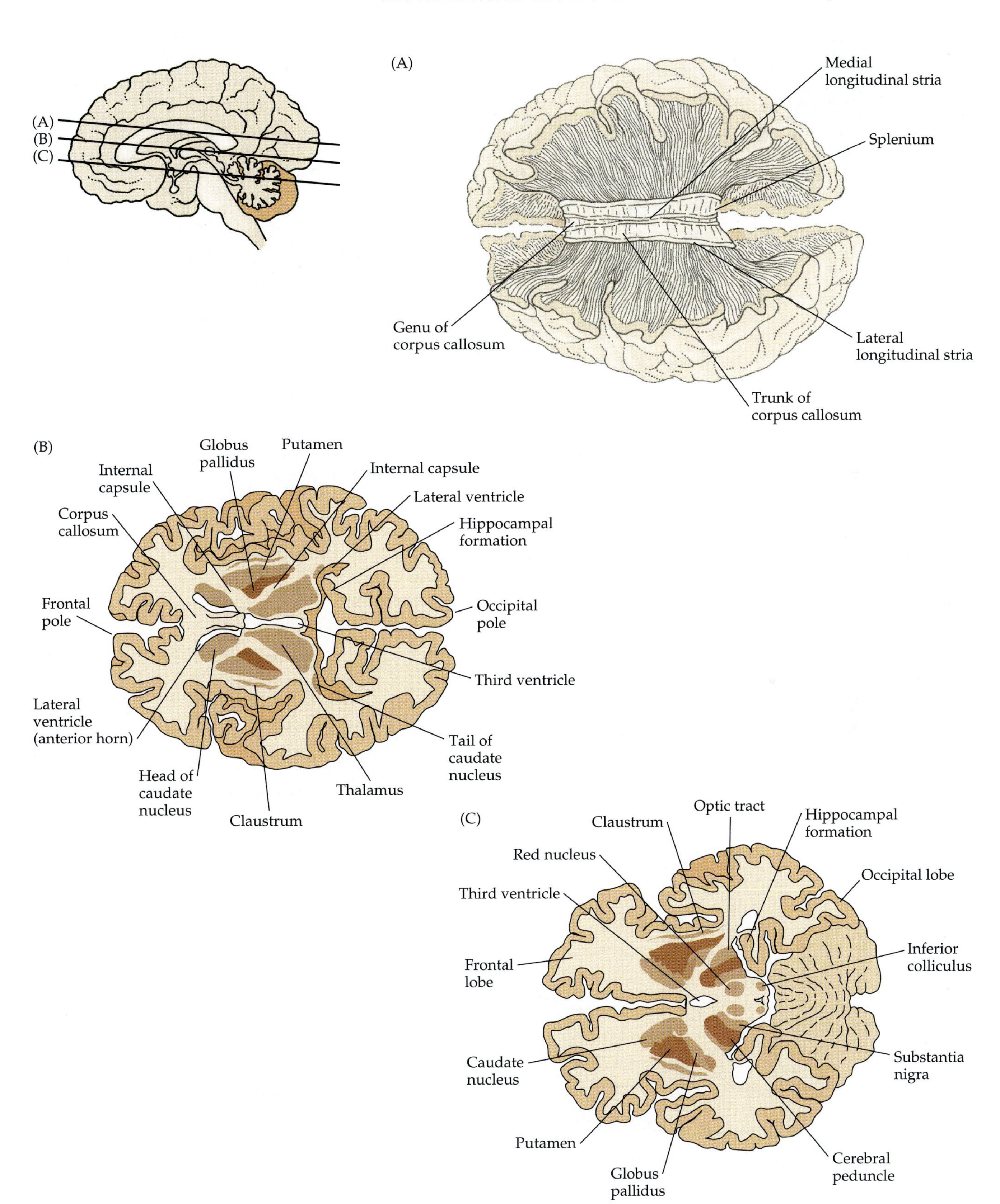

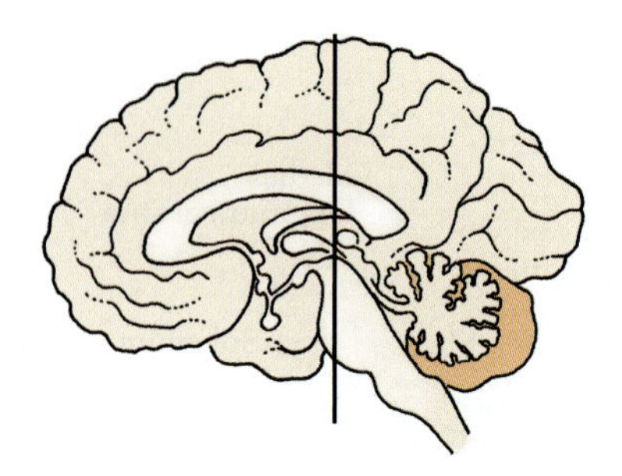

CORONAL SECTIONS

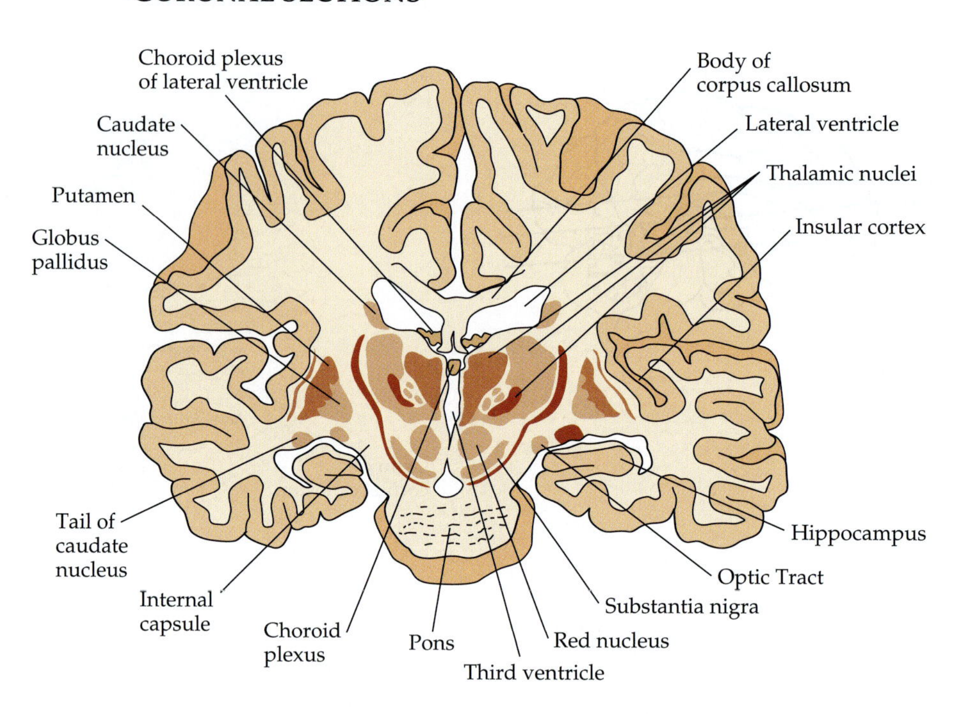

THE CEREBELLUM

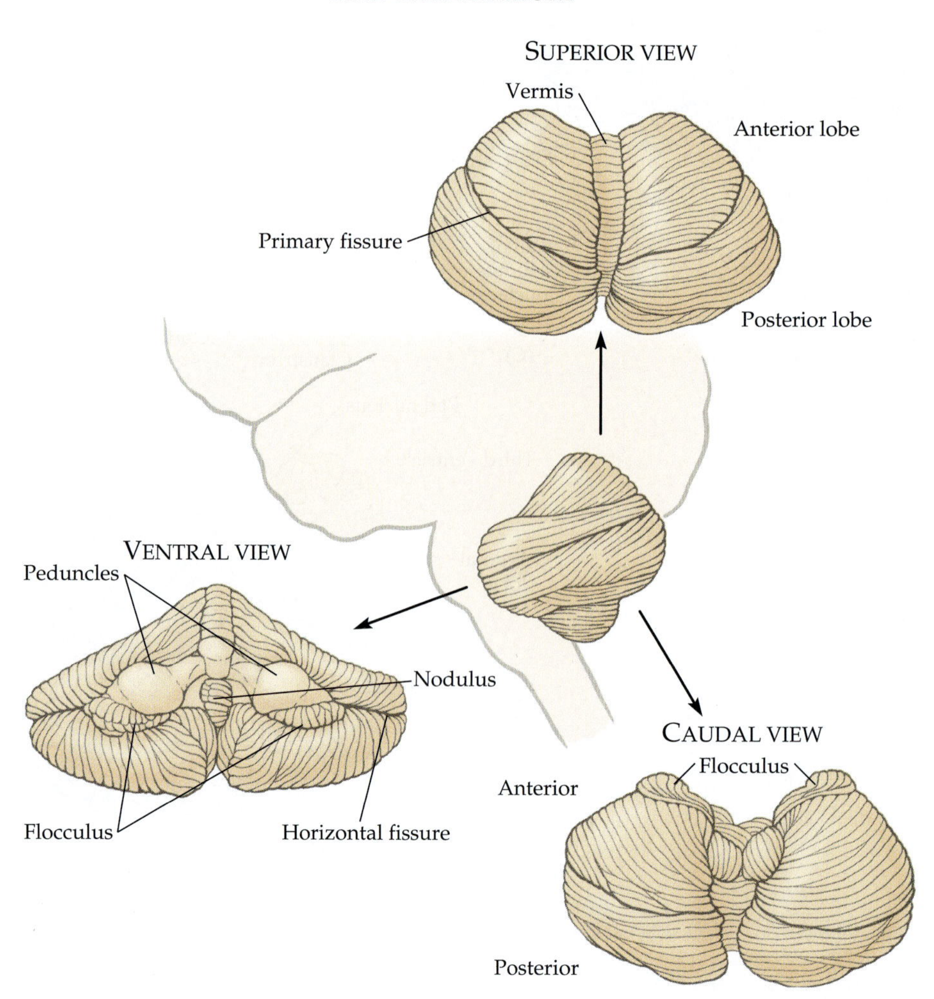

MAJOR SENSORY PATHWAYS

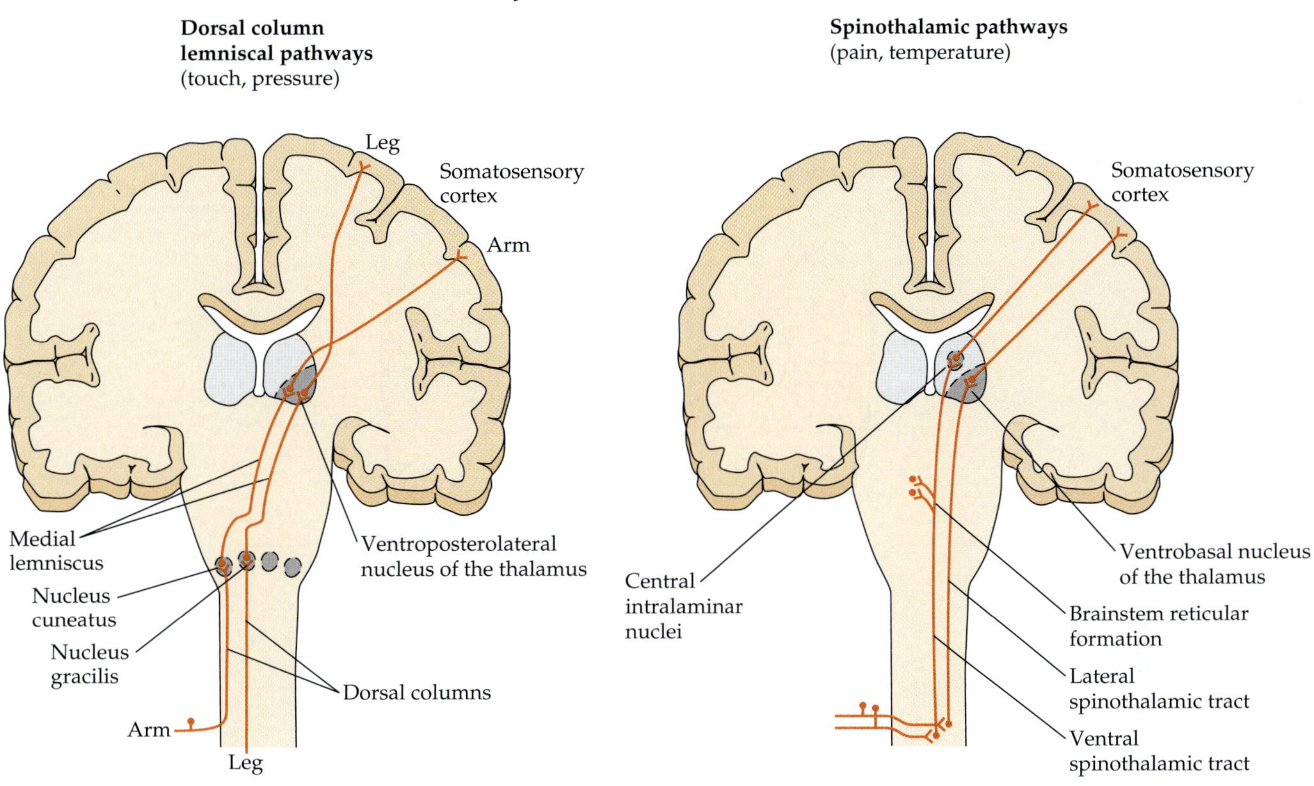

CROSS SECTION OF CERVICAL SPINAL CORD

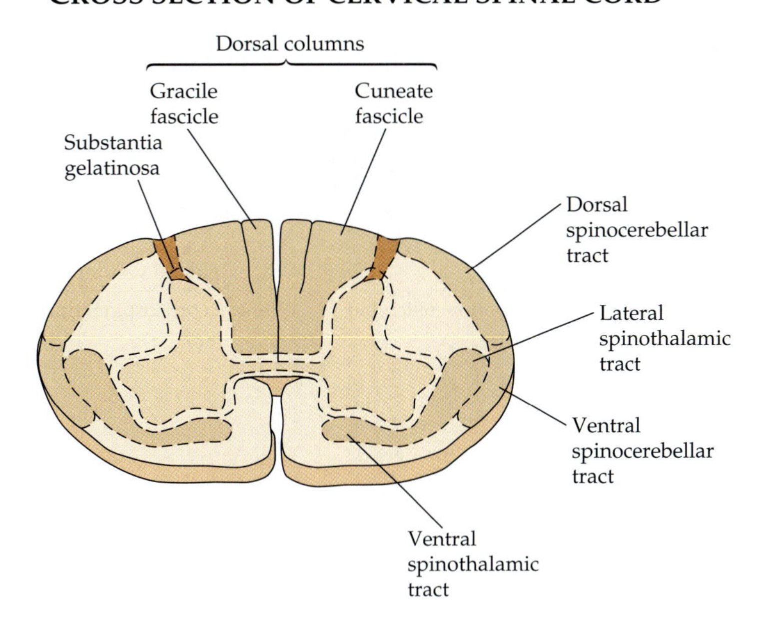

Major Motor Pathways

Tracts descending to the spinal cord

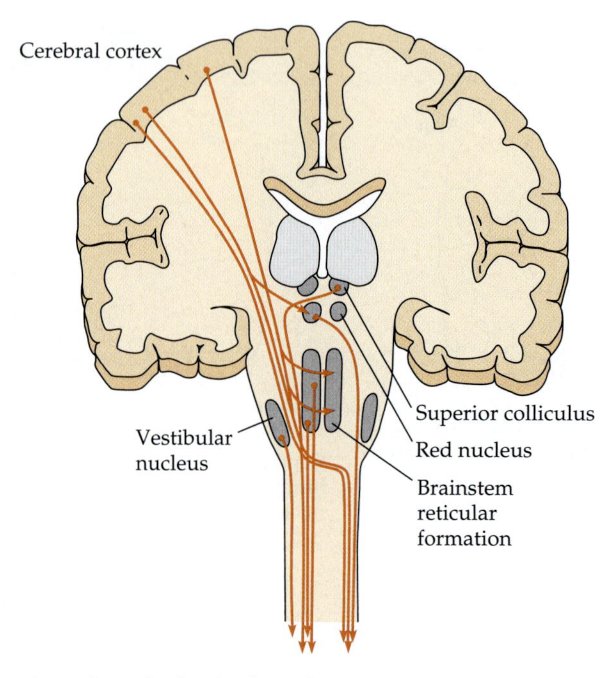

Cross section of cervical spinal cord

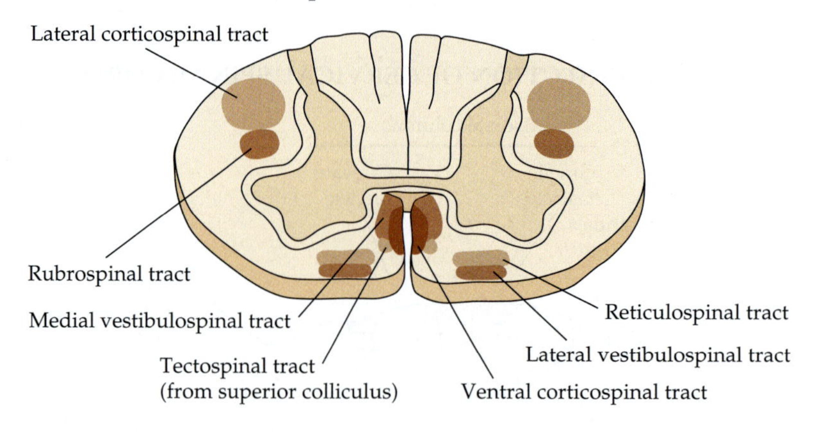

GLOSSARY

The definitions below apply to the terms used in the context of this book. Thus, **excitation**, **adaptation**, and **inhibition** all have additional meanings that are not included.

For structural formulae of transmitters see Appendix B.

For anatomical terms see Appendix C.

acetylcholine (ACh) Transmitter liberated by vertebrate motoneurons, preganglionic autonomic neurons, and in various central nervous system pathways.

acetylcholine receptor (ACh receptor) Membrane protein that binds ACh. Two different varieties:

nicotinic ACh receptor Activated by nicotine, consists of five polypeptide subunits that form a cation channel when activated.

muscarinic ACh receptor Activated by muscarine, contains a single protein molecule coupled by a G protein to one or more intracellular second messenger systems.

A channel A type of voltage-activated potassium channel.

action potential Brief regenerative, all-or-nothing electrical potential that propagates along an axon or muscle fiber.

activation 1. Initiation of an action potential. 2. Increase in the probability that an ion channel will open.

active transport Movement of ions or molecules against an electrochemical gradient.

primary active transport Utilizes metabolic energy.

secondary active transport Utilizes energy provided by the electrochemical gradient for another ion (usually sodium).

active zone Region in a presynaptic nerve terminal characterized by densely staining material on the cytoplasmic surface of the presynaptic membrane and a cluster of synaptic vesicles; believed to be the site of transmitter release.

adaptation Decline in response of a sensory neuron to a maintained stimulus.

adenosine 5-triphosphate (ATP) A common metabolite; hydrolysis of the terminal phosphoester linkage provides energy for many cellular reactions. Serves as phosphate donor in phosphorylation reactions. Also found in adrenergic and cholinergic synaptic vesicles; acts as a transmitter at synapses made by vertebrate sympathetic neurons.

adenylyl cyclase Enzyme catalyzing the synthesis of cyclic AMP from ATP.

adrenergic Referring to neurons releasing norepinephrine as a transmitter.

afferent Axon conducting impulses toward the central nervous system. Also called **primary afferent**.

after-hyperpolarization Slow hyperpolarizing potential seen in many neurons following a train of action potentials.

agonist A molecule that activates a receptor.

amphipathic Containing separate regions of hydrophilic and hydrophobic amino acid groups.

ampulla The sensory region of a semicircular canal in the vestibular apparatus.

anion A negatively charged ion.

antagonist A molecule that prevents activation of a receptor.

anterograde In the direction from the neuronal cell body toward the axon terminal. Compare with **retrograde**.

antibody An immunoglobulin molecule.

anticholinesterase Cholinesterase inhibitor (e.g., neostigmine, eserine); such agents prevent the hydrolysis of ACh and thereby allow its action to be prolonged.

astrocyte A class of glial cells found in the vertebrate CNS.

area centralis In the cat, the area of retina with highest acuity, containing cones.

autonomic nervous system Part of the vertebrate nervous system innervating viscera, skin, smooth muscle, glands, and heart, and consisting of two distinct divisions, **parasympathetic** and **sympathetic**.

axon The process or processes of a neuron conducting impulses, usually over long distances.

axon hillock Region of the cell body from which the axon originates; often the site of impulse initiation. See **initial segment**.

axonal transport Movement of material along an axon.

axoplasm Intracellular constituents of an axon.

axotomy Severing an axon.

basal lamina An extracellular, glycoprotein- and proteoglycan-containing matrix that ensheaths many tissues in the body, including nerves and muscle fibers.

basilar membrane The membrane in the cochlea upon which the hair cells sit, separating the **scala tympani** from the **scala media**.

biogenic amine A general term referring to any of several bioactive amines, including norepinephrine, epinephrine, dopamine, serotonin, and histamine.

bipolar cell Neuron with two major processes arising from the cell body; in the vertebrate retina, interposed between receptors and ganglion cells.

blobs Small, regularly spaced assemblies of neurons in the visual cortex of the monkey; they are stained with cytochrome oxidase and respond mainly to color stimuli.

blood–brain barrier Term denoting restricted access of substances to neurons and glial cells within the brain.

botulinum toxin A bacterial toxin that blocks release of transmitter from vertebrate motor nerve terminals. Also called botulin toxin.

bouton Small terminal expansion of the presynaptic nerve fiber at a synapse; site of transmitter release.

α-bungarotoxin Toxin from venom of the snake *Bungarus multicinctus*; binds to the nicotinic ACh receptor with high affinity.

calcium ATPase A molecule that transports calcium across a cell membrane against its electrochemical gradient, utilizing energy derived from the hydrolysis of ATP

capacitance of the membrane (C_m) Property of the cell membrane enabling electrical charge to be stored and separated and introducing distortion in the time course of passively conducted signals; usually measured in microfarads (μF).

carrier molecule A molecule involved in transporting ions or other molecules across cell membranes.

catecholamine A general term referring to molecules having both a catechol ring and an amino group, typically dopamine, norepinephrine, and epinephrine.

cation A positively charged ion.

cDNA (complementary DNA) DNA synthesized by reverse transcriptase using mRNA as a template.

centrifugal control Regulation of performance of peripheral sense organs by axons coming from the central nervous system.

cerebrospinal fluid (CSF) Clear liquid filling the ventricles of the brain and spaces between the meninges. See **subarachnoid space; ventricles.**

channel A pathway through a membrane allowing the passage of ions or molecules. All channels that have been characterized are aqueous pores formed by a single large protein or an assembly of polypeptide subunits.

charybdotoxin (CTX) A toxin obtained from scorpions that blocks potassium channels.

chimera An experimentally derived embryo or organ comprising cells arising from two or more genetically distinct sources.

cholinergic Neuron releasing acetylcholine as a transmitter.

cholinesterase An enzyme that hydrolyzes acetylcholine into acetic acid and choline.

choroid plexus Folded processes that are rich in blood vessels and project into ventricles of the brain and secrete cerebrospinal fluid.

chromaffin granules Large vesicles found in cells of the adrenal medulla, containing epinephrine (often norepinephrine as well), ATP, dopamine β-hydroxylase and chromogranins.

clone 1. All the progeny of a single cell. 2. To obtain a clone of cells containing a molecule of interest, such as a particular cDNA.

cochlea The bony canal containing the sensory apparatus for hearing.

conductance (g) Reciprocal of electrical resistance and thus a measure of the ability to conduct electricity; in cell membranes or ion channels, a measure of permeability to one or more ion species.

connexin One of a family of proteins that can assemble to form a connexon.

connexon A membrane channel bridging the space between two adjacent cells, connecting the cytoplasm of one to that of the other. See **gap junction.**

contralateral Relating to the opposite side of the body.

convergence The coming together of a group of presynaptic neurons to make synapses on one postsynaptic neuron.

cortical barrel Barrel-shaped aggregate of cortical neurons related to a specific peripheral sensory structure (e.g., a facial whisker).

cortical column An aggregate of cortical neurons, extending inward from the pial surface, that shares common properties (e.g., sensory modality, receptive field position, eye dominance, orientation, movement sensitivity).

coulomb Unit of electrical charge.

coupling potential A potential change in one cell produced by current spread from another through an electrical synapse.

crustacean Any member of the class Crustacea, which includes arthropods with hard shells, such as lobsters, crabs, barnacles, and shrimp.

curare A plant extract that blocks nicotinic ACh receptors.

dalton (Da) A unit of molecular mass, numerically equal to 1/12 the mass of a carbon atom; often expressed in thousands, or **kilodaltons (kDa).**

deactivation The response of a channel to a stimulus that reduces the probability that the channel will open.

delayed rectifier A type of potassium channel activated by membrane depolarization after a brief delay.

dendrite Process of a neuron specialized to act as a postsynaptic receptor region of a neuron.

depolarization Reduction in magnitude of the resting membrane potential toward zero.

depression Reduction of transmitter release from presynaptic terminals due to previous synaptic activity.

desensitization Reduction of the response of a receptor to a ligand after prolonged or repeated exposure.

diacylglycerol (DAG) An intracellular second messenger produced by phospholipase C–catalyzed hydrolysis of phospholipids. DAG activates protein kinase C.

divalent Having an electronic charge of +2 or –2.

divergence The branching of a nerve axon to form synapses with several other neurons.

domain A particular region of a polypeptide; for example, one of the four repeating regions of the voltage-activated sodium channel.

dopamine Transmitter liberated by neurons in some autonomic ganglia and in the central nervous system.

driving force The difference between the membrane potential and the equilibrium potential for movement of an ion species through a membrane channel (see electrochemical gradient).

efferent Axon conducting impulses outward from the central nervous system.

EGTA Ethylene glycol bis-(β-aminoethyl ether) N,N,N′N′-tetraacetic acid. A high-affinity calcium-binding compound.

electrical coupling The depolarization (or hyperpolarization) of one cell by current flow from another, usually through gap junctions.

electrochemical gradient The transmembrane difference in potential energy of an ion arising from the combined electrical and diffusional forces acting on it.

electroencephalogram (EEG) A recording of the electrical activity of the brain by external electrodes on the scalp.

electrogenic pump Active transport of ions across a cell membrane in which a net transfer of electrical charges contributes directly to the membrane potential.

electromyogram (EMG) A recording of muscular activity by external electrodes.

electroretinogram (ERG) Potential changes in response to light recorded by external electrodes on the eye.

electrotonic potentials Localized, graded, subthreshold potentials produced by artificially applied currents, and characterized by the passive electrical properties of cells.

endocytosis Process whereby membrane, together with some extracellular fluid, is internalized by the invagination and pinching off of a vesicle from the plasma membrane.

endothelial cells Layer of cells lining blood vessels.

end plate Postsynaptic area of vertebrate skeletal muscle fiber.

end plate potential (EPP) Postsynaptic potential change in a skeletal muscle fiber produced by ACh liberated from presynaptic terminals.

ependyma Layer of cells lining cerebral ventricles and central canal of spinal cord.

epinephrine (adrenaline) Hormone secreted by the adrenal medulla; certain of its actions resemble those of sympathetic nerves.

equilibrium potential Membrane potential at which there is no net passive movement of a permeant ion species into or out of a cell.

eserine An anticholinesterase; also known as physostigmine.

excitation Process tending to produce action potentials.

excitatory postsynaptic potential (EPSP) Depolarization of the postsynaptic membrane of a neuron produced by an excitatory transmitter released from presynaptic terminals.

exocytosis Process whereby synaptic vesicles fuse with presynaptic terminal membrane and empty transmitter molecules into the synaptic cleft.

explant A piece of tissue placed in culture.

extracellular matrix A scaffolding of large glycoproteins and proteoglycans that surrounds and separates cells or tissues.

extrafusal Muscle fibers making up the mass of a skeletal muscle (i.e., not within the sensory muscle spindles).

facilitation Increased release of transmitter from nerve terminals due to previous synaptic activity.

family A group of gene products with closely similar structure and function; for example, nicotinic ACh receptors. See **isotype**, **superfamily**.

farad (F) Unit of capacitance; more commonly used is **microfarad** ($\mu F = 10^{-6} F$).

faraday (*F*) The number of coulombs carried by 1 mole of a univalent ion (96,500).

fasciculation 1. The aggregation of neuronal processes to form a bundle. 2. The spontaneous contraction of muscle fibers in groups.

fetus A relatively late-stage mammalian embryo.

field axis orientation For simple cortical neurons in the visual system, the angle of the long axis of the receptive field (e.g., horizontal, vertical, oblique).

formants Frequency components of basic speech elements.

fovea Central depression in the retinal composed of slender cones; area of greatest visual resolution.

fusimotor Motoneurons supplying muscle fibers in a muscle spindle.

G proteins Receptor-coupled proteins that bind guanine nucleotides and activate intracellular messenger systems.

γ-aminobutyric acid (GABA) An inhibitory neurotransmitter.

γ-efferent fiber Small myelinated motor axon supplying intrafusal muscle fiber. See **fusimotor**.

ganglion A discrete collection of nerve cells.

gap junction The region of contact between two cells in which the space between adjacent membranes is reduced to about 2 nm and bridged by connecting channels. See **connexon**.

gate The mechanism whereby a channel is opened and closed.

gating current Movement of charge in a cell membrane associated with the opening or closing of a channel.

gene expression The transcription of DNA into mRNA and the translation of mRNA into protein.

glia See neuroglia.

glioblast A dividing cell, the progeny of which will develop into glial cells.

glutamate A transmitter liberated at many excitatory synapses in the vertebrate CNS.

glycine A transmitter liberated at many inhibitory synapses in the spinal cord and brain stem.

glycoprotein A protein containing carbohydrate residues.

golgi tendon organ Sensory element in muscle tendons activated by muscle stretch or contraction.

gray matter Part of the central nervous system composed predominantly of the cell bodies of neurons and fine terminals, as opposed to major axon tracts (**white matter**).

group I afferents Sensory fibers from muscle with conduction velocities in the range of 80–120 m/sec.

- **Group Ia** Arise from muscle spindles.
- **Group Ib** Arise from Golgi tendon organs.

group II afferents Sensory fibers from muscle spindles with conduction velocities in the range of 30–80 m/sec

growth cone The expanded tip of a growing axon.

hair cells Sensory cells in which bending of stereocilia ("hairs") causes a change in membrane potential; responsible for auditory transduction, transduction of vestibular stimuli, and vibratory transduction in lateral line organs of fish.

heterosynaptic Relating to the influence of activity at one synapse on the behavior of another synapse.

histamine A transmitter liberated by a small number of neurons in the vertebrate CNS.

homology The degree of identity of base pairs in two nucleotide sequences or of amino acids in two polypeptide sequences.

homosynaptic Relating to the influence of activity at a synapse on its own subsequent behavior.

horseradish peroxidase Enzyme used as a histochemical marker for tracing processes of neurons or spaces between cells.

hydropathy index A measure of insolubility of an amino acid, or amino acid sequence, in water, and hence its preference for a lipid environment.

hydrophilic Having a relatively high water solubility; polar.

hydrophobic Having a relatively low water solubility; nonpolar.

5-hydroxytryptamine (5-HT) See **serotonin.**

hyperpolarization Increase in membrane potential from the resting value, tending to reduce excitability.

impulse See **action potential.**

inactivation Reduction in conductance of a voltage-activated channel even though the activating voltage change is maintained.

inhibition Effect of one neuron upon another tending to prevent it from initiating impulses.

Postsynaptic inhibition Mediated through a permeability change in the postsynaptic cell, holding the membrane potential away from threshold.

Presynaptic inhibition Mediated by an inhibitory fiber ending on an excitatory terminal, reducing the release of transmitter.

initial segment The region of an axon close to the cell body; often the site of impulse initiation; see **axon hillock.**

input resistance (r_{input}) The resistance measured by injecting current into a cell or fiber; in a cylindrical fiber $r_{input} = 0.5\sqrt{(r_m r_i)}$.

in situ hybridization Technique for visualizing the distribution of mRNA for a particular protein by labeling the tissue with an antisense oligonucleotide probe.

inositol-1,4,5-trisphosphate (IP_3) An intracellular second messenger liberated by phospholipase C–catalyzed hydrolysis of phosphatidyl inositol. IP_3 triggers the release of calcium from intracellular stores.

integration The process whereby a neuron sums the various excitatory and inhibitory influences converging upon it and synthesizes a new output signal.

intercellular clefts Narrow fluid-filled spaces between membranes of adjacent cells; usually about 20 nm wide.

interneuron A neuron that is neither purely sensory nor purely motor but connects other neurons.

internode The myelinated portion of a nerve axon lying between two nodes of Ranvier.

intrafusal fiber Muscle fiber within a muscle spindle; its contraction initiates or modulates sensory discharge.

ion channel See **channel.**

ionophoresis Ejection of ions by passing current through a micropipette; used for applying charged molecules with a high degree of temporal and spatial resolution. Also spelled iontophoresis.

ipsilateral On the same side of the body.

inhibitory postsynaptic potential (IPSP) The potential change (usually hyperpolarizing) in a neuron produced by an inhibitory transmitter released from presynaptic terminals.

intracellular recording A recording of the membrane potential in an intact cell using a fine microelectrode inserted through the cell membrane.

invertebrate Any animal without a backbone.

in vitro Referring to a biological process studied outside an intact living organism.

in vivo Referring to a biological process studied within an intact living organism.

inward rectifier A type of potassium channel that allows potassium ions to move inward, but not outward, across the cell membrane.

isotypes Gene products of the same family, but with variations in amino acid sequence (e.g., voltage-activated sodium channels from brain and from muscle).

kilodalton (kDa) One thousand daltons.

laminin A prominent extracellular matrix glycoprotein; promotes neurite outgrowth in vitro.

lateral geniculate nucleus Small knee-shaped nucleus; part of posteroinferior aspect of thalamus acting as a relay in the visual pathway.

length constant, $\lambda = \sqrt{(r_m r_i)}$ Distance (usually in millimeters) over which a localized graded potential decreases to $1/e$ of its original size in an axon or a muscle fiber.

long-term potentiation (LTP) Increase in size of a synaptic potential lasting hours or more, produced by previous synaptic activation.

long-term depression (LTD) Decrease in size of a synaptic potential lasting hours or more, produced by previous synaptic activation.

magnocellular pathways In the visual system, large retinal ganglion cells and lateral geniculate cells that project to discrete cortical areas, particularly sensitive to movement and small changes in contrast.

Mauthner cell Large nerve cell in the mesencephalon of fishes and amphibians, up to 1 mm in length.

M channel A voltage-activated potassium channel that is inactivated by acetylcholine through muscarinic receptors

Meissner's corpuscle Rapidly adapting mechanoreceptor in superficial skin.

microglia Wandering, macrophage-like cells in the central nervous system that accumulate at sites of injury and scavenge debris.

microtubule A component of the cytoskeleton prominent in axons, formed by polymerization of tubulin monomers.

Merkel's disc Slowly adapting mechanoreceptor in superficial skin.

miniature end-plate potential (MEPP) Small depolarization at a neuromuscular synapse caused by spontaneous release of a single quantum of transmitter from the presynaptic terminal.

modality Class of sensation (e.g., touch, vision, olfaction).

monoclonal antibody An antibody molecule raised from a clone of transformed lymphocytes.

monovalent Having an electronic charge of +1 or –1.

monosynaptic A direct pathway from one neuron to the next, involving only one synapse.

motoneuron (motor neuron) A neuron that innervates muscle fibers.

α-motoneuron Supplies extrafusal muscle fibers.

γ-motoneuron Supplies intrafusal muscle fibers.

motor unit A single motoneuron and the muscle fibers it innervates.

MRI Magnetic resonance imaging; provides high-resolution pictures of brain structures. Formerly known as nuclear magnetic resonance imaging.

mRNA (messenger RNA) Polymer of ribonucleic acids transcribed from DNA that serves as a template for protein synthesis.

multimeric Composed of more than one polypeptide subunit (e.g., the **pentameric** acetylcholine receptor).

homomultimeric Composed of identical subunits.

heteromultimeric Composed of nonidentical subunits.

muscarinic See **acetylcholine receptor**.

muscle spindle Fusiform (spindle-shaped) structure in skeletal muscles containing small muscle fibers and sensory receptors activated by stretch.

mutagenesis Alteration of a gene to produce a product different from the naturally occurring, or "wild-type," variety.

myelin Fused membranes of Schwann cells or glial cells forming a high-resistance sheath surrounding an axon.

myoblast A dividing cell, the progeny of which develop into muscle cells.

myotube A developing muscle fiber formed by the fusion of myoblasts.

naloxone An agent that blocks opiate receptors specifically.

neostigmine An anticholinesterase; also known as prostigmine.

neurite Any neuronal process (axon or dendrite), typically used to refer to the processes of neurons in cell culture.

neuroblast A dividing cell, the progeny of which develop into neurons.

neuroglia Non-neuronal satellite cells associated with neurons. In the mammalian CNS, the main groupings are astrocytes and oligodendrocytes; in peripheral nerves the satellite cells are called Schwann cells.

neuromodulator A substance liberated from a neuron that modifies the efficacy of synaptic transmission.

neuropil A network of axons, dendrites, and synapses.

neurotransmitter See **transmitter**.

nicotinic See **acetylcholine receptor**.

nociceptive Responding to noxious (tissue-damaging or painful) stimulation.

node of Ranvier Localized area devoid of myelin occurring at intervals along a myelinated axon.

noise Fluctuations in membrane potential or current due to random opening and closing of ion channels.

norepinephrine (noradrenaline) Transmitter liberated by most sympathetic nerve terminals.

ocular dominance The greater effectiveness of one eye over the other for driving simple or complex cells in the visual cortex.

Ohm's law Relates current (I) to voltage (V) and resistance (R); $I = V/R$.

oligodendroglia Class of vertebrate CNS glial cells that forms myelin.

open channel block Block of ion flow though an open channel by a physical obstruction, such as a large molecule in the channel mouth.

opiate Term denoting products derived from the juice of opium poppy seed capsules.

opioid Any directly acting compound whose actions are similar to those of opiates and are specifically antagonized by naloxone.

optic chiasm The point of crossing, or decussation, of the optic nerves. In cats and primates, fibers arising from the medial part of the retina cross to supply the opposite lateral geniculate nucleus.

ouabain G-strophanthidin, a glycoside that specifically blocks sodium–potassium ATPase.

oval window The membranous partition between the **scala vestibuli** and the middle ear that receives auditory vibrations from the ear drum through bony interconnections.

overshoot Reversal of the membrane potential during the peak of the action potential.

Pacinian corpuscle A rapidly adapting mechanoreceptor sensitive to vibration; found in deep skin and other tissues.

parvocellular pathways In the visual system, smaller retinal ganglion cells and lateral geniculate cells that project to discrete areas of the visual cortex; concerned with color detection and fine discrimination.

parasympathetic nervous system A division of the autonomic nervous system arising from the cranial and sacral segments of the central nervous system.

patch clamp A technique whereby a small patch of membrane is sealed to the tip of a micropipette, enabling currents through single membrane channels to be recorded.

permeability Property of a membrane or channel allowing substances to pass into or out of the cell.

PET scan See **positron emission tomography**.

phagocytosis Endocytosis and degradation of foreign or degenerating material.

phenotype The physical characteristics of an animal.

phoneme Basic sound element of speech.

phospholipase C; phospholipase A2 Enzymes that hydrolyze phospholipids.

phosphorylation The covalent addition of one or more phosphate ions to a molecule, for example, a channel protein.

polar A molecule with separate positively charged and negatively charged regions. See **hydrophilic**.

polysynaptic A pathway involving a series of synaptic connections.

positron emission tomography A technique for mapping active areas of the brain. Glucose is labeled with isotopes that emit positrons; sites of glucose uptake are then located by detecting positron emissions.

postmitotic A cell that is no longer capable of undergoing cell division.

posttetanic potentiation (PTP) Increase in transmitter release from nerve terminals following a train of repetitive stimuli.

protease An enzyme that hydrolyzes protein molecules. Also spelled proteinase.

protein kinase An enzyme that phosphorylates proteins.

protein phosphatase An enzyme that cleaves phosphate residues from proteins.

pump Active transport mechanism.

pyramidal cell Any neuron with a long apical dendrite and shorter basal dendrites, a morphology characteristic of many cortical neurons.

quantal release Secretion of multimolecular packets (quanta) of transmitter by the presynaptic nerve terminal.

quantal size 1. The number of molecules of neurotransmitter in a quantum. 2. The change in postsynaptic potential or current produced by a quantum.

quantum content The number of quanta in a synaptic response.

receptive field The area of the periphery whose stimulation influences the firing of a neuron. For cells in the visual pathway, the receptive field refers to an area on the retina whose illumination influences the activity of a neuron.

receptor 1. A nerve terminal or accessory cell, associated with sensory transduction. 2. A molecule in the cell membrane that combines with a specific chemical substance.

Directly coupled (ionotropic) receptors Molecules that, when activated, form ion channels that span the membrane.

Indirectly coupled (metabotropic) receptors Molecules that activate G proteins, which, in turn, modify the activity of channels or pumps either directly or through a second messenger pathway.

receptor potential Graded, localized potential change in a sensory receptor initiated by the appropriate stimulus; the electrical sign of the transduction process.

reciprocal innervation Interconnections of neurons arranged so that pathways exciting one group of muscles inhibit the antagonistic motoneurons.

rectification The property of a membrane, or membrane channel, that allows it to conduct ionic current more readily in one direction than in the other.

reflex Involuntary movement or other response elicited by activation of sensory receptors, and involving conduction through one or more synapses in the CNS.

refractory period

absolute refractory period The time following an action potential during which a stimulus cannot elicit a second action potential.

relative refractory period The time following the absolute refractory period when the threshold for initiation of a second action potential is increased.

resistance of the membrane (R_m) Property of the cell membrane, measured in ohm cm^2, reflecting the difficulty ions encounter in moving across it. Inverse of conductance.

resting potential The steady electrical potential across the membrane in the quiescent state.

retinotectal Referring to the projection of retinal ganglion cells to neurons in the optic tectum.

retrograde In the direction from the axon terminal toward the cell body.

reversal potential The value of the membrane potential at which a chemical transmitter produces no change in potential.

Ringer's solution A saline solution containing sodium chloride, potassium chloride, and calcium chloride; named after Sidney Ringer.

RNA editing Replacement of one codon by another in messenger RNA after transcription.

round window The membranous partition between the **scala tympani** and the middle ear.

Ruffini's capsule Slowly adapting mechanoreceptor found deep in skin.

saccule The part of the vestibular apparatus that responds to vertical acceleration of the head.

saltatory conduction Conduction along a myelinated axon whereby the leading edge of the propagating action potential leaps from node to node.

scala tympani, scala media, scala vestibuli Fluid-filled compartments of the cochlea.

S channel A type of voltage-activated potassium channel inactivated by serotonin.

Schwann cell Satellite cell in the peripheral nervous system, responsible for making the myelin sheath.

second messenger A molecule forming part of a second messenger system.

second messenger system A series of molecular reactions inside a cell initiated by occupation of extracellular receptor sites and leading to a functional response, such as opening or closing of membrane channels.

semicircular canal Fluid-filled loop in the vestibular apparatus associated with detection of head rotation.

serotonin Also known as 5-hydroxytryptamine, or 5-HT; a neurotransmitter.

siemen (S) Unit of conductance; reciprocal of ohm.

size principle The orderly recruitment of motor units of increasing size as the strength of a muscle contraction increases.

site-directed mutagenesis The alteration of a gene to produce a product with one or more amino acid substitutions at specified locations.

sodium–potassium ATPase A molecule that transports sodium and potassium across a cell membrane against their electrochemical gradients, utilizing energy derived from the hydrolysis of ATP.

sodium–potassium exchange pump Sodium–potassium ATPase.

soma Cell body.

somatotopic Organized in an orderly manner in relation to an outline of the body, or a part of the body.

stereocilia Specialized microvilli of graded length projecting from the apical surface of a hair cell.

specific resistance 1. The resistance of one square centimeter of cell membrane 2. The resistance of a volume of cytoplasm one square centimeter in cross section and one centimeter in length.

striate cortex Also known as area 17 or visual I; primary visual region of occipital lobe marked by striation of Gennari, visible with the naked eye.

subarachnoid space The space filled by cerebrospinal fluid between the arachnoid and pia, two layers of connective tissue (meninges) surrounding the brain.

subunit The basic structural "building block" of a multimeric protein, such as a membrane channel; usually a single polypeptide.

summation Addition of synaptic potentials.

- **temporal summation** Addition in sequence, with each potential adding to the one preceding;
- **spatial summation** Addition of potentials arising in different parts of a cell; for example, of potentials spreading to the axon hillock from different branches of a dendritic tree.

superfamily A group of gene product families with similar structure and function (e.g., ligand-activated ion channels). See **family**.

supersensitivity An increase in responsiveness of target cells to chemical transmitters seen, for example, after denervation.

sympathetic nervous system A division of the autonomic nervous system arising from the thoracic and lumbar segments of the CNS.

synapse Site at which neurons make functional contact; a term coined by Sherrington.

synaptic cleft The space between the membranes of the pre- and postsynaptic cells at a chemical synapse across which transmitter diffuses.

synaptic delay The time between a presynaptic nerve impulse and a postsynaptic response.

synaptic vesicles Small membrane-enclosed sacs contained in presynaptic nerve terminals. Those with dense cores contain catecholamines and serotonin; clear vesicles are presumed to be the storage sites for other transmitters.

tectorial membrane Membrane in the cochlea in which hair cell cilia are imbedded.

tetraethylammonium (TEA) Quaternary ammonium compound that selectively blocks certain voltage- gated potassium channels in neurons and muscle fibers.

tetanus A train of action potentials; also the resulting sustained muscular contraction.

tetrodotoxin (TTX) Toxin from puffer fish that selectively blocks the voltage-activated sodium channel in neurons and muscle fibers.

threshold 1. Critical value of membrane potential or depolarization at which an impulse is initiated. 2. Minimal stimulus required for a sensation.

tight junction Site at which fusion occurs between the outer leaflets of membranes of adjacent cells, resulting in a five-layered junction. It is called a **macula occludens** if the area is a spot and a **zonula occludens** if the junction is a circumferential ring. Such complete junctions prevent the movement of substances through the extracellular space between the cells.

time constant (τ) A measure of the rate of buildup or decay of a localized graded potential; equal numerically to the product of the resistance and the capacitance of the membrane.

transcription Synthesis of mRNA using DNA as a template.

transducer Device for converting one form of energy into another (e.g., a microphone, photoelectric cell, loudspeaker, or light bulb).

translation Synthesis of protein using mRNA as a template.

transmitter Chemical substance liberated by a presynaptic nerve terminal causing an effect on the membrane of the postsynaptic cell, usually an increase in permeability to one or more ions.

transporter A protein mediating transport of ions or molecules across a cell membrane.

tuning The property of a receptor (e.g., a cochlear hair cell) that restricts its response to a specified frequency range.

undershoot Transient hyperpolarization following an action potential; caused by increased potassium conductance.

utricle Part of the vestibular apparatus that responds to horizontal acceleration of the head.

varicosity Swelling along an axon from which transmitter is liberated.

ventricles Cavities within the brain containing cerebrospinal fluid and lined by ependymal cells.

ventricular zone Region adjacent to the lumen of the neural tube (future ventricle) in the developing vertebrate neuroepithelium where cell proliferation occurs.

vertebrate An animal with a backbone.

voltage clamp Technique for displacing membrane potential abruptly to a desired value and keeping the potential constant while measuring currents across the cell membrane; devised by Cole and Marmont.

white matter The part of the CNS that appears white; consists of myelinated fiber tracts. See **gray matter**.

whole-cell patch recording Recording of membrane currents of an intact cell with a patch-clamp electrode, through an opening in the cell membrane.

BIBLIOGRAPHY

Abe, K., and Kimura, H. 1996. The possible role of hydrogen sulfide as an endogenous neuromodulator. *J. Neurosci.* 16: 1066–1071. [14]

Ache, B. W., and Zhainazarov, A. 1995. Dual second-messenger pathways in olfactory transduction. *Curr. Opin. Neurobiol.* 5: 461–466. [17]

Acklin, S. E. 1988. Electrical properties and anion permeability of doubly rectifying junctions in the leech central nervous system. *J. Exp. Biol.* 137: 1–11. [9]

Adams, D. J., Dwyer, T. M., and Hille, B. 1980. The permeability of endplate channels to monovalent and divalent metal cations. *J. Gen. Physiol.* 75: 493–510. [9]

Adams, P. R., and Brown, D. A. 1980. Lutenizing hormone-releasing factor and muscarinic agonists act on the same voltage-sensitive K^+-current in bullfrog sympathetic neurones. *Br. J. Pharmacol.* 68: 353–355. [16]

Adams, P. R., Brown, D. A., and Constanti, A. 1982. Pharmacological inhibition of the M-current. *J. Physiol.* 332: 223–262. [16]

Adams, P. R., Brown, D. A., and Jones, S. W. 1983. Substance P inhibits the M-current in bullfrog sympathetic neurones. *Br. J. Pharmacol.* 79: 330–333. [17]

Adams, R. D., Victor, M., and Ropper, A. H. 1997. *Principles of Neurology*, 6th Ed. McGraw-Hill, New York. [26]

Adler, E. M., Augustine, G. J., Duffy, S. N., and Charlton, M. P. 1991. Alien intracellular calcium chelators attenuate neurotransmitter release at the squid giant synapse. *J. Neurosci.* 11: 1496–1507. [11]

Adrian, E. D. 1946. *The Physical Background of Perception*. Clarendon, Oxford, England. [1, 22, 26]

Adrian, E. D. 1953. Sensory messages and sensation. The response of the olfactory organ to different smell. *Acta Physiol. Scand.* 29: 5–14. [17]

Adrian, E. D. 1959. *The Mechanism of Nervous Action*. University of Pennsylvania Press, Philadelphia. [22]

Adrian, E. D., and Zotterman, Y. 1926. The impulses produced by sensory nerve endings. Part II. The response of a single end organ. *J. Physiol.* 61: 151–171. [17]

Aghajanian, G. K. and Marek, G. J. 1999. Serotonin and hallucinogens. *Neuropsychopharmacology* 21 (2 suppl.): 16s–23s. [14]

Aguayo, A. J., Dickson, R., Trecarten, J., Attiwell, M., Bray, G. M., and Richardson, P. 1978. Ensheathment and myelination of regenerating PNS fibres by transplanted optic nerve glia. *Neurosci. Lett.* 9: 97–104. [24]

Ahlquist, R. P. 1948. A study of the adrenotropic receptors. *Am. J. Physiol.* 153: 586–600. [16]

Aigner, T. G. 1995. Pharmacology of memory: Cholinergic-glutamatergic interactions. *Curr. Opin. Neurobiol.* 5: 155–160. [14]

Akabas, M. H., Kauffman, C., Archdeacon, P., and Karlin, A. 1994. Identification of acetylcholine receptor channel-lining residues in the entire M2 segment of the α subunit. *Neuron* 13: 919–927. [3]

Akil, H., Owens, C., Gutstein, H., Taylor, L., Curran, E., and Watson, S. 1998. Endogenous opioids: Overview and current issues. *Drug Alcohol Depend.* 51: 127–140. [14]

Albillos, A., Dernick, G., Horstmann, H., Almers, W., Alvarez de Toledo, G., and Lindau, M. 1997. The exocytotic event in chromaffin cells revealed by patch amperometry. *Nature* 389: 509–512. [11]

Albin, R. L. 1995. The pathophysiology of chorea/ballism and parkinsonism. *Parkinsonism Rel. Disord.* 1: 3–11. [22]

Albin, R. L., Young, A. B., and Penney, J. B. 1995. The functional anatomy of disorders of the basal ganglia. *Trends Neurosci.* 18: 63–64. [22]

Albright, T. D. 1984. Direction and orientation selectivity of neurons in visual area MT of the macaque. *J. Neurophysiol.* 52: 1106–1130. [21]

Alcantara, S., Frisen, J., del Rio, J., Soriano, E., Barbacid, M., and Silos-Santiago, I. 1997. TrkB signaling is required for postnatal survival of CNS neurons and protects hippocampal and motor neurons from axotomy-induced cell death. *J. Neurosci.* 15: 3623–3633. [23]

Aldrich, R. W., and Stevens, C. F. 1983. Inactivation of open and closed sodium channels determined separately. *Cold Spring Harb. Symp. Quant. Biol.* 48: 147–153. [6]

Aldrich, R. W., and Stevens, C. F. 1987. Voltage-dependent gating of single sodium channels from mammalian neuroblastoma. *J. Neurosci.* 7: 418–431. [6]

Alexander, G. E., DeLong, M. R., and Strick, P. L. 1986. Parallel organization of functionally segregated circuits linking basal ganglia and cortex. *Annu. Rev. Neurosci.* 9: 357–381. [22]

Alkon, D. 1987. *Memory Traces in the Brain*. Cambridge University Press, Cambridge, England. [15]

Allen, R. D., Allen, N. S., and Travis, J. L. 1981. Video-enhanced differential interference contrast (AVEC-DIC) microscopy: A new method capable of analyzing microtubule-related movement in the reticulopodial network of *Allogromia laticollaris*. *Cell Motil.* 1: 291–302. [13]

Almers, W., Stanfield, P., and Stühmer, W. 1983. Lateral distribution of sodium and potassium channels in frog skeletal muscle: Measurements with a patch clamp technique. *J. Physiol.* 336: 261–284. [6]

Alper, S. L. 1991. The band-three related anion exchange (AE) gene family. *Ann. Rev. Physiol.* 53: 549–564. [4]

Altamirano, A. A., and Russell, J. M. 1987. Coupled Na/K/Cl efflux. "Reverse" unidirectional fluxes in squid giant axons. *J. Gen. Physiol.* 89: 669–686. [4]

Altiok, N., Altiok, S., and Changeux, J. P. 1997. Heregulin-stimulated acetylcholine receptor gene expression in muscle: Requirement for MAP kinase and evidence for a parallel inhibitory pathway independent of electrical activity. *EMBO J.* 16: 717–725. [24]

Altman, J., and Das, G. D. 1965. Autoradiographic and histological evidence of postnatal neurogenesis in rats. *J. Comp. Neurol.* 124: 319–335. [23]

Amara, S. G., and Kuhar, M. J. 1993. Neurotransmitter transporters: Recent progress. *Annu. Rev. Neurosci.* 16: 73–93. [13]

Andersen, P., Sundberg, S. H., Sveen, O., and Wigstrom H. 1977. Specific long-lasting potentiation of synaptic transmission in hippocampal slices. *Nature* 266: 736–737. [12]

Anderson, C. R., and Stevens, C. F. 1973. Voltage clamp analysis of acetylcholine-produced end-plate current fluctuations at frog neuromuscular junction. *J. Physiol.* 235: 665–691. [2, 9]

Anderson, D. J. 1993. Molecular control of cell fate in the neural crest: The sympathoadrenal lineage. *Annu. Rev. Neurosci.* 16: 129–158. [23]

Anderson, J. P., and Vilsen, B. 1995. Structure-function relationships of cation translocation by Ca^{2+}- and Na^+, K^+-ATPases studied by site-directed mutagenesis. *FEBS Lett.* 359: 101–106. [4]

Anderson, M. J., and Cohen, M. W. 1977. Nerve-induced and spontaneous redistribution of acetylcholine receptors on cultured muscle cells. *J. Physiol.* 268: 757–773. [23]

Angevine, J. B., and Sidman, R. L. 1961. Autoradiographic study of cell migration during histogenesis of cerebral cortex in the mouse. *Nature* 192: 766–768. [23]

Angleson, J. K., and Betz, W. J. 1997. Monitoring secretion in real time: Capacitance, amperometry and fluorescence compared. *Trends Neurosci.* 20: 281–287. [11]

Anglister, L., and McMahan, U. J. 1985. Basal lamina directs acetylcholinesterase accumulation at synaptic sites in regenerating muscle. *J. Cell Biol.* 101: 735–743. [24]

Anniko, M., and Wroblewski, R. 1986. Ionic environment of cochlear hair cells. *Hear. Res.* 22: 279–293. [18]

Anton, E. S., Kreidberg, J. A., and Rakic, P. 1999. Distinct functions of $\alpha3$ and αv integrin receptors in neuronal migration and laminar organization of the cerebral cortex. *Neuron* 22: 277–289. [23]

Antonini, A., and Stryker, M. P. 1993a. Development of individual geniculocortical arbors in cat striate cortex and effects of binocular impulse blockade. *J. Neurosci.* 13: 3549–3573. [25]

Antonini, A., and Stryker, M. P. 1993b. Rapid remodeling of axonal arbors in the visual cortex. *Science* 260: 1819–1821. [25]

Antonini, A., and Stryker, M. P. 1998. Effect of sensory disuse on geniculate afferents to cat visual cortex. *Vis. Neurosci.* 15: 401–409. [25]

Antonini, A., Gillespie, D. C., Crair, M. C., and Stryker, M. P. 1998. Morphology of single geniculocortical afferents and functional recovery of the visual cortex after reverse monocular deprivation in the kitten. *J. Neurosci.* 18: 9896–9909. [25]

Antonov, I., Kandel, E. R., and Hawkins, R. D. 1999. The contribution of facilitation of monosynaptic PAPs to dishabituation and sensitization of the *Aplysia* siphon withdrawal reflex. *J. Neurosci.* 19: 10438–10450. [15]

Apel, E. D., and Merlie, J. P. 1995. Assembly of the postsynaptic apparatus. *Curr. Opin. Neurobiol.* 5: 62–67. [13]

Appenzeller, O. 1990. *The Autonomic Nervous System: An Introduction to Basic and Clinical Concepts.* Elsevier, New York. [16]

Araque, A., Li, N., Doyle, R. T., and Haydon, P. G. 2000. SNARE protein-dependent glutamate release from astrocytes. *J. Neurosci.* 20: 666–673. [8]

Arbuthnott, E. R., Boyd, I. A., and Kalu, K.U. 1980. Ultrastructural dimensions of myelinated peripheral nerve fibres in the cat and their relation to conduction velocity. *J. Physiol.* 308: 125–127. [7]

Arechiga, H. 1993. Circadian rhythms. *Curr. Opin. Neurobiol.* 3: 1005–1010. [16]

Arechiga, H., Fernandez-Quiroz, F., Fernandez de Miguel, F., and Rodriguez-Sosa, L. 1993. The circadian system of crustaceans. *Chronobiol. Int.* 10: 1–19. [16]

Armstrong, C. M. 1981. Sodium channels and gating currents. *Physiol. Rev.* 61: 644–683. [6]

Armstrong, C. M., and Bezanilla, F. 1974. Charge movements associated with the opening and closing of the activation gates of sodium channels. *J. Gen. Physiol.* 63: 533–552. [6]

Armstrong, C. M., and Bezanilla, F. 1977. Inactivation of the sodium channel II. Gating current experiments. *J. Gen. Physiol.* 70: 567–590. [6]

Armstrong, C. M., and Hille, B. 1972. The inner quaternary ammonium ion receptor in potassium channels of the node of Ranvier. *J. Gen. Physiol.* 59: 388–400. [6]

Armstrong, C. M., and Hille, B. 1998. Voltage-gated ion channels and electrical excitability. *Neuron* 20: 371–380. [6]

Armstrong, C. M., Bezanilla, F., and Rojas, E. 1973. Destruction of sodium conductance inactivation in squid axons perfused with pronase. *J. Gen. Physiol.* 62: 375–391. [6]

Arnold, A. P., and Schlinger, B. A. 1993. Sexual differentiation of brain and behavior: The zebra finch is not just a flying rat. *Brain Behav. Evol.* 42: 231–241. [25]

Arshavsky, Y. I., Deliagina, T. G., and Orlovsky, G. N. 1997. Pattern generation. *Curr. Opin. Neurobiol.* 7: 781–789. [22]

Art, J. J., and Fettiplace, R. 1987. Variation of membrane properties in hair cells isolated from the turtle cochlea. *J. Physiol.* 385: 207–242. [18]

Art, J. J., Crawford, A. C., Fettiplace, R., and Fuchs, P. A. 1985. Efferent modulation of hair cell tuning in the cochlea of the turtle. *J. Physiol.* 360: 397–421. [18]

Art, J. J., Fettiplace, R., and Fuchs, P. A. 1984. Synaptic hyperpolarisation and inhibition of turtle cochlea hair cells. *J. Physiol.* 356: 525–550. [18]

Art, J. J., Fettiplace, R., and Wu, Y-C. 1995. The calcium-activated potassium channels of turtle hair cells. *J. Gen. Physiol.* 105: 49–72. [18]

Arthur, R. M., Pfeiffer, R. R., and Suga, N. 1971. Properties of 'two-tone inhibition' in primary auditory neurones. *J. Physiol.* 212: 593–609. [18]

Asada, H., Kawamura, Y., Maruyama, K., Kume, H., Ding, R. G., Kanbara, N., Kuzume, H., Sanbo, M., Yagi, T., and Obata, K. 1997. Cleft palate and decreased brain gamma-aminobutyric acid in mice lacking the 67-kDa isoform of glutamic acid decarboxylase. *Proc. Natl. Acad. Sci. USA* 94: 6496–6499. [13]

Asanuma, H. 1989. *The Motor Cortex.* Raven, New York. [22]

Asanuma, H., and Pavlides, C. 1997. Neurobiological basis of motor learning in mammals. *Neuroreport* 8: 1–4. [22]

Ashmore, J. F. 1987. A fast motile response in guinea-pig outer hair cells: The cellular basis of the cochlear amplifier. *J. Physiol.* 388: 323–347. [18]

Assad, J. A., Shepherd, G. M. G., and Corey, D. P. 1991. Tip-link integrity and mechanical transduction in vertebrate hair cells. *Neuron* 7: 985–994. [17]

Astion, M. L., Chavatal, A., and Orkand, R. K. 1991. Further studies of electogenic Na^+/HCO_3^- cotransport in glial cells of *Necturus* optic nerve: Regulation of pH. *Glia* 4: 461–468. [8]

Attaix, D., Taillandier, D., Temparis, S., Larbaud, D., Aurousseau, E., Combaret, L., and Voisin, L. 1994. Regulation of ATP-ubiquitin-dependent proteolysis in muscle wasting. *Reprod. Nutr. Dev.* 34: 583–597. [24]

Attwell, D., Barbour, B., and Szatkowski, M. 1993. Nonvesicular release of neurotransmitter. *Neuron* 11: 401–407. [4]

Atwood, H. L. (ed.). 1982. *Biology of Crustacea*, Vol. 3. Academic Press, New York. [15]

Atwood, H. L., and Morin, W. A. 1970. Neuromuscular and axoaxonal synapses of the crayfish opener muscle. *J. Ultrastruct. Res.* 32: 351–369. [9]

Auerbach, A., and Akk, G. 1998. Desensitization of mouse nicotinic acetylcholine receptor channels. *J. Gen. Physiol.* 112: 181–197. [9]

Avenet, P., and Lindemann, B. 1991. Noninvasive recording of receptor cell action potentials and sustained currents from single taste buds maintained in the tongue: The response to mucosal NaCl and amiloride. *J. Membr. Biol.* 124: 33–41. [17]

Axelrod, J. 1971. Noradrenaline: Fate and control of its biosynthesis. *Science* 173: 598–606. [13]

Axelsson, J., and Thesleff, S. 1959. A study of supersensitivity in denervated mammalian skeletal muscle. *J. Physiol.* 147: 178–193. [24]

Ayali, A., and Harris-Warrick, R. M. 1999. Monoamine control of the pacemaker kernel and cycle frequency in the lobster pyloric network. *J. Neurosci.* 19: 6712–6722. [15]

Azzopardi, P., Jones, K. E., and Cowey, A. 1999. Uneven mapping of magnocellular and parvocellular projections from the lateral geniculate nucleus to the striate cortex in the macaque monkey. *Vision Res.* 39: 2179–2189. [20]

Baas, P. W., and Brown, A. 1997. Slow axonal transport: The polymer transport model. *Trends Cell Biol.* 7: 380–384. [13]

Babcock, D. F., Herrington, J., Goodwin, P. C., Park, Y. B., and Hille, B. 1997. Mitochondrial participation in the intracellular Ca^{2+} network. *J. Cell Biol.* 136: 833–844. [4]

Baccus, S. A. 1998. Synaptic facilitation by reflected action potentials: Enhancement of transmission when nerve impulses reverse direction at axon branch points. *Proc. Natl. Acad. Sci. USA* 95: 8345–8350. [7, 15]

Bacskai, B. J., Wallen, P., Lev-Ram, V., Grillner, S., and Tsien, R. Y. 1995. Activity-related calcium dynamics in lamprey motoneurons as revealed by video-rate confocal microscopy. *Neuron* 14: 19–28. [10]

Bainton, C. R., Kirkwood, P. A., and Sears, T. A. 1978. On the transmission of the stimulating effects of carbon dioxide to the muscles of respiration *J. Physiol.* 280: 249–272. [22]

Bajjalieh, S. M. 1999. Synaptic vesicle docking and fusion. *Curr. Opin. Neurobiol.* 9: 321–328. [13]

Bakalyar, H. A., and Reed, R. R. 1990. Identification of a specialized adenylyl cyclase that may mediate odorant detection. *Science* 250: 1403–1406. [17]

Baker, F. H., Grigg, P., and van Noorden, G. K. 1974. Effects of visual deprivation and strabismus on the response of neurons in the visual cortex of the monkey, including studies on the striate and prestriate cortex in the normal animal. *Brain Res.* 66: 185–208. [25]

Baker, P. F., and Blaustein, M. P. 1968. Sodium-dependent uptake of calcium by crab nerve. *Biochim. Biophys. Acta* 150: 167–179. [4]

Baker, P. F., Blaustein, M. P., Hodgkin, A. L., and Steinhardt, R. A. 1969a. The influence of calcium on sodium efflux in squid axon. *J. Physiol.* 200: 431–458. [4]

Baker, P. F., Blaustein, M. P., Keynes, R. D., Manil, J., Shaw, T. I., and Steinhardt, R. A. 1969b. *J. Physiol.* 200: 459–496. [4]

Baker, P. F., Hodgkin, A. L., and Ridgeway, E. B. 1971. Depolarization and calcium entry in squid giant axons. *J. Physiol.* 218: 709–755. [4, 5]

Baker, P. F., Hodgkin, A. L., and Shaw, T. I. 1962. Replacement of the axoplasm of giant nerve fibres with artificial solutions. *J. Physiol.* 164: 330–354. [5]

Baker, R., and Llinás, R. 1971. Electrotonic coupling between neurons in the rat mesencephalic nucleus. *J. Physiol.* 212: 45–63. [9]

Baldessarini, R. J., and Tarsy, D. 1980. Dopamine and the pathophysiology of dyskinesias induced by antipsychotic drugs. *Annu. Rev. Neurosci.* 3: 23–41. [14]

Balice-Gordon, R. J., and Lichtman, J. W. 1993. In vivo observations of pre- and postsynaptic changes during the transition from multiple to single innervation at developing neuromuscular junctions. *J. Neurosci.* 13: 834–855. [23]

Balice-Gordon, R. J., and Lichtman, J. W. 1994. Long-term synapse loss induced by focal blockade of postsynaptic receptors. *Nature* 372: 519–524. [23]

Ballerini, L., Galante, M., Grandolfo, M., and Nistri, A. 1999. Generation of rhythmic patterns of activity by ventral interneurones in rat organotypic spinal slice culture. *J. Physiol.* 517: 459–475. [22]

Balslev, Y., Saunders, N. R., and Mollgard, K. 1997. Ontogenetic development of diffusional restriction to protein at the pial surface of the rat brain: An electron microscopical study. *J. Neurocytol.* 26: 133–148. [8]

Banner, L. R., and Patterson, P. H. 1994. Major changes in the expression of the mRNAs for cholinergic differentiation factor/leukemia inhibitory factor and its receptor after injury to adult peripheral nerves and ganglia. *Proc. Natl Acad. Sci. USA* 91: 7109–7113. [24]

Barbour, B., Keller, B. U., Llano, I., and Marty, A. 1994. Prolonged presence of glutamate during excitatory synaptic transmission to cerebellar Purkinje cells. *Neuron* 12: 1331–1343. [4]

Barbur, J. L., Harlow, A. J., and Plant, G. T. 1994. Insights into the different exploits of colour in the visual cortex. *Proc. R. Soc. Lond. B* 258: 327–334. [21]

Barchi, R. L. 1983. Protein components of the purified sodium channel from rat skeletal muscle sarcolemma. *J. Neurochem.* 40: 1377–1385. [3]

Barchi, R. L. 1997. Ion channel mutations and diseases of skeletal muscle. *Neurobiol. Dis.* 4: 254–264. [5]

Barde, Y-A. 1989. Trophic factors and neuronal survival. *Neuron* 2: 1525–1534. [23]

Barlow, H. B. 1953. Summation and inhibition in the frog's retina. *J. Physiol.* 119: 69–88. [19]

Barlow, H. B., Blakemore, C., and Pettigrew, J. D. 1967. The neural mechanism of binocular depth discrimination. *J. Physiol.* 193: 327–342. [21]

Barlow, H. B., Hill, R. M., and Levick, W. R. 1964. Retinal ganglion cells responding selectively to direction and speed of image motion in the rabbit. *J. Physiol.* 173: 377–407. [19]

Barnard, E. A., Webb, T. E., Simon, J., and Kunapuli, S. P. 1996. The diverse series of recombinant P2Y purinoceptors. *Ciba Found. Symp.* 198: 166–180. [14]

Baro, D. J., Levini, R. M., Kim, M. T., Willms, A. R., Lanning, C. C., Rodriguez, H. E., and Harris-Warrick, R. M. 1997. Quantitative single-cell-reverse transcription-PCR demonstrates that A-current magnitude varies as a linear function of shal gene expression in identified stomatogastric neurons. *J. Neurosci.* 17: 6597–6610. [22]

Barrionuevo, G., and Brown, T. H. 1983. Associative long-term potentiation in hippocampal slices. *Proc. Natl. Acad. Sci. USA* 80: 7347–7351. [12]

Barron, D. H., and Matthews, B. H. C. 1935. Intermittent conduction in the spinal cord. *J. Physiol.* 85: 73–103. [15]

Barry, M. F., Vickery, R. M., Bolsover, S. R., and Bindman, S. J. 1996. Intracellular studies of heterosynaptic long-term depression (LTD) in CA1 hippocampal slices. *Hippocampus* 6: 3–8. [12]

Bartlett, S. E., Reynolds, A. J., and Hendry, I. A. 1998. Retrograde axonal transport of neurotrophins: Differences between neuronal populations and implications for motor neuron disease. *Immunol. Cell Biol.* 76: 419–423. [13]

Bartol, T. M., Land, B. R., Salpeter, E. E., and Salpeter, M. M. 1991. Monte Carlo simulation of miniature endplate current generation in the vertebrate neuromuscular junction. *Biophys. J.* 59: 1290–1307. [13]

Bartos, M., Manor, Y., Nadim, F., Marder, E., and Nusbaum, M. P. 1999. Coordination of fast and slow rhythmic neuronal circuits. *J. Neurosci.* 19: 6650–6660. [15]

Basbaum, A. I., and Fields, H. L. 1979. The origin of descending pathways in the dorsolateral funiculus of the spinal cord of the cat and rat: Further studies on the anatomy of pain modulation. *J. Comp. Neurol.* 187: 513–522. [18]

Bawa, P., and Lemon, R. N. 1993. Recruitment of motor units in response to transcranial magnetic stimulation in man. *J. Physiol.* 471: 445–464. [22]

Baxter, D. A., Bittner, G. D., and Brown, T. H. 1985. Quantal mechanism of long-term synaptic potentiation. *Proc. Natl. Acad. Sci. USA* 82: 5978–5982. [12]

Bayliss, W. M., and Starling, E. H. 1902. The mechanism of pancreatic secretion. *J. Physiol.* 28: 325–353. [14]

Baylor, D. 1996. How photons start vision. *Proc. Natl. Acad. Sci. USA* 93: 540–565. [19]

Baylor, D. A. 1987. Photoreceptor signals and vision (Proctor Lecture). *Invest. Ophthalmol. Vis. Sci.* 28: 34–49. [19]

Baylor, D. A., and Burns, M. E. 1998. Control of rhodopsin activity in vision. *Eye* 12: 521–525. [19]

Baylor, D. A., and Fettiplace, R. 1977. Transmission from photoreceptors to ganglion cells in turtle retina. *J. Physiol.* 271: 391–424. [19]

Baylor, D. A., and Fuortes, M. G. F. 1970. Electrical responses of single cones in the retina of the turtle. *J. Physiol.* 207: 77–92. [5, 19]

Baylor, D. A., and Hodgkin, A. L. 1973. Detection and resolution of visual stimuli by turtle photoreceptors. *J. Physiol.* 234: 163–198. [19]

Baylor, D. A., and Nicholls, J. G. 1969. Chemical and electrical synaptic connexions between cutaneous mechanoreceptor neurones in the central nervous system of the leech. *J. Physiol.* 203: 591–609. [9]

Baylor, D. A., Fuortes, M. G. F., and O'Bryan, P. M. 1971. Receptive fields of cones in the retina of the turtle. *J. Physiol.* 214: 265–294. [19]

Baylor, D. A., Lamb, T. D., and Yau, K. W. 1979. The membrane current of single rod outer segments. *J. Physiol.* 288: 589–611. [19]

Baylor, D. A., Nunn, B. J., and Schnapf, J. L. 1984. The photocurrent, noise and spectral sensitivity of rods of the monkey *Macaca fascicularis. J. Physiol.* 357: 575–607. [19]

Bazan, N. G. 1999. Eicosanoids, platelet-activating factor and inflammation. In G. J. Siegel, B. W. Agranoff, R. W. Albers, S. K. Fisher, and M. D. Uhler (eds.), *Basic Neurochemistry: Molecular, Cellular and Medical Aspects*, 6th Ed. Lippincott-Raven, Philadelphia, pp. 731–741. [10]

Beadle, D. J., Lees, G., and Kater, S. B. 1988. *Cell Culture Approaches to Invertebrate Neuroscience.* Academic Press, London. [15]

Beam, K. G., Caldwell, J. H., and Campbell, D. T. 1985. Na channels in skeletal muscle concentrated near the neuromuscular junction. *Nature* 313: 588–590. [6]

Bean, B. P. 1981. Sodium channel inactivation in the crayfish giant axon. Must a channel open before inactivating? *Biophys. J.* 35: 595–614. [6]

Bekkers, J. M. 1994. Quantal analysis of synaptic transmission in the central nervous system. *Curr. Opin. Neurobiol.* 4: 360–365. [11]

Bekkers, J. M., and Stevens, C. F. 1990. Presynaptic mechanism for long-term potentiation in the hippocampus. *Nature* 346: 724–729. [12]

Bekoff, A. 1992. Neuroethological approaches to the study of motor development in chicks: Achievements and challenges. *J. Neurobiol.* 23: 1486–1505. [22]

Belkin, K. J., and Abrams, T. W. 1993. FMRFamide produces biphasic modulation of the LFS motor neurons in the neural circuit of the siphon withdrawal reflex of *Aplysia* by activating Na^+ and K^+ currents. *J. Neurosci.* 13: 5139–5152. [10]

Bell, J., Bolanowski, S., and Holmes, M. H. 1994. The structure and function of Pacinian corpuscles: A review. *Prog. Neurobiol.* 42: 79–128. [17]

Bellocchio, E. E., Hu, H., Pohorille, A., Chan, J., Pickel, V. M., and Edwards, R. H. 1998. The localization of the brain-specific inorganic phosphate transporter suggests a specific presynaptic role in glutaminergic transmission. *J. Neurosci.* 18: 8648–8659. [13]

Belmar, J., and Eyzaguirre, C. 1966. Pacemaker site of fibrillation potentials in denervated mammalian muscle. *J. Neurophysiol.* 29: 425–441. [24]

Benardete, E. A., Kaplan, E., and Knight, B. W. 1992. Contrast gain control in the primate retina: P cells are not X-like, some M cells are. *Vis. Neurosci.* 8: 483–486. [19]

Bennett, M. K., and Scheller, R. H. 1994. Molecular correlates of synaptic vesicle docking and fusion. *Curr. Opin. Neurobiol.* 4: 324–329. [13]

Bennett, M. V. 1997. Gap junctions as electrical synapses. *J. Neurocytol.* 26: 349–366. [9]

Bentley, D., and Caudy, M. 1983. Navigational substrates for peripheral pioneer growth cones: Limb-axis polarity cues, limb segment boundaries, and guidepost neurons. *Cold Spring Harb. Symp. Quant. Biol.* 48: 573–585. [23]

Berardi, N., and Maffei, L. 1999. From visual experience to visual function: Roles of neurotrophins. *J. Neurobiol.* 41: 119–126. [25]

Berg, D. K., and Hall, Z. W. 1975. Increased extrajunctional acetylcholine sensitivity produced by chronic postsynaptic neuromuscular blockade. *J. Physiol.* 244: 659–676. [24]

Bergmann, A., Agapite, J., and Steller, H. 1998. Mechanisms and control of programmed cell death in invertebrates. *Oncogene* 17: 3215–3223. [23]

Berlucchi, G., and Rizzolatti, G. 1968. Binocularly driven neurons in visual cortex of split-chiasm cats. *Science* 159: 308–310. [21]

Berman, D. M., and Gilman, A. G. 1998. Mammalian RGS proteins: Barbarians at the gate. *J. Biol. Chem.* 273: 1269–1272. [10]

Berne, R. M., and Levy, M. N. (eds.). 1988. *Physiology*, 2nd Ed. Mosby, St. Louis, MO. [18]

Bernstein, J. 1902. Untersuchungen zur Thermodynamik der bioelek-trischen Strome. *Pflügers Arch.* 92: 521–562. [5]

Bernstein, M., and Lichtman, J. W. 1999. Axonal atrophy: The retraction reaction. *Curr. Opin. Neurobiol.* 9: 364–370. [25]

Berretta, N., and Cherubini, E. 1998. A novel form of long-term depression in the CA1 area of the adult rat hippocampus independent of glutamate receptor activation. *Eur. J. Neurosci.* 10: 2957–2963. [12]

Berridge, M. J. 1998. Neuronal calcium signaling. *Neuron* 21: 13–26. [10]

Bettler, B., Kaupmann, K., and Bowery, N. 1998. GABAB receptors: Drugs meet clones. *Curr. Opin. Neurobiol.* 8: 345–350. [14]

Betz, H. 1990. Ligand-gated ion channels in the brain: The amino acid receptor superfamily. *Neuron* 5: 383–392. [14]

Betz, W. J., and Bewick, G. S. 1993. Optical monitoring of transmitter release and synaptic vesicle recycling at the frog neuromuscular junction. *J. Physiol.* 460: 287–309. [11, 13]

Betz, W. J., and Sakmann, B. 1973. Effects of proteolytic enzymes on function and structure of frog neuromuscular junctions. *J. Physiol.* 230: 673–688. [9]

Betz, W. J., Caldwell, J. H., and Ribchester, R. R. 1980. The effects of partial denervation at birth on the development of muscle fibers and motor units in rat lumbrical muscle. *J. Physiol.* 303: 265–279. [23]

Bevan, S., and Yeats, J. 1991. Protons activate a cation conductance in a sub-population of rat dorsal root ganglion neurones. *J. Physiol.* 433: 145–161. [17]

Bialek, W. 1987. Physical limits to sensation and perception. *Annu. Rev. Biophys. Biophys. Chem.* 16: 455–478. [17]

Biber, M. P., Kneisley, L. W., and LaVail, J. H. 1978. Cortical neurons projecting to the cervical and lumbar enlargements of the spinal cord in young and adult rhesus monkeys. *Exp. Neurol.* 59: 492–508. [22]

Bignami, A., and Dahl, D. 1974. Astrocyte-specific protein and neuroglial differentiation: An immunofluorescence study with antibodies to the glial fibrillary acidic protein. *J. Comp. Neurol.* 153: 27–38. [8]

Billups, B., and Attwell, D. 1996. Modulation of non-vesicular glutamate release by pH. *Nature* 379: 171–174. [8]

Bina, K. G., Park, M., and O'Dowd, D. K. 1998. Regulation of alpha7 nicotinic acetylcholine receptors in mouse somatosensory cortex following whisker removal at birth. *J. Comp. Neurol.* 397: 1–9. [18]

Bingman, V. P., and Benvenuti, S. 1996. Olfaction and the homing ability of pigeons in the southeastern United States. *J. Exp. Zool.* 276: 186–192. [17]

Birch, E. E., Swanson, W. H., Stager, D. R., Woody, M., and Everett, M. 1993. Outcome after very early treatment of dense congenital unilateral cataract. *Invest. Ophthalmol. Vis. Sci.* 34: 3687–3699. [25]

Birks, R. I., and MacIntosh, F. C. 1961. Acetylcholine metabolism of a sympathetic ganglion. *J. Biochem. Physiol.* 39: 787–827. [13]

Birks, R., Huxley, H. E., and Katz, B. 1960. The fine structure of the neuromuscular junction of the frog. *J. Physiol.* 150: 134–144. [11]

Birks, R., Katz, B., and Miledi, R. 1960. Physiological and structural changes at the amphibian myoneural junction in the course of nerve degeneration. *J. Physiol.* 150: 145–168. [11]

Birnbaumer, L., Abramowitz, J., and Brown, A. M. 1990. Receptor-effector coupling by G proteins. *Biochim. Biophys. Acta* 1031: 163–224. [10]

Bixby, J. L., and Van Essen, D. C. 1979. Competition between foreign and original nerves in adult mammalian skeletal muscle. *Nature* 282: 726–728. [24]

Bixby, J. L., Lilien, J., and Reichardt, L. F. 1988. Identification of the major proteins that promote neuronal process outgrowth on Schwann cells in vitro. *J. Cell Biol.* 107: 353–361. [23]

Björklund, A. 1991. Neural transplantation—An experimental tool with clinical possibilities. *Trends Neurosci.* 14: 319–322. [24]

Björklund, A., Dunnett, S. B., Stenevi, U., Lewis, N. E., and Iversen, S. D. 1980. Reinnervation of the denervated striatum by substantia nigra transplants: Functional consequences as revealed by pharmacological and sensorimotor testing. *Brain Res.* 199: 307–333. [24]

Black, I. B. 1978. Regulation of autonomic development. *Annu. Rev. Neurosci.* 1: 183–214. [23]

Black, J. A., and Waxman, S. G. 1988. The perinodal astrocyte. *Glia* 1: 169–183. [8]

Black, J. W. 1976. Ahlquist and the development of beta-adrenoceptor antagonists. *Postgrad. Med. J.* 52: 11–13. [16]

Black, J. W., and Prichard, B. N. 1973. Activation and blockade of beta adrenoceptors in common cardiac disorders. *Br. Med. Bull.* 29: 163–167. [16]

Blackman, J. G., and Purves, R. D. 1969. Intracellular recordings from the ganglia of the thoracic sympathetic chain of the guinea-pig. *J. Physiol.* 203: 173–198. [11]

Blackshaw, S. E. 1981. Morphology and distribution of touch cell terminals in the skin of the leech. *J. Physiol.* 320: 219–228. [15]

Blackshaw, S. E., and Nicholls, J. G. 1995. Neurobiology and development of the leech. *J. Neurobiol.* 27: 267–276. [15]

Blackshaw, S. E., and Thompson, S. W. 1988. Hyperpolarizing responses to stretch in sensory neurones innervating leech body wall muscle. *J. Physiol.* 396: 121–137. [15, 17]

Blackshaw, S. E., Nicholls, J. G., and Parnas, I. 1982. Expanded receptive fields of cutaneous mechanoreceptor cells after single neurone deletion in leech central nervous system. *J. Physiol.* 326: 261–268. [15, 24]

Blair, L. A. C., Levitan, E. S., Marshall, J., Dionne, V. E., and Barnard, E. A. 1988. Single subunits of the $GABA_A$ receptor form ion channels with properties of the native receptor. *Science* 242: 577–579. [14]

Blakemore, C., and Van Sluyters, R. C. 1974. Reversal of the physiological effects of monocular deprivation in kittens: Further evidence for a sensitive period. *J. Physiol.* 237: 195–216. [25]

Blanchet, C., Erostegui, C., Sugasawa, M., and Dulon, D. 1996. Acetylcholine-induced potassium current of guinea pig outer hair cells: Its dependence on a calcium influx through nicotinic-like receptors. *J. Neurosci.* 16: 2574–2584. [18]

Blasdel, G. G. 1989. Visualization of neuronal activity in monkey striate cortex. *Annu. Rev. Physiol.* 51: 561–581. [21]

Blasdel, G. G., and Lund, J. S. 1983. Termination of afferent axons in macaque striate cortex. *J. Neurosci.* 3: 1389–1413. [20]

Blasdel, G. G., Obermayer, K., and Kiorpes, L. 1995. Organization of ocular dominance and orientation columns in the striate cortex of neonatal macaque monkeys. *Vis. Neurosci.* 12: 589–603. [21]

Blauert, J. 1982. Binaural localization. *Scand. Audiol. Suppl.* 15: 7–26. [18]

Bleuel, A., de Gasparo, M., Whitebread, S., Puttner, I., and Monard, D. 1995. Regulation of protease nexin-1 expression in cultured schwann cells is mediated by angiotensin II receptors. *J. Neurosci.* 15: 759–761. [8]

Bliss, T. V. P., and Lømo, T. 1973. Long-lasting potentiation of synaptic transmission in the dentate of the anesthetized rabbit following stimulation of the perforant path. *J. Physiol.* 232: 331–356. [12]

Boadle-Biber, M. C. 1993. Regulation of serotonin synthesis. *Prog. Biophys. Mol. Biol.* 60: 1–15. [13]

Boekhoff, I., Tareilus, E., Strotmann, J., and Breer, H. 1990. Rapid activation of alternative second messenger pathways in olfactory cilia from rats by different odorants. *EMBO J.* 9: 2453–2458. [17]

Boivin, D. B., Duffy, J. F., Kronauer, R. E., and Czeisler, C. A. 1996. Dose response relationships for resetting of human circadian clock by light. *Nature* 379: 540–542. [16]

Boncinelli, E., Mallamaci, A., and Broccoli, V. 1998. Body plan genes and human malformation. *Adv. Genet.* 38: 1–29. [15, 23]

Bondy, C. A., Whitnall, M. H., Brady, L. S., and Gainer, H. 1989. Coexisting peptides in hypothalamic neuroendocrine systems: Some functional implications. *Cell. Mol. Neurobiol.* 9: 427–446. [13]

Bonhoeffer, T., and Grinvald, A. 1991. Iso-orientation domains in cat visual cortex are arranged in pin-wheel-like patterns. *Nature* 353: 429–431. [21]

Boring, E. G. 1942. *Sensation and Perception in the History of Experimental Psychology.* Appleton-Century, New York. [17]

Borjigin, J., Li, X., and Snyder, S. H. 1999. The pineal gland and melatonin: Molecular and pharmacologic regulation. *Annu. Rev. Pharmacol. Toxicol.* 39: 53–65. [16]

Bormann, J., and Feigenspan, A. 1995. $GABA_C$ receptors. *Trends Neurosci.* 18: 515–519. [3]

Borst, J. G. G., and Sakmann, B. 1996. Calcium influx and transmitter release in a fast CNS synapse. *Nature* 383: 431–434. [11]

Bosher, S. K., and Warrren, R. L. 1978. Very low calcium content of cochlear endolymph, an extracellular fluid. *Nature* 273: 377–378. [18]

Bosking, W. H., Zhang, Y., Schofield, B., and Fitzpatrick, D. 1997. Orientation selectivity and the arrangement of horizontal connections in tree shrew striate cortex. *J. Neurosci.* 17: 2112–2127. [21]

Bostock, H., and Sears, T. A. 1978. The internodal axon membrane: Electrical excitability and continuous conduction in segmental demyelination. *J. Physiol.* 280: 273–301. [7]

Bostock, H., Sears, T. A., and Sherratt, R. M. 1981. The effects of 4-aminopyridine and tetraethylammonium on normal and demyelinated mammalian nerve fibres. *J. Physiol.* 313: 301–315. [7]

Bothwell, M. 1995. Functional interactions of neurotrophins and neurotrophin receptors. *Annu. Rev. Neurosci.* 18: 223–253. [23]

Bouvier, M., Szatkowski, M., Amato, A., and Attwell, D. 1992. The glial cell glutamate uptake carrier countertransports pH-changing ions. *Nature* 360: 471–474. [4]

Bouzat, C., and Barrantes, F. J. 1996. Modulation of muscle nicotinic acetylcholine receptors by the glucocorticoid hydrocortisone. Possible allosteric mechanism of channel blockade. *J. Biol. Chem.* 271: 25835–25841. [9]

Bowe, M. A., and Fallon, J. R. 1995. The role of agrin in synapse formation. *Annu. Rev. Neurosci.* 18: 443–462. [23, 24]

Bowling, D. B., and Michael, C. R. 1980. Projection patterns of single physiologically characterized optic tract fibers in the cat. *Nature* 286: 899–902. [20]

Bowling, D., Nicholls, J. G., and Parnas, I. 1978. Destruction of a single cell in the central nervous system of the leech as a means of analyzing its connections and functional role. *J. Physiol.* 282: 169–180. [15, 24]

Boycott, B. B., and Dowling, J. E. 1969. Organization of primate retina: Light microscopy. *Philos. Trans. R. Soc. Lond. B* 255: 109–184. [19]

Boycott, B., and Wässle, H. 1999. Parallel processing in the mammalian retina (Proctor Lecture). *Invest. Ophthalmol. Vis. Sci.* 40: 1313–1327. [19]

Boyd, I. A., and Martin, A. R. 1956. The end-plate potential in mammalian muscle. *J. Physiol.* 132: 74–91. [9, 11]

Bracci, E., Beato, M., and Nistri, A. 1998. Extracellular K^+ induces locomotor-like patterns in the rat spinal cord in vibro: Comparison with NMDA or 5-HT induced activity. *J. Neurophysiol.* 79: 2643–2652. [22]

Brady, S. T., Lasek, R. J., and Allen, R. D. 1982. Fast axonal transport in extruded axoplasm from squid giant axon. *Science* 218: 1129–1131. [13]

Brainard, M. S., and Knudsen, E. I. 1998. Sensitive periods for visual calibration of the auditory space map in the barn owl optic tectum. *J. Neurosci.* 18: 3929–3942. [25]

Brandstatter, J. H., Koulen, P., and Wässle, H. 1988. Diversity of glutamate receptors in the mammalian retina. *Vision Res.* 38: 1385–1397. [19]

Brannstrom, T. 1993. Quantitative synaptology of functionally different types of cat medial gastrocnemius alpha-motoneurons. *J. Comp. Neurol.* 330: 439–454. [22]

Bray, G. M., Villegas-Perez, M. P., Vidal-Sanz, M., Carter, D. A., and Aguayo, A. J. 1991. Neuronal and nonneuronal influences on retinal ganglion cell survival, axonal regrowth, and connectivity after axotomy. *Ann. N.Y. Acad. Sci.* 633: 214–228. [24]

Bredt, D. S., and Snyder, S. H. 1989. Nitric oxide mediates glutamate-linked enhancement of cGMP levels in the cerebellum. *Proc. Natl. Acad. Sci. USA* 86: 9030–9033. [10]

Bredt, D. S., Glatt, C. E., Hwang, P. M., Fotuhi, M., Dawson, T. M., and Snyder, S. H. 1991. Nitric oxide synthase protein and mRNA are discretely localized in neuronal populations of the mammalian CNS together with NADPH diaphorase. *Neuron* 7: 615–624. [14]

Breer, H., Boekhoff, I., and Tareilus, E. 1990. Rapid kinetics of second messenger formation in olfactory transduction. *Nature* 345: 65–68. [17]

Bregman, B. S., Kunkel-Bagden, E., Schnell, L., Dai, H. N., Gao, D., and Schwab, M. E. 1995. Recovery from spinal cord injury mediated by antibodies to neurite growth inhibitors. *Nature* 378: 498–501. [24]

Breitwieser, G. E., and Szabo, G. 1985. Uncoupling of cardiac muscarinic and β-adrenergic receptors from ion channels by a guanine nucleotide analogue. *Nature* 317: 538–540. [10]

Brenner, H. R., and Johnson, E. W. 1976. Physiological and morphological effects of post-ganglionic axotomy on presynaptic nerve terminals. *J. Physiol.* 260: 143–158. [24]

Brenner, H. R., and Martin, A. R. 1976. Reduction in acetylcholine sensitivity of axotomized ciliary ganglion cells. *J. Physiol.* 260: 159–175. [24]

Brew, H., Gray, P. T., Mobbs, P., and Attwell, D. 1986. Endfeet of retina glial cells have higher densities of ion channels that mediate K^+ buffering. *Nature* 324: 466–468. [8]

Bridge, J. H. B., Smolley, J. R., and Spitzer, K. W. 1990. The relationship between charge movements associated with I_{Ca} and $I_{Na\text{-}Ca}$ in cardiac myocytes. *Science* 248: 376–378. [4]

Brightman, M. W., and Reese, T. S. 1969. Junctions between intimately apposed cell membranes in the vertebrate brain. *J. Cell Biol.* 40: 668–677. [8]

Brightman, M. W., Reese, T. S., and Feder, N. 1970. Assessment with the electron microscope of the permeability to peroxidase of cerebral endothelium in mice and sharks. In E. H. Thaysen (ed.), *Capillary Permeability* (Alfred Benzon Symposium II). Munskgaard, Copenhagen, Denmark. [8]

Brisben, A. J., Hsiao, S. S., and Johnson, K. O. 1999. Detection of vibration transmitted through an object grasped in the hand. *J. Neurophysiol.* 81: 1548–1558. [18]

Briscoe, J., Sussel, L., Serup, P., Hartigan-O'Connor, D., Jessell, T. M., Rubenstein, J. L. R., and Ericson, J. 1999. Homeobox gene Nkx2.2 and specificiation of neuronal identity by graded Sonic hedgehog signalling. *Nature* 398: 622–627. [23]

Brittis, P. A., Lemmon, V., Rutishauser, U., and Silver, J. 1995. Unique changes of ganglion cell growth cone behaviour following cell adhesion molecule perturbations: A time-lapse study of the living retina. *Mol. Cell. Neurosci.* 6: 433–449. [23]

Brittis, P. A., Silver, J., Walsh, F. S., and Doherty, P. 1996. FGF receptor function is required for the orderly projection of ganglion cell axons in the developing mammalian retina. *Mol. Cell. Neurosci.* 8: 120–128. [23]

Brivanlou, I. H., Warland, D. K., and Meister, M. 1998. Mechanisms of concerted firing among retinal ganglion cells. *Neuron* 20: 527–539. [9, 25]

Broadley, K. J. 1996. *Autonomic Pharmacology*. Taylor & Francis, London. [16]

Brodfuehrer, P. D., Debski, E. A., O'Gara, B. A., and Friesen, W. O. 1995. Neuronal control of leech swimming. *J. Neurobiol.* 27: 403–418. [15]

Brodin, L., Shupliakov, O., Pieribone, V. A., Hellgren, J., and Hill, R. H. 1994. The reticulospinal glutamate synapse in lamprey: Plasticity and presynaptic variability. *J. Neurophysiol.* 72: 592–604. [9]

Bronner-Fraser, M. 1985. Alterations in neural crest migration by a monoclonal antibody that affects cell adhesion. *J. Cell Biol.* 101: 610–617. [23]

Brooks, C. M. 1988. The history of thought concerning the hypothalamus and its functions. *Brain Res. Bull.* 20: 657–667. [16]

Brooks, V. B. 1956. An intracellular study of the action of repetitive nerve volleys and of botulinum toxin on miniature endplate potentials. *J. Physiol.* 134: 264–277. [11]

Brose, K., Bland, K. S., Wang, K. H., Arnott, D., Henzel, W., Goodman, C. S., Tessier-Lavigne, M., and Kidd, T. 1999. Slit proteins bind Robo receptors and have an evolutionarily conserved role in repulsive axon guidance. *Cell* 96: 795–806. [23]

Brown, A. G., and Fyffe, R. E. W. 1981. Direct observations on the contacts made between Ia afferent fibers and α-motoneurones in the cat's lumbosacral spinal cord. *J. Physiol.* 313: 121–140. [22]

Brown, A. M., and Birnbaumer, L. 1990. Ionic channels and their regulation by G protein subunits. *Annu. Rev. Physiol.* 52: 197–213. [10]

Brown, D. 1988. M-currents: An update. *Trends Neurosci.* 11: 294–299. [16]

Brown, D. A., Higashida, H., Noda, M., Ishizaka, N., Hashii, M., Hoshi, W., Yokoyama, S., Fukuda, K., Katayama, M., Nukada, T., Kameyama, K., Robbins, J., Marsh, S. J., and Selyanko, A. A. 1993. Coupling of muscarinic receptor subtypes to ion channels: Experiments on neuroblastoma hybrid cells. *Ann. N.Y. Acad. Sci.* 707: 237–258. [16]

Brown, G. L. 1937. The actions of acetylcholine on denervated mammalian and frog's muscle. *J. Physiol.* 89: 438–461. [24]

Brown, H. C., and Perry, V. H. 1998. Differential adhesion of macrophages to white and grey matter in an in vitro assay. *Glia* 23: 361–373. [8]

Brown, H. M., Ottoson, D., and Rydqvist, B. 1978. Crayfish stretch receptor: An investigation with voltage-clamp and ion-sensitive electrodes. *J. Physiol.* 284: 155–179. [17]

Brown, L. L., Schneider, J. S., and Lidsky, T. I. 1997. Sensory and cognitive functions of the basal ganglia. *Curr. Opin. Neurobiol.* 7: 157–163. [22]

Brown, M. C. 1987. Morphology of labeled afferent fibers in the guinea pig cochlea. *J. Comp. Neurol.* 260: 591–604. [18]

Brown, M. C., and Ironton, R. 1977. Motor neurone sprouting induced by prolonged tetrodotoxin block of nerve action potentials. *Nature* 265: 459–461. [24]

Brown, M. C., Holland, R. L., and Hopkins, W. G. 1981. Motor nerve sprouting. *Annu. Rev Neurosci.* 4: 17–42. [24]

Brown, M. C., Hopkins, W. G., and Keynes, R. J. 1982. Short- and long-term effects of paralysis on the motor innervation of two different neonatal mouse muscles. *J. Physiol.* 329: 439–450. [23]

Brown, M. C., Jansen, J. K. S., and Van Essen, D. 1976. Polyneuronal innervation of skeletal muscle in new-born rats and its elimination during maturation. *J. Physiol.* 261: 387–422. [23]

Brown, P. K., and Wald, G. 1963. Visual pigments in human and monkey retinas. *Nature* 200: 37–43. [19]

Brown, T. G. 1911. The intrinsic factor in the act of progression in the mammal. *Proc. R. Soc. Lond. B* 84: 308–319. [22]

Browne, D. L., Gancher, S. T., Nutt, J. G., Brunt, E. R., Smith, E. A., Kramer, P., and Litt, M. 1994. Episodic ataxia/myokymia syndrome is associated with point mutations in the human potassium channel gene, KCNA1. *Nature Genet.* 8: 136–140. [22]

Brownell, W. E., Bader, C. R., Bertrand, D., and de Ribaupierre, Y. 1985. Evoked mechanical responses of isolated cochlear outer hair cells. *Science* 227: 194–196. [18]

Bruce, E. N., and Cherniak, N. S. 1987. Central chemoreceptors. *J. Appl. Physiol.* 62: 389–402. [22]

Brundege, J. M., and Dunwiddie, T. V. 1997. Role of adenosine as a modulator of synaptic activity in the central nervous system. *Adv. Pharmacol.* 39: 353–391. [13, 14]

Brunet, L. J., Gold, G. H., and Ngai, J. 1996. General anosmia caused by a targeted disruption of the mouse olfactory cyclic nucleotide-gated cation channel. *Neuron* 17: 681–693. [17]

Bruns, D., and Jahn, R. 1995. Real-time measurement of transmitter release from single synaptic vesicles. *Nature* 377: 62–65. [15]

Brussaard, A. B. 1997. Antisense oligonucleotides induce functional deletion of ligand gated ion channels in cultured neurons and brain explants. *J. Neurosci. Methods* 71: 55–64. [14]

Buck, L., and Axel, R. 1991. A novel multigene family may encode odorant receptors: A molecular basis for odor recognition. *Cell* 65: 175–187. [17]

Bucossi, G., Nizzari, M., and Torre, V. 1997. Single-channel properties of ionic channels gated by cyclic nucleotides. *Biophys. J.* 72: 1165–1181. [19]

Buell, T. N., and Hafter, E. R. 1988. Discrimination of interaural differences of time in the envelopes of high-frequency signals: Integration times. *J. Acoust. Soc. Am.* 84: 2063–2066. [18]

Bullock, T. H., and Hagiwara, S. 1957. Intracellular recording from the giant synapse of the squid. *J. Gen. Physiol.* 40: 565–577. [11]

Bunge, R. P. 1968. Glial cells and the central myelin sheath. *Physiol. Rev.* 48: 197–251. [8]

Buonomano, D. V., and Merzenich, M. M. 1998. Cortical plasticity: From synapses to maps. *Annu. Rev. Neurosci.* 21: 149–186. [25]

Burden, S. J., Sargent, P. B., and McMahan, U. J. 1979. Acetylcholine receptors in regenerating muscle accumulate at original synaptic sites in the absence of the nerve. *J. Cell Biol.* 82: 412–425. [9, 24]

Burgess, D. L., and Noebels, J. L. 1999. Voltage-dependent calcium channel mutations in neurological disease. *Ann. N.Y. Acad. Sci.* 868: 199–212. [26]

Burgess, G. M., Mullaney, I., McNeill, M., Dunn, P. M., and Rang, H. P. 1989. Second messengers involved in the mechanism of action of bradykinin in sensory neurons in culture. *J. Neurosci.* 9: 3314–3325. [17]

Burgess, P. R., and Perl, E. R. 1967. Myelinated afferent fibres responding specifically to noxious stimulation of the skin. *J. Physiol.* 190: 541–562. [17]

Burgess, P. W., and Shallice, T. 1996. Bizarre responses, rule detection and frontal lobe lesions. *Cortex* 32: 241–259. [26]

Burgess, R. W., Nguyen, Q. T., Son, Y. J., Lichtman, J. W., and Sanes, J. R. 1999. Alternatively spliced isoforms of nerve- and muscle-derived agrin: Their roles at the neuromuscular junction. *Neuron* 23: 33–44. [23]

Burke, R. E., and Glenn, L. L. 1996. Horseradish peroxidase study of the spatial and electrotonic distribution of group Ia synapses on type-identified ankle extensor motoneurons in the cat. *J. Comp. Neurol.* 372: 465–485. [22]

Burnstock, G. 1993. Physiological and pathological roles of purines: An update. *Drug Dev. Res.* 28: 195–206. [16]

Burnstock, G. 1995. Noradrenaline and ATP: Cotransmitters and neuromodulators. *J. Physiol. Pharmacol.* 46: 365–384. [13, 16]

Burnstock, G. 1996. P2-purinoceptors: Historical perspective and classification. In *P2-Purinoceptors: Localization, Function and Transduction Mechanisms* (Ciba Foundation Symposium 198). Wiley, London, pp. 1–34. [3]

Burnstock, G. 1996. Purinoceptors: Ontogeny and phylogeny. *Drug Dev. Res.* 39: 204–242. [3]

Burnstock, G. 1997. The past, present and future of purine nulceotides as signalling molecules. *Neuropharmacology* 36: 1127–1139. [14]

Burnstock, G., and Holman, M. E. 1961. The transmission of excitation from autonomic nerve to smooth muscle. *J. Physiol.* 155: 115–133. [16]

Burnstock, G., and Wood, J. N. 1996. Purinergic receptors: Their role in nociception and primary afferent neurotransmission. *Curr. Opin. Neurobiol.* 6: 526–532. [14]

Bursztajn, S., Berman, S. A., and Gilbert, W. 1989. Differential expression of acetylcholine receptor mRNA in nuclei of cultured muscle cells. *Proc. Natl. Acad. Sci. USA* 86: 2928–2932. [24]

Butt, A. M., and Ransom, B. R. 1993. Morphology of astrocytes and oligodendrocytes during development in the intact rat optic nerve. *J. Comp. Neurol.* 338: 141–158. [8]

Buttner, N., Siegelbaum, S. A., and Volterra, A. 1989. Direct modulation of *Aplysia* S-K^+ channels by a 12-lipoxygenase metabolite of arachidonic acid. *Nature* 342: 553–555. [10]

Cabelli, R. J., Hohn, A., and Shatz, C. J. 1995. Inhibition of ocular dominance column formation by infusion of NT-4/5 or BDNF. *Science* 267: 1662–1666. [23]

Cain, D. P. 1998. Testing the NMDA, long-term potentiation, and cholinergic hypothesis of spatial learning. *Neurosci. Biobehav. Rev.* 22: 181–193. [12]

Calabrese, R. L. 1998. Cellular, synaptic, network, and modulatory mechanisms involved in rhythm generation. *Curr. Opin. Neurobiol.* 8: 710–717. [15]

Caldwell, J. H., and Daw, N. W. 1978. Effects of picrotoxin and strychnine on rabbit retinal ganglion cells: Changes in centre surround receptive fields. *J. Physiol.* 276: 299–310. [14]

Caldwell, P. C., Hodgkin, A. L., Keynes, R. D., and Shaw, T. L. 1960. The effects of injecting "energy-rich" phosphate compounds on the active transport of ions in the giant axons of *Loligo*. *J. Physiol.* 152: 561–590. [4]

Callaway, E. M. 1998. Local circuits in primary visual cortex of the macaque monkey. *Annu. Rev. Neurosci.* 21: 47–74. [20, 21]

Callaway, E. M., and Wiser, A. K. 1996. Contributions of individual layer 2-5 spiny neurons to local circuits in macaque primary visual cortex. *Vis. Neurosci.* 13: 907–922. [21]

Calvert, G. A., Bullmore, E. T., Brammer, M. J., Campbell, R., Williams, S. C., McGuire, P. K., Woodruff, P. W., Iversen, S. D., and David, A. S. 1997. Activation of auditory cortex during silent lipreading. *Science* 276: 593–596. [18]

Caminiti, R., Johnson, P. B., and Urbano, A. 1990. Making arm movements within different parts of space: Dynamic aspects in the primate motor cortex. *J. Neurosci.* 10: 2039–2058. [22]

Cammack, J. N., and Schwartz, E. A. 1993. Ions required for the electrogenic transport of GABA by horizontal cells of the catfish retina. *J. Physiol.* 472: 81–102. [4, 11]

Cammack, J. N., Rakhilin, S. V., and Schwartz, E. A. 1994. A GABA transporter operates asymmetrically and with variable stoichiometry. *Neuron* 13: 949–960. [11]

Campenot, R. B. 1977. Local control of neurite development by nerve growth factor. *Proc. Natl. Acad. Sci. USA* 74: 4516–4519. [23]

Campenot, R. B. 1982. Development of sympathetic neurons in compartmentalized cultures. II. Local control of neurite survival by nerve growth factor. *Dev. Biol.* 93: 13–21. [23]

Canessa, C. M., Schold, L., Buell, B., Thorens, I., Gautschi, I., Horisberger, J., and Rossier, B. 1994. Amiloride-sensitive epithelial Na^+ channel is made of three homologous subunits. *Nature* 367: 463–467. [17]

Cannon, S. C. 1996. Ion channel defects and aberrant excitability in myotonia and periodic paralysis. *Trends Neurosci.* 19: 3–10. [5]

Cannon, W. B., and Rosenbluth, A. 1949. *The Supersensitivity of Denervated Structures: Law of Denervation*. Macmillan, New York. [24]

Cantino, D., and Mugnani, E. 1975. The structural basis for electrotonic coupling in the avian ciliary ganglion: A study with thin sectioning and freeze-fracturing. *J. Neurocytol.* 4: 505–536. [7]

Cao, Y. Q., Mantyh, P. W., Carlson, E. J., Gillespie, A. M., Epstein, C. J., and Basbaum, A. I. 1998. Primary afferent tachykinins are required to experience moderate to intense pain. *Nature* 392: 390–394. [14]

Capecchi, M. R. 1997. Hox genes and mammalian development. *Cold Spring Harb. Symp. Quant. Biol.* 62: 273–281. [23]

Capsoni, S., Tongiorgi, E., Cattaneo, A., and Domenici, L. 1999. Dark rearing blocks the developmental down-regulation of brain-derived neurotrophic factor messenger RNA expression in layers IV and V of the rat visual cortex. *Neuroscience* 88: 393–403. [25]

Caputo, C., Bezanilla, F., and DiPolo, R. 1989. Currents related to the sodium-calcium exchange in squid giant axon. *Biochim. Biophys. Acta* 986: 250–256. [4]

Carafoli, E. 1994. Plasma membrane calcium ATPase: 15 years of work on the purified enzyme. *FASEB J.* 8: 993–1002. [4]

Carlson, M., Hubel, D. H., and Wiesel, T. N. 1986. Effects of monocular exposure to oriented lines on monkey striate cortex. *Brain Res.* 390: 71–81. [25]

Caroni, P., and Grandes, P. 1990. Nerve sprouting in innervated adult muscle induced by exposure to elevated levels of insulin-like growth factors. *J. Cell Biol.* 110: 1307–1317. [23]

Caroni, P., and Schwab, M. E. 1988. Two membrane protein fractions from rat central myelin with inhibitory properties for neurite growth and fibroblast spreading. *J. Cell Biol.* 106: 1281–1288. [8]

Caroni, P., Rotsler, S., Britt, J. C., and Brenner, H. R. 1993. Calcium influx and protein phosphorylation mediate the metabolic stabilization of synaptic acetylcholine receptors in muscle. *J. Neurosci.* 13: 1315–1325. [24]

Carrol, R. C., Lissin, D. V., von Zastrow, M., Nicoll, R. A., and Malenka, R. C. 1999. Rapid redistribution of glutamate receptors contributes to long-term depression in hippocampal cultures. *Nature Neurosci.* 2: 454–460. [12]

Casagrande, V. A. 1994. A third parallel visual pathway to primate area V1. *Trends Neurosci.* 17: 305–310. [20, 21]

Casey, P. J., Graziano, M. P., Freissmuth, M., and Gilman, A. G. 1988. Role of G proteins in transmembrane signaling. *Cold Spring Harb. Symp. Quant. Biol.* 53: 203–208. [10]

Cassel, J-C., and Jeitsch, H. 1995. Serotonergic modulation of cholinergic function in the central nervous system: Cognitive implications. *Neuroscience* 69: 1–41. [14]

Cassone, V. M. 1998. Melatonin's role in vertebrate circadian rhythms. *Chronobiol. Int.* 15: 457–473. [16]

Castro-Alamancos, M. A., and Calcagnotto, M. E. 1999. Presynaptic long-term potentiation in corticothalamic synapses. *J. Neurosci.* 19: 9090–9097. [12]

Catarsi, S., and Drapeau, P. 1996. Modulation and selection of neurotransmitter responses during synapse formation between identified leech neurons. *Cell Mol. Neurobiol.* 16: 699–713. [15]

Caterina, M. J., Schumacher, M. A., Tominaga, M., Rosen, T. A., Levine, J. D., and Julius, D. 1997. The capsaicin receptor: A heat-activated ion channel in the pain pathway. *Nature* 389: 816–824. [17]

Catterall, W. A. 1980. Neurotoxins that act on voltage-sensitive sodium channels in excitable membranes. *Ann. Rev. Pharmacol. Toxicol.* 20: 15–43. [6]

Caviness, V. S., Jr. 1982. Neocortical histogenesis in normal and reeler mice: A developmental study based upon [3H]-thymidine autoradiography. *Dev. Brain Res.* 4: 293–302. [23, 25]

Cazalets, J. R., Borde, M., and Clarac, F. 1996. The synaptic drive from the spinal locomotor network to motoneurons in the newborn rat. *J. Neurosci.* 16: 298–306. [22]

Ceccarelli, B., and Hurlbut, W. P. 1980. Vesicle hypothesis of the release of quanta of acetylcholine. *Physiol. Rev.* 60: 396–441. [11]

Cepko, C. L., Austin, C. P., Yang, X., Alexiades, M., and Ezzeddine, D. 1996. Cell fate determination in the vertebrate retina. *Proc. Natl. Acad. Sci. USA* 93: 589–595. [23]

Cervetto, L., Lagnado, L., Perry, R. J., Robinson, D. W., and McNaughton, P. A. 1989. Extrusion of calcium from rod outer segments is driven by both sodium and potassium gradients. *Nature* 337: 740–743. [4]

Cesare, P., and McNaughton, P. A. 1996. A novel heat-activated current in nociceptive neurones and its sensitization by bradykinin. *Proc. Natl. Acad. Sci. USA* 93: 15435–15439. [17]

Chapman, B., Stryker, M. P., and Bonhoeffer, T. 1996. Development of orientation preference maps in ferret primary visual cortex. *J. Neurosci.* 16: 6443–6453. [25]

Chau, C., Barbeau, H., and Rossignol, S. 1998. Effects of intrathecal alpha1- and alpha2 agonists and norepinephrine on locomotion in chronic spinal cats. *J. Neurophysiol.* 79: 2941–2963. [22]

Chaudhari, N., Landin, A. M., and Roper, S. D. 2000. A metabotropic glutamate receptor variant functions as a taste receptor. *Nature Neurosci.* 3: 113–119. [17]

Chaudhari, N., Yang, H., Lamp, C., Delay, E., Cartford, C., Than, T., and Roper, S. 1996. The taste of MSG: Membrane receptors in taste buds. *J. Neurosci.* 16: 3817–3826. [17]

Chen, A., Kumar, S. M., Sahley, C. L., and Muller, K. J. 2000. Nitric oxide influences injury-induced microglial migration and accumulation in the leech CNS. *J. Neurosci.* 20: 1036–1043. [8]

Chen, C. C., Akopian, A. N., Sivilotti, L., Colquhoun, D., Burnstock, G., and Wood, J. N. 1995a. A P2X purinoceptor expressed by a subset of sensory neurons. *Nature* 377: 428–431. [17]

Chen, J., Makino, C. L., Peachey, N. S., Baylor, D. A., and Simon, M. I. 1995b. Mechanisms of rhodopsin inactivation in vivo as revealed by a COOH-terminal truncation mutant. *Science* 267: 374–377. [19]

Chen, W., Zhu, X. H., Thulborn, K. R., and Ugurbil, K. 1999. Retinotopic mapping of lateral geniculate nucleus in humans using functional magnetic resonance imaging. *Proc. Natl. Acad. Sci. USA* 96: 2430–2434. [20]

Cheney, D. P., and Fetz, E. E. 1980. Functional classes of primate corticomotoneuronal cells and their relation to active force. *J. Neurophysiol.* 44: 773–791. [22]

Cheng, H., Lederer, M. R., Lederer, W. J., and Channell, M. B. 1996. Calcium sparks and $[Ca^{2+}]_i$ waves in cardiac myocytes. *Am. J. Physiol.* 270: C148–C159. [10]

Cherubini, E., Gaiarsa, J. L., and Ben-Ari, Y. 1991. GABA: An excitatory transmitter in early postnatal life. *Trends Neurosci.* 14: 515–519. [16]

Chiu, S. Y., and Ritchie, J. M. 1981. Evidence for the presence of potassium channels in the paranodal region of acutely demyelinated mammalian nerve fibres. *J. Physiol.* 313: 415–437. [7]

Chiu, S. Y., Ritchie, J. M., Rogart, R. B., and Stagg, D. 1979. A quantitative description of potassium currents in the paranodal region of acutely demyelinated mammalian nerve fibres. *J. Physiol.* 292: 149–166. [6]

Choi, D. W., Koh, J-Y., and Peters, S. 1988. Pharmacology of glutamate neurotoxicity in cortical cell culture: Attenuation by NMDA antagonists. *J. Neurosci.* 8: 185–196. [14]

Choi, K. L., Aldrich, R. W., and Yellen, G. 1991. Tetraethylammonium blockade distinguishes two inactivation states in voltage-gated K^+ channels. *Proc. Natl. Acad. Sci. USA* 88: 5092–5095. [6]

Chow, D. C., and Forte, J. G. 1995. Functional significance of the β-subunit for heterodimeric p-type ATPases. *J. Exp. Biol.* 198: 1–17. [4]

Christie, B. R., and Abraham, W. C. 1994. L-type voltage-sensitive calcium channel antagonists block heterosynaptic long-term facilitation depression in the dentate gyrus of anesthetized rats. *Neurosci. Lett.* 167: 41–45. [12]

Christofi, G., Nowicky, M. V., Bolsover, S. R., and Bindman, L. J. 1993. The postsynaptic induction of nonassociative long-term depression of excitatory synaptic transmission in rat hippocampal slices. *J. Neurophysiol.* 69: 219–229. [12]

Chua, M., and Betz, W. J. 1991. Characterisics of a non-selective cation channel in the surface membrane of adult rat skeletal muscle. *Biophys. J.* 59: 1251–1260. [5]

Clapham, D. E., and Neer, E. J. 1997. G protein βγ subunits. *Annu. Rev. Pharmacol. Toxicol.* 37: 167–203. [10]

Clarey, J. C., Barone, P., and Imig, T. J. 1994. Functional organization of sound direction and sound pressure level in primary auditory cortex of the cat. *J. Neurophysiol.* 72: 2383–2405. [18]

Classen, J., Liepert, J., Wise, S. P., Hallett, M., and Cohen, L. G. 1998. Rapid plasticity of human cortical movement representation induced by practice. *J. Neurophysiol.* 79: 1117–1123. [22]

Close, R. I. 1972. Dynamic properties of mammalian skeletal muscles. *Physiol. Rev.* 52: 129–197. [24]

Cochilla, A. J., Angleson, J. K., and Betz, W. J. 1999. Monitoring secretory membrane with FM1-43 fluorescence. *Annu. Rev. Neurosci.* 22: 1–10. [11]

Coggeshall, R. E., and Fawcett, D. W. 1964. The fine structure of the central nervous system of the leech, *Hirudo medicinalis. J. Neurophysiol.* 27: 229–289. [15]

Cohen, I., Rimer, M., Lømo, T., and McMahan, U. J. 1997. Agrin-induced postsynaptic-like apparatus in skeletal muscle fibers in vivo. *Mol. Cell. Neurosci.* 9: 237–253. [23]

Cohen, M. W. 1972. The development of neuromuscular connexions in the presence of D-tubocurarine. *Brain Res.* 41: 457–463. [24]

Cohen, M. W., Jones, O. T., and Angelides, K. J. 1991. Distribution of Ca^{2+} channels on frog motor nerve terminals revealed by fluorescent Ω-conotoxin. *J. Neurosci.* 11: 1032–1039. [11]

Cohen, S. 1959. Purification and metabolic effects of a nerve growth-promoting protein from snake venom. *J. Biol. Chem.* 234: 1129–1137. [23]

Cohen, S. 1960. Purification of a nerve-growth promoting protein from the mouse salivary gland and its neuro-cytotoxic antiserum. *Proc. Natl. Acad. Sci. USA* 46: 302–311. [23]

Colamarino, S. A., and Tessier-Lavigne, M. 1995. The axonal chemoattractant netrin-1 is also a chemorepellent for trochlear motor axons. *Cell* 81: 621–629. [23]

Colbert, C. M., Burger, B. S., and Levy, W. B. 1992. Longevity of synaptic depression in the hippocampal dentate gyrus. *Brain Res.* 571: 159–161. [12]

Cole, K. S. 1968. *Membranes, Ions and Impulses.* University of California Press, Berkeley. [6]

Coleridge, H. M., and Coleridge, J. C. 1994. Neural regulation of bronchial blood flow. *Respir. Physiol.* 98: 1–13. [16]

Coles, J. A., and Abbott, N. J. 1996. Signalling from neurones to glial cells in invertebrates. *Trends Neurosci.* 19: 358–362. [8]

Collett, M., Collett, T. S., and Wehner, R. 1999. Calibration of vector navigation in desert ants. *Curr. Biol.* 9: 1031–1034. [15]

Collett, M., Collett, T. S., Bisch, S., and Wehner, R. 1998. Local and global vectors in desert ant navigation. *Nature* 394: 269–272. [15]

Collett, T. S. 1996. Insect navigation en-route to the goal—Multiple strategies for the use of landmarks. *J. Exp. Biol.* 199: 227–235. [15]

Collett, T. S., and Baron, J. 1994. Biological compasses and the coordinate frame of landmark memories in honeybees. *Nature* 368: 137–140. [15]

Collingridge, G. L., Kehl, S. J., and McClennan, H. 1983. Excitatory amino acids in synaptic transmission in the Schaffer collateral-commissural pathway of the rat hippocampus. *J. Physiol.* 334: 33–46. [12]

Collins, W. F., Jr., Nelson, F. E., and Randt, C. T. 1960. Relation of peripheral nerve fiber size and sensation in man. *Arch. Neurol.* 3: 381–385. [18]

Colquhoun, L. M., and Patrick, J. W. 1997. Pharmacology of neuronal nicotinic acetylcholine receptor subtypes. *Adv. Pharmacol.* 39: 191–220. [3]

Comb, M., Hyman, S. E., and Goodman, H. M. 1987. Mechanisms of trans-synaptic regulation of gene expression. *Trends Neurosci.* 10: 473–478. [13]

Conover, J. C., and Yancopoulos, G. D. 1997. Neurotrophin regulation of the developing nervous system: Analyses of knockout mice. *Rev. Neurosci.* 8: 13–27. [23]

Constantine-Paton, M., Cline, H. T., and Debski, E. 1990. Patterned activity, synaptic convergence, and the NMDA receptor in developing visual pathways. *Annu. Rev. Neurosci.* 13: 129–154. [23]

Conti, F., and Stühmer, W. 1989. Quantal charge redistributions accompanying structural transitions of sodium channels. *Eur. Biophys. J.* 17: 53–59. [6]

Conti, F., Minelli, A., Debiasi, S., and Melone, M. 1997. Neuronal and glial localization of NMDA receptors in the cerebral-cortex. *Mol. Neurobiol.* 14: 1–18. [8]

Cook, O., Low, W., and Rahamimoff, H. 1998. Membrane topology of the rat brain Na^+-Ca^{2+} exchanger. *Biochim. Biophys. Acta* 1371: 40–52. [4]

Cook, S. P., Vulchanova, L., Hargreaves, K. M., Elde, R., and McCleskey, E. W. 1997. Distinct ATP receptors on pain-sensing and stretch-sensing neurons. *Nature* 387: 505–508. [17]

Coombs, J. S., Eccles, J. C., and Fatt, P. 1955a. The electrical properties of the motoneuron membrane. *J. Physiol.* 130: 291–325. [7]

Coombs, J. S., Eccles, J. C., and Fatt, P. 1955b. The specific ionic conductances and the ionic movements across the motoneuronal membrane that produce the inhibitory post-synaptic potential. *J. Physiol.* 130: 326–373. [9]

Cooper, E., Couturier, S., and Ballivet, M. 1991. Pentameric structure and subunit stoichiometry of a neuronal nicotinic acetylcholine receptor. *Nature* 350: 235–238. [3]

Cooper, J. R., Bloom, F. E., and Roth, R. H. 1996. *The Biochemical Basis of Neuropharmacology*, 7th Ed. Oxford University Press, New York. [4, 14, 16]

Cooper, N. G. F., and Steindler, D. A. 1986. Monoclonal antibody to glial fibrillary acidic protein reveals a parcellation of individual barrels in the early postnatal mouse somatosensory cortex. *Brain Res.* 380: 341–348. [8]

Coppola, D. M., White, L. E., Fitzpatrick, D., and Purves, D. 1998. Unequal representation of cardinal and oblique contours in ferret visual cortex. *Proc. Natl. Acad. Sci. USA* 95: 2621–2623. [21]

Corey, D. P., and Hudspeth, A. J. 1979. Ionic basis of the receptor potential in a vertebrate hair cell. *Nature* 281: 675–677. [17]

Corey, D. P., and Hudspeth, A. J. 1983. Kinetics of the receptor current in bullfrog saccular hair cells. *J. Neurosci.* 3: 962–976. [17]

Cormick, K. A., Isom, L. L., Ragsdale, D., Smith, D., Scheuer, T., and Catterall, W. A. 1998. Molecular determinants of Na^+ channel function in the extracellular domain of the beta1 subunit. *J. Biol. Chem.* 273: 3954–3962. [3]

Costa, E. 1991. The allosteric modulation of $GABA_A$ receptors. *Neuropsychopharmacology* 4: 225–235. [14]

Costa, E. 1998. From $GABA_A$ receptor diversity emerges a unified vision of GABAergic inhibition. *Annu. Rev. Pharmacol. Toxicol.* 38: 321–350. [14]

Cota, G., and Armstrong, C. M. 1989. Sodium channel gating in clonal pituitary cells. The inactivation state is not voltage dependent. *J. Gen. Physiol.* 94: 213–232. [6]

Courtney, S. M., and Ungerleider, L. G. 1997. What fRMI has taught us about human vision. *Curr. Opin. Neurobiol.* 7: 554–561. [21]

Couteaux, R., and Pecot-Déchavassine, M. 1970. Vésicules synaptiques et poches au niveau des zones actives de la jonction neuromusculaire. *C. R. Acad. Sci. (Paris)* 271: 2346–2349. [11]

Cowey, A., and Heywood, C. A. 1995. There's more to color than meets the eye. *Behav. Brain Res.* 71: 89–100. [21]

Cox, E. C., Muller, B., and Bonhoeffer, F. 1990. Axonal guidance in the chick visual system: Posterior tectal membranes induce collapse of growth cones from the temporal retina. *Neuron* 4: 31–47. [23]

Crago, P. E., Houk, J. C., and Rymer, W. Z. 1982. Sampling of total muscle force by tendon organs. *J. Neurophysiol.* 47: 1069–1083. [22]

Craig, A. M., Blackstone, C. D., Huganir, R. L., and Banker, G. 1994. Selective clustering of glutamate and γ-aminobutyric acid receptors opposite terminals releasing the corresponding transmitters. *Proc. Natl. Acad. Sci. USA* 91: 12373–12377. [23]

Crair, M. C., Gillespie, D. C., and Stryker, M. P. 1998. The role of visual experience in the development of columns in cat visual cortex. *Science* 279: 566–570. [25]

Crair, M. C., Ruthazer, E. S., Gillespie, D. C., and Stryker, M. P. 1997. Ocular dominance peaks at pinwheel center singularities of the orientation map in cat visual cortex. *J. Neurophysiol.* 77: 3381–3385. [21]

Crawford, A. C., and Fettiplace, R. 1981. An electrical tuning mechanism in turtle cochlear hair cells. *J. Physiol.* 312: 377–412. [18]

Crawford, A. C., and Fettiplace, R. 1985. The mechanical properties of ciliary bundles of turtle cochlear hair cells. *J. Physiol.* 364: 359–379. [17]

Crawford, A. C., Evans, M. G., and Fettiplace, R. 1989. Activation and adaptation of transducer currents in turtle hair cells. *J. Physiol.* 419: 405–434. [17]

Crawford, A. C., Evans, M. G., and Fettiplace, R. 1991. The actions of calcium on the mechanoelectrical transducer current of turtle hair cells. *J. Physiol.* 434: 369–398. [17]

Crepel, F., and Jaillard, D. 1991. Pairing of pre- and postsynaptic activities in cerebellar Purkinje cells induces long-term changes in synaptic efficacy in vitro. *J. Physiol.* 432: 123–141. [12]

Crick, F. 1999. The impact of molecular biology on neuroscience. *Philos. Trans. R. Soc. Lond. B* 354: 2021–2025. [26]

Critchlow, V., and von Euler, C. 1963. Intercostal muscle spindle activity and its γ–motor control. *J. Physiol.* 168: 820–847. [22]

Croner, L. J., and Kaplan, E. 1995. Receptive fields of P and M ganglion cells across the primate retina. *Vision Res.* 35: 7–24. [19]

Crosby, E. C., Humphrey, T., and Lauer, E. W. 1966. *Correlative Anatomy of the Nervous System*. Macmillan, New York. [22]

Crowley, C., Spencer, S. D., Nishimura, M. C., Chen, K. S., Pitts-Meek, S., Armanini, M. P., Ling, L. H., McMahon, S. B., Shelton, D. L., Levinson, A. D., and Phillips, H. S. 1994. Mice lacking nerve growth factor display perinatal loss of sensory and sympathetic neurons yet develop basal forebrain cholinergic neurons. *Cell* 76: 1001–1011. [23]

Cull-Candy, S. G., Miledi, R., and Parker, I. 1980. Single glutamate-activated channels recorded from locust muscle fibres with perfused patch-clamp electrodes. *J. Physiol.* 321: 195–210. [2]

Culman, J., and Unger, T. 1995. Central tachykinins: Mediators of defence reaction and stress reactions. *Can. J. Physiol. Pharmacol.* 73: 885–891. [14]

Cunningham, E. T., Jr., and Sawchenko, P. E. 1991. Reflex control of magnocellular vasopressin and oxytocin secretion. *Trends Neurosci.* 14: 406–411. [16]

Curtis, B. M., and Catterall, W. A. 1986. Reconstitution of the voltage-sensitive calcium channel purified from skeletal muscle transverse tubules. *Biochemistry* 25: 3077–3083. [10]

Curtis, H. J., and Cole, K. S. 1940. Membrane action potentials from squid giant axon. *J. Cell. Comp. Physiol.* 15: 147–157. [6]

Cynader, M., and Mitchell, D. E. 1980. Prolonged sensitivity to monocular deprivation in dark-reared cats. *J. Neurophysiol.* 43: 1026–1040. [25]

D'Arcangelo, G., Nakajima, K., Miyata, T., Ogawa, M., Mikoshiba, K., and Curran, T. 1997. Reelin is a secreted glycoprotein recognized by the CR-50 monoclonal antibody. *J. Neurosci.* 17: 23–31. [23]

Da Silva, K. M. C., Sayers, B. M., Sears, T. A., and Stagg, D. T. 1977. The changes in configuration of the rib cage and abdomen during breathing in the anaesthetized cat. *J. Physiol.* 266: 499–521. [22]

Dacey, D. M., and Lee, B. B. 1994. The blue-on opponent pathway in primate retina originates from a distinct bistratified ganglion cell type. *Nature* 367: 731–735. [20]

Dacheux, R. F., and Raviola, E. 1986. The rod pathway in the rabbit retina: A depolarizing bipolar and amacrine cell. *J. Neurosci.* 6: 331–345. [19]

Dale, H. H. 1953. *Adventures in Physiology*. Pergamon, London. [9]

Dale, H. H., Feldberg, W., and Vogt, M. 1936. Release of acetylcholine at voluntary motor nerve endings. *J. Physiol.* 86: 353–380. [9]

Damasio, A. R., and Damasio, H. 1983. The anatomic basis of pure alexia. *Neurology* 33: 1573–1583. [21]

Damasio, A. R., Tranel, D., and Damasio, H. 1990. Face agnosia and the neural substrates of memory. *Annu. Rev. Neurosci.* 13: 89–109. [21]

Dani, J. A., and Mayer, M. I. 1995. Structure and function of glutamate and nicotinic acetylcholine receptors. *Curr. Opin. Neurobiol.* 5: 310–317. [3]

Daniel, H., Levenes, C., and Crepel, F. 1998. Cellular mechanisms of cerebellar LTD. *Trends Neurosci.* 21: 401–407. [10, 12]

Daniel, P. M., and Whitteridge, D. 1961. The representation of the visual field on the cerebral in monkeys. *J. Physiol.* 159: 203–221. [20]

Dartnall, H. J. A., Bowmaker, J. K., and Molino, J. D. 1983. Human visual pigments: Microspectrophotometric results from the eyes of seven persons. *Proc. R. Soc. Lond. B* 220: 115–130. [19]

David, S., and Aguayo, A. J. 1981. Axonal elongation into peripheral nervous system "bridges" after central nervous system injury in adult rats. *Science* 214: 931–933. [24]

Davies, A. M., and Lumsden, A. 1990. Ontogeny of the somatosensory system: Origins and early development of primary sensory neurons. *Annu. Rev. Neurosci.* 13: 61–73. [23]

Davies, R. W., Gallagher, E. J., and Savioz, A. 1994. Reverse genetics of the mouse central nervous system: Targeted genetic analysis of neuropeptide function and reverse genetic screens for genes involved in human neurodegenerative disease. *Prog. Neurobiol.* 42: 319–331. [14]

Davies, S. J. A., Fitch, M. T., Memberg, S. P., Hall, A. K., Raisman, G., and Silver, J. 1997. Regeneration of adult axons in white matter tracts of the central nervous system. *Nature* 390: 680–683. [24]

Daw, N. W. 1968. Colour-coded ganglion cells in the goldfish retina: Extension of their receptive fields by means of new stimuli. *J. Physiol.* 197: 567–592. [21]

Daw, N. W. 1984. The psychology and physiology of colour vision. *Trends Neurosci.* 7: 330–335. [21]

Daw, N. W. 1998. Critical periods and amblyopia. *Arch. Ophthalmol.* 116: 502–505. [25]

Daw, N. W., Jensen, R. J., and Brunken, W. J. 1990. Rod pathways in mammalian retinae. *Trends Neurosci.* 13: 110–115. [19]

Daw, N. W., Reid, S. N., Wang, X. F., and Flavin, H. J. 1995. Factors that are critical for plasticity in the visual cortex. *Ciba Found. Symp.* 193: 258–276. [25]

Dawson, T., and Snyder, S. H. 1994. Gases as biological messengers: Nitric oxide and carbon monoxide in the brain. *J. Neurosci.* 14: 5147–5159. [14]

Dawson, V. L., and Dawson, T. M. 1998. Nitric oxide in neurodegeneration. *Prog. Brain Res.* 118: 215–229. [14]

De Boysson-Bardies, B., Halle, P., Sagart, L., and Durand, C. 1989. A crosslinguistic investigation of vowel formants in babbling. *J. Child Lang.* 16: 1–17. [18]

De Camilli, P., and Greengard, P. 1986. Synapsin I: A synaptic vesicle-associated neuronal phosphoprotein. *Biochem. Pharmacol.* 35: 4349–4357. [13]

De Felipe, C., Herrero, J. F., O'Brien, J. A., Palmer, J. A., Doyle, C. A., Smith, A. J., Laird, J. M., Belmonte, C., Cervero, F., and Hunt, S. P. 1998. Altered nociception, analgesia and aggression in mice lacking the receptor for substance P. *Nature* 392: 394–397. [14]

de Groat, W. C., Vizzard, M. A., Araki, I., and Roppolo, J. 1996. Spinal interneurons and preganglionic neurons in sacral autonomic reflex pathways. *Prog. Brain Res.* 107: 97–111. [16]

de la Torre, J. C. 1980. An improved approach to histofluorescence using the SPG method for tissue monoamines. *J. Neurosci. Methods* 3: 1–5. [14]

de la Villa, P., Kurahashi, T., and Kaneko, A. 1995. L-Glutamate-induced responses and cGMP-activated channels in three subtypes of retinal bipolar cells dissociated from the cat. *J. Neurosci.* 15: 3571–3582. [10]

De Leon, R. D., Hodgson, J. A., Roy, R. R., and Edgerton, V. R. 1998. Locomotor capacity attributable to step training versus spontaneous recovery after spinalization in adult cats. *J. Neurophysiol.* 79: 1329–1340. [22]

De Potter, W. P., Smith, A. D., and De Schaepdryver, A. F. 1970. Subcellular fractionation of splenic nerve: ATP, chromogranin A, and dopamine β-hydroxylase in noradrenergic vesicles. *Tissue Cell* 2: 529–546. [13]

De Renzi, E. 1982. *Disorders of Space Exploration and Cognition.* Wiley, New York. [22]

De Renzi, E. 1997. Prosopagnosia. In T. E. Feinberg and M. J. Farah (eds.), *Behavioral Neurology and Neuropsychology*. McGraw-Hill, New York, pp. 245–255. [21]

DeAngelis, G. C., Cumming, B. G., and Newsome, W. T. 1998. Cortical area MT and the perception of stereoscopic depth. *Nature* 394: 677–680. [21]

Dearry, A., Gingrich, J. A., Falardeau, P., Fremeau, R. T., Jr., Bates, M. D., and Caron, M. G. 1990. Molecular cloning and expression of the gene for a human D1 dopamine receptor. *Nature* 347: 72–76. [14]

Debanne, D., and Thompson, S. M. 1996. Associative long-term depression in the hippocampus in vitro. *Hippocampus* 6: 9–16. [12]

Debby-Brafman, A., Burstyn-Cohen, T., Klar, A., and Kalcheim, C. 1999. F-Spondin, expressed in somite regions avoided by neural crest cells, mediates inhibition of distinct somite domains to neural crest migration. *Neuron* 22: 475–488. [23]

DeChiara, T. M., Bowen, D. C., Valenzuela, D. M., Simmons, M. V., Poueymirou, W. T., Thomas, S., Kinetz, E., Compton, D. L., Rojas, E., Park, J. S., Smith, C., DiStephano, P. S., Glass, D. J., Burden, S. J., and Yancopoulos, G. D. 1996. The receptor tyrosine kinase MuSK is required for neuromuscular junction formation in vivo. *Cell* 85: 501–512. [23]

DeChiara, T. M., Vejsada, R., Poueymirou, W. T., Acheson, A., Suri, C., Conover, J. C., Friedman, B., McClain, J., Pan, L., Stahl, N., and Yancopolous, G. 1995. Mice lacking the CNTF receptor, unlike mice lacking CNTF, exhibit profound motor neuron deficits at birth. *Cell* 83: 313–322. [23]

DeKoninck, P., and Schulman, H. 1998. Sensitivity of Ca^{2+}/calmodulin-dependent protein kinase II to the frequency of Ca^{2+} oscillations. *Science* 279: 227–230. [10]

del Castillo, J., and Katz, B. 1954a. Quantal components of the end-plate potential. *J. Physiol.* 124: 560–573. [11]

del Castillo, J., and Katz, B. 1954c. Changes in end-plate activity produced by presynaptic polarization. *J. Physiol.* 124: 586–604. [11]

del Castillo, J., and Katz, B. 1955. On the localization of end-plate receptors. *J. Physiol.* 128: 157–181. [9]

del Castillo, J., and Stark, L. 1952. The effect of calcium ions on the motor end-plate potentials. *J. Physiol.* 116: 507–515. [11]

Del Río, J. A., Heimrich, B., Borrell, V., Förster, E., Drakew, A., Alcántara, S., Nakajima, K., Miyata, T., Ogawa, M., Mikoshiba, K., Derer, P., Frotscher, M., and Soriana, E. 1997. A role for Cajal-Retzius cells and *reelin* in the development of hippocampal connections. *Nature* 385: 70–74. [23]

Del Rio-Hortega, P. 1920. La microglia y su transformacion celulas en basoncito y cuerpos granulo-adiposos. *Trab. Lab. Invest. Biol. Madrid* 18: 37–82. [8]

Delany, K. R., and Zucker, R. S. 1990. Calcium released by photolysis of DM-nitrofen stimulates transmitter release at squid giant synapse. *J. Physiol.* 426: 473–498. [11]

del Castillo, J., and Katz, B. 1954b. Statistical factors involved in neuromuscular facilitation and depression. *J. Physiol.* 124: 574–585. [12]

Deller, T., and Frotscher, M. 1997. Lesion-induced plasticity of central neurons: Sprouting of single fibers in the rat hippocampus after unilateral entorhinal cortex lesion. *Prog. Neurobiol.* 53: 687–727. [24]

Deller, T., Drakew, A., and Frotscher, M. 1999. Different primary target cells are important for fiber lamination in the fascia denata: A lesson from reeler mutant mice. *Exp. Neurol.* 156: 239–253. [23]

Dellovade, T., Schwanzel-Fukuda, M., Gordan, J., and Pfaff, D. 1998. Aspects of GnRH neurobiology conserved across vertebrate forms. *Gen. Comp. Endocrinol.* 112: 276–282. [16]

Dennis, M., and Miledi, R. 1974a. Electrically induced release of acetylcholine from denervated Schwann cells. *J. Physiol.* 237: 431–452. [8]

Dennis, M. J., and Miledi, R. 1974b. Characteristics of transmitter release at regenerating frog neuromuscular junctions. *J. Physiol.* 239: 571–594. [11]

Denk, W., Holt, J. R., Shepherd, G. M. G., and Corey, D. P. 1995. Calcium imaging of single stereocilia in hair cells: Localization of transduction channels at both ends of tip links. *Neuron* 15: 1311–1321. [17]

Dennis, M. J., and Sargent, P. B. 1979. Loss of extrasynaptic acetylcholine sensitivity upon reinnervation of parasympathetic ganglion cells. *J. Physiol.* 289: 263–275. [24]

Dennis, M. J., and Yip, J. W. 1978. Formation and elimination of foreign synapses on adult salamander muscle. *J. Physiol.* 274: 299–310. [24]

Dennis, M. J., Harris, A. J., and Kuffler, S. W. 1971. Synaptic transmission and its duplication by focally applied acetylcholine in parasympathetic neurones of the heart of the frog. *Proc. R. Soc. Lond. B* 177: 509–539. [9]

Denzer, A. J., Schulthess, T., Fauser, C., Schumacher, B., Kammerer, R. A., Engel, J., and Ruegg, M. A. 1998. Electron microscopic structure of agrin and mapping of its binding site in laminin-1. *EMBO J.* 17: 335–343. [23]

Detrich, H. W., III, Westerfield, M., and Zon, L. I. 1999. Overview of the Zebrafish system. *Methods Cell Biol.* 59: 3–10. [23]

Devinsky, O., Morrell, M. J., and Vogt, B. A. 1995. Contributions of anterior cingulate cortex to behavior. *Brain* 118: 279–306. [22]

DeVries, S. H., and Schwartz, E. A. 1989. Modulation of an electrical synapse between solitary pairs of catfish horizontal cells by dopamine and second messengers. *J. Physiol.* 414: 351–375. [9]

DeYoe, E. A., Carman, G. J., Bandettini, P., Glickman, S., Wieser, J., Cox, R., Miller, D., and Neitz, J. 1996. Mapping striate and extrastriate visual areas in human cerebral cortex. *Proc. Natl. Acad. Sci. USA* 93: 2382–2386. [21]

DeYoe, E. A., Hickfield, S., Garren, H., and Van Essen, D. C. 1990. Antibody labeling of functional subdivisions in visual cortex: Cat-301 immunoreactivity in striate and extrastriate cortex of the macaque monkey. *Vis. Neurosci.* 5: 67–81. [20, 21]

Dhallan, R. S., Yau, K. W., Schrader, K. A., and Reed, R. R. 1990. Primary structure and functional expression of a cyclic nucleotide activated channel from olfactory neurons. *Nature* 347: 184–187. [17]

Diamond, J., and Miledi, R. 1962. A study of foetal and new-born muscle fibres. *J. Physiol.* 162: 393–408. [24]

Dickinson-Nelson, A., and Reese, T. S. 1983. Structural changes during transmitter release at synapses in the frog sympathetic ganglion. *J. Neurosci.* 3: 42–52. [11]

Dickson, B. J. 1998. Photoreceptor development: Breaking down the barriers. *Curr. Biol.* 8: R90–R92. [23]

Dietzel, I., Heinemann, U., and Lux, H. D. 1989. Relations between slow extracellular potential changes, glial potassium buffering, and electrolyte and cellular volume changes during neuronal hyperactivity in cat brain. *Glia* 2: 25–44. [8]

Dilger, J. P., Boguslavsky, R., Barann, M., Katz, T., and Vidal, A. M. 1997. Mechanisms of barbiturate inhibition of aceylcholine receptor channels. *J. Gen. Physiol.* 109: 401–414. [9]

Dilger, J. P., Liu, Y., and Vidal. A. M. 1995. Interactions of general anaesthetics with single acetylcholine receptor channels. *Eur. J. Anaesthesiol.* 12: 31–39. [9]

Ding, R., Asada, H., and Obata, K. 1998. Changes in extracellular glutamate and GABA levels in the hippocampal CA3 and CA1 areas and the induction of glutamic acid decarboxylase-67 in dentate granule cells of rats treated with kainic acid. *Brain Res.* 800: 105–113. [13]

Dionne, V. E., and Leibowitz, M. D. 1982. Acetylcholine receptor kinetics. A description from single-channel currents at the snake neuromuscular junction. *Biophys. J.* 39: 253–261. [9]

DiPaola, M., Czajkowski, C., and Karlin, A. 1989. The sidedness of the COOH terminus of the acetylcholine receptor δ subunit. *J. Biol. Chem.* 264: 15457–15463. [3]

DiPolo, R., and Beaugé, L. 1979. Physiological role of ATP-driven calcium pump in squid axons. *Nature* 278: 271–273. [4]

Dityatev, A. E., and Clamann, H. P. 1996. Reliability of spike propagation in arborizations of dorsal root fibers studied by analysis of postsynaptic potentials mediated by electrotonic coupling in the frog spinal cord. *J. Neurophysiol.* 76: 3451–3459. [9]

Dodd, J., and Jessell, T. M. 1988. Axon guidance and the patterning of neuronal projections in vertebrates. *Science* 242: 692–699. [23]

Dodge, F. A., Jr., and Rahamimoff, R. 1967. Cooperative action of calcium ions in transmitter release at the neuromuscular junction. *J. Physiol.* 193: 419–432. [11]

Doe, C. Q., and Skeath, J. B. 1996. Neurogenesis in the insect central nervous system. *Curr. Opin. Neurobiol.* 6: 18–24. [23]

Dolkart-Gorin, P., and Johnson, E. M. 1979. Experimental autoimmune model of nerve growth factor deprivation: Effects on developing peripheral sympathetic and sensory neurons. *Proc. Natl. Acad. Sci. USA* 76: 5382–5386. [23]

Dolphin, A. C. 1990. G protein modulation of calcium currents in neurons. *Annu. Rev. Physiol.* 52: 243–255. [10]

Domenici, M. R., Berretta, N., and Cherubini, E. 1998. Two distinct forms of long-term depression co-exist at the mossy fiber-CA3 synapse in the hippocampus during development. *Proc. Natl. Acad. Sci. USA* 95: 8310–8315. [12]

Donoghue, J. P., and Sanes, J. N. 1994. Motor areas of the cerebral cortex. *J. Clin. Neurophysiol.* 11: 382–396. [22]

Douglas, W. W. 1966. The mechanism of release of catecholamines from the adrenal medulla. *Pharmacol. Rev.* 18: 471–480. [16]

Douglas, W. W. 1980. Autocoids. In A. G. Gilman, L. S. Goodman, and A. Gilman (eds.), *Goodman and Gilman's The Pharmacological Basis of Therapeutics*. Macmillan, New York, pp. 608–618. [14]

Doupnik, C. A., Davidson, N., and Lester, H. A. 1995. The inward rectifier potassium channel family. *Curr. Opin. Neurobiol.* 5: 268–277. [3]

Dowdall, M. J., Boyne, A. F., and Whittaker, V. P. 1974. Adenosine triphosphate: A constituent of cholinergic synaptic vesicles. *Biochem. J.* 140: 1–12. [13]

Dowling, J. E. 1987. *The Retina: An Approachable Part of the Brain.* Harvard University Press, Cambridge, MA. [19]

Dowling, J. E., and Boycott, B. B. 1966. Organization of the primate retina: Electron microscopy. *Proc. R. Soc. Lond. B* 166: 80–111. [19]

Doyle, D. A., Cabral, J. M., Pfeutzner, A. K. Gulbis, J. M., Cohen, S. L., Chait, B. T., and MacKinnon, R. 1998. The structure of the potassium channel: Molecular basis of K^+ conductance and selectivity. *Science* 280: 69–77. [3]

Drachman, D. B. 1994. Myasthenia gravis. *New England J. Med.* 330: 1797–1810. [11]

Draguhn, A., Traub, R. D., Schmitz, D., and Jeffreys, J. G. 1998. Electrical coupling underlies high-frequency oscillations in the hippocampus in vitro. *Nature* 394: 189–192. [9]

Drescher, U., Kremoser, C., Handwerker, C., Loschinger, J., Noda, M., and Bonhoeffer, F. 1995. In vitro guidance of retinal ganglion cell axons by RAGS, a 25 kDa tectal protein related to ligands for Eph receptor tyrosine kinases. *Cell* 82: 359–370. [23]

Drew, C. A., Johnston, G. A., Weatherby, R. P. 1984. Bicuculline-insensitive GABA receptors: studies on the binding of (-)-baclofen to rat cerebellar membranes. *Neurosci. Lett.* 52: 317–321. [14]

Driver, J., and Halligan, P. W. 1991. Can visual neglect operate in object-centred co-ordinates? An affirmative single case study. *Cogn. Neuropsychol.* 8: 475–496. [26]

Droz, B., and Leblond, C. P. 1963. Axonal migration of proteins in the central nervous system and peripheral nerves as shown by radioautography. *J. Comp. Neurol.* 121: 325–346. [13]

Du Bois-Reymond, E. 1848. *Untersuchungen über thierische Electricität.* Reimer, Berlin. [9]

Duchen, L. W., and Strich, S. J. 1968. The effects of botulinum toxin on the pattern of innervation of skeletal muscle in the mouse. *Q. J. Exp. Physiol.* 53: 84–89. [24]

Duclert, A., and Changeux, J. P. 1995. Acetylcholine receptor gene expression at the developing neuromuscular junction. *Physiol. Rev.* 75: 339–368. [24]

Dudek, S. M., and Bear, M. F. 1992. Homosynaptic long-term depression in area CA1 of hippocampus and the effects of N-methyl-D-aspartate receptor blockade. *Proc. Natl. Acad. Sci. USA* 89: 4363–4367. [12]

Dudel, J., Adelsberger, H., and Heckmann, M. 1997. Neuromuscular glutamatergic and GABAergic channels. *Invertebr. Neurosci.* 3: 89–92. [9]

Dudel, J., and Kuffler, S. W. 1961. Presynaptic inhibition at the crayfish neuromuscular junction. *J. Physiol.* 155: 543–562. [9, 12]

Dumoulin, A., Rostaing, P., Bedet, C., Levi, S., Isambert, M. F., Henry, J. P., Triller, A., and Gasnier, B. 1999. Presence of the vesicular inhibitory amino acid transporter in GABAergic and glycinergic synaptic terminal boutons. *J. Cell Sci.* 112: 811–823. [13]

Dunant, Y., and Israel, M. 1998. In vitro reconstitution of neurotransmitter release. *Neurochem. Res.* 23: 709–718. [11]

Dunlap, K., and Fischbach, G. D. 1981. Neurotransmitters decrease the calcium conductance activated by depolarization of embryonic chick sensory neurones. *J. Physiol.* 317: 519–535. [10]

Dunlap, K., Holz, G. G., and Rane, S. G. 1987. G proteins as regulators of ion channel function. *Trends Neurosci.* 10: 244–247. [10]

Dunn, P. M., and Marshall, L. M. 1985. Lack of nicotinic supersensitivity in frog sympathetic neurones following denervation. *J. Physiol.* 363: 211–225. [24]

Dunnett, S. B., and Fibiger, H. C. 1993. Role of forebrain cholinergic systesm in learning and memory: Relevance to the cognitive deficits of aging and Alzheimer's dementia. *Prog. Brain Res.* 98: 413–420. [14]

Dunnett, S. B., Björklund, A., and Stenevi, U. 1983. Dopamine-rich transplants in experimental parkinsonism. *Trends Neurosci.* 6: 266–270. [24]

Dunnett, S. B., Kendall, A. L., Watts, C., and Torres, E. M. 1997. Neuronal cell transplantation for Parkinson's and Huntington's diseases. *Br. Med. Bull.* 53: 757–776. [14]

Dupin, E., Ziller, C., and Le Douarin, N. M. 1998. The avian embryo as a model in developmental studies: Chimeras and in vitro clonal analysis. *Curr. Top. Dev. Biol.* 36: 1–35. [23]

Dursteler, R. M., Wurtz, R. H., and Newsome, W. T. 1987. Directional pursuit deficits following lesions of the foveal representation within the superior temporal sulcus of the macaque monkey. *J. Neurophysiol.* 57: 1262–1287. [21]

Dwyer, T. M., Adams, D. J., and Hille, B. 1980. The permeability of the endplate channel to organic ions in frog muscle. *J. Gen. Physiol.* 75: 469–492. [3]

Eccles, J. C., and O'Connor, W. J. 1939. Responses which nerve impulses evoke in mammalian striated muscles. *J. Physiol.* 97: 44–102. [9]

Eccles, J. C., and Sherrington, C. S. 1930. Numbers and contraction-values of individual motor-units examined in some muscles of the limb. *Proc. R. Soc. Lond. B* 106: 326–357. [22]

Eccles, J. C., Eccles, R. M., and Fatt, P. 1956. Pharmacological investigations on a central synapse operated by acetylcholine. *J. Physiol.* 131: 154–169. [14]

Eccles, J. C., Eccles, R. M., and Magni, F. 1961. Central inhibitory action attributable to presynaptic depolarization produced by muscle afferent volleys. *J. Physiol.* 159: 147–166. [9]

Eccles, J. C., Katz, B. and Kuffler, S. W. 1942. Effects of eserine on neuromuscular transmission. *J. Neurophysiol.* 5: 211–230. [9]

Eden, G. F., VanMeter, J. W., Rumsey, J. M., Maisog, J. M., Woods, R. P., and Zeffiro, T. A. 1996. Abnormal processing of visual motion in dyslexia revealed by functional brain imaging. *Nature* 382: 66–69. [21]

Edes, I., and Kranias, E. G. 1995. Ca^{2+} ATPases. In N. Sperelakis (ed.), *Cell Physiology Source Book.* Academic Press, New York, pp. 156–165. [4]

Edwards, C. 1982. The selectivity of ion channels in nerve and muscle. *Neuroscience* 7: 1335–1366. [5]

Edwards, C., Ottoson, D., Rydqvist, B., and Swerup, C. 1981. The permeability of the transducer membrane of the crayfish stretch receptor to calcium and other divalent cations. *Neuroscience* 6: 1455–1460. [17]

Edwards, F. A., Gibb, A. J., and Colquhoun, D. 1992. ATP receptor mediated synaptic currents in the central nervous system. *Nature* 359: 144–147. [14]

Edwards, F. A., Konnerth, A., and Sakmann, B. 1990. Quantal analysis of inhibitory synaptic transmission in the dentate gyrus of rat hippocampal slices: A patch-clamp study. *J. Physiol.* 430: 213–249. [11]

Edwards, F. A., Robertson, S. J., and Gibb, A. J. 1997. Properties of ATP receptor mediated synaptic transmission in the rat medial habenula. *Neuropharmacology* 36: 1253–1268. [14]

Edwards, J. S. 1997. The evolution of insect flight: Implications for the evolution of the nervous system. *Brain Behav. Evol.* 50(1): 8–12. [15]

Egberongbe, Y. I., Gentleman, S. M., Falkai, P., Bogerts, B., Polak, J. M., and Roberts, G. W. 1994. The distribution of nitric oxide synthase immunoreactivity in the human brain. *Neuroscience* 59: 561–578. [14]

Egebjerg, J., Bettler, B., Hermans-Borgmeyer, I., and Heinemann, S. 1991. Cloning of the cDNA for a glutamate receptor subunit activated by Kainate but not by AMPA. *Nature* 351: 745–748. [3]

Ehrlich, J. S., Boulis, N. M., Karrer, T., and Sahley, C. L. 1992. Differential effects of serotonin depletion on sensitization and dishabituation in the leech, Hirudo medicinalis. *J. Neurobiol.* 23: 270–279. [15]

Ekstrom, J., Asztely, A., and Tobin, G. 1998. Parasympathetic non-adrenergic, non-cholinergic mechanisms in salivary glands and their role in reflex secretion. *Eur. J. Morphol.* 36 Suppl.: 208–212. [16]

Elbert, T., Pantev, C., Wienbruch, C., Rockstroh, B., and Taub, E. 1995. Increased cortical representation of the fingers of the left hand in string players. *Science* 270: 305–307. [18]

Eldridge, F. L. 1977. Maintenance of respiration by central neural feedback mechanisms. *Fed. Proc.* 36: 2400–2404. [22]

Elgersma, Y., and Silva, A. J. 1999. Molecular mechanisms of synaptic plasticity. *Curr. Opin. Neurobiol.* 9: 209–213. [12]

Elgoyhen, A. B., Johnson, D. S., Boulter, J., Vetter, D. E., and Heinemann, S. 1994. α9: An acetylcholine receptor with novel pharmacological properties expressed in rat cochlear hair cells. *Cell* 79: 705–715. [18]

Elliot, T. R. 1904. On the action of adrenalin. *J. Physiol.* 31: (Proc.) xx–xxi. [9, 13]

Elliott, E. J., and Muller, K. J. 1983. Sprouting and regeneration of sensory axons after destruction of ensheathing glial cells in the leech central nervous system. *J. Neurosci.* 3: 1994–2006. [15]

Elson, R. C., Huerta, R., Abarbanel, H. D. I., Rabinovich, M. I., and Selverston, A. I. 1999. Dynamic control of irregular bursting in an identified neuron of an oscillatory circuit. *J. Neurophysiol.* 82: 115–122. [15]

Engel, J. 1992. Laminins and other strange proteins. *Biochemistry* 31: 10643–10651. [23]

Engel, J. E., and Hoy, R. R. 1999. Experience-dependent modification of ultrasound auditory processing in a cricket escape response. *J. Exp. Biol.* 202: 2797–2806. [15]

Engel, S. A., Glover, G. H., and Wandell, B. A. 1997. Retinotopic organization in human visual cortex and the spatial precision of functional MRI. *Cerebral Cortex* 7: 181–192. [20]

Engert, F., and Bonhoeffer, T. 1999. Dendritic spine changes associated with hippocampal long-term synaptic plasticity. *Nature* 399: 66–70. [12]

England, J. D., Levinson, S. R., and Shrager, P. 1996. Immunocytochemical investigations of sodium channels along nodal and internodal portions of demyelinating axons. *Microsc. Res. Tech.* 34: 445–451. [7]

Enroth-Cugell, C., and Robson, J. G. 1966. The contrast sensitivity of retinal ganglion cells of the cat. *J. Physiol.* 187: 517–552. [19]

Enz, R., and Cutting, G. R. 1998. Molecular composition of $GABA_C$ receptors. *Vision Res.* 38: 1431–1441. [14]

Ericson, J., Muhr, J., Placzek, M., Lints, T., Jessell, T. M., and Edlund, T. 1995. Sonic hedgehog induces the differentiation of ventral forebrain neurons: A common signal for ventral patterning within the neural tube. *Cell* 81: 747–756. [23]

Erlander, M. G., Tillakaratne, N. J. K., Feldblum, S., Patel, N., and Tobin, A. J. 1991. Two genes encode distinct glutamate decarboxylases. *Neuron* 7: 91–100. [13]

Erxleben, C. 1989. Stretch-activated current through single ion channels in the abdominal stretch receptor organ of the crayfish. *J. Gen. Physiol.* 94: 1071–1083. [17]

Erxleben, C. F. 1993. Calcium influx through stretch-activated cation channels mediates adaptation by potassium current activation. *Neuroreport* 4: 616–618. [17]

Erxleben, C., and Kriebel, M. E. 1988. Subunit composition of the spontaneous miniature end-plate currents at the mouse neuromuscular junction. *J. Physiol.* 400: 659–676. [11]

Eugenin, J., and Nicholls, J. G. 1997. Chemosensory and cholinergic stimulation of fictive respiration in isolated CNS of neonatal opossum. *J. Physiol.* 501: 425–437. [22]

Evans, M. 1996. Acetylcholine activates two currents in guinea-pig outer hair cells. *J. Physiol.* 491: 563–578. [18]

Evarts, E. V. 1965. Relation of discharge frequency to conduction velocity in pyramidal neurons. *J. Neurophysiol.* 28: 216–228. [22]

Evarts, E. V. 1966. Pyramidal tract activity associated with a conditioned hand movement in the monkey. *J. Neurophysiol.* 29: 1011–1027. [22]

Evarts, E. V. 1968. Relation of pyramidal tract activity to force exerted during voluntary movement. *J. Neurophysiol.* 31: 14–27. [22]

Evers, J., Laser, M., Sun, Y-A., Xie, Z-P., and Poo, M-M. 1989. Studies of nerve-muscle interactions in *Xenopus* cell culture: Analysis of early synaptic surrents. *J. Neurosci.* 9: 1523–1539. [23]

Ewer, J., Rosbash, M., and Hall, J. C. 1998. An inducible promoter fused to the period gene in *Drosophila* conditionally rescues adult per-mutant arrhythmicity. *Nature* 333: 82–84. [16]

Eyzaguirre, C., and Kuffler, S. W. 1955. Processes of excitation in the dendrites and in the soma of single isolated sensory nerve cells of the lobster and crayfish. *J. Gen. Physiol.* 39: 87–119. [17]

Fain, G. L., Matthews, H. R., and Cornwall, M. C. 1996. Dark adaptation in vertebrate photoreceptors. *Trends Neurosci.* 19: 502–507. [19]

Falck, B., Hillarp, N-A., Thieme, G., and Thorp, A. 1962. Fluorescence of catecholamines and related compounds condensed with formaldehyde. *J. Histochem. Cytochem.* 10: 348–354. [14]

Fambrough, D. M. 1979. Control of acetylcholine receptors in skeletal muscle. *Physiol. Rev.* 59: 165–227. [24]

Fanselow, M. S., and Kim, J. J. 1994. Acquisition of contextual Pavlovian fear conditioning is blocked by application of an NMDA receptor antagonist D,L-2-amino-5-phosphonovaleric acid to the basolateral amygdala. *Behav. Neurosci.* 108: 210–212. [12]

Farah, M. J. 1990. *Visual Agnosia.* MIT Press, Cambridge, MA. [21]

Farbman, A. I. 1994. Developmental biology of olfactory sensory neurons. *Semin. Cell Biol.* 5: 3–10. [17]

Farinas, I., Yoshida, C. K., Backus, C. and Reichardt, L. F. 1996. Lack of neurotrophin-3 results in death of spinal sensory neurons and premature differentiation of their precursors. *Neuron* 17: 1068–1078. [23]

Fatt, P., and Ginsborg, B. L. 1958. The ionic requirements for the production of action potentials in crustacean muscle fibres. *J. Physiol.* 142: 516–543. [6]

Fatt, P., and Katz, B. 1951. An analysis of the end-plate potential recorded with an intra-cellular electrode. *J. Physiol.* 115: 320–370. [5, 9, 13]

Fatt, P., and Katz, B. 1952. Spontaneous subthreshold potentials at motor nerve endings. *J. Physiol.* 117: 109–128. [11]

Fatt, P., and Katz, B. 1953. The effect of inhibitory nerve impulses on a crustacean muscle fibre. *J. Physiol.* 121: 374–389. [9]

Fawcett, J. W., and Asher, P. A. 1999. The glial scar and central nervous system repair. *Brain Res. Bull.* 49: 377–391. [24]

Fawcett, J. W., and Keynes, R. J. 1990. Peripheral nerve regeneration. *Annu. Rev. Neurosci.* 13: 43–60. [24]

Fazeli, A., Dickinson, S. L., Hermiston, M. L., Tighe, R. V., Steen, R. G., Small, C. G., Stoeckli, S. T., Keino-Masu, K., Masu, M., Rayburn, H., Simons, J., Bronson, R. T., Gordon, J. I., Tessier-Lavigne, M., and Weinberg, R. A. 1997. Phenotype of mice lacking functional *Deleted in colorectal cancer* (*Dcc*) gene. *Nature* 386: 796–804. [23]

Fekete, D. M., Rouiller, E. M., Liberman, M. C., and Ryugo, D. K. 1984. The central projections of intracellularly labeled auditory nerve fibers in cats. *J. Comp. Neurol.* 229: 432–450. [18]

Feldberg, W. 1945. Present views of the mode of action of acetylcholine in the central nervous system. *Physiol. Rev.* 25: 596–642. [9]

Feldman, D. E., and Knudsen, E. I. 1997. An anatomical basis for visual calibration of the auditory space map in the barn owl's midbrain. *J. Neurosci.* 17: 6820–6837. [25]

Fernandez, J. M., Neher, E., and Gomperts, B. D. 1984. Capacitance measurements reveal stepwise fusion events in degranulating mast cells. *Nature* 312: 453–455. [11]

Fernandez-de-Miguel, F., and Drapeau, P. 1995. Synapse formation and function: Insights from identified leech neurons in culture. *J. Neurobiol.* 27: 367–379. [15]

Ferro-Novick, S., and Jahn, R. 1994. Vesicle fusion from yeast to man. *Nature* 370: 191–193. [13]

Ferster, D. 1981. A comparison of binocular depth mechanisms in areas 17 and 18 of the cat visual cortex. *J. Physiol.* 311: 623–655. [21]

Ferster, D. 1988. Spatially opponent excitation and inhibition in simple cells of the cat visual cortex. *J. Neurosci.* 8: 1172–1180. [20]

Ferster, D. and Miller, K. D. 2000. Neural mechanisms of orientation selectivity in the visual cortex. *Annu. Rev. Neurosci.* 23: 441–471. [20]

Ferster, D., Chung, S., and Wheat, H. 1996. Orientation selectivity of thalamic input to simple cells of cat visual cortex. *Nature* 380: 249–252. [20]

Fertuck, H. C., and Salpeter, M. M. 1974. Localization of acetylcholine receptor by ^{125}I-labeled α-bungarotoxin binding at mouse motor endplates. *Proc. Natl. Acad. Sci. USA* 71: 1376–1378. [9]

Fesce, R., Grohovaz, F., Valtorta, F., and Meldolesi, J. 1994. Neurotransmitter release: Fusion or "kiss and run"? *Trends Cell Biol.* 4: 1–4. [11]

Fesenko, E. E., Kolesnikov, S. S., and Lyubarsky, A. L. 1985. Induction by cyclic GMP of cationic conductance in plasma membrane of retinal rod outer segment. *Nature* 313: 310–313. [19]

Fettiplace, R. 1987. Electrical tuning of hair cells in the inner ear. *Trends Neurosci.* 10: 421–425. [18]

Fettiplace, R., and Fuchs, P. A. 1999. Mechanisms of hair cell tuning. *Annu. Rev. Physiol.* 61: 809–834. [18]

Fex, S., Sonessin, B., Thesleff, S., and Zelena, J. 1966. Nerve implants in botulinum poisoned mammalian muscle. *J. Physiol.* 184: 872–882. [24]

Fields, H. L. 1987. *Pain.* McGraw-Hill, New York. [18]

Fields, H. L., and Basbaum, A. I. 1978. Brain stem control of spinal pain transmission neurons. *Annu. Rev. Physiol.* 40: 217–248. [14, 18]

Fields, H. L., and Besson, J-M. (eds.). 1988. Pain modulation. *Prog. Brain Res.* 77. [18]

Fillenz, M. 1990. *Noradrenergic Neurons.* Cambridge University Press, Cambridge. [14]

Filoteo, A. G., Elwess, N. L., Enyedi, A., Caride, A., Aung, H. H., and Penniston, J. T. 1997. Plasma membrane Ca^{2+} pumps in rat brain. Patterns of alternative splices seen by isoform-specific antibodies. *J. Biol. Chem.* 272: 23741–23747. [4]

Fink, G. R., Frackowiak, R. S., Pietrzyk, U., and Passingham, R. E. 1997. Multiple nonprimary motor areas in the human cortex. *J. Neurophysiol.* 77: 2164–2174. [22]

Fink, M., Lesage, F., Duprat, F., Heurteaux, C., Reyes, R., Fosset, M., and Lazdunski, M. 1998. A neuronal two P domain K^+ channel stimulated by arachidonic acid and polyunsaturated fatty acids. *EMBO J.* 17: 3297–3308. [10]

Finn, J. T., Grunwald, M. E., and Yau, K-W. 1996. Cyclic nucleotide-gated ion channels: An extended family with diverse functions. *Ann. Rev. Physiol.* 58: 395–426. [3, 19]

Finn, J. T., Solessio, E. C., and Yau, K. W. 1997. A cGMP-gated cation channel in depolarizing photoreceptors of the lizard parietal eye. *Nature* 385: 815–819. [19]

Finn, J. T., Xiong, W. H., Solessio, E. C., and Yau, K. W. 1998. A cGMP-gated cation channel and phototransduction in depolarizing photoreceptors of the lizard parietal eye. *Vision Res.* 38: 1353–1357. [19]

Firestein, S., Shepherd, G. M., and Werblin, F. 1990. Time course of the membrane current underlying sensory transduction in salamander olfactory receptor neurones. *J. Physiol.* 430: 135–158. [17]

Fischbach, G. D., and Rosen, K. M. 1997. ARIA: A neuromuscular junction neuregulin. *Annu. Rev. Neurosci.* 20: 429–458. [23]

Fischer, B., and Poggio, G. F. 1979. Depth sensitivity of binocular cortical neurones of behaving monkeys. *Proc. R. Soc. Lond. B* 204: 409–414. [21]

Fischer, W., Bjorklund, A., Chen, K., and Gage, F. H. 1991. NGF improves spatial memory in aged rodents as a function of age. *J. Neurosci.* 11: 1889–1906. [23]

Fischer, W., Gage, F. H., and Bjorklund, A. 1989. Degenerative changes in forebrain cholinergic nuclei correlate with cognitive impairments in aged rats. *Eur. J. Neurosci.* 1: 34–45. [23]

Fisher, S. K., and Boycott, B. B. 1974. Synaptic connections made by horizontal cells within the outer plexiform layer of the retina of the cat and the rabbit. *Proc. R. Soc. Lond. B* 186: 317–331. [1]

Fitch, M. T., and Silver, J. 1999. Beyond the glial scar: Cellular and molecular mechanisms by which glial cells contribute to CNS regenerative failure. In M. H. Tuszynski and J. H. Kordower (eds.), *CNS Regeneration: Basic Science and Clinical Advances.* Academic Press, San Diego, CA, pp. 55–88. [24]

Fitzpatrick, D., Lund, J. S., and Blasdel, G. G. 1985. Intrinsic connections of macaque striate cortex: Afferent and efferent connections of lamina 4C. *J. Neurosci.* 5: 3329–3349. [21]

Fletcher, E. L., and Wassle, H. 1999. Indoleamine-accumulating amacrine cells are presynaptic to rod bipolar cells through GABA(C) receptors. *J. Comp. Neurol.* 413: 155–167. [14]

Flock, A. 1965. Transducing mechanisms in the lateral line canal organ receptors. *Cold Spring Harb. Symp. Quant. Biol.* 30: 133–145. [17]

Flock, A., Flock, B., and Murray, E. 1977. Studies on the sensory hairs of receptor cells in the inner ear. *Acta Otolaryngol. (Stockh.)* 83: 85–91. [17]

Flockerzi, V., Oeken, H-J., Hofmann, F., Pelzer, D., Cavalie', A., and Trautwein, W. 1986. Purified dihydropyridine-binding site from skeletal muscle t-tubules is a functional calcium channel. *Nature* 323: 66–68. [10]

Fonnum, F. 1984. Glutamate: A neurotransmitter in mammalian brain. *J. Neurochem.* 42: 1–11. [14]

Fontaine, B., and Changeux, J-P. 1989. Localization of nicotinic acetylcholine receptor α-subunit transcripts during myogenesis and motor endplate development in the chick. *J. Cell Biol.* 108: 1025–1037. [24]

Foote, S. L., Bloom, F. E., and Aston-Jones, G. 1983. Nucleus locus coeruleus: New evidence of anatomical and physiological specificity. *Physiol. Rev.* 63: 844–914. [14]

Forscher, P., and Smith, S. J. 1988. Actions of cytochalasins on the organization of actin filaments and microtubules in a neuronal growth cone. *J. Cell Biol.* 4: 1505–1516. [23]

Fox, P. T., Miezin, F. M., Allman, J. M., Van Essen, D. C., and Raichle, M. E. 1987. Retinotopic organization of human visual cortex mapped with positron-emission tomography. *J. Neurosci.* 7: 913–922. [20]

Fox, P. T., and Raichle, M. E. 1986. Focal physiological uncoupling of cerebral flood flow and oxidative metabolism during somatosensory stimulation in human subjects. *Proc. Natl. Acad. Sci. USA* 83: 1140–1144. [21]

Frade, J. M., and Barde, Y. A. 1998. Nerve growth factor: Two receptors, multiple functions. *BioEssays* 20: 137–145. [23]

Frade, J. M., Rodriguez-Tebar, A., and Barde, Y-A. 1996. Induction of cell death by endogenous nerve growth factor through its p75 receptor. *Nature* 383: 166–168. [23]

Francis, N. J., and Landis, S. C. 1999. Cellular and molecular determinants of sympathetic neuron development. *Annu. Rev. Neurosci.* 22: 541–566. [23]

Francois, J. 1979. Late results of congenital cataract surgery. *Ophthalmology* 86: 1586–1598. [25]

Frank, E., and Fischbach, G. D. 1979. Early events in neuromuscular junction formation *in vitro*: Induction of acetylcholine receptor clusters in the postsynaptic membrane and morphology of newly formed synapses. *J. Cell Biol.* 83: 143–158. [23, 24]

Frank, K., and Fuortes, M. G. F. 1957. Presynaptic and postsynaptic inhibition of monosynaptic reflexes. *Fed. Proc.* 16: 39–40. [9]

Frankenhaeuser, B., and Hodgkin, A. L. 1957. The actions of calcium on the electrical properties of squid axons. *J. Physiol.* 137: 218–244. [6]

Fraser, D. D., Hoehn, K., Weiss, S., and MacVicar, B. A. 1993. Arachidonic acid inhibits sodium currents and synaptic transmission in cultured striatal neurons. *Neuron* 11: 633–644. [10]

Fraser, S. E., Murray, B. A., Chuong, C-M., and Edelman, G. M. 1984. Alteration of the retinotectal map in *Xenopus* by antibodies to neural cell adhesion molecules. *Proc. Natl. Acad. Sci. USA* 81: 4222–4226. [23]

Frazer, A., and Hensler, J. G. 1999. Serotonin. In G. J. Siegel, B. W. Agranoff, R. W. Albers, S. K. Fisher, and M. D. Uhler (eds.), *Basic Neurochemistry: Molecular, Cellular, and Medical Aspects*, 6th Ed. Lippincott-Raven, Philadelphia, pp. 263–292. [13]

French, K. A., and Muller, K. J. 1986. Regeneration of a distinctive set of axosomatic contacts in the leech central nervous system. *J. Neurosci.* 6: 318–324. [15]

Friedlander, R. M., and Yuan, J. 1998. ICE, neuronal apoptosis and neurodegeneration. *Cell Death Differ.* 5: 823–831. [23]

Fritsch, G., and Hitzig, E. 1870. Ueber die electrische Erregbarkheit des Grosshirns. *Arch. Anat. Physiol. Wiss. Med.* 37: 300–332. [22]

Fuchs, P. A., and Getting, P. A. 1980. Ionic basis of presynaptic inhibitory potentials at crayfish claw opener. *J. Neurophysiol.* 43: 1547–1557. [9]

Fuchs, P. A., and Murrow, B. W. 1992a. Cholinergic inhibition of short (outer) hair cells of the chick's cochlea. *J. Neurosci.* 12: 800–809. [10, 18]

Fuchs, P. A., and Murrow, B. W. 1992b. A novel cholinergic receptor mediates inhibition of chick cochlear hair cells. *Proc. R. Soc. Lond. B* 248: 35–40. [18]

Fueshko, S., and Wray, S. 1994. LHRH cells migrate on peripherin fibes in embryonic olfactory explant cultures. An in vitro model for neurophilic neuronal migration. *Dev. Biol.* 166: 331–348. [16]

Fujisawa, H., and Kitsukawa, T. 1998. Receptors for collapsin/semaphorins. *Curr. Opin. Neurobiol.* 8: 587–592. [23]

Fujito, Y., and Aoki, M. 1995. Monosynaptic rebrospinal projections to distal forelimb motoneurons in the cat. *Exp. Brain Res.* 105: 181–190. [22]

Fukami, T. 1982. Further morphological and electrophysiological studies on snake muscle spindles. *J. Neurophysiol.* 47: 810–846. [22]

Fukami, Y., and Hunt, C. C. 1977. Structures in sensory region of snake spindles and their displacement during stretch. *J. Neurophysiol.* 40: 1121–1131. [17]

Fuller, J. L. 1967. Experimental deprivation and later behavior. *Science* 158: 1645–1652. [25]

Fuller, R. W. 1995. Serotonin uptake inhibitors: Uses in clinical therapy and in laboratory research. *Prog. Drug Res.* 45: 167–204. [13]

Fullerton, S. M., Strittmatter, W. J., and Matthew, W. D. 1998. Peripheral sensory nerve defects in apolipoprotein E knockout mice. *Exp. Neurol.* 153: 156–163. [24]

Fumagalli, G., Balbi, S., Cangiano, A., and Lømo, T. 1990. Regulation of turnover and number of acetylcholine receptors at neuromuscular junctions. *Neuron* 4: 563–569. [24]

Funke, K., and Worgotter, F. 1997. On the significance of temporally structured activity in the dorsal lateral geniculate nucleus (LGN). *Prog. Neurobiol.* 53: 67–119. [20]

Fuortes, M. G. F., and Poggio, G. F. 1963. Transient responses to sudden illumination in cells of the eye of *Limulus*. *J. Gen. Physiol.* 46: 435–452. [19]

Furchgott, R. F., and Zawadzki, J. V. 1980. The obligatory role of the endothelial cells in the relaxation of arterial smooth muscle by acetylcholine. *Nature* 288: 373–376. [10]

Furshpan, E. J., and Potter, D. D. 1959. Transmission at the giant motor synapses of the crayfish. *J. Physiol.* 145: 289–325. [9]

Furshpan, E. J., MacLeish, P. R., O'Lague, P. H., and Potter, D. D. 1976. Chemical transmission between rat sympathetic neurons and cardiac myocytes developing in microcultures: Evidence for cholinergic, adrenergic, and dual-function neurons. *Proc. Natl. Acad. Sci. USA* 73: 4225–4229. [23]

Gabella, G. 1976. *The Structure of the Autonomic Nervous System.* Chapman and Hall, London. [16]

Gage, F. H., Armstrong, D. M., Williams, L. R., and Varon, S. 1988. Morphologic response of axotomized septal neurons to nerve growth factor. *J. Comp. Neurol.* 269: 147–155. [23]

Gainer, H., Wolfe, S. A., Jr., Obaid, A. L., and Salzberg, B. M. 1986. Action potentials and frequency-dependent secretion in the mouse neurohypophysis. *Neuroendocrinology* 43: 557–563. [16]

Galambos, R. 1956. Suppression of auditory nerve fibers by stimulation of efferent fibers to the cochlea. *J. Neurophysiol.* 19: 424–437. [18]

Galzi, J-L., Revah, F., Bessis, A., and Changeux, J-P. 1991. Functional architecture of the nicotinic acetylcholine receptor: From electric organ to brain. *Annu. Rev. Pharmacol.* 31: 37–72. [9]

Gamlin, P. D., Zhang, H., Harlow, A., and Barbur, J. L. 1998. Pupil responses to stimulus color, structure and light flux increments in the rhesus monkey. *Vision Res.* 38: 3353–3358. [16]

Gan, W-B., and Lichtman, J. W. 1998. Synaptic segregation at the developing neuromuscular junction. *Science* 282: 1508–1511. [23]

Gantz, I., Schaffer, M., DelValle, J., Logsdon, C., Campbell, V., Uhler, M., and Yamada, T. 1991. Molecular cloning of a gene encoding the histamine H2 receptor. *Proc. Natl. Acad. Sci. USA* 88: 429–433. [14]

Garcia, U., Onetti, C., Valdiosera, R., and Aréchiga, H. 1994. Excitatory action of gamma-aminobutyric acid (GABA) on crustacean neurosecretory cells. *Cell. Mol. Neurobiol.* 14: 71–88. [15]

Garcia-Anoveros, J., and Corey, D. P. 1997. The molecules of mechanosensation. *Annu. Rev. Neurosci.* 20: 567–594. [17]

Garthwaite, J., Garthwaite, G., Palmer, R. M. J., and Moncada, S. 1989. NMDA receptor activation induces nitric oxide synthesis from arginine in rat brain slices. *Eur. J. Pharmacol.* 172: 413–416. [10]

Gasser, H. S., and Erlanger, J. 1927. The rôle played by the sizes of the constituent fibers of a nerve trunk in determining the form of its action potential wave. *Am. J. Physiol.* 80: 522–547. [7]

Gautam, M., Noakes, P. G., Moscoso, L., Rupp, F., Scheller, R. H., Merlie, J. P., and Sanes, J. R. 1996. Defective neuromuscular synaptogenesis in agrin-deficient mutant mice. *Cell* 85: 525–535. [23]

Gautam, M., Noakes, P. G., Mudd, J., Nichol, M., Chu, G. A., Sanes, J. R., and Merlie, J. P. 1995. Failure of postsynaptic specialization to develop at neuromuscular junctions of rapsyn-deficient mice. *Nature* 377: 232–236. [23]

Gautam, N., Downes, G. B., Yan, K. and Kisselev, O. 1998. The G-protein betagamma complex. *Cell. Signalling* 10: 447–455. [10]

Gauthier, I., Skudlarski, P., Gore, J. C., and Anderson, A. W. 2000. Expertise for cars and birds recruits brain areas involved in face recognition. *Nature Neurosci.* 3: 191–197. [21]

Gazzaniga, M. S. 1967. The split brain in man. *Sci. Am.* 217(8): 24–29. [21]

Gazzaniga, M. S. 1989. Organization of the human brain. *Science* 245: 947–952. [21]

Gazzaniga, M. S. 2000. Cerebral specialization and interhemispheric communication: does the corpus callosum enable the human condition? *Brain* 123: 1293–1326. [21]

Geleoc, G. S., Casalotti, S. O., Forge, A., and Ashmore, J. F. 1999. A sugar transporter as a candidate for the outer hair cell motor. *Nature Neurosci.* 2: 713–719. [18]

Gelperin, A. 1999. Oscillatory dynamics and information processing in olfactory systems. *J. Exp. Biol.* 18: 1855–1864. [15]

Georgopolous, A. P., Kettner, R. E., and Schwartz, A. B. 1988. Primate motor cortex and free arm movements to visual targets in three-dimensional space. II. Coding of directional movement by a neuronal population. *J. Neurosci.* 8: 2928–2937. [22]

Georgopoulos, A. P. 1991. Higher order motor control. *Annu. Rev. Neurosci.* 14: 361–377. [22]

Georgopoulos, A. P. 1994. New concepts in generation of movement. *Neuron* 13: 257–268. [22]

Georgopoulos, A. P., Schwartz, A. B., and Kettner, R. E. 1986. Neuronal population coding of movement direction. *Science* 233: 1416–1419. [22]

Georgopoulos, A. P., Taira, M., and Lukashin, A. 1993. Cognitive neurophysiology of the motor cortex. *Science* 260: 47–52. [22]

Gerfen, C. R. 1995. Dopamine receptor function in the basal ganglia. *Clin. Neuropharmacol.* 18: 162–177. [22]

Gerhardt, C. C., and van Heerikhuizen, H. 1997. Functional characteristics of heterologously expressed 5-HT receptors. *Eur. J. Pharmacol.* 334: 1–23. [14]

Gerst, J. E. 1999. SNAREs and SNARE regulators in membrane fusion and exocytosis. *Cell. Mol. Life Sci.* 55: 707–734. [13]

Getting, P. A. 1989. Emerging principles governing the operation of neural networks. *Annu. Rev. Neurosci.* 12: 185–204. [22]

Ghosh, A., and Greenberg, M. E. 1995. Calcium signaling in neurons: Molecular mechanisms and cellular consequences. *Science* 268: 239–247. [10]

Ghosh, A., Antonini, A., McConnell, S. K., and Shatz, C. J. 1990. Requirement for subplate neurons in the formation of thalamocortical connections. *Nature* 347: 179–181. [23]

Giancotti, F. G., and Ruoslahti, E. 1999. Integrin signaling. *Science* 285: 1028–1032. [23]

Gierdalski, M., Jablonska, B., Smith, A., Skangiel-Kramska, J., and Kossut, M. 1999. Deafferentation induced changes in GAD67 and GluR2 mRNA expression in mouse somatosensory cortex. *Brain Res. Mol. Brain Res.* 71: 111–119. [18]

Gilbert, C. D. 1977. Laminar differences in receptive field properties of cells in cat primary visual cortex. *J. Physiol.* 268: 391–421. [20]

Gilbert, C. D. 1983. Microcircuitry of the visual cortex. *Annu. Rev. Neurosci.* 6: 217–247. [14, 20]

Gilbert, C. D. 1998. Adult cortical dynamics. *Physiol. Rev.* 78: 467–485. [25]

Gilbert, C. D., and Wiesel, T. N. 1979. Morphology and intracortical projections of functionally characterised neurones in the cat visual cortex. *Nature* 280: 120–125. [20, 21]

Gilbert, C. D., and Wiesel, T. N. 1983. Cluster intrinsic connections in cat visual cortex. *J. Neurosci.* 3: 1116–1133. [21]

Gilbert, C. D., and Wiesel, T. N. 1989. Columnar specificity of intrinsic horizontal and corticocortical connections in cat visual cortex. *J. Neurosci.* 9: 2432–2442. [21]

Gilbert, S. F. 1994. *Developmental Biology*, 4th Ed. Sinauer Associates, Sunderland, MA. [23]

Gilbertson, T. A., Roper, S. D., and Kinnamon, S. C. 1993. Proton currents through amiloride-sensitive Na^+ channels in isolated hamster taste cells: Enhancement by vasopressin and cAMP. *Neuron* 10: 931–942. [17]

Gillen, C., Brill, S., Payne, J. A., and Forbush, B., III. 1996. Molecular cloning and functional expression of the K-Cl cotransporter from rabbit, rat, and human. A new member of the cation-chloride cotransporter family. *J. Biol. Chem.* 271: 16237–16244. [4]

Gillespie, P. G., Wagner, M. C., and Hudspeth, A. J. 1993. Identification of a 120 kd hair-bundle myosin located near stereociliary tips. *Neuron* 11: 581–594. [17]

Gilman, A. G. 1987. G proteins: Transducers of receptor-generated signals. *Ann. Rev. Biochem.* 56: 615–649. [10]

Gladden, M. H., Jankowska, E., and Czarkowska-Bauch, J. 1998. New observations on coupling between group II muscle afferents and feline gamma-motoneurones. *J. Physiol.* 512: 507–520. [22]

Glover, J. C., and Kramer, A. P. 1982. Serotonin analog selectively ablates identified neurons in the leech embryo. *Science* 216: 317–319. [15]

Glowatzki, E., Ruppersberg, J. P., Zenner, H. P., and Rusch, A. 1997. Mechanically and ATP-induced currents of mouse outer hair cells are independent and differentially blocked by d-tubocurarine. *Neuropharmacology* 36: 1269–1275. [14]

Glowatzki, E., Wild, K., Brandle, U., Fakler, G., Fakler, B., Zenner, H. P., and Ruppersberg, J. P. 1995. Cell-specific expression of the α9 n-ACh receptor subunit in auditory hair cells revealed by single cell RT-PCR. *Proc. R. Soc. Lond. B* 262: 141–147. [18]

Godecke, I., and Bonhoeffer, T. 1996. Development of identical orientation maps for two eyes without common visual experience. *Nature* 379: 251–254. [25]

Gold, J. I., and Knudsen, E. I. 1999. Hearing impairment induces frequency-specific adjustments in auditory spatial tuning in the optic tectum of young owls. *J. Neurophysiol.* 82: 2197–2209. [25]

Gold, M. R., and Martin, A. R. 1983a. Characteristics of inhibitory post-synaptic currents in brain-stem neurones of the lamprey. *J. Physiol.* 342: 85–98. [11]

Gold, M. R., and Martin, A. R. 1983b. Analysis of glycine-activated inhibitory post-synaptic channels in brain-stem neurones of the lamprey. *J. Physiol.* 342: 99–117. [2, 5, 9]

Gold, M. S., Reichling, D. B., Shuster, M. J., and Levine, J. D. 1996. Hyperalgesic agents increase a tetrodotoxin-resistant Na^+ current in nociceptors. *Proc. Natl. Acad. Sci. USA* 93: 1108–1112. [17]

Goldman, D. E. 1943. Potential, impedance and rectification in membranes. *J. Gen. Physiol.* 27: 37–60. [5]

Goldman-Rakic, P. S. 1999. The physiological approach: Functional architecture of working memory and disordered cognition in schizophrenia. *Biol. Psychiatry* 46: 650–661. [26]

Goldowitz, D. 1989. The *weaver* phenotype is due to intrinsic action of the mutant locus in granule cells: Evidence from homozygous *weaver* chimeras. *Neuron* 2: 1565–1575. [23]

Golgi, C. 1903. *Opera Omnia*, Vols. 1 and 2. Hoepli, Milan, Italy. [8]

Golovina, V. A., and Blaustein, M. P. 1997. Spatially and functionally distinct Ca^{2+} stores in sarcoplasmic and endoplasmic reticulum. *Science* 275: 1643–1648. [4]

Gomez, T. M., and Spitzer, N. C. 1999. In vivo regulation of axon extension and pathfinding by growth-cone calcium transients. *Nature* 397: 350–355. [23]

Goodenough, D. A., Goliger, J. A., and Paul, D. 1996. Connexins, connexons, and intercellular communication. *Annu. Rev. Biochem.* 65: 475–502. [7]

Goodman, C. S. 1996. Mechanisms and molecules that control growth cone guidance. *Annu. Rev. Neurosci.* 19: 341–377. [23]

Goodman, M., and Art, J. J. 1996. Variations in the ensemble of potassium currents underlying resonance in turtle hair cells. *J. Physiol.* 497: 395–412. [18]

Goodrum, J. F., Bouldin, T. W., Zhang, S. H., Maeda, N., and Popko, B. 1995. Nerve regeneration and cholesterol reutilization occur in the absence of apolipoproteins E and A-I in mice. *J. Neurochem.* 64: 408–416. [24]

Gorelova, N., and Reiner, P. B. 1996. Histamine depolarizes cholinergic septal neurons. *J. Neurophysiol.* 75: 707–714. [14]

Gossard, J. P., Brownstone, R. M., Barajon, I., and Hultborn, H. 1994. Transmission in a locomotor-related group Ib pathway from hindlimb extensor muscles in the cat. *Exp. Brain Res.* 98: 213–228. [22]

Gotti, C., Fornasari, D., and Clementi, F. 1997. Human neuronal nicotinic receptors. *Prog. Neurobiol.* 53: 199–237. [14]

Gould, A., Itasaki, N., and Krumlauf, R. 1998. Initiation of rhombomeric *Hoxb4* expression requires induction by somites and a retinoid pathway. *Neuron* 21: 39–51. [23]

Graba, Y., Aragnol, D., and Pradel, J. 1997. *Drosophila* Hox complex downstream targets and the function of homeotic genes. *BioEssays* 19: 379–388. [23]

Grady, E., Bohm, S., McConalogue, K., Garland, A., Ansel, J., Olerud, J., and Bunnett, N. 1997. Mechanisms attenuating cellular responses to neuropeptides: Extracellular degradation of ligands and desensitization of receptors. *J. Invest. Dermatol. Symp. Proc.* 2: 69–75. [13]

Grafstein, B. 1983. Chromatolysis reconsidered: A new view of the reaction of the nerve cell body to axon injury. In F. J. Seil (ed.), *Nerve, Organ, and Tissue Regeneration: Research Perspectives.* Academic Press, New York, pp. 37–50. [24]

Grafstein, B., and Forman, D. S. 1980. Intracellular transport in neurons. *Physiol. Rev.* 60: 1167–1283. [13]

Grassi, F., Epifano, O., Mileo, A. M., Barabino, B., and Eusebi, F. 1998. The open duration of fetal ACh receptor-channel changes during mouse muscle development. *J. Physiol.* 508: 393–400. [9]

Gray, C. M., Konig, P., Engel, A. K., and Singer, W. 1989. Oscillatory responses in cat visual cortex exhibit intercolumnar synchronization which reflects global stimulus properties. *Nature* 338: 334–337. [21]

Gray, R., Rajan, A. S., Radcliffe, K. A., Yakehiro, M., and Dani, J. A. 1996. Hippocampal synaptic transmission enhanced by low concentrations of nicotine. *Nature* 383: 713–716. [14]

Graybiel, A. M., Aosaki, T., Flaherty, A. W., and Kimura, M. 1994. The basal ganglia and adaptive motor control. *Science* 265: 1826–1831. [22]

Greene, L. A., and Shooter, E. M. 1980. The Nerve Growth Factor: Biochemistry, synthesis and mechanism of action. *Annu. Rev. Neurosci.* 3: 353–402. [23]

Greer, J. J., and Stein, R. B. 1990. Fusimotor control of muscle spindle sensitivity during respiration in the cat. *J. Physiol.* 422: 245–264. [22]

Gribkoff, V. K., et al. 1999. A reexamination of the role of GABA in the mammalian suprachiasmatic nucleus. *J. Biol. Rhythms.* 14: 126–130. [16]

Griffiths, T. D., Buchel, C., Frackowiak, R. S., and Patterson, R. D. 1998. Analysis of temporal structure in sound by the human brain. *Nature Neurosci.* 1: 422–427. [18]

Grillner, S. 1975. Locomotion in vertebrates: Central mechanisms and reflex interaction. *Physiol. Rev.* 55: 247–304. [22]

Grillner, S., Parker, D., and el Manira, A. 1998. Vertebrate locomotion—A lamprey perspective. *Ann. N.Y. Acad. Sci.* 860: 1–18. [22]

Grillner, S., Wallen, P., and Brodin, L. 1991. Neuronal network generating locomotor behavior in lamprey: Circuitry, transmitters, membrane properties and simulation. *Annu. Rev. Neurosci.* 14: 169–199. [22]

Grinnell, A. D. 1970. Electrical interaction between antidromically stimulated frog motoneurones and dorsal root afferents: Enhancement by gallamine and TEA. *J. Physiol.* 210: 17–43. [9]

Grinnell, A. D., and Rheuben, M. B. 1979. The physiology, pharmacology and trophic effectiveness of synapses formed by autonomic preganglionic nerves on frog skeletal muscles. *J. Physiol.* 289: 219–240. [24]

Grinvald, A., Lieke, E., Frostig, R. D., Gilbert, C. D., and Wiesel, T. N. 1986. Functional architecture of cortex revealed by optical imaging of intrinsic signals. *Nature* 324: 361–364. [21]

Groh, J. M., Born, R. T., and Newsome, W. T. 1997. How is a sensory map read out? Effects of microstimulation in visual area MT on saccades and smooth pursuit eye movements. *J. Neurosci.* 17: 4312–4330. [21, 22]

Grossman, Y., Parnas, I., and Spira, M. E. 1979. Ionic mechanisms involved in differential conduction of action potentials at high frequency in a branching axon. *J. Physiol.* 295: 307–322. [15]

Grumbacher-Reinert, S. 1989. Local influence of substrate molecules in determining distinctive growth patterns of identified neurons in culture. *Proc. Natl. Acad. Sci. USA* 86: 7270–7274. [23]

Grumbacher-Reinert, S., and Nicholls, J. 1992. Influence of substrate on retraction of neurites following electrical activity of leech Retzius cells in culture. *J. Exp. Biol.* 167: 1–14. [23]

Grusser, O. J., and Landis, T. 1991. *Visual Agnosias*. Vol. 12 of *Vision and Visual Dysfunction*. (J. R. Cronly-Dillon, ed.). CRC, Boca Raton, FL. [21]

Gu, X. 1991. Effect of conduction block at axon bifurcations on synaptic transmission to different postsynaptic neurones in the leech. *J. Physiol.* 441: 755–778. [15]

Gu, X. N., Macagno, E. R., and Muller, K. J. 1989. Laser microbeam axotomy and conduction block show that electrical transmission at a central synapse is distributed at multiple contacts. *J. Neurobiol.* 20: 422–434. [7]

Gu, X., Muller, K. J., and Young, S. R. 1991. Synaptic integration at a sensory-motor reflex in the leech. *J. Physiol.* 441: 733–754. [15]

Guénard, V., Xu, X. M., and Bunge, M. B. 1993. The use of Schwann cell transplantation to foster central nervous system repair. *Sem. Neurosci.* 5: 401–411. [24]

Guertin, P. A., and Hounsgaard, J. 1998. Chemical and electrical stimulation induce rhythmic motor activity in an in vitro preparation of the spinal cord from adult turtles. *Neurosci. Lett.* 245: 5–8. [22]

Guharay, R., and Sachs, F. 1984. Stretch-activated single ion channel currents in tissue-cultured embryonic chick skeletal muscle. *J. Physiol.* 352: 685–701. [17]

Guidry, G., and Landis, S. C. 1998. Target-dependent development of the vesicular acetylcholine transporter in rodent sweat gland innervation. *Dev. Biol.* 199: 175–184. [16]

Guillery, R. W. 1970. The laminar distribution of retinal fibers in the dorsal lateral geniculate nucleus of the rat: A new interpretation. *J. Comp. Neurol.* 138: 339–368. [20]

Guillery, R. W. 1974. Visual pathways in albinos. *Sci. Am.* 230(5): 44–54. [25]

Guillery, R. W., and Stelzner, D. J. 1970. The differential effects of unilateral lid closure upon the monocular and binocular segments of the dorsal lateral geniculate nucleus in the cat. *J. Comp. Neurol.* 139: 413–422. [25]

Gusella, J. F., and MacDonald, M. E. 1998. Huntingtin: A single bait hooks many species. *Curr. Opin. Neurobiol.* 8: 425–430. [22]

Guth, L. 1968. "Trophic" influences of nerve. *Physiol. Rev.* 48: 645–687. [24]

Gutkind, J. S. 1998. The pathways connecting G protein-coupled receptors to the nucleus through divergent mitogen-activated protein kinase cascades. *J. Biol. Chem.* 273: 1839–1842. [10]

Hacohen, N., Assad, J. A., Smith, W. J., and Corey, D. P. 1989. Regulation of tension on hair-cell transduction channels: Displacement and calcium dependence. *J. Neurosci.* 9: 3988–3997. [17]

Hagiwara, S. 1983. *Membrane Potential-Dependent Ion Channels in Cell Membrane. Phylogenetic and Developmental Approaches.* Raven, New York. [6]

Hagiwara, S., and Byerly, L. 1981. Calcium channel. *Annu. Rev. Neurosci.* 4: 69–125. [6]

Halder, G., Callaerts, P., and Gehring, W. J. 1995. Induction of ectopic eyes by targeted expression of the eyeless gene in *Drosophila*. *Science* 267: 1788–1792. [1, 15]

Hall, J. C. 1995. Tripping along the trail to the molecular mechanisms of biological clocks. *Trends Neurosci.* 18: 230–240. [16]

Hall, Z. W., Bownds, M. D., and Kravitz, E. A. 1970. The metabolism of γ-aminobutyric acid in the lobster nervous system. *J. Cell Biol.* 46: 290–299. [13]

Hall, Z. W., Hildebrand, J. G., and Kravitz, E. A. 1974. *Chemistry of Synaptic Transmission*. Chiron Press, Newton, MA. [13]

Halligan, P. W., and Marshall, J. C. 1991. Left neglect for near but not far space in man. *Nature* 350: 498–500. [22]

Hallworth, R., Evans, B. N., and Dallos, P. 1993. The location and mechanism of electromotility in guinea pig outer hair cells. *J. Neurophysiol.* 70: 549–558. [18]

Hamburger, V. 1939. Motor and sensory hyperplasia following limb-bud transplantations in chick embryos. *Physiol. Zool.* 12: 268–284. [23]

Hamill, M. B., and Koch, D. D. 1999. Pediatric cataracts. *Curr. Opin. Ophthalmol.* 10: 4–9. [25]

Hamill, O. P., and Sakmann, B. 1981. Multiple conductance states of single acetylcholine receptor channels in embryonic muscle cells. *Nature* 294: 462–464. [2]

Hamill, O. P., Marty, A., Neher, E., Sakmann, B. and Sigworth, J. 1981. Improved patch-clamp techniques for high-resolution current recording from cells and cell-free membrane patches. *Pflügers Arch.* 391: 85–100. [2]

Hamm, H. E. 1998. The many faces of G protein signaling. *J. Biol. Chem.* 273: 669–672. [10]

Hamon, M., Bourgoin, S., Artaud, F., and El Mestikawy, S. 1981. The respective roles of tryptophan uptake and tryptophan hydroxylase in the regulation of serotonin synthesis in the central nervous system. *J. Physiol. (Paris)* 77: 269–279. [13]

Hampson, E. C. G. M., Vaney, D. I., and Weiler, R. 1992. Dopaminergic modulation of gap junction permeability between amacrine cells in mammalian retina. *J. Neurosci.* 12: 4911–4922. [9]

Hantraye, P., Brouillet, E., Ferrante, R., Palfi, S., Dolan, R., Matthews, R. T., and Beal, M. F. 1996. Inhibition of neuronal nitric oxide synthase prevents MPTP-induced parkinsonism in baboons. *Nature Med.* 2: 1017–1021. [14]

Haque, N. S., Borghesani, P., and Isacson, O. 1997. Therapeutic strategies for Huntington's disease based on a molecular understanding of the disorder. *Mol. Med. Today* 3: 175–183. [26]

Hardwick, J. C., and Parsons, R. L. 1996. Activation of the protein phosphatase calcineurin during carbachol exposure decreases the extent of recovery from end-plate desensitization. *J. Neurophysiol.* 76: 3609–3616. [9]

Harik, S. I. 1984. Locus ceruleus lesion by local 6-hydroxydopamine infusion causes marked and speciic destruction of noradrenergic neurons, long-term depletion of norepinephrine and the enzymes that synthesize it, and enhanced dopaminergic mechanisms in the ipsilaterial cerebral cortex. *J. Neurosci.* 4: 699–707. [14]

Harlow, J. M. 1868. Recovery from passage of an iron bar through the head. *Publ. Mass. Med. Soc.* 2: 328–334. [26]

Harris, A. J., Kuffler, S. W., and Dennis, M. L. 1971. Differential chemosensitivity of synaptic and extrasynaptic areas on the neuronal surface membrane in parasympathetic neurones of the frog, tested by microapplication of acetylcholine. *Proc. R. Soc. Lond. B* 177: 541–553. [24]

Harris, G. W., and Naftolin, F. 1970. The hypothalamus and control of ovulation. *Br. Med. Bull.* 26: 3–9. [16]

Harris, G. W., and Ruf, K. B. 1970. Luteinizing hormone releasing factor in rat hypophysial portal blood collected during electrical stimulation of the hypothalamus. *J. Physiol.* 208: 243–250. [16]

Harris, G. W., Reed, M., and Fawcett, C. P. 1966. Hypothalamic releasing factors and the control of anterior pituitary function. *Br. Med. Bull.* 22: 266–272. [14]

Harris, J. A., Petersen, R. S., and Diamond, M. 1999. Distribution of tactile learning and its neural basis. *Proc. Natl. Acad. Sci.* 96: 7587–7591. [18]

Harris, J. A., Petersen, R. S., and Diamond, M. 1999. Distribution of tactile learning and its neural basis. *Proc. Natl. Acad. Sci. USA* 96: 7587–7591. [18]

Harris, K. M., and Landis, D. M. M. 1986. Membrane structure at synaptic junctions in area CA1 of the rat hippocampus. *Neuroscience* 19: 857–872. [11]

Harris, W. A., and Hartenstein, V. 1999. Cellular determination. In M. J. Zigmond, F. E. Bloom, S. C. Landis, J. L. Roberts, and L. R. Squire (eds.), *Fundamental Neuroscience.* Academic Press, New York, pp. 481–517. [23]

Hartinger, J., and Jahn, R. 1993. An anion binding site that regulates the glutamate transporter of synaptic vesicles. *J. Biol. Chem.* 268: 23122–23127. [4]

Hartline, H. K. 1940. The receptive fields of optic nerve fibers. *Am. J. Physiol.* 130: 690–699. [15, 19]

Hartshorn, W. A., and Catterall, W. A. 1984. The sodium channel from rat brain: Purification and subunit composition. *J. Biol. Chem.* 259: 1667–1675. [3]

Hartzell, H. C., Kuffler, S. W., and Yoshikami, D. 1975. Post-synaptic potentiation: Interaction between quanta of acetylcholine at the skeletal neuromuscular synapse. *J. Physiol.* 251: 427–463. [11]

Hata, Y., Tsumoto, T., and Stryker, M. P. 1999. Selective pruning of more active afferents when cat visual cortex is pharmacologically inhibited. *Neuron* 22: 375–381. [25]

Hatten, M. E. 1990. Riding the glial monorail: A common mechanism for glial-guided neuronal migration in different regions of the developing mammalian brain. *Trends Neurosci.* 13: 179–184. [8]

Hatten, M. E. 1999. Central nervous system neuronal migration. *Annu. Rev. Neurosci.* 22: 511–539. [8]

Hatten, M. E., Liem, R. K. H., and Mason, C. A. 1986. Weaver mouse cerebellar granule neurons fail to migrate on wild-type astroglial processes *in vitro*. *J. Neurosci.* 6: 2676–2683. [23]

Hawkins, R. D., Zhuo, M., and Arancio, O. 1994. Nitric oxide and carbon monoxide as possible retrograde messengers in hippocampal long-term potentiation. *J. Neurobiol.* 25: 652–665. [10]

Hebb, D. O. 1949. *The Organization of Behavior.* Wiley, New York. [12]

Hecht, S., Shlaer, S., and Pirenne, M. H. 1942. Energy, quanta and vision. *J. Gen. Physiol.* 25: 819–840. [19]

Heckmann, M., and Dudel, J. 1997. Desensitization and resensitization kinetics of glutamate receptor channels from *Drosophila* larval muscle. *Biophys. J.* 72: 2160–2169. [9]

Hediger, M. A., Kanai, Y., You, G., and Nussberger, S. 1995. Mammalian ion-coupled solute transporters. *J. Physiol.* 482P: 7S–17S. [4]

Heidelberger, R., and Matthews, G. 1992. Calcium influx and calcium current in single synaptic terminals of goldfish retinal bipolar neurons. *J. Physiol.* 447: 235–256. [11]

Heidelberger, R., Heinnemann, C., Neher, E., and Matthews, G. 1994. Calcium dependence of the rate of exocytosis in a synaptic terminal. *Nature* 371: 513–515. [11]

Heilbronn, E., Jarlebark, L., and Lawoko, G. 1995. Cholinergic and purinergic signalling in outer hair cells of mammalian cochlea. *Neurochem. Int.* 27: 301–311. [14]

Heiligenberg, W. 1989. Coding and processing electrosensory information in gymnotiform fish. *J. Exp. Biol.* 146: 255–275. [17]

Heinemann, S. H., Terlau, H., Stühmer, W., Imoto, K., and Numa, S. 1992. Calcium channel characteristics conferred on the sodium channel by single mutations. *Nature* 356: 441–443. [3]

Heist, E. K., and Schulman, H. 1998. The role of Ca^{2+}/calmodulin-dependent protein kinases within the nucleus. *Cell Calcium* 23: 103–114. [10]

Helmholtz, H. von. 1889. *Popular Scientific Lectures.* Longmans, London. [1]

Helmholtz, H. von. 1962/1927. *Helmholtz's Treatise on Physiological Optics.* (J. P. C. Southhall, ed.). Dover, New York. [1, 19, 21]

Hendrickson, A. E. 1985. Dots, stripes and columns in monkey visual cortex. *Trends Neurosci.* 8: 406–410. [21]

Hendrickson, A. E., Ogren, M. P., Vaughn, J. E., Barber, R. P., and Wu, J-Y. 1983. Light and electron microscope immunocytochemical localization of glutamic acid decarboxylase in monkey geniculate complex: Evidence for GABAergic neurons and synapses. *J. Neurosci.* 3: 1245–1262. [14]

Hendry, I. A., Stockel, K., Thoenen, H., and Iversen, L. L. 1974. The retrograde axonal transport of nerve growth factor. *Brain Res.* 68: 103–121. [23]

Hendry, S. H. C., and Calkins, D. J. 1998. Neuronal chemistry and functional organization in the primate visual system. *Trends Neurosci.* 21: 344–349. [20]

Hendry, S. H. C., and Yoshioka, T. 1994. A neurochemically distinct third channel in the macaque dorsal lateral geniculate nucleus. *Science* 264: 575–577. [20]

Hendry, S. H. C., Fuchs, J., deBlas, A. L., and Jones, E. G. 1990. Distribution and plasticity of immunocytochemically localized $GABA_A$ receptors in adult monkey visual cortex. *J. Neurosci.* 10: 2438–2450. [14]

Henkel, A. W., Lübke, J. and Betz, W. J. 1996. FM1-43 dye ultrastructural localization in and release from motor nerve terminals. *Proc. Natl. Acad. Sci. USA* 93: 1918–1923. [13]

Henkel, A. W., Simpson, L. L., Ridge, R. M., and Betz, W. J. 1996. Synaptic vesicle movements monitored by fluorescence recovery after photobleaching in nerve terminals stained with FM1-43. *J. Neurosci.* 16: 3960–3967. [13]

Henneman, E., Somjen, G., and Carpenter, D. O. 1965. Functional significance of cell size in spinal motoneurons. *J. Neurophysiol.* 28: 560–580. [22]

Henson, R., Shallice, T., and Dolan, R. 2000. Neuroimaging evidence for dissociable forms of repetition priming. *Science* 287: 1269–1272. [26]

Hepp-Reymond, M. C. 1988. In H. D. Seklis and J. Erwin (eds.), *Comparative Primate Biology*, Vol. 4. Liss, New York, pp. 501–624. [22]

Herbst, H., and Their, P. 1996. Different effects of visual deprivation on vasoactive intestinal polypeptide (VIP)-containing cells in the retinas of juvenile and adult rats. *Exp. Brain Res.* 111: 345–355. [19]

Hering, E. 1986. *Outline of a Theory of the Light Sense.* Harvard University Press, Cambridge, MA. [21]

Herlitze, S., Garcia, D. E., Mackie, K., Hille, B., Scheuer, T., and Catterall, W. A. 1996. Modulation of Ca^{2+} channels by G-protein beta gamma subunits. *Nature* 380: 258–262. [10]

Herlitze, S., Villarroel, A., Witzemann, V., Koenen, M., and Sakmann, B. 1996. Structural determinants of channel conductance in fetal and adult rat muscle acetylcholine receptors. *J. Physiol.* 492: 775–787. [9]

Herrada, G., and Dulac, C. 1997. A novel family of putative pheromone receptors in mammals with a topographically organized and sexually dimorphic distribution. *Cell* 90: 763–773. [17]

Herzog, E. D., Takahashi, J. S., and Block, G. I. D. 1998. Clock controls circadian period in isolated suprachiasmatic nucleus neurons. *Nature Neurosci.* 1: 708–713. [16]

Heuman, R. 1987. Regulation of the synthesis of nerve growth factor. *J. Exp. Biol.* 132: 133–150. [8]

Heumann, R., Korsching, S., Brandtlow, C., and Thoenen, H. 1987. Changes of nerve growth factor synthesis in nonneuronal cells in response to sciatic nerve transection. *J. Cell Biol.* 104: 1623–1631. [24]

Heuser, J. E. 1989. Review of electron microscopic evidence favouring vesicle exocytosis as the structural basis for quantal release during synaptic transmission. *Q. J. Exp. Physiol.* 74: 1051–1069. [9]

Heuser, J. E., and Reese, T. S. 1973. Evidence for recycling of synaptic vesicle membrane during transmitter release at the frog neuromuscular junction. *J. Cell Biol.* 57: 315–344. [11]

Heuser, J. E., and Reese, T. S. 1981. Structural changes after transmitter release at the frog neuromuscular junction. *J. Cell Biol.* 88: 564–580. [11]

Heuser, J. E., Reese, T. S., and Landis, D. M. D. 1974. Functional changes in frog neuromuscular junction studied with freeze-fracture. *J. Neurocytol.* 3: 109–131. [11]

Heuser, J. E., Reese, T. S., Dennis, M. J., Jan, Y., Jan, L., and Evans, L. 1979. Synaptic vesicle exocytosis captured by quick freezing and correlated with quantal transmitter release. *J. Cell Biol.* 81: 275–300. [11]

Heywood, C. A., Kentridge, R. W., and Cowey, A. 1998. Form and motion from colour in cerebral achromatopsia. *Exp. Brain Res.* 123: 145–153. [21]

Hickie, C., Cohen, L. B., and Balaban, P. M. 1997. The synapse between LE sensory neurons and gill motoneurons makes only a small contribution to the *Aplysia* gill-withdrawal reflex. *Eur. J. Neurosci.* 9: 627–636. [15]

Hiel, P., Rajan, R., and Irvine, D. 1994. Topographic representation of tone intensity along the isofrequency axis of cat primary auditory cortex. *Hear. Res.* 76: 188–202. [18]

Hilaire, G. G., Nicholls, J. G., and Sears, T. A. 1983. Central and proprioceptive influences on the activity of levator costae motoneurones in the cat. *J. Physiol.* 342: 527–548. [22]

Hilfiker, S., Pieribone, V. A., Czernik, A. J., Kao, H-T., Augustine, G. J., and Greengard, P. 1999. Synapsins as regulators of neurotransmitter release. *Philos. Trans. R. Soc. Lond. B* 354: 269–279. [13]

Hill, D. R., and Bowery, N. G. 1981. 3H-baclofen and 3H-GABA bind to bicuculline-insensitive GABA B sites in rat brain. *Nature* 290: 149–152. [14]

Hille, B. 1968. Charges and potentials at the nerve surface: Divalent ions and pH. *J. Gen. Physiol.* 51: 221–236. [6]

Hille, B. 1970. Ionic channels in nerve membranes. *Prog. Biophys. Mol. Biol.* 21: 1–32. [6]

Hille, B. 1992. *Ion Channels in Excitable Membranes*, 2nd Ed. Sinauer Associates, Sunderland, MA. [2, 3, 5, 6, 9]

Hille, B. 1994. Modulation of ion-channel function by G-protein-coupled receptors. *Trends Neurosci.* 17: 531–536. [10]

Hirano, T. 1990a. Depression and potentiation of the synaptic transmission between a granule cell and a Purkinje cell in rat cerebellar culture. *Neurosci. Lett.* 119: 141–144. [12]

Hirano, T. 1990b. Effects of postsynaptic depolarization in the induction of synaptic depression between a granule cell and a Purkinje cell in rat cerebellar culture. *Neurosci. Lett.* 119: 145–147. [12]

Hirning, L. D., Fox, A. P., McCleskey, E. W., Olivera, B. M., Thayer, S. A., Miller, R. J. and Tsien, R. W. 1988. Dominant role of N-type Ca^{2+} channels in evoked release of norepinephrine from sympathetic neurons. *Science* 239: 57–61. [10]

Hirokawa, N. 1998. Kinesin and dynein superfamily proteins and the mechanism of organelle transport. *Science* 279: 519–552. [13]

Hirokawa, N., Terada, S., Funakoshi, T., and Takeda, S. 1997. Slow axonal transport: The subunit transport model. *Trends Cell Biol.* 7: 382–388. [13]

Hirsch, J. A., Alonso, J-M., Reid, R. C., and Martinez, L. M. 1998a. Synaptic integration in striate cortical simple cells. *J. Neurosci.* 18: 9517–9528. [20]

Hirsch, J. A., Gallagher, C. A., Alonso, J. M., and Martinez, L. M. 1998b. Ascending projections of simple and complex cells in layer 6 of the cat striate cortex. *J. Neurosci.* 18: 8086–8094. [20]

Hoang, B., and Chiba, A. 1998. Genetic analysis of the role of integrin during axon guidance in *Drosophila*. *J. Neurosci.* 18: 7847–7855. [23]

Hochner, B., Parnas, H., and Parnas, I. 1989. Membrane depolarization evokes neurotransmitter release in the absence of calcium entry. *Nature* 342: 433–435. [11]

Hodgkin, A. L. 1954. A note on conduction velocity. *J. Physiol.* 125: 221–224. [7]

Hodgkin, A. L. 1964. *The Conduction of the Nervous Impulse*. Liverpool University Press, Liverpool, England. [1, 5] Hodgkin, A. L. 1973. Presidential address. *Proc. R. Soc. Lond. B* 183: 1–19. [5]

Hodgkin, A. L., and Huxley, A. F. 1939. Action potentials recorded from inside a nerve fibre. *Nature* 144: 710–711. [6]

Hodgkin, A. L., and Huxley, A. F. 1952a. Currents carried by sodium and potassium ion through the membrane of the giant axon of *Loligo*. *J. Physiol.* 116: 449–472. [6]

Hodgkin, A. L., and Huxley, A. F. 1952b. The components of the membrane conductance in the giant axon of *Loligo*. *J. Physiol.* 116: 473–496. [6]

Hodgkin, A. L., and Huxley, A. F. 1952c. The dual effect of membrane potential on sodium conductance in the giant axon of *Loligo*. *J. Physiol.* 116: 497–506. [6]

Hodgkin, A. L., and Huxley, A. F. 1952d. A quantitative description of membrane current and its application to conduction and excitation in nerve. *J. Physiol.* 117: 500–544. [6]

Hodgkin, A. L., and Katz, B. 1949. The effect of sodium ions on the electrical activity of the giant axon of the squid. *J. Physiol.* 108: 37–77. [5, 6]

Hodgkin, A. L., and Keynes, R. D. 1955a. Active transport of cations in giant axons from *Sepia* and *Loligo*. *J. Physiol.* 128: 28–60. [4]

Hodgkin, A. L., and Keynes, R. D. 1955b. The potassium permeability of a giant nerve fibre. *J. Physiol.* 128: 253–281. [5]

Hodgkin, A. L., and Keynes, R. D. 1956. Experiments on the injection of substances into squid giant axons by means of a microsyringe. *J. Physiol.* 131: 592–617. [5]

Hodgkin, A. L., and Rushton, W. A. H. 1946. The electrical constants of a crustacean nerve fibre. *Proc. R. Soc. Lond. B* 133: 444–479. [7]

Hodgkin, A. L., Huxley, A. F., and Katz, B. 1952. Measurement of current-voltage relations in the membrane of the giant axon of *Loligo*. *J. Physiol.* 116: 424–448. [6]

Hodgkin, J. 1999. Sex, cell death, and the genome of *C. elegans*. *Cell* 98(3): 277–280. [15]

Hofmann, F., Biel, M., and Flockerzi, V. 1994. Molecular basis for Ca^{2+} channel diversity. *Ann. Rev. Neurosci.* 17: 399–418. [3]

Hökfelt, T., Johansson, O., Llungdahl, A., Lundberg, M., and Schultzberg, M. 1980. Peptidergic neurones. *Nature* 284: 515–521. [16]

Holder, N., and Klein, R. 1999. Eph receptors and ephrins: Effectors of morphogenesis. *Development* 126: 2033–2044. [23]

Holland, R. L., and Brown, M. C. 1980. Postsynaptic transmission block can cause sprouting of a motor nerve. *Science* 207: 649–651. [24]

Hollmann, M., Maron, C., and Heinemann, S. 1994. N-Glycosylation site tagging suggests a three transmembrane domain topology for the glutamate receptor GluR1. *Neuron* 13: 1331–1343. [3]

Hollmann, M., and Heinemann, S. 1994. Cloned glutamate receptors. *Annu. Rev. Neurosci.* 17: 31–108. [14]

Hollyday, M., and Hamburger, V. 1976. Reduction of the naturally occurring motor neuron loss by enlargement of the periphery. *J. Comp. Neurol.* 170: 311–320. [23]

Holman, M. E., Coleman, H. A., Tonta, M. A., and Parkington, H. C. 1994. Synaptic transmission from splanchnic nerves to the adrenal medulla of guinea pigs. *J. Physiol.* 478: 115–124. [16]

Holmes, G. 1939. The cerebellum of man. *Brain* 62: 1–30. [22]

Holscher, C. 1999. Synaptic plasticity and learning and memory: LTP and beyond. *J. Neurosci. Res.* 58: 62–75. [12]

Holtzman, D. M., Kilbridge, J., Bredt, D. S., Black, S. M., Li, Y., Clary, D. O., Reichardt, L. F., and Mobley, W. C. 1994. NOS induction by NGF in basal forebrain cholinergic neurones: Evidence for regulation of brain NOS by a neurotrophin. *Neurobiol. Dis.* 1: 51–60. [14]

Homberg, U. and Hildebrand, J. G. 1989. Serotonin immunoreactivity in the optic lobes of the sphinx moth *Manduca sexta* and colocalization with FMRFamide and SCPB immunoreactivity. *J. Comp. Neurol.* 288: 243–253. [13]

Honig, M. C., Collins, W. F., and Mendell, L. 1983. Alpha-motoneuron epsps exhibit different frequency sensitivities to single Ia-afferent fiber stimulation. *J. Neurophysiol.* 49: 886–901. [22]

Hoover, J. E., and Strick, P. L. 1999. The organization of cerebellar and basal ganglia outputs to primary motor cortex as revealed by retrograde transneuronal transport of herpes simplex virus type 1. *J. Neurosci.* 19: 1446–1463. [22]

Hope, B. T., Michael, G. J., Knigge, K. M., and Vincent, S. R. 1991. Neuronal NADPH diaphorase is a nitric oxide synthase. *Proc. Natl. Acad. Sci. USA* 88: 2811–2814. [14]

Horellou, P., and Mallet, J. 1997. Gene therapy for Parkinson's disease. *Mol. Neurobiol.* 15: 241–256. [14]

Horn, R., and Marty, A. 1988. Muscarinic activation of ionic currents measured by a new whole-cell recording method. *J. Gen. Physiol.* 92: 145–159. [2]

Horton, J. C., and Hocking, D. R. 1996a. An adult-like pattern of ocular dominance columns in striate cortex of newborn monkeys prior to visual experience. *J. Neurosci.* 16: 1791–1807. [25]

Horton, J. C., and Hocking, D. R. 1996b. Pattern of ocular dominance columns in human striate cortex in strabismic amblyopia. *Vis. Neurosci.* 13: 787–795. [25]

Horton, J. C., and Hocking, D. R. 1997. Timing of the critical period for plasticity of ocular dominance columns in macaque striate cortex. *J. Neurosci.* 17: 3684–3709. [25]

Horton, J. C., and Hocking, D. R. 1998. Effect of early monocular enucleation upon ocular dominance columns and cytochrome oxidase activity in monkey and human visual cortex. *Vis. Neurosci.* 15: 289–303. [25]

Hoshi, T. W., Zagotta, W. N., and Aldrich, R. W. 1991. Two types of inactivation in *Shaker* K^+ channels: Effects of alterations in the carboxy terminal region. *Neuron* 7: 547–556. [6]

Hoshi, T., Zagotta, W. N., and Aldrich, R. W. 1990. Biophysical and molecular mechanisms of *Shaker* potassium channel inactivation. *Science* 250: 533–550. [6]

Housley, G. D. 1998. Extracellular nucleotide signaling in the inner ear. *Mol. Neurobiol.* 16: 21–48. [14]

Housley, G., and Ashmore, J. 1991. Direct measurement of the action of acetylcholine on isolated outer hair cells of the guinea pig cochlea. *Proc. R. Soc. Lond. B* 244: 161–167. [18]

Howard, J., and Hudspeth, A. J. 1988. Compliance of the hair bundle associated with gating of mechanoelectrical transduction channels in the bullfrog's saccular hair cell. *Neuron* 1: 189–199. [17]

Howard, J., Hudspeth, A. J., and Vale, R. D. 1989. Movement of microtubules by single kinesin molecules. *Nature* 342: 154–158. [13]

Howell, B. W., Hawkes, R., Soriano, P., and Cooper, J. A. 1997. Neuronal position in the developing brain is regulated by *mouse disabled-1*. *Nature* 389: 733–737. [23]

Hsiao, C. F., Trueblood, P. R., Levine, M. S., and Chandler, S. H. 1997. Multiple effects of serotonin on membrane properties of trigeminal motoneurons in vitro. *J. Neurophysiol.* 77: 2910–2924. [14]

Hsiao, S. S., Johnson, K. O., and Twombly, I. A. 1993. Roughness coding in the somatosensory system. *Acta Psychol. (Amst.)* 84: 53–67. [18]

Huang Y., Jellies, J., Johansen, K. M., and Johansen, J. 1998. Development and pathway formation of peripheral neurons during leech embryogenesis. *J. Comp. Neurol.* 397: 394–402. [15]

Huang, A. Y., and May, B. J. 1996. Sound orientation behavior in cats. II. Mid-frequency spectral cues for sound localization. *J. Acoust. Soc. Am.* 100: 1070–1080. [18]

Huang, C-L., Slesinger, P. A., Casey, P. J., Jan, Y. N., and Jan, L. Y. 1995. Evidence that direct binding of $G_{\beta\gamma}$ to the GIRK1 G protein-gated inwardly rectifying K^+ channel is important for channel activation. *Neuron* 15: 1133–1143. [10]

Huang, Z., Huang, P. L., Panahian, N., Dalkara, T., Fishman, M. C., and Moskowitz, M. A. 1994. Effects of cerebral ischemia in mice deficient in neuronal nitric oxide synthase. *Science* 265: 1883–1885. [14]

Hubel, D. H. 1982. Exploration of the primary visual cortex. *Nature* 299: 515–524. [20]

Hubel, D. H. 1988. *Eye, Brain and Vision.* Scientific American Library, New York. [1, 19, 20, 21, 25]

Hubel, D. H., and Livingstone, M. S. 1987. Segregation of form, color, and stereopsis in primate area 18. *J. Neurosci.* 7: 3378–3415. [21]

Hubel, D. H., and Livingstone, M. S. 1990. Color and contrast sensitivity in the lateral geniculate body and primary visual cortex of the macaque monkey. *J. Neurosci.* 10: 2223–2237. [21]

Hubel, D. H., and Wiesel, T. N. 1959. Receptive fields of single neurones in the cat's striate cortex. *J. Physiol.* 148: 574–591. [1, 20, 21]

Hubel, D. H., and Wiesel, T. N. 1961. Integrative action in the cat's lateral geniculate body. *J. Physiol.* 155: 385–398. [20]

Hubel, D. H., and Wiesel, T. N. 1962. Receptive fields, binocular interaction and functional architecture in the cat's visual cortex. *J. Physiol.* 160: 106–154. [20, 21]

Hubel, D. H., and Wiesel, T. N. 1963a. Receptive fields of cells in striate cortex of very young, visually inexperienced kittens. *J. Neurophysiol.* 26: 994–1002. [25]

Hubel, D. H., and Wiesel, T. N. 1963b. Shape and arrangement of columns in cat striate cortex. *J. Physiol.* 165: 559–568. [21]

Hubel, D. H., and Wiesel, T. N. 1965a. Receptive fields and functional architecture in two non-striate visual areas (18 and 19) of the cat. *J. Neurophysiol.* 28: 229–289. [20]

Hubel, D. H., and Wiesel, T. N. 1965b. Binocular interaction in striate cortex of kittens reared with artificial squint. *J. Neurophysiol.* 28: 1041–1059. [25]

Hubel, D. H., and Wiesel, T. N. 1967. Cortical and callosal connections concerned with the vertical meridian of visual field in the cat. *J. Neurophysiol.* 30: 1561–1573. [21]

Hubel, D. H., and Wiesel, T. N. 1968. Receptive fields and functional architecture of monkey striate cortex. *J. Physiol.* 195: 215–243. [20, 21]

Hubel, D. H., and Wiesel, T. N. 1970. The period of susceptibility to the physiological effects of unilateral eye closure in kittens. *J. Physiol.* 206: 419–436. [25]

Hubel, D. H., and Wiesel, T. N. 1971. Aberrant visual projections in the Siamese cat. *J. Physiol.* 218: 33–62. [25]

Hubel, D. H., and Wiesel, T. N. 1972. Laminar and columnar distribution of geniculo-cortical fibers in the macaque monkey. *J. Comp. Neurol.* 146: 421–450. [20, 21]

Hubel, D. H., and Wiesel, T. N. 1974. Sequence regularity and geometry of orientation columns in the monkey striate cortex. *J. Comp. Neurol.* 158: 267–294. [21]

Hubel, D. H., and Wiesel, T. N. 1977. Functional architecture of macaque monkey visual cortex (Ferrier Lecture). *Proc. R. Soc. Lond. B* 198: 1–59. [1, 21, 25]

Hubel, D. H., Wiesel, T. N., and LeVay, S. 1977. Plasticity of ocular dominance columns in monkey striate cortex. *Philos. Trans. R. Soc. Lond. B* 278: 377–409. [25]

Hubener, M., Shoham, D., Grinvald, A., and Bonhoeffer, T. 1997. Spatial relationships among three columnar systems in cat area 17. *J. Neurosci.* 17: 9270–9284. [21]

Hudspeth, A. J. 1982. Extracellular current flow and the site of transduction by vertebrate hair cells. *J. Neurosci.* 2: 1–10. [17]

Hudspeth, A. J., and Corey, D. P. 1977. Sensitivity, polarity and conductance change in the response of vertebrate hair cells to controlled mechanical stimuli. *Proc. Natl. Acad. Sci. USA* 74: 2407–2411. [17]

Hudspeth, A. J., and Gillespie, P. G. 1994. Pulling springs to tune transduction: Adaptation by hair cells. *Neuron* 12: 1–9. [17]

Hudspeth, A. J., and Jacobs, R. 1979. Stereocilia mediate transduction in vertebrate hair cells. *Proc. Natl. Acad. Sci. USA* 76: 1506–1509. [17]

Hudspeth, A. J., and Lewis, R. S. 1988. Kinetic analysis of voltage- and ion-dependent conductances in saccular hair cells of the bull-frog, *Rana catesbeiana. J. Physiol.* 400: 237–274. [18]

Hudspeth, A. J., Poo, M. M., and Stuart, A. E. 1977. Passive signal propagation and membrane properties in median photoreceptors of the giant barnacle. *J. Physiol.* 272: 25–43. [17]

Huganir, R. L., and Greengard, P. 1990. Regulation of neurotransmitter receptor desensitization by protein phosphorylation. *Neuron* 5: 555–567. [9]

Hughes J. (ed.). 1983. Opioid peptides. *Br. Med. Bull.* 39: 1–106. [18]

Hughes, J., Smith, T. W., Kosterlitz, H. W., Fothergill, L. A., Morgan, B. A., and Morris, H. R. 1975. Identification of two related pentapeptides from the brain with potent opiate agonist activity. *Nature* 258: 577–579. [14]

Humphrey, A. L., Sur, M., Uhlrich, D. J., and Sherman, S. M. 1985. Projection patterns of individual X- and Y-cell axons from the lateral geniculate nucleus to cortical area 17 in the cat. *J. Comp. Neurol.* 233: 159–189. [25]

Humphrey, D. R., and Reed, D. J. 1983. Separate cortical systems for the control of joint movement and joint stiffness: Reciprocal activation and co-activation of antagonist muscles. *Adv. Neurol.* 39: 347–372. [22]

Hunt, C. C. 1990. Mammalian muscle spindle: Peripheral mechanisms. *Physiol. Rev.* 70: 643–663. [17]

Hunt, C. C., Wilkerson, R. S., and Fukami, Y. 1978. Ionic basis of the receptor potential in primary endings of mammalian muscle spindles. *J. Gen. Physiol.* 71: 683–698. [17]

Huntington's Disease Collaborative Research Group. 1993. A novel gene containing a trinucleotide repeat that is expanded and unstable on Huntington's disease chromosomes. *Cell* 72: 971–983. [22]

Huxley, A. 1928. *Point Counter Point.* Harper Collins, New York. [18]

Huxley, A. F., and Stampfli, R. 1949. Evidence for saltatory conduction in peripheral myelinated nerve fibers. *J. Physiol.* 108: 315–339. [7]

Ichijo, H., and Bonhoeffer, F. 1998. Differential withdrawal of retinal axons induced by a secreted factor. *J. Neurosci.* 18: 5008–5018. [23]

Iggo, A. 1974. In J. I. Hubbard (ed.), *The Peripheral Nervous System.* Plenum, New York, pp. 347–404. [18]

Ignarro, L. J. 1990. Haem-dependent activation of guanylate cyclase and cyclic GMP formation by endogenous nitric oxide: A unique transduction mechanism for transcellular signaling. *Pharmacol. Toxicol.* 67: 1–7. [10]

Ikeda, S. R. 1996. Voltage-dependent modulation of N-type calcium channels by G-protein beta gamma subunits. *Nature* 380: 255–258. [10, 14]

Inagaki, C., Hara, M., and Zeng, H. T. 1996. A Cl^- pump in rat-brain neurons. *J. Exp. Zool.* 275: 262–268. [4]

Ingber, D. E. 1997. Tensegrity: The architectural basis of cellular mechanotransduction. *Annu. Rev. Physiol.* 59: 575–599. [17]

Inoue, S. 1981. Video image processing greatly enhances contrast, quality and speed in polarization-based microscopy. *J. Cell Biol.* 89: 346–356. [13]

Inui, A. 1999. Feeding and body-weight regulation by hypothalamic neuropeptides—Mediation of the actions of leptin. *Trends Neurosci.* 22: 62–67. [16]

Isaac, J. T., Nicoll, R. A., and Malenka, R. C. 1995. Evidence for silent synapses: Implications for the expression of LTP. *Neuron* 15: 427–434. [12]

Isaken, D. E., Liu, N. J., and Weisblat, D. A. 1999. Inductive regulation of cell fusion in leech. *Development* 126: 3381–3390. [23]

Isaksen, D. E., Liu, N. J., and Weisblat, D. A. 1999. Inductive regulation of cell fusion in leech. *Development* 126: 3381–3390. [23]

Isbister, C. M., and O'Connor, T. P. 1999. Filopodial adhesion does not predict growth cone steering events in vivo. *J. Neurosci.* 19: 2589–2600. [23]

Ito, M. 1984. *The Cerebellum and Neural Control.* Raven, New York. [22]

Ito, M., and Simpson, J. I. 1971. Discharges in Purkinje cell axons during climbing fiber activation. *Brain Res.* 31: 215–219. [22]

Ito, M., Sakurai, M., and Tongroach, P. 1982. Climbing fibre induced depression of both mossy fibre responsiveness and glutamate sensitivity of cerebellar Purkinje cells. *J. Physiol.* 324: 113–134. [12]

Iversen, L. L., Lee, C. M., Gilbert, R. F., Hunt, S., and Emson, P. C. 1980. Regulation of neuropeptide release. *Proc. R. Soc. Lond. B* 210: 91–111. [14]

Iwamura, U., Tanaka, M., Sakamoto, M., and Hikosaka, O. 1993. Rostrocaudal gradients in the neuronal receptive field complexity of postcentral gyrus. *Exp. Brain Res.* 92: 360–368. [18]

Izquierdo, I., and Medina, J. H. 1995. Correlation between the pharmacology of long-term potentiation and the pharmacology of memory. *Neurobiol. Learn. Mem.* 63: 19–32. [12]

Jacobs, B. L., and Fornal, C. A. 1991. Activity of brain serotonergic neurons in the behaving animal. *Pharmacol. Rev.* 43: 563–578. [14]

Jaffe, E. H., Marty, A., Schulte, A., and Chow, R. H. 1998. Extrasynaptic vesicular transmitter release from the somata of substantia nigra neurons in rat midbrain slices. *J. Neurosci.* 18: 3548–3553. [13]

Jafri, M. S., Moore, K. A., Taylor, G. E., and Weinreich, D. 1997. Histamine H1 receptor activation blocks two classes of potassium current, IK(rest) and IAHP, to excite ferret vagal afferents. *J. Physiol.* 503: 533–546. [14]

Jan, L. Y., and Jan, Y. N. 1997. Cloned potassium channels from eukaryotes and prokaryotes. *Annu. Rev. Neurosci.* 20: 91–123. [3]

Jan, Y. N., Jan, L. Y., and Kuffler, S. W. 1980. Further evidence for peptidergic transmission in sympathetic ganglia. *Proc. Natl. Acad. Sci. USA* 77: 5008–5012. [16]

Janig W., and McLachlan, E. M. 1987. Organization of lumbar spinal outflow to distal colon and pelvic organs. *Physiol. Rev.* 67: 1332–1404. [16]

Janig, W., and McLachlan, E. M. 1992. Characteristics of function-specific pathways in the sympathetic nervous system. *Trends Neurosci.* 15: 475–481. [16]

Jankowska, E., and Riddell, J. S. 1995. Interneurones mediating presynaptic inhibition of group II muscle afferents in the cat spinal cord. *J. Physiol.* 483: 461–471. [22]

Jansen, J. K. S., and Matthews, P. B. C. 1962. The central control of the dynamic response of muscle spindle receptors. *J. Physiol.* 161: 357–378. [17]

Jansen, J. K. S., Lømo, T., Nicholaysen, K., and Westgaard, R. H. 1973. Hyperinnervation of skeletal muscle fibers: Dependence on muscle activity. *Science* 181: 559–561. [24]

Jaramillo, F., and Hudspeth, A. J. 1991. Localization of the hair cell's transduction channels at the hair bundle's top by iontophoretic application of a channel blocker. *Neuron* 7: 409–420. [17]

Jasser, A., and Guth, P. S. 1973. The synthesis of acetylcholine by the olivo-cochlear bundle. *J. Neurochem.* 20: 45–53. [18]

Jeannerod, M., and Frak, V. 1999. Mental imaging of motor activity in humans. *Curr. Opin. Neurobiol.* 9: 735–739. [26]

Jenkinson, D. H., and Nicholls, J. G. 1961. Contractures and permeability changes produced by acetylcholine in depolarized denervated muscle. *J. Physiol.* 159: 111–127. [9]

Jentsch, T. J., Günther, W., Pusch, M., and Schwappach, B. 1995. Properties of voltage-gated chloride channels in the ClC gene family. *J. Physiol.* 482P: 19S–25S. [3]

Jentsch, T. J., Steinmeyer, K., and Schwarz, G. 1990. Primary structure of *Torpedo marmorata* chloride channel isolated by expression cloning in *Xenopus* oocytes. *Nature* 348: 510–514. [3]

Jessel, T. M., and Iversen, L. L. 1977. Opiate analgesics inhibit substance P release from rat trigeminal nucleus. *Nature* 268: 549–551. [14]

Ji, T. H., Grossmann, M., and Ji, I. 1998. G protein-coupled receptors. I. Diversity of receptor-ligand interactions. *J. Biol. Chem.* 273: 17299–17302. [10]

Jiang, G-J., Zidanic, M., Michaels, R., Michael, T., Griguer, C., and Fuchs, P. A. 1997. Cslo encodes Ca^{2+}-activated potassium channels in the chick's cochlea. *Proc. R. Soc. Lond. B* 264: 731–737. [18]

Jiang, W., Tremblay, F., and Chapman, C. E. 1997. Neuronal encoding of texture changes in the secondary somatosensory cortical areas of monkeys during passive texture discrimination. *J. Neurophysiol.* 77: 1656–1662. [18]

Jo, Y. H., and Schlichter, R. 1999. Synaptic corelease of ATP and GABA in cultured spinal neurons. *Nature Neurosci.* 2: 241–245. [13]

Joh, T. H., Park, D. H., and Reis, D. J. 1978. Direct phosphorylation of brain tyrosine hydroxylase by cyclic AMP-dependent protein kinase: Mechanism of enzyme activation. *Proc. Natl. Acad. Sci. USA* 75: 4744–4748. [13]

Johansson, C. B., Momma, S., Clarke, D. L., Risling, M., Lendahl, U., and Frisén, J. 1999. Identification of a neural stem cell in the adult mammalian central nervous system. *Cell* 96: 25–34. [23]

Johansson, R. S., and Vallbo, A. B. 1979. Detection of tactile stimuli. Thresholds of afferent units related to psychophysical thresholds in the human hand. *J. Physiol.* 297: 405–422. [18]

Johansson, R. S., and Vallbo, Å. B. 1983. Tactile sensory coding in the glabrous skin of the human hand. *Trends Neurosci.* 6: 27–32. [18]

Johnson, E. M., Jr., Taniuchi, M., and DiStefano, P. S. 1988. Expression and possible function of nerve growth factor receptors on Schwann cells. *Trends Neurosci.* 11: 299–304. [24]

Johnson, E. W., and Wernig, A. 1971. The binomial nature of transmitter release at the crayfish neuromuscular junction. *J. Physiol.* 218: 757–767. [11]

Johnson, F. H., Eyring, H., and Polissar, M. J. 1954. *The Kinetic Basis of Molecular Biology.* Wiley, New York. [2]

Johnson, G. V., and Guttmann, R. P. 1997. Calpains: Intact and active? *BioEssays* 19: 1011–1018. [10]

Johnson, J. W. and Ascher P. 1987. Glycine potentiates the NMDA response in cultured mouse brain neurons. *Nature* 325: 529–531. [14]

Johnson, J. W. and Ascher P. 1990. Voltage-dependent block by intracellular Mg^{2+} of N-methyl-D-aspartate-activated channels. *Biophys. J.* 57: 1085–90. [14]

Johnson, K. O., and Hsiao, S. S. 1992. Neural mechanisms of tactual form and texture perception. *Annu. Rev. Neurosci.* 15: 227–250. [18]

Johnson, K. O., Hsiao, S. S., and Twombly, I. A. 1995. Neural mechanisms of tactile form recognition. In M. Gazzaniga (ed.), *The Cognitive Sciences*. MIT Press, Cambridge, MA, pp. 253–268. [18]

Johnson, R. G., Jr. 1988. Accumulation of biological amines into chromaffin granules: A model of hormone and neurotransmitter transport. *Physiol. Rev.* 68: 232–307. [13]

Johnston, G. A. 1996. $GABA_A$ receptor pharmacology. *Pharmacol. Ther.* 69: 173–198. [14]

Jonas, P., Bischofberger, J., and Sandkuhler, J. 1998. Corelease of two fast neurotransmitters at a central synapse. *Science* 281: 419–424. [13, 14]

Jonas, P., Major, G., and Sakmann, B. 1993. Quantal components of unitary EPSCs at the mossy fibre synapse on CA3 pyramidal cells of rat hippocampus. *J. Physiol.* 472: 615–663. [11]

Jones, D. T., and Reed, R. R. 1989. Golf: An olfactory neuron specific-G protein involved in odorant signal transduction. *Science* 244: 790–796. [17]

Jones, E. G., and Wise, S. P. 1977. Size, laminar and columnar distribution of efferent cells in the sensorimotor cortex of monkeys. *J. Comp. Neurol.* 175: 391–438. [22]

Jones, E. M. C., Gray-Keller, M., and Fettiplace, R. 1999. The role of Ca^{2+}-activated K^+ channel spliced variants in the tonotopic organization of the turtle cochlea. *J. Physiol.* 518: 653–665. [18]

Jones, E. M. C., Laus, C., and Fettiplace, R. 1998. Identification of Ca^{2+}-activated K channel splice variants and their distribution in the turtle cochlea. *Proc. R. Soc. Lond. B* 265: 685–692. [18]

Jones, M. V., and Westbrook, G. L. 1996. The impact of receptor desensitization on fast synaptic transmission. *Trends Neurosci.* 19: 96–101. [9]

Jones, S. W., and Adams, P. R. 1987. The M-current and other potassium currents of vertebrate neurons. In L. K. Kaczmarek and I. B. Levitan (eds.), *Neuromodulation: The Biochemical Control of Neuronal Excitability*. Oxford University Press, New York, pp. 159–186. [16]

Jope, R. 1979. High-affinity choline uptake and acetylcholine production in the brain. Role in regulation of ACh synthesis. *Brain Res. Rev.* 1: 313–344. [13]

Jouvet, M. 1972. The role of monoamines and acetylcholine-containing neurons in the regulation of the sleep-waking cycle. *Ergeb. Physiol.* 64: 166–307. [14]

Junge, D. 1992. *Nerve and Muscle Excitation*, 3rd Ed. Sinauer Associates, Sunderland, MA. [5]

Kaas, J. H. 1993. The functional organization of somatosensory cortex in primates. *Ann. Anat.* 175: 509–518. [18]

Kaas, J. H. 1996. Theories of visual cortex organization in primates: Areas of the third level. *Prog. Brain Res.* 112: 213–221. [20]

Kaas, J. H., Hackett, T. A., and Tramo, M J. 1999. Auditory processing in primate cerebral cortex. *Curr. Opin. Neurobiol.* 9: 164–170. [18]

Kacharmina, J. E., Crino, P. B., and Eberwine, J. 1999. Preparation of cDNA from single cells and subcellular regions. *Methods Enzymol.* 303: 3–18. [14]

Kaczmarek, L. K., and Levitan, I. B. (eds.). 1987. *Neuromodulation: The Biochemical Control of Neuronal Excitability*. Oxford University Press, New York. [10]

Kalaska, J. F., Scott, S. H., Cisek, P., and Sergio, L. E. 1997. Cortical control of reaching movements. *Curr. Opin. Neurobiol.* 7: 849–859. [22]

Kallen, R. G., Sheng, Z-H., Yang, J., Chen, L.,Rogart, R. B., and Barchi, R. L. 1990. Primary structure and expression of a sodium channel characteristic of denervated and immature rat skeletal muscle. *Neuron* 4: 233–342. [3, 24]

Kallenberger, S., West, J. W., Catterall, W. A., and Scheuer, T. 1997a. Molecular analysis of potential hinge residues in the inactivation gate of brain type IIA Na^+ channels. *J. Gen. Physiol.* 109: 607–617. [6]

Kallenberger, S., West, J. W., Scheuer, T., and Catterall, W. A. 1997b. Molecular analysis of a putative inactivation particle in the inactivation gate of brain type IIA Na^+ channels. *J. Gen. Physiol.* 109: 589–605. [6]

Kalmidjn, A. J. 1982. Electric magnetic field detection in elasmobranch fishes. *Science* 218: 916–918. [17]

Kanai, Y. 1997. Family of neutral and acidic amino acid transporters: Molecular biology, physiology and medical implications. *Curr. Opin. Cell Biol.* 4: 565–572. [4]

Kandel, E. R. 1979. *Behavioral Biology of* Aplysia. W. H. Freeman, San Francisco. [15]

Kaneko, A. 1970. Physiological and morphological identification of horizontal, bipolar and amacrine cells in goldfish retina. *J. Physiol.* 207: 623–633. [19]

Kaneko, A. 1971. Electrical connexions between horizontal cells in the dogfish retina. *J. Physiol.* 213: 95–105. [19]

Kaneko, A., and Hashimoto, H. 1969. Electrophysiological study of single neurons in the inner nuclear layer of the carp retina. *Vision Res.* 9: 37–55. [1, 19]

Kaneko, A., and Tachibana, M. 1986. Effects of gamma-aminobutyric acid on isolated cone photoreceptors of the turtle retina. *J. Physiol.* 373: 443–461. [19]

Kaneko, A., Delavilla, P., Kurahashi, T., and Sasaki, T. 1994. Role of L-glutamate for formation of on-responses and off-responses in the retina. *Biomed. Res.* 15(Suppl. 1): 41–45. [19]

Kaneko, T., Akiyama, H., Nagatsu, I., and Muzuno, N. 1990. Immunohistochemical demonstration of glutaminase in catecholaminergic and serotonergic neurons of rat brain. *Brain Res.* 507: 151–154. [13]

Kaneko, T., Caria, M. A., and Asanuma, H. 1994. Information processing within the motor cortex. II. Intracortical connections between neurons receiving somatosensory cortical input and motor output neurons of the cortex. *J. Comp. Neurol.* 345: 172–184. [22]

Kanjhan, R., Housley, G. D., Burton, L. D., Christie, D. L., Kippenberger, A., Thorne, P. R., Luo, L., and Ryan, A. F. 1999. Distribution of the P2X2 receptor subunit of the ATP-gated ion channels in the rat central nervous system. *J. Comp. Neurol.* 407: 11–32. [14]

Kanner, B. I. 1994. Sodium-coupled neurotransmitter transport: Structure, function and regulation. *J. Exp. Biol.* 196: 237–249. [4]

Kanwisher, N., McDermott, J., and Chun, M. M. 1997. The fusiform face area: A module in human extrastriate cortex specialized for face perception. *J. Neurosci.* 17: 4302–4311. [21]

Kao, C. T. 1966. Tetrodotoxin, saxotoxin and their significance in the study of excitation phenomena. *Pharmacol. Rev.* 18: 977–1049. [6]

Kaplan, E., and Shapley, R. M. 1986. The primate retina contains two types of ganglion cells, with high and low contrast sensitivity. *Proc. Natl. Acad. Sci. USA* 83: 2755–2757. [19]

Kaplan, M. R., Meyer-Franke, A., Lambert, S., Bennett, V., Duncan, I. D., Levinson, S. R., and Barres, B. A. 1997. Induction of sodium channel clustering by oligodendrocytes. *Nature* 386: 724–728. [8]

Kaplan, M. R., Mount, D. B., Delpire, E., Gambo, G., and Hebert, S. C. 1996. Molecular mechanisms of NaCl cotransport. *Ann. Rev. Physiol.* 58: 649–668. [4]

Karni, A., Meyer, G., Jezzard, P., Adams, M. M., Turner, R., Underleiger, L. G. 1995. Functional MRI evidence for adult motor cortex plasticity during motor skill learning. *Nature* 377: 155–158. [22]

Karowski, C. J., Lu, H-K., and Newman, E. A. 1989. Spatial buffering of light-evoked potassium increases by retinal Müller (glial) cells. *Science* 224: 579–580. [8]

Karplus, M., and Petsko, G. A. 1990. Molecular dynamics simulations in biology. *Nature* 347: 631–639. [2]

Kasa, P., Rakonczay, Z., and Gulya, K. 1997. The cholinergic system in Alzheimer's disease. *Prog. Neurobiol.* 52: 511–535. [14]

Kasai, H. 1999. Comparative biology of Ca^{2+}-dependent exocytosis: Implications of kinetic diversity for secretory function. *Trends Neurosci.* 22: 88–93. [11]

Kasakov, L., Ellis, J., Kirkpatrick, K., Milner, P. and Burnstock, G. 1988. Direct evidence for concomitant release of noradrenaline, adenosine 5′-triphosphate and neuropeptide Y from sympathetic nerve supplying the guinea-pig vas deferens. *J. Auton. Nerv. Syst.* 22: 75–82. [16]

Kaspar, J., Schor, R. H., and Wilson, V. J. 1988. *J. Neurophysiol.* 60: 1765–1768. [22]

Katsuki, Y. 1961. Neural mechanisms of auditory sensation in cats. In W. A. Rosenblith (ed.), *Sensory Communication*. MIT Press, Cambridge, MA, pp. 561–583. [18]

Katz, B. 1950. Depolarization of sensory nerve terminal and the initiation of impulses in the muscle spindle. *J. Physiol.* 111: 261–282. [17]

Katz, B. 1966. *Nerve, Muscle, and Synapse*. McGraw-Hill, New York. [1]

Katz, B., and Miledi, R. 1964. The development of acetylcholine sensitivity in nerve-free segments of skeletal muscle. *J. Physiol.* 170: 389–396. [24]

Katz, B., and Miledi, R. 1965. The effect of temperature on the synaptic delay at the neuromuscular junction. *J. Physiol.* 181: 656–670. [11]

Katz, B., and Miledi, R. 1967a. The timing of calcium action during neuromuscular transmission. *J. Physiol.* 189: 535–544. [11]

Katz, B., and Miledi, R. 1967b. A study of synaptic transmission in the absence of nerve impulses. *J. Physiol.* 192: 407–436. [11]

Katz, B., and Miledi, R. 1968. The role of calcium in neuromuscular facilitation. *J. Physiol.* 195: 481–492. [12]

Katz, B., and Miledi, R. 1972. The statistical nature of the acetylcholine potential and its molecular components. *J. Physiol.* 224: 665–699. [2, 11]

Katz, B., and Miledi, R. 1973. The binding of acetylcholine to receptors and its removal from the synaptic cleft. *J. Physiol.* 231: 549–574. [13]

Katz, B., and Miledi, R. 1977. Transmitter leakage from motor nerve endings. *Proc. R. Soc. Lond. B* 196: 59–72. [11]

Katz, B., and Thesleff, S. 1957. A study of "desensitization" produced by acetylcholine at the motor endplate. *J. Physiol.* 138: 63–80. [9]

Katz, L. C., and Shatz, C. J. 1996. Synaptic activity and the construction of cortical circuits. *Science* 274: 1133–1138. [23]

Katz, L. C., Gilbert, C. D., and Wiesel, T. N. 1989. Local circuits and ocular dominance columns in monkey striate cortex. *J. Neurosci.* 9: 1389–1399. [21]

Kaufman, C. M., and Menaker, M. J. 1993. Effect of transplanting suprachiasmatic nuclei from donors of different ages into completely SCN lesioned hamsters. *Neural Transplant Plast.* 4: 257–265. [16]

Kaupmann, K., Huggel, K., Heid, J., Flor, P. J., Bischoff, S., Mickel, S. J., McMaster, G., Angst, C., Bittiger, H., Froestl, W., and Bettler, B. 1997. Expression cloning of $GABA_B$ receptors uncovers similarity to metabotropic glutamate receptors. *Nature* 386: 239–246. [14]

Kaupmann, K., Schuler, V., Mosbacher, J., Bischoff, S., Bittiger, H., Heid, J., Froestl, W., Leonhard, S., Pfaff, T., Karschin, A., and Bettler, B. 1998. Human gamma-aminobutyric acid type B receptors are differentially expressed and regulate inwardly rectifying K^+ channels. *Proc. Natl. Acad. Sci. USA* 95: 14991–14996. [14]

Kaupp, U. B. 1995. Family of cyclic-nucleotide gated ion channels. *Curr. Opin. Neurobiol.* 5: 434–442. [19]

Kavalali, E., Klingauf, J., and Tsien, R. W. 1999. Properties of fast endocytosis at hippocampal synapses. *Philos. Trans. R. Soc. Lond. B* 354: 337–346. [13]

Kawai, N., Yamagishi, S., Saito, M., and Furuya, K. 1983. Blockade of synaptic transmission in the squid giant synapse by a spider toxin (JSTX). *Brain Res.* 278: 346–349. [14]

Kawakami, K., Noguchi, S., Noda, M., Tqakahashi, H., Ohta, T., Kawamura, M., Nojmia, H., Hagano, K. Hirose, T. Inayama, S., Hayashida, H., Miyata, T., and Numa, S. 1985. Structure of α-subunit of *Torpedo californica* (Na^+-K^+)ATPase deduced from cDNA sequence. *Nature* 316: 733–736. [4]

Kawata, M., Yuri, K., and Sano, Y. 1991. Localization and regulation of mRNAs in the nervous tissue as revealed by in situ hybridization. *Comp. Biochem. Physiol. C* 98: 41–50. [14]

Keinän, K., Wisden, W., Sommer, B., Weiner, P., Herb, A., Verdoorn, T. A., Sakmann, B., and Seeberg, P. 1990. A family of AMPA-selective glutamate receptors. *Science* 249: 556–560. [3]

Keirstead, H. S., Dyer, J. K., Sholomenko, G. N., McGraw, J., Delaney, K. R., and Steeves, J. D. 1995. Axonal regeneration and physiological activity following transection and immunological disruption of myelin within the hatchling chick spinal cord. *J. Neurosci.* 15: 6963–6974. [24]

Keirstead, S. A., and Miller, R. F. 1997. Metabotropic glutamate receptor agonists evoke calcium waves in isolated Muller cells. *Glia* 21: 194–203. [8]

Kelly, D. R. 1996. When is a butterfly like an elephant? *Chem. Biol.* 8: 595–602. [17]

Kemp, D. T. 1978. Stimulated acoustic emissions from within the human auditory system. *J. Acoust. Soc. Am.* 64: 1386–1391. [17]

Kennedy, M. B. 1989. Regulation of neuronal function by calcium. *Trends Neurosci.* 12: 417–420. [10]

Kennedy, P. R. 1990. Corticospinal, rubrospinal and rubro-olivary projections: A unifying hypothesis. *Trends Neurosci.* 13: 474–479. [22]

Kennedy, T. E., Serafini, T., de la Torre, J. R., and Tessier-Lavigne, M. 1994. Netrins are diffusible chemotropic factors for commissural axons in the embryonic spinal cord. *Cell* 78: 425–435. [23]

Kettenmann, H., and Ransom, B. R. (eds.). 1995. *Neuroglia*. Oxford University Press, New York. [8]

Keynes, R. D. 1990. A series-parallel model of the voltage-gated sodium channel. *Proc. R. Soc. Lond. B* 240: 425–432. [6]

Keynes, R. D., and Elinder, F. 1998. Modelling the activation, opening, inactivation and reopening of the voltage-gated sodium channel. *Proc. R. Soc. Lond. B* 265: 263–270. [6]

Keynes, R. D., and Lewis, P. R. 1951. The sodium and potassium content of cephalopod nerve fibers. *J. Physiol.* 114: 151–182. [6]

Keynes, R. D., and Lumsden, A. 1990. Segmentation and the origin of regional diversity in the vertebrate central nervous system. *Neuron* 2: 1–9. [23]

Keynes, R. D., and Rojas, E. 1974. Kinetics and steady-state properties of the charged system controlling sodium conductance in the squid giant axon. *J. Physiol.* 239: 393–434. [6]

Kiang, N. Y., Rho, J. M., Northrop, C. C., Liberman, M. C., and Ryugo, D. K. 1982. Hair-cell innervation by spiral ganglion cells in adult cats. *Science* 217: 175–177. [18]

Kidd, T., Bland, K. S., and Goodman, C. S. 1999. Slit is the midline repellent for the robo receptor in *Drosophila*. *Cell* 96: 785–794. [23]

Kidd, T., Brose, K., Mitchell, K. J., Fetter, R. D., Tessier-Lavigne, M., Goodman, C. S., and Tear, G. 1998. Roundabout controls axon crossing of the CNS midline and defines a novel subfamily of evolutionarily conserved guidance receptors. *Cell* 92: 205–215. [23]

Kiehn, O., Harris-Warrick, R. M., Jordan, L. M., Hultborn, H., and Kudo, N. (eds.). 1998. *Neuronal Mechanisms for Generating Locomotor Activity* (Annals of the New York Academy of Sciences, vol. 860). New York Academy of Sciences, New York. [22]

Kier, C. K., Buchsbaum, G., and Sterling, P. 1995. How retinal microcircuits scale for ganglion cells of different size. *J. Neurosci.* 15: 7673–7683. [19]

Kikkawa, S., Nakagawa, M., Iwasa, T., Kaneko, A., and Tsuda, M. 1993. GTP-binding protein couples with metabotropic glutamate receptor in bovine retinal on-bipolar cell. *Biochem. Biophys. Res. Commun.* 195: 374–379. [19]

Kim, D. S., and Bonhoeffer, T. 1994. Reverse occlusion leads to a precise restoration of orientation preference maps in visual cortex. *Nature* 370: 370–372. [25]

Kinnamon, S. C., and Margolskee, R. F. 1996. Mechanisms of taste transduction. *Curr. Opin. Neurobiol.* 6: 506–513. [17]

Kinnamon, S. C., Dionne, V. E., and Beam, K. G. 1988. Apical localization of K^+ channels in taste cells provides the basis for sour taste transduction. *Proc. Natl. Acad. Sci. USA* 85: 7023–7027. [17]

Kirkwood, P. A., and Sears, T. A. 1974. Monosynaptic excitation of motoneurones from secondary endings of muscle spindles. *Nature* 252: 243–244. [22]

Kirkwood, P. A., and Sears, T. A. 1982. Excitatory postsynaptic potentials from single muscle spindle afferents in external intercostal motoneurones of the cat. *J. Physiol.* 322: 287–314. [22]

Kirn, J. R., and Nottebohm, F. 1993. Direct evidence for loss and replacement of projection neurons in adult canary brain. *J. Neurosci.* 13: 1654–1663. [23]

Kirsch, J. 1999. Assembly of signaling machinery at the postsynaptic membrane. *Curr. Opin. Neurobiol.* 9: 329–335. [13]

Kirsch, J., Wolters, I., Triller, A., and Betz, H. 1993. Gephyrin antisense oligonucleotides prevent glycine receptor clustering in spinal neurons. *Nature* 366: 745–748. [14]

Kirshner, N. 1969. Storage and secretion of adrenal catecholamines. *Adv. Biochem. Psychopharmacol.* 1: 71–89. [11]

Kisvardy, Z. F., Toth, E., Rausch, M., and Eysel, U. T. 1997. Orientation-specific relationship between populations of excitatory and inhibitory lateral connections in the visual cortex of the cat. *Cerebral Cortex* 7: 605–618. [21]

Kiyama, H., Sato, K., Kuba, T., and Tohyama, M. 1993. Sympathetic and parasympathetic ganglia express non-NMDA type glutamate receptors: Distinct receptor subunit composition in the principal and SIF cells. *Brain Res. Mol. Brain Res.* 19: 345–348. [16]

Kleene, S. J., and Gesteland, R. C. 1991. Calcium-activated chloride conductance in frog olfactory cilia. *J. Neurosci.* 11: 3624–3629. [17]

Kleinhaus, A. L., and Angstadt, J. D. 1995. Diversity and modulation of ionic conductances in leech neurons. *J. Neurobiol.* 27: 419–433. [15]

Klingauf, J., Kavalali, E. T., and Tsien, R. W. 1998. Kinetics and regulation of fast endocytosis at hippocampal synapses. *Nature* 394: 581–585. [11]

Kliot, M., and Shatz, C. J. 1985. Abnormal development of the retinogeniculate projection in Siamese cats. *J. Neurosci.* 5: 2641–2653. [25]

Klockgether, T., and Evert, B. 1998. Genes involved in hereditary ataxias. *Trends Neurosci.* 21: 413–418. [22]

Kneussel, M., and Betz, H. 2000. Receptors, gephyrin and gephyrin-associated proteins: novel insights into the assembly of inhibitory postsynaptic membrane specializations. *J. Physiol.* 525: 1–9. [13]

Knudsen, E. I. 1998. Capacity for plasticity in the adult owl auditory system expanded by juvenile experience. *Science* 279: 1531–1533. [25]

Knudsen, E. I. 1999a. Early experience and critical periods. In S. C. Landis, M. J. Zigmond, L. R. Squire, J. L. Roberts, and F. E. Bloom (eds.), *Fundamental Neuroscience*. Academic Press, New York, pp. 637–654. [25]

Knudsen, E. I. 1999b. Mechanisms of experience-dependent plasticity in the auditory localization pathway of the barn owl. *J. Comp. Physiol. [A]* 185: 305–321. [25]

Knudsen, E. I., and Knudsen, P. F. 1990. Sensitive and critical periods for visual calibration of sound localization by barn owls. *J. Neurosci.* 10: 222–232. [25]

Koehnle, T. J., and Brown, A. 1999. Slow axonal transport of neurofilament protein in cultured neurons. *J. Cell Biol.* 144: 447–458. [13]

Koh, D. S., Jonas, P., and Vogel, W. 1994. Na^+-activated K^+ channels localized in the nodal region of myelinated axons of *Xenopus*. *J. Physiol.* 479(Pt.2): 183–197. [6, 7]

Kohler, C., Swanson, L. W., Haglund, L., and Wu, J-Y. 1985. The cytoarchitecture, histochemistry and projections of the tuberomammillary nucleus in the rat. *Neuroscience* 16: 85–110. [14]

Köhler, M., Burnashev, N., Sakmann, B., and Seeburg, P. H. 1993. Determinants of Ca^{2+} permeability in both TM1 and TM2 of high-affinity kainate receptor channels: Diversity by RNA editing. *Neuron* 10: 491–500. [3]

Kokaia, Z., Bengzon, J., Metsis, M., Kokaia, M., Persson, H., and Lindvall, O. 1993. Coexpression of neurotrophins and their receptors in neurons of the central nervous system. *Proc. Natl. Acad. Sci. USA* 90: 6711–6715. [23]

Kolb, H. 1997. Amacrine cells of the mammalian retina: Neurocircuitry and functional roles. *Eye* 11: 904–923. [19]

Kolodkin, A. L. 1998. Semaphorin-mediated neuronal growth cone guidance. *Prog. Brain Res.* 117: 115–132. [23]

Komatsu, H. 1998. Mechanisms of color vision. *Curr. Opin. Neurobiol.* 8: 503–508. [21]

Konietzny, F., Perl, E. R., Trevino, D., Light, A., and Hensel, H. 1981. Sensory experiences in man evoked by intraneural electrical stimulation of intact cutaneous afferent fibers. *Exp. Brain Res.* 42: 219–222. [18]

Konnerth, A., Llano, I., and Armstrong, C. M. 1990. Synaptic currents in cerebellar Purkinje cells. *Proc. Natl. Acad. Sci. USA* 87: 2662–2665. [22]

Kopin, I. J. 1968. False adrenergic transmitters. *Annu. Rev. Pharmacol.* 8: 377–394. [13]

Kopin, I. J., Breese, G. R., Krauss, K. R., and Weise, V. K. 1968. Selective release of newly synthesized norepinephrine from the cat spleen during sympathetic nerve stimulation. *J. Pharmacol. Exp. Ther.* 161: 271–278. [13]

Kornack, D. R., and Rakic, P. 1999. Continuation of neurogenesis in the hippocampus of the adult macaque monkey. *Proc. Natl. Acad. Sci. USA* 96: 5768–5773. [23]

Korneev, S., Fedorov, A., Collins, R., Blackshaw, S. E., and Davies, J. A. 1997. A subtractive cDNA library from an identified regenerating neuron is enriched in sequences up-regulated during nerve regeneration. *Invertebr. Neurosci.* 3: 185–192. [24]

Kornhauser, J. M, Mayo, K. E., and Takahashi, J. S. 1996. Light, immediate-early genes, and circadian rhythms. *Behav. Genet.* 26: 221–240. [16]

Koulen, P., Brandstatter, J. H., Enz, R., Bormann, J., and Wässle, H. 1998. Synaptic clustering of GABA(C) receptor rho-subunits in the rat retina. *Eur. J. Neurosci.* 10: 115–127. [23]

Koutalos, Y., and Yau, K. W. 1996. Regulation of sensitivity in vertebrate rod photoreceptors by calcium. *Trends Neurosci.* 19: 73–81. [19]

Krauthammer, C. 2000. Restoration, reality and Christopher Reeve. *Time* February 14: 76. [26]

Kremer, E., and Lev-Tov, A. 1997. Localization of the spinal network associated with generation of hindlimb locomotion in the neonatal rat and organization of its transverse coupling system. *J. Neurophysiol.* 77: 1155–1170. [22]

Kreutzberg, G. W. 1996. Microglia: A sensor for pathological events in the CNS. *Trends Neurosci.* 19: 312–318. [8]

Krieger, D. T. 1983. Brain peptides: What, where and why? *Science* 222: 975–985. [14]

Kristan, W. B. 1983. The neurobiology of swimming in the leech. *Trends Neurosci.* 6: 84–88. [15]

Kristan, W. B., Lockery, S. R., and Lewis, J. E. 1995. Using reflexive behaviors of the medicinal leech to study information processing. *J. Neurobiol.* 27: 380–389. [15]

Krubitzer, L., Manger, P., Pettigrew, J., and Calford, M. 1995. Organization of somatosensory cortex in monotremes: In search of the prototypical plan. *J. Comp. Neurol.* 351: 261–306. [18]

Krug, M., Muller-Welde, P., Wagner, M., Ott, T., and Mathies, H. 1985. Functional plasticity in two afferent systems of the granule cells in the rat dentate area: Frequency-related changes, long-term potentiation and heterosynaptic depression. *Brain Res.* 360: 264–272. [12]

Kuffler, S. W. 1953. Discharge patterns and functional organization of the mammalian retina. *J. Neurophysiol.* 16: 37–68. [1, 15, 19, 20]

Kuffler, S. W. 1967. Neuroglial cells: Physiological properties and a potassium mediated effect of neuronal activity on the glial membrane potential. *Proc. R. Soc. Lond. B* 168: 1–21. [8]

Kuffler, S. W. 1980. Slow synaptic responses in autonomic ganglia and the pursuit of a peptidergic transmitter. *J. Exp. Biol.* 89: 257–286. [16]

Kuffler, S. W., and Eyzaguirre, C. 1955a. Processes of excitation in the dendrites and in the soma of single isolated sensory nerve cells of the lobster and crayfish. *J. Gen. Physiol.* 39: 87–119. [7]

Kuffler, S. W., and Eyzaguirre, C. 1955b. Synaptic inhibition in an isolated nerve cell. *J. Gen. Physiol.* 39: 155–184. [9]

Kuffler, S. W., and Nicholls, J. G. 1966. The physiology of neuroglial cells. *Ergeb. Physiol.* 57: 1–90. [8]

Kuffler, S. W., and Potter, D. D. 1964. Glia in the leech central nervous system: Physiological properties and neuron-glia relationship. *J. Neurophysiol.* 27: 290–320. [8, 15]

Kuffler, S. W., and Vaughan Williams, E. M. 1953. Small-nerve junctional potentials: The distribution of small motor nerves to frog skeletal muscle, and the membrane characteristics of the fibres they innervate. *J. Physiol.* 121: 289–317. [24]

Kuffler, S. W., and Yoshikami, D. 1975a. The distribution of acetylcholine sensitivity at the post-synaptic membrane of vertebrate skeletal twitch muscles: Iontophoretic mapping in the micron range. *J. Physiol.* 244: 703–730. [9, 11]

Kuffler, S. W., and Yoshikami, D. 1975b. The number of transmitter molecules in a quantum: An estimate from iontophoretic application of acetylcholine at the neuromuscular synapse. *J. Physiol.* 251: 465–482. [11]

Kuffler, S. W., Dennis, M. J., and Harris, A. J. 1971. The development of chemosensitivity in extrasynaptic areas of the neuronal surface after denervation of parasympathetic ganglion cells in the heart of the frog. *Proc. R. Soc. Lond. B* 177: 555–563. [24]

Kuffler, S. W., Hunt, C. C., and Quilliam, J. P. 1951. Function of medullated small-nerve fibers in mammalian ventral roots: Efferent muscle spindle innervation. *J. Neurophysiol.* 14: 29–51. [22]

Kuffler, S. W., Nicholls, J. G., and Orkand, R. K. 1966. Physiological properties of glial cells in the central nervous system of amphibia. *J. Neurophysiol.* 29: 768–787. [8]

Kuhar, M. J., De Souza, E. B., and Unnerstall, J. R. 1986. Neurotransmitter receptor mapping by autoradiography and other methods. *Annu. Rev. Neurosci.* 9: 27–59. [14]

Kuhse, J., Betz, H., and Kirsch, J. 1995. The inhibitory glycine receptor: Architecture, synaptic localization and molecular pathology of a post-synaptic ion-channel complex. *Curr. Opin. Neurobiol.* 5: 318–323. [3]

Kuljis, R. O., and Rakic, P. 1990. Hypercolumns in primate visual cortex can develop in the absence of cues from photoreceptors. *Proc. Natl. Acad. Sci. USA* 87: 5303–5306. [25]

Kuno, M. 1964a. Quantal components of excitatory synaptic potentials in spinal motoneurones. *J. Physiol.* 175: 81–99. [11]

Kuno, M. 1964b. Mechanism of facilitation and depression of the excitatory synaptic potential in spinal motoneurones. *J. Physiol.* 175: 100–112. [9, 12]

Kuno, M. 1971. Quantum aspects of central and ganglionic synaptic transmission in vertebrates. *Physiol. Rev.* 51: 647–678. [22]

Kupfermann, I. 1991. Functional studies of cotransmission. *Physiol. Rev.* 71: 683–732. [13]

Kurahashi, T., and Yau, K. Y. 1993. Co-existence of cationic and chloride components in odorant-induced current of vertebrate olfactory receptor cells. *Nature* 363: 71–74. [17]

Kuromi, H., and Kidokoro, Y. 1998. Two distinct pools of synaptic vesicles in single presynaptic boutons in a temperature-sensitive *Drosophila* mutant, *shibire. Neuron* 20: 917–925. [13]

Kuypers, H. G. J. M. 1981. In J. M. Brookhart and V. B. Mountcastle (eds.), *Handbook of Physiology*, Section 1: *The Nervous System*, Vol. 2: *Motor Control*, Part 2. American Physiological Society, Bethesda, MD. [22]

Kuypers, H. G. J. M., and Ugolini, G. 1990. Viruses as transneuronal tracers. *Trends Neurosci.* 13: 71–75. [13]

Kyte, J., and Doolittle, R. F. 1982. A simple method for displaying the hydrophobic character of a protein. *J. Mol. Biol.* 157: 105–132. [3]

Laake, J. H., Takumi, Y., Eidet, J., Torgner, I. A., Roberg, B., Kvamme, E., and Ottersen, O. P. 1999. Postembedding immunogold labelling reveals subcellular localization and pathway-specific enrichment of phosphate-activated glutaminase in rat cerebellum. *Neuroscience* 88: 1137–1151. [13]

Labhart, T. 1988. Polarization-opponent interneurons in the insect visual system. *Nature* 331: 435–437. [15]

Lachica, E. A., and Casagrande, V. A. 1992. Direct W-like geniculate projections to the cytochrome oxidase (CO) blobs in primate visual cortex: Axon morphology. *J. Comp. Neurol.* 319: 141–158. [21]

Lachica, E. A., Beck, P. D., and Casagrande, V. A. 1992. Parallel pathways in macaque striate cortex: Anatomically defined columns in layer III. *Proc. Natl. Acad. Sci. USA* 89: 3566–3570. [21]

Lam, D. M. 1997. *Invest. Ophthalmol. Vis. Sci.* 38: 553–556. [14]

Lam, D. M-K., and Ayoub, G. S. 1983. Biochemical and biophysical studies of isolated horizontal cells from the teleost retina. *Vision Res.* 23: 433–444. [14]

Lambright, D. G., Sondek, J., Bohm, A., Skiba, N. P., Hamm, H. E., and Sigler, P. B. 1996. The 2.0 crystal structure of a heterotrimeric G protein. *Nature* 379: 311–319. [10]

Lamme, V. A. F., Super, H., and Spekreijse, H. 1998. Feedforward, horizontal, and feedback processing in the visual cortex. *Curr. Opin. Neurobiol.* 8: 529–535. [20]

LaMotte, R. H., and Mountcastle, V. B. 1975. Capacities of humans and monkeys to discriminate between vibratory stimuli of different frequency and amplitude: A correlation between neuronal events and psychophysical measurements. *J. Neurophysiol.* 38: 539–559. [18]

Land, E. H. 1983. Recent advances in retinex theory and some implications for cortical computations: Color vision and the natural image. *Proc. Natl. Acad. Sci. USA* 80: 5163–5169. [21]

Land, E. H. 1986a. An alternative technique for the computation of the designator in the retinex theory of color vision. *Proc. Natl. Acad. Sci. USA* 83: 3078–3080. [21]

Land, E. H. 1986b. Recent advances in retinex theory. *Vision Res.* 26: 7–21. [21]

Land, M. F. 1993. News and views: Old twist in a new tale. *Nature* 363: 581–582. [15]

Lander, A. D. 1989. Understanding the molecules of neural cell contacts: Emerging patterns of structure and function. *Trends Neurosci.* 12: 189–195. [23]

Lang, E. J., Sugihara, I., Welsh, J. P., and Llinás, R. R. 1999. Patterns of spontaneous Purkinje cell complex spike activity in the awake rat. *J. Neurosci.* 19: 2728–2739. [22]

Lang, T., Wacker, I., Steyer, J., Kaether, C., Wunderlich, I., Soldati, T., Gerdes, H-H., and Almers, W. 1997. Ca^{2+}-triggered peptide secretion in single cells imaged with green fluorescent protein and evanescent-wave microscopy. *Neuron* 18: 857–863. [11]

Langer, G., Bronke, D., and Scheich, H. 1981. Neuronal discrimination of natural and synthetic vowels in field L of trained mynah birds. *Exp. Brain Res.* 43: 11–24. [18]

Langley, J. N. 1895. Note on regeneration of pre-ganglionic fibres of the sympathetic. *J. Physiol.* 18: 280–284. [23]

Langley, J. N. 1907. On the contraction of muscle, chiefly in relation to the presence of "receptive" substances. *J. Physiol.* 36: 347–384. [9]

Langley, J. N., and Anderson, H. K. 1892. The action of nicotin on the ciliary ganglion and on the endings of the third cranial nerve. *J. Physiol.* 13: 460–468. [9]

Langley, J. N., and Anderson, H. K. 1904. The union of different kinds of nerve fibres. *J. Physiol.* 31: 365–391. [24]

Lanno, I., Webb, C. K., and Bezanilla, F. 1988. Potassium conductance of the squid giant axon. Single channels studies. *J. Gen. Physiol.* 92: 179–196. [6]

Larkum, M. E., Zhu, J. J., and Sakmann, B. 1999. A new cellular mechanism for coupling inputs arriving at different cortical layers. *Nature* 398: 338–341. [7, 25]

Larsen, W. J., and Veenstra, R. D. 1998. Gap junction channels and biology. In N. Sperelakis (ed.), *Cell Physiology Source Book*, 2nd Ed. Academic Press, New York, pp. 467–480. [7]

Larsson, H. P., Baker, O. S., Dhillon, D. S., and Isacoff, E. Y. 1996. Transmembrane movement of the *Shaker* K^+ channel S4. *Neuron* 16: 387–397. [6]

Laskowski, M. B., and Sanes, J. R. 1988. Topographically selective reinnervation of adult mammalian skeletal muscles. *J. Neurosci.* 8: 3094–3099. [24]

Lassignal, N. L., and Martin, A. R. 1977. Effect of acetylcholine on postjunctional membrane permeability in eel electroplaque. *J. Gen. Physiol.* 70: 23–36. [9]

Lawrence, D. G., and Kuypers, H. G. J. M. 1968. The functional organization of the motor system in the monkey. I. The effects of bilateral pyramidal lesions. *Brain* 91: 1–14. [22]

Le Douarin, N. M. 1980. The ontogeny of the neural crest in avian embryo chimeras. *Nature* 286: 663–669. [23]

Le Douarin, N. M. 1986. Cell line segregation during peripheral nervous system ontogeny. *Science* 231: 1515–1522. [23]

LeBlanc, N., and Hume, J. R. 1990. Sodium-current induced release of calcium from cardiac sarcoplasmic reticulum. *Science* 248: 372–376. [4]

Ledent, C., Vaugeois, J. M., Schiffmann, S. N., Pedrazzini, T., El Yacoubi, M., Vanderhaeghen, J. J., Costentin, J., Heath, J. K., Vassart, G., and Parmentier, M. 1997. Aggressiveness, hypoalgesia and high blood pressure in mice lacking the adenosine A2a receptor. *Nature* 388: 674–678. [14]

Lee, H-K., Kameyama, K., Huganir, R. L., and Bear, M. F. 1998. NMDA induces long-term synaptic depression and dephosphorylation of the GluR1 subunit of AMPA receptors in hippocampus. *Neuron* 21: 1151–1162. [12]

Lee, K. J., and Jessell, T. M. 1999. The specification of dorsal cell fate in the vertebrate central nervous system. *Annu. Rev. Neurosci.* 22: 261–294. [23]

Lefkowitz, R. J. 1998. G protein-coupled receptors. III. New roles for receptor kinases and β-arrestins in receptor signaling and desensitization. *J. Biol. Chem.* 273: 18677–18680. [10]

Lehrer, M., and Collett, T. S. 1994. Approaching and departing bees learn different cues to the distance of a landmark. *J. Comp. Physiol. [A]* 175: 171–177. [15]

Leksell, L. 1945. The action potential and excitatory effects of the small ventral root fibres to skeletal muscle. *Acta Physiol. Scand.* 10(Suppl. 31): 1–84. [22]

Lemke, G. 1996. Neuregulins in development. *Mol. Cell. Neurosci.* 7: 247–262. [24]

Lena, C., and Changeux, J. P. 1997. Pathological mutations of nicotinic receptors and nicotine-based therapies for brain disorders. *Curr. Opin. Neurobiol.* 7: 674–682. [14]

Lennie, P., Krauskopf, J., and Sclar, G. 1990. Chromatic mechanisms in striate cortex of macaque. *J. Neurosci.* 10: 649–669. [21]

Leonard, J. P., and Wickelgren, W. O. 1986. Prolongation of calcium potentials by gamma-aminobutyric acid in primary sensory neurones of the lamprey. *J. Physiol.* 375: 481–497. [10]

Leonard, R. J., Labarca, C. G., Charnet, P., Davidson, N., and Lester, H. A. 1988. Evidence that the M2 membrane-spanning region lines the ion channel pore of a nicotinic receptor. *Science* 242: 1578–1581. [3]

Leonards, U., and Singer, W. 1997. Selective temporal interactions between processing streams with differential sensitivity for color and luminance contrast. *Vision Res.* 37: 1129–1140. [21]

Lepre, M., Fernandéz, J., and Nicholls, J. G. 1998. Re-establishment of direct synaptic connections between sensory axons and motoneurons after lesions of neonatal opossum CNS (*Monodelphis domestica*) in culture. *Eur. J. Neurosci.* 10: 2500–2510. [24]

Lester, H. A. 1992. The permeation pathway of neurotransmitter-gated ion channels. *Annu. Rev. Biophys. Biomol. Struct.* 21: 267–292. [9]

Letourneau, P. C. 1996. The cytoskeleton in nerve growth cone motility and axonal pathfinding. *Perspect. Dev. Neurobiol.* 4: 111–123. [23]

LeVay, S., and McConnell, S. K. 1982. On and off layers in the lateral geniculate nucleus of the mink. *Nature* 300: 350–351. [20]

LeVay, S., and Voight, T. 1988. Ocular dominance and disparity coding in cat visual cortex. *Vis. Neurosci.* 1: 395–414. [21]

LeVay, S., Connolly, M., Houde, J., and Van Essen, D. C. 1985. The complete pattern of ocular dominance stripes in the striate cortex and visual field of the macaque monkey. *J. Neurosci.* 5: 486–501. [20]

LeVay, S., Hubel, D. H., and Wiesel, T. N. 1975. The pattern of ocular dominance columns in macaque visual cortex revealed by a reduced silver stain. *J. Comp. Neurol.* 159: 559–576. [20]

LeVay, S., Stryker, M. P., and Shatz, C. J. 1978. Ocular dominance columns and their development in layer IV of the cat's visual cortex: A quantitative study. *J. Comp. Neurol.* 179: 223–244. [25]

LeVay, S., Wiesel, T. N., and Hubel, D. H. 1980. The development of ocular dominance columns in normal and visually deprived monkeys. *J. Comp. Neurol.* 191: 1–51. [25]

Levenes, C., Daniel, H., and Crepel, F. 1998. Long-term depression of synaptic transmission in the cerebellum: Cellular and molecular mechanisms revisited. *Prog. Neurobiol.* 55: 79–91. [12]

Levi-Montalcini, R. 1964. Growth-control of nerve cells by a protein factor and its antiserum. *Science* 143: 105–110. [23]

Levi-Montalcini, R. 1982. Developmental neurobiology and the natural history of nerve growth factor. *Annu. Rev. Neurosci.* 5: 341–362. [23]

Levi-Montalcini, R., and Angeletti, P. U. 1968. Nerve growth factor. *Physiol. Rev.* 48: 534–569. [23]

Levi-Montalcini, R., and Cohen, S. 1960. Effects of the extract of the mouse submaxillary salivary glands on the sympathetic system of mammals. *Ann. N.Y. Acad. Sci.* 85: 324–341. [23]

LeVine, H., III. 1999. Structural features of heterotrimeric G-protein-coupled receptors and their modulatory proteins. *Mol. Neurobiol.* 19: 111–149. [13]

Levinson, S. R. 1998. Structure and mechanism of voltage-gated ion channels. In N. Sperelakis (ed.), *Cell Physiology Source Book*, 2nd Ed. Academic Press, San Diego, CA, pp. 406-428. [3]

Levinson, S. R., and Meves, H. 1975. The binding of tritiated tetrodotoxin to squid giant axon. *Philos. Trans. R. Soc. Lond. B* 270: 349–352. [6]

Levitan, I. B., and Kaczmarek, L. K. 1997. *The Neuron: Cell and Molecular Biology*, 2nd Ed. Oxford University Press, New York. [10]

Lev-Tov, A., Miller, J. P., Burke, R. E., and Rall, W. 1983. Factors that control amplitude of EPSPs in dendritic neurons. *J. Neurophysiol.* 50: 399–412. [7]

Lewin, G. R., and Barde, Y-A. 1996. Physiology of the neurotrophins. *Annu. Rev. Neurosci.* 19: 289–317. [23]

Lewis, C., Neidhart, S., Holy, C., North, R. A., Buell, G., and Surprenant, A. 1995. Coexpression of P2X2 and P2X3 receptor subunits can account for ATP-gated currents in sensory neurons. *Nature* 377: 432–435. [17]

Lewis, J. E., and Kristan, W. B., Jr. 1998a. A neuronal network for computing population vectors in the leech. *Nature* 391: 76–79. [15]

Lewis, J. E., and Kristan, W. B., Jr. 1998b. Quantitative analysis of a directed behavior in the medicinal leech: Implications for organizing motor output. *J. Neurosci.* 18: 1571–1582. [15]

Lewis, R. S. 1999. Store-operated calcium channels. *Adv. Second Messenger Phosphoprotein Res.* 33: 279–307. [10]

Lewis, R. S., and Hudspeth, A. J. 1983. Voltage- and ion-dependent conductances in solitary vertebrate hair cells. *Nature* 304: 538–541. [18]

Li, C-L., and Jasper, H. H. 1953. Microelectrode studies of the electrical activity in the cerebral cortex of the cat. *J. Physiol.* 121: 117–140. [7]

Li, X-J., Blackshaw, S., and Snyder, S. H. 1994. Expression and localization of amiloride-sensitive sodium channel indicate a role for non-taste cells in taste perception. *Proc. Natl. Acad. Sci. USA* 91: 1814–1818. [17]

Li, Y., Field, P. M., and Raisman, G. 1998. Regeneration of adult corticospinal axons induced by transplanted olfactory ensheathing cells. *J. Neurosci.* 18: 10514–10524. [24]

Liao, D., Hessler, N. A., and Malinow, R. 1995. Activation of postsynaptically silent synapses during pairing-induced LTP in the CA1 region of hippocampal slice. *Nature* 375: 400–404. [12]

Liberman, M. C. 1988. Response properties of cochlear efferent neurons: Monaural vs. binaural stimulation and the effects of noise. *J. Neurophysiol.* 60: 1779–1798. [18]

Liberman, M. C., and Brown, M. C. 1986. Physiology and anatomy of single olivocochlear neurons in the cat. *Hear. Res.* 24: 17–36. [18]

Lieberman, A. R. 1971. The axon reaction: A review of the principal features of perikaryal responses to axon injury. *Int. Rev. Neurobiol.* 14: 49–124. [24]

Lim, N. F., Nowycky, M. C., and Bookman, R. J. 1990. Direct measurement of exocytosis and calcium currents in single vertebrate nerve terminals. *Nature* 344: 449–451. [11]

Lin, C-H., Thompson, C. A., and Forscher, P. 1994. Cytoskeletal reorganization underlying growth cone motility. *Curr. Opin. Neurobiol.* 4: 640–647. [23]

Lin, J. H., and Rydqvist, B. 1999. The mechanotransduction of the crayfish stretch receptor neurone can be differentially activated or inactivated by local anaesthetics. *Acta Physiol. Scand.* 166: 65–74. [17]

Lindemann, B. 1996. Taste reception. *Physiol. Rev.* 76: 718–766. [17]

Linden, D. J., and Conner, J. A. 1995. Long-term synaptic depression. *Annu. Rev. Neurosci.* 18: 319–357. [12]

Linden, D. J., Smeyne, M., and Connor, J. A. 1993. Induction of cerebellar long-term depression in culture requires postsynaptic action of sodium ions. *Neuron* 10: 1093–1100. [12]

Lindsay, R. M. 1996. Therapeutic potential of the neurotrophins and neurotrophin-CNTF combinations in peripheral neuropathies and motor neuron diseases. *Ciba Found. Symp.* 196: 39–48. [24]

Ling, G., and Gerard, R. W. 1949. The normal membrane potential of frog sartorius fibers. *J. Cell Comp. Physiol.* 34: 383–396. [2, 9]

Linial, M., Ilouz, N., and Parnas, H. 1997. Voltage-dependent interaction between the muscarinic ACh receptor and proteins of the exocytic machinery. *J. Physiol.* 504: 251–258. [11]

Linsker, R. 1989. From basic network principles to neural architecture: Emergence of orientation columns. *Proc. Natl. Acad. Sci. USA* 83: 8779–8783. [21]

Lipp, P., and Niggli, E. 1994. Sodium-current induced calcium signals in isolated guinea-pig ventricular myocytes. *J. Physiol.* 474: 439–446. [4]

Lipscombe, D., Kongsamut, S., and Tsien, R. W. 1989. α-Adrenergic inhibition of sympathetic neurotransmitter release mediated by modulation of N-type calcium-channel gating. *Nature* 340: 639–642. [10]

Lipscombe, D., Madison, D. V., Poenie, M., Reuter, H., Tsien, R. W., and Tsien, R. Y. 1988. Imaging of cytosolic Ca^{2+} transients arising from Ca^{2+} stores and Ca^{2+} channels in sympathetic neurons. *Neuron* 1: 355–365. [10]

Liu, Y., Holmgren, M., Jurman, M. E., and Yellen, G. 1997. Gated access to the pore of a voltage-dependent K^+ channel. *Neuron* 19: 175–184. [3]

Livesey, F. J., O'Brien, J. A., Li, M., Smith, A. G., Murphey, L. J., and Hunt, S. P. 1997. A Schwann cell mitogen accompanying regeneration of motor neurons. *Nature* 390: 614–618. [24]

Livingstone, M. S., and Hubel, D. 1988. Segregation of form, color, movement, and depth: Anatomy, physiology, and perception. *Science* 240: 740–749. [21]

Livingstone, M. S., and Hubel, D. H. 1984. Anatomy and physiology of a color system in the primate visual cortex. *J. Neurosci.* 4: 309–356. [21]

Livingstone, M. S., and Hubel, D. H. 1987a. Connections between layer 4B of area 17 and the thick cytochrome oxidase stripes of area 18 in the squirrel monkey. *J. Neurosci.* 7: 3371–3377. [21]

Livingstone, M. S., and Hubel, D. H. 1987b. Psychophysical evidence for separate channels for the perception of form, color, movement, and depth. *J. Neurosci.* 7: 3416–3468. [21]

Livingstone, M. S., Rosen, G. D., Drislane, F. W., and Galaburda, A. 1991. Physiological and anatomical evidence for a magnocellular defect in developmental dyslexia. *Proc. Natl. Acad. Sci. USA* 88: 7943–7497. [21]

Lledo, P. M., Zhang, X., Sudhof, T. C., Malenka, R. C., and Nicoll, R. A. 1998. Postsynaptic membrane fusion and long-term potentiation. *Science* 279: 339–403. [12]

Llinás, R. R. 1975. The cortex of the cerebellum. *Sci. Am.* 232(1): 56–71. [22]

Llinás, R. 1982. Calcium in synaptic transmission. *Sci. Am.* 247(4): 56–65. [11]

Llinás, R. 1985. Electrotonic transmission in the mammalian central nervous system. In M. V. L. Bennett and D. C. Spray (eds.), *Gap Junctions.* Cold Spring Harbor Laboratory, Cold Spring Harbor, NY, pp. 337–353. [9]

Llinás, R., and Sugimori, M. 1980. Electrophysiological properties of *in vitro* Purkinje cell dendrites in mammalian cerebellar slices. *J. Physiol.* 305: 197–213. [6, 7]

Llinás, R., McGuinness, T. L., Leonard, C. S., Sugimori, M., and Greengard, P. 1985. Intraterminal injection of synapsin I or calcium/calmodulin-dependent protein kinase II alters neurotransmitter release at the squid giant synapse. *Proc. Natl. Acad. Sci. USA* 82: 3035–3039. [13]

Llinás, R., Sugimori, M., and Silver, R. B. 1992. Microdomains of high calcium concentration in a presynaptic terminal. *Science* 256: 677–679. [11]

Llinás, R., Sugimori, M., and Silver, R. B. 1995. Time resolved calcium microdomains and synaptic transmission. *J. Physiol. (Paris)* 89: 77–81. [10]

Lloyd, D. P. C. 1943. Conduction and synaptic transmission of the reflex response to stretch in spinal cats. *J. Neurophysiol.* 6: 317–326. [22]

Lockery, S. R., and Kristan, W. B., Jr. 1990. Distributed processing of sensory information in the leech. II. Identification of interneurons contributing to the local bending reflex. *J. Neurosci.* 10: 1816–1829. [15]

Loewenstein, O., and Wersall, J. 1959. A functional interpretation of the electron-microscopic structure of sensory hairs in the cristae of the elasmobranch *Raja clavata* in terms of directional sensitivity. *Nature* 184: 1807–1808. [17]

Loewenstein, W. 1981. Junctional intercellular communication: The cell-to-cell membrane channel. *Physiol. Rev.* 61: 829–913. [7, 9]

Loewenstein, W. R. 1999. *The Touchstone of Life.* Oxford University Press, New York. [8, 9]

Loewenstein, W. R., and Mendelson, M. 1965. Components of adaptation in a Pacinian corpuscle. *J. Physiol.* 177: 377–397. [17]

Loewi, O. 1921. Über humorale Übertragbarkeit der Herznervenwirkung. *Pflügers Arch.* 189: 239–242. [9]

Lohr, C., and Deitmar, J. W. 1997. Intracellular Ca^{++} release mediated by metabotropic glutamate receptor activation in the leech giant glial cell. *J. Exp. Biol.* 200: 2565–2573. [8]

Lomeli, J., Quevedo, J., Linares, P., and Rudomin, P. 1998. Local control of information flow in segmental and ascending collaterals of single afferents. *Nature* 395: 600–604. [9]

Lømo, T., and Rosenthal, J. 1972. Control of ACh sensitivity by muscle activity in the rat. *J. Physiol.* 221: 493–513. [24]

Long, Q., Meng, A., Wang, H., Jessen, J. R., Farrell, M. J., and Lin, S. 1997. GATA-1 expression pattern can be recapitulated in living transgenic zebrafish using GFP reporter gene. *Development* 124: 4105–4111. [23]

Lonsbury-Martin, B. L., Martin, G. K., McCoy, M. J., and Whitehead, M. L. 1995. New approaches to the evaluation of the auditory system and a current analysis of otoacoustic emissions. *Otolaryngol. Head Neck Surg.* 112: 50–63. [17]

Lorenz, K. 1970. *Studies in Animal and Human Behavior,* Vols. 1 and 2. Harvard University Press, Cambridge, MA. [25]

Loring, R. H., and Zigmond, R. E. 1987. Ultrastructural distribution of $[^{125}I]$-toxin F binding sites on chick ciliary neurons: Synaptic localization of a toxin that blocks ganglionic nicotinic receptors. *J. Neurosci.* 7: 2153–2162. [24]

Lowel, S., Schmidt, K. E., Kim, D. S., Wolf, F., Hoffsummer, F., Singer, W., and Bonhoeffer, T. 1998. The layout of orientation and ocular dominance domains in area 17 of strabismic cats. *Eur. J. Neurosci.* 10: 2629–2643. [21]

Löwel, S., and Singer, W. 1992. Selection of intrinsic horizontal connections in the visual cortex by correlated neuronal activity. *Science* 255: 209–212. [25]

Lucas, S. M., and Binder, M. D. 1984. Topographic factors in distribution of homonymous group Ia-afferent input to cat medial gastrocnemius motoneurons. *J. Neurophysiol.* 51: 50–63. [22]

Luddens, H., and Wisden, W. 1991. Function and pharmacology of multiple $GABA_A$ receptor subunits. *Trends Pharmacol. Sci.* 12: 49–51. [14]

Ludwig, A., Flockerzi, V., and Hofmann, F. 1997. Regional expression and cellular localization of the alpha1 and beta subunit of high voltage-activated calcium channels in rat brain. *J. Neurosci.* 17: 1339–1349. [22]

Luebke, A. E., Dickerson, E. M., and Muller, K. J. 1993. *In situ* hybridization reveals changing distribution of laminin expression during neuronal development and after injury in the leech CNS. *Soc. Neurosci. Abstr.* 19: 1084. [8]

Luján, R., Nusser, Z., Roberts, J. D. B., Shigemoto, R., and Somogyi, P. 1996. Perisynaptic location of metabotropic glutamate receptors mGluR1 and mGluR5 on dendrites and dendritic spines in the rat hippocampus. *Eur. J. Neurosci.* 8: 1488–1500. [13]

Lukasiewicz, P. D. 1996. $GABA_C$ receptors in the vertebrate retina. *Mol. Neurobiol.* 12: 181–194. [14]

Lumpkin, E. A., and Hudspeth, A. J. 1995. Detection of Ca^{2+} entry through mechanosensitive channels localizes the site of mechanoelectrical transduction in hair cells. *Proc. Natl. Acad. Sci. USA* 92: 10297–10301. [17]

Lumsden, A. G. S., and Davies, A. M. 1986. Chemotropic effect of specific target epithelium in the developing mammalian nervous system. *Nature* 323: 538–539. [23]

Lumsden, A., and Krumlauf, R. 1996. Patterning the vertebrate neuroaxis. *Science* 274: 1109–1115. [23]

Lund, J. S. 1988. Anatomical organization of macaque monkey striate visual cortex. *Annu. Rev. Neurosci.* 11: 253–288. [20]

Lundberg, A. 1979. Multisensory control of spinal reflex pathways. *Prog. Brain Res.* 50: 11–28. [22]

Lundberg, A., and Quilisch, H. 1953. On the effect of calcium on presynaptic potentiation and depression at the neuromuscular junction. *Acta Physiol. Scand.* 30(Suppl. III): 121–129. [12]

Lundberg, J. M. 1996. Pharmacology of cotransmission in the autonomic nervous system: Integrative aspects of amines, neuropeptides, adenosine triphosphate, amino acids and nitric oxide. *Pharmacol. Rev.* 48: 113–178. [13]

Luo, Y., Shepherd, I., Li, J., Renzi, M. J., Chang, S., and Raper, J. A. 1995. A family of molecules related to collapsin in the embryonic chick nervous system. *Neuron* 14: 1131–1140. [23]

Luqmani, Y. A., Sudlow, G., and Whittaker, V. P. 1980. Homocholine and acetylhomocholine: False transmitters in the cholinergic electromotor system of *Torpedo*. *Neuroscience* 5: 153–160. [13]

Luscher, C., Jan, L. Y., Stoffel, M., Malenka, R. C., and Nicoll, R. A. 1997. G protein-coupled inwardly-rectifying K channels (GIRKs) mediate postsynaptic but not presynaptic transmitter actions in hippocampal neurons. *Neuron* 19: 687–695. [14]

Luskin, M. 1993. Restricted proliferation and migration of postnatally generated neurons derived from the forebrain subventricular zone. *Neuron* 11: 173–189. [23, 24]

Luskin, M. B., and Shatz, C. J. 1985a. Neurogenesis of the cat's primary visual cortex. *J. Comp. Neurol.* 242: 611–631. [23]

Luskin, M. B., and Shatz, C. J. 1985b. Studies of the earliest generated cells of the cat's visual cortex: Cogeneration of subplate and marginal zones. *J. Neurosci.* 5: 1062–1075. [23]

Luskin, M. B., Pearlman, A. L., and Sanes, J. R. 1988. Cell lineage in the cerebral cortex of the mouse studied in vivo and in vitro with a recombinant retrovirus. *Neuron* 1: 635–647. [8, 23]

Lynch, G. S., Dunwiddie, T., and Gribkoff, V. 1977. Heterosynaptic depression: A postsynaptic correlate of long-term potentiation. *Nature* 266: 737–739. [12]

Lynch, G., Larson, J., Kelso, S., Barrionuevo, G., and Schottler, F. 1983. Intracellular injections of EGTA block induction of hippocampal long-term potentiation. *Nature* 304: 719–721. [12]

Macagno, E. R., Muller, K. J., and Pitman, R. M. 1987. Conduction block silences parts of a chemical synapse in the leech central nervous system. *J. Physiol.* 387: 649–664. [15]

Macdonald, R. L., and Werz, M. A. 1986. Dynorphin A decreases voltage-dependent calcium conductance of mouse dorsal root ganglion neurones. *J. Physiol.* 377: 237–249. [14]

MacKenzie, A. B., Surprenant, A., and North, R. A. 1999. Functional and molecular diversity of purinergic ion channel receptors. *Ann. N.Y. Acad. Sci.* 868: 716–729. [14]

MacKinnon, R. 1991. Determination of the subunit stoichiometry of a voltage-activated potassium channel. *Nature* 350: 232–238. [3]

MacKinnon, R., Cohen, S. L., Kuo, A., Lee, A., and Chait, B. T. 1998. Structural conservation in prokaryotic and eukaryotic potassium channels. *Science* 280: 106–109. [3]

MacNeil, M. A., Heussy, J. K., Dacheux, R. F., Raviola, E., and Masland, R. H. 1999. The shapes and numbers of amacrine cells: Matching of photofilled with Golgi-stained cells in the rabbit retina and comparison with other mammalian species. *J. Comp. Neurol.* 413: 305–326. [19]

MacVicar, B. A., and Dudek, F. E. 1981. Electrotonic coupling between pyramidal cells: A direct demonstration in rat hippocampal slices. *Science* 213: 782–785. [9]

Madison, D. V., and Nicoll, R. A. 1986. Cyclic adenosine 3′,5′-monophosphate mediates beta-receptor actions of noradrenaline in rat hippocampal pyramidal cells. *J. Physiol.* 372: 245–259. [14]

Maeno, T., Edwards, C., and Anraku, M. 1977. Perrmeability of the endplate membrane activated by acetylcholine to some organic cations. *J. Neurobiol.* 8: 173–184. [3]

Maffei, L., and Galli-Resta, L. 1990. Correlation in the discharges of neighboring rat retinal ganglion cells during prenatal life. *Proc. Natl. Acad. Sci. USA* 87: 2861–2964. [19, 25]

Mager, S., Kleinberger-Doron, N., Keshet, G. I., Davidson, N., Kanner, B. I., and Lester, H. A. 1996. Ion binding and permeation at the GABA transporter GAT1. *J. Neurosci.* 16: 5405–5414. [4]

Maggio, J. E. 1988. Tachykinins. *Annu. Rev. Neurosci.* 11: 13–28. [14]

Magleby, K. L., and Pallotta, B. S. 1981. A study of desensitization of acetylcholine receptors using nerve-released transmitter in the frog. *J. Physiol.* 316: 225–250. [9]

Magleby, K. L., and Stevens, C. F. 1972. The effect of voltage on the time course of end-plate currents. *J. Physiol.* 223: 151–171. [9]

Magleby, K. L., and Weinstock, M. M. 1980. Nickel and calcium ions modify the characteristics of the acetylcholine receptor-channel complex at the frog neuromuscular junction. *J. Physiol.* 299: 203–218. [11]

Magleby, K. L., and Zengel, J. E. 1976. Augmentation: A process that acts to increase transmitter release at the frog neuromuscular junction. *J. Physiol.* 257: 449–470. [12]

Maienschein, V., and Zimmermann, H. 1996. Immunocytochemical localization of ecto-5'-nucleotidase in cultures of cerebellar granule cells. *Neuroscience* 70: 429–438. [13]

Mains, R. E., and Eipper, B. A. 1999. Peptides. In G. J. Siegel, B. W. Agranoff, R. W. Albers, S. K. Fisher, and M. D. Uhler (eds.), *Basic Neurochemistry: Molecular, Cellular, and Medical Aspects*, 6th Ed. Lippincott-Raven, Philadelphia, pp. 363–382. [13]

Mains, R. E., and Patterson, P. H. 1973. Primary cultures of dissociated sympathetic neurons. I. Establishment of long-term growth in culture and studies of differentiated properties. *J. Cell Biol.* 59: 329–345. [23]

Mains, R. E., Eipper, B. A., and Ling, N. 1977. Common precursor to corticotropins and endorphins. *Proc. Natl. Acad. Sci. USA* 74: 3014–3018. [14]

Maison, S. F. and Liberman, M. C. 2000. Predicting vulnerability to acoustic injury with a noninvasive assay of olivocochlear reflex strength. *J. Neurosci.* 20: 4701–4707. [18]

Majewski, H., and Iannazzo, L. 1998. Protein kinase C: A physiological mediator of enhanced transmitter output. *Prog. Neurobiol.* 55: 463–475. [10]

Makowski, L., Caspar, D. L., Phillips, W. C., and Goodenough, D. A. 1977. Gap junction structure. II. Analysis of the X-ray diffraction data. *J. Cell Biol.* 74: 629–645. [7]

Malach, R., Ebert, R., and Van Sluyters, R. C. 1984. Recovery from effects of brief monocular deprivation in the kitten. *J. Neurophysiol.* 51: 538–551. [25]

Malenka, R. C., and Nicoll, R. A. 1999. Long-term potentiation—A decade of progress? *Science* 285: 1870–1874. [12]

Malenka, R. C., Kauer, J. A., Perkel, D. J., Mank, M. D., Kelly, P. T. , Nicoll, R. A., and Waxham, M. N. 1989. An essential role for postsynaptic calmodulin and protein kinase activity in long-term potentiation. *Nature* 340: 554–557. [12]

Malenka, R. C., Kauer, J. A., Zucker, R. S., and Nicoll, R. A. 1988. Postsynaptic calcium is sufficient for potentiation of hippocampal synaptic transmission. *Science* 242: 81–84. [12]

Malinow, R., and Tsien, R. W. 1990. Presynaptic enhancement shown by whole-cell recordings of long-term potentiation in hippocampal slices. *Nature* 346: 177–180. [12]

Malinow, R., Schulman, H., and Tsien, R. W. 1989. Inhibition of postsynaptic PKC or CaMKII blocks induction but not expression of LTP. *Science* 245: 862–866. [12, 14]

Mallart, A., and Martin, A. R. 1967. Analysis of facilitation of transmitter release at the neuromuscular junction of the frog. *J. Physiol.* 193: 679–697. [12]

Mallart, A., and Martin, A. R. 1968. The relation between quantum content and facilitation at the neuromuscular junction of the frog. *J. Physiol.* 196: 593–604. [12]

Malonek, D., Tootell, R. B. H., and Grinvald, A. 1994. Optical imaging reveals the functional architecture of neurons processing shape and motion in owl monkey area MT. *Proc. R. Soc. Lond. B* 258: 109–119. [21]

Mandelkow, E., and Hoenger, A. 1999. Structures of kinesin and kinesin-microtubule interactions. *Curr. Opin. Cell Biol.* 11: 34–44. [13]

Maren, S., and Fanselow, M. S. 1995. Synaptic plasticity in the basolateral amygdala induced hippocampal formation stimulation *in vivo*. *J. Neurosci.* 15: 7548–7564. [12]

Maricq, A. V., Peterson, A. S., Brake, A. J., Myers, R. M., and Julius, D. 1991. Primary structure and functional expression of the 5HT3 receptor, a serotonin-gated ion channel. *Science* 254: 432–437. [3]

Markram, H., Lubke, J., Frotscher, M., and Sakmann, B. 1997. Regulation of synaptic efficacy by coincidence of postsynaptic APs and EPSPs. *Science* 275: 213–215. [7]

Marks, W. B., Dobelle, W. H., and MacNichol, E. F. 1964. Visual pigments of single primate cones. *Science* 143: 1181–1183. [19]

Marmont, G. 1940. Studies on the axon membrane. *J. Cell. Comp. Physiol.* 34: 351–382. [6]

Marrion, N. V. 1997. Control of M-current. *Annu. Rev. Physiol.* 59: 483–504. [16]

Marsden, C. D., Rothwell, J. C., and Day, B. L. 1984. The use of peripheral feedback in the control of movement. *Trends Neurosci.* 7: 253–257. [22]

Martin, A. R. 1990. Glycine- and GABA-activated chloride conductances in lamprey neurons. In O. P. Ottersen and J. Storm-Mathisen (eds.), *Glycine Neurotransmission*. Wiley, New York, pp. 171–191. [14]

Martin, A. R., and Dryer, S. E. 1989. Potassium channels activated by sodium. *Q. J. Exp. Physiol.* 74: 1033–1041. [6]

Martin, A. R., and Fuchs, P. A. 1992. The dependence of calcium-activated potassium currents on membrane potential. *Proc. R. Soc. Lond. B* 250: 71–76. [18]

Martin, A. R., and Levinson, S. R. 1985. Contribution of the Na^+-K^+pump to membrane potential in familial periodic paralysis. *Muscle Nerve* 8: 354–362. [5]

Martin, A. R., and Pilar, G. 1963. Dual mode of synaptic transmission in the avian ciliary ganglion. *J. Physiol.* 168: 443–463. [9, 11]

Martin, A. R., and Pilar, G. 1964a. Quantal components of the synaptic potential in the ciliary ganglion of the chick. *J. Physiol.* 175: 1–16. [11]

Martin, A. R., and Pilar, G. 1964b. Presynaptic and postsynaptic events during post-tetanic potentiation and facilitation in the avian ciliary ganglion. *J. Physiol.* 175: 16–30. [12]

Martin, A., Haxby, J. V., Lalonde, F. M., Wiggs, C. L., and Ungerleider, L. G. 1995. Discrete cortical regions associated with knowledge of color and knowledge of action. *Science* 270: 102–105. [21]

Martin, D. L. 1987. Regulatory properties of brain glutamate decarboxylase. *Cell. Mol. Neurobiol.* 7: 237–253. [13]

Martin, D. P., Schmidt, R. E., DiStefano, P. S., Lowry, O. H., Carter, J. G., and Johnson, E. M. 1988. Inhibitors of protein synthesis and RNA synthesis prevent neuronal death caused by nerve growth factor deprivation. *J. Cell Biol.* 106: 829–844. [23]

Martin, P. R., White, A. J., Goodchild, A. K., Wilder, H. D., and Sefton, A. E. 1997. Evidence that blue-on cells are part of the third geniculocortical pathway in primates. *Eur. J. Neurosci.* 9: 1536–1541. [20]

Martinez de la Escalera, G., Choi, A. L. H., and Weiner, R. L. 1992. Generation and synchronization of gonadotropin-releasing hormone (GnRH) pulses: Intrinsic properties of the GT1-1 GnRH neuronal cell line. *Proc. Natl. Acad. Sci. USA* 89: 1852–1855. [16]

Martinez, F. E., Crill, W. E., and Kennedy, T. T. 1971. Electrogenesis of cerebellar Purkinje cell responses in cats. *J. Neurophysiol.* 34: 348–356. [22]

Martinez, J. L., Jr., and Derrick, B. E. 1996. Long-term potentiation and learning. *Annu. Rev. Psychol.* 47: 173–203. [12]

Martinez-Serrano, A., and Bjorklund, A. 1997. Immortalized neural progenitor cells for CNS gene transfer and repair. *Trends Neurosci.* 20: 530–538. [24]

Martini, R., and Schachner, M. 1997. Molecular bases of myelin formation as revealed by investigations on mice deficient in glial cell surface molecules. *Glia* 19: 298–310. [8]

Marty, A. 1989. The physiological role of calcium-dependent channels. *Trends Neurosci.* 12: 420–424. [10]

Masland, R. H. 1988. Amacrine cells. *Trends Neurosci.* 11: 405–410. [19]

Mason, A., and Muller, K. J. 1996. Accurate synapse regeneration despite ablation of the distal axon segment. *Eur. J. Neurosci.* 8: 11–20. [15]

Masson, G. S., Busettini, C., and Miles, F. A. 1997. Vergence eye movements in response to binocular disparity without depth perception. *Nature* 389: 283–286. [21]

Masuda-Nakagawa, L. M., and Nicholls, J. G. 1991. Extracellular matrix molecules in development and regeneration of the leech CNS. *Philos. Trans. R. Soc. Lond. B* 331: 323–335. [8, 23]

Mathie, A., Wooltorton, J. R., and Watkins, C. S. 1998. Voltage-activated potassium channels in mammalian neurons and their block by novel pharmacological agents. *Gen. Pharmacol.* 30: 13–44. [6]

Matsuda, T., Wu, J-Y., and Roberts, E. 1973. Immunochemical studies on glutamic acid decarboxylase (EC 4.1.1.15) from mouse brain. *J. Neurochem.* 21: 159–166, 167–172. [14]

Matsui, K., Hosoi, N., and Tachibana, M. 1999. Active role of glutamate uptake in the synaptic transmission from nonspiking neurons. *J. Neurosci.* 19: 6755–6766. [8]

Matthews, B. H. C. 1931a. The response of a single end organ. *J. Physiol.* 71: 64–110. [17]

Matthews, B. H. C. 1931b. The response of a muscle spindle during active contraction of a muscle. *J. Physiol.* 72: 153–174. [17]

Matthews, B. H. C. 1933. Nerve endings in mammalian muscle. *J. Physiol.* 78: 1–53. [17]

Matthews, G., and Wickelgren, W. O. 1979. Glycine, GABA and synaptic inhibition of reticulospinal neurones of the lamprey. *J. Physiol.* 293: 393–414. [5, 14]

Matthews, M. R., and Nelson, V. H. 1975. Detachment of structurally intact nerve endings from chromatolytic neurones of rat superior cervical ganglion during the depression of synaptic transmission induced by postganglionic axotomy. *J. Physiol.* 245: 91–135. [24]

Matthews, P. B. C. 1964. Muscle spindles and their motor control. *Physiol. Rev.* 44: 219–288. [17]

Matthews, P. B. C. 1972. *Mammalian Muscle Receptors and Their Central Action*. Edward Arnold, London. [22]

Matthews, P. B. C. 1981. Evolving views on the internal operation and functional role of the muscle spindle. *J. Physiol.* 320: 1–30. [17]

Matthews, R. G., Hubbard, R., Brown, P. K., and Wald. G. 1963. Tautomeric forms of metarhodopsin. *J. Gen. Physiol.* 47: 215–240. [19]

Matthey, R. 1925. Récupération de la vue après résection des nerfs optiques chez le triton. *C. R. Soc. Biol.* 93: 904–906. [24]

Maturana, H. R., Lettvin, J. Y., McCulloch, W. S., and Pitts, W. H. 1960. Anatomy and physiology of vision in the frog (*Rana pipiens*). *J. Gen. Physiol.* 43: 129–175. [19]

Maue, R. A., and Dionne, V. E. 1987. Patch-clamp studies of isolated mouse olfactory receptor neurons. *J. Gen. Physiol.* 90: 95–125. [17]

Maunsell, J. H. R., and Van Essen, D. C. 1983. Functional properties of neurons in the middle temporal visual area (MT) of the macaque monkey. I. Selectivity for stimulus direction, speed and orientation. *J. Neurophysiol.* 49: 1127–1147. [21]

Maunsell, J. H., and Newsome, W. T. 1987. Visual processing in monkey extrastriate cortex. *Annu. Rev. Neurosci.* 10: 363–401. [20, 21]

May, B. J., and Huang, A. Y. 1996. Sound orientation behavior in cats. I. Localization of broadband noise. *J. Acoust. Soc. Am.* 100: 1059–1069. [18]

May, B. J., and McQuone, S. J. 1995. Effects of bilateral olivocochlear lesions on pure-tone intensity discrimination in cats. *Auditory Neurosci.* 1: 385–400. [18]

Mayer, M. L., and Vyklicky, L., Jr. 1989. The action of zinc on synaptic transmission and neuronal excitability in cultures of mouse hippocampus. *J. Physiol.* 415: 351–365. [14]

Mayer, M. L., and Westbrook, G. L. 1987. The physiology of excitatory amino acids in the vertebrate central nervous system. *Prog. Neurobiol.* 28: 197–276. [14]

Mayer, M. L., Westbrook, G. L., and Guthrie, P. B. 1984. Voltage-dependent block by Mg^{2+} of NMDA responses in spinal cord neurons. *Nature* 309: 261–263. [12]

Mayser, W., Schloss, P., and Betz, H. 1992. Primary structure and functional expression of a choline transporter expressed in the rat nervous system. *FEBS Lett.* 305: 31–36. [4]

McAllister, A. K., Katz, L. C., and Lo, D. C. 1999. Neurotrophins and synaptic plasticity. *Annu. Rev. Neurosci.* 22: 295–318. [23]

McConnell, S. K. 1988. Fates of visual cortical neurons in the ferret after isochronic and heterochronic transplantation. *J. Neurosci.* 8: 945–974. [23]

McConnell, S. K. 1995. Constructing the cerebral cortex: Neurogenesis and fate determination. *Neuron* 15: 761–768. [23]

McCrea, P. D., Popot, J-L., and Engleman, D. M. 1987. Transmembrane topography of the nicotinic acetylcholine receptor subunit. *EMBO J.* 6: 3619–3626. [3]

McDonald, N. Q., and Chao, M. V. 1995. Structural determinants of neurotrophin action. *J. Biol. Chem.* 270: 19669–19672. [23]

McDonald, T. F., Pelzer, S., Trautwein, W., and Pelzer, D. J. 1994. Regulation and modulation of calcium channels in cardiac, skeletal, and smooth muscle cells. *Physiol. Rev.* 74: 365–507. [10]

McEachern, A. E., Jacob, M. H., and Berg, D. K. 1989. Differential effects of nerve transection on the ACh and GABA receptors of chick ciliary ganglion neurons. *J. Neurosci.* 9: 3899–3907. [24]

McFarlane, S., Cornel, E., Amaya, E., and Holt, C. E. 1996. Inhibition of FGF receptor activity in retinal ganglion cell axons causes error in target recognition. *Neuron* 17: 245–254. [23]

McGehee, D. S., and Role, L. W. 1995. Physiological diversity of nicotinic acetylcholine receptors expressed by vertebrate neurons. *Ann. Rev. Physiol.* 57: 521–546. [3]

McGehee, D. S., Heath, M. J., Gelber, S., Devay, P., and Role, L. W. 1995. Nicotine enhancement of fast excitatory synaptic transmission in CNS by presynaptic receptors. *Science* 269: 1692–1696. [14]

McGlade-McCulloh, E., Morrissey, A. M., Norona, F., and Muller, K. J. 1989. Individual microglia move rapidly and directly to nerve lesions in the leech central nervous system. *Proc. Natl. Acad. Sci. USA* 86: 1093–1097. [8]

McIntire, S. L., Reimer, R. J., Schuske, K., Edwards, R. H., and Jorgensen, E. M. 1997. Identification and characterization of the vesicular GABA transporter. *Nature* 389: 870–876. [4]

McIntyre, A. K. 1980. Biological seismography. *Trends Neurosci.* 3: 202–205. [17]

McKay, R. 1997. Stem cells in the central nervous system. *Science* 276: 66–71. [14]

McKenna, M. P., and Raper, J. A. 1988. Growth cone behavior on gradients of substrate bound laminin. *Dev. Biol.* 130: 232–236. [23]

McKernan, M. G., and Shinnick-Gallagher, P. 1997. Fear conditioning induces a lasting potentiation of synaptic currents in vitro. *Nature* 390: 607–611. [12]

McKernan, R. M., and Whiting, P. J. 1996. Which $GABA_A$ receptor subtypes really occur in brain. *Trends Neurosci.* 19: 139–143. [3]

McLachlan, E. M. (ed.). 1995. *Autonomic Ganglia.* (Autonomic Nervous System, Vol. 6). Gordon and Breach, London. [16]

McLaughlin, S. K., McKinnon, P. J., and Margolskee, R. F. 1992. Gustducin is a taste-cell-specific G protein closely related to the transducins. *Nature* 357: 563–569. [17]

McMahan, U. J. 1990. The agrin hypothesis. *Cold Spring Harb. Symp. Quant. Biol.* 50: 407–418. [24]

McMahan, U. J., and Slater, C. R. 1984. The influence of basal lamina on the accumulation of acetylcholine receptors at synaptic sites in regenerating muscle. *J. Cell Biol.* 98: 1453–1473. [24]

McMahan, U. J., and Wallace, B. G. 1989. Molecules in basal lamina that direct the formation of synaptic specializations at neuromuscular junctions. *Dev. Neurosci.* 11: 227–247. [23, 24]

McMahan, U. J., Edgington, D. R., and Kuffler, D. P. 1980. Factors that influence regeneration of the neuromuscular junction. *J. Exp. Biol.* 89: 31–42. [24]

McMahan, U. J., Sanes, J. R., and Marshall, L. M. 1978. Cholinesterase is associated with the basal lamina at the neuromuscular junction. *Nature* 271: 172–174. [13]

McMahan, U. J., Spitzer, N. C., and Peper, K. 1972. Visual identification of nerve terminals in living isolated skeletal muscle. *Proc. R. Soc. Lond. B* 181: 421–430. [9]

McMahon, D. G. 1994. Modulation of electrical synaptic transmission in zebrafish retinal horizontal cells. *J. Neurosci.* 14: 1722–1734. [9]

McManaman, J. L., Haverkamp, L. J., and Oppenheim, R. W. 1991. Skeletal muscle proteins rescue motor neurons from cell death *in vivo*. *Adv. Neurol.* 56: 81–88. [23]

Meadows, J. C. 1974. The anatomical basis of prosopagnosia. *J. Neurol. Neurosurg. Psychiatry* 37: 489–501. [21]

Meech, R. W. 1974. The sensitivity of *Helix aspera* neurones to injected calcium ions. *J. Physiol.* 237: 259–277. [6]

Meier, T., and Wallace, B. G. 1998. Formation of the neuromuscular junction: Molecules and mechanisms. *BioEssays* 20: 819–829. [24]

Meissirel, C., Wikler, K. C., Chalupa, L. M., and Rakic, P. 1997. Early divergence of magnocellular and parvocellular functional subsystems in the embryonic primate visual system. *Proc. Natl. Acad. Sci. USA* 94: 5900–5905. [25]

Meister, M., and Berry, M. J., II 1999. The neural code of the retina. *Neuron* 22: 435–450. [19]

Meister, M., Lagnado, L., and Baylor, D. A. 1995. Concerted signaling by retinal ganglion cells. *Science* 270: 1207–1210. [19]

Meister, M., Wong, R. O., Baylor, D. A., and Shatz, C. J. 1991. Synchronous bursts of action potentials in ganglion cells of the developing mammalian retina. *Science* 252: 939–943. [25]

Mellitzer, G., Xu, Q., and Wilkinson, D. G. 1999. Eph receptors and ephrins restrict cell intermingling and communication. *Nature* 400: 77–81. [23]

Melzack, R., and Wall, P. D. 1965. Pain mechanisms: A new theory. *Science* 150: 971–979. [18]

Mendell, L. M., and Henneman, E. 1971. Terminals of single Ia fibers: Location, density, and distribution within a pool of 300 homonymous motoneurons. *J. Neurophysiol.* 34: 171–187. [22]

Menei, P., Montero-Menei, C., Whittemore, S. R., Bunge, R. P., and Bunge, M. B. 1998. Schwann cells genetically modified to secrete human BDNF promote enhanced axonal regrowth across transected adult rat spinal cord. *Eur. J. Neurosci.* 10: 607–621. [24]

Mengod, G., Charli, J. L., and Palacios, J. M. 1990. The use of in situ hybridization histochemistry for the study of neuropeptide gene expression in the human brain. *Cell Mol. Neurobiol.* 10: 113–126. [14]

Menini, A., Picco, C., and Firestein, S. 1995. Quantal-like current fluctuations induced by odorants in olfactory receptor cells. *Nature* 373: 435–437. [17]

Merigan, W. H., and Maunsell, J. H. R. 1993. How parallel are the primate visual pathways? *Annu. Rev. Neurosci.* 16: 369–402. [21]

Merlie, J. P., and Sanes, J. R. 1985. Concentration of acetylcholine receptor mRNA in synaptic regions of adult muscle fibers. *Nature* 317: 66–68. [24]

Merzenich, M. M., and Brugge, J. F. 1973. Representation of the cochlear partition on the superior temporal plane of the macaque monkey. *Brain Res.* 50: 275–296. [18]

Merzenich, M. M., Knight, P. L., and Roth, G. L. 1975. Representation of cochlea within primary auditory cortex in the cat. *J. Neurophysiol.* 38: 231–249. [18]

Messersmith, E. K., Leonardo, D., Shatz, C. J., Tessier-Lavigne, M., Goodwin, C. S., and Kolodkin, A. L. 1995. Semaphorin III can function as a selective chemorepellent to pattern sensory projections in the spinal cord. *Neuron* 14: 949–959. [23]

Meyer, G., Soria, J. M., Martinez-Galan, J. R., Martin-Clemente, B., and Fairen, A. 1998. Different origins and developmental histories of transient neurons in the marginal zone of the fetal and neonatal rat cortex. *J. Comp. Neurol.* 397: 493–518. [23]

Meyer, M., Matsuoka, I., Wetmore, C., Olson, L., and Thoenen, H. 1992. Enhanced synthesis of brain-derived neurotrophic factor in the lesioned peripheral nerve: Different mechanisms are responsible for the regulation of BDNF and NGF mRNA. *J. Cell Biol.* 119: 45–54. [24]

Meyer, T., and Stryer, L. 1991. Calcium spiking. *Annu. Rev. Biophys. Biophys. Chem.* 20: 153–174. [10]

Meyer-Lohmann, J., Hore, J., and Brooks, V. B. 1977. Cerebellar participation in generation of prompt arm movements. *J. Neurophysiol.* 40: 1038–1050. [22]

Miachon, S., Berod, A., Leger, L., Chat, M., Hartman, B., and Pujol, J. F. 1984. Identification of catecholamine cell bodies in the pons and pons-mesencephalon junction of the cat brain, using tyrosine hydroxylase and dopamine-β-hydroxylase immunohistochemistry. *Brain Res.* 305: 369–374. [14]

Michel, S., Guesz, M. E., Zaritsky, J. J., and Block, G. D. 1993. Circadian rhythm in membrane conductance expressed in isolated neurons. *Science* 259: 239–241. [22]

Middlebrooks, J. C., Dykes, R. W., and Merzenich, M. M. 1980. Binaural response-specific bands in primary auditory cortex (AI) of the cat: Topographical organization orthogonal to isofrequency contours. *Brain Res.* 181: 31–48. [18]

Middleton, R. E., Pheasant, D. J., and Miller, C. 1996. Homodimeric architecture of a ClC-type chloride ion channel. *Nature* 383: 337–340. [3]

Midtgaard, J., Lasser-Ross, N., and Ross, W. N. 1993. Spatial distribution of a Ca^{2+} influx in turtle Purkinje cell dendrites in vitro: Role of a transient outward current. *J. Neurophysiol.* 70: 2455–2469. [22]

Miele, M., Boutelle, M. G., and Fillenz, M. 1996. The source of physiologically stimulated glutamate efflux from the striatum of conscious rats. *J. Physiol.* 497: 745–751. [13]

Miledi, R. 1960a. The acetylcholine sensitivity of frog muscle fibers after complete or partial denervation. *J. Physiol.* 151: 1–23. [24]

Miledi, R. 1960b. Junctional and extra-junctional acetylcholine receptors in skeletal muscle fibres. *J. Physiol.* 151: 24–30. [9]

Miledi, R., Parker, I., and Sumikawa, K. 1983. Recording single γ-aminobutyrate- and acetylcholine-activated receptor channels translated by exogenous mRNA in *Xenopus* oocytes. *Proc. R. Soc. Lond. B* 218: 481–484. [3]

Miledi, R., Stefani, E., and Steinbach, A. B. 1971. Induction of the action potential mechanism in slow muscle fibres of the frog. *J. Physiol.* 217: 737–754. [24]

Miller, C. 1992. Hunting for the pore of voltage-gated channels. *Curr. Biol.* 2: 573–575. [3]

Miller, J. A., Agnew, W. S., and Levinson, S. R. 1983. Principal glycopeptide of the tetrodotoxin/saxitoxin binding protein from *Electrophorus electricus*: Isolation and partial chemical and physical characterization. *Biochemistry* 22: 462–470. [3, 6]

Miller, S. G., and Kennedy, M. B. 1986. Regulation of brain type II Ca^{2+}/calmodulin-dependent protein kinase by autophosphorylation: A Ca^{2+}-triggered switch. *Cell* 44: 861–870. [10]

Miller, T. M., and Heuser, J. E. 1984. Endocytosis of synaptic vesicle membrane at the frog neuromuscular junction. *J. Cell Biol.* 98: 685–698. [11, 13]

Ming, G-L., Song, H-J., Berninger, B., Holt, C. E., Tessier-Lavigne, M., and Poo, M-M. 1997. cAMP-dependent growth cone guidance by netrin-1. *Neuron* 19: 1225–1235. [23]

Mink, J. W., and Thach, W. T. 1991a. Basal ganglia motor control. I. Nonexclusive relation of pallidal discharge to five movement modes. *J. Neurophysiol.* 65: 273–300. [22]

Mink, J. W., and Thach, W. T. 1991b. Basal ganglia motor control. II. Late pallidal timing relative to movement onset and inconsistent pallidal coding of movement parameters. *J. Neurophysiol.* 65: 301–329. [22]

Mink, J. W., and Thach, W. T. 1991c. Basal ganglia motor control. III. Pallidal ablation: Normal reaction time, muscle cocontraction, and slow movement. *J. Neurophysiol.* 65: 330–351. [22]

Mink, J. W., and Thach, W. T. 1993. Basal ganglia intrinsic circuits and their role in behavior. *Curr. Opin. Neurobiol.* 3: 950–957. [22]

Miserendino, M. J., Sananes, C. B., Melia, K. R., and Davis, M. 1990. Blocking acquisition but not expression of conditioned fear-potentiated startle by NMDA antagonists in the amygdala. *Nature* 345: 716–718. [12]

Mishina, M., Takai, T., Imoto, K., Noda, M., Takahashi, T., Numa, S., Methfessel, C., and Sakmann, B. 1986. Molecular distinction between fetal and adult forms of muscle acetylcholine receptor. *Nature* 321: 406–411. [3, 9, 24]

Missale, C., Nash, S. R., Robinson, S. W., Jaber, M., and Caron, M. G. 1998. Dopamine receptors: From structure to function. *Physiol. Rev.* 78: 189–225. [14]

Mitz, A. R., and Wise, S. P. 1987. The somatotopic organization of the supplementary motor area: Intracortical microstimulation mapping. *J. Neurosci.* 7: 1010–1021. [22]

Miyakawa, H., Lev-Ram, V., Lasser-Ross, N., and Ross, W. N. 1992. Calcium transients evoked by climbing fiber and parallel fiber synaptic inputs in guinea pig cerebellar Purkinje neurons. *J. Neurophysiol.* 68: 1178–1189. [22]

Miyawaki, A., Llopis, J., Heim, R., McCaffery, J. M., Adams, J. A., Ikura, M., and Tsien, R. Y. 1997. Fluorescent indicators for Ca^{2+} based on green fluorescent proteins and calmodulin. *Nature* 388: 882–887. [4]

Miyazaki, S. 1995. Inositol trisphosphate receptor mediated spatiotemporal calcium signalling. *Curr. Opin. Cell Biol.* 7: 190–196. [10]

Mladinić, M., Becchetti, A., Didelon, F., Bradbury, A., and Cherubini, E. 1999. Low expression of the ClC-2 chloride channel during postnatal development: A mechanism for the paradoxical depolarizing action of GABA and glycine in the hippocampus. *Proc. R. Soc. Lond. B* 266: 1207–1213. [5, 9]

Modney, B. K., Sahley, C. L., and Muller, K. J. 1997. Regeneration of a central synapse restores nonassociative learning. *J. Neurosci.* 17: 6478–6482. [15]

Moeller, F. G., Dougherty, D. M., Swann, A. C., Collins, D., Davis, C. M., and Cherek, D. R. 1996. Tryptophan depletion and aggressive responding in healthy males. *Psychopharmacology* 126: 96–103. [13]

Mombaerts, P. 1999. Molecular biology of odorant receptors in vertebrates. *Annu. Rev. Neurosci.* 22: 487–509. [17]

Monsma, F. J., Jr., McVittie, L. D., Gerfen, C. R., Mahan, L. C., and Sibley, D. R. 1989. Multiple D2 dopamine receptors produced by alternative RNA splicing. *Nature* 342: 926–929. [14]

Moore, R. Y. 1997. Circadian rhythms: Basic neurobiology and clinical applications. *Annu. Rev. Med.* 48: 253–266. [16]

Morgan, C., de Groat, W. C., and Nadelhaft, I. 1986. The spinal distribution of sympathetic preganglionic and visceral primary afferent neurons that send axons into the hypogastric nerves of the cat. *J. Comp. Neurol.* 243: 23–40. [16]

Morgans, C. W., El Far, O., Berntson, A., Wässle, H., and Taylor, W. R. 1998. Calcium extrusion from mammalian photoreceptor terminals. *J. Neurosci.* 18: 2467–2474. [19]

Morita, K., and Barrett, E. F. 1990. Evidence for two calcium-dependent potassium conductances in lizard motor nerve terminals. *J. Neurosci.* 10: 2614–2625. [11]

Moriyoshi, K., Masu, M., Ishi, T, Shigemoto, R., Mizuno, N., and Nakanishi, S. 1991. Molecular cloning and characterization of the rat NMDA receptor. *Nature* 354: 31–37. [3]

Moriyoshi, K., Richards, L. J., Akazawa, C., O'Leary, D. D. M., and Nakanishi, S. 1996. Labeling neural cells using adenoviral gene transfer of membrane-targeted GFP. *Neuron* 16: 255–260. [23]

Morrison, A. D. 1998. 1 + 1 = r4 and much much more. *BioEssays* 20: 794–797. [23]

Mountcastle, V. 1998. Brain science at the century's ebb. *Daedalus* 127: 1–36. [26]

Mountcastle, V. B. 1957. Modality and topographic properties of single neurons of cat's somatic sensory cortex. *J. Neurophysiol.* 20: 408–434. [18, 20]

Mountcastle, V. B. 1995. The parietal system and some higher brain functions. *Cerebral Cortex* 5: 377–390. [18, 22]

Mountcastle, V. B. 1997. The columnar organization of the neocortex. *Brain* 120: 701–722. [18]

Mountcastle, V. B., and Powell, T. P. S. 1959. Neural mechanisms subserving cutaneous sensibility with special reference to the role of afferent inhibition in sensory perception and discrimination. *Bull. Johns Hopkins Hosp.* 105: 201–232. [18]

Mountcastle, V. B., Atluri, P. P., and Romo, R. 1992. Selective output-discriminative signals in the motor cortex of waking monkeys. *Cerebral Cortex* 2: 277–294. [22]

Mudge, A. W., Leeman, S. E., and Fischbach, G. D. 1979. Enkephalin inhibits release of substance P from sensory neurons in culture and decreases action potential duration. *Proc. Natl. Acad. Sci. USA* 76: 526–530. [10, 14]

Mueller, B. K. 1999. Growth cone guidance: First steps towards a deeper understanding. *Annu. Rev. Neurosci.* 22: 351–388. [23]

Muir, J. L. 1997. Acetylcholine, aging and Alzheimer's disease. *Pharmacol. Biochem. Behav.* 56: 687–696. [14]

Mulkey, R. M., and Malenka, R. C. 1992. Mechanisms underlying induction of homosynaptic long-term depression in area CA1 of the hippocampus. *Neuron* 9: 967–975. [12]

Mulkey, R. M., Herron, C. E., and Malenka, R. C. 1993. An essential role for protein phosphatases in hippocampal long-term depression. *Science* 261: 1051–1055. [12]

Muller, D., Joly, M., and Lynch, G. 1988. Contributions of quisqualate and NMDA receptors to the induction and expression of LTP. *Science* 242: 1694–1697. [12]

Muller, H. W., and Stoll, G. 1998. Nerve injury and regeneration: Basic insights and therapeutic interventions. *Curr. Opin. Neurol.* 11: 557–562. [24]

Muller, K. J. 1973. Photoreceptors in the crayfish compound eye: Electrical interactions between cells as related to polarized-light sensitivity. *J. Physiol.* 232: 573–595. [15]

Muller, K. J. 1981. Synapses and synaptic transmission. In K. J. Muller, J. G. Nicholls, and G. S. Stent (eds.), *Neurobiology of the Leech*. Cold Spring Harbor Laboratory, Cold Spring Harbor, NY, pp. 79–111. [15]

Muller, K. J., and Carbonetto, S. 1979. The morphological and physiological properties of a regenerating synapse in the C.N.S. of the leech. *J. Comp. Neurol.* 185: 485–516.

Muller, K. J., and McMahan, U. J. 1976. The shapes of sensory and motor neurones and the distribution of their synapses in ganglia of the leech: A study using intracellular injection of horseradish peroxidase. *Proc. R. Soc. Lond. B* 194: 481–499. [15]

Muller, K. J., and Nicholls, J. G. 1974. Different properties of synapses between a single sensory neurone and two different motor cells in the leech CNS. *J. Physiol.* 238: 357–369. [15]

Muller, K. J., Nicholls, J. G., and Stent, G. S. (eds.). 1981. *Neurobiology of the Leech*. Cold Spring Harbor Laboratory, Cold Spring Harbor, NY. [15]

Muller, M., and Wehner, R. 1994. The hidden spiral: Systematic search and path integration in desert ants, *Cataglyphis fortis*. *J. Comp. Physiol. [A]* 175: 525–530. [15]

Mullins, L. J. 1975. Ion selectivity of carriers and channels. *Biophys. J.* 15: 921–931. [3]

Mullins, L. J., and Noda, K. 1963. The influence of sodium-free solutions on membrane potential of frog muscle fibers. *J. Gen. Physiol.* 47: 117–132. [5]

Murashima, M., and Hirano, T. 1999. Entire course and distinct phases of day-lasting depression of miniature EPSC amplitudes in cultured Purkinje neurons. *J. Neurosci.* 19: 7326–7333. [12]

Murrell, J., Farlow, M., Ghetti, B., and Benson, M. D. 1991. A mutation in the amyloid precursor protein associated with hereditary Alzheimer's disease. *Science* 254: 97–99. [14]

Myers, R. D., Adell, A., and Lankford, M. F. 1998. Simultaneous comparison of cerebral dialysis and push-pull perfusion in the brain of rats: A critical review. *Neurosci. Biobehav. Rev.* 22: 371–387. [13]

Nadim, F., and Calabrese, R. L. 1997. A slow outward current activated by FMRFamide in heart interneurons of the medicinal leech. *J. Neurosci.* 17: 4461–4472. [15]

Naegele, J. R., and Barnstable, C. J. 1989. Molecular determinants of GABAergic local-circuit neurons in the visual cortex. *Trends Neurosci.* 12: 28–34. [14]

Nagai, T., McGeer, P. L., Araki, M., and McGeer, E. G. 1984. GABA-T intensive neurons in the rat brain. In A. Bjorklund, T. Hokfelt, and M. J. Kuhar (eds.), *Classical Transmitters and Transmitter Receptors in the CNS*. Vol. 3 of *Handbook of Chemical Neuroanatomy*. Elsevier, Amsterdam, pp. 247–272. [14]

Nagatsu, T. 1995. Tyrosine hydroxylase: Human isoforms, structure and regulation in physiology and pathology. *Essays Biochem.* 30: 15–35. [13]

Nagatsu, T., and Ichinose, H. 1999. Regulation of pteridine-requiring enzymes by the cofactor tetrahydrobiopterin. *Mol. Neurobiol.* 19: 79–96. [13]

Nakajima, S., and Onodera, K. 1969a. Membrane properties of the stretch receptor neurones of crayfish with particular referent to mechanisms of sensory adaptation. *J. Physiol.* 200: 161–185. [17]

Nakajima, S., and Onodera, K. 1969b. Adaptation of the generator potential in crayfish stretch receptors under constant length and constant tension. *J. Physiol.* 200: 187–204. [17]

Nakajima, S., and Takahashi, K. 1966. Post-tetanic hyperpolarization and electrogenic Na pump in stretch receptor neurone of crayfish. *J. Physiol.* 187: 105–127. [17]

Nakajima, Y., Tisdale, A. D., and Henkart, M. P. 1973. Presynaptic inhibition at inhibitory nerve terminals: A new synapse in the crayfish stretch receptor. *Proc. Natl. Acad. Sci. USA* 70: 2462–2466. [9]

Nakamura, H., and O'Leary, D. D. M. 1989. Inaccuracies in initial growth and arborization of chick retinotectal axons followed by course corrections and axon remodeling to develop topographic order. *J. Neurosci.* 9: 3776–3795. [23]

Nakamura, T., and Gold, G. H. 1987. A cyclic nucleotide-gated conductance in olfactory receptor cilia. *Nature* 532: 442–444. [17]

Nakanishi, S., Nakajima, Y., Masu, M., Ueda, Y., Nakahara, K., Watanabe, D., Yamaguchi, S., Kawabata, S., and Okada, M. 1998. Glutamate receptors: Brain function and signal transduction. *Brain Res. Rev.* 26: 230–235. [19]

Narahashi, T., Moore, J. W., and Scott, W. R. 1964. Tetrodotoxin blockage of sodium conductance increase in lobster giant axons. *J. Gen. Physiol.* 47: 965–974. [6]

Nastuk, W. L. 1953. Membrane potential changes at a single muscle end-plate produced by transitory application of acetylcholine with an electrically controlled microjet. *Fed. Proc.* 12: 102. [9]

Nathans, J. 1987. Molecular biology of visual pigments. *Annu. Rev. Neurosci.* 10: 163–194. [19]

Nathans, J. 1989. The genes for color vision. *Sci. Am.* 260(2): 42–49. [19]

Nathans, J. 1999. The evolution and physiology of human color vision: Insights from molecular genetic studies of visual pigments. *Neuron* 24: 299–312. [19]

Nathans, J., and Hogness, D. S. 1984. Isolation and nucleotide sequence of the gene encoding human rhodopsin. *Proc. Natl. Acad. Sci. USA* 81: 4851–4855. [19]

Navaratnam, D. S., Bell, T. J., Tu, T. D., Cohen, E. L., and Oberholtzer, J. C. 1997. Differential distribution of Ca^{2+}-activated K^+ channel splice variants among hair cells along the tonotopic axis of the chick cochlea. *Neuron* 19: 1077–1085. [18]

Neher, E., and Sakmann, B. 1976. Single channel currents recorded from membrane of denervated frog muscle fibres. *Nature* 260: 799–802. [9]

Neher, E., and Steinbach, J. H. 1978. Local anaesthetics transiently block currents through single acetylcholine receptor channels. *J. Physiol.* 277: 153–176. [9]

Neher, E., Sakmann, B., and Steinbach, J. H. 1978. The extracellular patch clamp: A method for resolving currents through individual open channels in biological membranes. *Pflügers Arch.* 375: 219–228. [2]

Neil, E. 1954. Reflexogenic areas of the circulation. *Arch. Middlesex Hosp.* 4: 16. (Modified from Berne, M., and Levy, M. N. 1990. *Principles of Physiology*. Wolfe, London.) [16]

Nelken, I., Rotman, Y., and Bar Yosef, O. 1999. Responses of auditory cortex neurons to structural features of natural sounds. *Nature* 397: 154–157. [18]

Nelson, B. H., and Weisblat, D. A. 1992. Cytoplasmic and cortical determinants interact to specify ectoderm and mesoderm in the leech embryo. *Development* 115: 103–115. [23]

Nelson, N., and Lill, H. 1994. Porters and neurotransmitter transporters. *J. Exp. Biol.* 196: 213–228. [4]

Neumann, H., Boucraut, J., Hahnel, C., Misgeld, T., and Wekerle, H. 1996. Neuronal control of MHC class II inducibility in rat astrocytes and microglia. *Eur. J. Neurosci.* 8: 2582–2590. [8]

Neumann, H., Misgeld, T., Matsumuro, K., and Wekerle, H. 1998. Neurotrophins inhibit major histocompatibility class II inducibility of microglia: Involvement of the P75 neurotrophin receptor. *Proc. Natl. Acad. Sci. USA* 95: 5779–5784. [8]

Neurobiology and development of the leech (Special Issue). 1995. *J. Neurobiol.* 27(Pt.3). [15]

Newman, E. A. 1986. High potassium conductance in astrocyte endfeet. *Science* 233: 453–454. [8]

Newman, E. A. 1987. Distribution of potassium conductance in mammalian Müller (glial) cells: A comparative study. *J. Neurosci.* 7: 2423–2432. [8]

Newman, E. A., and Zahs, K. R. 1997. Calcium waves in retinal glial cells. *Science* 275: 844–847. [8]

Newman, E. A., and Zahs, K. R. 1998. Modulation of neuronal activity by glial cells in the retina. *J. Neurosci.* 18: 4022–4028. [8]

Newman, E., and Reichenbach, A. 1996. The Muller cell: A functional element of the retina. *Trends Neurosci.* 19: 307–312. [8]

Newsome, W. T., and Pare, E. B. 1988. A selective impairment of motion perception following lesions of the middle temporal visual area (MT). *J. Neurosci.* 8: 2201–2211. [21]

Newsome, W. T., and Wurtz, R. H. 1988. Probing visual cortical function with discrete chemical lesions. *Trends Neurosci.* 11: 394–400. [21]

Neyton, J., and Trautmann, A. 1985. Single channel currents of an intercellular junction. *Nature* 317: 331–335. [7]

Nicholls, C. G., and Lopatin, A. N. 1997. Inward rectifier potassium channels. *Ann. Rev. Physiol.* 59: 171–191. [3]

Nicholls, J. G., and Baylor, D. A. 1968. Specific modalities and receptive fields of sensory neurons in the CNS of the leech. *J. Neurophysiol.* 31: 740–756. [5, 15]

Nicholls, J. G., and Purves, D. 1972. A comparison of chemical and electrical synaptic transmission between single sensory cells and a motoneurone in the central nervous system of the leech. *J. Physiol.* 225: 637–656. [9, 15]

Nicholls, J. G., and Wallace, B. G. 1978. Modulation of transmission at an inhibitory synapse in the central nervous system of the leech. *J. Physiol.* 281: 157–170. [9, 15]

Nicholls, J. G., Stewart, R. R., Erulkar, S. D., and Saunders, N. R. 1990. Reflexes, fictive respiration, and cell division in the brain and spinal cord of the newborn opossum, *Monodelphis domestica*, isolated and maintained *in vitro*. *J. Exp. Biol.* 152: 1–15. [22]

Nicholls, J., and Saunders, N. 1996. Regeneration of immature mammalian spinal cord after injury. *Trends Neurosci.* 19: 229–234. [24]

Nicoll, D. A., Longoni, S., and Phillipson, K. D. 1990. Molecular cloning and functional expression of the cardiac sarcolemmal Na^+-Ca^{2+} exchanger. *Science* 250: 562–565. [4]

Nicoll, R. A., Malenka, R. C., and Kauer J. A. 1990. Functional comparison of neurotransmitter receptor subtypes in mammalian central nervous system. *Physiol. Rev.* 70: 513–565. [14]

Nieuwenhuys, R., Donkelaar, H. J., and Nicholson, C. 1998. *The Central Nervous System of Vertebrates*. Springer, New York. [22]

Niu, L., Abood, L. G., and Hess, G. P. 1995. Cocaine: Mechanism of inhibition of a mouse acetylcholine receptor studied by a laser-pulse photolysis technique. *Proc. Natl. Acad. Sci. USA* 92: 12008–12012. [9]

Niu, L., and Hess, G. P. 1993. An acetylcholine receptor regulatory site in BC3H1 cells: Characterized by laser-pulse photolysis in the microsecond-to-millisecond time region. *Biochemistry* 32: 3831–3835. [9]

Nja, A., and Purves, D. 1978. The effects of nerve growth factor and its antiserum on synapses in the superior cervical ganglion of the guinea-pig. *J. Physiol.* 277: 55–75. [24]

Noakes, P. G., Gautam, M., Mudd, J., Sanes, J. R., and Merlie, J. P. 1995. Aberrant differentiation of neuromuscular junctions in mice lacking s-laminin/laminin $\beta 2$. *Nature* 374: 258–262. [23]

Noda, M., Shimizu, S., Tanabe, T., Takai, T., Kayano, T., Ikeda, T., Takahashi, H., Nakayama, H., Kanaoka, Y., Minamino, N., Kangawa, K., Matsuo, H., Raftery, M. A., Hirose, T., Inagama, S., Hayashida, H., Miyata, T., and Numa, S. 1984. Primary structure of *Electrophorus electricus* sodium channel deduced from cDNA sequence. *Nature* 312: 121–127. [3]

Noda, M., Takahashi, H., Tanabe, T., Toyosato, M., Furutani, Y., Hirose, T., Asai, M., Inayama, S., Miyata, T., and Numa, S. 1982. Primary structure of α-subunit precursor of *Torpedo californica* acetylcholine receptor deduced from cDNA sequence. *Nature* 299: 793–797. [3]

Noda, M., Takahashi, H., Tanabe, T., Toyosato, M., Kikyotani, S., Furutani, Y., Hirose, T., Takashima, H., Inayama, S., Miyata, T., and Numa, S. 1983a. Structural homology of *Torpedo californica* acetylcholine receptor subunits. *Nature* 302: 528–532. [3]

Noda, M., Takahashi, H., Tanabe, T., Toyosato, M., Kikyotani, S., Hirose, T., Asai, M., Takashima, H., Inayama, S., Miyata, T., and Numa, S. 1983b. Primary structure of β- and δ-subunit precursors of *Torpedo californica* acetylcholine receptor deduced from cDNA sequences. *Nature* 301: 251–255. [3]

Noguchi, S., Noda, M., Takahashi, H., Kawakami, K., Ohta, T., Nagano, K., Hirosi, T., Inayama, S., Kawamura, M., and Numa, S. 1986. Primary structure of the β-subunit of *Torpedo californica* (Na^+-K^+)-ATPase deduced from cDNA sequence. *FEBS Lett.* 196: 315–320. [4]

Nolte, J. 1988. *The Human Brain*, 2nd Ed. Mosby, St. Louis, MO. [23]

Norris, C. H., and Guth, P. S. 1973. The release of ACh by the crossed olivo-cochlear bundle. *Acta Otolaryngol.* 77: 318–326. [18]

Nottebohm, F. 1989. From bird song to neurogenesis. *Sci. Am.* 260(2): 74–79. [23]

Notterpek, L., Shooter, E. M., and Snipes, G. J. 1997. Up-regulation of the endosomal-lysosomal pathway in the trembler. *J. Neurosci.* 17: 4190–4200. [8]

Nowak, L., Bregestovski, P., Ascher, P., Herbet, A., and Prochiantz, A. 1984. Magnesium gates glutamate-activated channels in mouse central neurones. *Nature* 307: 462–465. [12]

Nowak, L., et al. 1984. Magnesium gates glutamate-activated channels in mouse central neurones. *Nature* 307: 463–465. [14]

Nudo, R. J., Milliken, G. W., Jenkins, W. M., and Merzenich, M. M. 1996. Use-dependent alterations of movement representations in primary motor cortex in adult squirrel monkeys. *J. Neurosci.* 16: 785–807. [22]

Numa, S., Noda, M., Takahashi, H., Tanabe, T., Toyosato, M., Furutani, Y., and Kikyotani, S. 1983. Molecular structure of the nicotinic acetylcholine receptor. *Cold Spring Harb. Symp. Quant. Biol.* 48: 57–69. [3]

Nusser, Z., Mulvihill, E., Streit, P., and Somogyi, P. 1994. Subsynaptic segregation of metabotropic and ionotropic glutamate receptors as revealed by immunogold localization. *Neuroscience* 61: 421–427. [13]

Nussinovitch, I., and Rahamimoff, R. 1988. Ionic basis of tetanic and post-tetanic potentiation at a mammalian neuromuscular junction. *J. Physiol.* 396: 435–455. [12]

O'Brien, R. A. D., Ostberg, A. J. C., and Vrbova, G. 1978. Observations on the elimination of polyneuronal innervation in developing mammalian skeletal muscle. *J. Physiol.* 282: 571–582. [23]

O'Brien, R. J., Lau, L. F., and Huganir, R. L. 1998. Molecular mechanisms of glutamate receptor clustering at excitatory synapses. *Curr. Opin. Neurobiol.* 8: 364–369. [13]

O'Leary, D. D. 1992. Development of connectional diversity and specificity in the mammalian brain by the pruning of colateral projections. *Curr. Opin. Neurobiol.* 2: 70–77. [23]

O'Leary, D. D. M., and Terashima, T. 1988. Cortical axons branch to multiple subcortical targets by interstitial axon budding: Implications for target recognition and "waiting periods." *Neuron* 1: 901–910. [23]

O'Leary, D. D. M., Borngasser, D. J., Fox, K., and Schlaggar, B. L. 1995. Plasticity in the development of neocortical areas. *Ciba Found. Symp.* 193: 214–230. [23]

O'Leary, D. D. M., Fawcett, J. W., and Cowan, W. M. 1986. Topographic targeting errors in the retinocollicular projection and their elimination by selective ganglion cell death. *J. Neurosci.* 6: 3692–3705. [23]

O'Leary, D. D., and Wilkinson, D. G. 1999. Eph receptors and ephrins in neural development. *Curr. Opin. Neurobiol.* 9: 65–73. [23]

O'Malley, J., Moore, C. T., and Salpeter, M. M. 1997. Stabilization of acetylcholine receptors by exogenous ATP and its reversal by cAMP and calcium. *J. Cell Biol.* 138: 159–165. [24]

Obaid, A. L., and Salzberg, B. M. 1996. Micromolar 4-aminopyridine enhances invasion of a vertebrate neurosecretory terminal arborization: Optical recording of action potential propagation using an ultrafast photodiode-MOSFET camera and a photodiode array. *J. Gen. Physiol.* 107: 353–368. [16]

Obaid, A. L., Koyano, T., Lindstrom, J., Sakai, T., and Salzberg, B. M. 1999. Spatiotemporal patterns of activity in an intact mammalian network with single-cell resolution: Optical studies of nicotinic activity in an enteric plexus. *J. Neurosci.* 19: 3073–3093. [16]

Obata, K. 1969. Gamma-aminobutyric acid in Purkinje cells and motoneurones. *Experientia* 25: 1283. [14]

Obata, K., Takeda, K., and Shinozaki, H. 1970. Further study on pharmacological properties of the cerebellar-induced inhibition of Deiters neurones. *Exp. Brain Res.* 11: 327–342. [14]

Obermayer, K., and Blasdel, G. G. 1993. Geometry of orientation and ocular dominance columns in monkey striate cortex. *J. Neurosci.* 13: 4114–4129. [21]

Ochoa, J., and Torebjork, E. 1989. Sensations evoked by intraneural microstimulation of C nociceptor fibres in human skin nerves. *J. Physiol.* 415: 583–599. [18]

Odette, L. L., and Newman, E. A. 1988. Model of potassium dynamics in the central nervous system. *Glia* 1: 198–210. [8]

Oheim, M., Loerke, D., Chow, R. H., amd Stühmer, W. 1999. Evanescent-wave microscopy: A new tool to gain insight into the control of transmitter release. *Philos. Trans. R. Soc. Lond. B* 354: 307–318. [11]

Ohmori, H. 1985. Mechanoelectrical transduction currents in isolated vestibular hair cells of the chick. *J. Physiol.* 359: 189–217. [17]

Okada, Y., Miyamoto, T., and Sato, T. 1994. Activation of a cation conductance by acetic acid in taste cells isolated from bullfrog. *J. Exp. Biol.* 187: 19–32. [17]

Olanow, C. W., and Tatton, W. G. 1999. Etiology and pathogenesis of Parkinson's disease. *Annu. Rev. Neurosci.* 22: 123–144. [22]

Olavarria, J. F., and Van Essen, D. C. 1997. The global pattern of cytochrome oxidase stripes in visual area V2 of the macaque monkey. *Cerebral Cortex* 7: 395–404. [21]

Oliet, S., Malenka, R. C., and Nicoll, R. A. 1996. Bidirectional control of quantal size by synaptic activity in the hippocampus. *Science* 271: 1294–1297. [12]

Olivera, B. M., Miljanich, G. P., Ramachandran, J., and Adams, M. E. 1994. Calcium channel diversity and neurotransmitter release: The omega-conotoxins and omega-agatoxins. *Annu. Rev. Biochem.* 63: 823–867. [11]

Olsen, R. W., and Leeb-Lundberg, F. 1981. Convulsant and anticonvulsant drug binding sites related to GABA-regulated chloride ion channels. In E. Costa, G. DiChiara, and G. L. Gessa (eds.), *GABA and Benzodiazepine Receptors.* Raven, New York, pp. 93–102. [14]

Onimaru, H. 1995. Studies of the respiratory center using isolated brainstem-spinal cord preparations. *Neurosci. Res.* 21: 183–190. [22]

Ophoff, R. A., Terwindt, G. M., Vergouwe, M. N., van Eijk, R., Oefner, P. J., Hoffman, S. M., Lamerdin, J. E., Mohrenweiser, H. W., Bulman, D. E., Ferrari, M., Haan, J., Lindhout, D., van Ommen, G. J., Hofker, M. H., Ferrari, M. D., and Frants, R. R. 1996. Familial hemiplegic migraine and episodic ataxia type-2 are caused by mutations in the Ca^{2+} channel gene CACNL1A4. *Cell* 87: 543–552. [22]

Oppenheim, R. W. 1991. Cell death during development of the nervous system. *Annu. Rev. Neurosci.* 14: 453–501. [23]

Oppenheim, R. W., Houenou, L. J., Johnson, J. E., Lin, L. F., Li, L., Lo, A. C., Newsome, A. L., Prevette, S. M., and Wang, S. 1995. Developing motor neurons rescued from programmed and axotomy-induced cell death by GDNF. *Nature* 373: 344–346. [23]

Orkand, R. K., Nicholls, J. G., and Kuffler, S. W. 1966. Effect of nerve impulses on the membrane potential of glial cells in the central nervous system of amphibia. *J. Neurophysiol.* 29: 788–806. [8]

Ortells, M. O., and Lunt, G. G. 1995. Evolutionary history of the ligand-gated ion channel superfamily of receptors. *Trends Neurosci.* 18: 121–127. [3]

Otsuka, M., Obata, K., Miyata, Y., and Tanaka, Y. 1971. Measurement of γ-aminobutyric acid in isolated nerve cells of cat central nervous system. *J. Neurochem.* 18: 287–295. [14]

Ottoson, D. 1956. Analysis of the electrical activity of the olfactory epithelium. *Acta Physiol. Scand.* 35(Suppl. 122): 1–83. [17]

Overton, E. 1902. Beiträge zur allgemeinen Muskel- und Nervenphysiologie. II. Über die Unentbehrlichkeit von Natrium- (oder Lithium-) Ionen für den Kontraktionsakt des Muskels. *Pflügers Arch.* 92: 346–386. [6]

Pace, U., Hansky, E., Salomon, Y., and Lancet, D. 1985. Odorant sensitive adenylate cyclase may mediate olfactory reception. *Nature* 316: 255–258. [17]

Paganetti, P., and Scheller, R. H. 1994. Proteolytic processing of the *Aplysia* A peptide precursor in AtT-20 cells. *Brain Res.* 633: 53–62. [13]

Palacín, M., Estévez, R., Bertran, J., and Zorzano, A. 1998. Molecular biology of mammalian plasma membrane amino acid transporters. *Physiol. Rev.* 78: 969–1054. [4, 13]

Palacios, J. M., Mengod, G., Vilaro, M. T., Wiederhold, K. H., Boddeke, H., Alvarez, F. J., Chinaglia, G., and Probst, A. 1990a. Cholinergic receptors in the rat and human brain: Microscopic visualization. *Prog. Brain Res.* 84: 243–253. [14]

Palacios, J. M., Waeber, C., Hoyer, D., and Mengod, G. 1990b. Distribution of serotonin receptors. *Ann. N.Y. Acad. Sci.* 600: 36–52. [14]

Palka, J., Whitlock, K. E., and Murray, M. A. 1992. Guidepost cells. *Curr. Opin. Neurobiol.* 2: 48–54. [23]

Pallis, C. A. 1955. Impaired identification of faces and places with agnosia for colors. *J. Neurol. Neurosurg. Psychiatry* 18: 218–224. [21]

Palmer, L. A., and Rosenquist, A. C. 1974. Visual receptive fields of single striate cortical units projecting to the superior colliculus in the cat. *Brain Res.* 67: 27–42. [20]

Palmer, R. M., Ferrige, A. G., and Moncada, S. 1987. Nitric oxide release accounts for the biological activity of endothelium-derived relaxing factor. *Nature* 327: 524–526. [14]

Palmer, T. M., and Stiles, G. L. 1995. Adenosine receptors. *Neuropharmacology* 34: 683–694. [14]

Pantev, C., Bertrand, O., Eulitz, C., Verkindt, C., Hampson, S., Schuirer, G., and Elbert, T. 1995. Specific tonotopic organizations of different areas of the human auditory cortex revealed by simultaneous magnetic and electric recordings. *Electroencephalogr. Clin. Neurophysiol.* 94: 26–40. [18]

Panula, P., Airaksinen, M. S., Pirvola, U., and Kotilainen, E. 1990. A histamine-containing neuronal system in human brain. *Neuroscience* 34: 127–132. [14]

Papazian, D. M., Timpe, L. C., Jan, Y. N., and Jan, L. Y. 1991. Alteration of voltage-dependence of Shaker potassium channel by mutations in the S4 sequence. *Nature* 349: 305–349. [6]

Papazian, D. M., Timpe, L. C., Jan, Y. N., and Jan, L. Y. 1987. Cloning and complementary DNA from *Shaker*, a putative potassium channel gene from *Drosophila. Science* 237: 749–753. [3]

Paradiso, K., and Brehm, P. 1998. Long-term desensitization of nicotinic acetylcholine receptors is regulated via protein kinase A-mediated phosphorylation. *J. Neurosci.* 18: 9227–9237. [9]

Pareek, S., Notterpek, L., Snipes, G. J., Naef, R., Sossin, W., Laliberte, J., Iacampo, S., Suter, U., Shooter, E. M., and Murphy, R. A. 1997. Neurons promote the translocation of peripheral myelin protein 22 into myelin. *J. Neurosci.* 17: 7754–7762. [8]

Park, H-J., Niedzielski, A. S., and Wenthold, R. J. 1997. Expression of the nicotinic acetylcholine receptor subunit, α9, in the guinea pig cochlea. *Hear. Res.* 112: 95–105. [18]

Park, J. H., Straub, V. A., and O'Shea, M. 1998. Anterograde signaling by nitric oxide: Characterization and in vitro reconstitution of an identified nitrergic synapse. *J. Neurosci.* 18: 5463–5476. [10]

Parnas, H., Parnas, I., and Segel, L. A. 1990. On the contribution of mathematical models to the understanding of neurotransmitter release. *Int. Rev. Neurobiol.* 32: 1–50 [12]

Parnas, H., Segel, L., Dudel, J., and Parnas, I. 2000. Autoreceptors, membrane potential and the regulation of transmitter release. *Trends Neurosci.* 23:60–68. [11]

Parnas, I. 1987. Strengthening of synaptic inputs after elimination of a single neurone innervating the same target. *J. Exp. Biol.* 132: 231–247. [24]

Parsons, S. M., Bahr, B. A., Gracz, L. M., Kaufman, R., Kornreich, W. D., Nilsson, L., and Rodgers, G. A. 1987. Acetylcholine transport: Fundamental properties and effects of pharmacological agents. *Ann. N.Y. Acad. Sci.* 493: 220–233. [13]

Partridge, L. D., and Thomas, R. C. 1976. The effects of lithium and sodium on potassium conductance in snail neurones. *J. Physiol.* 254: 551–563. [6]

Paschal, B. M., and Vallee, R. B. 1987. Retrograde transport by the microtubule associated protein MAP 1C. *Nature* 330: 181–183. [13]

Patacchini, R., and Maggi, C. A. 1995. Tachykinin receptors and receptor subtypes. *Arch. Int. Pharmacodyn. Ther.* 329: 161–184. [14]

Patlach, J. 1991. Molecular kinetics of voltage-dependent Na^+ channels. *Physiol. Rev.* 71: 1047–1080. [6]

Patterson, P. H., and Chun, L. L. Y. 1977. The induction of acetylcholine synthesis in primary cultures of dissociated rat sympathetic neurons. I. Effects of conditioned medium. *Dev. Biol.* 56: 263–280. [23]

Patuzzi, R. 1996. Cochlear micromechanics and macromechanics. In P. Dallos, A. N. Popper, and R. R. Fay (eds.), *The Cochlea.* Springer, New York, pp. 186–257. [18]

Paulson, H. L., and Fischbeck, K. H. 1996. Trinucleotide repeats in neurogenetic disorders. *Annu. Rev. Neurosci.* 19: 79–107. [22]

Paulson, O. B., and Newman, E. A. 1987. Does the release of potassium from astrocyte endfeet regulate cerebral blood flow? *Science* 237: 896–898. [8]

Payne, J. A. 1997. Functional characteristics of the neuronal-specific K-Cl cotransporter: Implication for $[K]_o$ regulation. *Am. J. Physiol.* 273: C1516–C1525. [4]

Payne, J. A., Stevenson, T. J., and Donaldson, L. F. 1996. Molecular characteristics of a K-Cl co-transporter in rat brain- a neuronal specific isoform. *J. Biol. Chem.* 271: 16245–16252. [4]

Payton, W. B. 1981. History of medicinal leeching and early medical references. In K. J. Muller, J. G. Nicholls, and G. S. Stent (eds.), *Neurobiology of the Leech.* Cold Spring Harbor Laboratory, Cold Spring Harbor, NY, pp. 27–34. [15]

Pearlman, A. L., Birch, J., and Meadows, J. C. 1979. Cerebral color blindness: An acquired defect in hue discrimination. *Ann. Neurol.* 5: 253–261. [21]

Pearson, K. 1976. The control of walking. *Sci. Am.* 235(6): 72–86. [22]

Pearson, K. G. 1995. Proprioceptive regulation of locomotion. *Curr. Opin. Neurobiol.* 5: 786–791. [15]

Pearson, K. G., and Iles, J. F. 1970. Discharge patterns of coxal levator and depressor motoneurons of the cockroach *Periplaneta americana*. *J. Exp. Biol.* 52: 139–165. [22]

Pearson, K. G., Misiaszek, J. E., and Fouad, K. 1998. Enhancement and resetting of locomotor activity by muscle afferents. *Ann. N.Y. Acad. Sci.* 860: 203–215. [22]

Pearson, K. G., Ramirez, J. M., and Jiang, W. 1992. Entrainment of the locomotor rhythm by group Ib afferents from ankle extensor muscles in spinal cats. *Exp. Brain Res.* 90: 557–566. [22]

Pei, X., Vidyasagar, T. R., Volgushev, M., and Creutzfeldt, O. D. 1994. Receptive field analysis and orientation selectivity of postsynaptic potentials of simple cells in cat visual cortex. *J. Neurosci.* 14: 7130–7140. [20]

Pellizzari, R., Rossetto, O., Schiavo, G., and Montecucco, C. 1999. Tetanus and botulinum neurotoxins: Mechanism of action and therapeutic uses. *Philos. Trans. R. Soc. Lond. B* 354: 259–268. [13]

Penfield, W. 1932. *Cytology and Cellular Pathology of the Nervous System*, Vol. 2. Hafner, New York. [8]

Penfield, W., and Rasmussen, T. 1950. *The Cerebral Cortex of Man. A Clinical Study of Localization of Function*. Macmillan, New York. [18, 22]

Penner, R., and Neher, E. 1988. The role of calcium in stimulus-secretion coupling in excitable and non-excitable cells. *J. Exp. Biol.* 139: 329–345. [11]

Penner, R., and Neher, E. 1989. The patch-clamp technique in the study of secretion. *Trends Neurosci.* 12: 159–163. [11]

Peper, K., and McMahan, U. J. 1972. Distribution of acetylcholine receptors in the vicinity of nerve terminals on skeletal muscle of the frog. *Proc. R. Soc. Lond. B* 181: 431–440. [9]

Peper, K., Dreyer, F., Sandri, C., Akert, K., and Moore, H. 1974. Structure and ultrastructure of the frog motor end-plate: A freeze-etching study. *Cell Tissue Res.* 149: 437–455. [11]

Pepperberg, D. R., Okajima, T. L., Wiggert, B., Ripps, H., Crouch, R. K., and Chader, G. J. 1993. Interphotoreceptor retinoid-binding protein (IRBP). Molecular biology and physiological role in the visual cycle of rhodopsin. *Mol. Neurobiol.* 7: 61–85. [19]

Perkins, G., Goodenough, D., and Sosinsky, G. 1997. Three dimensional structure of the gap junction connexon. *Biophys. J.* 72: 533–544. [7]

Perl, E. R. 1996. Cutaneous polymodal receptors: Characteristics and plasticity. *Prog. Brain Res.* 113: 21–37. [18]

Perl, E. R. 1999. Causalgia, pathological pain, and adrenergic receptors. *Proc. Natl. Acad. Sci. USA* 96: 7664–7667. [17]

Perschak, H., and Cuenod, M. 1990. In vivo release of endogenous glutamate and aspartate in the rat striatum during stimulation of the cortex. *Neuroscience* 35: 283–287. [13]

Persechini, A., Lynch, J. A., and Romoser, V. A. 1997. Novel fluorescent indicator proteins for monitoring free intracellular Ca^{2+}. *Cell Calcium* 22: 209–216. [4]

Pert, C. B., and Snyder, S. H. 1973. Opiate receptor: Demonstration in nervous tissue. *Science* 179: 1011–1014. [14]

Peters, A., Palay, S. L., and Webster, H. de F. 1991. *The Fine Structure of the Nervous System: Neurons and Their Supporting Cells*, 3rd Ed. Oxford University Press, New York. [8]

Petrof, B. J. 1998. The molecular basis of activity-induced muscle injury in Duchenne muscular dystrophy. *Mol. Cell. Biochem.* 179: 11–123. [13]

Pettersson, L. G., Lundberg, A., Alstermark, B., Isa, T., and Tantisira, B. 1997. Effect of spinal cord lesions on forelimb target-reaching and on visually guided switching of target-reaching in the cat. *Neurosci. Res.* 29: 241–256. [22]

Pevsner, J., Reed, R. R., Feinstein, P. G., and Snyder, S. H. 1988. Molecular cloning of odorant-binding protein: Member of a ligand carrier family. *Science* 241: 336–339. [17]

Pfaffinger, P. J., Martin, J. M., Hunter, D. D., Nathanson, N. M., and Hille, B. 1985. GTP-binding proteins couple cardiac muscarinic receptors to a K channel. *Nature* 317: 536–538. [10]

Pfrieger, F. W., and Barres, B. A. 1996. New views on synapse-glia interactions. *Curr. Opin. Neurobiol.* 6: 615–621. [8]

Phelan, K. A., and Hollyday, M. 1990. Axon guidance in muscleless chick wings: The role of muscle cells in motoneuronal pathway selection and muscle nerve formation. *J. Neurosci.* 10: 2699–2716. [23]

Philipson, K. D., Nicoll, D. A., Matsuoka, S., Hryshko, L. V., Levisky, D. O., and Weiss, J. N. 1996. Molecular regulation of the Na^+-Ca^{2+} exchanger. *Ann. N.Y. Acad. Sci.* 779: 20–28. [4]

Picard, N., and Strick, P. L. 1996. Motor areas of the medial wall: A review of their location and functional activation. *Cerebral Cortex* 6: 342–353. [22]

Picciotto, M. R., Zoli, M., Lena, C., Bessis, A., Lallemand, Y., Le Novere, N., Vincent, P., Merlo Pich, E., Brulet, P., and Changeux, J. P. 1995. Abnormal avoidance learning in mice lacking functional high-affinity nicotine receptor in the brain. *Nature* 374: 65–67. [14]

Pickles, J. O., Comis, S. D., and Osborne, M. P. 1984. Cross-links between stereocilia in the guinea pig organ of Corti, and their possible relation to sensory transduction. *Hear. Res.* 15: 103–112. [17]

Pieroni, J. P., and Byrne, J. H. 1992. Differential effects of serotonin, FMRFamide, and small cardioactive peptide on multiple, distributed processes modulating sensorimotor synaptic transmission in *Aplysia*. *J. Neurosci.* 12: 2633–2647. [10]

Piomelli, D. 1994. Eicosanoids in synaptic transmission. *Crit. Rev. Neurobiol.* 8: 65–83. [10]

Piomelli, D., Volterra, A., Dale, N., Siegelbaum, S. A., Kandel, E. R., Schwartz, J. H. and Belardetti, F. 1987. Lipoxygenase metabolites of arachidonic acid as second messengers for presynaptic inhibition of *Aplysia* sensory cells. *Nature* 328: 38–43. [10]

Pizzorusso, T., Fagiolini, M., Fabris, M., Ferrari, G., and Maffei, L. 1994. Schwann cells transplanted in the lateral ventricles prevent the functional and anatomical effects of monocular deprivation in the rat. *Proc. Natl. Acad. Sci. USA* 91: 2572–2576. [25]

Placzek, M., Yamada, T., Tessier-Lavigne, M., Jessell, T., and Dodd, J. 1991. Control of dorsoventral pattern in vertebrate neural development: Induction and polarizing properties of the floor plate. *Development* Suppl. 2: 105–122. [23]

Poggio, G. F., and Mountcastle, V. B. 1960. A study of the functional contributions of the lemniscal and spinothalamic systems to somatic sensibility. *Bull. Johns Hopkins Hosp.* 106: 266–316. [18]

Pohl, C. R., and Knobil, E. 1982. The role of the central nervous system in the control of ovarian function in higher primates. *Annu. Rev. Physiol.* 44: 583–593. [16]

Polley, D. B., Chen-Bee, C. H., and Frostig, R. D. 1999. Two directions of plasticity in the sensory-deprived adult cortex. *Neuron* 24: 623–637. [25]

Pons, T. P., Garraghty, P. E., and Mishkin, M. 1987. Physiological evidence for serial processing in somatosensory cortex. *Science* 237: 417–420. [18]

Pons, T. P., Garraghty, P. E., and Mishkin, M. 1992. Serial and parallel processing in somatosensory cortex of rhesus monkeys. *J. Neurophysiol.* 68: 518–527. [18]

Poritsky, R. 1969. Two- and three-dimensional ultrastructure of boutons and glial cells on the motoneuronal surface in the cat spinal cord. *J. Comp. Neurol.* 135: 423–452. [1]

Porter, C. W., and Barnard, E. A. 1975. The density of cholinergic receptors at the postsynaptic membrane: Ultrastructural studies in two mammalian species. *J. Membr. Biol.* 20: 31–49. [11]

Porter, J. T., and Mccarthy, K. D. 1998. Astrocytic neurotransmitter receptors in-situ and in-vivo. *Prog. Neurobiol.* 51: 439–455. [8]

Porter, L. L., and Sakamoto, K. 1988. Organization and synaptic relationships of the projection from the primary sensory to the primary motor cortex in the cat. *J. Comp. Neurol.* 271: 387–396. [22]

Potter, L. T. 1970. Synthesis, storage, and release of [^{14}C]acetylcholine in isolated rat diaphragm muscles. *J. Physiol.* 206: 145–166. [13]

Pourcho, R. G. 1996. Neurotransmitters in the retina. *Curr. Eye Res.* 15: 797–803. [19]

Powell, T. P. S., and Mountcastle, V. B. 1959. Some aspects of the functional organization of the cortex of the postcentral gyrus of the monkey: A correlation of findings obtained in a single unit analysis with cytoarchitecture. *Bull. Johns Hopkins Hosp.* 105: 133–162. [18]

Prell, G. D., and Green, J. P. 1986. Histamine as a neuroregulator. *Annu. Rev. Neurosci.* 9: 209–254. [14]

Pressley, T. A. 1996. Structure and function of the Na,K Pump: Ten years of molecular biology. *Miner. Electrolyte Metab.* 22: 264–271. [4]

Preuss, T. M., Stepniewska, I., and Kaas, J. H. 1996. Movement representation in the dorsal and ventral premotor areas of owl monkeys: A microstimulation study. *J. Comp. Neurol.* 371: 649–676. [22]

Pritchett, D. B., Sontheimer, H., Shivers, B. D. S., Ymer, S., Kettenmann, H., Schofield, P. R., and Seeburg, P. H. 1989. Importance of a novel $GABA_A$ receptor subunit for benzodiazepine pharmacology. *Nature* 338: 582–585. [14]

Prud'homme, M. J., Houdeau, E., Serghini, R., Tillet, Y., Schemann, M., and Rousseau, J. P. 1999. Small intensely fluorescent cells of the rat paracervical ganglion synthesize adrenaline, receive afferent innervation from postganglionic cholinergic neurones, and contain muscarinic receptors. *Brain Res.* 821: 141–149. [16]

Puce, A., Allison, T., Asgari, M., Gore, J. C., and McCarthy, G. 1996. Differential sensitivity of human visual cortex to faces, letterstrings and textures: A functional magnetic resonance imaging study. *J. Neurosci.* 16: 5205–5215. [21]

Pun, R. Y. K., and Lecar, H. 1998. Patch clamp techniques and analysis. In N. Sperelakis (ed.), *Cell Physiology Source Book*, 2nd Ed. Academic Press, San Diego, CA, pp. 391–405. [2]

Purves, D. 1975. Functional and structural changes in mammalian sympathetic neurones following interruption of their axons. *J. Physiol.* 252: 429–463. [24]

Purves, D., and Lichtman, J. W. 1983. Specific connections between nerve cells. *Annu. Rev. Physiol.* 45: 553–565. [23]

Purves, D., and Lichtman, J. W. 1985. *Principles of Neural Development.* Sinauer, Sunderland, MA. [24]

Purves, D., and Sakmann, B. 1974b. Membrane properties underlying spontaneous activity of denervated muscle fibers. *J. Physiol.* 239: 125–153. [24]

Purves, D., and Sakmann, B. 1974a. The effect of contractile activity on fibrillation and extrajunctional acetylcholine sensitivity in rat muscle maintained in organ culture. *J. Physiol.* 237: 157–182. [24]

Purves, D., Thompson, W., and Yip, J. W. 1981. Reinnervation of ganglia transplanted to the neck from different levels of the guinea pig sympathetic chain. *J. Physiol.* 313: 49–63. [24]

Putney, J. W., Jr., and McKay, R. R. 1999. Capacitative calcium entry channels. *BioEssays* 21: 38–46. [10]

Qian, H., Li, L., Chappell, R. L., and Ripps, H. 1997. GABA receptors of bipolar cells from the skate retina: Actions of zinc on GABA-mediated membrane currents. *J. Neurophysiol.* 78: 2402–2412. [14, 19]

Qian, H., Malchow, R. P., Chappell, R. L., and Ripps, H. 1996. Zinc enhances ionic currents induced in skate Müller (glial) cells by the inhibitory neurotransmitter GABA. *Proc. R. Soc. Lond. B* 263: 791–796. [8]

Quick, D. C., Kennedy, W. R., and Donaldson, L. 1979. Dimensions of myelinated nerve fibers near the motor and sensory terminals in cat tenuissimus muscles. *Neuroscience* 4: 1089–1096. [7]

Quilliam, T. A., and Armstrong, J. 1963. Mechanoreceptors. *Endeavour* 22: 55–60. [17]

Quiring, R., Walldorf, U., Kloter, U., and Gehring, W. J. 1994. Homology of the eyeless gene of *Drosophila* to the Small eye gene in mice and Aniridia in humans. *Science* 265: 785–789. [23]

Rabchevsky, A. G., and Streit, W. J. 1997. Grafting of cultured microglial cells into the lesioned spinal cord of adult rats enhances neurite outgrowth. *J. Neurosci. Res.* 47: 34–48. [24]

Rabow, L. E., Russek, S. J., and Farb, D. H. 1995. From ion currents to genomic analysis: Recent advances in $GABA_A$ receptor research. *Synapse* 21: 189–274. [14]

Radcliffe, K. A., Fisher, J. L., Gray, R., and Dani, J. A. 1999. Nicotinic modulation of glutamate and GABA synaptic transmission of hippocampal neurons. *Ann. N.Y. Acad. Sci.* 868: 591–610. [9]

Radel, J. D., Kustra, D. J., and Lund, R. D. 1995. The pupillary light response: Functional and anatomical interaction among inputs to the pretectum from transplanted retinae and host eyes. *Neuroscience* 68: 893–907. [24]

Raff, M. 1996. Neural development: Mysterious no more? *Science* 274: 1063. [26]

Raftery, M. A., Hunkapiller, M. W., Strader, C. D., and Hood, L. E. 1980. Acetylcholine receptor: Complex of homologous subunits. *Science* 208: 1454–1457. [3]

Rahamimoff, R., and Fernandez, J. M. 1997. Pre- and postfusion regulation of transmitter release. *Neuron* 18: 17–27. [11]

Rajan, R. 1995. Frequency and loss dependence of the protective effects of the olivocochlear pathways in cats. *J. Neurophysiol.* 74: 598–615. [18]

Rakic, P. 1974. Neurons in rhesus monkey visual cortex: Systematic relationship between time of origin and eventual disposition. *Science* 183: 425–427. [23]

Rakic, P. 1977. Prenatal development of the visual system in rhesus monkey. *Philos. Trans. R. Soc. Lond. B* 278: 245–260. [25]

Rakic, P. 1981. Neuronal-glial interaction during brain development. *Trends Neurosci.* 4: 184–187. [8, 23]

Rakic, P. 1986. Mechanism of ocular dominance segregation in the lateral geniculate nucleus: Competitive elimination hypothesis. *Trends Neurosci.* 9: 11–15. [25]

Rakic, P., and Caviness, V. S., Jr. 1995. Cortical development: View from neurological mutants two decades later. *Neuron* 14: 1101–1104. [25]

Raleigh, M. J., McGuire, M. T., Brammer, G. L., Pollack, D. B., and Yuwiler, A. 1991. Serotonergic mechanisms promote dominance acquisition in adult male vervet monkeys. *Brain Res.* 559: 181–190. [14]

Rall, W. 1967. Distinguishing theoretical synaptic potentials computed from different soma-dendritic distributions of synaptic input. *J. Neurophysiol.* 30: 1138–1168. [7]

Ralph, M. R., Foster, R., Davis, F. C., and Menaker, M. 1990. Transplanted suprachiasmatic nucleus determines circadian period. *Science* 247: 975–978. [16]

Ramachandran, R., Davis, K. A., and May, B. J. 1999. Single-unit responses in the inferior colliculus of decerebrate cats I. Classification based on frequency response maps. *J. Neurophysiol.* 82: 152–163. [18]

Ramanathan, K., Michael, T., Jiang, G-J., Hiel, H., and Fuchs, P. A. 1999. A molecular mechanism for electrical tuning of cochlear hair cells. *Science* 283: 215–217. [18]

Ramirez-Latorre, J., Yu, C. R., Qu, X., Perin, F., Karlin, A., and Role, L. 1996. Functional contributions of alpha5 subunit to neuronal acetylcholine receptor channels. *Nature* 380: 347–351. [14]

Ramón y Cajal, S. 1955. *Histologie du Système Nerveux*, Vol. 2. C.S.I.C., Madrid. [20]

Ramón y Cajal, S. 1995. *Histology of the Nervous System*, 2 vols. Oxford University Press, New York. [1, 8]

Ramón-Cueto, A., Plant, G., Avila, J., and Bunge, M. B. 1998. Long-distance axonal regeneration in the transected adult rat spinal cord is promoted by olfactory ensheathing glia transplants. *J. Neurosci.* 18: 3803–3815. [24]

Randall, A., and Tsien, R. W. 1995. Pharmacological dissection of multiple types of calcium channel currents in rat cerebellar granule neurons. *J. Neurosci.* 15: 2995–3012. [3]

Rang, H. P., Dale, M. M., and Ritter, J. M. 1999. *Pharmacology*, 4th Ed. Churchill Livingstone, Edinburgh, Scotland. [16]

Ransom, B. R. 1991. Vertebrate glial classification: Lineage and heterogeneity. *Ann. N.Y. Acad. Sci.* 633: 19–26. [8]

Ransom, B. R., and Fern, R. 1997. Does astrocytic glycogen benefit axon function and survival in CNS white matter during glucose deprivation? *Glia* 21: 134–141. [8]

Ransom, B. R., and Goldring, S. 1973. Slow depolarization in cells presumed to be glia in cerebral cortex of cat. *J. Neurophysiol.* 36: 869–878. [8]

Ransom, B. R., and Orkand, R. K. 1996. Glial-neuronal interaction in non-synaptic areas of the brain: Studies in the optic nerve. *Trends Neurosci.* 19: 352–358. [8]

Ransom, B. R., and Sontheimer, H. 1992. The neurophysiology of glial cells. *J. Clin. Neurophysiol.* 9: 224–251. [8]

Rao, A., Kim, E., Sheng, M., and Craig, A. M. 1998. Heterogeneity in the molecular composition of excitatory postsynaptic sites during development of hippocampal neurons in culture. *J. Neurosci.* 18: 1217–1229. [23]

Rao, S. M., Binder, J. R., Hammeke, T. A., Bandettini, P. A., Bobholz, J. A., Frost, J. A., Myklebust, B. M., Jacobson, R. D., and Hyde, J. S. 1995. Somatotopic mapping of the human primary motor cortex with functional magnetic resonance imaging. *Neurology* 45: 919–924. [22]

Rasband, M. N., Peles, E., Trimmer, J. S., Levinson, S. R., Lux, S. E., and Shrager, P. 1999. Dependence of nodal sodium channel clustering on paranodal axoglial contact in the developing CNS. *J. Neurosci.* 19: 7516–7528. [8]

Rasband, M. N., Trimmer, J. S., Schwartz, T. L., Levinson, S. R., Ellisman, M. H., Schachner, M., and Shrager, P. 1998. Potassium channel distribution, clustering, and function in remyelinating rat axons. *J. Neurosci.* 18: 36–47. [7]

Rasgado-Flores, H., Santiago, E. M., and Blaustein, M. P. 1989. Kinetics and stoichiometry of coupled Na efflux and Ca influx (Na/Ca exchange) in barnacle muscle cells. *J. Gen. Physiol.* 93: 1219–1241. [4]

Rasika, S., Alvarez-Buylla, A., and Nottebohm, F. 1999. BDNF mediates the effects of testosterone on the survival of new neurons in an adult brain. *Neuron* 22: 53–62. [23]

Rasmussen, G. L. 1946. The olivary peduncle and other fiber projections to the superior olivary complex. *J. Comp. Neurol.* 84: 141–220. [18]

Rasmusson, D. D. 1982. Reorganization of raccoon somatosensory cortex following removal of the fifth digit. *J. Comp. Neurol.* 205: 313–326. [18]

Rauschecker, J. P. 1998. Cortical processing of complex sounds. *Curr. Opin. Neurobiol.* 8: 516–521. [18]

Rauschecker, J. P., Tian, B., Pons, T., and Mishkin, M. 1997. Serial and parallel processing in the rhesus monkey auditory cortex. *J. Comp. Neurol.* 382: 89–103. [18]

Ravdin, P., and Axelrod, D. 1977. Fluorescent tetramethyl rhodamine derivatives of α-bungarotoxin: Preparation, separation, and characterization. *Anal. Biochem.* 80: 585–592. [9]

Raviola, E., and Wiesel, T. N. 1990. Neural control of eye growth and experimental myopia in primates. *Ciba Found. Symp.* 155: 22–38. [19]

Raybould, H. E., Holzer, P., Thiefin, G., Holzer, H. H., Yoneda, M., and Tache, Y. F. 1991. Vagal afferent innervation and regulation of gastric function. *Adv. Exp. Med. Biol.* 298: 109–127. [16]

Ready, D. 1989. A multifaceted approach to neural development. *Trends Neurosci.* 12: 102–110. [23]

Reale, R. A., and Imig, T. J. 1980. Tonotopic organization in auditory cortex of the cat. *J. Comp. Neurol.* 192: 265–291. [18]

Reddy, P. H., Williams, M., and Tagle, D. A. 1999. Recent advances in understanding the pathogenesis of Huntington's disease. *Trends Neurosci.* 22: 248–255. [22]

Redfern, P. A. 1970. Neuromuscular transmission in new-born rats. *J. Physiol.* 209: 701–709. [23]

Redman, R. S., and Silinsky, E. M. 1994. ATP released together with acetylcholine as the mediator of neuromuscular depression at frog motor nerve endings. *J. Physiol.* 477: 117–127. [12]

Redman, S. 1990. Quantal analysis of synaptic potentials in neurons of the central nervous system. *Physiol. Rev.* 70: 165–198. [11]

Reese, T. S., and Karnovsky, M. J. 1967. Fine structural localization of a blood-brain barrier to exogenous peroxidase. *J. Cell Biol.* 34: 207–217. [8]

Refinetti, R., and Menaker, M. 1992. The circadian rhythm of body temperature. *Physiol. Behav.* 51: 613–637. [16]

Reger, J. F. 1958. The fine structure of neuromuscular synapses of gastrocnemii from mouse and frog. *Anat. Rec.* 130: 7–23. [11]

Reichardt, L. F., and Tomaselli, K. J. 1991. Extracellular matrix molecules and their receptors: Functions in neural development. *Annu. Rev. Neurosci.* 14: 531–570. [23]

Reichelt, W., Hernandez, M., Damian, R. T., Kisaalita, W. S., and Jordan, B. L. 1997. Voltage- and GABA-evoked currents from Müller glial cells of the baboon retina. *Neuroreport* 8: 541–544. [8]

Reichert, H., and Boyan, G. 1997. Building a brain: Developmental insights in insects. *Trends Neurosci.* 20: 258–264. [15]

Reid, C. A., and Clements, J. D. 1999. Postsynaptic expression of long-term potentiation in the rat dentate gyrus demonstrated by variance – mean analysis. *J. Physiol.* 518(Pt.1): 121–130. [12]

Reiländer, H., Achilles, A., Freidel, U., Maul, G., Lottspeich, F., and Cook, N. J. 1992. Primary structure and functional expression of the Na/Ca,K-exchanger from bovine rod photoreceptors. *EMBO J.* 11: 1689–1695. [4]

Reimer, R. J., Fon, E. A., and Edwards, R. H. 1998. Vesicular neurotransmitter transport and the presynaptic regulation of quantal size. *Curr. Opin. Neurobiol.* 8: 405–412. [13]

Reiser, G., and Miledi, R. 1989. Changes in the properties of synaptic channels opened by acetylcholine in denervated frog muscle. *Brain Res.* 479: 83–97. [11]

Reiter, H. O., Waitzman, D. M., and Stryker, M. P. 1986. Cortical activity blockade prevents ocular dominance plasticity in the kitten visual cortex. *Exp. Brain Res.* 65: 182–188. [25]

Rekling, J. C., and Feldman, J. L. 1998. PreBotzinger complex and pacemaker neurons: Hypothesized site and kernel for respiratory rhythm generation. *Annu. Rev. Physiol.* 60: 385–405. [22]

Ressler, K. J., Sullivan, S. L., and Buck, L. B. 1993. A zonal organization of odorant receptor gene expression in the olfactory epithelium. *Cell* 73: 597–609. [17]

Restrepo, D., Miyamoto, T., Bryant, B. P., and Teeter, J. H. 1990. Odor stimuli trigger influx of calcium into olfactory neurons of the channel catfish. *Science* 249: 1166. [17]

Rettig, J., Sheng, Z-H., Kim, D. K., Hodson, C. D., Snutch, T. P., and Catterall, W. A. 1996. Isoform-specific interaction of the alpha1A subunits of brain Ca^{++} channels with the presynaptic proteins syntaxin and SNAP-25. *Proc. Natl. Acad. Sci. USA* 93: 7363–7368. [13]

Reuter, H. 1974. Localization of β adrenergic receptors, and effects of noradrenaline and cyclic nucleotides on action potentials, ionic currents and tension in mammalian cardiac muscle. *J. Physiol.* 242: 429–451. [10]

Reuter, H., and Seitz, N. 1968. The dependence of calcium efflux from cardiac muscle on temperature and external ion composition. *J. Physiol.* 195: 451–470. [4]

Reuter, H., Cachelin, A. B., DePeyer, J. E., and Kokubun, S. 1983. Modulation of calcium channels in cultured cardiac cells by isoproternenol and 8-bromo-cAMP. *Cold Spring Harb. Symp. Quant. Biol.* 48: 193–200. [10]

Reuveny, E., Slesinger, P. A., Inglese, J., Morales, J. M., Iñiguez-Lluhi, J. A., Lefkowitz, R. J., Bourne, H. R., Jan, Y. N., and Jan, L. Y. 1994. Activation of the cloned muscarinic potassium channel by G protein $\beta\gamma$ subunits. *Nature* 370: 143–146. [10]

Reyes, A., Lujan, R., Rozov, A., Burnashev, N., Somogyi, P., and Sakmann, B. 1998. Target-cell-specific facilitation and depression in neocortical circuits. *Nature Neurosci.* 1(4): 279–285. [15]

Reynolds, B. A., and Weiss, S. 1996. Clonal and population analyses demonstrate that an EGF-responsive mammalian embryonic precursor is a stem cell. *Dev. Biol.* 175: 1–13. [23]

Ribchester, R. R., and Taxt, T. 1983. Motor unit size and synaptic competition in rat lumbrical muscles reinnervated by active and inactive motor axons. *J. Physiol.* 344: 89–111. [23]

Ribchester, R. R., and Taxt, T. 1984. Repression of inactive motor nerve terminals in partially denervated rat muscle after regeneration of active motor axons. *J. Physiol.* 347: 497–511. [23]

Ribeiro, J. A., Cunha, R. A., Correia-de-Sa, P., and Sebastio, A. M. 1996. Purinergic regulation of acetylcholine release. *Prog. Brain Res.* 109: 231–241. [14]

Ricci, A. J., and Fettiplace, R. 1997. The effects of calcium buffering and cyclic AMP on mechano-electrical transduction in turtle auditory hair cells. *J. Physiol.* 501: 111–124. [17]

Ricci, A. J., Wu, Y-C., and Fettiplace, R. 1998. The endogenous calcium buffer and the time course of transducer adaptation in the auditory hair cells. *J. Neurosci.* 18: 8261–8277. [17]

Rice, J. J., May, B. J., Spirou, G. A., and Young, E. D. 1992. Pinna-based spectral cues for sound localization in cat. *Hear. Res.* 58: 132–152. [18]

Richardson, P. M., McGuinness, U. M., and Aguayo, A. J. 1980. Axons from CNS neurones regenerate into PNS grafts. *Nature* 284: 264–265. [24]

Ridley, R. M., and Baker, H. F. 1991. Can fetal neural transplants restore function in monkeys with lesion-induced behavioural deficits? *Trends Neurosci.* 14: 366–370. [24]

Rieke, F., and Baylor, D. A. 1998. Origin of reproducibility in the responses of retinal rods to single photons. *Biophys. J.* 75: 1836–1857. [19]

Riesen, A. H., and Aarons, L. 1959. Visual movement and intensity discrimination in cats after early deprivation of pattern vision. *J. Comp. Physiol. Psychol.* 52: 142–149. [25]

Rijntjes, M., Dettmers, C., Buchel, C., Kiebel, S., Frackowiak, R. S., and Weiller, C. 1999. A blueprint for movement: Functional and anatomical representations in the human motor system. *J. Neurosci.* 19: 8043–8048. [22]

Rissman, E. F. 1996. Behavioral regulation of gonadotropin releasing hormone. *Biol. Reprod.* 54: 413–419. [16]

Ritchie, A., and Fambrough, D. M. 1975. Ionic properties of the acetylcholine receptor in cultured rat myotubes. *J. Gen. Physiol.* 65: 751–767. [9]

Ritchie, J. M. 1986. Distribution of saxitoxin binding sites in mammalian neural tissue. *Ann. N.Y. Acad. Sci.* 479: 385–401. [6]

Ritchie, J. M. 1987. Voltage-gated cation and anion channels in mammalian Schwann cells and astrocytes. *J. Physiol. (Paris)* 82: 248–257. [8]

Ritchie, J. M., Black, J. A., Waxman, S. G., and Angelides, K. J. 1990. Sodium channels in the cytoplasm of Schwann cells. *Proc. Natl. Acad. Sci. USA* 87: 9290–9294. [8]

Rizzolatti, G., Fogassi, L., and Gallese, V. 1997. Parietal cortex: From sight to action. *Curr. Opin. Neurobiol.* 7: 562–567. [22]

Rizzolatti, G., Luppino, G., and Matelli, M. 1998. The organization of the cortical motor system: New concepts. *Electroencephalogr. Clin. Neurophysiol.* 106: 283–296. [22]

Roberts, A., and Bush, B. M. H. 1971. Coxal muscle receptors in the crab: The receptor current and some properties of the receptor nerve fibrers. *J. Exp. Biol.* 54: 515–524. [17]

Roberts, E. 1986. GABA: The road to neurotransmitter status. In R. W. Olsen and J. C. Venter (eds.), *Benzodiazepine/GABA Receptors and Chloride Channels: Structural and Functional Properties.* Alan R. Liss, New York, pp. 1–39. [13]

Robertson, D., Anderson, C-J., and Cole, K. S. 1987. Segregation of efferent projections to different turns of the guinea pig cochlea. *Hear. Res.* 25: 69–76. [18]

Robinson, D. A., and Fuchs, A. F. 1969. Eye movement evoked by stimulation of frontal eye fields. *J. Neurophysiol.* 32: 637–648. [21]

Robinson, M. S. 1994. The role of clathrin, adaptors, and dynamin in endocytosis. *Curr. Opin. Cell Biol.* 6: 538–544. [13]

Robinson, S. R., and Smotherman, W. P. 1992. Fundamental motor patterns of the mammalian fetus. *J. Neurobiol.* 23: 1574–1600. [22]

Robitaille, R., Adler, E. M., and Charlton, M. P. 1990. Strategic location of calcium channels at release sites of frog neuromuscular synapses. *Neuron* 5: 773–779. [11]

Roelink, H., Porter, J. A., Chiang, C., Tanabe, Y., Chang, D. T., Beachy, P. A., and Jessel, T. M. 1995. Floor plate and motor neuron induction by different concentrations of the amino-terminal cleavage product of sonic hedgehog autoproteolysis. *Cell* 81: 445–455. [23]

Rogan, M. T., and LeDoux, J. E. 1995. LTP is accompanied by a commensurate enhancement of auditory-evoked responses in a fear conditioning circuit. *Neuron* 15: 127–136. [12]

Rogan, M. T., Staubil, U. V., and LeDoux, J. E. 1997. Fear conditioning induces associative long-term potentiation in the amygdala. *Nature* 390: 604–607. [12]

Rogart, R. B., Cribs, L. L., Muglia, L. K., Kephart, D., and Kaiser, M. W. 1989. Molecular cloning of a putative tetrodotoxin-resistant heart Na^+ channel isoform. *Proc. Natl. Acad. Sci. USA* 86: 8170–8174. [3]

Rohrbough, J., and Spitzer, N. C. 1996. Regulation of intracellular Cl^- levels by Na^+-dependent Cl^- cotransport distinguishes depolarizing from hyperpolarizing GABA(A) receptor-mediated responses in spinal neurons. *J. Neurosci.* 16: 82–91. [4]

Rojas, L., and Orkand, R. K. 1999. K^+ channel density increases selectively in the endfoot of retinal glial cells during development of *Rana catesbiana*. *Glia* 25: 199–203. [8]

Romani, G. L., Williamson, S. J., and Kaufman, L. 1982. Tonotopic organization of the human auditory cortex. *Science* 216: 1339–1340. [18]

Rose, C. R., and Ransom, B. R. 1996. Intracellular sodium homeostasis in rat hippocampal astrocytes. *J. Physiol.* 491: 291–305. [8]

Rose, C. R., Ransom, B. R., and Waxman, S. G. 1997. Pharmacological characterization of Na^+ influx via voltage-gated Na+ channels in spinal cord astrocytes. *J. Neurophysiol.* 78: 3249–3258. [8]

Rosenblatt, K. P., Sun, Z-P., Heller, S., and Hudspeth, A. J. 1997. Distribution of Ca^{2+}-activated K^+ channel isoforms along the tonotopic gradient of the chicken's cochlea. *Neuron* 19: 1061–1075. [18]

Rosenthal, J. L. 1969. Post-tetanic potentiation at the neuromuscular junction of the frog. *J. Physiol.* 203: 121–133. [12]

Ross, E. M. 1989. Signal sorting and amplification through G protein-coupled receptors. *Neuron* 3: 141–152. [10]

Ross, W. N., and Werman, R. 1987. Mapping calcium transients in the dendrites of Purkinje cells form the guinea pig cerebellum in vitro. *J. Physiol.* 389: 319–336. [12]

Ross, W. N., Arechiga, H., and Nicholls, J. G. 1988. Influence of substrate on the distribution of calcium channels in identified leech neurons in culture. *Proc. Natl. Acad. Sci. USA* 85: 4075–4078. [10]

Ross, W. N., Lasser-Ross, N., and Werman, R. 1990. Spatial and temporal analysis of calcium-dependent electrical activity in guinea pig Purkinje cell dendrites. *Proc. R. Soc. Lond. B* 240: 173–185. [6]

Rossant, J. 1985. Interspecific cell markers and lineage in mammals. *Philos. Trans. R. Soc. Lond. B* 312: 91–100. [23]

Rossi, A., Zalaffi, A., and Decchi, B. 1996. Interaction of nociceptive and non-nociceptive cutaneous afferents from foot sole in common reflex pathways to tibialis anterior motoneurones in humans. *Brain Res.* 714: 76–86. [22]

Rothman, J. E. 1994. Mechanisms of intracellular protein transport. *Nature* 372: 55–63. [13]

Rothstein, J. D., Dykes-Hoberg, M., Pardo, C. A., Bristol, L. A., Jin, L., Kuncl, R. W., Kanai, Y., Hediger, M. A., Wang, Y., Schielke, J. P., and Welty, D. F. 1996. Knockout of glutamate transporters reveals a major role for astroglial transport in excitotoxicity and clearance of glutamate. *Neuron* 16: 675–686. [13]

Rothstein, J. D., Martin, L., Levey, A. I., Dykes-Hoberg, M., Jin, L., Wu, D., Nash, N., and Kuncl, R. W. 1994. Localization of neuronal and glial glutamate transporters. *Neuron* 13: 713–725. [13]

Rothwell, J. C. 1997. Techniques and mechanisms of action of transcranial stimulation of the human motor cortex. *J. Neurosci. Methods* 74: 113–122. [22]

Rotshenker, S. 1988. Multiple modes and sites for the induction of axonal growth. *Trends Neurosci.* 11: 363–366. [24]

Rotzler, S., Schramek, H., and Brenner, H. R. 1991. Metabolic stabilization of endplate acetylcholine receptors regulated by Ca^{2+} influx associated with muscle activity. *Nature* 349: 337–339. [24]

Rouiller, E. M., Moret, V., Tanne, J., and Boussaoud, D. 1996. Evidence for direct connections between the hand region of the supplementary motor area and motoneurons in the macaque monkey. *Eur. J. Neurosci.* 8: 1055–1059. [22]

Rovainen, C. M. 1967. Physiological and anatomical studies on large neurons of the central nervous system of the sea lamprey (*Petromyzon marinus*). II. Dorsal cells and giant interneurons. *J. Neurophysiol.* 30: 1024–1042. [9]

Rubenstein, J. L., and Beachy, P. A. 1998. Patterning of the embryonic forebrain. *Curr. Opin. Neurobiol.* 8: 18–26. [23]

Rubenstein, J. L., Shimamura, K., Martinez, S., and Puelles, L. 1998. Regionalization of the prosencephalic neural plate. *Annu. Rev. Neurosci.* 21: 445–477. [23]

Rubin, L. L., Barbu, K., Bard, F., Cannon, C., Hall, D. E., Horner, H., Janatpour, M., Liaw, C., Manning, K., Morales, J., Porter, S., Tanner, L., Tomaselli, K., and Yednock, T. 1992. Differentiation of brain endothelial cells in cell culture. *Ann. N.Y. Acad. Sci.* 633: 420–425. [8]

Rudomin, P. 1994. Segmental and descending control of the synaptic effectiveness of muscle afferents. *Prog. Brain Res.* 100: 97–104. [9]

Rudomin, P. and Schmidt, R. F. 1999. Presynaptic inhibition in the vertebrate spinal cord revisited. *Exp. Brain Res.* 129: 1–37. [9]

Rushton, W. A. H. 1951. A theory of the effects of fibre size in medullated nerve. *J. Physiol.* 115: 101–122. [7]

Russell, J. M. 1983. Cation-coupled chloride influx in squid axon. Role of potassium and stoichiometry of the transport process. *J. Gen. Physiol.* 81: 909–925. [4]

Ryan, S. G., Buckwalter, M. S., Lynch, J. W., Handford, C. A., Segura, L., Shiang, R., Wasmuth, J. J., Camper, S. A., Schofield, P., and O'Connell, P. 1994. A missense mutation in the gene encoding the alpha 1 subunit of the inhibitory glycine receptor in the spasmodic mouse. *Nature Genet.* 7: 131–135. [14]

Ryan, T. A., Reuter, H., and Smith, S. J. 1997. Optical detection of a quantal presynaptic membrane turnover. *Nature* 388: 478–482. [11]

Ryan, T. A., Reuter, H., Wendland, B., Schweizer, F. E., Tsein, R. W., and Smith, S. J. 1993. The kinetics of synaptic vesicle recycling measured at single presynaptic boutons. *Neuron* 11: 713–724. [11, 13]

Rydqvist, B., and Purali, N. 1993. Transducer properties of the rapidly adapting stretch receptor neurone in the crayfish (*Pacifastacus leniusculus*). *J. Physiol.* 469: 193–211. [17]

Sachs, F. 1988. Mechanical transduction in biological systems. *Crit. Rev. Biomed. Eng.* 16: 141–169. [17]

Sachs, M. B. 1984. Neural coding of complex sounds: Speech. *Annu. Rev. Physiol.* 46: 261–273. [18]

Safronov, V. B., and Vogel, W. 1995. Single voltage-activated Na^+ and K^+ channels in the somata of rat motoneurones. *J. Physiol.* 487: 91–106. [6]

Sagne, C., El Mestikaway, S., Isambert, M. F., Hamon, M., Henry, J. P., Giros, B., and Gasnier, B. 1997. Cloning of a functional vesicular GABA and glycine transporter by screening genome databases. *FEBS Lett.* 417: 177–183. [4, 13]

Sahley, C. L., Modney, B. K., Boulis, N. M., and Muller, K. J. 1994. The S cell: An interneuron essential for sensitization and full dishabituation of leech shortening. *J. Neurosci.* 14: 6715–6721. [15]

Sakai, K., Watanabe, E., Onoderaa, Y., Uchida, I., Kato, H., Yamamoto, E., Koizumi, H., and Miyashita, Y. 1995. Functional mapping of the human colour centre with echo-planar magnetic resonance imaging. *Proc. R. Soc. Lond. B* 261: 89–98. [21]

Sakamoto, T., Arissan, K., and Asanuma, H. 1989. Functional role of sensory cortex in learning motor skills in cats. *Brain Res.* 503: 258–264. [22]

Sakmann, B., Noma, A., and Trautwein, W. 1983. Acetylcholine activation of single muscarinic K^+ channels in isolated pacemaker cells of the mammalian heart. *Nature* 303: 250–253. [10]

Sakurai, M. 1987. Synaptic modification of parallel fibre-Purkinje cell transmission in *in vitro* cerebellar slices. *J. Physiol.* 394: 463–480. [12]

Sakurai, M. 1990. Calcium is an intracellular mediator of the climbing fiber induction of cerebellar long-term depression. *Proc. Natl. Acad. Sci. USA* 87: 3383–3385. [12]

Sala, C., O'Malley, J., Xu, R., Fumagalli, G., and Salpeter, M. M. 1997. ε Subunit-containing acetylcholine receptors in myotubes belong to the slowly degrading population. *J. Neurosci.* 17: 8937–8944. [24]

Salinas, E., and Romo, R. 1998. Conversion of sensory signals into motor commands in primary motor cortex. *J. Neurosci.* 18: 499–511. [22]

Salkoff, L., Baker, K., Butler, A., Covarrubius, M., Pak, M. D., and Wei, A. 1992. An essential 'set' of K^+ channels conserved in flies, mice and humans. *Trends Neurosci.* 15: 161–166. [3]

Salmons, S., and Sreter, F. A. 1975. The role of impulse activity in the transformation of skeletal muscle by cross innervation. *J. Anat.* 120: 412–415. [24]

Salpeter, M. M. (ed.). 1987a. *The Vertebrate Neuromuscular Junction.* Alan R. Liss, New York. [9, 23]

Salpeter, M. M. 1987b. Vertebrate neuromuscular junctions: General morphology, molecular organization, and functional consequences. In M. M. Salpeter (ed.), *The Vertebrate Neuromuscular Junction.* Alan R. Liss, New York, pp. 1–54. [9, 11, 13]

Salpeter, M. M., and Loring, R. H. 1985. Nicotinic acetylcholine receptors in vertebrate muscle: Properties, distribution and neural control. *Prog. Neurobiol.* 25: 297–325. [24]

Salpeter, M. M., and Marchaterre, M. 1992. Acetylcholine receptors in extrajunctional regions of innervated muscle have a slow degradation rate. *J. Neurosci.* 12: 35–38. [24]

Sandler, V. M., and Ross, W. N. 1999. Serotonin modulates spike backpropagation and associated $[Ca^{2+}]_i$ changes in the apical dendrites of hippocampal CA1 pyramidal neurons. *J. Neurophysiol.* 81: 216–224. [7]

Sandyk, R. 1992. L-Tryptophan in neuropsychiatric disorders: A review. *Int. J. Neurosci.* 67: 127–144. [13]

Sanes, J. N., Suner, S., and Donoghue, J. P. 1990. Dynamic organization of primary motor cortex output to target muscles in adult rats. I. Long-term patterns of reorganization following motor or mixed peripheral nerve lesions. *Exp. Brain Res.* 79: 479–491. [22]

Sanes, J. R., and Lichtman, J. W. 1999a. Can molecules explain long term potentiation? *Nature Neurosci.* 2: 597–604. [12]

Sanes, J. R., and Lichtman, J. W. 1999b. Development of the vertebrate neuromuscular junction. *Annu. Rev. Neurosci.* 22: 389–442. [13, 23, 24]

Sanes, J. R., Marshall, L. M., and McMahan, U. J. 1978. Reinnervation of muscle fiber basal lamina after removal of muscle fibers. *J. Cell Biol.* 78: 176–198. [24]

Sargent, P. B. 1993. The diversity of neuronal nicotinic acetylcholine receptors. *Annu. Rev. Neurosci.* 16: 403–443. [14, 16]

Sargent, P. B., and Pang, D. Z. 1989. Acetylcholine receptor-like molecules are found in both synaptic and extrasynaptic clusters on the surface of neurons in the frog cardiac ganglion. *J. Neurosci.* 9: 1062–1072. [24]

Sargent, P. B., Bryan, G. K., Streichert, L. C., and Garrett, E. N. 1991. Denervation does not alter the number of neuronal bungarotoxin binding sites on autonomic neurons in the frog cardiac ganglion. *J. Neurosci.* 11: 3610–3623. [24]

Sasai, Y. 1998. Identifying the missing links: Genes that connect neural induction and primary neurogenesis in vertebrate embryos. *Neuron* 21: 455–458. [23]

Sassoe-Pognetto, M., and Wassle, H. 1997. Synaptogenesis in the rat retina: Subcellular localization of glycine receptors, GABA(A) receptors, and the anchoring protein gephyrin. *J. Comp. Neurol.* 381: 158–174. [23]

Sato, H., Katsuyama, N., Tamura, H., Hata, Y., and Tsumoto, T. 1996. Mechanisms underlying orientation selectivity in the primary visual cortex of the macaque. *J. Physiol.* 494: 757–771. [20]

Sato, T., Kawada, T., Inagaki, M., Shishido, T., Takaki, H., Sugimachi, M., and Sunagawa, K. 1999. New analytic framework for understanding sympathetic baroreflex control of arterial pressure. *Am. J. Physiol.* 276: H2251–H2261. [16]

Sato, Y., Hirata, T., Ogawa, M., and Fujisawa, H. 1998. Requirement for early-generated neurons recognized by monoclonal antibody Lot1 in the formation of lateral olfactory tract. *J. Neurosci.* 18: 7800–7810. [23]

Saunders, N. R., Habgood, M. D., and Dziegielewska, K. M. 1999a. Barrier mechanisms in the brain. I. Adult brain. *Clin. Exp. Pharmacol. Physiol.* 26: 11–19. [8]

Saunders, N. R., Habgood, M. D., and Dziegielewska, K. M. 1999b. Barrier mechanisms in the brain. II. Immature brain. *Clin. Exp. Pharmacol. Physiol.* 26: 85–91. [8]

Saunders, N. R., Kitchener, P., Knott, G. W., Nicholls, J. G., Potter, A., and Smith, T. J. 1998. Development of walking, swimming and neuronal connections after complete spinal cord transection in the neonatal opossum, *Monodelphis domestica. J. Neurosci.* 18: 339–355. [24]

Saunders, N., and Dziegielewska, K. M. 1997. Barriers in the developing brain. *News Physiol. Sci.* 12: 21–31. [8]

Sauve, Y., Sawai, H., and Rasminsky, M. 1995. Functional synaptic connections made by regenerated retinal ganglion cell axons in the superior colliculus of adult hamsters. *J. Neurosci.* 15: 665–675. [24]

Savchenko, A., Barnes, S., and Kramer, R. H. 1997. Cyclic-nucleotide-gated channels mediate synaptic feedback by nitric oxide. *Nature* 390: 694–698. [10, 19]

Sawatari, A., and Callaway, E. M. 1996. Convergence of magno- and parvocellular pathways in layer 4B of macaque primary visual cortex. *Nature* 380: 442–446. [21]

Schalling, M., Stieg, P. E., Lindquist, C., Goldstein, M., and Hokfelt, T. 1989. Rapid increase in enzyme and peptide mRNA in sympathetic ganglia after electrical stimulation in humans. *Proc. Natl. Acad. Sci. USA* 86: 4302–4305. [13]

Schatzmann, H. J. 1989. The calcium pump of the surface membrane of the sarcoplasmic reticulum. *Ann. Rev. Physiol.* 51: 473–485. [4]

Scheutze, S. M., and Role, L. M. 1987. Developmental regulation of nicotinic acetylcholine receptors. *Annu. Rev. Neurosci.* 10: 403–457. [24]

Schikorski, T., and Stevens, C. F. 1997. Quantitative ultrastructural analysis of hippocampal excitatory synapses. *J. Neurosci.* 17: 5858–5867. [11]

Schiller, P. H., and Lee, K. 1991. The role of the primate extrastriate area V4 in vision. *Science* 251: 1251–1253. [21]

Schiller, P. H., and Malpeli, J. G. 1978. Functional specificity of lateral geniculate nucleus laminae of the rhesus monkey. *J. Neurophysiol.* 41: 788–797. [20]

Schlaggar, B. L., and O'Leary, D. D. M. 1991. Potential of visual cortex to develop an array of functional units unique to somatosensory cortex. *Science* 252: 1556–1560. [23]

Schmahmann, J. D., and Sherman, J. C. 1998. The cerebellar cognitive affective syndrome. *Brain* 121: 561–579. [22]

Schmidt, H. H., and Walter, U. 1994. NO at work. *Cell* 78: 919–925. [10]

Schmidt, K. E., Kim, D. S., Singer, W., Bonhoeffer, T., and Löwel, S. 1997. Functional specificity of long-range intrinsic and interhemispheric connections in the visual cortex of strabismic cats. *J. Neurosci.* 17: 5480–5492. [21]

Schmidt, R. F. 1971. Presynaptic inhibition in the vertebrate central nervous system. *Ergeb. Physiol.* 63: 20–101. [9]

Schmidt-Rose, Y., and Jentsch, T. J. 1997. Transmembrane topology of a CLC chloride channel. *Proc. Natl. Acad. Sci. USA* 94: 7633–7638. [3]

Schmied, A., Morin, D., Vedel, J. P., and Pagni, S. 1997. The "size principle" and synaptic effectiveness of muscle afferent projections to human extensor carpi radialis motoneurones during wrist extension. *Exp. Brain Res.* 113: 214–229. [22]

Schnapf, J. L., and Baylor, D. A. 1987. How photoreceptor cells respond to light. *Sci. Am.* 256(4): 40–47. [19]

Schnapf, J. L., Kraft, T. W., Nunn, B. J., and Baylor, D. A. 1988. Spectral sensitivity of primate photoreceptors. *Vis. Neurosci.* 1: 255–261. [19]

Schnapp, B. J., Vale, R. D., Sheetz, M. P., and Reese, T. S. 1985. Single microtubules from squid axoplasm support bidirectional movement of organelles. *Cell* 40: 455–462. [13]

Schnell, L., and Schwab, M. E. 1990. Axonal regeneration in the rat spinal cord produced by an antibody against myelin-associated neurite growth inhibitors. *Nature* 343: 269–272. [24]

Schnetkamp, P. P. 1995. Calcium homeostasis in vertebrate retinal rod outer segments. *Cell Calcium* 18: 322–330. [4]

Schnetkamp, P. P., Basu, D. K., and Szerencsei, R. T. 1989. Na^+-Ca^{2+} exchange in bovine rod outer segments requires and transports K^+. *Am. J. Physiol.* 257: C153–C157. [4]

Schofield, P. R., Darlison, M. G., Fujita, N., Burt, D. R., Stephenson, F. A., Rodriguez, H., Rhee, L. M., Ramchandran, J., Reale, V., Glencorse, T. A., Seeburg, P. H., and Barnard, E. A. 1987. Sequence and functional expression of the $GABA_A$ receptor shows a ligand-gated receptor superfamily. *Nature* 328: 221–227. [3]

Schramm, M., and Selinger, Z. 1984. Message transmission: Receptor controlled adenylate cyclase system. *Science* 225: 1350–1356. [10]

Schreiber, M. H., and Thach, W. T., Jr. 1985. Trained slow tracking. II. Bidirectional discharge patterns of cerebellar nuclear, motor cortex and spindle afferent neurons. *J. Neurophysiol.* 54: 1228–1270. [22]

Schuldiner, S., Shirvan, A., and Linial, M. 1995. Vesicular neurotransmitter transporters: From bacteria to humans. *Physiol. Rev.* 75: 369–392. [4, 13]

Schuldiner, S., Steiner-Mordoch, S., and Yelin, R. 1998. Molecular and biochemical studies of rat vesicular monamine transporter. *Adv. Pharmacol.* 42: 223–227. [4]

Schulman, H. 1995. Protein phosphorylation in neuronal plasticity and gene expression. *Curr. Opin. Neurobiol.* 5: 375–381. [12]

Schulman, H., and Hyman, S. E. 1999. Intracellular signaling. In M. J. Zigmond, F. E. Bloom, S. C. Landis, J. L. Roberts, and L. R. Squire (eds.), *Fundamental Neuroscience.* Academic Press, New York, pp. 269–316. [10]

Schultzberg, M., Hökfelt, T., and Lundberg, J. M. 1982. Coexistence of classical neurotransmitters and peptides in the central and peripheral nervous system. *Br. Med. Bull.* 38: 309–313. [13]

Schuman, E. V., and Madison, D. V. 1995. Nitric oxide and synaptic function. *Annu. Rev. Neurosci.* 17: 153–183. [14]

Schwab, M. E., and Bartholdi, D. 1996. Degeneration and regeneration of axons in the lesioned spinal cord. *Physiol. Rev.* 76: 319–370. [24]

Schwab, M. E., and Caroni, P. 1988. Oligodendrocytes and CNS myelin are nonpermissive substrates for neurite growth and fibroblast spreading in vitro. *J. Neurosci.* 8: 2381–2393. [24]

Schwartz, A. B., Kettner, R. E., and Georgopoulos, A. P. 1988. Primate motor cortex and free arm movement to visual targets in three-dimensional space. I. Relations between cell discharge and angle of movement. *J. Neurosci.* 8: 2913–2927. [22]

Schwartz, E. A. 1987. Depolarization without calcium can release γ-aminobutyric acid from a retinal neuron. *Science* 238: 350–355. [11, 19]

Schwarz, C., and Bolz, J. 1991. Functional specificity of a long-range horizontal connection in cat visual cortex: a cross-correlation study. *J. Neurosci.* 11: 2995–3007. [21]

Sclar, G., Maunsell, J. H. R., and Lennie, P. 1990. Coding of image contrast in central visual pathways of the macaque monkey. *Vision Res.* 30: 1–10. [21]

Scott, S. H., and Kalaska, J. F. 1995. Changes in motor cortex activity during reaching movements with similar hand paths but different arm postures. *J. Neurophysiol.* 73: 2563–2567. [22]

Seagar, M., Lévêque, C., Charvin, N., Marquèze, B., Martin-Moutot, N., Boudier, J. A., Boudier, J-L., Shoji-Kasai, Y., Sato, K., and Takahashi, M. 1999. Interactions between proteins implicated in exocytosis and voltage-gated calcium channels. *Philos. Trans. R. Soc. Lond. B* 354: 289–297. [13]

Sears, T. A. 1964a. Efferent discharges in alpha and fusimotor fibers of intercostal nerves of the cat. *J. Physiol.* 174: 295–315. [22]

Sears, T. A. 1964b. The slow potentials of thoracic respiratory motoneurones and their relation to breathing. *J. Physiol.* 175: 404–424. [22]

Seeburg, P. H. 1993. The molecular biology of the mammalian glutamate receptor channels. *Trends Neurosci.* 16: 359–365. [3]

Segal, R. A., and Greenberg, M. E. 1996. Intracellular signaling pathways activated by neurotrophic factors. *Annu. Rev. Neurosci.* 19: 463–489. [23]

Segovia, C. T., Salgado, Z. O., Clapp, C., and Martinez de la Escalera, G. 1996. The catecholaminergic stimulation of gonadotropin-releasing hormone release by GT1-1 cells does not involve phosphoinositide hydrolysis. *Life Sci.* 58: 1453–1459. [16]

Seilheimer, B., and Schachner, M. 1988. Studies of adhesion molecules mediating interactions between cells of peripheral nervous system indicate a major role for L1 in mediating sensory neuron growth on Schwann cells in culture. *J. Cell Biol.* 107: 341–351. [23]

Selkoe, D. J. 1991. The molecular pathology of Alzheimer's disease. *Neuron* 6: 487–498. [14]

Selverston, A., Elson, R., Rabinovich, M., Huerta, R., and Abarbanel, H. 1998. Basic principles for generating motor output in the stomatogastric ganglion. *Ann. N.Y. Acad. Sci.* 860: 35–50. [16]

Selyanko, A. A., and Brown, D. A. 1996. Intracellular calcium directly inhibits potassium M channels in excised membrane patches from rat sympathetic neurons. *Neuron* 16: 151–162. [16]

Selyanko, A. A., Hadley, J. K., Wood, I. C., Abogadie, F. C., Delmas, P., Buckley, N. J., London, B., and Brown, D. A. 1999. Two types of K(+) channel subunit, Erg1 and KCNQ2/3, contribute to the M-like current in a mammalian neuronal cell. *J. Neurosci.* 19: 7742–7756. [16]

Sendtner, M., Holtmann, B., and Hughes, R. A. 1996. The response of motoneurons to neurotrophins. *Neurochem. Res.* 21: 831–841. [23]

Sengpiel, F., Stawinski, P., and Bonhoeffer, T. 1999. Influence of experience on orientation maps in cat visual cortex. *Nature Neurosci.* 2: 727–732. [25]

Sengupta, P., and Bargmann, C. I. 1996. Cell fate specification and differentiation in the nervous system of *Caenorhabditis elegans*. *Dev. Genet.* 18: 73–80. [23]

Sereno, R. I., et al. 1995. Borders of multiple visual areas revealed by functional magnetic reonance imaging. *Science* 268: 889–893. [21]

Shain, D. H., Ramírez-Weber, F-A., Hsu, J., and Weisblat, D. A. 1998. Gangliogenesis in leech: Morphogenetic processes leading to segmentation in the central nervous system. *Dev. Genes Evol.* 208: 28–36. [23]

Shamma, S. A., Fleshman, J. W., Wiser, P. R., and Versnel, H. 1993. Organization of response areas in ferret primary auditory cortex. *J. Neurophysiol.* 69: 367–383. [18]

Shankland, M. 1995. Formation and specification of neurons during the development of the leech central nervous system. *J. Neurobiol.* 27: 294–309. [23]

Shapovalov, A. I., and Shiriaev, B. I. 1980. Dual mode of junctional transmission at synapses between single primary afferent fibres and motoneurones in the amphibian. *J. Physiol.* 306: 1–15. [9]

Shatz, C. J. 1977. Anatomy of interhemispheric connections in the visual system. *J. Comp. Neurol.* 173: 497–518. [21]

Shatz, C. J. 1996. Emergence of order in visual system development. *Proc. Natl. Acad. Sci. USA* 93: 602–608. [25]

Shatz, C. J., and Luskin, M. B. 1986. The relationship between the geniculocortical afferents and their cortical target cells during development of the cat's primary visual cortex. *J. Neurosci.* 6: 3655–3658. [23]

Sheetz, M. P. 1999. Motor and cargo interactions. *Eur. J. Biochem.* 262: 19–25. [13]

Shen, L., and Alexander, G. E. 1997. Neural correlates of a spatial sensory-to-motor transformation in primary motor cortex. *J. Neurophysiol.* 77: 1171–1194. [22]

Sheng, M. and Lee, S. H. 2000. Growth of the NMDA receptor industrial complex. *Nature Neurosci.* 3: 633–635. [13]

Sherrington, C. S. 1906. *The Integrative Action of the Nervous System*, 1961 Ed. Yale University Press, New Haven, CT. [1]

Sherrington, C. S. 1910. Flexor-reflex of the limb, crossed extension reflex, and reflex stepping and standing (cat and dog). *J. Physiol.* 40: 28–121. [22]

Sherrington, C. S. 1951. *Man on His Nature*. Cambridge University Press, Cambridge. [19]

Shi, S. H., Hayashi, Y., Petralia, R. S., Zaman, S. H., Wenthold, R. J., Svoboda, K., and Maninow, R. 1999. Rapid spine delivery and redistribution of AMPA receptors after synaptic NMDA receptor activation. *Science* 284: 1811–1816. [12]

Shiang, R., Ryan, S. G., Zhu, Y. Z., Hahn, A. F., O'Connell, P., and Wasmuth, J. J. 1993. Mutations in the alpha 1 subunit of the inhibitory glycine receptor cause the dominant neurologic disorder, hyperekplexia. *Nature Genet.* 5: 351–358. [14]

Shibuya, I., Tanaka, K., Hattori, Y., Uezono, Y., Harayama, N., Noguchi, J., Ueta, Y., Izumi, F., and Yamashita, H. 1999. Evidence that multiple P2X purinoceptors are functionally expressed in rat supraoptic neurones. *J. Physiol.* 514: 351–367. [14]

Shima, K., and Tanji, J. 1998. Role for cingulate motor area cells in voluntary movement selection based on reward. *Science* 282: 1335–1338. [22]

Shipp, S., and Zeki, S. 1985. Segregation of pathways leading from area V2 to areas V4 and V5 of macaque monkey visual cortex. *Nature* 315: 322–325. [20]

Shotwell, S. L., Jacobs, R., and Hudspeth A. J. 1981. Directional sensitivity of individual vertebrate hair cells to controlled deflection of their hair bundles. *Ann. N.Y. Acad. Sci.* 374: 1–10. [17]

Shrager, P., Chiu, S. Y., and Ritchie, J. M. 1985. Voltage-dependent sodium and potassium channels in mammalian cultured Schwann cells. *Proc. Natl. Acad. Sci. USA* 82: 948–952. [8]

Shute, C. C. D., and Lewis, P. R. 1963. Cholinesterase-containing systems of the brain of the rat. *Nature* 199: 1160–1164. [14]

Shyng, S.-L., and Salpeter, M. M. 1990. Effect of reinnervation on the degradation rate of junctional acetylcholine receptors synthesized in denervated skeletal muscles. *J. Neurosci.* 10: 3905–3915. [24]

Shyng, S.-L., Xu, R., and Salpeter, M. M. 1991. Cyclic AMP stabilizes the degradation of original junctional acetylcholine receptors in denervated muscle. *Neuron* 6: 469–475. [24]

Si, J., and Mei, L. 1999. ERK MAP kinase activation is required for acetylcholine receptor inducing activity-induced increase in all five acetylcholine receptor subunit mRNAs as well as synapse-specific expression of acetylcholine receptor varepsilon-transgene. *Brain Res. Mol. Brain Res.* 67: 18–27. [24]

Si, J., Luo, Z., and Mei, L. 1996. Induction of acetylcholine receptor gene expression by ARIA requires activation of mitogen-activated protein kinase. *J. Biol. Chem.* 271: 19752–19759. [24]

Sicard, G., and Holley, A. 1984. Receptor cell responses to odorants: Similarities and differences among odorants. *Brain Res.* 292: 283–296. [17]

Siegel, G. J., Agranoff, B. W., Albers, R. W., Fisher, S. K., and Uhler, M. D. (eds.). 1999. *Basic Neurochemistry: Molecular, Cellular, and Medical Aspects*, 6th Ed. Lippincott-Raven, Philadelphia. [13]

Sieghart, W. 1995. Structure and pharmacology of γ-aminobutyric acid receptor subtypes. *Pharmacol. Rev.* 47: 181–234. [3]

Sigworth, F. J. 1994. Voltage gating of ion channels. *Q. Rev. Biophys.* 27: 1–40. [6]

Sigworth, F. J., and Neher, E. 1980. Single Na^+ channel currents observed in cultured rat muscle cells. *Nature* 287: 447–449. [6]

Silinsky, E. M., and Redman, R. S. 1996. Synchronous release of ATP and neurotransmitter within milliseconds of a motor-nerve impulse in the frog. *J. Physiol.* 492: 815–822. [9, 11]

Sillito, A. M. 1979. Inhibitory mechanisms influencing complex cell orientation selectivity and their modification at high resting discharge levels. *J. Physiol.* 289: 33–53. [14, 20]

Sillito, A. M., Jones, H. E., Gerstein, G. L., and West, D. C. 1994. Feature-linked synchronization of thalamic relay cell firing induced by feedback from the visual cortex. *Nature* 369: 479–482. [20]

Silva, A. J., Stevens, C. F., Tonegawa, S., and Wang, Y. 1992. Deficient hippocampal long-term potentiation in alpha-calcium-calmodulin kinase II mutant mice. *Science* 257: 501–506. [12]

Simon, H., Hornbruch, A., and Lumsden, A. 1995. Independent assignment of antero-posterior and dorso-ventral positional values in the developing chick hindbrain. *Curr. Biol.* 5: 205–214. [23]

Simons, D. J., and Woolsey, T. A. 1979. Functional organization in mouse barrel cortex. *Brain Res.* 165: 327–332. [18]

Simpson, P. B., Challiss, R. A. J., and Nahorski, S. R. 1995. Neuronal Ca^{2+} stores: Activation and function. *Trends Neurosci.* 18: 299–306. [4]

Sims, T. J., Waxman, S. G., Black, J. A., and Gilmore, S. A. 1985. Perinodal astrocytic processes at nodes of Ranvier in developing normal and glial cell deficient rat spinal cord. *Brain Res.* 337: 321–331. [8]

Singer, W. 1995. Development and plasticity of cortical processing architectures. *Science* 270: 758–764. [25]

Skene, J. H. P., and Shooter, E. M. 1983. Denervated sheath cells secrete a new protein after nerve injury. *Proc. Natl. Acad. Sci. USA* 80: 4169–4173. [24]

Skou, J. C. 1957. The influence of some cations on an adenosine triphosphatase from peripheral nerves. *Biochim. Biophys. Acta* 23: 394–401. [4]

Skou, J. C. 1988. Overview: The Na,K pump. *Methods Enzymol.* 156: 1–25. [4]

Smeyne, R. J., Klein, R., Schnapp, A., Long, L. K., Bryant, S., Lewin, A., Lira, S. and Barbacid, M. 1994. Severe sensory and sympathetic neuropathies in mice carrying a disrupted Trk/NGF receptor gene. *Nature* 368: 246–249. [23]

Smith, A. D., de Potter, W. P., Moerman, E. J., and Schaepdryver, A. F. 1970. Release of dopamine β-hydroxylase and chromogranin A upon stimulation of the splenic nerve. *Tissue Cell* 2: 547–568. [11]

Smith, J. C., Ellengerger, J. J., Ballanyi, K., Richter, D. W., and Feldman, J. L. 1991. Pre-Botzinger complex: A brainstem region that may generate respiratory rhythm in mammals. *Science* 254: 726–729. [22]

Smith, P. J., Howes, E. A., and Treherne, J. E. 1987. Mechanisms of glial regeneration in an insect central nervous system. *J. Exp. Biol.* 132: 59–78. [8]

Smith, S. J. 1988. Neuronal cytomechanics: The actin-based motility of growth cones. *Science* 242: 708–715. [23]

Smith, S. J., and Thompson, S. H. 1987. Slow membrane currents in bursting pace-maker neurones of *Tritonia*. *J. Physiol.* 382: 425–448. [22]

Snyder, S. H., Jaffrey, S. R., amd Zakhary, R. 1998. Nitric oxide and carbon monoxide: Parallel roles as neural messengers. *Brain Res. Brain Res. Rev.* 26: 167–175. [10]

Sodickson, D. L., and Bean, B. P. 1996. $GABA_B$ receptor-activated inwardly rectifying potassium current in dissociated hippocampal CA3 neurons. *J. Neurosci.* 16: 6374–6385. [14]

Soejima, M., and Noma, A. 1984. Mode of regulation of the ACh-sensitive K-channel by the muscarinic receptor in rabbit atrial cells. *Pflügers Arch.* 400: 424–431. [10]

Sokoloff, L. 1977. Relation between physiological function and energy metabolism in the central nervous system. *J. Neurochem.* 29: 13–26. [21]

Sokolove, P. G., and Cooke, I. M. 1971. Inhibition of impulse activity in a sensory neuron by an electrogenic pump. *J. Gen. Physiol.* 57: 125–163. [17]

Solc, C. K., Derfler, B. H., Duyk, G. M., and Corey, D. P. 1994. Molecular cloning of myosins from the bullfrog saccular macula: A candidate for the adaptation motor. *Auditory Neurosci.* 1: 63–75. [17]

Sommer, B., Keinanen, K., Verdoorn, T. A., Wisden, W., Burnashev, N., Herb, A., Kohler, M., Takagi, T., Sakmann, B., and Seeburg, P. H. 1990. Flip and flop: A cell-specific functional switch in glutamate-operated channels of the CNS. *Science* 249: 1580–1585. [14]

Sommer, B., Kohler, M., Sprengle, R., and Seeburg, P. H. 1991. RNA editing in brain controls a determinant of ion flow in glutamate-gated channels. *Cell* 67: 11–19. [3]

Sompolinsky, H., and Shapley, R. 1997. New perspectives on the mechanisms for orientation selectivity. *Curr. Opin. Neurobiol.* 7: 514–522. [20]

Son, Y. J., and Thompson, W. J. 1995a. Schwann cell processes guide regeneration of peripheral axons. *Neuron* 14: 125–132. [8, 24]

Son, Y. J., and Thompson, W. J. 1995b. Nerve sprouting in muscle is induced and guided by processes extended by Schwann cells. *Neuron* 14: 133–141. [8, 24]

Son, Y. J., Trachtenberg, J. T., and Thompson, W. J. 1996. Schwann cells induce and guide sprouting and reinnervation of neuromuscular junctions. *Trends Neurosci.* 19: 280–285. [8]

Song, H-J., and Poo, M-M. 1999. Signal transduction underlying growth cone guidance by diffusible factors. *Curr. Opin. Neurobiol.* 9: 355–363. [23]

Song, H-J., Ming, G-L., and Poo, M-M. 1997. cAMP-induced switching in turning direction of nerve growth cones. *Nature* 388: 275–279. [23]

Song, H-J., Ming, G-L., He, Z., Lehmann, M., McKerracher, L., Tessier-Lavigne, M., and Poo, M-M. 1998. Conversion of neuronal growth cone responses from repulsion to attraction by cyclic nucleotides. *Science* 281: 1515–1518. [23]

Sorimachi, H., Ishiura, S., and Suzuki, K. 1997. Structure and physiological function of calpains. *Biochem. J.* 328: 721–732. [10]

Sossin, W. S., Fisher, J. M., and Scheller, R. H. 1989. Cellular and molecular biology of neuropeptide processing and packaging. *Neuron* 2: 1407–1417. [13]

Sotelo, C., Alvarado-Mallart, R. M., Frain, M., and Vernet, M. 1994. Molecular plasticity of adult Bergmann fibers is associated with radial migration of grafted Purkinje cells. *J. Neurosci.* 14: 124–133. [24]

Sotelo, C., and Alvarado-Mallart, R. M. 1991. The reconstruction of cerebellar circuits. *Trends Neurosci.* 14: 350–355. [24]

Sotelo, C., Llinás, R., and Baker, R. 1974. Structural study of inferior olivary nucleus of the cat: Morphological correlates of electrotonic coupling. *J. Neurophysiol.* 37: 541–559. [7]

Soto, F., Garcia-Guzman, M., and Stühmer, W. 1997. Cloned ligand-gated channels activated by extracellular ATP (P2X receptors). *J. Membr. Biol.* 160: 91–100. [3, 16]

Soucy, E., Wang, Y., Nirenberg, S., Nathans, J., and Meister, M. 1998. A novel signaling pathway from rod photoreceptors to ganglion cells in mammalian retina. *Neuron* 21: 481–493. [19]

Specht, S., and Grafstein, B. 1973. Accumulation of radioactive protein in mouse cerebral cortex after injection of [^{3}H]-fucose into the eye. *Exp. Neurol.* 41: 705–722. [20]

Spector, S. A. 1985. Trophic effect on the contractile and histochemical properties of rat soleus muscle. *J. Neurosci.* 5: 2189–2196. [24]

Sperry, R. W. 1970. Perception in the absence of neocortical commissures. *Proc. Res. Assoc. Nerv. Ment. Dis.* 48: 123–138. [21]

Spoendlin, H. 1972. Innervation densities of the cochlea. *Acta Otolaryngol.* 73: 235–248. [18]

Sretavan, D. W., Shatz, C. J., and Stryker, M. P. 1988. Modification of retinal ganglion cell axon morphology by prenatal infusion of tetrodotoxin. *Nature* 336: 468–471. [25]

Stadler, H., and Kiene, M-L. 1987. Synaptic vesicles in electromotoneurones. II. Heterogeneity of populations is expressed in uptake properties, exocytosis and insertion of a core proteoglycan into the extracellular matrix. *EMBO J.* 6: 2217–2221. [13]

Staley, K., Smith, R., Schaak, J., Wilcox, C., and Jentsch, T. J. 1996. Alteration of $GABA_A$ receptor function following gene transfer of the CLC-2 chloride channel. *Neuron* 17: 543–551. [5]

Staras, K., Kemenes, G., and Benjamin, P. R. 1999. Electrophysiological and behavioral analysis of lip touch as a component of the food stimulus in the snail *Lymnaea*. *J. Neurophysiol.* 81: 1261–1273. [15]

Starling, E. H. 1941. *Starling's Principles of Human Physiology*. (C. Lovatt Evans, ed.). Churchill, London. [16]

Stehle, J. H., Foulkes, N. S., Molina, C. A., Simonneaux, V., Pevet, P., and Sassone-Corsi, P. 1993. Adrenergic signals direct rhythmic expression of transcriptional repressor CREM in the pineal gland. *Nature* 365: 314–320. [14]

Stein, P. S., McCullough, M. L., and Currie, S. N. 1998. Spinal motor patterns in the turtle. *Ann. N.Y. Acad. Sci.* 860: 142–154. [22]

Steinbusch, H. W. M. 1984. Serotonin-immunoreactive neurons and their projections in the CNS. In A. Bjorklund, T. Hokfelt, and M. Kuhar (eds.), *Classical Transmitters and Transmitter Receptors in the CNS*. Vol. 3 of *Handbook of Chemical Neuroanatomy*. Elsevier, New York, pp. 68–125. [14]

Steinlein, O. K., Magnusson, A., Stoodt, J., Bertrand, S., Weiland, S., Berkovic, S. F., Nakken, K. O., Propping, P., and Bertrand, D. 1997. An insertion mutation of the CHRNA4 gene in a family with autosomal dominant nocturnal frontal lobe epilepsy. *Hum. Mol. Genet.* 6: 943–947. [14]

Steinlein, O. K., Mulley, J. C., Propping, P., Wallace, R. H., Phillips, H. A., Sutherland, G. R., Scheffer, I. E., and Berkovic, S. F. 1995. A missense mutation in the neuronal nicotinic acetylcholine receptor alpha 4 subunit is associated with autosomal dominant nocturnal frontal lobe epilepsy. *Nature Genet.* 11: 201–203. [14]

Stent, G. S., and Weisblat, D. 1982. The development of a simple nervous system. *Sci. Am.* 246(1): 136–146. [15]

Stent, G. S., Kristan, W. B., Jr., Torrence, S. A., French, K. A., and Weisblat, D. A. 1992. Development of the leech nervous system. *Int. Rev. Neurobiol.* 33: 109–193. [8, 23]

Sterling, P. 1983. Microcircuitry of the cat retina. *Annu. Rev. Neurosci.* 6: 149–185. [14]

Sterling, P. 1997. Retina. In G. M. Shepherd (ed.), *Synaptic Organization of the Brain*. Oxford University Press, New York, Chapter 6. [19]

Stern, K., and McClintock, M. K. 1998. Regulation of ovulation by human pheromones. *Nature* 392: 177–179. [17]

Sterr, A., Muller, M. M., Elbert, T., Rockstroh, B., Pantev, C., and Taub, E. 1998. Perceptual correlates of changes in cortical representation of fingers in blind multifinger Braille readers. *J. Neurosci.* 18: 4417–4423. [25]

Stevens, T. H., and Forgac, M. 1997. Structure, function and regulation of the vacuolar (H+)-ATPase. *Annu. Rev. Cell Dev. Biol.* 13: 779–808. [13]

Stewart, R. R., Adams, W. B., and Nicholls, J. G. 1989. Presynaptic calcium currents and facilitation of serotonin release at synapses between cultured leech neurones. *J. Exp. Biol.* 144: 1–12. [12]

Steyer, J. A., and Almers, W. 1999. Tracking single secretory granules in live chromaffin cells by evanescent-field fluorescence microscopy. *Biophys. J.* 76: 2262–2271. [11]

Steyer, J. A., Horstmann, H., and Almers, W. 1997. Transport, docking and exocytosis of single secretory granules in live chromaffin cells. *Nature* 388: 474–478. [11]

Stjärne, L., Greengard, P., Grillner, S. E., Hökfelt, T. G. M., and Ottoson, D. R. (eds.). 1994. *Molecular and Cellular Mechanisms of Neurotransmitter Release* (Advances in Second Messenger and Phosphoprotein Research, vol. 29). Raven, New York. [11]

Stoeckli, E. T., and Landmesser, L. T. 1995. Axonin-1, Nr-CAM, and Ng-CAM play different roles in the in vivo guidance of chick commissural neurons. *Neuron* 14: 1165–1179. [23]

Stoeckli, E. T., Sonderegger, P., Pollerberg, G. E., and Landmesser, L. T. 1997. Interference with axonin-1 and NrCAM interactions unmasks a floor-plate activity inhibitory for commissural axons. *Neuron* 18: 209–221. [23]

Stone, R. A., Laties, A. M., Raviola, E., and Wiesel, T. N. 1988. Increase in retinal vasoactive intestinal polypeptide after eyelid fusion in primates. *Proc. Natl. Acad. Sci. USA* 85: 257–260. [25]

Stratford, K. J., Tarczy-Hornoch, K., Martin, K. A. C., Bannister, N. J., and Jack, J. J. B. 1996. Excitatory synaptic inputs to spiny stellate cells in cat visual cortex. *Nature* 382: 258–261. [20]

Strecker, G. J., Wuarin, J. P., and Dudek, F. E. 1997. GABA A-mediated local synaptic pathways connect neurons in the rat suprachiasmatic nucleus. *J. Neurophysiol.* 78: 2217–2220. [16]

Streichert, L. C., and Sargent, P. B. 1992. The role of acetylcholinesterase in denervation supersensitivity in the frog cardiac ganglion. *J. Physiol.* 445: 249–260. [24]

Streisinger, G., Walker, C., Dower, N., Knauber, D., and Singer, F. 1981. Production of clones of homozygous diploid zebra fish (*Brachydanio rerio*). *Nature* 291: 293–296. [23]

Strettoi, E., and Masland, R. H. 1996. The number of unidentified amacrine cells in the mammalian retina. *Proc. Natl. Acad. Sci. USA* 93: 14906–14911. [19]

Stroud, R. M., and Finer-Moore, J. 1985. Acetylcholine receptor structure, function and evolution. *Annu. Rev. Cell Biol.* 1: 317–351. [3]

Stryer, L. 1987. The molecules of visual excitation. *Sci. Am.* 257(1): 42–50. [19]

Stryer, L. 1991–1992. Molecular mechanism of visual excitation. *Harvey Lect.* 87: 129–143. [19]

Stryer, L., and Bourne, H. R. 1986. G proteins: A family of signal transducers. *Annu. Rev. Cell Biol.* 2: 391–419. [19]

Stryker, M. P., and Harris, W. A. 1986. Binocular impulse blockade prevents the formation of ocular dominance columns in cat visual cortex. *J. Neurosci.* 6: 2117–2133. [25]

Stryker, M. P., and Zahs, K. R. 1983. On and off sublaminae in the lateral geniculate nucleus of the ferret. *J. Neurosci.* 10: 1943–1951. [20]

Stuart, A. E. 1970. Physiological and morphological properties of motoneurones in the central nervous system of the leech. *J. Physiol.* 209: 627–646. [15]

Stuart, G., Schiller, J., and Sakmann, B. 1997. Action potential initiation and propagation in rat neocortical pyramical neurons. *J. Physiol.* 505: 617–632. [7]

Stuart, G., Spruston, N., Sakmann, B. and Hausser, M. 1997. Action potential initiation and backpropagation in neurons of the mammalian CNS. *Trends Neurosci.* 20: 125–131. [7]

Stühmer, W., Conti, F., Suzuki, H., Wang, X., Noda, M., Yahagi, N., Kubo, H., and Numa, S. 1989. Structural parts involved in activation and inactivation of the sodium channel. *Nature* 239: 597–603. [6]

Suga, N., Zhang, Y., and Yan, J. 1997. Sharpening of frequency tuning by inhibition in the thalamic auditory nucleus of the mustached bat. *J. Neurophysiol.* 77: 2098–2114. [18]

Sugiura, Y., Woppmann, A., Miljanich, G. P., and Ko, C-P. 1995. A novel omega-conopeptide for the presynaptic localization of calcium channels at the mammalian neuromuscular junction. *J. Neurocytol.* 24: 15–27. [11]

Sulzer, D., Joyce, M. P., Lin, L., Geldwert, D., Haber, S. N., Hattori, T., and Rayport, S. 1998. Dopamine neurons make glutamatergic synapses in vitro. *J. Neurosci.* 18: 4588–4602. [13]

Sumikawa, K., and Miledi, R. 1988. Repression of nicotinic acetylcholine receptor expression by antisense RNAs and an oligonucleotide. *Proc. Natl. Acad. Sci. USA* 85: 1302–1306. [14]

Sumikawa, K., Parker, I., and Miledi, R. 1989. Expression of neurotransmitter receptors and voltage-activated channels from brain mRNA in *Xenopus* oocytes. *Methods Neurosci.* 1: 30–45. [3]

Sunahara, R. K., Niznik, H. B., Weiner, D. M., Stormann, T. M., Brann, M. R., Kennedy, J. L., Gelertner, J. E., Rozmahel, R., Yang, Y. L., et al. 1990. Human dopamine D1 receptor encoded by an intronless gene on chromosome 5. *Nature* 347: 80–83. [14]

Sunahara, R. K., Tesmer, J. J. G., Gilman, A. G., and Sprang, S. R. 1996. Crystal structure of the adenylyl cyclase activator $G_{s\alpha}$. *Science* 278: 1943–1947. [10]

Suter, D. M., Errante, L. D., Belotserkovsky, V., and Forscher, P. 1998. The Ig superfamily cell adhesion molecule, apCAM, mediates growth cone steering by substrate-cytoskeletal coupling. *J. Cell Biol.* 141: 227–240. [23]

Sutherland, E. W. 1972. Studies on the mechanism of hormone action. *Science* 177: 401–408. [10]

Sutter, M. L., and Schreiner, C. E. 1995. Topography of intensity tuning in cat primary auditory cortex: single-neuron versus multiple-neuron recordings. *J. Neurophysiol.* 73: 190–204. [18]

Svoboda, K., Helmchen, F., Denk, W., and Tank, D. W. 1999. Spread of dendritic excitation in layer 2/3 pyramidal neurons in rat barrel cortex in vivo. *Nature Neurosci.* 2: 65–73. [7]

Svoboda, K., Schmidt, C. F., Schnapp, B. J., and Block, S. M. 1993. Direct observation of kinesin stepping by optical trapping interferometry. *Nature* 365: 721–727. [13]

Swanson, L. W. 1976. The locus coeruleus: A cytoarchitectonic, Golgi, and immunohistochemical study in the albino rat. *Brain Res.* 110: 39–56. [14]

Swensen, K. I., Jordan, J. R., Beyer, E. C., and Paul, D. L. 1989. Formation of gap junctions by expression of connexins in *Xenopus* oocyte pairs. *Cell* 57: 145–155. [7]

Swindale, N. V., Matsubara, J. A., and Cynader, M. S. 1987. Surface organization of orientation and direction selectivity in cat area 18. *J. Neurosci.* 7: 1414–1427. [21]

Szabo, G., and Otero, A. S. 1990. G protein mediated regulation of K^+ channels in heart. *Annu. Rev. Physiol.* 52: 293–305. [10]

Szallasi, A., and Blumberg, P. M. 1996. Vanilloid receptors: New insights enhance potential as a therapeutic target. *Pain* 68: 195–208. [17]

Szatkowski, M., Barbour, B., and Attwell, D. 1990. Non-vesicular release of glutamate from glial cells by reversed electrogenic glutamate uptake. *Nature* 348: 443–446. [8]

Szczupak, L., Edgar, J., Peralta, M. L., and Kristan, W. B., Jr. 1998. Long-lasting depolarization of leech neurons mediated by receptors with a nicotinic binding site. *J. Exp. Biol.* 201: 1895–1906. [15]

Szentágothai, J. 1973. Neuronal and synaptic architecture of the lateral geniculate nucleus. In H. H. Kornhuker (ed.), *Central Visual Information*. Vol. 6 of *Handbook of Sensory Physiology*. Springer, Berlin, pp. 141–176. [20]

Takahashi, M., Billups, B., Rossi, D., Sarantis, M., Hamann, M., and Atwell, D. 1997a. The role of glutamate transporters in glutamate homeostasis in the brain. *J. Exp. Biol.* 200: 401–409. [4, 8]

Takahashi, M., Ishida, T., Traub, O., Corson, M. A., and Berk, B. C. 1997b. Mechanotransduction in endothelial cells: Temporal signaling events in response to shear stress. *J. Vasc. Res.* 34: 212–219. [17]

Takahashi, M., Sarantis, M., and Atwell, D. 1996. Postsynaptic glutamate uptake in rat cerebellar Purkinje cells. *J. Physiol.* 497: 523–530. [4]

Takai, T., Noda, M., Mishina, M., Shimizu, S., Furutani, Y., Kayano, T., Ikeda, T., Kubo, T., Takahashi, H., Takahashi, T., Kuno, M., and Numa, S. 1985. Cloning, sequencing, and expression of cDNA for a novel subunit of acetylcholine receptor from calf muscle. *Nature* 315: 761–764. [3]

Take, A. K., and Malpeli, J. G. 1998. Effects of focal inactivation of dorsal or ventral layers of the lateral geniculate nucleus on cats' ability to see and fixate small targets. *J. Neurophysiol.* 80: 2206–2209. [20]

Takechi, H., Onoe, H., Shizuno, H., Yoshikawa, E., Sadato, N., Tsukada, H., and Watanabe, Y. 1997. Mapping of cortical areas involved in color vision in non-human primates. *Neurosci. Lett.* 230: 17–20. [21]

Takeichi, M. 1995. Morphogenic roles of classic cadherins. *Curr. Opin. Cell Biol.* 7: 619–627. [23]

Takeuchi, A., and Takeuchi, N. 1959. Active phase of frog's end-plate potential. *J. Neurophysiol.* 22: 395–411. [9]

Takeuchi, A., and Takeuchi, N. 1960. On the permeability of the end-plate membrane during the action of transmitter. *J. Physiol.* 154: 52–67. [9]

Takeuchi, A., and Takeuchi, N. 1966. On the permeability of the presynaptic terminal of the crayfish neuromuscular junction during synaptic inhibition and the action of γ-aminobutyric acid. *J. Physiol.* 183: 433–449. [9]

Takeuchi, A., and Takeuchi, N. 1967. Anion permeability of the inhibitory post-synaptic membrane of the crayfish neuromuscular junction. *J. Physiol.* 191: 575–590. [9]

Takeuchi, N. 1963. Some properties of conductance changes at the end-plate membrane during the action of acetylcholine. *J. Physiol.* 167: 128–140. [9]

Takumi, Y,. Matsubara, A., Rinvik, E., and Otterson, O. P. 1999a. The arrangement of glutamate receptors in excitatory synapses. *Ann. N.Y. Acad. Sci.* 868: 474–481. [12]

Takumi, Y., Ramirez-Leon, V., Laake, P., Rinvik, E., and Otterson, O. P. 1999b. Different modes of expression of AMPA and NMDA receptors in hippocampal synapses. *Nature Neurosci.* 2: 618–624. [12]

Talbot, S. A., and Marshall, W. H. 1941. Physiological studies on neural mechanisms of visual localization and discrimination. *Am. J. Ophthalmol.* 24: 1255–1264. [20]

Tamir, H., Liu, K., Hsiung, S., Adlersberg, M., and Gershon, M. D. 1994. Serotonin binding protein: Synthesis, secretion, and recycling. *J. Neurochem.* 63: 97–107. [13]

Tan, J., Epema, A. H., and Voogd, J. 1995. Zonal organization of the flocculovestibular nucleus projection in the rabbit: A combined axonal tracing and acetylcholinesterase study. *J. Comp. Neurol.* 356: 51–71. [22]

Tanabe, T., Takashima, H., Mikami, A., Flockerzi, V., Takahashi, H., Kangawa, K., Kojima, M., Matsuo, H., Hirose, T., and Numa, S. 1987. Primary structure of receptors for calcium channel blockers from skeletal muscle. *Nature* 328: 313–318. [3]

Tanaka, K., Watase, K., Manabe, T., Yamada, K., Watanabe, M., Takahashi, K., Iwama, H., Nishikawa, T., Ichihara, N., Kikuchi, T., Okuyama, S., Kawashima, N., Hori, S., Takimoto, M., and Wada, K. 1997. Epilepsy and exacerbation of brain injury in mice lacking the glutamate transporter GLT-1. *Science* 276: 1699–1702. [8]

Tang, W-J., and Gilman, A. G. 1991. Type-specific regulation of adenylyl cyclase by G protein βγ subunits. *Science* 254: 1500–1503. [10]

Tang, Y-G., and Zucker, R. S. 1997. Mitochondrial involvement in post-tetanic potentiation of synaptic transmission. *Neuron* 18: 483–491. [12]

Tanji, J., and Wise, S. P. 1981. Submodality distribution in sensorimotor cortex of the unanesthetized monkey. *J. Neurophys.* 45: 467–481. [18]

Tank, D. W., Huganir, R. L., Greengard, P., and Webb, W. W. 1983. Patch-recorded single-channel currents of the purified and reconstituted *Torpedo* acetylcholine receptor. *Proc. Natl. Acad. Sci. USA* 80: 5129–5133. [3]

Tansey, M. G., Chu, G. C., and Merlie, J. P. 1996. ARIA/HRG regulates AChR ε subunit gene expression at the neuromuscular synapse via activation of phosphatidylinositol 3-kinase and Ras/MAPK pathway. *J. Cell Biol.* 134: 465–476. [24]

Tao-Cheng, J. H., Nagy, Z., and Brightman, M. W. 1987. Tight junctions of brain endothelium in vitro are enhanced by astroglia. *J. Neurosci.* 7: 3293–3299. [8]

Tao-Cheng, J. H., Nagy, Z., and Brightman, M. W. 1990. Astrocytic orthogonal arrays of intramembranous particle assemblies are modulated by brain endothelial cells in vitro. *J. Neurocytol.* 19: 143–153. [8]

Taraskevich, P. S., and Douglas, W. W. 1984. Electrical activity in adenohypophyseal cells and effects of hypophyseotropic substances. *Fed. Proc.* 43: 2373–2378. [16]

Tarozzo, G., De Andrea, M., Feuilloley, M., Vaudry, H., and Fasolo, A. 1998. Molecular and cellular guidance of neuronal migration in the developing olfactory system of rodents. *Ann. N.Y. Acad. Sci.* 839: 196–200. [16]

Tarozzo, G., Peretto, P., Biffo, S., Varga, Z., Nicholls, J., and Fasolo, A. 1995. Development and migration of olfactory neurones in the nervous system of the neonatal opossum. *Proc. R. Soc. Lond. B* 262: 95–101. [16]

Tasaki, I. 1959. Conduction of the nerve impulse. In J. Field (ed.), *Handbook of Physiology*, Section 1, Vol. 1, Chapter 3. American Physiological Society, Bethesda, MD, pp. 75–121. [7]

Taussig, R., and Gilman, A. G. 1995. Mammalian membrane-bound adenylyl cyclases. *J. Biol. Chem.* 270: 1–4. [10]

Teng, H., Cole, J. J., Roberts, R. L., and Wilkinson, R. S. 1999. Endocytic active zones: Hot spots for endocytosis in vertebrate neuromuscular terminals. *J. Neurosci.* 19: 4855–4866. [11, 13]

Teschemacher, H., Ophein, K. E., Cox, B. M., and Goldstein, A. 1975. A peptide-like substance from pituitary that acts like morphine. *Life Sci.* 16: 1771–1776. [14]

Tesmer, J. J. G., Sunahara, R. K., Gilman, A. G., and Sprang, S. R. 1997. Crystal structure of the catalytic domains of adenylyl cyclase in a complex with $G_{s\alpha}$⟨GTPγS. *Science* 278: 1907–1916. [10]

Tessier-Lavigne, M., Placzek, M., Lumsden, A. G. S., Dodd, J., and Jessell, T. M. 1988. Chemotropic guidance of developing axons in the mammalian central nervous system. *Nature* 336: 775–778. [23]

Teune, T. M., van der Burg, J., de Zeeuw, C. I., Voogd, J., and Ruigrok, T. J. 1998. Single Purkinje cell can innervate multiple classes of projection neurons in the cerebellar nuclei of the rat: A light microscopic and ultrastructural triple-tracer study in the rat. *J. Comp. Neurol.* 392: 164–178. [13]

Thach, W. T. 1978. Correlation of neural discharge with pattern and force of muscular activity, joint position, and direction of next intended movement in motor cortex and cerebellum. *J. Neurophysiol.* 41: 654–676. [22]

Thach, W. T. 1998. A role for the cerebellum in learning movement coordination. *Neurobiol. Learn. Mem.* 70: 177–188. [22]

Thach, W. T., Goodkin, H. G., and Keating, J. G. 1992. The cerebellum and the adaptive coordination of movement. *Annu. Rev. Neurosci.* 15: 403–442. [22]

Thanos, S., Bonhoeffer, F., and Rutishauser, U. 1984. Fiber-fiber interaction and tectal cues influence the development of the chick retinotectal projection. *Proc. Natl. Acad. Sci. USA* 81: 1906–1910. [23]

Thesleff, S. 1960. Supersensitivity of skeletal muscle produced by botulinum toxin. *J. Physiol.* 151: 598–607. [24]

Thoenen, H., Mueller, R. A., and Axelrod, J. 1969. Increased tyrosine hydroxylase activity after drug-induced alteration of sympathetic transmission. *Nature* 221: 1264. [13]

Thoenen, H., Otten, U., and Schwab, M. 1979. Orthograde and retrograde signals for the regulation of neuronal gene expression: The peripheral sympathetic nervous system as a model. In F. O. Schmitt and F. G. Worden (eds.), *The Neurosciences: Fourth Study Program*. MIT Press, Cambridge, MA, pp. 911–928. [13]

Thomas, R. C. 1969. Membrane currents and intracellular sodium changes in a snail neurone during extrusion of injected sodium. *J. Physiol.* 201: 495-514. [4]

Thomas, R. C. 1972. Intracellular sodium activity and the sodium pump in snail neurones. *J. Physiol.* 220: 55–71. [4]

Thomas, R. C. 1977. The role of bicarbonate, chloride and sodium ions in the regulation of intracellular pH in snail neurones. *J. Physiol.* 273: 317–338. [4]

Thompson, W. 1983. Synapse elimination in neonatal rat muscle is sensitive to pattern of muscle use. *Nature* 302: 614–616. [23]

Thompson, W. J., and Stent, G. S. 1976. Neuronal control of heartbeat in the medicinal leech. I. Generation of the vascular constriction rhythm by heart motor neurons. *J. Comp. Physiol.* 111: 309–333. [15]

Thompson, W., Kuffler, D. P., and Jansen, J. K. S. 1979. The effect of prolonged, reversible block of nerve impulses on the elimination of polyneuronal innervation of newborn rat skeletal muscle fibers. *Neuroscience* 4: 271–281. [23]

Thorogood, M. S., Almeida, V. W., and Brodfuehrer, P. D. 1999. Glutamate receptor 5/6/7-like and glutamate transporter-1-like immunoreactivity in the leech central nervous system. *J. Comp. Neurol.* 405: 334–344. [15]

Timpe, L. C., Schwartz, T. L., Tempel, B. L., Papazian, D. M., Jan, Y. N., and Jan, L. Y. 1988. Expression of functional potassium channels from *Shaker* cDNA in *Xenopus* oocytes. *Nature* 331: 143–145. [3]

Tischmeyer, W., and Grimm, R. 1999. Activation of immediate early genes and memory formation. *Cell. Mol. Life Sci.* 55: 564–574. [10]

Tobler, A. R., Notterpek, L., Naef, R., Taylor, V., Suter, U., and Shooter, E. M. 1999. Transport of Trembler-J mutant peripheral myelin protein 22 is blocked in the intermediate compartment and affects the transport of the wild-type protein by direct interaction. *J. Neurosci.* 19: 2027–2036. [8]

Tong, G., and Jahr, C. E. 1994. Block of glutamate transporters potentiates postsynaptic excitation. *Neuron* 13: 1195–1203. [4]

Tootell, R. B., Dale, A. M., Sereno, M. I., and Malach, R. 1996. New images from human visual cortex. *Trends Neurosci.* 19: 481–489. [21]

Tootell, R. B., Reppas, J. B., Kwong, K. K., Malach, R., Born, R. T., Brady, T. J., Rosen, B. R., and Belliveau, J. W. 1995. Functional analysis of human MT and related visual cortical areas using magnetic resonance imaging. *J. Neurosci.* 15: 3215–3230. [21]

Tootell, R. B., Silverman, M. S., and De Valois, R. L. 1981. Spatial frequency columns in primary visual cortex. *Science* 214: 813–815. [21]

Tootell, R. B., Silverman, M. S., Hamilton, S. L., De Valois, R. L., and Switkes, E. 1988. Functional anatomy of macaque striate cortex. III. Color. *J. Neurosci.* 8: 1569–1593. [21]

Torebjork, H. E., and Hallin, R. G. 1973. Perceptual changes accompanying controlled preferential blockade of A and C fibre responses in intact human skin nerve. *Exp. Brain Res.* 16: 321–332. [18]

Torre, V., Ashmore, J. F., Lamb, T. D., and Menini, A. 1995. Transduction and adaptation in sensory receptor cells. *J. Neurosci.* 15: 7757–7768. [19]

Tosini, G., and Menaker, M. 1996. Circadian-rhythms in cultured mammalian retina. *Science* 272: 419–421. [16]

Toyoshima, C., and Unwin, N. 1988. Ion channel of acetylcholine receptor reconstituted from images of postsynaptic membranes. *Nature* 336: 247–250. [3]

Treherne, J. M., Woodward, S. K. A., Varga, Z. M., Ritchie, J. M., and Nicholls, J. G. 1992. Restoration of conduction and growth of axons through injured spinal cord of neonatal opossum in culture. *Proc. Natl. Acad. Sci. USA* 89: 431–434. [24]

Trimmer, J. S., Cooperman, S. S., Tomiko, S. A., Zhou, J. Y., Crean, S. M., Boyle, M. B., Kallan, R. G., Sheng, Z. H., Barchi, R. L., Sigworth, F. J., Goodman, R. H., Agnew, W. S., and Mandel, G. 1989. Primary structure and functional expression of a mammalian skeletal muscle sodium channel. *Neuron* 3: 33–49. [3]

Trommsdorff, M., Gotthardt, M., Hiesberger, T., Shelton, J., Stockinger, W., Nimpf, J., Hammer, R. E., Richardson, J. A., and Herz, J. 1999. Reeler/Disabled-like disruption of neuronal migration in knockout mice lacking the VLDL receptor and ApoE receptor 2. *Cell* 97: 689–701. [23]

Truman, J. W., Thorn, R. S., and Robinow, S. 1992. Programmed neuronal death in insect development. *J. Neurobiol.* 23: 1295–1311. [23]

Trussell, L. 1997. Cellular mechanisms for preservation of timing in central auditory pathways. *Curr. Opin. Neurobiol.* 7: 487–492. [14]

Trussell, L. O. 1999. Synaptic mechanisms for coding timing in auditory neurons. *Annu. Rev. Physiol.* 61: 477–496. [9]

Ts'o, D. Y., and Gilbert, C. D. 1988. The organization of chromatic and spatial interactions in the primate striate cortex. *J. Neurosci.* 8: 1712–1727. [21]

Ts'o, D. Y., Frostig, R. D., Lieke, E. E., and Grinvald, A. 1990. Functional organization of primate visual cortex revealed by high resolution optical imaging. *Science* 249: 417–420. [21]

Ts'o, D. Y., Gilbert, C. D., and Wiesel, T. N. 1986. Relationships between horizontal interactions and functional architecture in cat striate cortex as revealed by cross-correlation analysis. *J. Neurosci.* 6: 1160–1170. [21]

Tsacopoulos, M., and Magistretti, P. J. 1996. Metabolic coupling between glia and neurons. *J. Neurosci.* 16: 877–885. [8]

Tsacopoulos, M., Poitry-Yamate, C. L., and Poitry, S. 1997. Ammonium and glutamate released by neurons are signals regulating the nutritive function of a glial-cell. *J. Neurosci.* 17: 2383–2390. [8]

Tsien, R. W. 1987. Calcium currents in heart cells and neurons. In L. K. Kaczmarek and I. B. Levitan (eds.), *Neuromodulation: The Biochemical Control of Neuronal Excitability*. Oxford University Press, New York, pp. 206–242. [10]

Tsien, R. W., and Tsien, R. Y. 1990. Calcium channels, stores, and oscillations. *Annu. Rev. Cell Biol.* 6: 715–760. [10]

Tsien, R. W., Bean, B. P., Hess, P., and Nowycky, M. 1983. Calcium channels: Mechanisms of β-adrenergic modulation and ion permeation. *Cold Spring Harb. Symp. Quant. Biol.* 48: 201–212. [10]

Tsien, R. Y. 1988. Fluorescent measurement and photochemical manipulation of cytosolic free calcium. *Trends Neurosci.* 11: 419–424. [4]

Tsubokawa, H., and Ross, W. N. 1996. IPSPs modulate spike backpropagation and associated $[Ca^{2+}]_i$ changes in the dendrites of hippocampal CA1 pyramidal neurons. *J. Neurophysiol.* 76: 2896–2906. [7]

Tsubokawa, H., and Ross, W. N. 1997. Muscarinic modulation of spike backpropagation in the apical dendrites of hippocampal CA1 pyramidal neurons. *J. Neurosci.* 17: 5782–5791. [15]

Tsuzuki, K., and Suga, N. 1988. Combination sensitive neurons in the ventroanterior area of the auditory cortex of the mustached bat. *J. Neurophysiol.* 60: 1908–1923. [18]

Tu, J. C., Xiao, B., Yuan, J. P., Lanahan, A. A., Leoffert, K., Li, M., Linden, D. J., and Worley, P. F. 1998. Homer binds a novel proline-rich motif and links group 1 metabotropic glutamate receptors with IP3 receptors. *Neuron* 21: 717–726. [13]

Tucker, T., and Fettiplace, R. 1995. Confocal imaging of calcium microdomains and calcium extrusion in turtle hair cells. *Neuron* 15: 1323–1335. [10]

Turman, A. B., Ferrington, D. G., Ghosh, S., Morley, J. W., and Rowe, M. J. 1992. Parallel processing of tactile information in the cerebral cortex of the cat: Effect of reversible inactivation of SI on responsiveness of SII neurons. *J. Neurophysiol.* 67: 411–429. [18]

Turner, A. J., and Tanzawa, K. 1997. Mammalian membrane metallopeptidases: NEP, ECE, KELL, and PEX. *FASEB J.* 11: 355–364. [13]

Turner, D. L., and Cepko, C. L. 1987. A common progenitor for neurons and glia persists in rat retina late in development. *Nature* 328: 131–136. [23]

Uĝurbil, K., Hu, X., Chen, W., Zhu, X-H., Kim, S-G., and Georgopoulos, A. 1999. Functional mapping in the human brain using high magnetic fields. *Philos. Trans. R. Soc. Lond. B* 354: 1195–1213. [1]

Ullian, E. M., McIntosh, J. M., and Sargent, P. B. 1997. Rapid synaptic transmission in the avian ciliary ganglion is mediated by two distinct classes of nicotinic receptors. *J. Neurosci.* 17: 7210–7219. [16]

Ungerleider, L. G., and Mishkin, M. 1982. Two cortical visual systems. In D. J. Ingle, R. J. W. Mansfield, M. S. Goodale (eds.), *The Analysis of Visual Behavior*. MIT Press, Cambridge, MA, pp. 549–586. [21]

Unwin, N. 1989. The structure of ion channels in membranes of excitable cells. *Neuron* 3: 665–676. [3]

Unwin, N. 1993. Nicotinic acetylcholine receptor at 9 Å resolution. *J. Mol. Biol.* 229: 1101–1124. [3]

Unwin, N. 1995. Acetylcholine receptor imaged in the open state. *Nature* 373: 37–43. [3]

Unwin, N., Toyoshima, C., and Kubalek, E. 1988. Arrangement of acetylcholine receptor subunits in the resting and desensitized states, determined by cryoelectron microscopy of crystallized *Torpedo* postsynaptic membranes. *J. Cell Biol.* 107: 1123–1138. [3]

Usherwood, P. N., and Machili, P. 1969. Chemical transmission at the insect excitatory neuromuscular synapse. *Nature* 210: 634–636. [14]

Vabnick, I., and Shrager, P. 1998. Ion channel redistribution and function during development of the myelinated axon. *J. Neurobiol.* 37: 80–96. [7]

Vafa, B., and Schofield, P. R. 1998. Heritable mutations in the glycine, $GABA_A$, and nicotinic acetylcholine receptors provide new insights into the ligand-gated ion channel superfamily. *Int. Rev. Neurobiol.* 42: 285–332. [14]

Vale, R. D., and Fletterick, R. J. 1997. The design plan of kinesin motors. *Annu. Rev. Cell Dev. Biol.* 13: 745–777. [13]

Vallbo, A. B. 1990. Muscle afferent responses to isometric contractions and relaxations in humans. *J. Neurophysiol.* 63: 1307–1313. [22]

Vallbo, A. B., and Hagbarth, K-E. 1968. Activity from skin mechanoreceptors recorded percutaneously in awake human subjects. *Exp. Neurol.* 21: 270–289. [18]

Vallee, R. B., and Bloom, G. S. 1991. Mechanisms of fast and slow axonal transport. *Annu. Rev. Neurosci.* 14: 59–92. [13]

Vallee, R. B., and Gee, M. A. 1998. Make room for dynein. *Trends Cell Biol.* 8: 490–494. [13]

Vallee, R. B., Shpetner, H. S., and Paschal, B. M. 1989. The role of dynein in retrograde axonal transport. *Trends Neurosci.* 12: 66–70. [13]

Valtorta, F., Jahn, R., Fesce, R., Greengard, P., and Ceccarelli, B. 1988. Synaptophysin (p38) at the frog neuromuscular junction: Its incorporation into the axolemma and recycling after intense quantal secretion. *J. Cell Biol.* 107: 2717–2727. [11]

van der Laan, L. J., De Groot, C. J., Elices, M. J., and Dijkstra, C. D. 1997. Extracellular matrix proteins expressed by human adult astrocytesin vivo and in vitro: An astrocyte surface protein containing the CS1 domain contributes to binding of lymphoblasts. *J. Neurosci. Res.* 50: 539–548. [8]

Van der Loos, H., and Woolsey, T. A. 1973. Somatosensory cortex: Structural alterations following early injury to sense organs. *Science* 179: 395–398. [18]

Van Essen, D. C., and Drury, H. A. 1997. Structural and functional analyses of human cerebral cortex using a surface-based atlas. *J. Neurosci.* 17: 7079–7102. [20]

Van Essen, D., and Jansen, J. K. 1974. Reinnervation of rat diaphragm during perfusion with α-bungarotoxin. *Acta Physiol. Scand.* 91: 571–573. [24]

Van Noorden, G. K. 1990. *Binocular Vision and Ocular Motility*. Mosby, St. Louis, MO. [25]

Vardi, N., Morigiwa, K., Wang, T. L., Shi, Y. J., and Sterling, P. 1998. Neurochemistry of the mammalian cone "synaptic complex." *Vision Res.* 38: 1359–1369. [19]

Varga, Z. M., Bandtlow, C. E., Erulkar, S. D., Schwab, M. E., and Nicholls, J. G. 1995. The critical period for repair of CNS of neonatal opossum (*Monodelphis domestica*) in culture: Correlation with development of glial cells, myelin and growth-inhibitory molecules. *Eur. J. Neurosci.* 7: 2119–2129. [24]

Varoqui, H., and Erickson, J. D. 1997. Vesicular neurotransmitter transporters. Potential sites for the regulation of synaptic function. *Mol. Neurobiol.* 15: 165–191. [4, 13]

Vassar, R., Ngai, J., and Axel, R. 1993. Spatial segregation of odorant receptor expression in the mammalian olfactory epithelium. *Cell* 74: 309–318. [17]

Vaucher, E., Linville, D., and Hamel, E. 1997. Cholinergic basal forebrain projections to nitric oxide synthase-containing neurons in the rat cerebral cortex. *Neuroscience* 79: 827–836. [14]

Vautrin, J., and Kriebel, M. E. 1991. Characteristics of slow-miniature end plate currents show a subunit composition. *Neuroscience* 41:71–88. [11]

Venable, N., Fernandez, V., Diaz, E., and Pinto-Hamuy, T. 1989. Effects of preweaning environmental enrichment on basilar dendrites of pyramidal neurons in occipital cortex: A Golgi study. *Brain Res. Dev. Brain Res.* 49: 140–144. [25]

Vergara, C., Latorre, R., Marrion, N. V., and Adelman, J. P. 1998. Calcium-activated potassium channels. *Curr. Opin. Neurobiol.* 8: 321–329. [6]

Verkhratsky, A., Orkand, R. K., and Kettenmann, H. 1998. Glial calcium: Homeostasis and signaling function. *Physiol. Rev.* 78: 99–139. [8]

Vertes, R. P. 1984. Brainstem control of the events of REM sleep. *Prog. Neurobiol.* 22: 241–288. [14]

Vidal-Sanz, M., Bray, G. M., and Aguayo, A. J. 1991. Regenerated synapses in the superior colliculus after the regrowth of retinal ganglion cell axons. *J. Neurocytol.* 20: 940–952. [24]

Villanueva, S., Fiedler, J., and Orrego, F. 1990. A study in rat brain cortex synaptic vesicles of endogenous ligands for N-methyl-D-aspartate receptors. *Neuroscience* 37: 23–30. [11]

Virchow, R. 1959. *Cellularpathologie* (F. Chance, trans.). Hirschwald, Berlin. [8]

Voderholzer, U., Hornyak, M., Thiel, B., Huwig-Poppe, C., Kiemen, A., Konig, A., Backhaus, J., Riemann, D., Berger, M., and Hohagen, F. 1998. Impact of experimentally induced serotonin deficiency by tryptophan depletion on sleep EEG in healthy subjects. *Neuropsychopharmacology* 18: 112–124. [13]

Vogel, H. J. 1994. Calmodulin: A versatile calcium mediator protein (Merck Frosst Award Lecture). *Biochem. Cell Biol.* 72: 357–376. [10]

Vollenweider, F. X., Cuenod, M., and Do, K. Q. 1990. Effect of climbing fiber deprivation on release of endogenous aspartate, glutamate, and homocysteate in slices of rat cerebellar hemispheres and vermis. *J. Neurochem.* 54: 1533–1540. [13]

von Bekesy, G. 1960. *Experiments in Hearing.* McGraw-Hill, New York. [18]

von Bernhardi, R., and Muller, K. J. 1995. Repair of the central nervous system: Lessons from lesions in leeches. *J. Neurobiol.* 27: 353–366. [24]

von Economo, G., and Koskinas, G. N. 1925. *Die Cytoarchiteck-tonik der Hirnrinde des erwachsenen Menschen.* Julius Springer, Heidelberg. [18]

von Euler, U. S. 1956. *Noradrenaline.* Charles Thomas, Springfield, IL. [13, 16]

von Euler, U. S., and Gaddum, J. H. 1931. Unidentified depressor substance in certain tissue extracts. *J. Physiol.* 72: 74–87. [14]

von Gersdorff, H., and Matthews, G. 1997. Depletion and replenishment of vesicle pools at a ribbon-type synaptic terminal. *J. Neurosci.* 17: 1919–1927. [11]

von Gersdorff, H., Vardi, E., Metthews, G., and Sterling, P. 1996. Evidence that vesicles on the synaptic ribbon of retinal bipolar neurons can be rapidly released. *Neuron* 16: 1221–1227. [19]

Voronin, L. L., Rossokhin, A. V., and Sokolov, M. V. 1999. Intracellular studies of the interaction between paired-pulse facilitation and the delayed phase of long-term potentiation in the hippocampus. *Neursoci. Behav. Physiol.* 29: 347–354. [12]

Vulchanova, L., Arvidsson, U., Riedl, M., Wang, J., Buell, G., Surprenant, A., North, R. A., and Elde, R. 1996. Differential distribution of two ATP-gated ion channel (P2X) receptors determined by immunocytochemistry. *Proc. Natl. Acad. Sci. USA* 93: 8063–8067. [14]

Vyskocil, F., Nikolsky, E., and Edwards, C. 1983. An analysis of mechanisms underlying the non-quantal release of acetylcholine at the mouse neuromuscular junction. *Neuroscience* 9: 429–435. [11]

Wada, H., Inagaki, N., Yamatodani, A., and Watanabe, T. 1991. Is the histaminergic neuron system a regulatory center for whole-brain activity? *Trends Neurosci.* 14: 415–418. [14]

Wagner, J. A., Carlson, S. S., and Kelly, R. B. 1978. Chemical and physical characterization of cholinergic synaptic vesicles. *Biochemistry* 17: 1199–1206. [11]

Wagner, S., Castel, M., Gainer, H., and Yarom, Y. 1997. GABA in the mammalian suprachiasmatic nucleus and its role in diurnal rhythmicity. *Nature* 387: 598–603. [16]

Wainer, B. H., Levey, A. I., Mufson, E. J., and Mesulam, M-M. 1984. Cholinergic systems in mammalian brain identified with antibodies against choline acetyltransferase. *Neurochem. Int.* 6: 163–182. [14]

Waldmann, R., Champigny, G., Bassilana, F., Heurteaux, C., and Lazdunski, M. 1997. A proton-gated cation channel involved in acid sensing. *Nature* 386: 173–177. [17]

Walker, D., and De Waard, M. 1998. Subunit interaction sites in voltage-dependent Ca^{2+} channels: Role in channel function. *Trends Neurosci.* 21: 148–154. [3]

Wallace, B. G., and Gillon, J. W. 1982. Characterization of acetylcholinesterase in individual neurons in the leech central nervous system. *J. Neurosci.* 2: 1108–1118. [14]

Walmsley, B., Alvarez, F. J., and Fyffe, R. E. W. 1998. Diversity of structure and function at mammalian central synapses. *Trends Neurosci.* 21: 81–88. [11]

Walro, J. M., and Kucera, J. 1999. Why adult mammalian intrafusal and extrafusal fibers contain different myosin heavy-chain isoforms. *Trends Neurosci.* 22: 180–184. [17, 22]

Walsh, C. A. 1999. Genetic malformations of the human cerebral cortex. *Neuron* 23: 19–29. [23]

Walsh, F. S., and Doherty, P. 1997. Neural cell adhesion molecules of the immunoglobulin superfamily: Role in axon growth and guidance. *Annu. Rev. Cell Dev. Biol.* 13: 425–456. [23]

Walter, J., Allsopp, T. E., and Bonhoeffer, F. 1990. A common denominator of growth cone guidance and collapse? *Trends Neurosci.* 13: 447–452. [23]

Walter, J., Henke-Fahle, S., and Bonhoeffer, F. 1987. Avoidance of posterior tectal membranes by temporal retinal axons. *Development* 101: 909–913. [23]

Walter, J., Kern-Veits, B., Huf, J., Stolze, B., and Bonhoeffer, F. 1987. Recognition of position-specific properties of tectal cell membranes by retinal axons *in vitro*. *Development* 101: 685–696. [23]

Walters, E. T., and Cohen, L. B. 1997. Functions of the LE sensory neurons in *Aplysia. Invertebr. Neurosci.* 3: 15–25. [15]

Wang, H. S., Pan, Z., Shi, W., Brown, B. S., Wymore, R. S., Cohen, I. S., Dixon, J. E., and McKinnon, D. 1998. KCNQ2 and KCNW potassium channel subunits: Molecular correlates of the M-channel. *Science* 282: 1890–1893. [16]

Wang, H., and Macagno, E. R. 1997. The establishment of peripheral sensory arbors in the leech: In vivo time-lapse studies reveal a highly dynamic process. *J. Neurosci.* 17: 2408–2419. [15]

Wang, H., Kunkel, D. D., Martin, T. M., Schwartzkroin, P. A., and Tempel, B. L. 1993. Heteromultimeric K^+ channels in terminal and juxtaparanodal regions of neurons. *Nature* 365: 75–79. [7]

Wang, R. 1998. Resurgence of carbon monoxide: An endogenous gaseous vasorelaxing factor. *Can. J. Physiol. Pharmacol.* 76: 1–15. [10]

Wang, T. F., and Guidotti, G. 1998. Widespread expression of ecto-apyrase (CD39) in the central nervous system. *Brain Res.* 790: 318–322. [13]

Wang, T. L., Guggino, W. B., and Cutting, G. R. 1994. A novel gamma-aminobutyric receptor subunit (rho2) cloned from human retina forms bicuculline-insensitive homooligomeric receptors in *Xenopus* oocytes. *J. Neurosci.* 14: 6524–6531. [14]

Wang, X., Merzenich, M. M., Beitel, R., and Schreiner, C. E. 1995. Representation of a species-specific vocalization in the primary auditory cortex of the common marmoset: Temporal and spectral characteristics. *J. Neurophysiol.* 74: 2685–2706. [18]

Warr, W. B. 1975. Olivocochlear and vestibular efferent neurons of the feline brain-stem: Their location, morphology and number determined by retrograde axonal transport and acetylcholinesterase histochemistry. *J. Comp. Neurol.* 161: 159–182. [18]

Wässle, H., Koulen, P., Brandstatter, J. H., Fletcher, E. L., and Becker, C. M. 1998. Glycine and GABA receptors in the mammalian retina. *Vision Res.* 38: 1411–1430. [14, 19, 23]

Wässle, H., and Chun, M. H. 1988. Dopaminergic and indoleamine-accumulating amacrine cells express GABA-like immunoreactivity in the cat retina. *J. Neurosci.* 8: 3383–3394. [14]

Waterhouse, B. D., Sessler, F. M., Liu, W., and Lin, C. S. 1991. Second messenger-mediated actions of norepinephrine on target neurons in central circuits: A new perspective on intracellular mechanisms and functional consequences. *Prog. Brain Res.* 88: 351–362. [14]

Watson, J. D., Myers, R., Frackowiak, R. S., Hajnal, J. V., Woods, R. P., Mazziotta, J. C., Shipp, S., and Zeki, S. 1993. Area V5 of the human brain: Evidence from a combined study using positron emission tomography and magnetic resonance imaging. *Cerebral Cortex* 3: 79–94. [21]

Watson, S. J., Akil, H., Richard, C. W., and Barchas, J. D. 1978. Evidence for two separate opiate peptide neuronal systems. *Nature* 275: 226–228. [14]

Waxman, S. G. 1997. Axon-glia interactions: Building a smart nerve fiber. *Curr. Biol.* 7: 406–410. [8]

Waxman, S. G., Black, J. A., Kocsis, J. D., and Ritchie, J. M. 1989. Low density of sodium channels support conduction in axons of neonatal rat optic nerve. *Proc. Natl. Acad. Sci. USA* 86: 1406–1410. [6]

Webb, T. E., Simon, J., Krishek, B. J., Bateson, A. N., Smart, T. G., King, B. F., Burnstock, G., and Barnard, E. A. 1993. Cloning and functional expression of a brain G-protein-coupled ATP receptor. *FEBS Lett.* 324: 219–225. [14]

Wehner, R. 1994b. The polarization-vision project: Championing organismic biology. *Fortschr. Zool.* 31: 11–53. [15]

Wehner, R. 1994a. Himmelsbild und Kompassauge—Neurobiologie eines Navigationssystems. *Verhand. Deutsch. Zool. Ges.* 87: 9–37. [15]

Wehner, R. 1996. Polarisationsmusteranalyse bei Insekten. *Nova Acta Leoplodina NF* 72: 159–183. [15]

Wehner, R. 1997. The ant's celestial compass system: Spectral and polarization channels. In M. Lehrer (ed.), *Orientation and Communication in Arthropods*, Birkhauser, Basel, Switzerland, pp. 145–185. [15]

Wehner, R., and Bernard, G. D. 1993. Photoreceptor twist: A solution to the false-color problem. *Proc. Natl. Acad. Sci. USA* 90: 4132–4135. [15]

Wehner, R., and Menzel, R. 1990. Do insects have cognitive maps? *Annu. Rev. Neurosci.* 13: 403–414. [15]

Wehner, R., Marsh, A. C., and Wehner, S. 1992. Desert ants on a thermal tightrope. *Nature* 586–587. [15]

Weiner, J. L., Gu, C., and Dunwiddie, T. V. 1997. Differential ethanol sensitivity of subpopulations of $GABA_A$ synapses onto rat hippocampal CA1 pyramidal neurons. *J. Neurophysiol.* 77: 1306–1312. [14]

Weiner, N., and Rabadjija, M. 1968. The effect of nerve stimulation on the synthesis and metabolism of norepinephrine from cat spleen during sympathetic nerve stimulation. *J. Pharmacol. Exp. Ther.* 160: 61–71. [13]

Weinreich, D. 1970. Ionic mechanisms of post-tetanic potentiation at the neuromuscular junction of the frog. *J. Physiol.* 212: 431–446. [12]

Weisblat, D. A., Huang, F. Z., Isaksen, D. E., Liu, N. J., and Chang, P. 1999. The other side of the embryo: An appreciation of the non-D quadrants in leech embryos. *Curr. Top. Dev. Biol.* 46: 105–132. [15]

Weiss, P. 1936. Selectivity controlling the central-peripheral relations in the nervous system. *Biol. Rev.* 11: 494–531. [23]

Weiss, P., and Hiscoe, H. B. 1948. Experiments of the mechanism of nerve growth. *J. Exp. Zool.* 107: 315–395. [13]

Weliky, M., and Katz, L. C. 1999. Correlational structure of spontaneous neuronal activity in the developing lateral geniculate nucleus in vivo. *Science* 285: 599–604. [20]

Weliky, M., Bosking, W. H., and Fitzpatrick, D. 1996. A systematic map of direction preference in primary visual cortex. *Nature* 379: 725–728. [21]

Welker, C. 1976. Microelectrode delineation of fine grain somatotopic organization of SM1 cerebral neocortex in albino rat. *J. Comp. Neurol.* 166: 173–190. [18]

Welker, E., Rao, S. B., Dorfl, J., Melzer, P., and van der Loos, H. 1992. Plasticity in the barrel cortex of the adult mouse: Effects of chronic stimulation upon deoxyglucose uptake in the behaving animal. *J. Neurosci.* 12: 153–170. [25]

Welsh, J. P., and Llinás, R. R. 1997. Some organizing principles for the control of movement based on olivocerebellar physiology. *Prog. Brain Res.* 114: 449–461. [22]

Werner, R., Miller, T., Azarnia, R., and Dahl, G. 1985. Translation and expression of cell-cell channel mRNA in *Xenopus* oocytes. *J. Membr. Biol.* 87: 253–268. [7]

Wernig, A. 1972. Changes in statistical parameters during facilitation at the crayfish neuromuscular junction. *J. Physiol.* 226: 751–759. [12]

Wernig, A., Muller, S., Nanassy, A., and Cagol, E. 1995. Laufband therapy based on 'rules of spinal locomotion' is effective in spinal cord injured persons. *Eur. J. Neurosci.* 7: 823–829. [22]

Werz, M. A., and Macdonald, R. L. 1983. Opioid peptides selective for mu- and delta-opiate receptors reduce calcium-dependent action potential duration by increasing potassium conductance. *Neurosci. Lett.* 42: 173–178. [14]

Wess, J. 1997. G-protein-coupled receptors: Molecular mechanisms involved in receptor activation and selectivity of G-protein recognition. *FASEB J.* 11: 346–354. [10]

Wess, J. 1998. Molecular basis of receptor/G-protein-coupling selectivity. *Pharmacol. Ther.* 80: 231–264. [10]

West, J. W., Patton, D. E., Scheuer, T., Wang, Y., Goldin, A. L., and Catterall, W. A. 1992. A cluster of hydrophobic amino acid residues required for fast Na+ channel inactivation. *Proc. Natl. Acad. Sci. USA* 89: 10910–10914. [6]

Weston, J. 1970. The migration and differentiation of neural crest cells. *Adv. Morphogenesis* 8: 41–114. [23]

Whim, M. D., Church, P. J., and Lloyd, P. E. 1993. Functional roles of peptide cotransmitters at neuromuscular synapses in *Aplysia*. *Mol. Neurobiol.* 7: 335–347. [13]

White, A. J., Wilder, H. D., Goodchild, A. K., Sefton, A. J., and Martin, P. R. 1998. Segregation of receptive field properties in the lateral geniculate nucleus of a New-World monkey, the marmoset *Callithrix jacchus*. *J. Neurophysiol.* 80: 2063–2076. [20]

White, S. R., Fung, S. J., Jackson, D. A., and Imel, K. M. 1996. Serotonin, norepinephrine and associated neuropeptides: Effects on somatic motoneuron excitability. *Prog. Brain Res.* 107: 183–199. [14]

Whitsel, B. L., Petrucelli, L. M., and Werner, G. 1969. Symmetry and connectivity in the map of the body surface in somatosensory area II of primates. *J. Neurophysiol.* 32: 170–183. [18]

Wichmann, T., and DeLong, M. R. 1996. Functional and pathophysiological models of the basal ganglia. *Curr. Opin. Neurobiol.* 6: 751–758. [22]

Wickelgren, W. O., Leonard, J. P., Grimes, M. J., and Clark, R. D. 1985. Ultrastructural correlates of transmitter release in presynaptic areas of lamprey reticulospinal axons. *J. Neurosci.* 5: 1188–1201. [11]

Wickman, K. D., Iñiguez-Lluhi, J. A., Davenport, P. A., Taussig, R., Krapivinsky, G. B., Linder, M. E., Gilman, A. G., and Clapham, D. E. 1994. Recombinant G-protein βγ-subunits activate the muscarinic-gated atrial potassium channel. *Nature* 368: 255–257. [10]

Wiederhold, M. L., and Kiang, N. Y. S. 1970. Effects of electric stimulation of the crossed olivocochlear bundle on single auditory nerve fibers of the cat. *J. Acoust. Soc. Am.* 48: 950–965. [18]

Wiesel, T. N. 1982. The postnatal development of the visual cortex and the influence of environment. *Nature* 299: 583–591. [25]

Wiesel, T. N., and Hubel, D. H. 1963a. Effects of visual deprivation on morphology and physiology of cells in the cat's lateral geniculate body. *J. Neurophysiol.* 26: 978–993. [25]

Wiesel, T. N., and Hubel, D. H. 1963b. Single-cell responses in striate cortex of kittens deprived of vision in one eye. *J. Neurophysiol.* 26: 1003–1017. [25]

Wiesel, T. N., and Hubel, D. H. 1965. Comparison of the effects of unilateral and bilateral eye closure on cortical unit responses in kittens. *J. Neurophysiol.* 28: 1029–1040. [25]

Wiesel, T. N., and Hubel, D. H. 1966. Spatial and chromatic interactions in the lateral geniculate body of the rhesus monkey. *J. Neurophysiol.* 29: 1115–1156. [21]

Wiesel, T. N., and Hubel, D. H. 1974. Ordered arrangement of orientation columns in monkeys lacking visual experience. *J. Comp. Neurol.* 158: 307–318. [25]

Wiesmann, C., Ultsch, M. H., Bass, S. H., and de Vos, A. M. 1999. Crystal structure of nerve growth factor in complex with the ligand-binding domain of the TrkA receptor. *Nature* 401: 184–188. [23]

Wigston, D. J., and Sanes, J. R. 1985. Selective reinnervation of intercostal muscles transplanted from different segmental levels to a common site. *J. Neurosci.* 5: 1208–1221. [24]

Willbold, E., Reinicke, M., Lance-Jones, C., Lagenaur, C., Lemmon, V., and Layer, P. G. 1995. Müller glia stabilizes cell columns during retinal development: Lateral cell migration but not neuropil growth is inhibited in mixed chick-quail retinospheroids. *Eur. J. Neurosci.* 7: 2277–2284. [8]

Willis, W. D., and Westlund, K. N. 1997. Neuroanatomy of the pain system and of the pathways that modulate pain. *J. Clin. Neurophysiol.* 14: 2–31. [18]

Wilson Horch, H. L., and Sargent, P. B. 1995. Perisynaptic surface distribution of multiple classes of nicotinic acetylcholine receptors on neurons in the chicken ciliary ganglion. *J. Neurosci.* 15: 7778–7795. [24]

Wilson Horch, H. L., and Sargent, P. B. 1996. Effects of denervation on acetylcholine receptor clusters on frog cardiac ganglion neurons as revealed by quantitative laser scanning confocal microscopy. *J. Neurosci.* 16: 1720–1729. [24]

Wilson, G. G., and Karlin, A. 1998. The location of the gate in the acetylcholine receptor channel. *Neuron* 20: 1269–1281. [3]

Winkler, J., Thal, L. J., Gage, F. H., and Fisher, L. J. 1998. Cholinergic strategies for Alzheimer's disease. *J. Mol. Med.* 76: 555–567. [23]

Winslow, R. L., and Sachs, M. B. 1987. Effect of electrical stimulation of the crossed olivocochlear bundle on auditory nerve response to tones in noise. *J. Neurophysiol.* 57: 1002–1021. [18]

Wise, D. S., Schoenborn, B. P., and Karlin, A. 1981. Structure of acetylcholine receptor dimer determined by neutron scattering and electron microscopy. *J. Biol. Chem.* 256: 4124–4126. [3]

Witzemann, V., Brenner, H-R., and Sakmann, B. 1991. Neural factors regulate AChR subunit mRNAs at rat neuromuscular synapses. *J. Cell Biol.* 114: 125–141. [24]

Wong, G. T., Gannon, K. S., and Margolskee, R. T. 1996. Transduction of bitter and sweet by gustducin. *Nature* 381: 796–800. [17]

Wong, L. A., and Gallagher, J. P. 1991. Pharmacology of nicotinic receptor-mediated inhibition in rat dorsolateral septal neurones. *J. Physiol.* 436: 325–346. [10]

Wong, R. O. 1999. Retinal waves and visual system development. *Annu. Rev. Neurosci.* 22: 29–47. [25]

Wong-Riley, M. 1989. Cytochrome oxidase: An endogenous metabolic marker for neuronal activity. *Trends Neurosci.* 12: 94–101. [21]

Woolsey, C. N. 1958. Organization of somatic sensory and motor areas of the cerebral cortex. In H. F. Harlow and C. N. Woolsey (eds.), *The Biological and Biochemical Bases of Behavior*. University of Wisconsin Press, Madison, pp. 63–82. [18]

Woolsey, C. N. 1972. Tonotopic organization of the auditory cortex. In M. B. Sachs (ed.), *Physiology of the Auditory System*. National Educational Consultants, Baltimore, MD, pp. 271–282. [18]

Woolsey, T. A., and Van der Loos, H. 1970. The structural organization of layer IV in the somatosensory region (SI) of mouse cerebral cortex: The description of a cortical field composed of discrete cytoarchitectonic units. *Brain Res.* 17: 205–242. [18]

Wray, S., Grant, P., and Gainer, H. 1989. Evidence that cells expressing luteinizing hormone-releasing hormone mRNA in mouse are derived from progenitor cells in the olfactory placode. *Proc. Natl. Acad. Sci. USA* 86: 8132–8136. [16, 23]

Wylie, C. (ed.). 1996. Zebrafish issue. *Development* 123: 1–481. [23]

Wylie, D. R., De Zeeuw, C. I., diGiorgi, P. L., and Simpson, J. I. 1994. Projections of individual Purkinje cells of identified zones in the ventral nodulus to the vestibular and cerebellar nuclei in the rabbit. *J. Comp. Neurol.* 349: 448–463. [22]

Xerri, C., Merzenich, M. M., Jenkins, W., and Santucci, S. 1999. Representational plasticity in cortical area 3b, paralleling tactual-motor skill acquisition in adult monkeys. *Cerebral Cortex* 9: 264–276. [18]

Xu, K., and Terakawa, S. 1999. Fenestration nodes and the wide submyelinic space form the basis for unusually fast impulse conduction of shrimp myelinated axons. *J. Exp. Biol.* 202: 1979–1989. [7]

Xu, R., and Salpeter, M. M. 1995. Protein kinase A regulates the degradation rate of Rs acetylcholine receptors. *J. Cell. Physiol.* 165: 30–39. [24]

Xu, X. M., Chen, A., Guénard, V., Kleitman, N., and Bunge, M. B. 1997. Bridging Schwann cell transplants promote axonal regeneration from both the rostral and caudal stumps of transected adult rat spinal cord. *J. Neurocytol.* 26: 1–16. [24]

Yabuta, N. H., and Callaway, E. M. 1998. Functional streams and local connections of layer 4C neurons in primary visual cortex of the macaque monkey. *J. Neurosci.* 18: 9489–9499. [21]

Yakel, J. L., and Jackson, M. B. 1988. 5HT3 receptors mediate rapid responses in cultured hippocampus and a clonal cell line. *Neuron* 1: 615–621. [3]

Yamada, T., Placzek, M., Tanaka, H., Dodd, J., and Jessell, T. M. 1991. Control of cell pattern in the developing nervous system: Polarizing activity of the floor plate and notochord. *Cell* 64: 635–647. [23]

Yamamori, T., Fukada, K., Aebersold, R., Korsching, S., Fann, M. J., and Patterson, P. H. 1989. The cholinergic neuronal differentiation factor from heart cells is identical to leukemia inhibitory factor. *Science* 246: 1412–1416. [23]

Yamashita, M., and Wässle, H. 1991. Responses of rod bipolar cells isolated from the rat retina to the glutamate agonist 2-amino-4-phosphonobutyric acid (APB). *J. Neurosci.* 11: 2372–2382. [19]

Yamashita, M., Fukui, H., Sugama, K., Horio, Y., Ito, S., Mizuguchi, H., and Wada H. 1991. Expression cloning of a cDNA encoding the bovine histamine H1 receptor. *Proc. Natl. Acad. Sci. USA* 88: 11515–11519. [14]

Yandava, B. G., Billinghurst, L. L., and Snyder, E. Y. 1999. "Global" cell replacement is feasible via neural stem cell transplantation: Evidence from the dysmyelinated *shiverer* mouse brain. *Proc. Natl. Acad. Sci. USA* 96: 7029–7034. [23]

Yang, N., George, A. L., and Horn, R. 1996. Molecular basis of charge movement in voltage-gated sodium channels. *Neuron* 16: 113–122. [6]

Yang, X. L., Gao, F., and Wu, S. M. 1999. Modulation of horizontal cell function by GABA(A) and GABA(C) receptors in dark- and light-adapted tiger salamander retina. *Vis. Neurosci.* 16: 967–979. [19]

Yau, K. W. 1976. Receptive fields, geometry and conduction block of sensory neurones in the central nervous system of the leech. *J. Physiol.* 263: 513–538. [7, 15]

Yau, K. W., and Chen, T. Y. 1995. Cyclic nucleotide-gated ion channels: An extended family with diverse functions. In R. A. North (ed.), *Handbook of Receptors and Channels: Ligand- and Voltage-Gated Ion Channels.* CRC Press, Boca Raton, FL, pp. 307–335. [19]

Yau, K. W., and Nakatani, K. 1985. Light-suppressible, cyclic GMP-sensitive conductance in the plasma membrane of the truncated rod outer segment. *Nature* 317: 252–255. [19]

Ye, W., Shimamura, K., Rubenstein, J. L., Hynes, M. A., and Rosenthal, A. 1998. FGF and Shh signals control dopaminergic and serotonergic cell fate in the anterior neural plate. *Cell* 93: 755–766. [23]

Yeckel, M. F., Kapur, A., and Johnston, D. 1999. Multiple forms of LTP in hippocampal CA3 neurons use a common postsynaptic mechanism. *Nature Neurosci.* 2: 525–633. [12]

Yellen, G. 1982. Single Ca^{+}-activated nonselective cation channels in neuroblastoma. *Nature* 296: 357–359. [5]

Yellen, G., Jurman, M. E., Abramson, T., and MacKinnon, R. 1991. Mutations affecting internal TEA blockade identify the probable pore-forming region of a K^{+} channel. *Science* 251: 939–942. [3]

Yool, A. J., and Schwartz, T. L. 1991. Alterations in ionic selectivity of a K^{+} channel by mutation of the H5 region. *Nature* 349: 700–704. [3]

Yoshioka, T., Dow, B. M., and Vautin, R. G. 1996. Neuronal mechanisms of color categorization in areas V1, V2 and V4 of macaque monkey visual cortex. *Behav. Brain Res.* 76: 51–70. [21]

Young, A. B., et al. 1995. Excitatory amino acid receptor distribution: quantitative autoradiographic ligand binding and *in situ* hybridization studies. In *Excitatory Amino Acids and Synaptic Transmission*. Academic Press, Harcourt, Brace and Co., New York, pp. 29–40. [14]

Young, E. D. 1984. Response characteristics of neurons of the cochlear nuclei. In C. Berlin (ed.), *Hearing Science*. College-Hill Press, San Diego, CA, pp. 423–460. [18]

Young, E. D., Spirou, G. A., Rice, J. J., and Voigt, H. F. 1992. Neural organization and responses to complex stimuli in the dorsal cochlear nucleus. *Philos. Trans. R. Soc. Lond. B* 336: 407–413. [18]

Young, J. Z. 1936. The giant nerve fibres and epistellar body of cephalopods. *Q. J. Microsc. Sci.* 78: 367–386. [5]

Yu, C., Brussaard, A. B., Yang, X., Listerud, M., and Role, L. W. 1993. Uptake of antisense oligonucleotides and functional block of acetylcholine receptor subunit gene expression in primary embryonic neurons. *Dev. Genet.* 14: 296–304. [14]

Yu, X., Nguyen, B., and Friesen, W. O. 1999. Sensory feedback can coordinate the swimming activity of the leech. *J. Neurosci.* 19: 4634–4643. [15]

Yurek, D. M., and Sladek, J. R., Jr. 1990. Dopamine cell replacement: Parkinson's disease. *Annu. Rev. Neurosci.* 13: 415–440. [14]

Z'Graggen, W. J., Metz, G. A. S., Kartje, G. L., Thallmair, M., and Schwab, M. E. 1998. Functional recovery and enhanced corticofugal plasticity after unilateral pyramidal tract lesion and blockade of myelin-associated neurite growth inhibitors in adult rats. *J. Neurosci.* 18: 4744–4757. [24]

Zafra, F., Aragon, C., and Gimenez, C. 1997. Molecular biology of glycinergic neurotransmission. *Mol. Neurobiol.* 14: 117–142. [14]

Zagotta, W. N., and Siegelbaum, S. A. 1996. Structure and function of cyclic-nucleotide-gated channels. *Ann. Rev. Neurosci.* 19: 235–263. [3]

Zagotta, W. N., Hoshi, T., and Aldrich, R. W. 1990. Restoration of inactivation in mutants of *Shaker* potassium channels by a peptide derived from ShB. *Science* 250: 568–571. [6]

Zagrebelsky, M., Buffo, A., Skerra, A., Schwab, M. E., Strata, P., and Rossi, F. 1998. Retrograde regulation of growth-associated gene expression in adult rat Purkinje cells by myelin-associated neurite growth inhibitory proteins. *J. Neurosci.* 18: 7912–7929. [24]

Zahs, K. R., and Newman, E. A. 1997. Asymmetric gap junctional coupling between glial-cells in the rat retina. *Glia* 20: 10–22. [8]

Zakrzewska, K. E., Sainsbury, A., Cusin, I., Rouru, J., Jeanrenaud, B., and Rohner-Jeanrenaud, F. 1999. Selective dependence of intracerebroventricular neuropeptide Y-elicited effects on central glucocorticoids. *Endocrinology* 140: 3183–3187. [13]

Zalutsky, R. A., and Nicoll, R. A. 1990. Comparison of two forms of long-term potentiation in hioppocampal neurons. *Science* 248: 1619–1624. [12]

Zehring, W. A., Wheeler, D. A., Reddy, P., Konopka, R. J., Kyriacou, C. P., Rosbash, M., and Hall, J. C. 1984. P-element transformation with period locus DNA restores rhythmicity to mutant, arrhythmic *Drosophila melanogaster*. *Cell* 39: 369–376. [16]

Zeki, S. 1990. Colour vision and functional specialisation in the visual cortex. *Disc. Neurosci.* 6: 1–64. [20, 21]

Zeki, S. 1999. Splendours and miseries of the brain. *Philos. Trans. R. Soc. Lond. B* 354: 2053–2065. [26]

Zeki, S. M. 1974. Functional organization of a visual area in the posterior bank of the superior temporal sulcus of the rhesus monkey. *J. Physiol.* 236: 549–573. [21]

Zeki, S., Watson, J. D., Lueck, C. J., Friston, K. J., Kennard, C., and Frackowiak, R. S. 1991. A direct demonstration of functional specialization in human visual cortex. *J. Neurosci.* 11: 641–649. [21]

Zengel, J. E., and Magleby, K. L. 1982. Augmentation and facilitation of transmitter release. A quantitative description at the frog neuromuscular junction. *J. Gen. Physiol.* 80: 582–611. [12]

Zhang, Z., Evans, R. L., and Culp, D. J. 1998. In vitro studies of the regulation of mucous gland secretion. *Eur. J. Morphol.* 36 Suppl.: 219–221. [16]

Zheng, C., Heintz, N., and Hatten, M. E. 1996. CNS gene encoding astrotactin, which supports neuronal migration along glial fibers. *Science* 272: 417–419. [8, 23]

Zheng, D., LaMantia, A. S., and Purves, D. 1991. Specialized vascularization of the primate visual cortex. *J. Neurosci.* 11: 2622–2629. [21]

Zheng, J., et al. 2000. Prestin is the motor protein of cochlear outer hair cells. *Nature* 405: 149–55. [18]

Zhou, W., Jiang, D., Raisman, G., and Zhou, C. 1998. Embryonic entorhinal transplants partially ameliorate the deficits in spatial memory in adult rats with entorhinal cortex lesions. *Brain Res.* 792: 97–104. [24]

Zhou, W., Raisman, G., and Zhou, C. 1998. Transplanted embryonic entorhinal neurons make functional synapses in adult host hippocampus. *Brain Res.* 788: 202–206. [24]

Zhou, Z. J. 1998. Direct participation of starburst amacrine cells in spontaneous rhythmic activities in the developing mammalian retina. *J. Neurosci.* 18: 4155–4165. [25]

Zhuchenko, O., Bailey, J., Bonnen, P., Ashizawa, T., Stockton, D. W., Amos, C., Dobyns, W. B., Subramony, S. H., Zoghbi, H. Y., and Lee, C. C. 1997. Autosomal dominant cerebellar ataxia (SCA6) associated with small polyglutamine expansions in the alpha 1A-voltage-dependent calcium channel. *Nature Genet.* 15: 62–69. [22]

Zigmond, M. J., Bloom, F. E., Landis, S. C., Roberts, J. L., and Squire, L. R. (eds.). 1999. *Fundamental Neuroscience.* Academic Press, New York. [23]

Zigmond, R. E., Schwarzchild, M. A., and Rittenhouse, A. R. 1989. Acute regulation of tyrosine hydroxylase by nerve activity and by neurotransmitters via phosphorylation. *Annu. Rev. Neurosci.* 12: 415–461. [13]

Zimmermann, H., and Braun, N. 1996. Extracellular metabolism of nucleotides in the nervous system. *J. Auton. Pharmacol.* 16: 397–400. [13]

Zimmermann, H., and Denston, C. R. 1977a. Recycling of synaptic vesicles in the cholinergic synapses of the *Torpedo* electric organ during induced transmitter release. *Neuroscience* 2: 695–714. [13]

Zimmermann, H., and Denston, C. R. 1977b. Separation of synaptic vesicles of different functional states from the cholinergic synapses of the *Torpedo* electric organ. *Neuroscience* 2: 715–730. [13]

Zipursky, S. L., and Rubin, G. M. 1994. Determination of neuronal cell fate: Lessons from the R7 neuron of *Drosophila. Annu. Rev. Neurosci.* 17: 373–397. [23]

Zollikofer, C., Wehner, R., and Fukushi, T. 1995. Optical scaling in conspecific *Cataglyphis* ants. *J. Exp. Biol.* 198: 1637–1646. [15]

Zucker, R. S. 1993. Calcium and transmitter release. *J. Physiol. (Paris)* 87: 25–36. [11]

INDEX

A

D

E

F

G

O

S

T

About the Book

Editor: Andrew D. Sinauer
Project Editor: Kerry Falvey
Production Manager: Christopher Small
Book Production: Jefferson Johnson, Joan Gemme, and Janice Holabird
Illustration Program: Imagineering Scientific and Technical Artworks, Inc.
Book Design: Jean Hammond
Cover Design: Jefferson Johnson
Color Separations: Burt Russell Litho, Inc.
Book and Cover Manufacture through: World Print Ltd.